Machining Fundamentals

From Basic to Advanced Techniques

by

JOHN R. WALKER

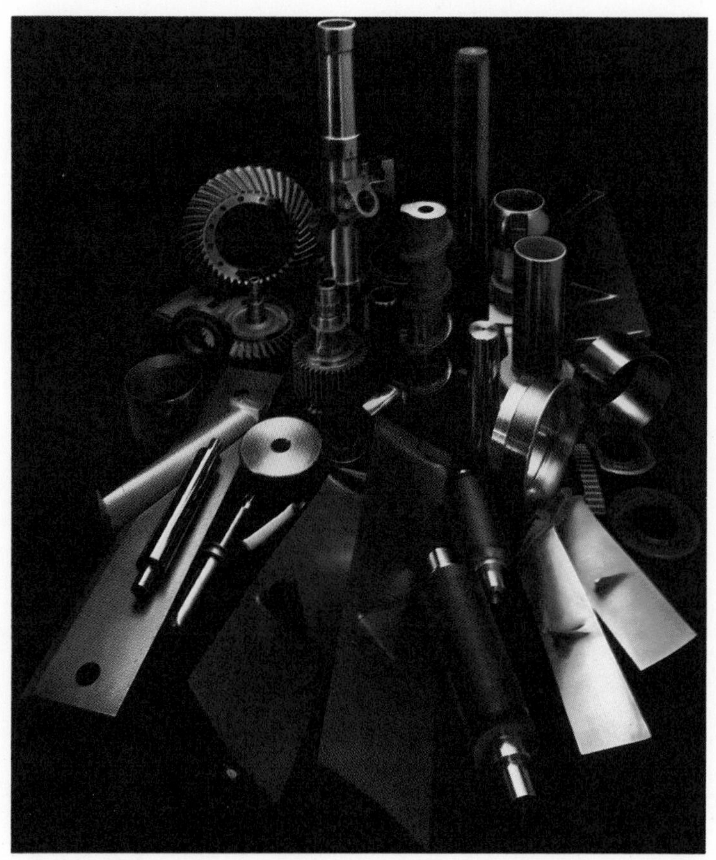

South Holland, Illinois

THE GOODHEART-WILLCOX COMPANY, INC.

Publishers

Library of Congress Catalog Card Number 88-33268
International Standard Book Number 0-87006-710-9

123456789-89-5432109

Library of Congress Cataloging in Publication Data

Walker, John R.
Machining fundamentals: fundamentals basic to industry/by John R. Walker.
p. cm.
Includes index.
ISBN 0-87006-710-9
1. Machine-shop practice. 2. Machining.
I. Title.
TJ1160.W25 1989
671.3'5--dc19 88-33268
CIP

INTRODUCTION

Machinists are highly skilled men and women. They use drawings, hand tools, precision measuring tools, drilling machines, grinders, lathes, milling machines, and other specialized machine tools to shape and finish metal and nonmetal parts. Machinists must have a sound understanding of basic and advanced machining technology, which includes:

1. A proficiency in safe machine tool operation (manual, automatic, and computer controlled).
2. A knowledge of the working properties of metals and nonmetals.
3. The academic skills (math, science, English, print reading, metallurgy, etc.) needed to make precision layouts and machine setups.

MACHINING FUNDAMENTALS provides an introduction to this important area of manufacturing technology. The text explains the ''How, Why, and When'' of the various machining operations, setups, and procedures. Through it, you will learn how machine tools operate and when to use one particular machine instead of another. In addition, the advantages and disadvantages of various machining techniques are discussed.

MACHINING FUNDAMENTALS details the many common methods of machining and shaping parts to meet given specifications. It also covers newer processes such as Laser Machining and Welding, Water Jet Cutting, High Energy Rate Forming (HERF), Cryogenics, Chipless Machining, Electrical Discharge Machining (EDM), Electro-Chemical Machining (ECM), Numerical Control (N/C and CNC), Robotics, and the importance of computers in the operation of most of these machining techniques.

MACHINING FUNDAMENTALS has many features that make it easy to read and understand. Learning objectives are presented at the beginning of each chapter. There are many photographs and line drawings, most made especially for the text, to help you visualize more clearly the machining operations and procedures.

Color is employed to emphasize safety precautions and to highlight important points in many illustrations. Technical terms are defined with bold, italic type. Review questions covering the contents of the chapter and supplemental activities are included at the end of each chapter. Metric terms are given after conventional values.

MACHINING FUNDAMENTALS will prove to be a valuable guide to anyone interested in machining because the procedures and techniques presented in the text are drawn from all areas of machining technology.

John R. Walker

CONTENTS

IMPORTANT SAFETY NOTICE

Work procedures and shop practices described in this book are effective, but general, methods of performing given operations. Use special tools and equipment as recommended. Carefully follow all safety warnings and cautions. Note that these warnings are not exhaustive. Proceed with care and under proper supervision to minimize the risk of personal injury or injury to others. Also follow specific equipment operating instructions.

High production, thru-feed grinder has precision air gauging system for continuous compensation and high accuracy. (SpeedFam Corp.)

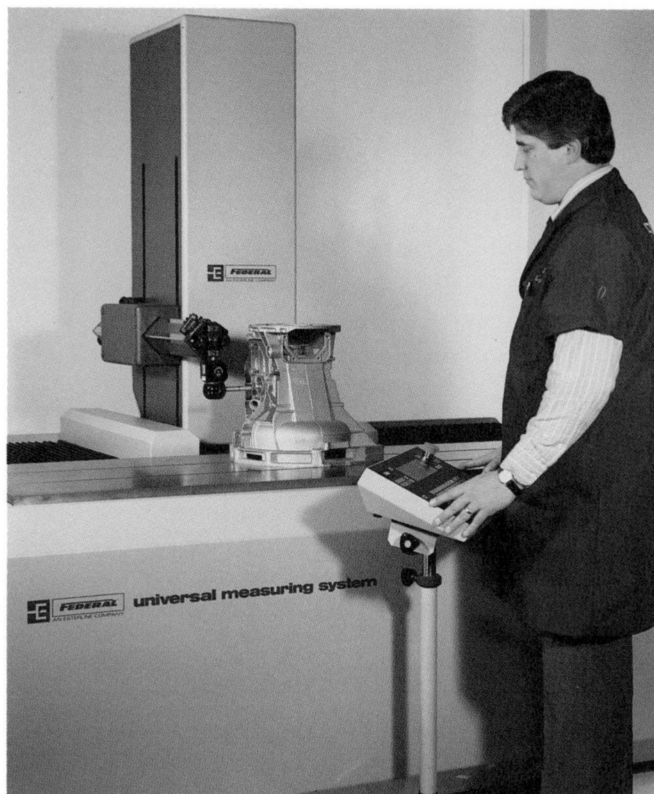

Computer-controlled measuring system is being used to check accuracy of machined transmission housing. (Federal)

Electric self-feeders can be used for automatic drilling, taping, deburring, spot-facing, and other operations. (Sugino)

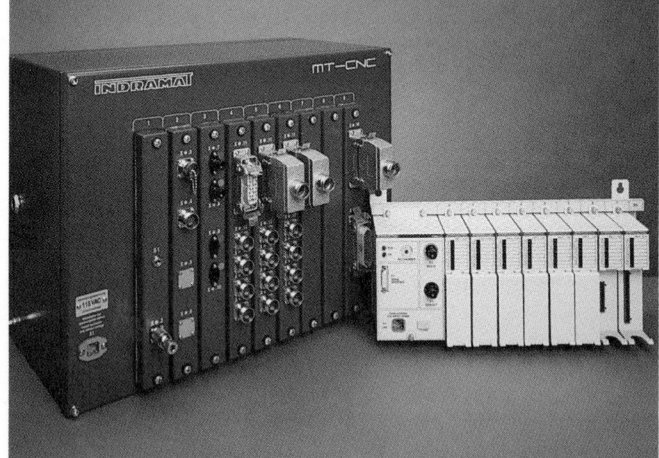

This is a computer numerical control or CNC unit that can be used for automatic tool management. (Indramat)

Machining technology has seen tremendous change in the past few years. Electronic devices and computers are now being used to increase production speed and precision. This text will detail these new techniques along with conventional methods to give you a solid background in machining fundamentals.

Chapter 1

OCCUPATIONS IN MACHINING TECHNOLOGY

After studying this chapter, you will be able to:

- List the requirements for the various machining technology occupations.
- Explain where to obtain information on occupations in machining technology.
- State what industry expects of an employee.
- Describe what an employee should expect from industry.
- Summarize the information given on a resume.

Many people want a career that is both challenging and interesting. Their philosophy is that a career is enjoyable; you only ''work'' when you dislike a job. Others are satisfied with whatever job comes along. In which of these categories are you?

If you are looking for a career that is challenging, interesting, and rewarding, the field of machining offers many opportunities. Whether you choose one of the machine shop areas or select a career in a related profession, you will fine that the study of MACHINING FUNDAMENTALS is basic to all of them. See Fig. 1-1.

No matter what choice is made, to be successful and advance in your career, a continuing program of education is usually necessary to keep up with technical progress.

Jobs in material machining fall into four general categories:

1. Semiskilled (not much training required).
2. Skilled (several years of training).
3. Technical (works with skilled and engineering staff).
4. Professional (college training essential).

SEMISKILLED WORKERS

Semiskilled workers are those who perform the basic operations that do NOT require a high degree of skill nor training, Fig. 1-2. Most of the work done by the semiskilled worker is routine and may be classified by the following general types:

1. Those who serve as helpers for skilled workers.
2. Those who operate machines and equipment used in making things. The machines are set up by skilled personnel.
3. Those who assemble the various manufactured parts into final products.

There is little chance for advancement out of semiskilled jobs without additional study and training. Most semiskilled work is found in production shops where there are great numbers of repeat operations.

In general, semiskilled workers are told what to do and how the work is to be done. They are usually the first to be let go or unemployed when there is a downturn in the economy.

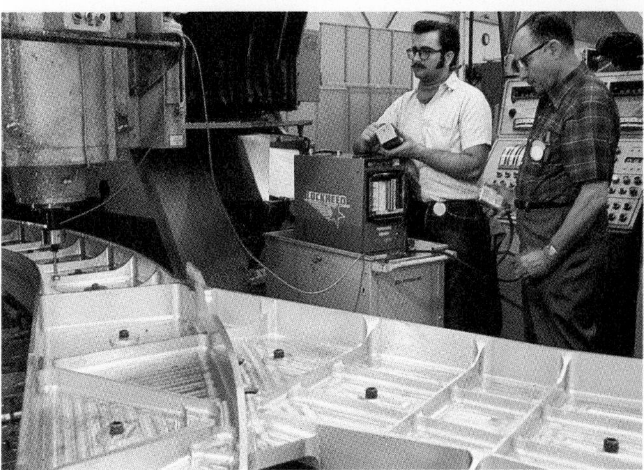

Fig. 1-1. The field of metal machining offers many opportunities, whether in the machine shop or in one of the related technical or professional areas of the technology. The study of MACHINING FUNDAMENTALS is basic to all of them. (Lockheed-California Company)

Fig. 1-2. The operator of this abrasive cutoff saw is classified as a semiskilled worker. The safe operation of the machine requires some training but it does not require a high degree of skill. (Norton Co.)

SKILLED WORKERS

Skilled workers have had training to do more complex tasks. They are found in all areas of material machining. Today, many of them received their training in ***apprentice programs*** (on-the-job training while working with skilled machinist), Fig. 1-3.

Four or more years of instruction under an experienced machinist is generally required. In addition to working in the shop, an apprentice usually studies related subjects, such as: math, science, English, print reading, metallurgy, safety, and production techniques. Upon completion of an apprentice program, the worker is capable of performing the precise work essential to the trade, Fig. 1-4.

Today however, the number of apprentice programs being offered is on the decline. Most workers now entering the field receive their training in the armed forces or in vocational/technical programs offered in high schools and community colleges. Many community college programs are offered in conjunction with local industry. Refer to Fig. 1-5.

Specialized machinists

There are several areas in which the machinist may specialize:

All-around machinist is a competent person who can set up and operate most types of machine tools. He or she is expected to plan and carry out all of the operations needed to machine a job, Fig. 1-6.

Fig. 1-3. The apprentice studies under an experienced machinist for a period of four or more years. The training program also includes the study of related subjects like math, English, science, etc.

Many all-around machinists work in ***job shops*** (shops where special and experimental work is machined, or where production runs are very small). In today's machine shop, this person must be familiar with computer-controlled machine tools and how they are programmed, Fig. 1-7.

Fig. 1-4. Upon completion of the apprentice program, the worker is capable of performing some of the precision work essential to the trade.

Fig. 1-5. The Army, and other branches of the Armed Forces, offer excellent opportunities for learning a trade. As a bonus, you get paid while you learn. (U.S. Army)

Fig. 1-6. The all-around machinist can set up and operate most types of machine tools. (Heidenhain Corp.)

Fig. 1-7. Some machinists specialize in computer-controlled machine tools. (Heidenhain Corp.)

A ***toolmaker,*** is a highly skilled person who specializes in producing the tools needed for machining operations.

1. ***Dies*** (special tools for shaping, forming, stamping, or cutting metal or other materials).
2. ***Jigs*** (devices that position work and guide cutting tools).
3. ***Fixtures*** (devices to hold work while it is machined).

These tools are necessary for modern mass production techniques. Toolmakers must have a broader background in machining operations and mathematics than most other skilled workers in the trade. See Fig. 1-8.

A ***diemaker*** is a toolmaker who specializes in making punches and dies needed to stamp out parts like auto body panels, electrical components, and similar products. He or she will also produce the dies for making ***extrusions*** (metal shaped by being pushed through an opening in a metal disc of proper configuration) and ***die castings*** (parts made by forcing molten metal into a mold). Like the toolmaker, a diemaker is a highly skilled machinist, Fig. 1-9.

The ***layout specialist*** is a person who interprets the drawings and uses precision measuring tools to mark off where metal must be removed by machining from castings, forgings, and metal stock. This person is a skilled machinist familiar with the operation and capabilities of machine tools. He or she is well trained in mathematics and print reading. Refer to Fig. 1-10.

Fig. 1-8. The toolmaker produces the tools and work-holding devices necessary for modern production techniques. They have a broader background in machining operations and mathematics than most other skilled workers in the trade. (Clausing)

Fig. 1-9. A few of the hundreds of die cast parts used by the auto industry. The molds in which they were cast were machined by diemakers. (Kelsey-Hayes)

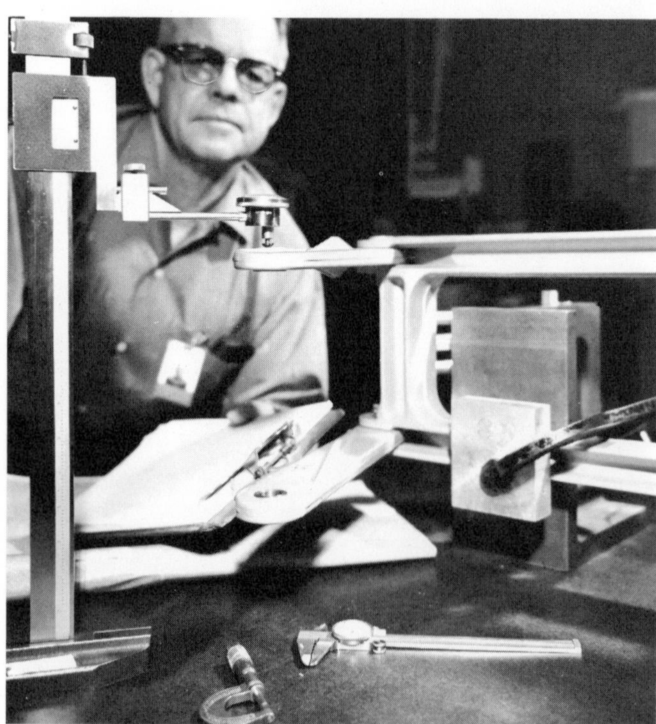

Fig. 1-10. Layout specialists interpret the drawings. Then, with precision measuring tools, they locate and mark off where material must be removed by machining from castings, forgings, and metal stock. This specialist is leveling an aircraft forging so that it can be laid out for machining.

A ***setup specialist*** is a person who ***sets up*** a machine tool (locate and position tools and work holding devices on machine) for use by a machine tool operator. This worker may also show the machine tool operator how to do the job and check the accuracy of the machined part, Fig. 1-11.

A ***part programmer*** inputs data into a computer-controlled machine tool for machining a product. Computer-controlled machine tools are revolutionizing the field of material machining. However, computers have no inherent intelligence and cannot think or exercise judgment. They must be ***programmed*** (instructed) by a highly skilled part programmer who studies the drawings and determines the sequences, tools, and motions the machine tool must utilize to machine the part, Fig. 1-12.

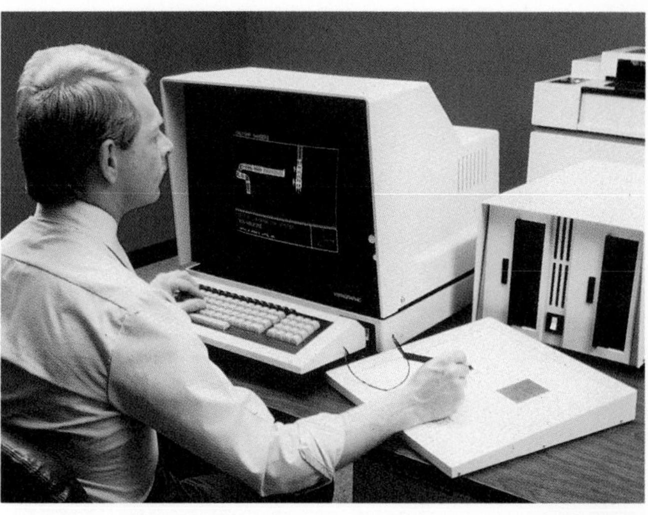

Fig. 1-12. This programmer is preparing the information (tools, tool paths, machining sequence, etc.) for a computer-controlled machine tool. This information or computer data is called a program and will use the computer to control the machine during the entire machining operation of a specific part. (California Computer Products, CALCOMP)

Fig. 1-11. This turret lathe has been prepared for operation by a setup specialist. After thoroughly checking the completed part to be sure it will meet manufacturing specifications, the machine tool will be turned over to a machine operator. (Clausing)

To perform this task, a part progrmamer must have a background that includes the following:

1. Formal training in computer technology as it relates to machine tool operation.
2. Experience at reading and interpreting drawings.
3. Be well versed in machining technology and procedures.
4. A working knowledge of cutting speeds and feeds for various tools and materials.
5. An extensive training in mathematics.

Many community colleges and vo-tech centers offer programs in machine tool computer programming to qualified persons, generally skilled machinists or other persons with extensive machine tool experience.

A ***supervisor*** or ***manager*** is usually a skilled machinist who has been promoted to a position of more responsiblity. This person will direct other workers in the shop and is responsible for meeting production "deadlines" and for keeping work quality high. In many shops, a manager may also be responsible for "firing" incompetent workers, Fig. 1-13.

TECHNICIANS

The ***technician*** is a member of the production team and operates in the realm between the shop and engineering departments, Fig. 1-14. The position is an outgrowth of today's highly technological and scientific world. The job usually requires at least two years of college, with a program of study centered on math, science, English, computer science, quality control, manufacturing, and produc-

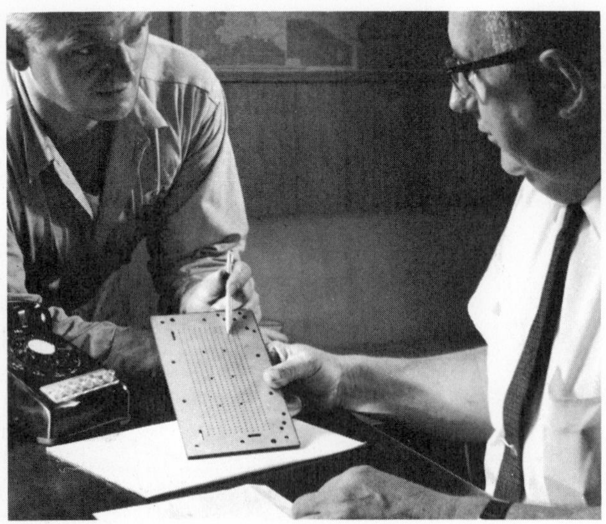

Fig. 1-13. The supervisor or manager of the production department works very closely with the engineering, metallurigical, and other staff to make sure that they satisfy the customer's specifications.

Fig. 1-15. Some technicians are responsible for the repair and maintenance of computer-controlled devices like welding robots. These robots are welding frames for minivans. (Ford Motor Co.)

Fig. 1-14. Modern quality control techniques require highly skilled technicians. Many technicians also assist engineers in developing, constructing, and testing experimental devices. (TEKTRONIX, Inc.)

tion processes. Many state and community colleges offer two year technical programs.

The technician assists the engineer by testing various experimental devices and machines, compiling statistics, making cost estimates, and preparing technical reports. Many inspection and quality control programs are managed by technicians. Technicians also repair and maintain computer-controlled machine tools and robots, Fig. 1-15.

THE PROFESSIONS

The professions offer many excellent opportunities in the field of metalworking.

Teaching is one of the most satisfying of the professions; it is a field that students too often overlook, Fig. 1-16. Teachers of industrial arts, industrial technology, industrial education, and vocational-technical education are in a fortunate position. Teaching is a challenging profession that offers a freedom NOT found in most other professions. It is not an overcrowded profession and it appears there will be a demand for teachers for many years to come. See Fig. 1-17.

To teach machining, four years of college training are usually needed. While industrial experience is ordinarily NOT required, it will prove very helpful.

Engineering is a fast growing and challenging profession. Engineers use mathematics, science, and a knowledge of manufacturing to develop new products and processes for industry, Fig. 1-18.

A bachelor's degree is usually the minimum requirement for entering the engineering profession. However, some men and women have been able to enter the profession without a degree after several years experience as machinists, drafters, or engineering technicians. They are usually required, however, to take some college level training.

The ***industrial engineer*** is primarily concerned with the safest and most efficient use of machines, materials, and personnel, Fig. 1-19. In some instances, he or she may be responsible for the design of special machinery and equipment to be utilized in manufacturing operations.

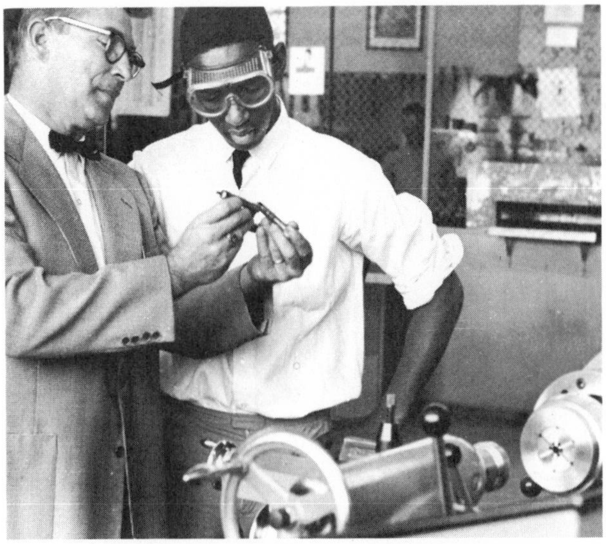

Fig. 1-16. The teaching profession is often overlooked by the student. Many skilled educators will be needed in machining technology if the United States is to maintain its position as a world leader in the industry.

Fig. 1-18. Engineering is a challenging profession. Here an engineer checks out an experimental scramjet engine capable of sustaining flight at speeds in excess of Mach 6.0. It is fueled with liquid hydrogen, which also cools the engine's combustion chamber components to keep them from melting. (NASA)

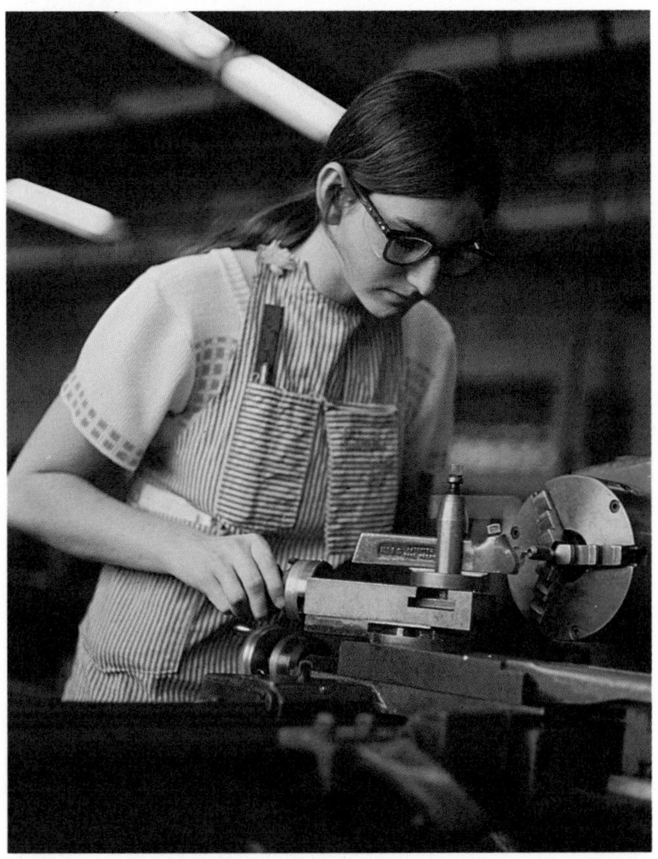

Fig. 1-17. This young lady may someday teach machine technology. During four or more years of training, she will learn all phases of machine tool operation and computer programming as it relates to machine technology.

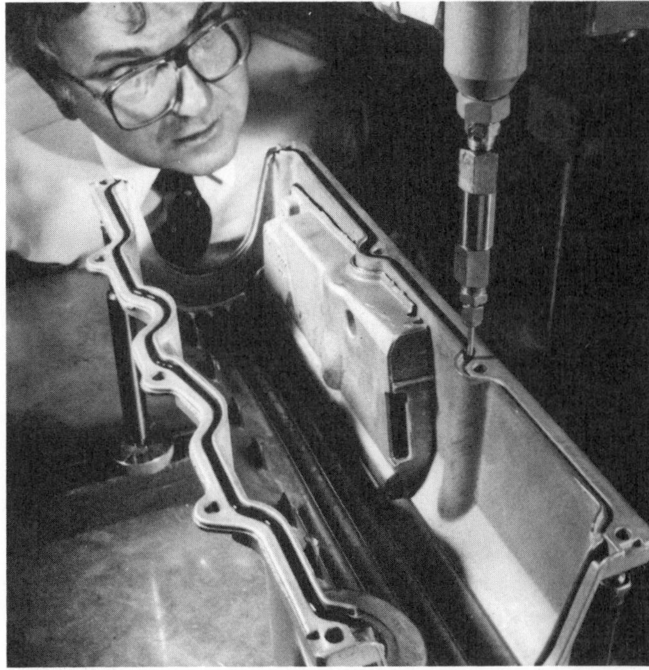

Fig. 1-19. Industrial engineers have many responsibilities. Here a computer controlled machine is applying a sealant bead. The chemical sealant will form the gasket for this engine cover. The machine, now in operation, produces a 60 percent reduction in material costs compared to traditional liquid-cork composition gasketing systems. (General Electric Corp.)

A ***mechanical engineer*** is normally responsible for the design and development of new machines, devices, and ideas, Fig. 1-20. This engineer is also involved with the redesign and improvements of existing equipment. Some mechanical engineers specialize in various areas of transportation (ships, transit vehicles, etc.).

The ***tool and manufacturing engineer*** often works with the other engineers, Fig. 1-21. A principle concern of the mechanical engineer is the design and development of the original or prototype model. When this model has been thoroughly tested and has met design requirements, the product is turned over to the tool and manufacturing engineer to devise methods and means required to manufacture and assemble the item.

A ***metallurgical engineer*** is involved in the development and testing of metals that are used in products and manufacturing processes. See Fig. 1-22.

OBTAINING INFORMATION ON MACHINING OCCUPATIONS

There are many sources of occupational information. The closest are the school's career center, and instructors of industrial/technical education.

State employment services are also excellent sources for getting information on local and state employment opportunities, as are the various trade unions concerned with the metalworking trades.

The field and regional offices of the Bureau of Apprenticeship, United States Department of Labor, may also be contacted for information on apprenticeship programs in your area.

Information on technical occupations is also available from community colleges. Many of them offer Associate Degrees in technical areas.

WHAT TO EXPECT WHEN ENTERING THE WORLD OF WORK

You will be very disappointed if you think that graduation means the end of training and study. Modern technology means constantly developing new ideas, materials, processes, and manufacturing techniques. In turn, this change is creating occupations that did not previously exist. It has been said that the average graduate will be employed in five different jobs during their lifetime, and three of them do NOT now exist!

To hold your job and advance in it, you will have to study to keep up-to-date with the knowledge and new skills that advanced technology demands. You must START WORK ON TIME, and be at WORK EVERY DAY. Sick leave is to help workers and should not be abused.

Industry also expects a fair day's work for a fair day's pay. High manufacturing costs and competi-

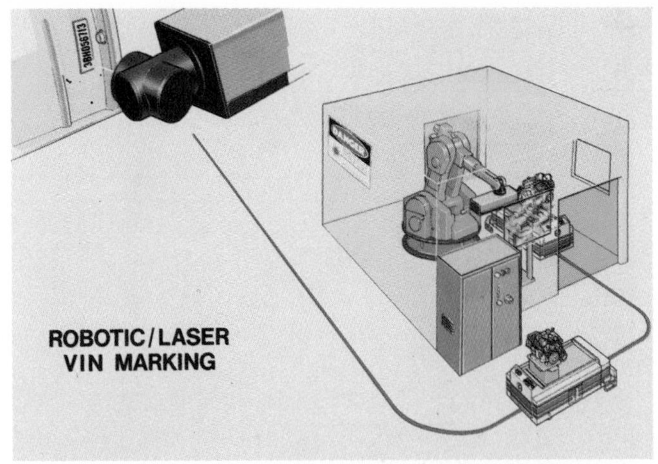

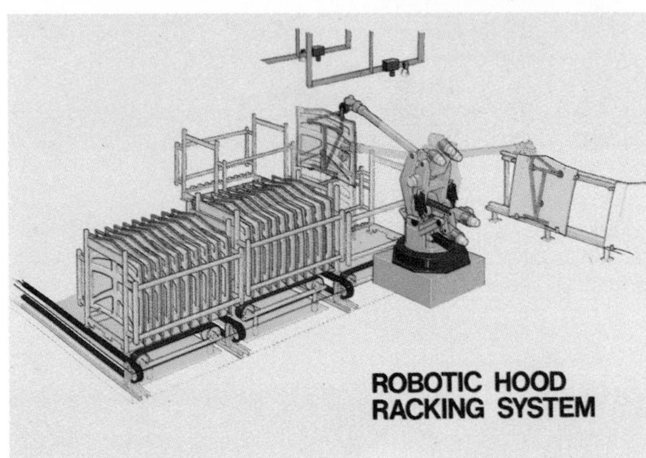

Fig. 1-20. Mechanical engineers are responsible for the design and development of new machines, devices, and ideas. These robots, used by the auto industry, were designed by mechanical engineers. (Oldsmobile Div., GMC)

Fig. 1-21. Tool and manufacturing engineers devise methods of producing complex products like the F-16. They must also plan techniques that will allow major modifications while the plane is in production. (General Dynamics)

Fig. 1-22. An experimental melt is poured from a 300-pound induction furnace at Bethlehem Steel Corporation's Homer Research Laboratory. Metallurgical engineers are responsible for testing and developing metals that will eventually be used in manufacturing products. (Bethlehem Steel Corp.)

tion with foreign-made products have made this a real necessity.

Do your assigned work, and NEVER knowingly turn out a piece of substandard or faulty work. Take pride in what you do! YOU must assume the responsibility for your actions. Industry is always on the lookout for bright young people who are not afraid to work and assume responsibility.

WHAT AN EMPLOYEE SHOULD EXPECT FROM INDUSTRY

You have some idea of what industry expects from an employee. However, are you aware of what an employee should expect from industry? These expectations are over and above salary and fringe benefits.

The following are but a few of the general conditions you might look for when selecting a place of employment.

Check that a relatively safe and clean work area is provided. Obviously, some areas can never be made as safe as others. For example, tapping a blast furnace is inherently more dangerous than working on a small lathe, or drill press.

Inspect whether work areas are adequately lighted, heated, and ventilated. Are noxious fumes and dust particles filtered from the air?

Look to see whether safety clothing and equipment are available for all dangerous work. Safety items such as goggles, hearing protectors, and steel-tipped safety shoes may be provided free or at minimum cost.

Are all necessary precautions observed when hazardous materials are being worked?

Check for a preventative safety program and whether safety regulations and precautions are rigorously enforced.

What other safety conditions do you think should be available from an employer?

HOW TO GET A JOB

Securing your first job after graduation will be a very important task. To be successful, you will have to spend as much time looking for this position as you would working at a regular job. In addition, there are several other things that you can do to make this task easier.

You will have to decide what type of work you would like to do. Most schools and state employment services administer tests that will help you determine the area(s) of employment where you will have a good chance of succeeding.

Answering the following questions will give you additional help:

1. What can I do with some degree of success?
2. What have I done that others have commended me for doing well?
3. What are the things I really like to do?
4. What are the things I do NOT like to do?
5. What jobs have I held? Why did I leave them?
6. What skills have I acquired while in school?

You will probably have two or more areas of interest. After listing them, start gathering information on these areas of interst. Use as many different sources as time permits. This may include reading, talking with persons doing this type of work, and visiting industry.

If time permits, plan your educational program to prepare for entry into a specific job, or for advanced schooling if you are near graduation.

The next problem is how do you go about getting that job? Jobs are always available. Workers get promoted, they retire, some quit, die, or get fired. Technological progress also creates new jobs. However, you must "track" these jobs down. There is no easy way to get a challenging job.

Concentrate on getting the job. Make your initial request for a job IN PERSON. Always be specific on the type of job you are after. Make sure you are qualified for that job. Never ask for any job or say—"What openings do you have?"

Resume

To speed the tedious task of filling out job applications, prepare a job resume in advance. A ***resume*** is a summary of your educational and employment background. It is submitted when applying for a job. It will assure uniform information with little chance for confusing responses.

Your resume should include:

1. Your full NAME.
2. Full NAME, and PHONE NUMBER. Do not forget the zip code.
3. PLACE and DATE OF BIRTH. For some jobs, it may be necessary to include a certified copy of your birth certificate. Have a copy available in advance.
4. Your SOCIAL SECURITY NUMBER.
5. LOCATION OF EMPLOYMENT or the places you have worked. Start with the most recent place of employment. Include the following items under each place of employment:
   a. Name and address.
   b. Dates employed.
   c. Immediate supervisor's name.
   d. Salary.
   e. Reason for leaving.
6. SCHOOLING and special training. Include dates attending.
7. The EQUIPMENT you can safely operate.
8. Names and addresses of REFERENCES. Do not include relatives unless you have worked for them. Make sure you secure permission before using a person for a reference!

Last, but not least, know where to look for a job. Look over the classified section of local newspapers each day. Talk with friends and relatives that are employed. They may be aware of job openings at other places of employment before the jobs become official and are advertised.

New offices and factories usually indicate possible job openings. It would also be to your advantage to prepare a list of desirable employers in your community and visit their employment offices. Plan these visits on a routine basis when jobs are not readily available. The employment office will then know you are interested in working for their firm and may give you preference.

Dress appropriately. Job hunting is not the time to wear old clothes or torn and beat-up shoes. Be clean and have a neat haircut.

One thing you must remember. The job will NOT come to you, YOU must search for it!

TEST YOUR KNOWLEDGE—Chapter 1

Please do not write in the text. Place your answers on a separate sheet of paper.

1. List the four general categories into which metalworking occupations fall.
2. ________ workers are those who perform operations that do NOT require a high degree of skill or training.
3. The ________ worker usually starts as an apprentice.
4. Since the number of apprentice programs is on the decline, where can this training now be obtained?
5. Describe what an all-around machinist is expected to do.
6. What does a layout specialist do?
7. To perform their job properly, a part programmer should have the following background: (List five.)
8. What are some of the duties of a technician?
9. List the areas of study included in a technician's educational program.
10. List three sources of information on metalworking occupations.
11. What does industry expect from you when on the job?
12. What is a job resume?
13. Why should a job resume be prepared in advance?
14. Explain why you think references are important.

RESEARCH AND DEVELOPMENT

1. Invite a speaker from the local Government Employment Service Office (it may be listed under different names in various parts of the country) to discuss local and national employment opportunities in the metalworking industries.
2. Prepare a chart that shows the hourly salaries of skilled and technical machine shop employees at local industries.
3. Ask someone who has graduated from your shop class and who has completed an apprentice program to describe their experiences while in training.
4. Prepare a bulletin board display around a local apprentice program. List the training requirements on a year-to-year basis.
5. Make a study of the help wanted column in the daily papers for a period of two weeks. Prepare a list of machine shop and related positions available, the salaries offered, and the requirements for securing the jobs. How often are additional benefits such as health insurance, etc., mentioned in the ads?
6. Summarize the information on machine shop and related occupations described in the occupational outlook handbook (a government publication) and make it available to the class.
7. Contact the International Association of Machinists for information on machinist trades apprentice programs.
8. Contact engineering societies and request speakers and/or films to address your class on their professions.

Chapter 2

UNDERSTANDING DRAWINGS

After studying this chapter, you will be able to:

- Read drawings that are dimensioned in fractional and/or decimal inches, and in metrics.
- Explain the information found on a typical drawing.
- Describe how detail, subassembly, and assembly drawings differ.
- Point out why drawings are numbered.

Many products manufactured today are an assembly of parts supplied by a number of different industries. These industries may be in distant geographic locations, Fig. 2-1.

Fig. 2-1. Thousands of drawings were needed to construct this experimental helicopter. Standards and specifications had to be precise because parts were manufactured in several geographic locations. (Sikorsky Aircraft, United Technologies)

It would not be possible for industry to manufacture a complex product without using drawings. ***Drawings*** show the craftworker what to make and the standards that must be followed so the various parts will fit together properly. The resulting parts will also be interchangeable with similar components on equipment already in service.

Drawings may range from a simple freehand sketch, Fig. 2-2, to many thousands of detailed drawings for a complex job or product, Fig. 2-3.

Symbols, lines, and figures are employed to give drawings meaning, Fig. 2-4. They have been standardized so they mean the same thing wherever drawings are made and used.

These symbols, lines, and figures have been devised by the American National Standards Institute, better known as ANSI. The symbols, lines, and figures on drawings are known as the "LANGUAGE OF INDUSTRY."

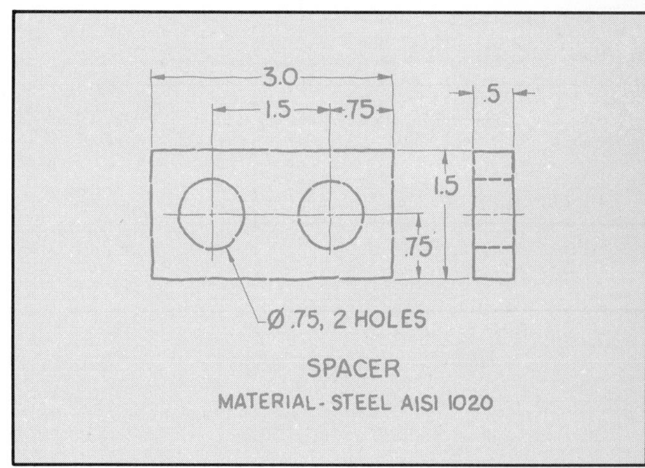

Fig. 2-2. Some drawings are as simple as this freehand sketch.

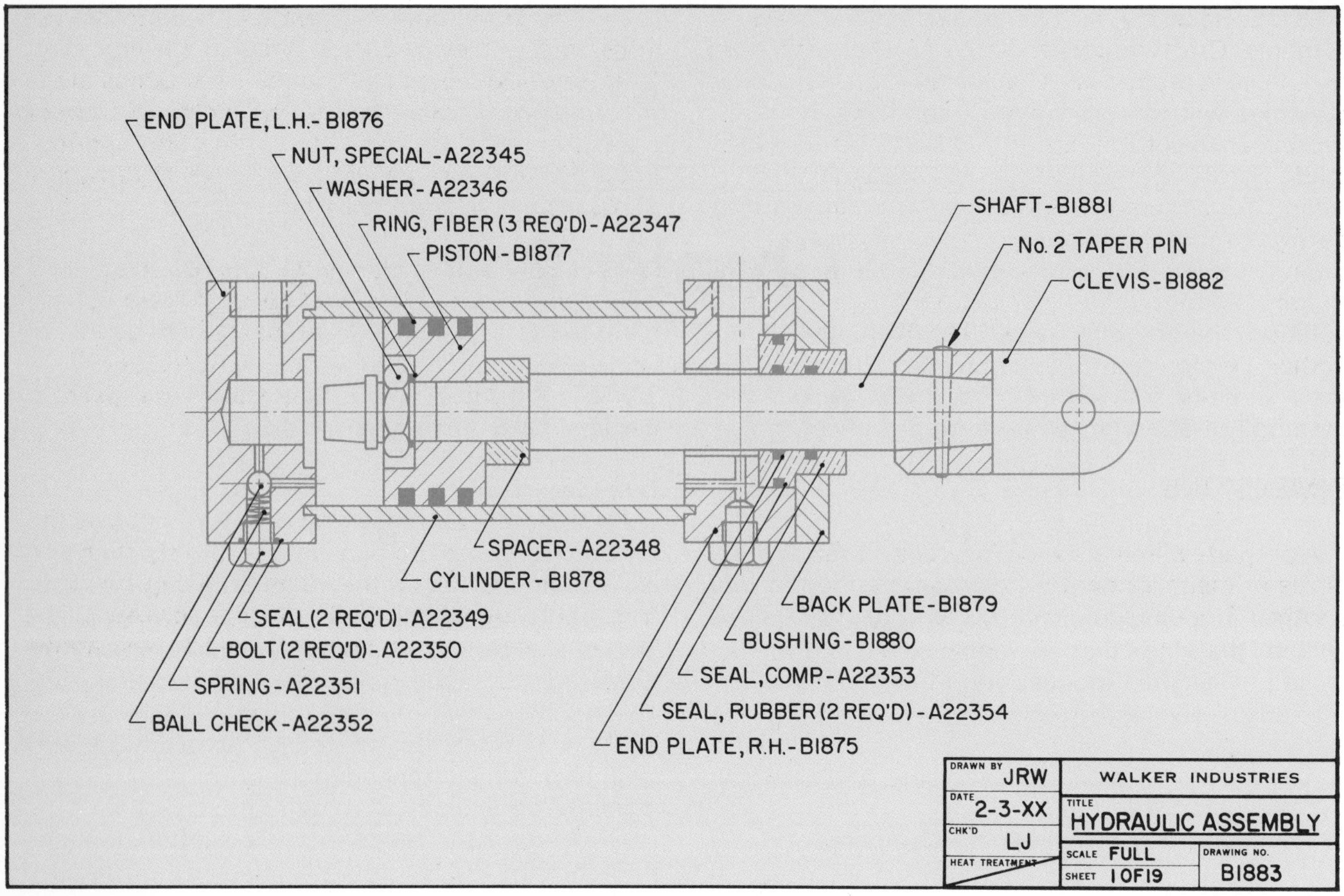

END PLATE, L.H.- B1876
NUT, SPECIAL-A22345
WASHER- A22346
RING, FIBER (3 REQ'D) - A22347
PISTON - B1877
SHAFT - B1881
No. 2 TAPER PIN
CLEVIS - B1882
SPACER - A22348
CYLINDER - B1878
SEAL (2 REQ'D) - A22349
BOLT (2 REQ'D) - A22350
SPRING - A22351
BALL CHECK - A22352
BACK PLATE - B1879
BUSHING - B1880
SEAL, COMP. - A22353
SEAL, RUBBER (2 REQ'D) - A22354
END PLATE, R.H. - B1875

DRAWN BY JRW	WALKER INDUSTRIES	
DATE 2-3-XX	TITLE HYDRAULIC ASSEMBLY	
CHK'D LJ		
HEAT TREATMENT	SCALE FULL	DRAWING NO.
	SHEET 1 OF 19	B1883

Fig. 2-3. Each manufactured product or unit may require dozens of drawings, one for each part, even the smallest screw, washer, or pin.

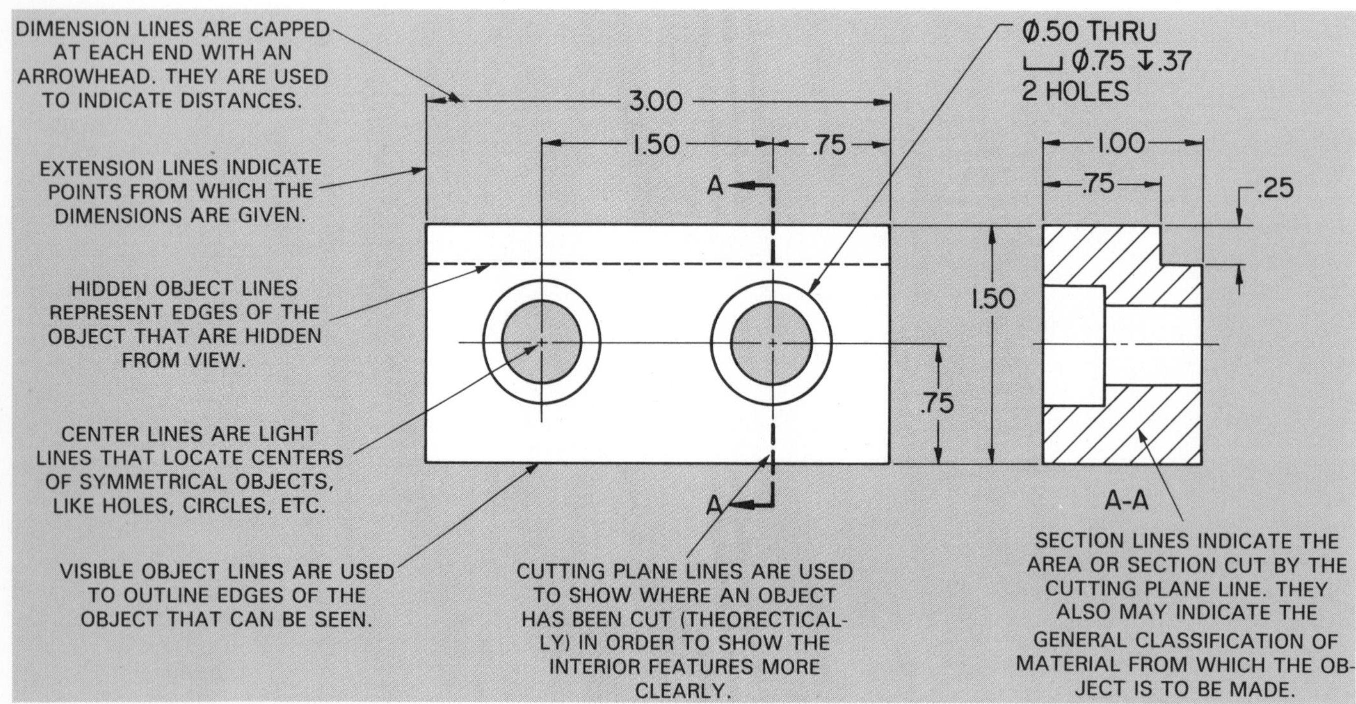

Fig. 2-4. These are a few of the many types of lines, symbols, and figures used to give a drawing exact meaning.

NOTE: Recently, ANSI changed several drawing symbols. Craftworkers must be familiar with past and present practices because only recently made drawings will contain the new standards. It will be too expensive to revise the millions of drawings made before the new standards were devised.

Fig. 2-5 shows past and present practices that pertain to metalworking.

Lines are used to draw views that are necessary to fully describe the object to be manufactured. In addition, the drawing usually includes other information needed to make the product. Details often show threads, for example. Fig. 2-6 shows several methods of showing threads on a drawing.

DIMENSIONS

A properly made drawing includes all ***dimensions*** (sizes or measurements), in proper relation to one another. The dimensions are needed to produce the part or object.

Until recently, drawings were only dimensioned in decimal and/or fractional parts of the inch, Fig. 2-7. However, some of America's industries are in the process of converting to the metric system of measurement. During this transition period, metalworkers will have to understand drawings dimensioned in more than one manner.

Dimensional and/or fractional dimensioning

Drawings using ***fractional dimensioning*** usually show objects that do NOT require a high degree of accuracy in their manufacture. GREATER PRECISION is indicated when dimensions are given in ***decimal*** parts of the inch.

Dual dimensioning

Dual dimensioning is a system that employs the English fraction of an inch and/or decimal inch and metric dimensions on the same drawing, Fig. 2-8. If the drawing is intended primarily for use in the United States, the decimal inch will appear above

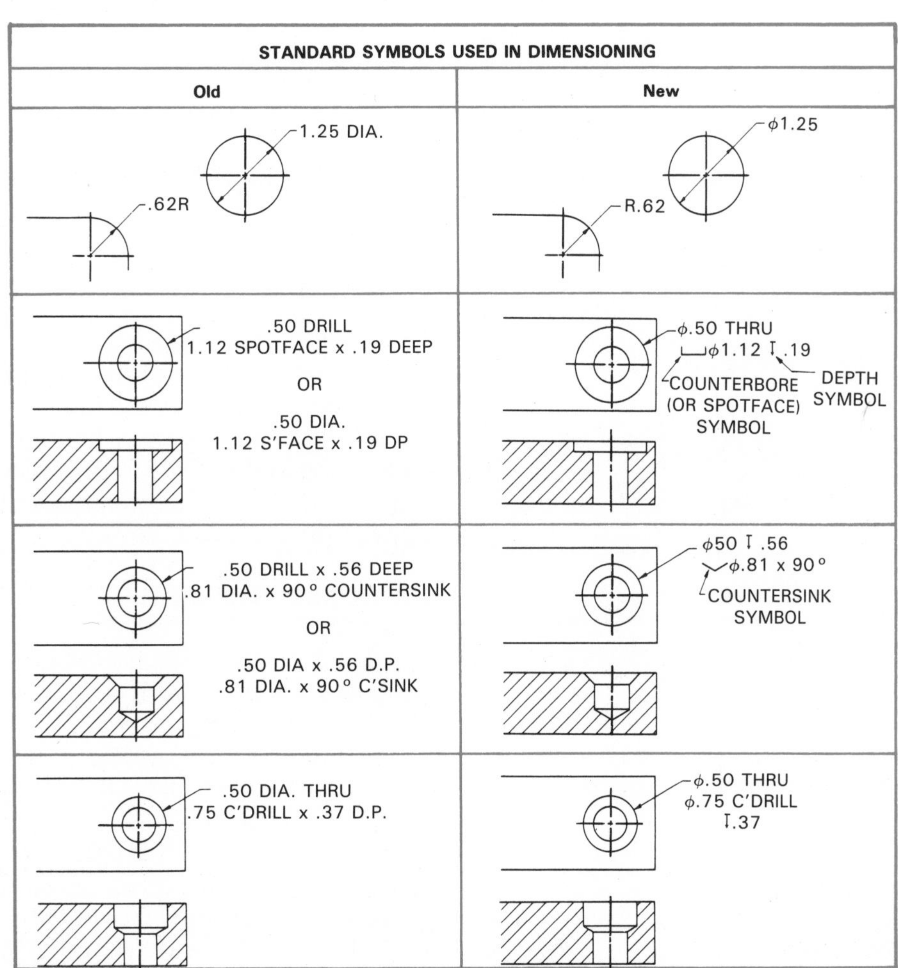

Fig. 2-5. Recently, ANSI recommended certain changes in specifying circles, arcs, and hole sizes. Since it will be many years before the new standards are found on all drawings and prints, both old and new methods are shown above. Study and compare the examples.

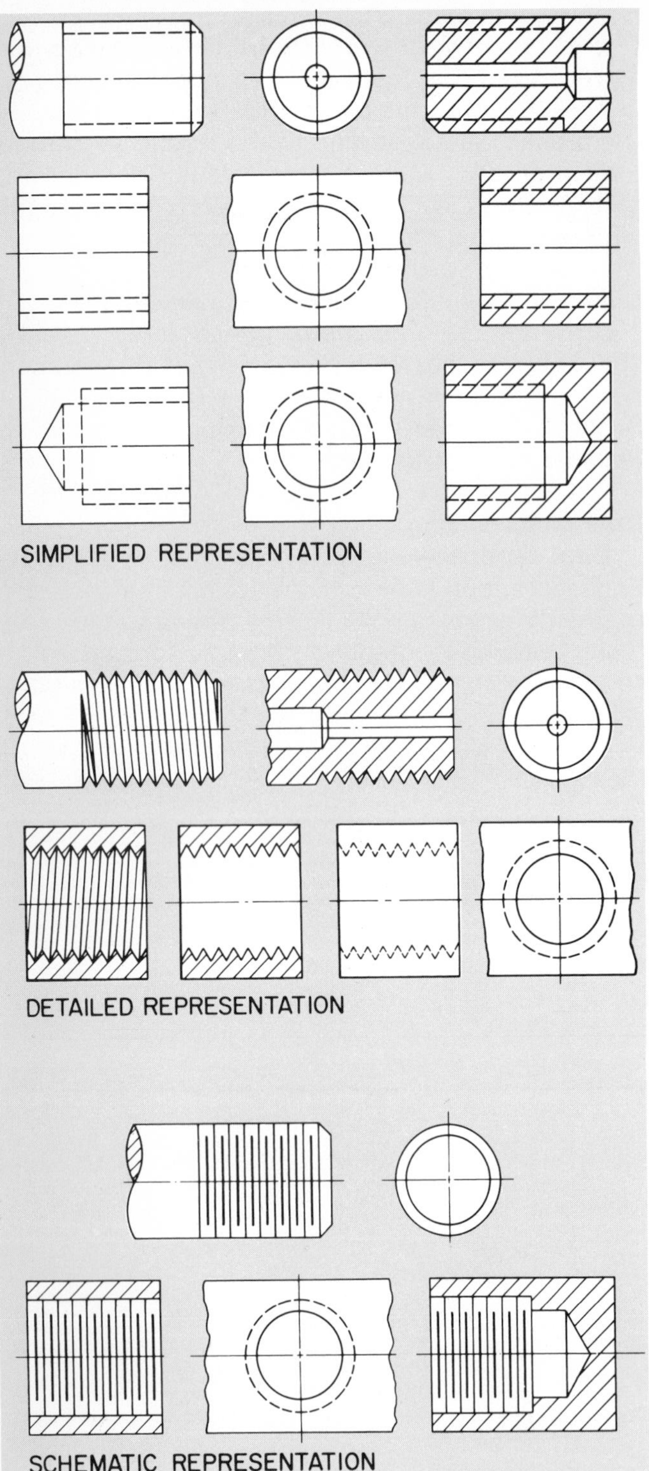

Fig. 2-6. Methods employed to depict threads on drawings. Only one type will be found on a drawing. The simplified version is the most common type.

the metric dimension, as in A, Fig. 2-9. The reverse is true if the drawing is to be used in a metric oriented country, as in B. Some American companies place the metric dimension within brackets, as in C.

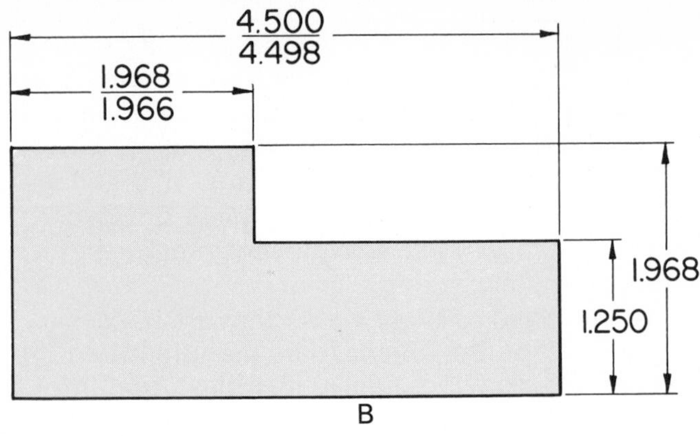

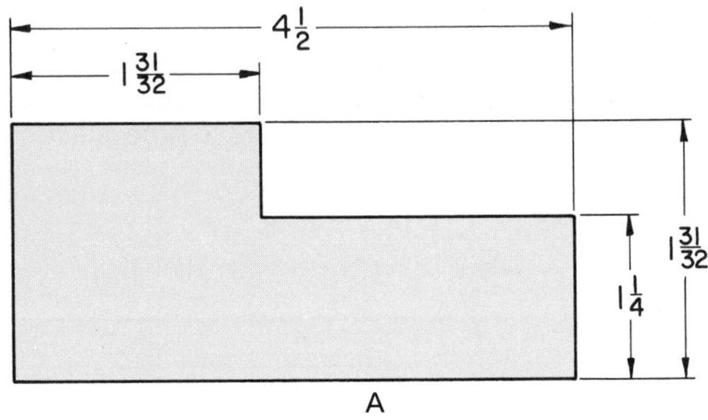

Fig. 2-7. A—Fractional dimensions do not require tolerances closer than ± 1/64 in. B—Decimal dimensions indicate tolerances of ± 0.001 in. unless otherwise specified.

Metric dimensioning

With ***metric dimensioning,*** all of the dimensions on the drawing are in the metric system, usually in millimeters. However, until there is full conversion to metrics in the United States, a ***conversion chart*** (equivalents for millimeter and inch dimensions) will appear on the drawing, Fig. 2-10.

INFORMATION INCLUDED ON DRAWINGS

Drawings will have ***additional information*** to inform the machinist of the type material to be used, surface finish required, tolerances, etc. It is important for you to be familiar with this information.

Materials to be used

The general classification of ***materials*** to be used in the manufacture of an object may be indicated by the TYPE OF SECTION LINE on the drawing or plan, Fig. 2-11. Exact material specification is included in a section of the title block, Fig. 2-12A. Sometimes, it may be found in the ***NOTES*** shown elsewhere on the drawing.

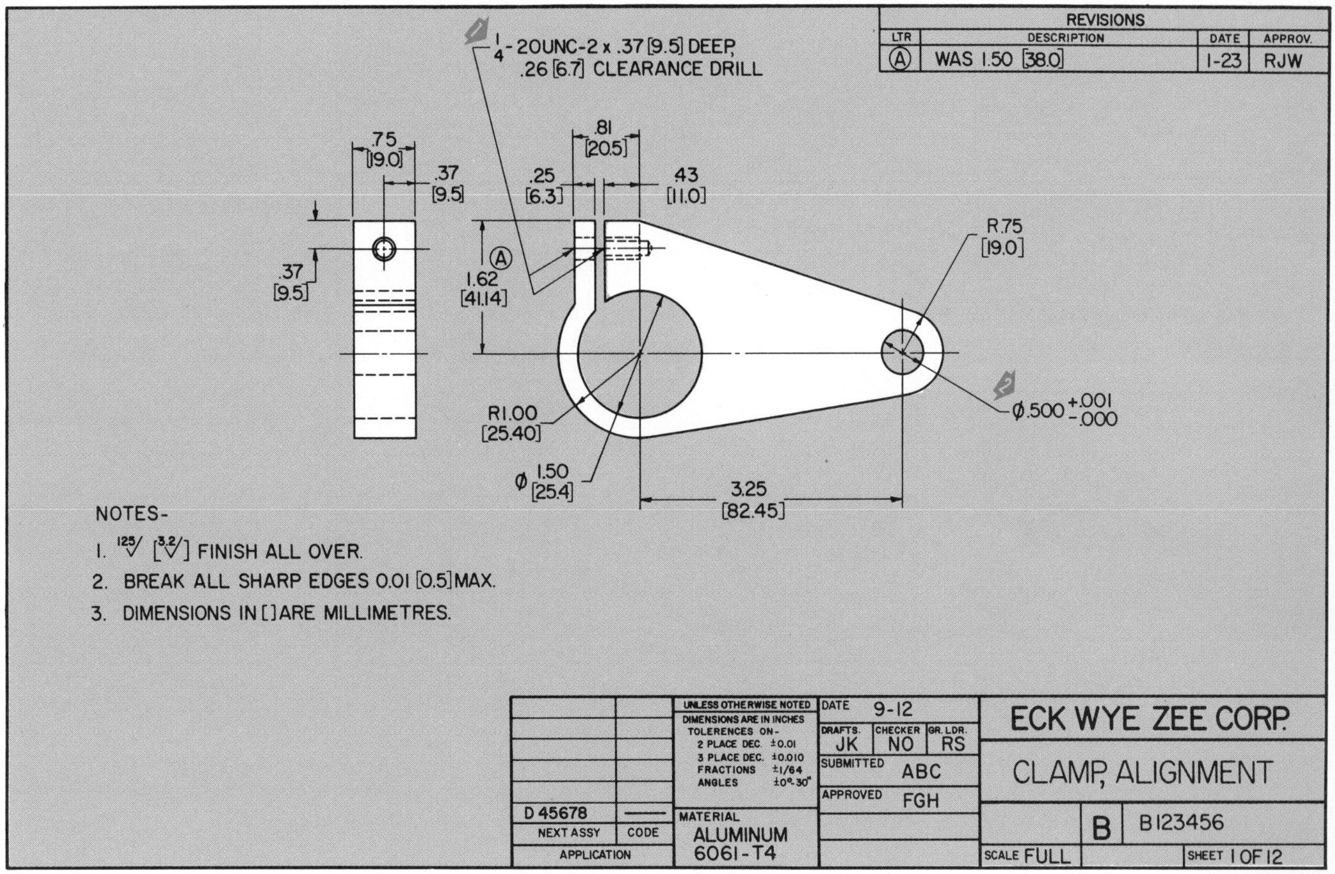

REVISIONS
LTR | DESCRIPTION | DATE | APPROV.
A | WAS 1.50 [38.0] | 1-23 | RJW
1/4-20UNC-2 x .37 [9.5] DEEP, .26 [6.7] CLEARANCE DRILL
.75 [19.0]
.37 [9.5]
.81 [20.5]
.25 [6.3]
.43 [11.0]
R.75 [19.0]
.37 [9.5]
1.62 [41.14]
R1.00 [25.40]
⌀ 1.50 [25.4]
⌀.500 +.001 -.000
3.25 [82.45]
NOTES-
1. 125/ [3.2/] FINISH ALL OVER.
2. BREAK ALL SHARP EDGES 0.01 [0.5] MAX.
3. DIMENSIONS IN [ ] ARE MILLIMETRES.
UNLESS OTHERWISE NOTED DIMENSIONS ARE IN INCHES TOLERENCES ON- 2 PLACE DEC. ±0.01 3 PLACE DEC. ±0.010 FRACTIONS ±1/64 ANGLES ±0°-30'
DATE 9-12
DRAFTS. JK CHECKER NO GR. LDR. RS
SUBMITTED ABC
APPROVED FGH
ECK WYE ZEE CORP.
CLAMP, ALIGNMENT
D 45678
NEXT ASSY | CODE
APPLICATION
MATERIAL ALUMINUM 6061-T4
B B123456
SCALE FULL
SHEET 1 OF 12

Fig. 2-8. Dual dimensioned drawing: 1—Thread size has not been given a metric size. There is no metric thread equal to this size fractional thread. 2—There is no metric size reamer equal to this size.

1.233
31.318
NOTE THAT SHOULD APPEAR ON DRAWING-
INCH
MILLIMETER
or
INCH/MILLIMETER
A
31.318
1.233
NOTE THAT SHOULD APPEAR ON DRAWING-
MILLIMETER
INCH
or
MILLIMETER/INCH
B
1.233 [31.318]
NOTE THAT SHOULD APPEAR ON DRAWING-
DIMENSIONS IN [ ] ARE MILLIMETERS
C

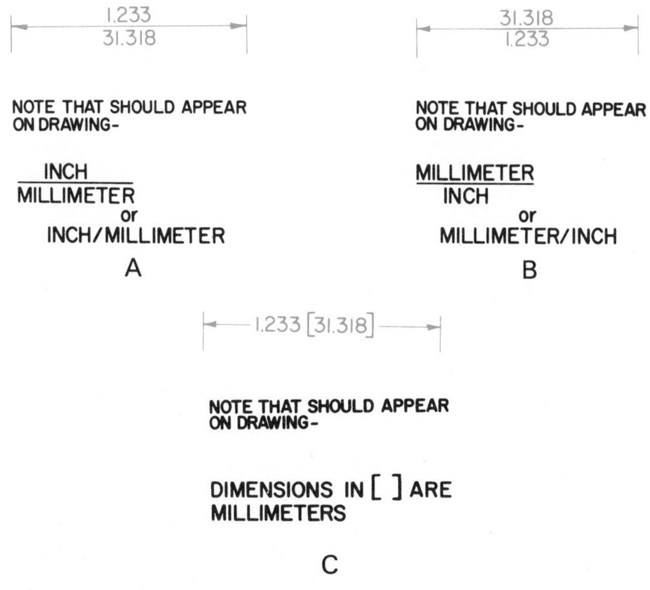

Fig. 2-9. Study how inch and metric dimensions are indicated on a dual dimensioned drawing. A—When drawing is to be used in the United States. B—When drawing is to be used primarily in a metric country and the United States. C—Brackets are sometimes used to indicate the metric equivalent on a drawing to be used in the United States.

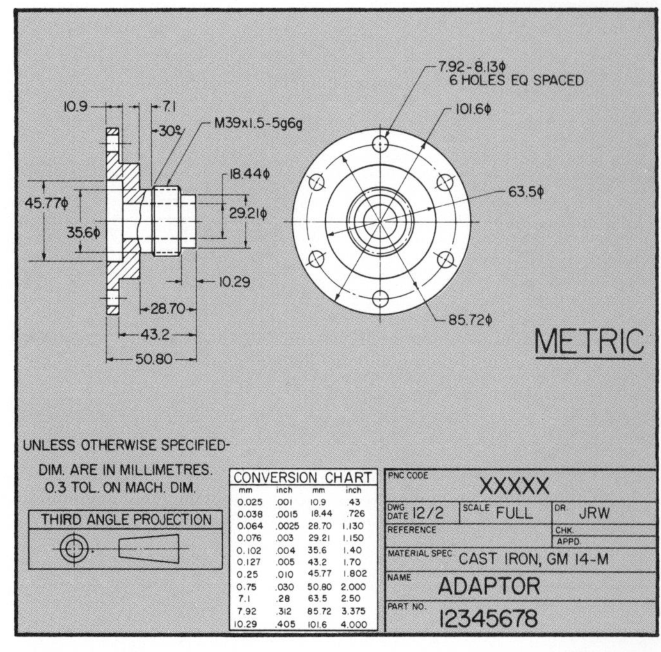

10.9
7.1
30°
M39x1.5-5g6g
7.92-8.13⌀ 6 HOLES EQ SPACED
101.6⌀
18.44⌀
63.5⌀
45.77⌀
35.6⌀
29.21⌀
10.29
28.70
43.2
50.80
85.72⌀
METRIC
UNLESS OTHERWISE SPECIFIED-
DIM. ARE IN MILLIMETRES.
0.3 TOL. ON MACH. DIM.
THIRD ANGLE PROJECTION
CONVERSION CHART
mm inch mm inch
0.025 .001 10.9 .43
0.038 .0015 18.44 .726
0.064 .0025 28.70 1.130
0.076 .003 29.21 1.150
0.102 .004 35.6 1.40
0.127 .005 43.2 1.70
0.25 .010 45.77 1.802
0.75 .030 50.80 2.000
7.1 .28 63.5 2.50
7.92 .312 85.72 3.375
10.29 .405 101.6 4.000
PNC CODE XXXXX
DWG DATE 12/2 SCALE FULL DR. JRW
REFERENCE
CHK.
APPD.
MATERIAL SPEC. CAST IRON, GM 14-M
NAME ADAPTOR
PART NO. 12345678

Fig. 2-10. A drawing utilizing SI or metric units of measurement. Until it is no longer considered necessary or when the United States has fully converted to metric measurement, a conversion chart will usually appear on the drawing.

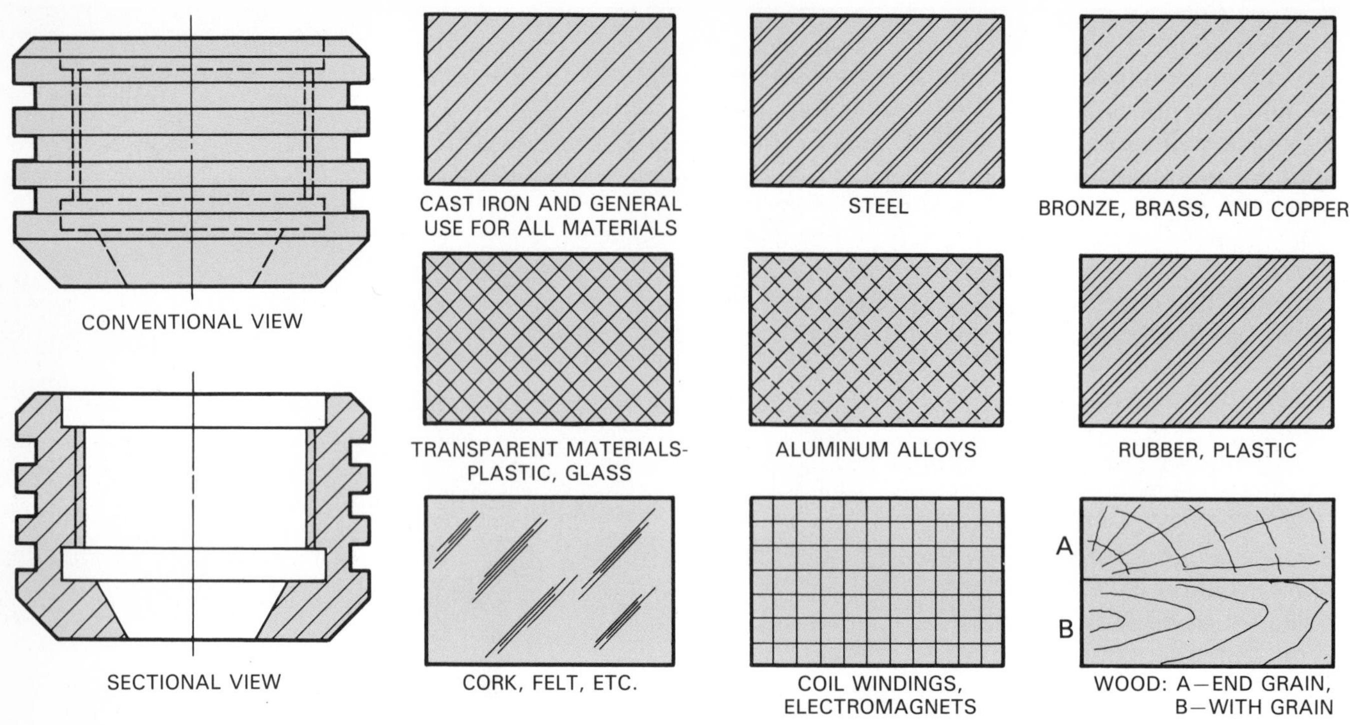

Fig. 2-11. Sectional views make a drawing easier to understand because internal details are shown more clearly. Section blocks illustrate a few materials and how they are shown in section. However, many drawings showing views in section use general section lining that is similar to cast iron section lining.

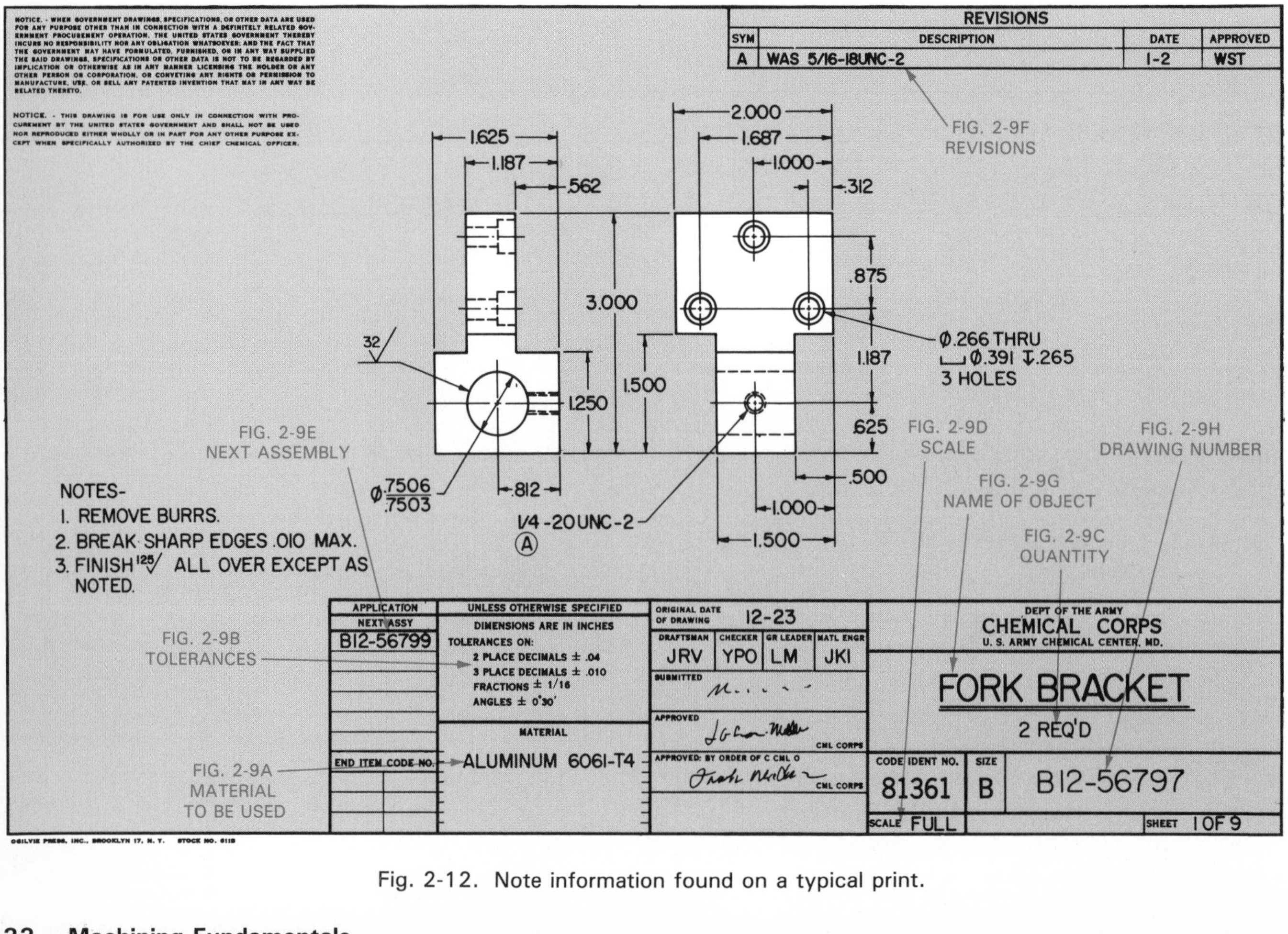

Fig. 2-12. Note information found on a typical print.

Surface finishes required

The quality of the ***surface finish*** (degree of surface smoothness) is important in the manufacture of many products. The smoothness of the bore of an engine cylinder is an example. Usually, the more superior or smoother the finish of a machined surface, the more expensive it is to manufacture.

In the past, symbols were employed to indicate that a surface was to be machined, Fig. 2-13. They may still be found on some older drawings. With so many machining techniques now in use, symbols such as these do NOT indicate, in sufficient detail, the quality of the surface finish required on a part.

The method presently used provides more complete surface information. Shown in Fig. 2-14, a CHECK MARK and NUMBER are utilized to indicate surface roughness in ***microinches*** (0.000001 in.) or ***micrometer*** (0.000 001 m). A microinch is one-millionth of an inch and may be abbreviated μ in. A micrometer is one-millionth of a meter and is abbreviated μ m.

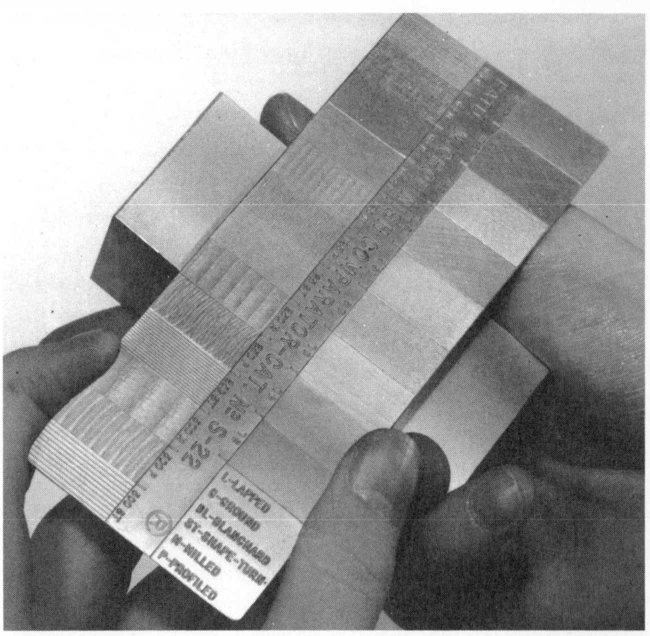

Fig. 2-15. Surface Roughness Comparison Standards being used to check whether the milled surface meets the required specifications.

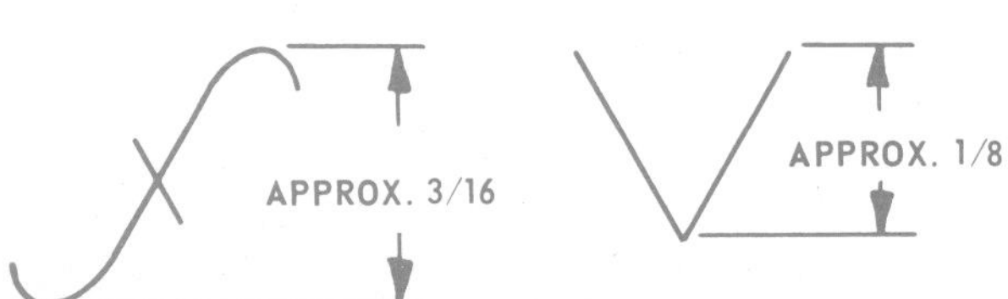

Fig. 2-13. These are old style finish marks. They do not indicate the degree of smoothness required other than that the surface is machined. Finish marks of these types are still found on older drawings.

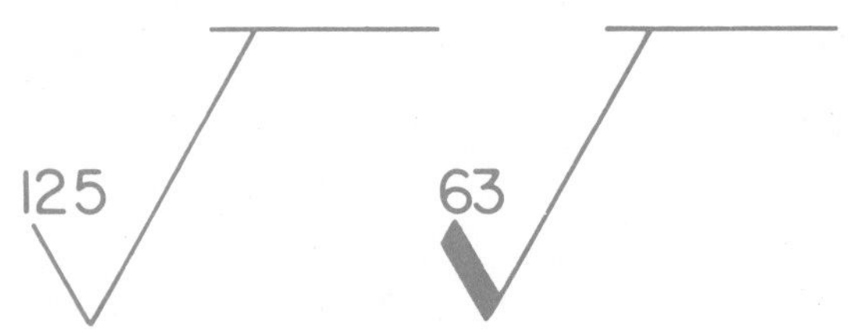

Fig. 2-14. These are finish marks now used. The number indicates the degree of smoothness in microinches—the larger the number, the rougher the finish.

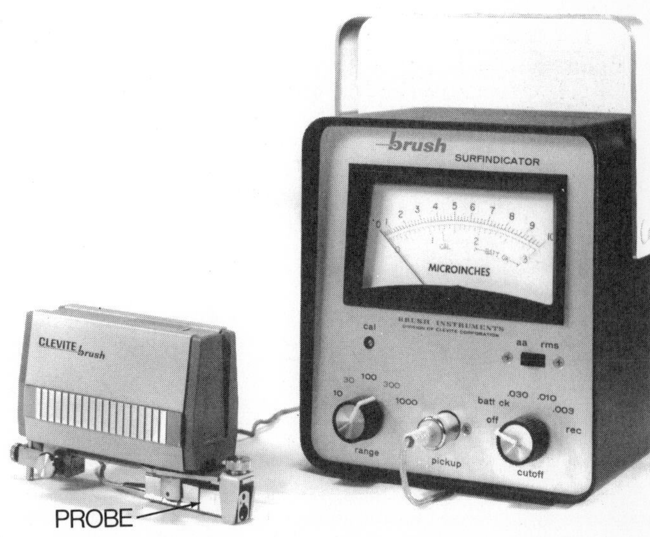

Fig. 2-16. Surface roughness is best determined with a profilimeter. The probe on the unit is moved across the work surface and measures it electronically. A direct reading of surface roughness is given on the dial or on digital readout. (Clevite Corp.)

A machinist can compare some surface finishes to required specifications by employing a ***surface roughness comparison standard*** as a guide, Fig. 2-15. If the surface finish is critical, as it is in some jet engine components, the surface finish is measured electronically with a device called a ***profilimeter,*** Fig. 2-16.

Tolerances

The control of dimensions to achieve interchangeable manufacturing is known as ***tolerancing.*** It controls the size of the features of a part.

NOTE: Because of the nature of this text, this

chapter will contain only information on general tolerancing. To guarantee interchangeability of parts, industry is adopting a system of ***geometrical dimensioning and tolerancing.*** However, it is necessary to completely understand general tolerancing before you can be expected to work with geometrical dimensioning and tolerancing.

Tolerances are allowances, either oversize or undersize, permitted when machining or making a part. See Fig. 2-12B. Acceptable tolerances may be shown on drawings in several different ways.

When the dimension is given in inches and fractions of an inch, unless otherwise indicated, the permissible tolerances can be assumed to be ± 1/64 in. The symbol ± means that the machined surface can be ***plus*** (larger) or ***minus*** (smaller) by 1/64 in. and still be acceptable. When the tolerance is plus AND minus (both directions), it is called a ***bilateral tolerance.***

If it is permissible to machine the part larger, but not smaller, the dimension on the drawing might read $2\ 1/2\ ^{+\ 1/64}$. If only a minus tolerance is permitted, the dimension might read $2\ 1/2\ ^{-\ 1/64}$. When the tolerance is plus OR minus (one direction), it is called a ***unilateral dimension.***

Dimensions shown as inches and decimals usually indicate that the work must be machined more precisely. Two methods of showning these tolerances are in general use. Under normal conditions, unless otherwise indicated, the tolerances can be assumed to be ± 0.001 in.

A PLUS tolerance could be shown as:

$$2.500\ ^{+.001} \text{ or } \frac{2.501}{2.500}.$$

A MINUS tolerance could be shown as:

$$2.500\ ^{-.001} \text{ or } \frac{2.500}{2.499}.$$

The dimensions show that the part can be used as long as the machined dimensions measure within these limits.

Metric tolerances are presented in the same way as decimal tolerances, Fig. 2-17.

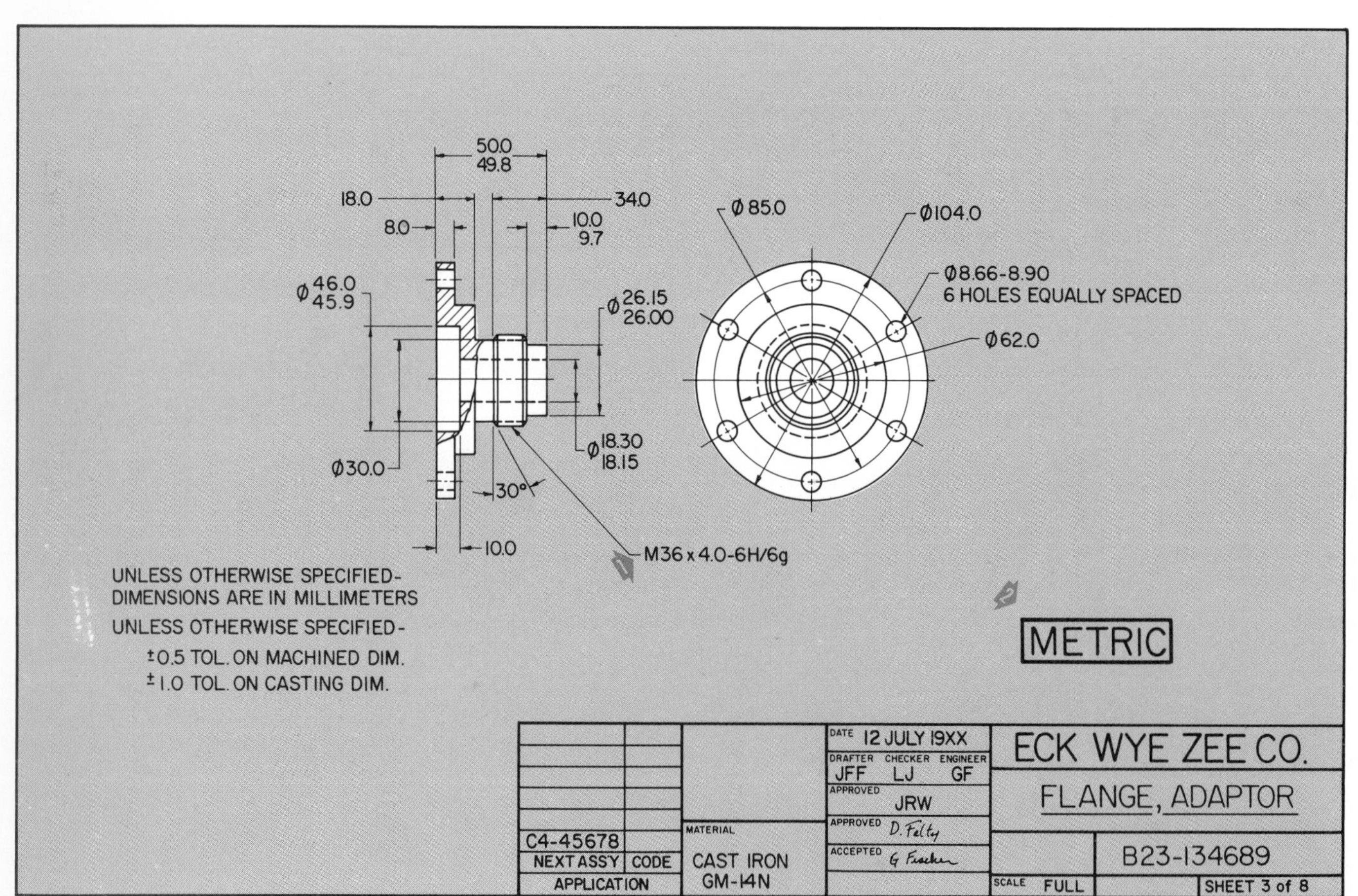

Fig. 2-17. This is a metric dimensioned detail drawing. 1—Note that metric thread specifications are different from the more familiar UNC (coarse) and UNF (fine) series threads. The letter "M" denotes the thread system symbol for standard metric screw threads. The 36 indicates the nominal thread diameter in millimeters. The 4.0 denotes thread pitch in millimeters. The 6H and 6g are tolerance class designations. 2—To avoid possible misunderstanding, metric is shown on the drawing in large letters.

Quantity of units

Also shown on the drawing is the ***number of units*** (number of parts) needed in each assembly, Fig. 2-12C. A work order, included with the job information received by the shop, will give the total number of units to be manufactured. This facilitates ordering the necessary materials, and will help in determining the most economical way to manufacture the pieces.

Scale of drawing

Drawings made other than actual size (1:1) are called ***scale drawings.*** The scale is usually shown in a section of the title block, Fig. 12-12D. A drawing made one-half size would be shown by the figure 1:2 (one to two). A figure of 2:1 (two to one) would mean that the drawing is twice the actual size of the part.

Assembly or subassembly

Assembly or ***subassembly*** information is necessary to correctly fit the various parts together. The term ***application*** is sometimes used in place of term NEXT ASSEMBLY, Fig. 2-12E.

Revisions

Revisions indicate what changes were made on the original drawing and when they were made on the drawing, Fig. 2-12F.

Name of the object to be made

A portion of the title block provides this information. It tells the machinist the correct name of the piece, Fig. 2-12G.

TYPES OF PRINTS

Seldom are the original drawings used in the shop because they might be lost, damaged, or destroyed. On many jobs, several sets of plans are required.

There are several methods of duplicating the original drawings:

1. ***Blueprints.*** The term "blueprint" is often employed to refer to all types of prints. However, the process is seldom used today because of the time required to make the print. A real ***blueprint*** had white lines on a blue background.
2. ***Diazo process.*** These are direct ***positive copies*** (dark lines on a white background) of the original drawing. They are often referred to as ***whiteprints.***
3. ***Xerography (electrostatic) process.*** Pronounced ze-rog'-ra'-fee, the process makes an exact copy of the original drawing. The print can be enlarged or reduced in size if necessary. Full color copies can be made on some xerographic machines.
4. ***Microfilm process.*** This is a technique where the original drawing is reduced by photographic means. Finished negatives can be stored in roll form, or on cards adaptable to computerized storage and retrieval. To produce a working print, the microfilm image is retrieved from files and enlarged, called ***blow-backs,*** on photographic paper. The print is discarded or destroyed when it is no longer needed. Microfilms can also be viewed on a ***reader*** (machine for making enlarged projection on display screen), Fig. 2-18.

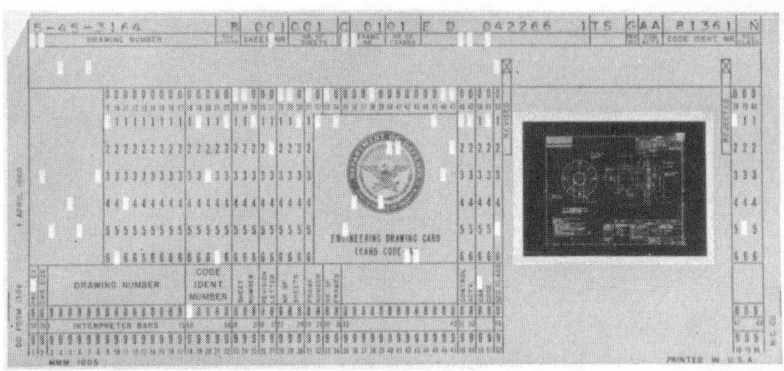

A

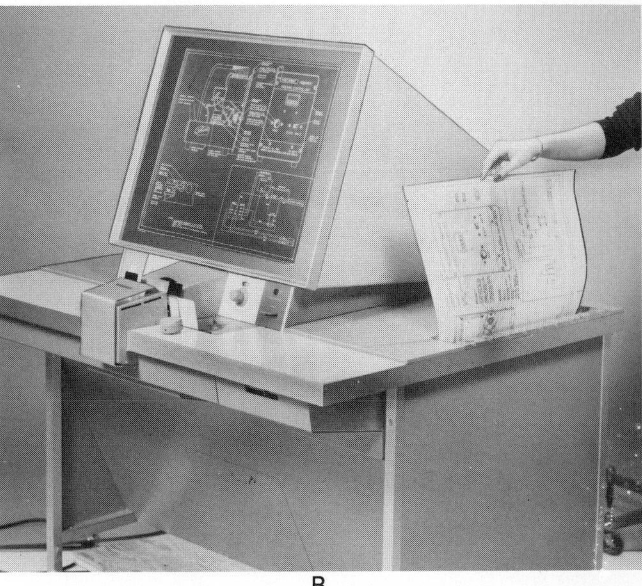

B

Fig. 2-18. A—Small negative on the microfilm aperture card is enlarged by a photographic process to the desired print size on a microfilm reader/printer. The enlarged print can be verified or confirmed on the view screen. B—A microfilm reader/printer. (RECORDAK)

5. ***Computer generated.*** Prints called ***hardcopy,*** are generated on a ***plotter*** (automatic drafting machine) from information stored electronically in computer memory, Fig. 2-19. This same

Fig. 2-19. Left—Many companies now use Computer Aided Design and Drafting (CAD or CADD) techniques to prepare detail drawings. (Lockheed-California) Right—Prints, called hardcopy, are generated on a plotter from CAD developed information. (Hewlett-Packard Marketing Communications)

information can also be used to control machine tools using ***CAM*** (computer aided manufacturing). When used, the overall manufacturing technique is called CIM (computer integrated manufacturing). The prints are generated by a process called ***CAD*** (computer aided drafting.).

More information on the computer in manufacturing will be included in later chapters.

TYPES OF DRAWINGS USED IN THE SHOP

The ***working drawings,*** also called ***prints*** tell the craftworker what to make and establish the standards for the product. There are two major kinds of working drawings:

1. The ***detail drawing*** includes a drawing of the part, usually multiview, with dimensions and other information for making the part, Fig. 2-20.
2. The ***assembly drawing*** shows where and how the parts, described on detail drawing, fit into the completed assembly, Fig. 2-21.

On large or complex products, ***subassembly drawings,*** are used to show the assembly of a small portion of the completed object, Fig. 2-22.

Some assembly and subassembly drawings are shown as ***exploded pictorial drawings*** (drawing with parts separated but in proper relationship). One is shown in Fig. 2-23.

The detail drawing, in most instances, provides information on only one item. However, if the mechanism is small in size or if it is composed of only a few parts, the detail and assembly drawings may appear on the same sheet, Fig. 2-24.

PARTS LIST

Parts are identified by ***circled numbers*** and/or by giving them on a title note. Some drawings will also include a ***parts list*** or ***bill of materials*** that summarize what is given on all of the drawings. See Fig. 2-25 and Fig. 2-26.

DRAWING SIZES

Most firms centralize the preparation and storage of drawings in the engineering department. Generally, engineers and drafters prepare their drawings on standard size sheets, thereby simplifying stocking, handling, and storage of the completed drawings.

Standard sizes for drawing sheets include:

ENGLISH MEASURE SHEET SIZES
A size = 8 1/2 x 11 in.
B size = 11 x 17 in.
C size = 17 x 22 in.
D size = 22 x 34 in.
E size = 34 x 44 in.

METRIC MEASURE SHEET SIZES
A4 size = 210.0 x 297.0 mm
A3 size = 297.0 x 420.0 mm
A2 size = 420.0 x 594.0 mm
A1 size = 594.0 x 841.0 mm
A0 size = 841.0 x 1 189.0 mm

Also, for convenience in filing and locating drawings in storage, industry gives each plate or drawing an ***identifying number,*** Fig. 2-12H.

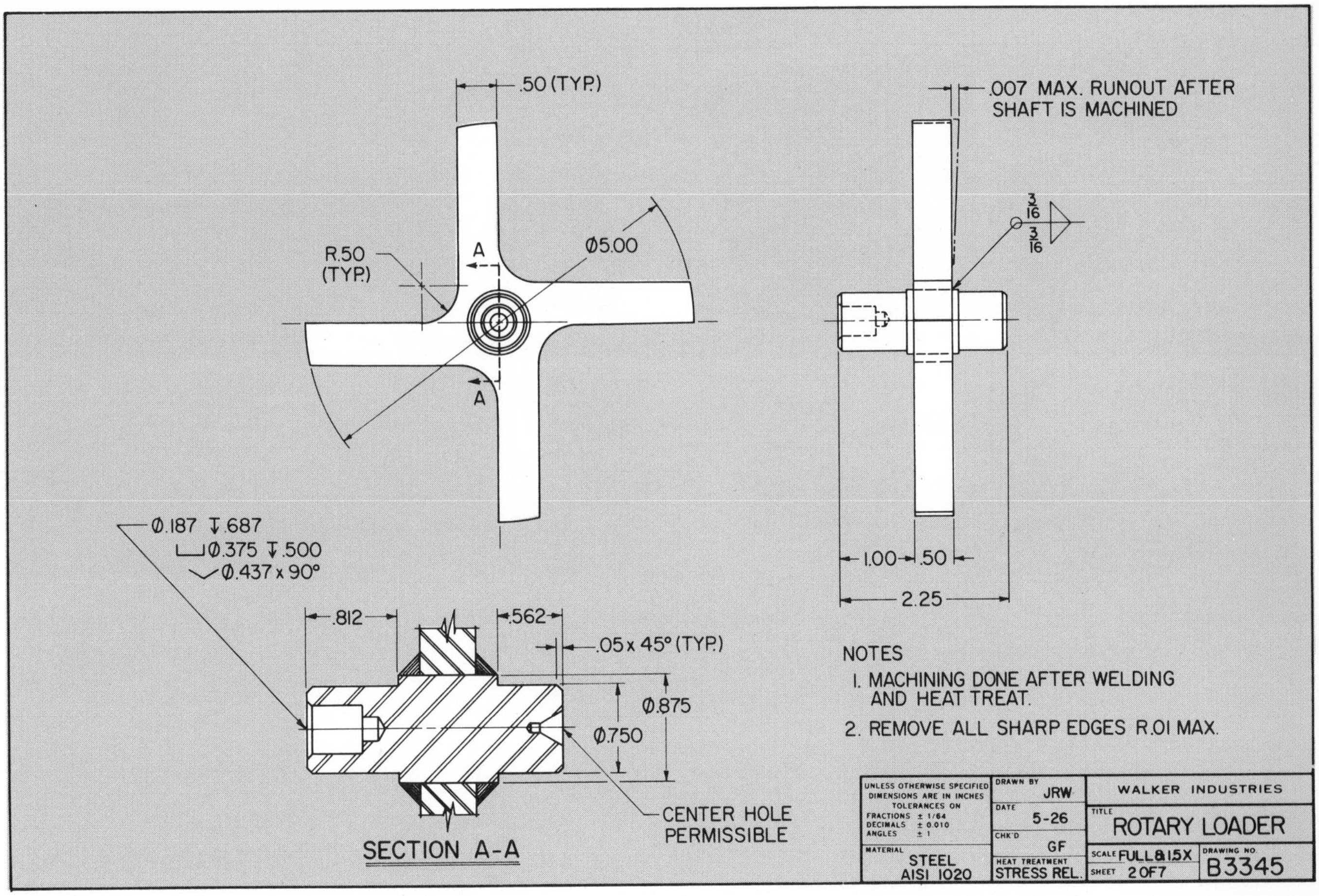

Fig. 2-20. Detail drawing contains all of the information needed to produce the part.

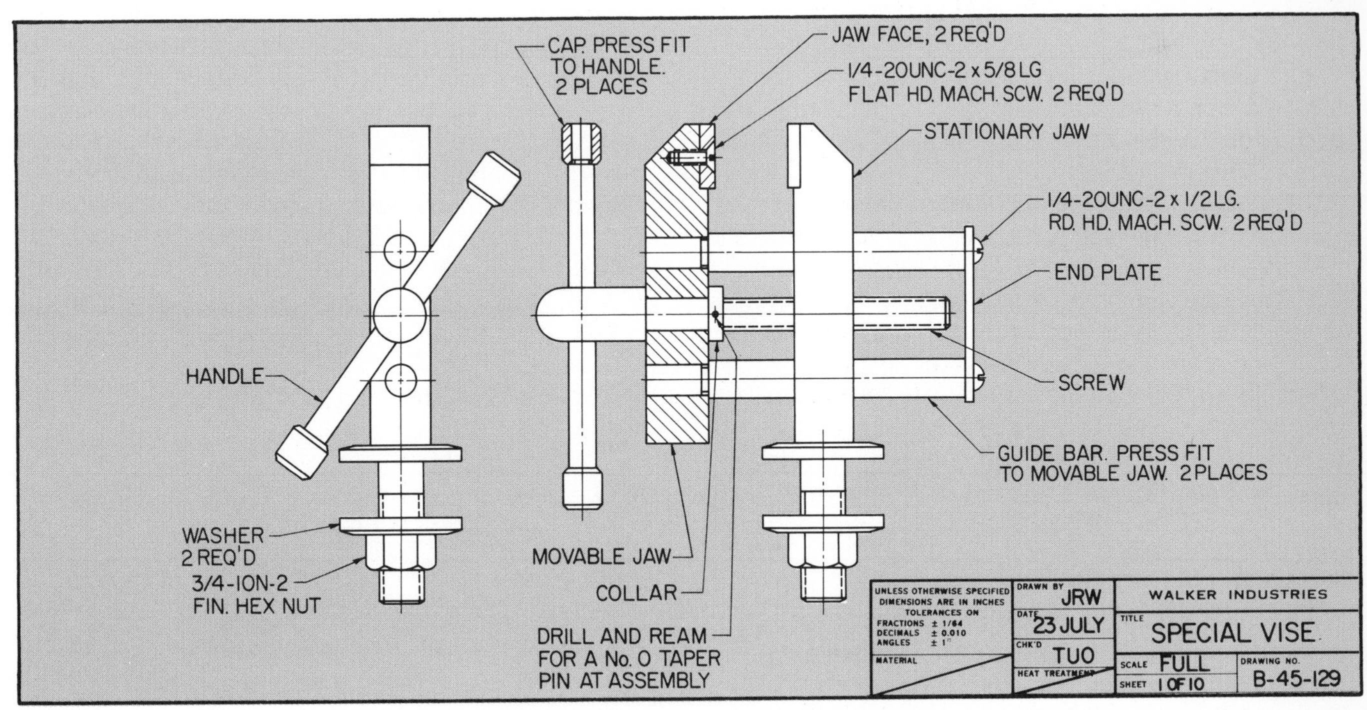

Fig. 2-21. This is an assembly drawing.

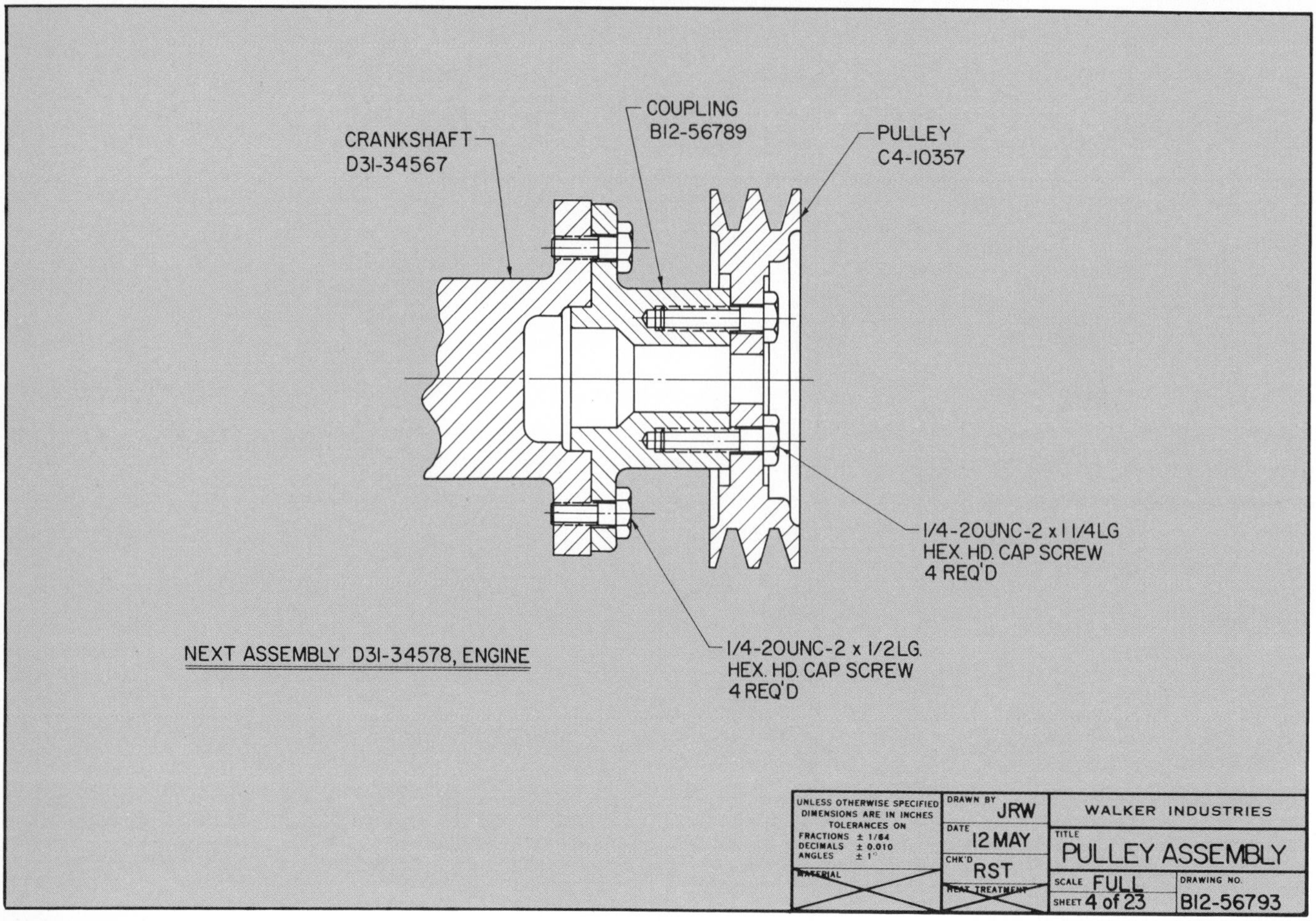

CRANKSHAFT
D31-34567
COUPLING
B12-56789
PULLEY
C4-10357
1/4-20UNC-2 x 1 1/4LG
HEX. HD. CAP SCREW
4 REQ'D
NEXT ASSEMBLY D31-34578, ENGINE
1/4-20UNC-2 x 1/2LG.
HEX. HD. CAP SCREW
4 REQ'D
UNLESS OTHERWISE SPECIFIED DIMENSIONS ARE IN INCHES TOLERANCES ON FRACTIONS ± 1/64 DECIMALS ± 0.010 ANGLES ± 1°
MATERIAL
DRAWN BY JRW
DATE 12 MAY
CHK'D RST
HEAT TREATMENT
WALKER INDUSTRIES
TITLE PULLEY ASSEMBLY
SCALE FULL
SHEET 4 of 23
DRAWING NO. B12-56793

Fig. 2-22. Subassembly drawing contains the assembly of only a portion of the entire product.

5 2
B
1
2
4
6
3
SUPPORT ASM
1 575510 - SPRING
2 585792 - FLAT WASHER
3 9422299 - NUT
4 395586 - PIN
5 9422277 - NUT
6 1363524 - BUMPER
7 395578 - ARM ASM
8 391350 - PAD
A 7
8
EXISTING PART
B LUBRICATE WITH 9985038 GREASE
A INSTALL WITH IDENTIFICATION (LETTERS) FACING OUTWARD TOWARD LEFT HAND SIDE OF CAR
1 35 - 45 LB-FT
2 16 - 26 LB-FT

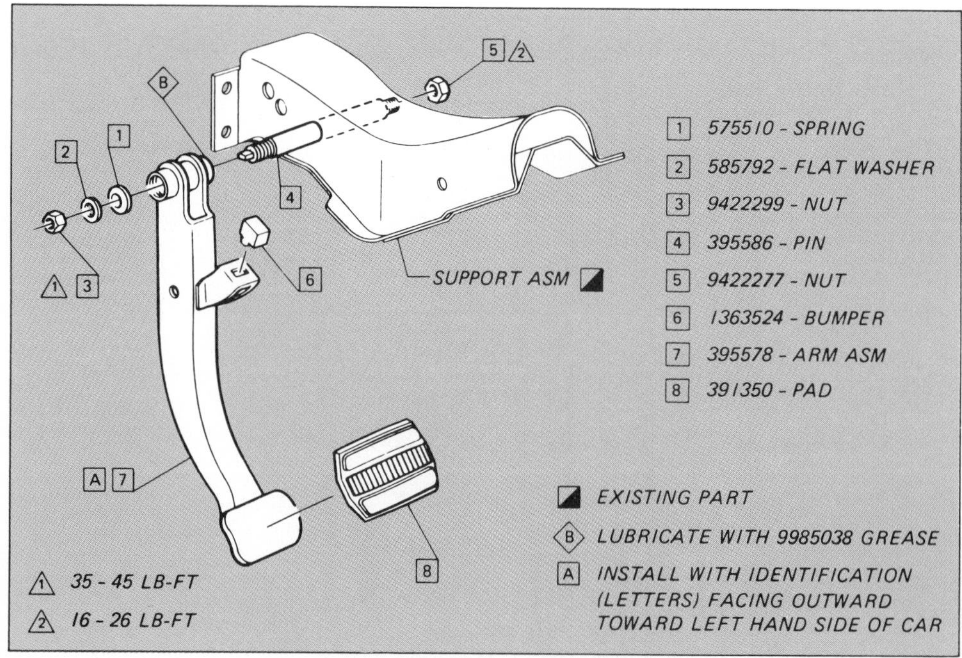

Fig. 2-23. Pictorial type subassembly drawing is often used with semiskilled workers who have received a minimum of training in print reading. (General Motors Corp.)

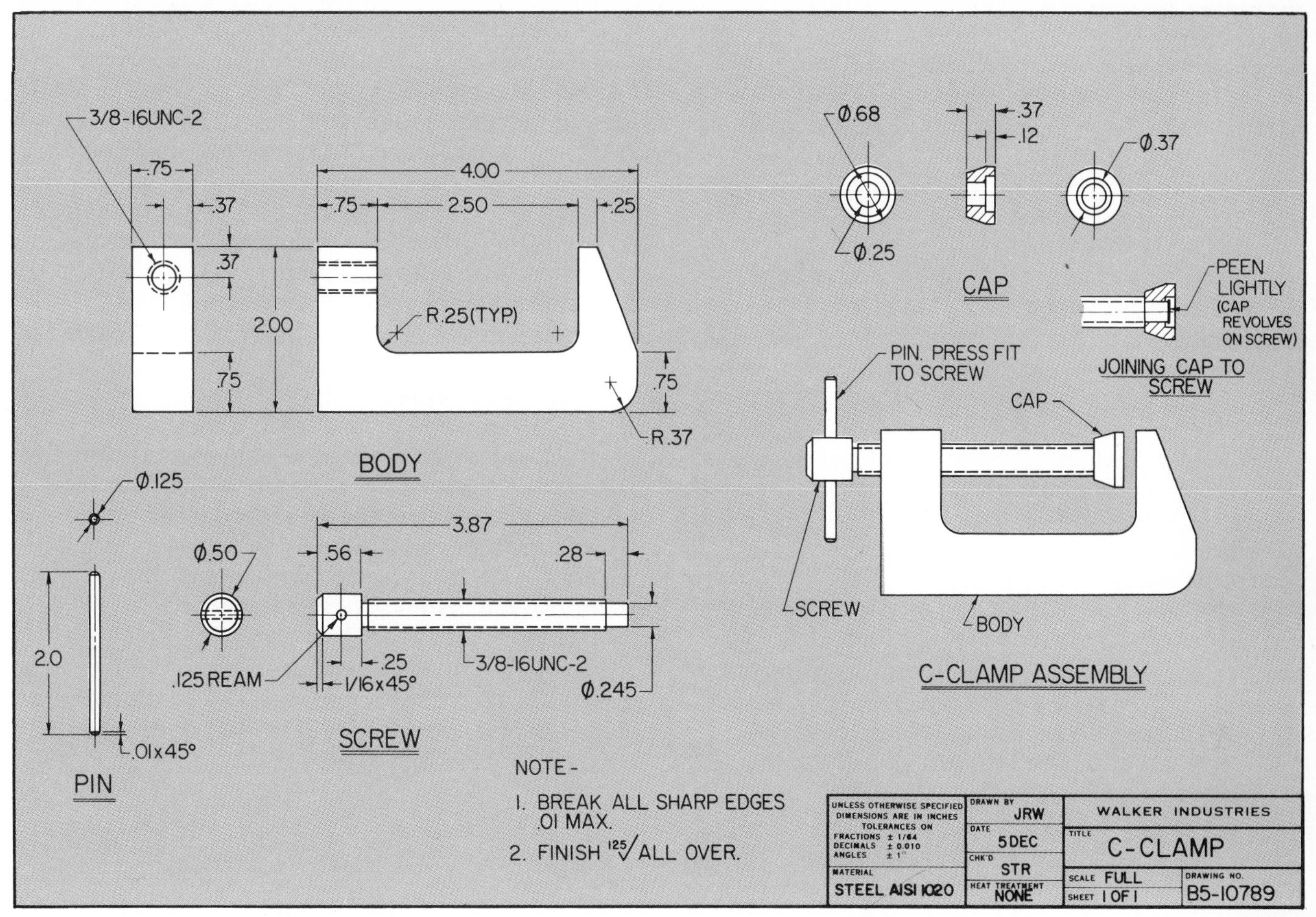

Fig. 2-24. Note detail and assembly drawing on same sheet.

PARTS LIST		
No.	*Name*	*Quan.*
1	CRANKCASE	1
2	CRANKSHAFT	1
3	CRANKCASE COVER	1
4	CYLINDER	2
5	PISTON	2

Fig. 2-25. This is a typical, but partial, parts list.

<table>
<tr><td>A1776</td><td>NUT</td><td>BRASS</td><td>6</td></tr>
<tr><td>A1985</td><td>BOLT</td><td>BRASS</td><td>6</td></tr>
<tr><td>B1765</td><td>PLATE</td><td>ALUMINUM</td><td>2</td></tr>
<tr><td>B1767</td><td>CYLINDER</td><td>CAST IRON</td><td>2</td></tr>
<tr><td>Pt. No.</td><td>Name</td><td>Material</td><td>Quan.</td></tr>
<tr><td colspan="4">BILL OF MATERIALS</td></tr>
</table>

Fig. 2-26. Note example of a partial bill of materials.

TEST YOUR KNOWLEDGE—Chapter 2

Please do not write in the text. Place your answers on a separate sheet of paper.

1. Drawings are used to:
   a. Show, in multiview, what an object looks like before it is made.
   b. Standardize parts.
   c. Show what to make and the sizes to make it.
   d. All of the above.
   e. None of the above.
2. The symbols, lines, and figures that make up a drawing are frequently called the ________ ________ ________.
3. A microinch is ________ of an inch.
4. A micrometer is ________ of a meter.
5. How can surface roughness of a machined part be checked against specifications on the drawing? How can it be measured electronically?
6. When tolerances are plus AND minus, it is called a ________ tolerance.
7. When tolerances are plus OR minus, it is called a ________ tolerance.

8. Tolerances are:
   a. The different materials that can be used.
   b. Allowances in either oversize or undersize that a part can be made and still be acceptable.
   c. Dimensions.
   d. All of the above.
   e. None of the above.
9. Drawings made other than actual size are called ________ ________.
10. A subassembly drawing differs from an assembly drawing by:
   a. Showing only a small portion of the complete object.
   b. Making it possible to use smaller drawings.
   c. Showing the object without all needed dimensions.
   d. All of the above.
   e. None of the above.
11. Why are prints used in place of the original drawings?
12. The craftworker is given all of the information needed to make a part on a ________ drawing.
13. What does an assembly drawing show?
14. Why are standard size drawing sheets used?

RESEARCH AND DEVELOPMENT

1. Make a tracing and reproduce it by the diazo and electrostatic processes.
2. Prepare a display utilizing the microfilming technique of print reproduction. Include prints, samples of film cards and photographs or magazine advertisements illustrating the equipment used to make them.
3. Secure sample prints from industry.
4. Secure prints produced by the CAD (Computer Aided Design) technique.
5. Prepare a display panel that shows a simple project, from print to finished product.
6. Prepare projectuals (transparencies) for the overhead projector that show the title block, parts list, and material list from an actual industrial drawing. Use them to explain or describe an industrial drawing to the class. If possible, borrow a sample of the part shown on the drawing.
7. Contact a local industry and borrow prints of a simple assembly. If possible, also secure a sample of the object shown on the print. Develop a display around them.

Chapter 3

SHOP SAFETY

After studying this chapter, you will be able to:

- Give reasons why shop safety is important.
- Explain why it is important to develop safe work habits.
- Recognize and correct unsafe work practices.
- Apply safe work practices when employed in a machine shop.

Shop safety is NOT something to be studied at the start of a training program and then forgotten; most accidents are caused by carelessness or by breaking safety rules. Remember this when your instructor insists on safe work practices. If you are deligent and follow instructions with care, machining operations can be safe and enjoyable. Safe work practices should become a force of habit!

Since it is NOT possible to include every safety precaution, the safety practices in this chapter are general. Safety precautions for specific tools and machines are described in the text where they apply, along with the description and operation of the equipment. Refer to Fig. 3-1.

Study all safety rules carefully and constantly apply them. When in doubt about any task, get help! DO NOT take chances!

SAFETY FOR THE SHOP

Keep the ***shop clean.*** Metal scraps should be placed in the scrap bin. Never allow them to remain on the bench or floor.

Exercise extreme care when machining unfamiliar materials. For example, ***magnesium chips*** burn with great intensity under certain conditions. Applying water to the burning magnesium chips only intensifies the fire. Machining equipment can be damaged beyond repair and *very* serious burns can result.

Inhaling fumes or dust from some of the newer space age and exotic materials can cause serious ***respiratory ailments.*** DO NOT machine materials until you KNOW WHAT IT IS and how it should be handled.

An approved type respirator and special protective clothing must be worn when machining some materials. Machines must be fitted with effective vacuum systems as needed.

Fig. 3-1. The pilot of this Navy aircraft wouldn't think of taking off even though the aircraft was flown earlier in the day, until all of its systems were in safe operating condition. You should do the same before operating any type of equipment. (Grumman Aerospace Corp.)

The shop is a place to work, not play. It is NOT a place for ''horseplay.'' A ''joker'' in a machine shop is a ''walking hazard'' to everyone. Daydreaming also increases your chances of injury.

If you have been ill and are on ***medication,*** check with your doctor or school clinic to determine whether it is safe for you to operate machinery. For example, many cold remedies recommend that you do NOT OPERATE MACHINERY while taking the medication because of possible drowiness.

Avoid using ***compressed air*** to remove chips and cutting oil from machines. The flying chips may cause serious eye injuries. Vaporized oil may also ignite and result in painful burns and property damage.

Oily rags must be placed in approved ***safety containers*** (metal can with metal lid). Rags or waste used to clean machines become imbedded with metal slivers. Be sure to place them where they will NOT be used again and dispose of them daily. This will minimize the possibility of ***spontaneous combustion*** (ignition by rapid oxidation or burning of oil without heat from an external source).

Keep ***hand tools*** in good condition and store them in such a way that people cannot be injured when a tool is taken from the tool panel or storage rack.

Care must be taken when handling long sections of metal stock. One example, accidently contacting a light fixture may cause severe electrical burns and even ***electrocution*** (electric current passes through body tissues). An electric shock is like being ''hit by a truck.''

Secure help when moving heavy machine accessories or large pieces of metal stock. ***Back injuries*** are usually long term injuries!

Dress properly. Avoid wearing loose fitting sweaters or clothing. A snug fitting shop coat or apron can be worn to protect your street clothes, Fig. 3-2. Rings and other jewelry should be removed. Keep sleeves rolled up. Remove your tie if you wear one. If clothing or jewelry gets caught in machinery, severe injuries or even death can result.

Have ***adequate ventilation*** for jobs where dust and fumes are a hazard. Return oils and solvents to proper storage. Wipe up spilled oil and solvent right away.

Wear appropriate ***safety gear,*** Fig. 3-3. In noisy areas, use a hearing protector. Disposable plastic gloves will protect your hands when handling oils, cutting fluids, and solvents. Wear a dust mask when machining produces airborne

Fig. 3-2. This student is properly dressed for the job he is doing. He is wearing approved eye protection, and a snug fitting apron. The machine has been carefully checked out and lubricated before it was operated.

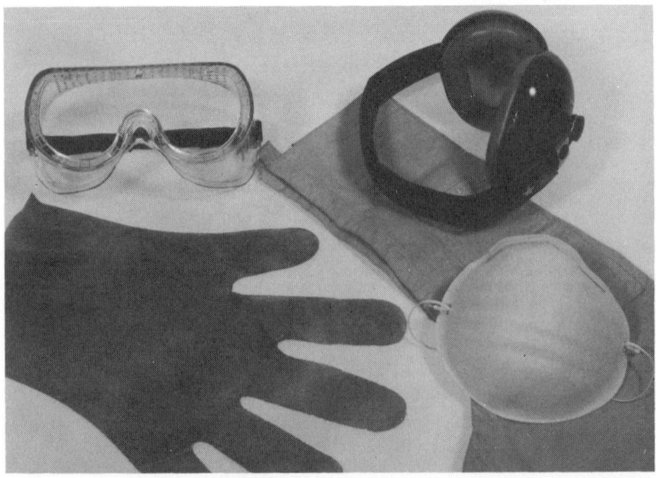

Fig. 3-3. Wear appropriate safety gear. Shown are approved eye protection, apron to protect clothing, plastic gloves for handling oils and solvents, hearing protector, and dust mask.

particles (machining sand castings, plastics, some grinding operations, etc.).

Always ***protect your eyes.*** Eyesight that has been damaged or destroyed cannot be replaced. Wear OSHA (Occupational Safety and Health Administration) approved safety glasses, goggles, or face shields. It is good practice to have your own personal safety glasses. The cost is reasonable. Your instructor can help you determine the style best suited for your needs.

If you wear glasses, special safety lenses are available that can be ground to your prescription. Your eye doctor or optician can help get them.

Take no chances! Wear eye protection whenever you are in the shop. Special eye protection should be used when conditions call for them, Fig. 3-4.

Know your job! It is foolish, and often disastrous, to operate machines without first receiving proper instructions. Get additional help if you are NOT sure what must be done or how a task should be performed.

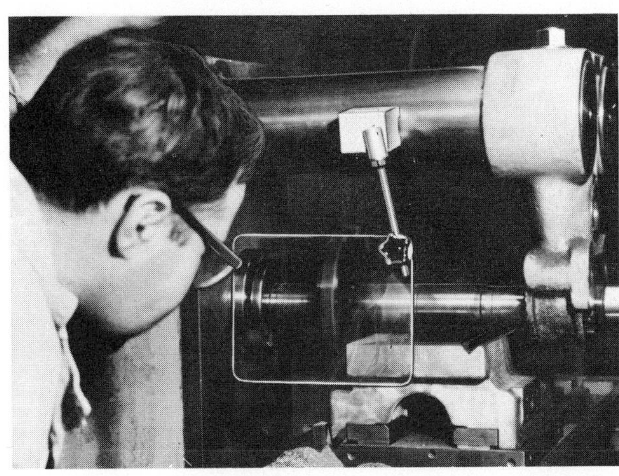

Fig. 3-4. Protection from hot chips is provided by the installation of a portable magnetic safety shield. (Enco Mfg. Co.)

GENERAL MACHINE SAFETY

1. Avoid operating a machine until all guards are in place!
2. Stop your machine to make adjustments or measurements! Resist the urge to touch a surface that has been machined while the machine is running. Severe lacerations can result.
3. Keep the floor around your machine clear of oil, chips, and metal scrap.
4. It is NOT considered good practice to talk to anyone while you are operating a machine. You might become distracted and injure yourself, or someone else.
5. NEVER attempt to remove chips or cuttings with your hands or while the machine is operating. Use a brush, Fig. 3-5. Pliers are one of the safest ways to remove long, stringy chips from a lathe. Better still, learn how to grind the cutting tool to break chips off shorter. This is explained later in the text.
6. Secure prompt medical attention for any cut, bruise, scratch, burn or other injury. No matter how minor the injury may appear, report it to your instructor!

GENERAL TOOL SAFETY

1. Never carry sharp pointed tools in your pockets! When using these tools, lay them on the bench in such a way that you will NOT injure yourself when you reach for them, Fig. 3-6.
2. Make sure tools are properly sharpened, are in good condition, and fitted with suitable handles.

Fig. 3-5. Use a brush to remove accumulated chips—NOT your hands!

IN CASE OF FIRE!

Know what to do in case of a fire! Be familiar with the location of fire exits and how they are opened. Be aware of alternate escape routes.

In some situations, students are trained in the use of fire extinguishers by the local fire department. If you are one of these students, know where the fire extinguishers are located. If you have NOT received such training, get out of the fire area immediately.

SPECIAL SAFETY NOTE: ***Think before acting!*** It costs nothing and you may be saved from painful injury that could result in a permanent disability. If you think it tiring to sit through a one hour class in school, think what it would be like to spend your entire life in a wheelchair!

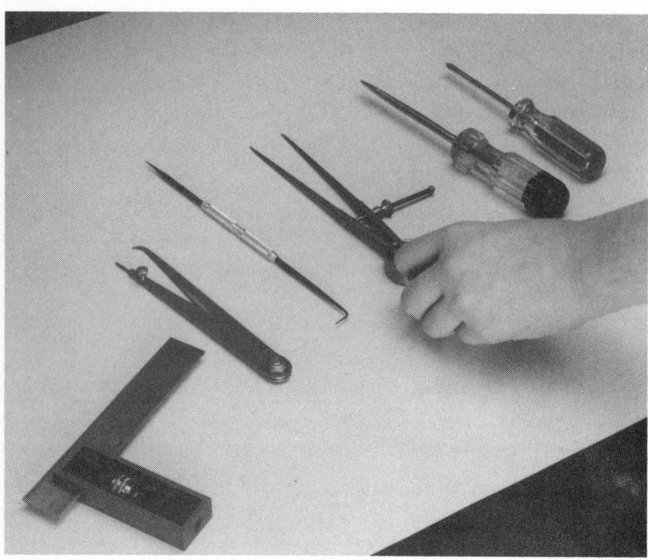

Fig. 3-6. Place sharp pointed tools on bench in such a manner that they will not injure you when reaching to pick them up.

TEST YOUR KNOWLEDGE—Chapter 3

Please do not write in this text. Place your answers on a separate sheet of paper.

1. Why is shop safety so important?
2. Most shop accidents are caused by _______.
3. Safety glasses should be worn:
   a. Most of the time.
   b. Only when working on machines.
   c. The entire time you are in the shop.
   d. None of the above.
4. Oily rags should be placed in a safety container to prevent _______ _______.
5. Why should compressed air NOT be used to clean chips from machine tools?
6. Never attempt to operate a machine until _______________.
7. Always stop machine tools before making _______ and _______.
8. Use a _______ to remove chips and shavings, NOT your _______.
9. When working in an area contaminated with dust or solvent fumes, be sure there is _______ _______. A _______ _______ should also be worn when working in a dusty area.
10. Secure prompt _______ for any cut, bruise, scratch, or burn.
11. Get help when moving _______________.

RESEARCH AND DEVELOPMENT

1. Ask your fire department to demonstrate the proper use of various types of fire extinguishers and what to do in case of a fire in the shop.
2. Invite a safety expert from a local industry or a supplier of safety equipment to evalute your present shop safety program and make recommendations, if necessary, to improve it.
3. Work with school officials to see what protective eye wear is available for each student. Prepare a table of estimated needs, cost per year, and cost in succeeding years.
4. Develop and produce a series of colorful safety posters for the shop.
5. Prepare visual aids on safe work habits to be observed when using hand and machine tools.
6. Secure samples of safety posters and other safety program features used by local industry.
7. Design and construct a bulletin board on eye safety.
8. Contact the NATIONAL SOCIETY FOR THE PREVENTION OF BLINDNESS, 1790 Broadway, New York, New York 10019, for information on the Wise Owl Club of America, sponsored by that organization.
9. Show a film or TV program on eye safety.
10. Produce a series of 35 mm slides or a TV program on safe work habits to be observed when using the lathe, drill press, grinder, or vertical milling machine.

Chapter 4

MEASUREMENT

After studying this chapter, you will be able to:

- Measure to 1/64 in. (0.5 mm) with a steel rule.
- Measure to 0.0001 in. (0.002 mm) using a Vernier micrometer caliper.
- Measure to 0.001 in. (0.02 mm) using Vernier measuring tools.
- Measure angles to 0°5′ using a universal Vernier bevel.
- Identify and use the various types of gages found in a machine shop.
- Use a dial indicator.
- Employ the various helper measuring tools found in a machine shop.

Without some form of accurate measurement, modern industry could not exist. The science that deals with systems of measurement is called ***metrology***.

Today, industry makes measurements accurately to one-millionth (0.000001) of the inch. This is known as a ***microinch***. One-millionth of a meter (0.000001) is called a ***micrometer*** and is pronounced mi-kro-me-ter.

If a microinch were as thick as a dime, an inch would be as high as FOUR Empire State Buildings. An engineer, with tongue in cheek, once estimated that a steel railroad rail, supported on both ends, would sag one-millionth of an inch when a "fat horsefly" landed on it in the middle.

In addition to using English units of measure (inch, foot, etc.), American industry is gradually converting to metric units of measure (millimeter, meter, etc.). To this end, the United States has adopted the modern form of the metric system known as the ***International System of Units*** (abbreviated SI).

All of the familiar measuring tools are available in metric units, Fig. 4-1. The metric (millimeter) rule is compared with the conventional fractional and decimal rule in Fig. 4-2. Metric-based measuring tools will offer no problems. As a matter of fact, many think they are easier to read than inch-based measuring tools.

Regardless of how fine industry can measure, the job at hand is to first learn to read a rule to 1/64 inch and 0.5 millimeter (mm). Then you can pro-

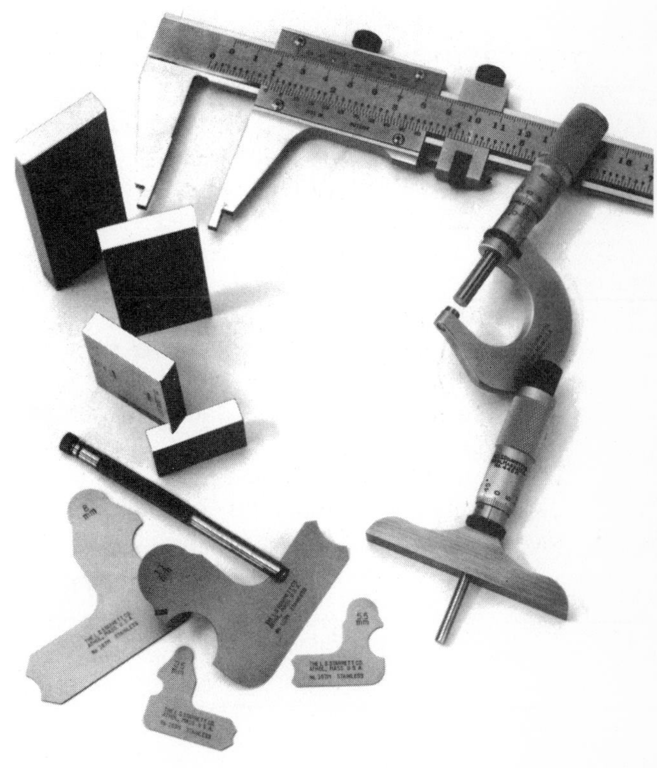

Fig. 4-1. These are but a few of the many measuring tools available in both inch and metric base systems. You will have to know how to read these and many others if you work in industry. (L.S. Starrett Co.)

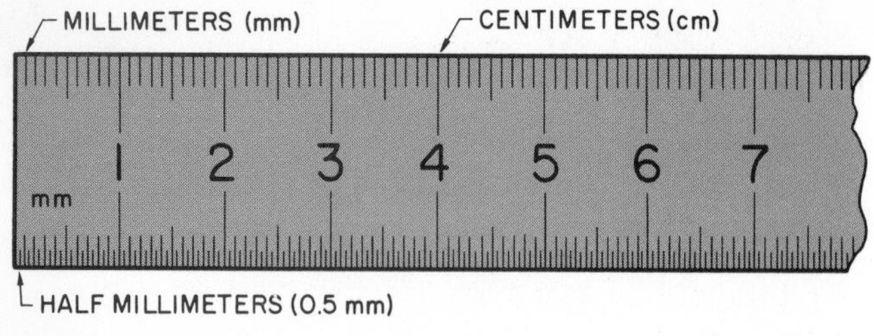

METRIC

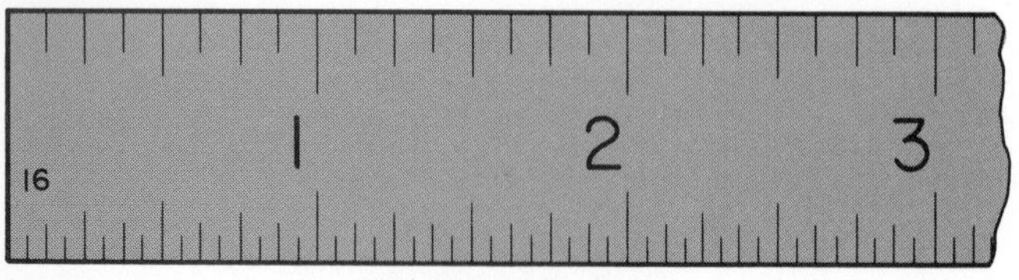

FRACTIONAL INCH

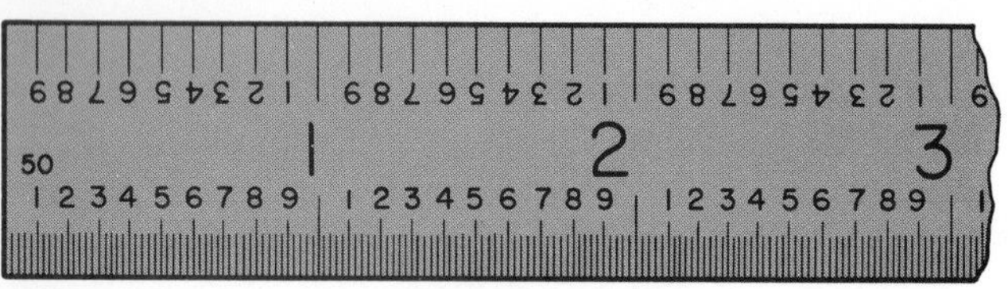

DECIMAL INCH

Fig. 4-2. Compare metric or millimeter graduated rule with the more familiar fractional and decimal rules.

gress through 1/1000 (0.001) in. and 1/100 (0.01) mm by micrometer and Vernier type measuring tools. Finally, you can progress to 1/10,000 (0.0001) in. and 1/500 or 2/1000 (0.002) mm by the Vernier scale on some micrometers.

THE RULE

The ***steel rule,*** often incorrectly referred to as a scale, is the simplest of the measuring tools found in the shop. See Fig. 4-2 for the three basic types of rule graduations. A few of the many rule styles are shown in Figs. 4-3 through 4-12. Study them closely.

Reading the rule (English units)

A careful study of the enlarged rule section will show the different fractional divisions of the inch from 1/8 to 1/64 in., Fig. 4-13. The lines representing the divisions are called ***graduations.*** On many rules, every fourth graduation is numbered on the 1/32 edge, and every eighth graduation on the 1/64 edge.

The best way to learn to read the rule is to:

1. Become thoroughly familiar with making 1/8 and 1/16 measurements.
2. Do the same with the 1/32 and 1/64 measurements.
3. Practice until you become proficient enough to read the measurements accurately and quickly.

Some steel rules (inch-based) are graduated in 10ths, 20ths, 50ths, and 100ths. Additional practice will be necessary to read these rules accurately and quickly.

Fractional measurements are ALWAYS reduced to the lowest terms. A measurement of 14/16 is 7/8; 2/8 is 1/4; etc.

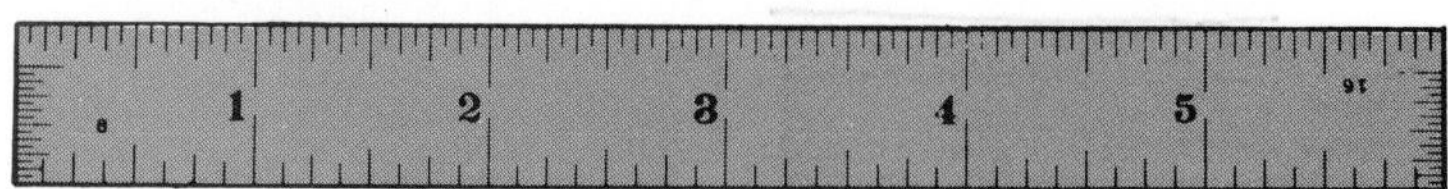

Fig. 4-3. This is a 6 in. steel pocket rule.

Fig. 4-4. Narrow rule can be for measuring in narrow slots, grooves, etc.

Fig. 4-5. This is a flexible steel rule.

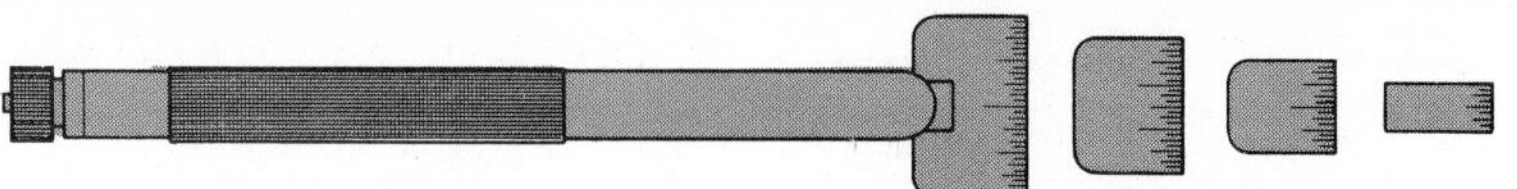

Fig. 4-6. Note narrow rules with holder. The four small rule sections are interchangeable in the holder, and can be set at various angles. Rules are graduated in 1/32 and 1/64 in. divisions. A metric version with the rules graduated in 1.0 and 0.5 mm divisions is also available.

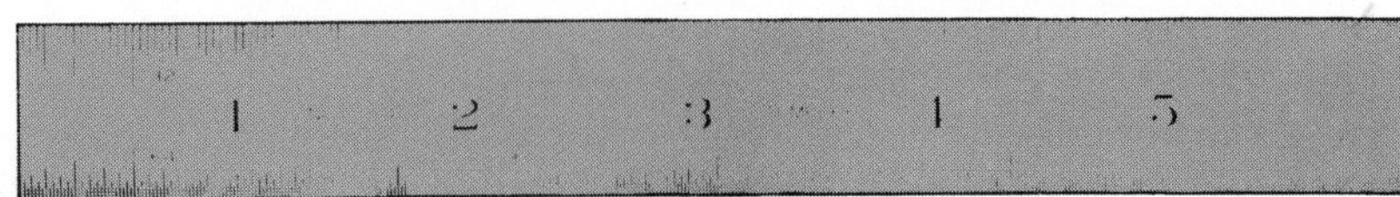

Fig. 4-7. Rule, with beveled edge, will put graduations closer to work surface, thus reducing the possibility for error.

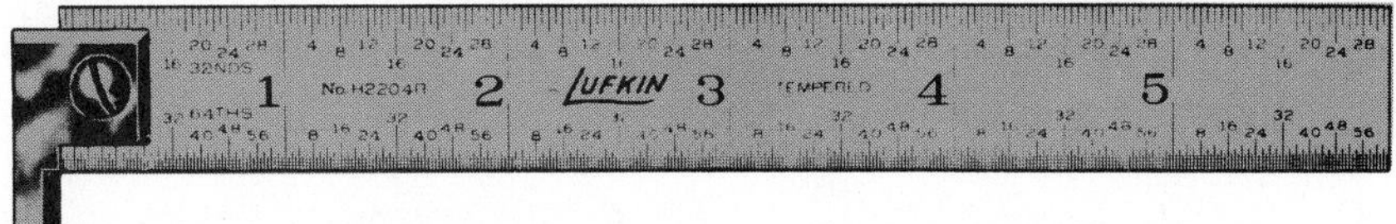

Fig. 4-8. Note hook rule with a fixed hook.

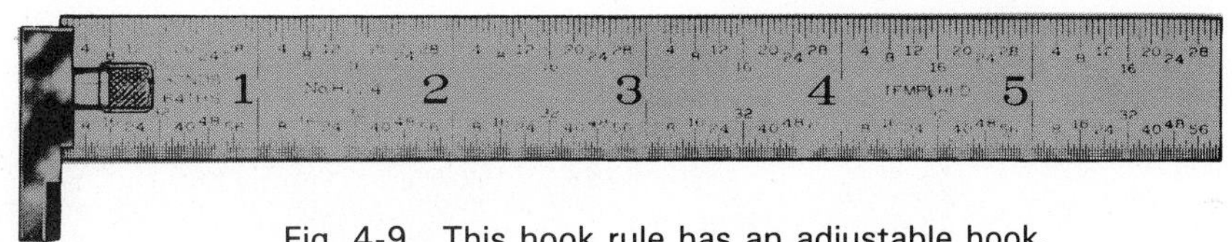

Fig. 4-9. This hook rule has an adjustable hook.

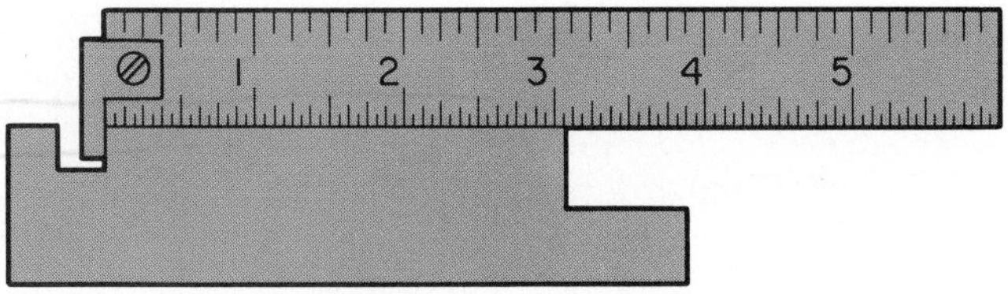

Fig. 4-10. Study how the hook rule is used.

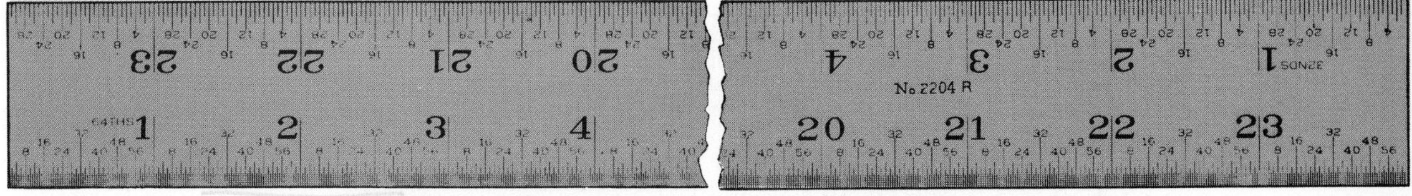

Fig. 4-11. Note how 12 in. steel rule has numbered graduations. This rule has No. 4 graduations (see table). The metric version is 300 mm long with 1.0 and 0.5 mm graduations on both sides. Every 10 mm is numbered.

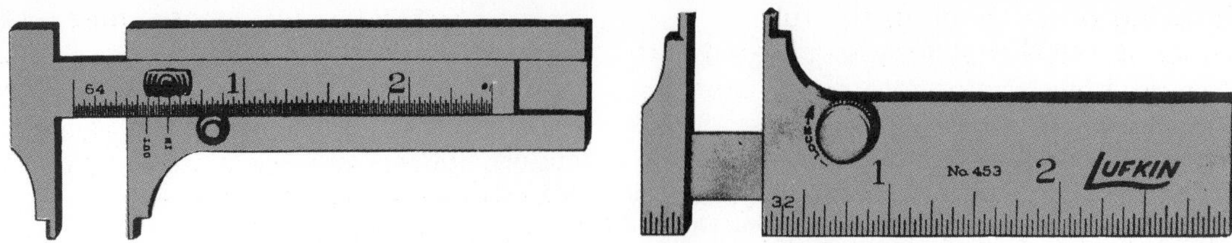

Fig. 4-12. A slide caliper rule is fitted to the work and locked. Outside dimensions are read at the point marked "out." Inside dimensions are read at the point marked "in."

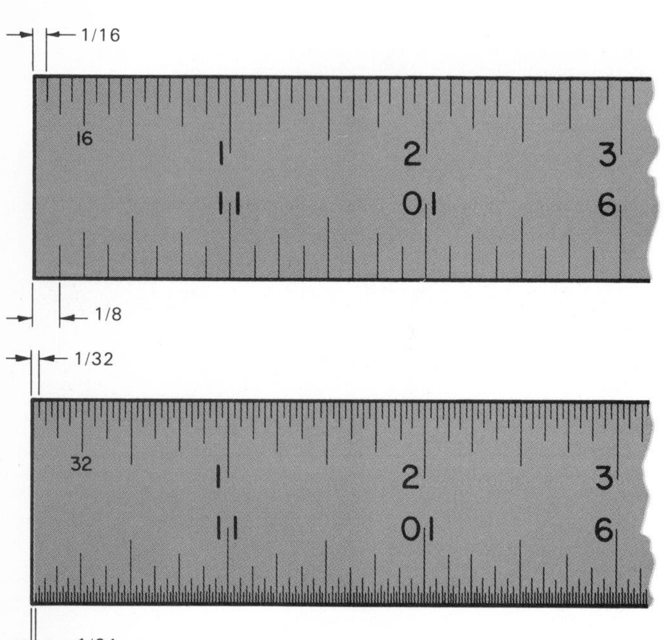

Fig. 4-13. These are fractional measurements found on a No. 4 rule.

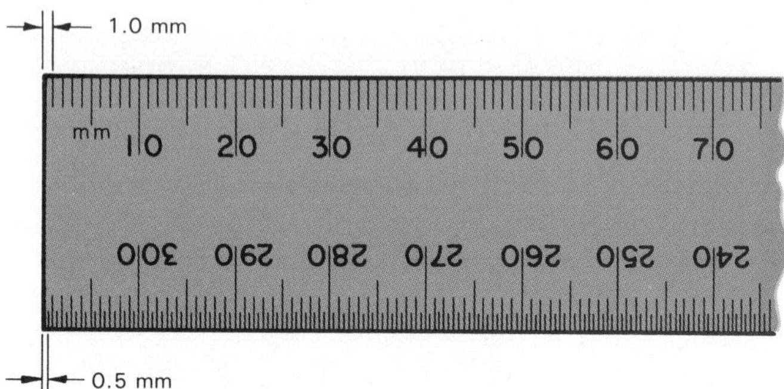

Fig. 4-14. Most metric rules are graduated in millimeters and 0.5 millimeters. They are available in 150 mm, 300 mm, 600 mm, and 1200 mm lengths.

Reading the rule (Metric units)

Most metric rules are divided into millimeters or 0.5 millimeters. They are numbered every 10 mm. See Fig. 4-14. The distance of the measurement is determined by counting the number of millimeters. For example, a measurement of 11 millimeters is written 11.0 mm. A measurement of 15 1/2 millimeters is written 15.5 mm.

How many measurements in Fig. 4-15 can you read?

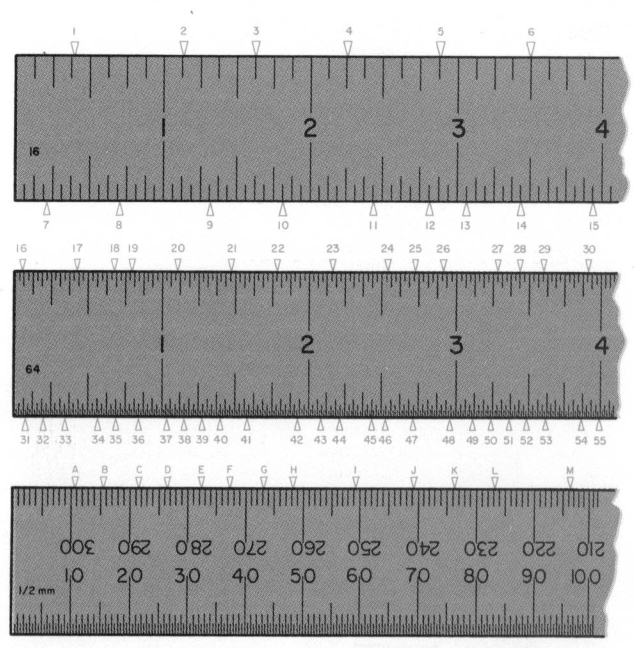

Fig. 4-15. Study and determine the measurements indicated on each rule.

Care of the rule

The steel rule is precision made and, like all tools, the quality of service depends upon the care it receives. Here are a few suggestions:

1. Use a screwdriver to loosen and tighten screws and to open paint cans. Be careful not to bend your rule.
2. Keep the rule clear of moving machinery. Never use it to clean metal chips as they form on the cutting tool. This will not only ruin the rule, but will prove extremely dangerous.

3. Avoid laying other tools on the rule.
4. Steel rules will NOT rust if they are wiped with an oily cloth before being returned to storage.
5. An occasional cleaning with steel wool will keep the graduations legible.
6. Make it a practice to make measurements and tool settings from the 1 in. line (10 mm line on metric rules), or other major graduations, rather than from the end of the rule.
7. Store rules separately. Do not throw them in a drawer with other tools.
8. Use the rule with care so the ends do not become nicked and worn.
9. Use the correct rule for the job being done.
10. Coat the tool with wax or rust preventative if the rule is to be stored for a prolonged period.

THE MICROMETER CALIPER

A Frenchman, Jean Palmer, devised and patented a measuring tool that made use of a screw thread, making it possible to read measurements quickly and accurately without calculations. It incorporated a series of engraved lines on the sleeve and around the thimble. The device, Fig. 4-16, called "Systeme Palmer," is the basis for the modern micrometer caliper, Fig. 4-17.

The ***micrometer caliper,*** also known as a "mike," is a precision measuring tool capable of measuring to 1/1000 (0.001) in. or 1/100 (0.01) mm. When fitted with a Vernier scale, it will read to 1/10,000 (0.0001) in. or 2/1000 (0.002) mm.

While manufactured in sizes up to 60 in. and 1500 mm, the spindle movement is limited to 1 in. and 25 mm. Only the tool frame is enlarged.

Types of micrometers

Micrometers are made in a large variety of models. A few of the more commonly used are:

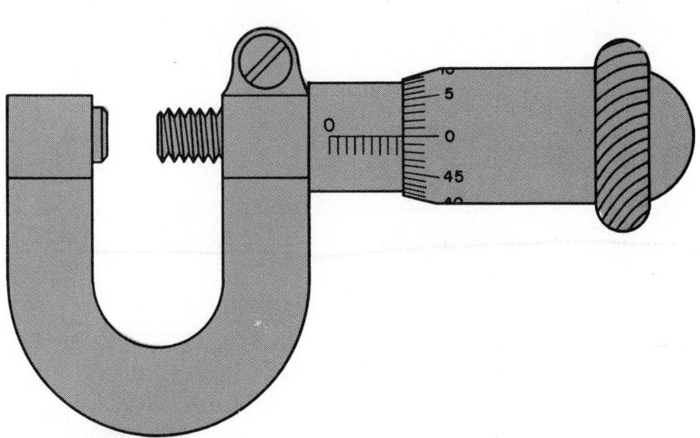

Fig. 4-16. A drawing of Systeme Palmer, the modern micrometer caliper operates on the same principle as this 1848 measuring tool.

An ***outside micrometer*** measures external diameters and thicknesses, Fig. 4-18.

The ***inside micrometer*** measures internal diameters of cylinders and rings, widths of slots, setting gages, etc. There are two general styles: the ***conventional inside micrometer,*** whose range is extended by fitting longer rods to the micrometer head, Fig. 4-19; and the ***jaw-type inside micrometer*** whose range is limited to 1 in. or 25 mm. Note that the scale is graduated from right to left, Fig. 4-20.

A ***direct reading micrometer*** is read directly from the numbers appearing in the readout opening in the

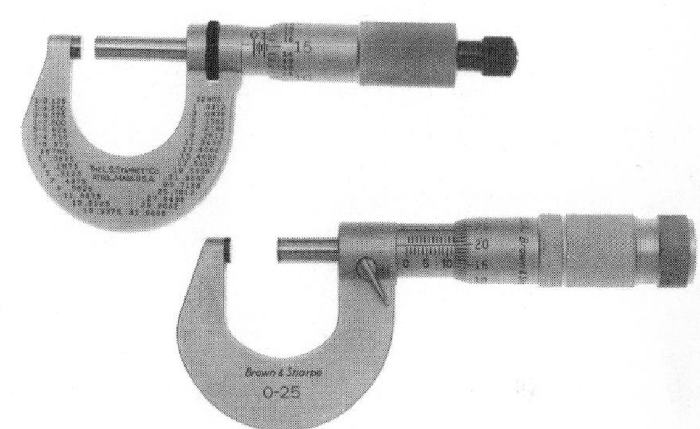

Fig. 4-17. Inch based and metric based micrometer calipers are both available in a large range of sizes.

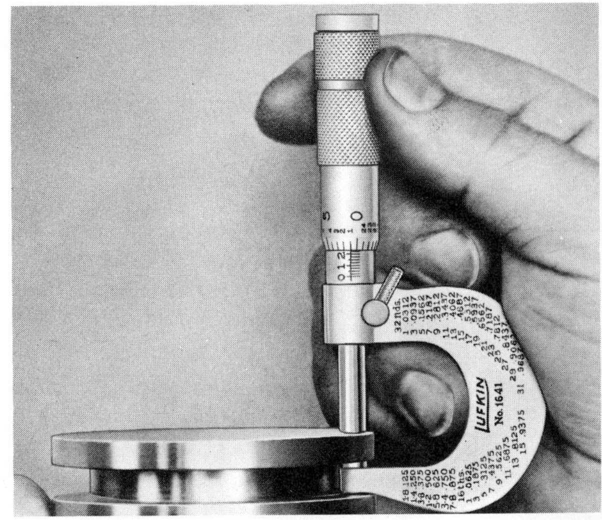

Fig. 4-18. Note how machinest is holding the 0 to 1 in. outside micrometer.

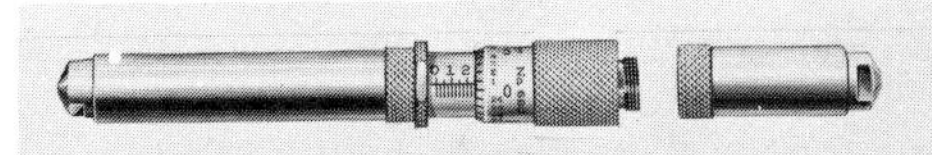

Fig. 4-19. This is a conventional inside micrometer.

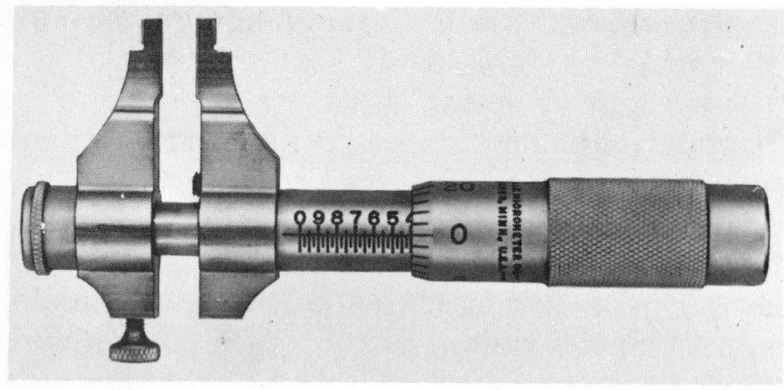

Fig. 4-20. One jaw-type inside caliper, note how divisions on sleeve are numbered. They are in the reverse order of conventional outside micrometer. (Scherr-Tumico)

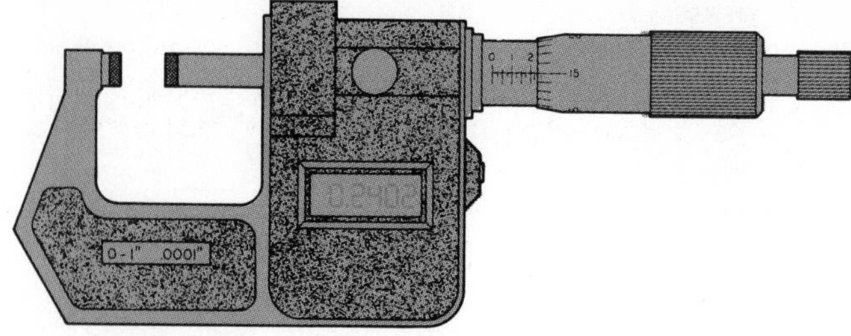

Fig. 4-22. A digital readout micrometer is an electronic measuring tool that can convert from readings in inches to millimeters with a push of a button.

frame, Fig. 4-22. An ***electronic digital readout micrometer*** senses the spindle position on the work and indicates the measurement on the digital display. Refer to Fig. 4-22.

A ***micrometer depth gage,*** will measure the depths of holes, slots, projections, etc., Fig. 4-23. The measuring range can be increased by changing to spindles of longer lengths. Note that measurements are read from RIGHT TO LEFT.

A ***screw thread micrometer*** has a pointed spindle and a double "V" anvil, both correctly shaped to contact the screw thread, Fig. 4-24. It measures the pitch diameter of the thread, which equals the outside diameter of the thread, minus the depth of one thread. Since each thread micrometer is designed to measure only a limited number of threads per inch, a set of thread micrometers are necessary to measure a full range of thread pitches.

A ***chamfer micrometer,*** will accurately measure countersunk holes and other chamfer type measurements. With fastener tolerances so critical on some aerospace and other applications, it is important that countersunk holes and tapers on fasteners meet specifications. A chamfer micrometer makes it possible to check these critical areas, Fig. 4-25.

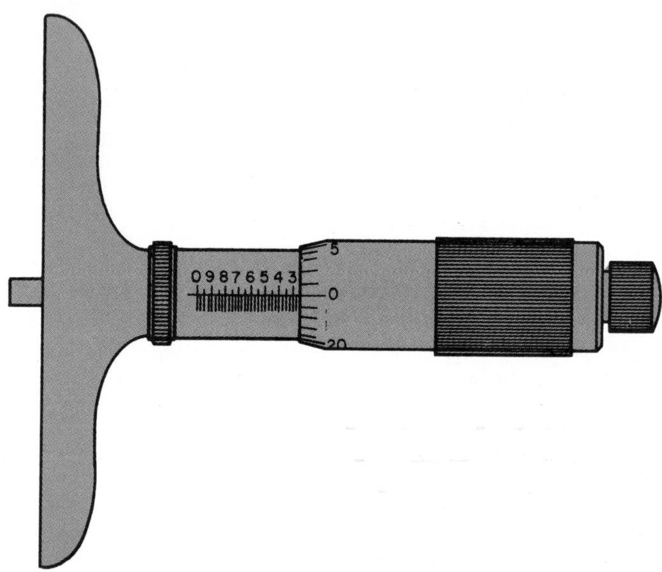

Fig. 4-23. Note micrometer depth gage.

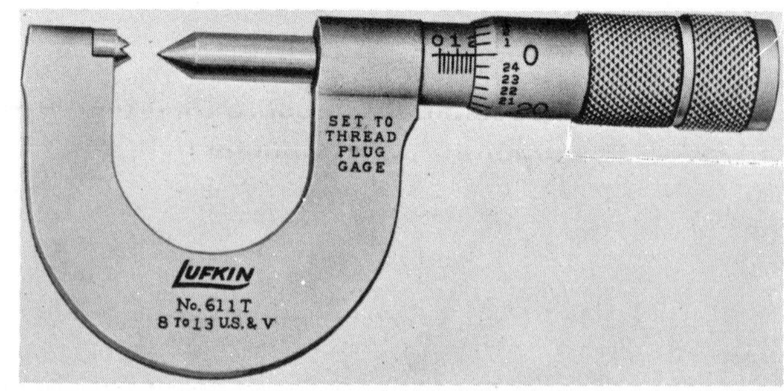

Fig. 4-24. Note that this screw thread micrometer can be used to measure 8 to 13 threads per inch.

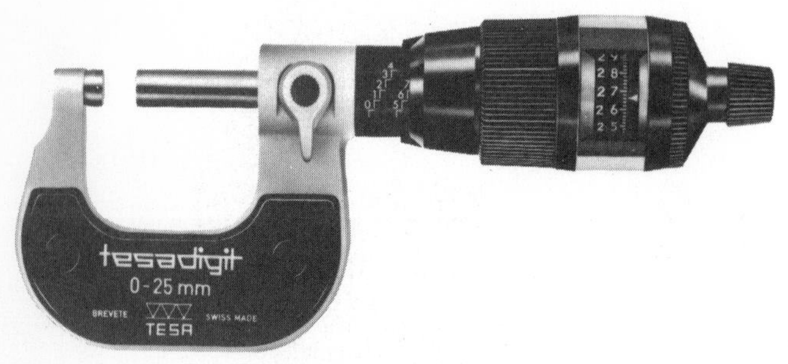

Fig. 4-21. Note one type of direct reading micrometer. To read this metric based micrometer, add the total reading in millimeters visible on the sleeve to the reading in hundredths of a millimeter on the thimble. The reading here is 7.266 mm. (Brown and Sharp Mfg. Co.)

Special micrometers are available for unique applications, Fig. 4-26. For example, many cutting tools have cutting edges that are uneven in number.

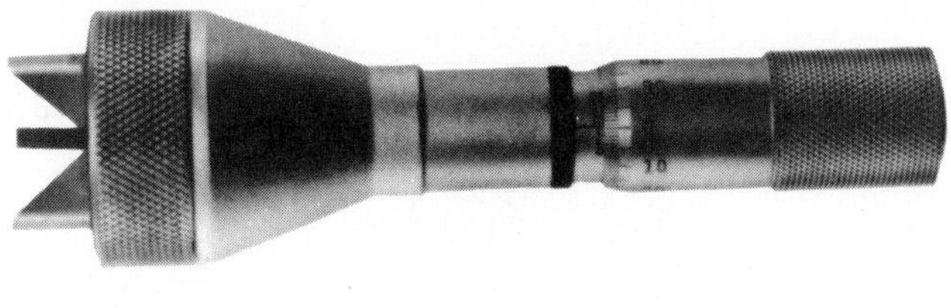

Fig. 4-25. Chamfer micrometer is for direct measuring of chamfers, countersinks, and top diameter of tapered holes. (Swiss Precision Instruments, Inc.)

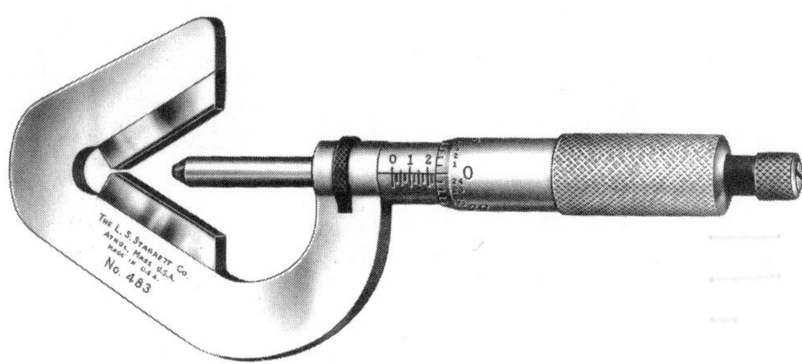

Fig. 4-26. This is one of many specially designed micrometer-type measuring tools. This particular model can measure the diameter of odd fluted taps, reamers, and milling cutters. It can also be used to check out-of-roundness that cannot be checked with a conventional micrometer.

This makes it impossible to measure their diameter with a conventional "mike." Special micrometers have been devised to handle this and other measurement problems.

Reading an inch-based micrometer

The principle of the micrometer is based on a very accurately made screw thread that rotates in a fixed nut. The screw thread is ground on the ***spindle*** and is attached to the ***thimble.*** The spindle advances or recedes from the ***anvil*** by rotating the thimble. The threaded section has 40 threads per inch; therefore, each revolution of the spindle moves the spindle 1/40 (0.025) in. A sectional view of a micrometer is shown in Fig. 4-27.

The line engraved lengthwise on the sleeve is divided into 40 equal parts per inch which corresponds to the number of threads per inch on the spindle. Each vertical line equals 1/40 or 0.025 in. Every fourth division is numbered 1, 2, 3, etc., representing 0.100 in., 0.200 in., etc.

The beveled edge of the thimble is divided into 25 equal parts around its circumference. Each division equals 1/1000 (0.001) in. On some micrometers, every division is numbered, while every fifth is numbered on others.

The "mike" is read by recording the highest number on the sleeve (1 = 0.100, 2 = 0.200, etc.). To this number, add the number of vertical lines visible between the number and thimble edge (1 = 0.025, 2 = 0.050, etc.). To this total, add the number of thousandths indicated by the line that coincides with the horizontal sleeve line. See Fig. 4-28.

Micrometer example 1

The reading is composed of:

4 large graduations or 4 × 0.100 = 0.400
2 small graduations or 2 × 0.025 = 0.050
and 8 graduations on the thimble or 8 × 0.001 = 0.008
Total mike reading = 0.458 in.

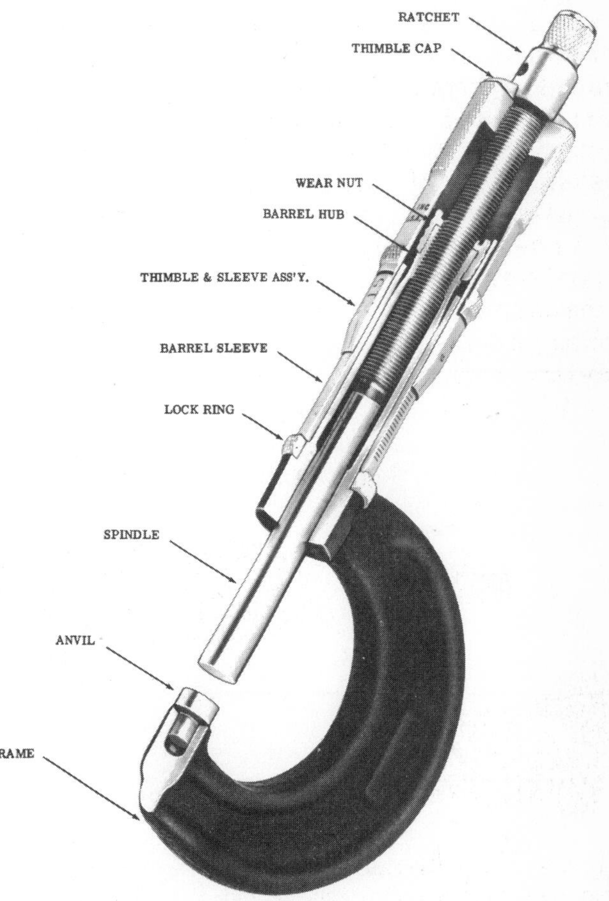

Fig. 4-27. Study sectional view of micrometer caliper. (Scherr-Tumico)

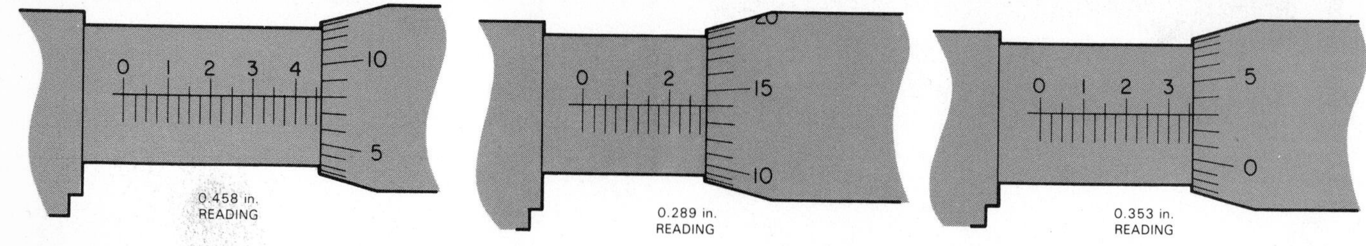

Fig. 4-28. Study micrometer caliper readings—Examples 1, 2, and 3.

Micrometer example 2

The reading is composed of:
2 large graduations or 2 × 0.200 = 0.200
3 small graduations or 3 × 0.025 = 0.075
and 14 graduations on the
thimble or 14 × 0.001 = 0.014
Total mike reading = 0.289 in.

Micrometer example 3

The reading is composed of:
3 large graduations or 3 × 0.100 = 0.300
2 small graduations or 2 × 0.025 = 0.050
and 3 graduations on the
thimble or 3 × 0.001 = 0.003
Total mike reading = 0.353 in.

Reading a Vernier micrometer

On occasion, it is necessary to measure finer than 1/1000 (0.001) in. When this situation is encountered, the Vernier micrometer caliper is employed. This micrometer has a third scale AROUND THE SLEEVE that will furnish the 1/10,000 (0.0001) in. reading without estimating or guessing. See Fig. 4-29.

The Vernier scale has 11 parallel lines that occupy the same space as 10 lines on the thimble. The lines around the sleeve are numbered 1 to 10. The difference between the spaces on the sleeve and those on the thimble is one-tenth of a thousandth of an inch.

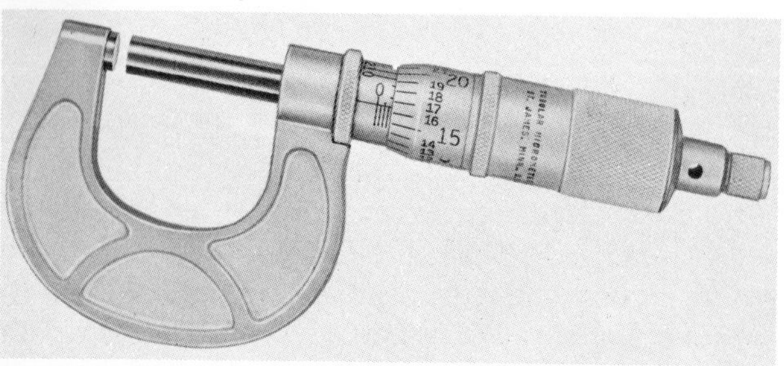

Fig. 4-29. This is a Vernier micrometer caliper.

To read, first obtain the thousandths reading, then observe which of the lines on the Vernier scale coincides (lines up) with a line on the thimble. Only one of them can line up. If the line is 1, add 0.0001 to the reading, if line 2, add 0.0002 to the reading, etc. See Fig. 4-30.

Reading a metric-based micrometer

The metric-based micrometer, Fig. 4-31, is read as shown in Fig. 4-32. If you are able to read the conventional inch-based micrometer, the change over to reading the metric-based tool will offer no difficulties.

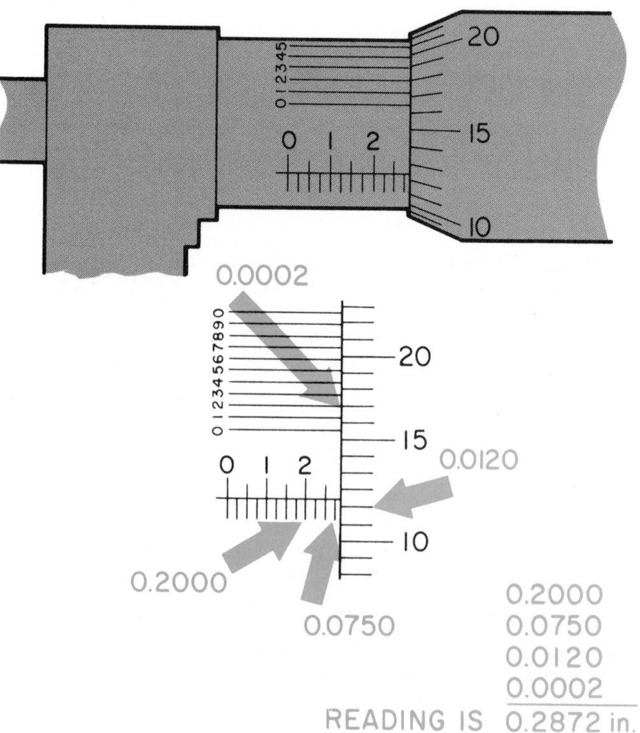

Fig. 4-30. Note how to read an inch based Vernier micrometer caliper. Add the total reading in thousandths, then observe which of the lines on the Vernier scale coincides with a line on the thimble. Only one of them can align. In this case, it is line 2, so 0.0002 will be added to the reading.

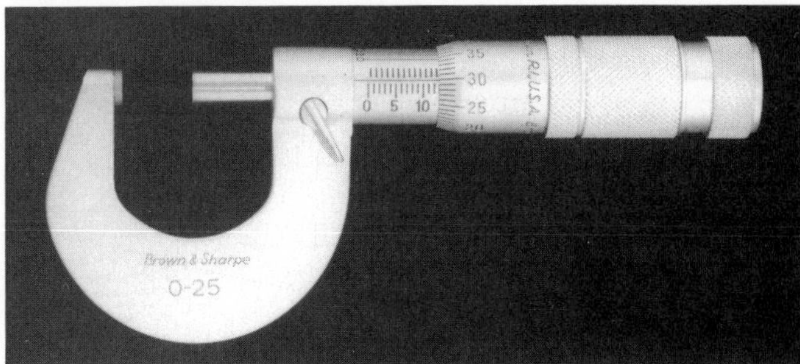

Fig. 4-31. This is a metric based micrometer caliper.

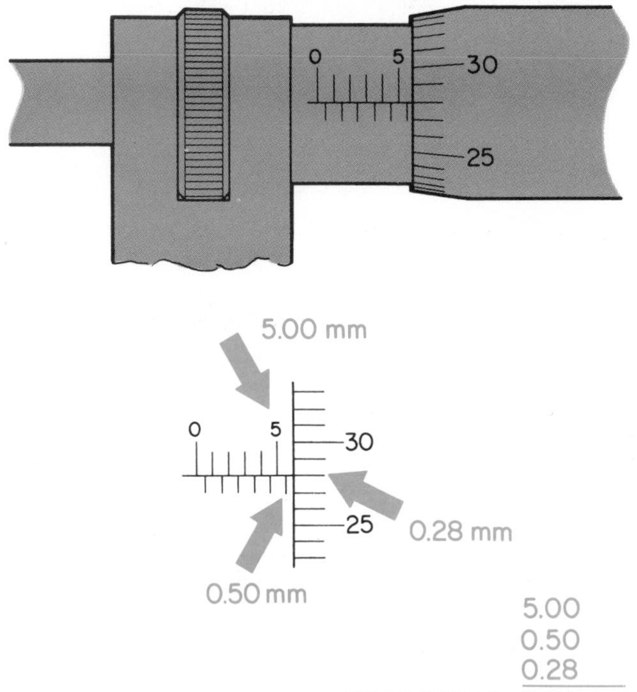

Fig. 4-32. To read a metric micrometer, add the total reading in millimeters visible on the sleeve to the reading of hundredths of a millimeter, indicated by the graduation on the thimble. Note that the thimble reading coincides with the longitudinal line on the micrometer sleeve.

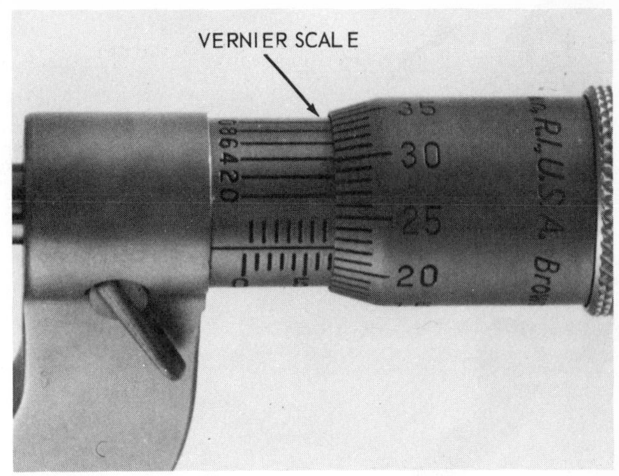

Fig. 4-33. Study Vernier scale on a metric based micrometer.

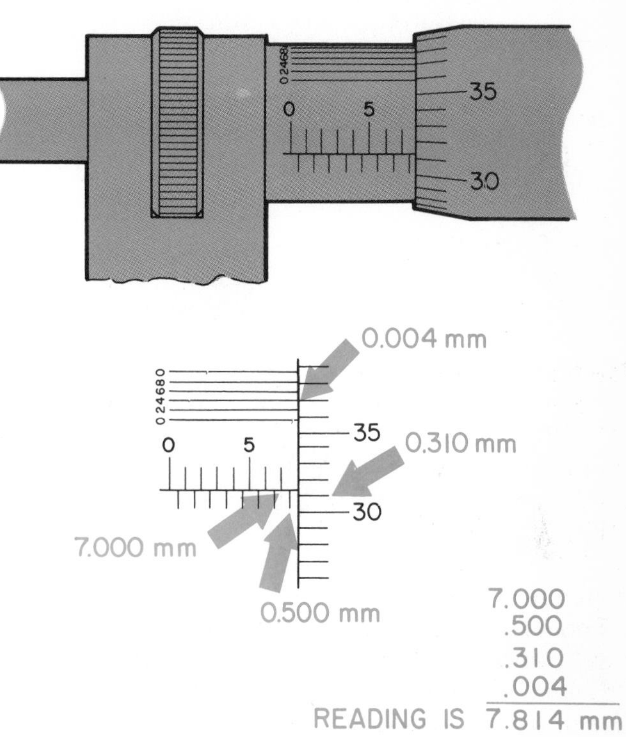

Fig. 4-34. Note reading a metric based Vernier micrometer caliper. To the regular reading in hundredths of a millimeter (0.01), add the reading from the Vernier scale that coincides with a line on the thimble. Each line on the Vernier scale is equal to two thousandths of a millimeter (0.002 mm).

Reading a metric Vernier micrometer

Metric Vernier micrometers are used like those graduated in hundredths of a millimeter (0.01 mm), Fig. 4-33. However, using the Vernier scale on the sleeve, an additional reading of two-thousandths of a millimeter can be obtained, Fig. 4-34.

Using the micrometer

The proper way to hold a micrometer when making a measurement is shown in Fig. 4-35. The work is placed into position, and the thimble rotated until the part is clamped LIGHTLY between the anvil and spindle. Guard against excessive pressure. It will cause an erroneous reading.

The correct contact pressure will be applied if a "mike" with a ***ratchet stop*** is used, Fig. 4-36. This device is used to rotate the spindle. When the pressure reaches a predetermined amount, the ratchet stop slips and prevents further turning of the spindle. Uniform contact pressure with the work

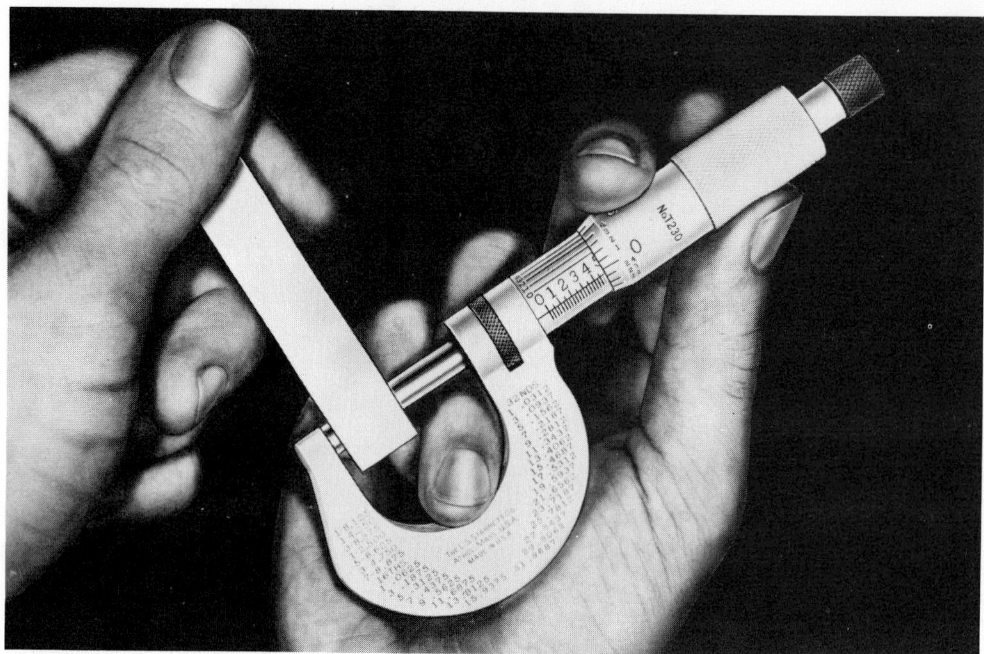

This is the correct ways to hold a 0 — 1 in. (0 — 25 mm) micrometer caliper for making measurements.

This is the proper way to hold a micrometer when making a measurement on work mounted in a machine.

This is an example of making a measurement over a large piece. The micrometer frame is hollow to reduce weight. Store large micrometers like this model out of direct sunlight and away from any heat source. The heat will affect their accuracy.

Fig. 4-35. Note ways of using micrometers. (L.S. Starrett Co. and Lufkin)

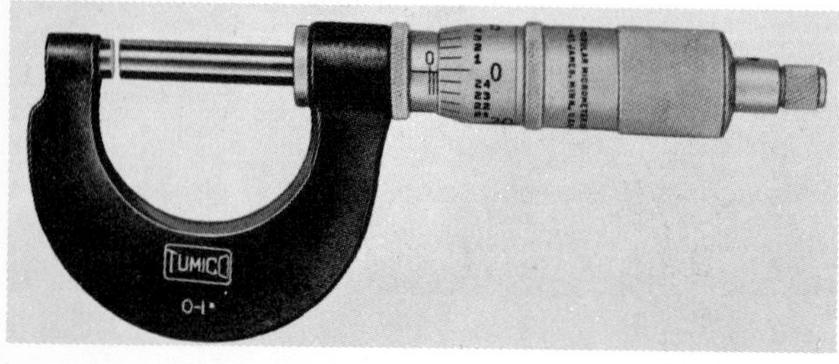

Fig. 4-36. This is a micrometer with a ratchet stop.

is assured even if different people use the same micrometer.

Some micrometers are fitted with a ***friction thimble,*** Fig. 4-18. It is a friction control built into the upper section of the thimble. It produces the same results as the ratchet stop but permits one hand use of the micrometer.

When several identical parts are to be gaged, lock the spindle into place with the ***lock ring,*** Fig. 4-27. Gaging parts with a micrometer locked at the proper setting is an easy way to determine whether the pieces are oversize, correct size, or undersize.

Reading an inside micrometer

To get a correct reading with an inside micrometer, it is important that the tool be HELD SQUARE across the diameter of the work. It must be positioned so that it will measure across the diameter on exact center, Fig, 4-37.

Measurement is made by holding one end of the tool in place and then "feeling" for the MAXIMUM possible setting by moving the other end from left to right, and then in and out of the opening. The measurement is made when no left or right movement is felt, and a slight drag is noticeable on the in and out swing. It may be necessary to take several readings and average them.

Reading a micrometer depth gage

Be sure to read a micrometer depth gage correctly, Fig. 4-38. Unlike an outside micrometer, the graduations on this measuring tool are in REVERSE ORDER. That is: they read 0, 9, 8, 7, 6, etc. The graduations UNDER the thimble must be read rather than those that are exposed.

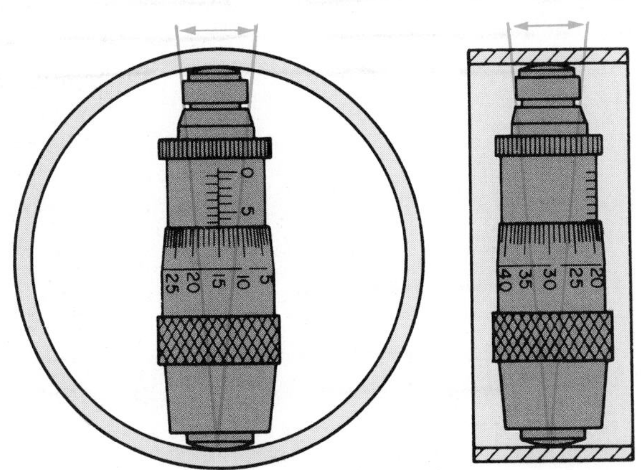

Fig. 4-37. Study how inside micrometer is used. Extension rods can be added to extend the tool's measuring range.

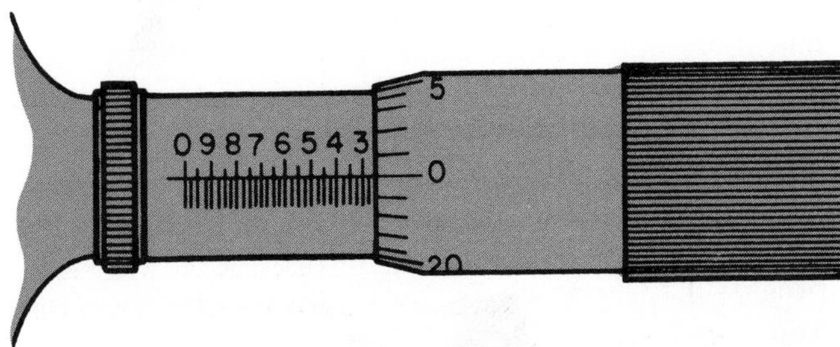

Fig. 4-38. Note micrometer depth gage. Remember, when making measurements with a depth micrometer, the graduations are in reverse order.

CARE OF A MICROMETER

Micrometers must be handled with care or their accuracy will be destroyed. The following techniques are recommended:

1. Place the micrometer on the work carefully so the faces of the anvil and spindle will not be damaged. The same applies when removing the tool after a measurement has been made on a part.
2. Keep the micrometer clean. Wipe it with a slightly oiled cloth to prevent rust and tarnish. A drop of light oil on the screw thread will keep the tool operating smoothly.
3. Avoid "springing" micrometers not fitted with a ratchet stop or friction thimble by applying too much pressure when making a measurement. Micrometer accuracy could be affected.
4. Clean the anvil and spindle faces before use. This can be done with a soft cloth or by LIGHTLY closing the jaws on a clean piece of paper and drawing the paper out.
5. Check for accuracy by closing the spindle gently on the anvil and note whether the zero line on the thimble coincides with the zero on the sleeve. If they are not aligned, make adjustments by following the manufacturer's recommendations for this operation.
6. Avoid placing a micrometer where it may fall on the floor, or have other tools placed on it.
7. If the micrometer must be opened or closed a considerable distance, do NOT "twirl" the frame; gently roll the thimble on the palm of the hand. See Figs. 4-39 and 4-40.
8. Never attempt to make a micrometer reading until the machine has come to a complete stop.
9. Clean and oil the tool if it is to be stored for some time. If possible, place the micrometer in a small box for protection.

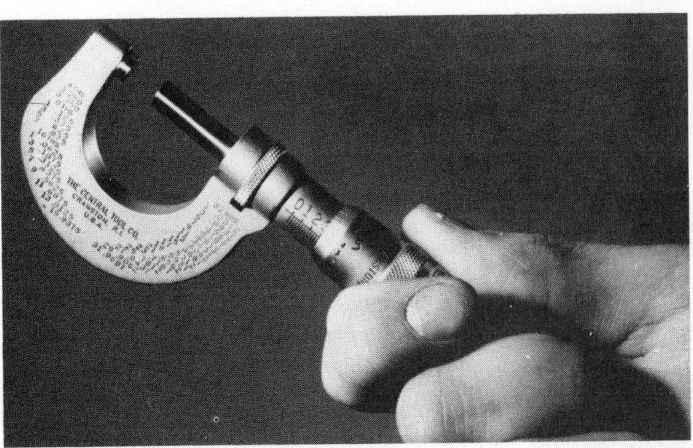

Fig. 4-39. Micrometer accuracy can be destroyed by "twirling" the tool to open or close it.

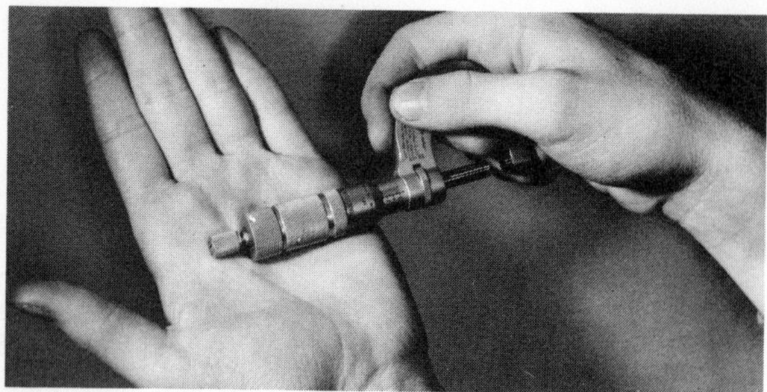

Fig. 4-40. Roll the micrometer thimble on the palm of your hand if the instrument must be opened or closed a considerable distance.

VERNIER MEASURING TOOLS

The Vernier principle of measuring was named for its inventor, Pierre Vernier (1580-1637), a French mathematician.

The ***Vernier caliper,*** unlike the micrometer caliper, can make both inside and outside measurements, Figs. 4-41 through 4-43. The design of the tool permits measurements to be made over a large range of sizes. It is manufactured as a standard item in 6 in., 12 in., 24 in., 36 in., and 48 in. lengths. Metric Vernier calipers are available in 150 mm, 300 mm, and 600 mm lengths. The 6 in. and 12 in. and 150 mm and 300 mm sizes are most commonly used.

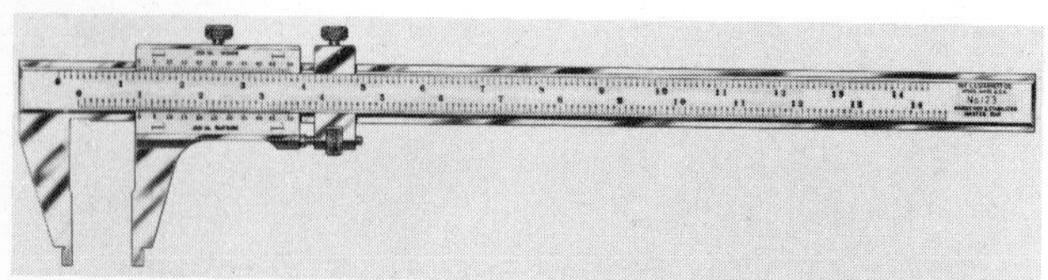

Fig. 4-41. A Vernier caliper can make accurate measurements to 1/1000 (0.001) in. and 1/50 (0.02) mm.

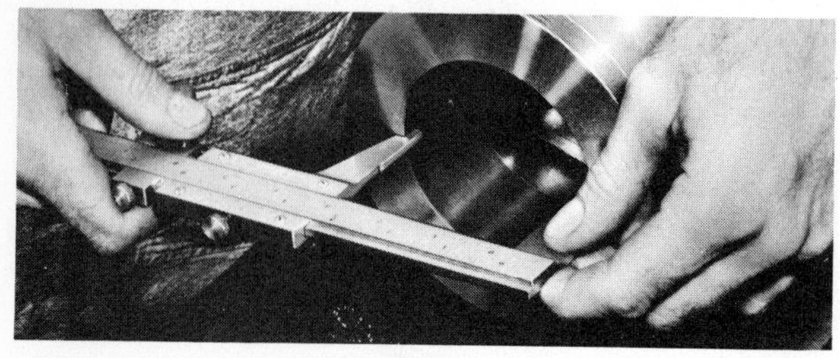

Fig. 4-42. Machinist is making an internal measurement with a Vernier caliper. (L.S. Starrett Co.)

Fig. 4-43. Note proper method for making an external measurement with a Vernier caliper. Machinist is making the final adjustment with the adjusting nut. (L.S. Starrett Co.)

The Vernier caliper can make accurate measurements to 1/1000 (0.001) in. and 1/50 or 2/100 (0.02) mm.

The following measuring tools also utilize the Vernier principle:

The ***Vernier height gage*** is designed for use in tool rooms and inspection departments, for layout work, checking hole and/or pin location, jig and fixture work, etc. See Figs. 4-44 and 4-45.

The ***Vernier depth gage*** is ideal for measuring the depth of holes, slots, and recesses. It is ordinarily fitted with a 6 in. or 12 in. and 150 mm or 300 mm blades. Refer to Fig. 4-46.

A ***gear tooth Vernier caliper,*** is needed to measure gear teeth, form and threading tools, Figs. 4-47 and 4-48.

A ***universal Vernier bevel protractor*** is designed for the precision layout and measurement of angles. Angles are measured in degrees, minutes, and seconds, Fig. 4-49.

Vernier measuring tools, with the exception of the Vernier bevel protractor, consist of a graduated beam with fixed jaw or base, and a Vernier slide assembly. The ***Vernier slide assembly*** is composed of a movable jaw or scribe, Vernier plate, and clamping screws. The slide moves as a unit along the beam.

Unlike other Vernier measuring tools, the beam of the Vernier caliper is graduated on both sides. One side is for making OUTSIDE measurements, the other for INSIDE measurements.

Many of the newer Vernier measuring tools are graduated to make both inch and millimeter measurements.

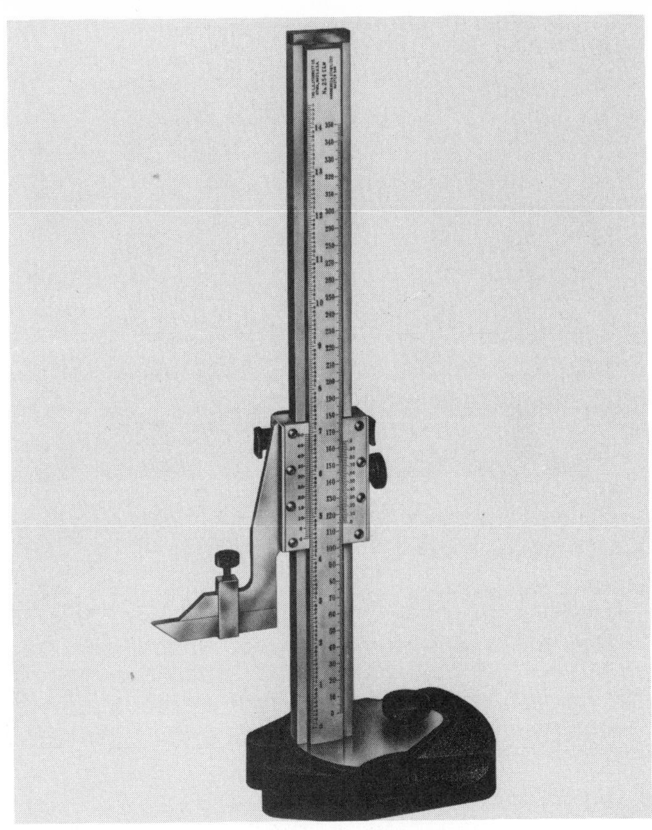

Fig. 4-44. This is a conventional Vernier height gage. One edge is graduated in inches and the other in millimeters. (L.S. Starrett Co.)

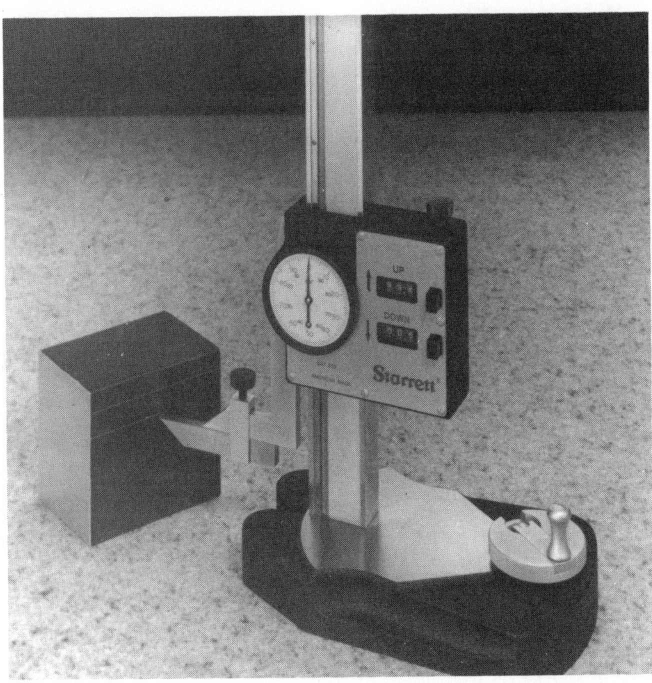

Fig. 4-45. This is an electronic digital readout Vernier height gage, capable of making inch and millimeter measurements. (L.S. Starrett Co.)

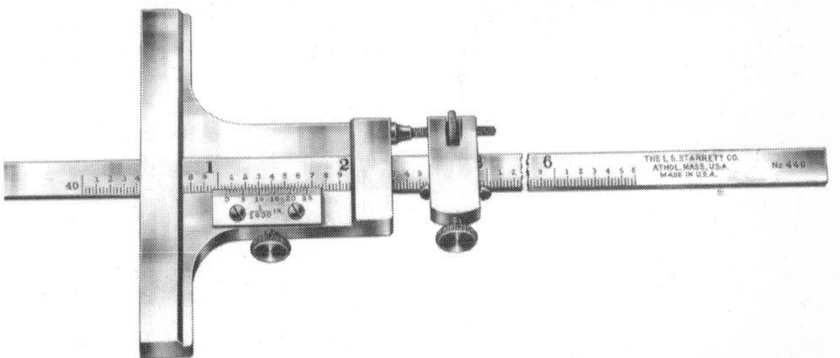

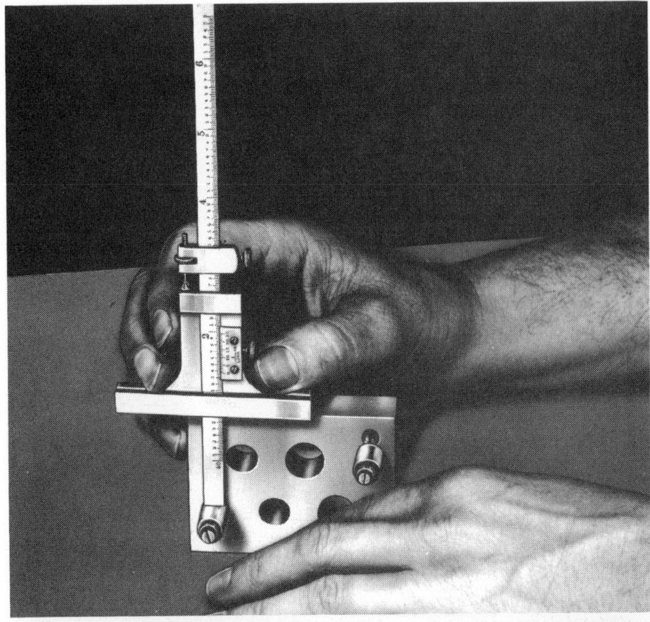

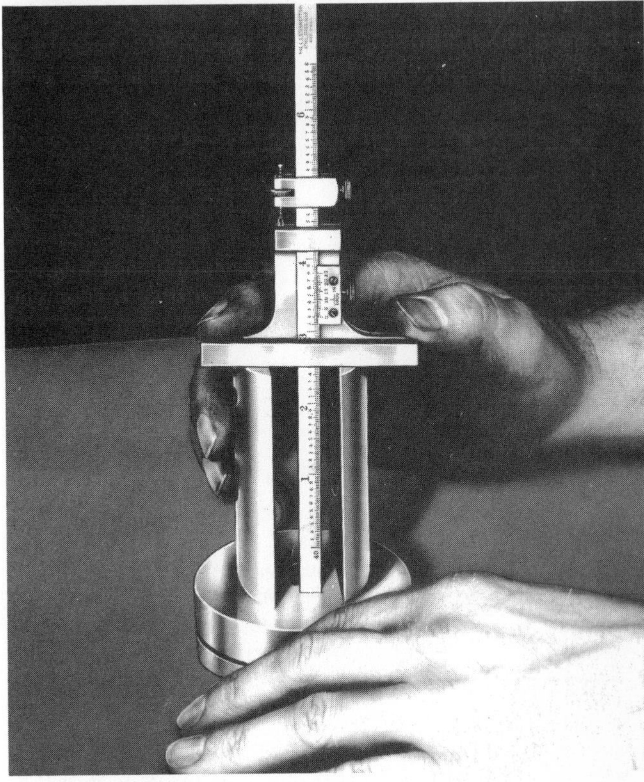

Fig. 4-46. Note Vernier depth gage and several ways it is used. (L.S. Starrett Co.)

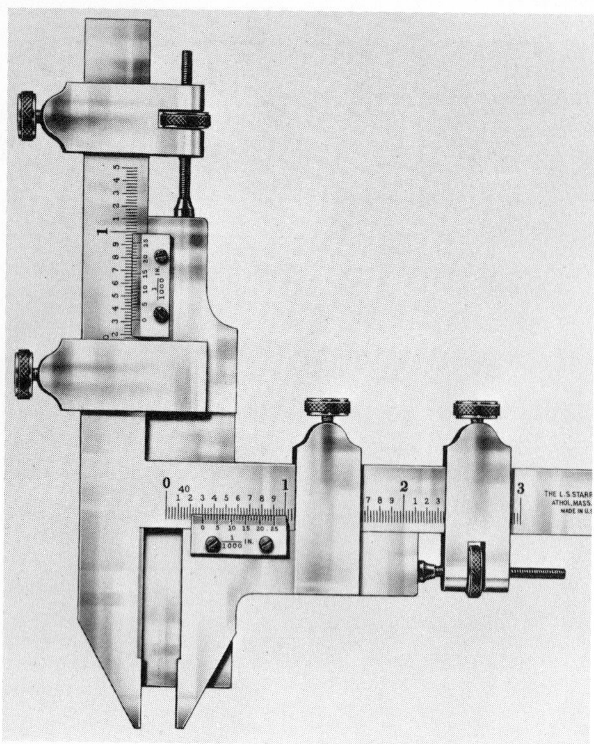

Fig. 4-47. This is a gear tooth Vernier caliper. (L.S. Starrett Co.)

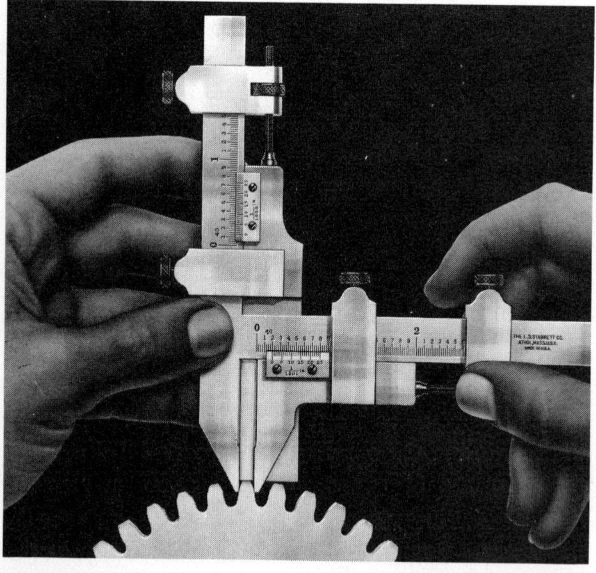

Fig. 4-48. Study how to measure gear tooth with gear tooth Vernier caliper. Tool is read in the same manner as a Vernier caliper and height gage.

Reading an inch-based Vernier scale

These measuring tools are available with either 25 or 50 division Vernier plates. Both plates can be read to 0.001 in.

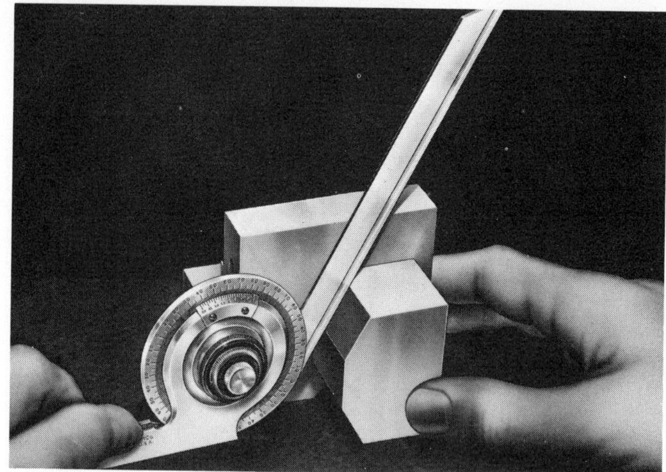

Fig. 4-49. Note universal Vernier bevel protractor. (L.S. Starrett Co.)

Every inch section on the beam of measuring tools using the 25 division Vernier plate is graduated into forty (40) equal parts. Each graduation is 1/40 or 0.025 in. Every fourth division, representing 1/10 or 0.100 in., is numbered.

There are 25 divisions on the Vernier plate. Every fifth line is numbered: 5, 10, 15, 20, and 25. The 25 divisions occupy the same space as 24 divisions on the beam. This slight difference, equal to 0.001 (1/1000) in. per division, is the basis of the Vernier principle of measuring.

To read a 25 division Vernier plate measuring tool, note how many inches (1, 2, 3, etc.), tenths (0.100, 0.200, etc.), and fortieths (0.025, 0.050, or 0.075) there are between the "0" on the Vernier scale and the "0" line on the beam; then add them.

Now count the number of graduations (each graduation equals 0.001 or 1/1000 in.) that lie between the "0" line on the Vernier plate and the line that coincides (corresponds exactly) with a line on the beam. Only one line will coincide. Add this to the above total for the reading.

Vernier 25 division example (See Fig. 4-50.)

The reading is composed of:

The "0" line on the Vernier plate is between 2 and 3 on the beam	=	2.000
Plus three 0.100 (1/10) graduations	=	0.030
Plus two 0.025 (1/40) graduations	=	0.050
Plus eighteen 0.001 (1/1000) graduations	=	0.018
Total reading	=	2.368 in.

On the 50 division Vernier plate, every second graduation between the inch lines is numbered, and equals 1/10 or 0.100 in. The unnumbered graduations equal 1/20 or 0.050 in.

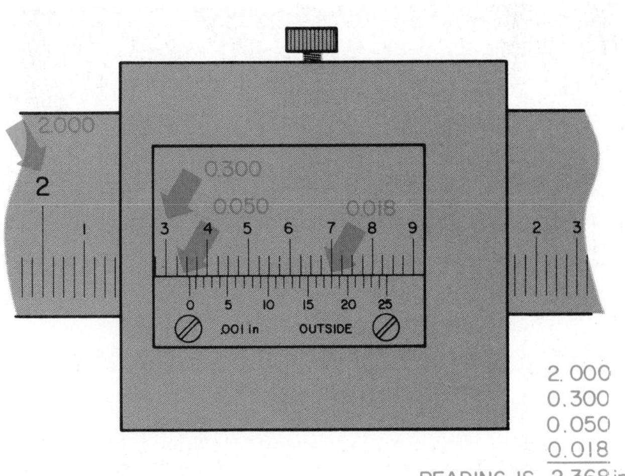

Fig. 4-50. Study how to read a 25 division Vernier scale.

The Vernier plate is graduated into 50 parts, each representing 0.001 (1/1000) in. Every fifth line is numbered: 5, 10, 15, 20 . . . 40, 45, and 50.

To read a 50 division Vernier measuring tool, first count how many inches, tenths (0.100), twentieths (0.050) there are between the "0" line on the beam, and the "0" line on the Vernier plate. Then add them.

Next, count the number of 0.001 graduations on the Vernier plate from its "0" line to the line that coincides with a line on the beam. Add this to the above total. This is the reading.

Vernier 50 division example

The reading is composed of:

The "0" line on the Vernier plate is between 2 and 3 on the beam	=	2.000
Plus two 0.100 (1/10) graduations	=	0.200
Plus one 0.050 (1/20) graduations	=	0.050
Plus fifteen 0.001 (1/1000) graduations	=	0.015
Total reading	=	2.265 in.

Reading a metric-based Vernier scale

The same principles are used in reading metric Vernier measuring tools as those for English measure. However, the readings on the Vernier scale are obtained in 0.02 mm. A 25 division Vernier scale is explained in Fig. 4-52, while a 50 division scale is described in Fig. 4-53.

Using the Vernier caliper

As with any precision tool, a Vernier caliper must NOT be forced on the work. Slide the Vernier assembly until the jaws almost contact the section being measured. Lock the clamping screw. Make the tool adjustment with the fine adjusting nut. The jaws must contact the work firmly but not tightly.

2.000
0.200
0.050
0.015
READING IS 2.265 in.

Fig. 4-51. Study how to read a 50 division Vernier scale.

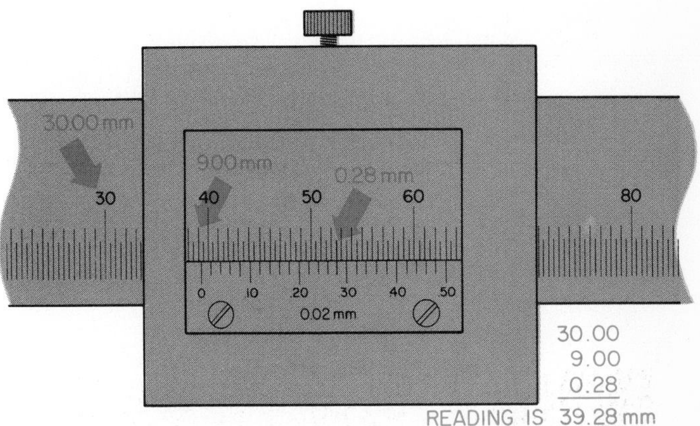

Fig. 4-52. Study how to read a 25 division metric based Vernier scale. Readings on scale are obtained in two hundredths of a millimeter (0.02 mm).

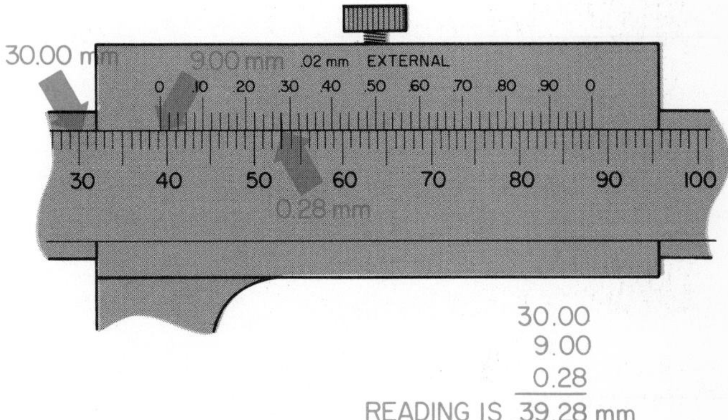

Fig. 4-53. Study how to read a 50 division metric based Vernier scale. Each division also equals two hundredths of a millimeter (0.02 mm).

Lock the slide on the beam. Carefully remove the tool from the work and make your reading.

Points permitting accurate divider and trammel point settings, for precise layout work, are located on the outside measuring scale and on the slide assembly.

Two other measuring tools are also used like the Vernier caliper. They are:

Electronic digital readout calipers are direct reading, and may be switchable to give both inch or metric measure. They are battery powered and very convenient, Fig. 4-54.

Dial calipers can be used to make outside, inside, and depth measurements, Fig. 4-55. The dial reads directly to 0.001 in. or 0.02 mm and is easily adjustable for setting. A lock permits the tool to be employed for repetitive measuring.

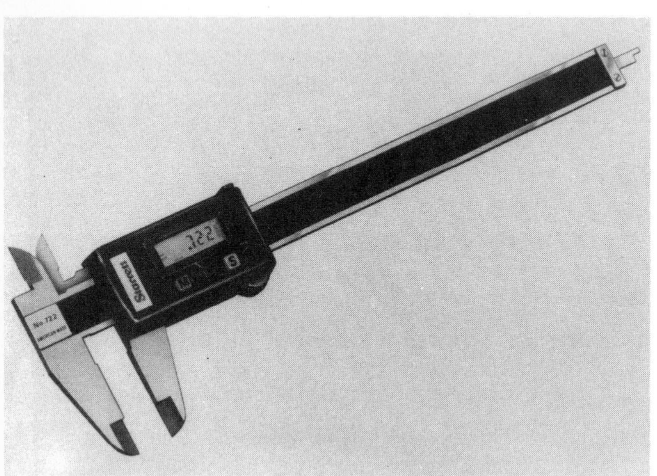

Fig. 4-54. Note electronic digital readout caliper. (L.S. Starrett Co.)

Fig. 4-55. Dial calipers are available in 0 to 6 in. and 0 to 12 in. range, and 0 to 150 mm range. (L.S. Starrett Co.)

Universal Vernier bevel protractor

There are many times when angles must be measured with great accuracy. A ***universal Vernier bevel protractor*** can measure angles accurately to 1/12 degree or 5 minutes, Fig. 4-56. A quick review of the circle, angles, and units of measurement associated with them will help in understanding how to read this instrument.

1. ***Degree***—A circle, no matter what size, contains 360 degrees. This is normally written 360°. Angles are also measured by degrees.
2. ***Minute***—If a degree were divided into 60 equal parts, each part would be one (1) minute. The minute is utilized to represent a fractional part of a degree. It would be written like this: 0° 0′.
3. ***Second***—Very accurate work requires that the minute be divided into smaller units known as seconds. There are 60 seconds in one minute. An angular measurement written in degrees, minutes, and seconds would be 36° 18′ 22″. This would read "36 degrees, 18 minutes, and 22 seconds."

The ***universal bevel protractor*** is a finely made tool with a dial graduated into degrees, a base or stock, and a sliding blade that can be extended in either direction or set at any angle to the stock. The blade can be locked against the dial by tightening the blade clamp nut. The blade and dial can be rotated as a unit to any desired position, and locked by tightening the dial clamp nut.

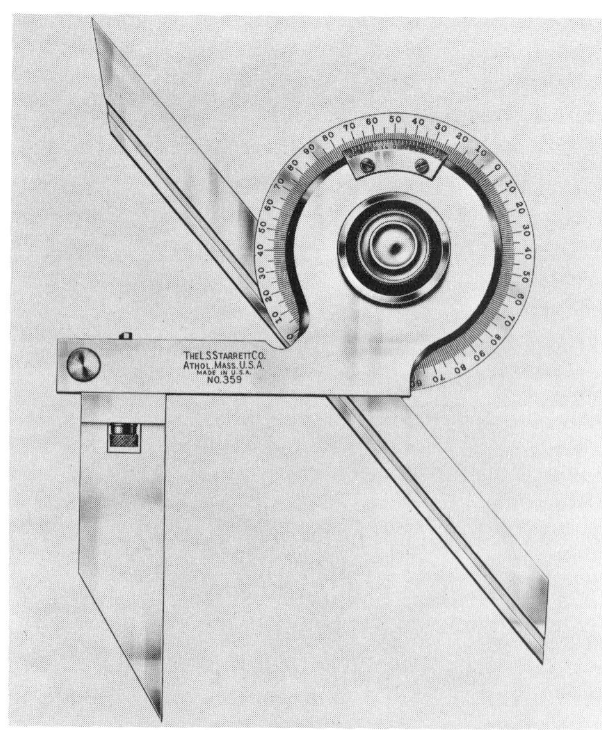

Fig. 4-56. Note this Vernier protractor.

The protractor dial, graduated into 360 degrees reads 0-90 degrees and 90-0 degrees. Every ten degrees is numbered, and each 5 degrees is indicated by a fine line longer than those on either side. The Vernier scale is divided into twelve equal parts on each side of the "0." Every third graduation is numbered 0, 15, 30, 45, and 60, representing minutes. Each division equals 5 minutes. Since each degree is divided into 60 minutes, each division is equal to 5/60 of a degree.

To read the protractor, note the number of degrees that can be read up to the "0" on the Vernier plate. To this, add the number of minutes indicated by the line beyond the "0" on the Vernier plate that aligns exactly with a line on the dial.

Protractor example (See Fig. 4-57.)

The reading is composed of:

The "0" is slightly beyond 50 = 50° 00′

The line indicating 20 minutes is aligned with a line on the dial = 20′

Total reading = 50° 20′

Care of Vernier tools

Reasonable care in handling these expensive tools will ensure their accuracy.

1. Wipe with a soft lint-free cloth before using. This will prevent dirt and grit from being "ground in," which would eventually destroy the accuracy of the tool.
2. Store the tool in the case designed to hold it.
3. Never force the tool when making measurements.
4. Use a magnifying glass or a jeweler's lope to make Vernier readings. Hold the tool so the light is reflected on the scale.
5. Hold the tool as little as possible. Sweat and body acids cause rapid rusting and staining.
6. Periodically check for accuracy. Use a measuring standard, Jo block, or ground parallel.
7. Wipe the tool with a lightly oiled, soft cloth after use and before storage. Return the tool to the manufacturer for adjustments and repairs.
8. Lay Vernier heights gages on their side when not in use at the surface place or layout table. There will then be no danger that they will get knocked over and damaged.

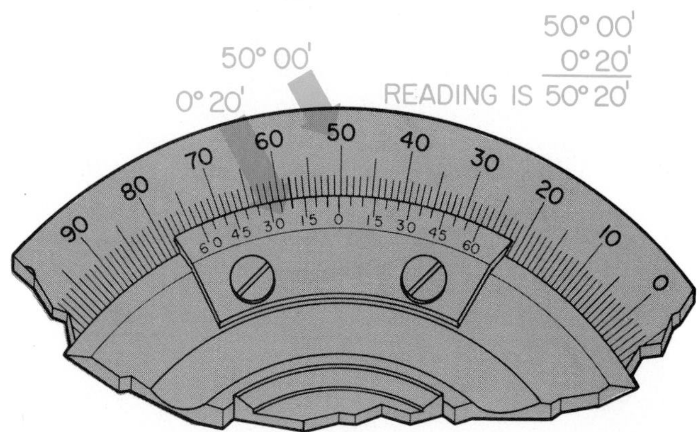

Fig. 4-57. Study how to read a universal Vernier bevel protractor.

GAGES

It is impractical to check every dimension on every manufactured part with conventional measuring tools. For rapid checking, ***plug, ring,*** and ***snap gages, precision gage blocks, dial indicators*** and other ***electronic, optical, air type*** and ***laser gages*** are employed. These gaging devices can quickly determine whether the dimensions of a manufactured part are within specified limits or tolerances.

Gaging, which is the term used when checking parts with various gages, differs somehwat from measuring. ***Measuring*** requires the skillful use of precision measuring tools to determine the exact geometric size of the piece. Gaging, on the other hand, simply shows whether the piece is made within the specified tolerances.

When great numbers of an item, with several critical dimensions, are manufactured, it may NOT be possible to check each piece. It therefore becomes necessary to decide how many pieces, picked at random, must be checked to assure satisfactory quality and adherence to specifications. This technique is called ***statistical quality control.***

Several types of gages have been developed. Each has been devised to do a specific job.

Remember! Always handle gages carefully. If dropped or mishandled, their accuracy could be affected. Gages provide a method of checking your work and are very important tools.

Plug gage

Plug gages are used to check whether hole diameters are within specified tolerances. The ***double end cylindrical plug gage,*** has two gaging members known as GO and NO-GO plugs, Fig. 4-58. The ***GO plug*** should enter the hole with little or no interference. The ***NO-GO plug*** should NOT enter if the opening is made to specifications.

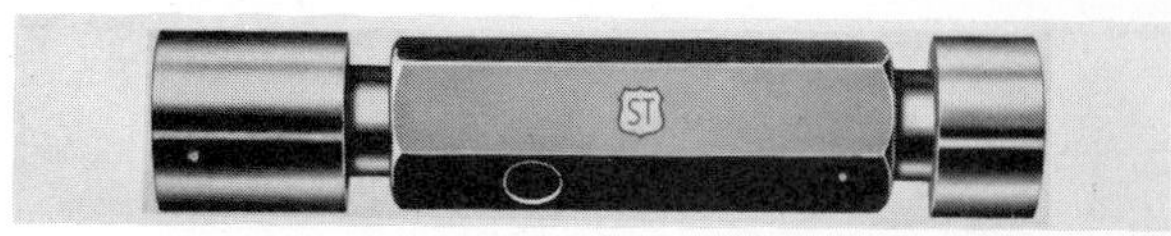

Fig. 4-58. This is a double end cylindrical plug gage. (Standard Tool Co.)

The GO plug is made LONGER to distinguish it from the NO-GO plug. A ***progressive*** or ***step plug gage,*** is able to check the GO and NO-GO dimensions in one motion. See Fig. 4-59.

Ring gage

External diameters are checked with ***ring gages.*** The GO and NO-GO ring gage are separate units, and can be distinguished from each other by a groove cut on the knurled outer surface of the NO-GO gage. Refer to Fig. 4-60.

On ring gages, the gage tolerance is OPPOSITE to that applied to the plug gage. The opening of the GO gage is larger than the NO-GO gage.

Fig. 4-59. This is a progressive or step plug gage (Standard Tool Co.)

Fig. 4-60. Ring gages, the larger sizes are cut away to reduce weight. (Standard Tool Co.)

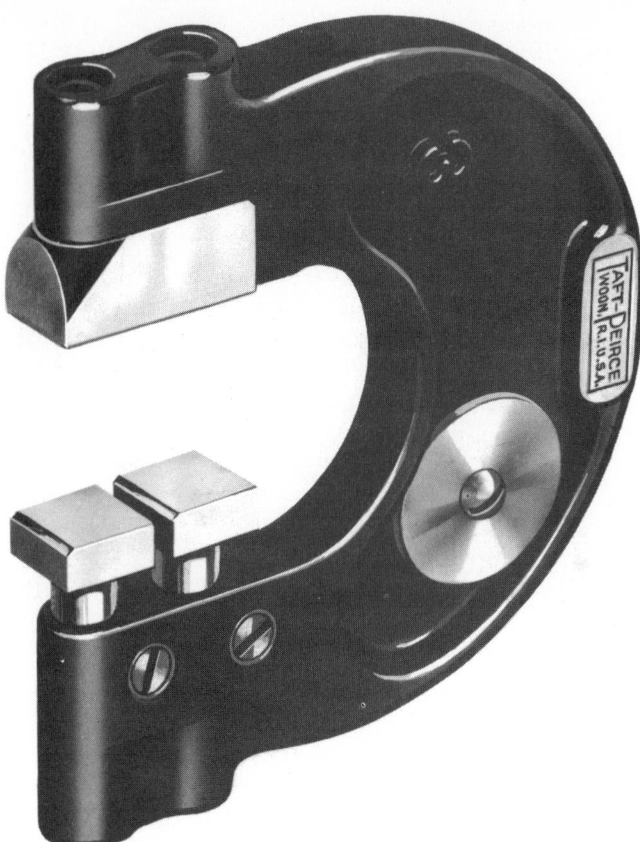

Fig. 4-61. This is an adjustable type snap gage. (Taft-Pierce Co.)

Snap gage

A ***snap gage*** functions much the same as a ring gage. It is made in three general types:

1. The ***adjustable snap gage,*** Fig. 4-61, which can be adjusted through a range of sizes.
2. The ***nonadjustable snap gage,*** Fig. 4-62, which is made for one specific size.
3. The ***dial indicator snap gage,*** Fig. 4-63, on which plus or minus tolerances are read directly from the indicator. The dial face has a double row of graduations reading in opposite directions from zero. MINUS graduations are in red and PLUS graduations are in black. They are available for gaging outside diameters from 0 – 2 in., 2 – 4 in., 4 – 6 in., and 6 – 8 in.

Thread gages

Gages similar to those just described are used to check screw thread fits and tolerances and are known as ***thread plug gages,*** Fig. 4-64 and Fig. 4-65, ***thread ring gages,*** Fig. 4-66, and ***thread roll snap gages,*** Fig. 4-67.

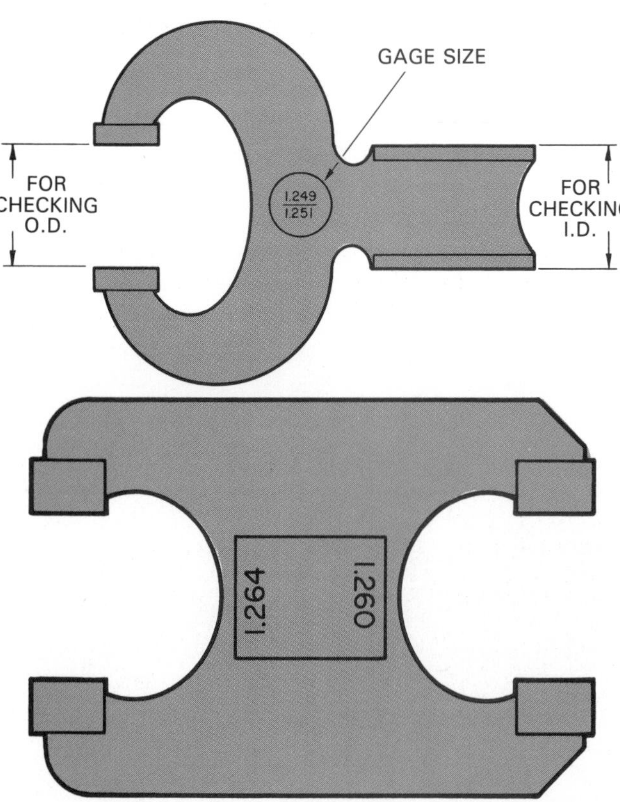

Fig. 4-62. Top—Combination internal-external nonadjustable snap gage. Bottom—Nonadjustable type external snap gage.

Fig. 4-63. This is a dial indicator snap gage. (L.S. Starrett Co.)

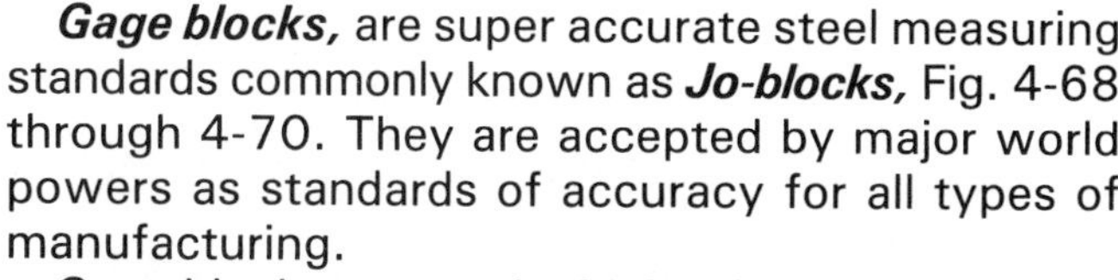

Gage blocks, are super accurate steel measuring standards commonly known as ***Jo-blocks,*** Fig. 4-68 through 4-70. They are accepted by major world powers as standards of accuracy for all types of manufacturing.

Gage blocks are used widely to check and verify the accuracy of master gages; as working gages for toolroom work; and for laying out and setting up work for machining where extreme accuracy is required.

Gage blocks can be purchased in various combinations or sets ranging from a few carefully selected blocks that meet conditions found in many shops, to a complete set of 121 blocks, Fig. 4-71.

The new Federal Accuracy Grades for gage blocks are shown in Fig. 4-72.

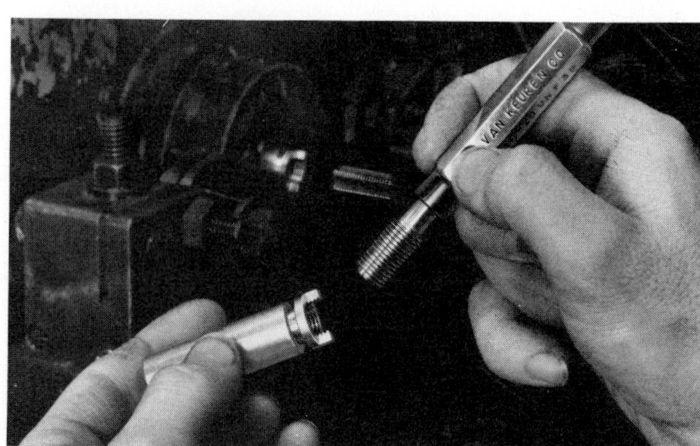

Fig. 4-64. Machinist is using a thread plug gage to check a job.

Fig. 4-65. Note thread plug gage. (Standard Tool Co.)

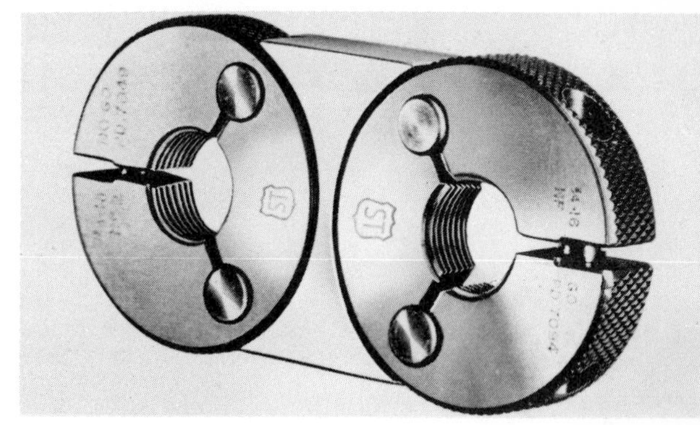

Fig. 4-66. Note thread ring gage. (Standard Tool Co.)

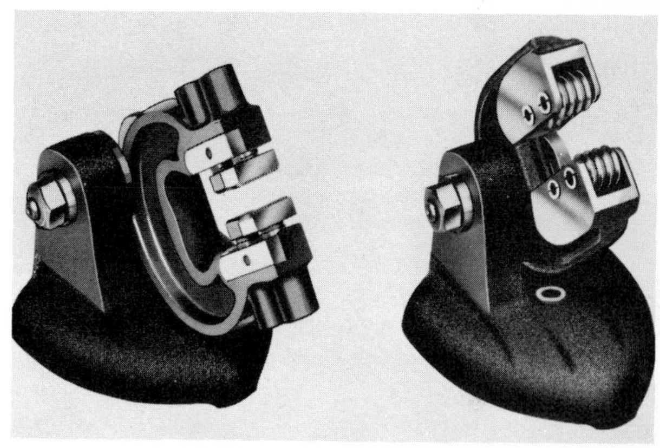

Fig. 4-67. This is a Go NO-GO thread snap gage. (Taft-Pierce Co.)

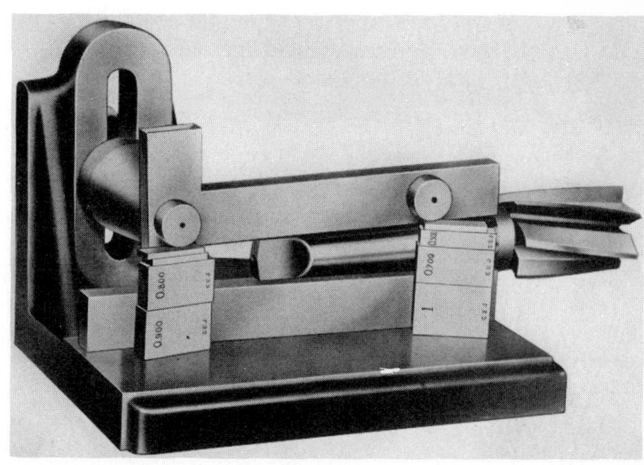

Fig. 4-68. This is an example of measuring a taper with Johannson sine bar and precision gage blocks. (C.E. Johannson & Co.)

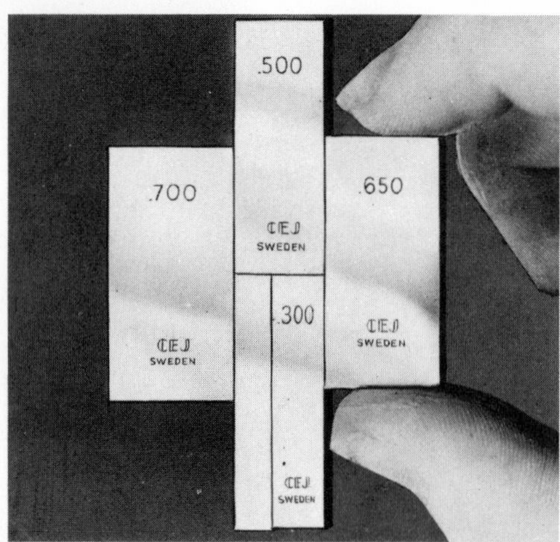

Fig. 4-69. Gage blocks are so accurately made that clean blocks will adhere to one another with considerable pressure when they are "wrung" together. Two or more smaller blocks can be assembled into a larger unit and still maintain the accuracy of the single unit. (C.E. Johannson & Co.)

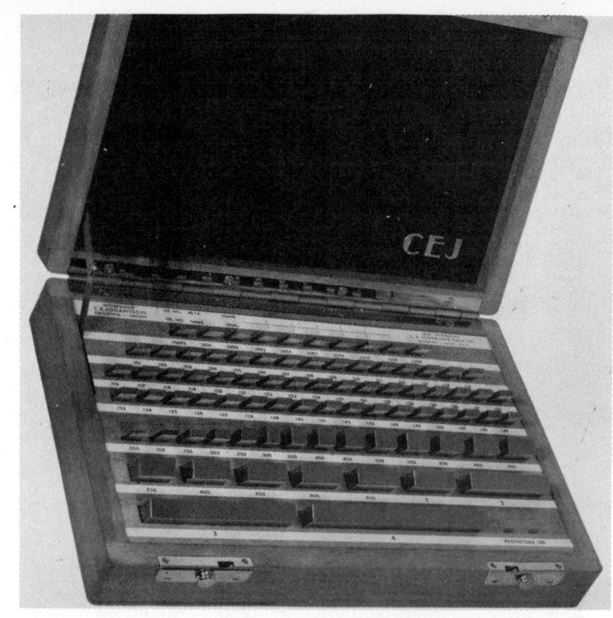

Fig. 4-71. This is a typical set of jo-blocks or gage blocks. (C.E. Johannson & Co)

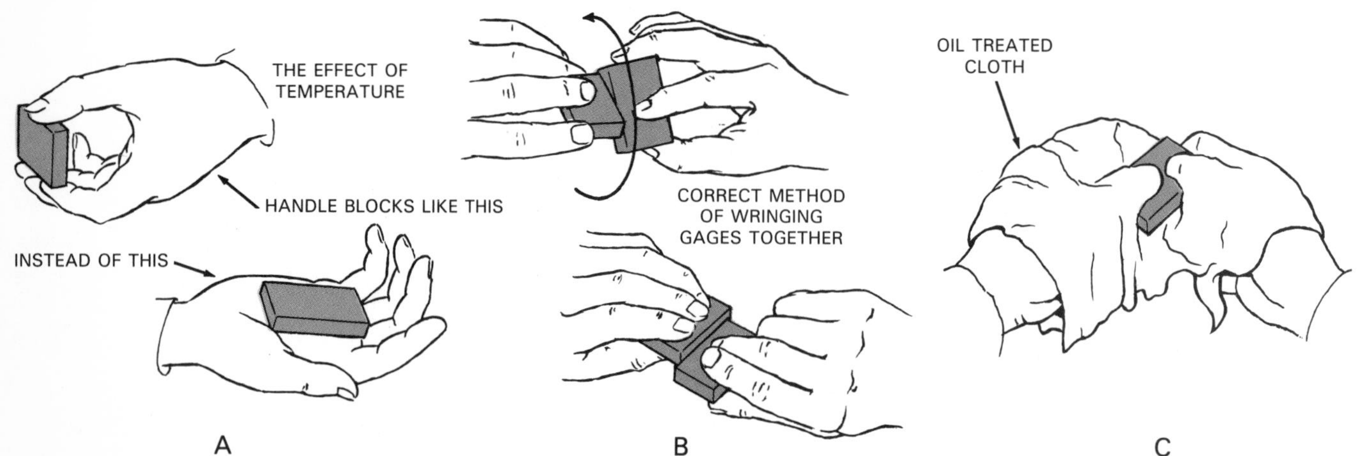

Fig. 4-70. Study proper care of gage blocks: A—Improper handling of gage blocks causes temperature changes. This can expand or contract metal in blocks making errors of measurement. For the most accurate results, measuring should be made in air conditioned, temperature controlled gaging rooms. Handle the blocks only when they must be moved; then only with the tips of the fingers as illustrated, and for the briefest period possible. B—When wringing gage blocks together to build up to desired size, wipe the gages and then carefully slide them together. They should then take hold and adhere to each other strongly. Do not leave them together for extended periods because of the possibility of the contacting surfaces corroding. C—Wipe blocks with soft cloth or chamois campered with an oil recommended by the gage manufacturer before storing. (Webber Gage Div., L.S. Starrett Co.)

FEDERAL ACCURACY GRADES			
		Tolerance	
Accuracy Grade	**Former Designation**	**English System (Inch)**	**Metric System (Millimeter)**
0.5	AAA	±.000001"	±.00003 mm
1	AA	±.000002"	±.00005 mm
2	A+	+.000004" −.000002"	+.0001 mm −.00005 mm
3	A&B	+.000006" −.000002"	+.00015 mm −.00005 mm

Reference Temperature: 68 °F (20 °C)
One Inch = 25.4 millimeters exactly

Fig. 4-72. Note Federal Accuracy Grades for gage blocks.

DIAL INDICATORS

Industry is constantly searching for ways to reduce costs, yet maintain quality. Inspection has always been a costly part of manufacturing. To speed up this phase of production, without sacrificing accuracy, dial indicators and electronic gaging are receiving increased attention.

Much use is made of ***dial indicators*** for centering and aligning work on machine tools, Fig. 4-73, checking for eccentricity, Fig. 4-74 and Fig. 4-75, and visual inspection of work.

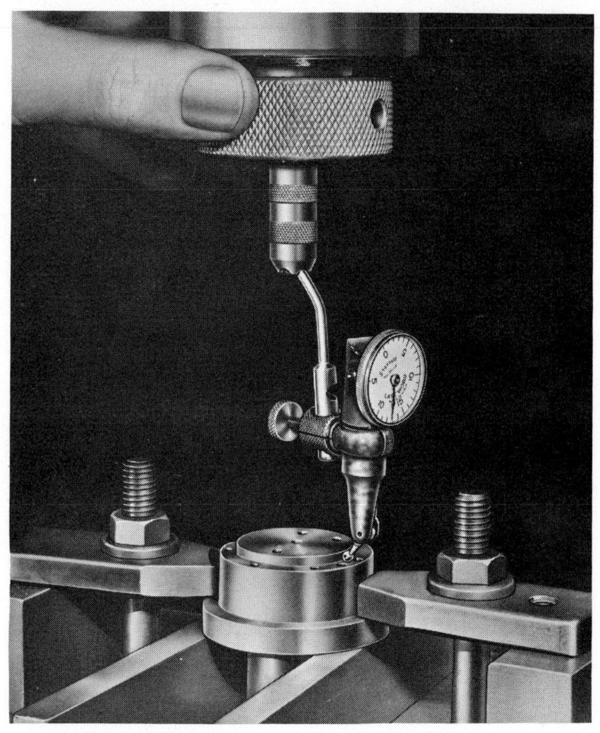

Fig. 4-73. Machinst is centering work on a vertical milling machine with a dial indicator. (L.S. Starrett Co.)

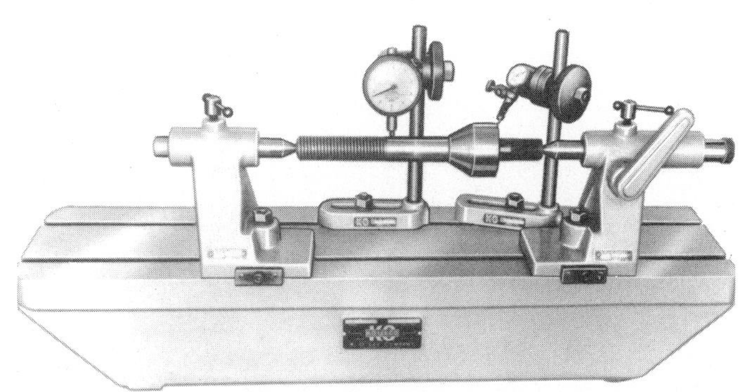

Fig. 4-74. Dial indicators are being used to check runout of a shaft. The bench center and surface plate, on which shaft is mounted, provides an accurate means for inspecting all types of work. (K.O. Lee)

Fig. 4-75. Checking drill press column and quill bearing surfaces for alignment with a dial indicator. Note air gage in background for checking hole diameters. (Clausing)

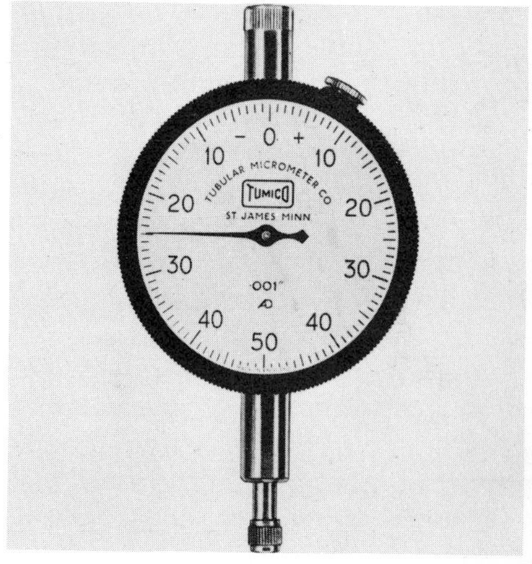

Fig. 4-76. This is a balanced type dial indicator face. (Scherr-Tumico)

Dial indicators are made like fine watches with shockproof movements and jeweled bearings. They are either of the ***balanced type,*** Fig. 4-76, where the figures read in BOTH directions from "0," and the ***continuous type,*** Fig. 4-77, that reads from "0" in a clockwise direction.

Dial faces are available in a wide range of graduations. They usually read in 1/1000 (0.001) in./ 1/100 (0.01) mm or 1/10.000 (0.0001) in./ 2/1000 (0.002) mm increments. Dial indicators must be mounted to rigid holding devices.

How to use a dial indicator

The hand on the dial is actuated by a sliding plunger. Place the plunger lightly against the work until the hand moves. The dial face is turned until the "0" line coincides with the hand. As the work or unit touching the plunger is slowly moved, the indicator hand will measure this movement. For example, it might show the difference between the high and low points or the total run-out of the piece in a lathe. When machining, adjustments are made until there is little or no indicator movement.

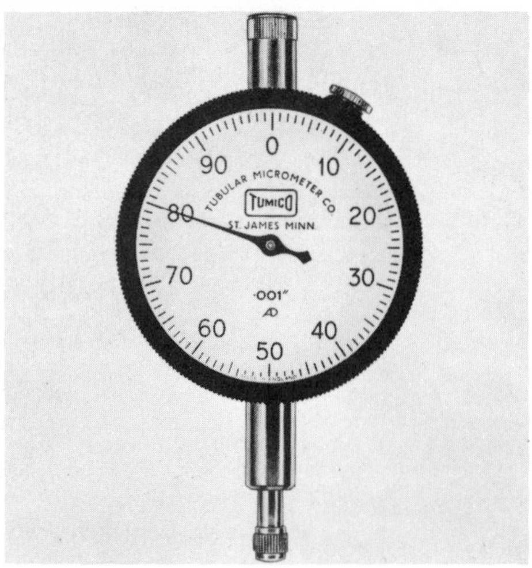

Fig. 4-77. Note continuous type dial indicator face. (Scherr-Tumico)

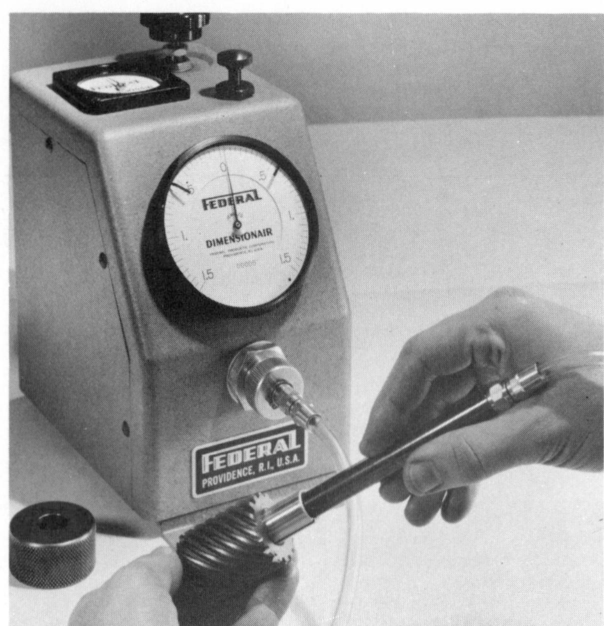

Fig. 4-78. An air gage is slid into hole. Amount of air leakage out of hole determines measurement. Note graduations on dial face. Knurled device, to the left, is the ring gage used to set and check gage accuracy. (Federal Products Corp.)

OTHER GAGING TOOLS

Industry makes wide use of other types of gaging tools. Most of them are for special purposes and are NOT usually found in a school shop. However, since you might have to use them in industry, it is important to learn about such tools.

Air gage

An ***air gage*** uses air pressure to measure hole sizes and is especially helpful when measuring deep internal bores, Figs. 4-78 and 4-79. There is no actual contact between the measuring plug or gage and wall of the bore being measured. The bore measurement depends on the air leakage between the plug and the hole wall. Pressure is built up and the measurement of the back pressure gives an accurate measurement of the hole size. A larger bore diameter would allow more air leakage and vice versa.

Change in pressure (air leakage) is measured by a dial type indicator, a cork floating on the air stream, or by a ***manometer*** (U-tube in which height of fluid in tube indicates pressure).

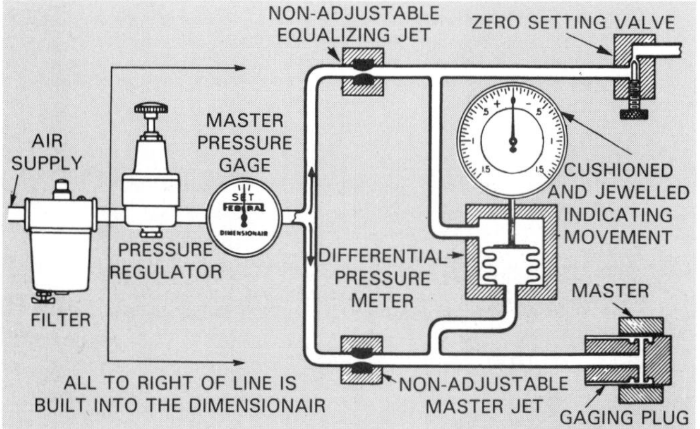

Fig. 4-79. Diagram illustrates how air gage operates.

Electronic gage

Another gage that can make extremely close measurements is the ***electronic gage,*** Fig. 4-80. It is a comparison type gage and must be calibrated by means of master gauge blocks.

Laser gaging

The ***laser*** is a device that produces a very narrow beam of extremely intense light that can be utilized for communication, medical, and industrial applications. LASER is the abbreviation for ***L***ight ***A***mplification by ***S***timulated ***E***mission of ***R***adiation.

The laser is another area of technology that has moved from the laboratory into the shop. When employed for inspection purposes, it can check the accuracy of critical areas in machined parts quickly and accurately. Refer to Fig. 4-81.

Optical comparator

The ***optical comparator*** uses magnification as a means for production inspection, Fig. 4-82. An enlarged image of the part is projected upon a screen for inspection. The part image is superimposed upon an enlarged accurate drawing of the correct shape and size. The comparison is made visually. Variations as small as 0.0005 in. (0.012 mm) can be noted by a skilled operator, Fig. 4-83.

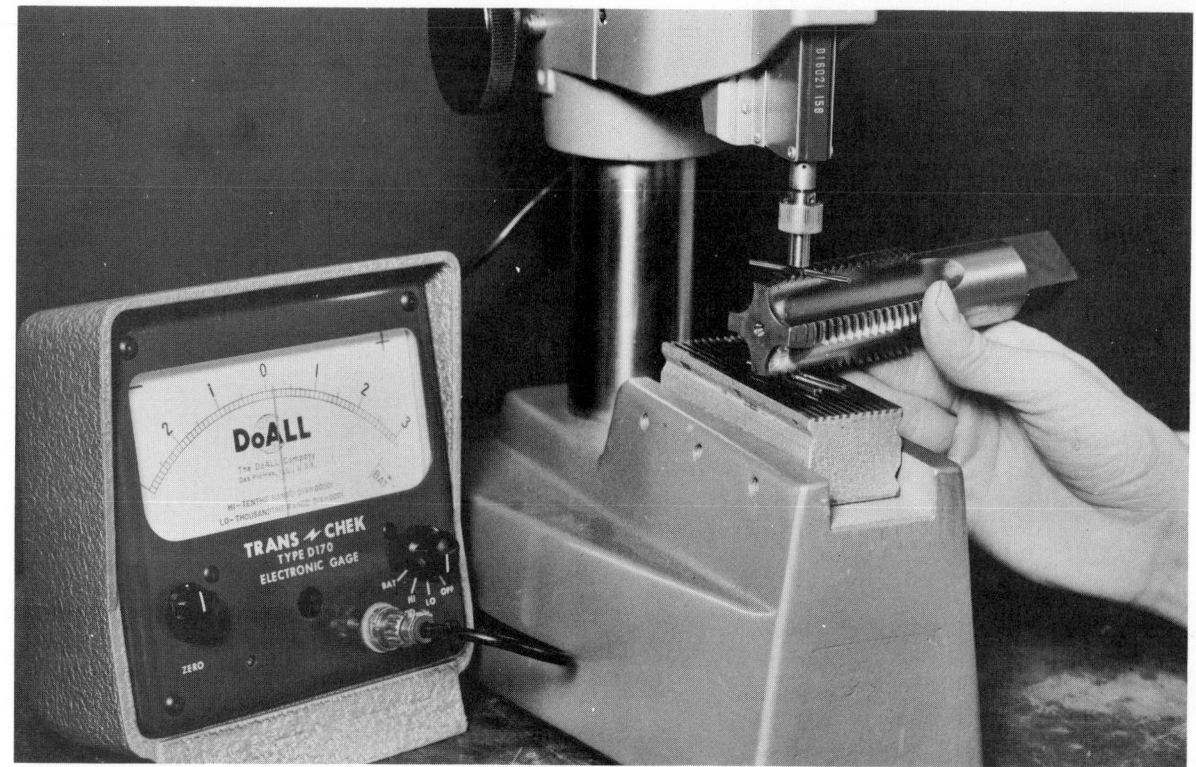

Fig. 4-80. Electronic comparator is being used to check pitch diameter of a tap by means of the 3-wire measuring technique. (DoALL Co.)

Fig. 4-81. Laser is being used to inspect a part from a car's automatic transmission. Manually, one person could inspect no more than four units an hour. The laser can inspect the part as fast as machine is turned on—120 or more an hour. (Ford Motor Co.)

Fig. 4-82. This 50-power optical comparator permits a fast check of the tooth formation on a tap.

Optical flats

Optical flats are precise measuring instruments that use light waves as a measuring standard, Fig. 4-84. The flats are made of quartz with one face ground and polished to optical flatness. When this face is placed on a machined surface and a special light passed through it, light bands appear on the surface, Fig. 4-85. The shape of these bands in-

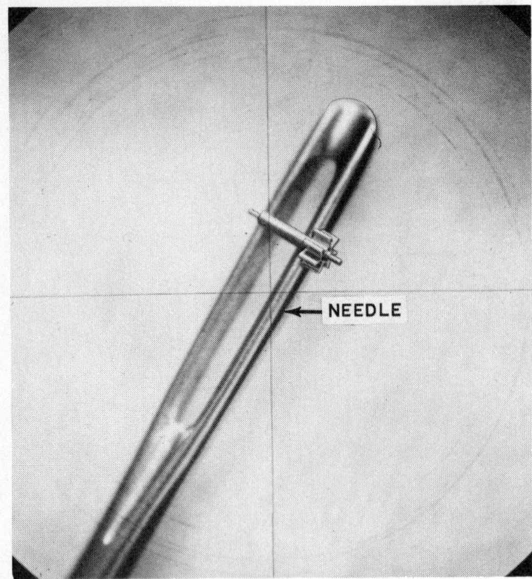

Fig. 4-83. Optical comparator can be used to check the accuracy of tiny components, like this watch part. Note that watch gear was not being inspected when this photo was taken. (Hamilton Watch Co.)

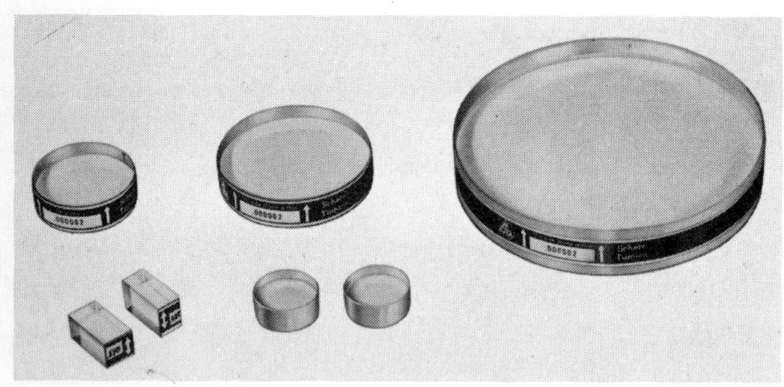

Fig. 4-84. These are optical flats.

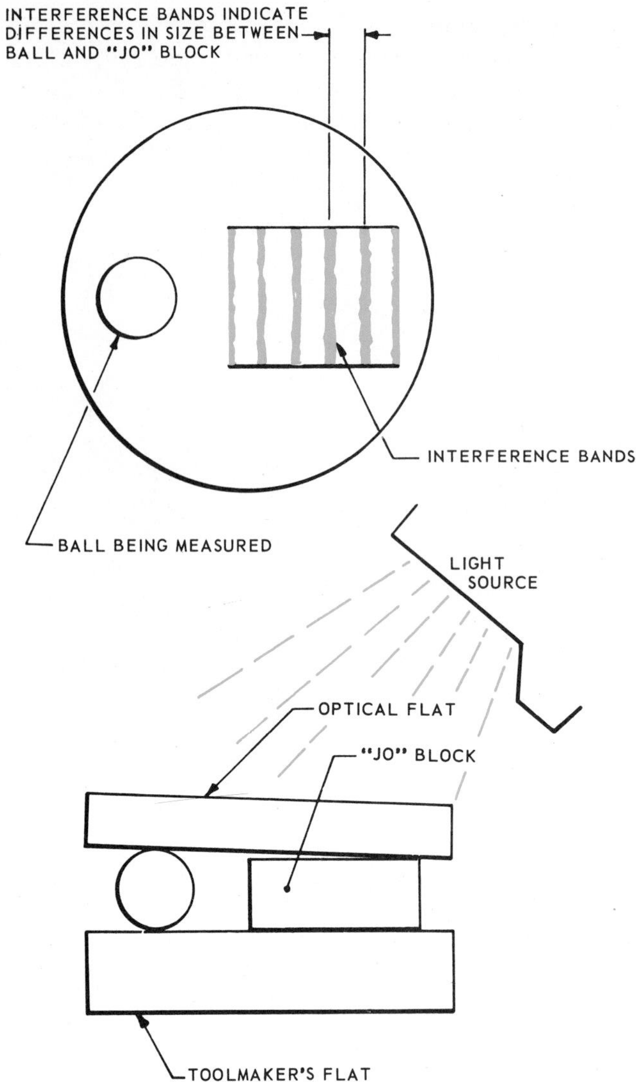

Fig. 4-85. Study principle of the optical flat.

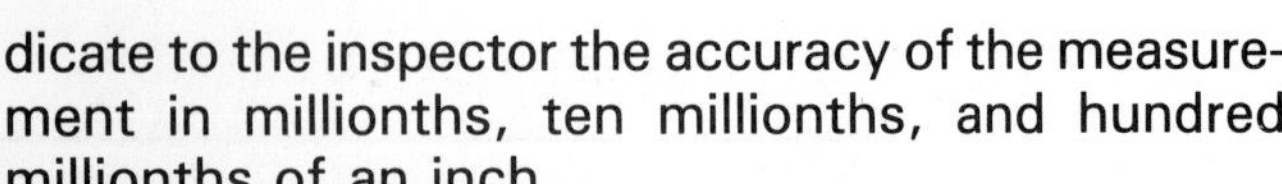

dicate to the inspector the accuracy of the measurement in millionths, ten millionths, and hundred millionths of an inch.

MISCELLANEOUS MEASURING TOOLS

There are some measuring tools that do not fall under a specific category, yet are frequently found in the shop.

Thickness (feeler) gage

Thickness gages, are pieces or leafs of metal manufactured to precise thicknesses, Fig. 4-86. Various thicknesses, 0.001 to 0.030 in. (0.03 to 0.50 mm), are available in short lengths and long 25 ft. (7.6 m) rolls. For convenience, several blade thicknesses are provided and marked at regular intervals. Thickness gages are made of tempered steel and are usually 1/2 in. (12.7 mm) wide.

Thickness gages are ideal for measuring narrow slots, setting small gaps and clearances, determining fit between mating surfaces, and for checking flatness of parts in straightening operations.

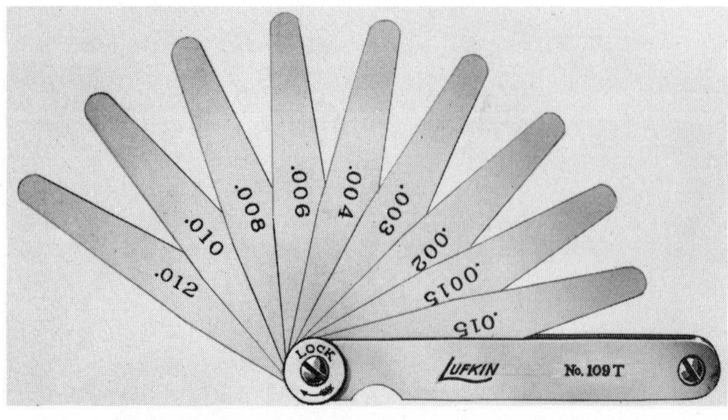

Fig. 4-86. These are thickness or feeler gages.

Screw pitch gage

Screw pitch gages are employed to determine the pitch or number of threads per inch, Fig. 4-87. Each blade is stamped with the pitch or number of threads per inch. Screw pitch gages are available in English and metric thread sizes.

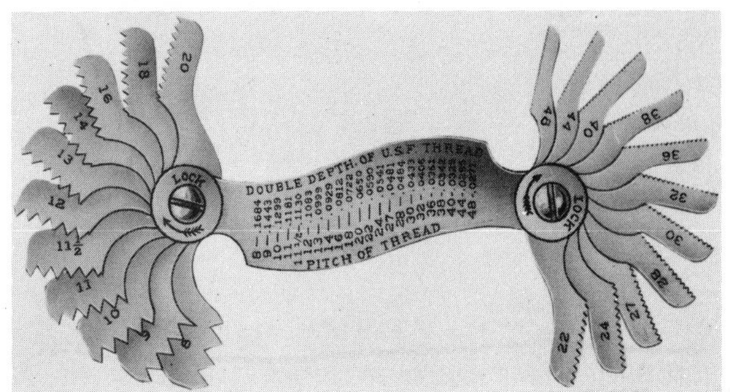

Fig. 4-87. Screw pitch gages are made for both inch based and metric threads.

Fillet and radius gage

The thin steel blades of a ***fillet and radius gage,*** are used to check concave and convex radii on corners or against shoulders, for layout work and inspection, or as a template when grinding form cutting tools. See Figs. 4-88 and 4-89.

A holder, Fig. 4-90, is especially helpful for checking radii in hard to reach locations. The gages increase in radius by 1/64 in. (0.5 mm) increments.

Drill rod

Drill rods are steel rods manufactured to close tolerances to twist drill diameters. They are useful for inspecting hole alignment and location, checking hole diameters in the same manner as a plug gage, etc. Drill rod is available in both inch and metric sizes.

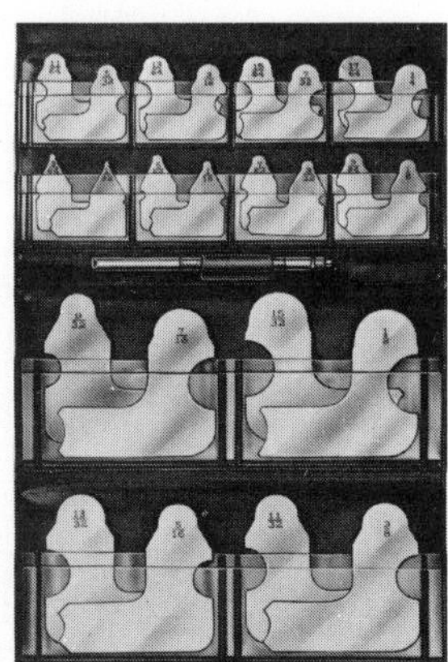

Fig. 4-88. This is a set of radius and fillet gages.

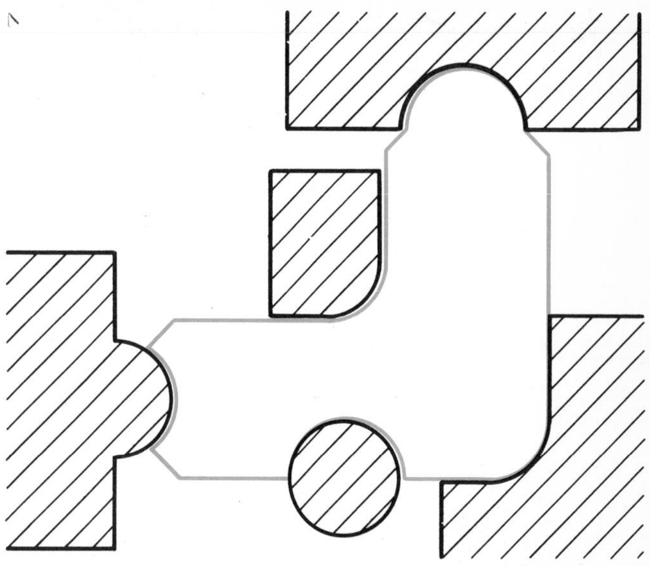

Fig. 4-89. Note several uses of the radius gage.

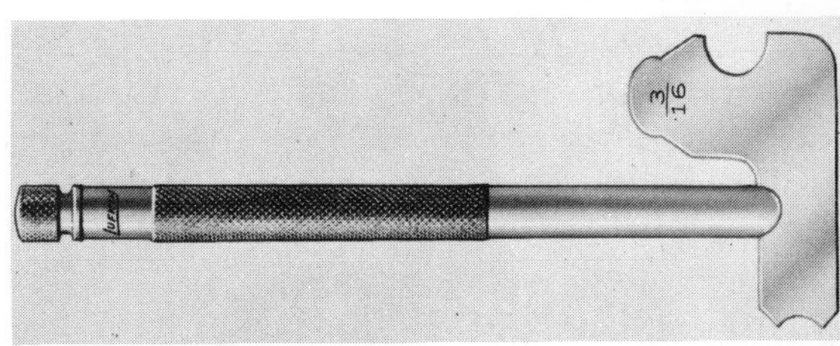

Fig. 4-90. This is radius gage and holder. Holder is made so that it permits gage to be set at different angles. (Lufkin Rule Co.)

HELPER TYPE MEASURING TOOLS

Some measuring tools are NOT direct reading and require the help of a rule, micrometer, or Vernier caliper to determine the size of the measurement taken. These are called ***helper measuring tools.***

Calipers

External measurements of 1/64 in. (0.4 mm) can be made with ***outside calipers,*** Fig. 4-91. A caliper does NOT have a dial nor scale which shows a measurement; it must be compared with a steel rule.

Round stock is measured by setting the caliper square with the work and moving the caliper legs down on the stock. Adjust the tool until the caliper point bears lightly on the center line of the stock. Caliper weight should cause the caliper to slip over the diameter. Hold the caliper next to the rule to make the reading, Fig. 4-92.

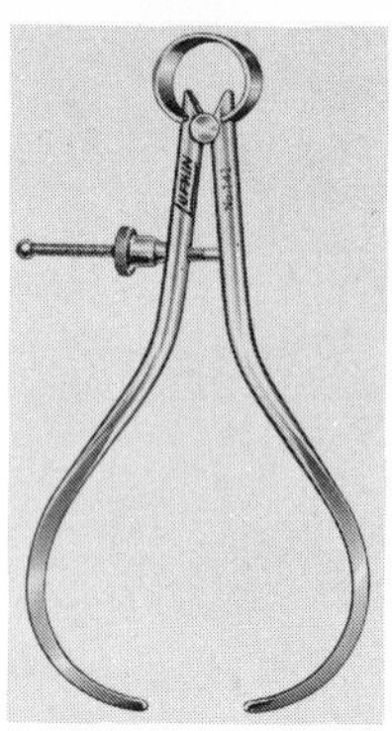

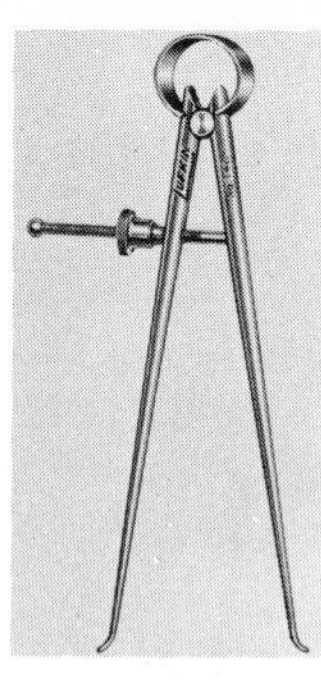

Fig. 4-91. A—Spring-joint outside caliper. B—Spring-joint inside caliper. (Lufkin Rule Co.)

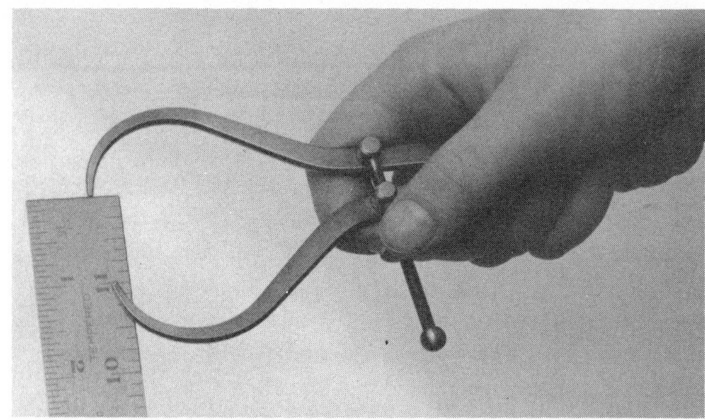

Fig. 4-92. Setting an outside caliper.

An *inside caliper* is used for making internal measurements where 1/64 in. (0.4 mm) accuracy is acceptable, Fig. 4-93.

Hole diameter can be measured by setting the caliper to approximate size, and inserting the legs into the opening, Fig. 4-94. Hold one leg firmly against the hole wall, and adjust the thumbscrew until the other leg lightly touches the wall exactly opposite the first leg. The legs should "drag" slightly when moved in and out, or from side to side. Read hole size as shown in Fig. 4-95.

Considerable skill is required to make accurate measurements with a caliper. Much depends upon the machinist's sense of touch. With practice, measurements with accuracy within 0.003 to 0.005 in. (0.07 to 0.12 mm) can be made. However, a micrometer or Vernier caliper is preferred and must be utilized when greater accuracy is required.

Telescoping gage

The *telescoping gage* is intended for use with the micrometer to determine internal dimensions, Fig. 4-96. To use such a gage, compress the contact legs. The legs telescope within one another under spring tension. Insert the gage into the hole and allow the legs to expand, Fig. 4-97. After the proper fitting is obtained, lock the contacts into position. Remove the gage from the hole and make your reading with a micrometer.

Six telescoping gages, with a measuring capacity of 5/16 to 6 in. (8.0 to 150.0 mm) comprise a set. See Fig. 4-98.

Small hole gage

The *small hole gage* permits measuring smaller openings than is possible with a telescoping gage, Fig. 4-99. The contacts are designed to allow accurate measurement of shallow grooves, and small

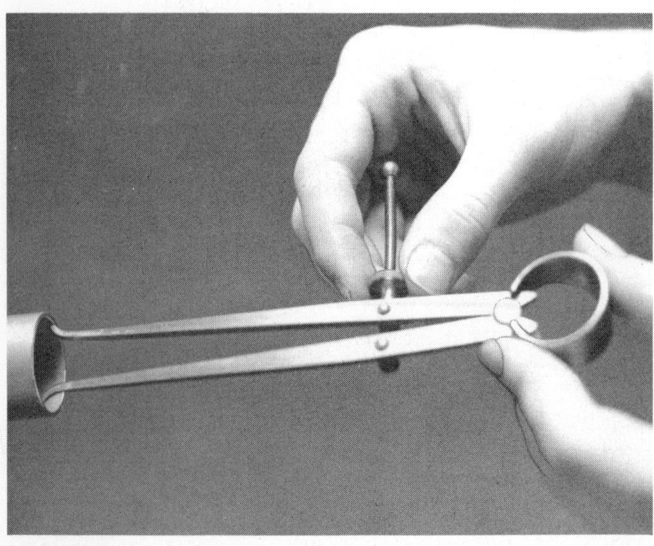

Fig. 4-93. Practice is needed to measure with an inside caliper.

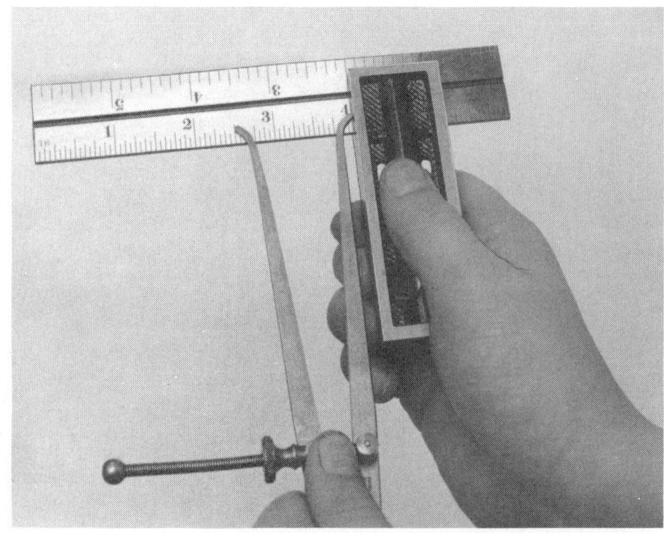

Fig. 4-94. Read inside caliper by holding it on ruler scale.

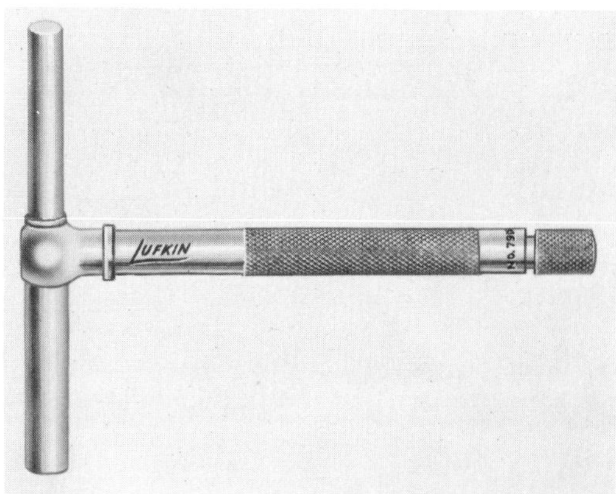

Fig. 4-95. Telescoping gage is spring loaded and will lock into place. (Lufkin Rule Co.)

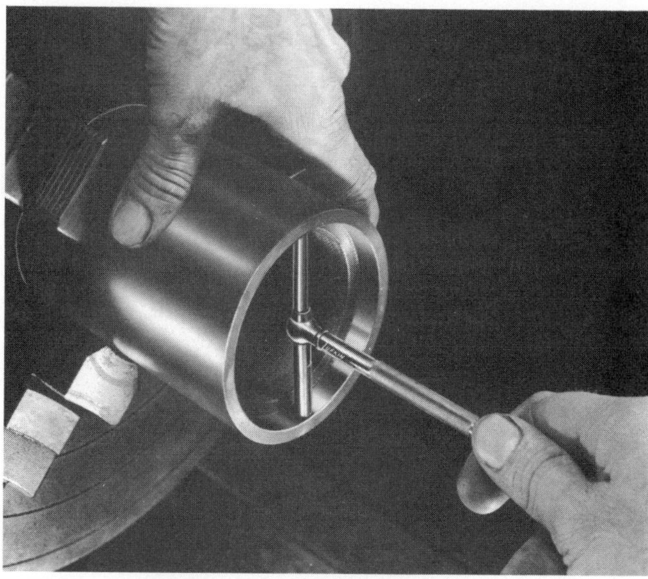

Fig. 4-96. Machinist is fitting a telescoping gage into a hole.

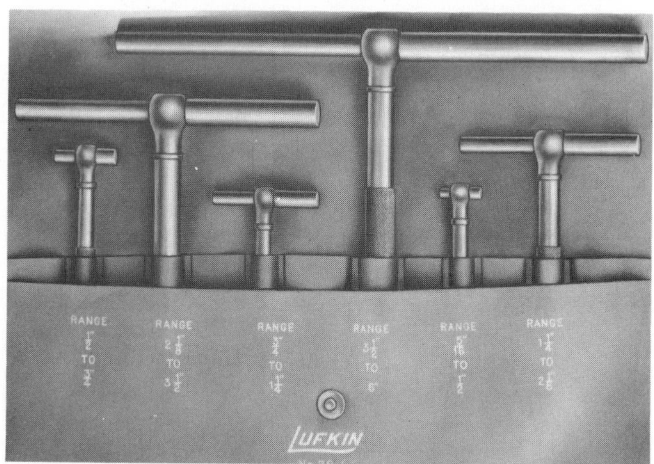

Fig. 4-97. This is a typical set of telescoping gages.

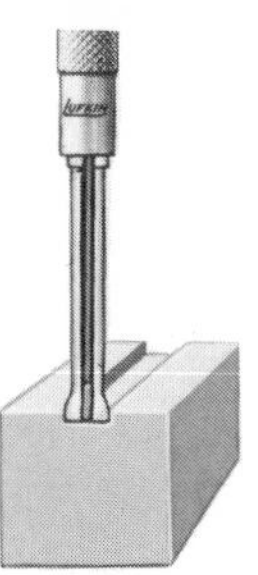

Fig. 4-98. Note how to measure a shallow groove with a small hole gage. (Lufkin Rule Co.)

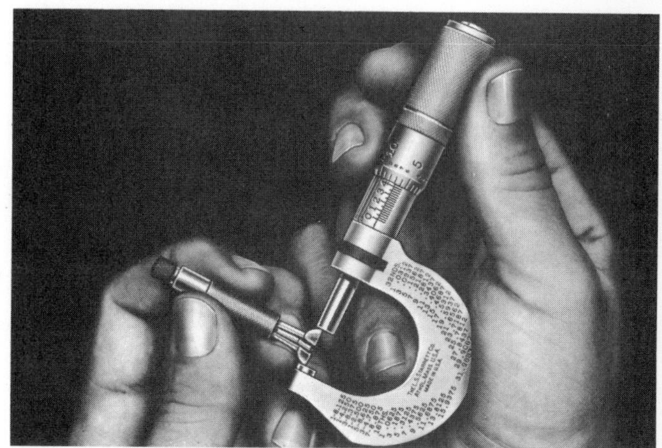

Fig. 4-99. Note correct way to measure a small hole gage with a micrometer. (L.S. Starrett Co.)

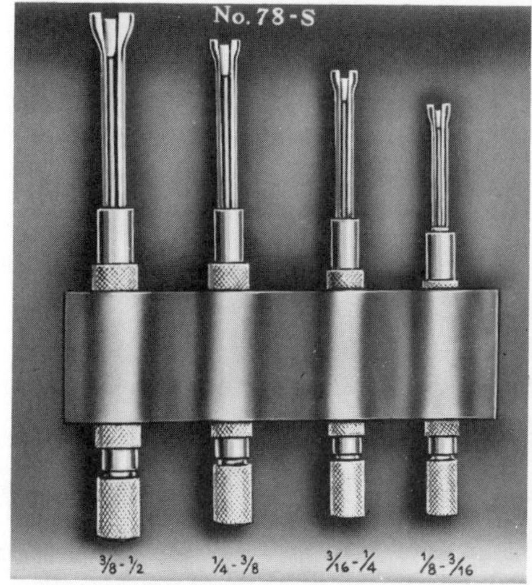

Fig. 4-100. This small hole gage set has a range of 1/8 to 1/2 in. or 3.5 to 12.5 mm.

diameter holes. They are adjusted to size by the knurled knob at the end of the handle. Measurement is made over the contacts with a micrometer, Fig. 4-100. A set consists of four gages with a range of 1/8 to 1/2 in. (3.0 to 12.5 mm), Fig. 4-100.

TEST YOUR KNOWLEDGE—Chapter 4

Please do not write in the text. Place your answers on a separate sheet of paper.

1. Make reading from the rules.

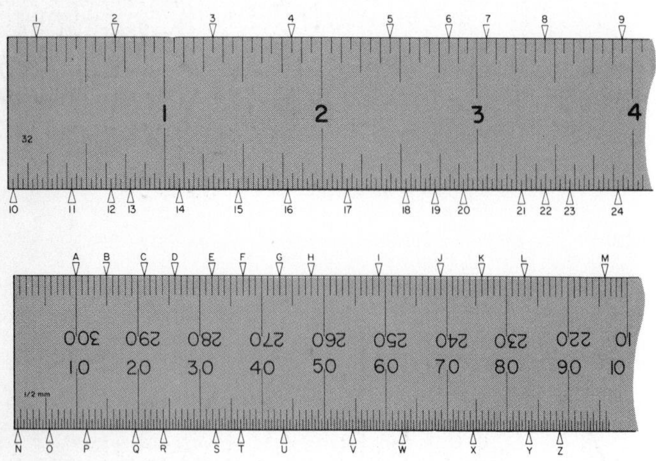

2. Make readings from the Vernier scale shown below.

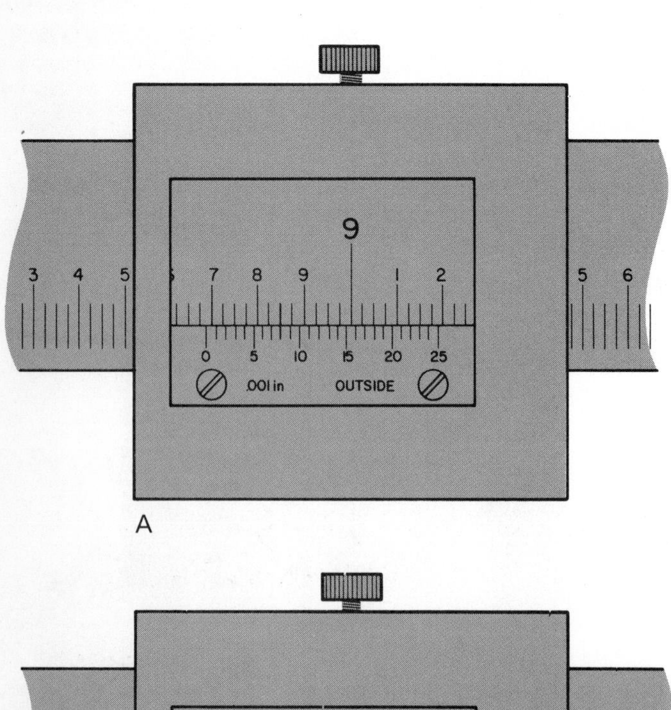

A

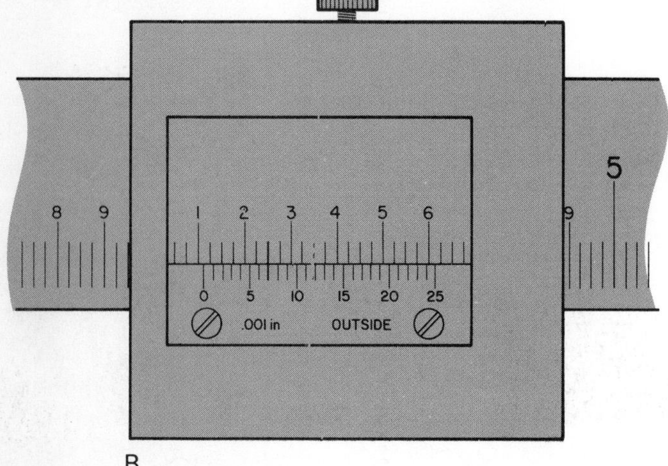

B

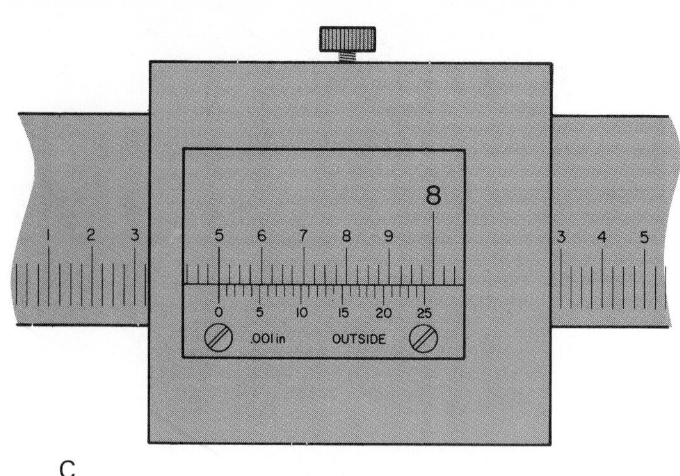

C

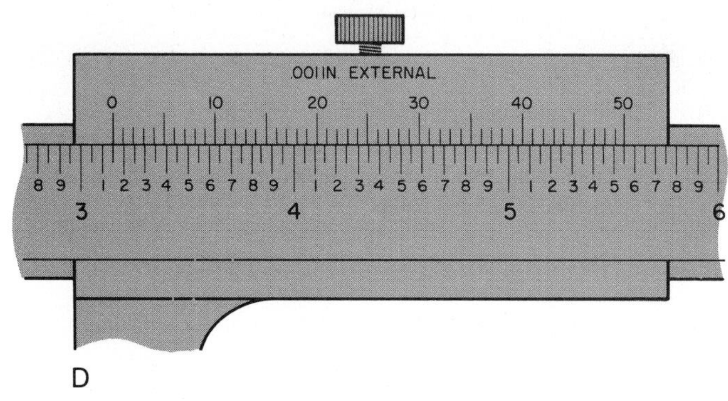

D

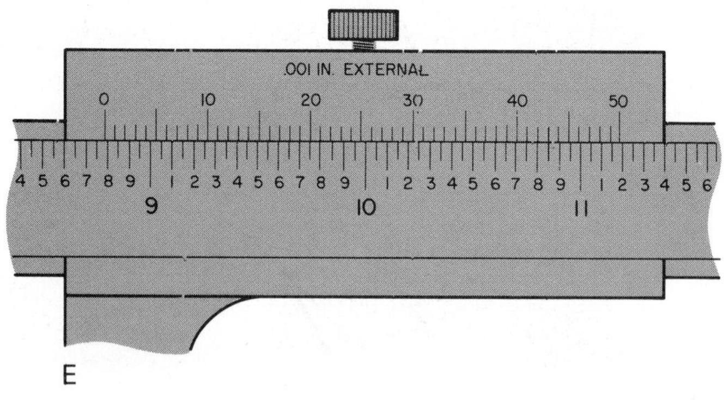

E

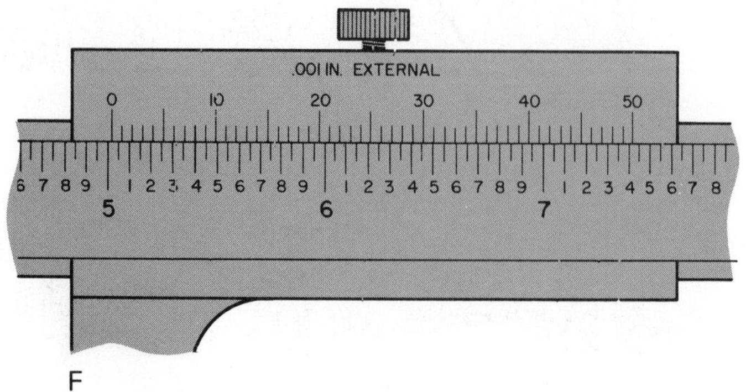

F

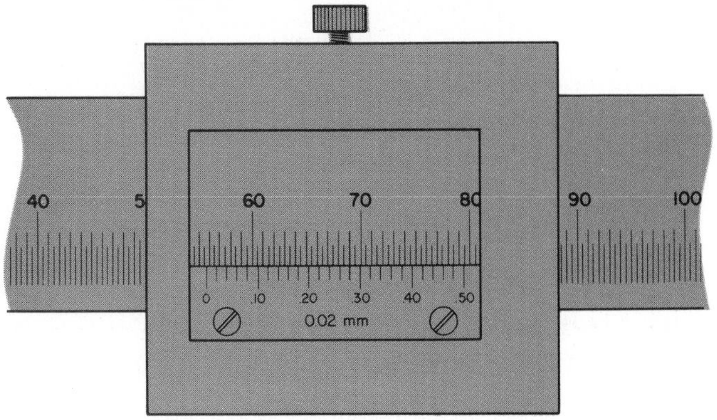

G

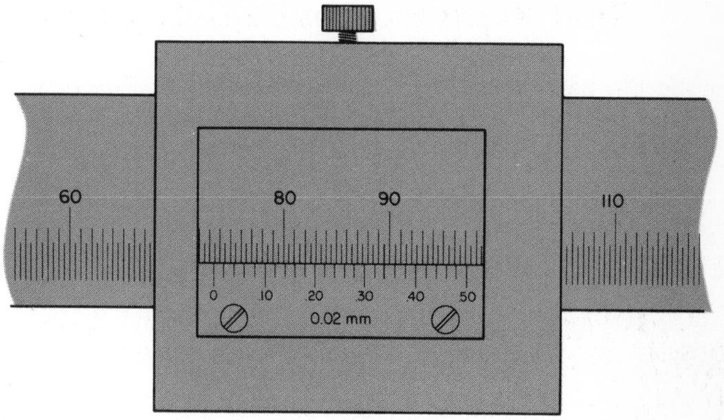

H

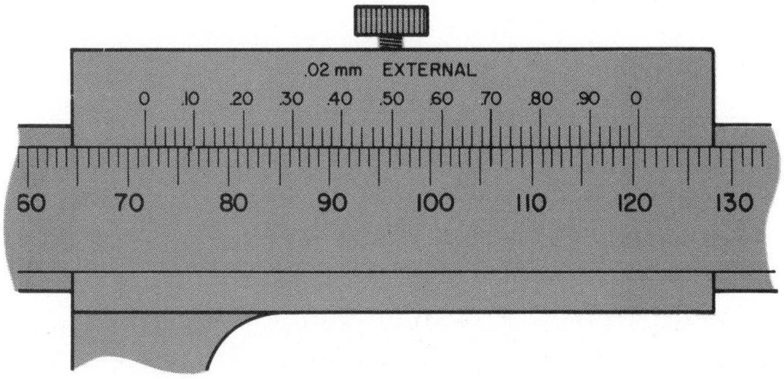

I

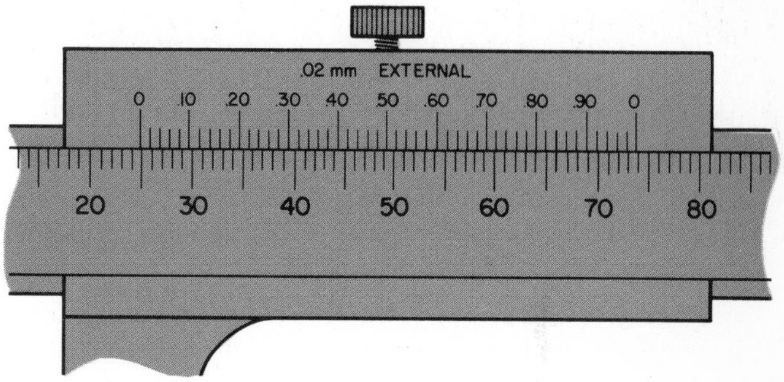

J

3. Make readings from the micrometer illustrations.

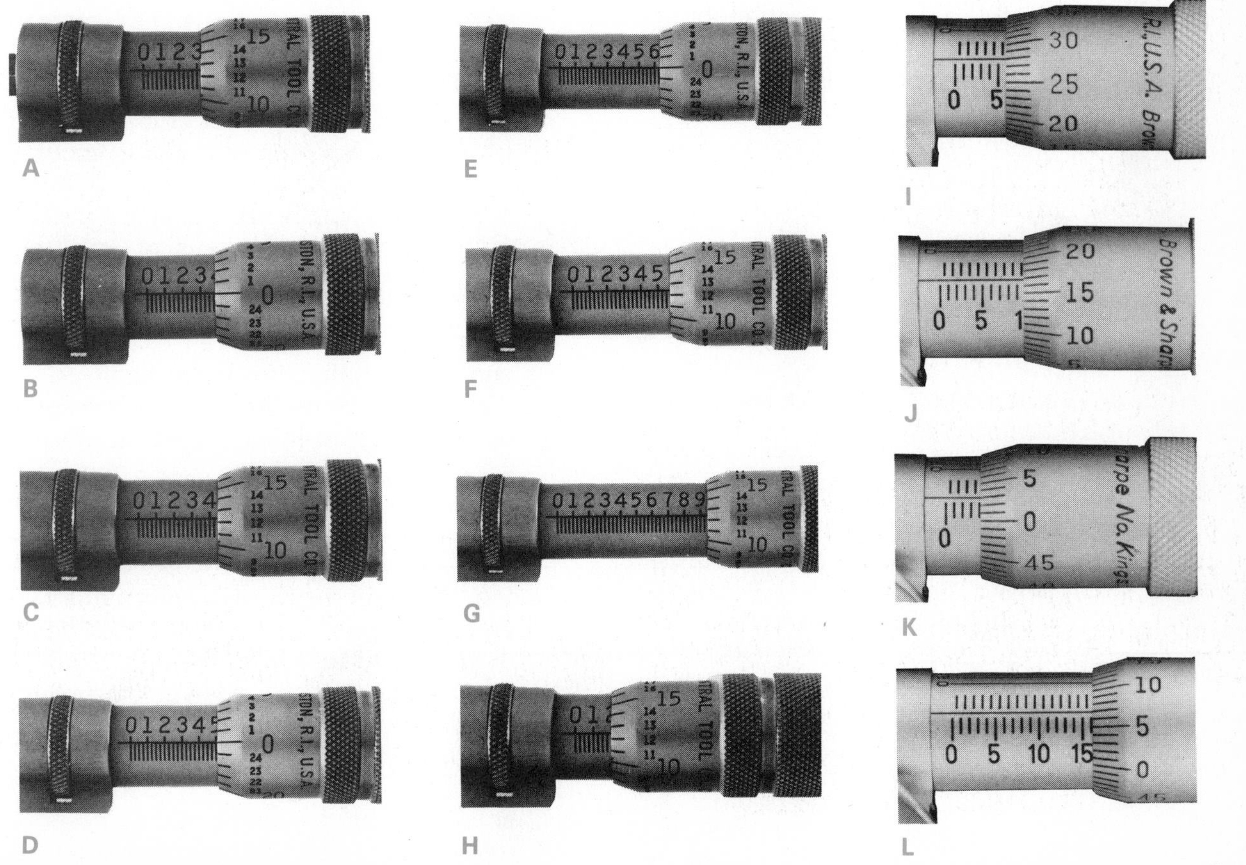

4. The micrometer has been nicknamed _______.

Answer the following questions as they pertain to measurement.

5. A millionth part of a standard inch is known as a _______.
6. One millionth part of a meter is known as a _______.
7. The micrometer is capable of measuring accurately to the _______ and _______ part of standard inch and metric versions to _______ and _______ millimeters.
8. The Vernier caliper has several advantages over the micrometer. List two of them.
9. A Vernier caliper can measure to the _______ part of the inch and the metric version to _______ millimeters.
10. List six (6) precautions that must be observed when using a micrometer or Vernier caliper.
11. The Vernier type tool for measuring angles is called a _______ _______ _______ _______.
12. How does a double end cylindrical plug gage differ from a progressive or step plug gage?
13. The ring gage is used to check whether _______ are within the specified _______ range.
14. Gage blocks are often referred to as _______ blocks.
15. An air gage type measuring tool employs air pressure to measure deep internal openings. It operates on the principle of:
    a. Air pressure leakage between the plug and hole walls.
    b. The amount of air pressure needed to insert the tool properly in the hole.
    c. Amount of air pressure needed to eject the gage from the hole.
    d. All of the above.
    e. None of the above.
16. The dial indicator is available in two basic types. List them.
17. What are some uses for the dial indicator?
18. Name the measuring device that employs light waves as a measuring standard.
19. The _______ _______ is used for production inspection. An enlarged image of the part is projected on a screen where it is superimposed upon an accurate drawing.
20. The pitch of a thread can be determined with a _______ _______ _______.
21. Of what use are fillet and radius gages?
22. What are helper type measuring tools?
23. How is a telescoping gage used?

RESEARCH AND DEVELOPMENT

1. Make a large working model of the hub and thimble of a micrometer. Use different size cardboard mailing tubes.
2. Develop a working model of a Vernier caliper. Make it large enough to be used to instruct the class. The model may be of a 25 division scale or a 50 division scale.
3. Make an enlarged section of a No. 4 rule at least ten times actual size. Use bass wood, plywood, or hardboard.
4. Prepare a transparency of a No. 4 rule using several overlays that can be used on an overhead projector to teach beginners how to read a rule.
5. Design and make a series of posters showing how to read a metric micrometer.
6. Prepare a display of various types of gages. Secure samples of work checked by gage type measuring tools.
7. Arrange for someone to demonstrate how optical flats are used. Use a film or video presentation if the actual equipment cannot be secured.
8. Have a quality control expert, employed by a local industry, describe their job and the specialized measuring tools employed. If such a person cannot be secured, make arrangements to visit such a department at a local industry. Do not go to the meeting "cold." Prepare questions in advance that the class would like to have answered or explained.
9. Prepare a research paper on how temperature changes can affect measuring accuracy.

   Prepare 1.000 in. (25.00 mm) long pieces of aluminum, brass, steel, plastic, and cast iron with exactly the same size cross section. Record their exact lengths at room temperature with a Vernier micrometer caliper.

   Place the sections in a freezer for 24 hours and quickly measure them again.

   Record your findings. Place the sections in boiling water or in a heat treating furnace for 15 minutes at 200 deg. F (93 deg. C). Quickly measure them again. Record your findings.

   Record a graph that will show how sizes varied under the three conditions of temperature. Using this information, have a class discussion how products can be affected by great changes in temperature and how industry takes into account this problem when certain products (aircraft, auto engines, etc.) are designed.
10. Prepare a paper on early measuring tools and some of the problems encountered before measuring standards were established.

Chapter 5

LAYOUT WORK

After studying this chapter, you will be able to:

- Explain why layouts are needed.
- Identify common layout tools.
- Use layout tools safely.
- Make basic layouts.
- List safety rules for layout work.

Laying out is the term that describes the locating and marking of lines, circles, arcs, and points for drilling holes or making cuts. These lines and reference points on the metal show the machinist where to machine.

The tools for this work are known as ***layout tools.*** Many common hand tools fall into this category. The accuracy of the job will depend upon the proper and careful application of these tools.

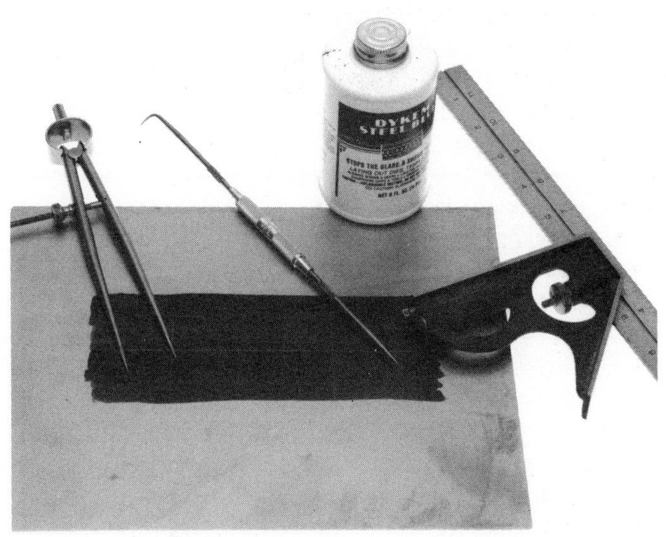

Fig. 5-1. These are a few of the tools needed for a simple layout.

MAKING LINES ON METAL

The shiny finish of most metals would make it difficult to distinguish or see any layout lines. For this reason, some coating *must be* placed on the metal before layout. See Fig. 5-1.

Layout dye

Layout dye, is probably the easiest to use of the many coatings for making layout lines stand out better, Fig. 5-2. This blue-colored fluid, when applied to the metal, offers an excellent contrast between the metal and the layout lines. All dirt, grease and oil must be removed before applying the dye, otherwise, it will NOT adhere properly.

In a pinch, layout fluid can be made by dissolving the coating on spirit duplicator carbons in alcohol. Chalk will also work on hot finished steel as a layout background. A pencil should not be used because it will rub off and would mark too wide.

Scriber

A layout, to be accurate, requires fine lines that must be scribed or scratched into the metal. A ***scriber*** will produce these lines, Figs. 5-2 and 5-3. The point is made of hardened steel, and is kept needle sharp by frequent honing on a fine oilstone. Many styles of scribers are available.

CAUTION! *Never carry a scriber in your pocket.* It will puncture the skin easily.

Divider

Where the scriber is used to draw straight, gradually curved lines, circles and arcs are drawn with the ***divider,*** Fig. 5-4. It is essential that both legs of the tool be equal in length and kept pointed. Measured distances can be laid out with a divider, Fig. 5-5. To set the tool to the correct distance, set

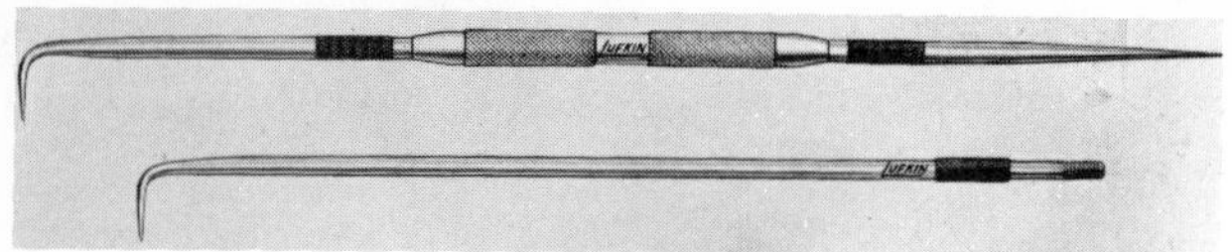

Fig. 5-2. Scribers are used to mark on parts during layout. (Lufkin Rule Co.)

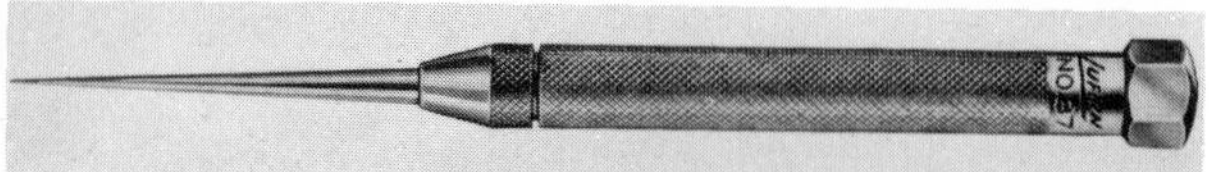

Fig. 5-3. This pocket scriber has point that can be removed, reversed, and stored in handle when tool is not being used.

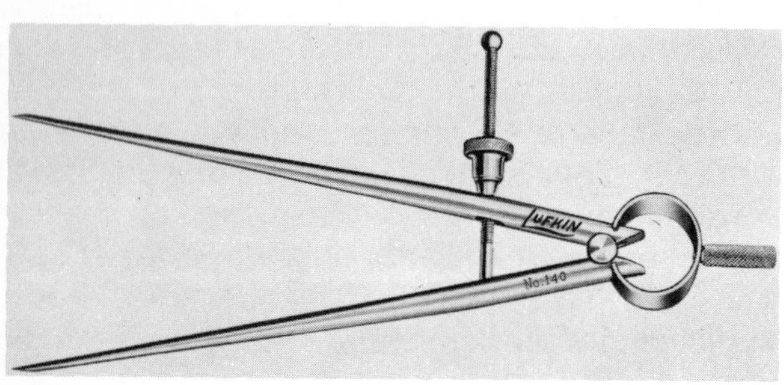

Fig. 5-4. Divider is needed for marking lines, arcs, and circles.

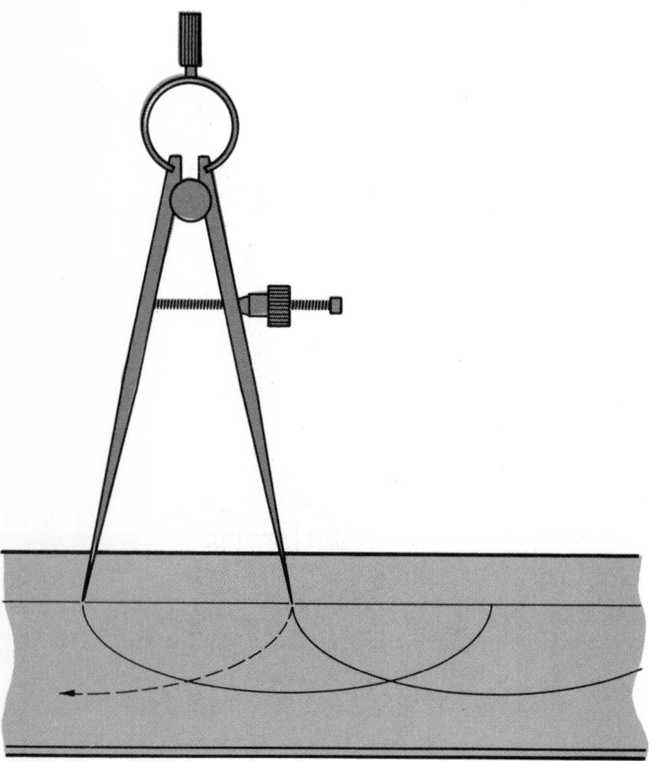

Fig. 5-5. Note how to mark equal spaces with a divider.

one point in the inch or centimeter mark of a steel rule, and open the divider until the other leg is set to the proper measurement, Fig. 5-6.

Circles and arcs TOO LARGE to be made with a divider are drawn with a ***trammel,*** Fig. 5-7. This consists of a long thin rod, called a ***beam,*** on which two ***sliding heads*** with scriber points are mounted. One head is equipped with an adjusting screw. ***Extension rods*** can be added to the beam to increase the capacity of the tool.

The ***hermaphrodite caliper*** is a layout tool which has one leg shaped like a caliper, and the other pointed like a divider, Fig. 5-8. Lines parallel to the edge of the material, either straight or curved, can be drawn with the tool, Fig. 5-9. It can also be used to locate the center of irregularly shaped stock.

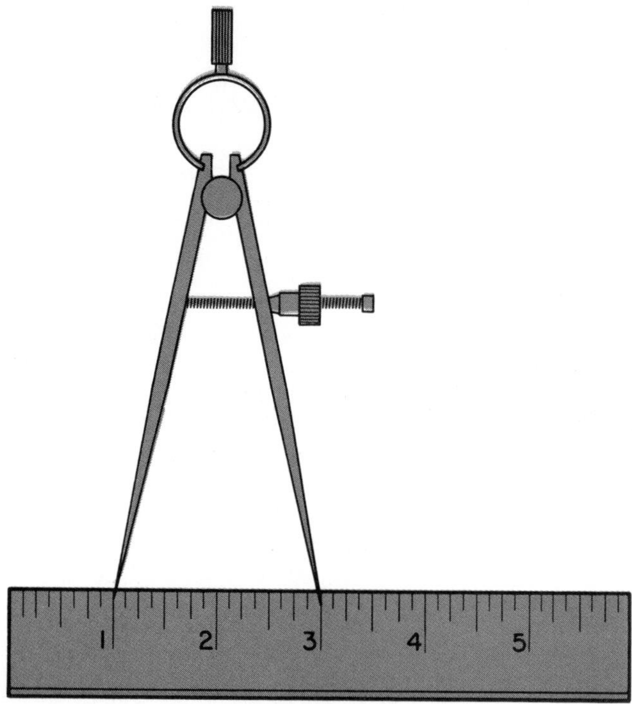

Fig. 5-6. To set a divider to desired dimension, place it on ruler as shown.

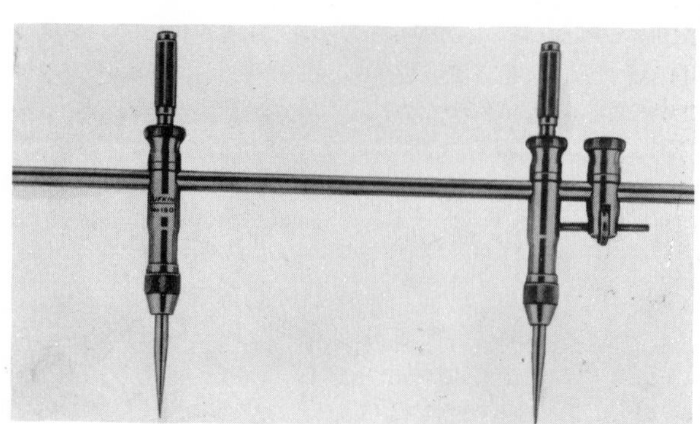

Fig. 5-7. The trammel is used to draw large circles and arcs.

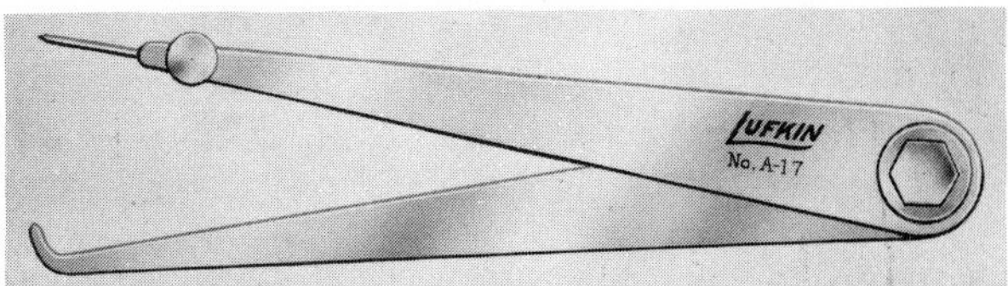

Fig. 5-8. Hermaphrodite caliper has blunt end for sliding surface and point for scribing.

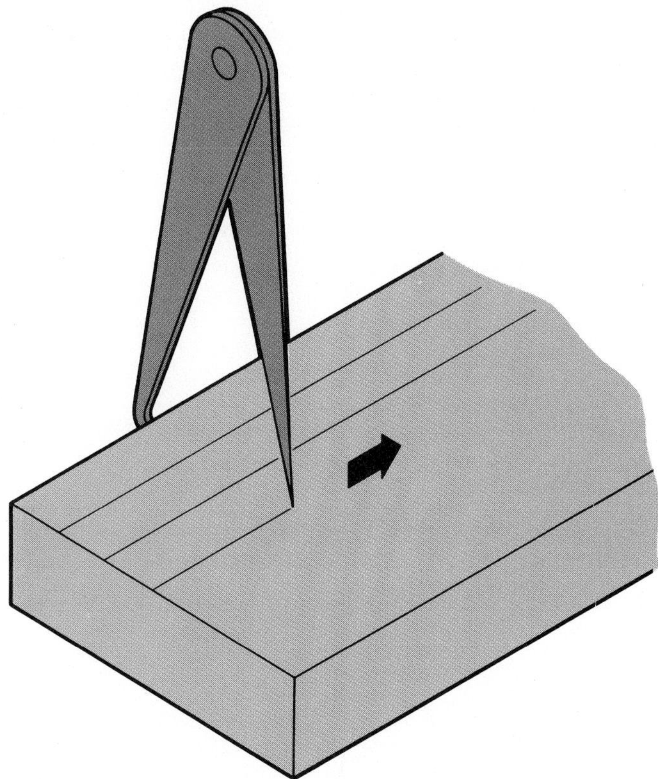
Fig. 5-9. Note how you can scribe lines parallel to an edge with hermaphrodite caliper.

Surface gage

A ***surface gage,*** has many uses, but is most frequently employed for layout work, Fig. 5-10. It consists of a ***base, spindle,*** and ***scriber.*** An adjusting screw is fitted for making fine adjustments. The scriber is mounted in such a manner that it can be pivoted into any position. A surface gage can be utilized to scribe lines, at a given height and parallel to the surface, Fig. 5-11. A V-slot in the base permits the tool to be also employed on a curve surface.

To check whether a part is parallel to a given surface, fit the surface gage with a dial indicator. The indicator is then set to the required dimension with the aid of gage blocks. The tool is then moved back and forth along the work, Fig. 5-12.

Precision layout tools are required when drawings call for positions to be located to within 0.001 in. (0.02 mm). See Figs. 5-13 through 5-15.

Surface plate

Every linear measurement depends upon an accurate reference surface from which final dimensions are taken. ***Surface plates*** provide this reference surface or plane for work inspection and for layout prior to machining.

Surface plates can be purchased in sizes up to 72 in. by 144 in. (1800 mm by 3600 mm) and in various grades. ***Surface plate grade differences*** are given in degrees of flatness:

1. Grade AA for laboratory.
2. Grade A for inspection.
3. Grade B for toolroom and layout applications.

Surface plates are made from semisteel, Fig. 5-16, or granite, Fig. 5-17. However, most surface plates made today are granite. It is more stable and not affected as much by temperature changes.

Surface plates are primarily for layout and inspection work. They should NEVER be used for any job that might mar or nick the surface.

When square reference surfaces are needed, the ***right angle plate,*** is used, Fig. 5-18. The plates can be placed in about any position with the work clamped to the face for layout, measurement, or inspection.

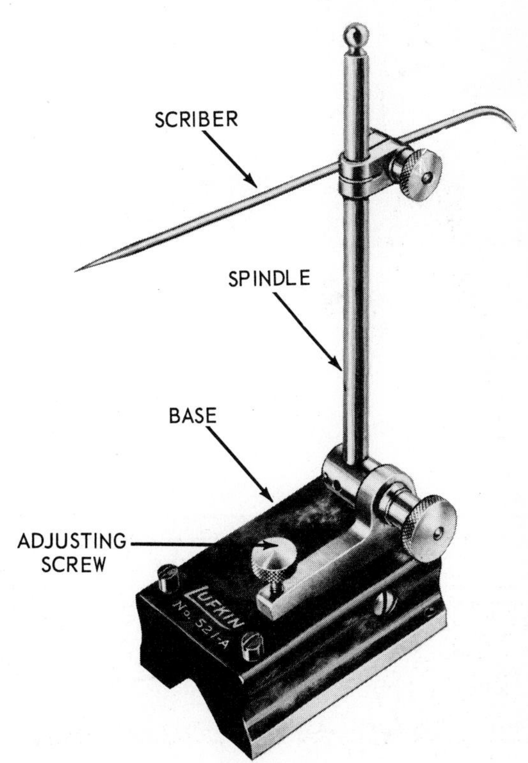

Fig. 5-10. Study parts of surface gage.

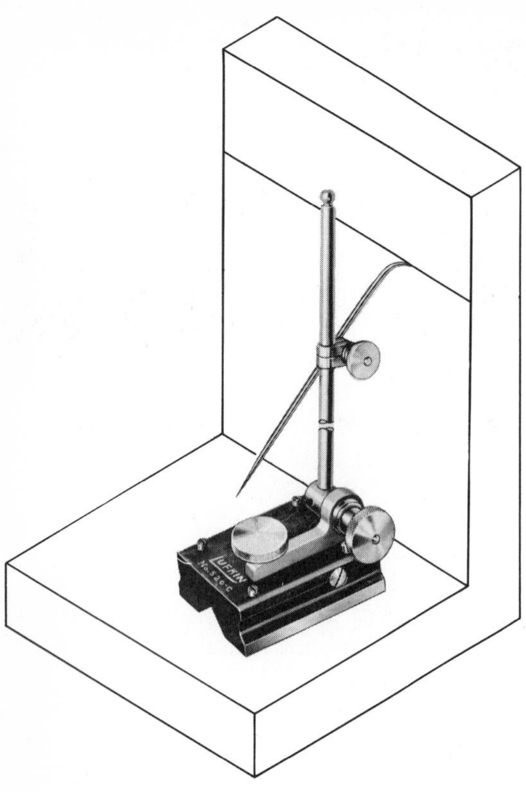

Fig. 5-11. Carefully slide surface gage to scribe lines parallel to base. Handle gauge carefully because sharp points can cause injury.

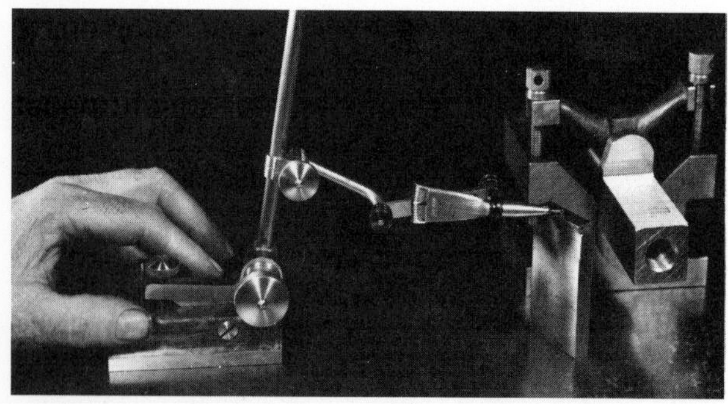

Fig. 5-12. Machinist is setting indicator, mounted on surface gage, using gage blocks. (Lufkin Rule Co.)

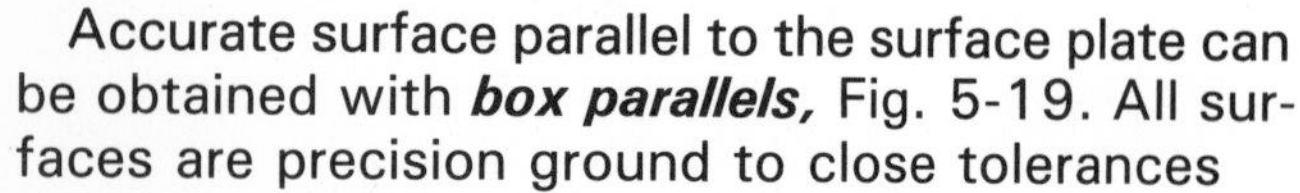

Accurate surface parallel to the surface plate can be obtained with ***box parallels,*** Fig. 5-19. All surfaces are precision ground to close tolerances

V-blocks

V-blocks support round work for layout and inspection, Fig. 5-20. They are furnished in matched pairs with surfaces that are ground square to close tolerances. Ribs are cast into the body for strength, weight reduction, and to provide clamping surfaces.

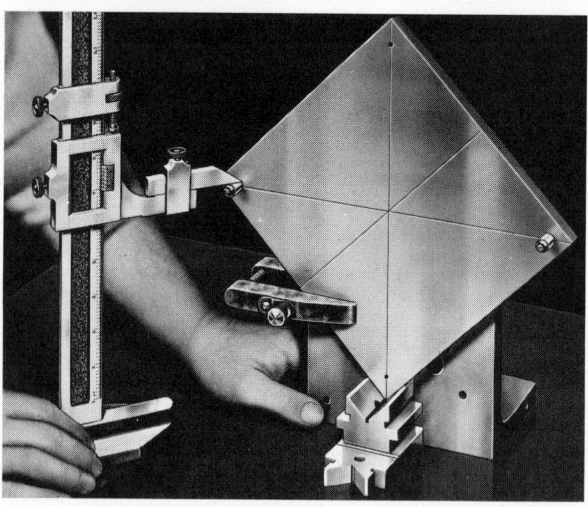

Fig. 5-13. Machinist is using Vernier height gage. Note that a V-block and angle plate support the job during layout. (L.S. Starrett Co.)

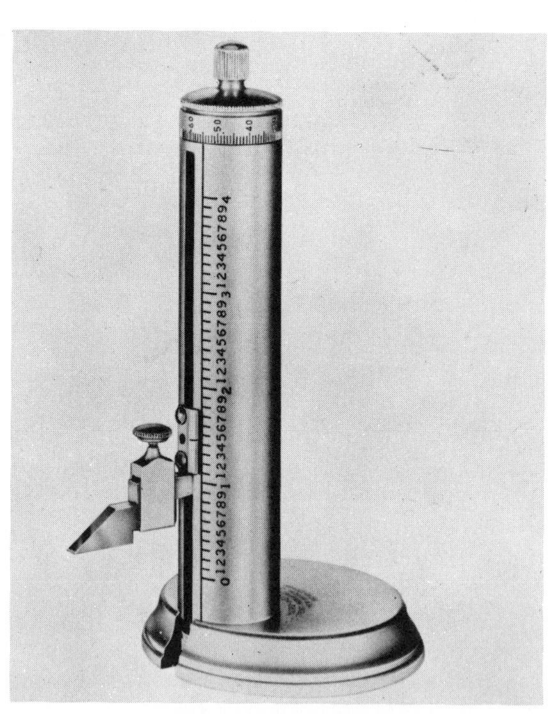

Fig. 5-14. This is another type precision height gage. It is available in standard and metric measure. It operates like a micrometer. (H.B. Tools)

Straightedge

Long flat surfaces are checked for accuracy with a ***straightedge,*** Fig. 5-21. This tool is also used for laying out long straight lines. Straightedges can be made from steel or granite, steel being more common.

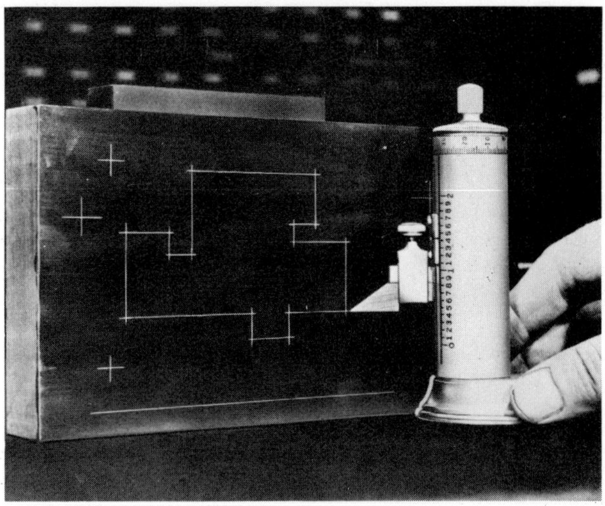

Fig. 5-15. Height gage is being used to scribe lines.

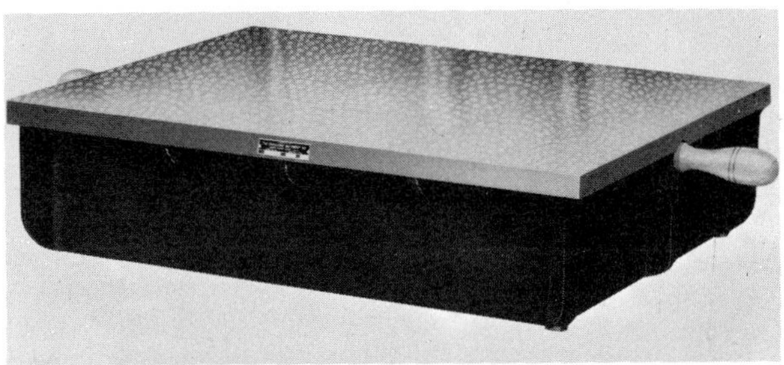

Fig. 5-16. This is a semisteel surface plate. (Challenge Machinery Co.)

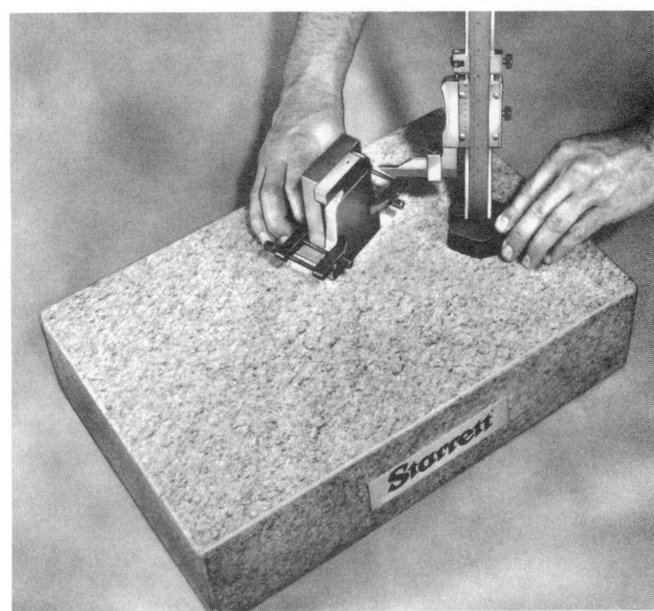

Fig. 5-17. This is a granite surface plate. Most surface plates are now being made from pink or black granite. (L.S. Starrett Co.)

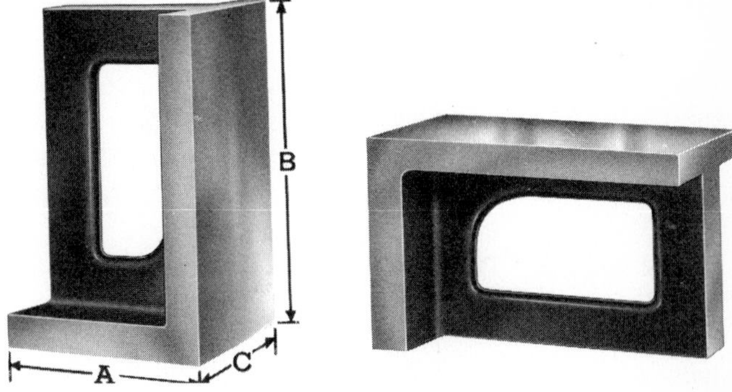

Fig. 5-18. These are right angle plates.

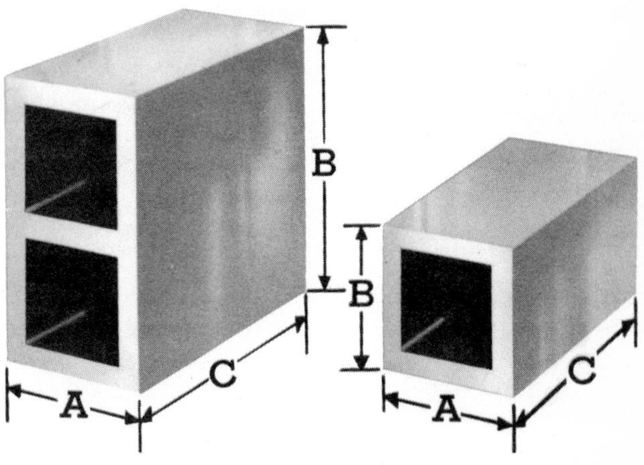

Fig. 5-19. These are box parallels.

Fig. 5-20. V-blocks can be used to hold round stock for layout and mesurement work.

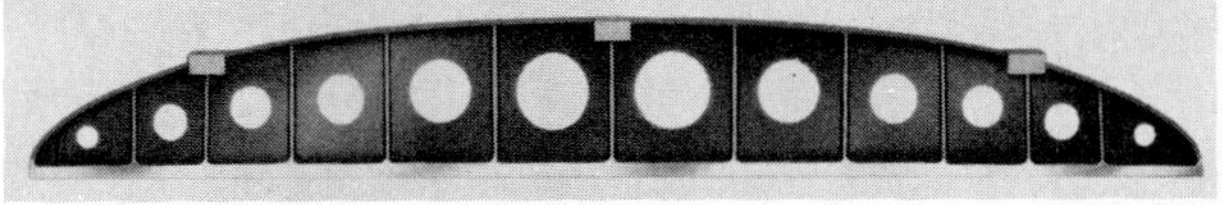

Fig. 5-21. Steel straightedge is very common but they are also available in granite.

SQUARES

The ***square*** is employed to check 90 deg. (square) angles. The tool is also used for laying out lines that must be at right angles to a given edge or parallel to another edge. Some simple machine setups can be made quickly and easily with a square.

Many different types of squares are available. A few of the most common are:

The ***hardened steel square*** is recommended when extreme accuracy is required, Fig. 5-22. It has true right angles, both inside and outside. It is accurately ground and lapped for straightness and parallelism. The tool comes in sizes up to 36 in. (914 mm).

Handle a square with care. The blade is mounted solidly to the beam but if the tool is dropped, the blade can be "sprung," ruining the square.

A ***double square*** is more practical for many jobs than the solid square because the sliding blade is adjustable and interchangeable with other blades, Fig. 5-23. The tool should NOT be used where great precision is required. The bevel blade has one angle for checking ***octagons*** (45 deg. angles), and another one for checking ***hexagons*** (60 deg. angles).

Fig. 5-22. Hardened steel square is handy during layout work.

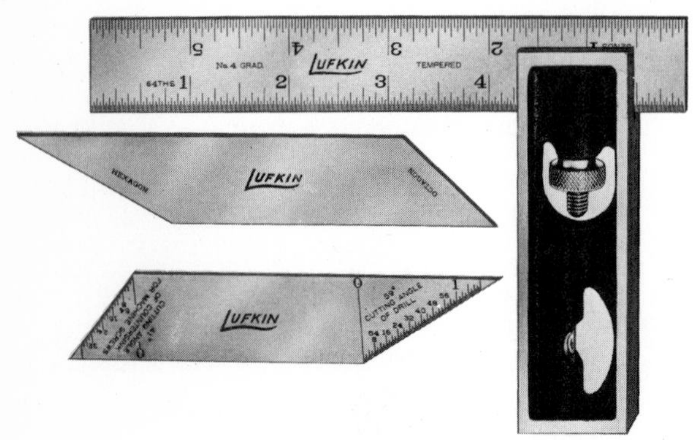

Fig. 5-23. This is a double square. (Lufkin Rule Co.)

A ***drill grinding blade*** is also available for this type square. One end is beveled to 59 deg. for drill grinding and the other end is at 41 deg. for checking the cutting angles of machine screw countersinks. Both ends are graduated for measuring the length of the cutting lips, to assure that the cutting tools are sharpened on center.

A complete ***combination set*** consists of a hardened blade (#4 graduated rule), square head, center head, and bevel protractor. The blade fits all three heads. Combination sets are adaptable to a large variety of operations, making them especially valuable in the shop, Fig. 5-24.

The ***square head,*** having one 45 deg. edge, makes it possible for the tool to serve as a miter square. By projecting the blade the desired distance below the edge, it serves as a depth gage, Fig. 5-25. The spirit level, fitted in one edge, allows it to be used as a simple level.

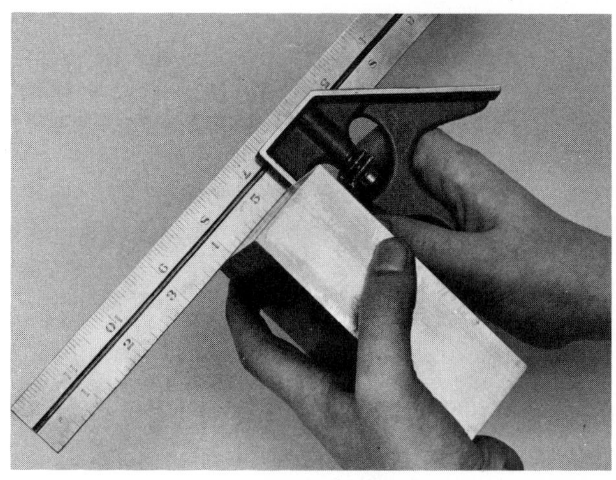

Fig. 5-24. Combination set will perform various layout tasks.

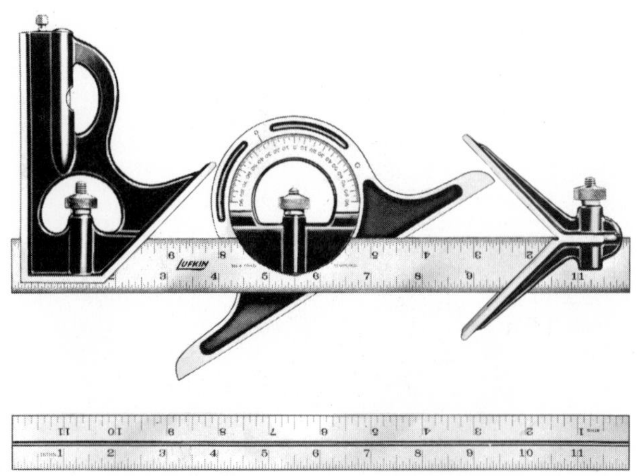

Fig. 5-25. The combination set can be used to check squareness and to measure like a depth gage.

With the rule properly inserted, the ***center head*** can be used to quickly find the center of round stock. This is illustrated in Fig. 5-26.

The ***protractor head*** can be rotated through 180 deg. and is graduated accordingly. The head can be locked with a locking nut, making it possible to accurately determine or scribe angles, Fig. 5-27. The head also has a level built in, making it possible to use it as a level for positioning angles for inspection, layout, or machining.

MEASURING ANGLES

In addition to the protractor head of the combination set, other angle measuring tools are employed in layout work. The accuracy required by a job will determine which tool must be used.

When angles do NOT have to be laid out or checked to extreme accuracy, a ***plain protractor*** will prove satisfactory, Fig. 5-28. The head is graduated from 0 to 180 deg. in both directions for easy reading of angles.

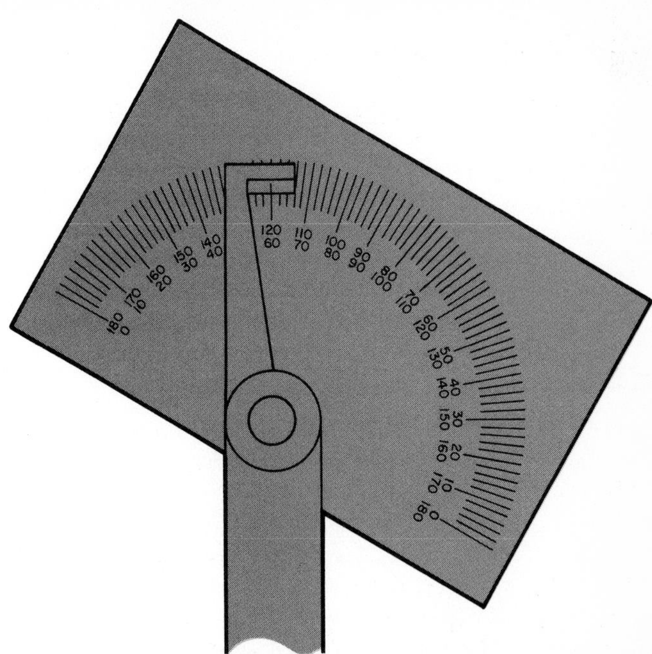

Fig. 5-28. Plain steel protractor will show angles with moderate precision.

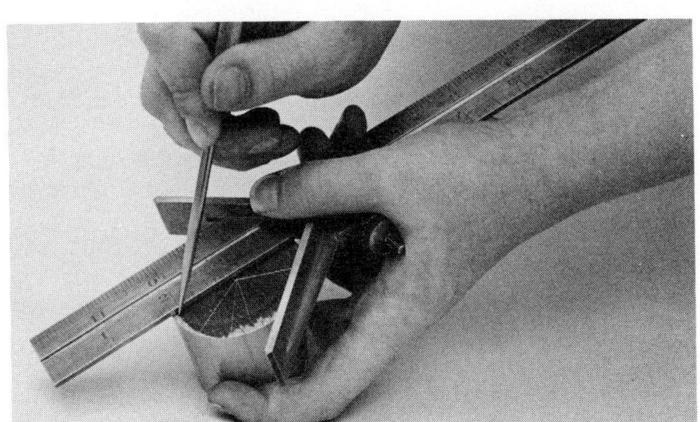

Fig. 5-26. Machinist is using a center head and rule to locate center of a piece of round stock.

Fig. 5-27. Angular settings on layouts can be made with the protractor head and rule of the combination set.

The ***protractor depth gage*** is suitable for checking angles and measuring slot depths, Fig. 5-29.

A universal bevel is useful for checking, laying out, and transferring angles, Fig. 5-30. Both blade and stock are slotted, making it possible to adjust the blade into any desired position. A thumbscrew locks it tightly in place.

When a job requires extreme accuracy, the machinist uses the ***Vernier protractor,*** Fig. 5-31. With this tool, angles of 1/12 of a degree (5 minutes) can be accurately measured.

SIMPLE LAYOUT STEPS

Each layout job has its pecularities and requires some planning before the operation can be started. Fig. 5-32 shows a typical job.

1. Carefully study the drawings.

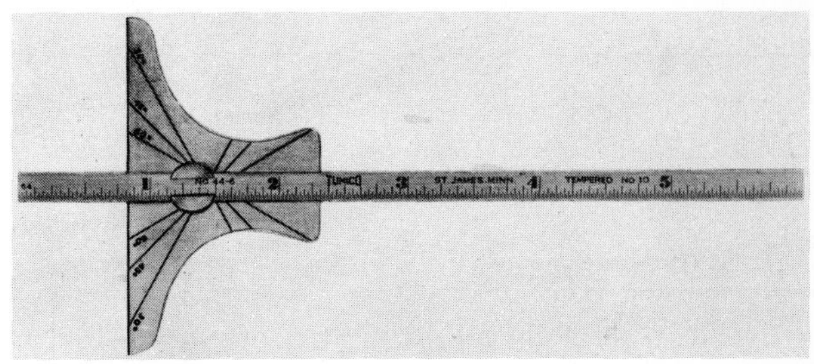

Fig. 5-29. This is a protractor depth gage. (Scherr-Tumico)

Fig. 5-30. Universal bevel can be locked at various angles.

Fig. 5-31. More precise angular measurements are made with a Vernier protractor. (L.S. Starrett Co.)

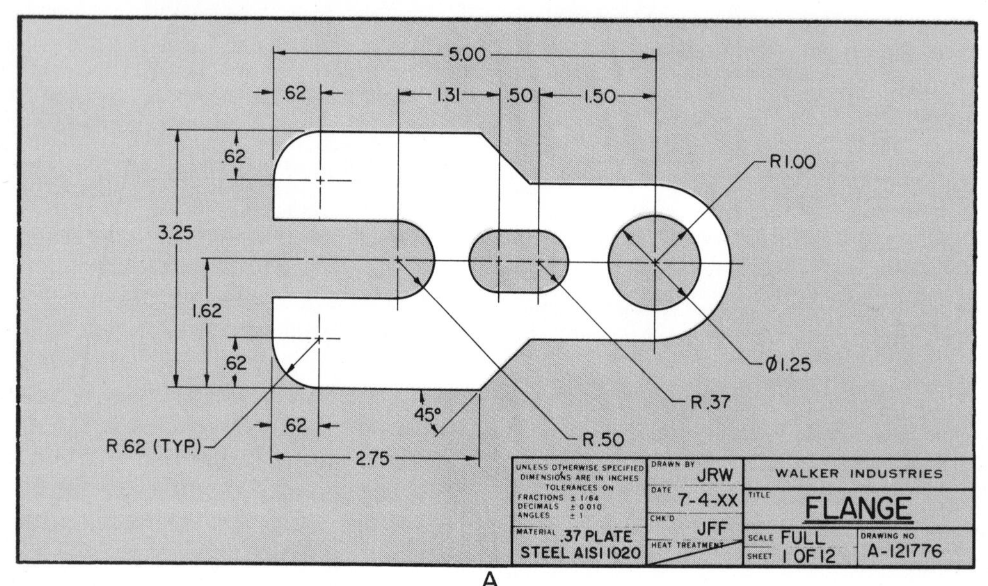

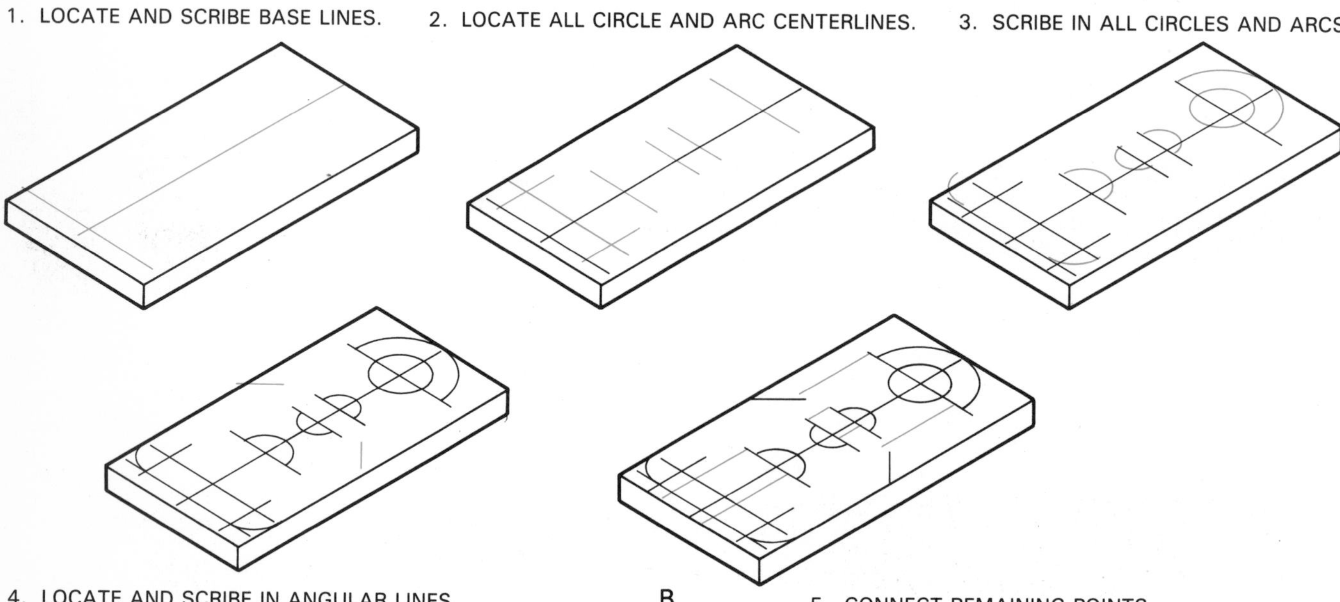

Fig. 5-32. Compare part drawing with steps in laying out a job. A—Plans for part. B—Layout steps.

2. Cut stock to size and remove all burrs and sharp edges.
3. Clean all dirt, grease and oil from the work surface. Apply layout dye.
4. Locate and scribe a ***reference*** or ***base line;*** you will make all measurements from this line. If the material has one true edge, it can be used in place of the base line.
5. Locate and center points of all circles and arcs.
6. Use a ***prick punch*** to mark the point where centerlines intersect, Fig. 5-33. The sharp point (30 to 60 deg.) of this punch makes it easy to locate this position. After the prick punch mark has been checked and found on center, it is enlarged slightly with the ***center punch,*** Fig. 5-33.
7. With dividers or trammel, scribe in all circles and arcs.
8. If angular lines are necessary, use the proper protractor type layout tool. You can also locate the correct points by measuring and connecting them using a rule or straightedge and a scribe.
9. Scribe in all other internal openings.
10. Lines should be clean and sharp. Any double or sloppy line work should be removed by cleaning it off with a solvent. Then apply another coat of dye before the line is rescribed.

LAYOUT SAFETY

1. Never carry an open scriber, divider, trammel, or hermaphrodite caliper in your pocket.
2. Always cover all sharp points with a cork when the tool is not being used.
3. Wear goggles when grinding points of scriber type tools.
4. Get help when you must move heavy angle plates, large V-blocks, etc.
5. Remove all burrs and sharp edges from stock before starting layout work.

TEST YOUR KNOWLEDGE—Chapter 5

Please do not write in the text. Place your answers on a separate sheet of paper.

1. What is used to make layout lines stand out and be easier to see?
2. Why are layout lines used?
3. Straight layout lines are drawn with a ________.
4. Circles and arcs are drawn on meal with a ________.
5. When circles and arcs are too large to be drawn with the tool listed in #4, they are made with a ________.
6. Why should a pencil NOT be used to make layout lines on metal?
7. A ________ ________ is the flat granite or steel surface used for layout and inspection work.
8. What layout operations can be performed with a combination set?
9. Round stock is usually supported on ________ for layout and inspection.
10. Long flat surfaces can be checked for trueness with a ________.
11. The center of round stock can be found quickly with the ________ ________ and rule of the combination set.
12. Angular lines, that must be very accurate, should be laid out with a ________ ________.
13. The ________ punch has a sharper point than the ________ punch.
14. List three safety precautions that should be observed when doing layout work.

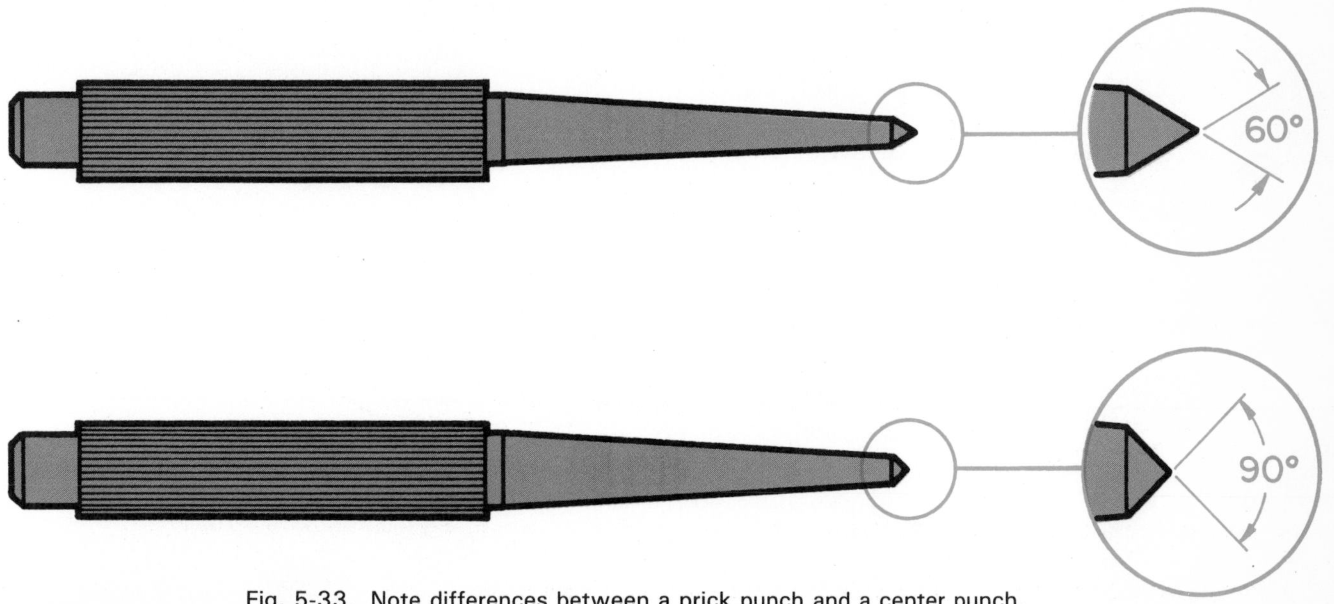

Fig. 5-33. Note differences between a prick punch and a center punch.

RESEARCH AND DEVELOPMENT

1. Make a display panel showing samples of the various kinds of layout fluids used by industry. Use a clear plastic spray to prevent the scribed lines from rusting and the coatings from rubbing off.
2. Prepare a sample that will show a good layout job. Develop it into a bulletin board display. Use colored twine or yarn running from the sample to printed notations explaining the various aspects that indicate a good layout job.
3. Write a paper on how surface plates are made. Secure literature from the various manufacturers to illustrate the paper. Also include:
   a. How surface plate grades are determined.
   b. Why cast iron, steel, and granite are used to make them rather than other materials.
   c. How to take care of the surface (maintain accuracy, keep it clean, etc.)
4. Prepare a series of overhead projector transparencies, 35 mm slides, or a video to show the correct sequence for producing a good layout job.

Chapter 6

HAND TOOLS

After studying this chapter, you will be able to:
- Identify the most commonly used machine shop hand tools.
- Select the proper hand tool for the job.
- Maintain hand tools properly.
- Explain how to use hand tools safely.

The correct selection and use of hand tools will help you do the job safely, and with a minimum expenditure of time. When used incorrectly, a hand tool can be damaged; but more importantly, you or someone else may be injured. It is to your advantage to learn to properly work with hand tools.

CLAMPING DEVICES

Clamping devices are employed to hold and/or position material while it is being worked. Several types are used in machining.

Vises

The ***machinist's*** or ***bench vise*** is for numerous holding tasks. It should be mounted far enough out on the bench edge to permit clamping long work in a vertical position. It may be a solid base vise, Fig. 6-1, or a swivel base type, Fig. 6-2, which allows the vise to be rotated.

Small precision parts may be held in a ***small bench vise,*** Fig. 6-3. This type vise can be rotated and tilted to any desired position. ***Vise size*** is determined by the width of the jaws, Fig. 6-4.

A vise's clamping action is obtained from a heavy screw turned by a handle. The handle is long enough to apply ample pressure for any work that will fit the vise. Under NO CIRCUMSTANCES should the

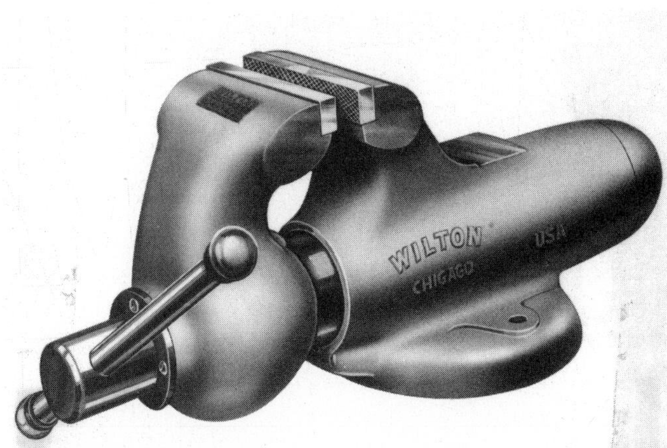

Fig. 6-1. Note construction of solid base machinist's or bench vise. (Wilton Tool Mfg., Inc.)

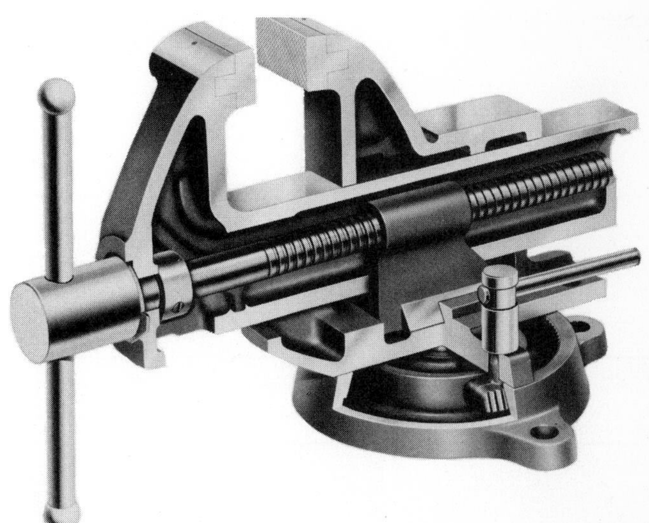

Fig. 6-2. Study cutaway of swivel base vise. Base is made in two parts so vise body can be rotated to any desired position. (Columbian Vise and Mfg. Co.)

Fig. 6-3. This is a small vise used by the toolmaker. It can be rotated and pivoted to secure desired working position. (Wilton Tool Mfg., Inc.)

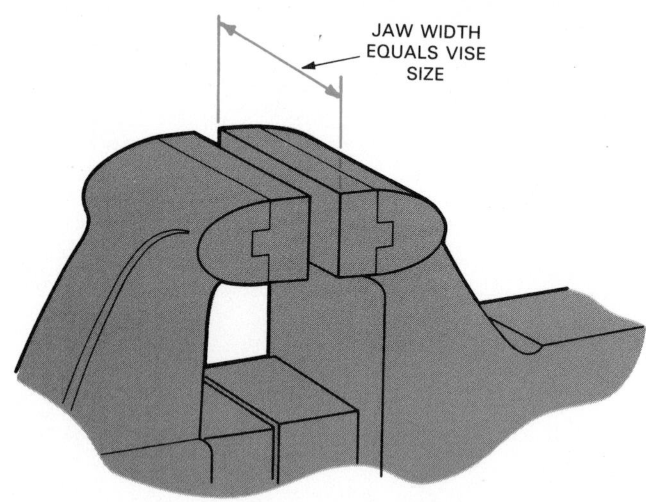

Fig. 6-4. Vise size is determined by width of vise jaws.

vise handle to hammered tight, nor should additional pressure be applied using a length of pipe for leverage.

Vise jaws are hardened, Fig. 6-5. Unless covered with soft copper, brass, or aluminum ***caps,*** the jaws should NOT be employed to clamp work that would be damaged or marred by the jaw serrations.

SAFETY NOTE: When clamping a job in a vise, avoid letting the vise handle or work project into the aisle, Fig. 6-6.

Clamps

The ***C-clamp*** and the ***parallel clamp*** hold parts together while they are worked on. The C-clamp is made in many sizes, Fig. 6-7. Jaw opening determines ***clamp size.*** See Fig. 6-8.

A parallel clamp is ideal for holding small work. For maximum clamping action, the jaw faces must be parallel. Strips of paper, the width of the clamp jaw, placed between the work and the jaws will improve clamping action.

PLIERS

The combination or slip-joint pliers are widely used for holding operation, Fig. 6-9. The slip-joint permits the pliers to be opened wider at the hinge pin to grip larger size work. They are made in 5, 6, 8, and 10 in. sizes. The ***plier size*** indicates the overall length of the tool.

Some combination pliers are made with cutting edges for clipping wire and small metal sections to needed lengths. The better grade pliers are forged.

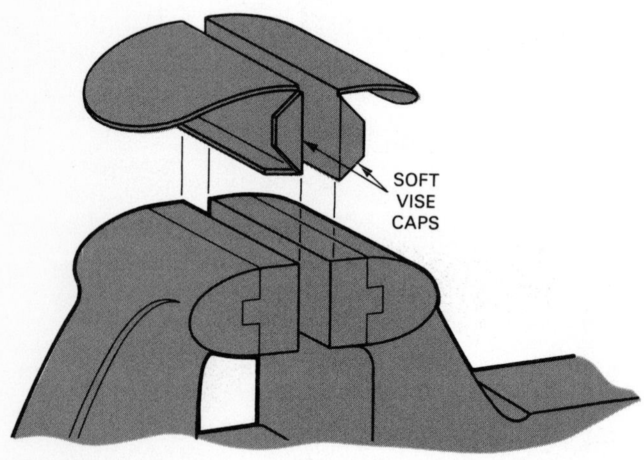

Fig. 6-5. Caps made of copper, lead, or aluminum are slipped over hardened vise jaws to protect work from becoming marred or damaged by jaw serrations.

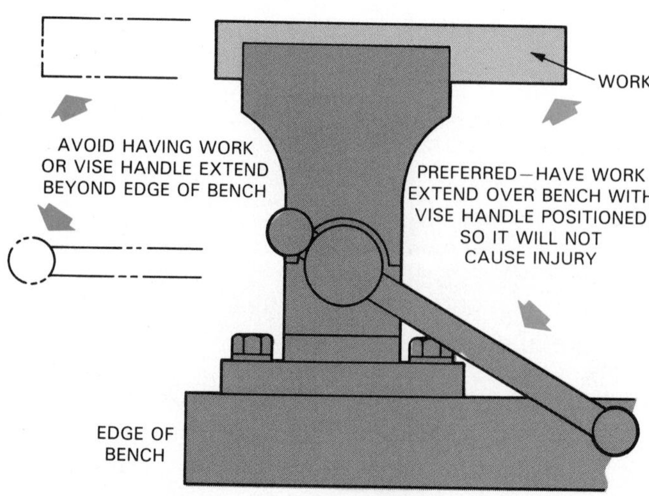

Fig. 6-6. CAUTION: To prevent injury, avoid letting the vise handle or work project into the aisle.

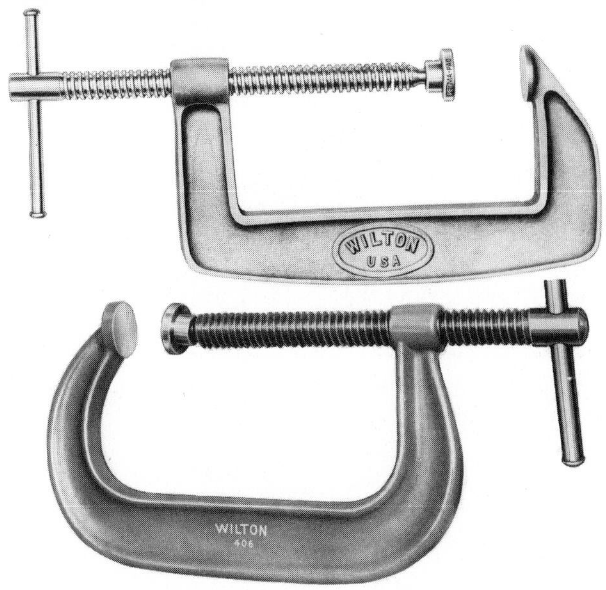

Fig. 6-7. These are C-clamps. (Wilton Tool Mfg., Inc.)

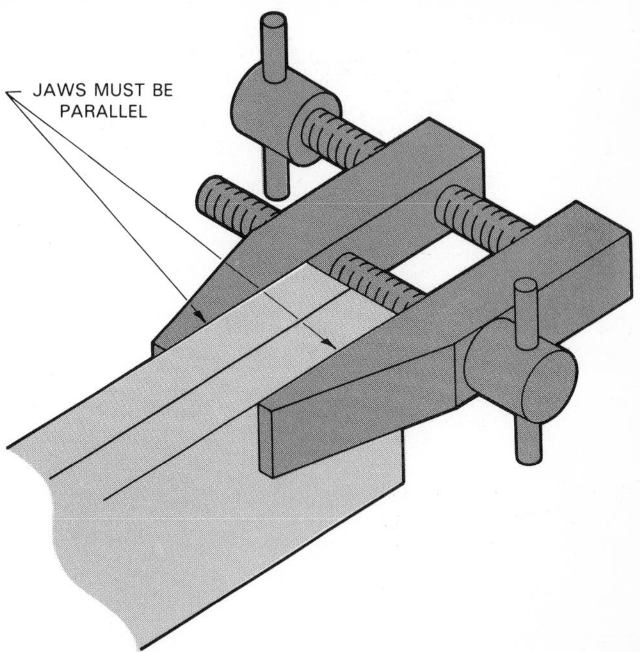

Fig. 6-8. For maximum clamping action with a parallel clamp, the jaws must be adjusted until parallel.

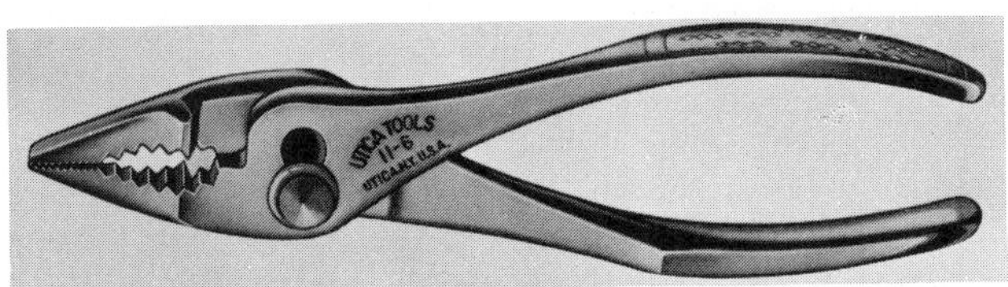

Fig. 6-9. Combination or slip-joint pliers are for holding tasks. (Utica Tools)

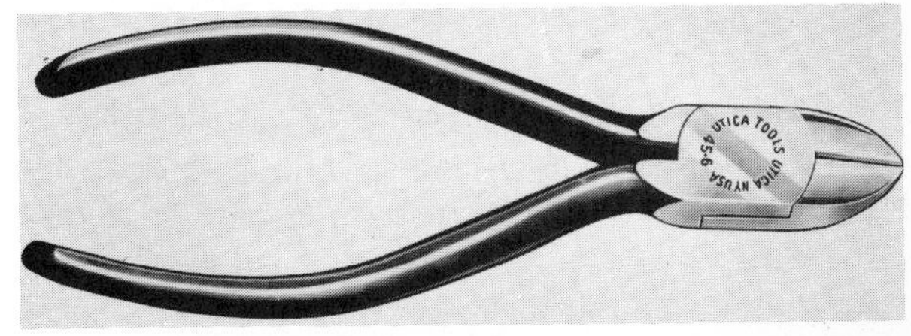

Fig. 6-10. Diagonal pliers will cut flush with a surface.

Diagonal pliers are another widely utilized tool for light cutting tasks, Fig. 6-10. The cutting edges are at an angle to permit the pliers to CUT FLUSH with the work surface. Diagonal pliers are made in 4, 5, 6, and 7 in. lengths.

Side-cutting pliers are capable of cutting heavier wire and pins, Fig. 6-11. Some of them have a wire stripping groove and insulated handles. They are made in 6, 7, and 8 in. lengths.

Round-nose pliers are helpful when forming wire and light metal, Fig. 6-12. Their jaws are smooth and will NOT marr metal being worked. Round-nose pliers are available in 4, 4 1/2, 5, and 6 in. sizes.

Needle-nose pliers, both straight, Fig. 6-13, and curved-nose, Fig. 6-14, are handy when work space is limited and for holding small work. They will reach into cramped places.

Tongue and groove pliers have aligned teeth for flexibility in gripping different size work, Fig. 6-15. Jaw opening size can be adjusted easily. The 6 in. size usually has five adjustments while the larger 16 in. size has eleven adjustments. They are made in many different sizes.

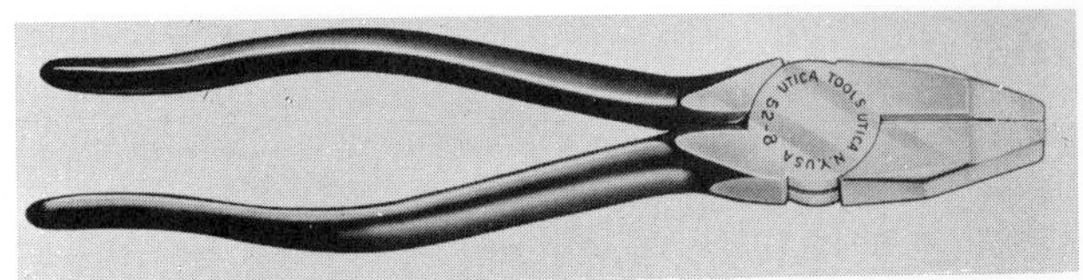

Fig. 6-11. Side-cutting pliers have square jaws for holding and cutting jaws.

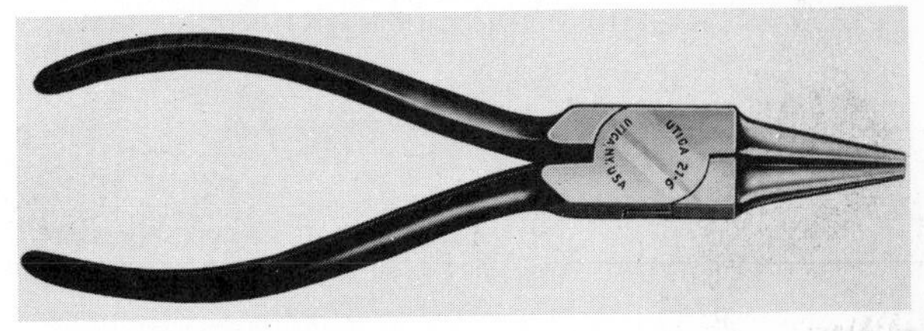

Fig. 6-12. Note round-nose pliers.

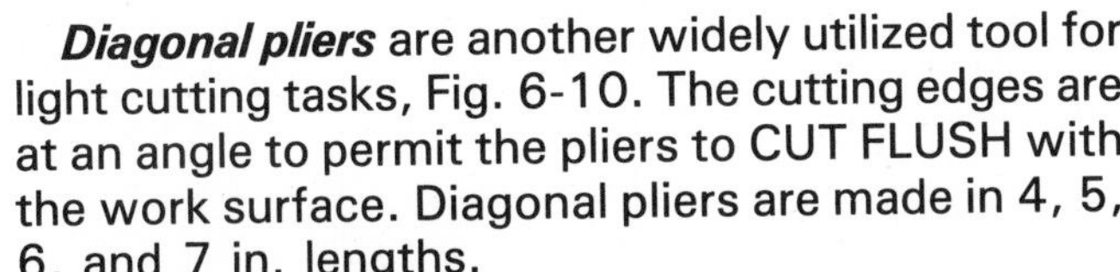

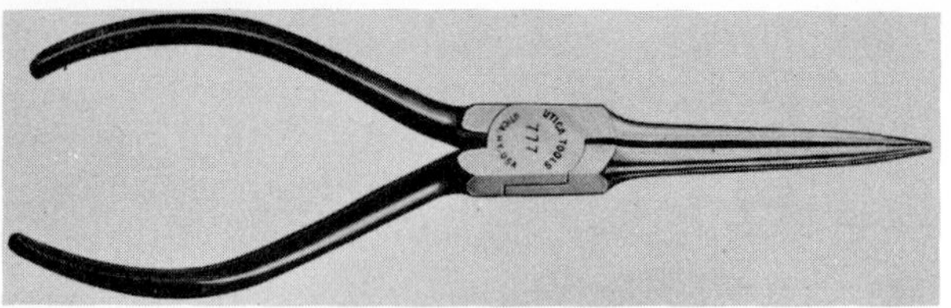

Fig. 6-13. Straight needle-nose pliers are for grasping smaller hard to reach objects.

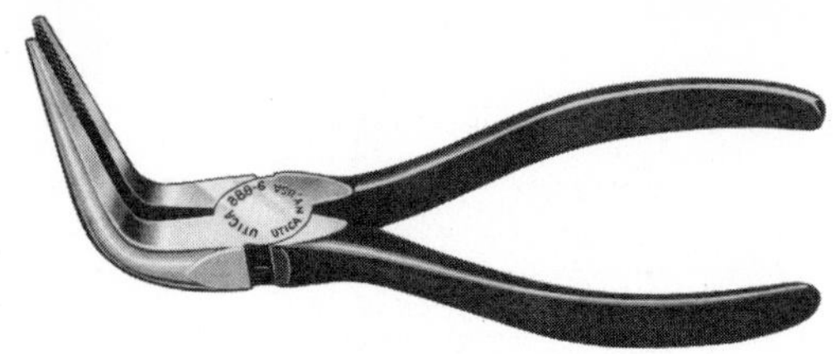

Fig. 6-14. Curved needle-nose pliers can sometimes be handy.

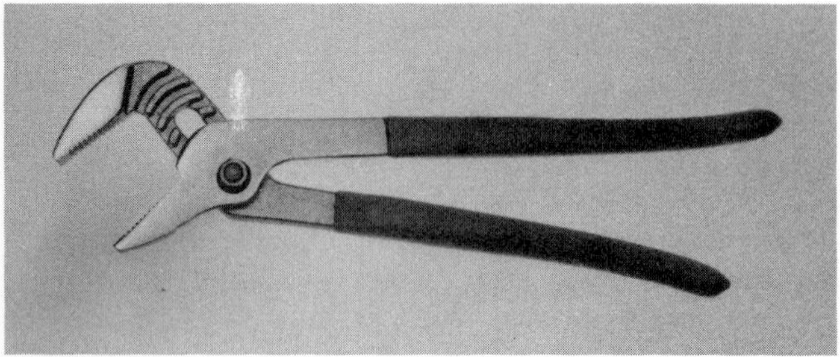

Fig. 6-15. Tongue and groove pliers will expand to hold large objects.

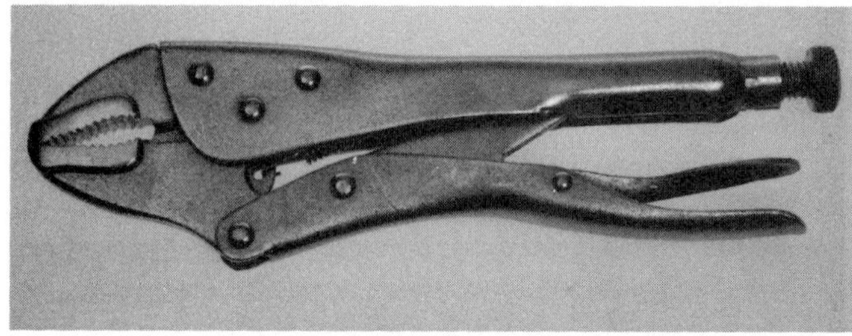

Fig. 6-16. These adjustable clamping pliers can be locked on work. They are known by many names: Vise Grip® pliers, adjustable locking pliers, etc.

Adjustable clamping pliers are a relatively new addition to the pliers family, Fig. 6-16. Jaw opening can be adjusted through a range of sizes. After fitting on the work, a squeeze of the hand can lock the jaws onto the work with more than 2000 lbs. of pressure. Jaw pressure can be relieved by opening the quick release on the handle. These pliers are made in many sizes with straight, curved, or long-nose jaws. They are known by several names, including: Vise Grips® , Tag-L-Lock Pliers® , and Locking Pliers.

Care of pliers

Like other tools, pliers will give long, useful service if a few simple precautions are taken:

1. NEVER use pliers as a substitute for a wrench.
2. Pliers with cutters will deform or break if used to cut metal sizes that are too large, or work that has been heat-treated. Breakage will also occur if additional leverage is applied to the handles.
3. Clean and oil pliers occasionally.
4. Store pliers in a clean, dry place. Avoid throwing them in a drawer or tool box with other tools.
5. Use pliers large enough for the job.

WRENCHES

Wrenches comprise a family of tools designed to assemble and disassemble many types of threaded fasteners. They are made in a vast number of types and sizes. Only the most commonly used wrenches will be covered.

Torque limiting wrenches

Torque is the amount of turning or twisting force applied to a threaded fastener or part. It is measured in force units of ***foot-pounds*** (ft.-lbs.) or in ***Newton-meters*** (N•m) for metrics. Torque is the product of the force applied, times the length of the lever arm. See Figs. 6-17 and 6-18.

A ***torque limiting wrench*** allows you to measure the tightening of a threaded fastener in foot-pounds or Newton-meters. This provides maximum holding power, without danger of the fastener or part fail-

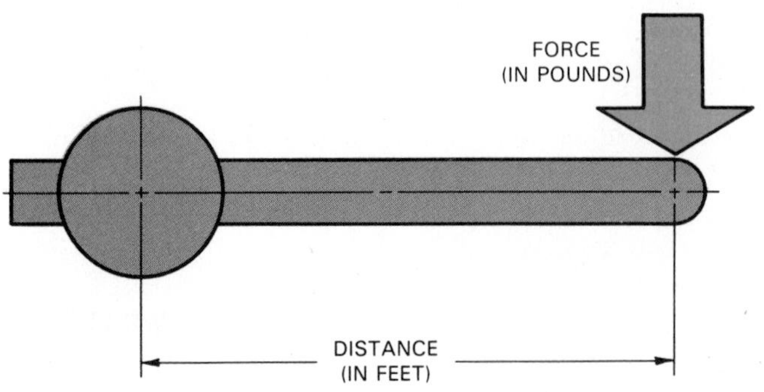

TORQUE = FORCE (in pounds) × DISTANCE (in feet)
= foot-pounds
= ft.-lbs.

Fig. 6-17. Note how torque is measured using standard inch-foot pound measurement.

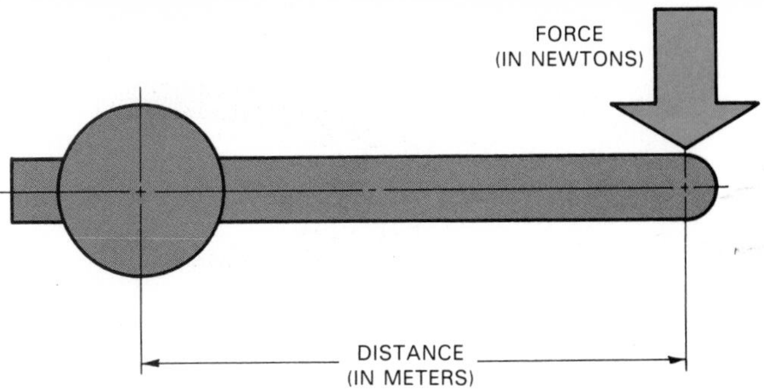

TORQUE = FORCE (in newtons) × DISTANCE (in meters)
= newton-meters
= N•m
(The newton (N) is that force which applied to a mass of 1 kilogram, gives it an acceleration of 1 meter per second squared.)

Fig. 6-18. Study how torque values in SI metric measure are given in Newton-meters (N•m).

ing, or causing the work to warp or spring out of shape. See Fig. 6-19.

There are many types of torque limiting wrenches, Fig. 6-20. It is possible to obtain torque wrenches that are direct reading, or that feature a sensory signaling mechanism (clicking sound or momentary release) when a present torque is reached.

The right and wrong methods of gripping the wrench handle are shown in Fig. 6-21. Under NO condition should the handle be lengthened for additional leverage. These tools are designed to take a specific maximum force load. Any force over this amount will destroy the accuracy of the wrench.

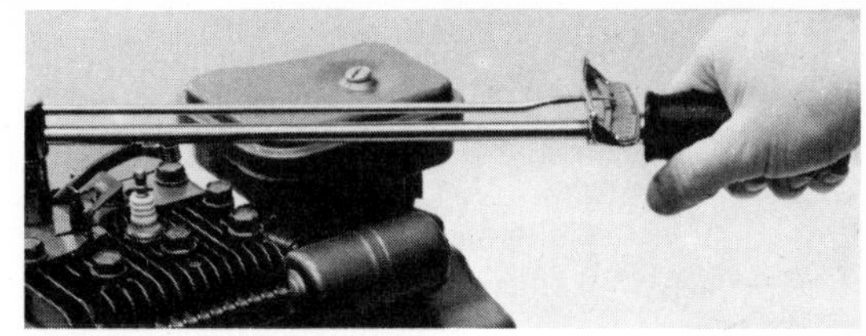

Fig. 6-19. Torque limiting wrenches are employed when fsteners must be tightened to within certain limits to prevent undue stresses and strains from developing in the part.

Torque limiting wrenches will give accurate measurements whether they are pushed or pulled. However, to prevent hand injury, the preferred method is to PULL on the wrench handle.

Adjustable wrenches

The term ''adjustable wrench'' is a misnomer (not named properly). Other wrenches, such as the ''monkey wrench'' and pipe wrench, are also adjustable. However, the wrench that is somewhat like an open-end wrench, but with an adjustable jaw, is commonly referred to as an ***''adjustable wrench,''*** Fig. 6-22.

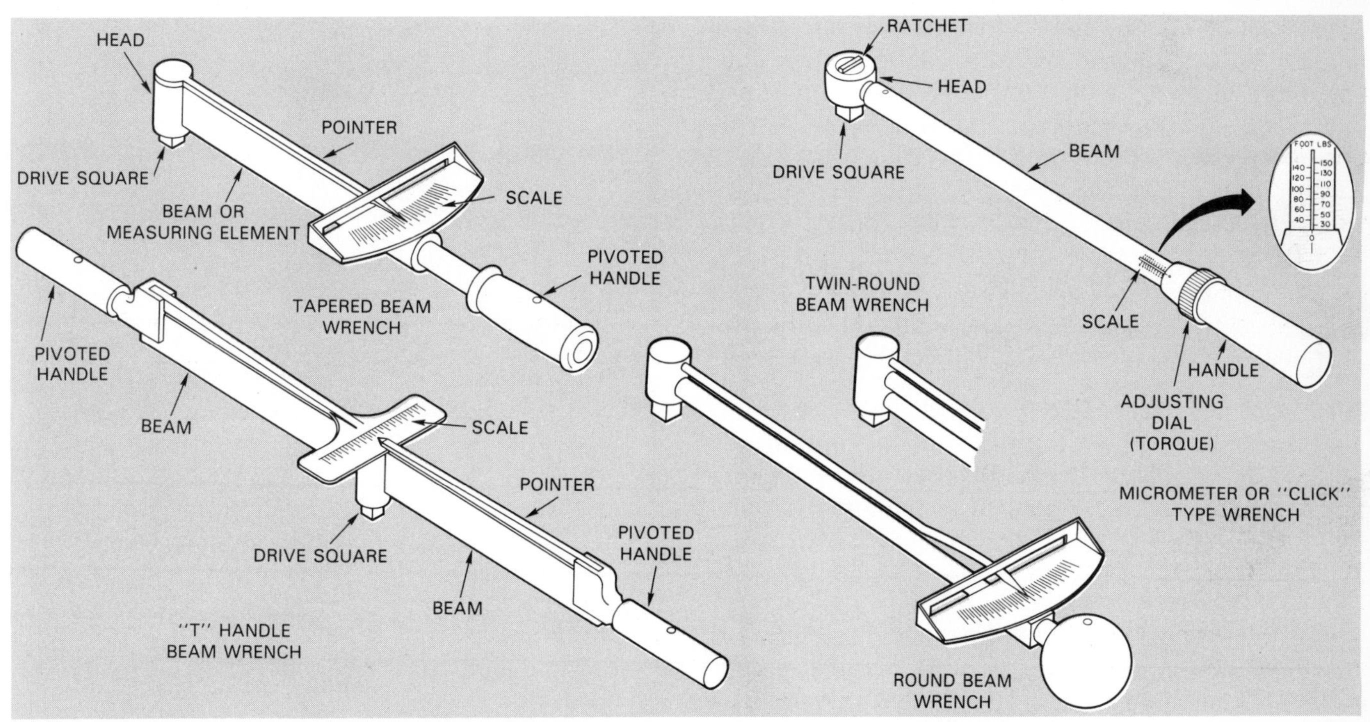

Fig. 6-20. Study several types of torque limiting wrenches.

Fig. 6-21. Note right and wrong ways to apply pressure to a torque limiting wrench handle.

As the name implies, the wrench can be adjusted to fit a range of bolt head and nut sizes. Although convenient at times, the adjustable wrench is NOT intended to take the place of open-end, box, and socket wrenches.

Three important points must be remembered when using the adjustable wrench:

1. The wrench should be placed on the bolt head or nut so the movable jaw FACES THE DIRECTION the fastener is to be rotated, Fig. 6-23.
2. Adjust the thumbscrew so the jaws fit the bolt head or nut SNUGLY, Fig. 6-24.
3. Avoid placing an extension on the wrench handle for additional leverage. Never hammer on the handle to loosen a stubborn fastener. Use the SMALLEST wrench that will fit the fastener on which you are working. This will minimize the possibility of twisting the fastener off.

SAFETY NOTE! Pushing on any wrench is normally considered dangerous. When the fastener loosens unexpectedly or fails, you will almost invariably strike and injure your knuckles on the work. This operation is commonly known as ''knuckle dusting.'' Always PULL ON a wrench; Do NOT PUSH!

The ***pipe wrench*** is a wrench that will grip round stock, Fig. 6-25. However, the jaws always leave marks on the work. Avoid using a pipe wrench on bolt heads or nuts unless they cannot be turned with another type of wrench. For instance, you might need a pipe wrench to remove a bolt if the corners of its head have been rounded.

Open-end wrenches

Open-end wrenches are usually double ended, with two different size openings, Fig. 6-26. They

Fig. 6-22. Adjustabale wrench is handy when full wrench set is not available. (Snap-On Tool Corp.)

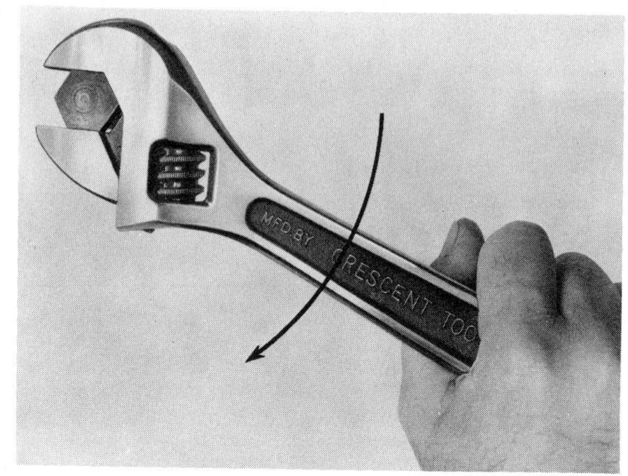

Fig. 6-23. The movable jaw should always face the direction of rotation. (Crescent Tool Co.)

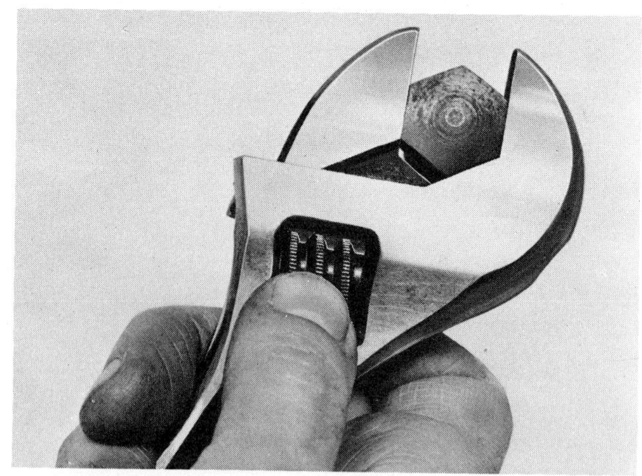
Fig. 6-24. A wrench must fit the nut or bolt snugly.

Fig. 6-25. Pipe wrench has jaws that will grasp round objects.

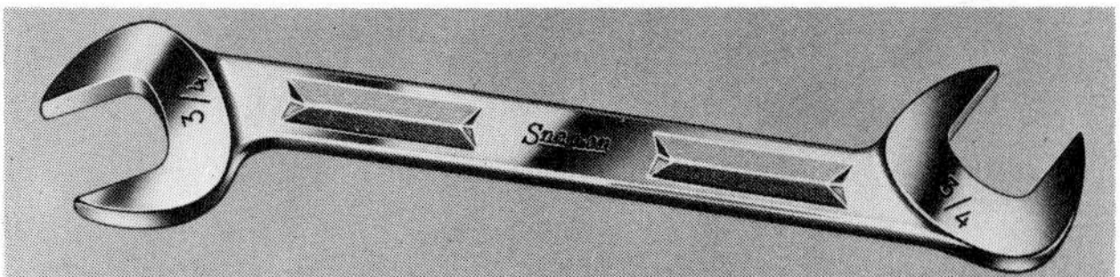

Fig. 6-26. Open-end wrench is acceptable when torque applied is low. (Snap-On Tool Corp.)

are made about 0.005 in. (0.13 mm) oversize to permit them to easily slip on bolt heads and nuts of the wrench size. Openings are angled with the wrench body so they can be applied in close quarters. Standard and metric size open-end renches are available.

Box wrenches

The body or jaw of the ***box wrench*** completely surrounds the bolt head or nut, Fig. 6-27. A properly fitted box wrench will not normally slip. It is preferred for many jobs. Box wrenches are available in the same sizes as open-end wrenches and with straight and offset handles.

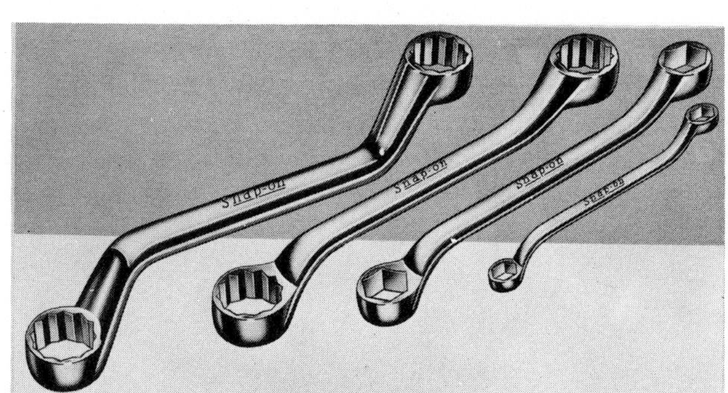

Fig. 6-27. Box wrenches can handle more torque than open-end wrench. (Snap-On Tool Corp.)

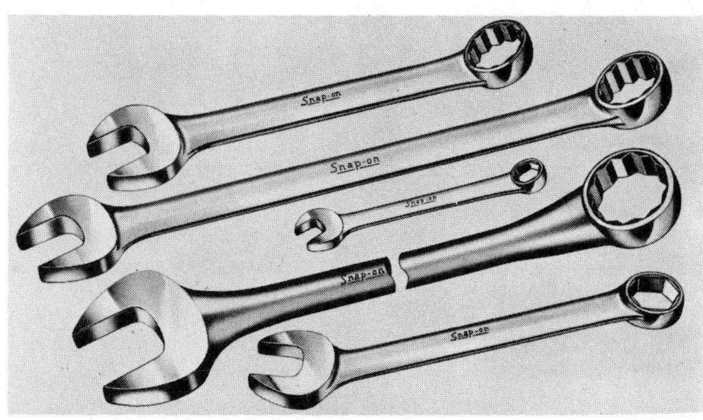

Fig. 6-28. Combination wrenches are handy because of two end configurations. (Snap-On Tool Corp.)

Combination open-end and box wrenches

A ***combination open-end and box wrench*** has an open-end wrench at one end and a box wrench at the other end of the handle. They are made in standard and metric sizes, Fig. 6-28.

Socket wrenches

Socket wrenches are box-like and are made with a detachable tool head-socket that fits many types of ***handles*** (either solid bar or ratchet type), Fig. 6-29. A typical socket wrench set contains various handles and a wide range of socket sizes. Many sets include both standard and metric sockets. Various types of socket openings are shown in Fig. 6-30.

Spanner wrenches

Spanner wrenches are special wrenches with drive lugs; they are usually furnished with machine tools and attachments. Spanner wrenches are designed to turn flush and recessed type threaded fittings. The fittings have slots or holes to receive the wrench end.

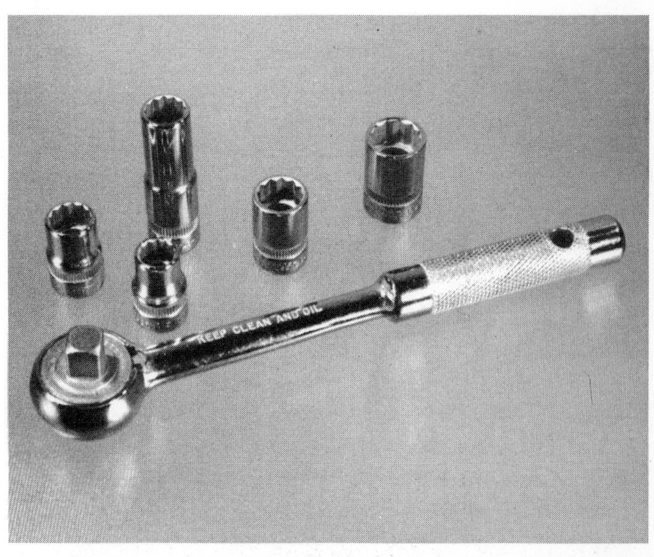

Fig. 6-29. Note a typical socket wrench and sockets. The wrench has a right- and left-hand ratchet mechanism.

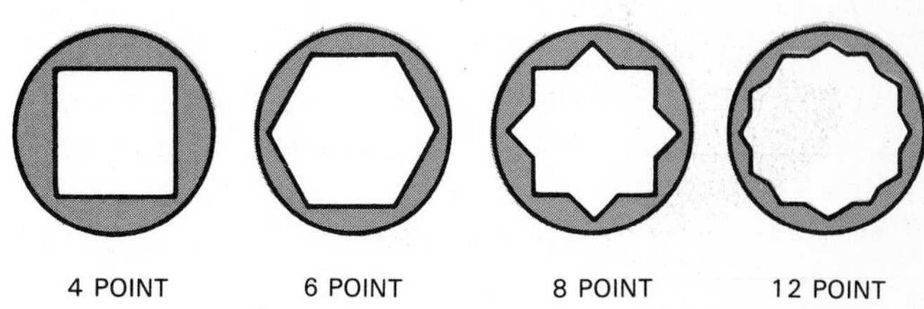

Fig. 6-30. Study types of socket openings available. Which socket opening can be used with both square and hex head fasteners?

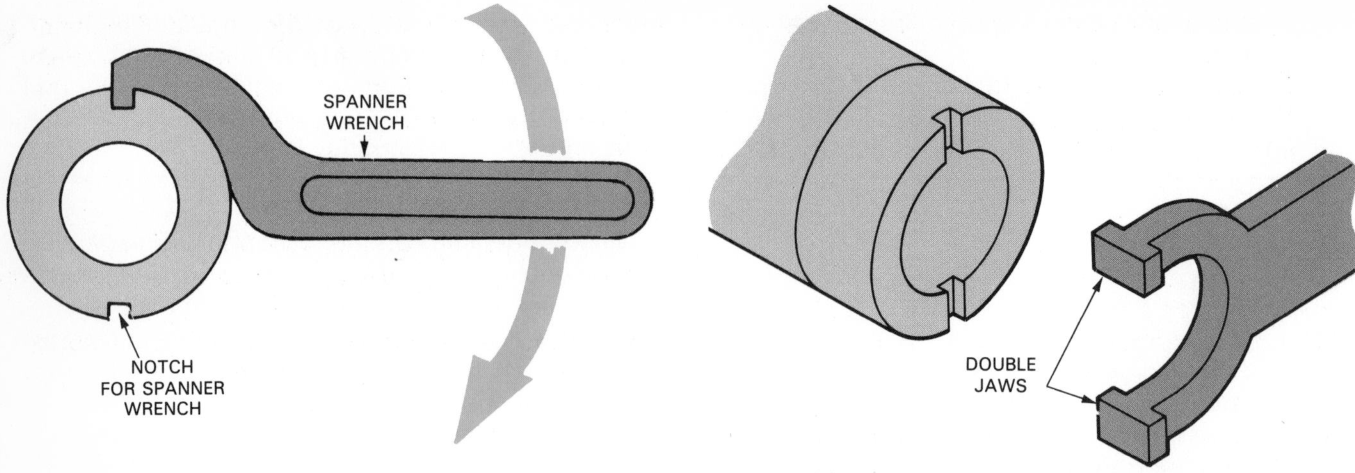

Fig. 6-31. Note hook spanner wrench. Some can be adjusted to fit different size fasteners.

Fig. 6-32. Note end spanner wrench.

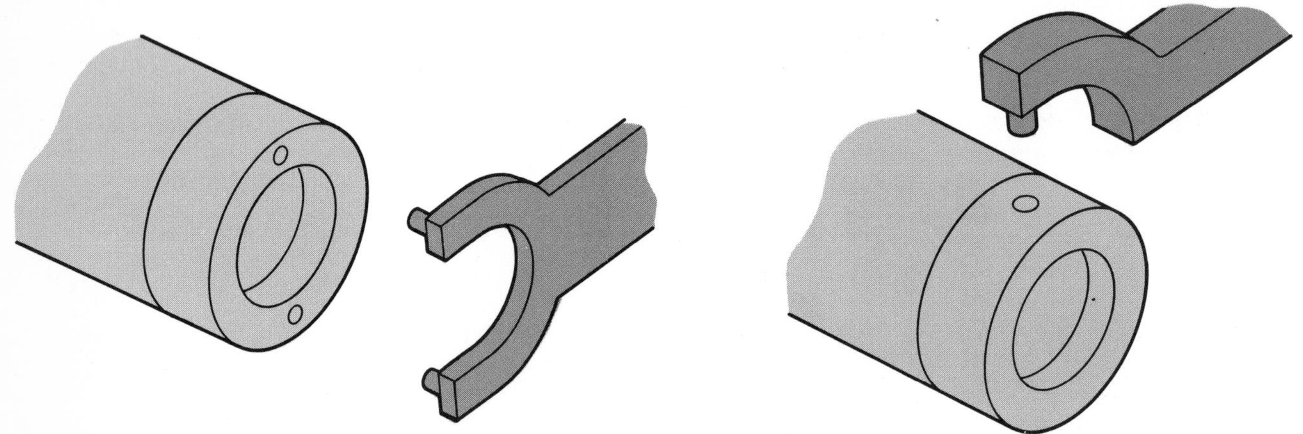

Fig. 6-33. Note pin spanner wrenches.

A ***hook spanner*** is equipped with a single lug that is placed in a slot or notch cut in the fitting, Fig. 6-31.

An ***end spanner*** has lugs on both faces of the wrench for better access to the fitting. The lugs fit notches or slots machined into the face of the fitting. Refer to Fig. 6-32.

On ***pin spanner wrenches,*** the lugs are replaced with pins that fit into holes on the fitting, rather than into notches, Fig. 6-33.

Allen wrenches

The wrench, used with socket headed fasteners, is more commonly known as an ***allen wrench,*** Fig. 6-34. It is manufactured in many sizes to fit fasteners of various standard and metric dimensions.

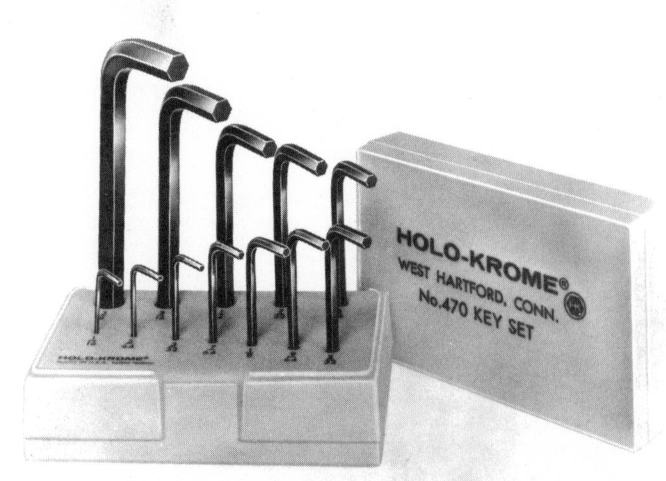

Fig. 6-34. Allen wrenches are used with socket head fasteners. They are made in both inch and metric sizes. (Holo-Krome)

General safety rules for wrenches

1. Always PULL on a wrench. You have more

control over the tool and there is less change of injury.

2. Select a wrench that fits properly. A loose fitting wrench, or one with worn jaws, may slip and cause injury. It can also round off and ruin the bolt or nut on which it is being used.
3. Never hammer on a wrench to loosen a stubborn fastener, unless the tool has been designed for this purpose.
4. It is a dangerous practice to lengthen a wrench handle for additional leverage. Use a larger wrench.
5. When employing a wrench, clean grease and oil from the handle and the floor in the work area. This will reduce the possibility of your hands or feet slipping.
6. Do NOT use a wrench on moving machinery.
7. Use the smallest adjustable wrench that will fit the fastener on which you are working. This will minimize and possibility of twisting the fastener off.

SCREWDRIVERS

Screwdrivers are manufactured with many different tip shapes, Fig. 6-35. Each shape has been designed for a particular type of fastener. The standard and Phillips type screwdrivers are familiar to all shop workers. The other shapes may not be as well known.

A ***standard screwdriver*** has a flattened wedge-shaped tip that fits into the slot in a screw head. This tool is made in 3 to 12 in. lengths. The shank diameter, and the width and thickness of the tip are proportional with the length. ***Screwdriver length*** is measured from blade tip of the handle. The blade is heat-treated to provide the necessary hardness and toughness to withstand the twisting pressures.

A few of the standard screwdriver tips are shown in Fig. 6-36.

The ***double-end offset screwdriver*** can work where there is NOT enough space to use a conventional straight shank tool. See Fig. 6-36A.

The conventional ***straight shank screwdriver*** is widely used for a variety of work, Fig. 6-36B.

The ***electrician's screwdriver*** has a long thin blade and an insulated handle, Fig. 6-36C. The long thin blade will reach into small areas.

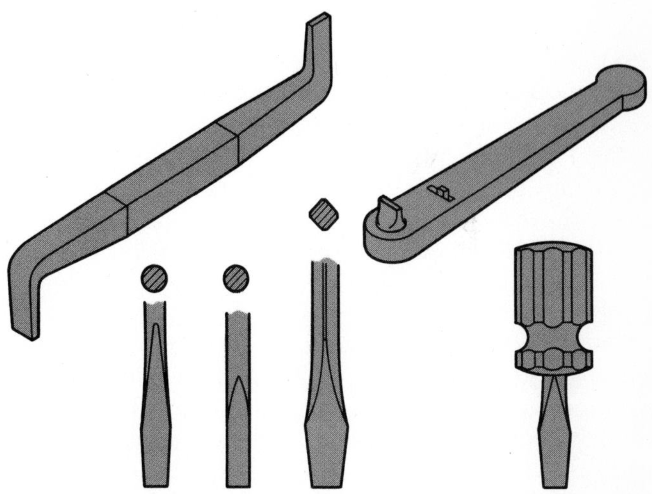

Fig. 6-36. Study styles and types of standard screwdrivers. A—Double end offset. B—Conventional straight shank. C—Electrician's. D—Heavy-duty. E—Stubby or close quarter. F—Ratchet type offset..

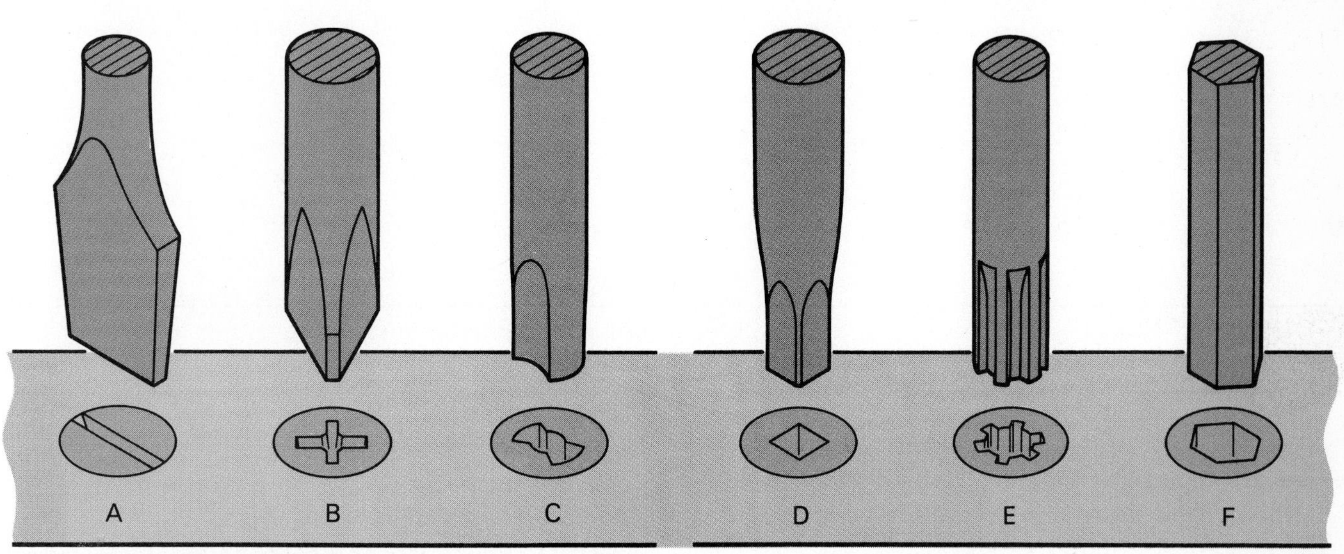

Fig. 6-35. Study screwdriver tips. A—Standard. B—Phillips. C—Clutch. D—Square. E—Torx. F—Hex.

A ***heavy-duty screwdriver*** has a heavy, square shank that permits a wrench to be applied for driving or removing large or stubborn screws, Fig. 6-36D.

Stubby or ***close quarter screwdrivers*** are designed for use where work space is limited, Fig. 6-36E.

The ***ratchet screwdriver*** moves the screw on the power stroke, but NOT on the return stroke. It can be set for right- or left-hand operation, Fig. 6-36F.

The ***Phillips screwdriver*** has an X-shaped tip for use with Phillips recessed head screws. Four sizes (#1, #2, #3, and #4) will handle the full range of this type fastener. They are manufactured in the same general styles as the standard screwdriver.

The ***Pozidriv® screwdriver*** tip is similar in appearance to the Phillips type tip but has a slightly different tip shape. This type has been designed for POSIDRIV® screws used extensively in the aircraft, automotive, electronic, and appliance industries. The tip of this screwdriver has a black oxide finish to distinguish it from a Phillips type tool. Note that a Phillips tip will damage the opening in the head of the Pozidriv® screw.

Clutch head, Robertson, Torx®, and ***hex screwdrivers*** are used for special industrial and security applications.

Using a screwdriver

Always select the correct size screwdriver for the screw being handled, Fig. 6-37. A poor fit damages the screw slot and frequently the tool's tip. Damaged screw heads are dangerous, and often difficult to drive or remove. They should be replaced.

When driving or removing screws, hold the screwdriver square with the fastener. Guide the tip with your free hand.

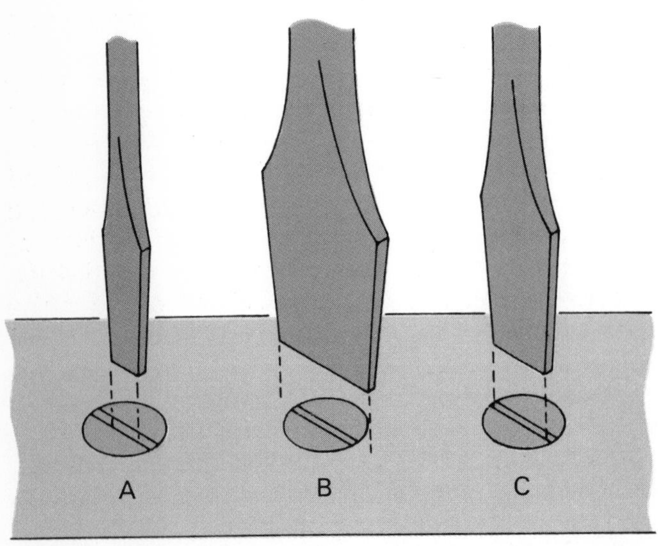

Fig. 6-37. Use correct tip for job being done. Tip A is too narrow and will damage the screw head. Tip B is too wide and will damage the work. Tip C is the correct width.

A worn screwdriver tip, Fig. 6-38, left, must be reground. A fine grinding wheel and light pressure is required. Check the tip during the grinding operation by fitting it to a screw slot. A properly ground tip, Fig. 6-38, right, fits snugly into the slot and holds the head firmly in the slot.

CAUTION! Avoid overheating the tip during the grinding operation. It will destroy the tool.

Safety precautions for screwdrivers

1. A screwdriver is NOT a substitute for a chisel, nor is it made to be hammered on, or used as a pry bar.
2. Wear goggles when regrinding screwdriver tips.
3. Screws with burred heads are dangerous and should be replaced or the burrs should be removed with a file or abrasive cloth.
4. ALWAYS turn electric power off when working on electrical equipment. The screwdriver should also have an insulated handle specifically designed for electrical work.
5. Avoid carrying a screwdriver in your pocket. It is a dangerous practice that can cause injury to you or to someone else. It can also damage your clothes.

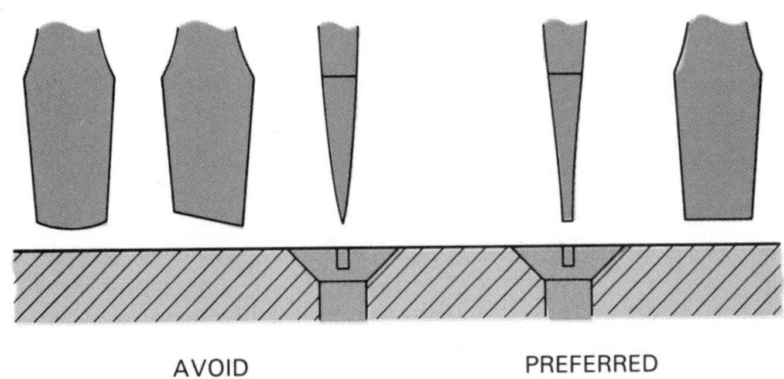

Fig. 6-38. Tips on the left are to be avoided. They are worn or improperly sharpened. Tip to right is ground correctly. Note that the sides are concave; this holds tip in slot when pressure is applied.

STRIKING TOOLS

The machinist's ball-peen hammer is the most commonly used shop hammer, Fig. 6-39. It has a hardened striking face and is employed for all general purpose work that requires a hammer.

Ball-peen ***hammer sizes*** are classified according to the weight of the head, without the handle. They are available in weights of 2, 4, 8, and 12 ounces, and 1, 1 1/2, 2, and 3 pounds.

Soft-face hammers or ***mallets*** permit heavy blows to be struck without part or surface damage. A steel

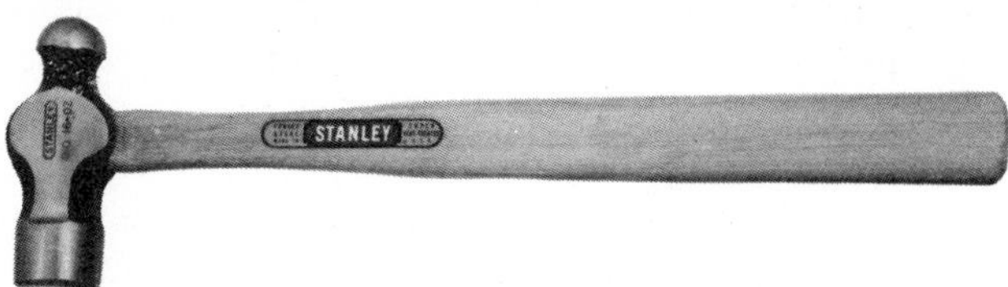

Fig. 6-39. Ball-peen hammer is most common type.

face hammer would damage or mar the work surface on the part.

Soft-face hammers are especially useful for setting work tightly on parallels (steel bars) when mounting material in a vise for machining, Fig. 6-40.

Soft-face hammers are made of many different materials: copper, brass, lead, rawhide, and plastic. They range in weight from a few ounces to several pounds.

Safety precautions for striking tools

1. NEVER strike two hammers together. The faces are very hard and the blow might cause a chip to break off and fly out at high speed.
2. Do NOT use a hammer unless the head is on tightly and the handle is in good condition.
3. Knuckles can be injured if you ''choke up'' too far on the handle when striking a blow.
4. Unless a blow is struck squarely, the hammer may glance off of the work and injure you, or someone working nearby.
5. Place a hammer on the bench carefully. A falling hammer can cause a painful foot injury, or damage precision tools on the bench.

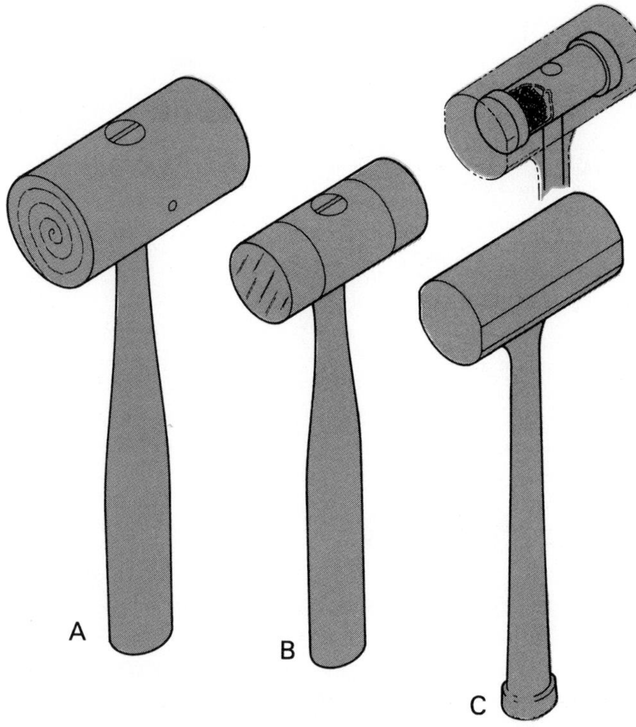

Fig. 6-40. There are several types of non-marring soft face hammers and mallets available for shop use. Shown are: A—Rawhide mallet. B—Plastic face hammer. C—''Dead blow'' hammer. Dead blow hammer has tiny steel shots encased in plastic. This provides the striking power but it will not rebound back as will other mallets and soft face hammers.

HAND CUTTING TOOLS

Not all cutting in metalworking is done by machine. There are several basic hand tools that are cutting implements. These tools, when in good condition, sharp, and properly handled, are safe to use.

Chisels

Chisels are tools used mostly to cut cold metal, hence the term ''cold'' chisels. The four chisels illustrated in Fig. 6-41 are the most common types. The general term ***cold chisel*** is used when referring to these chisels. Other chisels in this category are variations or combinations of these four chisels.

The work to be cut will determine how the chisel should be sharpened. A chisel with a slightly curved cutting edge will work better when cutting on a flat plate, Fig. 6-42A. A curved edge will help prevent the chisel from cutting unwanted grooves in the surrounding metal, as when shearing rivet heads. If it is to be used to shear metal held in a vise, the cutting edge should be straight, Fig. 6-42B.

The chisel is frequently employed to chip surplus metal from castings. Chipping is started by holding the chisel at an angle, as shown in Fig. 6-43A. The angle must be great enough to cause the cutting edge to enter the metal.

After the chisel cut has been started and the proper depth reached, the chisel angle can be decreased enough to keep the cutting action at the proper depth, Fig. 6-43B. Cut depth can be reduced by decreasing the chisel angle. However, decrease the angle too much and it will ride on the heel of the cutting edge and lift out of the cut, Fig. 6-43C.

Grip the metal so the layout line is just below the vise jaws when shearing it to a line. This will leave sufficient metal to finish by filing or grinding. When cutting, it is usually best to hold the metal in a vise WITHOUT using jaw caps. This provides a better shearing action between the vise jaws and chisel. Advance the chisel after each blow so the cutting is done by the center of the cutting edge.

A chisel is an ideal tool for removing rivets. The head can be sheared off and the rivet punched out. A variation of the conventional cold chisel for removing rivet heads is called a ***''rivet buster,''*** Fig. 6-44.

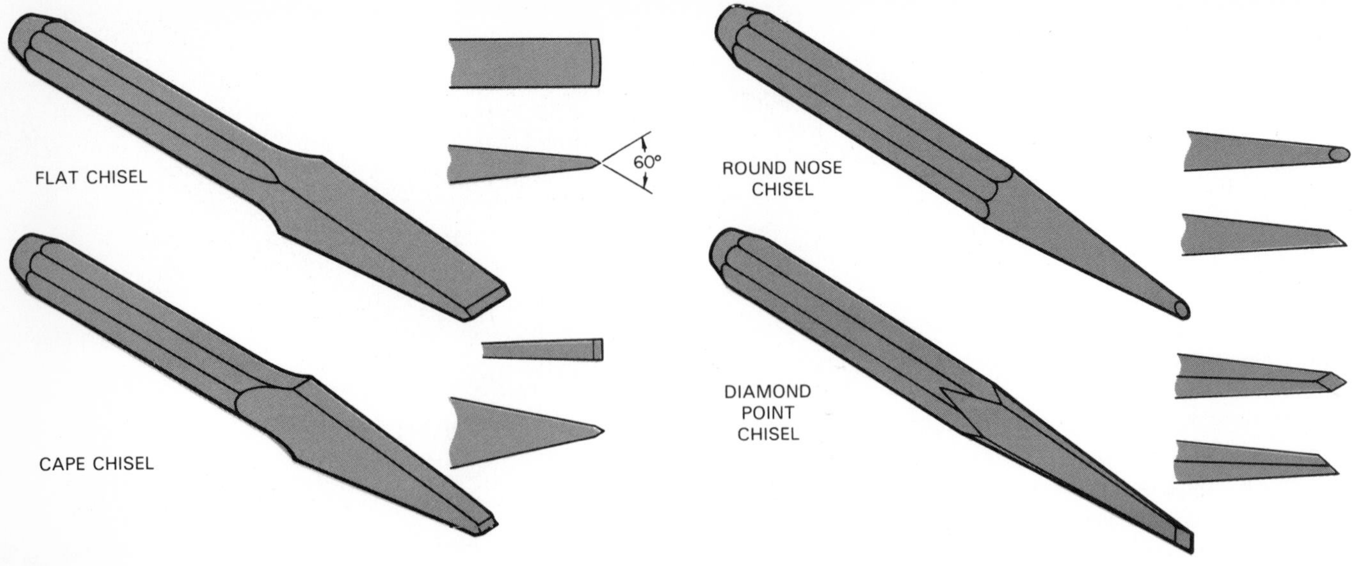

Fig. 6-41. Study cold chisel types. Flat chisel is used for general cutting and chipping work. Cape chisel has a narrower cutting edge than flat chisel and is used to cut grooves. Round nose chisel can cut radii and round grooves. Diamond point chisel is principally used for squaring corners.

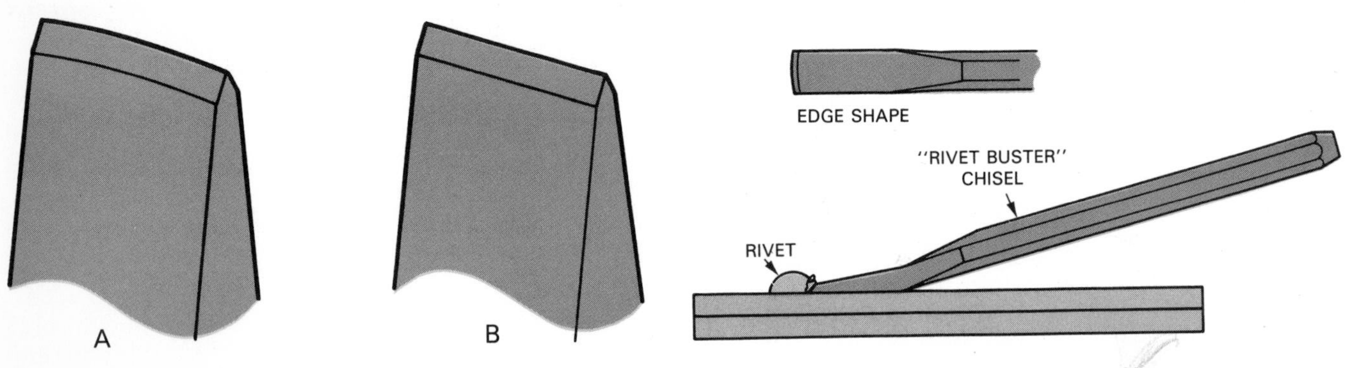

Fig. 6-42. The work to be done determines how a chisel should be sharpened. A—Slightly rounded for cutting on flat plates. B—Flat for shearing.

Fig. 6-44. This chisel, a variation of the flat chisel, is often referred to as a "river buster." Note drawing that shows how it is sharpened.

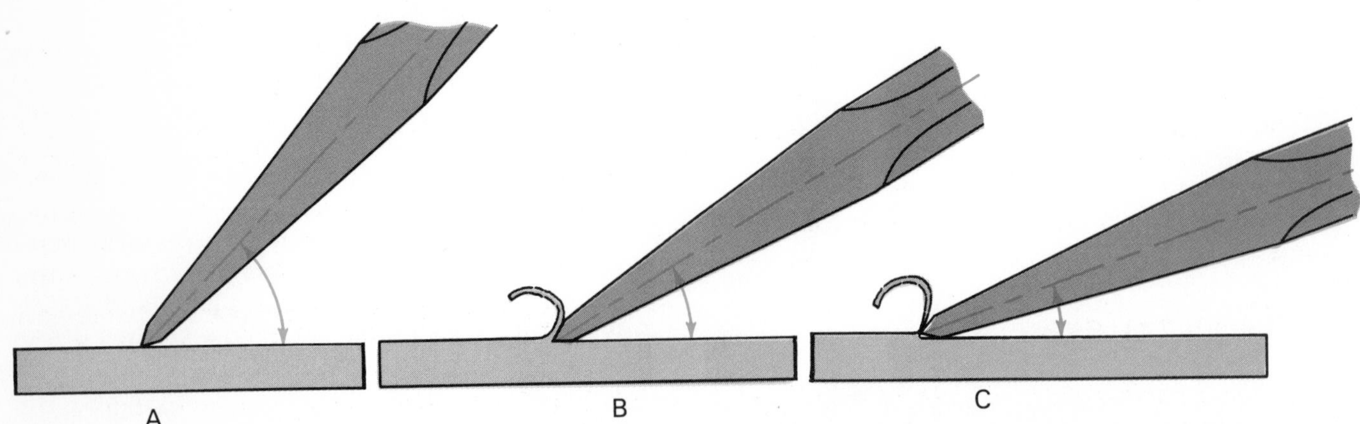

Fig. 6-43. Note proper chisel angles for various cutting operations. A—Starting the cut. B—Maintianing cut at desired depth. C—Reducing the cutting angle too much will cause chisel to lift out of cut.

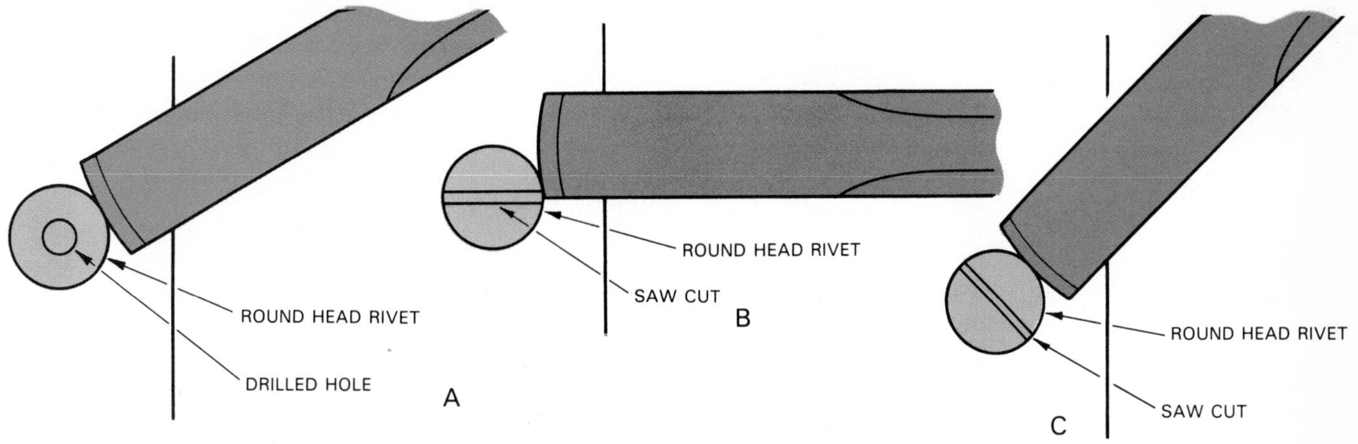

Fig. 6-45. Recommended practice for removing rivet heads. A—When there is not enough room to swing hammer with sufficient force. B and C—When rivet heads are too large to be removed at one time.

When there is NOT enough room to swing a hammer with sufficient force to cut a rivet, drill a hole about the size of the rivet body and almost through the head, Fig. 6-45A. The head can then be removed easily with a chisel.

If the head is so large that the entire head cannot be removed in one piece, saw the head almost through and cut away half of the head at a time. Fig. 6-45B and C shows how this may be done. Rivets can also be removed with a cape chisel, Fig. 6-46.

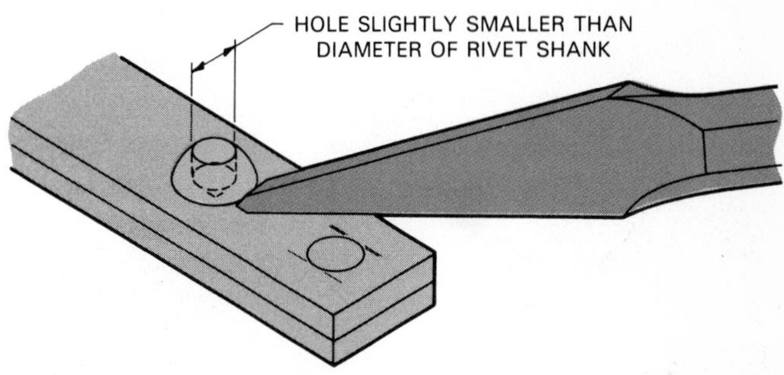

Fig. 6-46. A cape chisel may also be used to remove rivets.

WARNING! There are few things more dangerous than jagged metal knocked or chipped off of a ***mushroomed chisel head*** (head smashed and enlarged), Fig. 6-47A. Remove this hazardous condition by grinding before someone is injured, Fig. 6-47B.

Chisel safety

1. Flying chips are dangerous! Wear safety goggles, and erect a shield when cutting with a chisel. They will protect you and people working nearby.
2. Hold a chisel so that if you miss with the hammer it will not strike and injure your hand. If available, use a chisel holder.
3. Remove any chisel head mushrooming by grinding before it becomes dangerous.
4. Edges cut on metal with a chisel are sharp and can cause bad cuts. Remove them by grinding or filing.

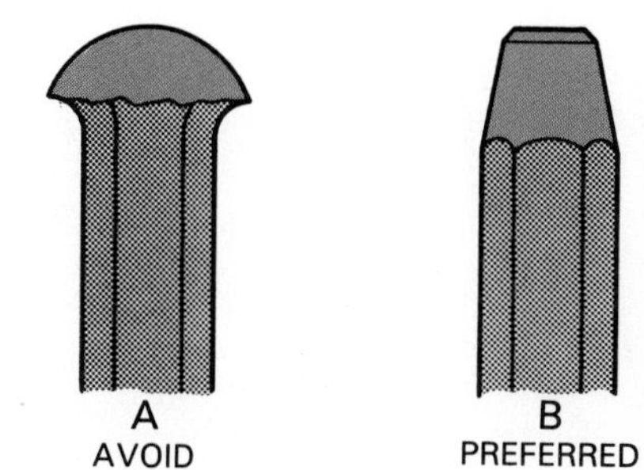

Fig. 6-47. A—Chisel has a dangerous mushroomed head. B—Safe chisel head. A mushroomed head should be ground to a safe condition before flying chips injure someone.

SAWING METAL BY HAND

The typical ***hacksaw*** is composed of a frame with a handle and a replaceable blade, Fig. 6-48. Almost all hacksaws made today are adjustable to accommodate several different blade lengths. They are also made so the blade can be installed in either a vertical or horizontal position, Figs. 6-49 and 6-50.

When placing a blade in the saw frame, make sure the frame is adjusted for the blade length being inserted. There should be sufficient adjustment remaining to permit tightening the blade until it "pings" when snapped with your finger.

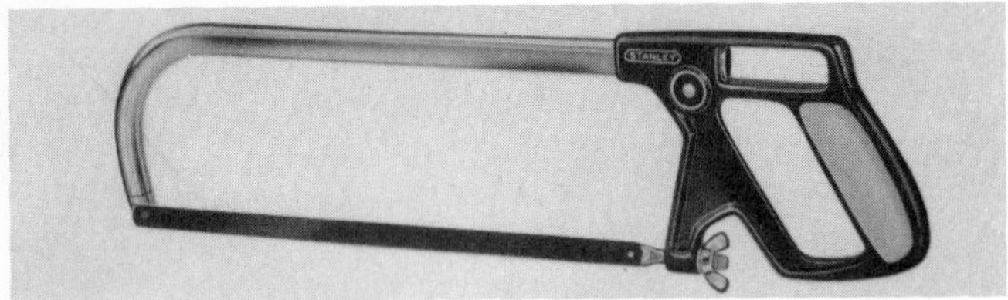

Fig. 6-48. This is a typical hacksaw. (Stanley Tool Co.)

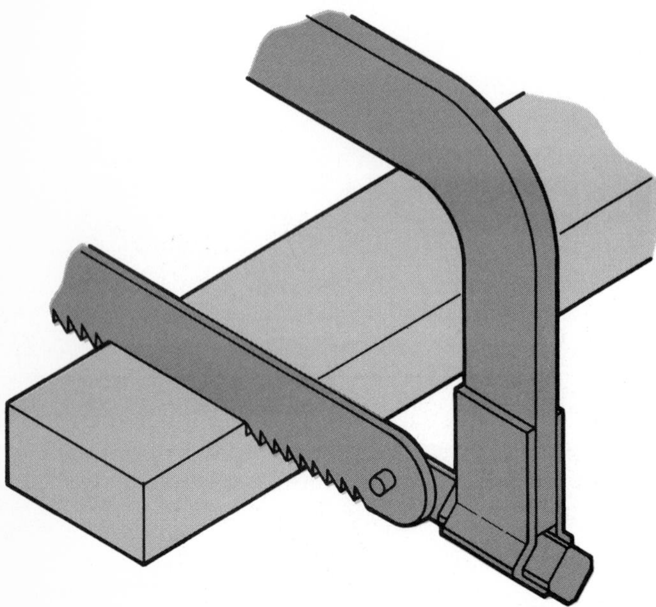

Fig. 6-49. This blade is set in a conventional vertical position.

The hacksaw blade must be positioned with the teeth pointing AWAY from the HANDLE, Fig. 6-51. Frequently, a new blade must be retightened after a few strokes because it will stretch slightly from the heat produced while cutting.

Holding the work

The work must be held securely, with the point to be cut as close to the vise as practical. This eliminates ''chatter'' and vibration that dull the saw teeth.

Fig. 6-52 offers suggestions for holding irregular shaped work. Note that the work is clamped so the cut is started on a flat side rather than on a corner or edge. This lessens the possibility of ruining the teeth or breaking the blade.

Starting a cut

When starting a cut to a marked line, it is best to notch the work with a file, Fig. 6-53. You can also use the thumb of the left hand to guide the blade until it starts the cut, Fig. 6-54. Some hacksaw blades are manufactured with very fine teeth at the FRONT to make starting a cut easier. Use enough pressure so the blade will begin to cut immediately.

Hand cutting metal

Grasp the hacksaw firmly by the handle and front of the frame. Apply enough pressure on the forward stroke to make the teeth cut. Insufficient pressure will permit the teeth to slide over the material and

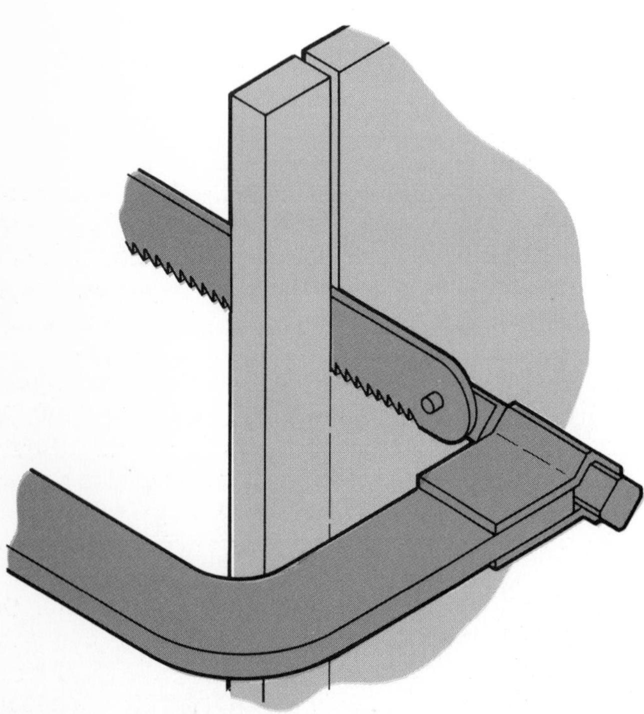

Fig. 6-50. This blade is positioned in a horizontal orientation.

Fig. 6-51. A hacksaw blade must be inserted with the teeth pointing to cut on the forward stroke of the hacksaw. The teeth point away from the handle.

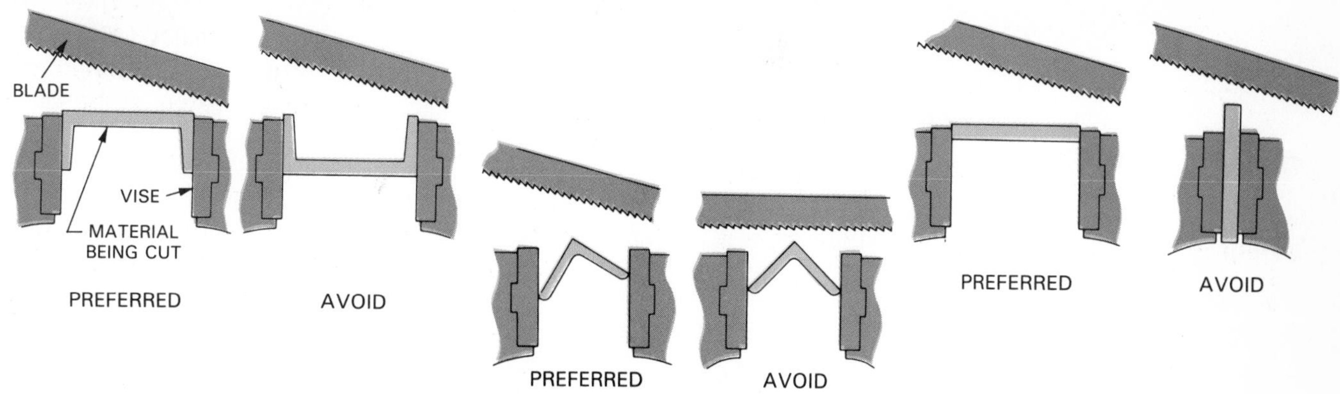

Fig. 6-52. Study how to hold irregular shaped work for sawing.

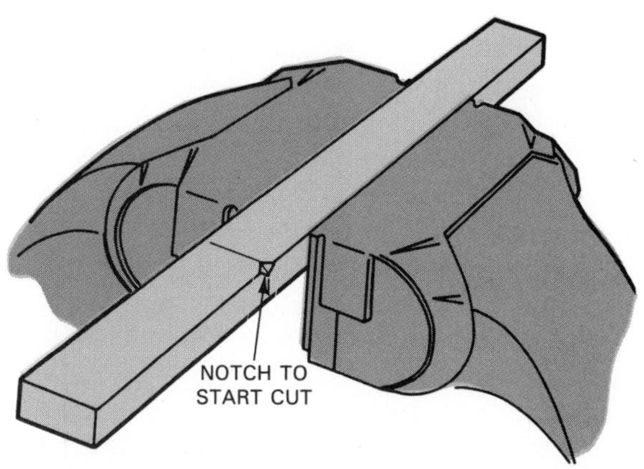

Fig. 6-53. Notching or nicking the edge of a piece to be cut with a hacksaw permits easier starting of the cut.

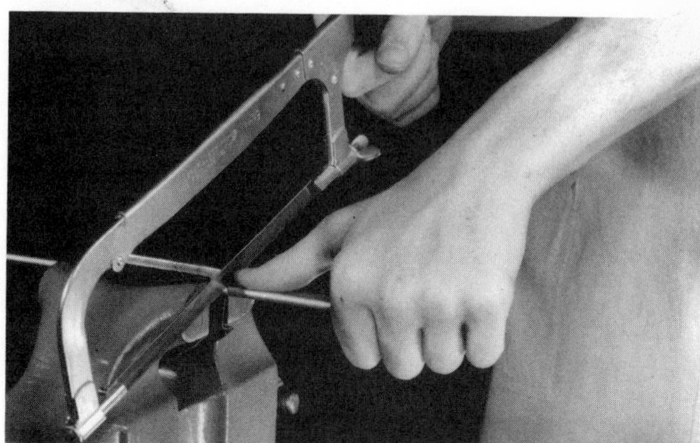

Fig. 6-54. Be careful when using thumb to guide blade until cut is started.

dull the teeth. Also, lift the saw slightly on the return stroke.

Cut the the FULL LENGTH of the blade and make about 40 to 50 strokes per minute. More strokes per minute may generate enough heat to draw the blade temper and dull the teeth. Keep the blade moving in a straight line. Avoid any twisting or binding which can break or bend the blade.

Dulling or breaking a hacksaw blade

If a cut is started with an old blade and the blade breaks or dulls, do NOT continue in the same cut with a new blade. As a blade become dull, the ***kerf*** (slot made by blade) becomes narrower. To continue the cut in the slot with a new blade will usually cause the blade to bind and be ruined in the first few strokes. If possible, rotate the work and start a new cut on the other side.

Finishing a cut

Saw carefully when the blade has cut almost through the material. Support the stock being cut off with your free hand to prevent it from dropping when the cut is completed.

Saw blades

All hacksaw blades are heat treated to provide the hardness and toughness needed to cut metal. The ***flexible back blade*** has only the teeth hardened. The ***all-hard blade*** is hardened throughout, except the hardness is reduced near the end holes to reduce the possibility or breakage at these points.

Flexible back blades are best for sawing soft materials or materials with thin cross sections. An all-hard blade does NOT buckle when heavy pressure is applied. It is best for cutting hard metals.

The number of teeth per inch on the blade has an important bearing on the shape and kind of material to be cut, and upon blade life. TWO or THREE TEETH should be cutting at all times, otherwise, the teeth will straddle the section being cut and snap off when cutting pressure is applied, Fig. 6-55.

The ***set*** of the blade provides the necessary clearance, and prevents the blade from binding in

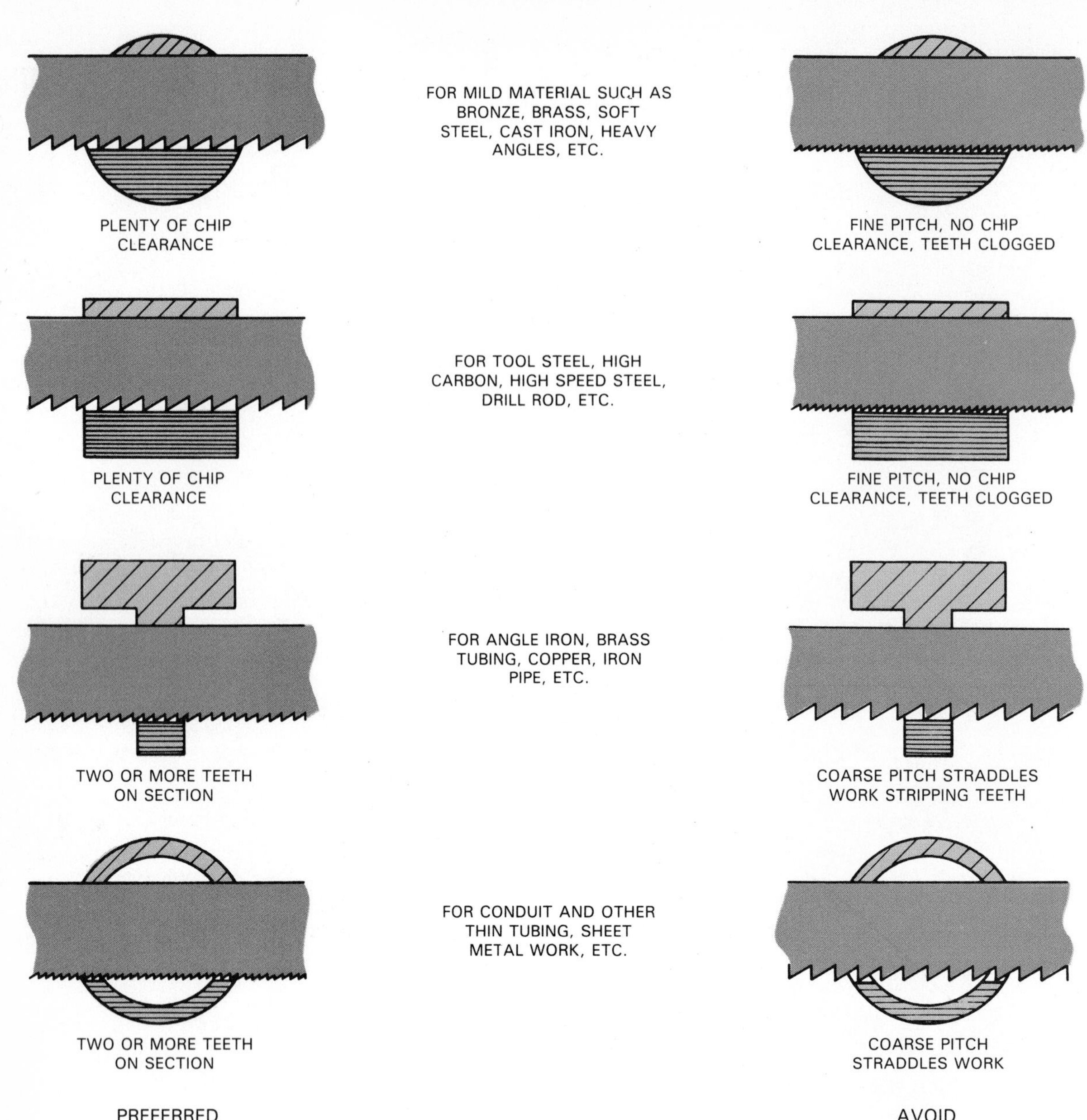

Fig. 6-55. The proper hacksaw blade should be used for each job to assure long blade life and rapid cutting action. Study recommendations.

the cut. A blade may have one of three sets: ***raker, alternate,*** or ***undulating.*** These are in Fig. 6-56.

Unusual cutting situations

Cutting soft metal tubing can be a problem. The blade may bind and tear the tubing or the tubing may flatten. This can be eliminated by inserting a wood dowel of the proper size into the tubing. Then cut through the tubing and the dowel, Figs. 6-56 and 6-58.

Cutting a long narrow strip can be done by setting the blade at right angles to the frame. Make the cut in the usual way, as in Fig. 6-59. Strips of any width, up to the capacity of the saw frame, can be made in this manner.

Thin metal can be cut more easily by putting it between two pieces of wood, and cutting through both of them, Fig. 6-60.

Hacksaw safety

1. Do not test the sharpness of a blade by running your fingers across its teeth.
2. Store saws so you will NOT accidently reach in and grasp the teeth when you pick up a hacksaw.
3. The burr formed on the cut surface is sharp and can cause a serious cut. Do NOT brush away chips with your hand; use a brush.
4. All-hard blades can shatter and produce flying chips. WEAR GOGGLES!

5. Be sure the blade is properly tensioned. If the blade should break while you are on the cutting stroke, your hand may strike the work and cause a painful injury.

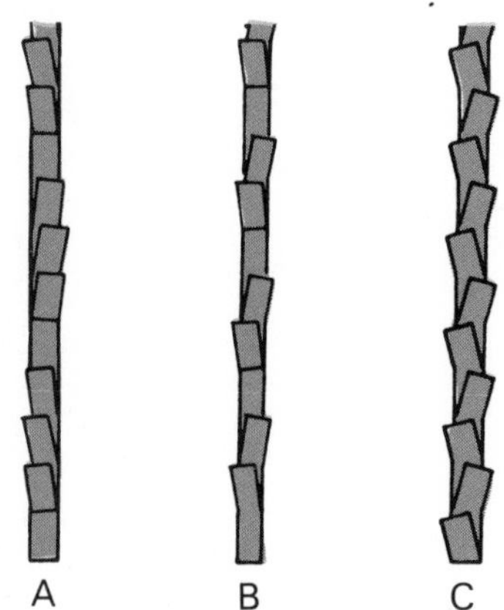

Fig. 6-56. Types of sets in hacksaw teeth. A—Undulated. B—Raker. C—Alternate.

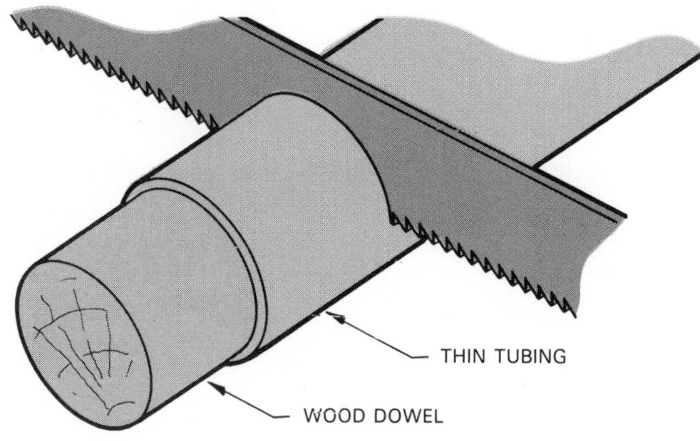

Fig. 6-57. A snug fitting dowel slid into thin wall tubing will make cutting easier.

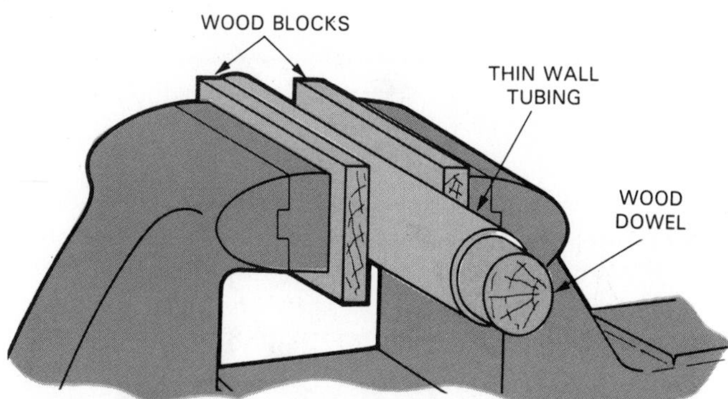

Fig. 6-58. Place soft wood blocks between vise jaws and work to prevent marring and damaging to the exterior surface of metal objects. CAUTION: Avoid applying too much clamping pressure to hold thin wall tubing when it is being cut. The tubing will collapse easily.

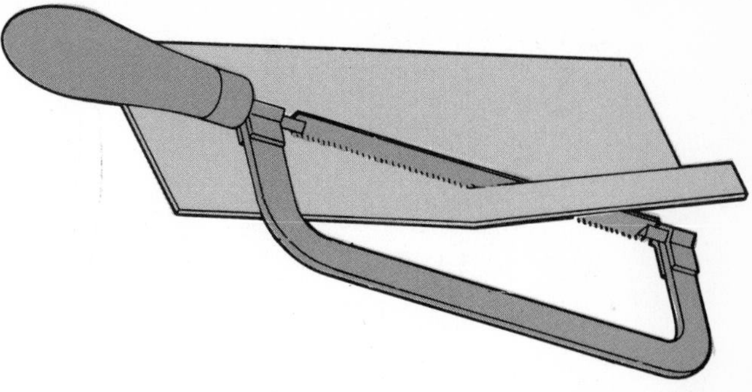

Fig. 6-59. Blade can be pivoted to a horizontal position for cutting long narrow strips. Best results can be obtained if strip is bent up slightly, as shown, during the sawing operation.

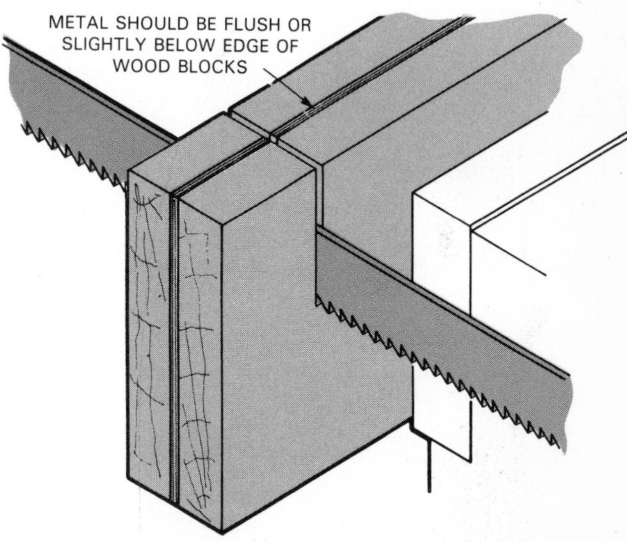

Fig. 6-60. "Sandwich" thin metal between two piece of wood to make cutting easier and more precise.

FILES

A ***file*** is used for hand smoothing and shaping operations. The modern file is made from high grade carbon steel and is heat treated to provide the necessary hardness and toughness.

In the first production step to manufacture a file, the blank is cut to the approximate shape and size. The tang and point are formed next. Then the blank is annealed and straightened. The point and tang are trimmed after the sides and faces have been ground and the teeth cut. After another straightening, it is heat treated, cleaned, and oiled. Tests are made continually to assure a quality tool.

Fig. 6-61 shows the steps in the manufacture of a modern file.

File classifications

Files are classified by their shape. The shape is the general outline and cross section. The outline is either tapered or blunt, Fig. 6-62.

Fig. 6-61. Note how a file is manufactured. A—Steel bar cut to correct length. B—Bar forged to shape. C—Blank after it has been annealed. D—Annealed blank straightened and ground smooth. E—Teeth cut on blank. F—Blank trimmed and coated for heat treatment. G—Completed file cleaned and inspected. (Nicholson File Co.)

Fig. 6-62. These are blunt and tapered files.

Files are also classified according to the cut of the teeth: ***single-cut, double-cut, rasp,*** and ***curved tooth,*** Fig. 6-63, and to the coarseness of the teeth: ***rough, coarse, bastard, second-cut, smooth,*** and ***dead smooth.***

File care

DANGER! A file should NEVER be used without a handle. It is too easy to drive the unprotected tang into your hand.

Fit the handle on the file by drilling a hole in the handle equal in diameter to the width of the tang at its midpoint, Fig. 6-64. Mate the file and handle

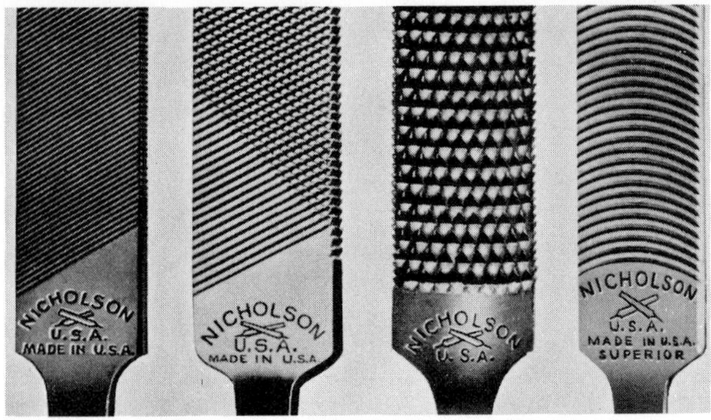

Fig. 6-63. Study single-cut, double-cut, rasp, and curved-tooth files.

by placing the tank into the hole then sharply striking the handle on a solid surface.

Store files so that they are always separated, Fig. 6-65. NEVER throw files together in a drawer, or store them in a damp place.

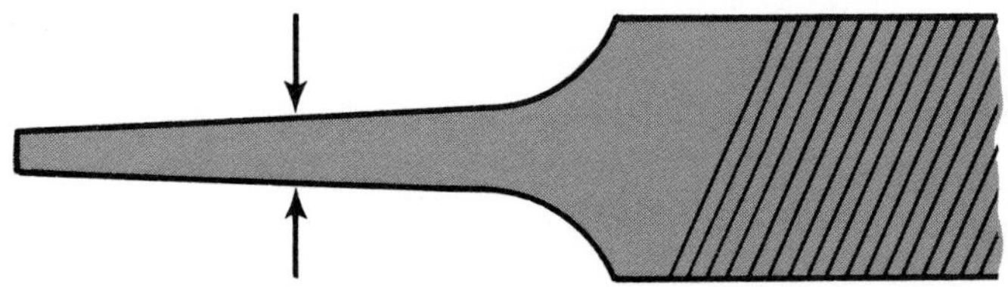

Fig. 6-64. File handle hole should be equal in diameter to width of file tang at point indicated.

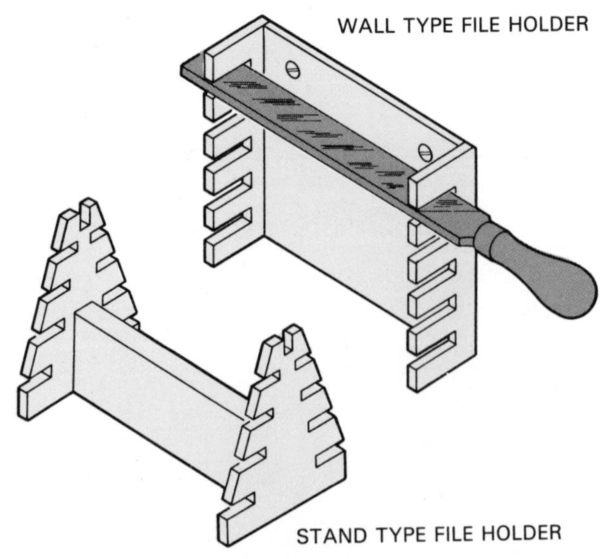

Fig. 6-65. Storing files properly greatly extends their useful lives.

Clean files frequently with a ***file card*** or ***brush,*** Fig. 6-66. Some soft metals cause ***pinning,*** that is, the teeth become loaded or clogged with some of the material the file has removed. Pinning causes gouging and scratching on the work surface. The particles can be removed from the file with a pick or scorer. A ***file card*** combines the card, brush, and pick for file cleaning.

Fig. 6-67. These are a few of the many hundred different kinds of files. (Nicholson File Co.)

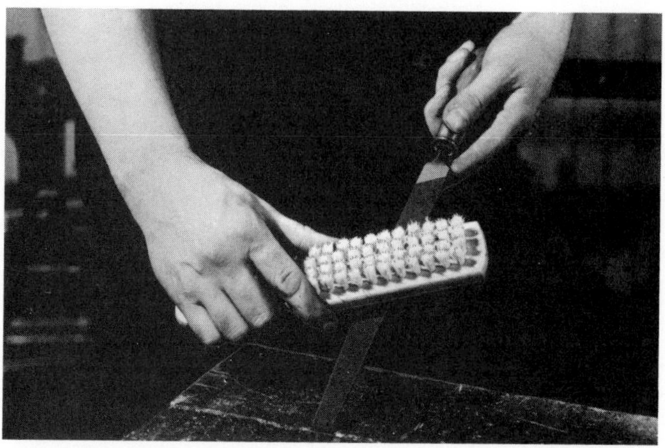

Fig. 6-66. Machinist is using file card to clean file.

File selection

There is almost no limit to the number of different kinds, shapes, and cuts of files that are manufactured, Fig. 6-67. Only the general classification of files will be covered.

Files have three distinct characteristics: length, kind, and cut. The ***file length*** is always measured from the heel to the point, Fig. 6-68. The tang is NOT included in the measurement.

The ***file type*** refers to its SHAPE, such as flat, mill, half-round, square, etc. The ***file cut*** indicates the relative COARSENESS of the teeth.

Single-cut files are usually used to produce a smooth surface finish. They require only light pressure to cut.

Double-cut files remove metal much faster than single-cut files. They require heavier pressure and they produce a rougher surface finish.

Rasps are best for working wood or other soft materials where a large amount of stock must be removed in a hurry.

A ***curved-tooth file*** is used to file flat surfaces of aluminum and steel sheet.

Some files have ***safe edges*** which denotes that the file has one or both edges without teeth, Fig. 6-69. This permits filing corners without danger to the portion of the work that is NOT to be filed.

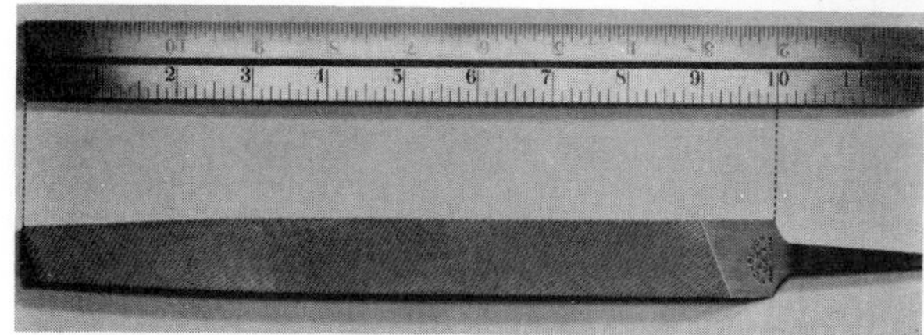

Fig. 6-68. Note how the file is measured. (Nicholson File Co.)

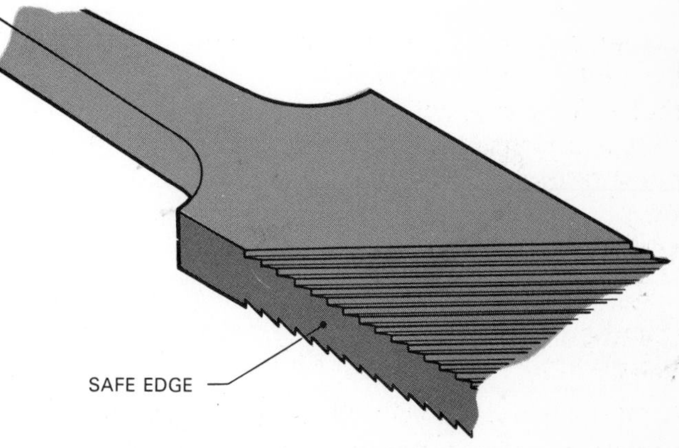

Fig. 6-69. Safe edge of a file does not have teeth.

Many factors must be considered in selecting the file if maximum cutting efficiency is to be attained:

1. The nature of the work (flat, concave, convex, notched, etc.).
2. Kind of material.
3. Amount of material to be removed.
4. Surface finish and accuracy demanded.

Of the many file shapes available, the most commonly used are flat, pillar, square, 3-square, knife, half-round, round, and crossing, Fig. 6-70. Each shape is available in many sizes and degrees of coarseness: rough, coarse, bastard, second-cut, smooth, and dead smooth, Fig. 6-71. Note that a rough cut, small file (4 in.) may be as fine as a large (16 in.) second-cut file.

Kinds of files

The vast variety of files fall into five general groups:

1. The ***machinist file*** is used whenever metal must be removed rapidly, and the finish is of secondary importance. It is made in a large range of shapes and sizes, and is double-cut.
2. The ***mill file*** is a single-cut and tapers the last third of the length towards the point. It is suitable for general filing when a smooth finish is required. A mill file works well for draw-filing, lathe work, and working on brass and bronze.
3. ***Swiss pattern*** and ***jewelers's files*** are manufactured in over a hundred different shapes. They are used primarily by tool and die makers, jewelers, and others who do precision filing.
4. The ***rasp*** has teeth that are individually formed and disconnected from each other. It is used for relatively soft materials (plastic for example) when large quantities of the material must be removed.
5. The groups of ***special purpose files*** include those specifically designed to cut one type of metal. The long-angle lathe file, Fig. 6-72, that does an efficient filing job on the lathe, falls into this category.

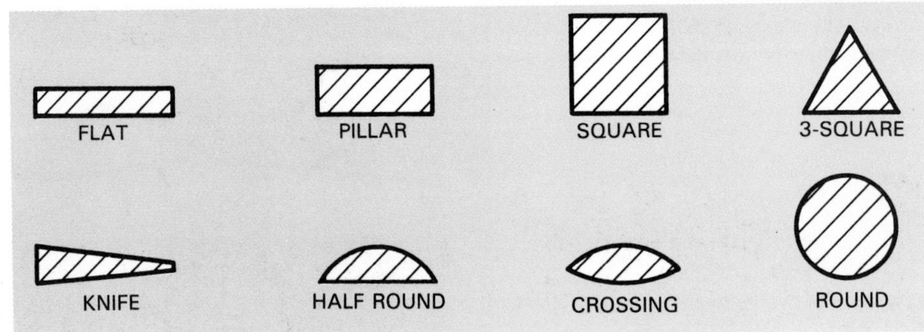

Fig. 6-70. Study cross-sectional views of most widely used files.

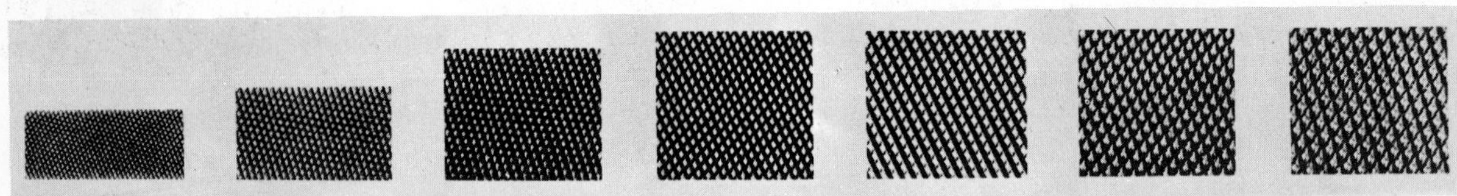

Fig. 6-71. Note range in coarseness of a typical machinist flat bastard file. File sizes range from 4 in. (100.0 mm) to 16 in. (400.0 mm). (Nicholson File Co.)

Using a file

Efficient filing requires that the work be held solidly. Where practical, hold the work at about elbow height for general filing, Fig. 6-73. Mount it slightly lower if large quantities of metal must be removed by heavy filing.

Straight or ***cross filing*** consists of pushing the file lengthwise, straight ahead or at a slight angle,

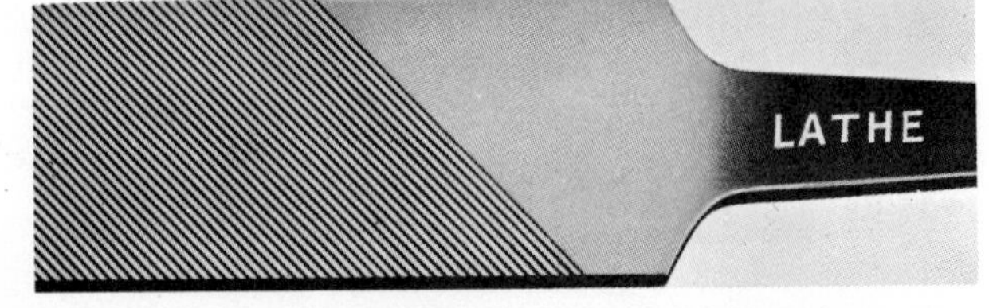

Fig. 6-72. This is a long angle lathe file.

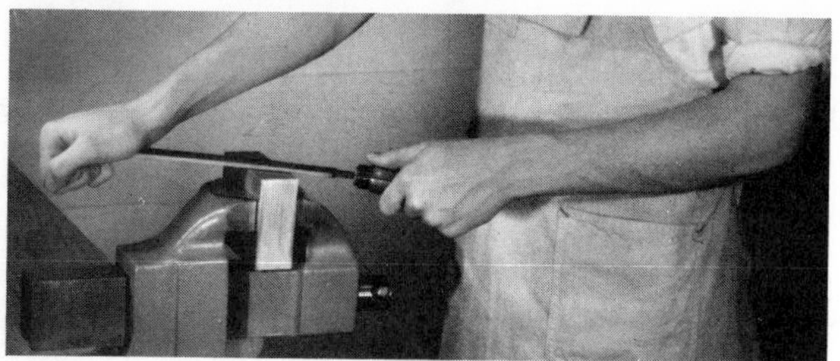

Fig. 6-73. Mount work at elbow height for general filing.

across the work. Grasp the file as shown in Fig. 6-74. Heavy-duty filing requires heavy pressure and can best be done if the file is held as in Fig. 6-75.

Files can be RUINED by using either TOO MUCH pressure or TOO LITTLE pressure on the cutting stroke. Apply just enough pressure to permit the file to cut on the entire forward stroke. Too little pressure allows the file to slide over the work and it becomes dull. Too much pressure ''overloads'' the file and causes the teeth to clog and chip.

Lift the file fromthe work on the reverse stroke except when filing soft metal. Then the pressure on the return stroke should be no more than the weight of the file.

Draw filing, when properly done, produces a finer finish than straight filing. Hold the file as shown in Fig. 6-76. Do NOT use a short angle file for draw filing, as there is the likelihood of scoring or scratching instead of shaving and shearing, as the file should while being pushed and pulled across the metal. Use a double-cut file to ''rough down'' the surface and a single-cut file to produce the final finish.

File safety

1. A file should NEVER be used without a handle. Painful injuries may reuslt!
2. Clean files with a file card, NOT your hand. The chips can penetrate your hand and cause a painful infection.
3. Files are very brittle. They will break if used to pry!
4. Clean the surface being filed with a piece of cloth. Short burrs are formed in filing and can cause serious cuts.
5. Never hammer on or with a file. It may shatter and chips can fly in all directions.
6. Avoid cleaning a file by slapping it on the bench. It may shatter.

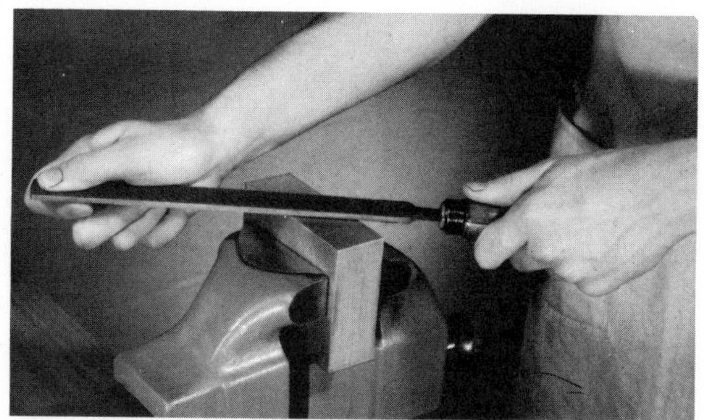

Fig. 6-74. Note proper way to hold a file for straight or cross filing.

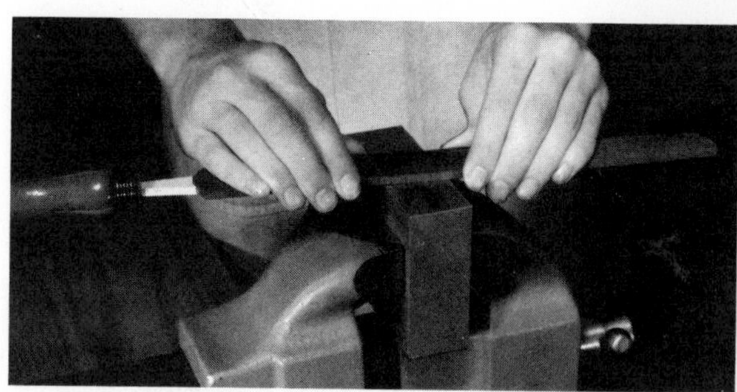

Fig. 6-76. Draw filing improves the surface finish when done properly.

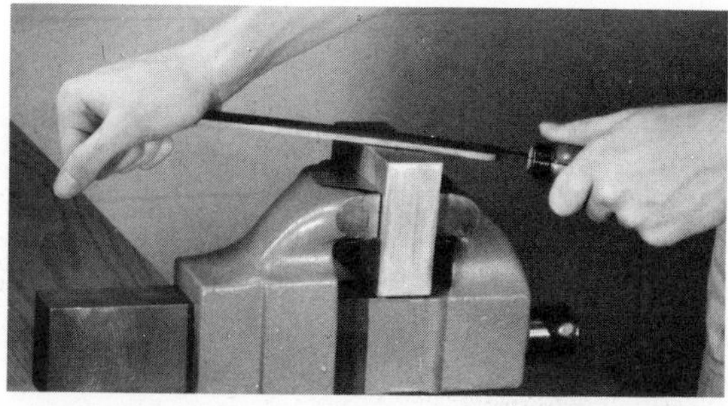

Fig. 6-75. Some additional pressure is required when a considerable quantity of metal must be removed.

HAND REAMING

A drill does NOT produce a smooth or accurte enough hole for a precision fit. Reaming is the operation that will produce smoothness and accuracy. Ordinarily, only final sizing of a hole is done by ***hand reaming.***

Hand reamer

A ***hand reamer*** has a square on the shank end that is suitable for holding it in a tap wrench, Fig. 6-77. The reamer may be made of high speed steel

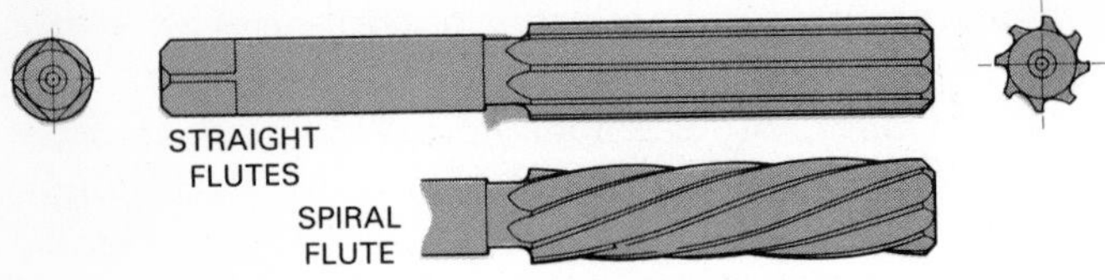

Fig. 6-77. This is a hand reamer.

or carbon steel, and is available in sizes from 1/8 to 1 1/2 in. (3.175 to 38.1 mm). The cutting end is ground with a slight taper to provide easy starting in the hole.

Straight fluted reamers are suitable for most work. However, when reaming a hole with a keyway or other interruption, it is better to have a ***spiral fluted reamer.***

When preparing a piece to be reamed by hand, 0.005 to 0.010 in. (0.15 to 0.25 mm) of stock should be left in the hole for removal by the reaming tool.

An ***expansion hand reamer*** is used when a hole must be cut a few thousandths over nominal size for fitting purposes, Fig. 6-78. Slots are cut into the center of the tool. The center opening is machined on a slight taper. The reamer is expanded by tightening a taper screw into this opening. The amount of expandion is limited and the reamer may be broken if expanded too much.

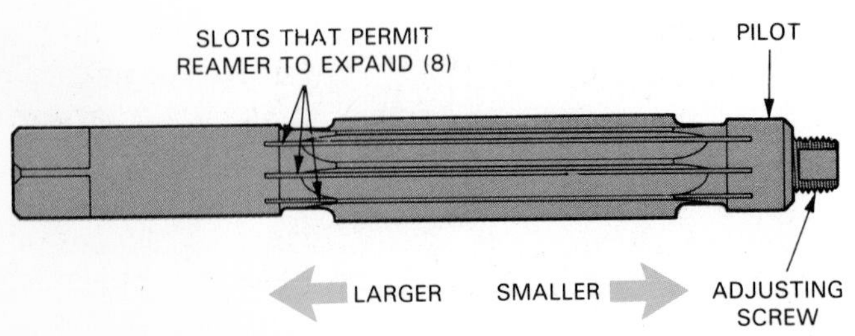

Fig. 6-78. Note parts of expansion hand reamer.

Do NOT use an expansion reamer in place of a solid reamer unless absolutely necessary because of the danger of producing OVERSIZE HOLES.

The ***adjustable hand reamer*** is threaded its entire length and fitted with tapered slots to receive the adjustable blades, Fig. 6-79. The blades are tapered along one edge to correspond with the taper slots in the reamer body so that the cutting edges of the blades are parallel.

Reamer diameter is set by loosening one adjusting nut and tightening the other. The blades can be moved in either direction. This type reamer is manufactured in sizes ranging from 3/8 to 3 1/2 in. (9.5 to 85.0 mm) and each reamer has sufficient adjustment to increase reamer diameter to the next larger reamer size.

The ***taper reamer*** will finish a taper hole accurately and with a smooth finish for taper pins, Fig. 6-80. Because of the long cutting edges, taper reamers are somewhat difficult to operate.

To provide for easier removal of surplus metal, a ***roughing reamer*** is first rotated into the hole. This reamer is slightly SMALLER (0.010 in. or 0.25 mm) than the finish reamer. It also has left-hand spiral grooves ground along the cutting edges to break up the chips.

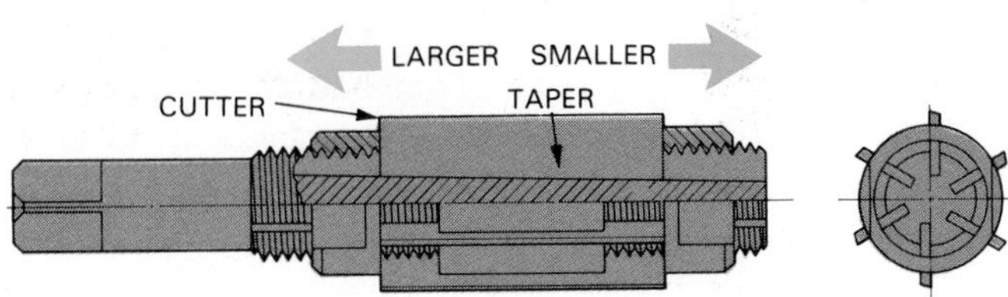

Fig. 6-79. Adjustable hand reamer can be set for odd sizes.

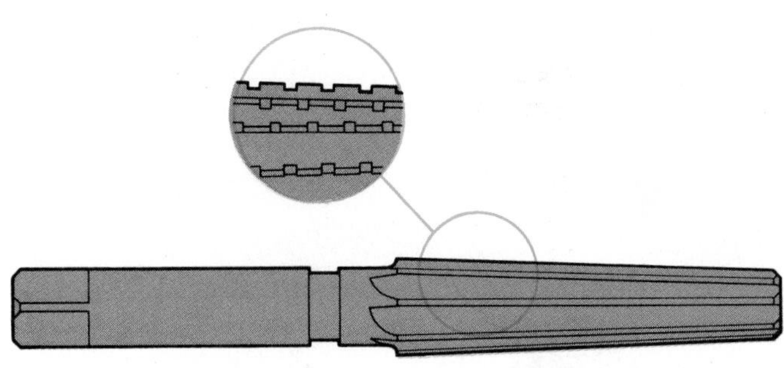

Fig. 6-80. This is a taper hand reamer. Inset shows how cutting edges are notched on roughing taper reamer.

Using a hand reamer

A two-handle tap wrench is commonly used because it permits an even application of pressure on the reamer. It is virtually impossible to secure a satisfactory hole using an adjustable wrench.

To start reaming, rotate the reamer slowly to allow it to align with the hole. It is desirable to check whether the reamer has started square at several points around its circumference, Fig. 6-81.

Feed should be steady and rapid. Keep the reamer cutting, or it will start to "chatter," producing a series of tool marks in the surface of the hole. This could also cause the hole to be out-of-round.

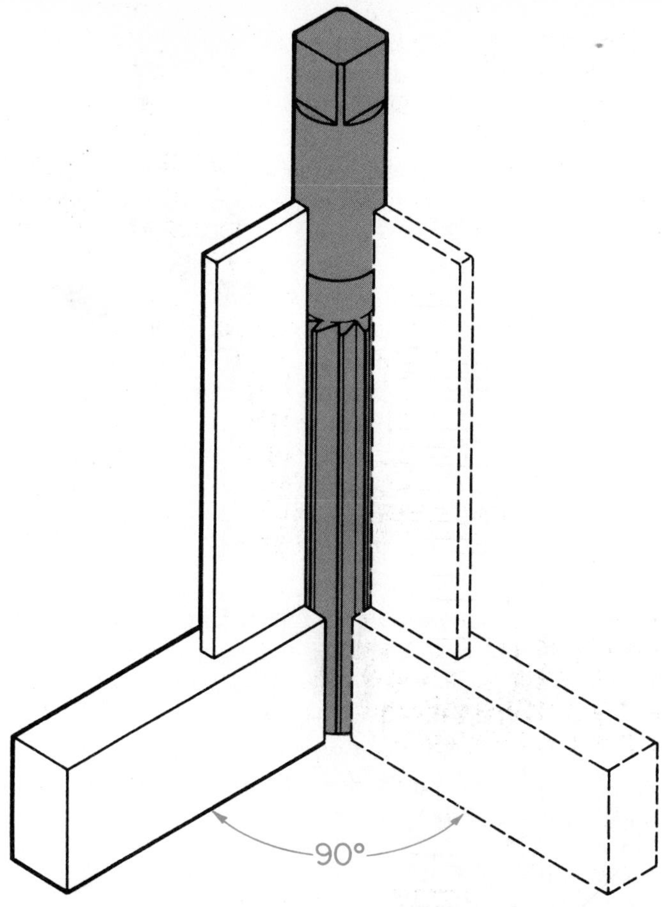

Fig. 6-81. Always make sure reamer is square with work.

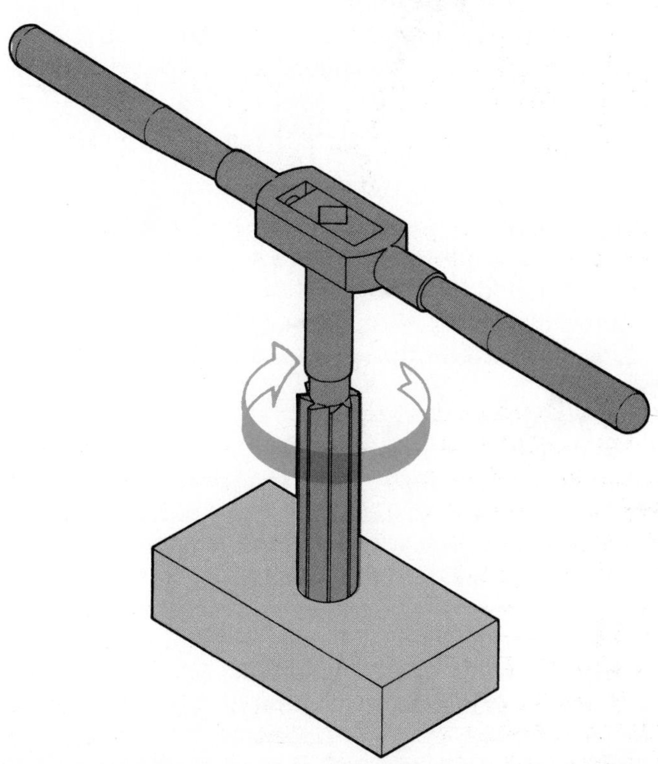
Fig. 6-82. Always turn a hand reamer in a clockwise direction.

Turning pressure is applied evenly with both hands, and ALWAYS in a clockwise direction, Fig. 6-82. NEVER turn a reamer in a counterclockwise direction because the cutting edges will be dulled.

Feed the reamer deeply enough into the hole to take care of the starting taper. Cutting fluid, to be applied, will depend upon the metal being reamed.

Hand reaming safety

1. Remove all burrs from reamed holes.
2. NEVER use your hands to remove chips and cutting fluid from the reamer. Use a piece of cotton waste.
3. Store reamers carefully so they do NOT touch one another. Reamers should never be stored loose or thrown in a drawer with other tools.
4. Clamp work solidly before starting to ream.
5. Avoid removing chips and cutting fluid or cleaning a reamed hole with compressed air.

HAND THREADING

Threaded sections have many applications in our everyday life. A ***thread*** is a spiral or helical ridge found on nuts and bolts. When required on a job, threads are indicated on the plans and drawings in a special way, Fig. 6-83. They are specified by diameter and number of threads per inch. Metric threads are specified by diameter and thread pitch is given in mm.

The ***American National Thread System*** was adopted in 1911. It is the common thread form used in the United States and is characterized by the 60 deg. angle formed by the sides of the thread.

The ***National Coarse*** (NC) is for general purpose work. ***National Fine*** (NF) is for precision assemblies. These are the most widely used thread groups in the American National Thread Series. The NF group has MORE threads per inch for a given diameter than the NC group.

A considerable amount of confusion resulted during World War II from the many different forms and kinds of threads used by the Allies. As a result, the powers that make up NATO (North Atlantic Treaty Organization) adopted a standard thread form. It is referred to as the ***Unified System,*** Figs. 6-84 and 6-85. It is very similar to the American National Thread System. It differs only in the thread shape. The thread root is rounded and the crest may be flat or rounded. The threads are identified by UNF and UNC (Unified National Fine and Unified National Coarse). Fasteners using this thread series are interchangeable with fasteners using the American National thread.

There are several other thread groups in addition to those just described. Included are the ***Unified***

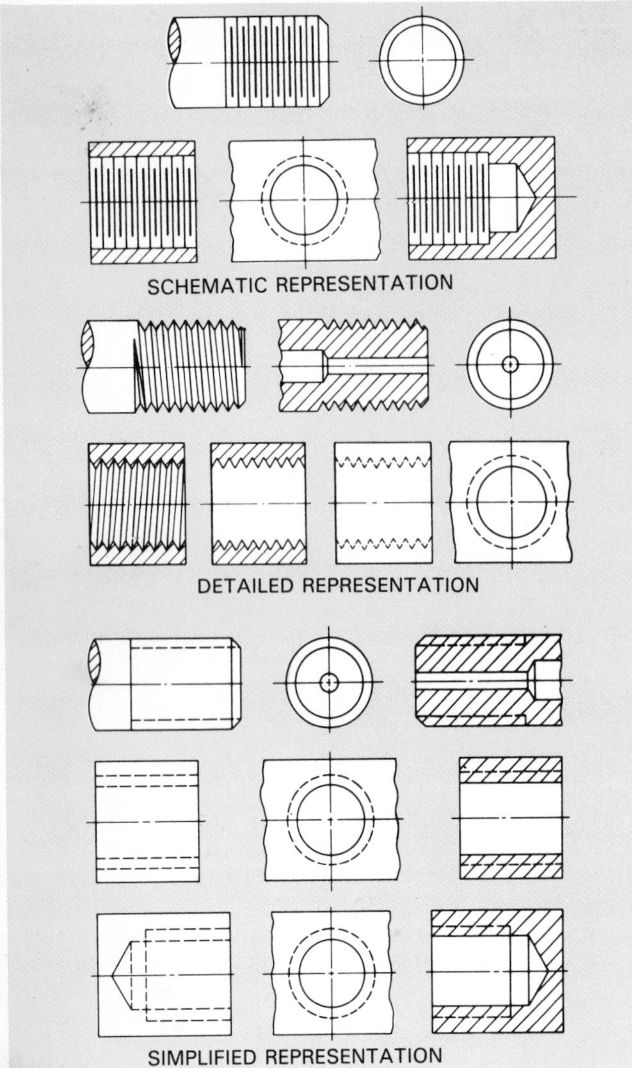

Fig. 6-83. Study methods used to visually depict threads on drawings. Only one type will be used on a drawing.

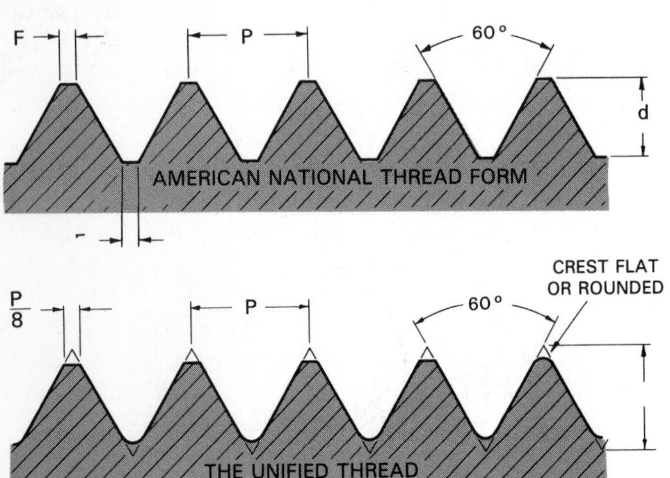

Fig. 6-84. Drawings illustrate difference betwen American National Thread form and Unified Thread form.

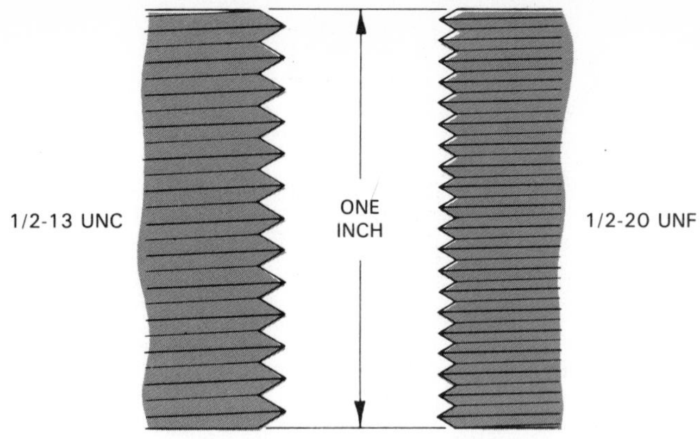

Fig. 6-85. Note comparison of Unified Coarse (UNC), and Unified Fine (UNF) threads. Both have same geometric shape.

National Extra Fine (UNEF), ***Unified National*** 8 ***Series,*** and ***Unified National*** 12 ***Series.*** The 8 Series has 8 threads per inch and is used on diameters ranging from 1 to 6 in. in 1/8 and 1/4 in. increments. The 12 Series has 12 threads per inch and is used on diameters ranging from 1 to 6 in.

Metric unit threads have the same shape as the Unified Thread but are specified in a different manner, Fig. 6-86. Metric threads and Unified National Series threads are NOT interchangeable. See Fig. 6-87.

Thread size

Threads of the Unified National system smaller than 1/4 in. diameter are NOT measured as fractional sizes. They are given by ***number sizes*** that range from #0 (approximately 1/16 or 0.060 in. diameter) to #12 (just under 1/4 or 0.216 in. diameter). Both UNC and UNF series are available.

Care must be taken so the number denoting the thread diameter and the number of threads are NOT mistaken for a fraction. For example: a #8-32 UNC thread would be a thread that as a #8 (0.164 in.) diameter and 32 threads per inch, NOT an 8/32 (1/4) in. diameter thread with a UNC series thread.

Cutting threads

Because thread dimensions have been standardized, the use of ***taps*** to cut INTERNAL THREADS, and ***dies*** to cut EXTERNAL THREADS have become universal practice whenever threads are to be cut by hand. See Fig. 6-88.

Internal threads

Internal threads are made with a tap, Fig. 6-89. Taps are made of carbon steel or high speed steel (HSS) and are carefully heat treated for long life. Taps are quite brittle and are easily broken if NOT handled properly.

ISO METRIC THREAD SERIES

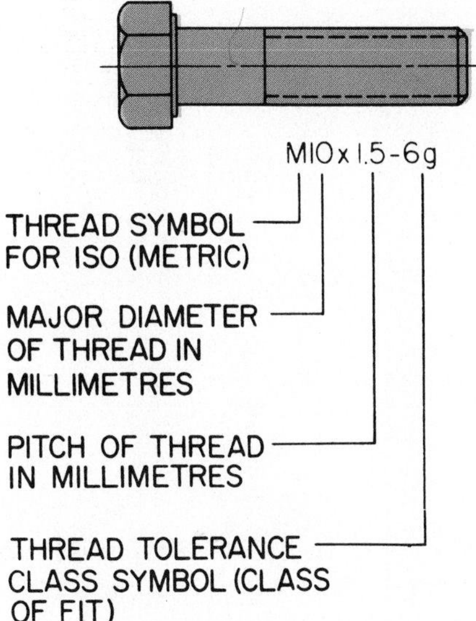

UNIFIED NATIONAL COARSE THREAD SERIES

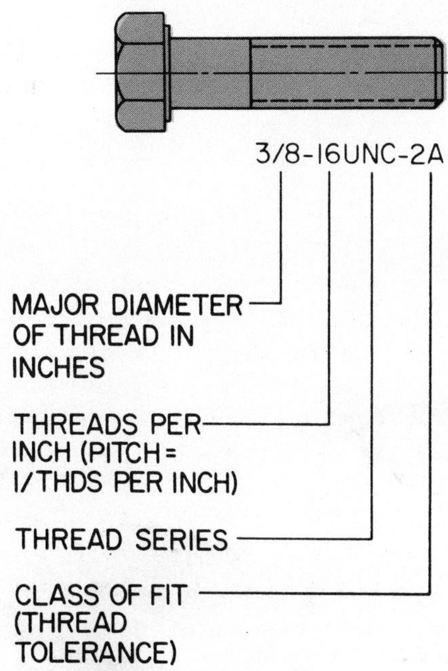

Fig. 6-86. Study how thread size is noted and what each term means.

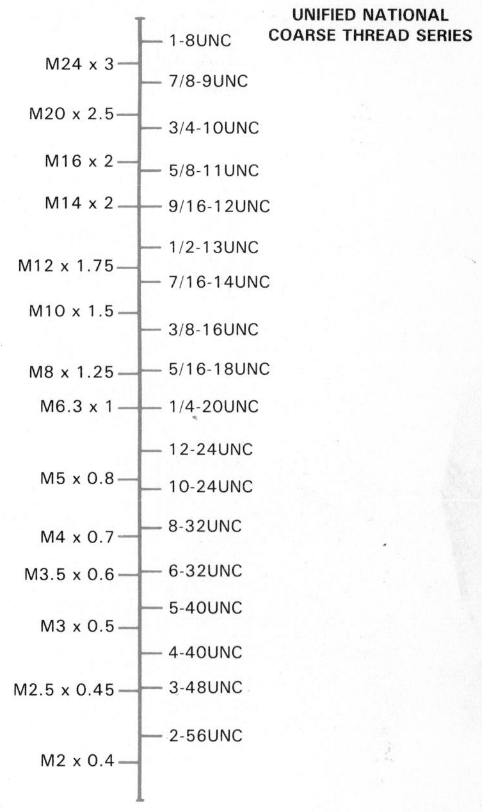

ISO and Unified National Thread Series ARE NOT INTERCHANGEABLE

Fig. 6-87. While ISO metric threads may appear to be similar in diameter to the Unified National thread series, they are NOT interchangeable.

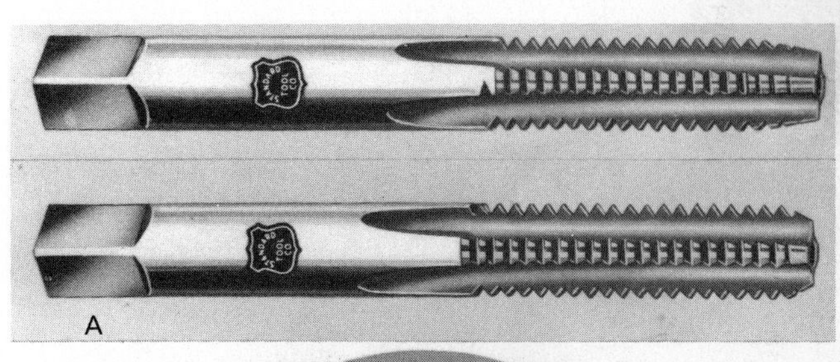

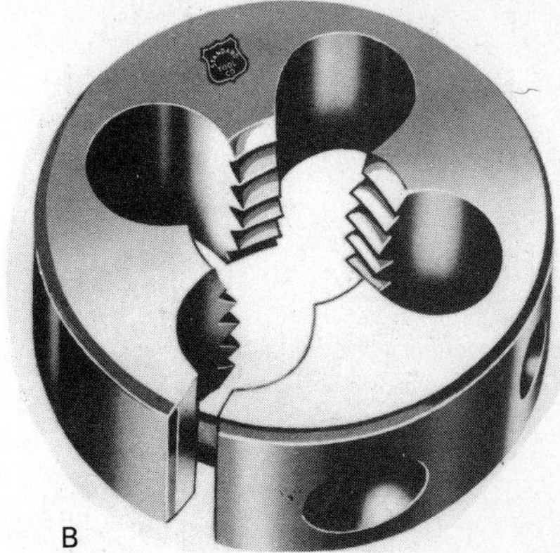

Fig. 6-88. A—Tap is for cutting internal threads. B—Die is for cutting external threads. (Standard Tool Co.)

To meet demands for varying degrees of thread accuracy, it became necessary for industry to adopt standard working tolerances for threads. Working tolerances for threads have been divided into ***classes of fits,*** which are indicated by the last number on the thread description (1/2-13UNC-2).

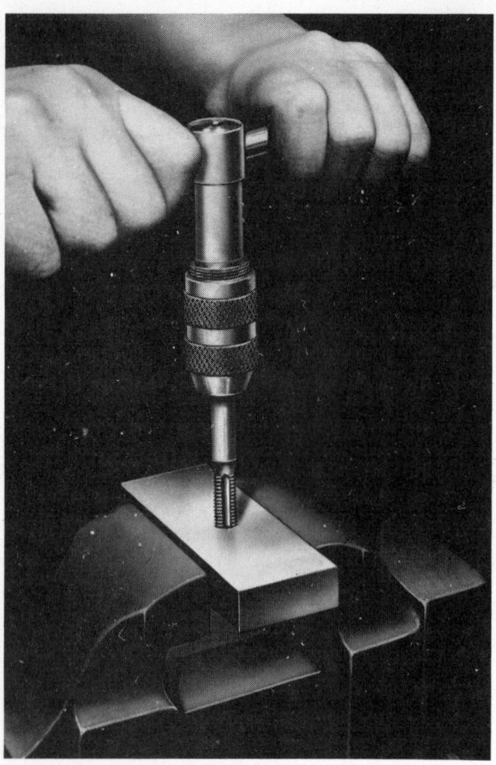

Fig. 6-89. Machinist is cutting internal threads with a tap.

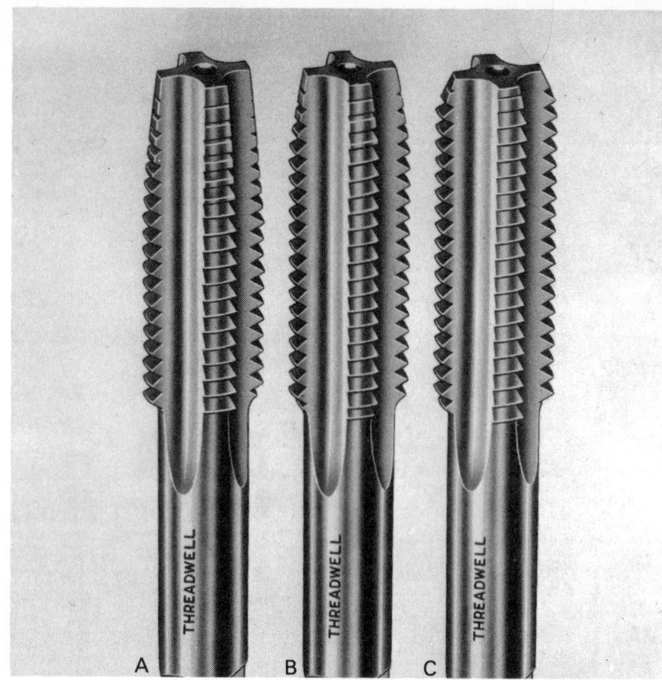

Fig. 6-90. Standard hand taps are manufactured in sets of three. A—Taper for starting thread. B—Plug for continuing thread after taper tap has cut as far into hole as it can. C—Bottoming for continuing threads to bottom of a blind hole. (Threadwell Manufacturing Co.)

Fits for inch-based threads are:

Class 1 = Loose fit.
Class 2 = Free fit.
Class 3 = Medium fit.
Class 4 = Close fit.

Under revised ISO standards, there will be two classes of thread tolerances for external threads, namely 6g for ***general purpose*** and 5g6g for ***close tolerance threads.*** There will be only one tolerance class for internal threads — 6H.

Note that a LOWER CASE LETTER indicates the tolerance on a bolt and a CAPITAL LETTER is used for the nut.

Taps

Standard hand taps are made in sets of three, known as taper, plug, and bottoming taps. These are shown in Fig. 6-90.

Threads are started with a ***taper tap.*** It is tapered back from the end 6 to 10 threads before full thread diameter is reached.

The ***plug tap*** is used after the taper tap has cut threads as far into the hole as possible. It tapers back 3 or 4 threads before full thread diameter is reached.

Threads are cut to the bottom of a ***blind hole*** (hole does NOT go through part) with a ***bottom tap.*** It is only necessary to use the full set of taps when a blind hole is to be tapped, Fig. 6-91.

Fig. 6-91. Cutaway of metal block illustrates three types of threaded holes. Left. Open or through hole. Center. Blind hole that has been drilled deeper than desired threads. Right. Blind hole with threads tapped to bottom.

Another tap used in the shop is a pipe tape, Fig. 6-92. A ***pipe tap*** cuts a tapered thread so there is a ''wedging'' action set up to make a leak-tight joint. The fraction that indicates pipe tap size may be confusing at first because it indicates pipe size and NOT the thread diameter.

A ***pipe thread*** is indicated by NPT (National Pipe Thread) and the threads taper 3/4 in. per foot of length.

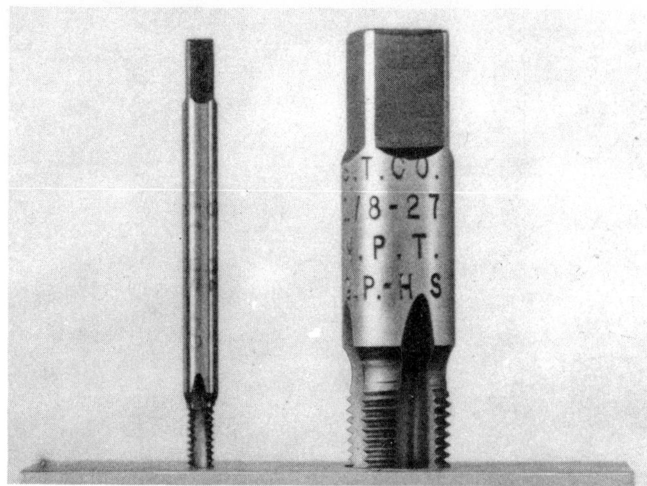

Fig. 6-92. Photo showing difference between one-eighth standard thread tap and one-eighth pipe thread tap.

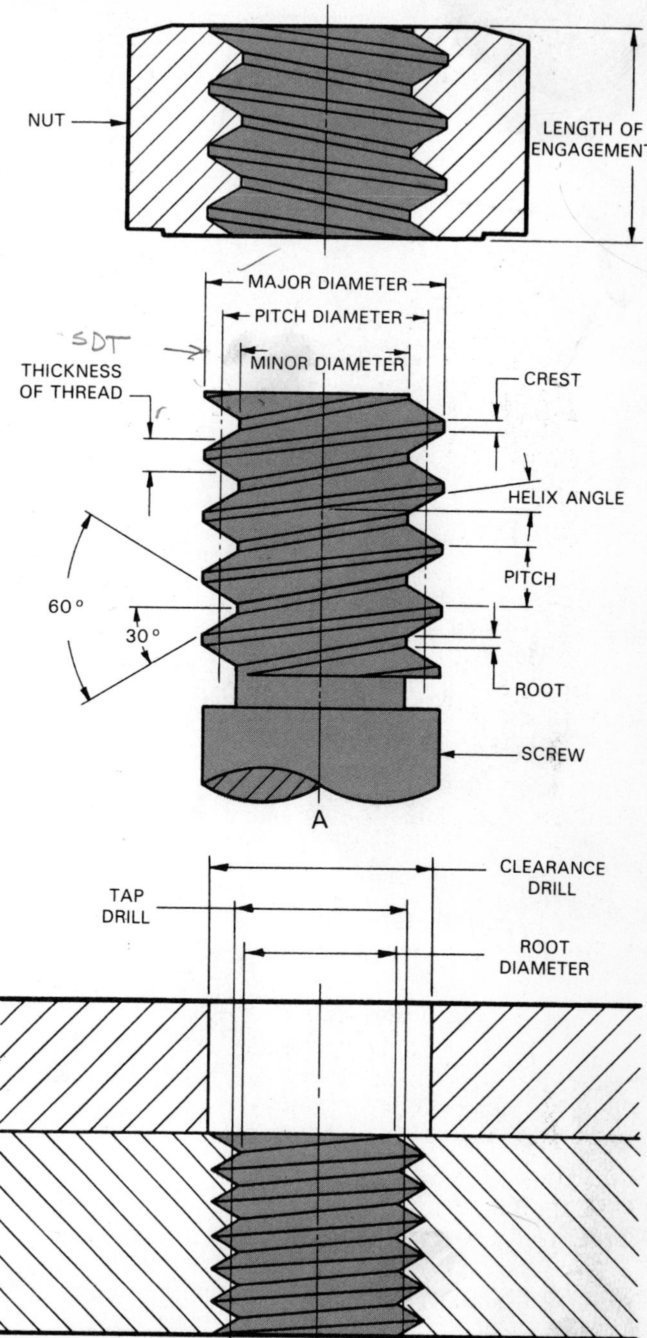

Fig. 6-93. A—Nomenclature of a thread. B—Cross section of a typical part at a point where a bolt will be used to clamp two pieces together. The clearance drill permits bolt or threaded section to enter without binding.

Tap drill

The drill used to make the hole prior to tapping is called a ***tap drill.*** Theoretically, it should be equal in diameter to the minor diameter of the screw that will be fitted into the tapped hole, Fig. 6-93. However, this situation would cause the tap to cut a full thread. The pressure required to rotate the tap would be so great that tap breakage could occur. Full depth threads are not necessary because three-quarter depth threads are strong enough that the fastener usually breaks before the threads strip.

Drill sizes can be secured from a ***tap drill chart,*** Figs. 6-94 and 6-95.

Tap wrenches

Two types of tap wrenches are available. The type to be employed will depend upon tap size. A ***T-handle tap wrench*** should be used with all small taps, Fig. 6-96. It allows a more sensitive "feel" when tapping.

The ***hand tap wrench*** is best suited for large taps where more leverage is required, Fig. 6-97.

When tapping by hand, the chief requirement is to make sure the tap is started straight, and remains square during the entire tapping operation, Fig. 6-98. The tap must be BACKED OFF a half of a turn every one or two cutting turns. This will break the chips free and allow them to drop through the tap flutes. Backing off prevents them from jamming the tap and damaging the threads.

Some machinists, when tapping blind holes, will place a piece of grease pencil, wax crayon, or a dab of grease in the holes. As the tap cuts the threads, the grease is forced up and out of the hole carrying the chips along.

Use a cutting fluid designed for the particular metal you are tapping.

Care in tapping

Considerable care must be exercised when tapping:

1. Use the correct size tap drill. Secure this information from a tap drill chart.
2. Use a sharp tap and apply sufficient quantities of cutting fluid. With some cutting fluids, the area is flooded with fluid; with others, a few

NATIONAL COARSE, AND NATIONAL FINE THREADS AND TAP DRILLS								
Size	Threads Per Inch	Major Dia.	Minor Dia.	Pitch Dia.	Tap Drill 75% Thread	Decimal Equiv-alent	Clearance Drill	Decimal Equiv-alent
2	56	.0860	.0628	.0744	50	.0700	42	.0935
	64	.0860	.0657	.0759	50	.0700	42	.0935
3	48	.099	.0719	.0855	47	.0785	36	.1065
	56	.099	.0758	.0874	45	.0820	36	.1065
4	40	.112	.0795	.0958	43	.0890	31	.1200
	48	.112	.0849	.0985	42	.0935	31	.1200
6	32	.138	.0974	.1177	36	.1065	26	.1470
	40	.138	.1055	.1218	33	.1130	26	.1470
8	32	.164	.1234	.1437	29	.1360	17	.1730
	36	.164	.1279	.1460	29	.1360	17	.1730
10	24	.190	.1359	.1629	25	.1495	8	.1990
	32	.190	.1494	.1697	21	.1590	8	.1990
12	24	.216	.1619	.1889	16	.1770	1	.2280
	28	.216	.1696	.1928	14	.1820	2	.2210
1/4	20	.250	.1850	.2175	7	.2010	G	.2610
	28	.250	.2036	.2268	3	.2130	G	.2610
5/16	18	.3125	.2403	.2764	F	.2570	21/64	.3281
	24	.3125	.2584	.2854	I	.2720	21/64	.3281
3/8	16	.3750	.2938	.3344	5/16	.3125	25/64	.3906
	24	.3750	.3209	.3479	Q	.3320	25/64	.3906
7/16	14	.4375	.3447	.3911	U	.3680	15/32	.4687
	20	.4375	.3725	.4050	25/64	.3906	29/64	.4531
1/2	13	.5000	.4001	.4500	27/64	.4219	17/32	.5312
	20	.5000	.4350	.4675	29/64	.4531	33/64	.5156
9/16	12	.5625	.4542	.5084	31/64	.4844	19/32	.5937
	18	.5625	.4903	.5264	33/64	.5156	37/64	.5781
5/8	11	.6250	.5069	.5660	17/32	.5312	21/32	.6562
	18	.6250	.5528	.5889	37/64	.5781	41/64	.6406
3/4	10	.7500	.6201	.6850	21/32	.6562	25/32	.7812
	16	.7500	.6688	.7094	11/16	.6875	49/64	.7656
7/8	9	.8750	.7307	.8028	49/64	.7656	29/32	.9062
	14	.8750	.7822	.8286	13/16	.8125	57/64	.8906
1	8	1.0000	.8376	.9188	7/8	.8750	1- 1/32	1.0312
	14	1.0000	.9072	.9536	15/16	.9375	1- 1/64	1.0156
1-1/8	7	1.1250	.9394	1.0322	63/64	.9844	1- 5/32	1.1562
	12	1.1250	1.0167	1.0709	1- 3/64	1.0469	1- 5/32	1.1562
1-1/4	7	1.2500	1.0644	1.1572	1- 7/64	1.1094	1- 9/32	1.2812
	12	1.2500	1.1417	1.1959	1-11/64	1.1719	1- 9/32	1.2812
1-1/2	6	1.5000	1.2835	1.3917	1-11/32	1.3437	1-17/32	1.5312
	12	1.5000	1.3917	1.4459	1-27/64	1.4219	1-17/32	1.5312

Fig. 6-94. Study thread and tap drill chart for Unified National threads.

NOMINAL SIZE	INTERNAL THREAD Minor Diameter		TAP DRILL DIAMETER
	Max.	Min.	
M1.6x0.35	1.321	1.221	1.25
M2x0.4	1.679	1.567	1.6
M2.5x0.45	2.138	2.013	2.05
M3x0.5	2.599	2.459	2.5
M3.5x0.6	3.010	2.850	2.9
M4x0.7	3.422	3.242	3.3
M5x0.8	4.334	4.134	4.2
M6.3x1	5.553	5.217	5.3
M8x1.25	6.912	6.647	6.8
M10x1.5	8.676	8.376	8.5
M12x1.75	10.441	10.106	10.2
M14x2	12.210	11.835	12.0

NOMINAL SIZE	INTERNAL THREAD Minor Diameter		TAP DRILL DIAMETER
	Max.	Min.	
M16x2	14.210	13.835	14.0
M20x2.5	17.744	17.294	17.5
M24x3	21.252	20.752	21.0
M30x3.5	26.771	26.211	26.5
M36x4	32.370	31.670	32.0
M42x4.5	37.799	37.129	37.5
M48x5	43.297	42.587	43.0
M56x5.5	50.796	50.046	50.5
M64x6	58.305	57.505	58.0
M72x6	66.305	65.505	66.0
M80x6	74.305	73.505	74.0
M90x6	84.305	83.505	84.0
M100x6	94.305	93.505	94.0

Fig. 6-95. Study thread and tap drill chart for metric threads.

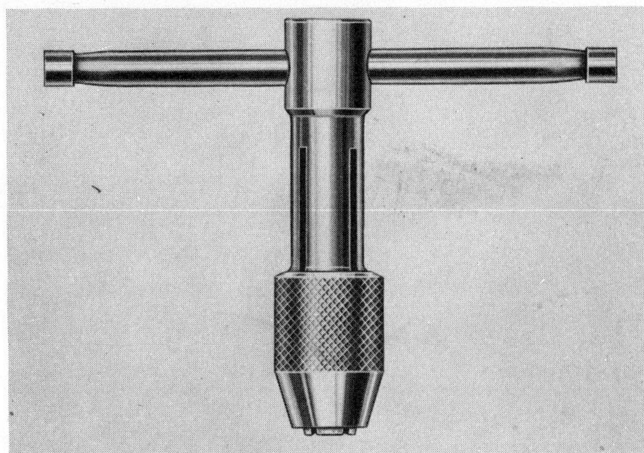

Fig. 6-96. Note T-Handle tap wrench. (Threadwell Manufacturing Co.)

Fig. 6-97. Note hand tap wrench. (Threadwell Manufacturing Co.)

Fig. 6-98. A tap must be started squarely with the hole. A quick way to check this is to use a machinist's square.

drops are sufficient. Read the container label.

3. Start the taper tap square.
4. Do NOT force the tap to cut. Remove the chips using a piece of cloth or cotton waste, NOT your fingers.
5. Avoid running a tap to the bottom of a blind hole and continuing to apply pressure. Do NOT allow the hole to fill with chips and jam the tap. Both conditions can cause the tap to break, especially small taps.
6. Remove burrs on the tapped hole with a smooth file.

Broken taps

Taps sometime break off in a hole. Several tools and techniques have been developed for removing them without damaging the threads already cut. Remember that these methods do NOT always work and the part may have to be discarded. Avoid tap breakage!

Many times a tap will shatter in the hole. It may then be possible to remove the fragments with a pointed tool such as a scribe.

Broken carbon steel taps can sometimes be removed from steel if the work can be heated to annealing temperature, and then drilled out. This CANNOT be done with high speed steel taps. If the high speed steel tap is large enough, it can be ground out with a hand grinder.

A ***tap extractor*** can be employed at times to remove a broken tap, Fig. 6-99. Penetrating oil should be applied and allowed to "soak in" for a short time before the fingers of the tap extractor are fitted into the flutes of the broken tap. The collar on the extractor is slipped down flush with the work surface. A tap wrench is then fitted on the extractor.

The tap extractor is then carefully twisted back and forth to loosen the tap segments. After the broken parts have been loosened, it is a simple matter to remove them.

A ***tap disintegrator*** is used in some shops to remove broken taps. This device makes use of an electric arc to cause the tap to disintegrate. If used properly, it will break up the tap without affecting the metal surrounding the broken tap.

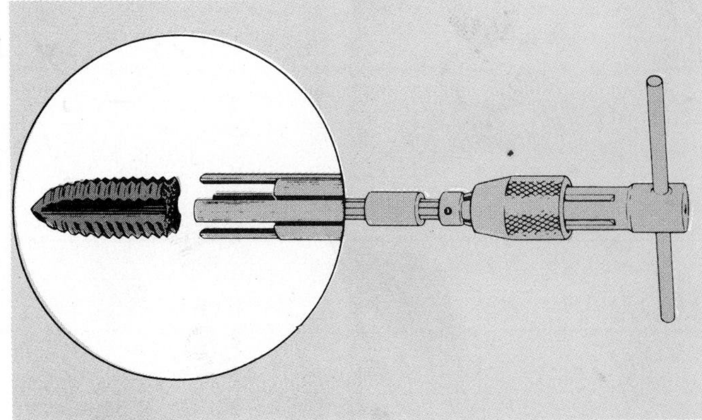

Fig. 6-99. Tap extractor will help remove a broken tap. Close-up shows fingers of extractor and how they fit into flutes of broken tap. It does not always work, however.

Cutting external threads

External threads are cut with a die. ***Solid dies*** are NOT adjustable and for that reason are NOT often used, Fig. 6-100. An ***adjustable die,*** Fig. 6-101, and the ***two-part adjustable die,*** Fig. 6-102, are preferred. The latter has a wide range of adjustment and is fitted with guides to keep it true and square on the work. Dies are available for cutting more standard threads.

Fig. 6-100. These are solid dies used to cut external threads by hand. They cannot be adjusted.

Fig. 6-101. Note split in adjustable die.

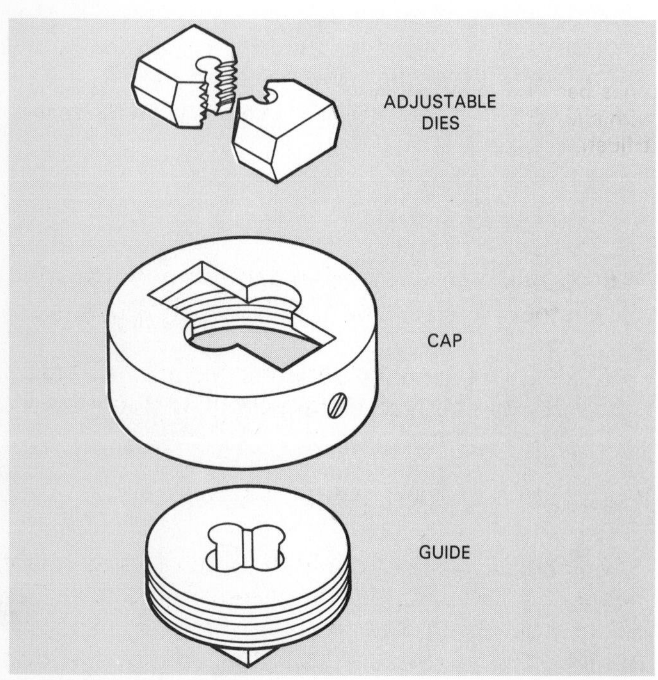

Fig. 6-102. Note construction of two-part adjustable die.

Die stocks

A ***die stock*** holds the die and provides leverage for turning the die on the work. See Fig. 6-103.

When cutting external threads, it is necessary to remember:

1. Material diameter is the SAME SIZE as the desired thread diameter. That is, 1/2-13UNC threads are cut on a 1/2 in. diameter shaft. What diameter shaft would be needed to cut 1/2-20UNF threads?
2. Mount work solidly in a vise.
3. Set the die to the proper size. Make trial cuts on a piece of scrap until the proper adjustment is found.
4. Grind a small chamfer on the shaft end, Fig. 6-104. This permits a die to start easily.
5. Start the cut with the tapered end of the die.
6. Back off the die every one or two turns to break the chips.

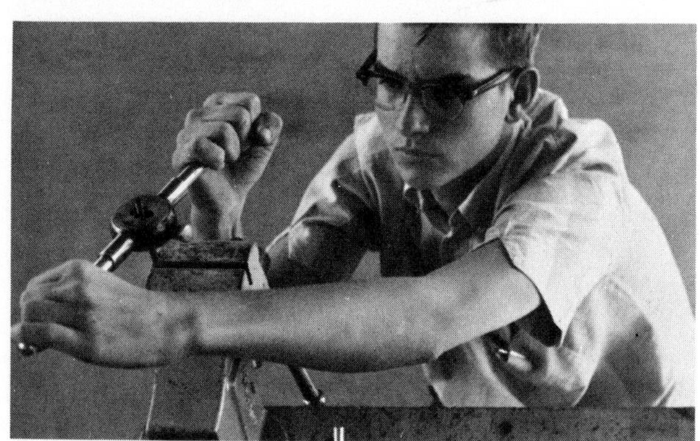

Fig. 6-103 The die is held in a die stock.

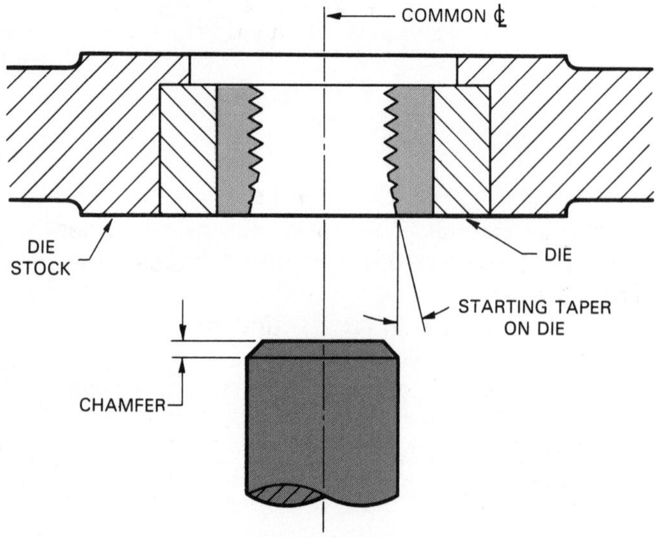

Fig. 6-104. A die will start easier if a small chamfer is cut or ground on the end of the shaft to be threaded. Section through die and die stock shows proper way to start threads.

7. Use cutting oil. Place a paper towel down over the work to absorb excess cutting oil. The towel will also prevent the oil from getting on the floor.
8. Remove any burrs from the finished thread with a fine cut file.

Threading to a shoulder

When a thread must be cut by hand to a shoulder, start and run the threads as far as possible in the usual manner, Fig. 6-105. Remove the die. Turn it over with the guides up. Run the threads down again to the shoulder. Remember to never try this operation without first starting the threads in the usual manner.

care NOT to endanger persons working in the area near you!

2. Chips produced by hand threading are sharp. Use a brush or piece of cloth, NOT your hand, to remove them!
3. Newly cut external threads are very sharp. Again, use a brush or cloth to clean them.
4. Wash your hands after using cutting fluids or oils! Some cause skin rash. This can develop into a serious skin disorder if the oils are left on hands for extended periods.
5. Have cuts treated by a qualified person. Infections can occur when cuts and other injuries are NOT properly treated.

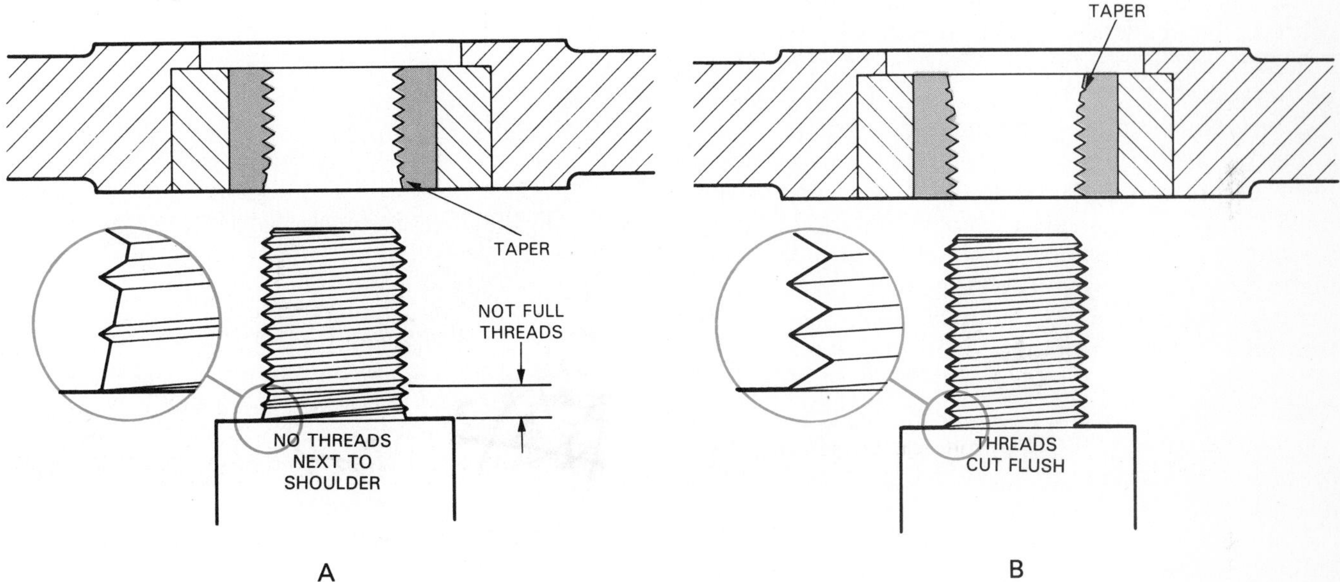

Fig. 6-105. Note how to cut threads to a shoulder. After die has been run down as far as possible, the die is reversed. When rotated down the shaft, it will cut threads almost flush with shoulder. A—Running die down normally. B—Reversing die to cut flush.

Problems in cutting external threads

Ragged threads is the most common problem encountered when cutting external threads with a die. They are caused by:

1. Applying little or no cutting oil.
2. Dull die cutters.
3. Stock too large for the threads being cut.
4. Die NOT started square.
5. One set of cutters could be upside down when using a two part die.

Hand threading safety

1. If a tap or threaded piece must be cleaned of chips with compressed air, protect your eyes from flying chips by wearing goggles. Take

6. Be sure the die is clamped firmly in the die stock. It NOT, it can fall from the holder and cause injury.
7. Broken taps have very sharp edges and are very dangerous. Handle them as you would broken glass!

HAND POLISHIHG WITH ABRASIVES

An ***abrasive*** is commonly thought of as any hard substance that will wear away another material. The substance, grain size, backing material, and the manner the abrasive is bonded to the backing material determines the performance and efficiency of an abrasive.

Abrasive materials

Emery is a natural abrasive. It is black in color and cuts slowly, with a tendency to polish.

Aluminum oxide has replaced emery as an abrasive when large quantities of metal must be removed. It is a manufactured abrasive that works best on high carbon and alloy steels. Aluminum oxide, when designed for use on metal, has a grain shape that is NOT as sharp as that made for woodworking.

Silicon carbide is the hardest and sharpest of the ***synthetic*** (manufactured) abrasives. It is ideal for "sanding" metals like cast iron, bronze, and aluminum.

Silicon carbide is greenish black in color. It is superior to aluminum oxide in its ability to cut fast under light pressure.

Crocus may be synthetic or natural iron oxide. It is bright red in color, very soft, and is used for cleaning and polishing when a minimum of stock is to be removed.

Diamonds are the hardest natural substance known. However, they can also be manufactured. Synthetic diamonds have no value as gems and are used almost exclusively for polishing and grinding by industry. Diamond dust polishing compound is made by crushing synthetic diamonds. It is the only abrasive hard enough to polish the newer heat treated, exotic alloy steels used by industry.

Grain size

The table in Fig. 6-106 shows a comparison of grain size and indicates how the various abrasives are graded.

TECHNICAL GRADES		SIMPLIFIED GRADES	OTHER GRADES
Mesh	Aluminum Oxide Silicon Carbide	Emery	Emery Polishing
600			4/0
			3/0
500			2/0
400	10/0		0
360			
320	9/0		1/2
280	8/0		
240	7/0		1 G
220	6/0		2
180	5/0		3
150	4/0	Fine	
120	3/0		
100	2/0	Medium	
80	0	Coarse	
60	1/2		
50	1	Extra Coarse	
40	1 1/2		
36	2		
30	2 1/2		
24	3		
20	3 1/2		
16	4		
12	4 1/2		

Fig. 6-106. Study abrasive Comparative Grading Chart. (Coated Abrasive Manufacturers Institute)

Coated abrasives

A ***coated abrasive*** is cloth or paper with abrasive grains bonded to one surface. Because of its flexibility, cloth is used as a backing material for abrasives found in metalworking or machining. It is available in 9 in. x 11 in. (210 mm x 280 mm) sheets. It is also available in rolls starting at 1/2 in. (12.5 mm) in width, and is called ***abrasive cloth.***

Using abrasive cloth

1. Abrasive cloth is quite expensive. Use only what you need. Tear the correct amount from the sheet or roll.
2. Do NOT discard abrasive cloth unless it is completely worthless. USED CLOTH is excellent for polishing.
3. If a job has been filed properly, only a fine-grain cloth will be needed to polish the surface. However, if scratches are deep, start the polishing operation by using a coarse-grain cloth first. Change to a medium-grain cloth next and finally a fine-grain abrasive. A few drops of oil will speed the operation. For a high polish, leave the oil on the surface after the scratches have been removed. Reverse the cloth and rub the smooth backing over the work.
4. Abrasive cloth must be supported to work efficiently. To get this support, wrap it around a block of wood or file, Fig. 6-107. Apply pressure and rub the abrasive back and forth in a straight line, parallel to the long side of the work.
5. Avoid using abrasives on MACHINED SURFACES.

Safety when hand polishing with abrasives

1. Avoid rubbing your fingers or hand over polished surfaces or surfaces yet to be polished. Burrs on the edges of the metal can cause painful cuts.
2. Wash your hands thoroughly after polishing operations.
3. Treat all cuts immediately, no matter how small!
4. Place all oily rags in a closed metal container. Never put them in your apron/shop coat or in a locker!
5. Wipe up any oil dropped on the floor during polishing operations.
6. If the lathe must be used for polishing, make sure the machine is protected from the abrasive grains that fall from the cloth during polishing. Stop the machine when inspecting your work.

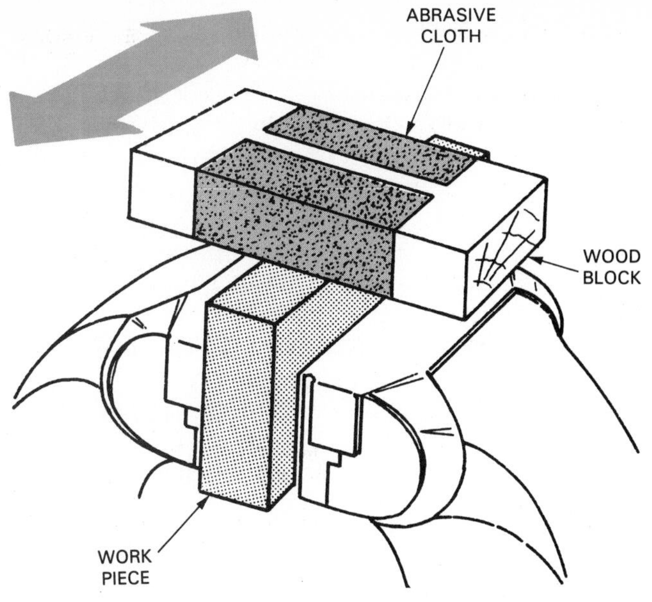

Fig. 6-107. Abrasive cloth should be supported with a block of wood or a file. Avoid supporting it with your fingers.

TEST YOUR KNOWLEDGE—Chapter 6

Please do not write in the text. Place your answers on a separate sheet of paper.

1. List two variations of the machinist's vise.
2. How is vise size determined?
3. Work held in a vise can be protected from damage by the jaw serrations if _______ are placed over the jaws.
4. To prevent injuries, what should be avoided when mounting work in a vise?
5. Work is often held together with a _______ _______ and/or _______ while being machined or worked on.
6. How do combination pliers have an advantage over many other types of pliers?
7. Why are the cutting edges on diagonal pliers set at an angle?
8. List three ways of extending the working life of pliers.
9. What are adjustable clamping pliers?
10. Of what use are torque limiting wrenches?
11. Do torque limiting wrenches give a more accurate reading when they are pushed or when they are pulled?
12. Several different wrenches can be classified as adjustable wrenches. Name three.
13. List three points that should be observed when using an adjustable wrench.
14. Round work can be gripped with a _______ wrench. Its main disadvantage is that the jaws will probably _______ the work.
15. Describe socket wrenches.
16. What wrenches are employed to turn flush and recessed types of threaded fasteners? The fasteners have slots or holes to receive the wrench lugs.
17. Rather than lengthen the wrench handle for additional leverage, it is better to use a _______ wrench.
18. List five safety precautions that should be observed when using a wrench.
19. What is the difference between a standard screwdriver tip and a Phillips screwdriver tip?

Match each phrase in the left column with the correct term in the right column.

20. Has a flattened wedge shaped tip.
21. Is short and used when space is limited.
22. Has a square shank to permit additional force to be applied with a wrench.
23. Useful when handling small screws.
24. Tip is similar to tip of a Phillips head screwdriver.
25. Has an insulated handle.

a. Stubby.
b. Offset.
c. Electricians.
d. Ratchet.
e. Standard.
f. Heavy-duty.
g. POZIDRIV® .
h. Automatic.
i. Hex type.

26. List three safety precautions that should be observed when using a screwdriver.
27. How is the size of a ball peen hammer determined?
28. Why are soft-face hammers and mallets used in place of a ball-peen hammer?
29. List three safety precautions that should be observed when using striking tools.
30. There are few things more dangerous than a chisel with a head that has become _______ from use. This danger can be removed by _______.
31. The chisel is an ideal tool for _______ _______.
32. List the four general types of cold chisels.
33. The standard hacksaw is designed to accommodate _______ _______ _______.
34. A hacksaw cuts best at about _______ to _______ strokes per minute.
35. Why should work be mounted solidly and close to the vise before hacksawing?
36. Should a blade break or dull before completing a cut, do NOT continue in the same cut with a new blade. Why?
37. The number of teeth per inch on a hacksaw blade has an important bearing on the shape and kind of metal being cut. At least _______ to _______ should be cutting at all times, otherwise _______.
38. What is the best way to hold thin metal for hacksawing?
39. What is the best way to hold thin wall tubing for hacksawing?

40. Files are cleaned with a _______ _______. Never with _______ _______.
41. Files are classified according to the cut of their teeth. List the four cuts.
42. What are the most commonly used file shapes?
43. List three safety precautions that should be observed when files are used.
44. When is reaming done?
45. How much stock should be left in a hole for hand reaming?
46. A _______ is used to cut internal threads. External threads are cut with a _______.
47. The hole to be tapped must be:
    a. The same diameter as the desired thread.
    b. A few thousandths larger than the desired thread.
    c. A few thousandths (0.003–0.004 in.) smaller than the threads.
    d. All of the above.
    e. None of the above.
48. The drill, used to make the hole prior to threading, is called a _______ _______.
49. How does the UNC thread series differ from the UNF thread series?
50. List the correct sequence taps should be used to tap a blind hole.
51. How much larger or smaller must a shaft be to cut external threads?
52. Taps are turned in with a _______ _______. A _______ _______ is used with dies.
53. What is an abrasive?
54. _______ surfaces are never polished with an abrasive.

RESEARCH AND DEVELOPMENT

1. Industry makes considerable use of the pneumatic chisel. Secure information on this tool for a bulletin board display and, if possible, borrow an example of the actual tool for examination from a local industry.
2. Design a safety poster than shows the correct way to use a chisel.
3. Secure samples of the various types of hacksaw blades used for hand sawing. Prepare them as a bulletin board display.
4. Design and produce a series of safety posters on the file that illustrate the following unsafe practices:
    a. Using a file as a pry.
    b. File used without a handle.
    c. File used as a hammer.
5. Design a panel that shows the file in various stages of manufacturer. Secure actual examples.
6. Inspect the files in your shop. Clean them; repair or replace damaged file handles. Make a new file rack if the present rack is badly worn or not suitable.
7. Examine the screwdrivers in your shop. Repair and/or regrind the tools when examination indicates they are in need of repair.
8. There are many other types of wrenches not covered in this unit. Prepare a paper featuring these wrenches. Include drawings. Reproduce the report for distribution to the class.
9. Give a demonstration on the proper way to use a torque limiting wrench.
10. Contact various tool manufacturers for information on how wrenches are manufactured. Prepare a bulletin board display with the material.
11. Repair and lubricate all adjustable wrenches in the shop.
12. Make a safety poster illustrating the proper way to use a wrench.
13. Prepare a sample block of metal that can be used to show the difference between a drilled hole and a reamed hole.
14. Develop and construct displays that show:
    a. Samples of various abrasive materials.
    b. A flow chart showing how synthetic abrasives are manufactured. If possible, secure samples of the raw materials.
    c. Metal samples in various stages of polishing. Spray them with lacquer or acrylic plastic to prevent rust.
15. Set up an experiment to determine what abrasive materials are best for aluminum, brass, cast iron, and tool steel. The investigation should include the quantity of material removed within a specified period of time; surface finish of the completed piece; degree of clogging, if any, of the abrasive cloth; and the effect lubricating oil has on the surface finish. Abrasives of similar grade value must be used if tests are to be valid.
16. Give a demonstration on various ways broken taps can be removed. Industry often uses a technique that erodes the tap electrically, permitting the parts to be removed easily. Secure information on this process for a bulletin board display.
17. Prepare a study on the accuracy of hand reamers. Make sample holes and measure them to determine whether they are within acceptable limits. Does the application of cutting fluid affect the size of a reamed hole?
18. Demonstrate the proper way to tap a blind hole.
19. Demonstrate the correct way to run a thread down to a shoulder.

Chapter 7

DRILLS AND DRILLING MACHINES

After studying this chapter, you will be able to:

- Select and safely use the correct drills and drilling machine for a given job.
- Make safe setups on a drill press.
- List the various drill series.
- Sharpen a twist drill.
- Explain the safety rules that pertain to drilling operations.

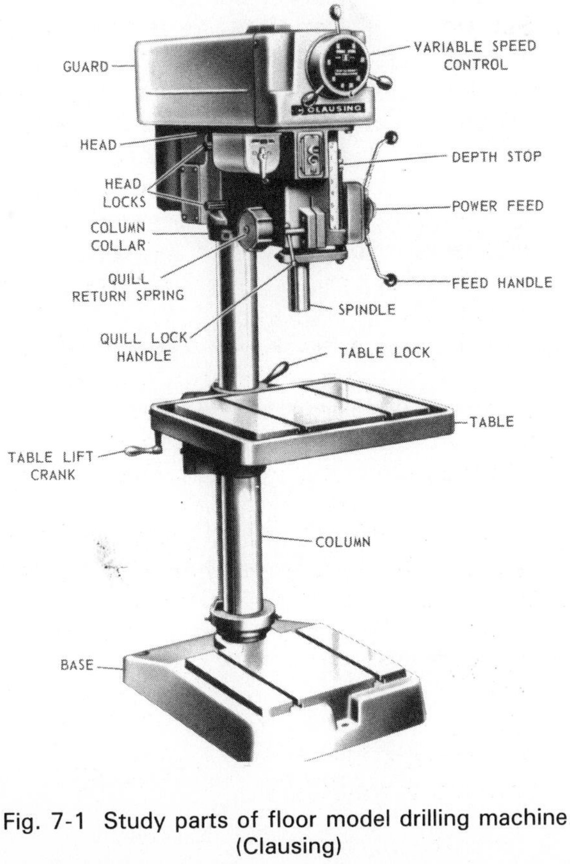

GUARD
VARIABLE SPEED CONTROL
HEAD
DEPTH STOP
HEAD LOCKS
POWER FEED
COLUMN COLLAR
QUILL RETURN SPRING
FEED HANDLE
SPINDLE
QUILL LOCK HANDLE
TABLE LOCK
TABLE
TABLE LIFT CRANK
COLUMN
BASE

Fig. 7-1 Study parts of floor model drilling machine. (Clausing)

The drill press, such as shown in Fig. 7-1, is probably the best known of the machine tools. A ***machine tool*** is a power-driven machine that holds the material and cutting tool and brings them together so the material is drilled, cut, shaved or ground.

A ***drill press*** is primarily used for cutting ROUND HOLES. It can also be used for many different machining operations, such as:

1. Reaming, Fig. 7-2.
2. Counterboring, Fig. 7-3.
3. Spotfacing, Fig. 7-4.

A drill press operates by rotating a cutting tool, or ***drill,*** against the material with sufficient pressure to cause the drill to penetrate the material. See Figs. 7-5 and 7-6.

SAFETY NOTE! NEVER attempt to operate a drilling machine while your senses are impaired by medication or other substances.

Fig. 7-2. Reaming is being done on this drill press. (Clausing)

Fig. 7-3. Counterboring is done to prepare a hole to receive a fillister or socket head screw. (Clausing)

Fig. 7-4. Spotfacing is machining a surface to permit a nut or bolt head to bear uniformly over its entire head. (Clausing)

Drill press size is determined by the LARGEST DIAMETER circular piece that can be drilled on center, Fig. 7-7. A 17 in. drill press can drill to the center of a 17 in. diameter piece. The center line of the drill is 8 1/2 in. from the column.

DRILLS

Common drills are known as ***twist drills*** because most are made by forging or milling rough flutes, and then twisting them to a spiral shape. After twisting, the drills are milled and ground to approximate size, Fig. 7-8. Then, they are heat treated and ground to exact size.

Most drills are made of ***high speed steel*** (HS or HSS) or ***carbon steel.*** High speed steel drills can be operated at much higher cutting speeds than carbon steel drills without danger of burning and drill damage.

Fig. 7-5. Drilling is the operation most often performed on a drill press. (Clausing)

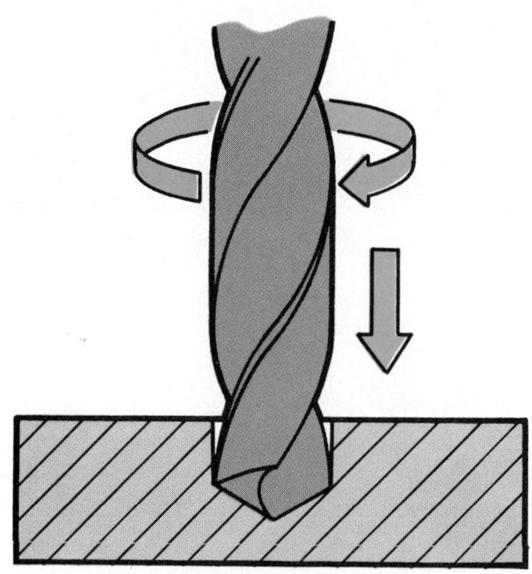

Fig. 7-6. Note how a drill works.

Types of drills

Industry utilizes special drills to improve the accuracy of a drilled hole, to speed production, and to improve drilling efficiency.

The ***straight flute gundrill*** is designed for ferrous and nonferrous metals, Fig. 7-9. It is usually fitted with a carbide cutting tip.

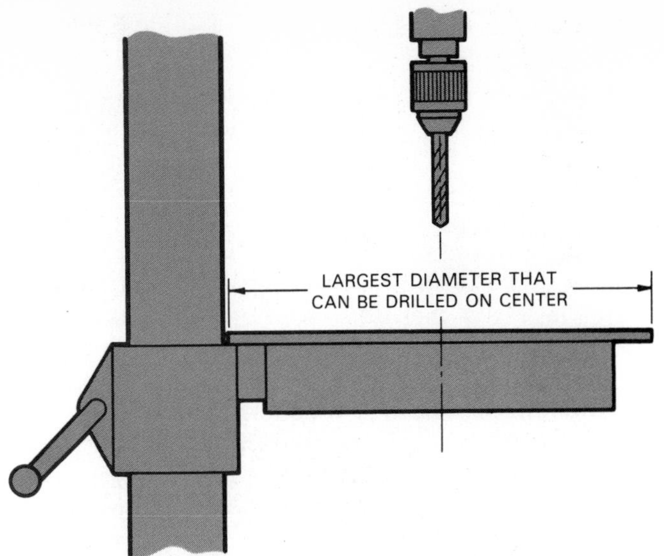

Fig. 7-7. Study how a drill press is measured.

Fig. 7-8. This is a closeup of flute milling operation in manufacture of a large drill. (Chicago-Latrobe)

An ***oil hole drill*** has coolant holes through the body to permit fluid or air to be forced to the point to remove heat, Fig. 7-10. The pressure of the fluid or air also ejects the chips from the hole while drilling.

Three- and four-flute core drills are used to enlarge core holes in casting. See Fig. 7-11.

Special ***step drills*** permit the elimination of one or more drilling operations in production work. Fig. 7-12 illustrates this type drill.

Combination drill and reamer drills also speeds up production by eliminating one operation, Fig. 7-13.

Spade drills have replaceable cutting tips, usually of tungsten carbide, Fig. 7-14. They are available in sizes from 1 in. (25.0 mm) to 5 in. (125.0 mm). They are less expensive than twist drills of the same size.

Most drills are available with straight or taper shanks and with tungsten carbide tips. Coating drills with Titanium Nitride greatly increases tool life.

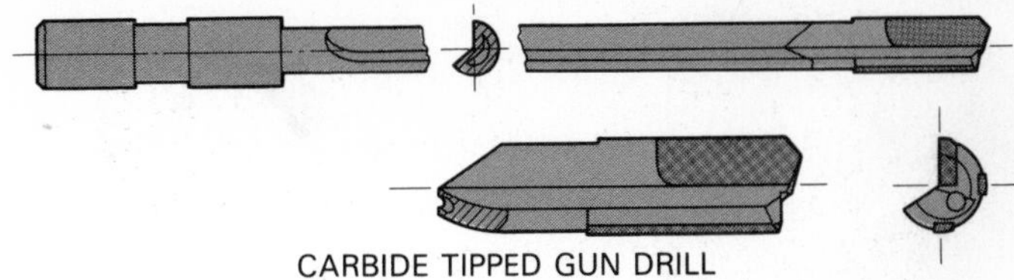

Fig. 7-9. This is the straight flute drill. Tip is shown in larger scale. Shaded portion is tungsten carbide. Large area does cutting; smaller sections act as wear surfaces. This type is also known as a gundrill.

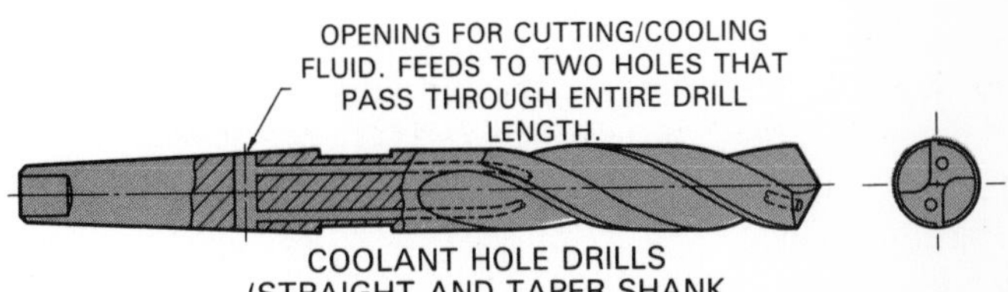

Fig. 7-10. Note oil hole drill.

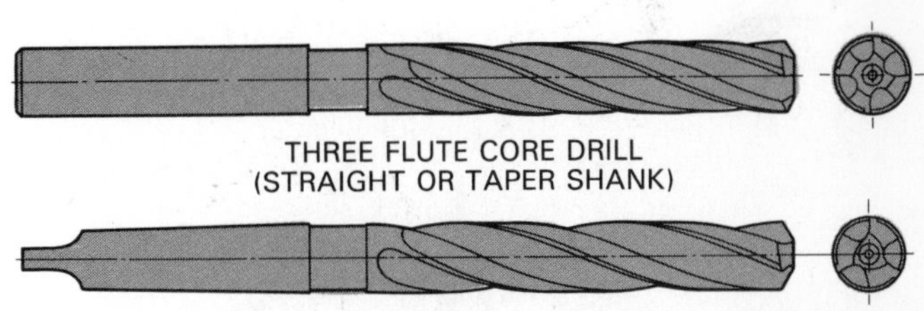

Fig. 7-11. Note three- and four-flute core drills.

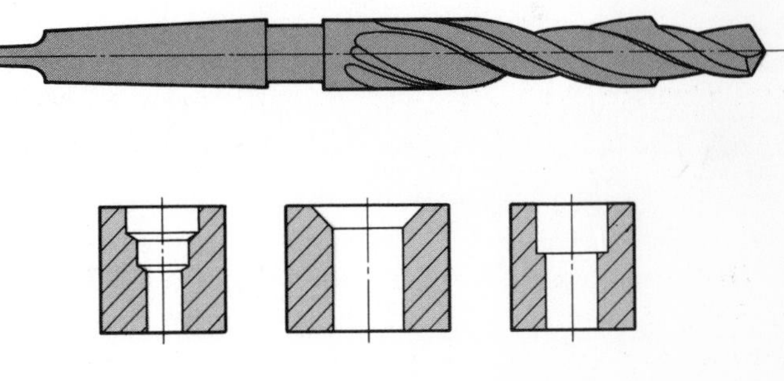

Fig. 7-12. This is a step drill.

COMBINATION DRILL AND REAMER
(STRAIGHT AND TAPER SHANK)

Fig. 7-13. This is a combination drill and reamer.

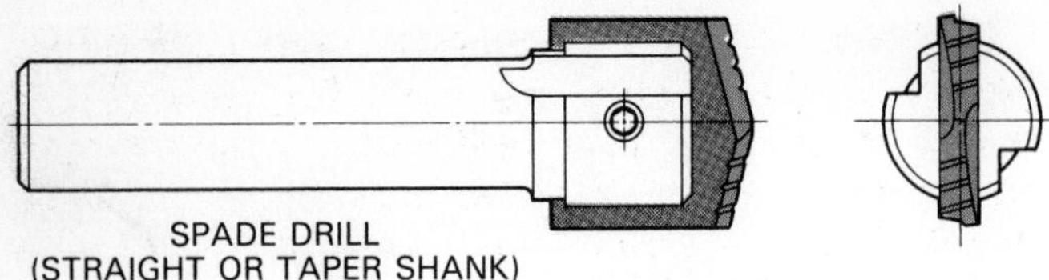

Fig. 7-14. This spade drill has cutting tip that is interchangeable with other size tips.

DRILL SIZE

Drill sizes are expressed by the following series:

1. ***Numbers*** — #80 to #1 (0.0135 in. to 0.288 in. diameters).
2. ***Letters*** — A to Z (0.234 in. to 0.413 in. diameters).
3. ***Inches and fractions*** — 1/64 in. to 3 1/2 in. diameters.
5. ***Metric*** — 3.0 mm to 76.0 mm (0.118 in. to 2.992 in.) diameters.

The drill size chart will give an idea of this vast array of drill sizes, Fig. 7-15.

Drill measurements

Most drills, with the exception of small number drills, have the diameter stamped on the shank. These figures frequently become obliterated (obscured) and it is impossible to determine the drill's diameter without measuring.

When a micrometer is used for measuring, the measurement is made across the DRILL MARGINS. However, if the drill is WORN, the measurement is made on the SHANK at the end of the flutes. See Fig. 7-16 for both techniques.

Inch	Mm.	Wire Gage	Decimals of an Inch
		80	.0135
		79	.0145
1/64			.0156
	.4		.0157
		78	.0160
		77	.0180
	.5		.0197
		76	.0200
		75	.0210
	.55		.0217
		74	.0225
	.6		.0236
		73	.0240
		72	.0250
	.65		.0256
		71	.0260
	.7		.0276
		70	.0280
		69	.0293
	.75		.0295
		68	.0310
1/32			.0313
	.8		.0315
		67	.0320
		66	.0330
	.85		.0335
		65	.0350
	.9		.0354
		64	.0360
		63	.0370
	.95		.0374
		62	.0380
		61	.0390
	1		.0394
		60	.0400
		59	.0410
	1.05		.0413
		58	.0420
		57	.0430
	1.1		.0433
	1.15		.0453
		56	.0465

Inch	Mm.	Wire Gage	Decimals of an Inch
3/64			.0469
	1.2		.0472
	1.25		.0492
	1.3		.0512
		55	.0520
	1.35		.0531
		54	.0550
	1.4		.0551
	1.45		.0571
	1.5		.0591
		53	.0595
	1.55		.0610
1/16			.0625
	1.6		.0630
		52	.0635
	1.65		.0650
	1.7		.0669
		51	.0670
	1.75		.0689
		50	.0700
	1.8		.0709
	1.85		.0728
		49	.0730
	1.9		.0748
		48	.0760
	1.95		.0768
5/64			.0781
		47	.0785
	2		.0787
	2.05		.0807
		46	.0810
		45	.0820
	2.1		.0827
	2.15		.0846
		45	.0860
	2.2		.0866
	2.25		.0886
		43	.0890
	2.3		.0906
	2.35		.0925
		42	.0935

Inch	Mm.	Wire Gage	Decimals of an Inch
3/32			.0938
	2.4		.0945
		41	.0960
	2.45		.0966
		40	.0980
	2.5		.0984
		39	.0995
		38	.1015
	2.6		.1024
		37	.1040
	2.7		.1063
		36	.1065
	2.75		.1083
7/64			.1094
		35	.1100
	2.8		.1102
		34	.1110
		33	.1130
	2.9		.1142
		32	.1160
	3		.1181
		31	.1200
	3.1		.1220
1/8			.1250
	3.2		.1260
	3.25		.1280
		30	.1285
	3.3		.1299
	3.4		.1339
		29	.1360
	3.5		.1378
		28	.1405
9/64			.1406
	3.6		.1417
		27	.1440
	3.7		.1457
		26	.1470
	3.75		.1476
		25	.1495
	3.8		.1496
		24	.1520
	3.9		.1535
		23	.1540

Inch	Mm.	Wire Gage	Decimals of an Inch
5/32			.1563
		22	.1570
	4		.1575
		21	.1590
		20	.1610
	4.1		.1614
	4.2		.1654
		19	.1660
	4.25		.1673
	4.3		.1693
		18	.1695
11/64			.1719
		17	.1730
	4.4		.1732
		16	.1770
	4.5		.1772
		15	.1800
	4.6		.1811
		14	.1820
		13	.1850
	4.7		.1850
	4.75		.1870
3/16			.1875
	4.8		.1890
		12	.1890
		11	.1910
	4.9		.1929
		10	.1935
		9	.1960
	5		.1969
		8	.1990
	5.1		.2008
		7	.2010
13/64			.2031
		6	.2040
	5.2		.2047
		5	.2055
	5.25		.2067
	5.3		.2087
		4	.2090
	5.4		.2126
		3	.2130
	5.5		.2165
7/32			.2188
	5.6		.2205
		2	.2210
	5.7		.2244
	5.75		.2264
		1	.2280
	5.8		.2283

Fig. 7-15. Decimal equivalents of drill sizes.

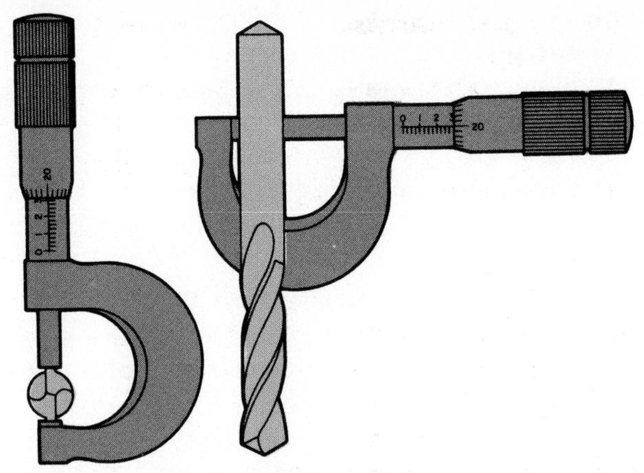

Fig. 7-16. Note how to measure drill diameter with a micrometer.

Diameters can also be checked with a ***drill gage,*** Fig. 7-17. Drill gages are made for various drill series; however, 1/2 in. drills are the largest that can be checked by this method in the fractional series. New drills are checked at the points; worked drills at the end of the flutes.

Remember! ALWAYS check drill diameter before using it. Using the wrong size drill can be a very expensive and time-consuming mistake!

PARTS OF A DRILL

The twist drill has been scientifically designed to be an efficient cutting tool. It is composed of three principle parts: point, shank, and body, Fig. 7-18.

Inch	Mm.	Letter Sizes	Decimals of an Inch
	5.9		.2323
		A	.2340
15/64			.2344
	6		.2362
		B	.2380
	6.1		.2402
		C	.2420
	6.2		.2441
		D	.2460
	6.25		.2461
	6.3		.2480
1/4		E	.2500
	6.4		.2520
	6.5		.2559
		F	.2570
	6.6		.2598
		G	.2610
	6.7		.2638
17/64			.2656
	6.75		.2657
		H	.2660
	6.8		.2677
	6.9		.2717
		I	.2720
	7		.2756
		J	.2770
	7.1		.2795
		K	.2810
9/32			.2812
	7.2		.2835
	7.25		.2854
	7.3		.2874
		L	.2900
	7.4		.2913
		M	.2950
	7.5		.2953
19/64			.2969
	7.6		.2992
		N	.3020
	7.7		.3031
	7.75		.3051
	7.8		.3071
	7.9		.3110
5/16			.3125
	8		.3150
		O	.3160
	8.1		.3189
	8.2		.3228
		P	.3230
	8.25		.3248
	8.3		.3268

Inch	Mm.	Letter Sizes	Decimals of an Inch
21/64			.3281
	8.4		.3307
		Q	.3320
	8.5		.3346
	8.6		.3386
		R	.3390
	8.7		.3425
11/32			.3438
	8.75		.3345
	8.8		.3465
		S	.3480
	8.9		.3504
	9		.3543
		T	.3580
	9.1		.3583
23/64			.3594
	9.2		.3622
	9.25		.3642
	9.3		.3661
		U	.3680
	9.4		.3701
	9.5		.3740
3/8			.3750
		V	.3770
	9.6		.3780
	9.7		.3819
	9.75		.3839
	9.8		.3858
		W	.3860
	9.9		.3898
25/64			.3906
	10		.3937
		X	39.70
		Y	.4040
13/32			.4063
		Z	.4130
	10.5		.4134
27/64			.4219
	11		.4331
7/16			.4375
	11.5		.4528
29/64			.4531
15/32			.4688
	12		.4724
31/64			.4844
	12.5		.4921
1/2			.5000
	13		.5118
33/64			.5156
17/32			.5313
	13.5		.5315

Inch	Mm.	Decimals of an Inch
35/64		.5469
	14	.5512
9/16		.5625
	14.5	.5709
37/64		.5781
	15	.5906
19/32		.5938
39/64		.6094
	15.5	.6102
5/8		.6250
	16	.6299
41/64		.6406
	16.5	.6496
21/32		.6563
	17	.6693
43/64		.6719
11/16		.6875
	17.5	.6890
45/64		.7031
	18	.7087
23/32		.7188
	18.5	.7283
47/64		.7344
	19	.7480
3/4		.7500
49/64		.7656
	19.5	.7677
25/32		.7812
	20	.7874
51/64		.7969
	20.5	.8071
13/16		.8125
	21	.8268
53/64		.8281
27/32		.8438
	21.5	.8465
55/64		.8594
	22	.8661
7/8		.8750
	22.5	.8858
57/64		.8906
	23	.9055
29/32		.9063
59/64		.9219
	23.5	.9252
15/16		.9375
	24	.9449
61/64		.9531
	24.5	.9646
31/32		.9688
	25	.9843
63/64		.9844

Inch	Mm.	Decimals of an Inch
1		1.0000
	25.5	1.0039
1 1/64		1.0156
	26	1.0236
1 1/32		1.0313
	26.5	1.0433
1 3/64		1.0469
1 1/16		1.0625
	27	1.0630
1 5/64		1.0781
	27.5	1.0827
1 3/32		1.0938
	28	1.1024
1 7/64		1.1094
	28.5	1.1220
1 1/8		1.1250
1 9/64		1.1406
	29	1.1417
1 5/32		1.1562
	29.5	1.1614
1 11/64		1.1719
	30	1.1811
1 3/16		1.1875
	30.5	1.2008
1 13/64		1.2031
1 7/32		1.2188
	31	1.2205
1 15/64		1.2344
	31.5	1.2402
1 1/4		1.2500
	32	1.2598
1 17/64		1.2656
	32.5	1.2795
1 9/32		1.2813
1 19/64		1.2969
	33	1.2992
1 5/16		1.3125
	33.5	1.3189
1 21/64		1.3281
	34	1.3386
1 11/32		1.3438
	34.5	1.3583
1 23/64		1.3594
1 3/8		1.3750
	35	1.3780
1 25/64		1.3906
	35.5	1.3976
1 13/32		1.4063
	36	1.4173
1 27/64		1.4219
	36.5	1.4370

Fig. 7-15. (Continued). Decimal equivalents of drill sizes.

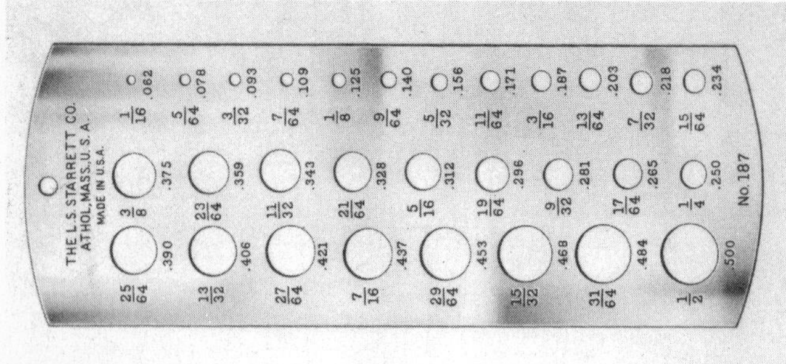

Fig. 7-17. Drill gage is used to measure fractional size drills. Similar gages are available for measuring letter, number, and millimeter size drills.

Shank

The ***shank*** is the portion of the drill that mounts into the chuck or spindle. Twist drills are made with shanks that are either straight or tapered, Fig. 7-19. ***Straight shank drills*** are used with a chuck. ***Tapered shank drills*** have self-holding tapers (No. 1 to No. 5 Morse taper) that fit directly into the drill press spindle.

A ***tang*** is on the taper shank; it fits into a slot in the spindle, sleeve, or socket, and assists in driving the tool. The tang also offers a means of separating the taper from the holding device (spindle, sleeve, or socket).

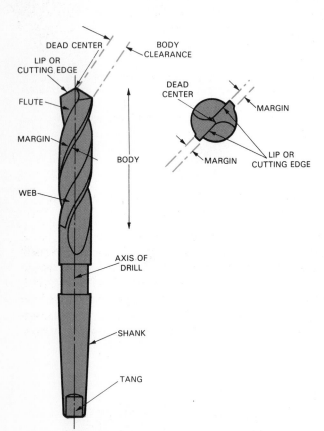

Fig. 7-18. Study parts of a twist drill.

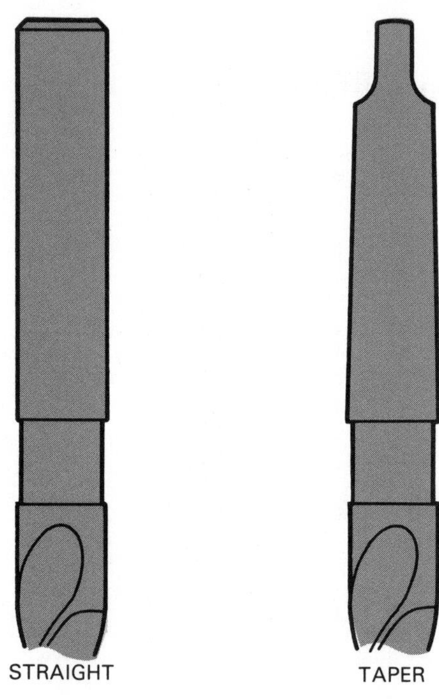

Fig. 7-19. Note types of drill shanks.

Point

The ***point*** is the cone-shaped end that does the cutting. The point consists of the following:

1. ***Dead center*** refers to the sharp edge at the extreme tip of the drill. This should always be in the EXACT CENTER of the drill AXIS.
2. The ***lips*** are the cutting edges of the drill.
3. The ***heel*** is the portion of the point back from the lips or cutting edges.
4. ***Lip clearance*** is the amount the surface of the point is relieved back from the lips.

Body

The ***body*** is the portion of the drill between the point and the shank. It consists of:

1. FLUTES are two or more spiral grooves that run the length of the drill body. The flutes do four things:
   a. Help form the cutting edges of the drill point.
   b. Curl the chip tightly for easier removal.
   c. Form channels through which the chips can escape as the hole is drilled.
   d. Allow coolant and lubricant to get down to the cutting edges.
2. The ***margin*** is the narrow strip extending back the entire length of the drill body.
3. ***Body clearance*** refers to the part of the drill body

that has been reduced in order to lower friction between the drill and the wall of the hole.

4. The ***web*** is the metal column that separates the flutes. It gradually increases in thickness toward the shank for added strength.

HOLDING DRILLS IN DRILL PRESS

A drill is held in the drill press by either of the following methods:

1. ***Chuck:*** Movable jaw mechanism for drills with straight shanks. See Fig. 7-20.
2. Tapered spindle: Tapered opening for drills with taper shanks, Fig. 7-21. Drill chucks with taper shanks make it possible to use straight shank drills when the drill press is fitted with a taper spindle opening.

Fig. 7-20. Drill chuck can be tightened with key to hold drill.

When a chuck is employed, the drill is inserted and the chuck jaws are TIGHTENED BY HAND. After a quick flip of the switch to determine that the chuck is centered and running true, tighten the chuck with a ***chuck key.***

DANGER! Always remove the key from the chuck before turning on the drill press. It could hit something or fly out with considerable force.

Taper shank drills must be wiped clean before inserting the shank into the spindle. Nicks in the shank must be removed with an oilstone; otherwise, the shank will NOT seat properly.

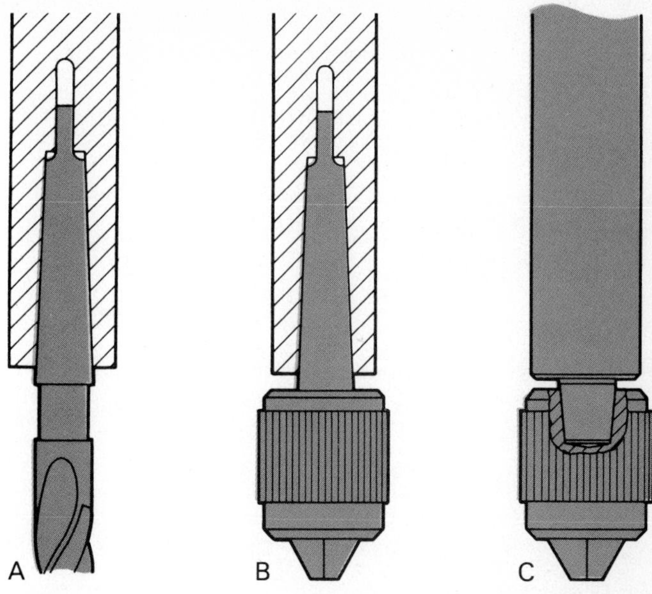

Fig. 7-21. A and B—These are taper shank tools used in a drill press spindle with a tapered opening. C—This is a solid spindle with a short external taper that only fits a drill chuck. The chuck is attached permanently to spindle.

SAFETY NOTE! Never attempt to use a taper shank drill mounted in a drill chuck.

Most drill press spindles are made with a No. 2 or No. 3 ***Morse Taper*** (often shown as MT). A drill with a shank smaller than the spindle taper must be enlarged to fit with a ***sleeve,*** Fig. 7-22. Drills with shanks larger than the spindle opening can often be used with a ***socket,*** Fig. 7-23. The taper opening in the socket is larger than the taper on its shank.

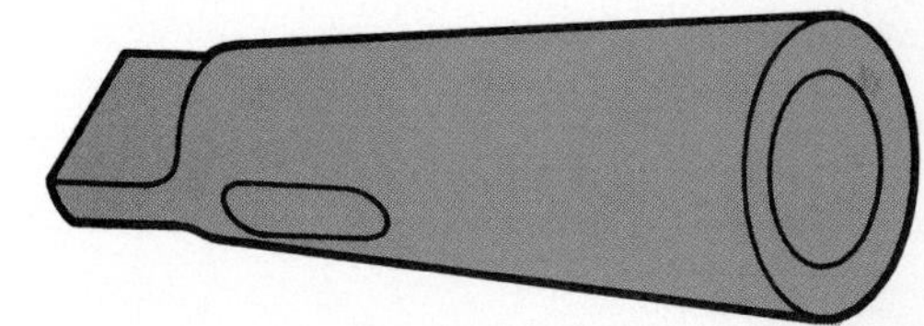

Fig. 7-22. This is a drill sleeve.

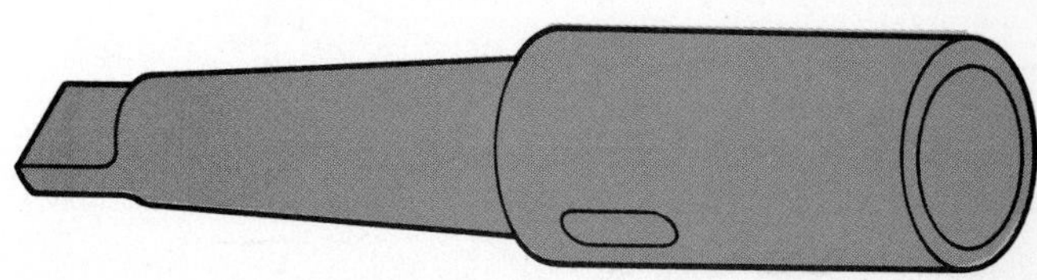

Fig. 7-23. This is a drill socket.

WARNING! A socket should only be used in an emergency when a larger drill press is NOT available. It is dangerous to overstress a drill press by using a drill larger than the machine's rated capacity.

Sleeves, sockets, and taper shank drills are separated with a ***drift,*** Fig. 7-24. To use a drift, insert it in the slot with the round edge up, Fig. 7-25. A sharp rap with a lead hammer will cause them to separate.

NOTE! A file tang should never be used in place of a drift. It will damage the drill shank and machine spindle. Then other drills will NOT fit properly. The file could also shatter!

Hold the drill when removing it from the spindle to prevent it from falling to the floor. Wrap a piece of clean cloth around the drill to protect your hand from metal chips and painful cuts. Dropping the drill may damage the drill point. If the drill is large enough, it could also injure your foot.

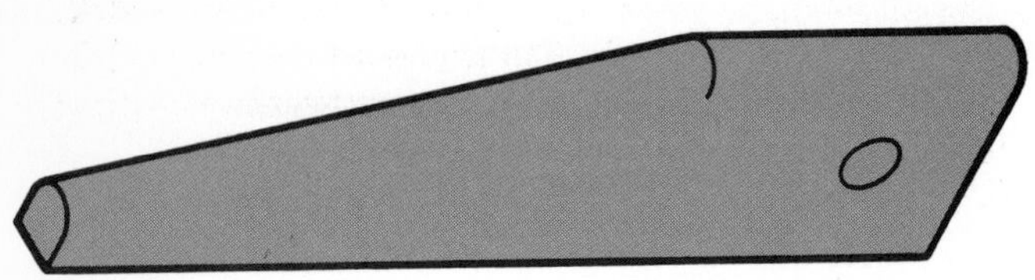

Fig. 7-24. This is a drift.

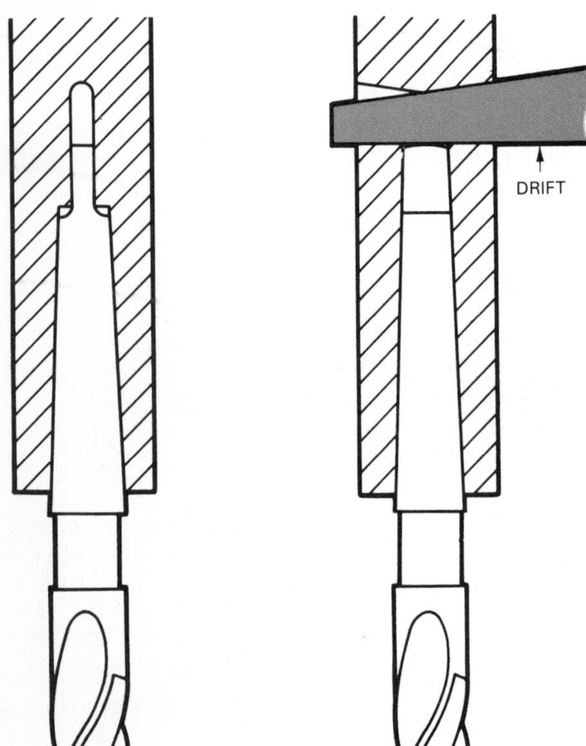

Fig. 7-25. A—Drill is locked in spindle. B—Note how to remove taper shank tool from drill spindle with a drift. Never use the tang of a file to do this job!

WORK HOLDING DEVICES

Work must be mounted solidly on the drilling machine. When mounted improperly, it may spring and/or move causing drill damage or breakage.

DANGER! Serious injury can result from work that becomes loose and spins about on a drill press. This dangerous situation is nicknamed a "merry-go-round."

The machinist has the following devices at his disposal for holding work:

Vises

Vises are widely employed to hold work, Fig. 7-26. For best results, the vise must be bolted to the drill table.

Parallels are often used to level and raise the work above the vise base, Fig. 7-27. This will permit the drill to come through the work and not damage the vise.

Seat the work on the parallels by tightening the vise and tapping the work with a mallet until the parallels do NOT move. Loose parallels indicate that the work is NOT seated properly.

An ***angular vise*** permits angular drilling without tilting the drill press table. See Figs. 7-28 and 7-29.

Fig. 7-26. Note typical vises. Above. Swivel vise base is designed to permit vise to swivel through 180 deg. Below. Quick acting vise will slide and lock on parts more quickly. (L-W Chuck Co.)

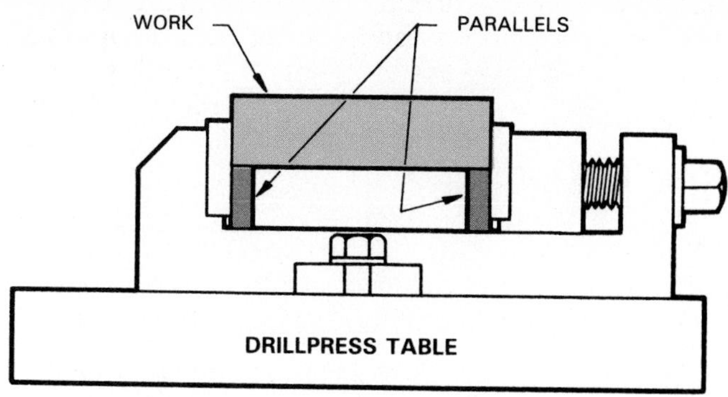

Fig. 7-27. Parallels are often used to raise work above vise base. This will prevent drill from cutting into vise as it goes through work.

Fig. 7-30. A Cross-slide permits rapid alignment of work for drilling.

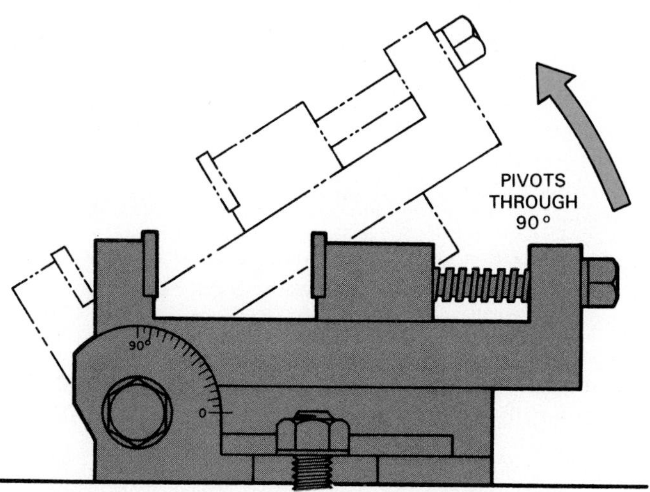

Fig. 7-28. Angular vise can be adjusted through 90 deg. to permit drilling on an angle without tilting entire vise or drill table.

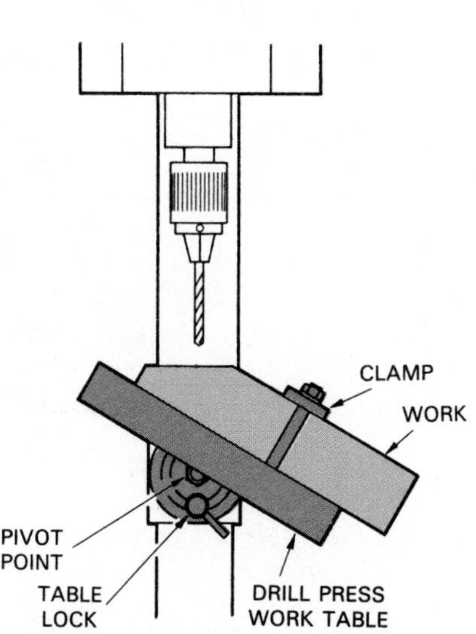

Fig. 7-29. Angular drilling can also be done by tilting drill table. Be sure table is locked tightly before starting to drill.

A ***cross-slide*** permits rapid alignment of the work. Some cross-slides are fitted with a vise, Fig. 7-30. Others have a series of tapped holes for mounting a vise or other work holding devices.

V-blocks

V-blocks support round work for drilling, Fig. 7-31. They are made in many sizes. Some are fitted with clamps to hold the work. Larger sizes must be clamped with the work, Fig. 7-32.

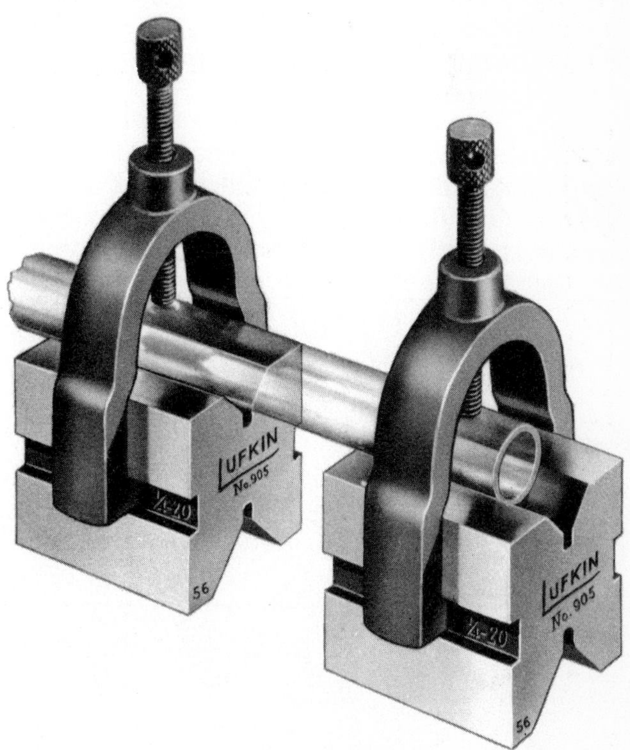

Fig. 7-31. V-blocks supporting round work for drilling.

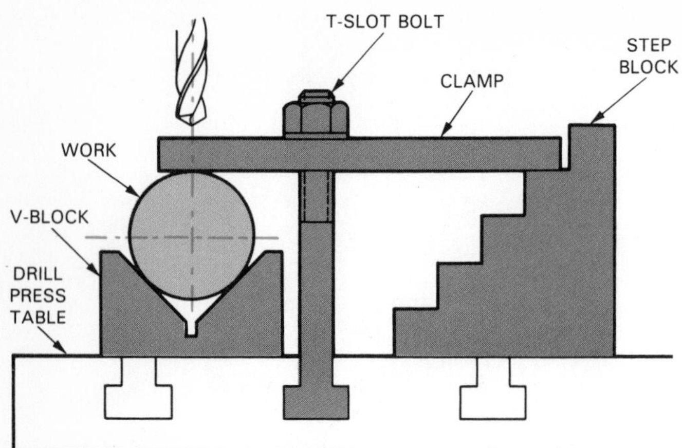

Fig. 7-32. Study one way of clamping large diameter round stock for drilling. Always check that drill will clear V-block when it comes through work.

T-bolts

T-bolts fit into the drill press table slots and fasten work or clamping devices to the machine, Fig. 7-33. A washer should always be used between the nut and holding device. For convenience, having an assortment of different length T-bolts is desirable. To reduce the chance of a setup working loose, place the bolts as CLOSE to the work as possible. See Fig. 7-34.

Strap clamps

Strap clamps make the clamping operation easier if a good assortment is available, Fig. 7-35. The enlongated slot permits some adjustment without removing the washer and nut.

A ***U-strap clamp*** is employed when the clamp must bridge the work. It can straddle the drill and NOT interfere with the drilling operation.

The small round section that projects from the ***finger clamp*** permits the use of small holes or openings in the work to be utilized for clamping.

Fig. 7-33. Note a few of the many types and sizes of T-bolts available.

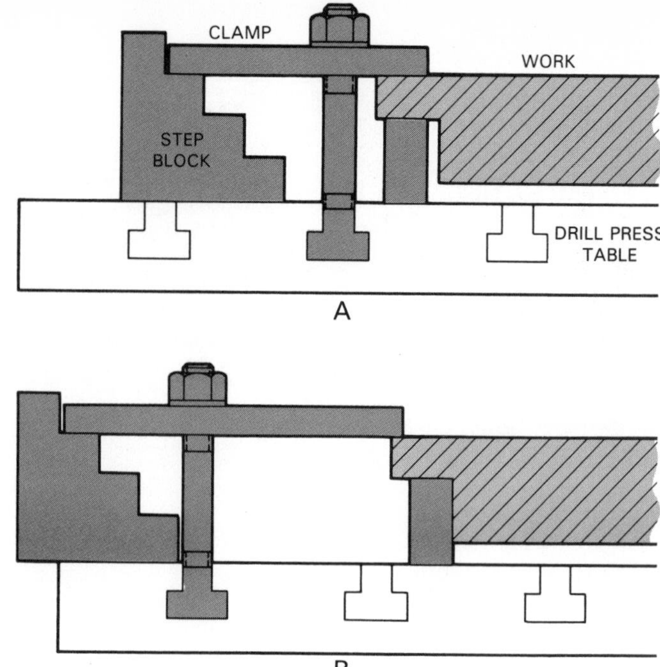

Fig. 7-34. Study examples of clamping techniques. A—Correct clamping technique. Note that clamp is parallel to work. Clamp slippage can be reduced by placing a piece of paper between work and clamp. B—Incorrect clamping technique. T-bolt is too far from work. This allows clamp to spring under pressure. Note that clamp barely engages work.

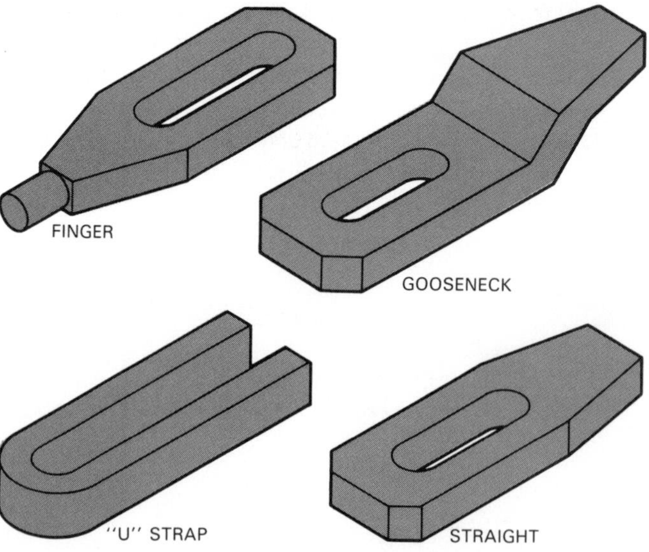

Fig. 7-35. Study types of strap clamps.

Use a strip of copper or aluminum to protect a machined surface that must be clamped.

Step blocks

A ***step block*** supports the strap clamp opposite the work, Fig. 7-36. The steps allow the ad-

Fig. 7-36. Step blocks are used to support strap clamps.

justments necessary to keep the strap parallel with the work.

Angle plate

The ***angle plate*** is often used when work must be clamped to a support. The angle plate is then bolted to the machine table, Figs. 7-37 and 7-38.

Drill jig

A ***drill jig*** permits holes to be drilled in a number of identical pieces, Fig. 7-39. It is a clamping device that supports and locks the work in proper position. With the use of drill bushings, it guides the drill to the correct location. This makes it unnecessary to layout each individual piece for drilling. Quantity production requires the use of jigs.

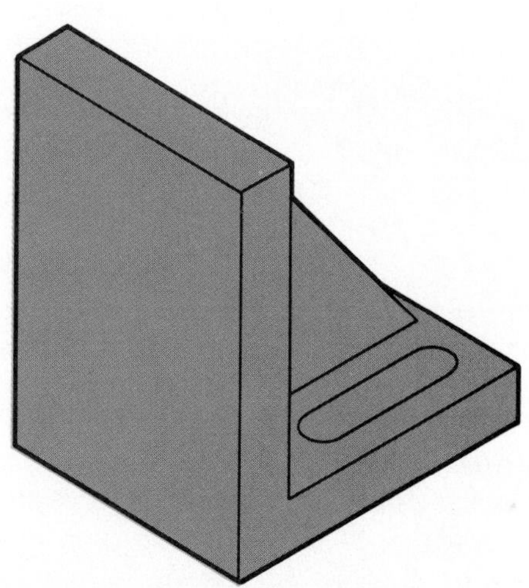

Fig. 7-37. Angle plate is used to support work.

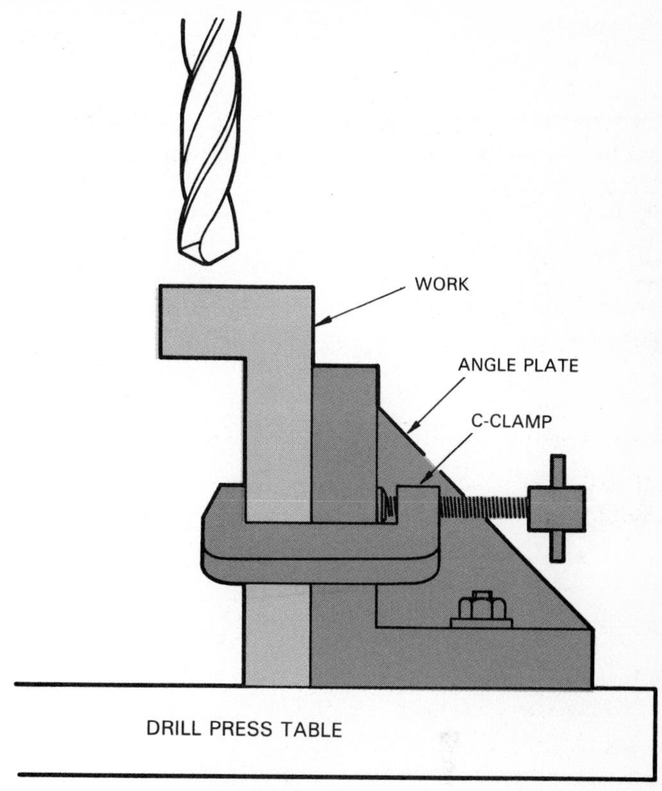

Fig. 7-38. Work must sometimes be mounted against an angle plate for adequate support for drilling.

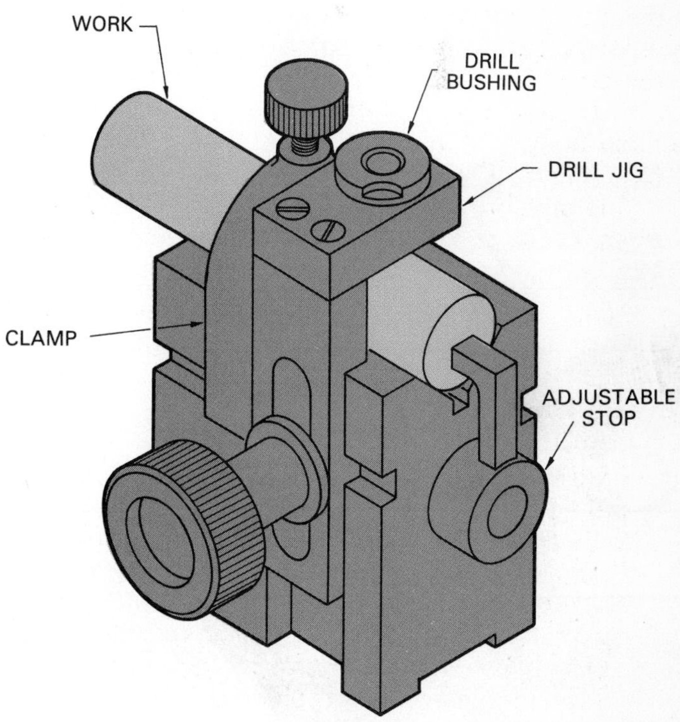

Fig. 7-39. This is a typical drill jig for holding round stock for drilling through center.

Parallels

Parallels are helpful for raising work in a vise so drilling can be better observed. Parallels can be made from stock steel bars or from special steel that is heat treated, Fig. 7-40. Heat treated parallels are ground to size. Parallels must be located so the drill does NOT contact them when the drill breaks through the work.

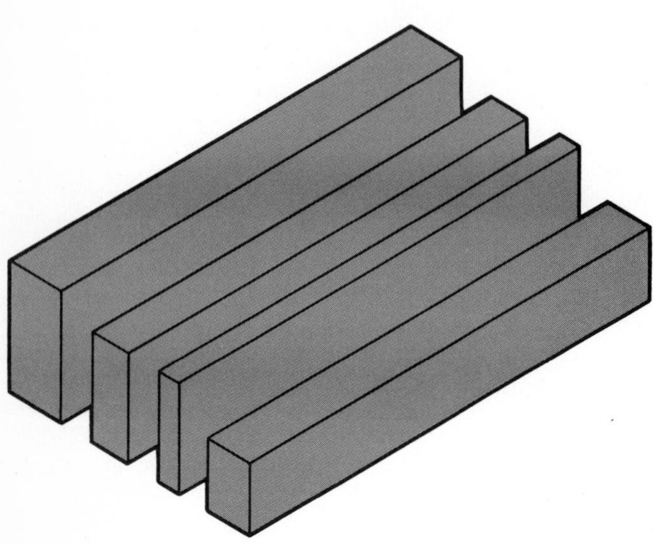

Fig. 7-40. Steel parallels are available in a large variety of sizes.

CUTTING SPEEDS AND FEEDS

The ***cutting speed*** is the speed the drill rotates. The ***feed*** is the distance the drill is moved into the work with each revolution. Both are important considerations because they determine the time required to produce the hole.

Drill cutting speed, also known as ***peripheral speed,*** does NOT refer to the ***revolutions per minute*** (RPM) of the drill, but rather to the distance drill CUTTING EDGE circumference travels per minute.

Feed

Contrary to popular belief, the spiral shape of the drill flutes does NOT cause the drill to pull itself into the work. Constant pressure must be applied and maintained to advance the drill point at a given rate. This advance is called "feed" and is measured in decimal fractions of an inch or in millimeters.

Because so many variables affect results, there can be no hard and fast rule to determine the EXACT cutting speed and feed for a given material. For this reason, the ***drill speed and feed table*** indicates only recommended speeds and feeds, Fig. 7-41. They are a starting point and can be increased or decreased for optimum cutting.

Feed CANNOT be controlled accurately on a hand fed drill press. A machinist must become aware of the cutting characteristics (such as uniform chips) that indicate whether the drill is being fed at the correct rate.

A feed that is TOO LIGHT will cause the drill to scrape and "chatter" and dull rapidly. Chipped cutting edges, drill breakage, and the drill heating up despite the application of coolant usually indicate that the rate is TOO GREAT.

Speed conversion

A problem arises in setting a drill press to the correct speed because its speed is given in revolutions per minute (RPM), while recommended drill cutting speed (CS) is given in feet or meters per minute (FPM/MPM).

The simple formula (RPM = CS divided by 0.2500) will determine the RPM to operate any diameter drill (D) at any specified speed.

Drill speed problem: At what speed (RPM) must a drill rotate when drilling aluminum with a 1/2 in. diameter high-speed steel drill?

To solve this problem:

1. Refer to the speed and feed table, Fig. 7-41. It will give the recommended cutting speed for aluminum (250 FPM).
2. Convert drill diameter (1/2 in.) to decimal fraction (0.500).
3. Write down the formula (RPM $= \frac{CS}{0.250D}$).
4. Substitute values and solve the formula:

$$RPM = \frac{250}{0.250 \times 0.500} = \frac{250}{0.125} = 2000 \text{ RPM}$$

Metric problems would be solved in a similar manner; however, the following formula would be used:

$$RPM = \frac{\text{CS in MPM (meters per minute)} \times 1000}{\text{D (in mm)} \times \pi}$$

MPM = Meters per minute

$\pi = 3.14$

Drill press speed control mechanisms

With some drill presses, it is possible to dial the desired RPM, Fig. 7-42. However, on most conventional drilling machines, it is NOT possible to set the machine at the exact speed desired. The machinist must settle for a speed nearest the desired RPM.

The number of speed settings is limited by the number of pulleys in the drive mechanism, Fig. 7-43. An engraved metal chart or decal showing spindle speeds at various bet settings is attached

A

MATERIAL	CUTTING FLUID	SPEED FEET PER MINUTE	FEEDS PER REVOLUTION Over .040 Diameter** Under 1/8	1/8 to 1/4	1/4 to 1/2	1/2 to 1	Over 1
Aluminum & Aluminum Alloys	Sol. Oil, Ker. & Lard Oil, Lt. Oil	200-300	.0015	.003	.006	.010	.012
Aluminum & Bronze	Sol. Oil, Ker. & Lard Oil, Lt. Oil	50-100	.0015	.003	.006	.010	.012
Brass, Free Machining	Dry, Sol. Oil, Ker. & Lard Oil, Lt. Min. Oil	150-300	.0025	.005	.010	.020	.025
Bronze, Common	Dry, Sol. Oil, Lard Oil, Min. Oil	200-250	.0025	.005	.010	.020	.025
Bronze, Soft and Medium Hard	Min. Oil with 5%–15% Lard Oil	70-300	.0025	.005	.010	.020	.025
Bronze, Phosphor, 1/2 hard	Dry, Sol. Oil, Lard Oil, Min. Oil	110-180	.0015	.003	.006	.010	.012
Bronze, Phosphor, Soft	Dry, Sol. Oil, Lard Oil, Min. Oil	200-250	.0025	.005	.010	.020	.025
Cast Iron, Soft	Dry or Airjet	100-150	.0025	.005	.010	.020	.025
Cast Iron, Medium	Dry or Airjet	70-120	.0015	.003	.006	.010	.012
Cast Iron, Hard	Dry or Airjet	30-100	.001	.002	.003	.005	.006
*Cast Iron, Chilled	Dry or Airjet	10-25	.001	.002	.003	.005	.006
*Cast Steel	Soluble Oil, Sulphurized Oil. Min. Oil	30-60	.001	.002	.003	.005	.006
Copper	Dry, Soluble Oil, Lard Oil, Min. Oil	70-300	.001	.002	.003	.005	.006
Magnesium & Magnesium Alloys	Mineral Seal Oil	200-400	.0025	.005	.010	.020	.025
Manganese Copper, 30% Mn.	Soluble Oil, Sulphurized Oil	10-25	.001	.002	.003	.005	.006
Malleable Iron	Dry, Soluble Oil, Soda Water, Min. Oil	60-100	.0025	.005	.010	.020	.025
Monel Metal	Sol. Oil, Sulphurized Oil, Lard Oil	30-50	.0015	.003	.006	.010	.012
Nickel, Pure	Sulphurized Oil	60-100	.001	.002	.003	.005	.006
Nickel, Steel 3-1/2%	Sulphurized Oil	40-80	.001	.002	.003	.005	.006
Plastics, Thermosetting	Dry or Airjet	100-300	.0015	.003	.006	.010	.012
Plastics, Thermoplastic	Soluble Oil, Soapy Water	100-300	.0015	.003	.006	.010	.012
Rubber, Hard	Dry or Airjet	100-300	.001	.002	.003	.005	.006
Spring Steel	Soluble Oil, Sulphurized Oil	10-25	.001	.002	.003	.005	.006
Stainless Steel, Free Mach'g.	Soluble Oil, Sulphurized Oil	60-100	.0025	.005	.010	.020	.025
Stainless Steel, Tough Mach'g.	Soluble Oil, Sulphurized Oil	20-27	.0025	.005	.010	.020	.025
Steel, Free Machin'g SAE 1100	Soluble Oil, Sulphurized Oil	70-120	.0015	.003	.006	.010	.012
Steel, SAE-AISI, 1000-1025	Soluble Oil, Sulphurized Oil	60-100	.0015	.003	.006	.010	.012
Steel, .30-.60% CARB., SAE 1000-9000							
Annealed 150-225 Brinn.	Soluble Oil, Sulphurized Oil	50-70	.0015	.003	.006	.010	.012
Heat Treated 225-283 Brinn.	Sulphurized Oil	30-60	.0025	.005	.010	.020	.025
Steel, Tool Hi. Car. & Hi. Speed	Sulphurized Oil	25-50	.0025	.005	.010	.020	.025
Titanium	Highly Activated Sulphurized Oil	15-20	.0025	.005	.010	.020	.025
Zinc, Alloy	Soluble Oil, Kerosene & Lard Oil	200-250	.0015	.003	.006	.010	.012

*Use Specially Constructed Heavy Duty Drills
**For drill under .040, feeds should be adjusted to produce chips and not powder with ability to dispose of same without packing.

B

DRILL SPEEDS IN R. P. M.

Diameter of Drill	Soft Metals 300 F.P.M.	Plastics and Hard Rubber 200 F.P.M.	Annealed Cast Iron 140 F.P.M.	Mild Steel 100 F.P.M.	Malleable Iron 90 F.P.M.	Hard Cast Iron 80 F.P.M.	Tool or Hard Steel 60 F.P.M.	Alloy Steel Cast Steel 40 F.P.M.
1/16 (No. 53 to 80)	18320	12217	8554	6111	5500	4889	3667	2445
3/32 (No. 42 to 52)	12212	8142	5702	4071	3666	3258	2442	1649
1/8 (No. 31 to 41)	9160	6112	4278	3056	2750	2445	1833	1222
5/32 (No. 23 to 30)	7328	4888	3420	2444	2198	1954	1465	977
3/16 (No. 13 to 22)	6106	4075	2852	2037	1833	1630	1222	815
7/32 (No. 1 to 12)	5234	3490	2444	1745	1575	1396	1047	698
1/4 (A to E)	4575	3055	2139	1527	1375	1222	917	611
9/32 (G to K)	4071	2712	1900	1356	1222	1084	814	542
5/16 (L, M, N)	3660	2445	1711	1222	1100	978	733	489
11/32 (O to R)	3330	2220	1554	1110	1000	888	666	444
3/8 (S, T, U)	3050	2037	1426	1018	917	815	611	407
13/32 (V to Z)	2818	1878	1316	939	846	752	563	376
7/16	2614	1746	1222	873	786	698	524	349
15/32	2442	1628	1140	814	732	652	488	326
1/2	2287	1528	1070	764	688	611	458	306
9/16	2035	1357	950	678	611	543	407	271
5/8	1830	1222	856	611	550	489	367	244
11/16	1665	1110	777	555	500	444	333	222
3/4	1525	1018	713	509	458	407	306	204

Figures are for High-Speed Drills. The speed of Carbon Drills should be reduced one-half. Use drill speed nearest to figure given.

Fig. 7-41. A—Study how to use drill speed and feed table. B—Drill speeds in rpm table is handy. (Chicago-Latrobe)

to many machines. If not available, information on spindle speeds can be found in the operator's manual, or it can be calculated if motor speed and pulley diameters are known.

CUTTING COMPOUNDS

Drilling at the recommended cutting speeds and feeds generates considerable heat at the cutting

Fig. 7-42. Split pulley speed control mechanism allows speed adjustment by turning dial in front. (Clausing)

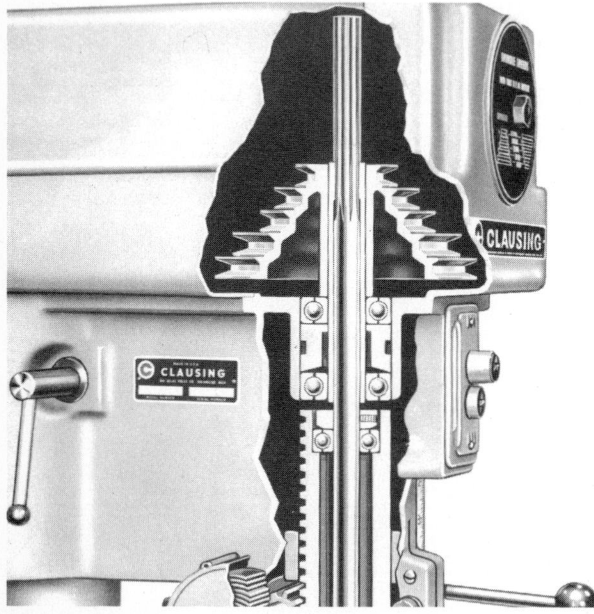

Fig. 7-43. With step pulley speed control, belt is transferred to different pulley ratios to change speed. (Clausing)

point. This heat must be dissipated (carried away) as fast as it is generated or it will destroy the drill's temper and cause it to dull rapidly.

Cutting compounds are applied to absorb the heat and lubricate. They cool the cutting tool, serve as a lubricant to reduce friction at the cutting edges, and minimize the tendency for the chips to weld to the lips. Cutting compounds also improve hole finish and aid in the rapid removal of chips from the hole.

There are many kinds of cutting fluids and compounds.

Many cutting compounds must be applied liberally. However, newer compounds must be applied sparingly. More is NOT necessarily better. Read instructions on the container for the compounds being employed.

Avoid using cutting fluids and compounds when drilling cast iron or other brittle materials because they tend to cause the chips to pack and glaze the opening. Compressed air, used with care, will work when drilling these materials.

WARNING! Always wash your hands thoroughly with soap and warm water after using cutting compounds and fluids. Before using the cutting compound or fluid, check with the compound container to determine what should be done if you get any in an eye.

SHARPENING DRILLS

A drill becomes dull with use and must be resharpened. Continued use of a dull drill may result in its breakage or cause it to burn as it is forced into the metal. Improper sharpening will cause the same problems.

Remove the entire point if it is badly worn or if the margins are burned, chipped, or worn off near the point. If, by accident, the drill becomes overheated during grinding, do NOT plunge it into water. Allow it to cool in still air. The shock of sudden cooling may cause it to crack.

Three factors must be considered when repointing a drill: lip clearance, length and angle of lips, proper location of dead center.

Lip clearance: The two cutting edges or lips are comparable to chisels, Fig. 7-44. To cut effective-

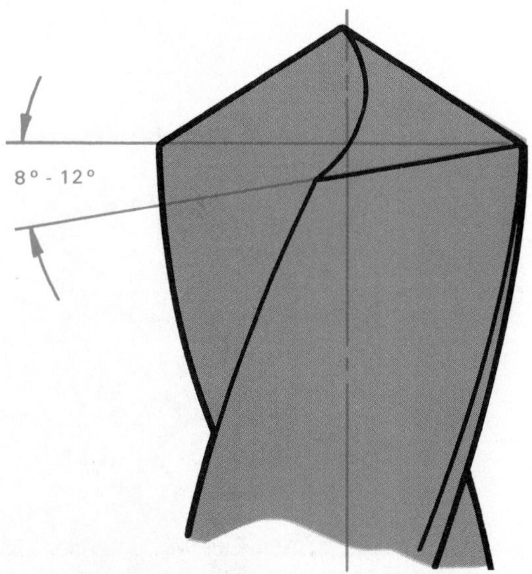

Fig. 7-44. Lip clearance of 8 - 12 deg. is satisfactory for most drilling.

ly, the heel or that part of the point back of the cutting edge must be relieved. Without this ***lip clearance,*** it would be impossible for the lips to cut. If there is TOO MUCH CLEARANCE, the cutting edges will be weakened. TOO LITTLE CLEARANCE results in the drill point merely rubbing without penetration into the material.

Gradually increase lip clearance towards the center until the line across dead center stands at an angle of 120 to 135 deg. with the cutting edge. See Fig. 7-45.

Length and angle of lips: The material to be drilled determines the proper point angle, Fig. 7-46. The angles, in relation to the axis, must be the same (59 deg.), has been found to be satisfactory for most metals. If the angles are unequal, only one lip will cut and the hole will be oversize.

Proper location of dead center: Drill dead center must be accurate. Equal angles, but lips of different lengths will result in oversize holes. The resulting "wobble" places tremendous pressures on the drill press spindle and bearings. See Fig. 7-47.

A combination of both faults can result in a broken drill. If the drill is very large, permanent damage to the drilling machine can result. The hole produced will be oversize and often out-of-round. Refer to Fig. 7-48.

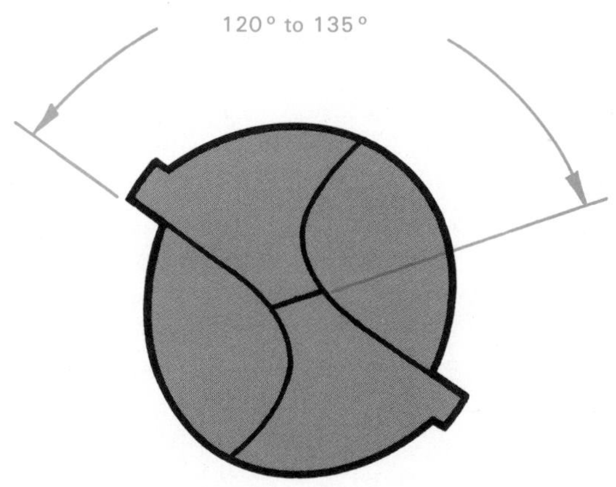

Fig. 7-45. Note proper angle of drill dead center.

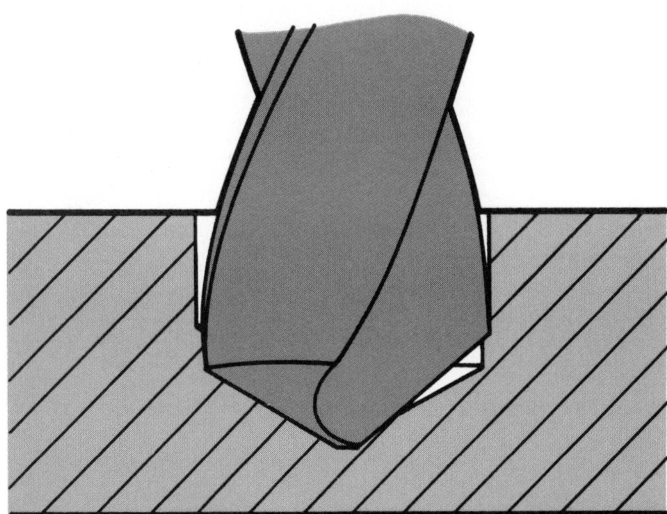

Fig. 7-46. Unequal drill point angles will prevent proper drilled hole.

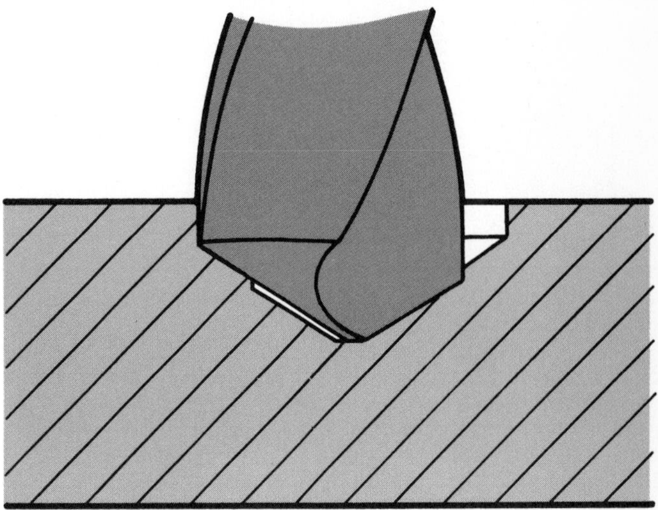

Fig. 7-47. This drill point is ground off center.

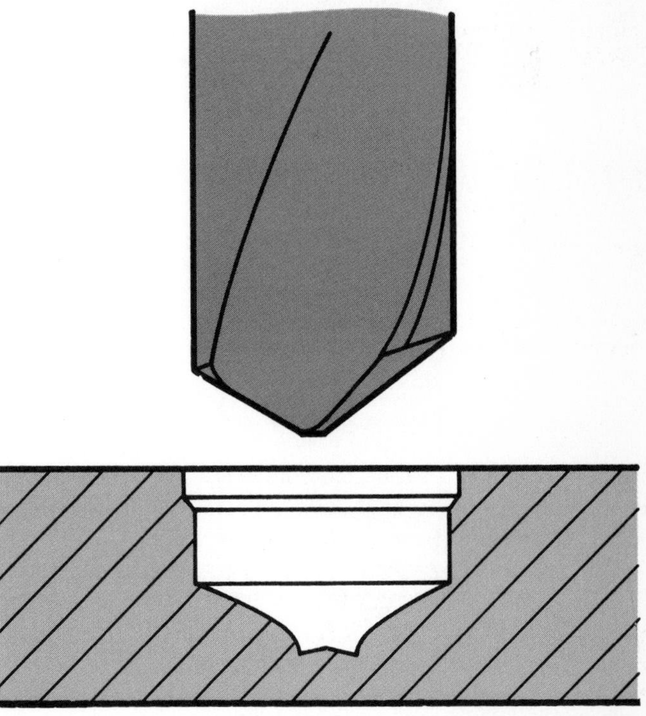

Fig. 7-48. This drill point has unequal point angles and the drill is sharpened off center.

The web of a drill increases in thickness towards the shank, Fig. 7-49. When a drill has been shortened by repeated grindings, the web must be thinned to minimize the pressure required to make the drill penetrate the material. The thinning must be done equally on both sides of the web and care must be taken to insure that the web is centered.

The ***drill point gage*** is the tool most frequently used to check a drill point during the sharpening operation. Its use is shown in Fig. 7-50.

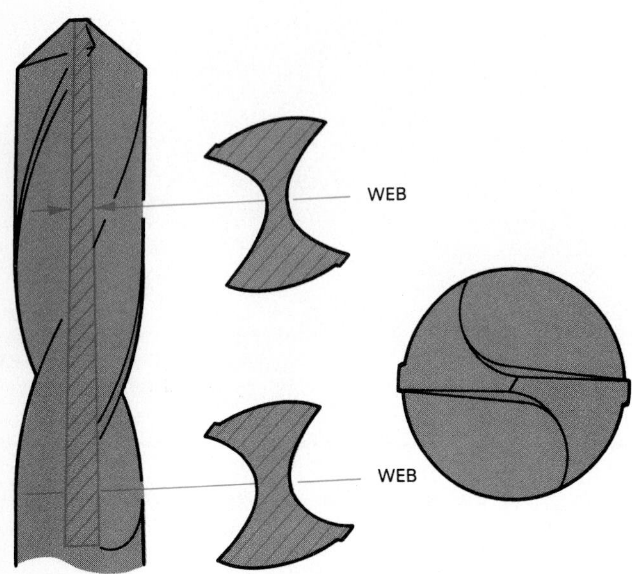

Fig. 7-49. Note web of a drill, and how drill point is relieved by grinding.

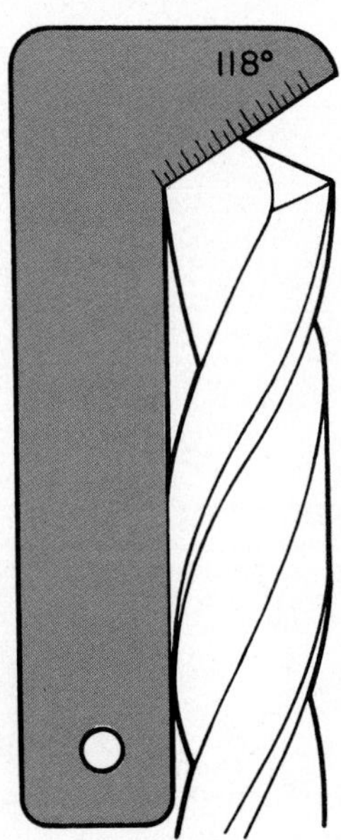

Fig. 7-50. Using a drill point gage will help assure proper drill sharpening.

Drill sharpening procedures

Use a coarse grinding wheel for roughing out the drill point if a large quantity of metal must be removed. Complete the operation on a fine wheel.

Many hand sharpening techniques have been developed. The following is suggested.

1. Grasp the drill shank with the right hand and the rest of the drill with the left hand. See Fig. 7-51.
2. Place your left-hand fingers that are supporting the drill on the grinder tool rest. The tool rest should be slightly below center (about 1 in. down on a 7 in. dia. wheel for example).
3. Stand so the center line of the drill will be at a 59 deg. angle with relation to the center line of the wheel, Fig. 7-52. Lightly touch the drill lip to the wheel in approximately a horizontal position.
4. Use the left hand as a pivot point and slowly lower the shank with the right hand. Increase pressure as the heel is reached to insure proper clearance.

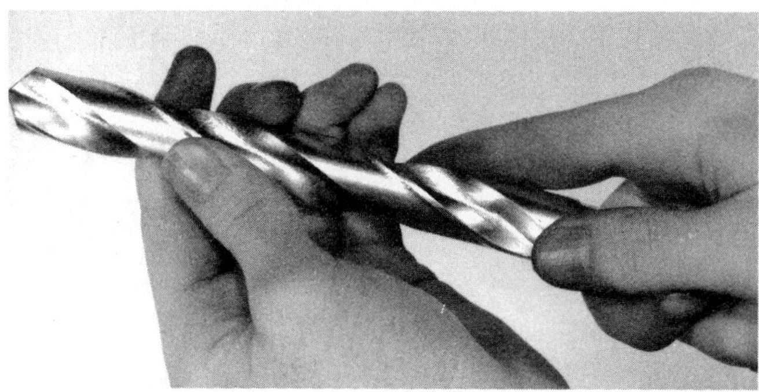

Fig. 7-51. This is one way recommended to hold a drill when it is being sharpened.

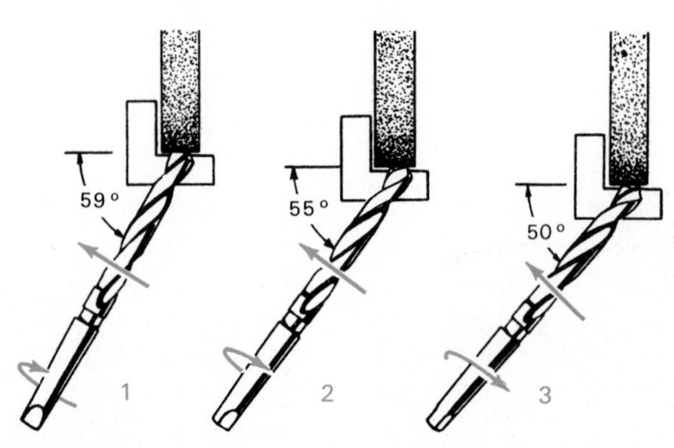

Fig. 7-52. Here is one technique that may be followed when sharpening a drill. The actual grinding of drill point consists of three definite motions of drill shank while point is held lightly against rotating wheel. These three motions are: (1) to the left. (2) clockwise rotation, (3) downward.

5. Repeat the operation on each lip until the drill is sharpened. DO NOT QUENCH high speed steel drills in water to cool them. Allow them to cool in calm air!
6. Check the drill tip frequently with a drill point gage to assure a correctly sharpened drill.

Sharpening a drill is not as difficult as it may first appear. However, before attempting to sharpen a drill, secure a properly sharpened drill and run through the motions explained above. When you have acquired sufficient skill, sharpen a dull drill.

To test, drill a hole in soft metal and observe the chip formation. When properly sharpened, chips will come out of the flutes in curled spirals of equal size and length. Tightness of the chip spiral is governed by the rake angle, Fig. 7-53.

The standard drill point has a tendency to stick in the hole when used to drill brass. When brass is drilled, sharpen the drill as shown in Fig. 7-54.

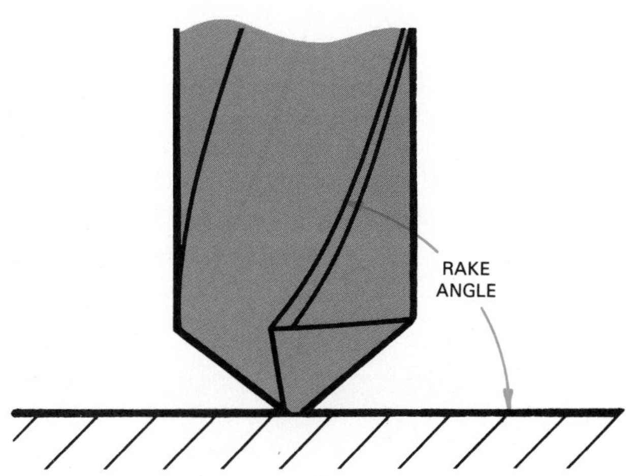

Fig. 7-53. Note rake angle of drill for ordinary work.

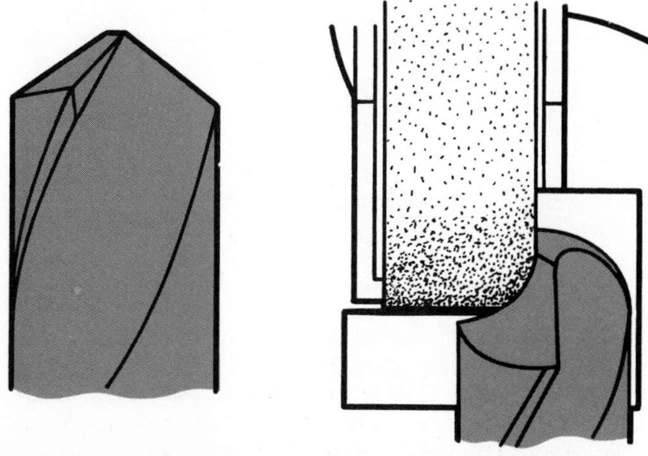

Fig. 7-54. Note modified rake angle for drilling brass.

Drill grinding attachments.

A drill sharpening device is shown in Fig. 7-55. An attachment for conventional grinders is shown in Fig. 7-56. In the machine shop where a high degree of hole accuracy is required and a large amount of sharpening must be done, these devices are a must.

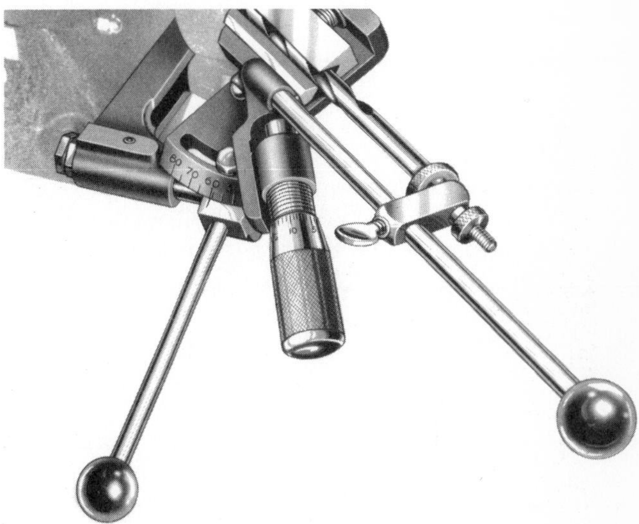

Fig. 7-55. A drill sharpening grinder makes drill sharpening easy. (Darex)

Fig. 7-56. A drill sharpening attachment mounted on a conventional bench grinder provides efficient, yet economical, way for sharpening drills. (Clausing)

DRILLING HOLES

Accurate drilling is dependent upon observing a few simple rules:

1. Carefully study the drawing to determine hole locations. Layout the position and mark the intersecting lines with a prick punch.
2. Secure a drill of the size indicated and check it for size.
3. Mount work solidly on the machine. NEVER try to hold it by hand. It may be whipped out of your hand and cause a "merry-go-round." The workpiece could then whip into your hand and cause serious injuries.
4. Insert a ***wiggler*** or ***center finder*** in the drill chuck to position the point to be drilled directly under the chuck or spindle. Turn on the power and center the wiggler point with your fingers. Position the work until the revolving centered wiggler point does not "wiggle" when it is lightly dropped into the punched hole location and removed. If there is any point movement, additional alignment is necessary because the work is NOT positioned properly, Fig. 7-57.
5. Remove the wiggler and insert a center drill. Hand tighten it, Fig. 7-58. Check to be sure it runs true. If it does, tighten the chuck with a chuck key. Remember to REMOVE THE KEY before starting the machine!
6. After center drilling, replace the tool with the required drill. Hand tighten it in the chuck. Turn on the machine. If it does NOT run true, the drill may be bent or placed in the chuck off center. Also check that it will drill to the required depth.
7. Calculate the correct cutting speed, and feed if you plan to use a power feed. Adjust the machine to operate as closely as possible to this speed.
8. Turn on the power and apply cutting fluid. Start the cut. Even pressure on the feed handle will keep the drill cutting freely.
9. Watch for the following signs that indicate a poorly cutting drill:
   a. A dull drill will squeak and overheat. Chips will be rough and blue, and cause the machine to slow down. Small drills will break.
   b. Infrequently, a chip will get under the dead center and act as a bearing preventing the drill from cutting. Remove it by raising and lowering the drill several times.
   c. Chips packed in the flutes cause the drill to bind and slow the machine or cause the drill to break. Remove the drill from the hole and clean it with a brush that has been dipped in cutting fluid. Do NOT use cutting fluid when drilling cast iron!

Fig. 7-57. It is difficult to align a drill with center lines by eye. To assist in this job, center finder or "wiggler" is used.

Fig. 7-58. After alignment with the "wiggler," center drilling will assure that drill will make hole where specified.

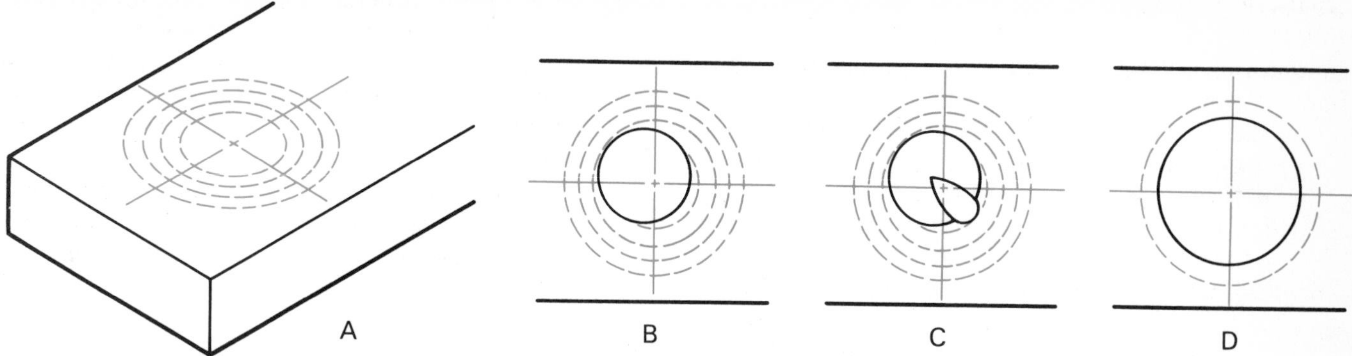

Fig. 7-59. Note how to bring a drill back on center. A—Proof circles. B—Drill has been made off center (exaggerated). C—Groove cut to bring drill back on center. D—Drill back on center. This operation will only work if drill has NOT started to cut to its full diameter.

10. Clear chips and apply cutting fluid as needed.
11. The most critical time of the drilling operation occurs when the drill starts to break through the work. Ease up on feed pressure at this point to prevent the drill from "digging in."
12. Remove the drill from the hole and turn off the power. NEVER try to stop the chuck with YOUR HAND. Clear the chips with a brush, NOT your hand. Unclamp the work and use a file to remove all burrs.
13. Clean chips and cutting fluid from the machine. Wipe it down with a soft cloth. Return equipment to storage after cleaning.

Observe extreme care in positioning the piece for drilling. A poorly planned setup may permit the drill to cut into the vise or drill table when it breaks through the work.

If a hole must be located precisely, certain additional precautions can be taken to insure that the hole will be drilled where it is supposed to be drilled. After the center point has been determined, a series of circles, called ***proof circles,*** are scribed, Fig. 7-59A. They will serve as reference points to help check whether the drill remains on center as it starts to penetrate the material.

Even when work is properly centered, the drill may "drift" when starting a hole. Various factors can cause this: hard spots in the metal, an improperly sharpened drill, etc. The drill cannot be brought back on center by moving the work because it will still try to follow the original hole. This condition, Fig. 7-59B, must be corrected before the full diameter of the drill is reached.

The drill is brought back on center by using a round nose cape chisel to cut a groove on the side of the hole where the drill must be drawn, Fig. 7-59C. This groove will "pull" the drill point to cut in that direction and recenter.

Repeat the operation until the hole is concentric with or centered in the proof circles, Fig. 7-59D.

Drilling larger holes

Drills larger than 1/2 in. (12.5 mm) diameter require considerable power and pressure to get started. Even then they may run off center. The pressure can be greatly reduced and accuracy improved by first drilling a ***pilot*** or ***lead hole*** which is smaller in diameter. See Fig. 7-60.

The small pilot hole permits pressure to come directly on the cutting edges of the large drill causing it to drill faster. Diameter of the pilot hole should be as large as, or slightly larger than, the width of the dead center.

Drilling round stock

Holes are more difficult to drill in the curved surface of round stock. Many difficulties can be

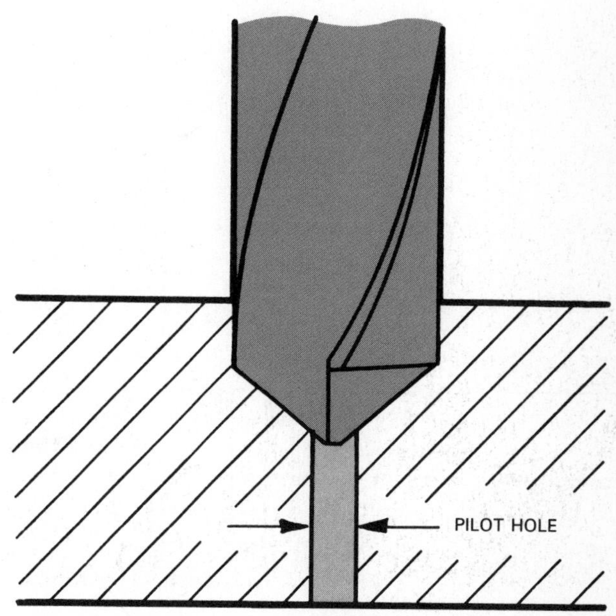

Fig. 7-60. Pilot or lead hole will make drilling large hole much easier.

eliminated by holding the round material in a V-block, Fig. 7-61. A V-block can be held in a vise or clamped directly to the table. Refer to Fig. 7-31.

To center round stock in a V-block:

1. Locate hole position on the stock. Prick punch the intersection of the layout lines. Place the stock in a V-block. Make certain that the drill, if the hole is to go through the piece, will clear the V-block. Also be sure there is ample clearance between the clamp and drill chuck.
2. To align the hole for drilling through exact center, place the work and V-block on the drill press table, or on a surface plate. Rotate the punch mark until it is upright. Place a steel square on the flat surface with the blade against the round stock as shown in Fig. 7-62. Measure from the square blade to the punch mark, and rotate the stock until the measurement is the same when taken from both positions.
3. From this point, the drilling sequence is identical to that previously described.

If a large number of indentical parts must be drilled, it may be desirable to make a drill jig as shown in Fig. 7-63. The drill jig automatically positions and centers each piece for drilling.

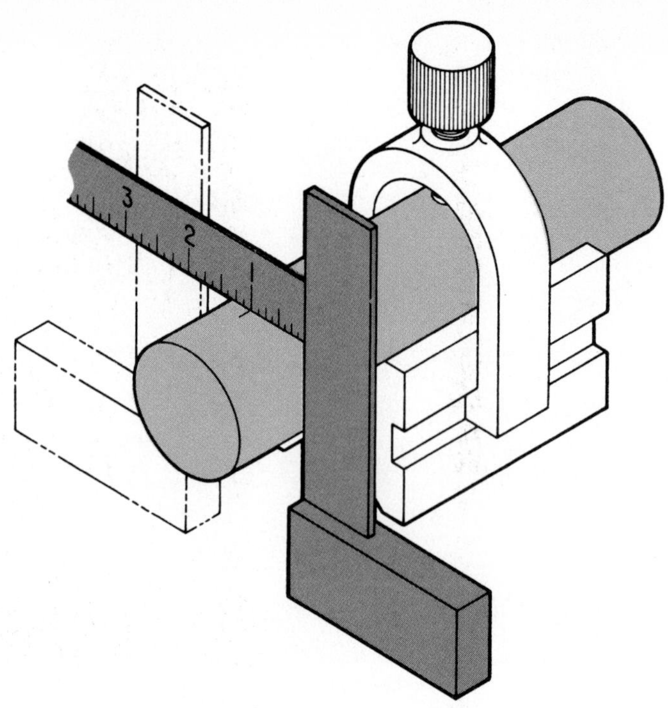

Fig. 7-62. Using square to center round stock in a V-block.

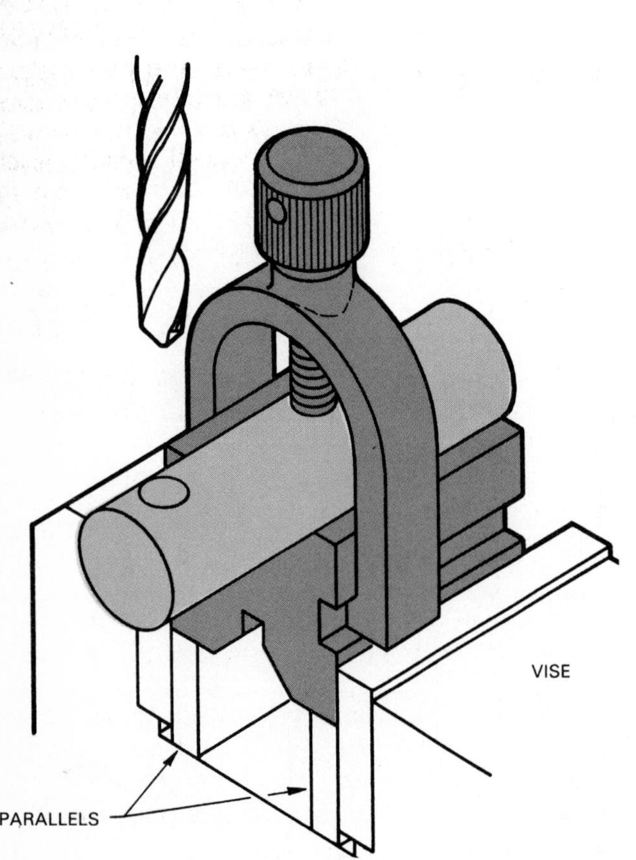

Fig. 7-61. V-block eliminates many difficulties when drilling round stock. Be sure drill will clear V-block when it comes through material.

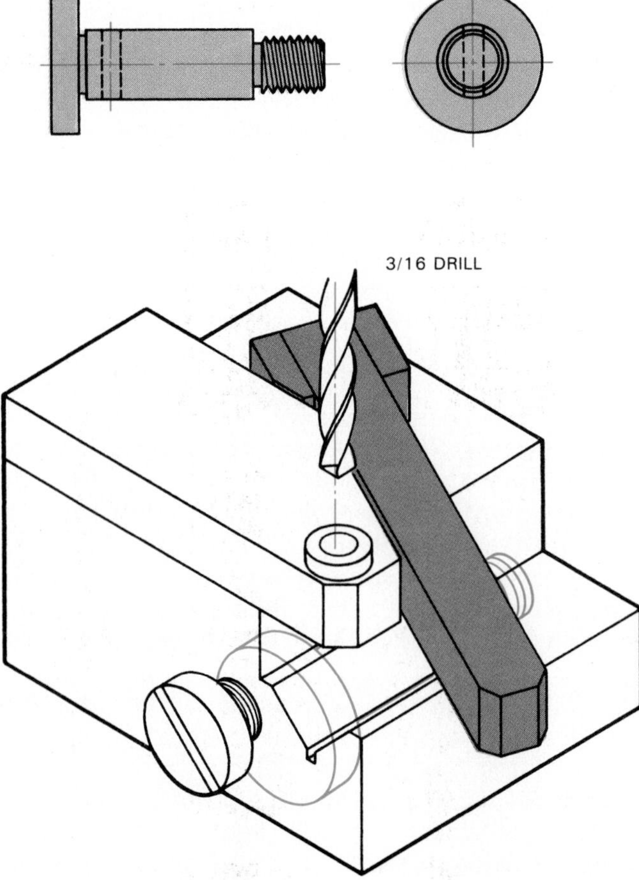

Fig. 7-63. A typical drill jig has arm that lifts to allow easy insertion and removal of bolt.

Blind holes

A ***blind hole*** is a hole that is NOT drilled all the way through the work. ***Hole depth*** is measured by the distance the full hole diameter goes into the work, Fig. 7-64. Using a drill press fitted with a ***depth stop*** or ***gage*** is the quickest means of securing proper depth when drilling blind holes, Fig. 7-65.

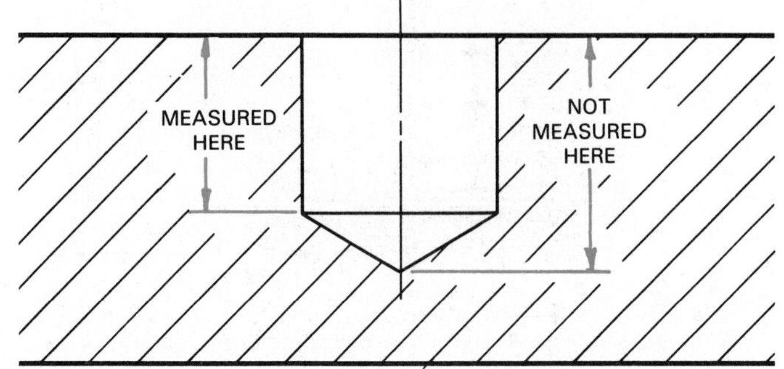

Fig. 7-64. Note how depth of a blind hole is measured.

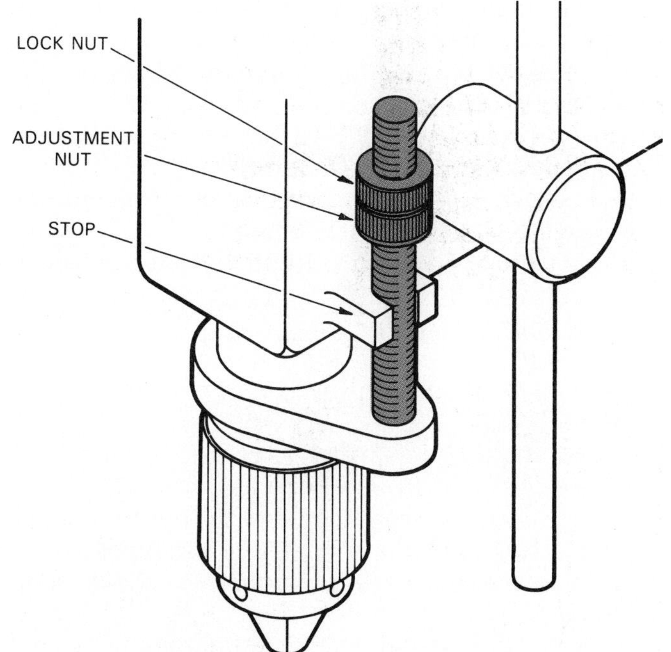

Fig. 7-65. Depth gage attachment fitted to most drill presses provides easy adjustment of how far drill moves into or through work.

OTHER DRILL PRESS OPERATIONS

Operations other than drilling can also be done on a drill press. As with drilling, they require a thorough knowledge of the machine and the cutting tools.

Reaming

Reaming is the operation that produces holes that are extremely accurate in diameter and have an exceptionally fine surface finish.

Machine reamers are made in a variety of sizes and styles. They are usually manufactured from high speed steel. Some are fitted with carbide cutting edges. Descriptions of a few of the more common machine reamers follow.

Jobber's reamer

The ***jobber's reamer*** is identical to the hand reamer except that a taper shank is available and the tool is designed for machine operation, Fig. 7-66.

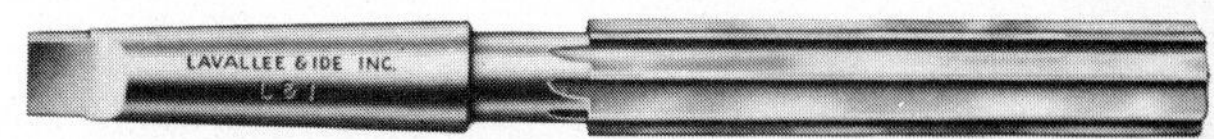

Fig. 7-66. Jobber's reamer is designed for machine operation.

Chucking reamer

A ***chucking reamer*** is manufactured with both straight and taper shanks. It is similar to the jobber's reamer but flutes are shorter and deeper. It is available with straight or spiral flutes, Fig. 7-67.

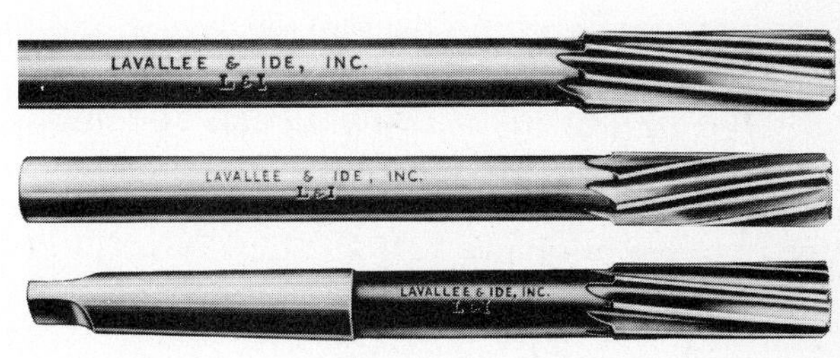

Fig. 7-67. Above—Chucking reamer, straight shank, R.H. spiral, R.H. cut. Center—Chucking reamer, straight shank, L.H. spiral, R.H. cut. Below—Chucking reamer, taper shank.

Rose chucking reamer

The ***rose chucking reamer*** is designed to cut on its end, Fig. 7-68. The flutes provide chip clearance and are ground to act only as guides. This type reamer is best used when considerable metal must be removed and the finish is NOT too critical.

Expansion chucking reamer

An ***expansion chucking reamer*** is available with either a straight or taper shank, has straight flutes, Fig. 7-69. Slots are cut into the body to permit the

Fig. 7-68. Note rose chucking reamer.

Fig. 7-69. Note expansion chucking reamer. Above—Taper shank. Below—Straight shank.

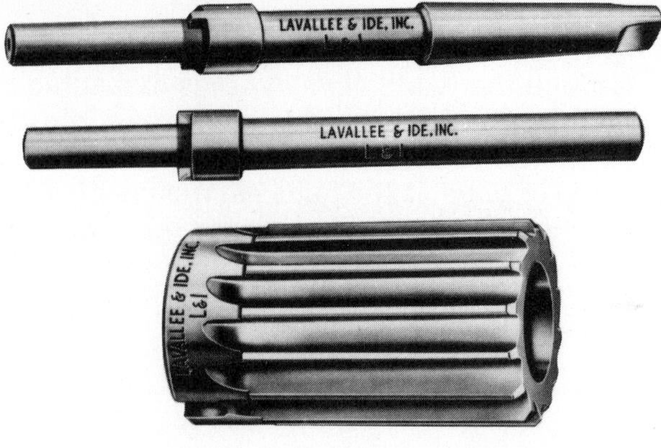

Fig. 7-71. Above—Arbors for shell reamer. Below—Shell reamer.

reamer to expand when an adjusting screw in the end is tightened.

A regular expansion reamer has several drawbacks. The slots, necessary for the reamer to expand, reduce tool rigidity. This diminishes accuracy and surface finish. Also, as it is expanded, cutting edge clearance is reduced which creates a "drag" which often causes tool chatter and a resulting decrease in finish quality.

A recently designed expansion reamer is solid and provides rigidity and accuracy not possible with the conventional expansion reamer, Fig. 7-70. To expand this type, a tapered plug is forced into the reamer end. The tool body expands well beyond the tip, and assures uniform parallel expansion the full length of the carbide cutting lips. Clearance is automatically provided. The plug can be removed for shimming to a larger size. However, once expanded, reamer diameter CANNOT be reduced, other than by grinding.

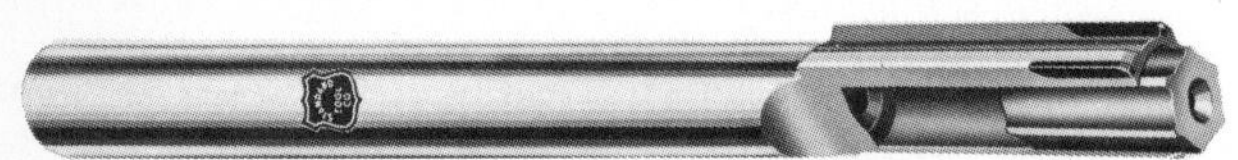

Fig. 7-70. This solid type expansion reamer has cutting edges that are tungsten carbide tipped for extended cutting life. (Standard Tool Co.)

Shell reamer

A ***shell reamer*** is mounted on a special arbor that can be used with several reamer sizes, Fig. 7-71. It can have straight or spiral flutes and is also made in the rose style. The arbor shank may be straight or tapered. A hole in the reamer is tapered to fit the arbor which is fitted with drive lugs.

Using machine reamers

Reamers are precision tools. They are also expensive. The quality of the finish and accuracy of the reamed hole will depend on how the tool is used.

1. Mount the reamer solidly.
2. Allow enough material in the drilled hole to permit the reamer to cut rather than burnish (smooth and polish). The following allowances are recommended:
   a. To 1/4 in. (6.3 mm) diameter allow 0.010 in. (0.25 mm).
   b. 1/4 to 1/2 in. (6.3 to 12.5 mm) diameters allow 0.015 in. (0.4 mm).
   c. 1/2 to 1.0 in. (12.5 to 25.0 mm) diameters allow 0.020 in. (0.5 mm).
   d. 1.0 to 1.5 in. (25.0 to 38.0 mm) diameters allow 0.025 in. (0.6 mm).
3. Use sharp reamers.
4. Cutting speed for a high speed steel reamer should be about two-thirds that of a similar size drill.
5. Feed should be as much as possible while giving a good finish and accurate hole size.
6. Carefully check reamer diameter before use. If the hole diameter is critical, drill and ream a hole in a piece of similar material to check tool accuracy.
7. When NOT being used, reamers should be stored in separate containers or storage compartments. This will minimize the danger of chipping or dulling the cutting edges.
8. Use an ample supply of cutting fluid.
9. Remove the reamer from the hole BEFORE stopping the machine.

COUNTERSINKING

Countersinking is the operation that cuts a chamfer in a hole to permit a flat-headed fastener to be inserted with the head flush to the surface,

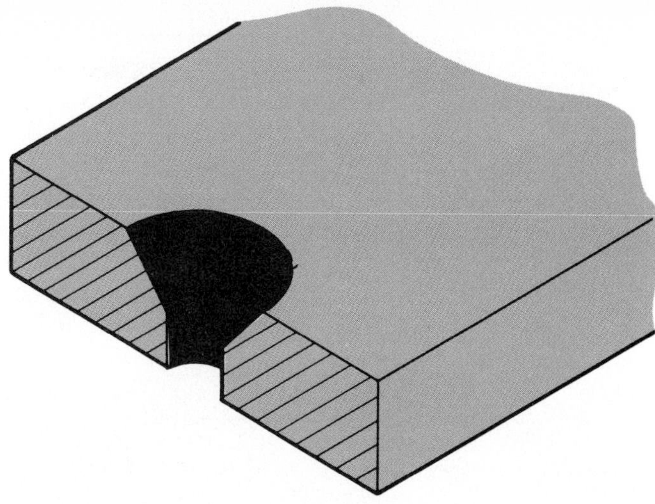

Fig. 7-72. This is a cross section of hole that has been countersunk.

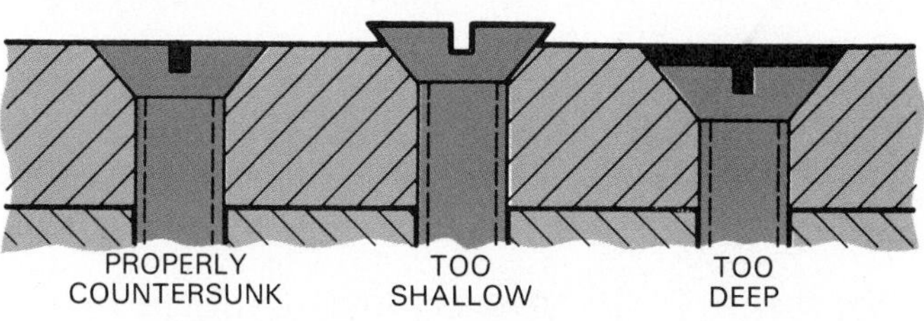

Fig. 7-73. Note correctly and incorrectly countersunk holes.

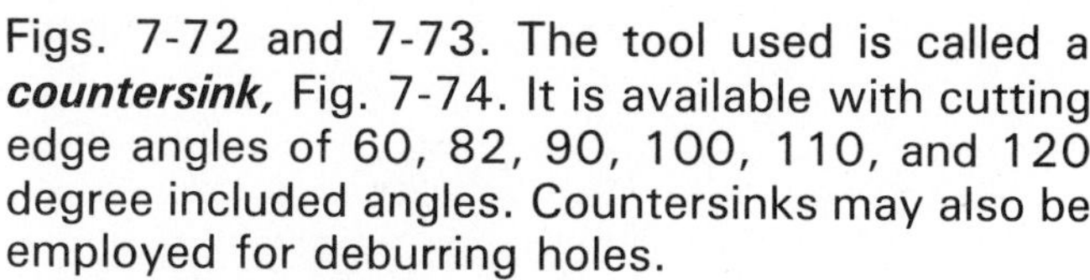

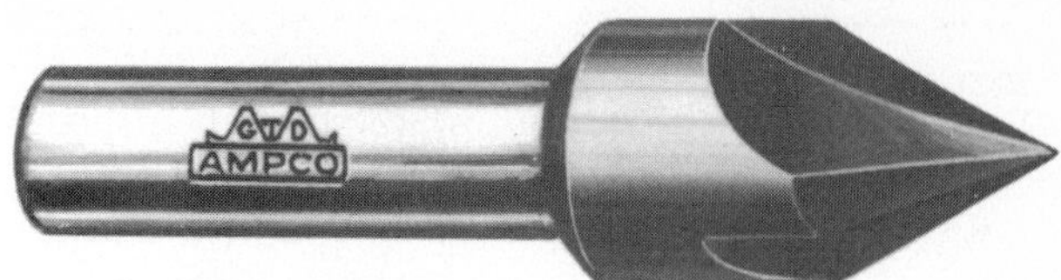

Fig. 7-74. Countersinks come in various sizes. (Greenfield Tap & Die)

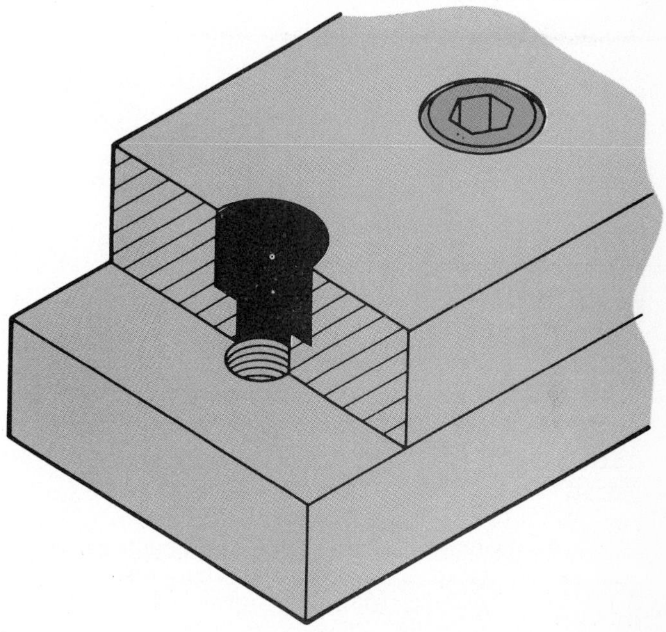

Fig. 7-75. Note sectional view of hole that has been drilled and counterbored to receive a socket head screw.

Figs. 7-72 and 7-73. The tool used is called a ***countersink,*** Fig. 7-74. It is available with cutting edge angles of 60, 82, 90, 100, 110, and 120 degree included angles. Countersinks may also be employed for deburring holes.

Using a countersink

1. The cutting speed should be about one-half that recommended for a similar size drill. This will minimize the possibility of ''chatter.''
2. Feed the tool into the work until the chamber is large enough for the fastener head to be flush.
3. Use the depth stop on the drill press if a number of similar holes must be countersunk.

COUNTERBORING

The heads of fillister and socket head screws are usually set BELOW the work surface. A ***counterbore*** is employed to enlarge the drilled hole to proper depth and machine a square shoulder on the bottom to secure maximum clamping action from the fastener, Fig. 7-75.

The counterbore tool has a guide, called a ***pilot,*** which keeps it positioned correctly in the hole.

Solid counterbores are available. However, counterbores with interachangeable pilots and cutters are commonly used, Fig. 7-76. They can be changed easily from one size cutter or pilot to another size.

A drop of oil on the pilot will prevent it from binding in the drilled hole.

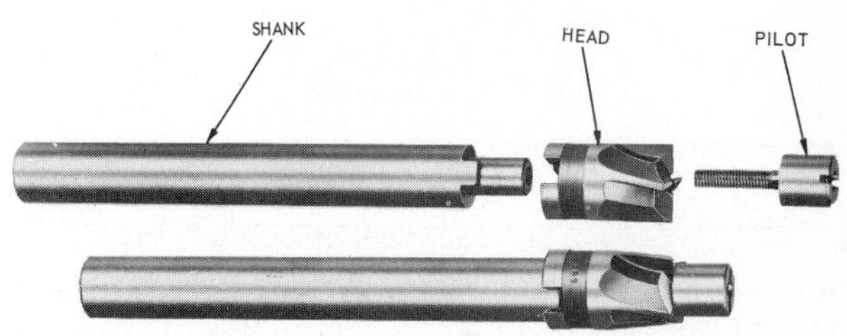

Fig. 7-76. This is straight shank interchangeable counterbore. Heads can be quickly changed.

SPOTFACING

Spotfacing is the term applied when a circular spot is machined on a rough surface (casting, forging, etc.) to furnish a bearing surface for the head of bolt, washer, or nut. A counterbore may be used for spotfacing, although a special tool manufactured for inverted spotfacing is available, Fig. 7-77.

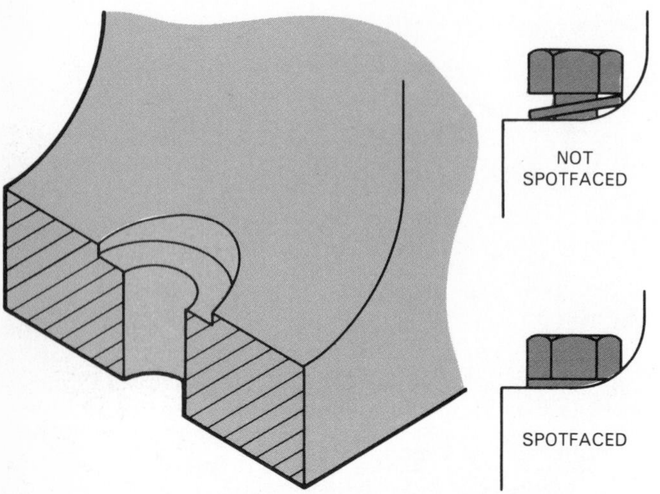

Fig. 7-77. Sectional view of a casting with mounting hole that has been spotfaced. Smaller drawings show a side view of casting before and after spotfacing. Note bolt head cannot be drawn down tightly until mounting hole is spotfaced.

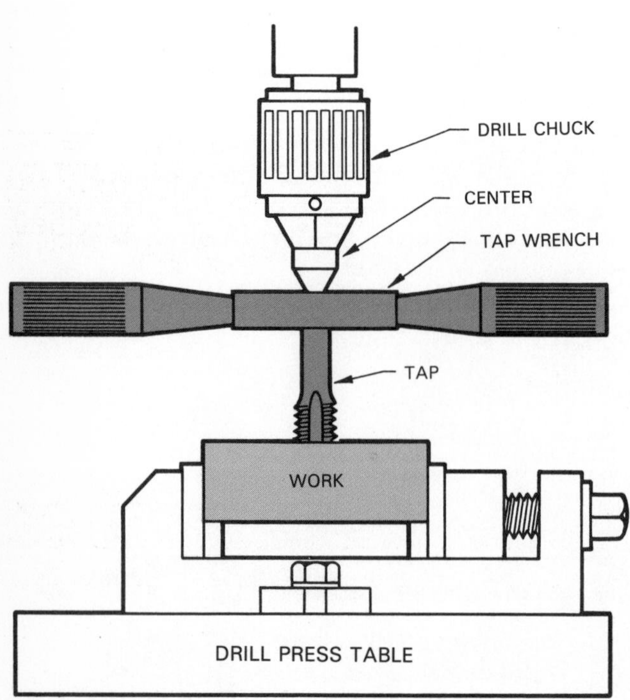

Fig. 7-78. Study setup for hand tapping on a drill press.

TAPPING

Tapping may be done by hand on a drill press by:

1. Drilling a correct size hole for the tap, Fig. 7-78.
2. With the work clamped in the machine, insert a small 60 deg. center in the chuck. The ***center*** holds the tap vertically.
3. Place the center point in the tap's center hole.
4. Feed the tap into the work by holding down on the feed handle and turning the tap with a TAP WRENCH.

DANGER! NEVER insert a tap into the drill chuck and attempt to use drill press POWER to run the tap into the work. The tap will shatter when power is applied. Turn the tap by hand!

Tapping can only be done with power through the use of a ***tapping attachment,*** Fig. 7-79. This device fits the standard drill press. It has reducing gears that slows the tap to about one-third drill press speed. A table with the attachment provides recommended spindle speeds for tapping.

A clutch arrangement drives the tap until it reaches the predetermined depth, at which time the tap stops rotation. Raising the feed handle causes the tap to reverse direction and back out of the hole.

POLISHING, GRINDING, AND BORING

Polishing, grinding and boring can be done with limited success on a standard drill press. However, such machines do NOT always have the necessary

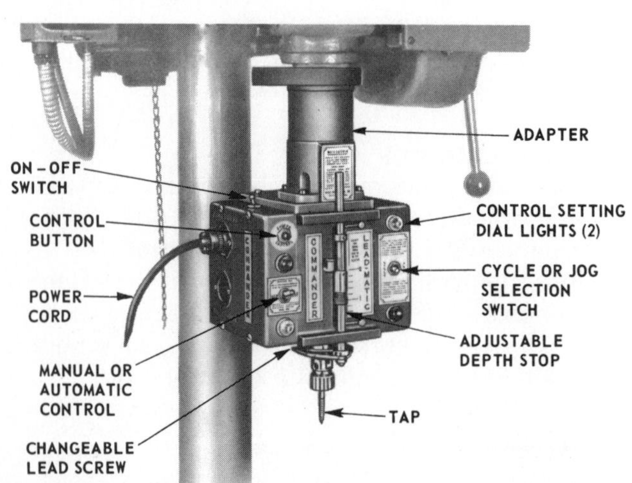

Fig. 7-79. Note tapping attachment on this drill press. Cycle or jog tapping action is selected by moving the selector switch. CYCLE: A touch on the control button sends the tap through the entire tapping reversing cycle. JOG: A touch on the control button will start the tap; releasing the button reverses the tap. This provides full control of setup and a thread-at-a-time tapping when working through hard or stringy materials. (Commander Mfg. Co.)

rigidity or means to make fine depth adjustments to do them as well as they can be done on more specialized equipment.

INDUSTRIAL APPLICATIONS

Specialized drilling machines enable industry to drill holes as small as 1/10,000 (0.0001) in. or 0.0025 mm in diameter, to as large as 3 1/2 in. or 88.9 mm in diameter.

The machine that drills these small holes is called an ultrasensitive microscopic precision drilling machine, Figs. 7-80 and 7-81. Theoretically, it could drill 25 holes in the diameter of one "human hair." However, ***laser drilling machines*** and ***electron beam drilling machines*** are replacing this precision technique of drilling (cutting) these small diameter holes.

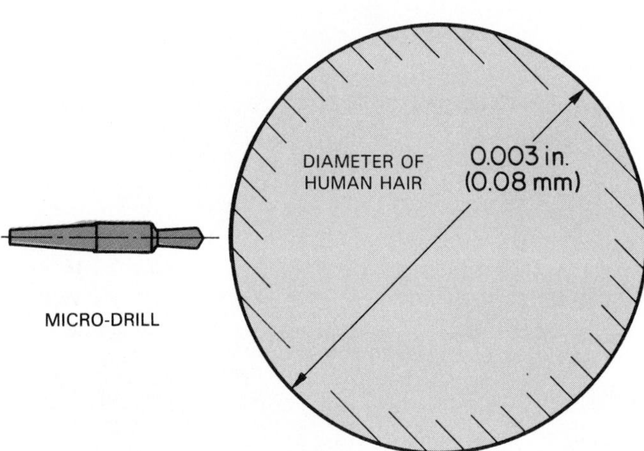

Fig. 7-80. Compare size of micro-drill with human hair.

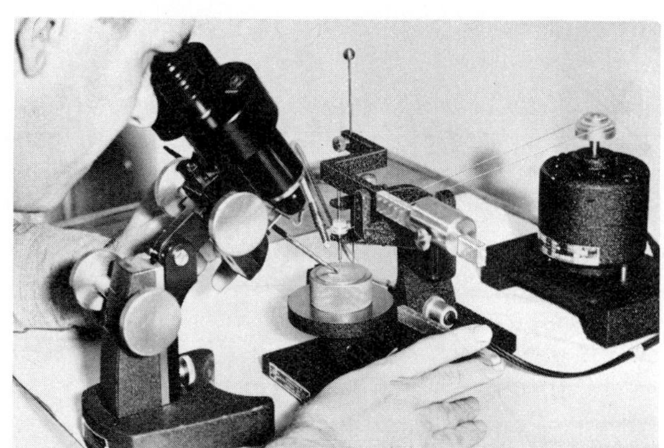

Fig. 7-81. A skilled operator using a micro-drilling machine. The drilling operation is so small that a microscope must be used to see what is being done. Laser and electron beam techniques are replacing this technique. (National Jet Co.)

The ***radial drill press*** is at the other end of the drilling machine scale and can handle very large work. The drill head is mounted in such a manner that it can be moved back and forth on an arm that extends from the massive machine column. The arm can be moved up and down and pivoted on the column, Fig. 7-82.

Often, a large pit is located along one side of the machine to permit the positioning of large, odd-shaped work. The pit is covered when NOT in use. It is NOT uncommon to drill holes 3 1/2 in. or approximately 90 mm in diameter with this machine.

A

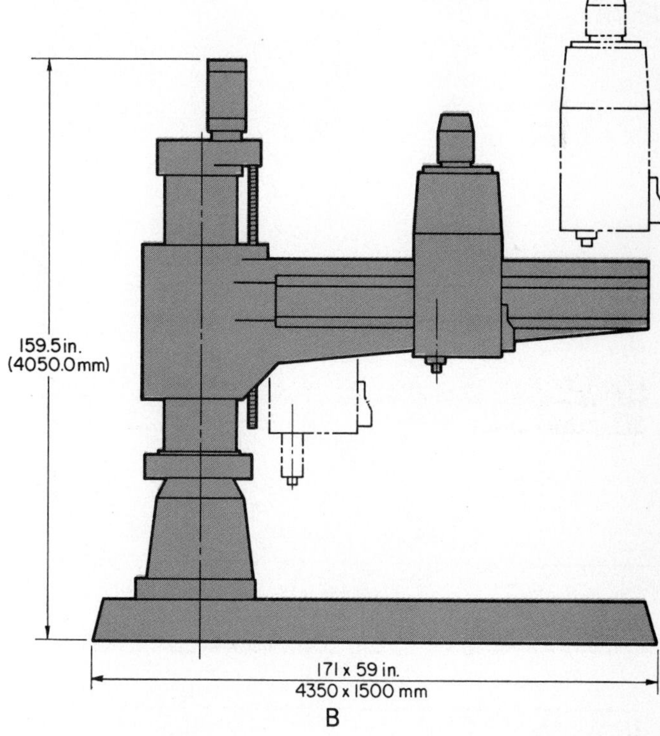

B

Fig. 7-82. A—Radial drill press is drilling large hole in part. B—Construction of radial drill press. (National Machine Tool Builders Assoc.)

A smaller ***bench type radial drill press*** is also available. One is illustrated in Fig. 7-83.

In between these two extremes are many types of drilling machines. The ***electric hand drill,*** the ***high speed sensitive drill press, tape controlled turret head*** and ***deep hole drilling machines,*** the ***gang*** and ***multiple spindle*** and familiar ***floor*** and ***bench drilling machines*** are in Fig. 7-84 through Fig. 7-90.

Some drilling machines are designed for special applications. What is perhaps the largest drilling machine ever constructed is shown in Fig. 7-91. This 2 1/2 story tall machine precisely drills and mills 39 mounting pads that locate and support the exterior body panels of the Pontiac automobile. The drilling/milling is done simultaneously with a tolerance of 0.039 to 0.059 in. (1.0 to 1.5 mm). Sensors tell the operator when a tool breaks or dulls, or when they begin to dull. The machine is completely automated.

Robotic drilling machines are the latest in automatic drilling technology, Fig. 7-92. To put the machine to work, the operator notifies a minicomputer, which has available parts programs on a disk or tape. Next the operator loads the part onto the work station fitted with the required fixtures. Some machines are loaded by robots which also unload the part after it has been drilled.

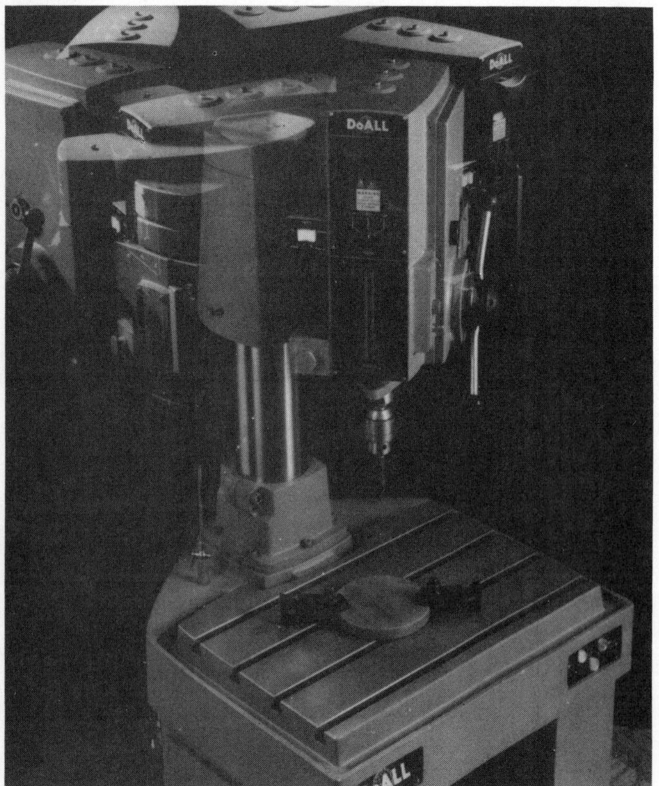

Fig. 7-83. Bench type radial drilling machine will handle large jobs. (DoALL)

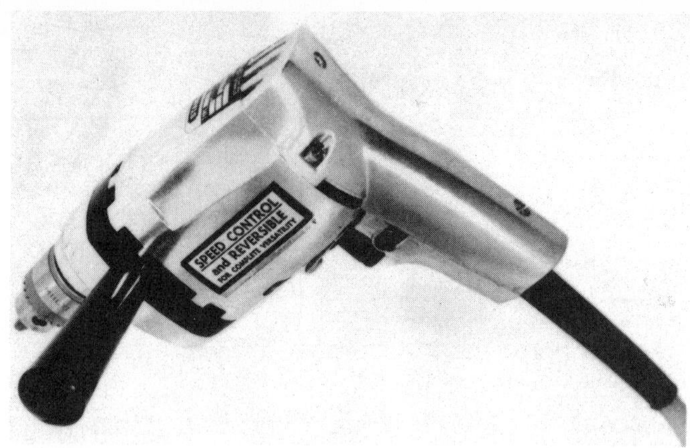

Fig. 7-84. Portable electric drill is manufactured in a large range of sizes and types. Some are battery powered. (Stanley Tool Co.)

Fig. 7-85. Note high speed sensitive drill presses. (Dumore Co.)

When the technician notifies the master computer that the part is ready for drilling, the robot, equipped with the correct tool, moves along the track to a starting position at the work station. The robot moves through the prescribed drilling, countersinking, spotfacing and/or counterboring operations, changing tools as needed. Upon completing the job, the robot moves to the next position in the work cell to begin another production sequence.

DRILL PRESS SAFETY

1. Remove neckties and tuck in loose clothing so there is no chance of them becoming entangled in the rotating drill!

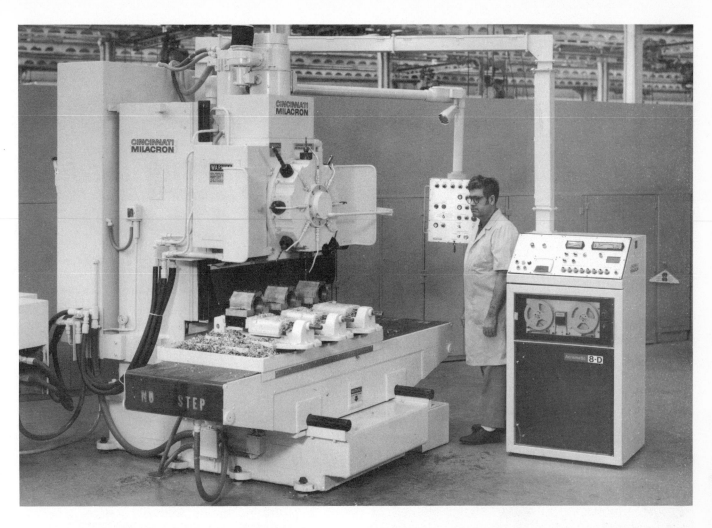

Fig. 7-86. This computer-controlled machining center is set up for drilling operations. No drill jigs are necessary as parts are positioned electronically. (Cincinnati Milacron)

Fig. 7-87. Gang of four drilling machines work together. Each machine is fitted with a different cutting tool. Work held in a drill jig moves from position to position as each operation is performed. (Clausing)

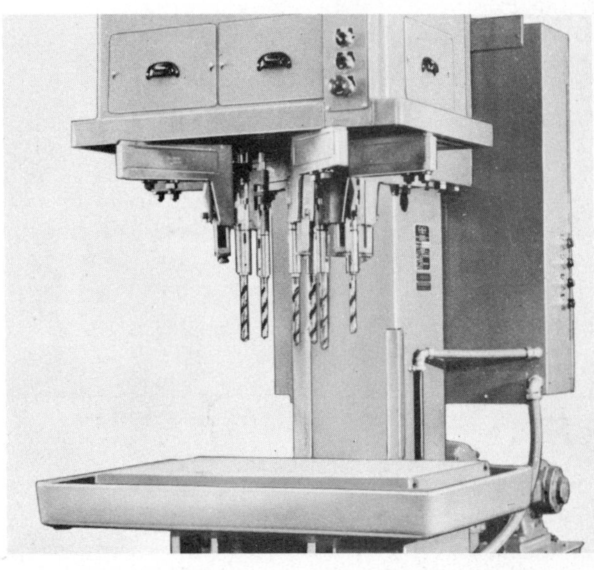

Fig. 7-88. Heavy-duty multiple spindle drill press has drilling heads that can be positioned to meet various drilling requirements. (Deka Drill Corp.)

Fig. 7-89. Specially designed portable drill is set up here to drill a tire mold. (Firestone Tire & Rubber Co.)

2. Check out the machine! Are all guards in place? Do switches work? Does the machine operate properly? Are the tools sharpened for the material being worked?
3. Clamp the work solidly. Do NOT hold work with your hand. A "merry-go-round" can inflict serious and painful injuries.
4. Wear goggles!
5. Place a piece of wood under drills being removed from the machine. Small drills can be damaged when dropped and the larger tools can injure you if dropped.
6. Use sharp tools.
7. Clean chips from the work with a brush, NOT your hands!
8. Treat cuts and scratches immediately!
9. Always remove the key from the chuck BEFORE turning on the power.
10. Let the drill spindle stop on its own after turning off the power. Do NOT attempt to stop it with your hand!
11. Keep the work area clear of chips. Place them in an appropriate container and NOT brush them onto the floor.
12. Wipe up all cutting fluid that spills on the floor right away.
13. Avoid trying to clean the tapered opening in the spindle while it is rotating.
14. After using a drill, wipe it clean of chips and cutting fluid. Replace the tool to proper storage.

Fig. 7-90. This deep hole drilling machine is tape controlled. (Pontiac Div., GMC)

15. Place all oily and dirty waste in a closed container when the job is finished!

TEST YOUR KNOWLEDGE—Chapter 7

Please do not write in the book. Place your answers on a separate piece of paper.

1. A twist drill works by:
   a. Being forced into material.
   b. Rotating against material and being pulled through by spiral of flutes.
   c. Rotating against material with sufficient pressure to cause it to penetrate material.
   d. All of the above.
   e. None of the above.
2. How is drill press size determined?
3. Drills are made from:
   a. High speed steel.
   b. Carbon steel.
   c. Both of the above.
   d. None of the above.
4. Drill sizes are expressed by what four series.
5. What are two techniques used to determine a drill's size?
6. List the two types of drill shanks.
7. _______ shank drills are used with a chuck.
8. _______ shank drills fit directly into the drill press spindle.
9. The spiral grooves that run the length of the drill body are called _______.
10. The spiral grooves in a drill body are used to:
    a. Help form the cutting edge of the drill point.
    b. Curl chips for easier removal.
    c. Form channels through which the chips can escape from the hole.
    d. All of the above.
    e. None of the above.

A

B

Fig. 7-91. A—One of largest drilling/milling machines made stands 2 1/2 stories high. It simultaneously drills and mills 30 pads on space frame chassis that supports and positions external plastic body panels on car. It performs all 30 drilling/milling operations in less than a minute. B—Another view of the huge drilling/milling machine.

Fig. 7-92. Robot is automated; routing and drilling cell moves between work stations on rails. Use of track for traversing motion simplified programming required to keep tools normal to the work. (Grumman Aerospace Corp.)

11. Name the device employed to enlarge a taper shank drill so it will fit the spindle opening.
12. The device used to permit a drill with a taper shank too large to fit the spindle opening is called a ________.
13. What is the name of the tool used to separate a taper shank drill from the above devices?
14. Cutting fluids are used to:
    a. Cool the drill.
    b. Improve the finish of a drilled hole.
    c. Aid in the removal of chips.
    d. All of the above.
    e. None of the above.
15. List the three factors that must be considered when repointing a drill.
16. What occurs when the cutting lips of a drill are NOT sharpened to the same lengths?
17. The ________ ________ ________ should be used frequently when sharpening a drill to assure a correctly sharpened drill.
18. The included angle of a drill point sharpened for general drilling is ______ degrees.
19. What coolant should be used when drilling cast iron?
20. Large drills require a considerable amount of power and pressure to get started. They also have a tendency to drift off center. These conditions can be minimized by first drilling a ________ or ________ hole. This hole should be as large as, or slightly larger than, the width of the ________ ________ of the drill point.
21. What is a blind hole?
22. How is the depth of a drilled hole measured?
23. The ________ ________ is almost identical to the hand reamer except that the shank has been designed for machine use.
24. The ________ ________ is ideal for finishing holes that must be a few thousandths of an inch over standard size.
25. How should a reamer be removed from a finished hole?
26. The cutting speed for a high speed reamer is approximately ________ that for a similar sized drill.
27. What is the name of the operation employed to cut a chamfer in a hole to receive a flat head screw?
28. The operation used to prepare a hole for a fillister or socket screw is called ________.
29. ________ is the operation that machines a circular spot on a rough surface for the head of a bolt or nut.

RESEARCH AND DEVELOPMENT

1. Drills are expensive. Keep a record of drills broken in the shop during a semester and what caused the breakage. Make recommendations that will reduce drill breakage and damage.
2. Make a series of safety posters on the use of a drill press.
3. Prepare a research paper on early drilling devices. Include sketches. You may want to reproduce this report on a copier, or make a series of transparencies of the early tools for the overhead projector.
4. Make a large teaching aid showing the various parts of the drill.
5. Develop a research project to investigate the effects of various types of cutting fluids and compounds on drilling, quality of finished hole, etc. Also include one hole drilled in the same material without coolant/cutting fluid.
6. Prepare a teaching aid that will show examples of a drilled hole, reamed hole, countersinking, spot facing, and counterboring.
7. Borrow drill jigs from a local industry. Describe to the class how they are used.
8. Demonstrate one of the following lessons:
    a. Centering round stock in a V-block.
    b. The proper way to use a wiggler.
    c. Sharpening a twist drill.
    d. Examples of several ways to clamp work safely on a drill press table.

Chapter 8

OFFHAND GRINDING

After studying this chapter, you will be able to:
- Identify the various types of offhand grinders.
- Dress and true a grinding wheel.
- Prepare a grinder for safe operation.
- Use an offhand grinder safely.
- List safety rules for offhand grinding.

Grinding is an operation that removes material by rotating an abrasive wheel or belt against the work, Fig. 8-1. It is usually employed for:
1. Sharpening tools.
2. Removing material too hard to be machined by other techniques.
3. Cleaning parting lines from castings and forgings.
4. Finishing and polishing molds used in die casting of metals and injection molding of plastics.

DANGER! Never attempt to operate offhand grinding machines while your senses are impaired by medication or other substances. They can remove flesh and bone in seconds.

ABRASIVE BELT GRINDERS

Abrasive belt grinding machines are heavy-duty versions of the belt and disc sanders found in woodworking, Fig. 8-2. A wide variety of abrasive belts permit these machine tools to be used for grinding to a line, deburring, contouring, sharpening tools, and finishing cast and forged parts, Fig. 8-3.

BENCH AND PEDESTAL GRINDER

The bench grinder and pedestal grinder are the simplest and most widely utilized grinding machines. The grinding done on them and abrasive belt grinders is called ***offhand grinding;*** work that does NOT require great accuracy is held in your HANDS and manipulated until ground to the desired shape.

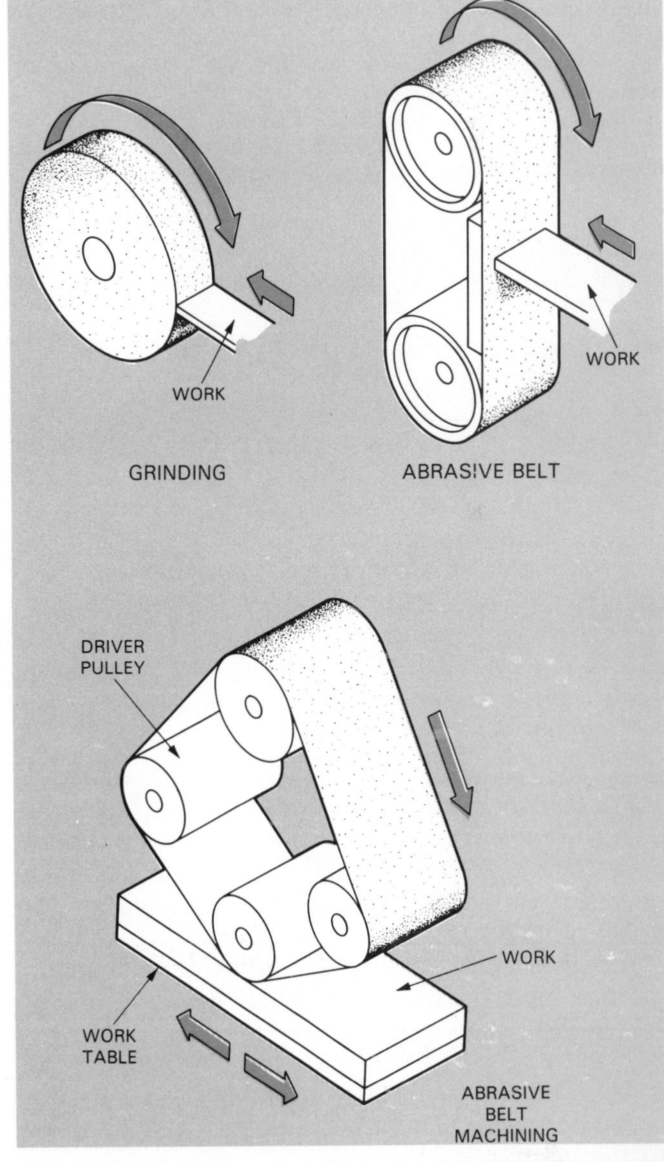

Fig. 8-1. Note principles of how typical grinding machines work.

A—Wet type abrasive belt platen grinder sprays water on abrasive belt to cool the work. It can be quickly changed from vertical to horizontal. (Hammond Machinery Inc.)

B—Flexible belt abrasive grinder adapts to three-dimensional contours. (Rockwell International, Power Tool Div.)

C—Abrasive belt grinder for contact wheel and platen grinding. (Hammond Machinery, Inc.)

Fig. 8-2. These are abrasive belt grinding machines.

The ***bench grinder*** is a grinder that is fitted to a bench or table, Fig. 8-4. The grinding wheels mount directly onto the motor shaft. One wheel is usually ***coarse*** for roughing, the other is usually ***fine*** for finish grinding.

A ***pedestal grinder*** is usually larger than the bench grinder, and is equipped with a ***pedestal*** (base) fastened to the floor.

The ***dry type pedestal grinder*** has NO provisions for cooling the work during grinding other than a water container. The tool or part can be dipped into the water. See Fig. 8-5.

With the ***wet type pedestal grinder,*** a coolant system is built into the grinder and keeps the wheels constantly flooded with fluid. The coolant washes away particles of loose abrasive and metal and keeps the work cool. Cooling prevents localized heat build-up which can ruin tools and ''burn'' areas of other types of work.

Bench and pedestal grinders can be dangerous if the operator is not careful.

A grinder must NEVER be used unless fitted with ***eye shields*** and the operator is wearing goggles.

The ***tool rest*** is provided to support the work while being ground. It is recommended that the rests be adjusted to within 1/16 in. (1.5 mm) of the wheels,

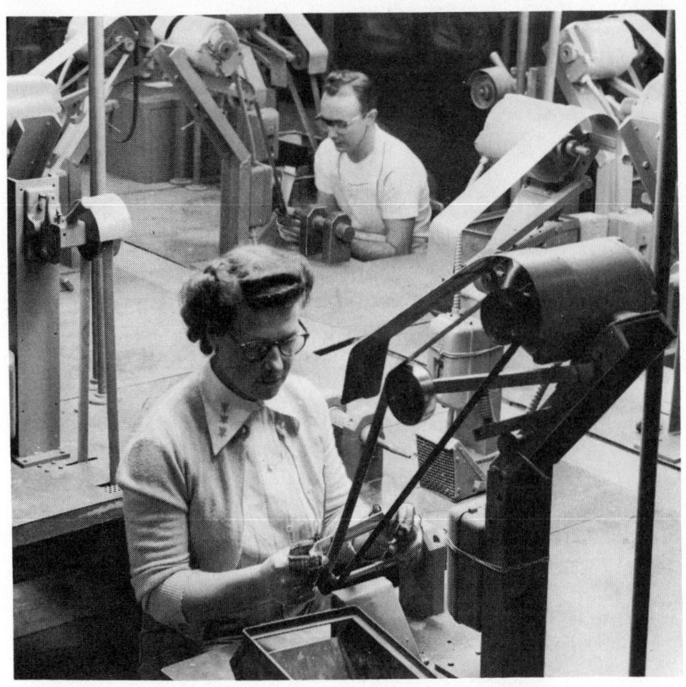

Fig. 8-3. Jet engine turbine blades are being finished by abrasive belt grinding. (Westinghouse Electric Corp.)

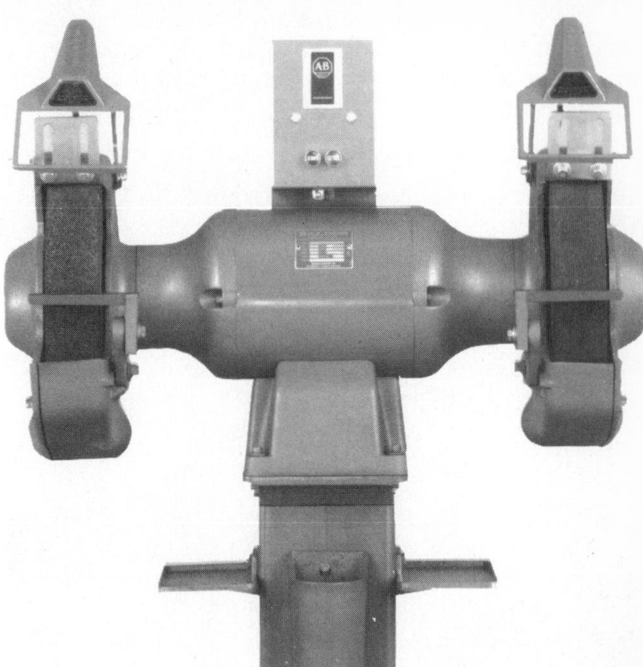

Fig. 8-5. This pedestal grinder has roughing wheel on left and finishing wheel on right. (Baldor)

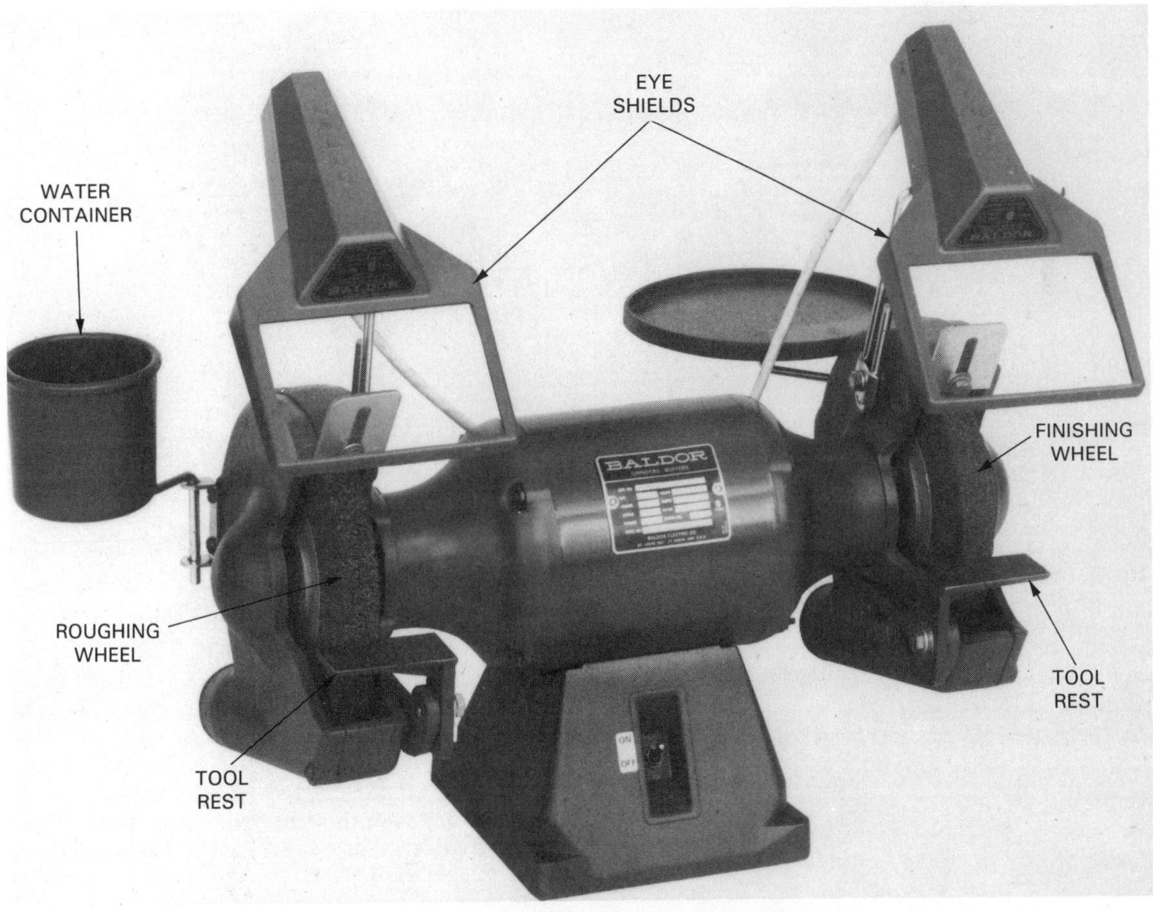

Fig. 8-4. Bench grinder can be used for many tasks. Never operate a bench or pedestal grinder unless all safety devices are in place and in sound condition. Also wear eye protection! (Baldor)

Fig. 8-7. This will prevent work from being wedged between the rest and the wheel. Turn the wheel by hand after adjusting the rest, to be sure there is sufficient clearance.

WARNING! Do NOT make adjustments while the grinding wheels are revolving and make sure the tool rest is adjusted properly.

Fig. 8-6. This pedestal grinder has built-in coolant system. (Hammond Machinery Inc.)

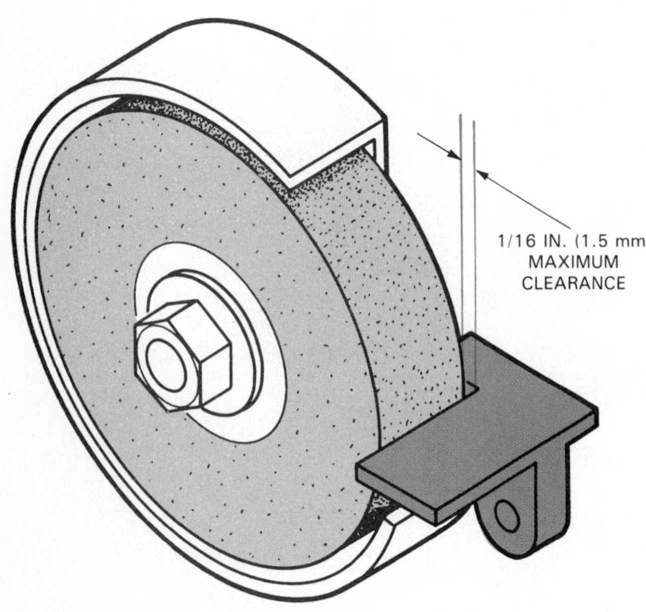

Fig. 8-7. Note properly spaced tool rest.

GRINDING WHEELS

Grinding wheels can be another source of danger and should be examined frequently for ***concentricity*** (run true), roundness, cracks, etc. A new wheel can be tested by suspending it on a string or wire, or holding it lightly with one finger. Then tap the side of the wheel lightly with a metal rod or screwdriver handle. A solid wheel will give off a clear ''RINGING SOUND.'' A wheel that does NOT give off a clear sound should be assumed to have a fault and be destroyed. Because it is NOT possible to use this technique to check the wheels each time the grinder is used, never stand in front of a spinning wheel.

Wheel dresser

The grinder wheels must run true and be balanced on the shaft. A ***wheel dresser*** will true the wheel and remove any glaze that may have formed during grinding operations, Figs. 8-8 through 8-10. The

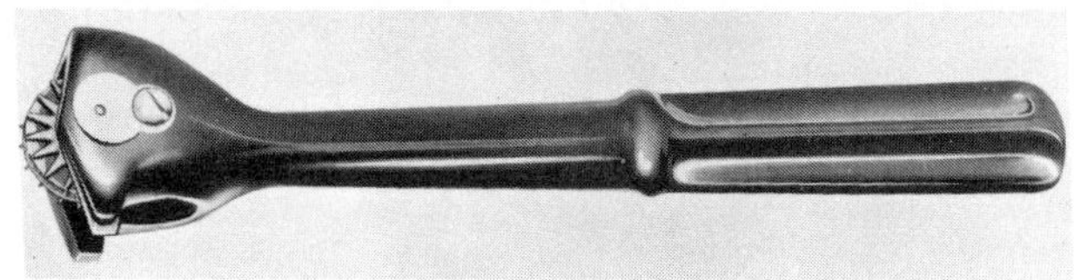

Fig. 8-8. A mechanical wheel dresser will true stone face.

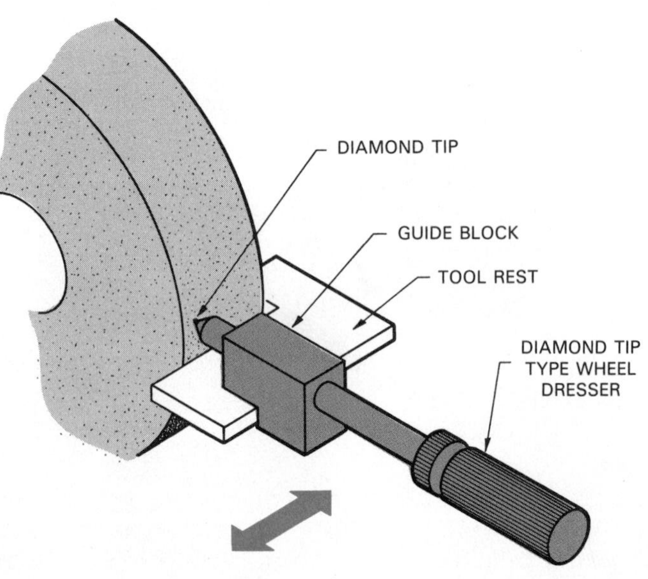

Fig. 8-9. Industrial diamonds are also used to dress and true grinding wheels. The wheel dresser illustrated trues wheel by being guided along the edge of the tool rest. A screw arrangement permits diamond to be adjusted into wheel. This type wheel dresser can be employed on grinders with a slotted tool rest by using a guide block that fits into slot or by clamping tool to protractor guide.

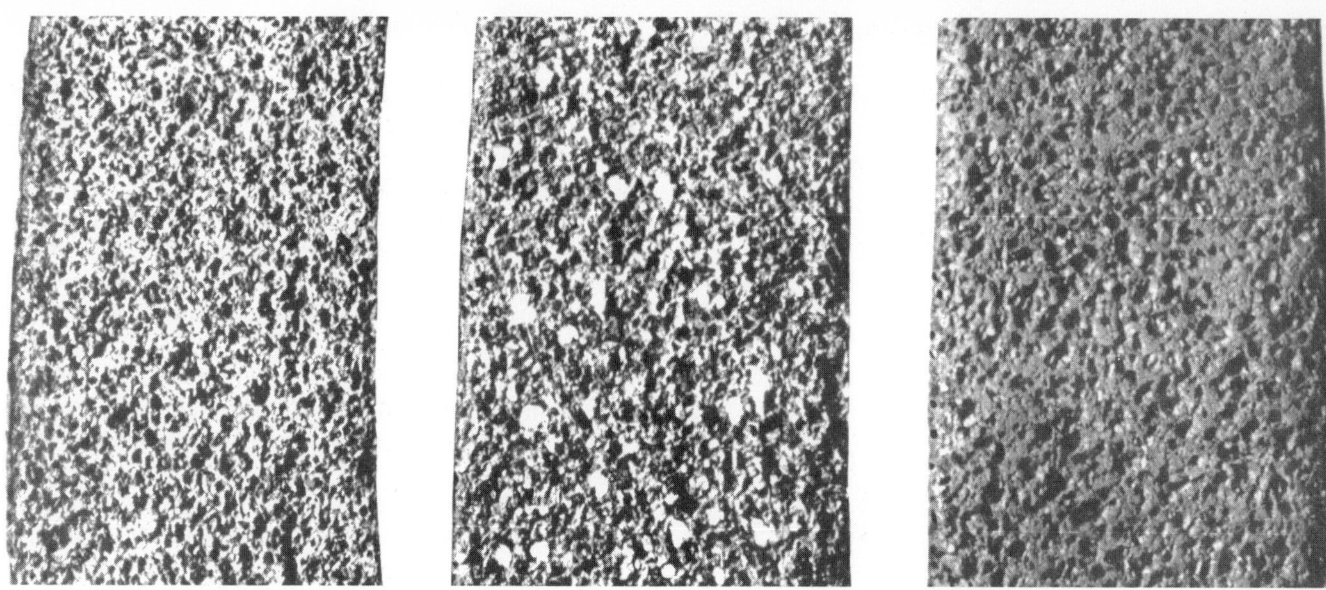

Fig. 8-10. Study grinding wheels in various conditions: A—Properly dressed. B—Loaded. C—Glazed. (Norton Co.)

wheel dresser is supported on the tool rest and is held firmly against the wheel with both hands, Fig. 8-11. It is moved back and forth across the wheel face to remove a thin layer of stone.

Grinding rules

To obtain maximum efficiency from a grinder, the following recommendations should be observed.

1. Use the face of a wheel, NOT the sides.
2. Move the work back and forth across the wheel face. Even wear will result and it will prevent grooves from forming on the face of the wheel.
3. Keep the wheel dressed and the tool rests properly adjusted.
4. Soft metals, like aluminum, brass and copper, tend to ***load*** (clog) grinding wheels. When possible, these metals should be ground on an abrasive belt grinder. If such a machine tool is NOT available, and the grinder is to be used primarily for tool grinding, it is suggested that another grinder be secured for this type of work.

USING DRY TYPE GRINDER

After examining the grinder and making the necessary adjustments, turn on the machine. It is assumed that you wear safety glasses whever you are in the shop. Remember to stand to one side until the grinder has reached operating speed.

Place the work on the tool rest and slowly push it against the grinding wheel. It too much pressure is applied, the work will begin to ''burn'' or discolor. Overheating can be minimized by dipping the work into the water container from time to time. Care must be taken when grinding edge tools because excessive heat will ''draw'' (remove) the ***temper*** (hardness) and ruin the tool.

Keep the work moving across the wheel face to prevent the formation of grooves or ridges. Dress and retrue the wheel as needed for maximum efficiency.

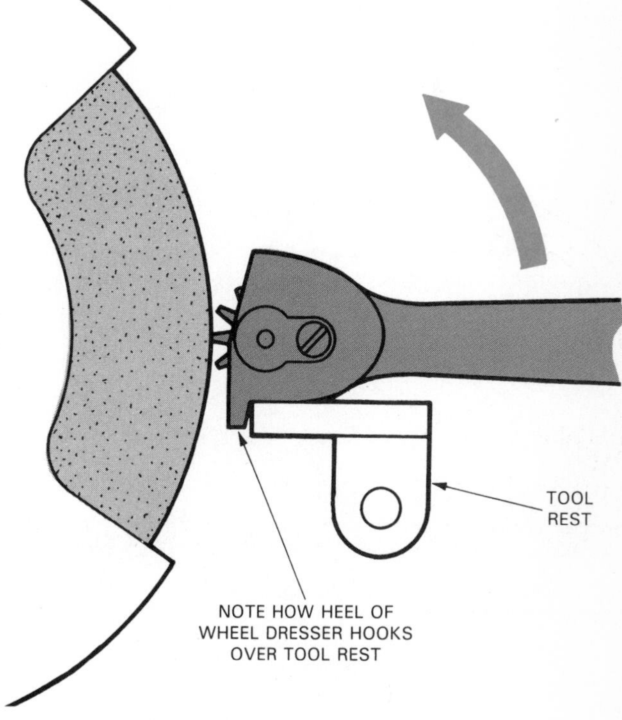
TOOL REST
NOTE HOW HEEL OF WHEEL DRESSER HOOKS OVER TOOL REST

Fig. 8-11. Note how to use a mechanical type wheel dresser. Move it back and forth over face of stone.

DANGER! Sections of cloth should never be used to hold work while it is being ground. Serious injuries will result if the cloth is pulled into the wheel. Hold the work, especially small lathe cutter bits, in hand vises designed for that purpose, Fig. 8-12.

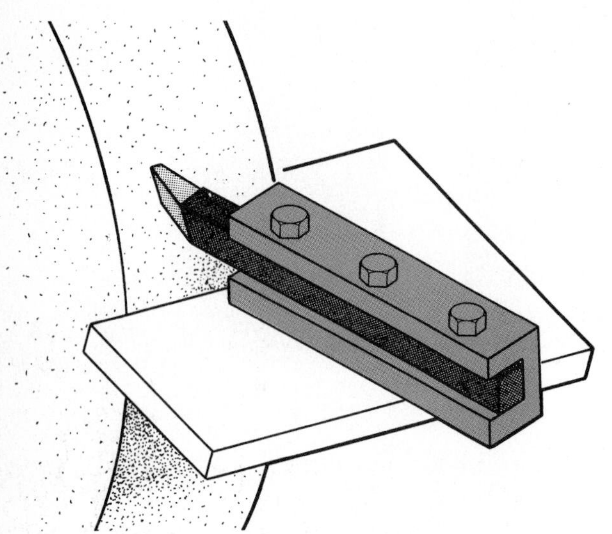

Fig. 8-12. This is one type of hand vise for holding cutter bits while they are sharpened.

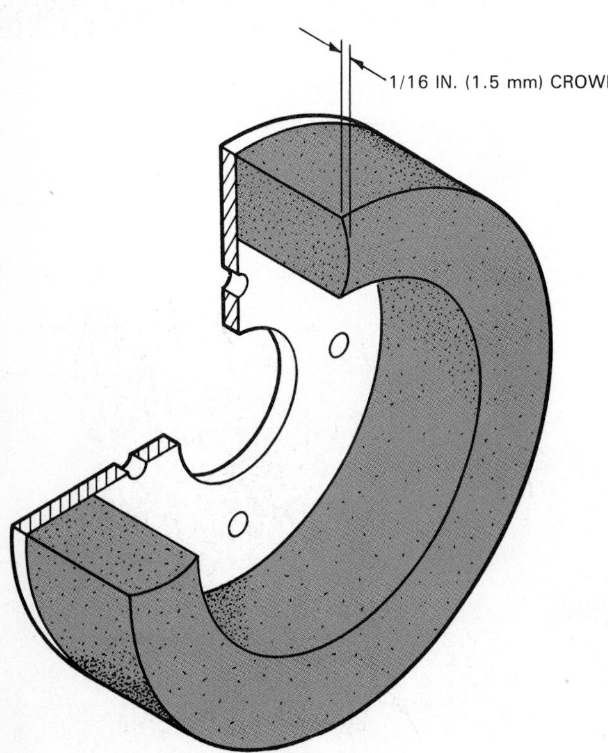

Fig. 8-13. The slight crown, on the wheel face, minimizes amount of contact between work and wheel. It reduces possibility of heat destroying carbide tip of a tool.

USING WET TYPE GRINDER

The wet type grinder is primarily used to grind carbide tipped tools. Since carbide type tools are often brazed onto a steel shank, both steel and carbide must be ground away when these tools are sharpened. It is recommended that aluminum oxide wheels be employed for grinding the STEEL SHANK; and that silicon carbide or diamond impregnated wheels be used for grinding the CARBIDE TIP.

A flat face is required on the wheel face on the dry type grinder but a slightly crowned wheel face is recommended when grinding carbide tipped tools, Fig. 8-13. The crown minimizes the contact between the wheel and the work. This reduces the possibility of the tip being damaged or destroyed by excessive heat.

The coolant attachment must be adjusted to keep a full flow of liquid directed on the tool at all times.

Adjust the tool table rest to obtain the correct clearance angle, Fig. 8-14. A protractor guide is helpful when compound clearance angles are required, Fig. 8-15.

Use the entire face of the wheel. Keep the tool in continuous motion to minimize wheel wear. Dress and retrue the wheel as needed.

FLEXIBLE SHAFT HAND GRINDERS

Flexible shaft hand grinders perform many grinding jobs from light deburring to polishing operations,

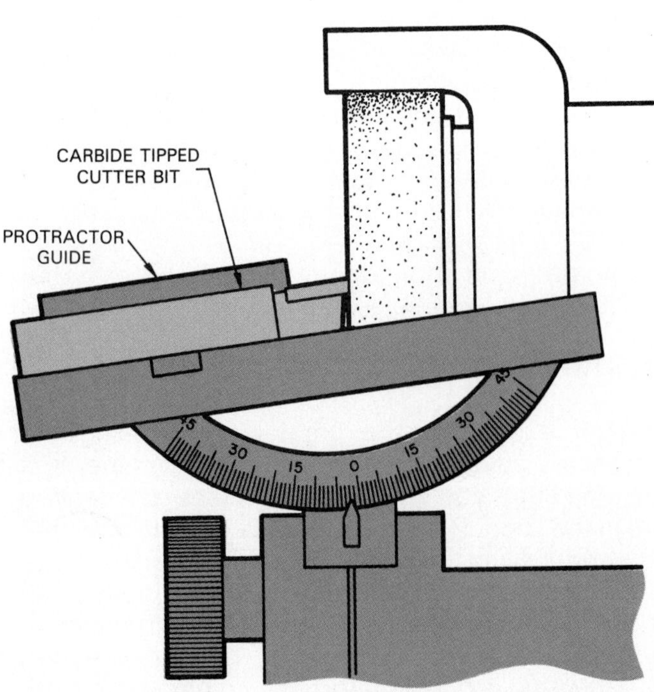

Fig. 8-14. Tool rest can be adjusted to any desired clearance angle with this setup.

Fig. 8-15. Using the protractor guide when grinding a compound clearance angle on a carbide tool.

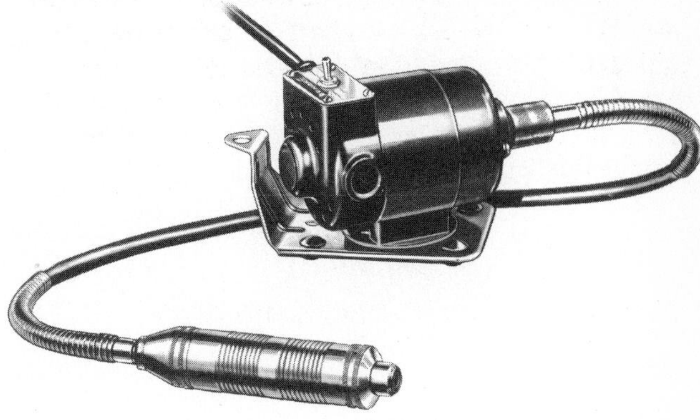

Fig. 8-16. A flexible shaft hand grinding unit is helpful for working in recessed areas on parts. (Dumore Co.)

Fig. 8-16. They are used extensively on jobs such as finishing dies, Fig. 8-17. Flexible shaft hand grinders can be electrically powered or air powered. Refer to, Fig. 8-18.

ABRASIVE BELT AND GRINDER SAFETY

1. Always wear goggles or a face shield when performing grinding operations even though the machines are fitted with eye shields.
2. Always check a grinding wheel for soundness before putting it on the grinder. Destroy wheels that are NOT sound or that have a badly worn center hole.
3. Allow the wheels or belt to stop completely before attempting to make machine adjustments.
4. Because it is NOT possible to check grinding wheels each time the machine is used, it is considered good practice to stand to one side of the machine during operation. This will keep you clear of flying fragments if the wheel shatters!
5. Be sure all wheel guards and safety devices are in place before attempting to use a grinder or abrasive belt machine.
6. If the grinding operation is to be performed dry, do NOT forget to hook up any exhaust attachments before starting.
7. Check the machine thoroughly before using it. Lubricate the machine only as recommended by the manufacturer.
8. Keep your hands clear of the rotating wheels. It is a cutting tool and can cause serious injuries!
9. Injuries caused by rotating grinding wheels are especially prone to infection. Have them, no matter how small, treated immediately.

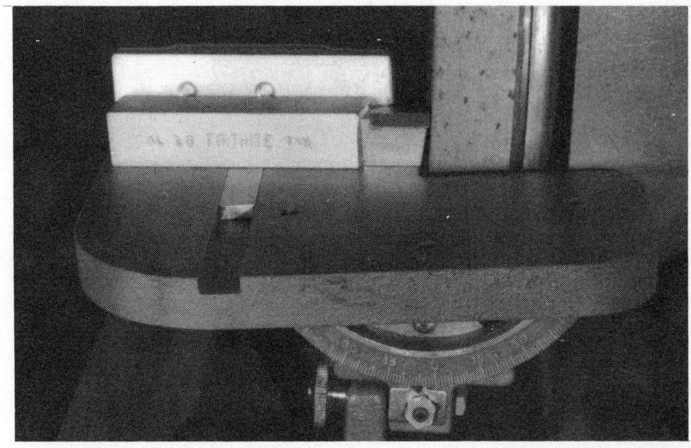

Fig. 8-17. Machinist is finishing a die with a flexible shaft hand grinder. (Dumore Co.)

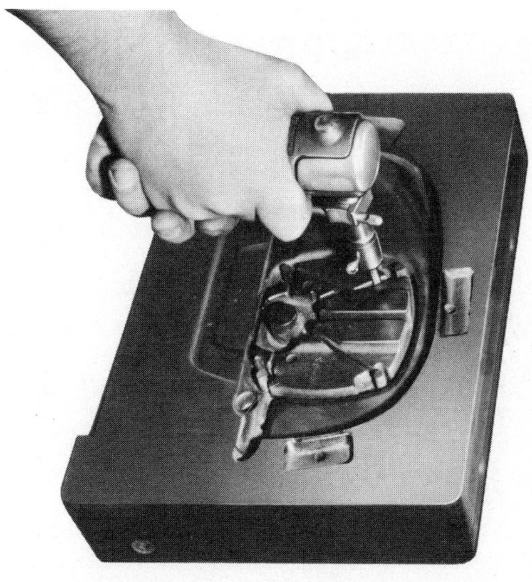

Fig. 8-18. A reciprocating type hand grinder is being used to polish a die casting mold. The abrasive stone can be replaced with various shaped files and hones. It is air powered. (Diamond Tool Div., Engis Corp.)

10. Under no condition should work be held with a cloth to protect your hands from the heat generated by the grinding operation. Hold the work in a clamp or hand vise!
11. Make sure the tool rest is properly adjusted.
12. Avoid work pressure on the side of the grinding wheel and do NOT use a wheel that is glazed or loaded with metal.
13. NEVER operate a grinding wheel at speeds higher than that recommended by the manufacturer! See Fig. 8-19.

Fig. 8-19. NEVER operate a grinding wheel at speeds greater than those listed on wheel.

TEST YOUR KNOWLEDGE—Chapter 8

Please do not write in the text. Place your answers on a separate sheet of paper.

1. Describe the grinding operation.
2. How do abrasive belt grinders differ from abrasive wheel grinders?
3. Bench and pedestal grinders are used to do _______ grinding.
4. The grinding technique referred to in the preceding statement is so named because:
   a. It can only do external work.
   b. Work is too hard to be machined by other methods.
   c. Work is manipulated with fingers until desired shape is obtained.
   d. All of the above.
   e. None of the above.
5. Name two types of pedestal grinders. How do they differ?
6. The tool rest should be about _____ in. or _____ mm away from the grinding wheel or belt for safety. This prevents the possibility of work being _______ between the tool _______ and _______.
7. How can grinding wheel soundness be checked?
8. As the wheel cannot be checked each time the grinder is used, it is recommended that the operator:
   a. Not use the grinder.
   b. Check with the instructor whether the wheel is sound.
   c. Stand to one side of the grinder when using the machine.
   d. All of the above.
   e. None of the above.
9. Work will _______ if it is forced against the wheel with too much pressure.
10. Carbide tipped tools are usually sharpened on a _______ type grinder.
11. The face of wheel on a wet type grinder is _______ slightly. Why is this done?
12. Never mount a grinding wheel on a grinder without _______________.
13. List four safety precautions to be observed when operating an abrasive wheel and belt type grinder.

RESEARCH AND DEVELOPMENT

1. How are natural sandstone grinding wheels made? Secure samples of natural sandstone and compare them with manufactured abrasives. What, in your opinion, makes the manufactured abrasive superior to the natural product?
2. Research the term MOH SCALE. Prepare a chart showing the common abrasives in order of their hardness. Secure samples to mount on the chart.
3. Grinders used by 18th century workers required them to work in an unusual position. Prepare a brief presentation for the class on how they worked, some of the health problems they had because of their job, and some of the articles they produced. Make drawings to show during the presentation. They may be made as transparencies for the overhead projector.
4. Try to borrow an old fashioned natural sandstone grinding machine. Demonstrate its use to the class.
5. True and dress the grinding wheels used on the machines in the shop.
6. Secure a list of the various grades and types of grinding wheels available from a typical abrasive manufacturer.

Chapter 9

SAWING AND CUTOFF MACHINES

After studying this chapter, you will be able to:

- Identify the various types of sawing and cutoff machines.
- Select the correct machine for the job to be done.
- Mount a blade and prepare the machine for use.
- Position the work for cutting most efficiently.
- Safely operate sawing and cutoff machines.

The first step in machining most metalworking jobs is to cut the stock to the desired length. This can be done using power saws.

There are three principal types of ***metal cutting saws.*** One makes use of a reciprocating or back and forth cutting action, Fig. 9-1. The blade is similar to that found on a hand hacksaw, only larger and heavier, Figs. 9-2 and 9-3. Another type makes use of a continuous or band type blade, that moves in only one direction, Fig. 9-4 and Fig. 9-5. The third kind has a circular blade (either abrasive, toothed, or friction) that rotates into the work, Fig. 9-6.

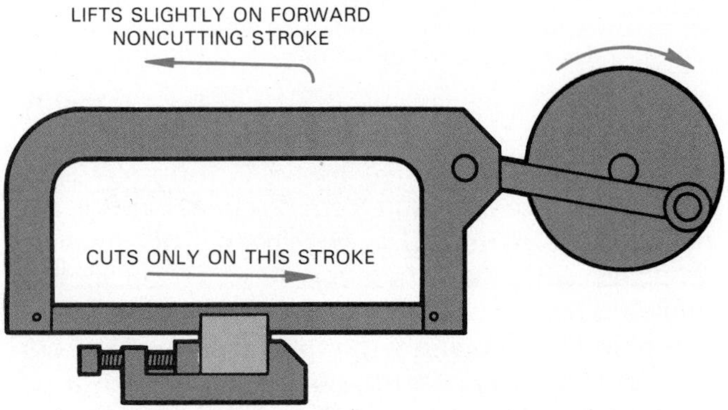

LIFTS SLIGHTLY ON FORWARD NONCUTTING STROKE

CUTS ONLY ON THIS STROKE

Fig. 9-1. Reciprocating power hacksaws cut on back stroke. A cam lifts blade from metal on noncutting forward stroke.

DANGER! Never attempt to operate a sawing or cutoff machine while your senses are impaired by medication or other substances.

RECIPROCATING POWER HACKSAW

A ***reciprocating hacksaw blade*** moves back and forth across the work, but cuts only on the back stroke. There are several types of feeds available.

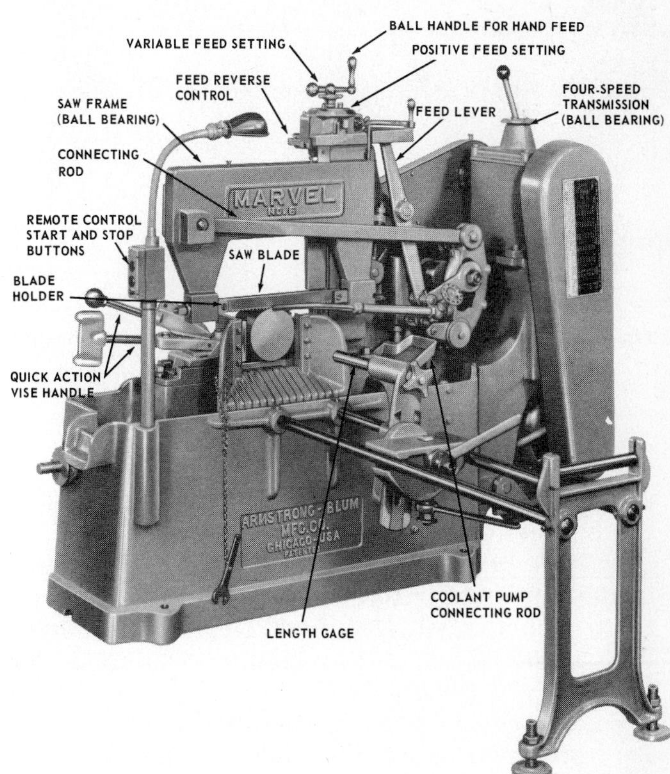

Fig. 9-2. Study parts of industrial type reciprocating power hacksaw. (Armstrong-Blum Mfg. Co.)

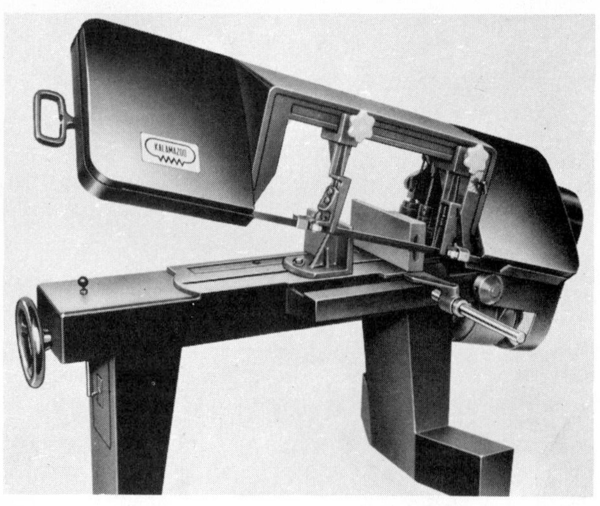

Fig. 9-3. This is a light duty, band type cutting saw. (Kalamazoo Tank & Silo Co.)

Fig. 9-4. Industrial type horizontal cutoff machine is another saw type. (DoALL Co.)

Positive feed produces an exact depth of cut on each stroke. The pressure on the blade varies with the number of teeth in contact with the work.

Definite pressure feed yields a pressure on the blade that is uniform regardless of the number of teeth in contact with the work. Depth of cut varies with the number of teeth in contact. This condition prevails with ***gravity feed.***

Feed can be adjusted to meet varying conditions. For best performance, the blade and feed must be selected to permit high-speed cutting and heavy-feed pressure with minimum blade bending and breakage.

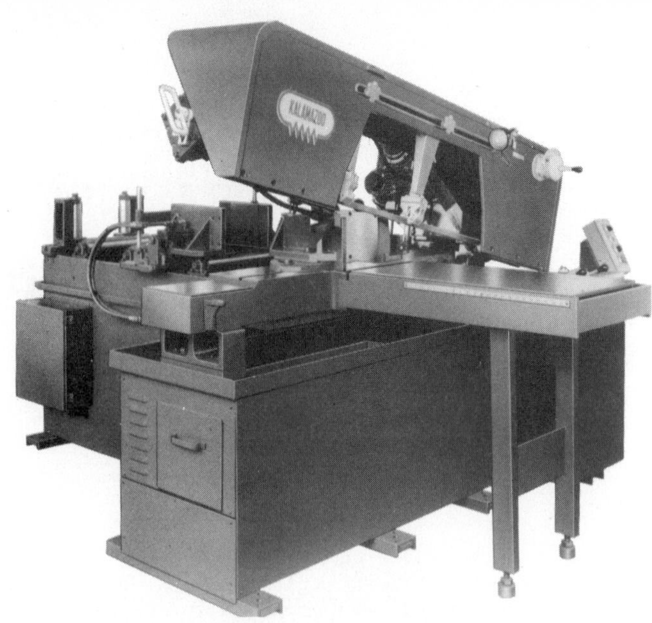

Fig. 9-5. Horizontal bank cutoff machine is fitted with automatic stock feed, blade tensioning, and variable speed drive to permit high speed production. (Kalamazoo Tank & Silo Co.)

Fig. 9-6. Abrasive cutoff machine is frequently used to cut space age metals that would be difficult to cut by conventional methods. Machine illustrated is removing spures and gates from jet engine castings. (Westinghouse)

Standard reciprocating metal cutting saws are available in sizes ranging from 6 x 6 in. (150 x 150 mm) to 24 x 24 in. (900 x 900 mm). The saws can be fitted with many accessories. These include quick acting vises, power stock movement, power clamping of work, and automatic cycling of the cutting operation. The latter moves the work out the required distance, clamps it, and makes the cut automatically. The cycle is repeated upon the completion of the cut, Fig. 9-7.

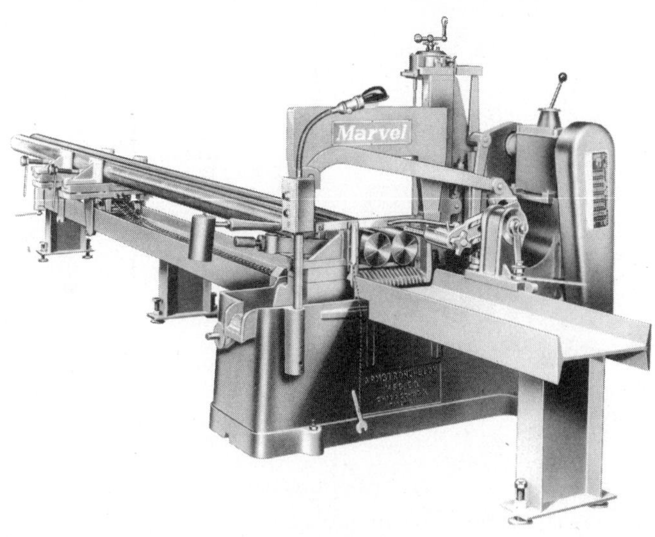

Fig. 9-7. With automatic cycling, work is placed on loading track and length gage is set. Machine will then make the cut, raise the blade, open the vise, feed the bars forward to gage setting, close the vise, and again begin to cut. (Armstrong-Blum Mfg. Co.)

High-speed cutting usually requires a coolant. Coolant is necessary on most materials EXCEPT cast iron and some brass alloys to reduce friction, blade wear, and chip clogged teeth.

Swivel vise permits angular cuts to be made quickly. Refer to Fig. 9-8.

Selecting a power hacksaw blade

Proper blade selection is important. The ***three-tooth rule*** still applies—at least three teeth in contact with the work. Large sections and soft materials require a COARSE TOOTH BLADE. Small or thin work and hard materials require a FINER TOOTH BLADE.

For best cutting action, apply HEAVY—FEED PRESSURE on hard materials and large work, and LIGHT-FEED PRESSURE on soft materials and work with small cross sections, Fig. 9-9.

Blades are made in two principal types: flexible back, and all hard. The choice depends upon use.

Flexible back blades should be utilized where safety requirements demand a shatterproof blade, or for cutting odd-shaped work when there is a possibility of it working loose in the vise.

For a majority of cutting jobs, the ***all-hard blade*** is first choice for straight accurate cutting under a variety of conditions.

WARNING! Care must be exercised when starting a cut with an all-hard blade to prevent it from dropping on the work when cutting starts. If this happens, the blade usually shatters and flying blade sections can cause serious injuries!

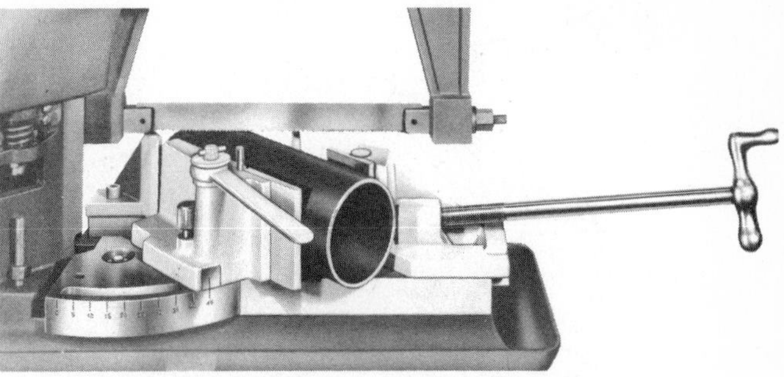

Fig. 9-8. Swivel vise permits angular cuts.

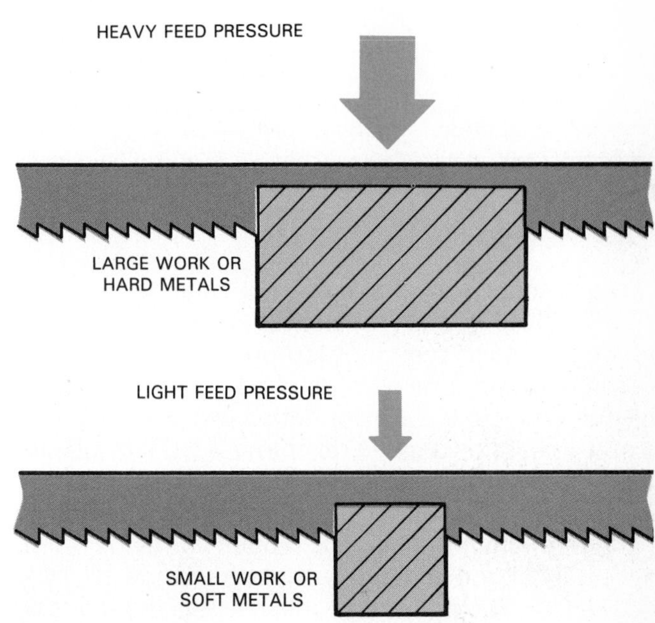

Fig. 9-9. Apply heavy feed pressure on large work with hard materials and light pressure on small work with soft metals.

Blades are made from tungsten and molybdenum steels, with tungsten carbide teeth on alloy steel backs. The following "rule-of-thumb" can be followed for selecting the correct blade:

1. Use a 4-tooth blade for cutting large sections or readily machined metals.
2. Use a 6-tooth blade for cutting harder alloys and miscellaneous cutting.
3. Use 10- and 14-tooth blades primarily on light-duty machines where work is limited to small sections and moderate and light-feed pressures.

Mounting a power hacksaw blade

The blade must be mounted to cut on the power or back stroke, Fig. 9-10, and to lie perfectly flat against the mounting plates. A blade must also be properly tensioned if long life and accurate cuts are to be achieved.

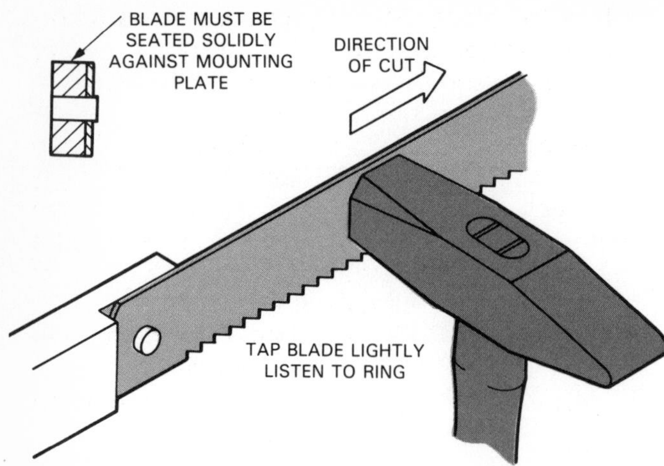

Fig. 9-10. When adjusting blade on reciprocating power hacksaw, make sure it is perfectly flat against mounting plate before tensioning. Tighten blade until a low toned musical ring is heard when blade is lightly tapped with a small hammer. Since blades have a tendency to stretch slightly after making a few cuts, it should be checked and retensioned if needed.

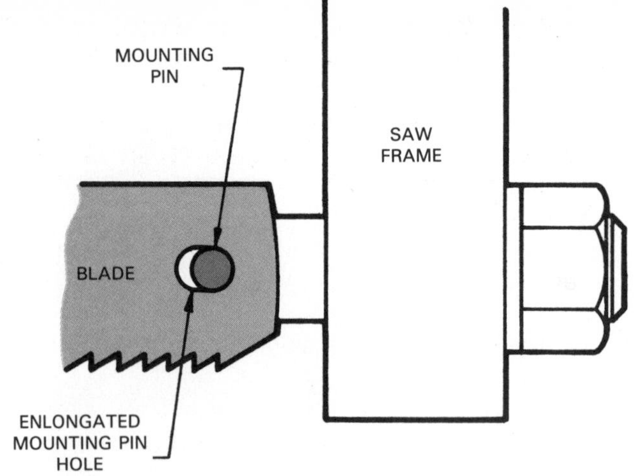

Fig. 9-11. Pin hole on a properly tensioned saw blade will be slightly enlongated or egg shaped instead of round.

Many techniques have been developed for properly mounting and tensioning blades. For best results, consult the manufacturer's literature. The recommended pressure can be secured with a torque wrench. If the information (proper torque for a given blade on a given machine) is NOT available, the following methods will work:

1. Tighten the blade until a low musical ring is heard when the blade is tapped lightly. A "high pitch" tone indicates that the blade is TOO TIGHT. A "dull thud" means that the blade is TOO LOOSE.
2. The shape of the blade pin hole can serve as an indicator as to whether a blade is tensioned properly. When properly tensioned, the pin holes will become slightly enlongated (sort of egg-shaped) instead of round, Fig. 9-11.

The blade will become more firmly seated after the first few cuts and it will stretch or elongate slightly. This will require ***blade retensioning*** (retightening) before further cutting.

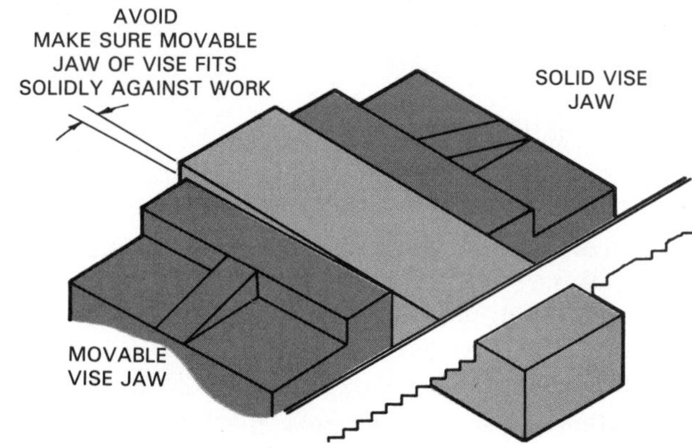

Fig. 9-12. Work must be clamped solidly; otherwise, work will twist and bind, ruining blade in first few seconds of use.

Cutting with power hacksaw

Measure off the distance to be cut. Allow ample material for facing, etc., if the work order does not provide length information. Mark the stock, and mount the work firmly on the machine, Fig. 9-12.

If several sections of the same length are to be cut, use a ***stop gage,*** Fig. 9-13. Apply ample supply of coolant if the machine is designed for it.

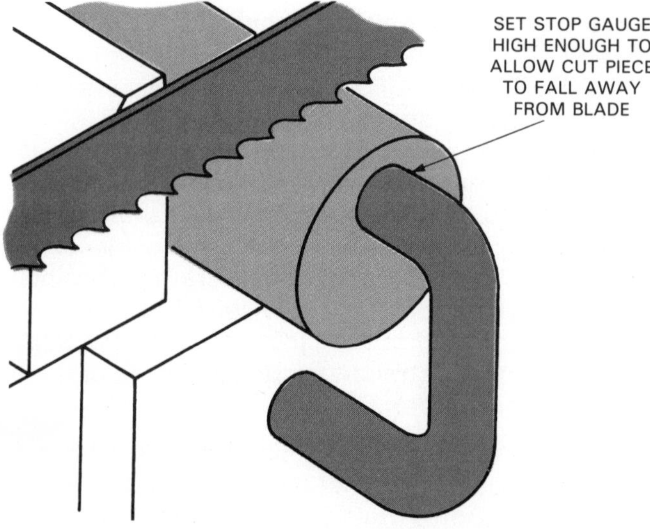

Fig. 9-13. Stop gage is used when several pieces of same length must be cut. Adjust it high to permit work to fall free when completely cut.

Power hacksawing

Most sawing problems can be overcome or prevented by careful planning and observing a few precautions or rules.

1. ***Broken blades*** usually happen when the blade is dropped on the work. A loose blade or excessive feed can also cause the blade to fracture. Loose work can be a cause of blade damage, so can making a cut on a corner or sharp edge where the three-tooth rule should NOT be observed.

   These problems can be avoided by exercising care when making the machine ready for cutting.
2. ***Crooked cuts*** are usually the result of a worn blade.

   Remember to reverse the work after replacing a blade; and start a new cut on the opposite side. See Fig. 9-14.

   A loose blade or the blade rubbing on the vise or clamping fixture will cause the same problem. It can also be caused by excessive blade pressure on the work or by a worn saw guide.
3. ***Blade pin holes breaking out*** can be caused by dirty mounting plates or too much tension on the blade. Worn mounting plates can cause a blade to twist and strain in such a way that the pin hole will break out instead of breaking the blade elsewhere in its length.
4. ***Premature blade teeth wear*** is evident when the teeth become rounded and dull quickly. INSUFFICIENT FEED PRESSURE, indicated by light powdery chips is one of the major causes of this condition. EXCESSIVE PRESSURE, indicated by burned chips, can cause the same problem.

   Insufficient pressure can be corrected by increasing cutting pressure until a FULL curled chip is produced. If too much pressure is the culprit, reduce feed pressure until a full curled chip is formed.

   Too little coolant or a poorly adjusted machine can cause rapid wear. Correct by following the manufacturer's handbook recommendations.
5. ***Teeth strip off*** results when the blade teeth snap off of the blade. Starting a cut on a sharp corner is a major cause of this problem. A proper machine setup that will present a flatter starting surface, will greatly reduce teeth stripping. Be sure the work is clamped securely as this can also cause the teeth to strip, Fig. 9-15.

   Check the manufacturer's blade chart to determine the proper blade for the job to be done. A blade with teeth TOO FINE will clog or ''load'' and jam, causing the teeth to shear off. A blade TOO COARSE (less than three teeth cutting) will cause the same problems.

   Make sure the blade is properly mounted and cutting on the power stroke.

HORIZONTAL BAND SAW

The horizontal hand saw is frequently referred to as a cutoff machine. It offers three advantages over the reciprocating hacksaw, Fig. 9-16.

1. A band saw will usually cut with more PRECISION. The blade can be guided more accurately than the blade on the reciprocating type saw, and can utilize a finer blade for a given piece of material. It is common practice to cut directly on the line when band sawing.
3. A band saw produces LITTLE WASTE. The small cross section of a band saw blade makes smaller and fewer chips for a given length or thickness of material, Fig. 9-17.

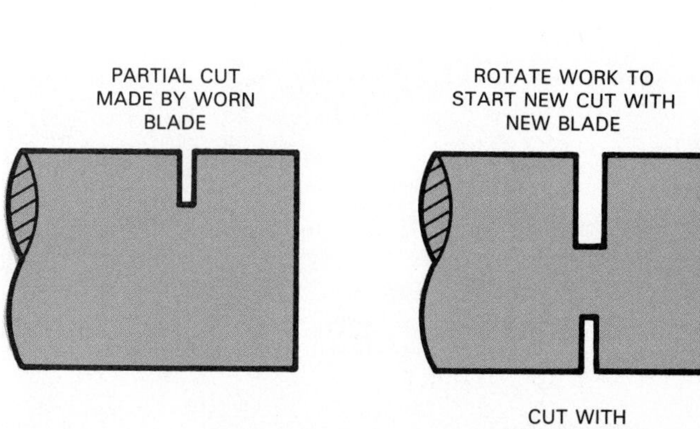

Fig. 9-14. Never attempt to start a new blade in a cut started by a worn blade. Reverse work and start another cut on opposite side to align and cut through to old cut.

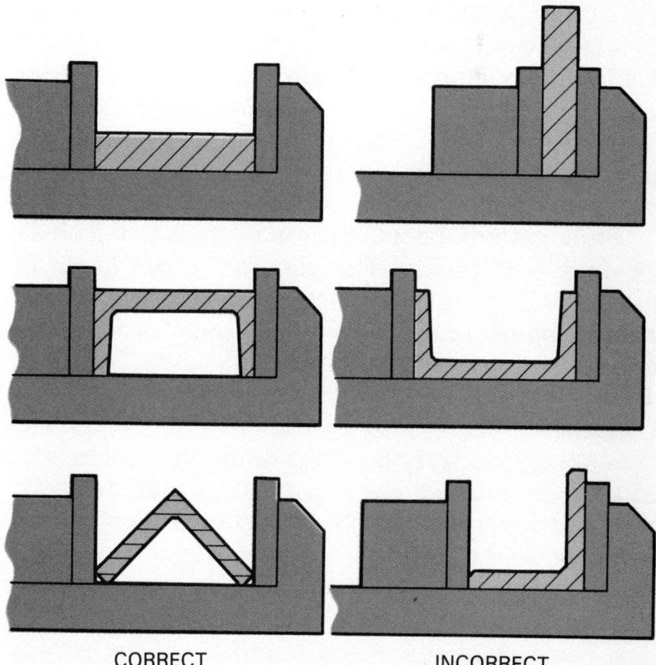

Fig. 9-15. Note recommended ways to hold sharp cornered work for cutting. A carefully planned setup will assure that at least three teeth on blade will be cutting most of the time, greatly extending blade life.

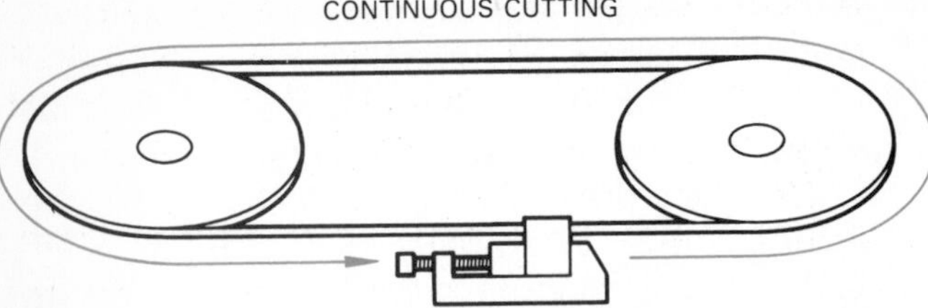

Fig. 9-16. A band type power saw offers three advantages over a reciprocating power hacksaw; it is faster, more precise, and produces less waste.

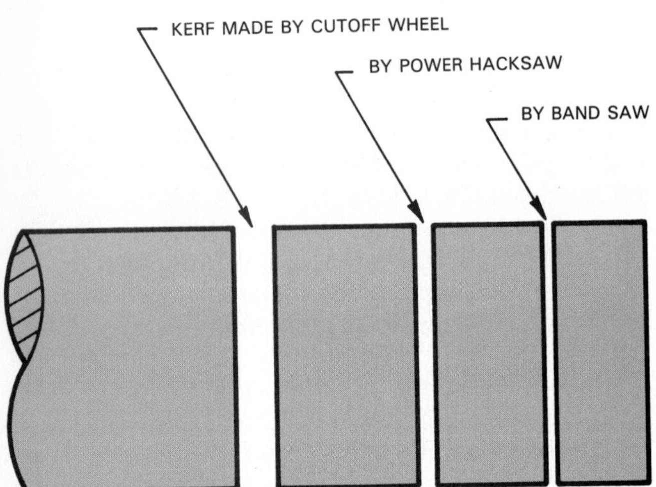

Fig. 9-17. Study differences in amount of metal converted to chips (waste) by each cutoff machine type.

Selecting band saw blade

Band saw blades are made with the teeth raker set or wavy set, Fig. 9-18. In addition to these sets, most blade manufacturers make blades that are variations and modifications of these sets. The ***raker set*** is preferred for general use.

Tooth pattern determines the efficiency of a blade in various materials. The ***standard tooth blade*** is best suited for cutting most ferrous metals. A ***skip tooth blade*** pattern is preferred for cutting aluminum, magnesium, copper, and brass. The ***hook tooth*** is recommended for most nonferrous metallic materials. See Fig. 9-19.

For best results, consult the blade manufacturer's chart or manual for the proper blade (set, pattern, and number of teeth per inch) for the particular material being cut.

Installing a band saw blade.

A blade must be installed carefully, if the saw is to work at top efficiency. Blade guides should be adjusted to provide adequate blade support, Fig. 9-20. Proper blade support is needed to cut true and square with the holding device.

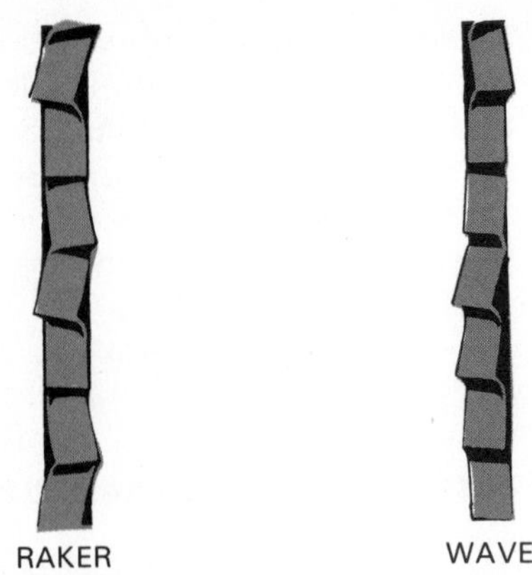

Fig. 9-18. Saw blade teeth are commonly made with teeth raker set or wavy set. Raker set teeth are preferred for general use and for cutting large solid sections or thick plate.

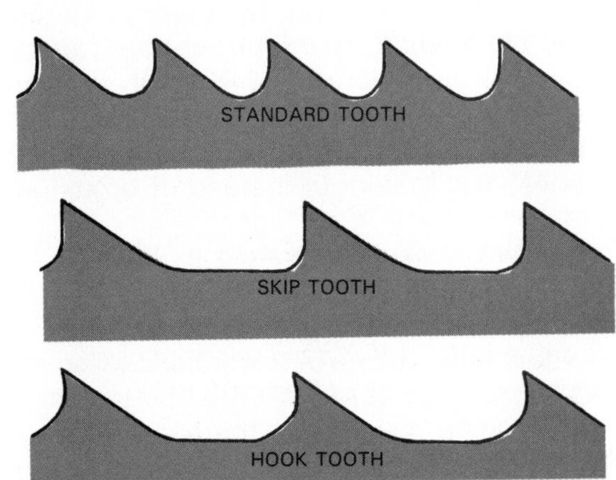

Fig. 9-19. Standard tooth blades, with their well rounded gullets, are usually best for most ferrous materials, hard bronzes, and brasses. Skip tooth blades provide for more chip clearance without weakening blade body. This one is desirable for aluminum, magnesium, copper, and soft brasses. Hook tooth blades offer two advantages over skip tooth blade: blade design makes it feed easier and its chip breaker design prevents "gumming" up.

Fig. 9-20. Adjust blade guides to provide adequate blade support; otherwise, blade will not cut true. (Kalamazoo Tank & Silo Co.)

Follow the manufacturer's instructions for adjusting blade tension. Improper blade tension ruins blades and can cause early failure of bearings in the drive and idler wheels.

Cutting problems encountered with the band saw are similar to those of the reciprocating type hacksaw. Most problems are caused by poor machine condition. They can be kept to a minimum if a routine maintenance program is followed on a regular basis. This typically includes checking guide alignment, wheel alignment, feed pressure, and hydraulic systems.

CIRCULAR METAL-CUTTING SAWS

Metal-cutting circular saws are found in many areas of metalworking. They are primarily production machines and are divided into three classifications:

1. An ***abrasive cutoff saw*** cuts material by means of a rapidly revolving, thin abrasive wheel, Fig. 9-21. Most materials, glass, ceramics, and a long list of metals can be cut to close tolerances. Hardened steel does NOT have to be annealed to be cut.

   Abrasive cutting falls into two classifications – dry and wet. ***Wet abrasive cutting,*** while NOT quite as rapid as dry cutting, produces a finer surface finish and permits cutting to close tolerances. The cuts are burn free and have little or no burrs. ***Dry abrasive cutting*** does NOT use a coolant and is for rapid, less critical cutting.

   Special heat-resistant abrasive wheels are available for high-speed cutoff of hot stock.
2. A ***cold circular saw*** makes use of a circular blade, Fig. 9-22. The toothed blade is capable of producing very accurate cuts. The larger machines can rapidly sever metals up to 27 in. (675 mm) in diameter.
3. A ***friction saw*** blade may or may not have teeth. The saw operates at very high speeds (20-25,000 surface feet per minute or 6 100-7 600 m per minute) and actually burns or melts its way through the metal.

   If teeth are on the friction saw blade, their primary use is to carry oxygen to the cut. These machines find many applications in steel mills to cut ***billets*** (sections of semifinished iron and steel) while they are still "red hot."

POWER SAW SAFETY

1. Get help when lifting and cutting heavy material.
2. Clean oil, grease, and coolant from the floor around the work area.
3. Burrs on cut pieces are sharp. Use special care when handling pieces with burrs until you can remove them.
4. Do NOT clean chips from the machine with your hands. A brush is much safer and does a better job. Stop the machine to clean it!
5. Follow the manufacturer's instructions for tensioning a blade. Too much tension could shatter the blade causing sharp particles of flying steel!
6. Keep your hands out of the way of moving parts.
7. Stop the machine before making adjustments.

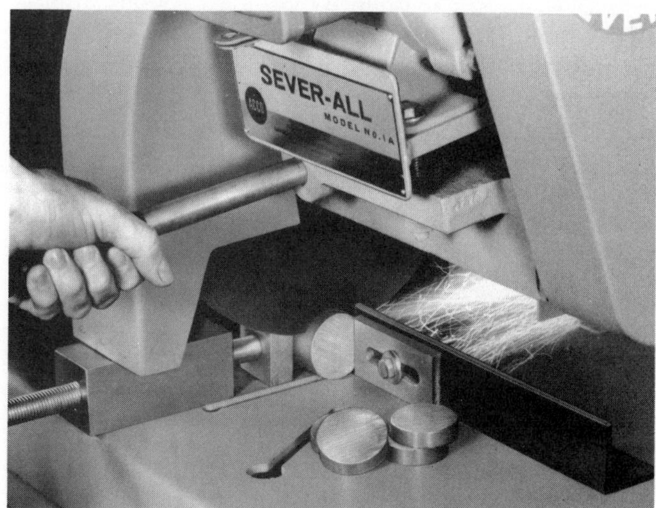

Fig. 9-21. Wear dust mask when cutting with dry type abrasive cutoff saw if machine is not fitted with a vacuum system.

Fig. 9-22. This is an automated cold circular saw. (Motch & Merryweather Machinery Co.)

8. Handle band saw blades with extreme care. They are long and springy and can uncoil at high speed to cause serious cuts.
9. Have all cuts, bruises, and scratches, even minor ones, treated immediately.
10. Do NOT operate a machine unless it is in good condition and all guards are in place.

TEST YOUR KNOWLEDGE—Chapter 9

Please do not write in the text. Place your answers on a separate sheet of paper.

1. List the three basic types of metal cutting saws.
2. The _______ type saw has a back and forth cutting action. However, it only cuts on the _______ or power stroke.
3. What does the "three-tooth rule" for sawing mean?
4. If the machine is equipped for coolant, coolant should be used when power sawing most metals, except _______ _______ and _______ _______, to reduce friction, blade wear, and chip clogging.
5. Blades for the hacksaw are manufactured in two principle types. List them.
6. The following "rule-of-thumb" should be followed for selecting the correct blade:
   a. _______ _______ per inch for cutting large sections or readily machined materials.
   b. _______ _______ per inch for cutting harder alloys and miscellaneous cutting.
   c. _______ and _______ per inch for cutting on the majority of light-duty machines where work is limited to small sections and moderate to light-feed pressures.
7. Many techniques have been devices to put the proper tension on power hacksaw blades. List the three most commonly used.
8. When is a stop gage used?
9. What are three advantages the continuous band sawing machine offers over other types of power saws?
10. Band saw blades are made with the teeth _______ set of _______ set.
11. The tooth pattern of a blade determines the efficiency of a blade in various materials.
    a. The _______ tooth is best suited for cutting most ferrous metals.
    b. The _______ tooth pattern is preferred for cutting aluminum, magnesium, copper, and soft brass.
    c. The _______ tooth is recommended for most nonferrous metallic materials.
12. List the three circular type metal cutting saws.
13. List five safety precautions to be observed when operating a power saw.

RESEARCH AND DEVELOPMENT

1. Make a display panel that includes samples of the different band saw blades used by industry. Label them according to their recommended use. Arrange to provide a magnifying glass so the blade teeth can be examined in detail.
2. Secure samples of abrasive cutoff wheels and of the material they are best suited to cut. Prepare a bulletin board display using these wheels and samples, and manufacturers brochures of abrasive cutoff machines.
3. Prepare large scale models of the following blade types from suitable aluminum sheet: raker and wavy set teeth, standard tooth blade, skip tooth blade, and hook tooth blade. Mount them in hardwood stands for easy examination.
4. Design and produce a series of safety panels pertaining to power sawing.
5. If the power saw in your shop has seen extensive service, use the machine's service manual to overhaul the machine. Contact the manufacturer for a service manual and parts list, if needed. If time permits, paint the machine according to "color dynamics" specifications.
6. Demonstrate to the class the various methods used to properly tension a blade on a power saw.
7. Contact a saw blade manufacture and request a chart showing the best type blade for various sawing situations. Mount it near the saw so students will have ready access to it.

Chapter 10

THE LATHE

After studying this chapter, you will be able to:

- Describe how a lathe operates.
- Identify the various parts of a lathe.
- Safely set up and operate a lathe using various work holding devices.
- Sharpen lathe cutting tools.
- Safely perform cutting, drilling, and boring operations on a lathe.
- Cut thread on a lathe.

The ***lathe*** operates on the principle of the work being rotated against the edge of a cutting tool, Fig. 10-1. It is one of the oldest and most important machine tools. The cutting tool is controllable and can be moved lengthwise on the lathe bed and into any desired angle across the revolving work. Refer to Fig. 10-2.

SAFETY NOTE! Never attempt to operate a lathe while your senses are impaired by medication or other substances.

LATHE SIZE

Lathe size is determined by the SWING and LENGTH OF THE BED, Fig. 10-3. The ***swing*** indicates the largest diameter that can be turned over the ***ways*** (flat or V-shaped bearing surface that aligns and guides movable part of machine). ***Bed length*** is entire length of the ways.

Bed length must not be mistaken for the maximum length of the work that can be turned between centers. The longest piece that can be turned

Fig. 10-1. Metal cutting lathe is one of the most versatile machine tools made. (Clausing Machine Tools)

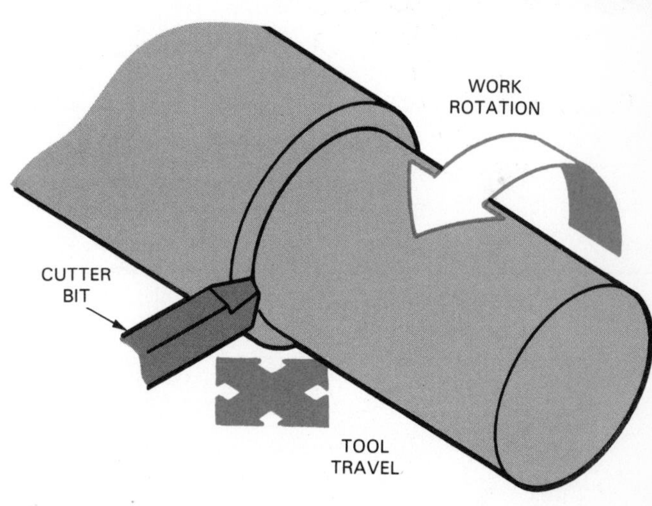

Fig. 10-2. Note operating principle of the lathe.

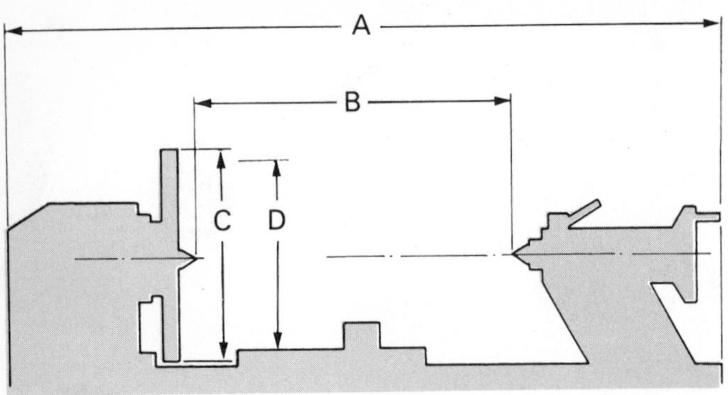

Fig. 10-3. Study how the lathe is measured. A—Length of bed. B—Distance between centers. C—Diameter of work that can be turned over ways. D—Diameter of work that can be turned over cross slide.

is equal to the length of the bed MINUS the distance taken up by the headstock and tailstock. See measurement B in Fig. 10-3.

As an example, consider the capacity and clearance of a modern 13 in. by 6 ft. (325 mm by 1800 mm) lathe:

Swing over bed . . . . 13 in. (325mm)
Swing over cross slide . . . . 8 3/4 in. (218 mm)
Bed length . . . . 72 in. (1800 mm)
Distance between centers . . . 50 in. (1240 mm)

MAJOR PARTS OF A LATHE

The chief function of any lathe, no matter how complex it may appear to be, is to rotate the work against a controllable cutting tool. Each of the lathe parts in Fig. 10-4 falls into one of the following categories:

1. Driving the lathe.
2. Holding and rotating the work.
3. Holding, moving, and guiding the cutting tool.

Driving the lathe

Power is transmitted to the various drive mechanisms by belt drive and/or gear train. See Figs. 10-5 through 10-7. Spindle speed can be varied by:

1. Moving the drive belt to another pulley ratio (seldom used today).
2. Shifting to a different gear ratio.
3. Adjusting a split pulley to another position.

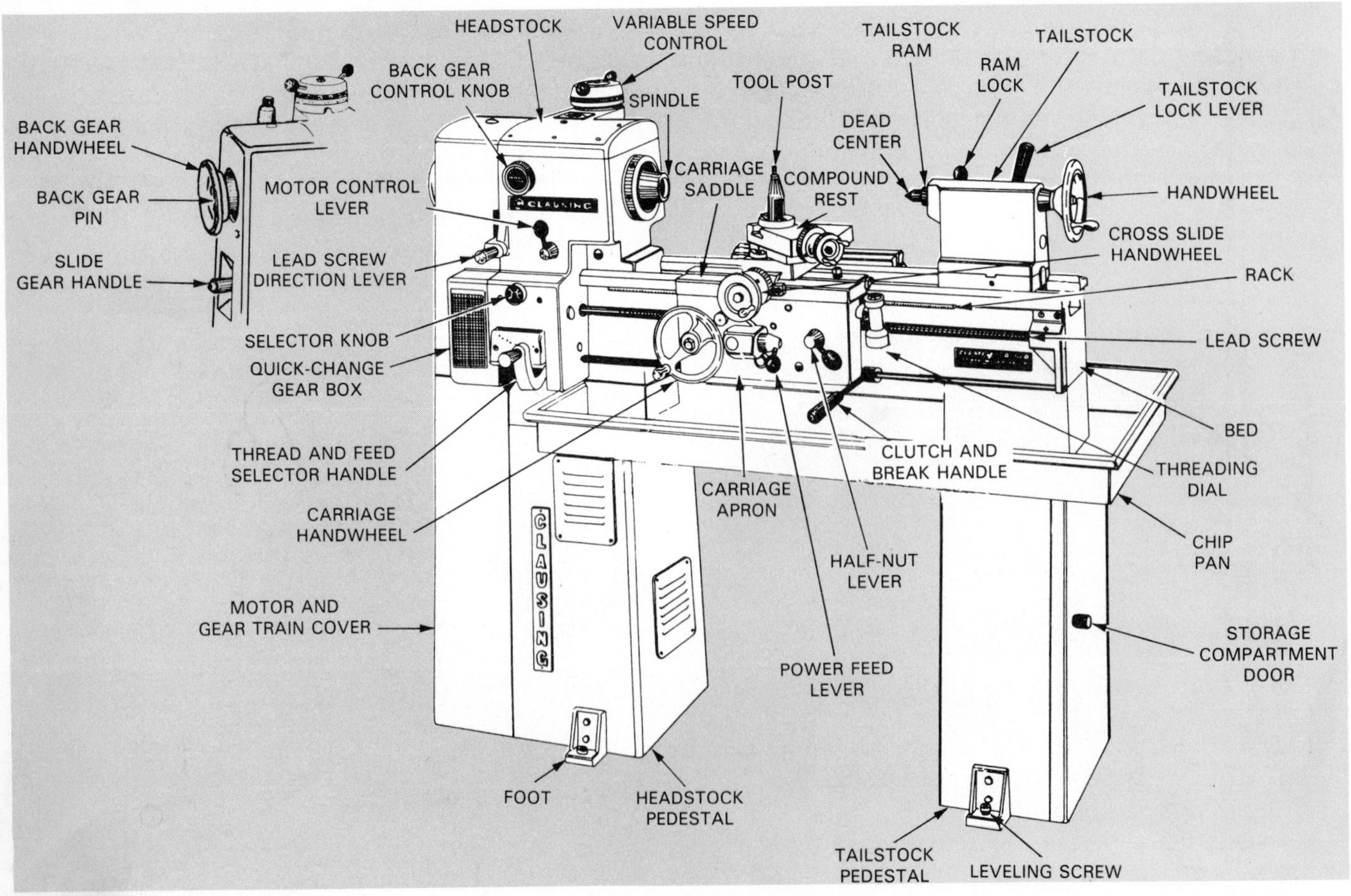

Fig. 10-4. Study engine lathe and its major parts. (Clausing Machine Tools)

Fig. 10-5. Note shifting levers on geared head lathe. Spindle speed is increased or decreased by shifting to different gear ratios. (LeBlond Makino Machine Tool Co.)

Fig. 10-6. Spindle speed is controlled on this lathe by an automatic transmission. Desired speed is dialed in. (LeBlond Makino Machine Tool Co.)

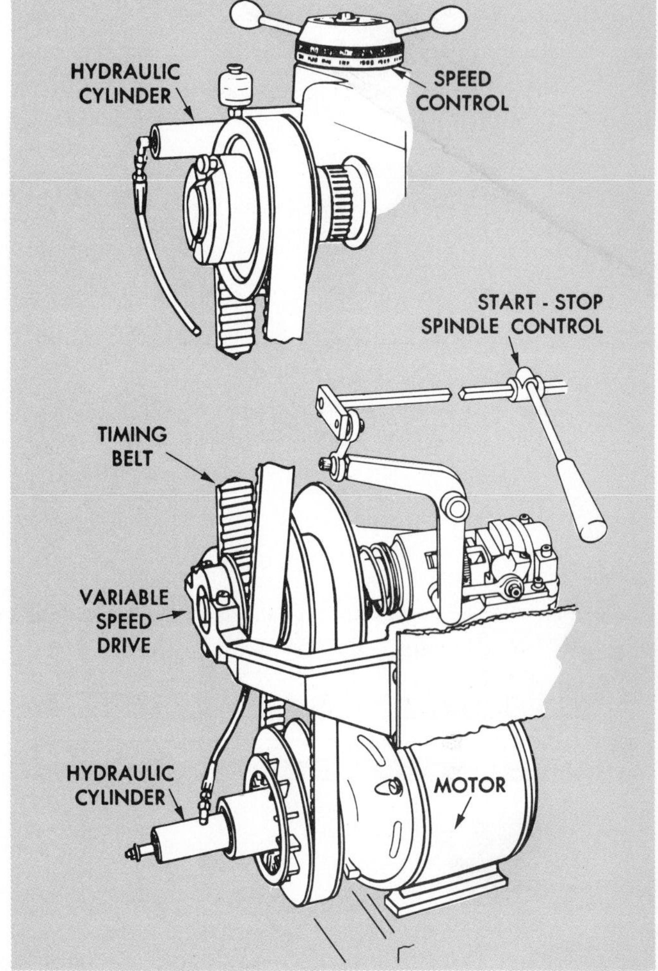

Fig. 10-7. A split pulley, hydraulically actuated from top of machine (speed control), is used to control spindle speeds on many lathes. (Clausing Machine Tools)

4. Controlling the speed hydraulically.

Slower speeds with greater power are obtained on some machines by engaging a ***back gear.*** See Fig. 10-8.

CAUTION! Do not engage the back gear while the lathe spindle is rotating.

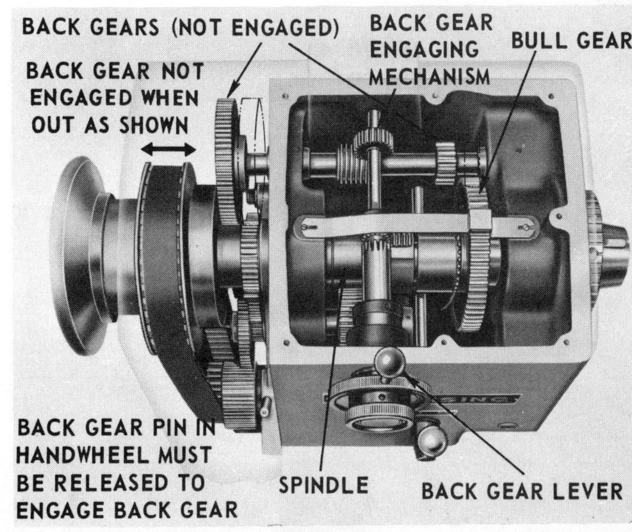

Fig. 10-8. Back gear mechanism is clearly shown in this view of a lathe headstock. Direct drive is disengaged before back gear slides into position. Do NOT engage back gear while lathe is running. (Clausing Machine Tools)

Holding and rotating the work

The ***headstock*** contains the spindle to which the various work holding attachments are fitted, Fig. 10-9. The ***spindle*** revolves in heavy/duty bearings and is rotated by belts, gears, or a combination of both. It is hollow with the front tapered internally to receive tools and attachments with taper shanks, Fig. 10-10. The hole permits long stock to be turned without dangerous overhang, Fig. 10-11. It also allows use of a ***knockout bar*** to remove taper shank tools, Fig. 10-12.

Externally, on the front end, a spindle is threaded or fitted with one of two types of tapered spin-

Fig. 10-9. Lathe headstock is driving end of lathe. (Clausing Machine Tools)

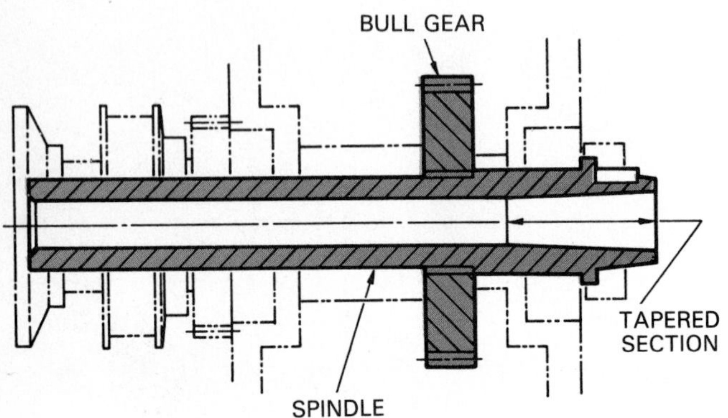

Fig. 10-10. Note construction of lathe spindle. Refer to Fig. 10-8 for photo of spindle in lathe headstock.

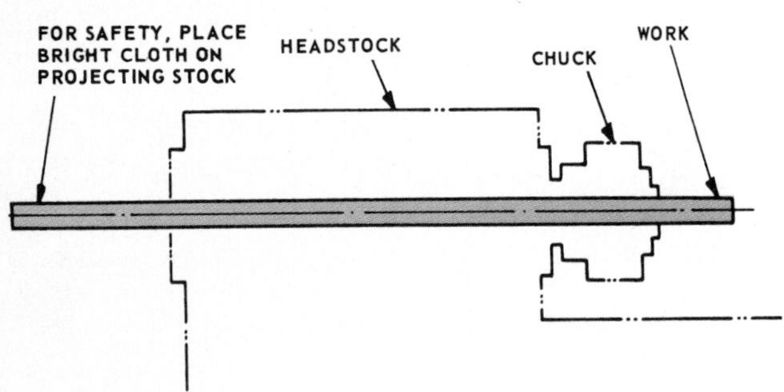

Fig. 10-11. Hollow spindle permits long stock to be turned without dangerous overhang. Some sort of flag should be tied to portion of stock that projects from rear of spindle to prevent others from running into it accidently.

dle noses to receive work holding attachments.

A ***threaded spindle nose*** permits mounting an attachment by screwing it directly on the threads until it seats on the spindle flange. Threaded spindle noses are seldom used on modern lathes. Refer to Fig. 10-13.

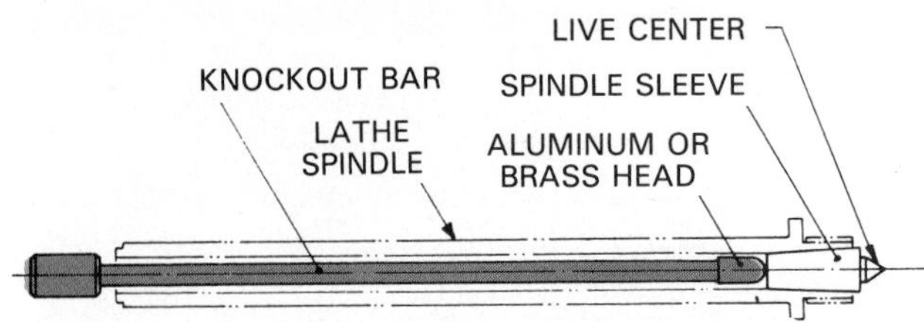

Fig. 10-12. A knockout bar is used to tap tapered shank lathe accessories from spindle.

Fig. 10-13. This is a threaded type spindle nose. (South Bend Lathe)

The ***cam-lock spindle nose*** has a short taper that fits into a tapered recess on the back of the work holding attachment, Fig. 10-14. A series of cam-locking studs, located on the back of the attachment, are inserted into holes in the spindle nose. The studs are locked by tightening the cams located around the spindle nose.

A ***long taper key spindle*** is fitted with a long taper and key that fits into a corresponding taper and keyway in the back of the work holding device, Fig. 10-15. Mounting is done by rotating the key until it is on the top. The keyway in the back of the work holding attachment is slid over the key to support the device until the threaded spindle collar can be engaged with the threaded section on the back of the chuck, faceplate, etc., and tightened.

NOTE: Attachment points on the spindle nose and work holding attachment must be cleaned carefully before mounting the device.

Work is held in the lathe by a chuck, faceplate, collet or between centers. These attachments will be described in detail later in this chapter.

The outer end of the work is often supported by the ***tailstock,*** Fig. 10-16. It can be adjusted along

Fig. 10-14. This is a cam-lock type spindle nose. (Clausing Machine Tools)

Fig. 10-15. This is a long taper key spindle nose. (Clausing Machine Tools)

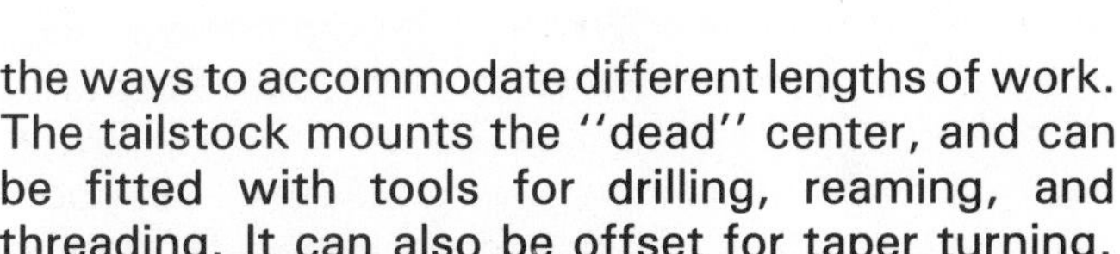

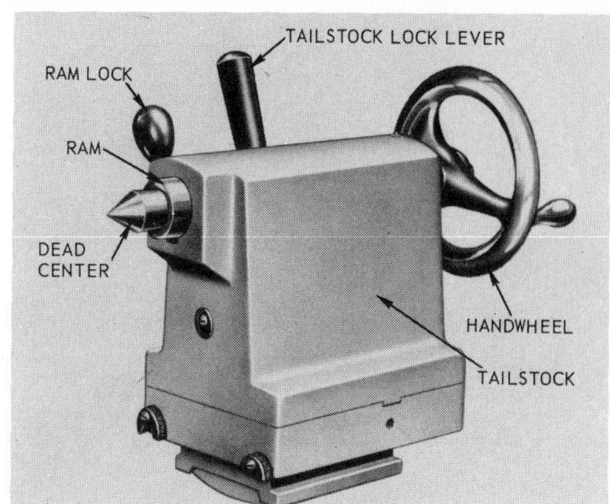

Fig. 10-16. Note parts of tailstock. (Clausing Machine Tools)

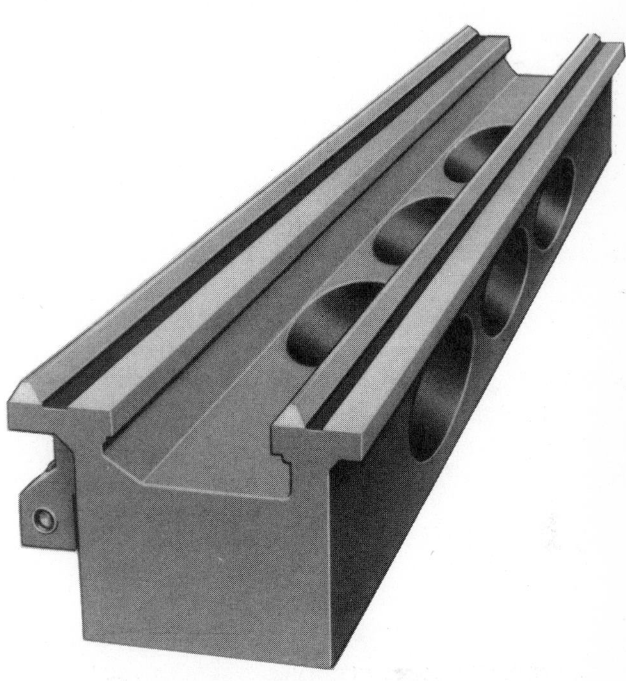

Fig. 10-17. Bed is the foundation of the lathe.

the ways to accommodate different lengths of work. The tailstock mounts the "dead" center, and can be fitted with tools for drilling, reaming, and threading. It can also be offset for taper turning.

The tailstock is locked on the ways by tightening a ***clamp bolt nut*** or ***binding lever.*** The tailstock spindle is positioned by rotating the handwheel and can be locked in position by tightening a binding lever.

Holding, moving and guiding the cutting tool

The ***bed*** is the foundation of a lathe, Fig. 10-17. All other parts are fitted to it. Ways are integral with the bed. The V-shape maintains precise alignment of the headstock and tailstock, and serves as rails to guide the travel of the carriage. The ***cutting tool*** is mounted on the carriage, Fig. 10-18.

The ***carriage*** controls and supports the cutting tool and is composed of:

1. The ***saddle*** is fitted to and slides along the ways.
2. The ***apron*** contains the drive mechanism to move the carriage along the ways using hand or power feed.
3. The ***cross slide*** permits ***traverse*** tool movement (movement toward or away from the operator).
4. The ***compound rest*** permits angular tool movement.
5. The ***tool rest*** mounts the cutting tool.

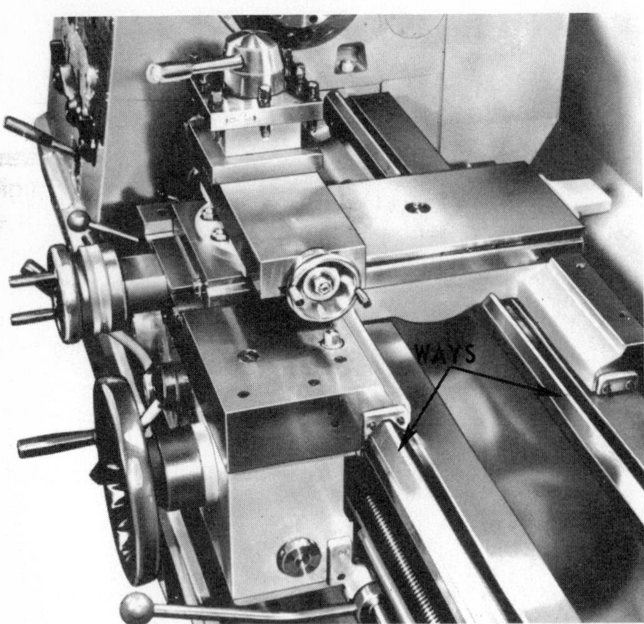

Fig. 10-18. V-shaped ways guide carriage. Cutting tool is mounted on carriage.

Fig. 10-19. Feed mechanism moves carriage along ways.

Power is transmitted to the carriage through the ***feed mechanism,*** which is located at the left of the lathe, Fig. 10-19. Power is transmitted through a train of gears to the ***quick change gear box*** which regulates the amount of tool travel per revolution of the spindle, Fig. 10-20. The gear train also contains gears for reversing tool travel.

The quick change gear box is arranged between the spindle and the lead screw. It contains gears of various ratios which makes it possible to machine various pitches of screw threads without changing loose gears. ***Longitudinal*** (back-and-forth) travel and cross (in-and-out) travel is controlled in the same manner.

An ***index plate*** provides instructions on how to set the lathe shift levers for various thread cutting and feed combinations, Fig. 10-21. It is located on the face of the gear box. The ***large numbers*** on the

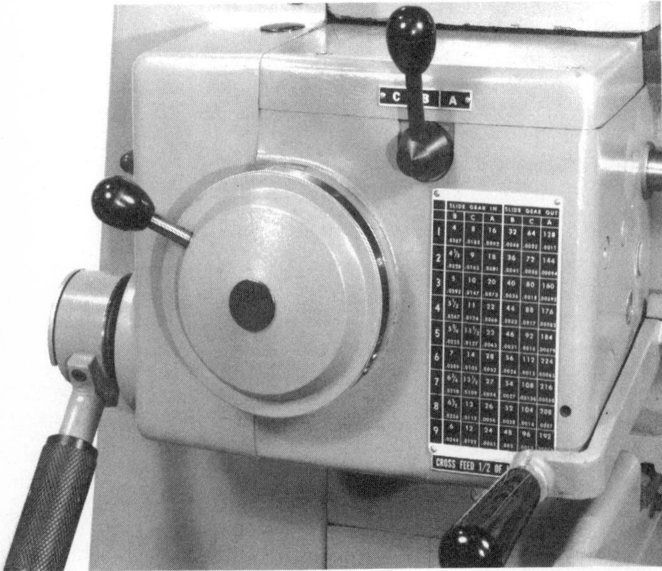

Fig. 10-20. Left—One type of quick-change gear box. Right—Quick-change gear box for cutting both inch and metric size threads. (Clausing Machine Tools)

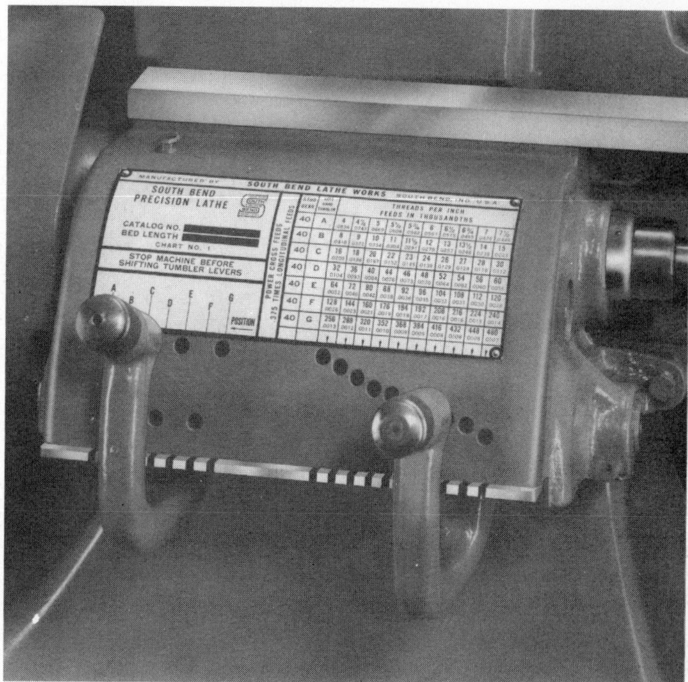

Fig. 10-21. Left—Index plate showing lever position for inch and metric feeds and threads. (Clausing Machine Tools) Right—Index plate for inch feeds and threads. (South Bend Lathe)

index plate indicate the number of threads that can be cut per inch or pitch of metric threads. The ***smaller figures*** indicate the carriage longitudinal movement, in thousandths of an inch or in mm for each spindle revolution.

The ***lead screw*** transmits power to the carriage through a gearing and clutch arrangement in the carriage apron, Fig. 10-22. ***Feed change levers*** on the apron control the operation of power longitudinal feed and power cross feed, Fig. 10-23.

When placed in neutral, the ***half-nuts*** may be engaged for thread cutting. The gear arrangement makes it possible to engage power feed and half-nuts simultaneously. The half-nuts are engaged ONLY for thread cutting and are NOT used as "automatic" feed for regular turning.

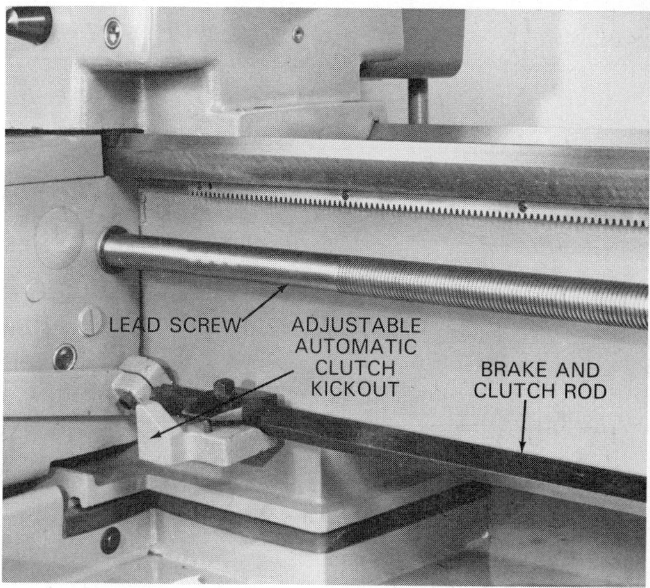

Fig. 10-22. The lead screw.

PREPARING LATHE FOR OPERATION

Before an aircraft is permitted to takeoff, the pilot or crew must go through a check-out procedure to determine whether the engines, controls, and safety features are in first-class operating condition. The same applies to the operation of a machine tool such as a lathe. The operator should inspect the machine for safe and proper operation. They should:

1. Clean and lubricate the machine. Use lubricant types and grades specified by the manufacturer. Many recommend a specific lubricating sequence to reduce any possibility of missing a vital lube point.
2. Be sure all guards are in position and locked in place.
3. Turn the spindle over by hand to be sure it is NOT locked nor engaged in back gear (unless you intend to use back gear).
4. Move the carriage along the ways, Fig. 10-24. There should be no binding.
5. Check cross slide movement. Adjust ***gibs*** if

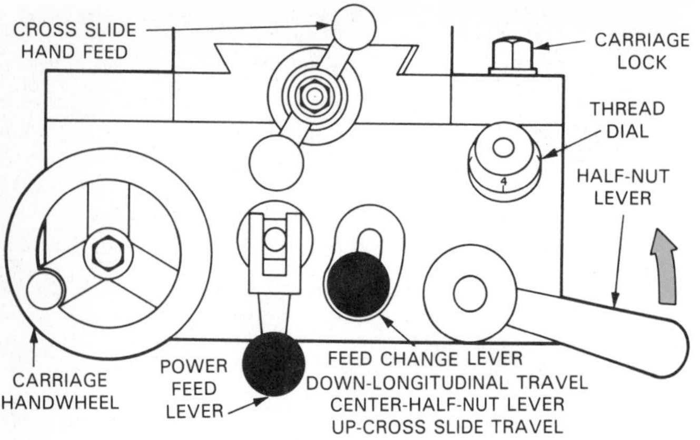

Fig. 10-23. Feed change levers on apron control operation of power feed. Half-nut is engaged only for thread cutting, sometimes called thread chasing.

Fig. 10-25. If there is too much play in unit, adjust gibs according to instructions in manufacturer's handbook. (Clausing Machine Tools)

Fig. 10-24. Carriage should be checked for binding by moving unit along ways. Correct any binding before attempting to use lathe. (Clausing Machine Tools)

Fig. 10-26. Witness lines on tailstock indicate whether tailstock is aligned properly with headstock.

there is too much play in the cross slide. See Fig. 10-25.

6. Mount the desired work holding attachment. Clean the nose with a soft brush. A threaded nose spindle should have a drop of lube oil before attaching the chuck or faceplate.
7. Adjust the drive mechanism for the desired speed and feed.
8. If the tailstock is used, check it for alignment, Fig. 10-26.
9. Clamp the cutter bit into an appropriate tool holder and mount it in the tool post. Do NOT permit excessive compound rest overhang as this often causes tool ''chatter'' and results in a poorly machined surface, Fig. 10-27.
10. Mount the work. Check for adequate clearance between the work and the various machine parts.

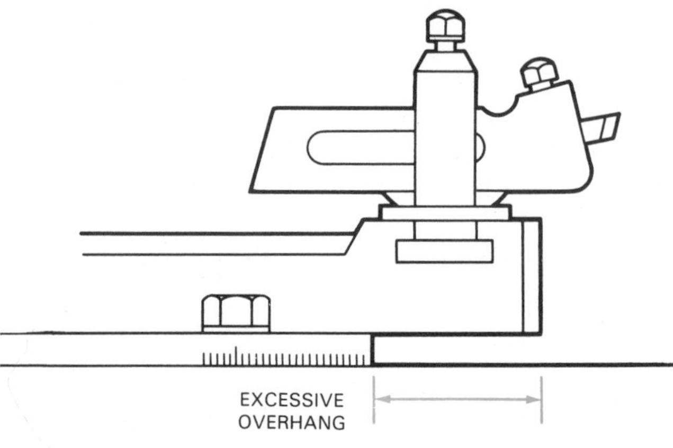

Fig. 10-27. Excessive compound rest overhang usually causes tool ''chatter'' and poorly machined surface.

In addition to the above procedures, the operator must take some precautions. Sleeves should be rolled up and rings, jewelry, and necktie or necklace removed.

A ***lathe board*** will aid in organizing and holding the tools and measuring instruments needed for the job. They must NOT be placed on the lathe ways or carriage, Fig. 10-28.

Fig. 10-28. Lathe board (school shop assembled) keeps tools within easy reach and away from chips and turnings.

CLEANING THE LATHE

To maintain the accuracy built into a lathe, it must be cleaned after each work period. Use a 2 in. paint brush, NOT a dust brush, to remove the accumulated chips.

WARNING! Lathe chips are sharp; do NOT remove them with your hands.

Wipe all painted surfaces with a soft cloth. To complete the job, move the tailstock to the extreme right on the ways. Use a soft cloth to remove the remaining chips, oil, and dirt from the machined surfaces.

DANGER! An air hose should NEVER be used to remove chips. The flying particles might injure you or a nearby person.

A light coating of machine oil on the machined surfaces will prevent rust until the next time the machine will be used. The lead screw occasionally needs cleaning. Adjust the screw to rotate at a slow speed, place a heavy cord around it and start the machine, Fig. 10-29. With the lead screw revolving, permit the cord to feed along the thread. Hold the cord just tightly enough to remove the accumulated dirt.

Fig. 10-29. Machinist is cleaning lead screw with a piece of cord. Do NOT wrap cord around your hand!

CAUTION! NEVER wrap the cord around your hands. The cord could catch and seriously injure your hand.

LATHE SAFETY

1. No attempt should be made to operate a lathe until you know the proper procedures and have been checked out on its safe operation by your instructor.
2. Dress appropriately! Remove necktie, necklace, wrist watch, rings and other jewelry, and loose fitting sweater. Wear an apron or a properly fitted shop coat. Safety glasses are a must!
3. Clamp all work solidly! Use the correct size tool and work holding device for the job. Get help when handling large sections of metal and heavy chucks and attachments.
4. Check work frequently when it is being machined between centers. The work expands as it heats up and could damage the tailstock center.
5. Be sure all guards are in place before attempting to operate the machine, Fig. 10-30.
6. Turn the faceplate or chuck by hand to be sure there is NO binding or danger of the work striking any part of the lathe.

Fig. 10-30. Remember! Never attempt to operate a machine until ALL guards and safety devices are in place and working properly.

Fig. 10-31. Remove stringy chips with pliers; NOT your hand.

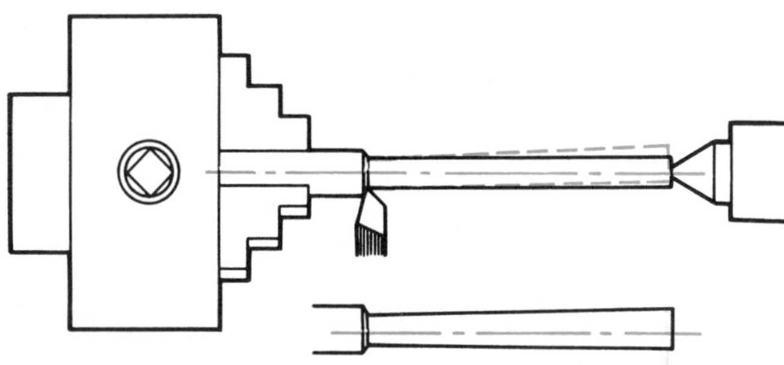

Fig. 10-32. If not properly supported by tailstock center, small diameter work will spring away from cutting tool and will be machined on a slight taper.

7. Keep the machine clear of tools!
8. Stop the machine before making measurements and adjustments.
9. Remember—chips are sharp! Do NOT try to remove them with your hands when they become "stringy" and build up on the tool post. Stop the machine and remove them with pliers, Fig. 10-31.
10. Do NOT permit small diameter work to project too far from the chuck without support from the tailstock. Without support, the work will be tapered, or worse, spring up over the cutting tool and/or break. See Fig. 10-32.
11. Be careful NOT to run the cutting tool into the chuck or dog. Check any readjustment of work or tool for ample clearance when the cutter has been moved left to the farthest point that will be machined.
12. Stop the machine before attempting to wipe down a machine surface. Keep the coolant brush clear of the work when knurling.
13. Before repositioning or removing work from the lathe, move the cutting tool clear of the work area. This will prevent accidental cuts from the cutter bit.
14. Avoid talking to anyone while running a lathe! Do NOT permit anyone to fool around with the machine while you are operating it. You are the only one who should turn the machine on or off, or make adjustments to the lathe.
15. If the lathe has a threaded spindle nose, never attempt to run the chuck on or off the spindle using power. It is also dangerous practice to stop such a lathe by reversing the direction of rotation. The chuck could spin off and cause serious injury to you. There is also the danger of damaging the machine.
16. You should always be aware of the direction of travel and speed of the carriage before engaging the half-nuts or automatic feed.
17. Always remove the key from the chuck. Make it a habit NEVER to let go of the key until it is out of the chuck and clear of the work area.
18. Tools must NOT be placed on the lathe ways. Use a tool board or place them on the lathe tray, Fig. 10-33.
19. When filing on the lathe, be sure the file has a securely fitting handle.

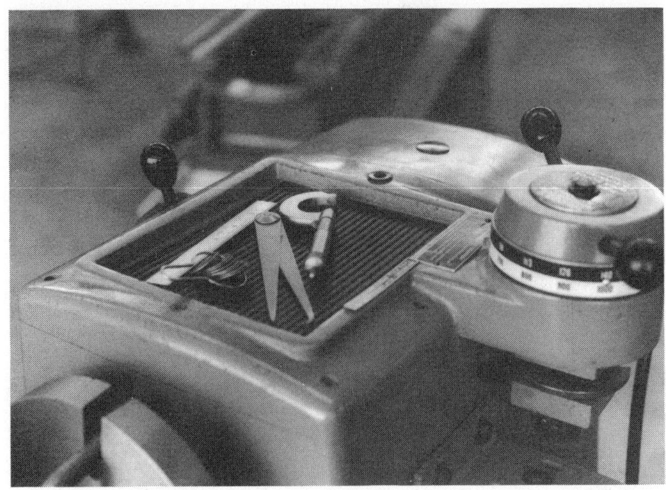

Fig. 10-33. Keep tools on a lathe tray; never on ways or in chip pan.

20. Stop the machine immediately if some odd sounding noise or vibration develops during operation. If you cannot locate the trouble, get help from your instructor. Under no condition should the machine be operated until the trouble has been corrected.
21. Remove sharp edges and burrs from work before removing it from the machine.
22. Plan your work thoroughly before starting. Have all needed tools on hand.
23. Use care when cleaning the lathe. Chips sometimes stick in recesses. Remove them with a brush or short stick. NEVER clean a machine tool with compressed air, or a floor dust brush.

Fig. 10-34. Note straight, left-hand, and right-hand toolholders. (J.H.Williams and Co.)

CUTTING TOOLS AND TOOL HOLDERS

To operate a lathe efficiently, the machinist must have a thorough knowledge of cutting tools and know how they must be shaped to machine various materials. The ***cutting tool*** is held in contact with the revolving work to machine material off of the work. In all probability, you will use a cutting tool made from high-speed steel (HSS).

The square cutter bit body is inserted in a lathe ***toolholder.*** Toolholders are made in straight, right-hand, and left-hand shapes, Fig. 10-34. To tell the difference between right-hand and left-hand toolholders, hold the head of the tool in your hand and note the direction the shank points. The shank of the right-hand holder points to the right, the left-hand toolholders points to the left. A ***turret holder*** may also be utilized, as in Fig. 10-35.

Cutting tool shapes

The parts of the cutter bit are shown in Fig. 10-36. Study them.

Most cutter bits are ground to cut in one direction only. Some cutting tools for general purpose turning are shown in Fig. 10-37.

To get best performance, the tool bit must have a keen, properly shaped cutting edge. The shape depends on the type of work (roughing or finishing) and the metal to be machined.

Left-cut and right-cut roughing tools

The left-cut roughing tool cuts most efficiently when it travels from left to right, Fig. 10-38A. The

Fig. 10-35. This turret type toolholder has four cutter bits. They can be brought into cutting position by loosening the lock (handle) and pivoting desired cutter bit into cutting position before locking it in place. (ENCO Mfg. Co.)

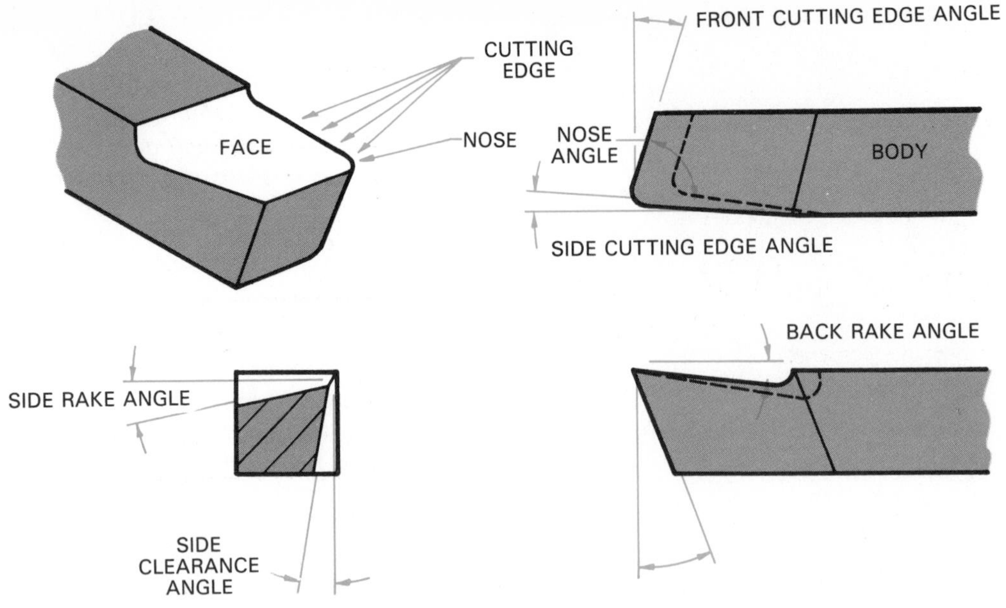

Fig. 10-36. Study parts of the cutter bit.

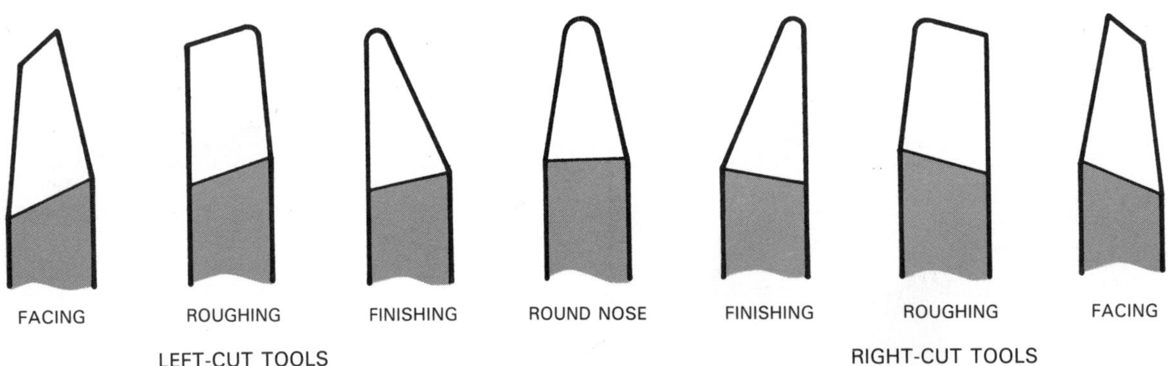

Fig. 10-37. Note standard cutting tool shapes.

right-cut roughing tool operates just the opposite, right to left, Fig. 10-38B.

Tool shape (shape of cutting tip), straight cutting edge with a small rounded nose, permits deep cuts at heavy feeds. The slight ***side relief*** provides ample support to the cutting edge.

Finishing tool

The nose of a ***finishing tool*** is more rounded, Fig. 10-39. Such a tool will produce a smooth finish, if the cutting edge is honed with a fine oil stone after grinding. A light cut and a fine feed is used with the tool.

Facing tool

The ***facing tool*** is ground to prevent interference with the tailstock center, Fig. 10-40. The tool point is set at a slight angle to the work face with the point leading slightly.

Round nose tool

A ***round nose tool*** is designed for lighter turning and is ground flat on the face (without back or side rake) to permit cutting in either direction, Fig. 10-41. A slight variation with a negative rake on the face is excellent for machining brass. Refer to Fig. 10-42.

ALUMINUM requires a different tool shape from those previously described, Fig. 10-43. The tool is set slightly above center to reduce any tendency to ''chatter'' (vibrate rapidly). The tool designs illustrated are typical of cutting tools used to machine aluminum alloys.

Grinding high speed cutter bits

When first attempting to grind a cutter bit, it may be best if you first practice on square sections of cold finished steel rod. You may also want to use chalk or bluing and draw the desired tool shape on

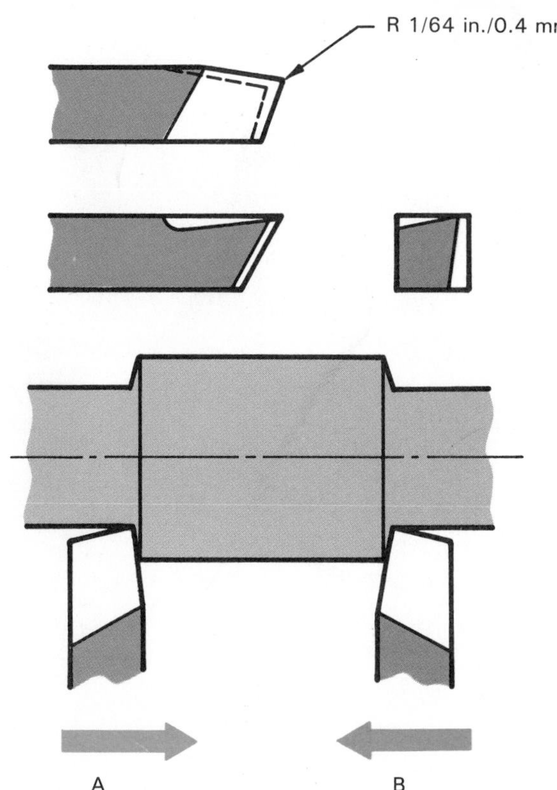

Fig. 10-38. Roughing tool is for rapid material removal.

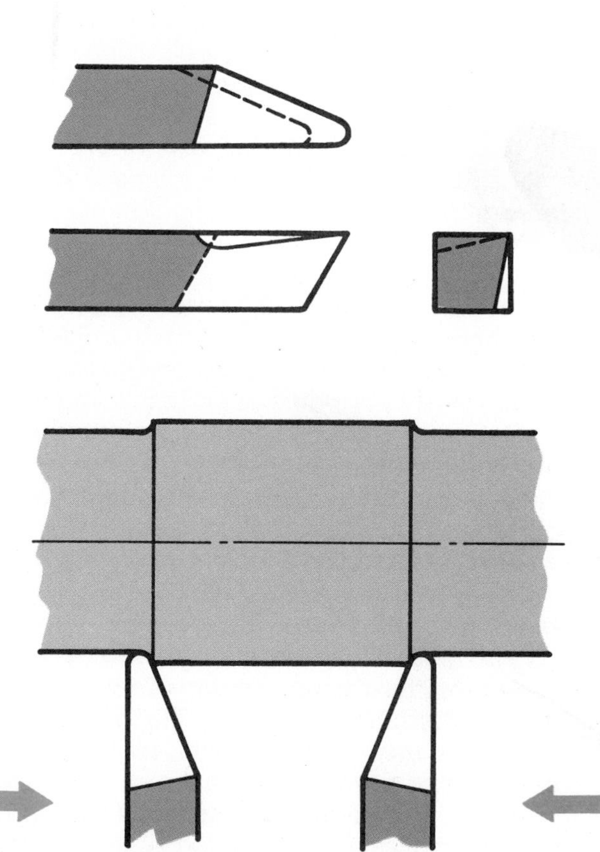
Fig. 10-39. Finishing tool will produce smooth surface.

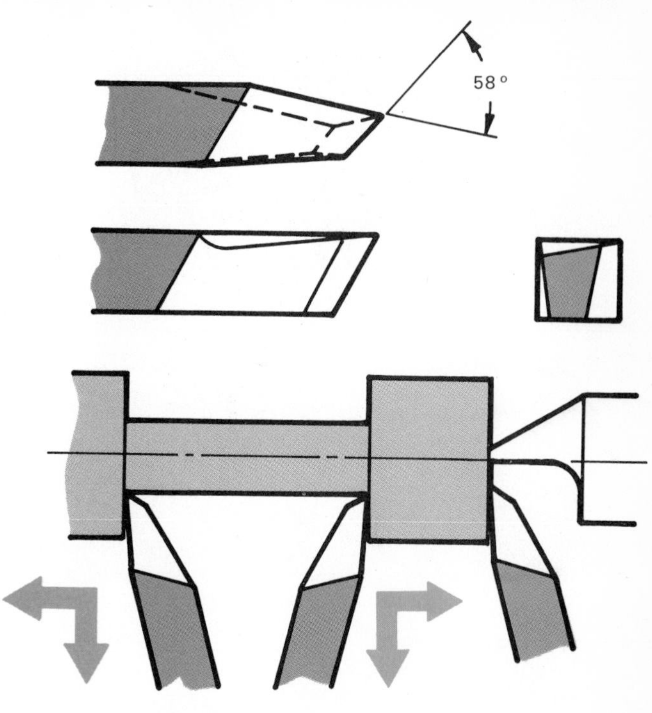

Fig. 10-40. Facing tool will machine surfaces perpendicular to spindle center line.

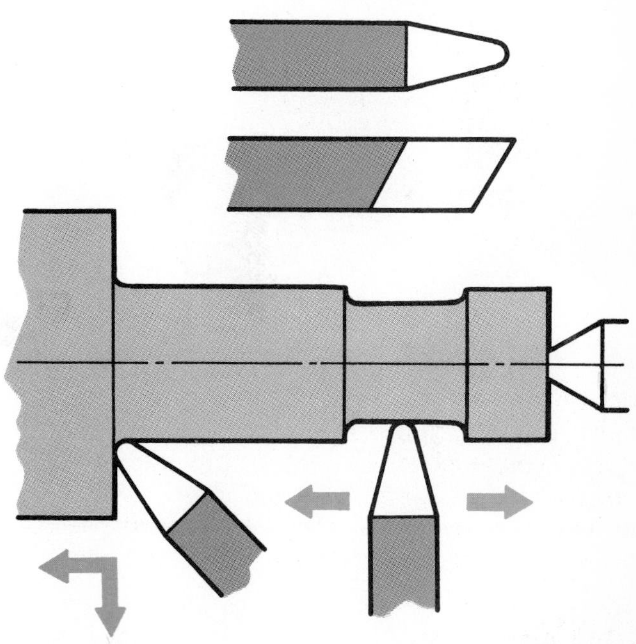
Fig. 10-41. Round nose tool will produce fillets.

the front portion of the blank, Fig. 10-44. The lines will serve as guides for grinding.

Shown in Fig. 10-45 is the recommended grinding sequence. Side clearance, top clearance, and end relief may be checked with a clearance and cutting angle gage, Fig. 10-46.

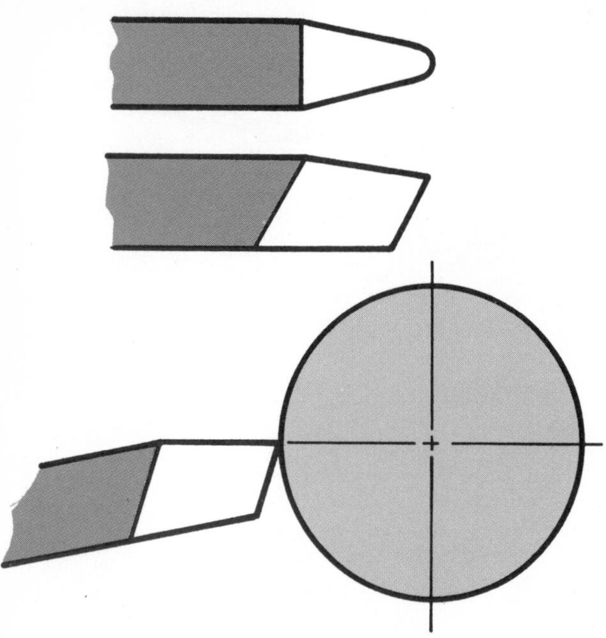

Fig. 10-42. This tool is for brass.

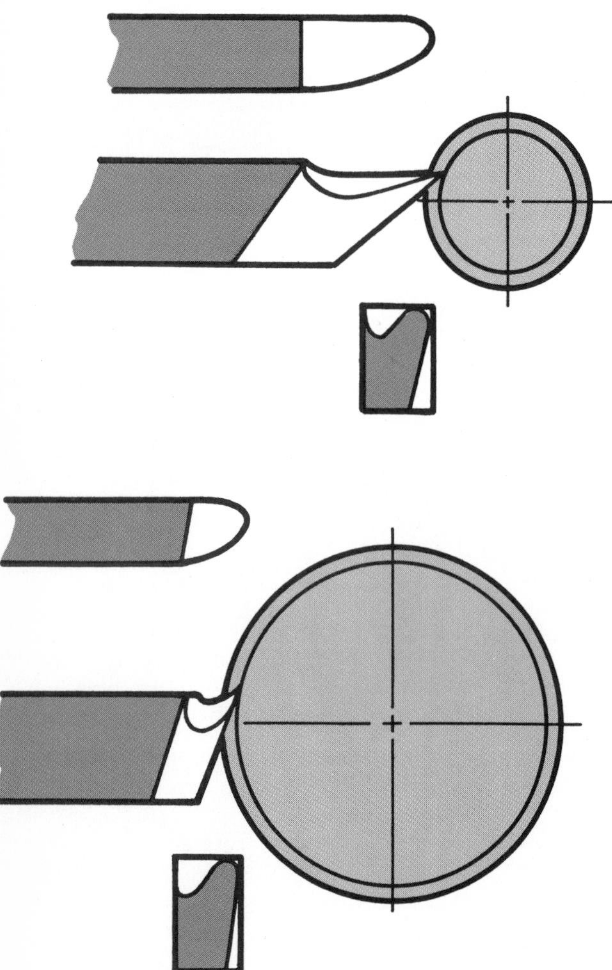

Fig. 10-43. These tools are for aluminum.

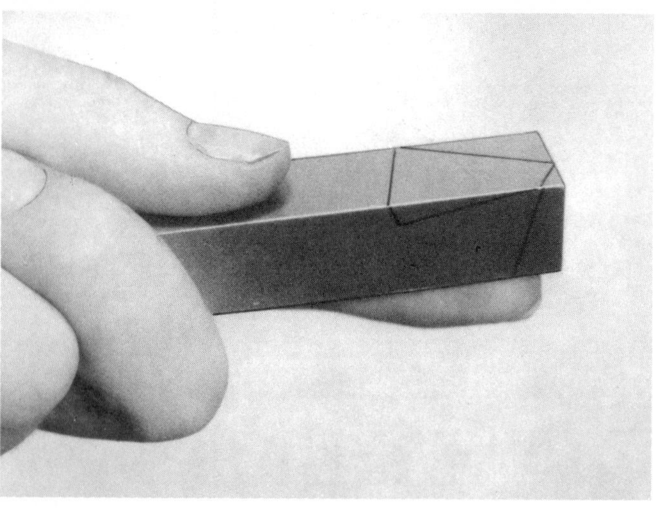

Fig. 10-44. Cutter blank has been laid out or marked in preparation for grinding.

Cemented carbide cutting tools

Cutting speeds can be increased 300 to 400 percent using ***carbide cutting tools.*** Powders of tungsten, carbon, and cobalt are molded into ***tool blanks*** (unmachined cutting tools) and heated to extremely high temperatures, Fig. 10-47. The result is one of the hardest metals made by humans. The hardness and strength of the blank can be changed by varying the amount of cobalt used to ***cement*** (bind) together the tungsten carbide particles. The tool blank is then securely brazed onto a prepared shank, Fig. 10-48.

For best results, cemented carbides should be sharpened on a special silicon carbide or ***diamond charged grinding wheel*** (grinding wheel with diamond dust or chips embedded in it).

Cutting tools designed for machining steel are chamfered 0.003 to 0.002 in. (0.050 to 0.075 mm) by honing lightly with a silicon carbide or diamond hone, Fig. 10-49. If NOT honed, the irregular edge produced by grinding will crumble when used. Honing, if done properly, does NOT interfere with the cutting action.

Note that the clearance angles of the carbide tools described are NOT as great as those required for high speed steel cutting tools.

Carbide tipped straight turning tools

The cutting tools shown in Fig. 10-50 are general purpose tools for facing, turning, and boring. The shape permits machining to a square shoulder.

The ***square nose*** carbide tipped tool is a general purpose tool with many applications, Fig. 10-51.

The carbide tipped ***threading tool*** (Style E), Fig. 10-52, has a 60 deg. included angle that conforms with the Unified National 60 deg. included angle thread. It is used for V-grooming and chamfering.

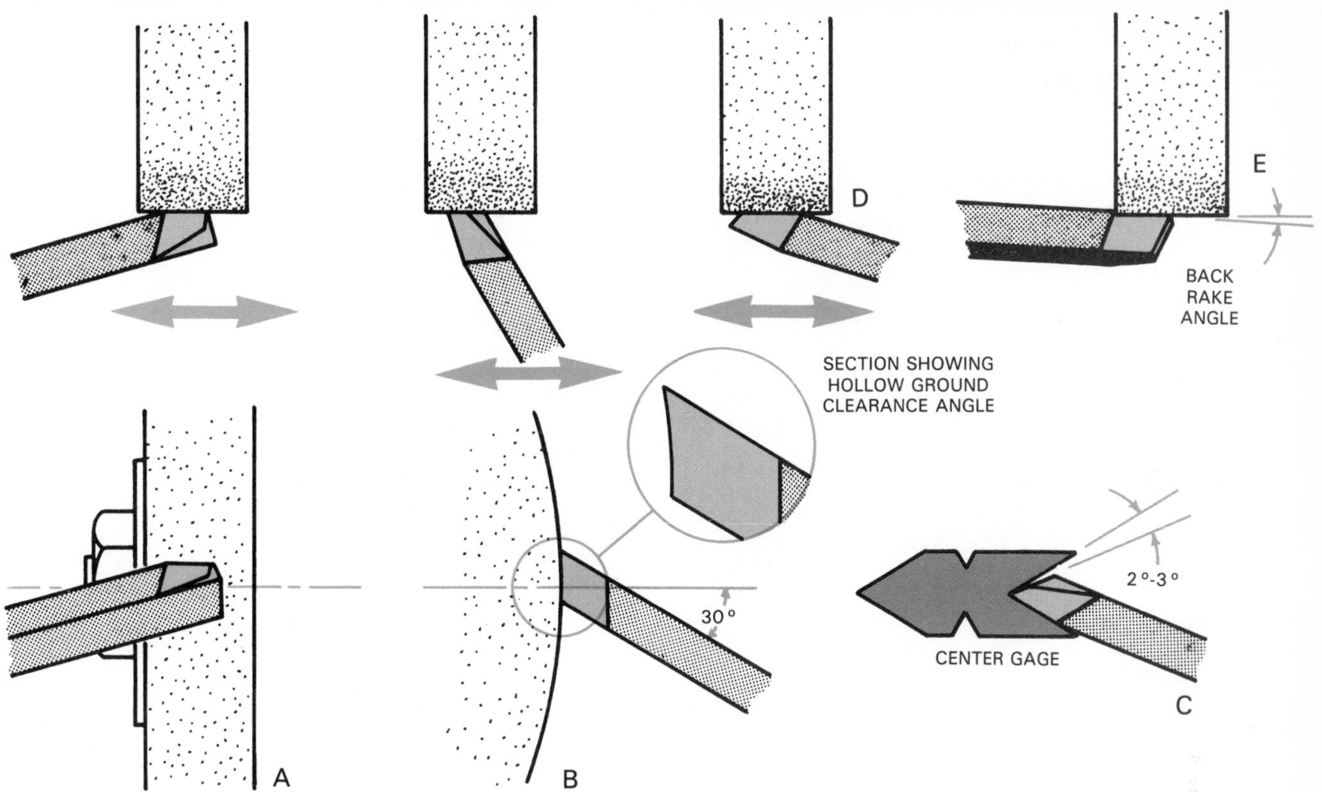

Fig. 10-45. Study cutter bit grinding sequence. A—Two views showing how to position a cutter bit blank on grinding wheel to shape side clearance angle/side cutting edge angle. B—Shaping end clearance angle/front cutting edge angle. C—Center gage is checking nose angle. D—Grinding other side clearance angle, when required. E—Grinding back/side rake angles. Accuracy of clearance angles can be checked with cutter bit gage.

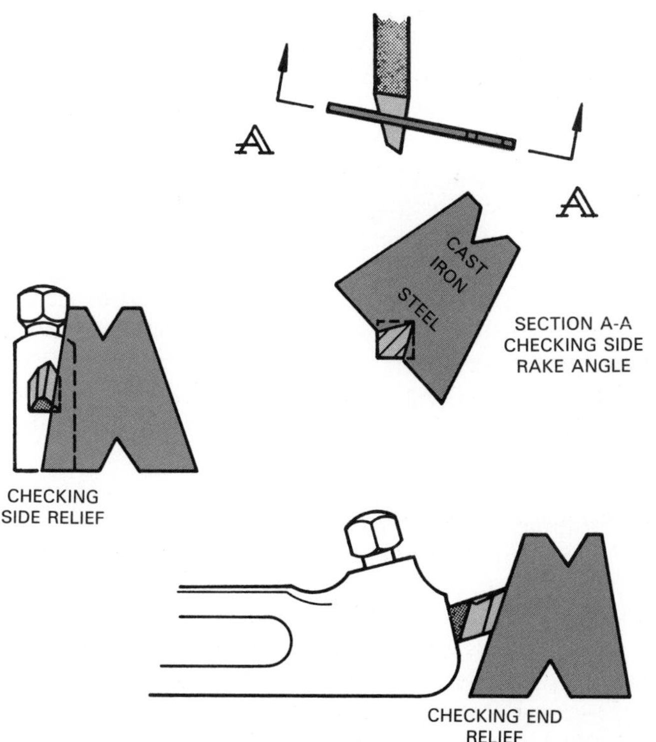

RAKE AND CLEARANCE ANGLE FOR LATHE TOOLS
(High Speed Steel)

	Cast Iron	Low Carbon Steel	High Carbon Steel
Back Rake	6– 8°	8–12°	4– 6°
Side Rake	10–12°	14–18°	8–10°
Clearance *	6– 9°	8–10°	6– 8°
	Alloy Steels	**Soft Brass**	**Aluminum**
Back Rake	5– 8°	0– 2°	25–50°
Side Rake	10–15°	0– 2°	10–20°
Clearance *	6– 8°	10–15°	7–10°
	Copper		
Back Rake	10–12°		
Side Rake	20–25°		
Clearance *	6– 8°		

* The end and side clearance angles are usually the same.

Fig. 10-46. A—Cutter bit gage is used to check accuracy after grinding cutter tip. B—Note rake and clearance angles for lathe tools to machine different metals.

Fig. 10-47. Tungsten carbide tool blanks are available in a wide selection of shapes, sizes, and degrees of hardness.

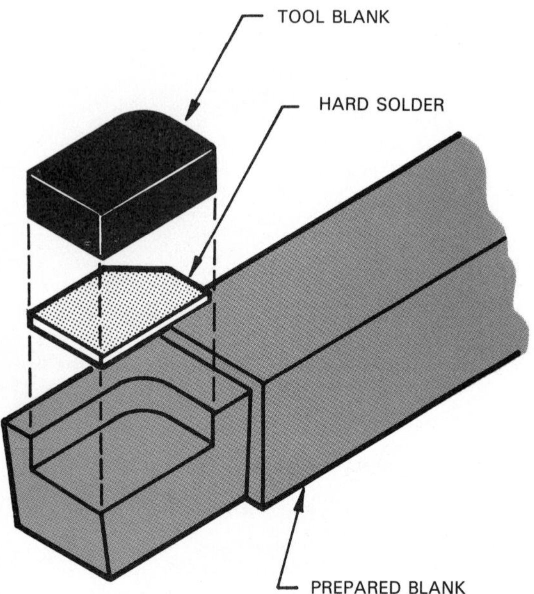

Fig. 10-48. Prepared shank is ready to have carbide tool blank brazed in place. Brazing must be done properly or blank will not be attached solidly, causing it to wear rapidly.

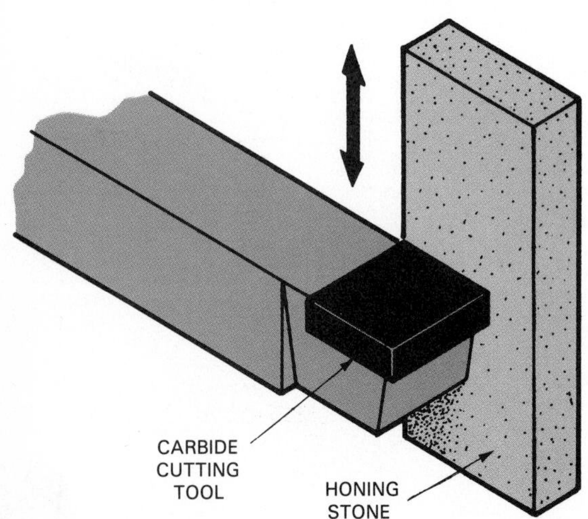

Fig. 10-49. Honing newly ground edge of a carbide type cutting tool will extend its life appreciably.

Indexable throwaway insert cutting tools

Throwaway insert cutting tools of carbide and sintered oxides (often referred to as ceramics, cermets, etc.) are made in a number of standard shapes, Fig. 10-53. They clamp into special holders, Fig. 10-54. As an edge becomes dull, the next edge is rotated into position until all edges are dulled. Since it is more costly to resharpen an insert than replace it, they are discarded after all edges have been dulled.

The extreme harness and heat resistance of ***ceramic cutting tools*** make them ideal cutting tools. However, they are brittle and must be rigidly supported. And while they are more expensive than cemented carbide tools, they perform better than a cemented carbide tool by a ratio of 5 to 1.

Other types of cutting tools

Diamonds, both natural and manufactured, are employed as single point cutting tools on materials whose hardness or abrasive qualtites make them difficult to machine with other types of cutting tools. These diamonds are known as ***industrial diamonds.***

CUTTING SPEEDS AND FEEDS

The matter of cutting speed and feed is most important since they govern the length of time required to machine the work and the quality of the surface finish.

Cutting speed indicates the distance in feet or meters per minute the work moves past the cutting tool. Measuring is done on the circumference of the work.

To explain this differently, if a lathe were to cut one long chip, the length of that chip in feet or meters cut in one minute would be the cutting speed of the lathe. Cutting speed is NOT the revolutions per minute (RPM) of the lathe.

Feed is the distance the cutter moves LENGTHWISE along the lathe bed during ONE revolution of the work.

Several factors that must be considered when determining the correct cutting speed and amount of feed are:

1. Cutting tool material.
2. Kind of material being machined.
3. Finish desired.
4. Condition of lathe.
5. Rigidity of work.
6. Kind of coolant, if any, being used.
7. Shape of material being machined.
8. Depth of cut.

If the machining is done with TOO SLOW of a cutting speed, extra time will be needed to complete the job. If speed is TOO HIGH, the cutting tool dulls rapidly and the finish will be substandard.

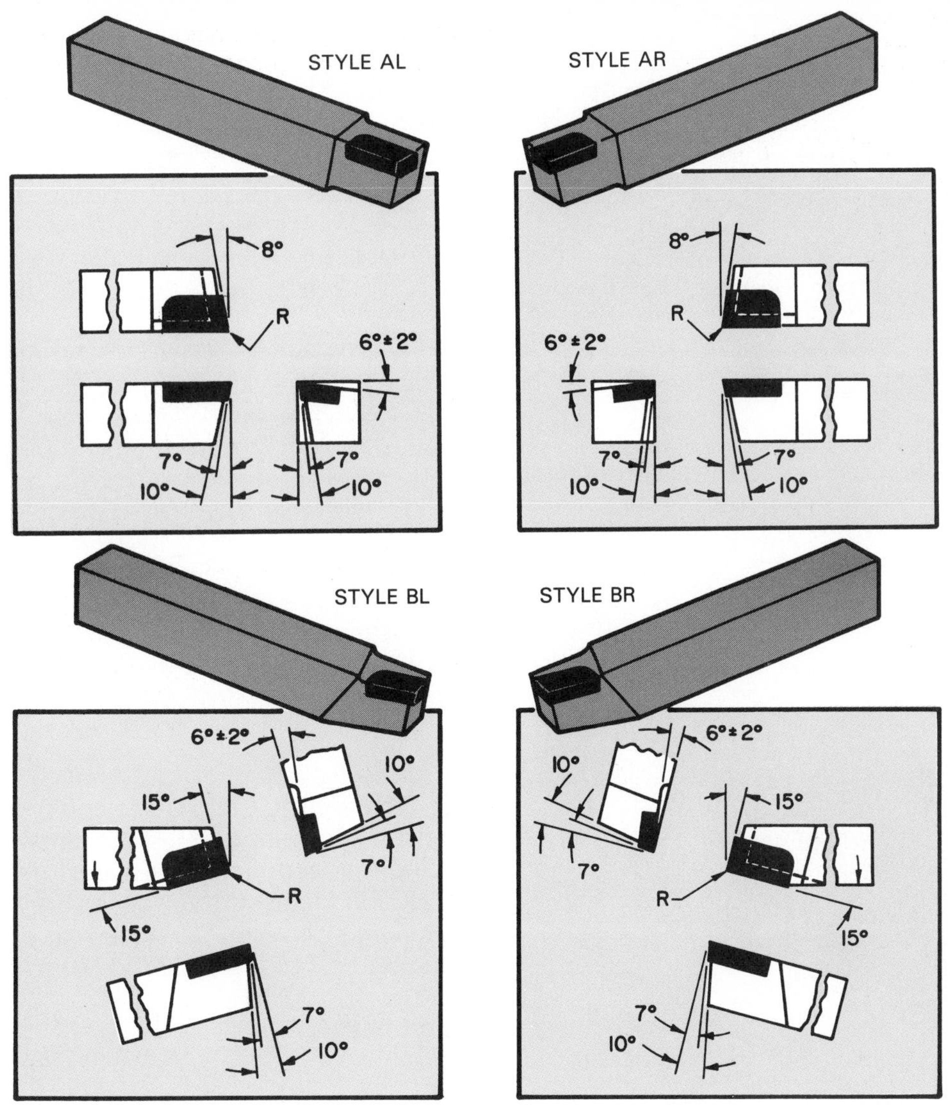

Fig. 10-50. Study typical standard cemented-carbide single-point tools. (Carboly Co.)

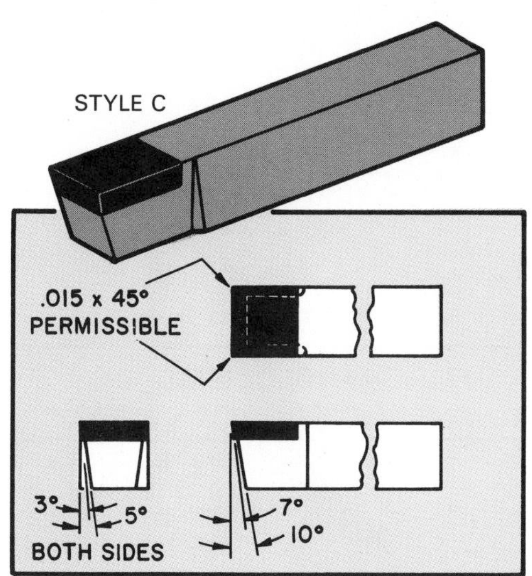

Fig. 10-51. Study typical square-nose, cemented-carbide single-point tool.

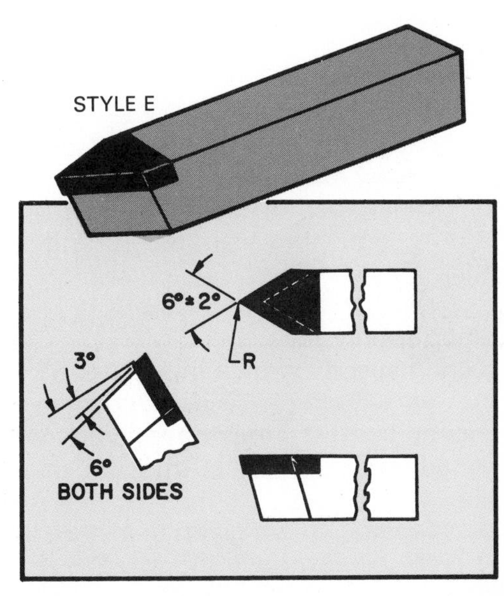

Fig. 10-52. Study carbide tipped threading tool.

Fig. 10-53. These are a few of the many different types of throwaway insert tools of carbide and sintered oxides, often referred to as ceramics, cermets, etc.). (Sandvik, Coromant Co.)

MATERIAL TO BE CUT	ROUGHING CUT 0.01 to 0.020 in. 0.25 to 0.50 mm Feed		FINISHING CUT 0.001 to 0.020 in. 0.050 to 0.24 mm Feed	
	ft/min	m/min	ft/min	m/min
Cast Iron	70	20	120	36
Steel				
Low Carbon	130	40	160	56
Med Carbon	90	27	100	30
High Carbon	50	15	65	20
Tool Steel				
(annealed)	50	15	65	20
Brass—Yellow	160	56	220	67
Bronze	90	27	100	30
Aluminum*	600	183	1000	300

The speeds for rough turning are offered as a starting point. It should be all the machine and work will withstand. The finishing feed depends upon the finish quality desired.

*The speeds for turning aluminum will vary greatly according to the alloy being machined. The softer alloys can be turned at speeds upwards of 1600 ft/min (488 m/min) roughing to 3500 ft/min (106 m/min) finishing. High silicon alloys require a lower cutting speed.

Fig. 10-55. Study suggested cutting speeds and feeds for turning with high speed steel tools.

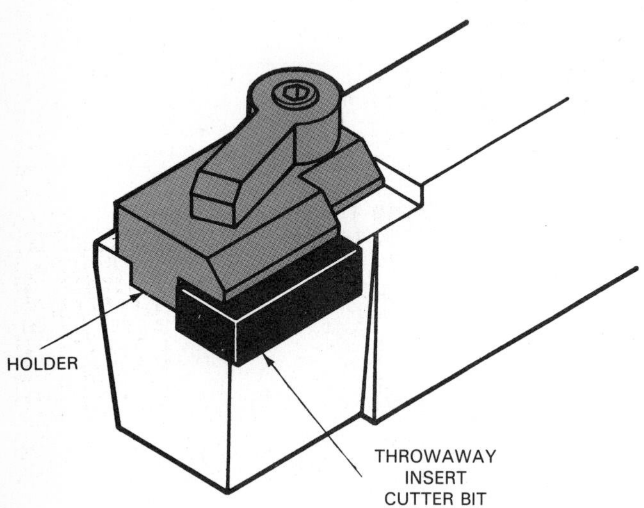

Fig. 10-54. Note tip of a typical throwaway insert carbide cutter bit and its holder. This is only one of many styles available.

The ***speed and feed chart*** takes the many factors into consideration and is recommended for use with high-speed cutter bits, Fig. 10-55. They can be increased 50 percent if a coolant is used, and 300 to 400 if a cemented carbide cutting tool is employed.

Calculating cutting speeds

The cutting speeds shown in Fig. 10-55 are offered only as a starting point. Depending upon machine condition,they may have to be increased or decreased until optimum cutting conditions are obtained.

Cutting speeds (CS) are given in ***feet per minute*** (FPM) or ***meters per minute*** (MPM). Speed of the work (spindle speed) is given in ***revolutions per minute*** (RPM). Thus, the ***peripheral speed*** (speed at circumference or outside edge) or CS of the work must be converted to RPM to determine the required spindle speed. The following formulas are used:

NOTE: Since cutting speeds are approximate, the following formulas have been simplified (π has been rounded off to 3 from 3.1416) to aid calculations.

Inch based:

$$RPM = \frac{CS \times 4}{D}$$

Where RPM = Revolutions per minute

CS = Cutting speed recommended for the particular material being machined (steel, aluminum, etc.) in feet per minute.

D = Diameter of work in inches. Convert all fractions to decimals.

Example: What spindle speed is required to finish turn 4 in. diameter aluminum alloy?

$$RPM = \frac{CS \times 4}{D}$$

$$= \frac{1000 \times 4}{4}$$

$$= 1000 \text{ RPM}$$

CS = Table recommends a cutting speed of 1000 FPM for finish turning aluminum alloy.

D = 4 in.

Adjust the spindle speed to as close to this speed as possible. Increase or decrease speed as needed to obtain desired surface finish.

Metric based:

$$RPM = \frac{CS \times 1000}{D \times 3}$$

Where RPM = Revolutions per minute
CS = Cutting speed recommended for particular material being machined (steel, aluminum, etc.) in meters (m) per minute (m/pm).
D = Diameter of work in millimeters (mm).

Example: What spindle speed is required to finish turn 100 mm diamter aluminum alloy?

$$RPM = \frac{CS \times 1000}{D \times 3}$$

$$= \frac{300 \times 1000}{100 \times 3}$$

$$= 1000 \text{ RPM}$$

CS = Table recommends a cutting speed of 300 m/pm for finish turning aluminum alloy.
D = 100 mm

Adjust the spindle to as close to this speed as possible. Increase or decrease speed as needed to obtain desired surface finish.

Roughing cuts

Roughing cuts are taken to reduce the work diameter to approximate size. The work is left 1/32 in. (0.08 mm) oversize for finish turning.

As the finish obtained on the roughing cut is of little importance, the highest speed and coarsest feed, consistent with safety and accuracy, is used.

Finishing cuts

The ***finish cut*** brings the work to the required diameter and surface finish. A high spindle speed,with a sharp tool and fine feed are employed.

Depth of cut

The ***depth of cut*** refers to the distance the cutter has been fed into the work surface. The depth of cut, like feed, varies greatly with lathe condition, material hardness, speed, feed, and amount of material to be removed, and whether it is to be a roughing or finishing cut.

Cut depth can be set accurately with a ***micrometer dial*** on both the cross slide and compound rest, Fig. 10-56.

The micrometer dial is usually graduated in 0.001 in. or 0.02 mm which means that a movement of ONE graduation feeds the cutting tool 0.001 in. or 0.02 mm into the piece. However, it must be remembered that material is removed around the periphery (outside edge) of the work at double the depth adjustment. For example, the diameter is reduced 0.002 in. or 0.04 mm for each 0.001 in. (0.02 mm) of infeed, Fig. 10-57. This must NOT be forgotten or TWICE as much material as specified will be removed.

Fig. 10-56. Left. Inch based micrometer dials on cross slide and compound rest handwheels. Right. Combination inch-metric graduated micrometer dials on cross slide and compound rest handwheels. (Clausing Machine Tools)

Some lathes, however, have a micrometer dial that is setup so that the number of graduations the cutter is fed into the work equals the amount the work diameter will be reduced. That is, if the cutter is fed in 0.005 in. (0.10 mm) or 5 graduations, the work diameter will be reduced 0.005 in. (0.10 mm). Check the lathe you will be using to be sure which system it employs.

A common mistake is to remove too little material at too slow a speed. Cuts as deep as 0.125 in. (3.00 mm) can be handled by light lathes with cuts of 0.250 in. (6.00 mm) and deeper with heavier machines without overtaxing the lathe.

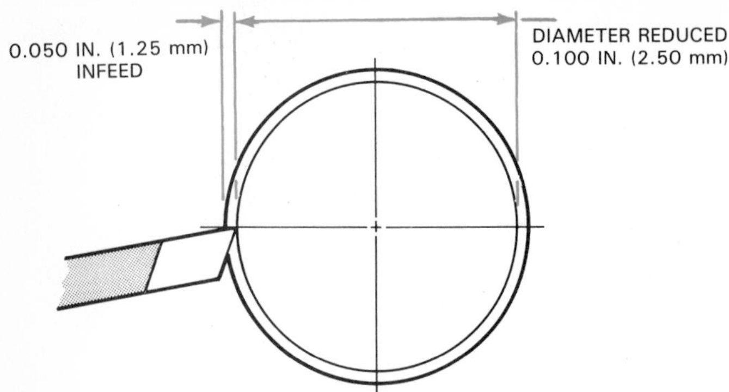

Fig. 10-57. Remember! Material is removed on each cut at two times infeed distance.

Fig. 10-59. Here work is held in chuck for machining. (Clausing Machine Tools)

WORK HOLDING ATTACHMENTS

One factor that makes the lathe such a versatile machine tool is the great variety of ways work may be mounted in or on a lathe. The most common way is to mount the work so it revolves permitting the cutting tool to move across the rotating surface. Large and/or odd-shaped pieces are sometimes mounted on the carriage and machined with a cutting tool mounted in the rotating spindle.

Most work, however, is machined while supported by one of the following methods:

1. Between centers using a faceplate and dog, Fig. 10-58.
2. Held in a chuck, Fig. 10-59.
   a. 3-jaw universal chuck.
   b. 4-jaw independent chuck.
   c. Jacobs type chuck.
3. Held in a collet, Fig. 10-60.
4. Bolted to the faceplate, Fig. 10-61.

Fig. 10-60. Note work held in a collet.

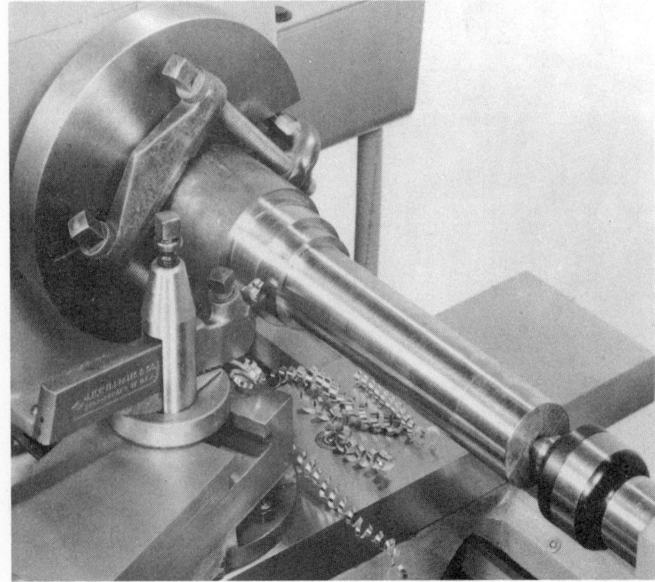

Fig. 10-58. This machining work is being done mounted between centers.

Fig. 10-61. Work has been bolted to faceplate for machining.

TURNING WORK BETWEEN CENTERS

Considerable lathe work is done with the work supported between centers. For this operation, a faceplate is attached to the spindle nose, Fig. 10-62. A sleeve and live center are inserted into the spindle opening, Fig. 10-63.

A ***dead center*** (center which does not revolve), Fig. 10-64, or heavy duty ***ball bearing center,*** Fig. 10-65, is fitted into the tailstock spindle. The ends of the stock are drilled to fit the center points.

A lathe dog, Fig. 10-66, is clamped to one end of the material, Fig. 10-66. Three types of ***lathe dogs*** are shown in Fig. 10-67:

1. ***Bent-tail standard dog*** has setscrew exposed.
2. ***Bent-tail safety dog*** has setscrew recessed. This type dog is usually preferred to the standard lathe dog.
3. ***Clamp-type dog*** is for turning square or rectagular shape work.

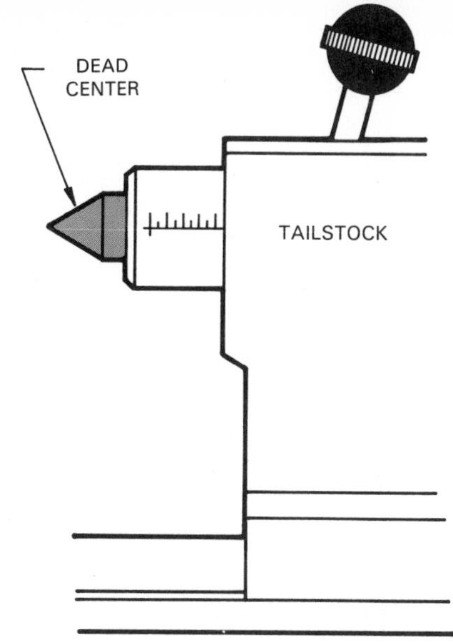

Fig. 10-64. Dead center does not rotate and is fitted to tailstock spindle.

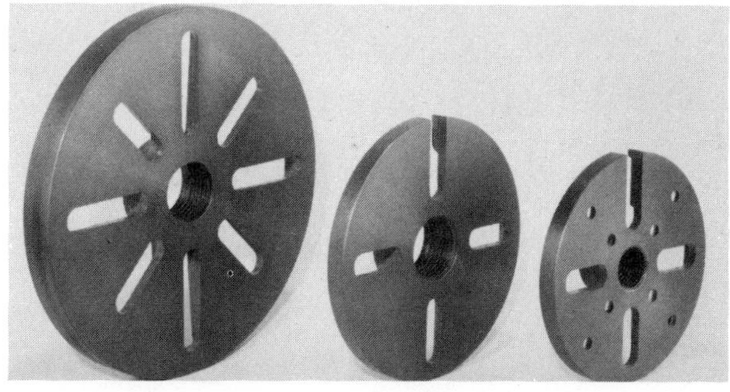

Fig. 10-62. Lathe faceplates come in various sizes.

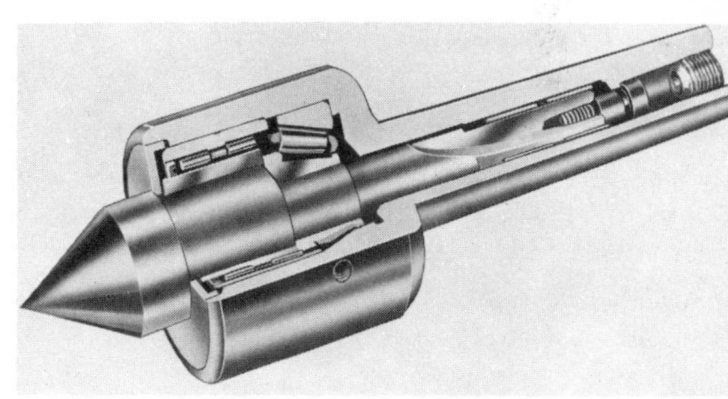

Fig. 10-65. Cutaway shows construction of heavy-duty ball bearing center.

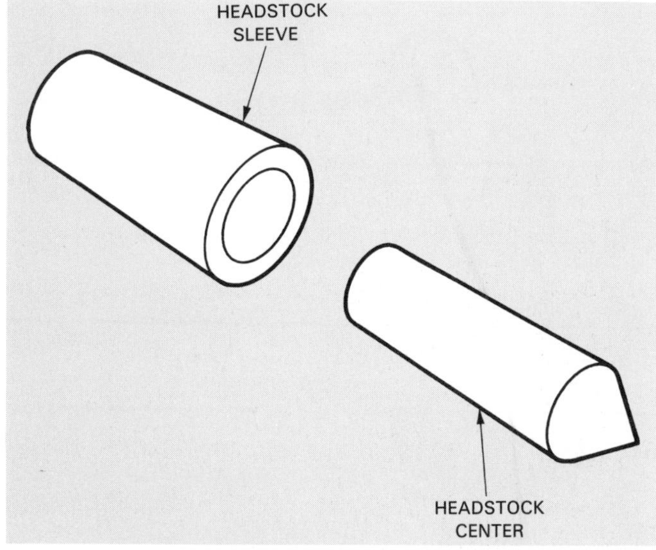

Fig. 10-63. Note sleeve and headstock center.

Fig. 10-66. Machining work is mounted between centers.

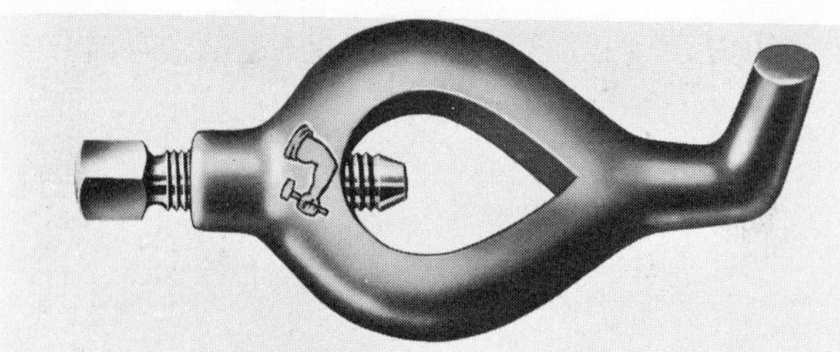

A—Bent tail standard lathe dog.

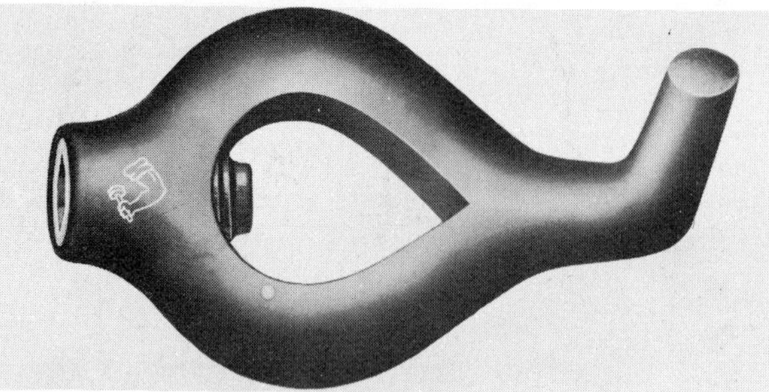

B—Bent tail safety lathe dog.

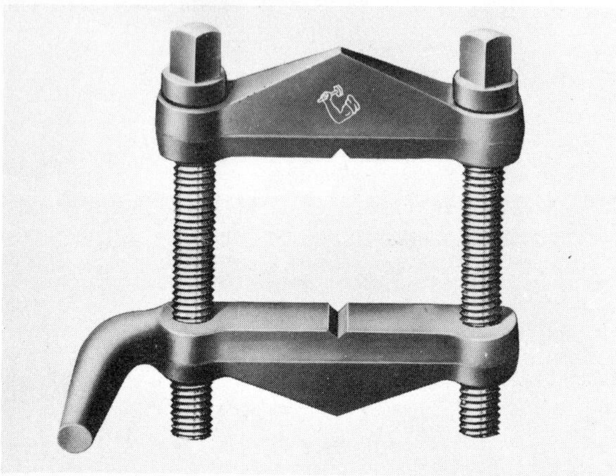

C—Clamp type lathe dog.

Fig. 10-67. Study variations of lathe dogs. (Armstrong Bros. Tool Co.)

Drilling center holes

Before work can be mounted between centers, it is necessary to locate and drill ***center holes*** in each end of the stock, Fig. 10-68. Fig. 10-69 illustrates several methods for locating the center of round stock.

Center holes are usually drilled with a ***combination drill and countersink,*** Fig. 10-70. The ***drill angle*** is identical to that of the center point. The ***straight drill*** provides clearance for the center point and serves as a reservoir for a lubricant. To select the correct size center drill, refer to Fig. 10-71.

The center holes can be drilled on:

1. A drill press, Fig. 10-72.
2. The lathe with the work centered in the chuck, Fig. 10-73.

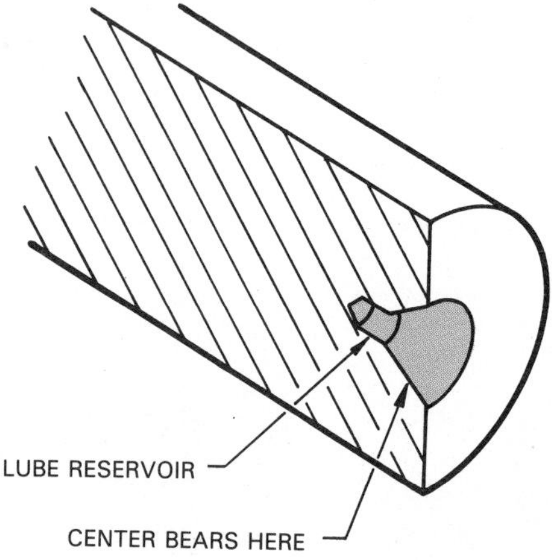

Fig. 10-68. Tail center rides in center hole. A supply of lubricant is placed in reservoir; this expands and lubes center as metals heats up.

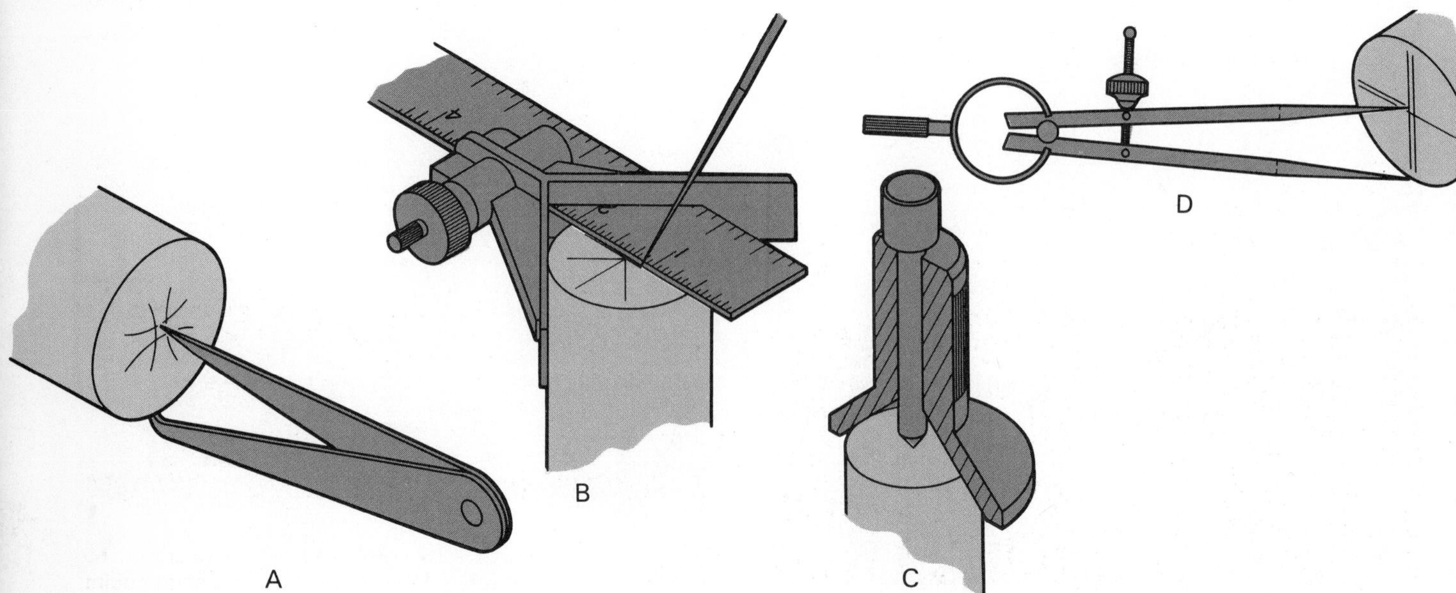

Fig. 10-69. Here are several ways to locate center of round stock. A—With hermaphrodite caliper. B—With center head and rule of combination set is recommended. C—With bell center punch. D—With dividers.

Fig. 10-70. Plain type combination drill and countersink will make hole and countersink it in one operation. (Standard Tool Co.)

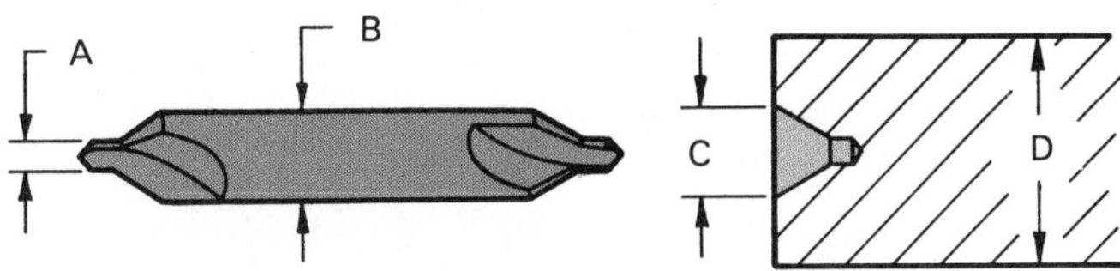

COMBINATION DRILL AND COUNTERSINK NO.	A	B	C	D
1	1/16	13/64	1/8	32/16 TO 5/16
2	3/32	3/10	3/16	3/8 TO 1
3	1/8	3/10	1/4	1 1/4 TO 2
4	5/32	7/16	5/16	2 1/4 TO 4

Fig. 10-71. Study center drill chart.

Fig. 10-72. Center holes can be drilled on drill press. Mount work in a V-block for support.

Fig. 10-73. Some work can be held in a lathe chuck for center drilling.

3. The lathe with the center drill held in the headstock, Fig. 10-74.

The center holes should be drilled deep enough to provide adequate support, Fig. 10-75.

Checking center alignment

Accurate turning between centers requires centers that run true and are in precise alignment. As the work must be reversed to machine its entire length, care must be taken to make the live center run true. If they do NOT, the diameters will be ***eccentric*** (not aligned on the same center line), Fig. 10-76. This can be prevented by truing the live center. A light truing cut can be made if the center is NOT hardened. A tool post grinder will be needed to true it.

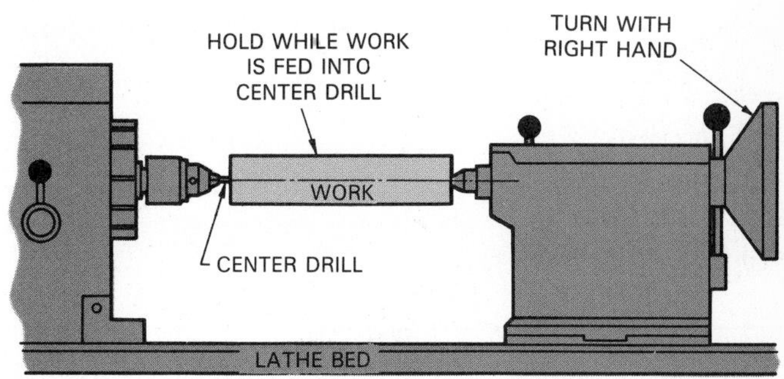

Fig. 10-74. Center holes can be drilled in large stock by mounting a Jacobs chuck in the headstock. Locate center point of each end and center punch. Support one end on tail center and feed other into center drill mounted in the Jacobs chuck. Repeat operation on second end.

A tapered piece will result if the centers are NOT aligned. Approximate alignment can be determined by:

1. Checking centers visually by bringing the points together, Fig. 10-77.
2. Checking the witness lines on the base of the tailstock for alignment, Fig. 10-78.

A more precise method for checking alignment is needed if close tolerance work is to be done:

1. Make a light trial cut across a few inches of the material. Check the diameter at each end of the cut with a micrometer, Fig. 10-79. The centers are aligned if the readings are identical.
2. Use a steel test bar and dial indicator, Fig. 10-80. Mount the test bar between centers and position the dial indicator in the tool post and at right angles to the work. Move the indicator contact point against the test bar until a reading is shown. Move the indicator along the test bar. If the readings remain constant, the centers are aligned.

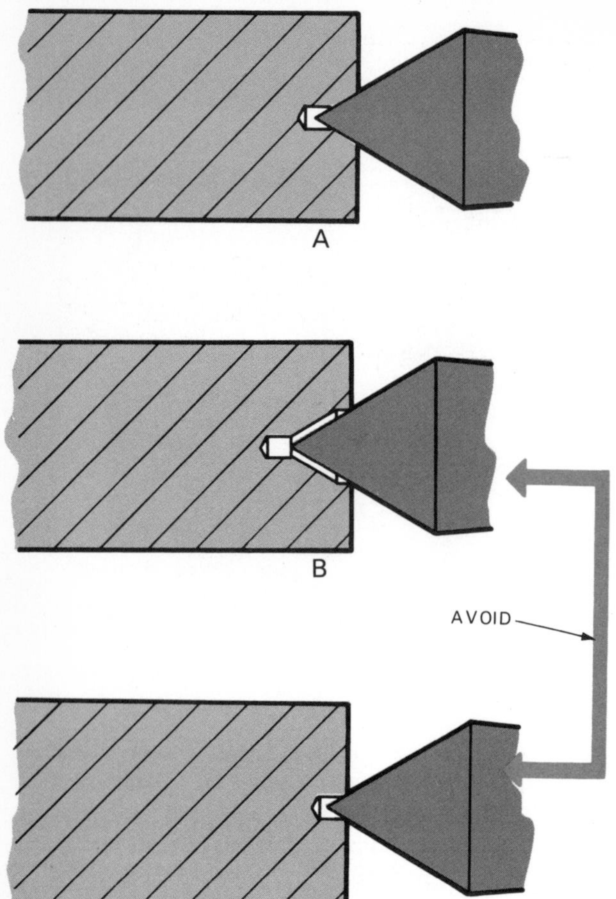

Fig. 10-75. Note correct and incorrect center holes. A—Properly drilled center hole. B—Hole drilled too deep. C—Hole not drilled deep enough. Not enough support, center point, if dead center, will burn off.

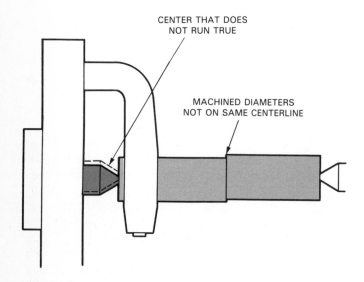

Fig. 10-76. Eccentric diameters will result if live center does NOT run true. Piece must be reversed in dog so stock can be machined its entire length.

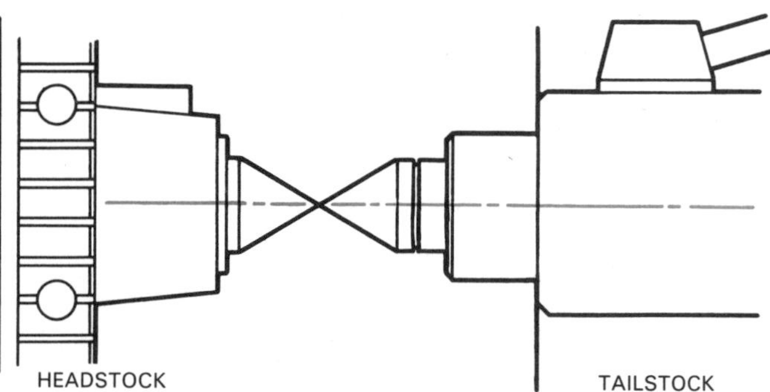

Fig. 10-77. You can check center alignment by bringing center points together. View is looking down on top of centers.

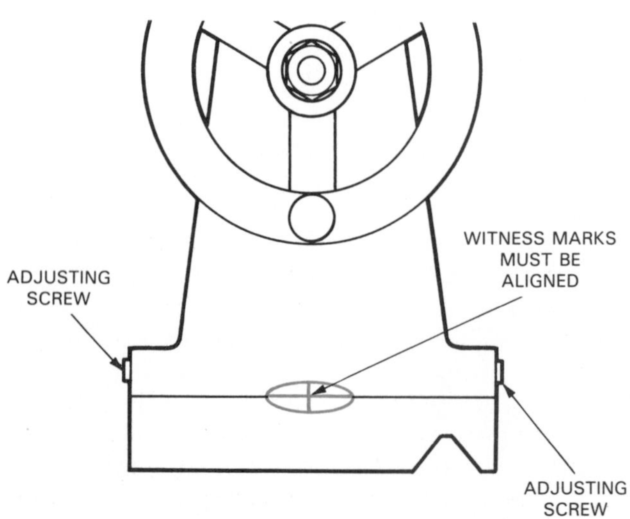

Fig. 10-78. Alignment of centers can also be detemined by checking witness marks on base of tailstock.

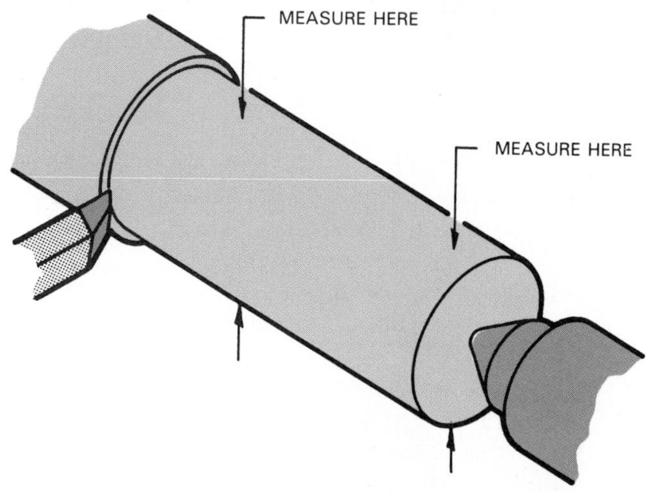

Fig. 10-79. Make a light cut on stock and measure diameter at two points to check alignment. Measurements must be equal.

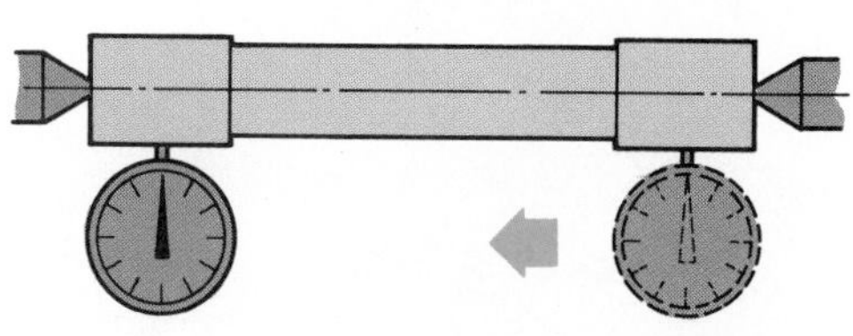

Fig. 10-80. Test bar and dial indicator is efficient way to check center alignment.

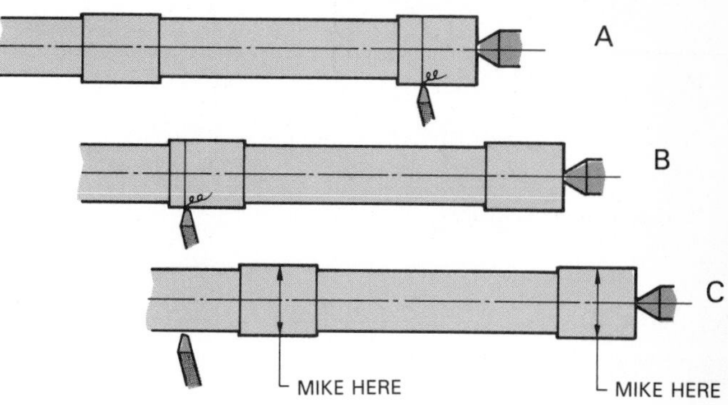

Fig. 10-81. A—Machine two shoulders on a test piece. B—Keep same tool setting and make a cut on both shoulders. C—Measure resulting diameters.

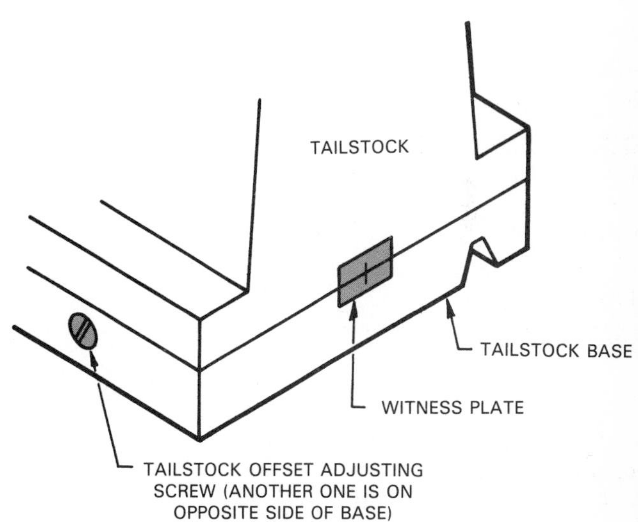

Fig. 10-82. An adjusting screw is located on BOTH sides of tailstock base. They are used to set over or shift tailstock for taper turning.

3. Machine a section of scrap, as shown in Fig. 10-81. Set the cross feed screw to make a light cut at the right end of the piece. With the same tool setting, move it to the left and continue the cut. Identical micrometer readings indicate center alignment.

Adjusting screws on the base of the tailstock (one on each side) are used to align the centers if checks indicate this is needed. Make adjustments gradually. See Fig. 10-82.

Mounting work between centers

Clamp a dog to one end of the work. Place a ***lubricant*** (white lead, graphite and oil, or a commercial center lubricant) in the center hole on the other end. Mount the piece on the centers and adjust the tail center until the work is snug. It will "clatter" if too loose, and will score or burn the center point if adjusted too tightly.

Check the adjustment from time to time as the heat generated by the machining process causes the work to expand. Using a ball bearing center will reduce or eliminate many problems when working between centers.

Use a different faceplace if the dog tail binds on the faceplate slot, Fig. 10-83. This can cause the work to be pulled off center. When machined, this will produce a surface that is NOT concentric with the center hole.

Fasing work held between centers

Facing is the operation that machines the end of the work square and reduces it to a specific length. At times, considerable material must be removed. In this situation, it is best to leave the work longer than finished size and drill deeper center holes for better support during the roughing operation.

Face the work to length before starting the finish cut. A ***right-cut facing tool*** will be needed. The 58 deg. point provides a slight clearance between the center point and the work face, Fig. 10-84. Do NOT damage the cutting tool point by running it into the center. A ***half center*** makes the operation easier, but is only used for facing as it does not provide

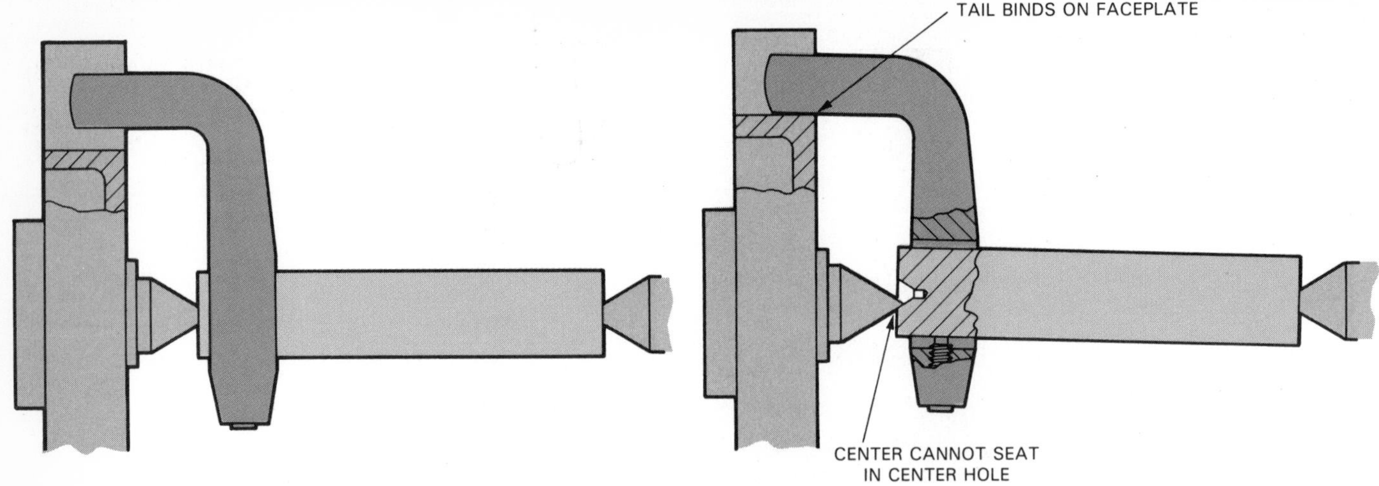
TAIL BINDS ON FACEPLATE
CENTER CANNOT SEAT
IN CENTER HOLE

Fig. 10-83. Diameter of turned surface will NOT be concentric with center holes if center hole is NOT seated properly on headstock center. A binding lathe dog is a common reason for this problem.

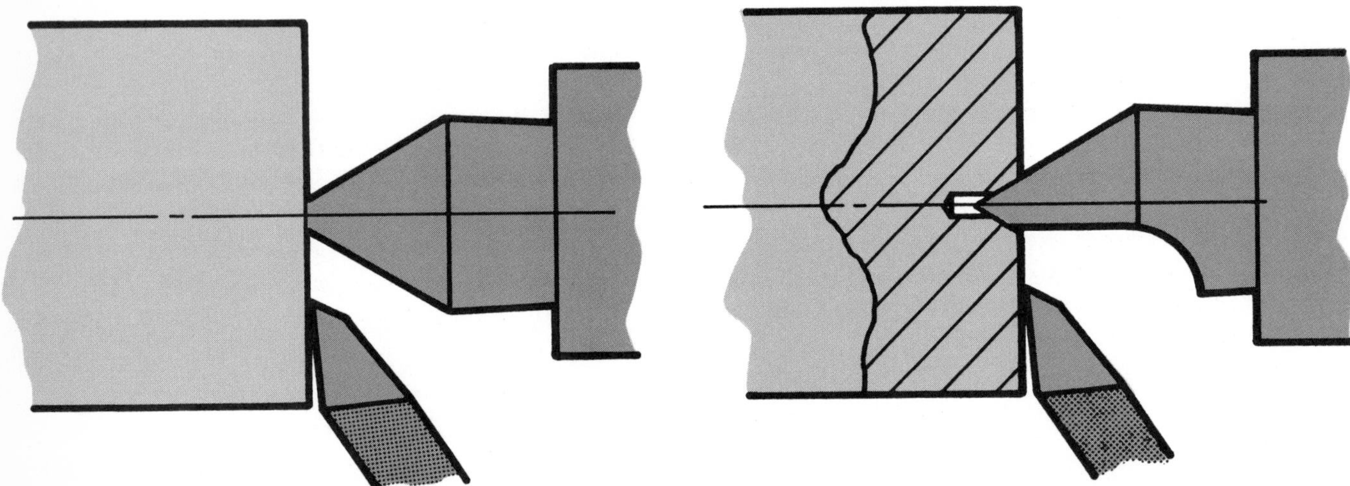

Fig. 10-84. Note relationship of cutter bit to work face when making a facing cut.

Fig. 10-85. Using half center will give more clearance when facing end of stock.

an adequate bearing surface and will not hold the lubricant. See Fig. 10-85.

Facing to length

It is standard practice to cut stock slightly longer than needed to permit its ends to be machined square. A steel rule may be employed if the dimension is NOT critical. For more accuracy, a Vernier caliper or large micrometer may be used. The difference between the rough length and the required length is the amount of material that must be removed.

Set the compound at 30 deg., Fig. 10-86. Bring the cutting tool up until it just touches the surface to be machined and lock the carriage. Remove material from each end of the stock until the specified length is attained.

Rough turning between centers

Rough turning is the operation whereby excess material is cut away rapidly with little regard for the quality of surface finish. The diameter is reduced to within 1/32 in. (0.8 mm) of required size by employing deep cuts and coarse feeds.

Set the compound at 30 deg. from a right angle to the work, Fig. 10-87. This will permit the tool to cut as close as possible to the left end of the work without the dog striking the compound rest.

WARNING! The maximum distance compound can be fed towards dog or chuck without striking them should always be checked BEFORE STARTING the lathe.

Use a left-hand toolholder. Position tool post as far to left in compound rest T-slot as possible. Avoid too much tool overhang, Fig. 10-88.

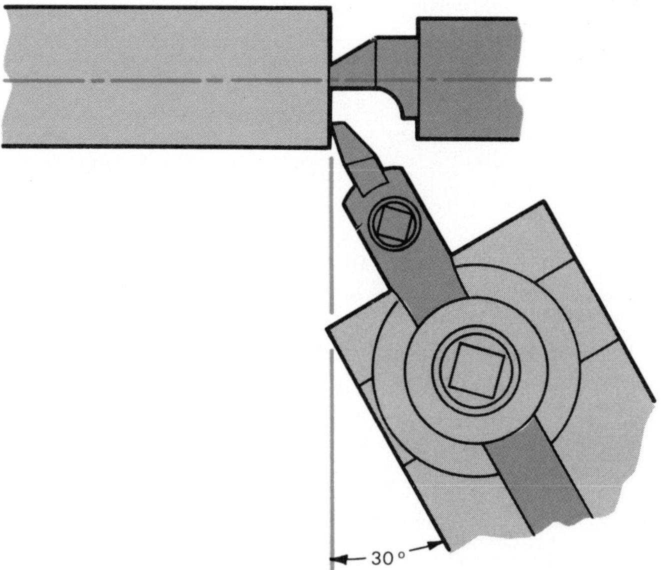

Fig. 10-86. This is recommended compound setting when facing stock to length.

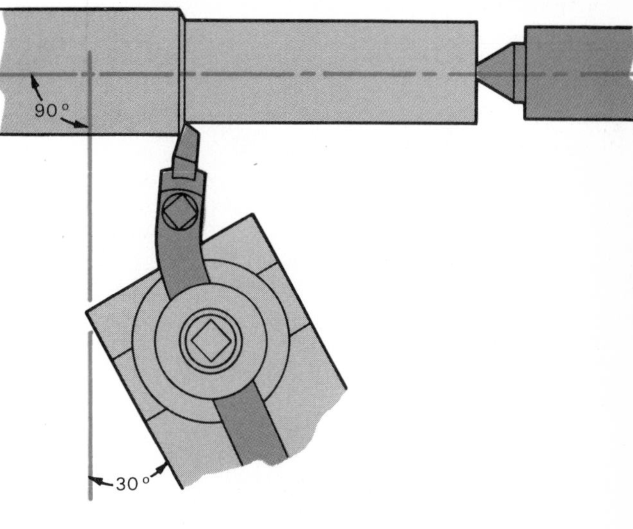

Fig. 10-87. Here is compound setting for rough turning.

Locate the cutting edge of the tool about 1/16 in. (1.5 mm) above work center for EACH INCH of diameter, Fig. 10-89. It can be set by comparing it with the tail center point or with an index line scribed on the tailstock ram of some lathes.

The toolholder must be positioned correctly; otherwise, the heavy side pressure developed during the machining will cause it to turn in the tool post and force the cutting tool deeper into the work, Fig. 10-90. Correctly positioned, the cutting tool will pivot away from the work, Fig. 10-91.

Make a trial cut to true up the stock. Measure the resulting diameter. The difference between the diameter and the required rough diameter is TWICE the distance the tool must be fed into the work. If the piece is greatly oversize, it will be necessary to make two or more cuts to bring it to size.

When depth of cut has been determined, engage the power feed. Observe the condition of the chips. They should be in SMALL sections and slightly BLUE in color. Long, stringy chips indicate a cutting tool that is NOT properly sharpened. Remove these chips with pliers after stopping the machine. Measure work diameter after each cut to prevent excess metal removal.

DANGER! Stop the machine before making measurements or cleaning out chips.

If a ball bearing center is NOT used, lubricate the tail center frequently. STOP THE MACHINE immediately if the center heats up and starts to smoke or ''squeal.''

Finish turning

The work is still oversize after rough turning. It must be machined to the specified diameter and to a smooth surface finish with ***finish turning.***

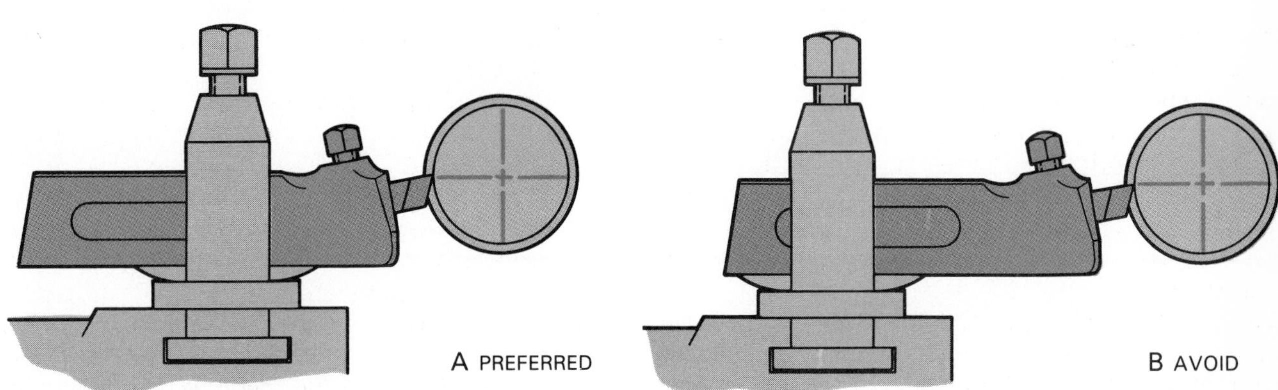

Fig. 10-88. Note correct and incorrect mounting of tool holder. A—Tool holder and center bit in proper position. B—Too much overhang, tool will ''chatter'' and produce a rough machined surface.

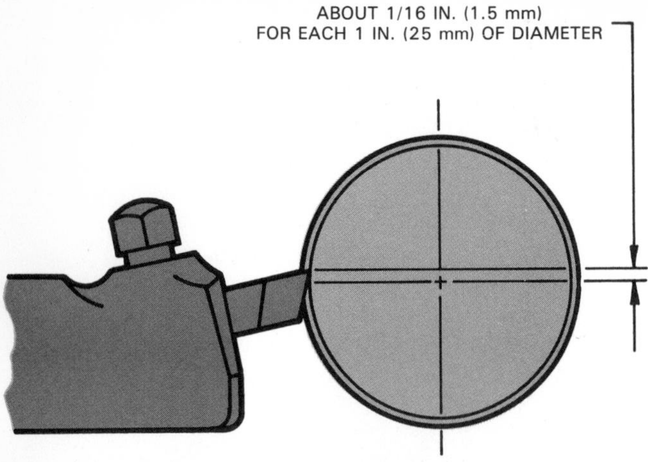

Fig. 10-89. Set point of cutter bit about 1/16 in. (1.5 mm) above center for each inch (25 mm) of diameter when rough turning. Remember to lower point as work diameter decreases in size.

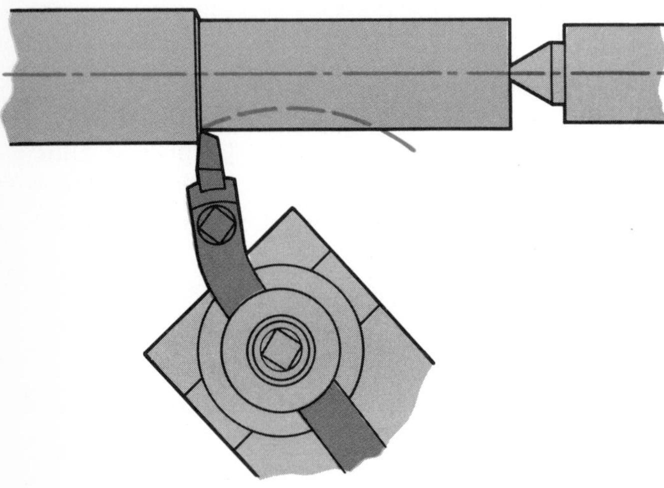

Fig. 10-90. An incorrectly positioned tool will cut deeper and deeper into work if it slips in tool post.

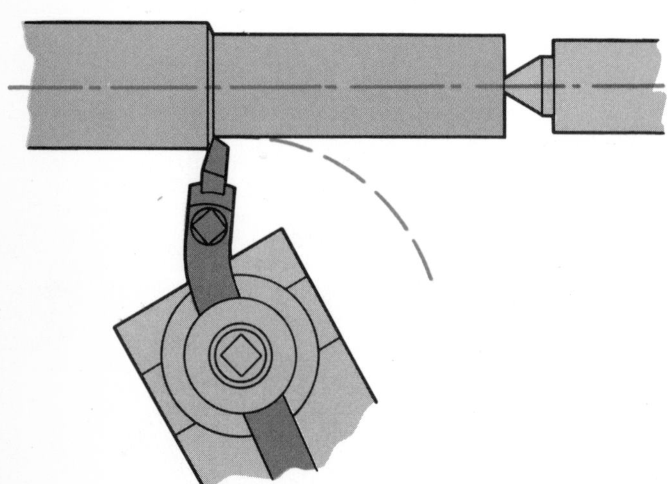

Fig. 10-91. A correctly positioned tool will swing clear of work if tool holder slips in tool post.

A right-cut finishing tool is fitted into the toolholder, Fig. 10-92. Position it on center and check for adequate clearance between the compound rest and the revolving lathe dog.

Adjust the lathe for a faster spindle speed and a fine feed. Run the cutting tool into the work until a light cut is being made; then engage the power feed. After a sufficient distance has been machined, disengage the power feed and stop the lathe.

SAFETY NOTE! Under NO condition should a lathe be reversed to brake it to a stop!

Do NOT interfere with the cross slide setting. ''Mike'' the diameter of the machined area. The difference between the specified diameter and the measurement is the amount of material that must be removed. Move the cutting tool clear of the work and feed it in ONE-HALF the amount that must be removed. For example, if the diameter is 0.008 in. (0.20 mm) OVERSIZE, tool infeed will be 0.004 in. (0.10 mm). Make another cut about 1/2 in. (13 mm) wide and again measure to make sure the correct diameter will be machined.

When reversing the work to permit machining its entire length, protect the section under the lathe dog setscrew by inserting a piece of soft aluminum or copper sheet, Fig. 10-93.

All rough and finish machining should be done TOWARDS THE HEADSTOCK (right to left) because the headstock offers a more solid base than the tailstock.

Turning to a shoulder

Up to this point, only plain turning in which the entire length of the piece has been machined to a specified diameter has been described. However, it is frequently necessary to machine a piece to several different diameters.

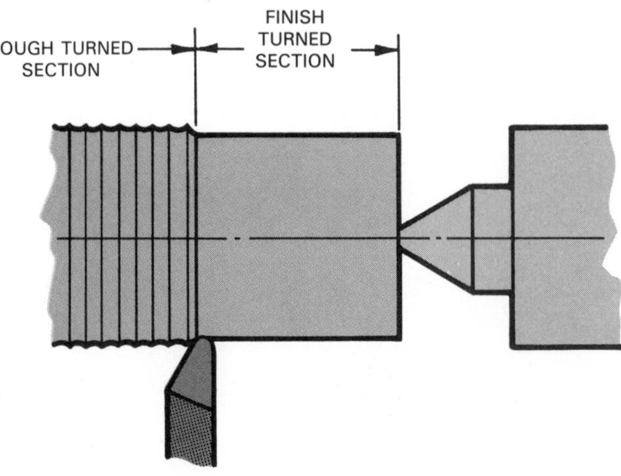

Fig. 10-92. Note exaggerated drawing of rough and finished turned sections while using right-cut finishing tool.

Locate the points to which the different diameters are to be cut. Scribe the lines with a hermaphrodite caliper which has been set to the required length, Fig. 10-94.

Maching is done as previously described with the exception of cutting the shoulder, Fig. 10-95. This is the point where the diameters change and one of the following types of ***shoulders*** is specified:

1. Square shoulder.
2. Angular shoulder.
3. Filleted shoulder.
4. Undercut shoulder.

A ***right-cut tool*** is used to make the square and angular type shoulders, Figs. 10-96 and 10-97.

A ***round nose tool*** is ground to the required radius using a fillet or radius gage to check radius accuracy for machining a filleted shoulder. See Fig. 10-98.

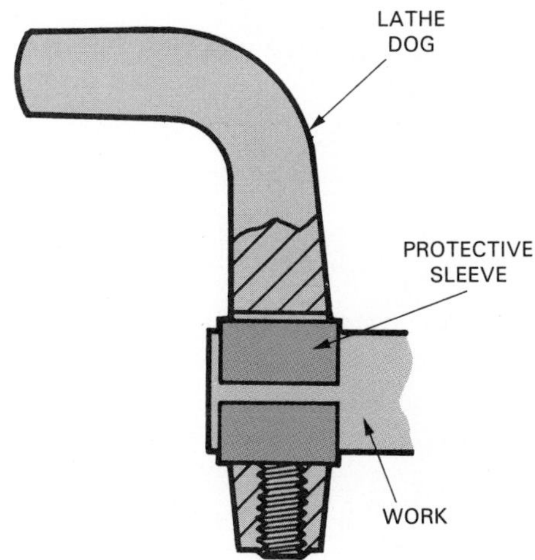

Fig. 10-93. Finished surface of work is protected from lathe dog screw by a section of soft metal.

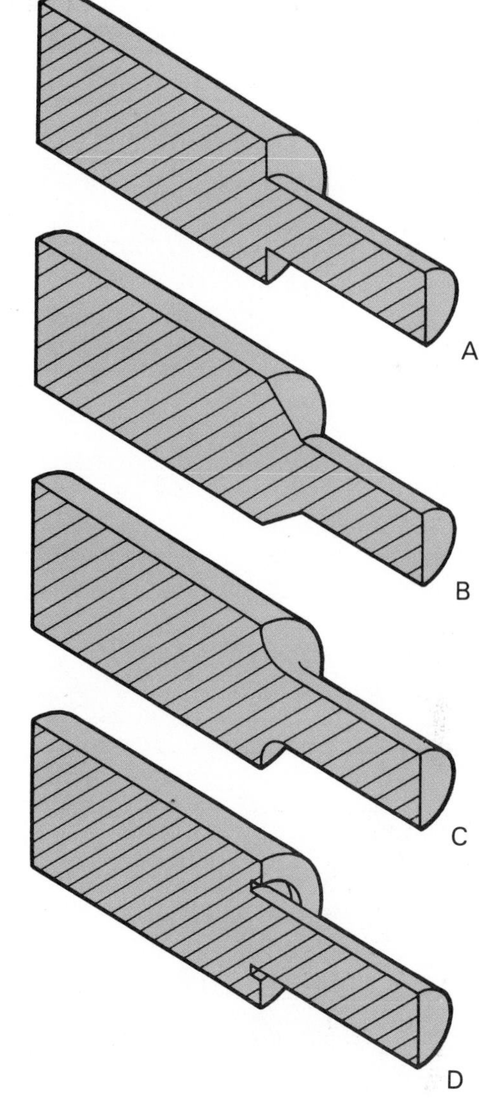

Fig. 10-95. Note four kinds of shoulders. A—Square. B—Angular. C-Filleted. D—Undercut.

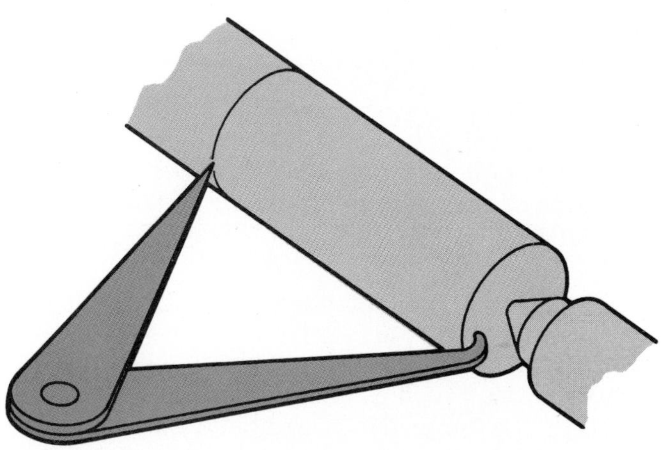

Fig. 10-94. Locating reference points on work with a hermaphrodite caliper.

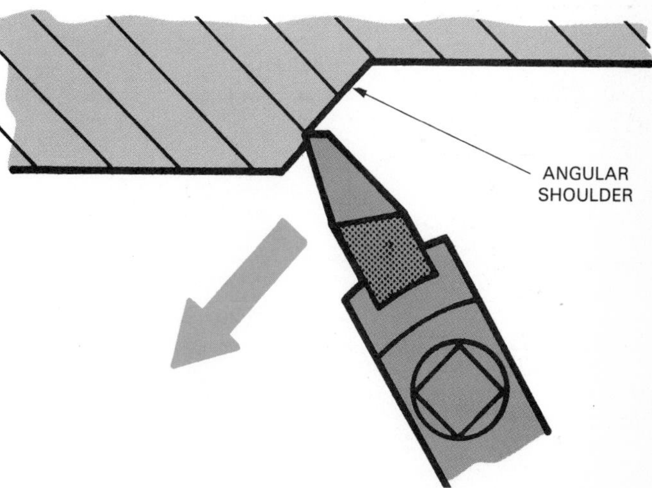

Fig. 10-96. To machine an angular shoulder, cut is made from smaller diameter to larger diameter.

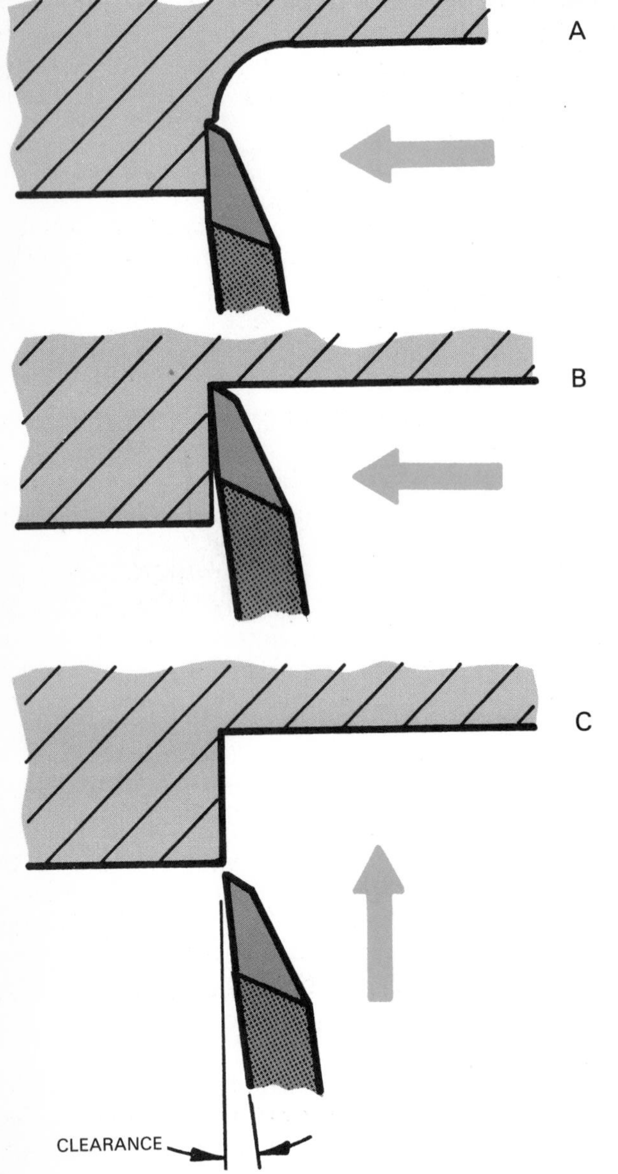

Fig. 10-97. Note machining sequence for cutting a square shoulder. A—First cut. B—Second cut. C—Facing cut.

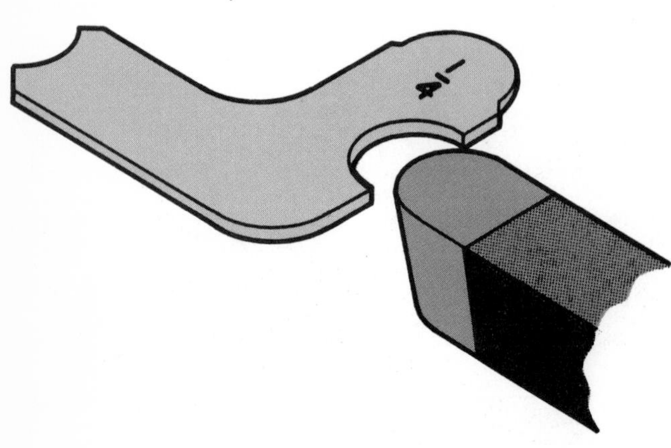

Fig. 10-98. Fillet gage can be used to check radius ground on a cutter bit.

Grooving or necking operations

It is sometimes necessary to cut a ***groove*** or ***neck*** on a shaft to terminate a thread, or to provide adequate clearance for mating pieces, Fig. 10-99. As any recess cut into a surface has a tendency to weaken a shaft, it is better to make the groove ROUND, rather than square.

The tool is set on center and fed in until it just touches the work surface. Set the cross feed micrometer dial to zero and feed the tool in the required number of thousandths/millimeters for the specified depth. Square grooves can be machined with a ***parting tool.***

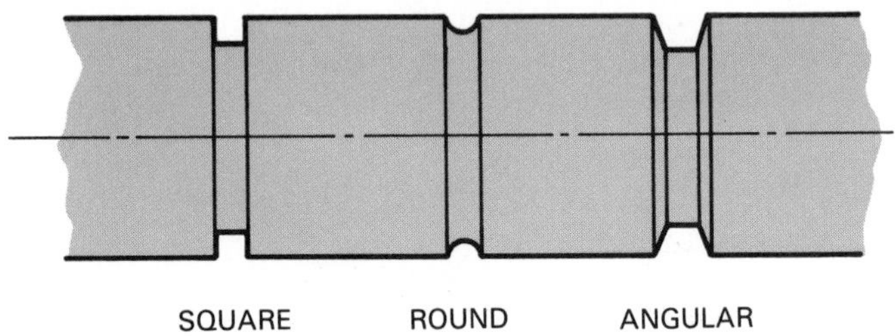

Fig. 10-99. A typical groove or neck is cut to one of these shapes.

USING LATHE CHUCKS

The chuck is another device for holding work in a lathe. ***Chucking*** is the most rapid method of mounting work and for that reason is widely preferred. Other operations, such as drilling, boring, reaming, and internal threading, can be done while the work is held in a chuck. Additional support can be obtained for the piece by supporting the free end with the tailstock center.

Chucks most commonly used are:

1. 3-jaw universal chuck.
2. 4-jaw independent chuck.
3. Jacobs chuck.
4. Draw-in collet chuck.

3-jaw universal chuck

The ***3-jaw universal chuck*** is designed to permit all jaws to operate at one time, Fig. 10-100. It will automatically center round or hexagonal shaped stock.

Two sets of jaws are supplied with each universal chuck because they CANNOT be reversed like the jaws on a 4-jaw independent chuck. One set of jaws is used to hold large diameter work, Fig. 10-101. The other set is to hold small diameter work, Fig. 10-102.

The jaws are numbered 1, 2, and 3 as are the slots in which they are fitted. The ***jaw number***

Fig. 10-100. The 3-jaw universal chuck is common. (L.W. Chuck Co.)

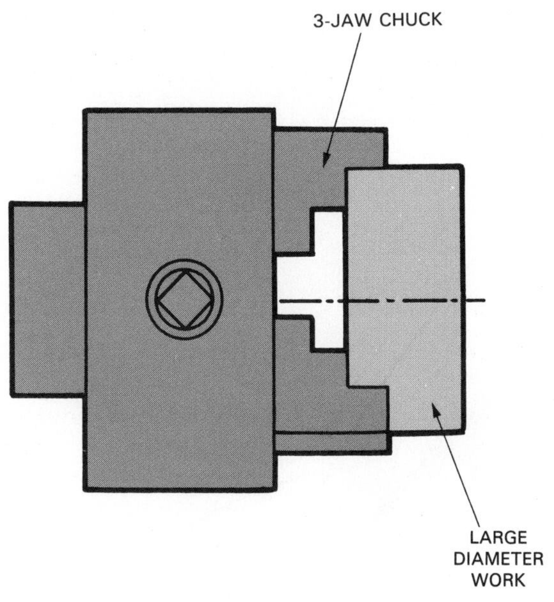

Fig. 10-101. One set of jaws, supplied with a 3-jaw universal chuck, is for mounting large diameter work.

MUST correspond with the ***slot number*** if the work is to be centered. The jaws are made for a particular chuck and are NOT interchangeable with other chucks. Make sure the chuck and jaws have the same serial number!

Installing chuck jaws

Clean the jaws, jaw slots, and ***scroll*** (spiral thread seen in the jaw slots). Turn the scroll until the first thread does NOT quite show in jaw slot #1. Slide jaw #1 into the slot as far as it will go. Now, turn the scroll until the spiral engages with the first tooth on the bottom of the jaw. Repeat the operation at slot #2 and slot #3, making sure the proper jaws are inserted.

Remember! Make it a habit that whenever you let go of a lathe chuck key, the chuck key is on the tool tray or lathe board.

Universal chuck jaws lose their centering accuracy as the scroll wears. The accuracy is also affected when too much pressure is used to mount the work, or by gripping work too near the front of the jaws, Fig. 10-103.

4-jaw independent chuck

Each jaw of the 4-jaw independent chuck operates individually, Fig. 10-104. This permits square, rectangular, and odd-shaped work to be centerd, Fig. 10-105. The jaws can be removed from the slot and reversed. The reversing feature permits the jaws to be used to hold large diameter work in one position and smaller diameter work when reversed, Fig. 10-106.

The most accurate technique for centering round work in this type chuck makes use of a dial indicator. The piece is centered approximately. The concentric rings on the chuck face serve as guide. A dial indicator is mounted in the tool post, Fig. 10-107. The jaws are then adjusted until the indicator needle does NOT fluctuate (move back and

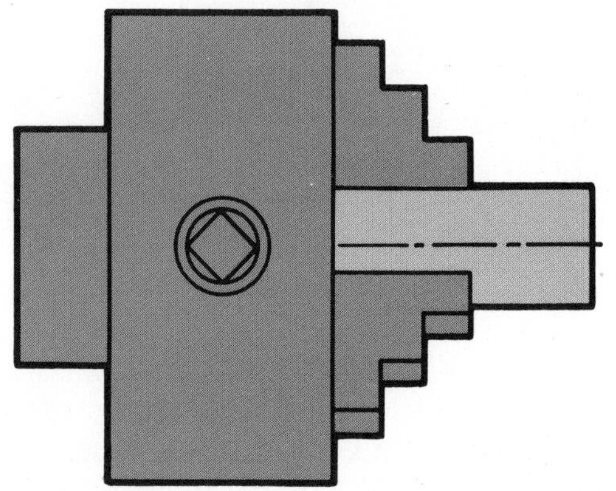

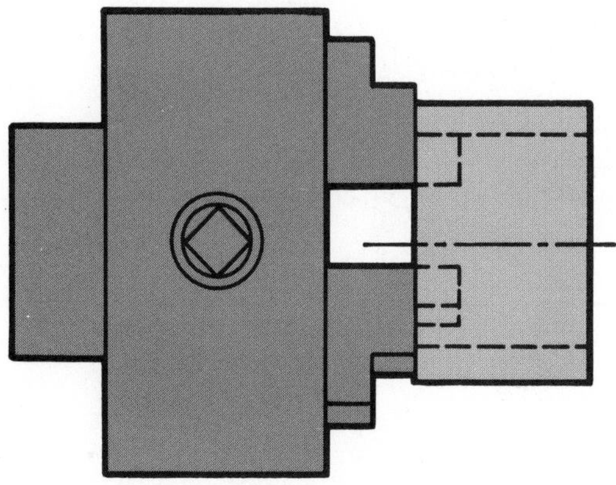

Fig. 10-102. Note methods of holding work using second set of jaws supplied with 3-jaw universal chuck.

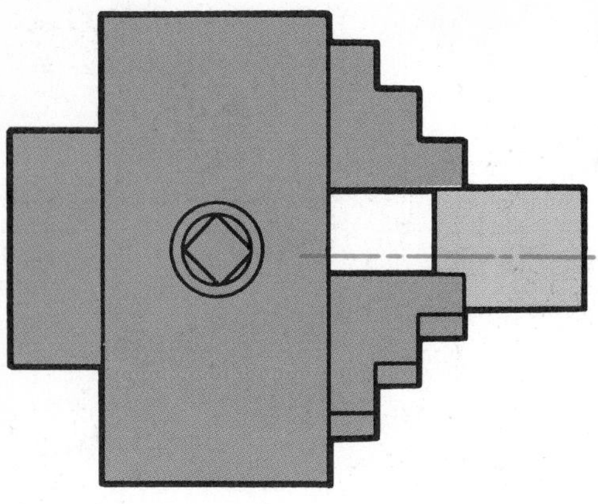

Fig. 10-103. Avoid gripping work near front of jaws. It could fly out and cause injuries.

Fig. 10-104. This is a 4-jaw independent chuck. (L.W. Chuck Co.)

A

B

C

D

Fig. 10-105. Study how irregular shaped work can be held in 4-jaw independent chuck. A—Position work between two aluminum sections. Work is held in place in a wooden form. B—Sections are clamped together. A final check is made to be sure work is positioned properly. C—A low melting temperature metal is poured into spaces and allowed to harden. Clamps and form are removed. D—Note completed holding fixture and piece that is to be machined while held in it.

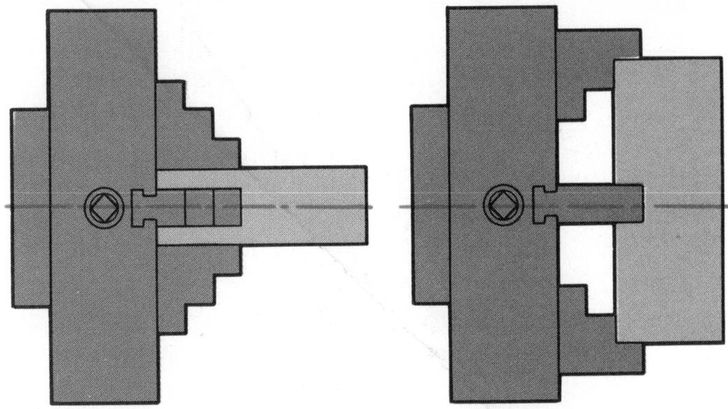

Fig. 10-106. Reversing feature of jaws in a 4-jaw independent chuck makes it possible to turn work having extreme differences in diameter without difficulty.

Fig. 10-107. Machinist is centering work in 4-jaw chuck using a dial indicator. Machine is a small modelmaker's lathe.

forth) when the work is rotated by hand. Tighten all jaws securely after the piece has been centered.

SAFETY NOTE! NEVER turn on the lathe until checking that you did NOT accidently leave the chuck key in the chuck.

Chalk can also be used to position the work on approximate center, Fig. 10-108. Rotate the work while bring the chalk into contact with it. Loosen the jaws opposite the chalk marks slightly. Then tighten the jaws on the side where the chalk mark appears. Continue the operation until the work is centered. A cutting tool may be used if the work is oversize enough.

Fig. 10-108. You can use chalk to center work in a 4-jaw chuck. Chalk mark indicates "high point." Loosen jaw opposite chalk mark; then tighthen jaw on chalk side. Use gradual adjustments rather than large movement.

Avoid trying to center stock in one or two adjustments, but rather in several small ***increments*** (very small steps). When making the final small adjustment, it may be necessary to loosen the jaw on the low side and retighten it, after which the high side is given a final tightening. This last method for making final adjustment applies, in particular, when centering work with a dial indicator.

Jacobs chuck

When small diameter work is to be turned, such as screws, pins, etc., the ***Jacobs chuck*** can be utilized, Fig. 10-109. This chuck is better suited for work of this nature than the larger universal or independent chuck.

A ***standard Jacobs chuck*** is normally fitted in tailstock for drilling. It can be mounted by fitting it in the spindle sleeve and then placing the unit in the headstock spindle. Wipe the chuck shank, sleeve, and headstock spindle hole with a clean soft cloth before they are fitted together and into the headstock spindle.

A ***headstock spindle Jacobs chuck*** is similar to the standard Jacobs chuck but is designed to fit directly on a threaded spindle nose, Fig. 10-110. The chuck has the advantage of NOT interfering with the compound rest, making it possible to work very close to the chuck.

Draw-in collet chuck

The ***draw-in collet chuck*** is another work holding device for securing work small enough to pass through the lathe spindle, Fig. 10-111.

Collets are accurately made sleeves; one end is threaded and the other split into three even sections. The slots are cut slightly more than half the length of the collet and permit the jaws to spring in and clamp the work.

The standard collet has a circular hole for round stock, but collets for holding square, hexagonal, and octagonal material are available, Fig. 10-112.

The chief advantage of collets lies in their ability to center work automatically and maintain accuracy over long periods of hard usage. They have the disadvantage of being expensive because a separate collet is needed for each different size or stock shape.

A collet chuck using steel segments, bonded to rubber is shown in Fig. 10-113. An advantage of

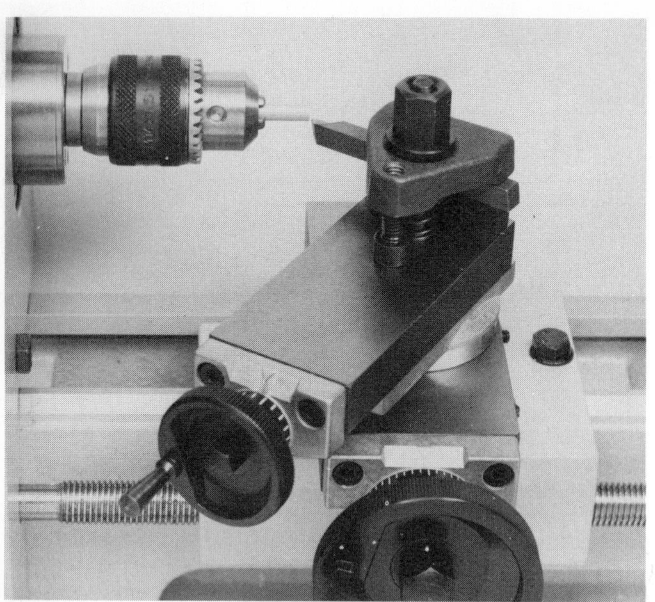

Fig. 10-109. Note turning of small diameter work in a Jacobs chuck.

Fig. 10-110. This is a headstock spindle Jacobs chuck. (The Jacobs Mfg. Co.)

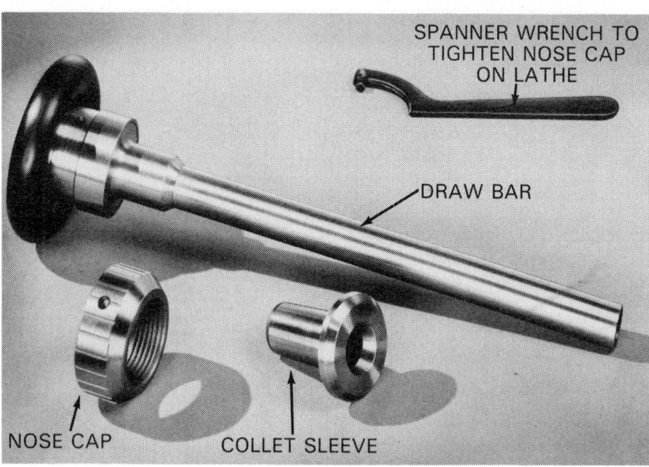

Fig. 10-112. Study units that make up a collet chuck. (South Bend Lathe, Inc.)

Fig. 10-111. Note design of collet chuck.

Fig. 10-113. This is a Rubber-Flex collet chuck in use. (The Jacobs Mfg. Co.)

this chuck is that each collet has a range of 0.100 in. (2.5 mm), rather than being a single size, like steel collets. However, they are only available for round work.

Mounting and removing chucks

If NOT installed on the spindle nose correctly, the accuracy of a chuck is affected. Remove the center and sleeve, if they are in place, by holding them with one hand and tapping them loose with a knockout bar. Carefully wipe the spindle end clean of chips and dirt. Apply a few drops of spindle oil. Clean the portion of the chuck that fits on the spindle.

On a chuck fitted to a threaded spindle nose, clean the threads with a ***spring cleaner.*** Refer to Fig. 10-114.

With the tapered key spindle nose, rotate the spindle until the key is in the up position. Slide the chuck on and tighten the threaded ring. Pins on the cam-lock spindle are fitted into place and locked.

Fitting a chuck onto a threaded spindle nose requires a different technique. Hold the chuck against the spindle nose with the right hand and turn the spindle with the left hand. Screw the chuck on until it fits firmly against the shoulder.

CAUTION! Do NOT spin the chuck on rapidly or use power. Release belt tension if possible, to eliminate any chance of power being transferred to the spindle.

During installation, place a board upon the ways under the chuck for protection of your hand and machine ways, Fig. 10-115.

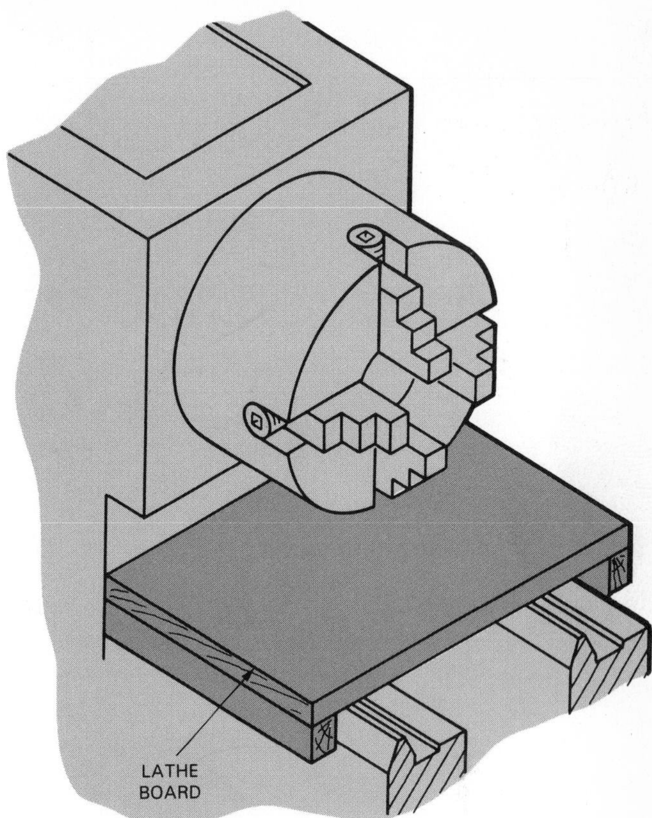

Fig. 10-115. A board placed under a chuck will protect your hand and the lathe should you accidently drop the chuck when mounting or removing it.

Removing chuck from threaded spindle

There are several accepted methods of removing chucks from a threaded lathe headstock spindle. The first step in any method, regardless of type of spindle nose, is to place wooden cradle across the ways beneath the chuck, Fig. 10-116. Then use one of the following techniques:

1. Lock the spindle in back gear and use a chuck key to apply leverage, Fig. 10-117.
2. Place a suitable size adjustable wrench on one jaw and apply pressure to the wrench. See Fig. 10-118.
3. If neither of the above two methods work, place a block of wood between the rear lathe ways and a chuck jaw. Engage the back gear and give the drive pulley a quick rearward turn. Refer to Fig. 10-119.

Fig. 10-116. A wooden cradle placed under a chuck will make removing chuck from a threaded spindle safer and easier.

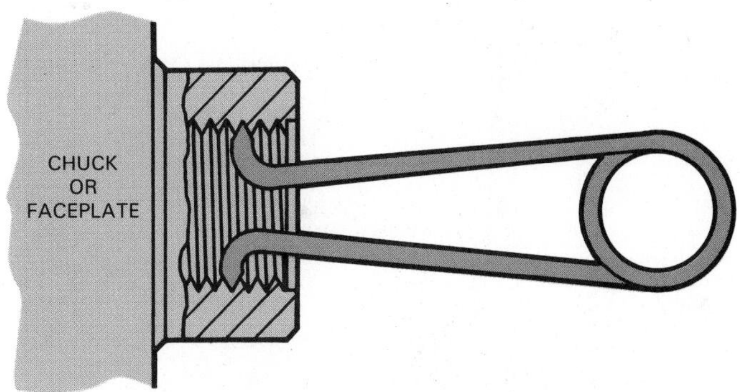

Fig. 10-114. Spring cleaner is used for cleaning threads in chuck before mounting it on a threaded spindle nose.

Fig. 10-117. Chuck wrench can sometimes be used to loosen chuck on a threaded spindle nose for removal.

Fig. 10-118. An adjustable wrench fitted to one of the jaws may be used to loosen a stubborn chuck on a threaded spindle nose.

Removing chuck from other spindle noses

Little difficulty should be encountered when removing a chuck from tapered and cam-lock spindle noses. For tapered spindle noses, after locking the spindle in back gear, place the appropriate spanner wrench in the locking ring. Give it a tap or two with a leather or plastic mallet. Turn the ring until the chuck is released.

WARNING! Place a wooden cradle under the chuck before attempting to remove it from the spindle. Removal will be easier and hand injuries will be avoided.

FACING STOCK HELD IN CHUCK

A round nose cutting tool, in a straight toolholder, is used to face stock held in a chuck. The compound is pivoted 30 deg. to the right. The toolholder is set to less than 90 deg. to face the work. The cutter bit is exactly on center. The carriage is then moved into position and locked to the way, Fig. 10-120.

A facing cut can be made in either direction. The tool may be started in the center and fed out, or, the reverse may be done. The usual practice is to start from the center and feed outward. If the material is over 1 1/2 in. (38.0 mm) in diameter, automatic feed may be employed.

With the cutting tool on center a smooth face will result from the cut. A ***rounded "nubbin"*** (remaining piece of unmachined face material) will result if the tool is slightly above center. A ***square shoulder*** "nubbin" indicates that the cutter is below center. Reposition the tool and repeat the operation if either condition is seen. See Fig. 10-121.

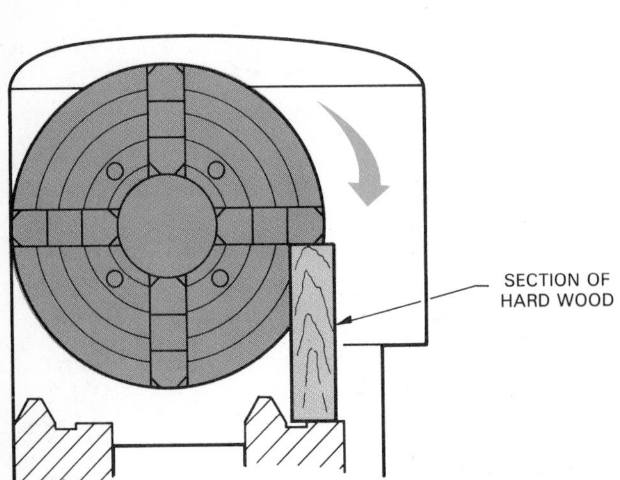

Fig. 10-119. Reversing chuck against a block of wood is often used to loosen a stubborn chuck on a threaded spindle nose.

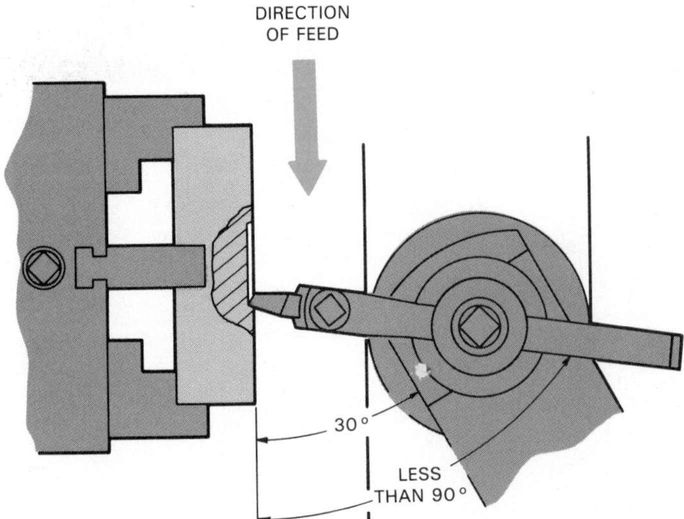

Fig. 10-120. Tool and tool holder are in the correct position for facing.

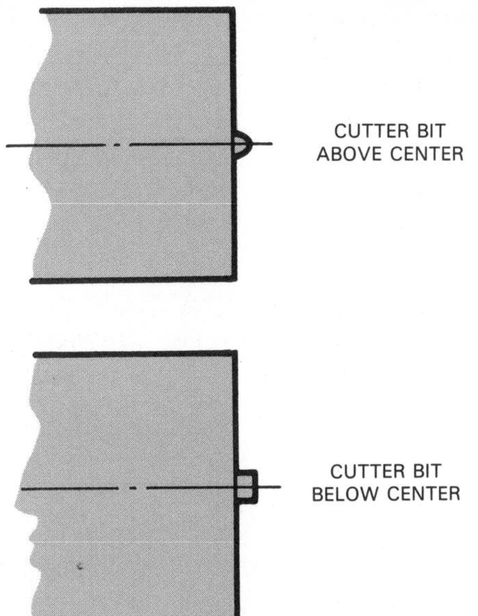

Fig. 10-121. If cutter bit is above or below center, a small ''nubbin'' of metal is formed. The ''nubbin'' shape indicates tool position.

PLAIN TURNING AND TURNING TO A SHOULDER

Work mounted in a chuck is machined in the same manner as if it were between centers. To prevent ''springing'' (flexing) while it is being machined, long work should be center drilled and supported with a tailstock center, Fig. 10-122.

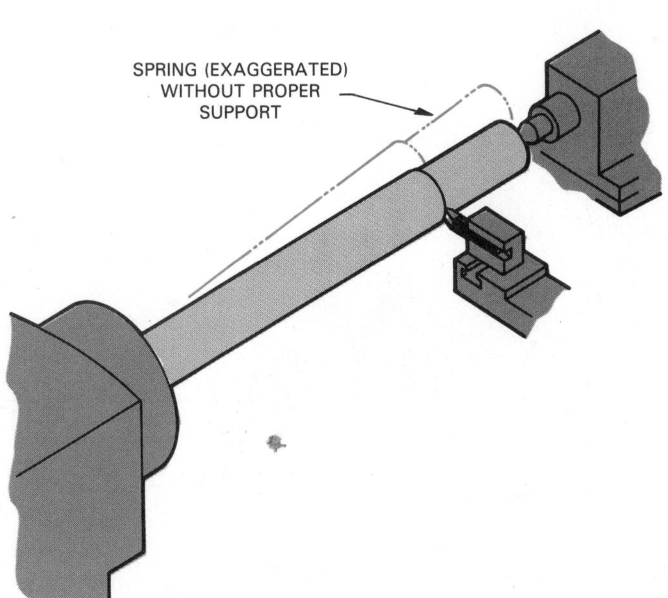

Fig. 10-122. Long work must be supported with tailstock center.

PARTING OPERATIONS

Parting is the operation of cutting off material after is has been machined, Fig. 10-123. This is one of the more difficult operations performed on a lathe.

The cutting tool must be ground with the correct clearance (front, side and end) and held in a straight or offset toolholder, Figs. 10-124 and 10-125. Grind a ***concave*** rake on top of the cutter to reduce chip width, thus preventing it from ***seizing*** (binding) in the groove.

Keep the tool sharp. This will permit easy penetration into the work. If NOT kept sharp, the tool may slip, and as pressure builds up, it may dig in suddenly and break the tool.

The cutoff blade is set at EXACTLY 90 deg. to the work surface, Fig. 10-126. The cutting edge

Fig. 10-123. Parting is one of the most difficult jobs performed on the lathe.

Fig. 10-124. Concave rake ground into top of parting tool will reduce width of chip and prevent binding.

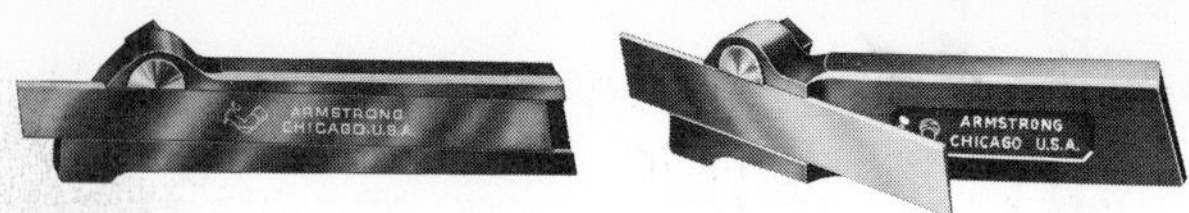

Fig. 10-125. These are parting tool holders. (Armstrong Bros. Tool Co.)

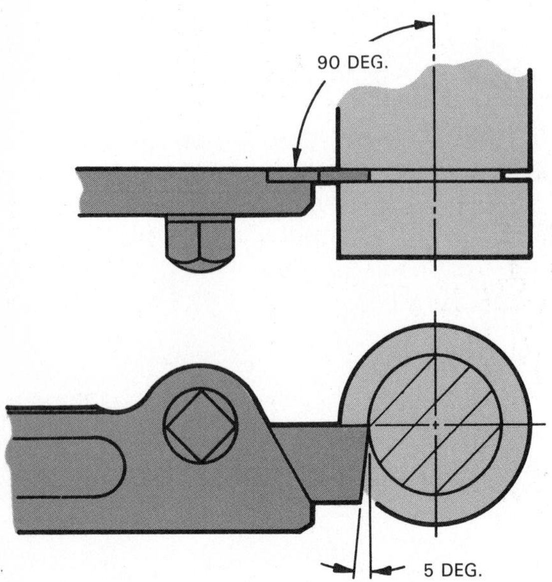

Fig. 10-126. Parting tool is in position for cutting off stock.

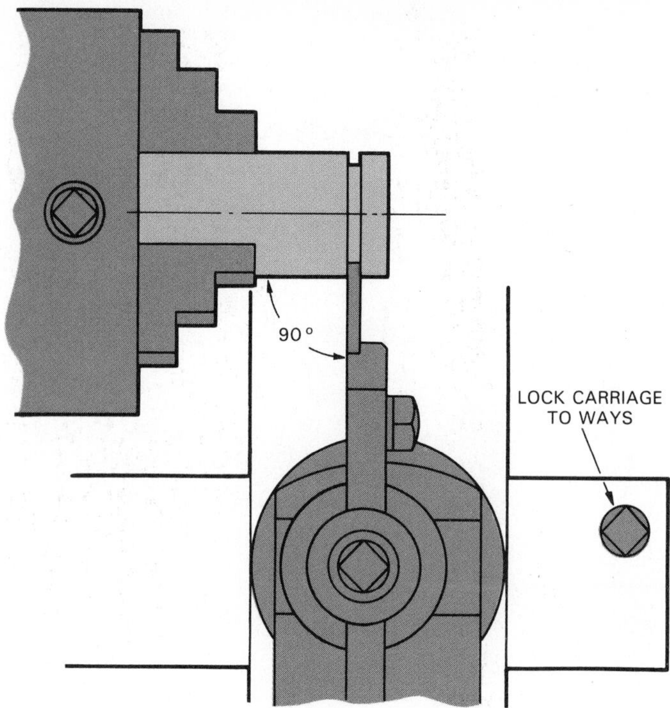

Fig. 10-127. Work is held close in chuck for parting operation. Parting tool blade is set at a 90° angle to cut, and carriage is locked to ways.

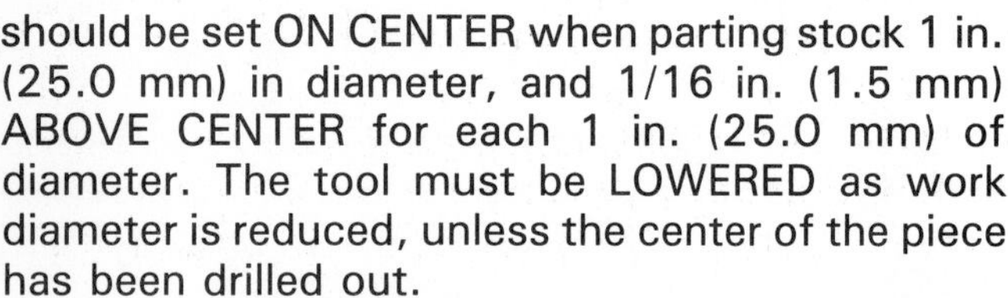

should be set ON CENTER when parting stock 1 in. (25.0 mm) in diameter, and 1/16 in. (1.5 mm) ABOVE CENTER for each 1 in. (25.0 mm) of diameter. The tool must be LOWERED as work diameter is reduced, unless the center of the piece has been drilled out.

Spindle speed is about one-third that employed for conventional turning. The compound and cross slide must be tightened to prevent play. Feed should be ample to provide a continuous chip. If feed is too slow, ***''hogging''*** (cutter digs in and takes a very heavy cut) can result. The tool will NOT cut continuously, but will ride on surface of metal for a revolution or two, then bite suddenly.

If the machine is in good condition, automatic cross-feed may be employed.

CAUTION! Do NOT forget to lock the carriage to the ways during a parting operation.

When parting, apply ample quantities of cutting fluid. Whenever possible, hold the work ''close'' in the chuck and, if necessary, use an offset toolholder, Fig. 10-127.

WARNING! NEVER attempt to cut off work held BETWEEN CENTERS. Serious trouble will be encountered, Fig. 10-128.

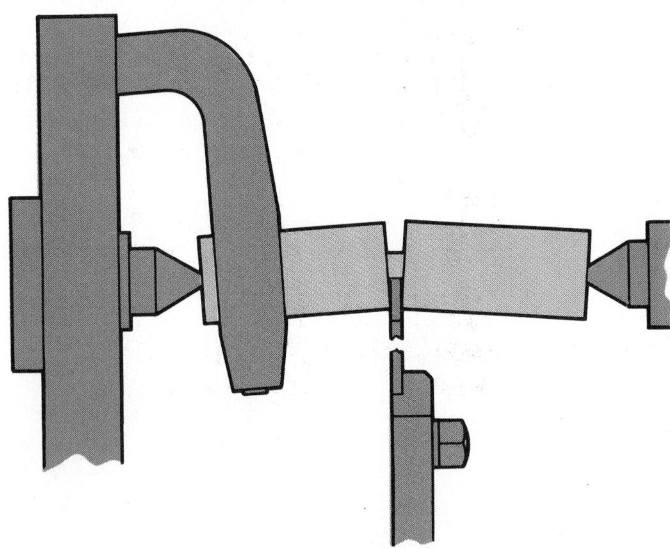

Fig. 10-128. Work cannot be parted safely while being held between centers.

TAPER TURNING

A section of material is considered to be ***tapered*** when it increases or decreases in diameter at a uniform rate, Fig. 10-129. A cone is an example of a taper. The ''wedging'' action of a taper makes it ideal as a means for driving drills, milling arbors,

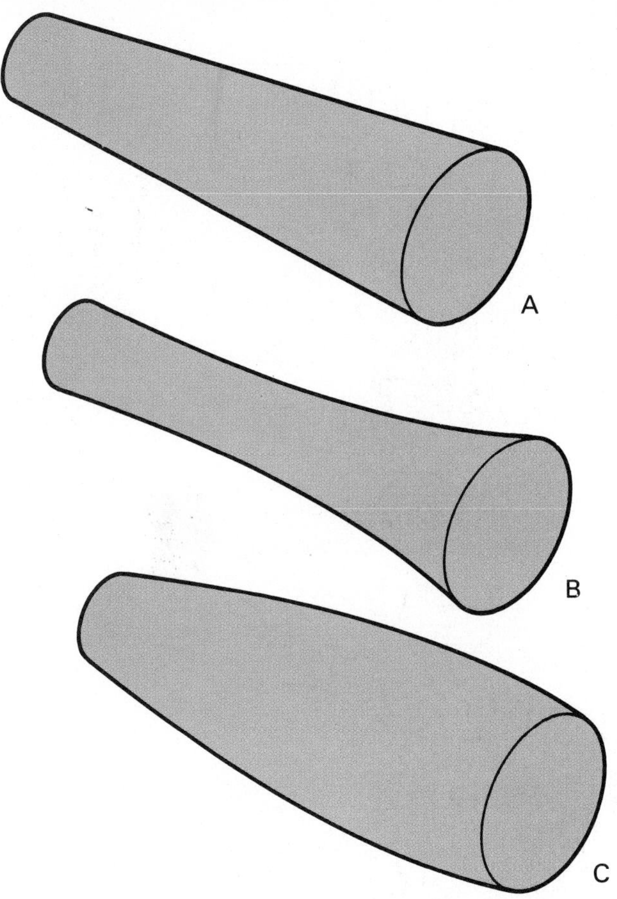

Fig. 10-129. A—Taper increases or decreases in diameter at a uniform rate. B and C—These are NOT tapered; they are "bell shaped."

end mills, centers, etc. In addition, it can be assembled and disassembled easily, and will align itself in a similary tapered hole automatically each time. Taper can be given in taper per inch, taper per foot, degrees, mm per 25.00 mm in length, or as a ratio, Fig. 10-130.

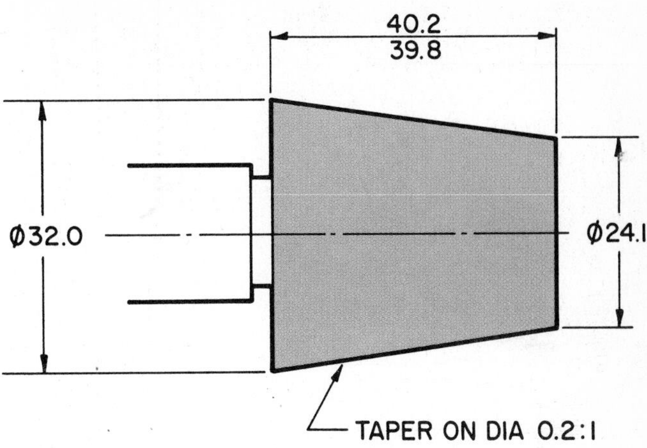

Fig. 10-130. Taper may be given as a ratio (shown above), taper per inch, taper per foot, degrees, or in millimeters per 25 millimeters.

There are five principle methods of machining tapers on a lathe. Each has its advantages and disadvantages. See Fig. 10-131.

Taper turning with compound rest

The compound rest method of turning tapers is the easiest, Figs. 10-132 and 10-133. Internal and external tapers can be machined, Fig. 10-134. However, taper length is limited by the movement of the compound rest. Because the compound rest base is graduated in degrees, the taper must be converted to degrees, Fig. 10-135. The table in Fig. 10-136 may be used.

A careful study of the plans will show whether the angle is given from center, or is the included angle, Fig. 10-137. If an included angle is given, it must be divided by two (2) to obtain the angle from the center line.

With the lathe center line representing 0 deg., pivot the compound to the desired angle and lock

	WAYS OF MACHINING TAPERS	
METHOD	ADVANTAGES AND DISADVANTAGES	INFORMATION NEEDED
1. Compound	Length of taper limited. Will cut external and internal taper.	Must know the taper angle.
2. Offset tailstock	External taper only. Must work between centers.	Taper per inch or taper per foot.
3. Taper attachment	Best method to use.	Angle or taper per inch or foot.
4. Tool bit	Very short taper.	Taper angle.
5. Reamer	Internal only.	Taper number.

Fig. 10-131. Study ways to machine tapers.

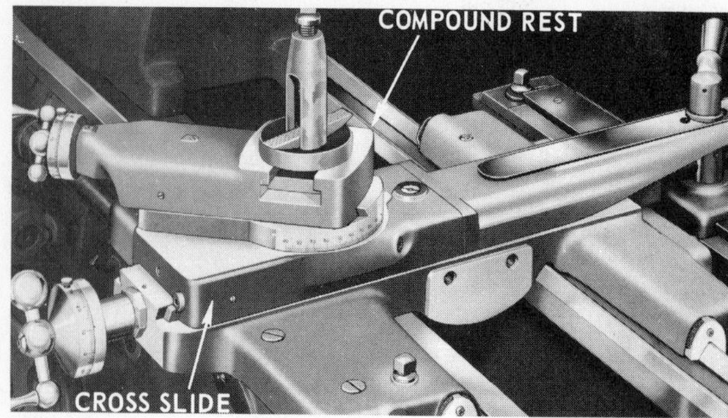

Fig. 10-132. Tapers can be cut using compound rest.

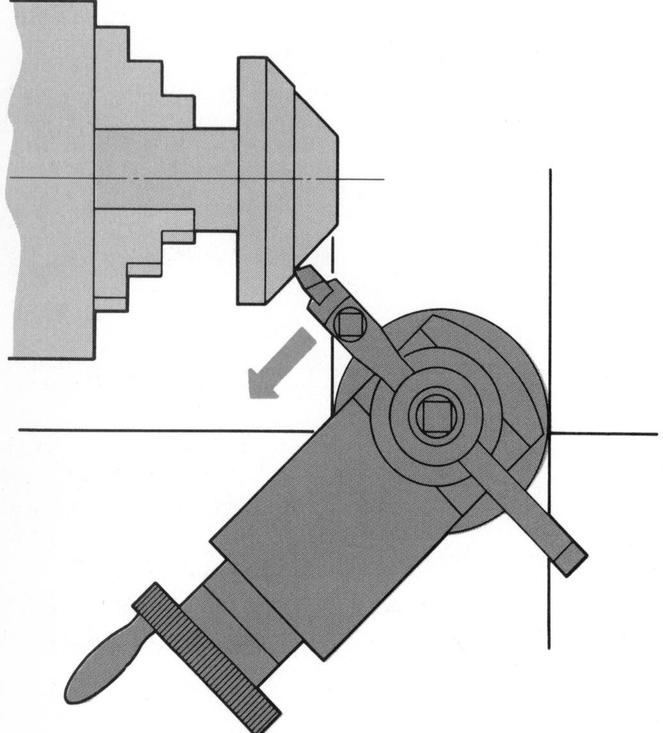

Fig. 10-133. When turning a taper using compound rest, note cut is made from small diameter to large diameter.

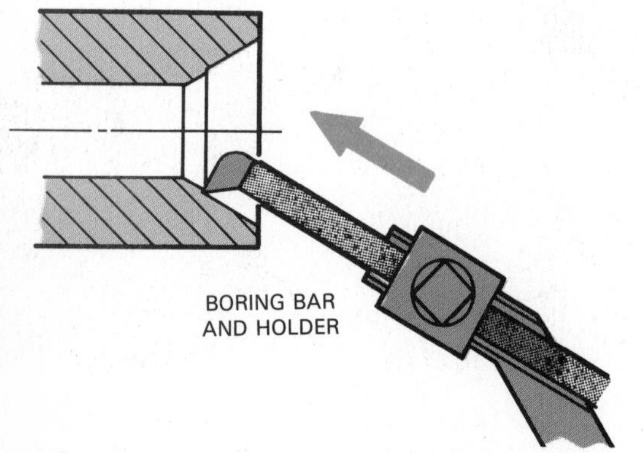

Fig. 10-134. Internal taper is being turned with compound rest.

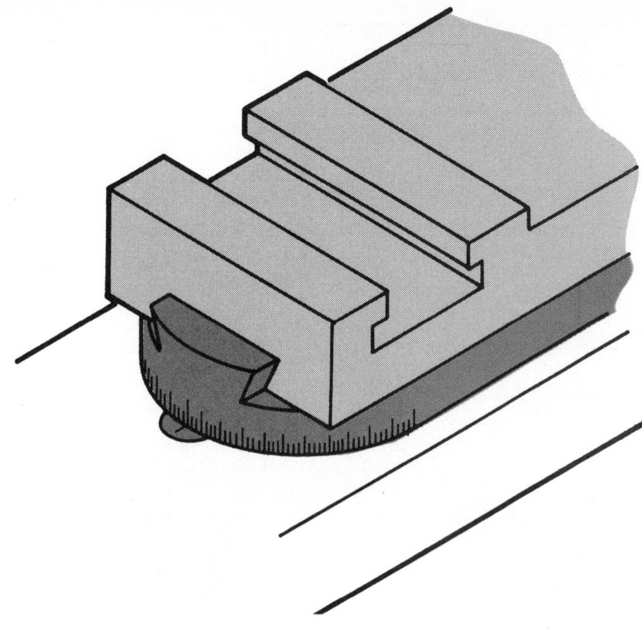

Fig. 10-135. Base of compound rest is divided into degrees to aid in positioning.

TAPER PER FOOT WITH CORRESPONDING ANGLES		
TAPER PER FOOT	INCLUDED ANGLE	ANGLE WITH CENTER LINE
1/16	0° 17' 53"	0° 8' 57"
1/8	0° 35' 47"	0° 17' 54"
3/16	0° 53' 44"	0° 26' 52"
1/4	1° 11' 38"	0° 35' 49"
5/16	1° 29' 31"	0° 44' 46"
3/8	1° 47' 25"	0° 53' 42"
7/16	2° 5' 18"	1° 2' 39"
1/2	2° 23' 12"	1° 11' 36"
9/16	2° 41' 7"	1° 20' 34"
5/8	2° 58' 3"	1° 29' 31"
11/16	3° 16' 56"	1° 38' 28"
3/4	3° 34' 48"	1° 47' 24"
13/16	3° 52' 42"	1° 56' 21"
7/8	4° 10' 32"	2° 5' 16"
15/16	4° 28' 26"	2° 14' 13"
1	4° 46' 19"	2° 23' 10"

Fig. 10-136. Chart gives taper per foot with corresponding angles.

it in position. It is the usual practice to turn a taper from the small diameter to the large diameter. See Fig. 10-138.

As will be the case when turning ALL tapers, the cutting tool must be set on exact center. A toolholder should be selected that will provide ample clearance.

To machine a taper, bring the cutting tool into position with the work and lock the carriage to prevent it from shifting during the turning operation.

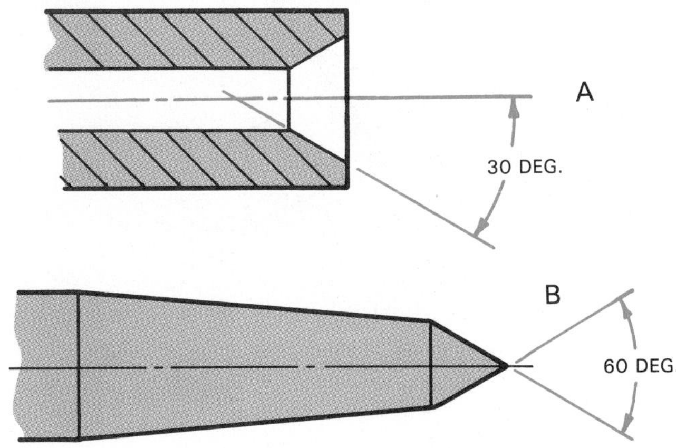

Fig. 10-137. Note methods for measuring angles. A—Angle measured from center. B—Included angle measured.

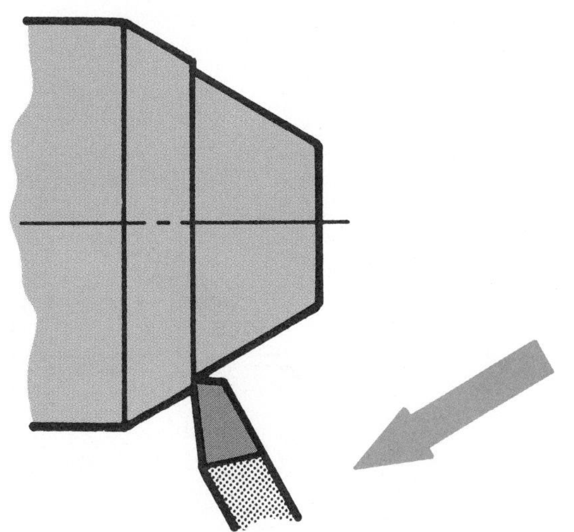

Fig. 10-138. It is usual practice to turn a taper from small diameter to large diameter.

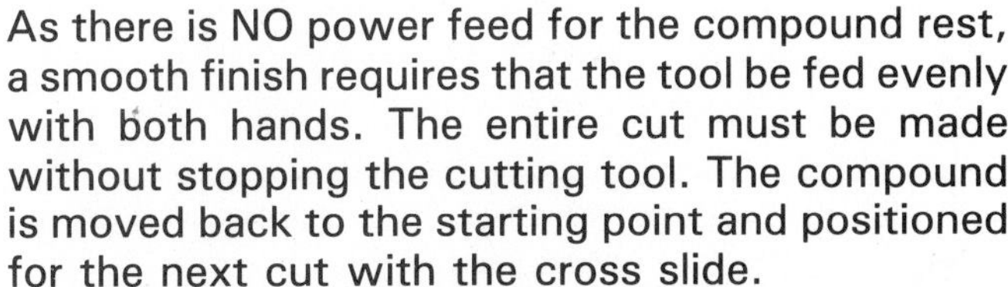

As there is NO power feed for the compound rest, a smooth finish requires that the tool be fed evenly with both hands. The entire cut must be made without stopping the cutting tool. The compound is moved back to the starting point and positioned for the next cut with the cross slide.

The work can be mounted between centers, or held in a chuck, when tapers are cut with a compound rest. A suitable boring bar is needed when machining internal tapers. Some internal tapers are finished to size with a taper reamer.

Taper turning by offset tailstock method

The ***offset tailstock method,*** also known as the ***tailstock setover method,*** is also employed for taper turning, Fig. 10-139. Jobs that can be turned between centers may be taper turned by this technique. Only EXTERNAL tapers can be machined.

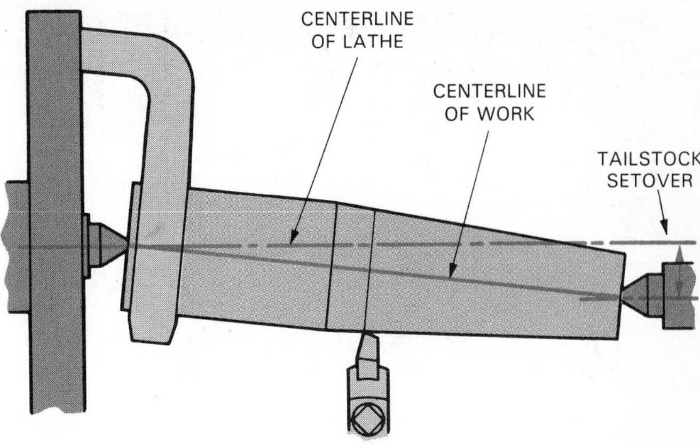

Fig. 10-139. Study how to machine a taper by offset tailstock technique.

Most lathe tailstocks are constructed in two parts which permits the dead center to be moved off center, Fig. 10-140. This is accomplished by loosening the anchorbolt that locks the tailstock to the ways and making the proper adjustments with screws on the tailstock. After the setover has been made, the screws are drawn up snug, but NOT tight.

CALCULATING TAILSTOCK SETOVER

Taper turning by this technique is NOT a precise method and requires some "trial and error" adjustments to produce an accurate tapered section. The approximate setover can be calculated when certain basic information is known.

Offset must be calculated for each job because the length of the piece plays an important part in

Fig. 10-140. Tailstock, on most lathes, is constructed in two parts permitting the section mounting the center to be moved off center. Distance off center can be checked by spacing of witness lines. (Rockwell International)

the calculations. When the length of the pieces vary, different tapers will be produced with the same tailstock offset, Fig. 10-141.

The following terms are used with calculating tailstock setover, Fig. 10-142.

TPI = Taper Per Inch
TPF = Taper Per Foot
D = Diameter at large end
d = Diameter at small end
ℓ = Length of taper
L = Total length of piece

Calculating setover when taper per inch is known

Information needed: TPI = Taper Per Inch
L = Total length of piece

FORMULA USED: $\text{Offset} = \frac{L \times TPI}{2}$

EXAMPLE: What will be the tailstock setover for the following job?

Taper Per Inch = 0.0125

Total length of piece = 8.000

$$\text{Offset} = \frac{L \times TPI}{2}$$

$$= \frac{8.000 \times 0.125}{2}$$

$$= 0.500 \text{ in.}$$

NOTE: The same procedure would be followed when using metrics. However, all dimensions must be in millimeters.

Calculating setover when taper per foot is known

When Taper Per Foot (TPF) is known, it must be converted to Taper Per Inch (TPI). The following formula takes this into account:

$$\text{Offset} = \frac{TPF \times L}{24}$$

Calculating setover when dimensions of tapered sections are known but TPI or TF is NOT given

Quite often plans do NOT specify TPI or TPF but do give other pertinent information. Calculations will be easier if all fractions are converted to decimals. All dimensions must either be in inches or in millimeters.

Information needed: D = Diameter at large end
d = Diameter at small end
ℓ = Length of taper
L = Total length of piece

FORMULA USED: $\text{Offset} = \frac{L \times (D - d)}{2\ell}$

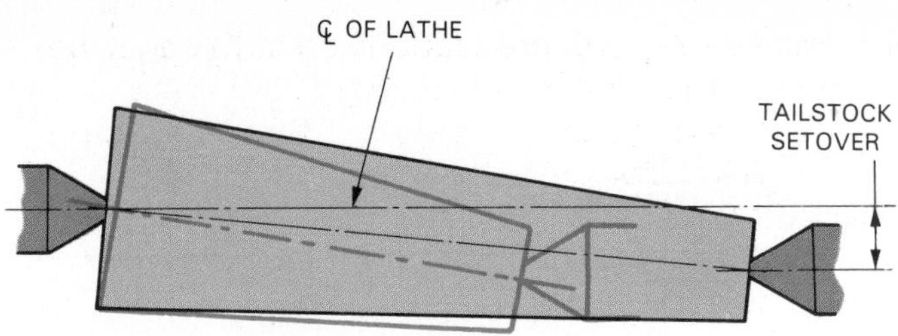

Fig. 10-141. Length of work causes taper to vary even though tailstock offset remains the same.

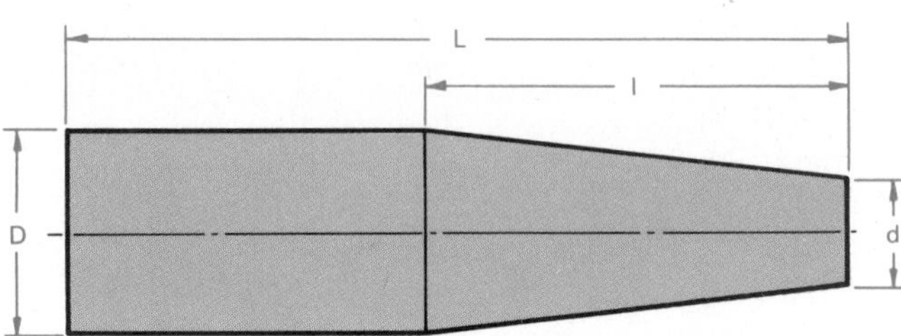

Fig. 10-142. Note basic taper information. D = Diameter at large end of taper; d = Diameter at small end of taper; ℓ = Length of taper; L = Total length of piece.

EXAMPLE: Calculate the tailstock setover for the following job.

D = 1.250 in.
d = 0.875 in.
ℓ = 3.000 in.
L = 9.000 in.

$$\text{Offset} = \frac{L \times (D - d)}{2\ell}$$

$$= \frac{9.000 \times (1.250 - 0.875)}{2 \times 3.000}$$

$$= \frac{9.000 \times 0.375}{6}$$

$$= 0.562 \text{ in.}$$

Calculating setover when taper is given in degrees

The space available in this text does NOT permit the introduction of a course in basic trigonometry. Trigonometry is necessary to make these calculations. However, any good machinist's handbook will provide this information. At least one such book should be part of a machinist's tool box.

MEASURING TAILSTOCK SETOVER

When an ample tolerance is allowed, (±0.015 in. or 0.05 mm), the setover can be measured with a rule:

1. Place a rule with graduations on both edges between the center points, Fig. 10-143.
2. Measure the distance between the two witness marks on the tailstock base, Fig. 10-144.

Accurate work requires more care in making the tailstock setover. An additional factor enters into the calculations—the distance the center point enters the piece. Typically, 1/4 in. (6.5 mm) is ample and it must be subtracted from the total length of the piece.

Use the appropriate method to calculate the offset. An accurate setover may be made using the MICROMETER COLLAR on the lathe cross slide. See Fig. 10-145.

1. Clamp the toolholder in a reverse position in the tool post.
2. Turn the cross slide screw back to remove all play.
3. Turn in the compound until the toolholder can

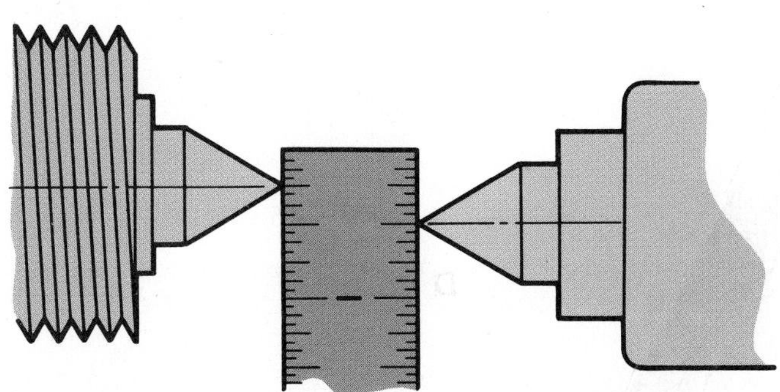

Fig. 10-143. Approximate tailstock setover can be determined by measuring distance between center points.

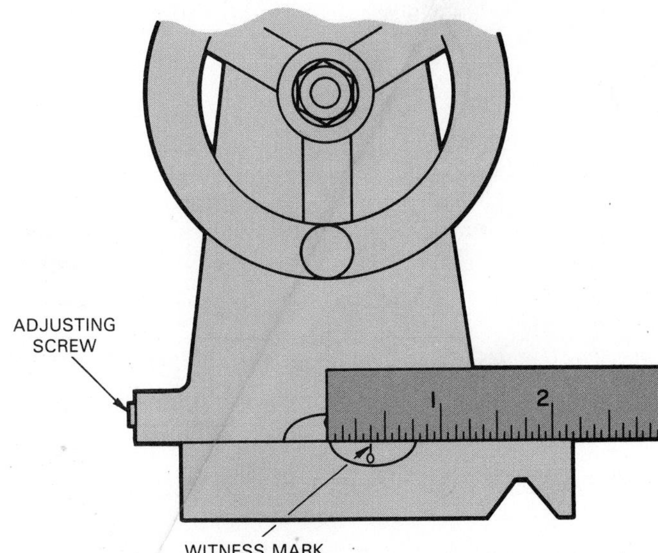

Fig. 10-144. Approximate setover can also be determined by measuring distance between witness lines with a rule.

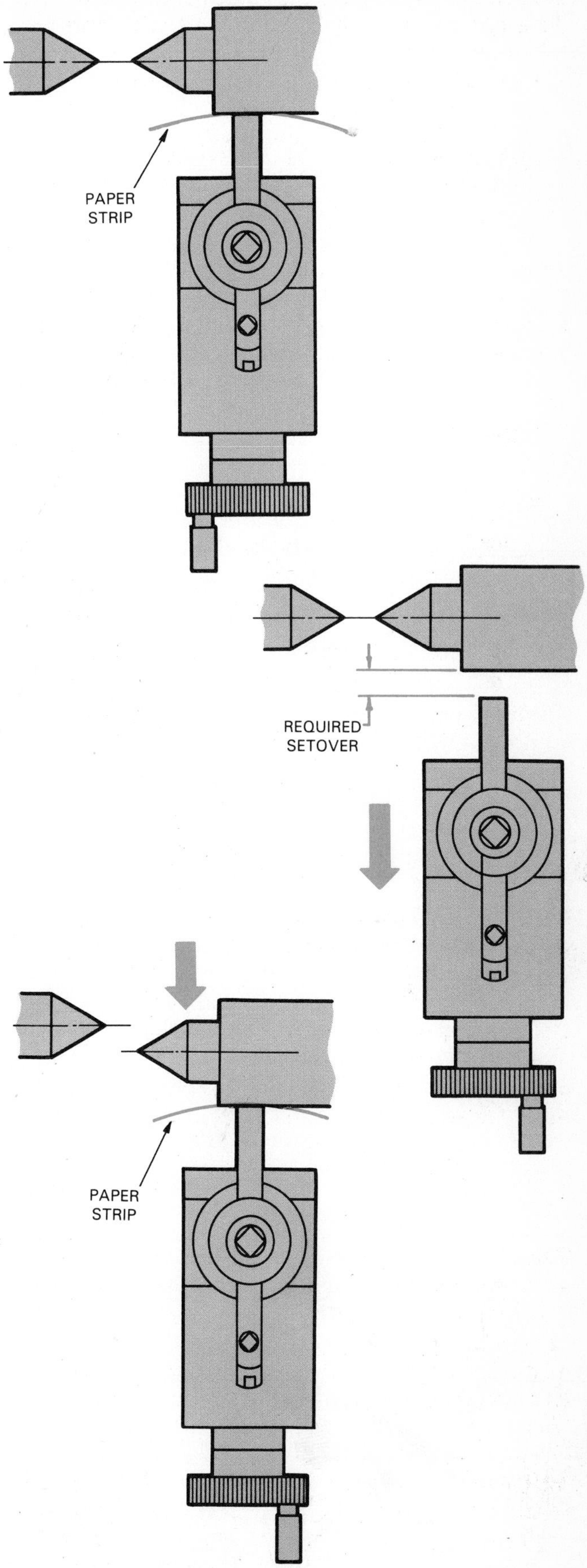

Fig. 10-145. Study how to use micrometer collar of compound rest to make setover measurement.

be felt with a piece of paper between the toolholder and tailstock spindle.

4. Use the micrometer collar and turn out the cross slide screw the distance the tailstock is to be set over.
5. Move the tailstock over until the spindle touches paper in same manner described in Step #3.
6. Check the setting again after "snugging up" the adjusting screws.

A DIAL INDICATOR can be employed in place of the toolholder and paper strip, Fig. 10-146.

1. Mount the dial indicator in the tool post.
2. Position it with the cross slide until the indicator reads zero when in contact with the tailstock spindle. There should be NO "play" in the cross slide.
3. Set the tailstock over the required distance using the dial indicator to make the measurement.
4. Recheck the reading after "snugging up" the adjusting screws. Make additional adjustments if a deviation in the indicator reading occurs.

CUTTING A TAPER

When cutting a taper, additional strain is imposed on the centers because they are out-of-line and do NOT bear true in the center holes, Fig. 10-147.

A bell type center drill offers some advantage in reducing strain.

Some machinists prefer a center with a ball tip to produce an improved bearing surface, Fig. 10-148. Because the pressures imposed are uneven, the work is more apt to heat up than when doing conventional turning between centers. It must be checked frequently for binding.

Make the cuts as in conventional turning. However, cutting should start at the small end of the taper.

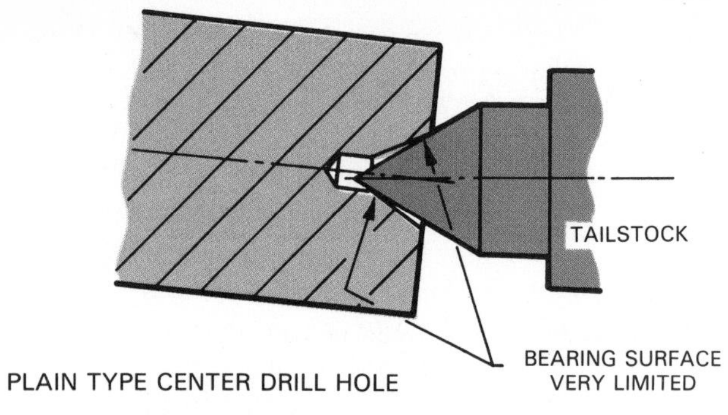

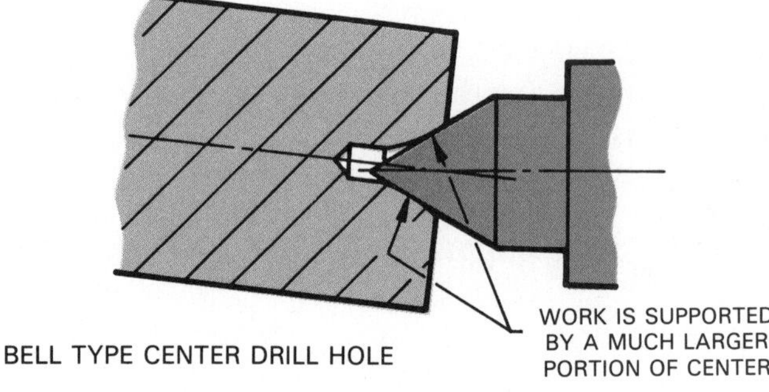

Fig. 10-147. Taper turning, by offset tailstock method, is hard on tail center because center point does not bear evenly in center hole. A center hole drilled with a bell type center drill reduces this problem.

Turning a taper with a taper attachment

A ***taper attachment*** is a guide that can be attached to most lathes, Figs. 10-149 and 10-150. It is an accurate way to cut tapers and offers advantages over other methods of machining tapers:

1. Internal and external tapers can be cut. An accurate fit is assured for mating parts.
2. Work can be held by any conventional means.
3. The lathe does NOT have to be altered. The machine can be used for straight turning by lock-

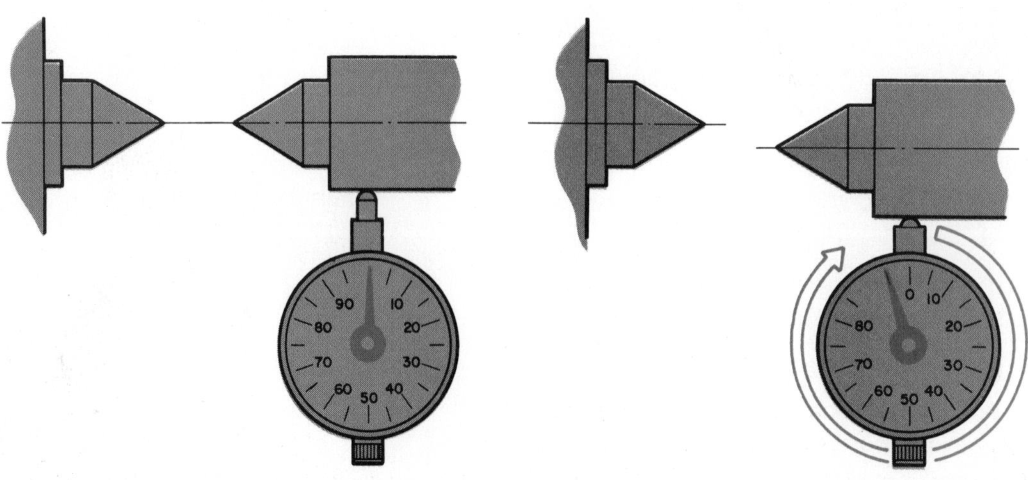

Fig. 10-146. Dial indicator can also be used to measure amount of setover.

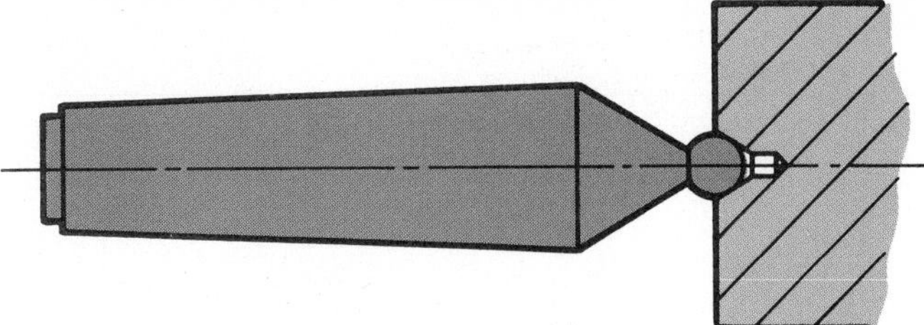

Fig. 10-148. A ball-tipped center lessens pressure on tail center when turning tapers.

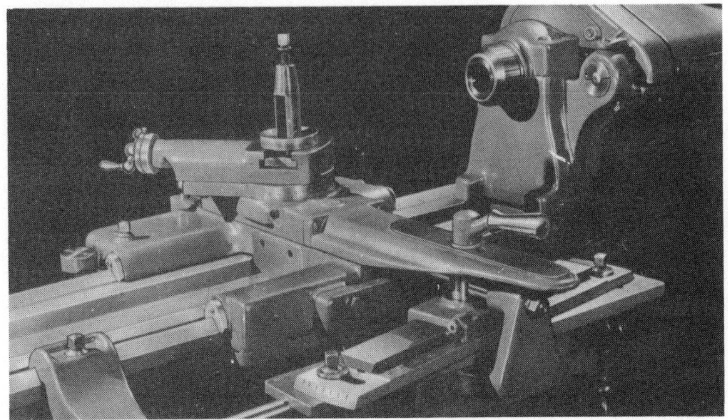

Fig. 10-149. This lathe is fitted with plain taper attachment. (South Bend Lathe, Inc.)

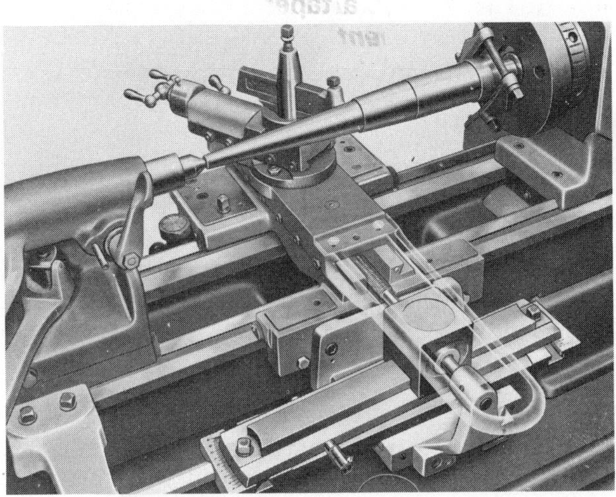

Fig. 10-150. Note lathe fitted with telescopic taper attachment. (Clausing Machine Tools)

ing the taper attachment out. No realignment of the lathe is necessary.

4. Once the attachment has been set, the taper can be machined on material of various lengths.
5. One end of the taper attachment ***swivel bar*** is graduated in total taper in inches per foot. The other end is graduated to indicate the included angle of the taper in degrees.

Types of taper attachments

There are two types of taper attachments: plain taper attachment and telescopic taper attachment.

The ***plain taper attachment*** requires the cross slide screw to be disengaged from the cross slide feed nut. The cutting tool must be advanced by the compound rest feed screw.

The ***telescopic taper attachment*** is made in such a manner that it is NOT necessary to disconnect the cross slide feed nut to use the taper attachment. The tool can be advanced into the work with the cross slide screw in the usual manner.

Setting a taper attachment

1. Study the plans and, if necessary, calculate the taper. Set the swivel bar as specified from the calculations.
2. Mount the work in the machine.
3. Slide the taper attachment unit to a position that will permit the cutting tool to travel the full length of the taper and lock it to the ways.
4. Move the carriage to the right until the cutting tool is about one inch (25 mm) away from the end of the work. This will permit any play to be taken up before the tool starts to cut.
5. Tighten the binding screw that engages the cross slide feed to the taper attachment, if the machine is fitted with a plain taper attachment.
6. Oil the bearing surfaces of the taper attachment and make a trial cut. If necessary, readjust until the taper is being cut to specifications. Complete the cutting operation.

Turning a taper with a square nose tool

This taper technique, using a square nose tool, is limited to the production of SHORT TAPERS, Fig. 10-151. The cutter bit is ground with a square nose

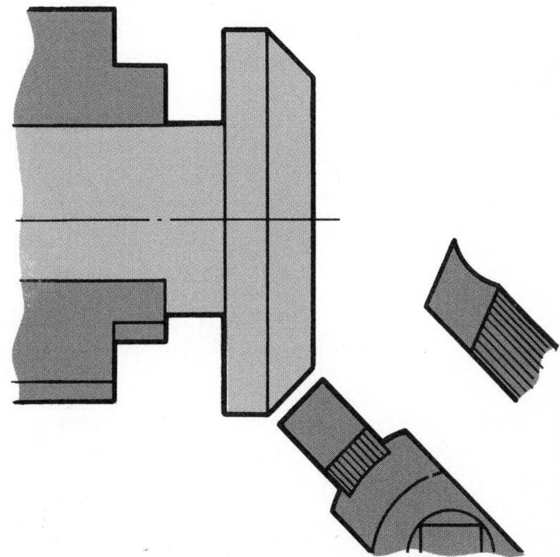

Fig. 10-151 Short taper is being turned with a square nose tool.

and set to the correct angle with the protractor head and blade of a combination set.

The tool is positioned on center and fed into the revolving work. "Chatter" can be minimized by running the work at a slow spindle speed. The carriage must be locked to the ways.

CAUTION! Before using any of the taper turning techniques on work mounted between centers, it is very important that the centers be *"zeroed in"* (put in perfect alignment). Then the necessary adjustments can be made (setting over tailstock, adjusting taper attachment, etc.).

MEASURING TAPERS

There are two basic methods of testing the accuracy of machined tapers. One is a ***comparison method;*** the other involves ***direct taper measurement.***

Measuring tapers by comparison

Taper plug and ***ring gages*** serve two purposes, Fig. 10-152. They measure the basic diameter of the taper as well as the angle of slope. The angle is checked by applying ***bluing*** (usually Prussian blue) to the machined surface or plug gage. Insert the blued section into the mating part and slowly rotate it. If the bluing rubs off EVENLY, it indicates that the taper is correct. If it rubs off UNEVENLY, it will show where the taper is off and will indicate what machine adjustments are needed, Fig. 10-153.

Gages are also provided with notches for the specified tolerance in taper diameter. The indentations include GO and NO GO positions, Fig. 10-154.

Direct measurement of tapers

A ***taper test gage*** is sometimes employed to check taper accuracy, Fig. 10-155. It consists of a base with two adjustable straight edges. Slots in the straight edges permit adapting the gage to check different tapers. The taper is set by means of two discs of known size which are located the correct distance apart.

Another technique for checking and/or measuring tapers is to set the tapered section on a surface plate. Two gage blocks or ground parallels of the same height are placed on opposite sides of the taper. Two cylindrical rods (sections of drill rod are satisfactory) of the same diameter are placed on the blocks. The distance across the rods is measured with a micrometer. See Fig. 10-156.

A second reading is made 1, 3, or 6 in. (25, 75, or 150 mm) above the first reading. The rods are the same diameter as those used to make the first reading. The taper per foot can be determined by subtracting to find the difference between the two measurements. Then multiply it by twelve (12) if

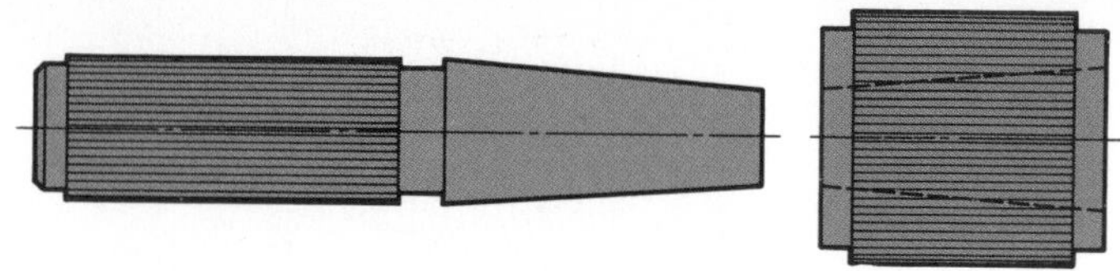

Fig. 10-152. Left—Plug gage. Right—Ring gage.

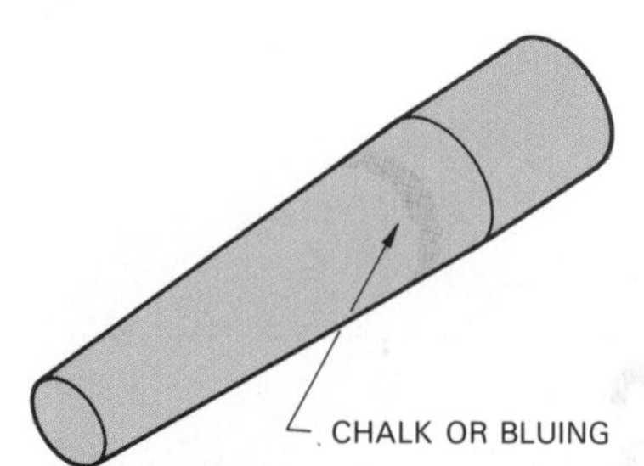

Fig. 10-153. When chalk or bluing does not rub off evenly, it indicates that taper does not fit properly and additional machine adjustments will have to be made.

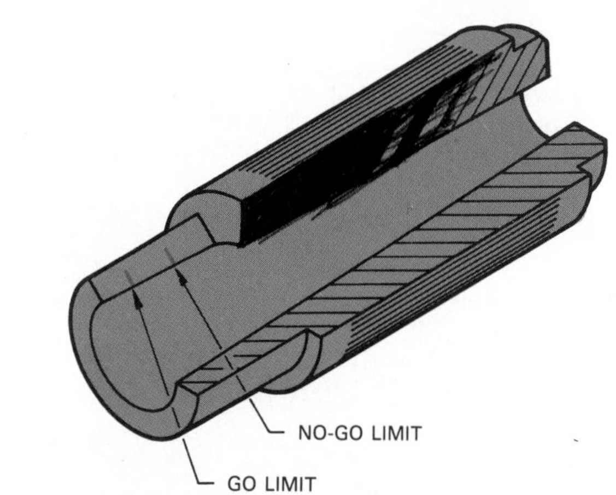

Fig. 10-154. Note close-up of a typical GO and NO-GO ring gage for measuring tapers.

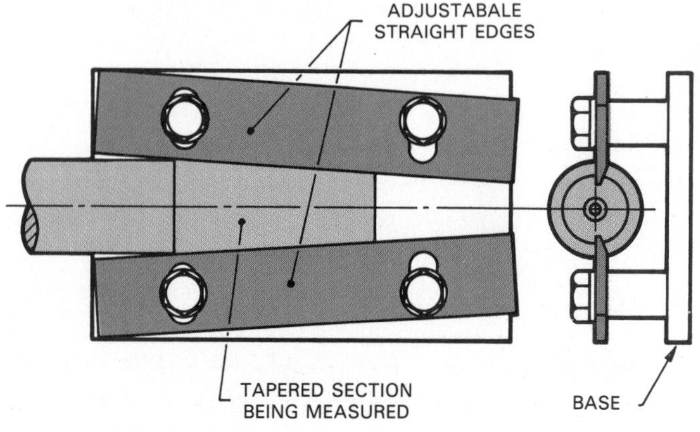

Fig. 10-155. Taper test gage can be set for different tapers.

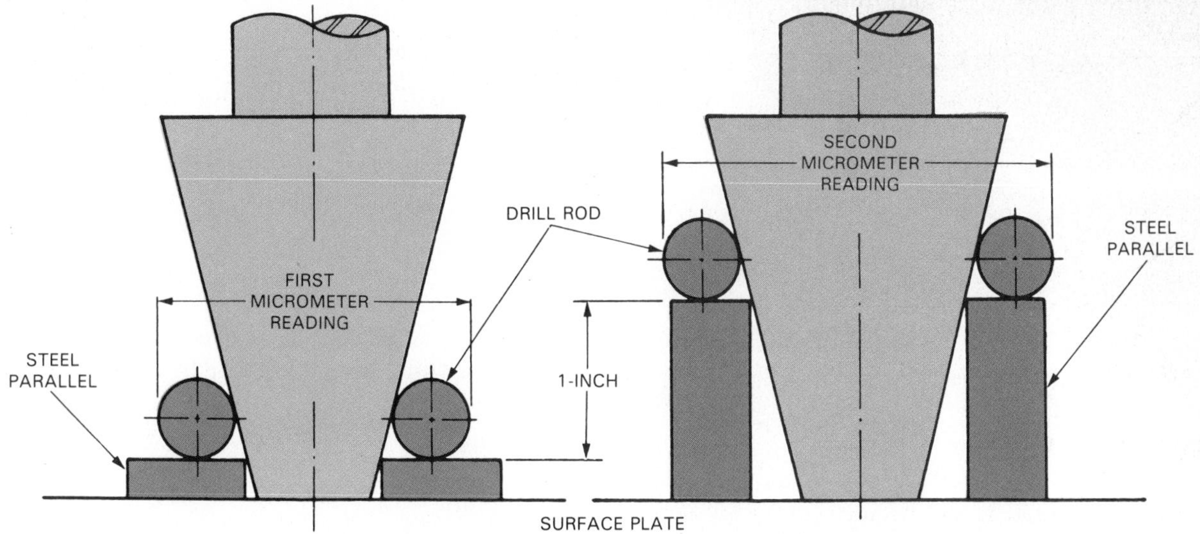

Fig. 10-156. Study method for measuring a taper using parallels, drill rod, micrometers, and a surface plate.

the readings were made 1 in. apart, by four (4) if they were made 3 in. apart, and by two (2) if they were made 6 in. apart.

A ***sine bar*** consists of a very accurately machined bar with edges that are parallel, Fig. 10-157. When used in conjunction with gage blocks and sine tables, angles can be measured with precision.

CUTTING SCREW THREADS ON THE LATHE

Screw threads are utilized for many reasons. The more important applications are:

1. Making adjustments (cross feed on lathe).
2. Assembling parts (nuts, bolts, and screws).
3. Transmitting motion (lead screw on lathe).
4. Applying pressure (clamps).
5. Making measurements (micrometer).

Screw thread forms

The first machine cut screw threads were square, but since that time, many different thread forms have been developed including American National, Unified, Sharp V, Square, Acme, Worm threads, etc. Each thread form has a particular use and formula for calculating its shape and size. See Figs. 10-158 through 10-161.

The following terms relate to ***screw threads,*** Fig. 10-162:

1. ***External thread*** means the threads are cut on the outside surface of piece.
2. ***Internal thread*** means threads are cut on the inside surface of piece.
3. ***Major diameter*** is the largest diameter of thread.
4. ***Minor diameter*** is the smallest diameter of thread.
5. ***Pitch diameter*** is the diameter of an imaginary cylinder that would pass through threads at such points to make width of thread and width of the spaces at these points EQUAL.
6. ***Pitch*** is the distance from one thread point to the next thread point measured parallel to the thread axis. Pitch of inch based threads is equal to one (1) OVER the number of threads per inch.
7. ***Lead*** is the distance a nut will travel in one complete revolution of the screw. On a single thread, the lead and pitch are the same. Multiple thread screws have been developed to secure an increase in lead without weakening the thread. See Fig. 10-163.

Preparing to cut sharp V-threads on a lathe

Sharpen the cutting tool to the correct shape including the correct clearance. The top is ground flat with no side or back rake, Fig. 10-164.

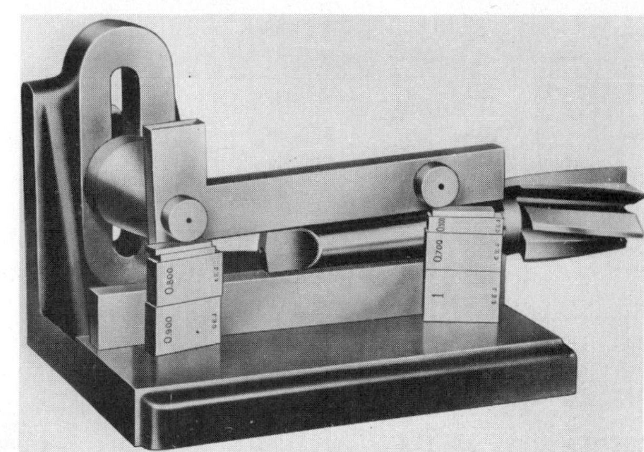

Fig. 10-157. A sine bar and precision gage blocks can also be used to measure a taper. (C.E. Johannson and Co.)

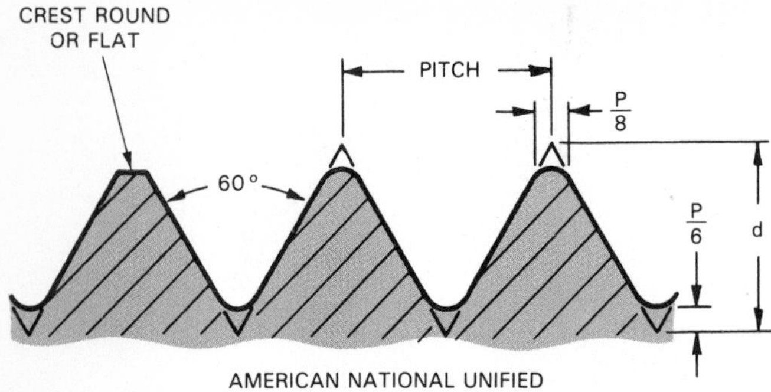

Fig. 10-158. Note Unified Thread form is interchangeable with American National Threads. N = Number of threads per inch; P = Pitch = $\frac{1}{N}$; d = Depth of thread = $\frac{0.866}{N}$.

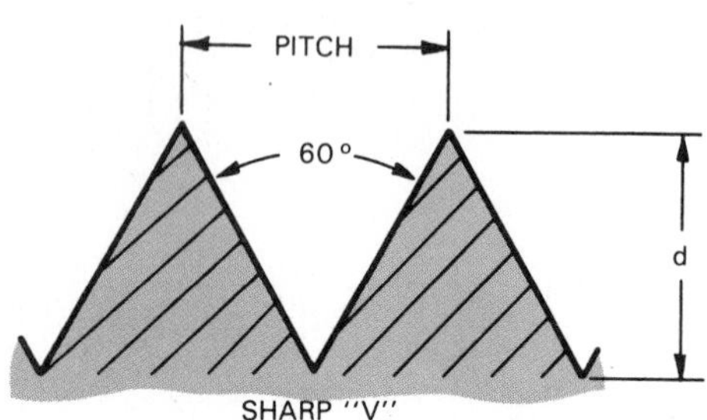

Fig. 10-159. Compare sharp "V" thread form. N = Number of threads per inch; P = Pitch = $\frac{1}{N}$; d = Depth of thread = $\frac{0.866}{N}$.

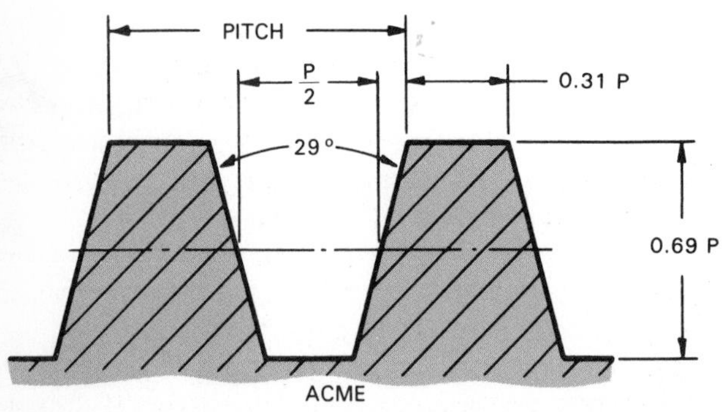

Fig. 10-160. Note Acme thread form. N = Number of threads per inch; P = Pitch = $\frac{1}{N}$; d = Depth of thread = $\frac{P}{2}$ + 0.010; Flat = 0.371P; Root = 0.371P − 0.0052.

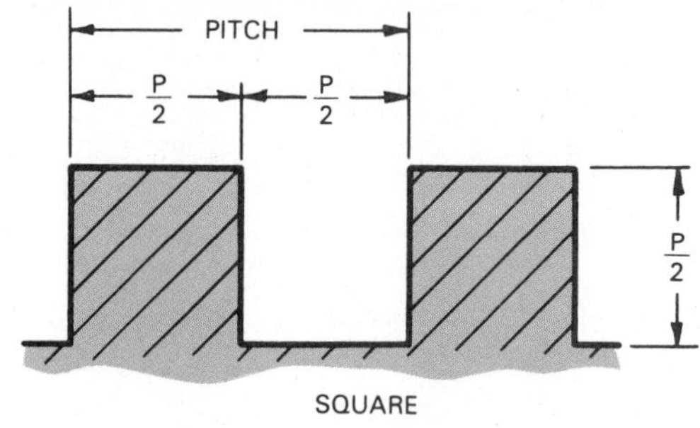

Fig. 10-161. Compare square thread form with previous types. N = Number of threads per inch. P = Pitch = $\frac{1}{N}$. Depth of thread = $\frac{P}{2}$. Flat or space = $\frac{P}{2}$.

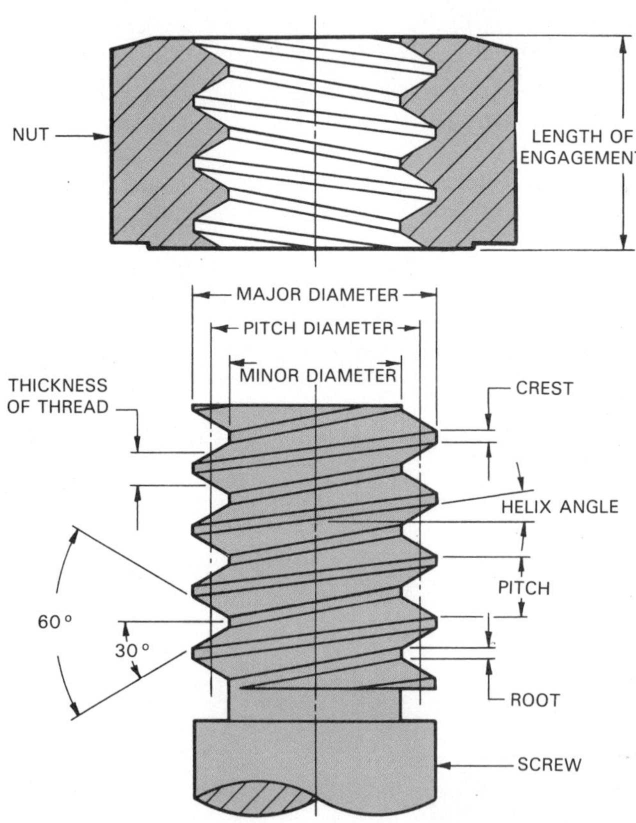

Fig. 10-162. Study nomenclature of a thread.

A ***center gage*** is used for grinding and setting the tool bit in position, Fig. 10-165. It is often referred to as a ***fish tail.***

The work is set up in the same manner as for straight turning. If mounted between centers, the centers must be precisely aligned; otherwise, a tapered thread will be produced. Then the thread will NOT be operable unless the threads are cut excessively deep at one end. The work must also run true with no "wobble." There must be NO play in the dog tail in the faceplate slot.

A groove is frequently cut at the point where the thread is to terminate, Fig. 10-166. The ***thread end groove*** is cut equal to the minor diameter of the thread and serves two purposes:

1. Provides a place to stop the threading tool at the end of its cut.
2. Permits a nut to be run up to the end of the thread.

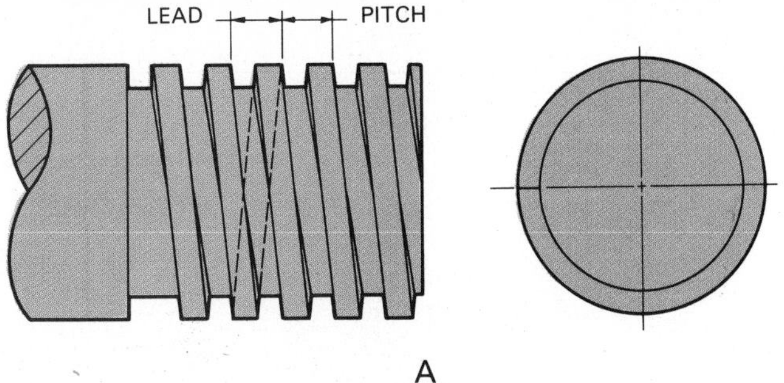

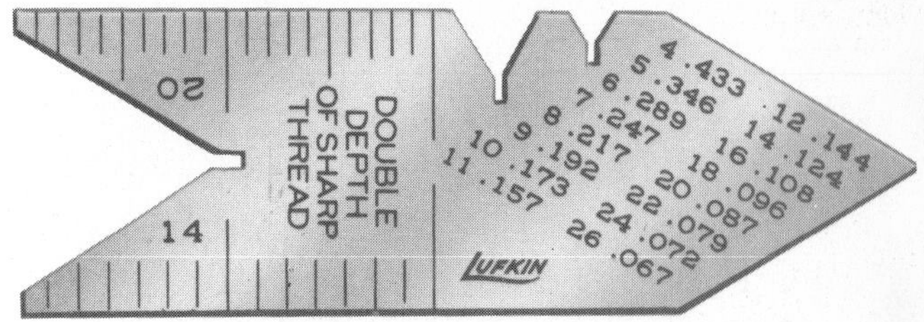

Fig. 10-165. A center gage can also be called a "fish tail." (The Lufkin Tool Co.)

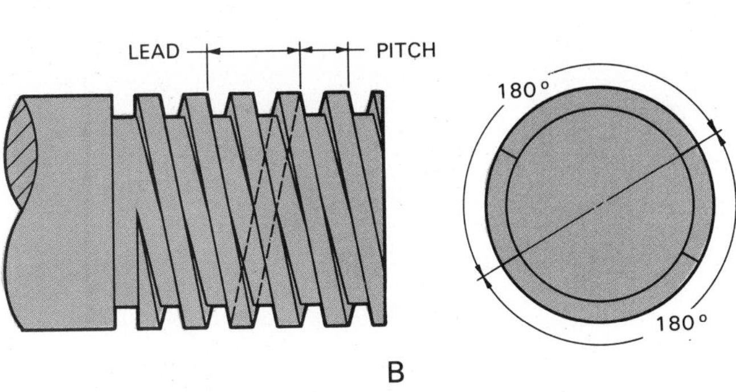

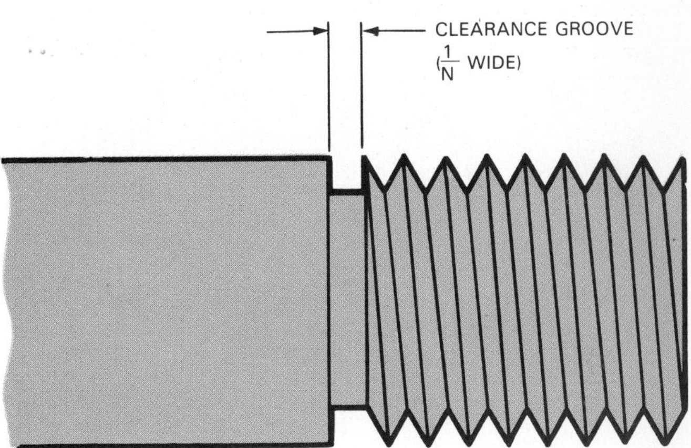

Fig. 10-166. Note typical technique for terminating a screw thread.

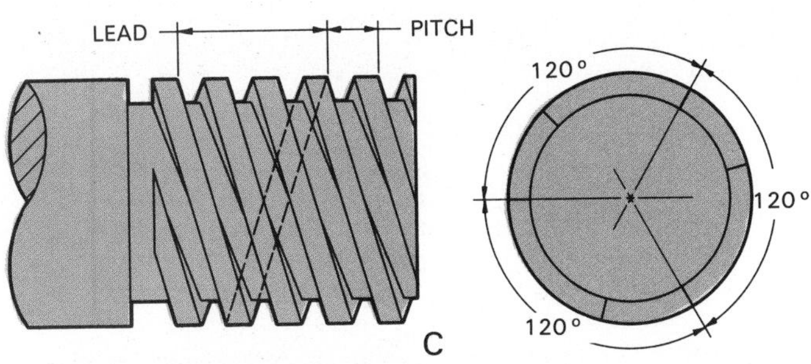

Fig. 10-163. Note difference between lead and pitch. A—Single thread screw, the pitch and lead are equal. B—Double thread screw, the lead is twice the pitch. C—Triple thread screw, the lead is three times the pitch.

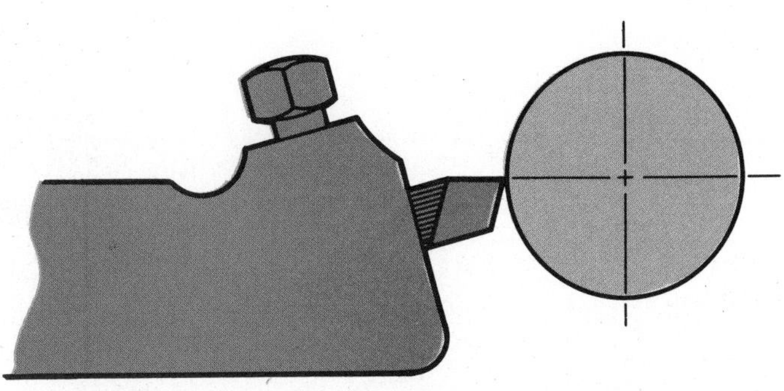

Fig. 10-164. This cutting tool is properly ground for machining sharp "V" threads. The tool is set on center as shown.

Several methods may be employed to terminate a thread, Fig. 10-167. Ordinarily, the beginner should use a groove until sufficient experience has been gained. However, the design of some parts does NOT permit a groove to be used. In which case, the threads must be terminated by another method. They require perfect coordination and very rapid operation of the cross slide to get the tool out of position at the end of the cut.

The gear box is adjusted to cut the correct number of threads. Make apron adjustments to permit the half-nuts to be engaged.

After the proper apron and gear adjustments have been made, pivot the compound rest to 29 deg. to the right, Fig. 10-168. Then set the threading tool in place.

It is essential that the tool be set on center with the tool axis at 90 deg. to the center line of the work. This is done with a center gage. Place the gage against the work while the tool is set into a V, Fig. 10-169. Tool height can be set by using the center line scribed on the tailstock spindle or with the center point.

The compound is set at 29 deg. to permit the tool to shear the chip better than if it were fed straight into the work, Fig. 10-170. Since the angle of the tool is 30 deg. and it is fed in at an angle of 29 deg.,

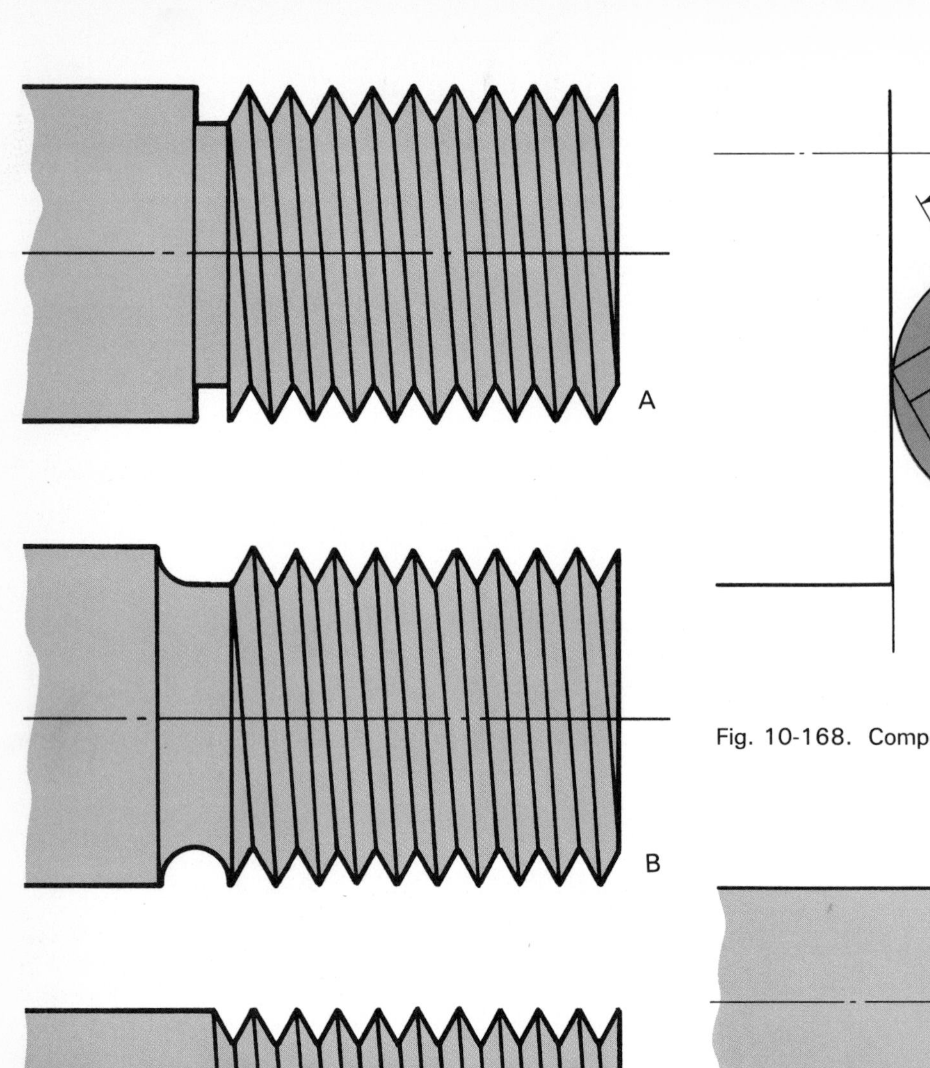

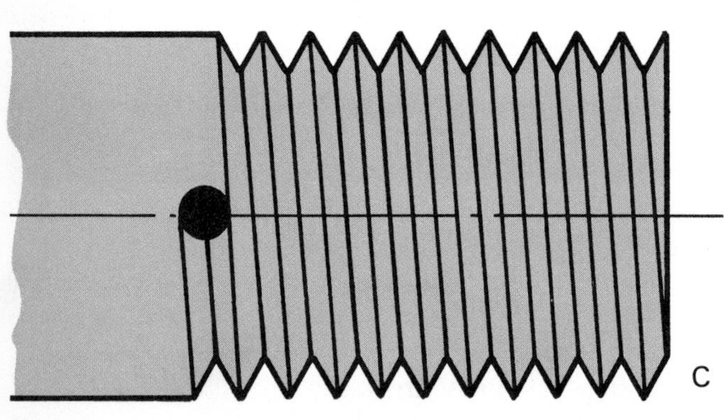

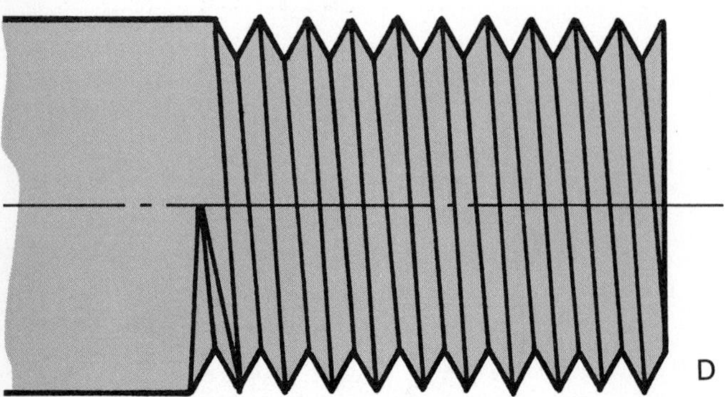

Fig. 10-167. Study methods for terminating a screw thread. A—Square groove. B—Round groove. C—Small shallow hole. D—Tool withdrawn from thread at end of cut.

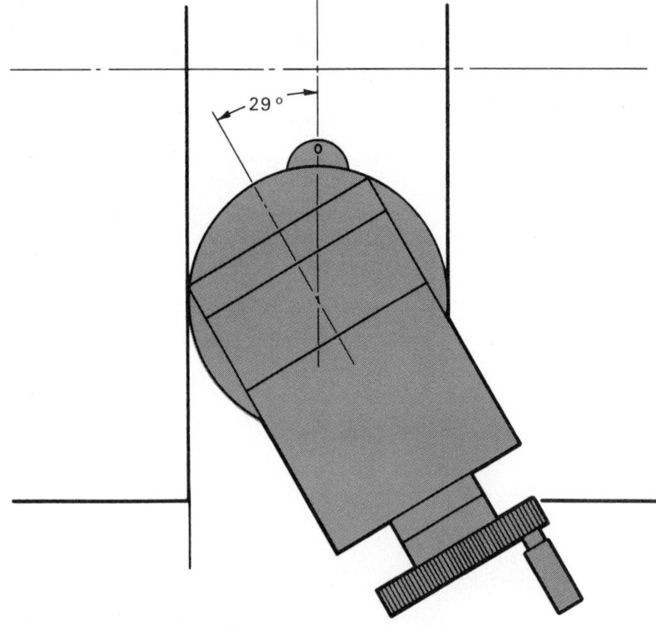

Fig. 10-168. Compound is set up for machining right-hand external threads.

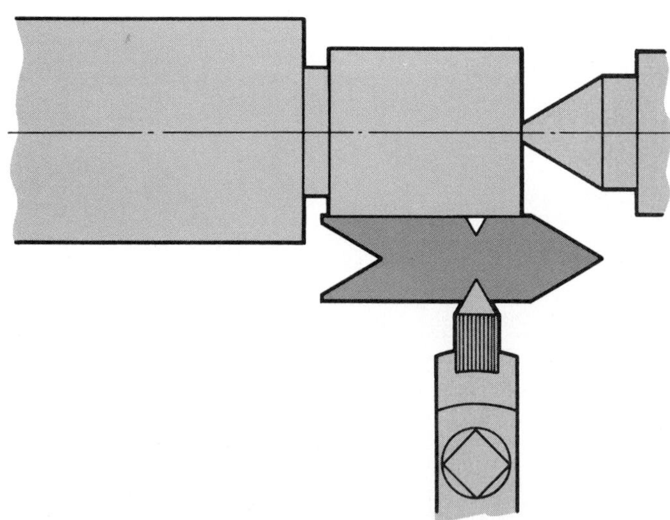

Fig. 10-169. Center gage is being used to position cutting tool for machining threads.

the slight shaving action that results will produce a smooth finish on the right side of the thread. At the same time, not enough metal is removed to interfere with the main chip that is removed by the left edge of the tool.

As the tool must be removed from the work after each cut and repositioned before the next cut can be started, a ***thread cutting stop*** may be used, Fig. 10-171. After the point of the tool is set to just touch the work, lock the stop to the saddle dovetail with the adjusting screw just bearing on the stop.

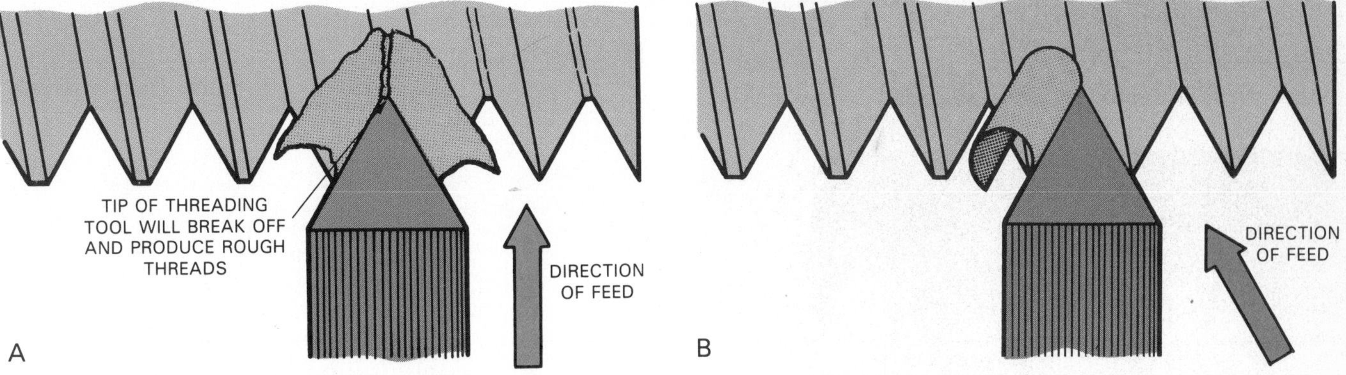

Fig. 10-170. Study cutting action of tool. A—When fed straight in, note that both edges are cutting and weakest part of tool, the point, is doing hardest work. B—When fed in at 29° angle, note that only one edge is cutting and cutting load is distributed across edge evenly.

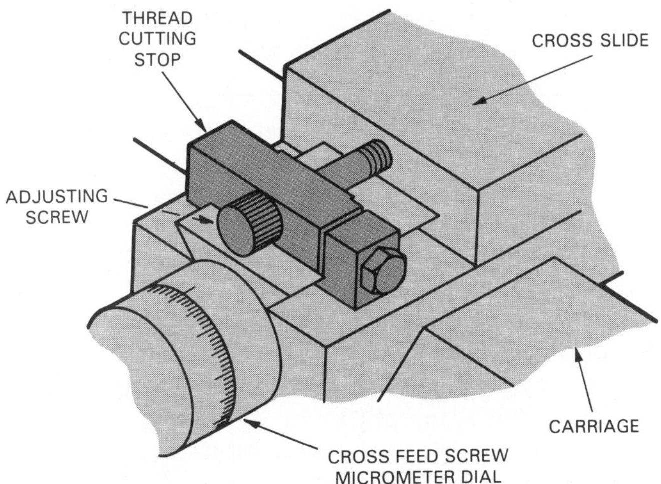

Fig. 10-171. After adjustment, thread cutting stop will let you start next cut in same location.

After a cutting pass has been made, move the tool back from the work with the cross slide screw. Move the carriage back to start another cut. Feed the tool into the work until the adjusting screw again bears against the thread cutting stop. By turning the compound rest in 0.002 to 0.005 in. (0.05 to 0.12 mm), the tool will be positioned for the next cut.

A thread dial is fitted to the carriage of most lathes, Fig. 10-172. It meshes with the lead screw. The ***thread dial*** is used to indicate when to engage the half-nuts to permit the tool to follow exactly in the original cut. The thread dial eliminates the need to reverse the spindle rotation to bring the tool back to the starting point after each cut.

The face of the thread dial rotates when the half-nuts are NOT engaged, Fig. 10-173. When the desired graduation moves into alignment with the index line, the half-nuts can be engaged.

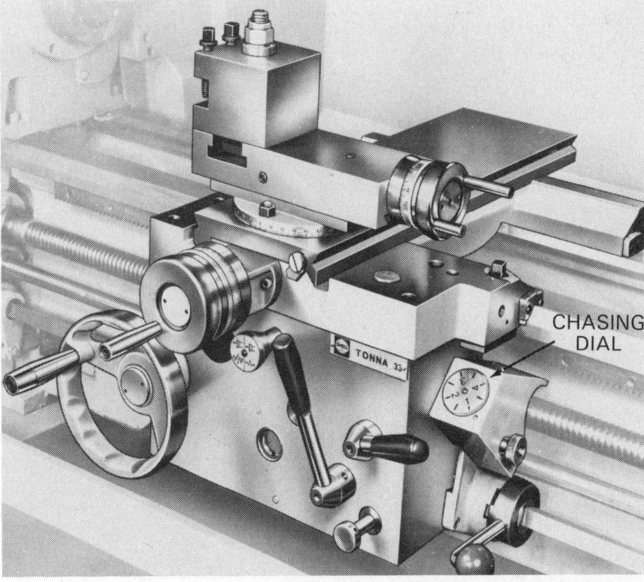

Fig. 10-172. Left—Thread dial for cutting inch base threads. Right—Thread dial for cutting both inch and metric base threads. Housing contains a series of gears. Gear selection depends upon threads being cut. (Clausing Machine Tool)

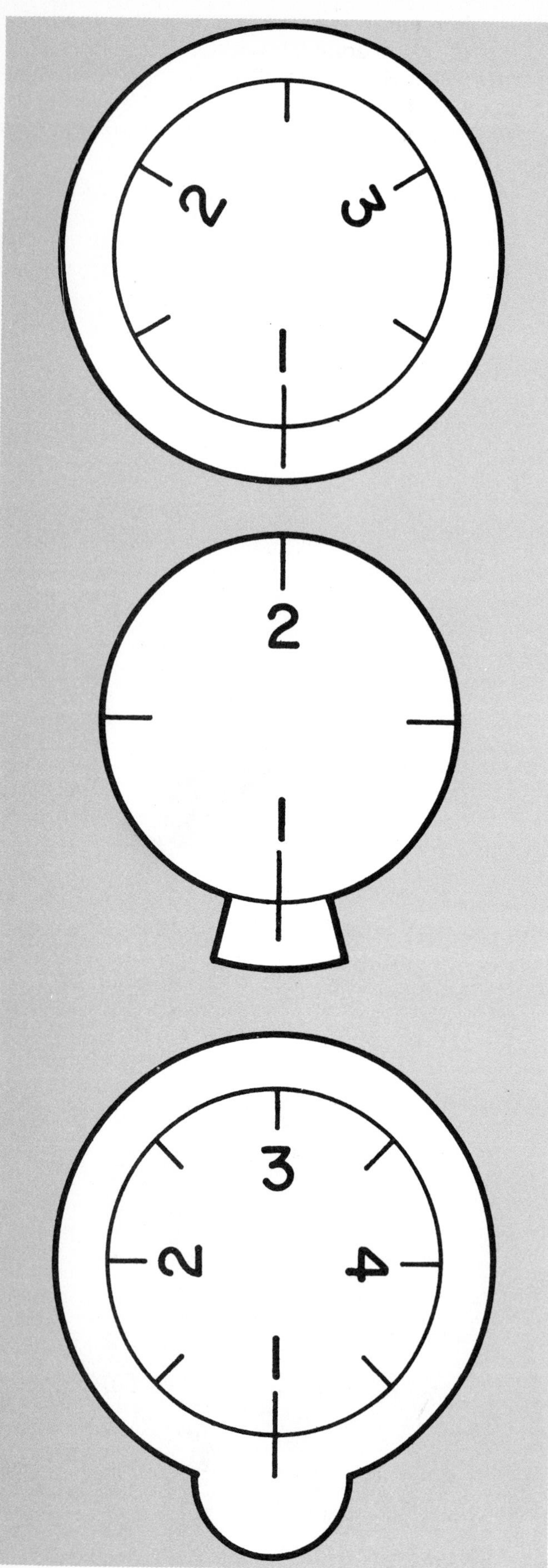

Fig. 10-173. These are typical thread dial faces.

The thread dial is used as follows for all inch based threads:

1. For all EVEN NUMBERED threads, close the half-nuts at any line on the dial.
2. For all ODD NUMBERED threads, close the half-nuts at any numbered line on the dial.
3. For all threads involving ONE-HALF OF A THREAD IN EACH INCH (such as 11 1/2), close the half-nuts at any odd numbered line.
4. For all threads involving ONE-FOURTH OF A THREAD IN EACH INCH (such as 4 3/4), return to the original starting line before closing the half-nuts.

NOTE! On lathes converted to metrics, the thread dial CANNOT be used. When thread cutting, once the half-nuts have been closed, they must NOT be opened until the thread is completely cut. The motor must be reversed after each cut to return the tool to the starting position.

The thread dial can be used on lathes with full metric capabilities. However, the thread dial varies with the lathe manufacturer and must be considered individually. To be sure of correct thread dial procedure, consult the manufacturer's handbook for the machine.

Making the cut

Set the spindle speed to about one-quarter that used for conventional turning. Feed the tool in until it just touches the work. After this, it is moved beyond the right end of the work and adjusted to take a 0.002 in. (0.05 mm) cut.

Turn on the power and engage the half-nuts when indicated by the thread dial. This cut is made to check whether the lathe is cutting the correct threads. Thread pitch can be checked with a rule or with a screw pitch gage, Fig. 10-174. When everything checks, additional cuts in 0.005 in. (0.12 mm) increments are made until the thread is almost to size. The last few cuts should be NO MORE than 0.002 in. (0.05 mm) deep. Note that all cutting tool advances are made with the compound rest feed screw.

A liberal application of lard or cutting oil, before each cut, will be helpful in obtaining a smooth finish.

Resetting tool in thread

It is sometimes necessary to replace a broken cutting tool, or to resharpen it for the finish cuts. After replacing the tool, it must be realigned with the portion of the thread already cut. This can be done as follows.

1. Set the tool on center and position it with a center gage.
2. Engage the half-nuts at the proper thread dial graduation.

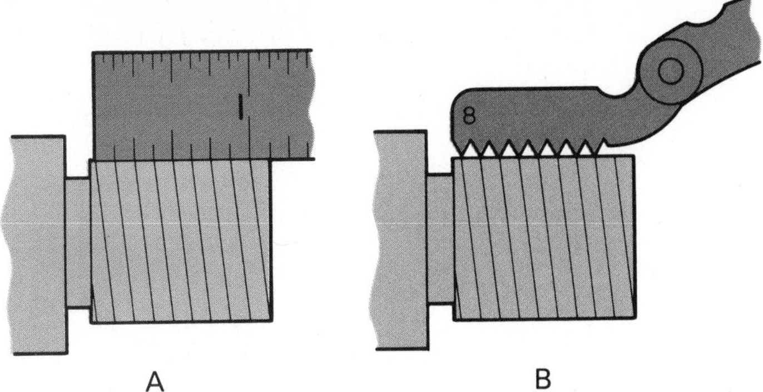

Fig. 10-174. Always check thread pitch after first light cut has been made. A—With a rule. B—With a screw pitch gage.

3. Set the tool back from the work and rotate the spindle until the tool reaches a position about halfway down the threaded section.
4. Using the compound rest screw and the cross slide screw, align the tool in the existing thread. Reset the thread cutting stop after the tool has been aligned.

Measuring threads

Measure the thread at frequent intervals during the machining operation to assure accuracy. The easiest way to check thread size is to fit them into a threaded hole or nut. If it does NOT fit, it is too large and further machining is necessary. This technique is not very accurate but is usually satisfactory when close tolerances are NOT specified.

A ***thread micrometer*** makes quick, accurate thread measurements, Fig. 10-175. It has a pointed spindle and a double-V anvil to engage the thread. The micrometer reading given is the true pitch diameter which equals the outside diameter of the screw MINUS the depth of one thread. Each micrometer is designed to read a limited number of thread pitches and is available in both inch and millimeter measurements.

The ***three-wire method of measuring threads*** has proven to be quite satisfactory, Fig. 10-176. A micrometer measurement is made over three wires of a specific diameter that are fitted into the threads. Shown in Fig. 10-177, the formula will provide the information necessary to calculate the correct measurement over the wires.

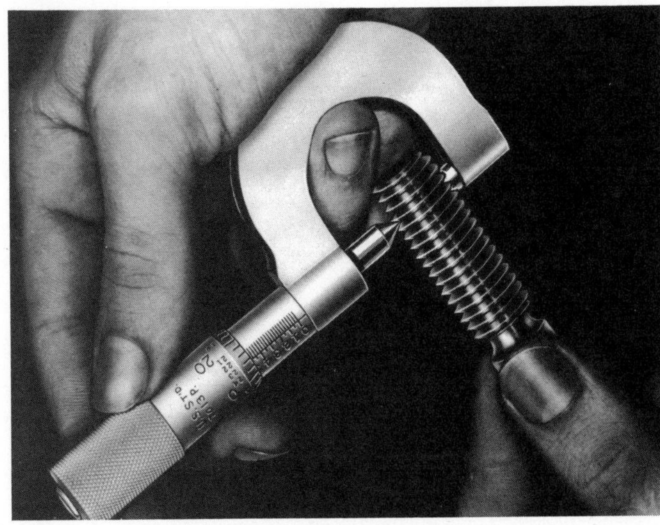

Fig. 10-175. Thread micrometer will check cut threads precisely. (L.S. Starrett Co.)

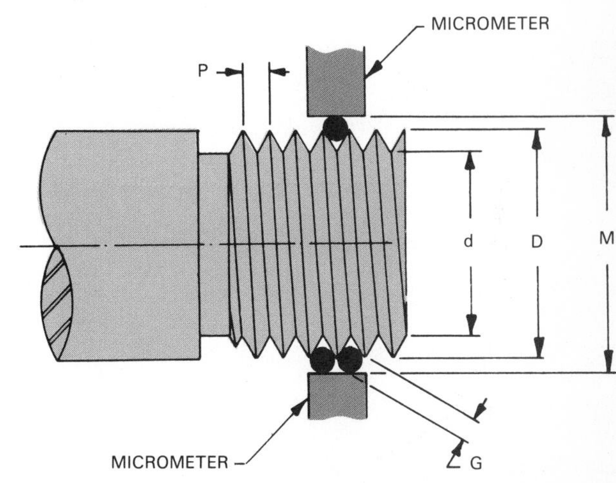

Fig. 10-176. Study three-wire method of measuring screw threads. Compare it to formula in Fig. 10-177.

$$M = D + 3G - \frac{1.5155}{N}$$

Where: M = Measurement over the wires
D = Major diameter of thread
d = Minor diameter of thread
G = Diameter of wires
P = Pitch = $\frac{1}{N}$
N = Number of threads per inch

The smallest wire size that may be used for a given thread—

$$G = \frac{0.560}{N}$$

The largest wire size that can be used for a given thread—

$$G = \frac{0.900}{N}$$

The three-wire formula will work only if "G" is no larger or smaller than the sizes determined above. Any wire diameter between the two extremes may be used. All wires must be the same diameter.

Fig. 10-177. Three-wire thread measuring formula is calculated by plugging in values.

Cutting left-hand threads

Left-hand threads are cut in basically the same manner as right-hand threads. The major differences involve pivoting the compound to the LEFT and changing lead screw rotation so the carriage travels towards the tailstock (left to right), Fig. 10-178.

Cutting Acme threads

The ***Acme thread*** has the top and bottom of the thread flat but the sides have a 29 deg. included angle. It was originally developed to replace the square thread. Its advantages are its strength and ease with which it can be cut compared to the square thread. The thread form is employed in machine tools for precise control of component movement.

The ***Acme screw thread gage*** is the standard for grinding and setting Acme thread cutting tools, Fig. 10-179. Tool angle is ground to fit a V in the thread gage. The width of the flat section varies with the pitch of the thread. This width is obtained by grinding back the tool point until it fits into the notch appropriate for the thread being cut.

In cutting the threads, the groove is usually roughed out with a square nosed tool to approximate depth, then finished with an Acme shaped tool. The compound is set to 14 deg. and the tool is set up using the thread gage, Fig. 10-180. Other than this, Acme threads are cut in the same manner as the sharp V thread.

Cutting internal threads

Internal threads are made on the lathe with a conventional boring bar and a cutting tool sharpened to the proper shape. Refer to Fig. 10-181.

Before internal threads can be machined, the work must be prepared. A hole is drilled and bored to correct size for the thread's minor diameter. A recess is machined with a square nosed tool at the

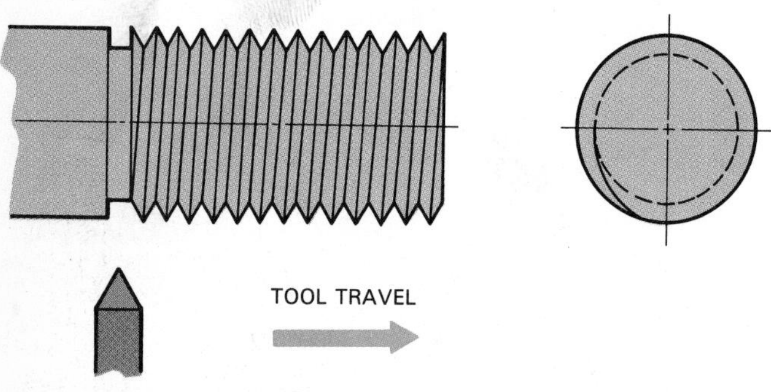

Fig. 10-178. Note cutting action for a left-hand thread.

Fig. 10-179. Acme screw thread gage and tool setup gage will allow you to check lathe settings.

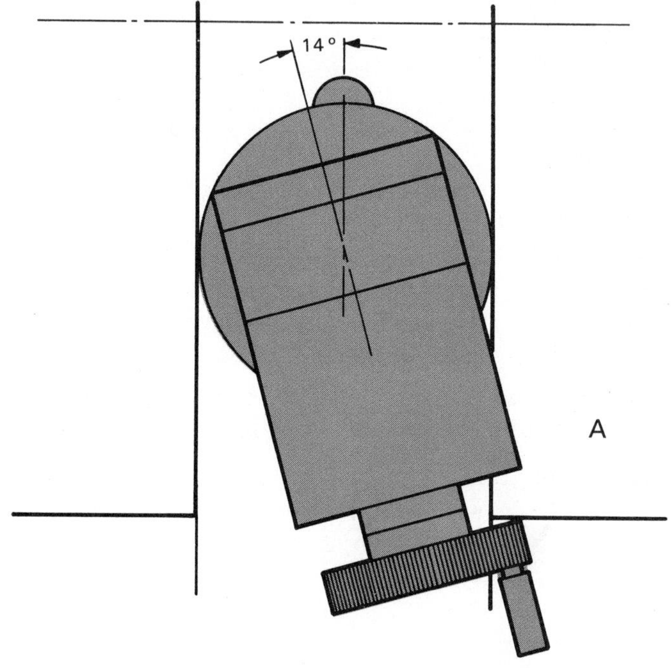

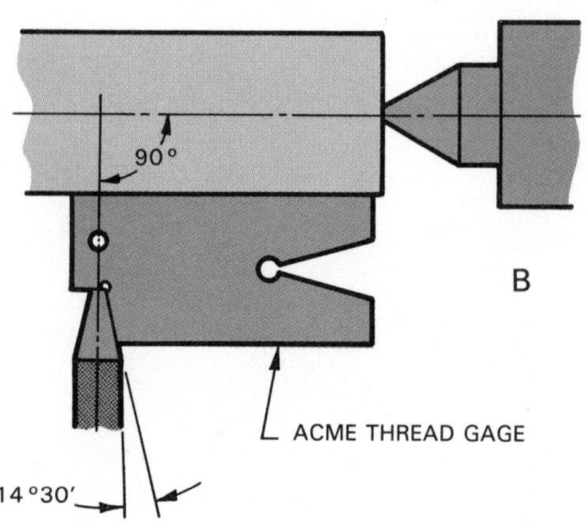

Fig. 10-180. A—This compound setting is for machining Acme threads. B—Cutting tool is positioned with an Acme thread gage.

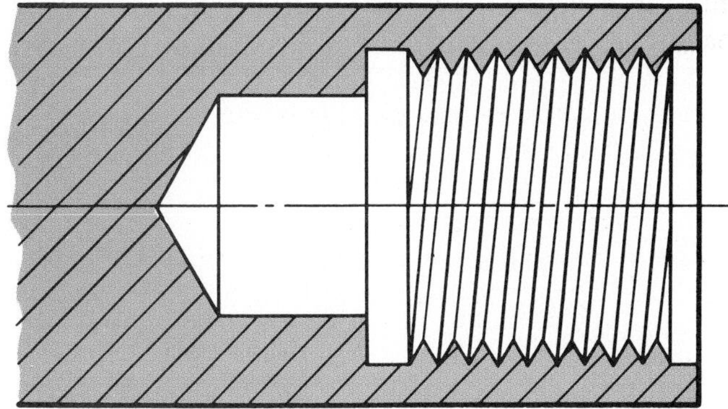

Fig. 10-181. Note internal screw threads.

point where the thread terminates, Fig. 10-182. The diameter of the recess is equal to the major diameter of the thread.

Pivot the compound rest 29 deg. to the LEFT for cutting right-hand internal threads, Fig. 10-183. Mount the tool on center and align it with a center gage, Fig. 10-184.

Bring the tool up until it just touches the work surface. Adjust the micrometer collar on the cross slide to zero (0) with the tool in position. Using the compound rest screw, adjust the cutter to make a cut of 0.002 in. (0.05 mm).

CAUTION! Remember that tool infeed and removal from the cut are the reverse of those used for cutting external threads.

A problem may arise in trying to determine when the tool has traveled far enough into the hole so the

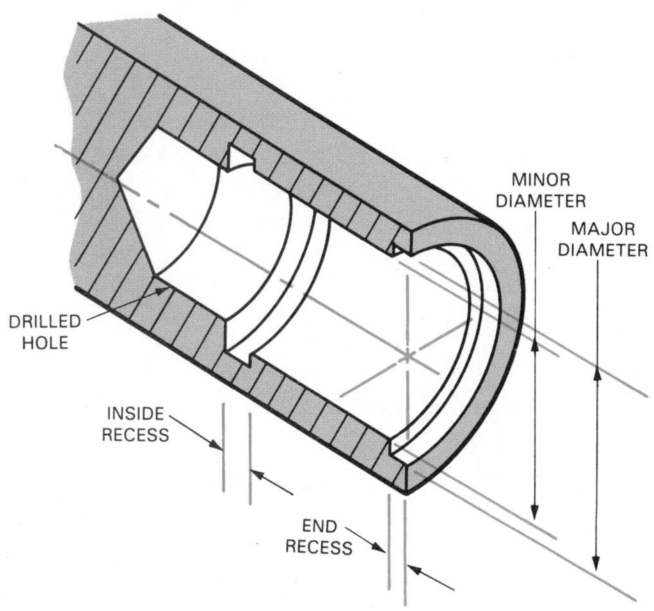

Fig. 10-182. Opening for internal screw threads has been machined.

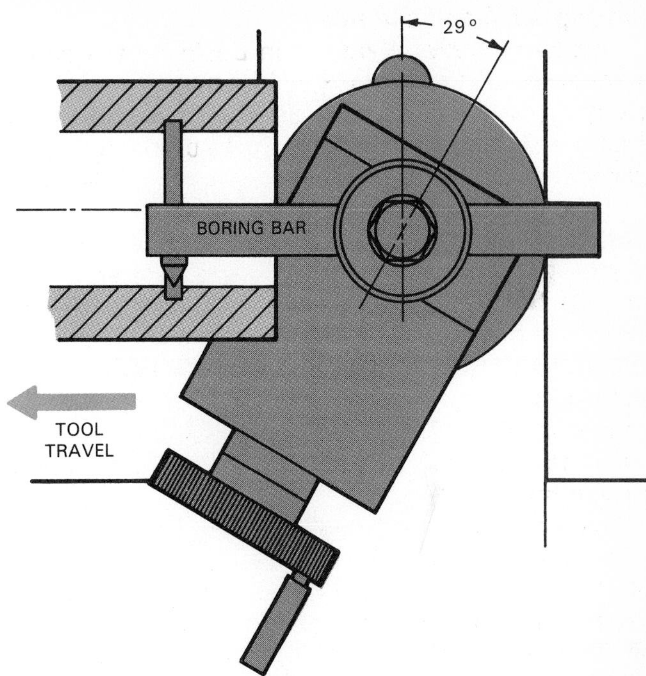

Fig. 10-183. This is compound setting for internal right-hand screw threads.

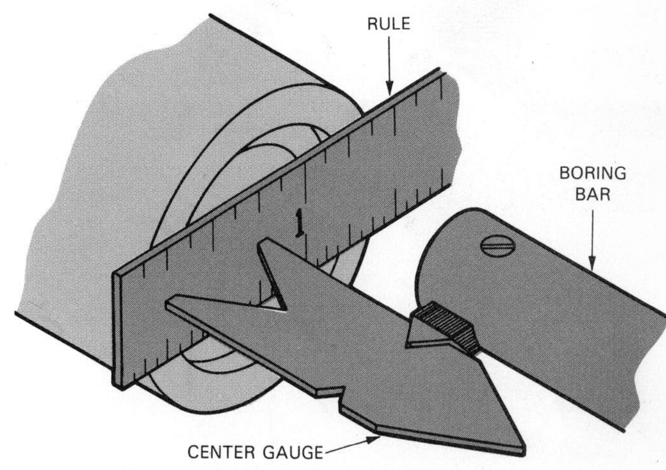

Fig. 10-184. Note how to position cutting tool for machining internal screw threads.

half-nuts can be disengaged. One method makes use of a lightly scribed line in a blued area on the flat way of the bed. The tool will have advanced far enough when the carriage reaches this point.

Another technique allows you to start at the back of the hole when cutting internal threads. Pivot the compound rest 29 deg. to the RIGHT. Place the threading tool to the rear of the boring bar with the cutting edge up. See Fig. 10-185.

The lathe spindle is run in reverse. To prevent the tool from being placed too far into the hole to start the cut, mount a micrometer carriage stop on the

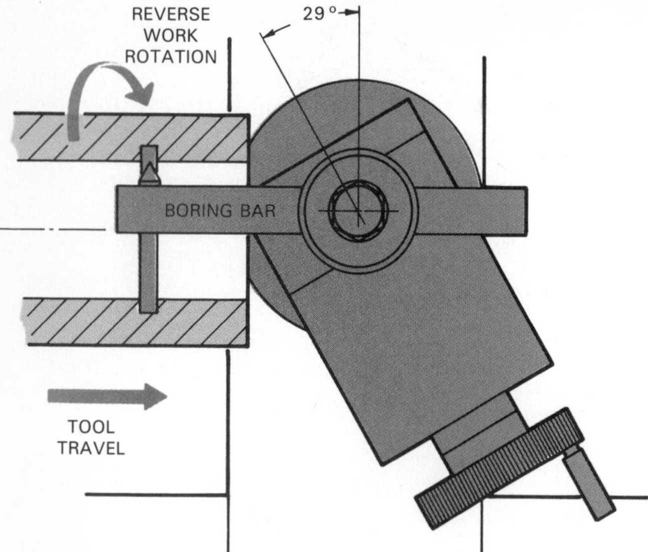

Fig. 10-185. This is alternate setup technique for cutting internal right-hand threads. Note that work rotates opposite that of normal turning operations.

ways, Fig. 10-186. The carriage is returned until it touches the stop. Follow the same general procedure for cutting the threads as previously described.

Continue making additional cuts until the threads are finished. Lighter cuts must be taken than those when machining external threads, because the toolholder is NOT as rigid. Keep the surface flooded with cutting fluid.

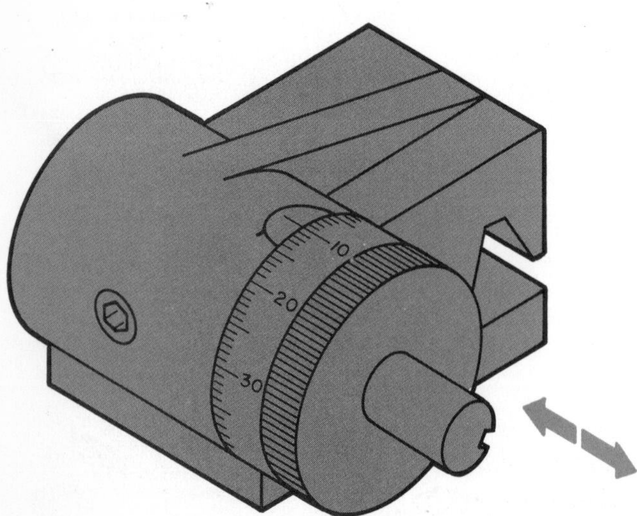

Fig. 10-186. When using alternate technique for cutting internal right-hand threads, mount a micrometer carriage stop on ways. Adjust it to prevent tool from being placed too far into hole when starting each cut.

Cutting threads on a taper

Tappered threads must be cut, at times to obtain a fluid or gas tight joint. When this situation arises, the threading tool must be positioned in relation to the center line of the taper rather than in relation to the taper, Fig. 10-187.

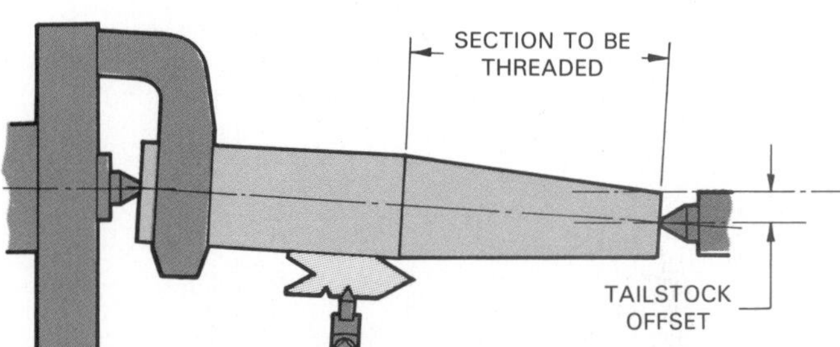

Fig. 10-187. Study how to position tool for machining screw threads on a taper. Tool is NOT positioned on taper when setting tool.

BORING ON A LATHE

Boring is an internal machining operation where a single point cutting tool is used to enlarge a hole, Fig. 10-188. Boring may be employed to enlarge a hole to a specified size where a drill or reamer will NOT do the job. When properly set up, it produces a hole that is concentric with the outside diameter of the work.

While the machining technique remains essentially the same as for external turning, several conditions are encountered that could cause difficulties. When boring on a lathe, you must make allowances for the following:

1. Movement of the cross slide screw is reversed.
2. The machinist must work by ''feel'' as the cutting action cannot always be observed.
3. Additional front clearance must be ground on the cutting tool to avoid rubbing, Fig. 10-189. Otherwise, the shape of the cutting tool is identical with that for external turning.
4. Boring deep or small diameter holes requires a long, slender boring bar. The overhang makes the tool more likely to spring away from the surface being machined. It is also necessary to take several light cuts INSTEAD OF one heavy cut to remove the same amount of material.
5. Internal measuring tools are more difficult to use for some people that those for making external measurements.

Boring the hole

The hole size to be bored determines the type and size boring bar required, Figs. 10-190 to 10-194.

Fig. 10-188. Boring or machining internal surfaces is sometimes done on lathe.

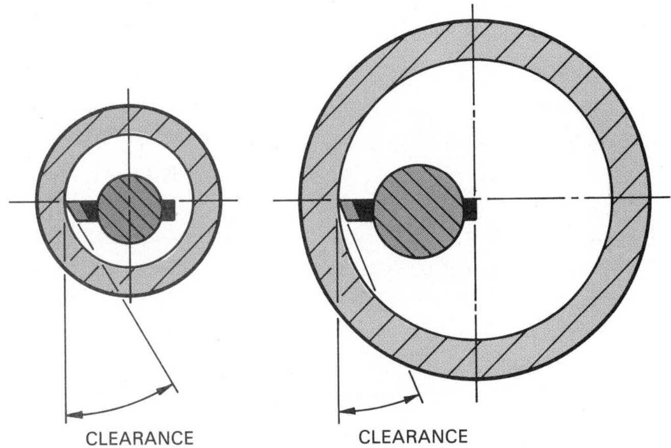

Fig. 10-189. Tool used to bore small diameters requires more front clearance to prevent rubbing.

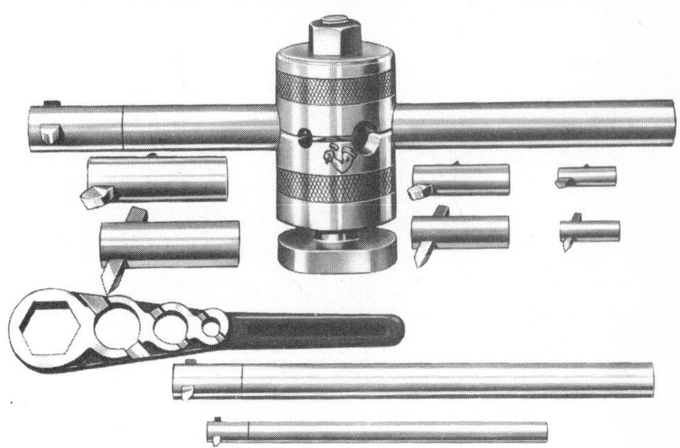

Fig. 10-190. Interchangeable type boring bar permits machinist to use most rigid bar for job. Body of this model replaces tool post. (Armstrong Bros. Tool Co.)

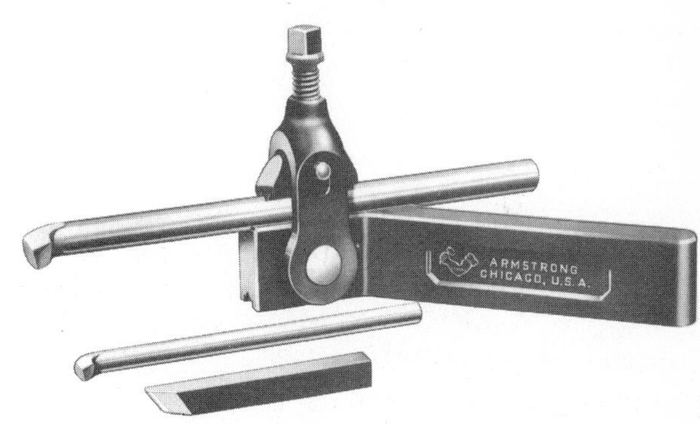

Fig. 10-191. Note boring toolholder and boring bar for light internal machining operations. (Armstrong Bros. Tool Co.)

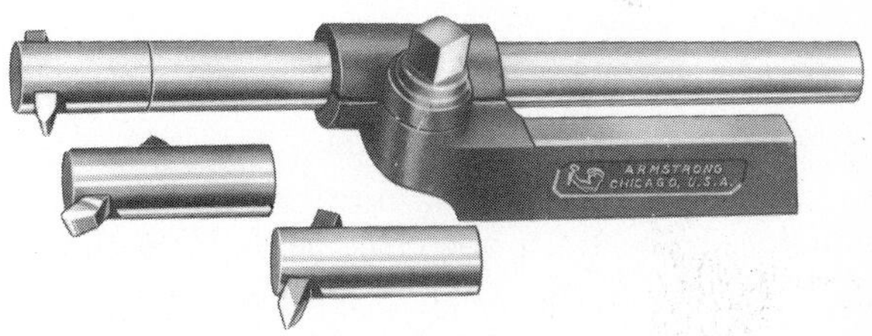

Fig. 10-192. This boring toolholder and bar has interchangeable end pieces. (Armstrong Bros. Tool Co.)

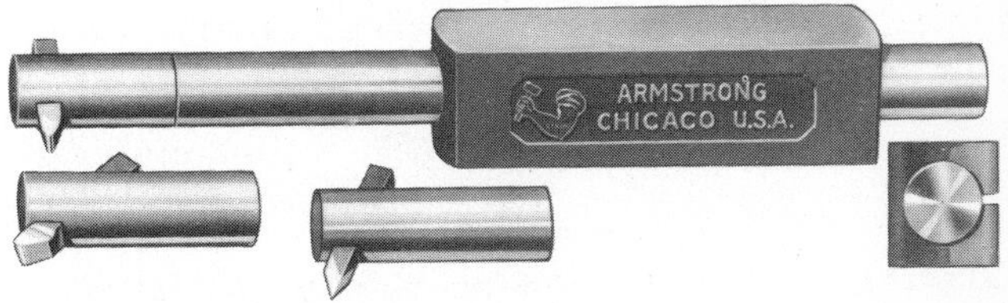

Fig. 10-193. With spring type boring toolholder, boring bar is held in holder by tool post screw pressure. (Armstrong Bros. Tool Co.)

Always use the largest bar possible to give maximum tool support. The bar should extend from the holder only far enough to permit the tool to cut to the required hole depth, Fig. 10-195.

The boring bar is set on center or slightly below center, with the boring bar parallel to tool travel, Fig. 10-196. Check for adequate clearance when the tool is at maximum depth in the hole.

Make a light cut in the same manner as in external machining. When the cut is completed, stop the machine. Set the cross slide micrometer dial to zero (0) and back the tool away from the work. Remove the boring bar from the hole.

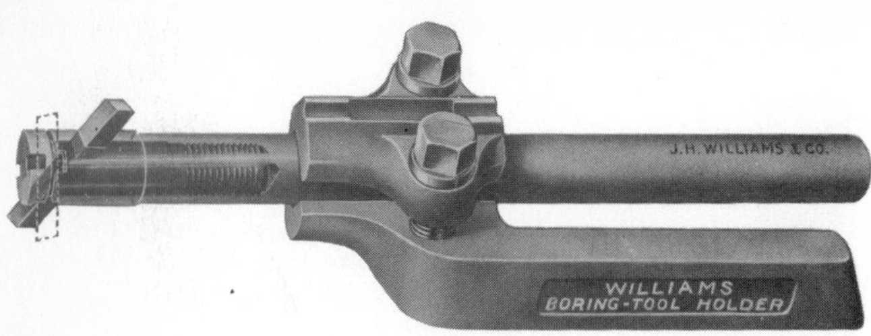

Fig. 10-194. With clamp type boring bar toolholder, boring bars of different diameters may be used. (Armstrong Bros. Tool Co.)

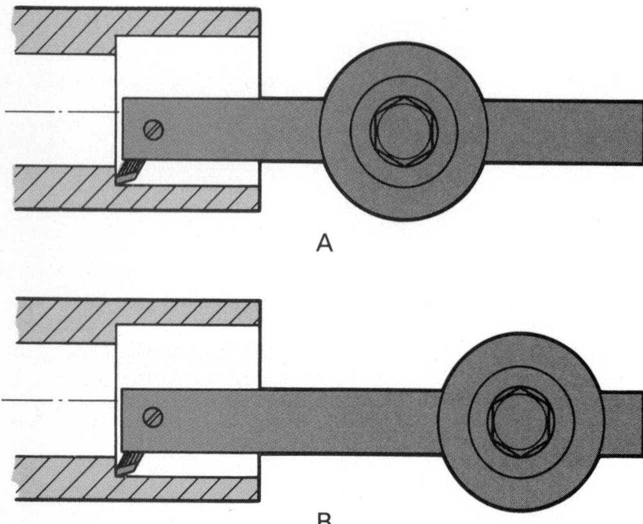

Fig. 10-195. Keep cutting tool as close to tool post as possible for maximum tool support. A—Properly positioned boring bar. B—Boring bar projects too far from tool post. Vibration and "chatter" could produce a rough machined surface.

Check hole diameter with a telescoping gage and micrometer or an inside micrometer. After checking hole accuracy, bring the cross slide back to zero (0), and advance the tool to make another cut (amount of infeed will be determined by boring bar in use and material being bored). Make additional cuts, checking hole size frequently, until the desired diameter is attained.

When making the final cut, it may be necessary to reverse tool travel after reaching the desired depth. Let the tool feed out of the hole without changing the tool setting. This will compensate for any tool spring.

CAUTION! Reverse carriage feed for the above operation. Do NOT reverse spindle rotation!

When boring holes with long, slender boring bars, it may be necessary to run the tool into the hole without changing its setting every two or so cuts to compensate for tool spring.

Because of the slender nature of some boring bars, "chatter" is more likely to occur than when doing external work. This can usually be eliminated by:

1. Using a slower spindle speed.
2. Reducing tool overhang.
3. Grinding a smaller radius on the cutting tool nose.
4. Placing a weight on the back overhang of the boring bar.
5. Placing the tool slightly below center.

DRILLING AND REAMING ON A LATHE

The lathe can perform many operations other than turning. It can sometimes be needed to drill or ream holes.

Drilling on a lathe

When a hole is cut in solid stock, the usual practice is to hold it in a suitable chuck and mount the drill in the tailstock, Fig. 10-197. For holes that are 1/2 in. (12.5 mm) in diameter, a straight shank drill

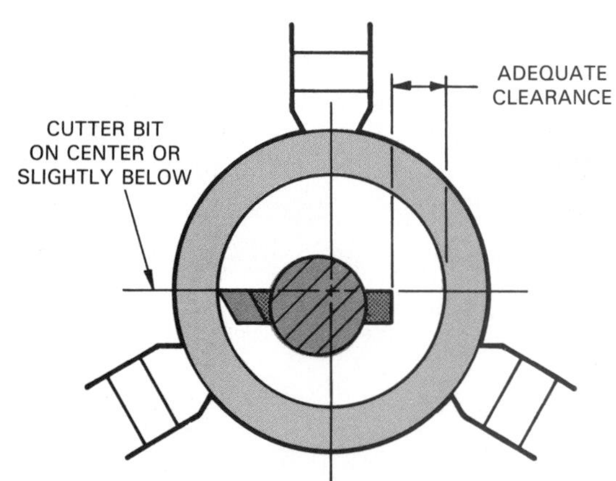

Fig. 10-196. Tool is set on, or slightly below, center when boring. Also check for adequate clearance between boring bar and hole.

Fig. 10-197. Drills that are larger than 1/2 in. (12.5 mm) in diameter are usually fitted with a self-holding taper that fits into tailstock spindle.

is placed in Jacobs chuck fitted in the tailstock spindle, Fig. 10-198. Holes larger than 1/2 in. (12.5 mm) diameter are made with taper shank drills.

Drills with taper shanks too large to be fitted in the tailstock can be used, with EXTREME CARE, if mounted as in Fig. 10-199. A dog is fitted to the neck of the drill. The tool is set up to permit the tailstock center to press into the center hole in the drill tang.

The drill's cutting point bears against the rotating work and is prevented from revolving by the dog held stationary by the tool post. The tailstock center keeps the drill aligned and enables it to be fed into the material by the tailstock hand wheel.

A ***commercial drill holder*** makes it unnecessary to use the make-shift lathe dog setup. See Figs. 10-200 and 10-201.

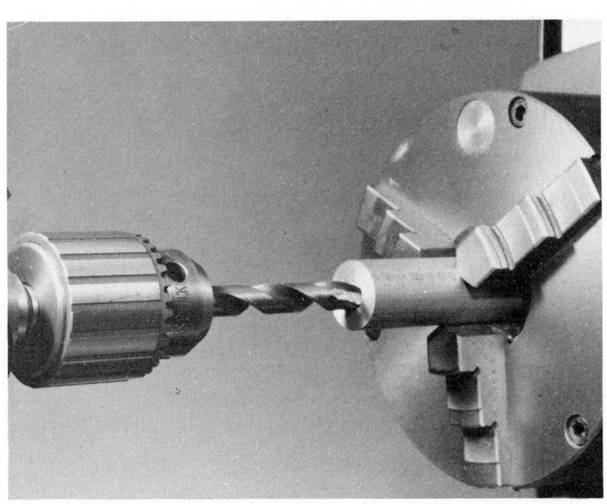

Fig. 10-198. Drilling is being done with straight shank drill held in Jacobs chuck.

Fig. 10-199. Note setup employed when drill shank is too large to be fitted in tailstock. Lathe dog, with its tail supported by compound, prevents it from revolving during drilling operation. CAUTION: This type drilling requires care to prevent drill from slipping from tailstock center when full drill diameter breaks through work.

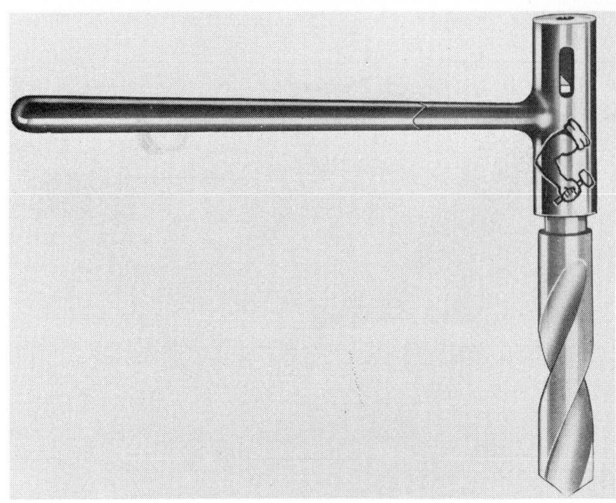

Fig. 10-200. Large taper shank drills can be used on lathe by fitting them in this commercial drill holder. (Armstrong Bros. Tool Co.)

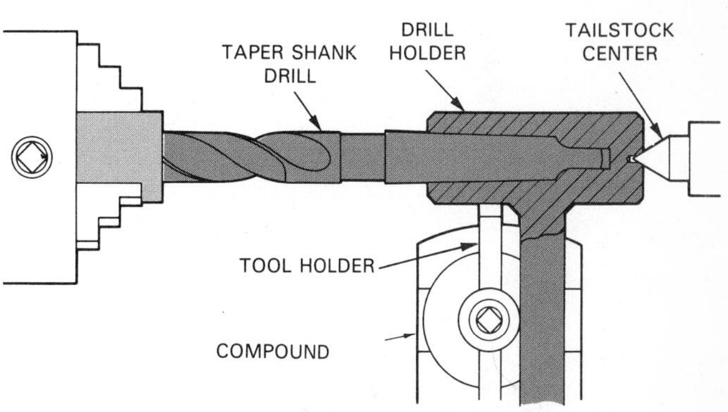

Fig. 10-201. This illustration shows how commercial drill holder is used.

Greater accuracy requires a centered starting point for the drill. A starting point made with a combination drill and countersink is adequate for most jobs, Fig. 10-202.

Holes over 1/2 in. (12.5 mm) diameter require a pilot hole. This hole should have a diameter equal to width of larger drill's dead center, Fig. 10-203.

Ample clearance must be provided in back of work to permit drill to break through without striking the chuck or headstock spindle, Fig. 10-204.

Reaming on a lathe

A hole is ***reamed*** to produce accuracy in diameter and finish, Fig. 10-205. The hole is first drilled, allowing stock for reaming.

The allowance for reaming is dependent upon hole size:

1. With a hole size to 1/4 in. (6.5 mm) diameter, allow 0.010 in. (0.25 mm) of material for reaming.

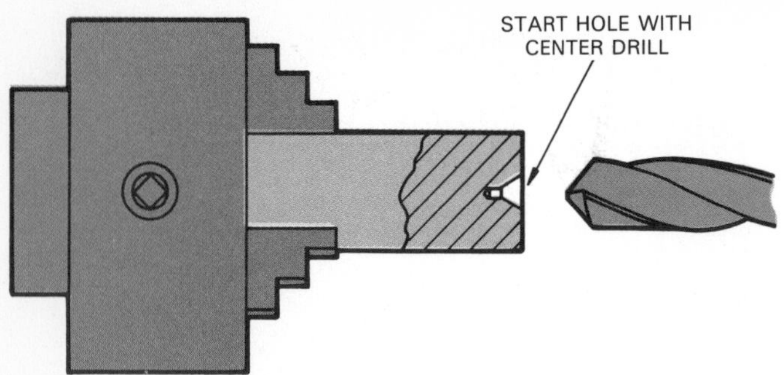

Fig. 10-202. Drill will cut exactly on center if hole is first started with a center drill.

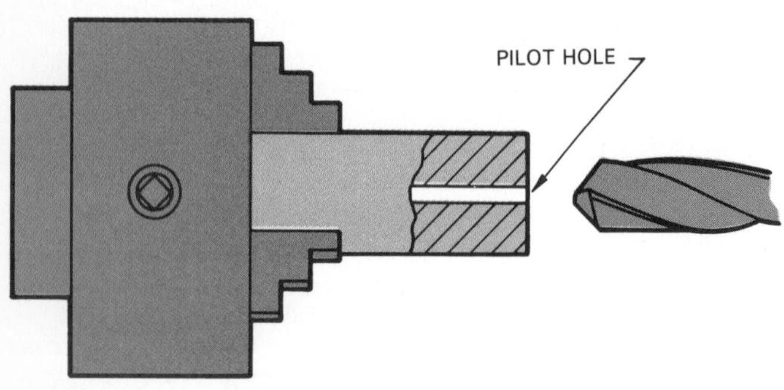

Fig. 10-203. Holes larger than 1.2 in. (12.5 mm) diameter require a pilot hole.

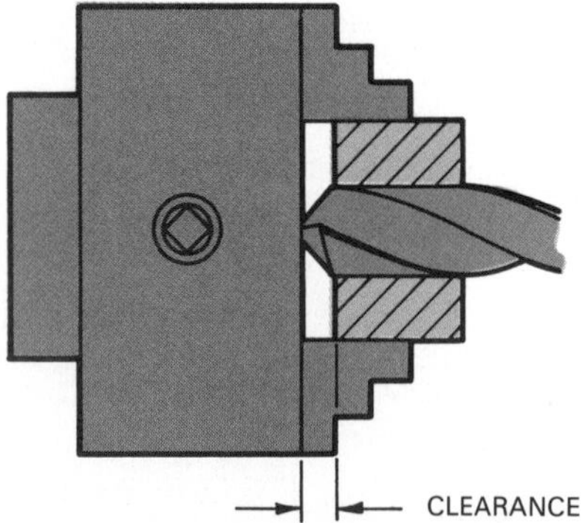

Fig. 10-204. There must be enough clearance, between back of work and chuck face, to permit drill to come through without damaging chuck.

2. With a hole size from 1/4 in. (6.5 mm) to 1/2 in. (12.5 mm) diameter, allow 0.015 in. (0.4 mm).
3. With a hole size from 1/2 in. (12.5 mm) to 1.0 in. (25.0 mm) diameter, allow 0.020 in. (0.5 mm).
4. With a hole size from 1.0 in. (25.0 mm) to 1.5 in. (37.5 mm) diameter, allow 0.025 in. (0.6 mm).
5. With a hole size above 1.5 in. (37.5 mm) diameter, allow 0.030 in. (0.8 mm) for reaming.

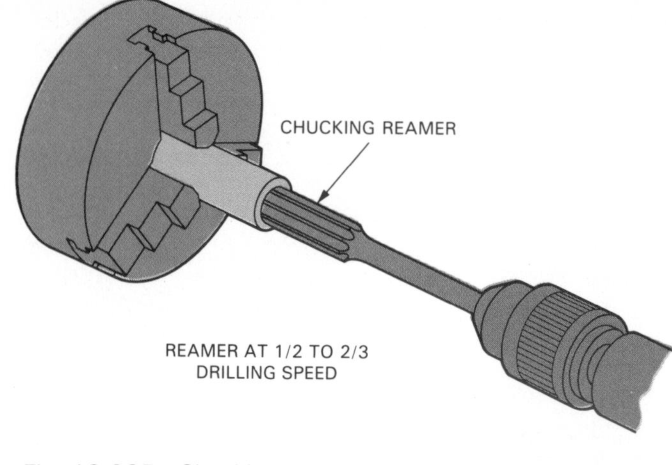

Fig. 10-205. Chucking reamer can be mounted in a Jacobs chuck to finish hole.

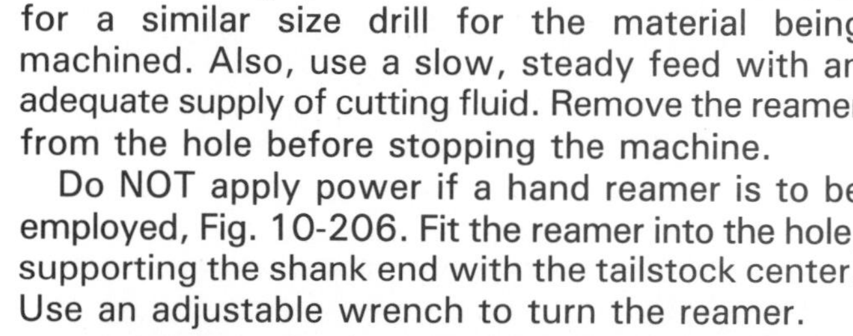

Use a cutting speed that is about two-thirds that for a similar size drill for the material being machined. Also, use a slow, steady feed with an adequate supply of cutting fluid. Remove the reamer from the hole before stopping the machine.

Do NOT apply power if a hand reamer is to be employed, Fig. 10-206. Fit the reamer into the hole, supporting the shank end with the tailstock center. Use an adjustable wrench to turn the reamer.

CAUTION! Continue to rotate the reamer CLOCKWISE as it is removed from the hole. AVOID turning it counterclockwise as it will ruin the tool's cutting edges.

KNURLING ON A LATHE

Knurling is the process of forming horizontal or diamond-shaped ***serrations*** (raised grooves or teeth)

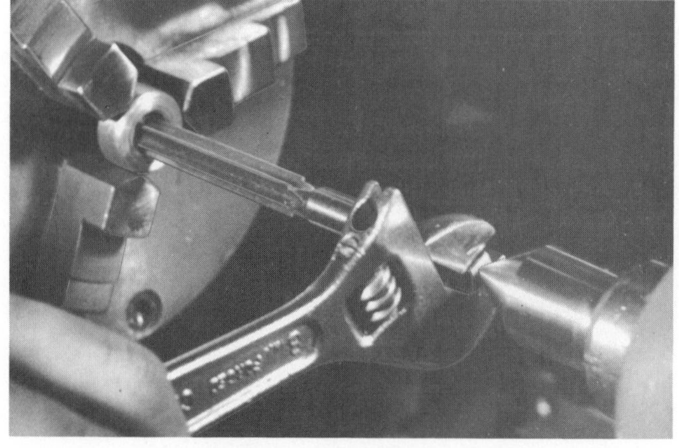

Fig. 10-206. Never turn on the power when using a hand reamer on the lathe.

on the circumference of the work to provide a gripping surface, Fig. 10-207. It is done with a ***knurling tool*** mounted in the tool post,Fig. 10-208. The knurled pattern is raised by rolling the knurls against the metal. This displaces the metal into the required pattern.

Angular knurls raise a diamond pattern, Fig. 10-209, while a ***straight knurl*** produces a straight pattern, Fig. 10-210, along the length of the work. The patterns can be produced in coarse, medium, and fine pitch.

Knurling procedure

If a knurling tool setup is NOT made properly, the knurls will NOT track and will quickly dull. The following procedure is recommended:

1. Mark off section to be knurled.
2. Adjust the lathe to a slow back geared speed and a fairly rapid feed.
3. Place the knurling tool in the tool post. Bring it up to the work. Both wheels must bear evenly on the work with the faces parallel with the center line of the piece.
4. Start the lathe and slowly force the knurls into the work surface until a pattern begins to form. Tool travel should be towards the headstock whenever possible. Engage the automatic feed and let the tool travel across the work.

Fig. 10-207. Knurling rollers are forming serrations on part.

Fig. 10-208. Knurling tool will mount in lathe. (Armstrong Bros. Tool Co.)

Fig. 10-209. Compare diamond knurl in coarse, medium, and fine patterns.

Fig. 10-210. Compare straight knurl in coarse, medium, and fine pitch.

5. When it reaches the proper position, reverse spindle rotation and allow the tool to move back across the work to the starting point. Apply additional pressure to force the knurls deeper into the work.
6. Repeat the operation until a satisfactory knurl is formed. Flood the work with cutting fluid.

Knurling difficulties

Knurling problems can arise and destroy the work is NOT handled properly. A ***double cut knurl*** occurs when one wheel makes twice as many ridges as the other. It is a common problem, Fig. 10-211.

A double knurl is usually caused by one wheel being dull. Raising or lowering the knurling tool to put more pressure on the dull wheel will frequently eliminate the trouble. Pivoting the tool slightly to allow the right side of the wheels to apply more pressure may also help.

Considerable side pressures are developed during the knurling operation. Watch the tool carefully. Do NOT permit the work to slip into the chuck or loosen on the tailstock center. If a ball bearing center is NOT used, keep the tail center well lubricated.

Knurling is done before turning a shaft to a smaller diameter, Fig. 10-212. If done after the smaller diameter has been machined, the work will spring away from the tool, giving the surface a superficial (light, nonpenetrating) knurl.

Avoid applying too much pressure to the knurling tool.The work surface becomes hardened during the operation and the knurled section could "flake" off. High pressure also tends to bend the shaft.

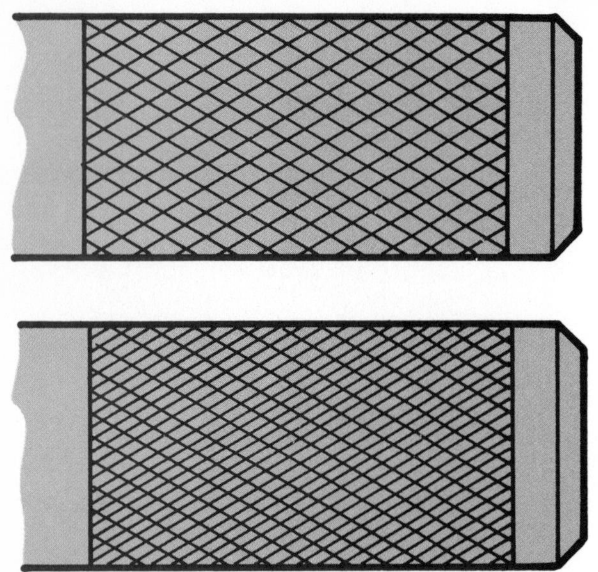

Fig. 10-211. Above—This is correctly made diamond knurl pattern. Below—Diamond knurl pattern where one knurl wheel was slightly above or below center.

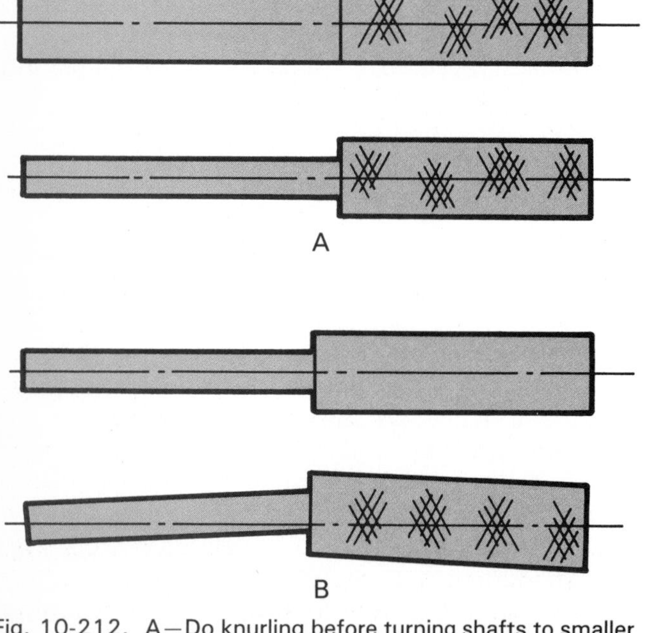

Fig. 10-212. A—Do knurling before turning shafts to smaller diameters. B—If knurled after being turned to smaller diameter, shaft may take on a permanent bend or break and receive only a superficial (very light) pattern.

WARNING! Never stop the lathe with the knurls engaged in the work. The piece will take on a permanent bend.

FILING AND POLISHING ON A LATHE

Lathe filing is done to remove burrs, round off sharp edges, and to blend-in form cut outlines. A file is NOT intended to replace a properly sharpened cutting tool and should NOT be used to improve the surface finish on a turned section.

Filing on a lathe

When filing on a lathe, avoid holding the tool stationary against the work. Keep it moving across the area being filed. If held in one position, the file will "load" and score the surface of the work.

An ordinary mill file will produce satisfactory results. However, a ***long angle lathe file*** produces superior results, Fig. 10-213.

Operate the lathe at high spindle speed and apply long, even strokes. Release pressure on the return stroke. Out-of-round work will result when uneven pressure is applied. Clean the file often.

As simple as filing on a lathe may appear, if can be DANGEROUS if a few precautions are not observed.

1. Move the carriage out of the way and remove the tool post.
2. Use the left-hand method of filing, Fig, 10-214. It involves holding the file handle in the left hand. The right hand is then well clear of the revolving chuck or faceplate. AVOID the right-hand method, Fig. 10-215. This technique places your left arm over the chuck or faceplate.

Polishing on a lathe

Polishing is sometimes done on a lathe. A strip of abrasive cloth, suitable for the material to be

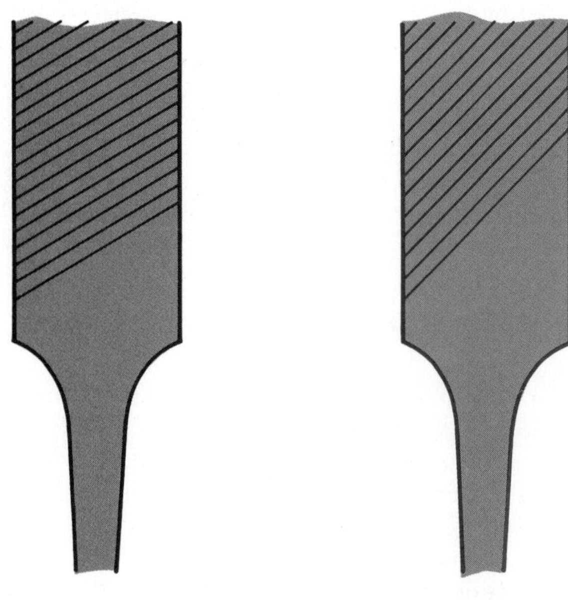

Fig. 10-213. Note how standard mill file differs from long angle lathe file.

Fig. 10-214. Left-hand method of filing on the lathe is preferred. How does it differ from the right-hand method shown in Fig. 10-215?

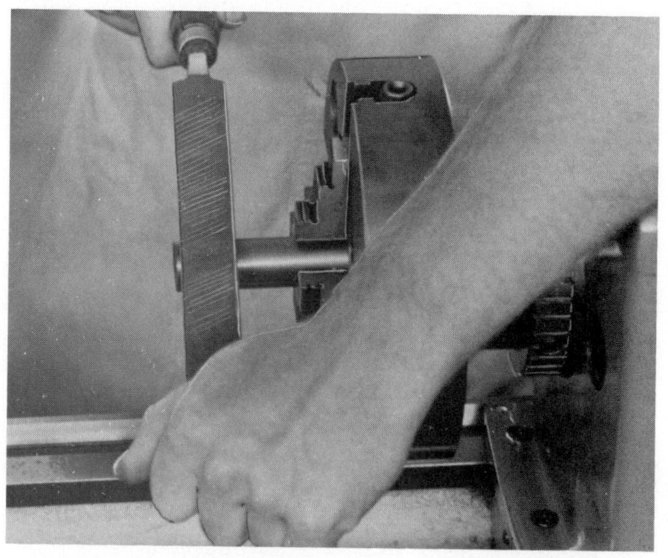

Fig. 10-215. Note how the left hand and arm must be over revolving chuck with right-hand method of filing on the lathe.

polished, is cut to length. It is then grasped between your fingers and held across the work, Fig. 10-216. If more pressure is required, mount the abrasive cloth on a strip of wood or on a file, Fig. 10-217. Use a high spindle speed for polishing.

The finer the abrasive used, the finer the resulting finish. A few drops of machine oil on the abrasive will improve the finish. For the final polish, remove the abrasive cloth from the job and reverse it so the cloth backing is in contact with the work surface.

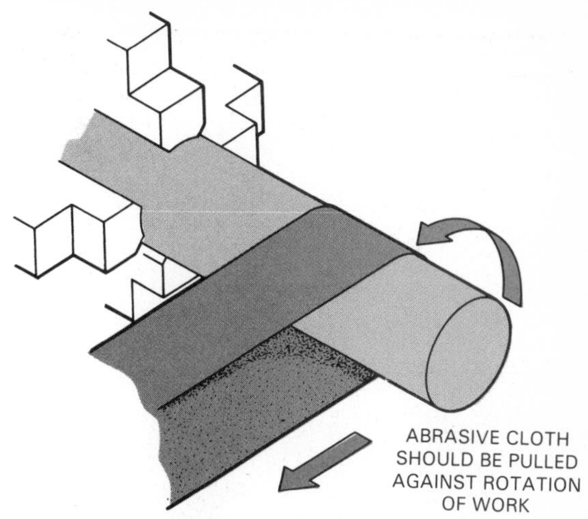

Fig. 10-216. Polishing with abrasive cloth held in hands. Keep hands away from revolving chuck or dog.

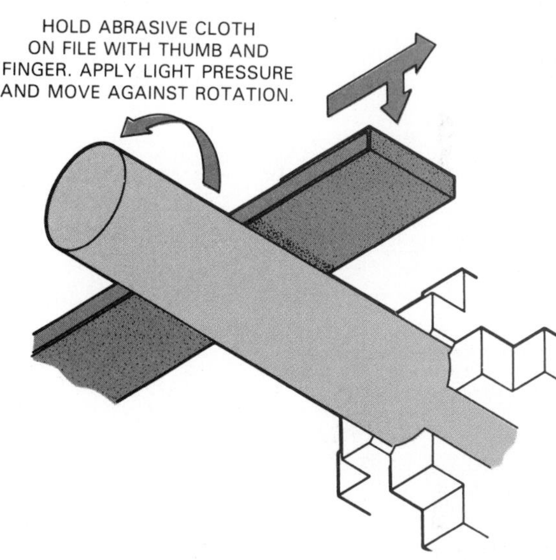

HOLD ABRASIVE CLOTH
ON FILE WITH THUMB AND
FINGER. APPLY LIGHT PRESSURE
AND MOVE AGAINST ROTATION.

Fig. 10-217. More pressure can be applied if abrasive cloth is supported by a file or block of wood.

Like filing, polishing is NOT a substitute for a properly sharpened tool bit.

Note! Clean the lathe thoroughly after polishing. The abrasive particles from the cloth will cause rapid wear of the machine's moving parts.

STEADY AND FOLLOWER RESTS

The ***steady rest*** and ***follower rest*** are needed for additional support when the work is long and slender to prevent it from springing or bending away from the cutting tool. The support is also needed to reduce "chattering" when long shafts are machined.

The steady rest, sometimes called a ***center rest,*** is bolted directly to the ways, Fig. 10-218. It is pro-

Fig. 10-218. The steady or center rest is needed when nature of work prevents it from being mounted between centers.

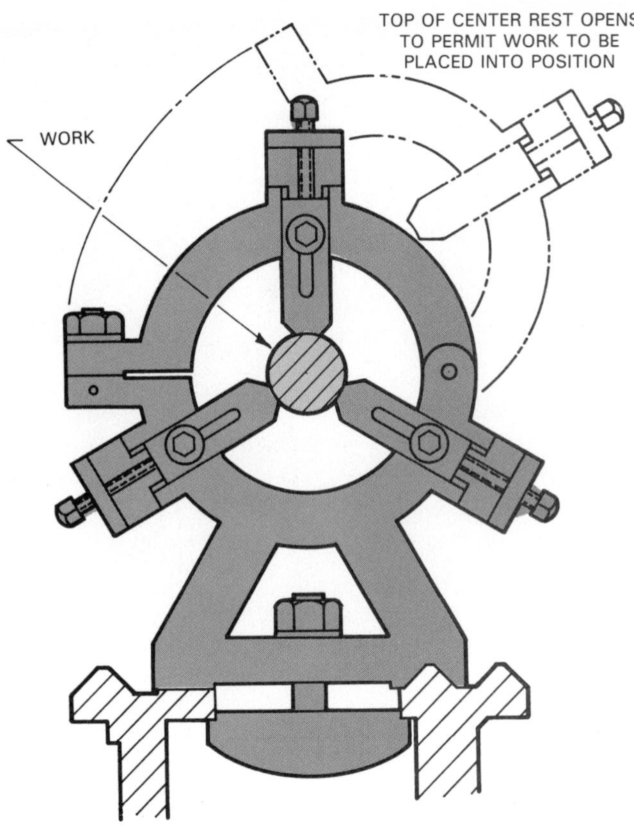

Fig. 10-219. Top of center rest swings open to permit easy installation of work. Care must be taken to have work accurately centered.

vided with three adjustable jaws, each with individual locking screws. The upper portion of the attachment is fitted with a single jaw. It can be opened to permit the work to be placed in position, Fig. 10-219.

Steady rest setup

To set up a steady rest:

1. Bolt the attachment to the ways at the desired position.
2. Back off all jaws and open the upper section.
3. Mount the work between centers or in a chuck and support the free end with the tailstock center.
4. Lower and lock the upper segment in place.
5. Adjust the jaws up to the work and lock them into position. The jaws act as bearing surfaces where they contact the work. They must be well lubricated.
6. A ***cat head*** is employed if the shaft being machined is unsuitable as a bearing surface (rough surface, out-of-round, square, etc.), Fig. 10-220. Care must be taken to center the shaft within the cat head.
7. When machining at the end of a long shaft and it is NOT possible to support it with the tailstock, center the work in the chuck. Adjust the center rest to the work as close to the chuck as possible before moving it to a point where the support is needed.
8. The same technique is employed when drilling, reaming, tapping, etc., on the end of a long shaft.

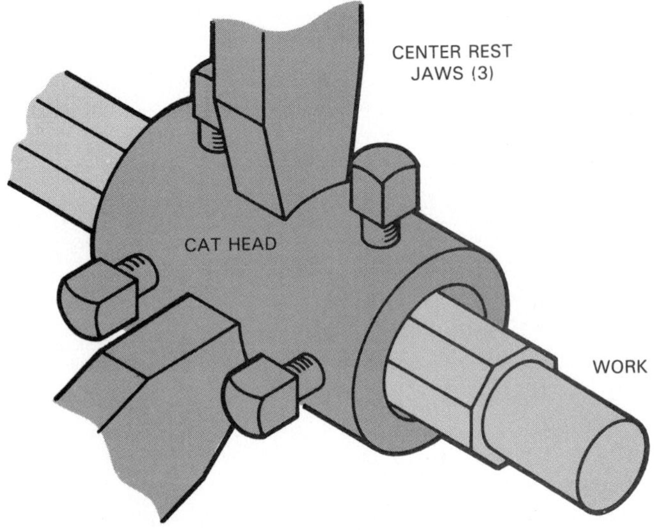

Fig. 10-220. Cat head will provide bearing surface when needed.

Follower rest setup

The follower rest, Fig. 10-221, operates on the same principle as and is used similarly to the steady rest, Fig. 10-221. The follower rest differs slightly

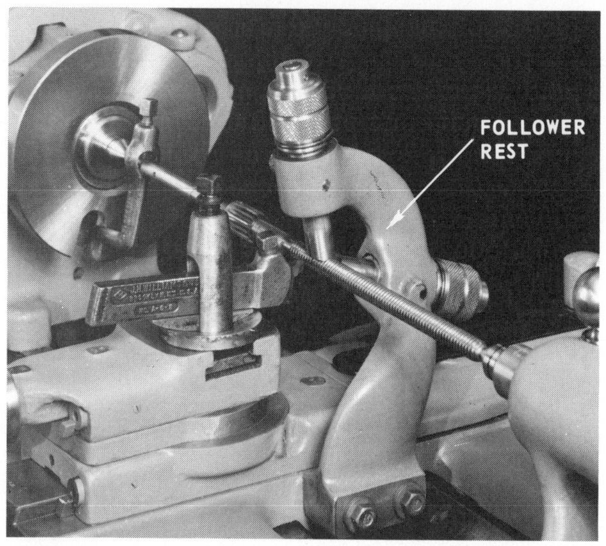

Fig. 10-221. Follower rest is being used to support a long slender shaft while threads are being machined on it. (South Bend Lathe, Inc.)

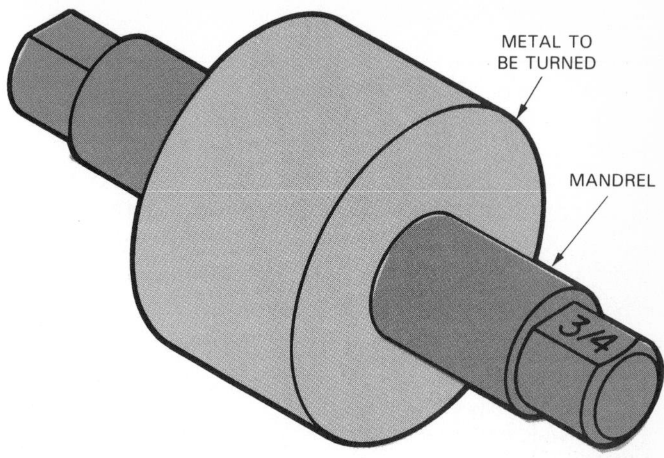

Fig. 10-222. This lathe mandrel has work mounted on it.

in that it supports directly in back of the cutting tool and follows along during the cut.

The follower rest bolts directly to the carriage and the jaws adjust in the same way as on the steady rest. Note that the jaws must be readjusted after each cut.

MANDRELS

At times, it is necessary to machine the outside diameter of a piece concentric with a hole that has been previously bored or reamed. This can be a simple operation if the material can be held in the lathe by conventional means. There are, however, times when the material cannot be gripped solidly to permit accurate machining. In such cases, the work is mounted on a ***mandrel*** and turned between centers, Fig. 10-222.

A ***solid mandrel*** is made from a section of hardened steel that has been machined with a slight taper (0.0005 in. per inch), Fig. 10-223A. They are made in standard sizes starting at 1/8 in. in diameter. The size is stamped on the large end. The other end is slightly smaller than specified size to permit easy installation in the work.

An ***expansion mandrel*** permits work with openings that vary from standard sizes to be turned, Fig. 10-223B. The shaft and sleeve are machined with corresponding tapers from hardened steel. The sleeve is slotted so it can expand when forced on the tapered shaft.

A ***gang mandrel*** is helpful when many pieces of the same configuration must be turned, Fig. 10-223C. Several pieces are mounted on the mandrel and locked in place by tightening the nut.

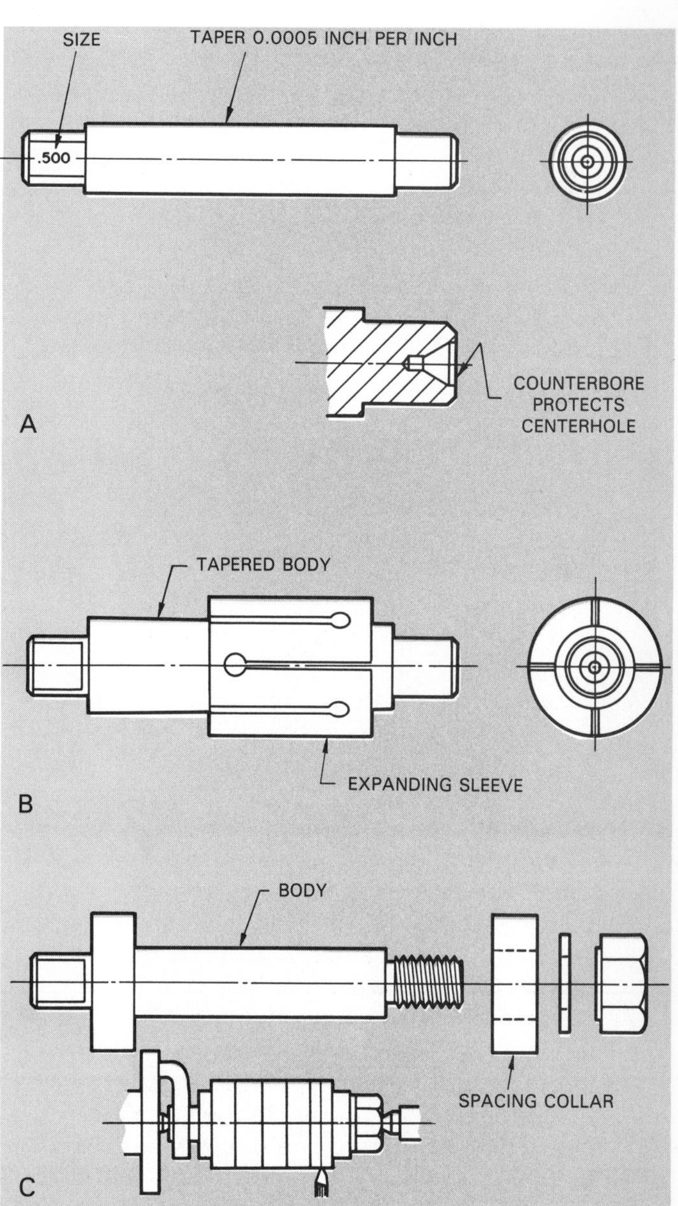

Fig. 10-223. Note different kinds of mandrels. A—Solid type. B—Expansion type. C—Gang mandrel.

Installing a mandrel

Work is pressed on a mandrel with an ***arbor press,*** Fig. 10-224. The work must first be checked for burrs and cleaned. Lubricate the work with a light oil to prevent it from "freezing" on the mandrel.

The mandrel is mounted between centers and driven by a lathe dog. Use care so the tool does NOT come into contact with the mandrel during the machining operation. In an emergency, a mandrel can be machined from a section of mild steel.

GRINDING ON THE LATHE

The ***tool post grinder*** permits the lathe to be used for internal and external grinding, Fig. 10-225. With a few simple attachments, it is possible to sharpen reamers and milling cutters on the lathe. You can also grind shafts and true lathe centers.

As steel parts sometimes warp during heat treatment, it is common to machine the piece to within 0.010 to 0.015 in. (0.2 to 0.3 mm) of finished size. After heat treatment, the metal is mounted on the lathe for grinding to finished size. A light grinding cut is made on each pass. When done properly, a very smooth finish results.

Preparing a lathe for grinding

As particles of the grinding wheel wear away during the grinding operation, it is important that the lathe be protected from them. Cover the bed, cross slide, and other parts with canvas or heavy craft paper for protection against the resulting abrasive dust and grit, Fig. 10-226.

WARNING! When placing protective covering on the lathe, be sure covering material cannot become entangled in the lead screw or other moving parts.

It is also good practice to place a small tray of water or oil just below the grinding wheel to collect as much grit and dust as possible. The abrasive particles can cause excessive wear should they get into moving parts.

Fig. 10-225. This is a typical tool post grinder. (The Dumore Co.)

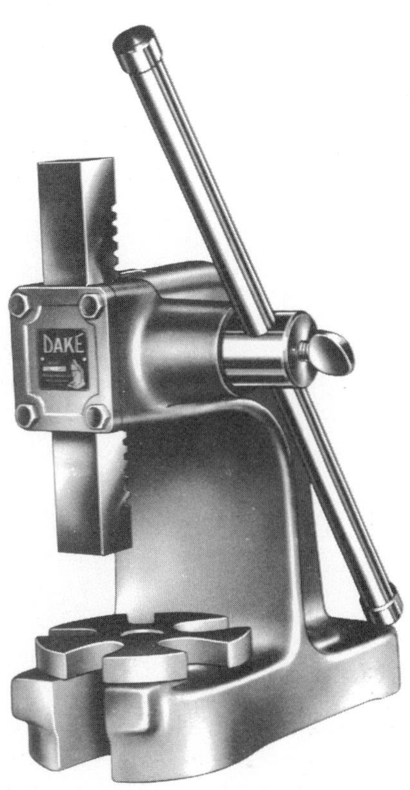

Fig. 10-224. Arbor press is used to press arbor into work.

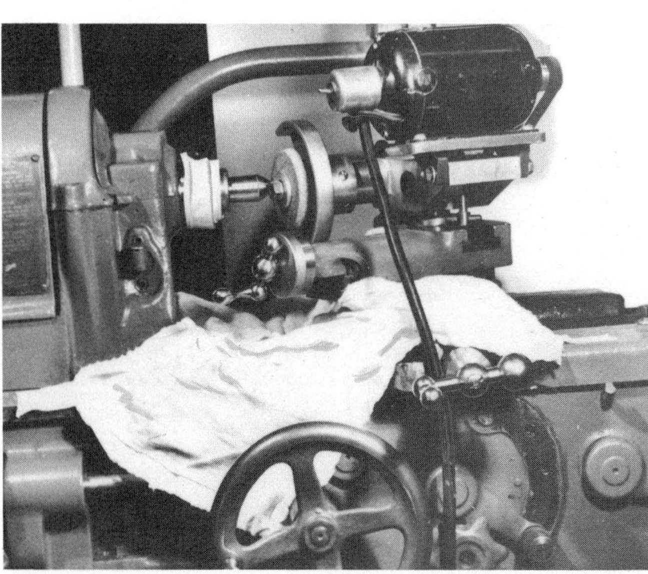

Fig. 10-226. Ways and other bearing surfaces of lathe should be carefully protected with canvas or heavy paper before performing grinding operations.

Preparing the grinder

Select the grinding wheel best suited for the job. It must be balanced and run true if a smooth, accurately sized job is to be obtained.

A ***diamond wheel dresser*** consisting of a steel shank, is used for the trueing operation, Fig. 10-227. It is mounted solidly to the lathe, on center or slightly below the center of the grinding wheel, Fig. 10-228. The rotating wheel is moved back and forth across the diamond, removing about 0.001 in. (0.02 mm) on each pass. Remove only enough to true the wheel.

External grinding

External grinding, Fig. 10-229, is done to finish the exterior surface of the piece, Fig. 10-229. The following steps are recommended to complete the job with the least amount of difficulty:

1. Mount the work solidly in the lathe. Provide adequate clearance.
2. Adjust lathe spindle speed for 80-100 rpm, and a feed of 0.005 - 0.007 in. (0.12 - 0.17 mm).
3. Turn on power for lathe and grinder. The work turns into the grinding wheel, Fig. 10-230.

Fig. 10-229. Top—Tool post grinder is finishing a shaft mounted between centers. Note cloth over lathe to catch grit. Bottom—During face grinding on this lathe, protective cloth has been removed to take photo.

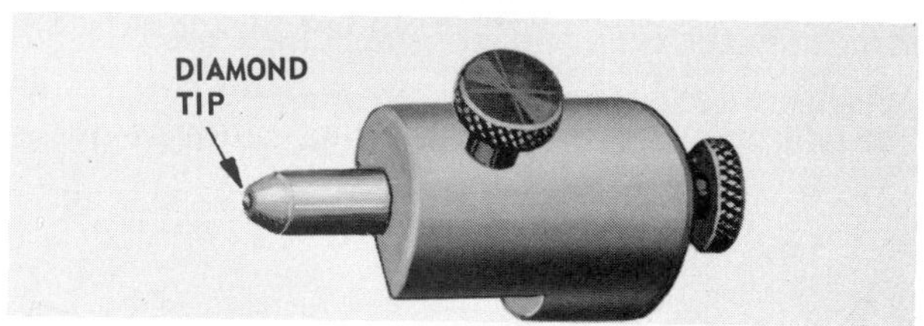

Fig. 10-227. Diamond wheel dresser is used to true tool post grinder wheels. (Black and Decker)

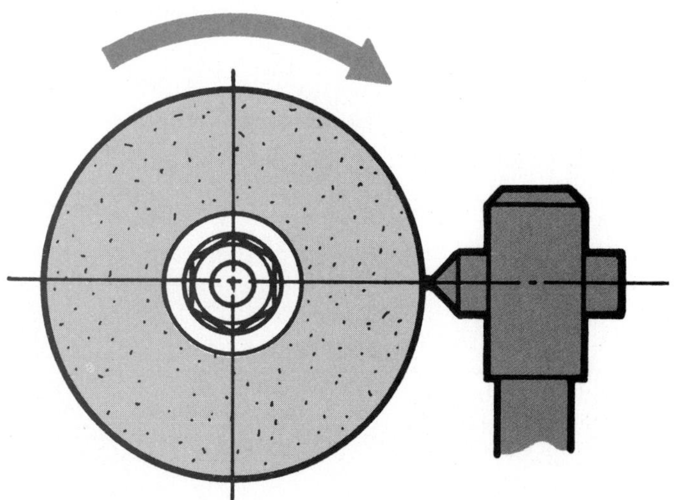

Fig. 10-228. If needed, true the wheel on a tool post grinder with a diamond dressing tool before grinding on lathe.

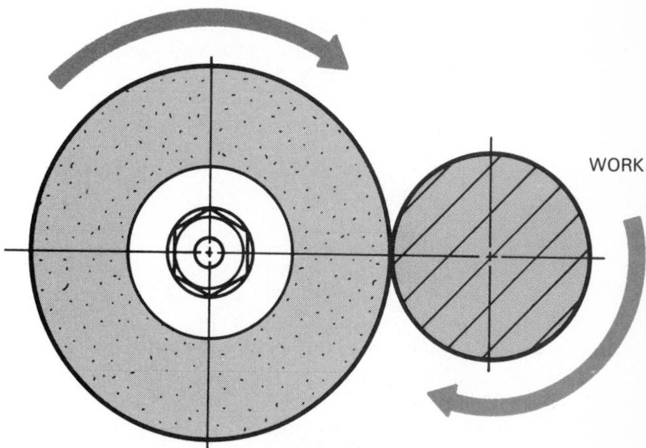

Fig. 10-230. With external grinding, the work turns INTO grinding wheel.

4. Feed grinding wheel into the work until it just begins to ''spark.''
5. Engage automatic longitudinal feed.
6. Check work diameter frequently with a micrometer. Use light cuts; otherwise, the piece might overheat and warp.

7. Redress the grinding wheel before making the final pass over the work. Allow the work to "spark out" (grinding wheel no longer cuts).

Internal grinding

Internal grinding is done in much the same manner as external grinding but on the inside of the work, Fig. 10-231. The work and grinding wheel must rotate in opposite directions, Fig. 10-232. Because the ***quill*** (shaft for mounting the grinding wheel for internal work) is quite slender, use very light cuts and slow feeds to prevent the hole from "bell mouthing," Fig. 10-233. For the same reason, it is suggested that the grinding wheel be allowed to "spark out" on the last cut.

Fig. 10-231. Note internal grinding operation. (The Dumore Co.)

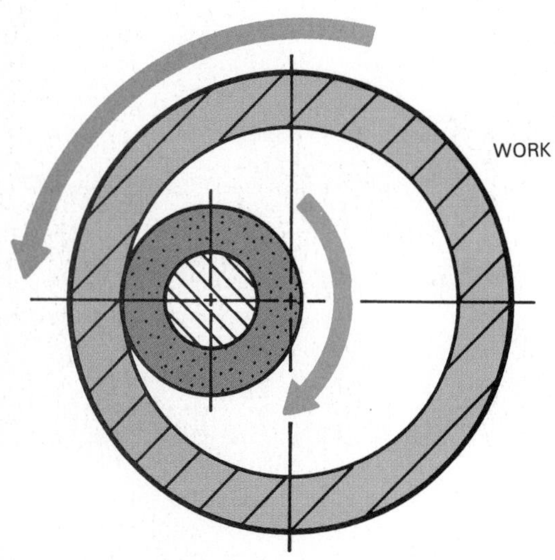

Fig. 10-232. With internal grinding, work and grinding wheel turn in opposite directions.

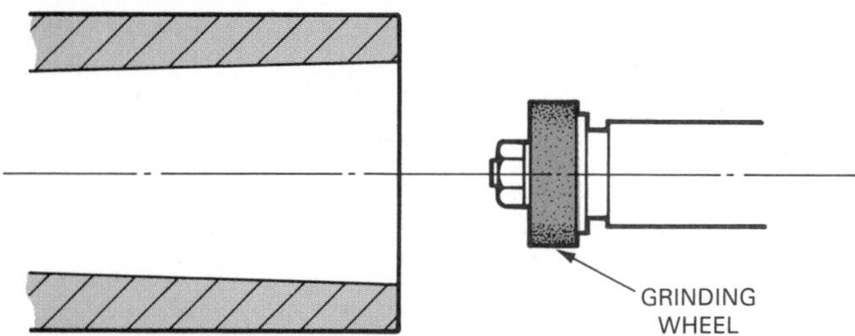

Fig. 10-233. Bell mouthing (hole larger at its mouth) is caused by taking too deep of a cut with grinder, or grinding with a feed that is too rapid.

MILLING ON A LATHE

Some lathes can be fitted with a vertical milling attachment, Fig. 10-234. Such machines are primarily designed for home work shops but are often used in model and experimental shops. A vise is mounted to the cross slide which also provides traverse (in-out) movement while longitudinal (back and forth) feed is furnished by the carriage.

A special milling attachment is available for some lathes. It permits limited milling operations to be performed, Fig. 10-235.

The cutter is mounted on an arbor or fitted into the headstock. Cutter depth is controlled by the adjusting screw on the device. Cutter movement is controlled by carriage and cross slide movements.

SPECIAL LATHE ATTACHMENTS

A ***tracing*** or ***duplicating unit*** should be utilized when several identical pieces must be produced. The duplicating unit improves the quality of the part because each is an exact duplicate of the master template. However, COMPUTER CONTROLLED

Fig. 10-234. This lathe is fitted with milling/drilling head. Work holding devices are mounted on cross slide.

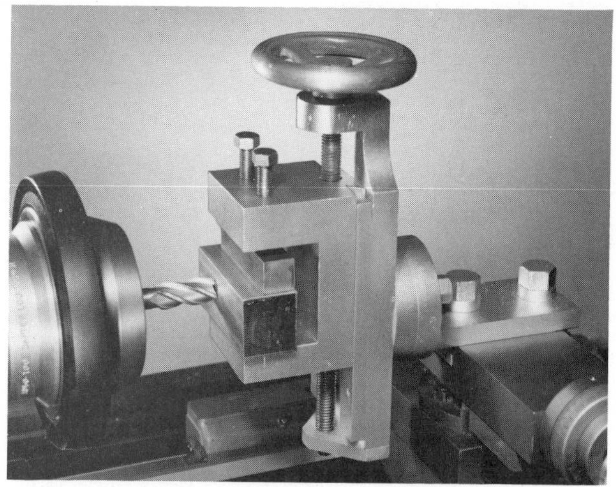

Fig. 10-235. Milling can be done on a lathe.

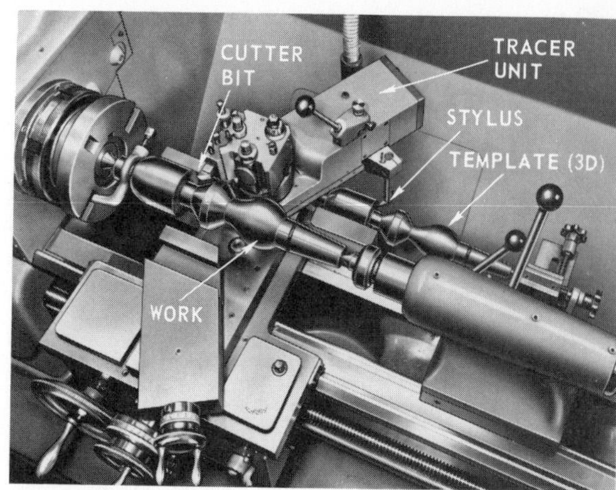

Fig. 10-237. This duplicating unit makes use of three-dimensional templates. (Clausing Machine Tools)

LATHES are now doing the work formerly done by these units when CNC lathes are available.

Several types of duplicating or tracing units are available. One type makes use of flat templates, Fig. 10-236. Another type employs a three-dimensional template or pattern, Fig. 10-237. Most units are hydraulically operated, Fig. 10-238.

INDUSTRIAL APPLICATIONS OF THE LATHE

Industry makes wide use of variations of the basic lathe, Fig. 10-239. The super-precision tool room

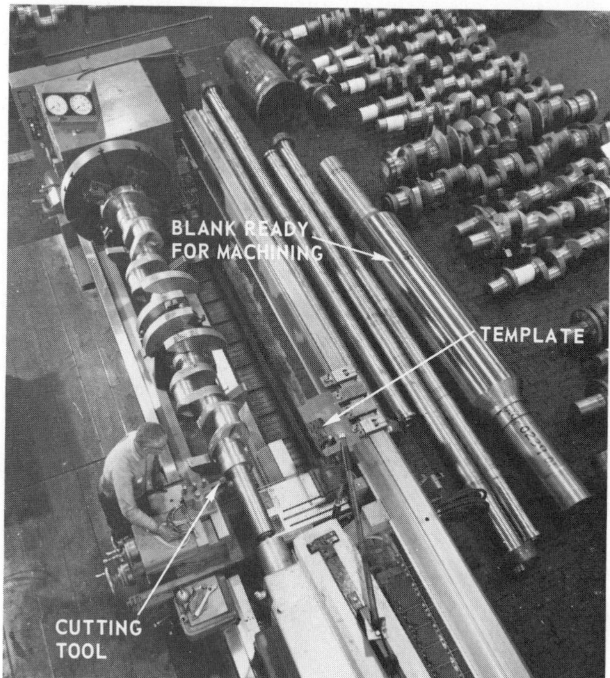

Fig. 10-236. Flat template is used to guide cutting tool as it machines bearing area of a crankshaft for a large diesel engine.

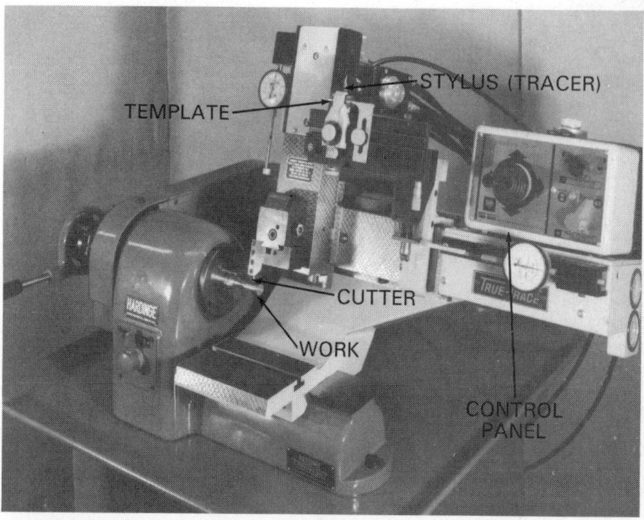

A

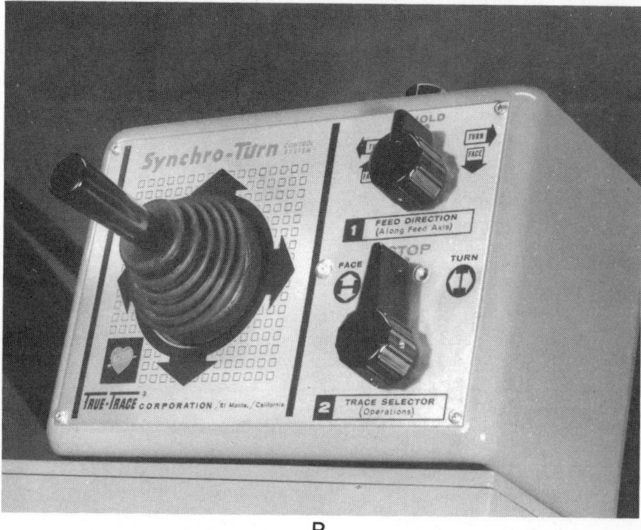

B

Fig. 10-238. A—A tracing unit is mounted on lathe. B—Schematic illustration shows basic relationship of lathe spindle with tracing attachment. Standard carriage is locked in a fixed position when unit is in operation. (True-Trace Corp.)

Fig. 10-239. Note typical engine lathe. (Harrison-Rem Sales, Inc.)

Fig. 10-240. This is a super precision toolroom lathe. (Hardinge Bros., Inc.)

Fig. 10-241. This is a manually operated turret lathe. (Clausing Machine Tools)

lathe is required to meet the close tolerances and fine surface finish specifications of tool rooms, model shops, research and development laboratories and departments. See Fig. 10-240.

For limited production runs, 100 to 5000 pieces, the manually operated ***turret lathe*** is used, Fig. 10-241. This is a conventional lathe equipped with a six-sided tool holder called a ***turret,*** to which a number of different cutting tools are fitted, Fig. 10-242. Stops control the length of tool travel and rotate the turret to bring the next cutting tool into position automatically.

A ***cross-slide*** unit is used for turning, facing, forming, and cutting off operations, Fig. 10-243. Turret lathes range in size from the small ***precision instrument turret lathe*** to the versatile ***automatic turret lathe.*** Refer to Figs. 10-244 to 10-246.

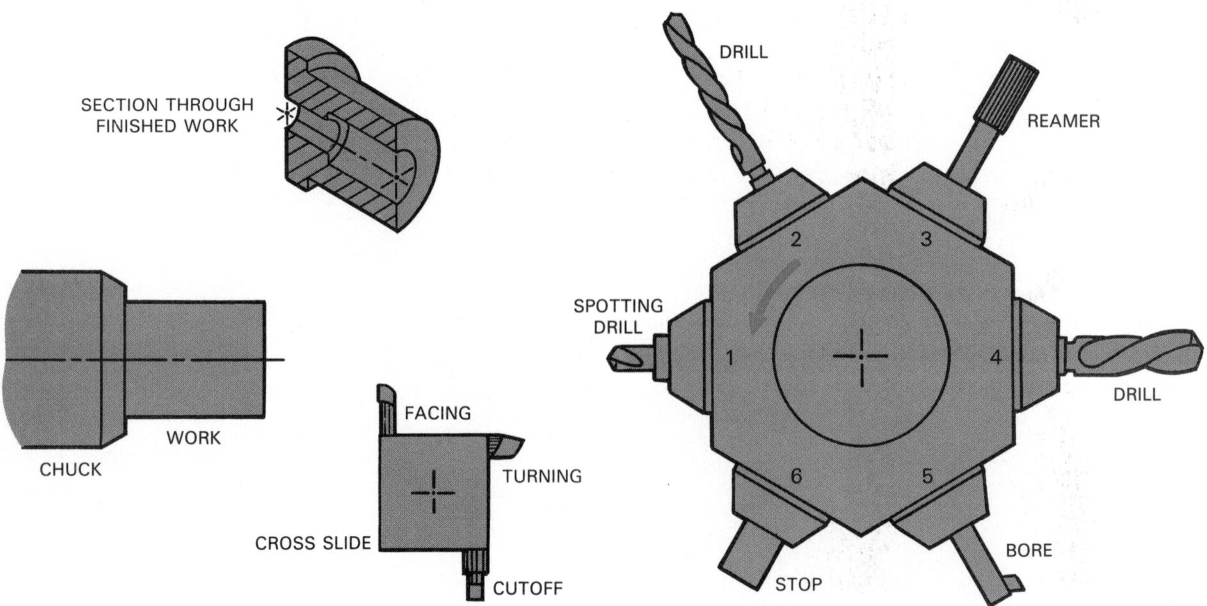

Fig. 10-242. Study turret in relation to other parts of a lathe. Turret rotates to bring tool (drill, reamer, etc.) into position. Depth tool cuts is controlled by stops.

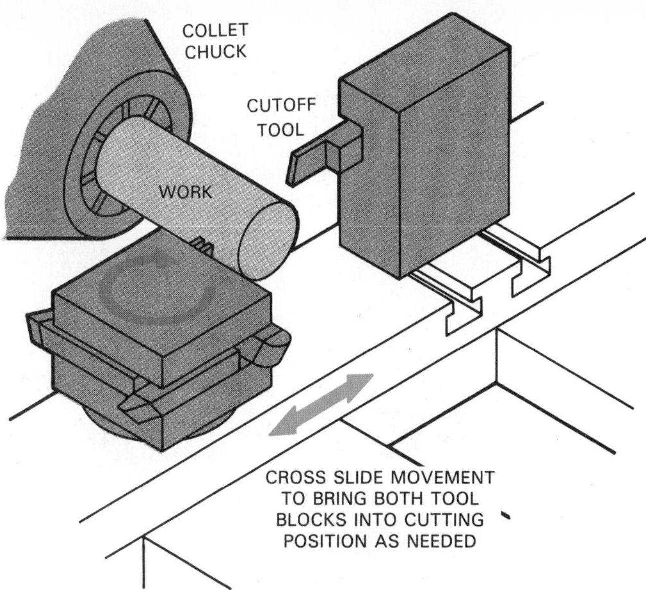

Fig. 10-243. Cross slide on turret lathe is similar to cross slide on a conventional lathe. However, it is fitted with several cutting tools that can be brought into position as needed.

Fig. 10-244. Machinist is using precision instrument turret lathe with magnifying lens. (Louis Levin and Son, Inc.)

Fig. 10-246. This is a larger, more versatile turret lathe. (Warner & Swasey Co.)

The ***screw machine,*** another variation of the lathe, was developed for the high-speed production of large numbers of precision parts. The machine is designed to perform a maximum number of operations, either simultaneously, or in a very rapid sequence. See Figs. 10-247 and 10-248.

Work too large or too heavy to be machined in a horizontal position, can be turned on a ***vertical boring machine,*** Fig. 10-249. These huge machines, known as ***boring mills,*** are capable of turning and boring work with diameters up to 40 ft. (12 m).

Conventional metalworking lathes are manufactured in a large range of sizes from the tiny jeweler's lathe to large machines that turn forming rolls for the steel industry, Fig. 10-250.

Numerically controlled (N/C) and ***computer numerically controlled*** (CNC) lathes and turning machines are finding increased usage by industry. When properly programmed, these machine tools are capable of producing complex work with a high degree of accuracy and repeatability.

Note! A detailed description of N/C and CNC can be found in later chapters.

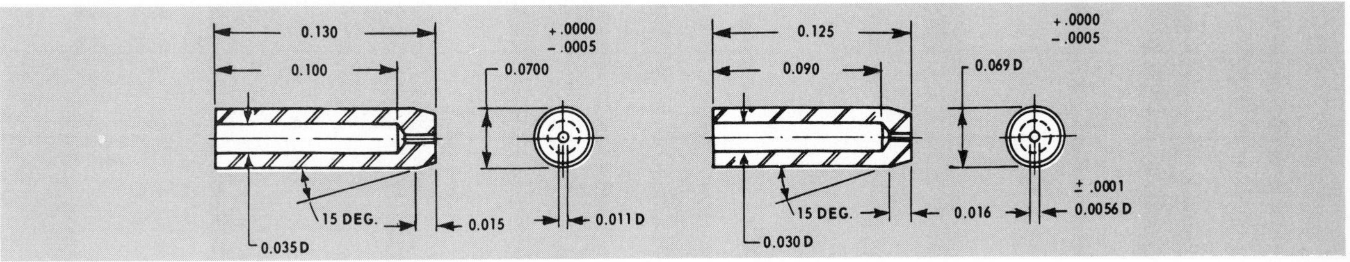

Fig. 10-245. Parts shown here were machined from stainless steel and produced in quantity on turret lathe illustrated in Fig. 10-244. Note small dimensions.

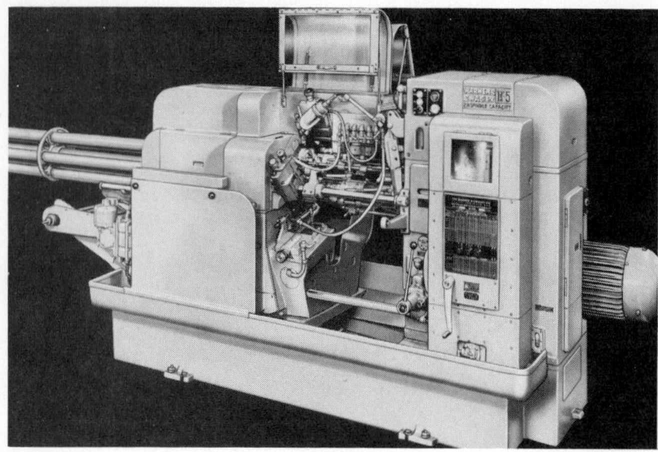

Fig. 10-247. Multiple spindle automatic screw machine is for precision high speed production. (Warner & Swasey Co.)

small

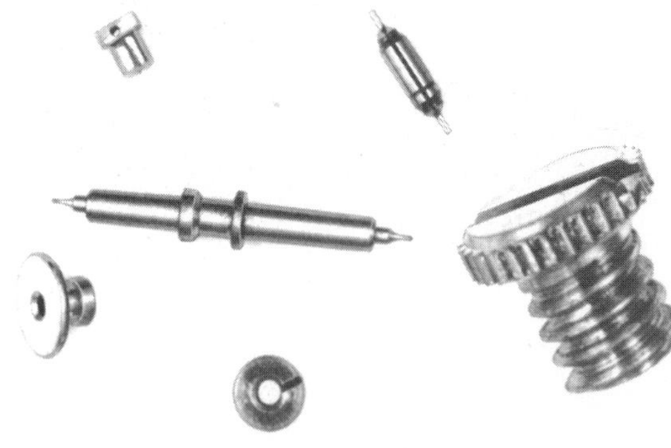

parts

Fig. 10-248. Work range of screw machine family is considerable. Watch parts illustrated above were made on one of the smaller machines. The lettering "small parts" was typed on a standard typewriter. The company advertises that they can make precision parts as small as 0.003 in. dia. x 0.010 in. long (0.08 mm dia. x 0.25 mm long), with a tolerance of ±0.0002 in. (±0.005 mm) from carbon steel, stainless steel, alloy steel, and most nonferrous metals. (Hamilton Watch Co.)

N/C and CNC lathes and turning machines include:

1. Small lathes used primarily by training purposes, Fig. 10-251.
2. Conventional lathes with two-axis CNC movement, Fig. 10-252.
3. Vertical spindle computer guided lathes, Fig. 10-253.
4. Turning centers with automatic tool changing, Fig. 10-254.
5. Completely automated turning centers with robot loading/unloading and automatic gaging and tool monitoring, Figs. 10-255 and 10-256.

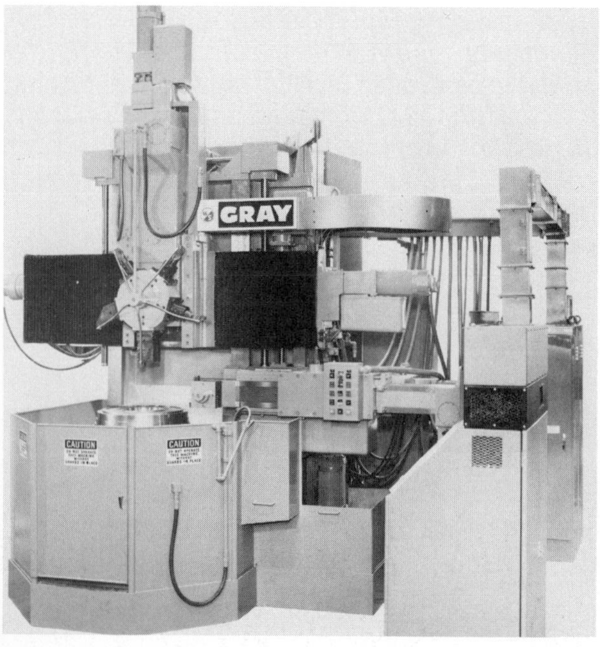

Fig. 10-249. Left. Note that the work is mounted horizontally on this typical boring mill. (The G.A. Gray Co.) Right. This 17 ft. (5.1 m) vertical boring mill features high cutting speeds, infinite speed variations while the machine operates, and CNC control. (Simmons Machine Tool Corp.)

Fig. 10-250. Some idea of the size of this large, modern lathe can be gained by comparing it with machinist in photo.

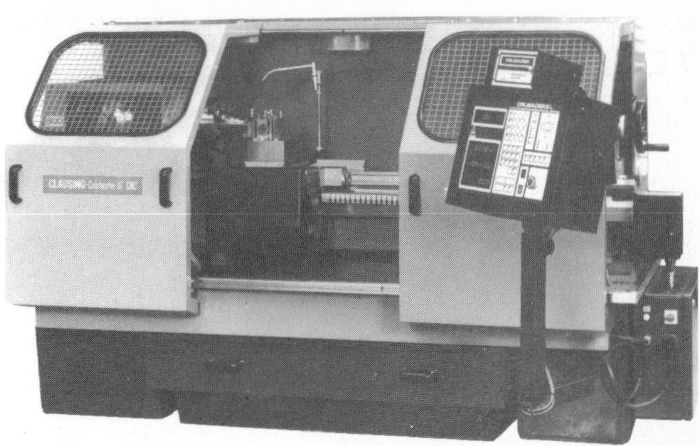

Fig. 10-252. This lathe has two-axis CNC controls.

Fig. 10-251. Instructor is programming CNC training lathe. Its computer activated controls are same as those on full size industrial machines. Machine tools such as this permit CNC training programs to be given without tying up expensive full size machines. (South Bend Lathe, Inc.)

Fig. 10-253. Vertical spindle CNC lathe can be programmed to do automatic machining. (Wickman Bennett Machine Tools Inc.)

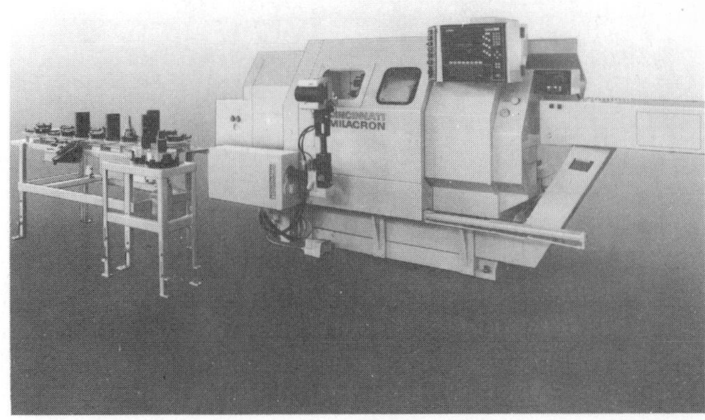

Fig. 10-254. This CNC turning center has automatic parts handling robot. (Cincinnati Milacron)

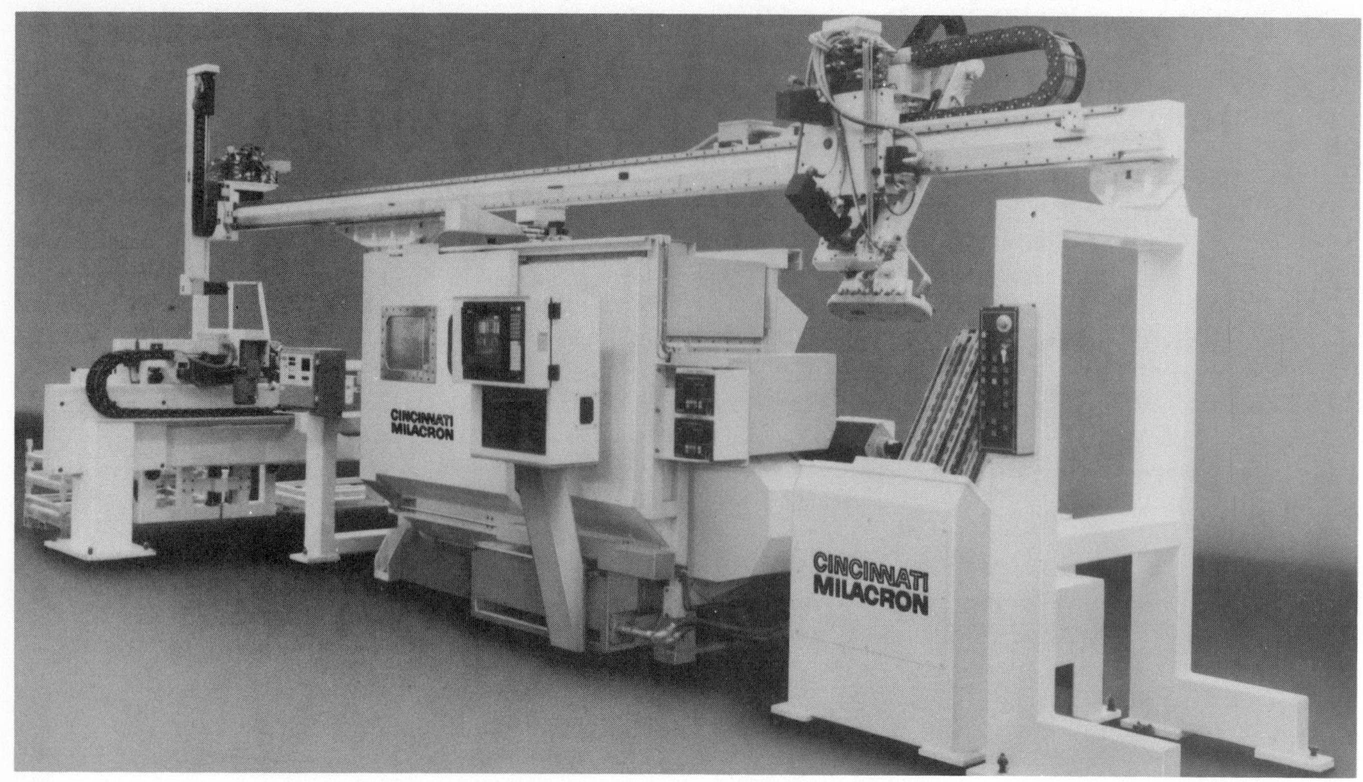

Fig. 10-255. Flexible CNC turning cell changes tools and parts automatically. It also maintains part size by tool-tip probing and laser gaging, and monitors its own performance automatically under computer control. Cell is served by an overhead gantry loader for changing parts and tools. (Cincinnati Milacron)

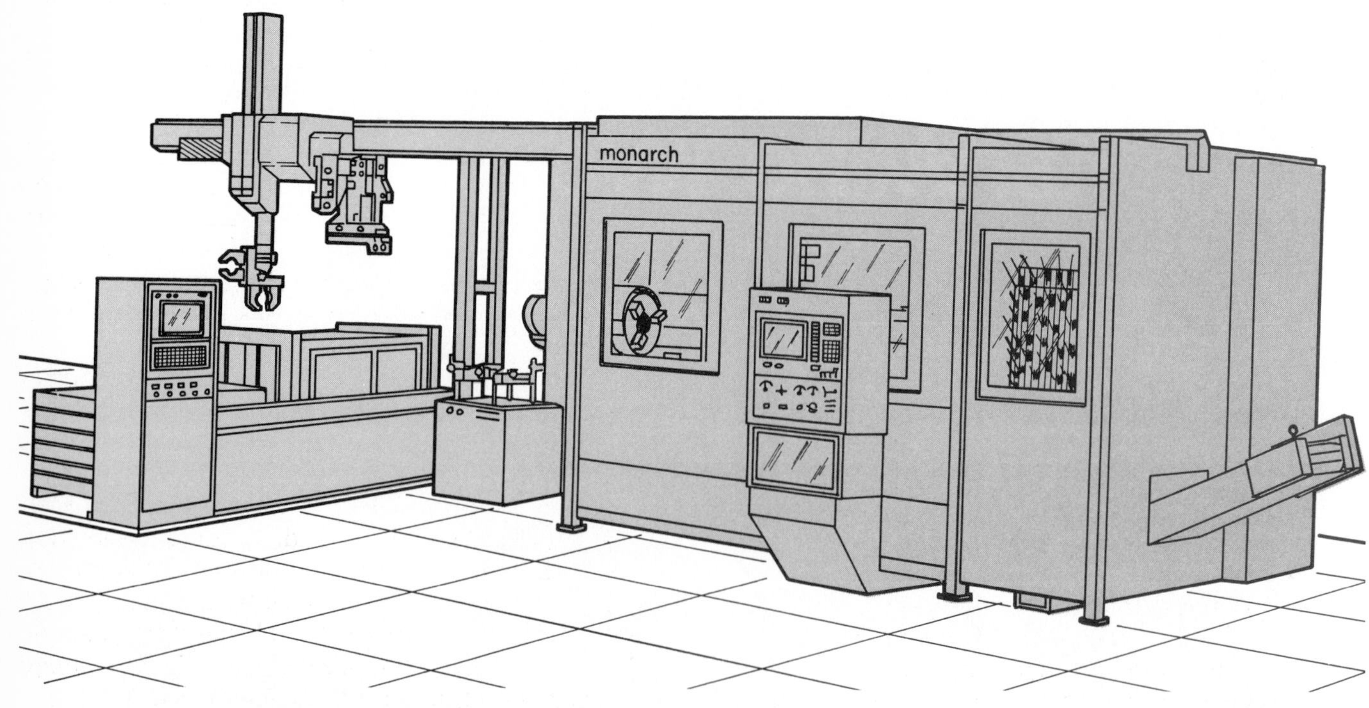

Fig. 10-256. This stand-alone turning cell is capable of automatic machining, part and tool changing, and palletization (stacking finished part) without operator intervention. Part and tool probes automatically monitor machining and determine when tools need to be changed. Accuracy and repeatability is enhanced with a post machining gaging system. Finished parts are unloaded by robot, placed in gage and then returned to pallet. Measured part data is fed back into the machine's CNC control where offsets (tool movement) are automatically updated to maintain consistent part accuracy. (Monarch Machine Tool Co.)

TEST YOUR KNOWLEDGE—Chapter 10

Please do not write in the text. Place your answers on a separate sheet of paper.

1. The lathe operates on the principle of:
   a. The cutter revolving against the work.
   b. The cutting tool, being controllable, can be moved vertically across the work.
   c. The work rotating against the cutting tool which is controllable.
   d. All of the above.
   e. None of the above.
2. The size of a lathe is determined by the ________ and the ________ of the ________.
3. The largest piece that can be turned between centers is equal to:
   a. The length of the bed minus the space taken up by the headstock.
   b. The length of the bed minus the space taken up by the tailstock.
   c. The length of the bed minus the space taken up by the headstock and the tailstock.
   d. All of the above.
   e. None of the above.
4. Into which of the following categories do the various parts of the lathe fall:
   a. Driving the lathe.
   b. Holding and rotating the work.
   c. Holding, moving, and guiding the cutting tool.
   d. All of the above.
   e. None of the above.
5. Explain the purpose of ways on the lathe bed.
6. Power is transmitted to the carriage through the feed mechanism to the quick change gear box which regulates the amount of ________ ________ per ________ ________.
7. The carriage supports and controls the cutting tool. Describe each of the following parts:
   a. Saddle.
   b. Cross slide.
   c. Compound rest.
   d. Tool rest.
8. Accumulated metal chips and dirt are cleaned from the lathe with a ________; NEVER with ________ ________.
9. Which of the following actions are considered dangerous when operating a lathe?
   a. Wearing eye protection.
   b. Wearing necktie and jewelry.
   c. Measure with work rotating.
   d. Operating lathe with most guards in place.
   e. Using compressed air to clean machine.
10. Most cutter bits used in the school shop are made of ________ ________ ________.
11. Cutting speeds can be increased 300 to 400 percent by using ________ ________ ________ tools.
12. What does cutting speed indicate?
13. ________ is used to indicate the distance the cutter moves in one revolution of the work.
14. Calculate the cutting speeds for the following metals. The information furnished is sufficient to do so.
    a. Formula: $RPM = \frac{CS \times 4}{D}$
    b. CS = Cutting speed recommended for material being machined.
    c. D = Diameter of work in inches.
    PROBLEM A: What is the spindle speed (rpm) required to finish turn 2 1/2 in. diameter aluminum alloy? A rate of 1000 fpm is the recommended speed for finish turning the material.
    PROBLEM B: What is the spindle speed (rpm) required to rough turn 1 in. diameter tool steel? 50 fpm is the recommended speed for rough turning the material.
15. Calculating the cutting for metric size material requires a slightly different formula.
    a. Formula: $RPM = \frac{CS \times 1000}{D \times 3}$
    b. CS = Cutting speed recommended for particular material bieng machined (steel, aluminum, etc.) in meters (m) per minute.
    c. D = Diameter of work in millimeters (mm).
    PROBLEM: What spindle speed is required to finish turn 200 mm diameter aluminum alloy? Recommended cutting speed for the material is 300 m/pm.
16. Most work is machined while supported by one of four methods. List them.
17. Sketch a correctly drilled center hole.
18. A tapered piece will result if centers are NOT aligned and the work is turned between centers. Approximate alignment can be determined by two methods. What are they?
19. Describe one method for checking center alignment if close tolerance work is to be done between centers.
20. It is often necessary to turn to a shoulder or to a point where the diameters of the work change. One of four types of shoulders will be specified. Make a sketch of each.
    a. Square shoulder. b. Angular shoulder.
    c. Filleted shoulder. d. Undercut shoulder.
21. What are the two lathe chucks most commonly used? Describe the characteristics of each.
22. When using the cutoff tool, the spindle speed of the machine is about ________ the speed used for conventional turning.
23. List six simple rules for parting (cutoff) operations.
24. There are four ways of machining tapers on a lathe. List them, with their advantages and disadvantages.

25. When is a section of material considered tapered?

26. Machine adjustments must be calculated for each tapering job. The information given below will enable you to calculate the necessary tailstock setover for the problems given.

FORMULA: When taper per inch is known,

$$\text{Offset} = \frac{L \times TPI}{2}$$

When taper per foot is known,

$$\text{Offset} = \frac{L \times TPF}{24}$$

When dimensions of tapered section are known but TPI or TPF is not given,

$$\text{Offset} = \frac{L \times (D - d)}{2 \times \ell}$$

WHERE: TPI = Taper Per Inch

TPF = Taper Per Foot

D = Diameter at large end of taper

d = Diameter at small end of taper

ℓ = Length of taper

L = Total length of piece

NOTE: The formulas, except for the TPF formula, can be used when dimensions are in mm.

PROBLEM A: What will be the tailstock setover for the following job?
Taper Per Inch = 0.125 in.
Total length of piece = 4.000 in.

PROBLEM B: What will be the tailstock setover for the following job?
D = 2.50 in.
d = 1.75 in.
ℓ = 6.00 in.
L = 9.00 in.

PROBLEM C: What will be the tailstock setover for the following job?
D = 45.0 mm
d = 25.0 mm
ℓ = 175.0 mm
L = 275.0 mm

27. Screw threads are used for many reasons. List five or more important uses.

The following questions are of the matching type. Place the letter of the correct explanation on your work sheet.

28. External thread.
29. Internal thread.
30. Major diameter.
31. Minor diameter.
32. Pitch diameter.
33. Pitch.
34. Lead.

a. Smallest diameter of thread.
b. Largest diameter of thread.
c. Distance from one point on a thread to a corresponding point on next thread.
d. Cut on outside surface of piece.
e. Diameter of imaginary cylinder that would pass through threads at such points as to make width of thread and width of space at these points equal.
f. Cut on inside surface of piece.
g. Distance a nut will travel in one complete revolution of screw.

35. A groove is cut at the point where a thread is to terminate. It is cut to the depth of the thread and serves to:
a. Provide a place to stop the threading tool after it makes a cut.
b. Permits a nut to be run up to the end of the thread.
c. Terminate the thread.
d. All of the above.
e. None of the above.

36. The tip of a cutting tool to cut a sharp V thread is sharpened using a _______ _______ to check that it is the correct shape. This tool is frequently called a _______ _______.

37. The _______ _______ is fitted to many lathe carriages. It meshes with the lead screw and is used to indicate when to engage the half-nuts to permit the thread cutting tool to follow exactly in the original cut.

38. The compound rest is set at _______ when cutting threads to permit the cutting tool to shear the material better than if it were fed straight into the work.

39. The three-wire thread measuring formula for inch based threads is:

$$M = D + 3G - \frac{1.5155}{N}$$

WHERE: G = Wire diameter.
D = Major diameter of thread (Convert to decimal size).
M = Measurement over the wires.
N = Number of threads per inch.

PROBLEM: Calculate the correct measurement over the wires for the following threads. Use the wire size given in the problem.

a. 1/2-20UNF
(wire size 0.032 in.)
b. 1/4-20UNC
(wire size 0.032 in.)
c. 3/8-16 UNC
(wire size 0.045 in.)
d. 7/16-14UNC
(wire size 0.060 in.)

40. Drills, used on the lathe, are fitted with ________ shanks or ________ shanks.
41. Boring is:
    a. A drilling operation.
    b. An internal machining operation whereby a single point cutting tool is employed to enlarge a hole.
    c. An external machining operation whereby a single point cutting tool is employed to reduce the diameter of a hole.
    d. All of the above.
    e. None of the above.
42. When is reaming done?
43. The process of forming horizontal or diamond shaped serrations on the circumference of the work is called ________. It is commonly done to provide a ________ ________.
44. When is filing on the lathe usually done?
45. Polishing is an operation used to produce a ________ ________ on the work.
46. When is a steady rest used?
47. What is the difference between a steady rest and a follower rest?
48. There are times when a shaft is unsuitable as a bearing surface and cannot be used with a steady rest. When this occurs, a ________ can be employed so the shaft can be supported with the steady rest.
49. What is a mandrel? When is it used?
50. A mandrel is usually pressed into the work with a ________ ________.
51. Internal and external grinding can be done on a lathe with a ________ ________ grinder.
52. What should be done to protect the lathe from the abrasive particles that wear away from the grinding wheel?
    a. Use a nonabrasive grinding wheel.
    b. Cover the bed and moving parts with a heavy cloth.
    c. Use a soft abrasive grinding wheel.
    d. All of the above.
    e. None of the above.
53. There are many safety precautions that must be observed when operating a lathe. List what YOU consider five (5) of the most important.

RESEARCH AND DEVELOPMENT

1. Make large scale wooden models of the basic cutting tool shapes. They should be cutaway models to permit the various clearance angles to be easily observed.
2. The invention of the lathe is lost in history. The first lathe was believed to have been a "tree" lathe. Work was mounted between two trees and a limber branch fitted with a section of cord provided the power. Prepare a paper on the known history of the lathe. Illustrate it with transparencies designed for the overhead projector.
3. Prepare a comparison test using carbon steel, high speed steel, and cemented carbide cutting tools. Make the tests on mild steel (annealed), tool steel (heat treated), and aluminum alloy. Employ the recommended cutting speeds and feeds. Make a graph that will show the times needed by the various cutting tools to perform an identical machining operation. Also indicate surface finish quality.
4. Develop and produce a series of posters on lathe safety.
5. Make a display board showing large scale models of sharp V, square, and Acme screw threads.
6. Visit a local industry that uses automatic screw machines, turret lathes, and/or CNC lathes and turning centers in manufacturing their products. Prepare a short paper describing your impressions of these machines in action.
7. Borrow samples of products made on automatic screw machines and turret lathes. What characteristics, if any, do they have in common?
8. Invite the Director of the local State or Federal Employment Agency to give a class lecture on work opportunities for the machinist in your community.
9. Study the help wanted columns in your local papers. Collect the advertisements in the metalworking trades. What type jobs are available? How do the various employers advertise? Do they offer fringe benefits? Are they equal opportunity employers? Do they pay a premium for second and third shift workers? Make this study every three months and prepare a chart that will show whether the demand increases or decreases during the various seasons.

10. Secure a selection of job applications. Study them carefully and discuss with the class the information they request. When can be done to make applying for a job easier and simpler?
11. Develop a paper, with illustrations if possible, on how the first screw threads were made. What does term ''chasing'' mean in reference to first threads cut on a metal lathe?
12. Demonstrate to the class the proper technique of machining screw threads. Illustrate how the tool can be repositioned after being resharpened and how to use the 3-wire thread measuring method of measuring threads.
13. Prepare a lesson on milling on the lathe.
14. Develop a program for a CNC lathe.
15. Demonstrate the operation of a CNC lathe.
16. Develop a research project to investigate the effects of cutting fluids upon the quality of the surface finish of turned work. Prepare a paper on your findings.
17. Show a film or video tape on the operation of a CNC lathe or turning center.
18. Prepare a display case or dislay of the various products made in your machine shop. Show several of them in various stages of construction.

Chapter 11

PLANING MACHINES

After studying this chapter, you will be able to:

- Identify the various types of planing machines.
- Explain how planing machines operate.
- Describe the industrial applications of planing machines.

Planing machines are designed to machine horizontal, vertical, and/or angular planed flat surfaces. These machines are classified into several categories.

SHAPER

The ***shaper*** has a single point cutting tool that moves back and forth over the work. Being too slow for modern production techniques, most work formerly done on a shaper is now performed by other machine tools, Fig. 11-1. However, the shaper is still found in some jobbing and specialty machine shops because it is easily and inexpensively tooled for some one-of-a-kind jobs.

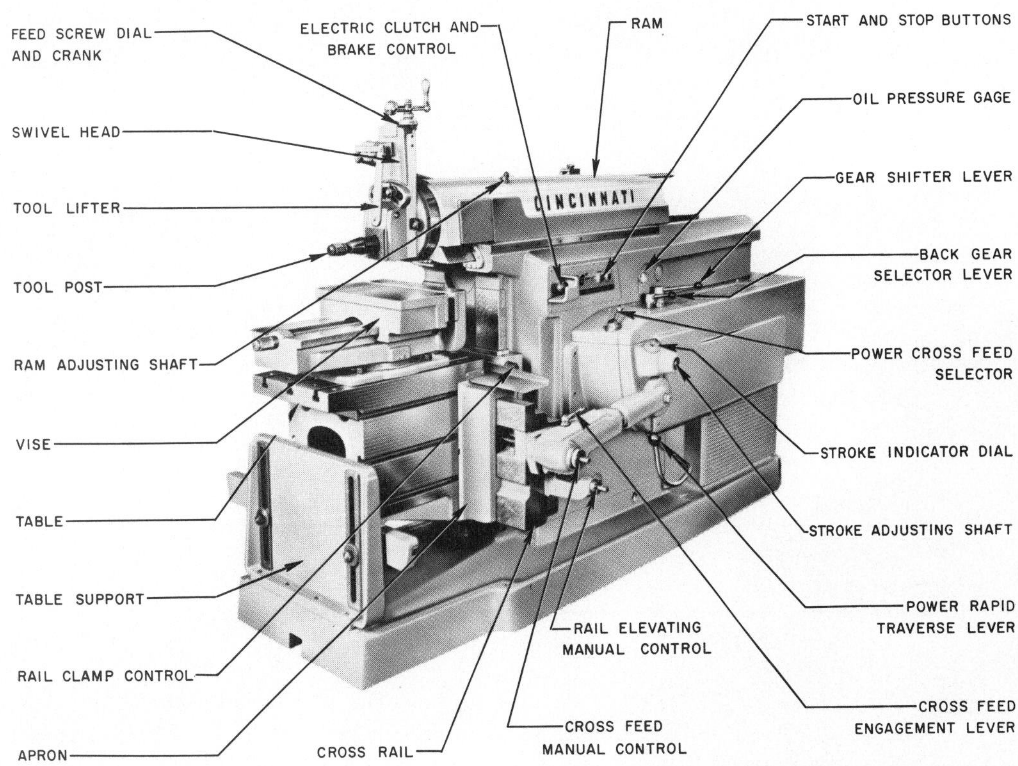

Fig. 11-1. Study various parts of shaper. (The Cincinnati Shaper Co.)

On a shaper, the work is stationary and the single point cutting tool moves against it, Fig. 11-2. While primarily used to machine flat surfaces, a skillful machinist can manipulate it to cut curved and irregular shapes, slots, grooves, and keyways. Work is usually mounted in a vise, Fig. 11-3.

As with any machine tool, carefully examine a shaper to be sure it is in safe operating condition. The machine should also be lubricated according to the manufacturer's specifications.

SAFETY NOTE! Never attempt to operate these machines while your senses are impaired by medication or other substances.

SHAPER SIZE

Shaper size is determined by the maximum length of material the tool can machine in one setup. See Fig. 11-4. For example, a 20 in. (500 mm) shaper has a ***stroke*** (distance cutting tool travels) that is sufficient to machine a surface 20 in. (500 mm) long.

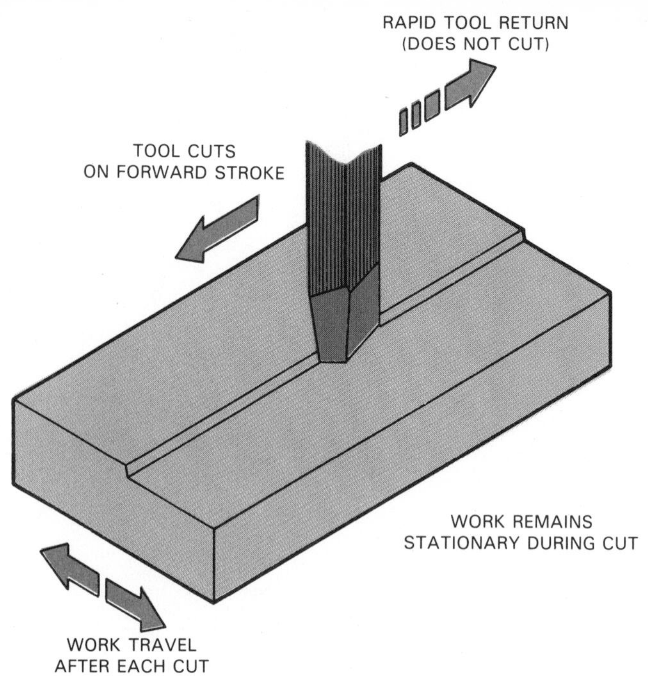

Fig. 11-2. Diagram shows how a shaper operates. Work is stationary and cutting tool moves against it.

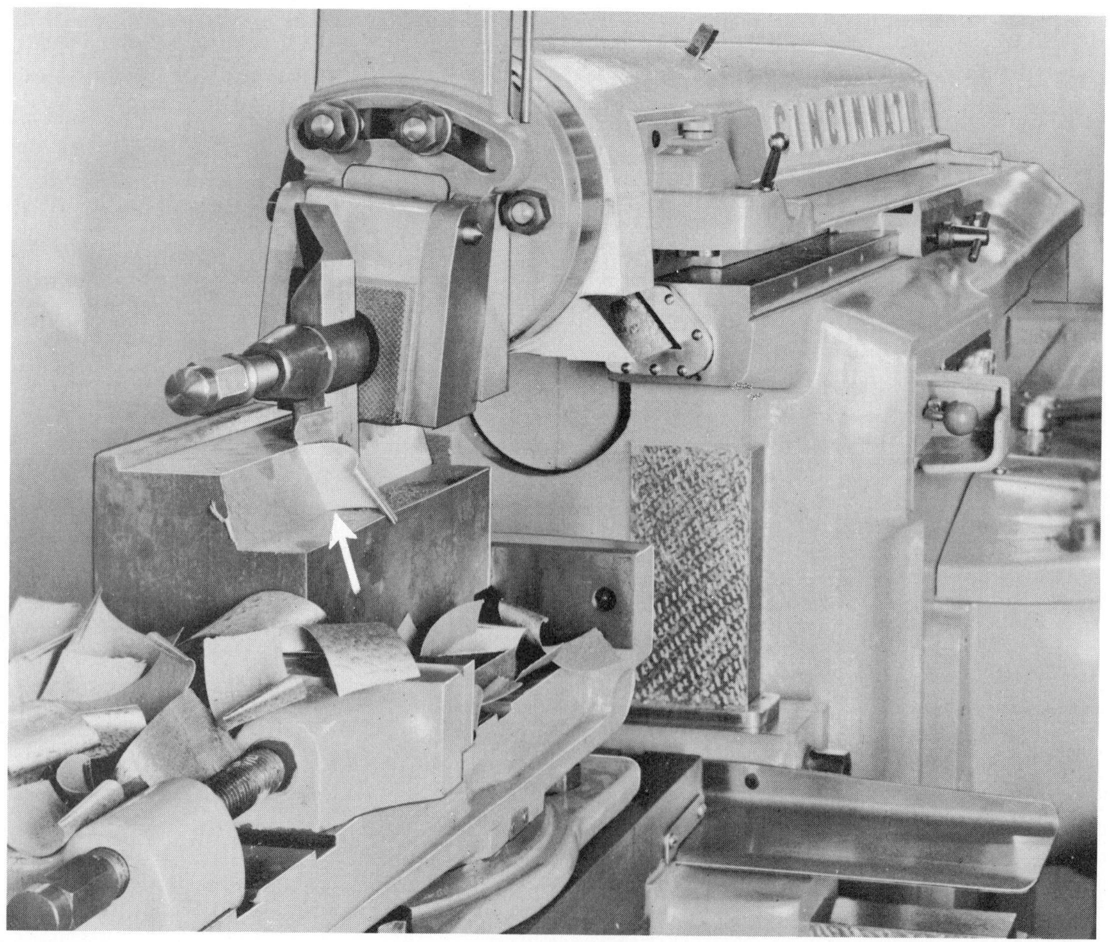

Fig. 11-3. With work held in a vise, shaper is making a cut that is 2 in. (50 mm) deep and 1/32 in. (0.8 mm) thick. (The Cincinnati Shaper Co.)

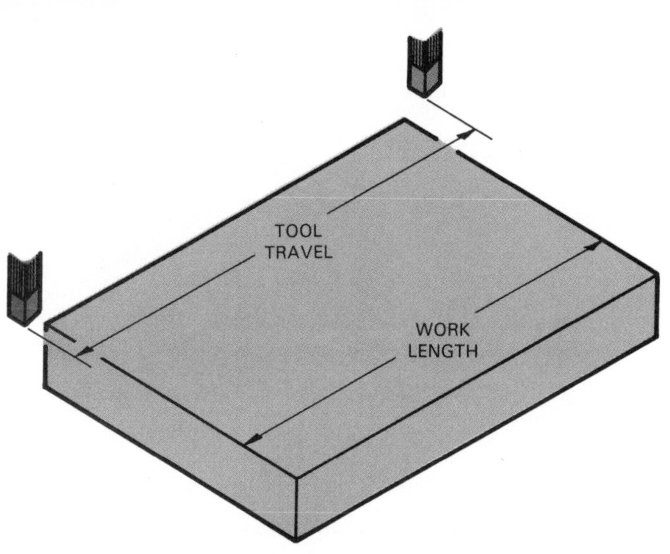

Fig. 11-4. Shaper size is determined by maximum work length that can be machined in one pass.

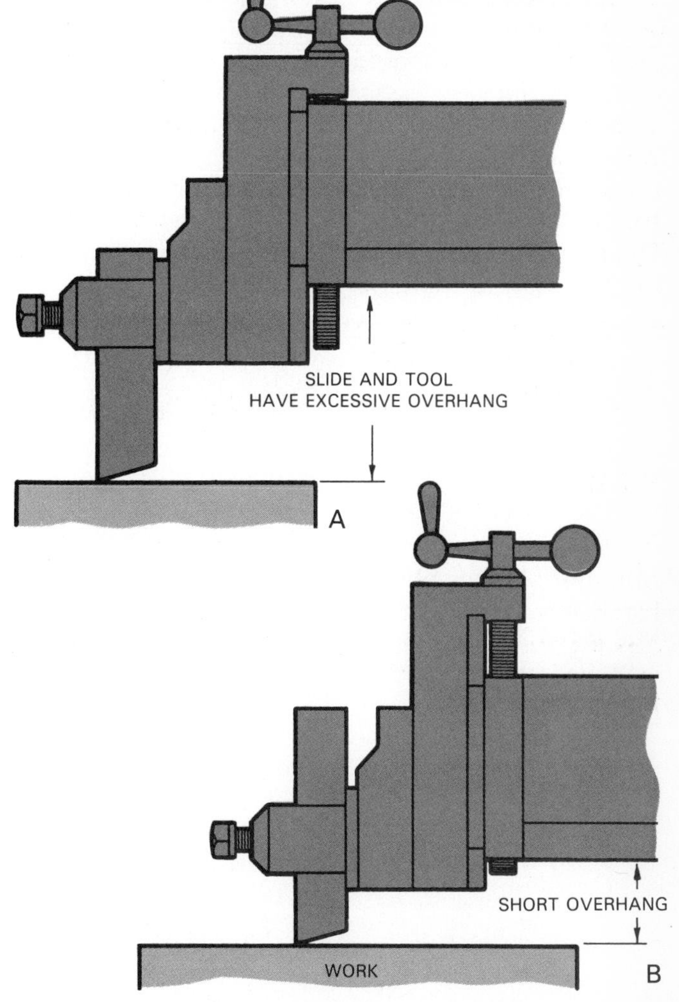

Fig. 11-6. Avoid excessive overhang of slide and/or cutting tool. A—Excessive overhang causes chatter, producing a rough finish. B—Keep slide up and a short grip on tool for increased rigidity.

Mounting work on shaper

The position of the vise is important when using a shaper. It should be positioned so the machining can be done in the shortest possible time. Assuming that the cutter is making the same number of strokes per minute, the setup shown in Fig. 11-5A will permit the work to be done in about a third of the time needed to machine the work in the setup in Fig. 11-5B.

Avoid excessive tool and slide overhang, Fig. 11-6. The resulting "chatter" will cause a very rough finish.

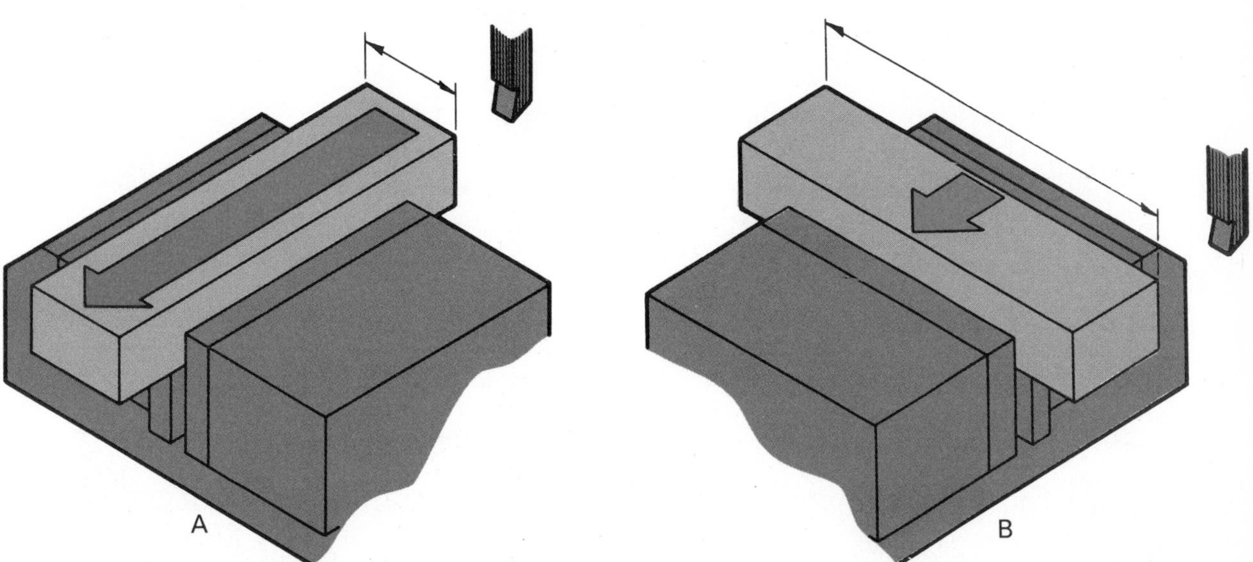

Fig. 11-5. Position work in vise so it can be machined in shortest amount of time. A—Proper setup. B—This setup will take approximately two times longer to machine than other setup.

The cutting stroke of the machine should be adjusted to work as shown in Fig. 11-7. Make the adjustments as recommended by the machine's manufacturer.

Shaper cutting speed and feed

The ***shaper speed*** is the number of cutting strokes the ram makes per minute. The ***shaper feed*** is the distance the work travels or moves after each cutting stroke. Generally, the following should be observed:

1. The harder the metal or the deeper the cut, the SLOWER the cutting speed.
2. The softer the metal or the lighter the cut, the FASTER the cutting speed.
3. Coarse feed, deep cut, and slow cutting speed for the roughing cut.
4. Fine feed, light cut, and fast cutting speed for the finishing cut.

Cutting tool shape is determined by the material being machined and the degree of finish desired. The tool shapes shown in Fig. 11-8 are recommended for mild steel.

SLOTTER

The chief difference between a shaper and a ***slotter*** is the direction of the cutting action, Fig. 11-9.

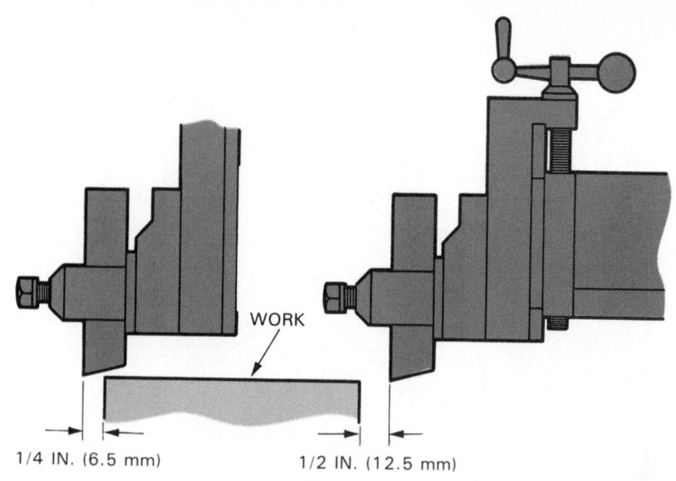

Fig. 11-7. The 1/4 in. (6.5 mm) allowance at end of stroke provides ample chip clearance, while 1/2 in. (12.5 mm) allowance permits cutter to drop back into cutting position for next stroke.

The slotting machine is classified as a VERTICAL SHAPER. It is used to cut slots, keyways (both internal and external), and to machine internal and external gears, Fig. 11-10.

The vertical shaper and the manner in which it cuts are shown in Figs. 11-11 through 11-13.

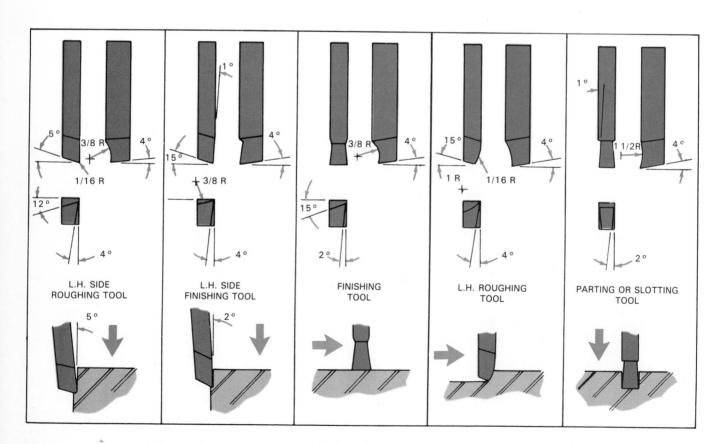

Fig. 11-8. Note cutting tool shapes recommended for machining mild steel.

Fig. 11-9. This is a modern slotter or vertical shaper. (Ex-Cell-O Manufacturing Systems Co.)

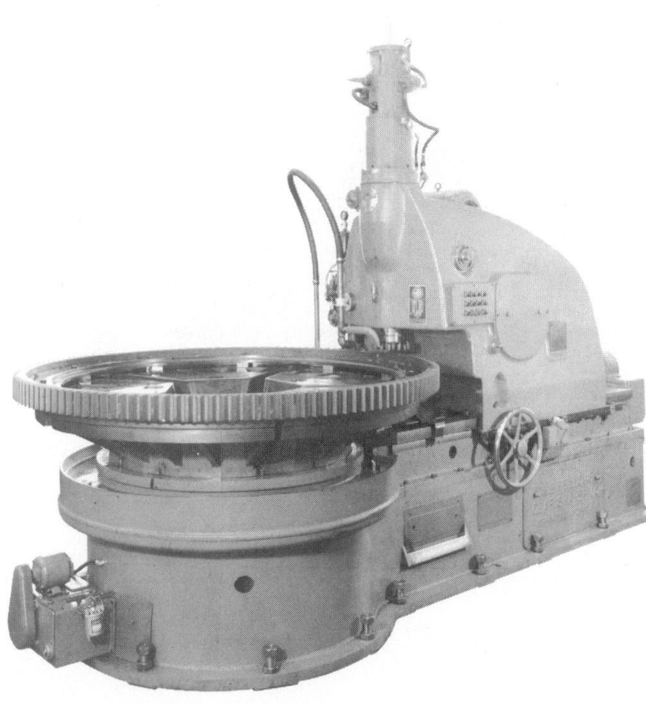

Fig. 11-10. Vertical gear shaper is cutting a 120 in. (3050 mm) external gear. (Fellows Gear Shaper Co.)

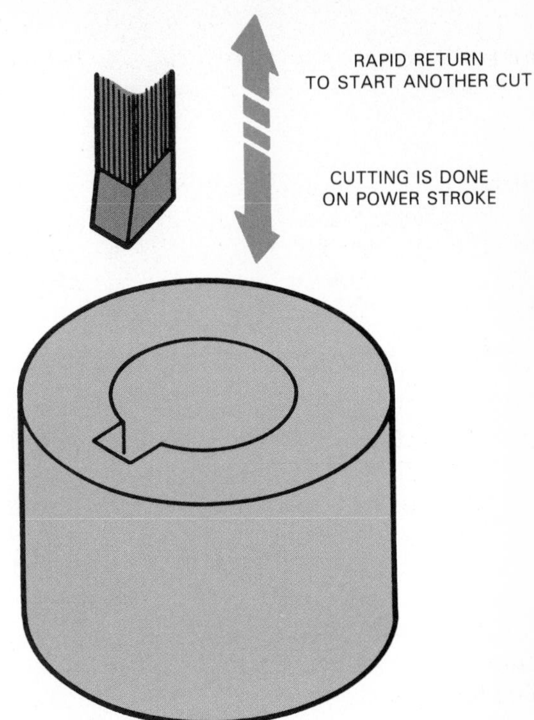

Fig. 11-11. Diagram shows slotter or vertical shaper operation. Slotter, unlike shaper, moves vertically rather than horizontally and work is held stationary.

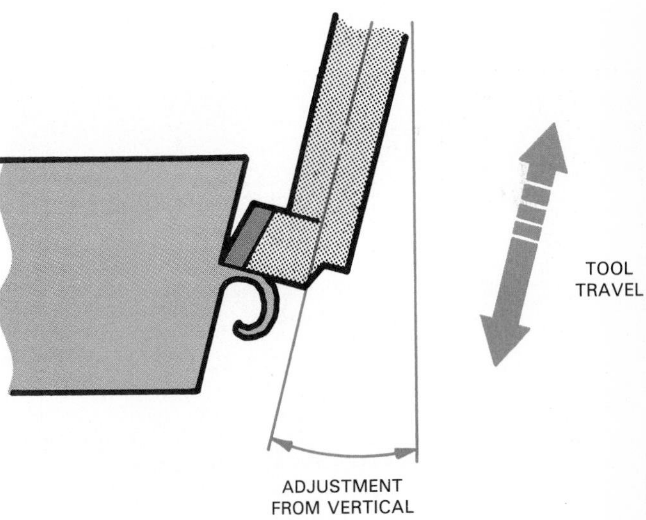

Fig. 11-12. Cutting head of a slotter can also be adjusted to make angular cuts.

PLANER

The ***planer*** differs from the shaper in that the WORK travels back and forth while the cutter remains stationary, Figs. 11-14 and 11-15. A planer can handle work that is too large to be machined on most milling machines. Planers are large pieces

Fig. 11-13. Job is being machined on a slotter.

Fig. 11-14. This 144 in. by 126 in. by 40 ft. (3.6 m x 3.2 m x 12.2 m) double housing planer has two cutting heads. People add perspective to machine's size. (G.A. Gray Co.)

Fig. 11-15. This is a 42 in. by 42 in. by 10 ft. (1.05 m by 1.05 m by 3.0 m) open side hydraulic planer. (Ex-Cell-O Manufacturing Systems Co.)

of equipment. Some are large enough to handle and machine work up to 20 ft. (6.1 m) wide and twice as long, Fig. 11-16.

BROACHING

Broaching is similar to shaping, but instead of a single cutting tool advancing slightly after each stroke, the broach is a long tool with MANY teeth. See Fig. 11-17.

Broaching machines are designed to push or pull this multitooth cutting tool across the work, Fig.

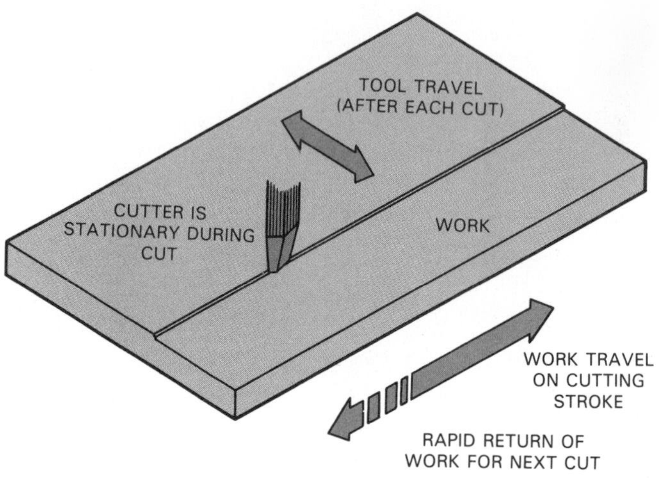

Fig. 11-16. Diagram shows how a planer works. The tool(s) remain stationary while work moves against it.

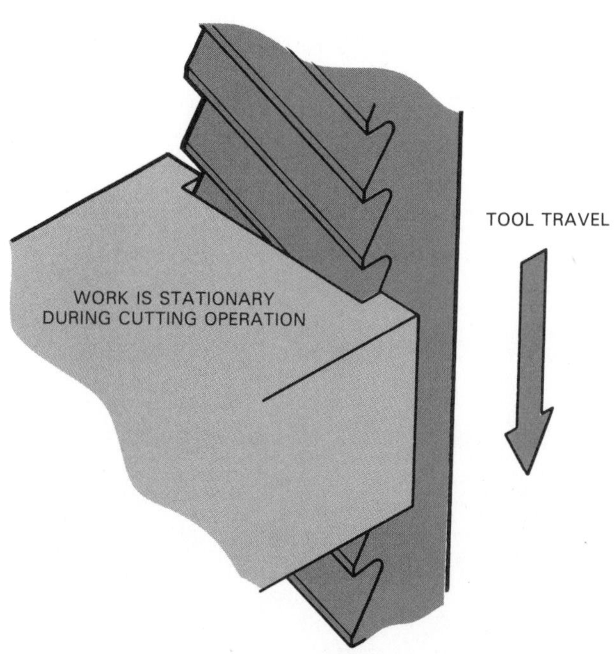

Fig. 11-17. Diagram shows how a broach operates. A multitooth cutting tool moves against work. Operation may be on a vertical or horizontal plane.

11-18. Each tooth on the ***broach*** (cutting tool) removes only a small portion of the material being machined, Fig. 11-19.

Roughing, semifinishing, and finishing teeth are usually on the same broach, Fig. 11-20. The machining operation can be completed in a single pass.

When properly employed, broaching can remove material faster than any other machining technique. Small parts can often be stacked and shaped in a single pass of the tool, Fig. 10-21. Larger units, such as auto engine blocks, may require several passes to machine all surfaces of the part.

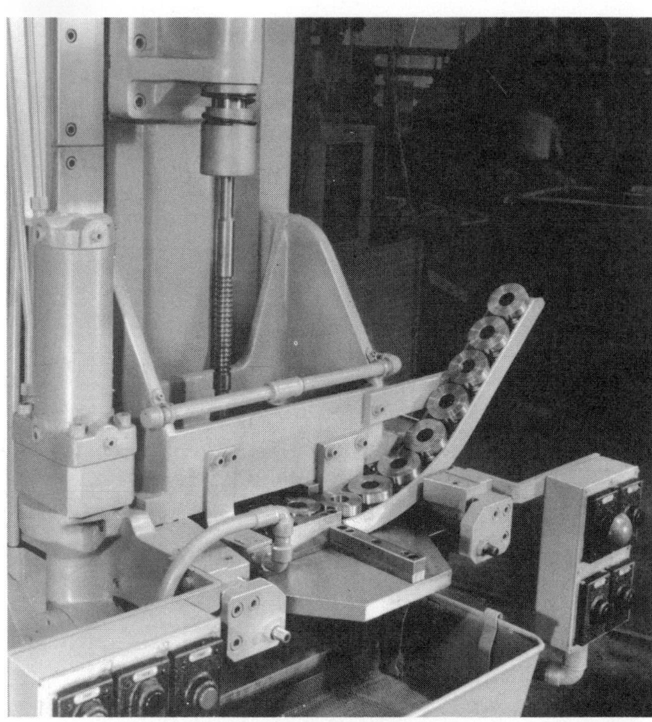

Fig. 11-18. Work is moving into position on modern broaching machine. (Sunstrand Corp.)

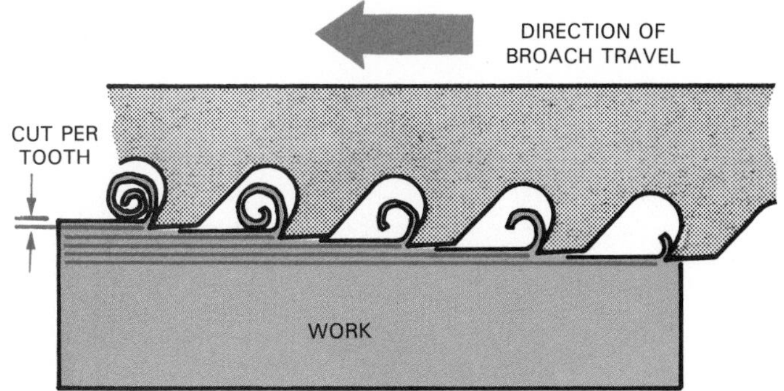

Fig. 11-19. Each tooth, on a broaching tool, removes only a small portion of material being machined.

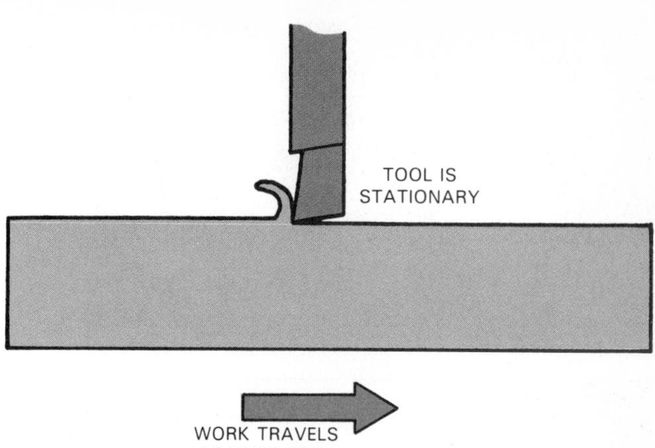

Fig. 11-20. Drawing shows a greatly shortened section of internal broaching tool and a cross section of splines it cuts. Pilot guides cutter in work. Each cutting tooth increases slightly in size until specified size is attained.

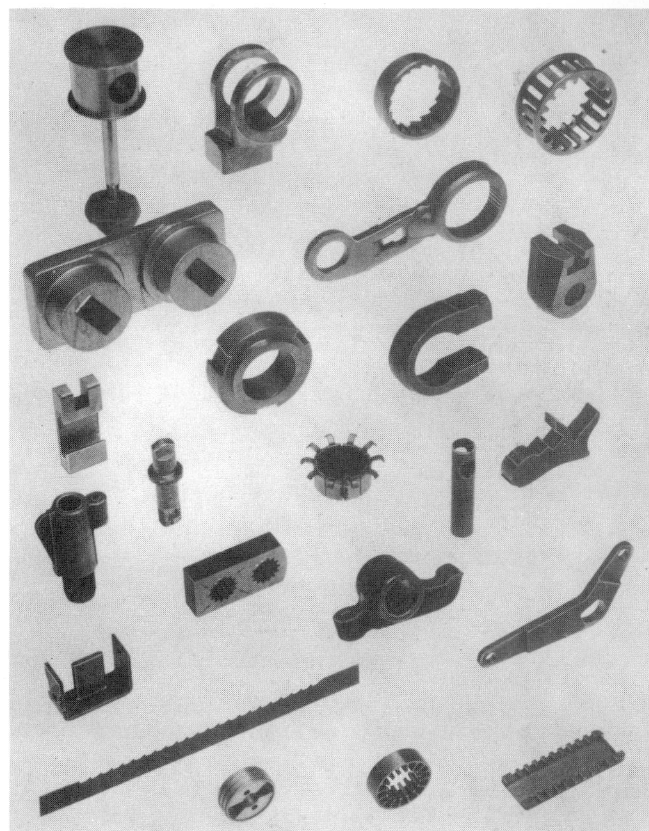

Fig. 11-21. These are some typical parts machined by broaching. (LaPoint Machine Tool Co.)

Consistently close tolerances can be maintained by broaching. While surface finishes produced are smooth compared to many other machining processes, they can be further improved by providing ***burnishing*** (noncutting) elements to the finishing end of the broach.

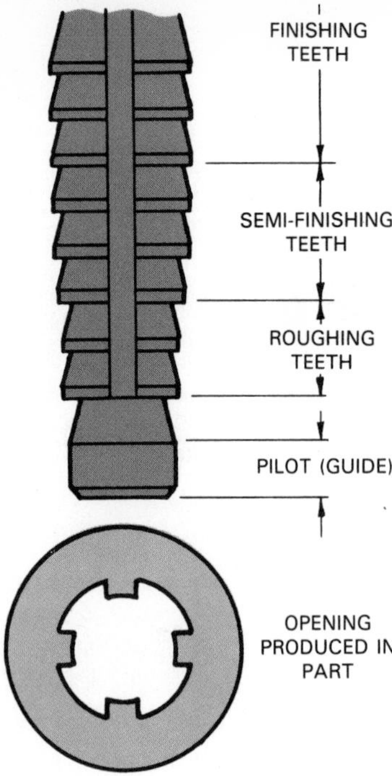

Fig. 11-22. Note typical broaching tool.

Broaching machines are available in vertical and horizontal configurations and in many sizes. Some are fitted with dual rams. Work requiring multiple broaching passes can be transferred from one ram to the other to reduce handling time, Fig. 11-22.

TEST YOUR KNOWLEDGE—Chapter 11

Please do not write in the text. Place your answers on a separate sheet of paper.

1. The shaper is a machine used to machine __________ surfaces.
2. How is shaper size determined?
3. The cutting tool on the shaper:
    a. Is stationary and the work moves against it.
    b. Moves across the work which is stationary.
    c. Is moved across work which, in turn, moves at a slower speed in opposite direction.
    d. All of the above.
    e. None of the above.
4. With a harder metal or a deeper cut, the cutting speed should be __________.
5. With softer metal or a lighter cut, the cutting speed should be __________.
6. Use __________ feed, __________ cut, and __________ cutting speed for the roughing cut.
7. Use __________ feed, __________ cut, and __________ cutting speed for the finishing cut.
8. The vertical shaper is also known as a __________.
9. When is a planer needed to machine work?
10. The cutting tool on a planer:
    a. Is stationary and the work moves against it.
    b. Moves across the work which is stationary.
    c. Is pulled or pushed across the work.
    d. All of the above.
    e. None of the above.
11. How does broaching differ from other planing machines?
12. What is unique about the cutting tool used on a broaching machine?

RESEARCH AND DEVELOPMENT

1. Prepare a term paper on the history and development of planing machines.
2. If your shop has a shaper, overhaul it. Repaint the machine after all mechanical repairs are made.
3. Construct a model of a broaching tool for machining a flat surface.
4. Secure samples of work that has been shaped by broaching.

Chapter 12

MILLING MACHINES

After studying this chapter, you will be able to:

- Describe how milling machines operate.
- Identify the various types of milling machines.
- Select the proper cutter for the job to be done.
- Set up and safely operate horizontal and vertical milling machines.
- Calculate cutting speeds and feeds.
- Perform various cutting, drilling, and boring operations on a milling machine.
- Make the needed calculations and cut spur gears.
- Point out safety precautions that must be observed when operating a milling machine.

A ***milling machine*** rotates a multitooth cutter into the work. A wide variety of cutting operations can be performed on milling machines. They are capable of machining flat or contoured surfaces, slots, grooves, recesses, threads, gears, spirals, and other configurations.

There are more variations of milling machines available than any other family of machine tools. Milling machines are well suited for computer controlled operations, Fig. 12-1. As mentioned, metal is removed in milling by means of turning a multitooth cutter into the work, Fig. 12-2. Each tooth of the cutter removes a small individual chip of material.

Work may be clamped directly to the machine table, held in a fixture, or mounted in or on one of the numerous work-holding devices available for milling machines.

SAFETY NOTE! Never attempt to operate a milling machine while your senses are impaired by medication or other substances.

TYPES OF MILLING MACHINES

It is difficult to classify the various categories of milling machines because their designs tend to merge with one another. However, for practical purposes, milling machines may be grouped into two large families:

1. Fixed-bed type.
2. Column and knee type.

Both groups are made with horizontal or vertical spindles. On a ***horizontal spindle milling machine,*** the cutter(s) is fitted onto an arbor mounted in the machine on an axis parallel with the work table. Refer to Fig. 12-3.

The cutter on a ***vertical spindle milling machine*** is normally perpendicular or at a right angle to the work table, Fig. 12-4. However, on many vertical

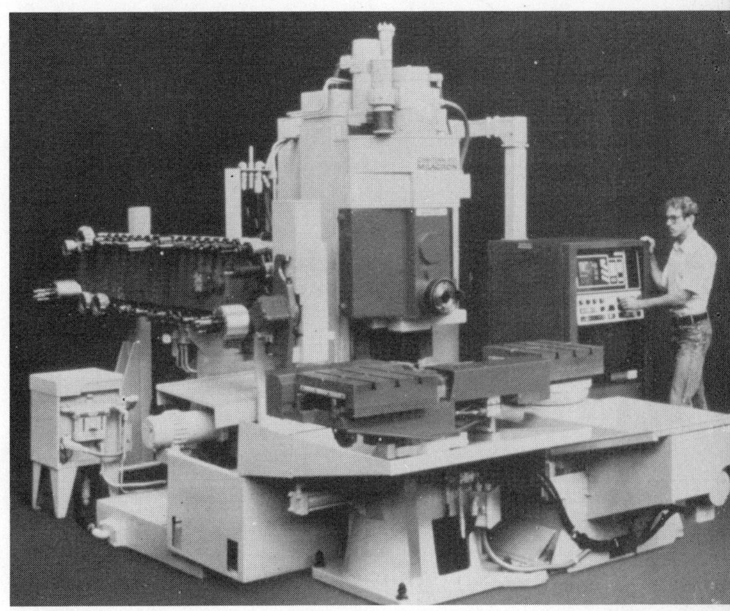

Fig. 12-1. Computer numerical controlled (CNC) milling machine is latest in technology. It is equipped with automatic tool changing and offers milling, drilling, boring, and reaming capabilities. It is easily programmed for machining a single part or for production work. (Cincinnati Milacron)

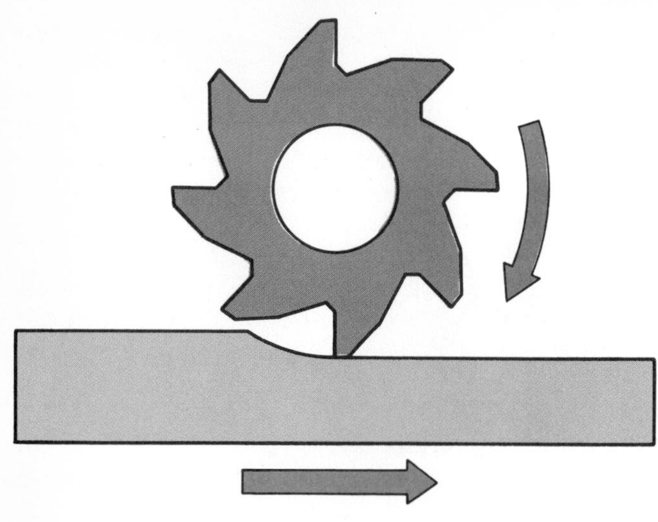

Fig. 12-2. Note how milling machine works.

Fig. 12-4. Cutter on a vertical milling machine is perpendicular to surface being machined. (Heidenhain Corp.)

spindle machines, it can be tilted to perform angular cutting operations.

Fixed-bed milling machines

Fixed-bed or ***bed-type milling machines*** are characterized by very rigid work table construction and support, Fig. 12-5. The work table moves only in a ***longitudinal*** direction (back and forth/X-axis) and can vary in length from 3 to 30 ft. (0.9 to 9.0 m). ***Vertical*** (up and down/Z-axis) and ***cross*** (in and out/Y-axis) movements are obtained by cutter head movements.

The type of bed permits heavy cutting on large, heavy work, Fig. 12-6.

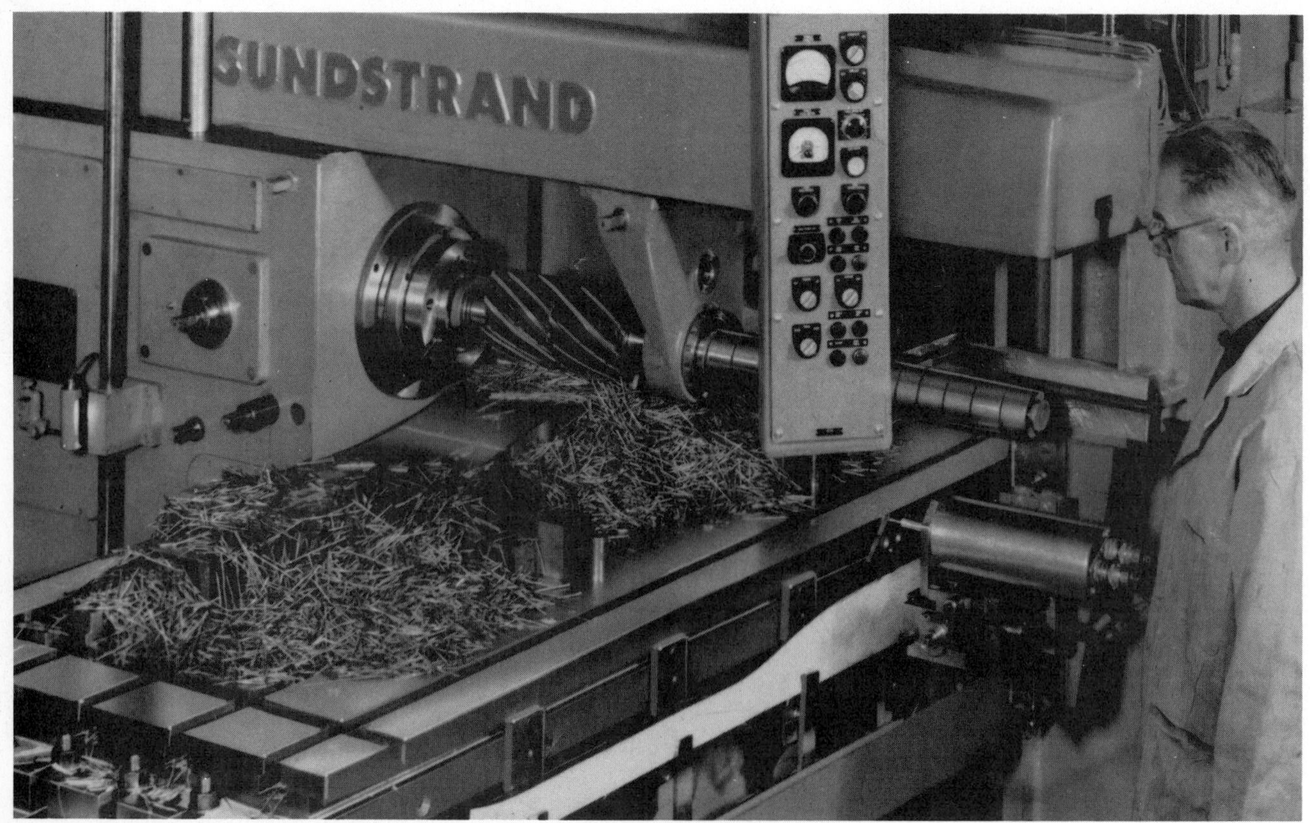

Fig. 12-3. This 50 hp horizontal milling machine can make a cut 1/8 in. deep by 12 in. wide. Note alignment of arbor on which cutter is mounted. (National Machine Tool Builders Assoc.)

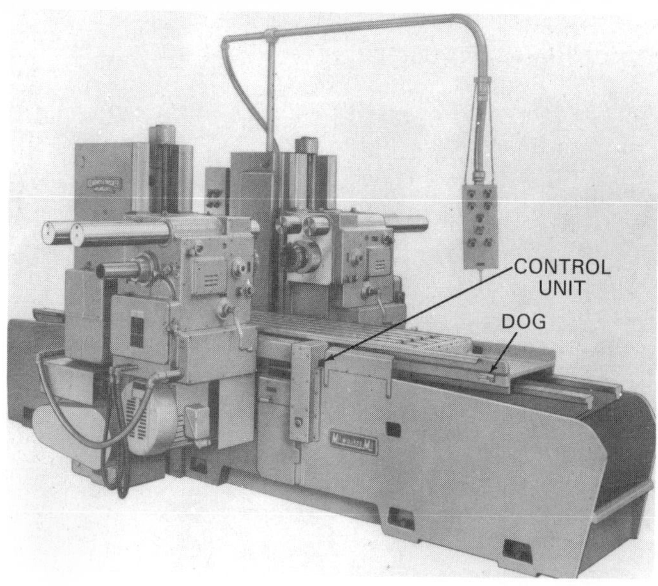

Fig. 12-5. Duplex type fixed bed milling machine makes use of two cutting heads. (Kearney & Trecker Corp.)

Bed-type milling machines can be further classified as horizontal, vertical, or planer type machines. Since primarily production type machines, they are given more detailed coverage under INDUSTRIAL APPLICATIONS OF THE MILLING MACHINE at the end of this chapter.

Column and knee milling machines

The ***column and knee type milling machine*** is so named because the parts that provide movement to the work consist of a ***column*** that supports and guides the knee in vertical movement (up and down/Z-axis). The ***knee*** supports the mechanism for obtaining traverse (in and out/Y-axis) and longitudinal (back and forth/X-axis) table movements. See Fig. 12-7.

There are three basic types of knee and column type milling machines:

1. Plain (horizontal spindle) milling machine.
2. Universal (horizontal spindle) milling machine.
3. Vertical spindle milling machine.

Fig. 12-6. Large, multi-spindle, 3-axis, fixed bed milling machine is shaping Boeing 727 upper wing skins. (Grumman Aerospace Corp.)

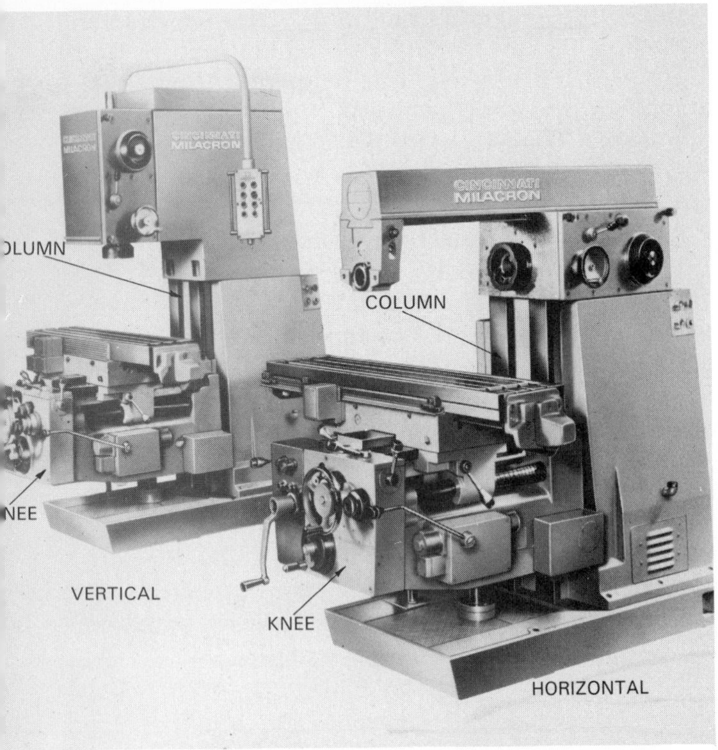

Fig. 12-7. These are column and knee type milling machines. (Cincinnati Milacron)

Fig. 12-8. Note table construction on plain type milling machine. (Greaves Machine Tool Div., J.A. Fay & Egan Co.)

Plain milling machine

The work table on the ***plain milling machine*** has three movements: VERTICAL, CROSS, and LONGITUDINAL (X-, Y-, and Z-axes), Figs. 12-8 and 12-9. The cutter spindle projects horizontally from the column.

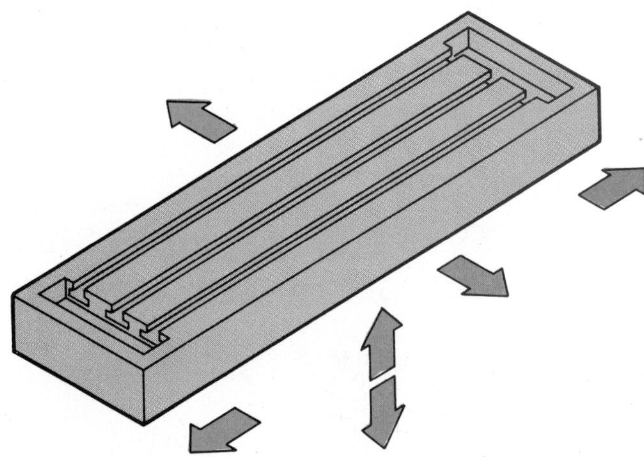

Fig. 12-9. Study table movements of plain type milling machine.

Universal milling machine

A ***universal milling machine*** is similar to the plain milling machine but the table has a fourth movement, Fig. 12-10. On this type of machine, the table can be swiveled on the saddle through an angle of 45 degrees or more, Fig. 12-11. This makes it possible to produce spiral gears, spiral splines, and similar work, Fig. 12-12.

Vertical spindle milling machine

A ***vertical spindle milling machine*** differs from the plain and universal machines by having the cutter spindle in a vertical position, at a right angle to the top of the work table. The cutter head can be raised and lowered by hand or by power feed, Fig. 12-13.

This type milling machine is best suited for an end mill or face mill cutter. Vertical mills include swivel head, sliding head, and rotary head types.

A ***swivel head milling machine*** is the type usually found in vocational-technical school training programs. The spindle can be swiveled for angular cuts.

Fig. 12-10. This is a universal type milling machine. (Greaves Machine Tool Div., J.A. Fay & Egan Co.)

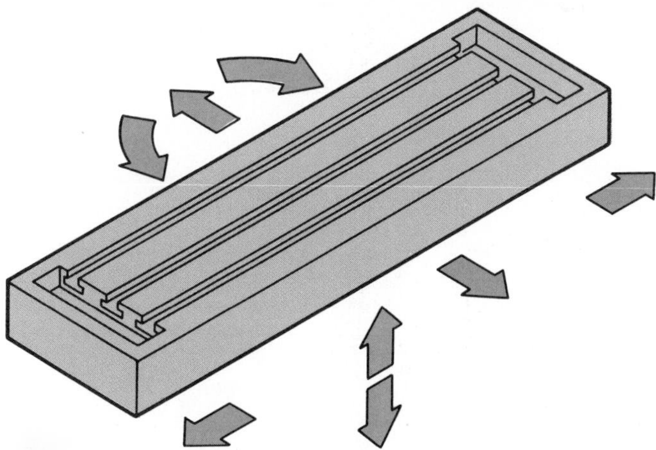

Fig. 12-11. Note table movements of universal type milling machine.

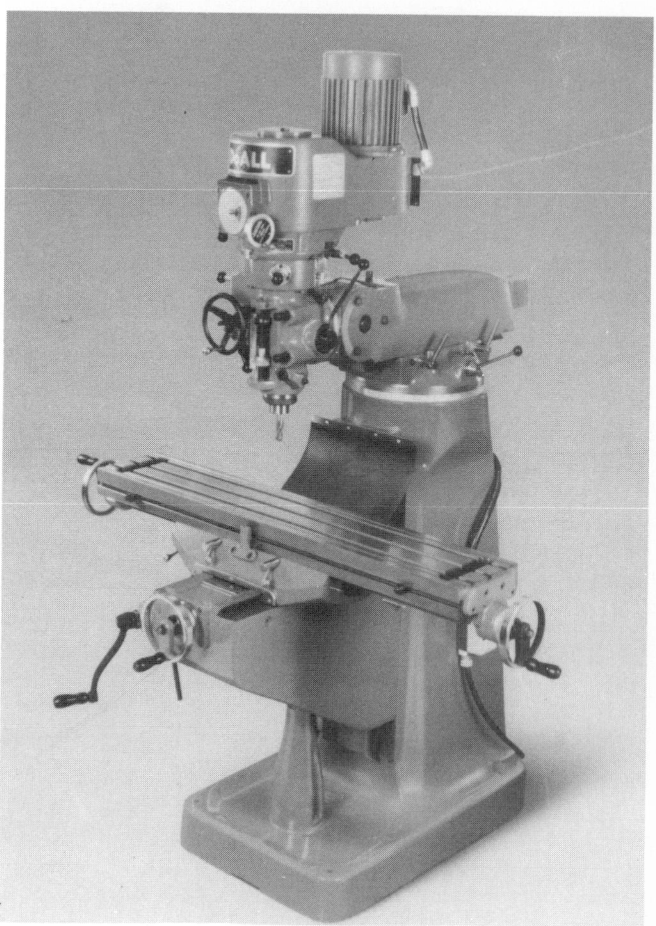

Fig. 12-13. The vertical milling machine. (DoALL Co.)

On the ***sliding head milling machine,*** the spindle head is fixed in a vertical position. The head can be moved in a vertical direction by hand or under power, Fig. 12-14.

The spindle on the ***rotary head milling machine*** can be moved vertically and in circular arcs of adjustable radii about a vertical center line. It can be adjusted manually or under power feed, Figs. 12-15 and 12-16.

Methods of milling machine control

The method employed to control table movement is another way of classifying milling machines, and

Fig. 12-12. Note how table is swiveled at an angle to cutting tool. This feature makes it possible to machine spiral gears like one being cut.

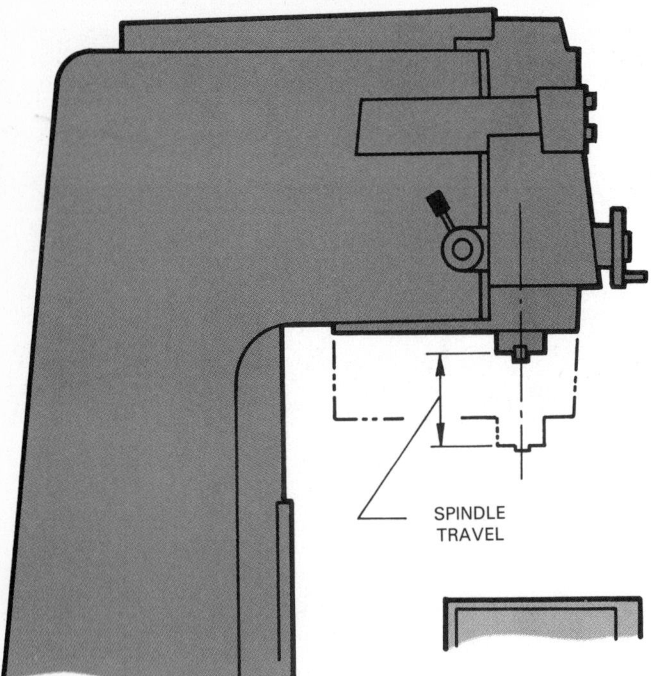

Fig. 12-14. Spindle head is fixed in a vertical position on sliding head milling machine. Entire head is moved to make cutting adjustments.

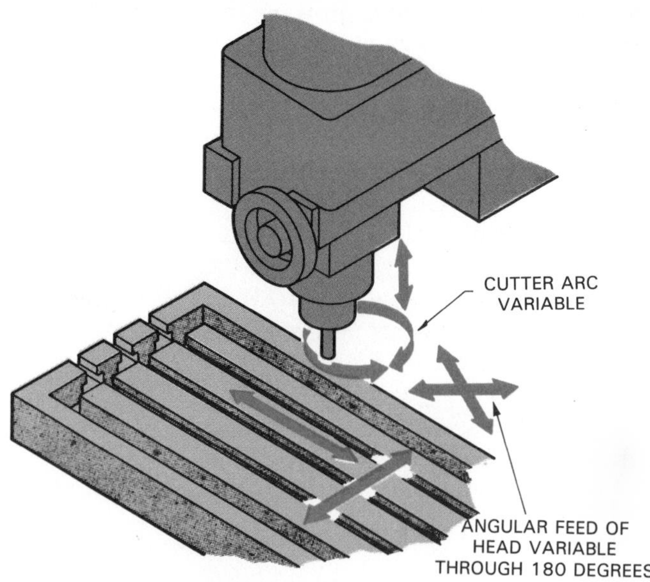

Fig. 12-16. Note movements possible on rotary head vertical milling machine.

Fig. 12-15. This is a rotary head vertical milling machine. (Kearney & Trecker Corp.)

all machine tools in general. Basically, there are four methods of control:

1. ***Manual***—All movements are made by hand lever control.
2. ***Semi-automatic***—Movements are controlled by hand and/or power feeds.
3. ***Fully automatic***—A complex hydraulic feed arrangement follows two or three dimensional templates to guide the cutter(s) automatically. Specifications can also be programmed on magnetic or perforated tape which guides the cutter(s) through the required machining operations.
4. ***Computerized*** (CNC)—Machining coordinates are entered into a master computer or computer on the machine using a special programming language. Computer instructions electronically guide the cutter(s) through the required machining sequence.

Small machines, such as the ***bench type milling machine,*** Fig. 12-17, have only longitudinal table

Fig. 12-17. Light plain type milling machine is sometimes referred to as a bench mill.

movement fitted with power feed, Fig. 12-18. Larger machines have automatic feed or power feed on all table movements, Fig. 12-19.

Table movement (feed) can be engaged at cutting speed; however, there is a ***rapid traverse feed*** which gives fast power movement in any direction of feed engagement. This permits work to be positioned at several times the fastest feed indicated on the feed chart. It operates by positioning the automatic power feed control lever to give the desired directional movement and activating the rapid traverse lever, Fig. 12-20.

WARNING! Never activate rapid traverse with the cutter in a cut.

Fig. 12-20. Machinist is engaging rapid traverse on a #2 plain milling machine. The #2 denotes machine size.

Fig. 12-18. Note table feed mechanism on this light milling machine.

Adjusting milling cutting speed and feed

Depending upon the milling machine make, ***cutting speed*** (cutter rpm) and ***rate of feed*** (table movement speed) may be changed by:

1. Shifting V-belts, Fig. 12-21.
2. Adjusting variable speed pulleys, Fig. 12-22.
3. Utilizing a quick change gear box and shifting or dialing to the required speed and feed setting, Figs. 12-23 and 12-24. On some machines, speed and feed changes are made hydraulically or electronically.

Always make the changes as specified in the instruction manual for the specific machine being used!

DANGER! Make sure the machine has come to a complete stop before attempting to adjust V-belts.

Fig. 12-19. These are typical table controls. (Cincinnati Lathe & Tool Co.)

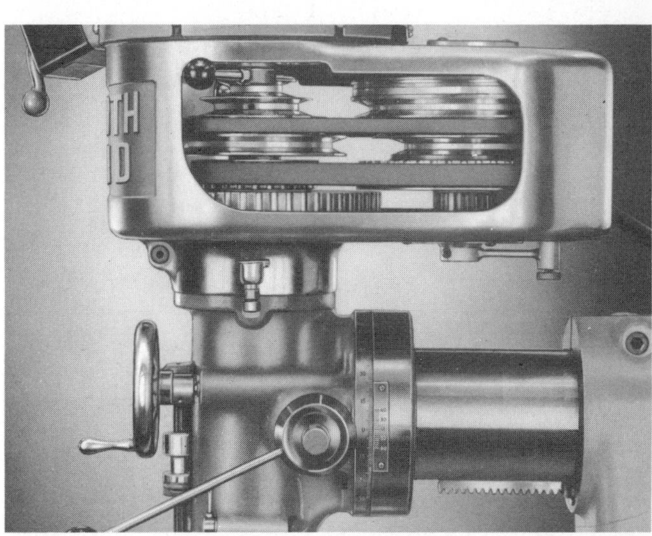

Fig. 12-21. Spindle speeds on a V-belt drive are changed by using various pulley ratios. (South Bend, Inc.)

Fig. 12-22. Spindle speed on this machine is controlled by adjusting a split pulley. Pulley can be spread open or narrowed in width to control where belt rides in pulley. This is done by dialing speed selector to proper position. (Cincinnati Lathe & Tool Co.)

Fig. 12-23. Note dialing mechanism used to shift gears and change spindle speed.

Fig. 12-24. This is another type dialing mechanism to change amount of feed.

MILLING SAFETY PRACTICES

Milling machines, like all machine tools, should be cleaned after each work session. A medium width paint brush may be used to remove accumulated chips, Fig. 12-25.

CHIPS are RAZOR SHARP; do NOT use your hand to remove them. NEVER remove chips with compressed air. The flying chips may injure you or a nearby person.

If cutting oil was used, the oily mist produced by the compressed air is highly flammable. If ignited by an open flame, it can produce explosive results. Finish by wiping down the machine with a soft cloth.

The following procedures are suggested for the safe operation of a milling machine.

1. Become thoroughly familiar with the milling machine before attempting to operate it. When in doubt, obtain additional instructions.
2. Wear appropriate clothing and approved safety glasses!
3. Stop the machine before attempting to make adjustments or measurements!
4. Get help to move any heavy machine attachment, such as a vise, dividing head, rotary table, or large work.
5. Stop the machine before trying to remove accumulated chips.
6. Never reach over or near a rotating cutter!
7. Be sure the work holding device is mounted solidly to the table, and the work is held

Fig. 12-25. Use a brush to remove metal chips; NEVER use your hand!

firmly. Spring or vibration in the work can cause thin cutters to jam and shatter!

8. Avoid talking with anyone while operating a machine tool, nor allow anyone to turn your machine on for you.
9. Keep the floor around your machine clear of chips and wipe up spilled cutting fluid immediately! Place sawdust or special oil obsorbing compound on slippery floors.
10. Be thoroughly familiar with the placement of the machine's STOP switch or lever, Fig. 12-26.
11. Treat any small cuts and skin punctures as potential infections! Clean them thoroughly. Apply antiseptic and cover injury with a bandage.
    Report any injury, no matter how minor, to your instructor or supervisor.
12. Have work clothes laundered frequently. Greasy clothing is a fire hazard.
13. Put all oily rags, used to wipe down machines, in an approved metal container that can be closed tightly.
14. Never "fool around" when operating a milling machine! Keep your mind on the job and be ready for any emergency!
15. Be sure all power to the machine is turned off before opening or removing guards and covers.
16. Take care to prevent running the saddle or work into the column.
17. Use a section of heavy cloth or gloves for protection when handling milling cutters. Avoid using your bare hands!

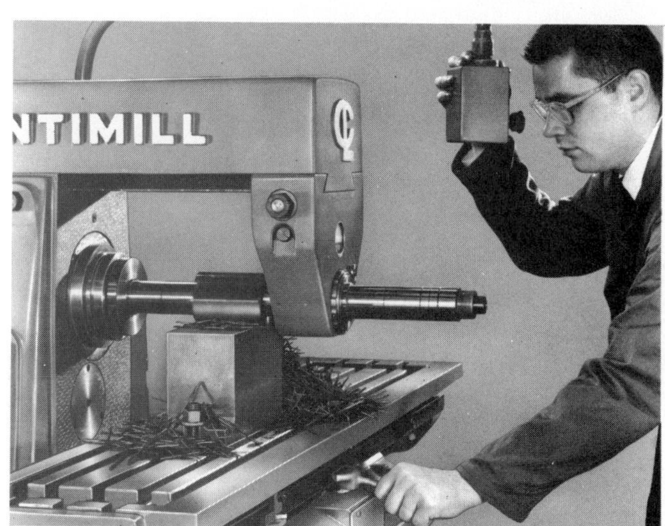

Fig. 12-26. Make it a "first point of business" to know the location of the STOP LEVER. WEAR safety glasses and STOP the machine before attempting to make measurements, adjustments, or to clear away chips. (Cincinnati Lathe & Tool Co.)

MILLING OPERATIONS

There are two main categories of milling operations:

1. ***Face milling*** is done when the surface being machined is PARALLEL with the CUTTER FACE, Figs. 12-27 and 12-28. Large, flat surfaces are machined with this technique.
2. ***Peripheral milling*** is done when the surface being machined is PARALLEL with the PERIPHERY of the cutter, Figs. 12-29 and 12-30.

MILLING CUTTERS

The typical ***milling cutter*** is circular in shape with a number of cutting edges (teeth) located around

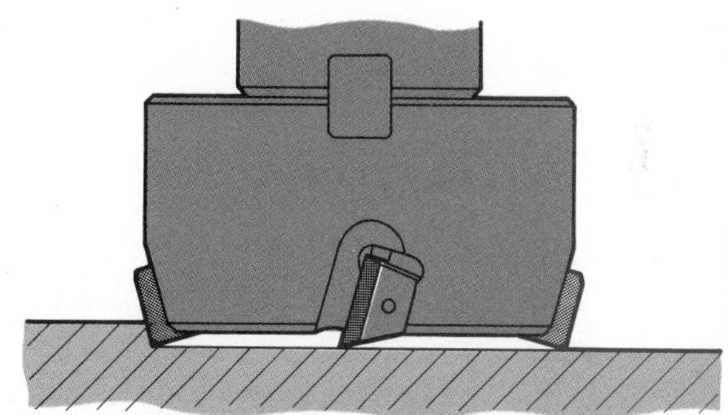

Fig. 12-27. With face milling, surface being machined is parallel with cutter face.

Fig. 12-28. Note example of face milling. (Lovejoy Tool Co., Inc.)

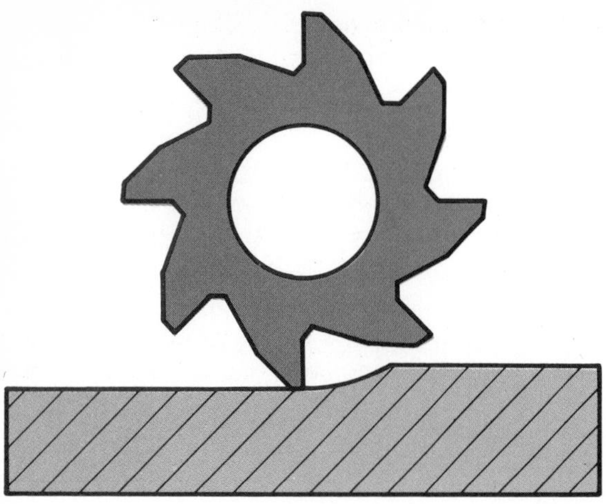

Fig. 12-29. With peripherial milling, surface being machined is parallel with periphery of cutter.

the circumference. A milling cutter is manufactured in a large number of stock shapes, sizes, and kinds, because it cannot be economically ground for a particular job as can a lathe cutter bit. See Fig. 12-31.

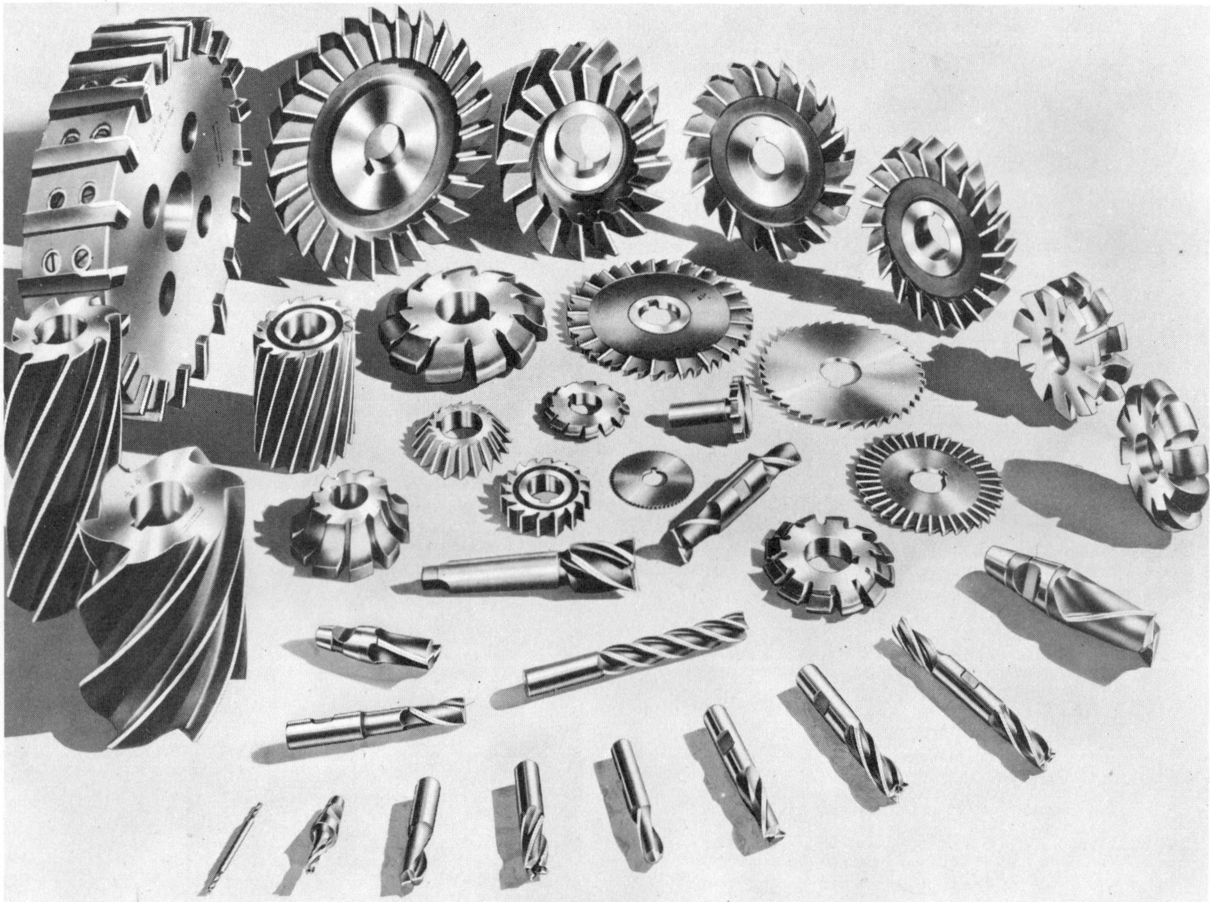

Fig. 12-30. Note example of peripherial milling.

Fig. 12-31. There are many variations of standard milling cutters. (Brown & Sharpe Mfg. Co.)

Types of milling cutters

There are two general types of milling cutters:

1. A ***solid cutter*** has the shank and body made in one piece, Fig. 12-32.
2. The ***inserted tooth cutter*** has teeth made of special cutting material which are brazed or clamped in place, Fig. 12-33. Worn and broken teeth can be replaced easily instead of discarding the entire cutter.

Fig. 12-32. With the exception of the cutter in the back row, these are solid milling cutters. (Brown & Sharpe Mfg. Co.)

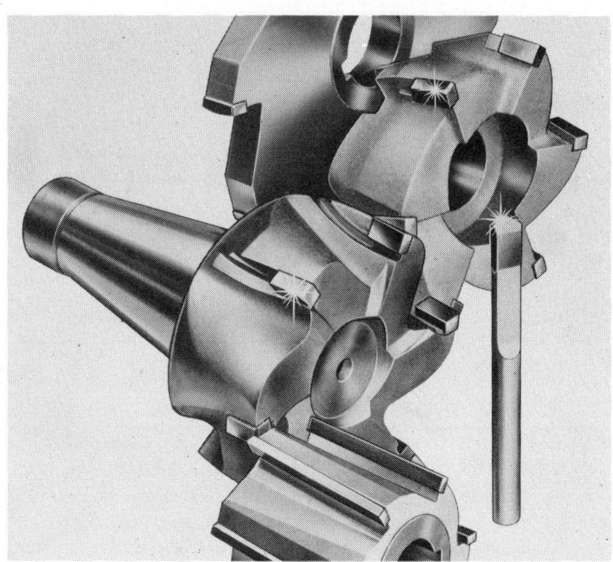

Fig. 12-33. Cutting edges of these inserted tooth cutters are brazed to cutter body. (Brown & Sharpe Mfg. Co.)

How milling cutters are classified

Milling cutters are frequently classified by the method used to mount them on the machine:

1. ***Arbor cutters*** have a suitable hole for mounting to an arbor, Fig. 12-34.
2. ***Shank cutters*** are fitted with either a straight or taper shank that is an integral part of the cutter, Fig. 12-35. They are held in the machine by collets or special sleeves.
3. ***Facing cutters*** can be mounted directly to a machine's spindle nose or on a stub arbor. See Fig. 12-36.

Fig. 12-34. Note hole in arbor type milling cutter.

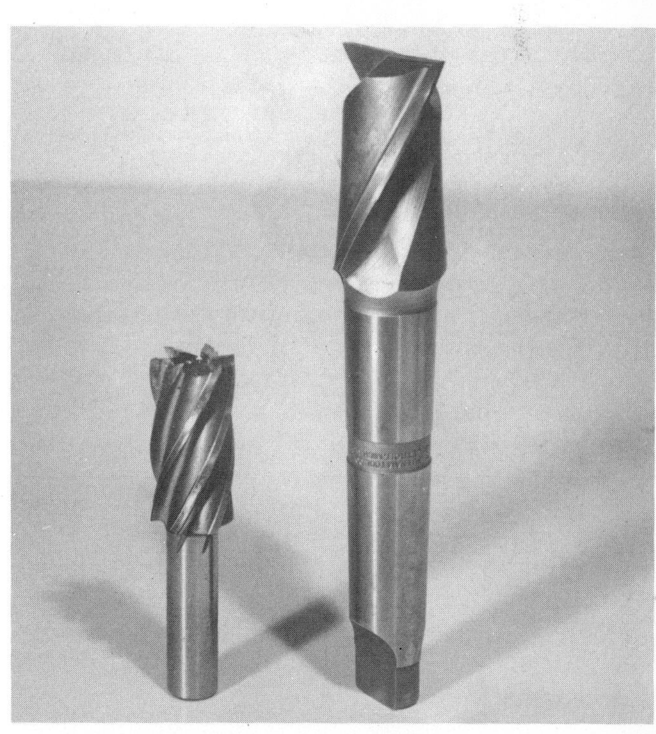

Fig. 12-35. These are shank type milling cutters.

Fig. 12-36. Note construction of facing type milling cutter. (Standard Tool Co.)

Milling cutter material

Considering the wide range of materials that must be machined, the ideal milling cutter should have:

1. ***High abrasion resistance*** so the cutting edges do not wear away rapidly due to the abrasive nature of some materials.
2. ***Red hardness*** so the cutting edges are NOT affected by the terrific heat generated by many machining operations.
3. ***Edge toughness*** so the cutting edges do NOT readily break down due to the loads imposed upon them by the cutting operation.

As no material can meet these requirements in all situations, cutters are made from materials that are, by necessity, a compromise.

High-speed steels (HSS) are the most versatile of the cutter materials. Cutters made from high-speed steels are excellent for general purpose work or where vibration and chatter are problems. They are preferred for use on machines of low power.

HSS milling cutters can be improved by the application of surface lubricating treatments, surface hardeneing treatments, or by having coatings such as chromium, tungsten, or tungsten carbide applied to the cutting surfaces by one of several processes. While these treated tools cost 2-6 times as much as conventional HSS tools, they may last 5-10 percent longer or provide 50-100 percent higher metal removal rates with the same tool life.

Cemented tungsten carbides include a broad family of hard metals. They are produced by powdered metallurgy techniques and have qualities that make them suitable for metalcutting tools. Cemented carbides can, in general, be operated at speeds 3-10 times faster then conventional HSS cutting tools.

Normally, only the cutting tips are made of cemented carbides, rather than the entire cutter. They are brazed or clamped to the cutter body. See Figs. 12-37 and 12-38.

Some of the newer inserted tooth cutters make use of indexable type inserts. The insert has several cutting edges at various corners. When they

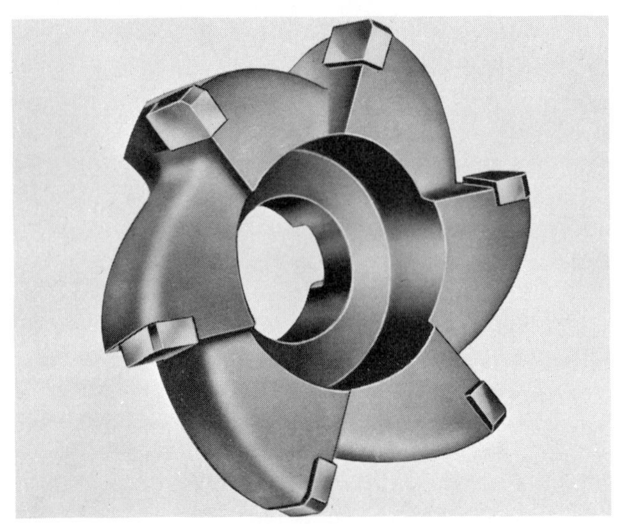

Fig. 12-37. Teeth on this inserted tooth milling cutter are brazed to cutter body. (Brown & Sharpe Mfg. Co.)

Fig. 12-38. This inserted tooth cutter has teeth that are clamped to cutter body. It is composed of the following parts: 1—Cutter body is made of hardened alloy steel. 2—Replaceable hardened front edge that clamps tooth to cutter body. 3—Blade pocket. 4—Solid carbide replaceable anvil to protect cutter body. 5—Replaceable rest button to support cutter tooth. 6—Solid carbide cutter tooth that can be indexed four times. (Standard Tool Co.)

become dull, the inserts are indexed or repositioned so a new cutting edge contacts the metal.

Cemented carbide cutters are excellent for long production runs and for milling materials with a scale-like surface (cast iron, cast steel, bronze, etc.)

TYPES AND USES OF MILLING CUTTERS

The following are the more commonly used milling cutters with a summary of the work to which they are best suited.

End mills

End milling cutters are designed for machining slots, keyways, pockets, and similar work, Fig. 12-39. The cutting edges are on the circumference and end, Fig. 12-40. Solid end mills may have straight or helical flutes, Figs. 12-41 and 12-42, and straight or taper shanks, Fig. 12-43. Straight shank end mills are available in single and double end styles, Figs. 12-44 and 12-45.

The term ***hand*** is used to describe the direction of cutter rotation (right- or left-hand) and the helix of the flutes, Fig. 12-46. A ***right-hand cutter*** rotates counterclockwise, and a ***left-hand cutter*** rotates clockwise when viewed from the cutting end.

Ball nose end mills are utilized for tracer milling, computer controlled contour milling, die sinking, fillet milling, and other radius work. A cut with a depth equal to one-half the end mill diameter can generally be taken in solid stock. Refer to Figs. 12-47 through 12-49.

Several end mill styles are available:

1. A ***two-flute end mill*** can be fed into the work like a drill. There are two cutting edges on the circumference with the end teeth cut to the center, Fig. 12-50.

Fig. 12-39. Note slot being end milled.

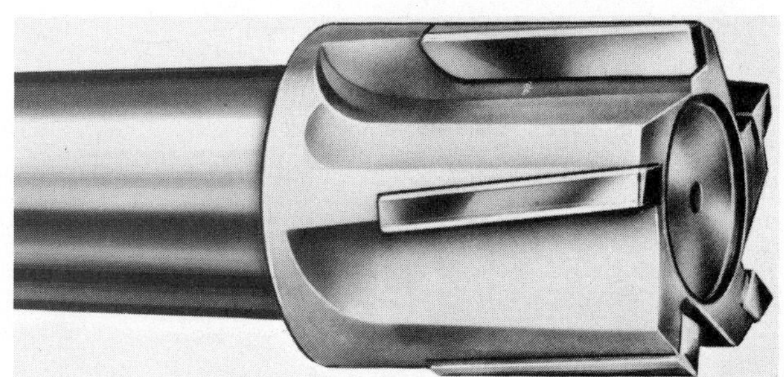

Fig. 12-40. This inserted tooth type end mill has cutting surfaces on circumference and face. (Brown & Sharpe Mfg. Co.)

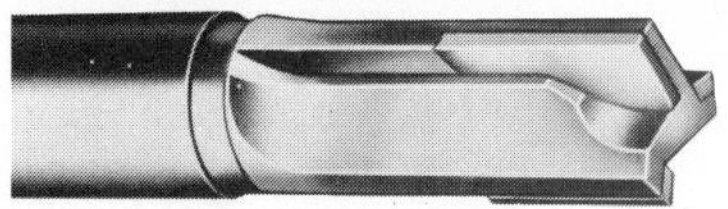

Fig. 12-41. Note inserted tooth type straight flute end mill. (Brown & Sharpe Mfg. Co.)

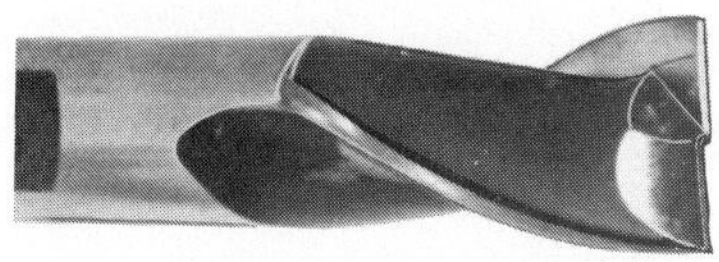

Fig. 12-42. This is a helical fluted end mill.

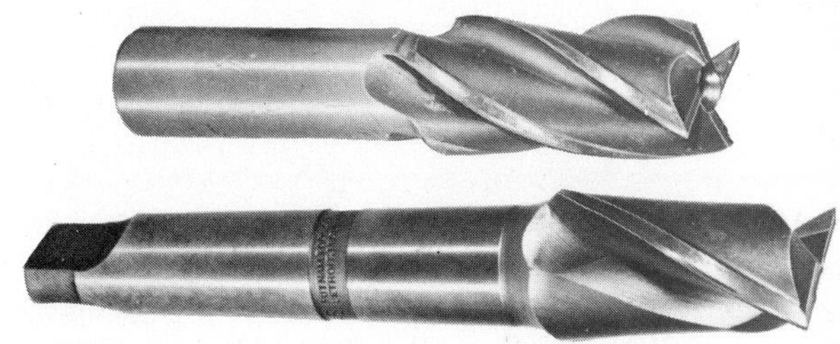

Fig. 12-43. Above—Straight shank end mill. Below—Taper shank end mill.

Fig. 12-44. Note single end style end mill.

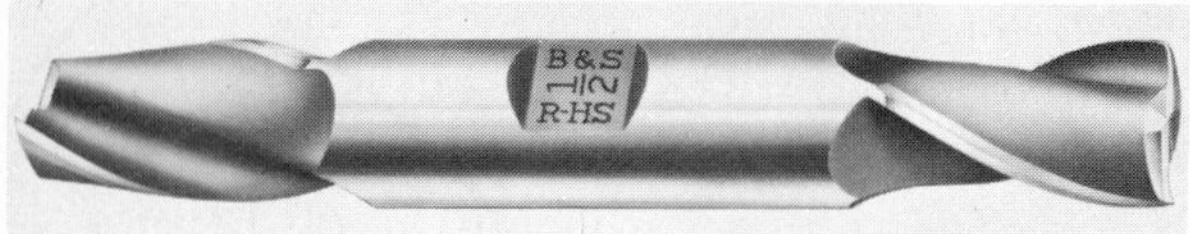

Fig. 12-45. Note double end style end mill.

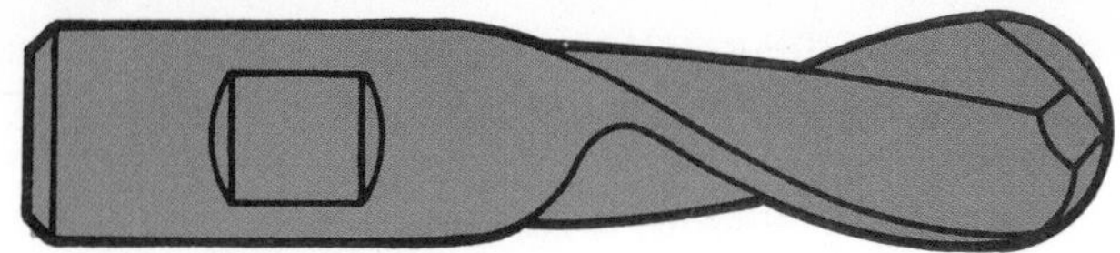

Fig. 12-47. Ball nosed end mill has rounded tip.

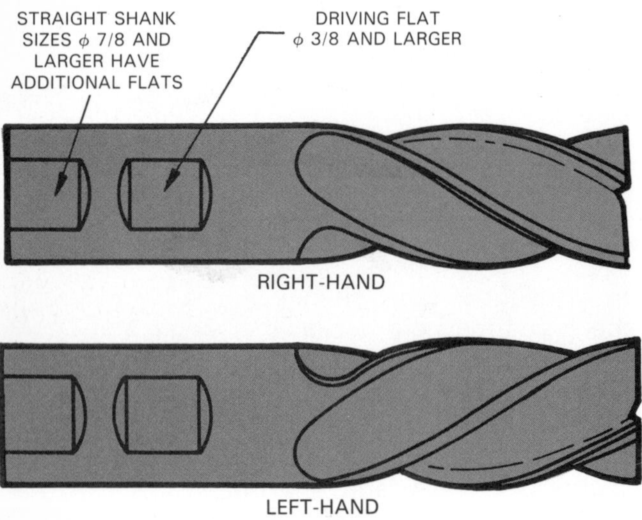

Fig. 12-46. Cutter is right-hand if it rotates counterclockwise when viewed from cutting end. It is left-hand if rotation is clockwise.

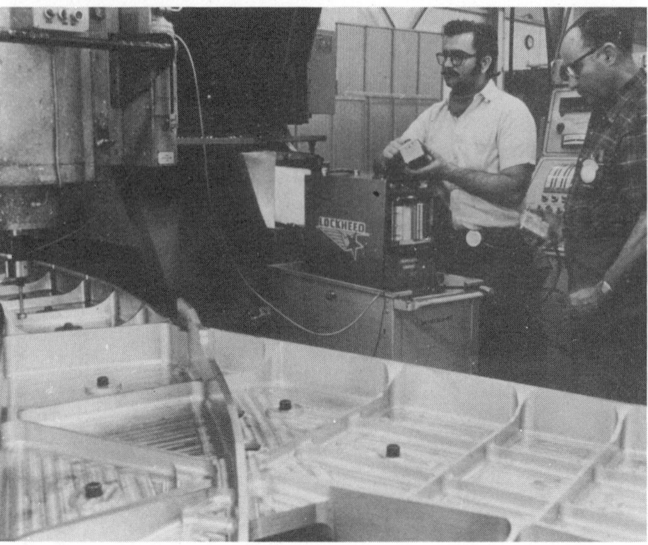

Fig. 12-49. Machinists are contouring a large aerospace member on a CNC milling machine. (Lockheed-California)

Fig. 12-48. Tracer milling will follow master shape to machine duplicate shape. (Bridgeport Machines)

Fig. 12-50. Two-flute end mill can be fed into work like a drill.

2. The ***multi-flute end mill*** can be run at the same speed and feed as a comparable two lip end mill, but it has a longer cutting life and will produce a better finish. It is recommended for conventional milling where ***plunge cutting*** (going into work like a twist drill) is NOT necessary. Refer to Fig. 12-51.
3. A ***shell end mill*** has teeth similar to the multi-flute end mill but is mounted on a stub arbor, Fig. 12-52. The cutter is designed for both face and end milling. Shell end mills are made with right-hand cut, right-hand helix, or with left-hand cut, left-hand helix.

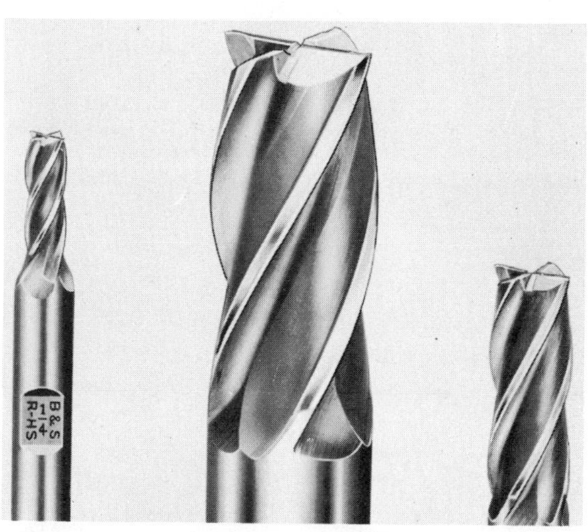

Fig. 12-51. These are multi-flute end mills.

Fig. 12-52. Shell end mill is being used in vertical mill.

Face milling cutters

Face milling cutters are intended for machining large flat surfaces parallel to the face of the cutter, Fig. 12-53. The teeth are designed to make the roughing and finishing cuts in one operation. Because of their size and cost, most face milling cutters have inserted cutting edges.

A ***fly cutter*** is a single point (cutting tool) face mill. One is shown in Fig. 12-54.

Arbor milling cutters

The more common arbor milling cutters and the work for which they are best adapted include:

Plain milling cutter

Plain milling cutters are cylindrical, with teeth located around the circumference. Plain milling cutters less than 3/4 in. (20 mm) are made with straight teeth. Wider plain cutters, called ***slab cutters*** are made with helical teeth designed to cut with a shearing action. This reduces the tendency for the cutter to chatter.

Fig. 12-53. Note inserted tooth type face mill. (Brown & Sharpe Mfg. Co.)

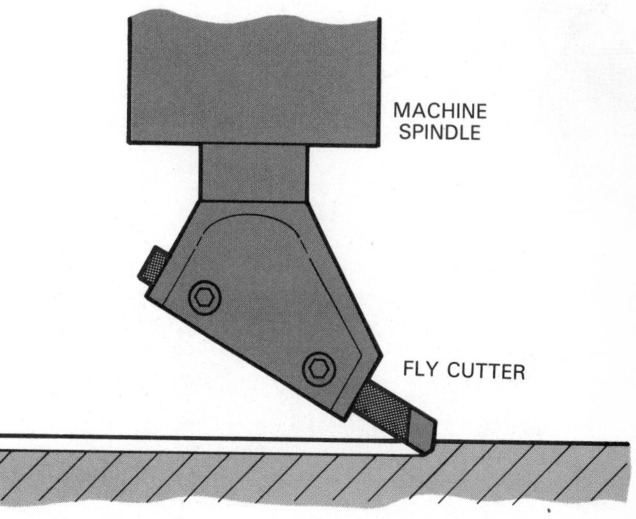

Fig. 12-54. Fly cutter is a single point cutting tool used for face milling.

1. The ***light-duty plain milling cutter*** is used chiefly for light slabbing cuts and shallow slots. Refer to Fig. 12-55.
2. The ***heavy-duty plain milling cuter*** is recommended for heavy cuts where considerable material must be removed. There are fewer teeth than on a comparable light-duty cutter, Fig. 12-56. The cutting edges are better supported and chip spaces are ample to increase the larger volume of chips.
3. A ***helical plain milling cutter*** has fewer teeth than either of the two previously mentioned cutters, Fig. 12-57. They can be run at high speeds and produce exceptionally smooth finishes.

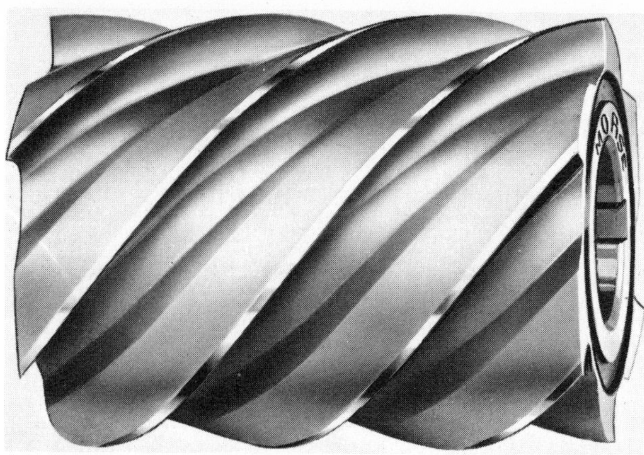

Fig. 12-57. Note helical plain milling cutter. (Morse Cutting Tools)

Side milling cutter

Cutting edges are located on the circumference and on one or both sides of ***side milling cutters.*** They are made in solid form or with inserted teeth.

1. ***Plain side milling cutter*** teeth are on the circumference and on both sides of this type cutter, Fig. 12-58. It is recommended for side cutting, straddle milling, and slotting. Plain side milling cutters are available in diameters ranging from 2 in./50 mm to 8 in./200 mm, and in widths from 3/16 in./5 mm to 1 in./25mm.

Fig. 12-58. This is a plain side milling cutter.

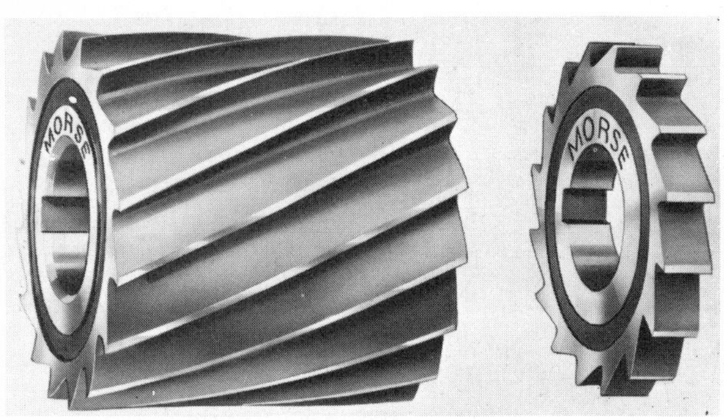

Fig. 12-55. These are light-duty plain milling cutters.

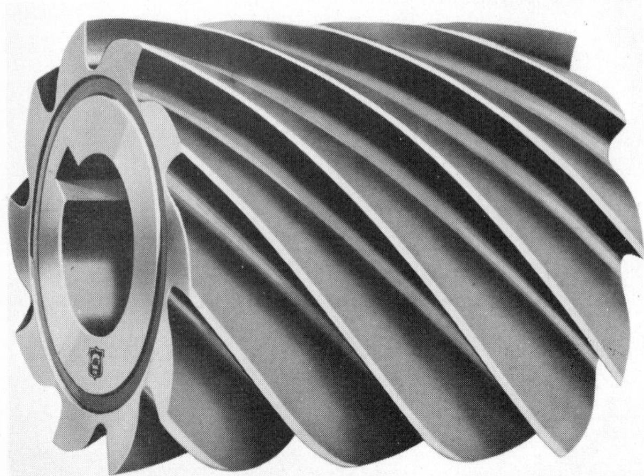

Fig. 12-56. Note heavy-duty plain milling cutter. (Standard Tool Co.)

2. A ***staggered-tooth side milling cutter*** has alternate right-hand and left-hand helical teeth. They aid in reducing chatter while providing adequate chip clearance for higher operating speeds and feeds than are possible with the plain side milling cutter. This type cutter is especially good for matching deep slots, Fig. 12-59.
3. The ***half side milling cutter*** has helical teeth on the circumference but side teeth on only ONE SIDE, Fig. 12-60. It is made as a right-hand or left-hand cutter and is recommended for heavy straddle milling and milling to a shoulder.
4. The ***interlocking side milling cutter*** is ideally suited for milling slots, bosses, and making other types of cuts that must be held to extremely close tolerances. The unit is made as two cutters with interlocking teeth that can be adjusted to the required width by using spacers or collars, Fig. 12-61. The alternating right and left shearing action eliminates side pressures, producing a good surface finish.

Fig. 12-59. Note staggered-tooth side milling cutter.

Fig. 12-60. These are half side milling cutters. (Standard Tool Co.)

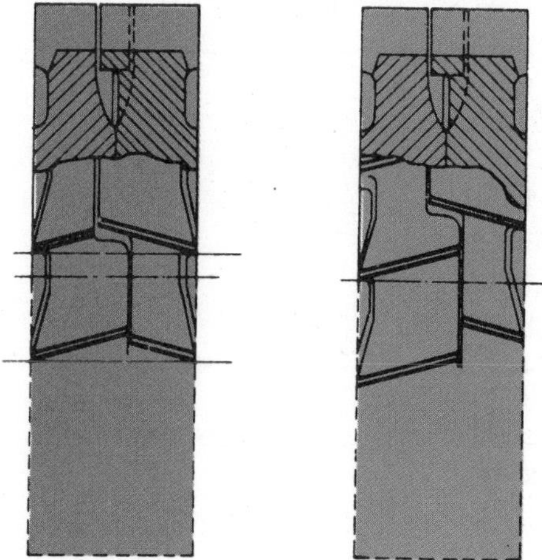

Fig. 12-61. Study tooth pattern on interlocking side milling cutters. (Morse Tool Co.)

Angle cutters

Angle cutters differ from other cutters in that the cutting edges are neither parallel, nor at right angles to the cutter axis.

1. On ***single-angle milling cutter,*** the teeth are on the angular face and on the side adjacent to the large diameter, Fig. 12-62. Single angle cuttes are made in both right-hand and left-hand cut, with included angles of 45 and 60 degs.
2. The ***double-angle milling cutter*** is used for milling threads, notches, serrations, and similar work. Double angle cutters are manufactured with included angles of 45, 60, and 90 degs. Other angles can be special ordered, Fig. 12-63.

Metal slitting saws

Metal slitting saws are thin milling cutters that resemble circular saw blades. They are employed for narrow slotting and cutoff operations. Slitting saws are available in diameters as small as 2 1/2 in. (60 mm) and as large as 8 in. (200 mm).

1. The ***plain metal slitting saw*** is essentially a thin plain milling cutter. It is used for ordinary slotting and cutoff operations, Fig. 12-64. Both sides are ground concave for clearance. The hub is the same thickness as the cutting edge. It is stocked in thicknesses ranging from 1/32 in. (0.8 mm) to 3/16 in. (5 mm).
2. A ***side chip clearance slitting saw*** is similar to the plain side milling cutter, Fig. 12-65. It is especially suitable for deep slotting and sawing because of its ample chip clearance.

Fig. 12-62. Note single-angle milling cutter.

Fig. 12-63. This is a double-angle milling cutter.

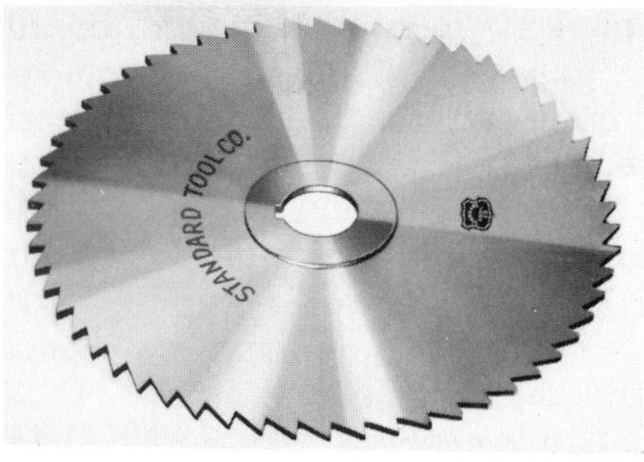

Fig. 12-64. Plain metal slitting saw is for slotting and cutoff operations.

Fig. 12-66. Concave milling cutter will produce accurate shape along cut.

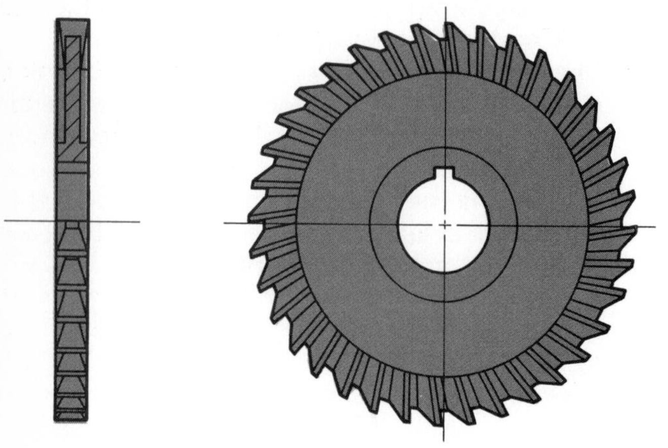

Fig. 12-65. Study construction of side chip clearance slitting saw.

Fig. 12-67. Convex milling cutter will produce curved slot.

Fig. 12-68. Corner rounding milling cutter is available as left- and right-hand cut.

Fig. 12-69. Note shape of gear cutter. (Standard Tool Co.)

Formed milling cutters

Formed milling cutters are employed to accurately duplicate a required contour. A wide range of shapes can be machined with standard cutters available. Included in this cutter classification are the ***concave cutter,*** Fig. 12-66, ***convex cutter,*** Fig. 12-67, ***corner rounding cutter,*** Fig. 12-68, and ***gear cutter,*** Fig. 12-69.

Miscellaneous milling cutters

Included in this category are cutters that do NOT fit into any of the previously mentioned groups.

1. The ***T-slot milling cutter*** has cutting edges for milling the bottoms of T-slots after cutting with an end mill or side cutter, Fig. 12-70.
2. A ***Woodruff keyseat cutter*** is used to mill the circular keyseat for Woodruff keys, Fig. 12-71.
3. A ***dovetail cutter*** mills dovetail type ways and it is used in much the same manner as the T-slot cutter. See Fig. 12-72.

Care of milling cutters

Milling cutters are expensive and easily damaged if care is not taken in use and storage. The following recommendations will help extend cutter life:

1. Use sharp cutting tools! Machining with dull tools results in low quality work and it eventually damages the cutting edges beyond salvage by grinding.
2. Tools must be properly supported and the work held rigidly, Fig. 12-73.
3. Use the correct cutting speed and feed for the material being machined.
4. An ample supply of cutting fluid is essential.
5. Employ the correct cutter for the job.
6. Store cutters in individual compartments or on wooden pegs. They should never come in contact with other cutters or tools, Fig. 12-74.
7. Clean cutters before storing them. If they are to be stored for any length of time, it is best to give them a light protective coating of oil.
8. Never hammer a cutter onto an arbor! Examine the arbor for nicks or burrs if the cutter does not slip onto the arbor easily. Do NOT forget to key the cutter to the arbor.
9. Place a board under the end mill when removing it from a vertical milling machine. This will prevent cutter damage if it is dropped accidently. Protect your hand with a heavy cloth or gloves. See Fig. 12-75.

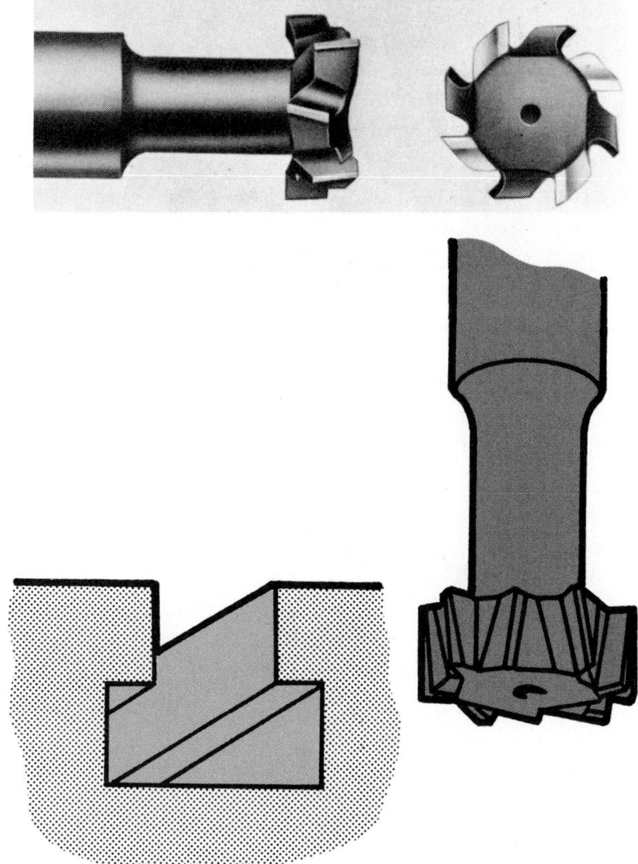

Fig. 12-70. Example of cut made with T-slot milling cutter. (Morse Cutting Tools)

Fig. 12-71. Woodruff keyseat cutter is for circular keyways.

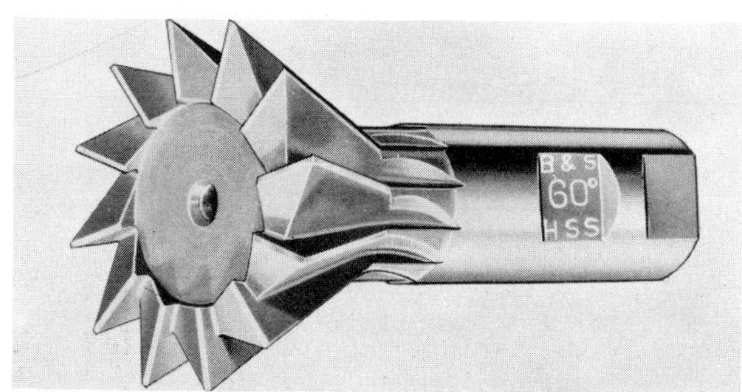

Fig. 12-72. Dovetail cutter will make slot with angled sides.

Fig. 12-73. Rigidly supported cutters will permit heavier cuts and prolong cutter life. An example supply of cutting fluid is also essential.

Fig. 12-74. Store cutters so they cannot come into contact with other cutters.

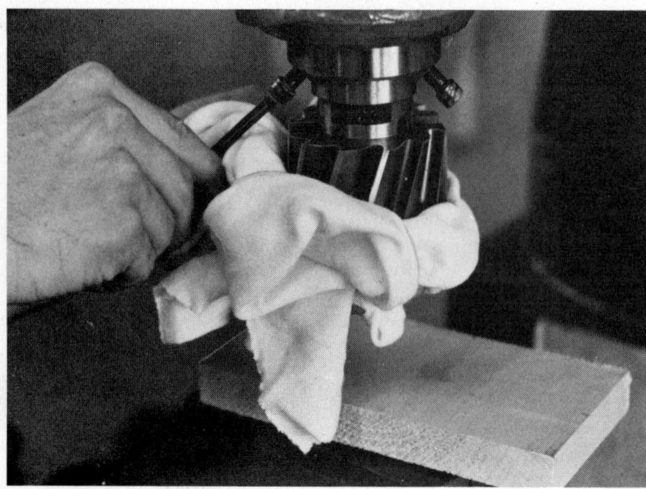

Fig. 12-75. When removing or replacing a cutter on a milling machine, protect your hand with a heavy cloth. On vertical milling machines, also place a section of wood under the cutter.

METHODS OF MILLING

Milling operations can be classified into one of two distinct methods:

1. With ***conventional*** or ***up-milling,*** the work is fed INTO the rotation of the cutter, Fig. 12-76. the chip is at minimum thickness at the start of the cut. The cut is so light that the cutter has a tendency to slide over the work until sufficient pressure is built up to cause the teeth to bite into the material. This alternate sliding to start, followed by the sudden breakthrough as the tooth completes the cut, leaves marks so familiar on many milled surfaces. The marks and ridges can be kept to a minimum by keeping the table gibs properly adjusted.
2. With ***climb*** or ***down-milling,*** the work moves in the same direction as cutter rotation, Fig. 12-77. Full engagement of the tooth is instantaneous. The sliding action of conventional milling is eliminated, resulting in a better finish and longer tool life.

The main advantage of climb milling is the tendency of the cutter to press the work down on the work table or holding device.

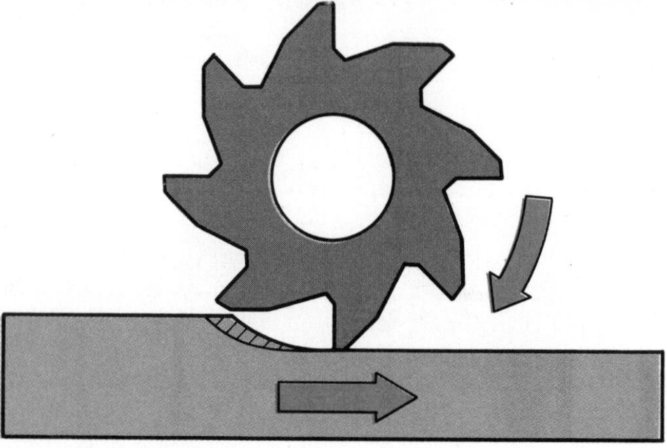

Fig. 12-76. Study movement with conventional milling.

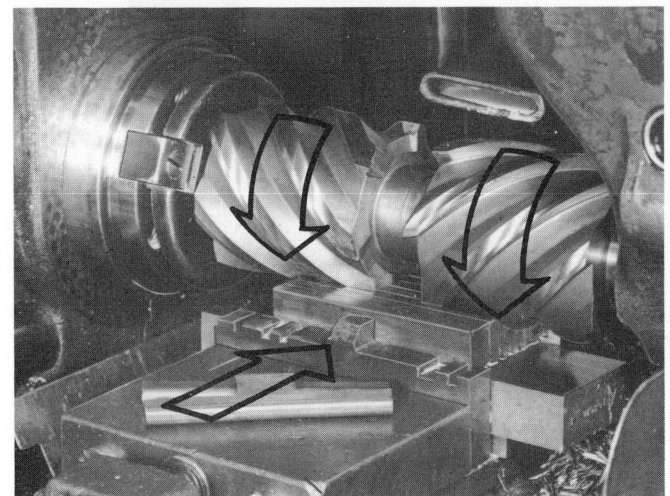

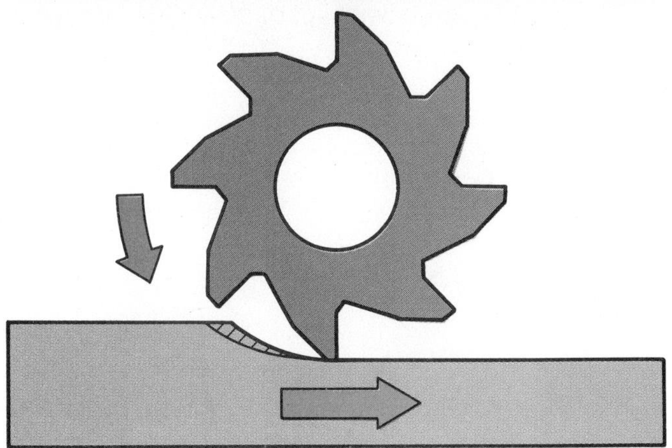

Fig. 12-77. Study movement with climb milling.

Climb milling is NOT recommended on LIGHT MACHINES nor on large OLDER MACHINES that are NOT in top condition, or fitted with an antiback-lash device to take up play. There is danger of a serious accident if there is play in the table, or if the work or holding device is NOT mounted securely.

HOLDING AND DRIVING CUTTERS

The ***arbor*** is the most common method employed to hold and drive cutters. It is made in a number of sizes and styles, Fig. 12-78. In addition to arbors with ***self-holding tapers*** for use on some small hand milling machines and older models of larger millers, there are three basic arbor styles in general use:

1. ***Style A*** is fitted with a small pilot end that runs in a bronze bearing in the arbor support, Fig. 12-79. This style is best used when maximum arbor support clearance is required.
2. ***Style B*** is characterized by the large bearing collar that can be positioned on any part of the arbor, Fig. 12-80. This feature makes it possible to mount the bearing support as close to the cutter as possible for maximum cutter support. This permits heavy cuts.
3. ***Style C*** is used to hold smaller sizes of shell end and face milling cutters that cannot be mounted directly to the spindle nose. See Fig. 12-81.

In general, use the SHORTEST ARBOR possible which permits adequate clearance between the arbor support and the work.

Style A and style B arbors have a keyway milled their entire length permitting a key to be employed to prevent the cutter from revolving on the arbor. See Fig. 12-82.

Spacing collars allow the cutter to be positioned on the arbor, Fig. 12-83. Accurately made in a number of widths, collars permit two or more cutters to be precisely spaced for ***gang*** and ***saddle milling.*** A left-hand threaded nut tightens the cutter and collars on the arbor. The nut should NOT be tightened directly against the bearing collar on the style B arbor because the bearing may be damaged.

Fig. 12-78. This arbor has self-holding taper.

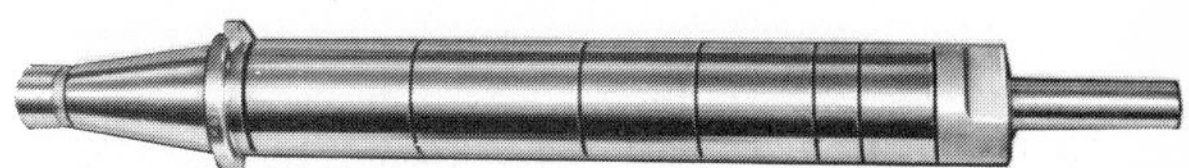

Fig. 12-79. Note style A arbor.

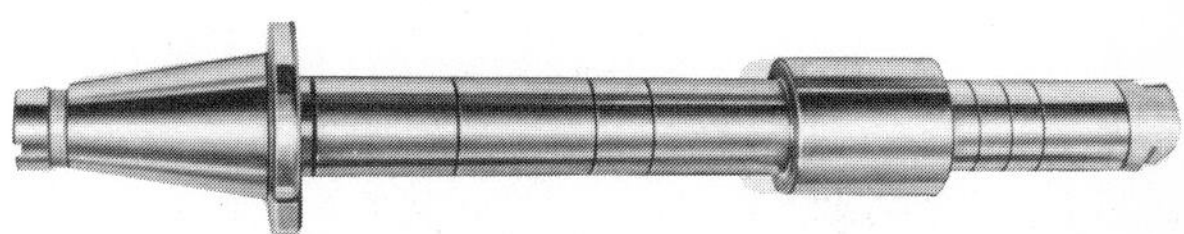

Fig. 12-80. Note style B arbor.

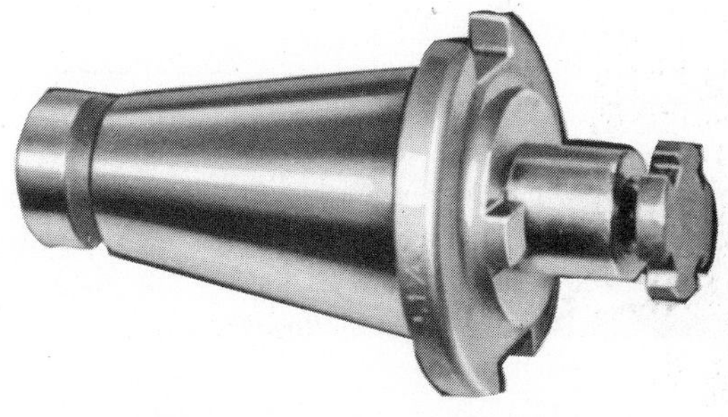

Fig. 12-81. Note style C arbor.

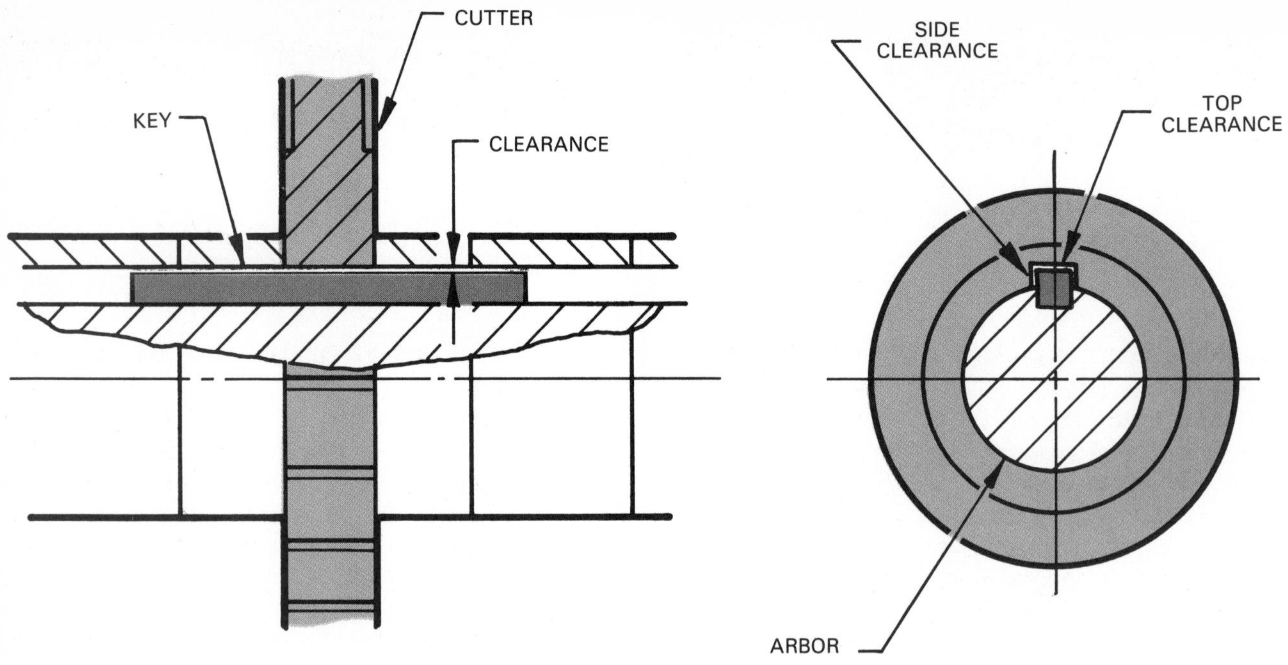

Fig. 12-82. Key cutter to arbor to prevent it from slipping during cutting operation.

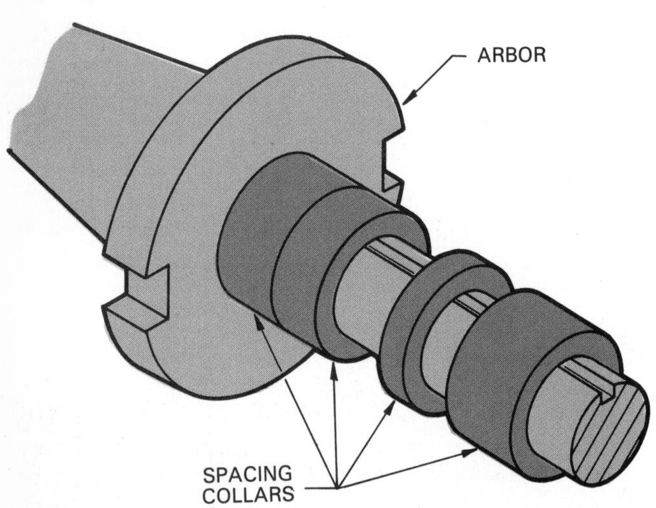

Fig. 12-83. Spacing collars are manufactured in many different widths and will position one or more cutters.

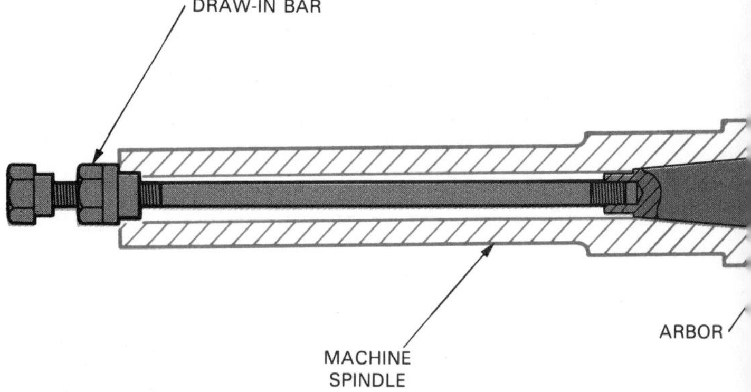

Fig. 12-84. Study draw-in bar action. Avoid operating a milling machine if arbor is not held in place with a draw-in bar.

A ***draw-in bar*** is used on most vertical and horizontal milling machines, Fig. 12-84. It fits through the spindle and screws into the arbor or collet to hold it firmly on the spindle.

Drive keys, on the nose of the spindle, fit into corresponding slots on the arbor, collet, or collet holder to provide positive (nonslip) drive, Fig. 12-85.

End mills may be mounted in ***spring collets, adapters, shell end mill holders,*** and ***stub arbors*** depending upon the type of work to be done.

Spring collets accommodate straight shank end mills and drills, Fig. 12-86. Some collets must be

Fig. 12-85. Clearly shown are spindle drive keys, style A arbor support, and style B arbor support. (Cincinnati Lathe & Tool Co.)

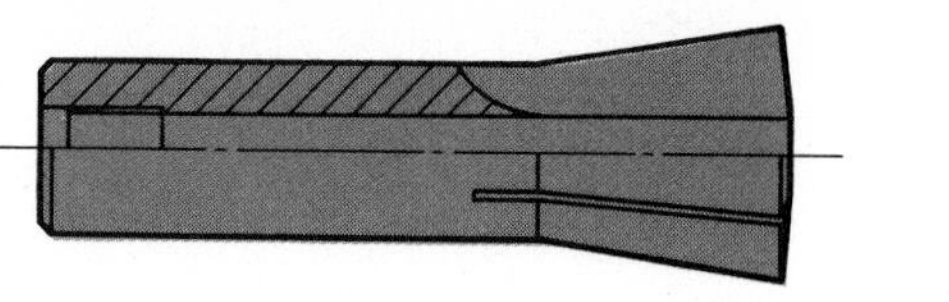

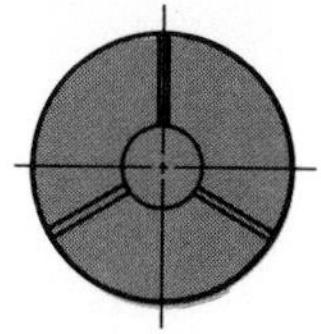

Fig. 12-86. This is a spring collet (R-8 taper type).

ARBOR NUT

STUB ARBOR

SPACERS

Fig. 12-90. Note stub arbor with an R-8 taper.

fitted in a ***collet chuck,*** Fig. 12-87. ***Adapters*** are used for taper shank end mills and drills, Fig. 12-88.

Shell end mill holders enable shell end mills, for face and side milling, to be fitted to a vertical milling machine, Fig. 12-89.

Stub arbors are short arbors that permit various side cutters, slitting saws, formed cutters, and angle cutters to be used on a vertical mill, Fig. 12-90.

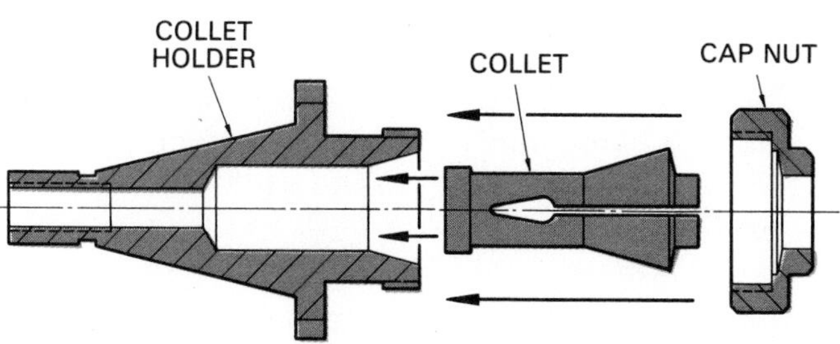

Fig. 12-87. Note collet chuck and collet.

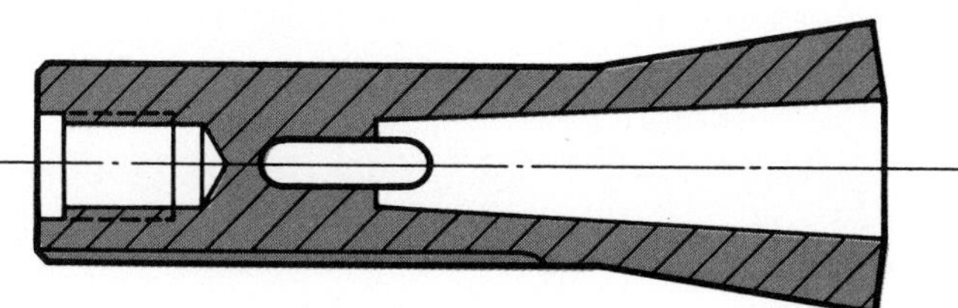

Fig. 12-88. This adapter is used with taper shank cutting tools.

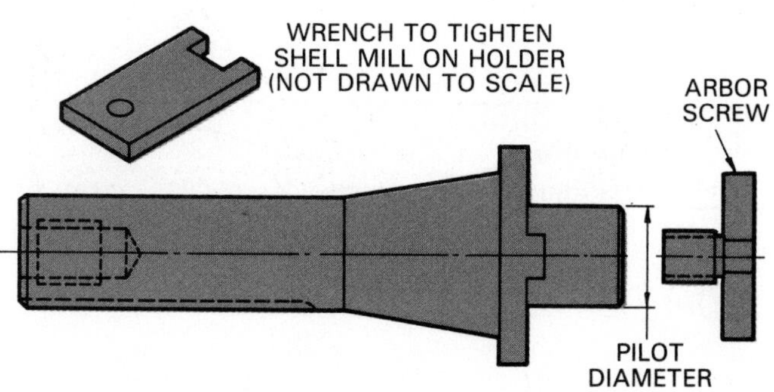

Fig. 12-89. Study this shell end mill holder (R-8 taper type).

Care of cutter holding and driving devices

To maintain precision and accuracy during a milling operation, care must be taken to prevent damage to cutter holding and driving devices.

1. Keep the taper end of the arbor clean and free of nicks. Same applies to spindle taper.
2. Clean and lubricate the bearing sleeve before placing the arbor support on it. Also make sure the bearing sleeve fits snugly.
3. Clean the spacing collars before slipping them on an arbor. Otherwise, cutter run-out will occur making it difficult to make an accurate cut. Refer to Fig. 12-91.
4. Store arbors separately and in a vertical position.
5. Never loosen or tighten an arbor nut unless the arbor support is locked in place, Fig. 12-92.
6. Use a wrench of the correct size, Fig. 12-93. Make sure at least FOUR THREADS are engaged before tightening the arbor nut.

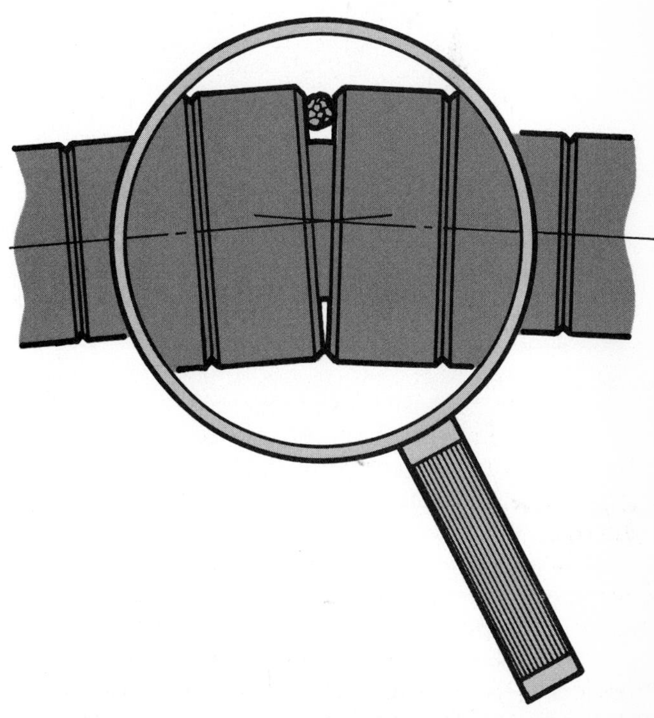

Fig. 12-91. A chip between spacing collars will cause arbor to be sprung out of true. Bend in arbor is exaggerated in this drawing.

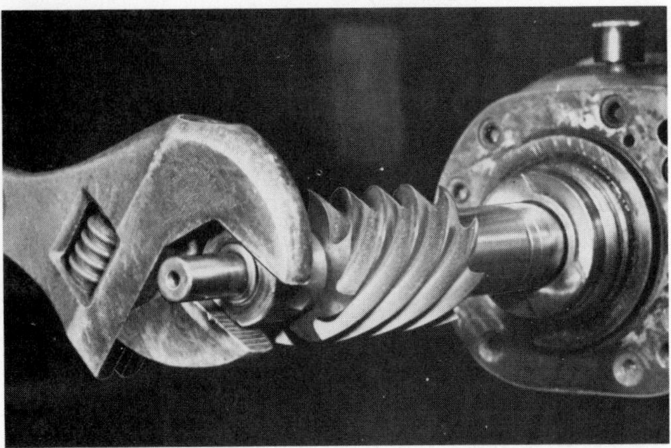

Fig. 12-92. Do not tighten or loosen an arbor nut without arbor support in place because this could spring arbor so that it will not run true.

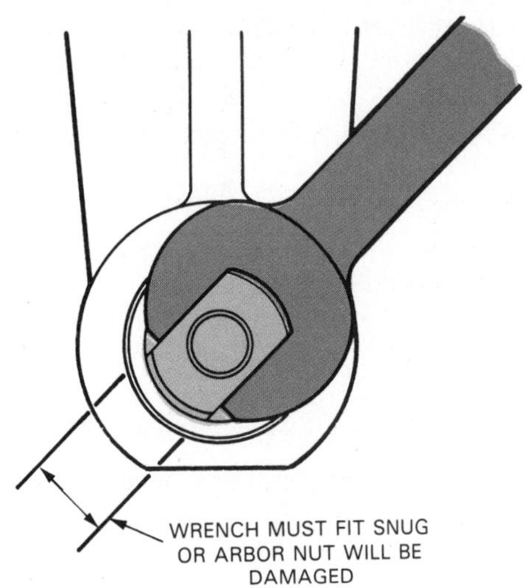
WRENCH MUST FIT SNUG OR ARBOR NUT WILL BE DAMAGED

Fig. 12-93. Use a wrench of correct type and size to loosen an arbor nut.

7. Cutters should NOT be forced on an arbor. Check to see what is making them difficult to slide on and correct any problem.
8. Key all cutters to the arbor.
9. Avoid tightening an arbor nut by striking the wrench with a hammer or mallet. This can crack the nut and/or distort the threads.
10. To remove an arbor or adapter from the machine:
    a. Loosen the draw-in bar nut a *few* turns. Do NOT remove it from the arbor completely.
    b. Tap the draw-in bar with a lead hammer to loosen the arbor in the spindle.
    c. Hold the loosened arbor with one hand and unscrew the draw-in bar with the other.
    d. Remove the arbor from the spindle. Clean and store it properly.

MILLING CUTTING SPEEDS AND FEEDS

The time required to complete a milling operation and the quality of the finish is almost completely governed by the cutting speed and feed of the cutter.

Milling cutting speed refers to the distance, measured in feet or meters, a point (tooth) on the cutter circumference moves in one minute. It is expressed in feet per minute (FPM) or meters per minute (MPM). Milling cutting speed is directly dependent on the revolutions per minute (RPM) of the cutter.

Milling feed is the rate at which work moves into the cutter. It is given in feed per tooth per revolution (FTR). Proper feed selection is probably the most difficult setting for a machinist to determine. In view of the many variables (width of cut, depth of cut, machine condition, cutter sharpness, etc.), feed should be as coarse as possible and consistent with the desired finish.

Calculating cutting speeds and feeds

Considering the previously mentioned variables, the speeds listed in Fig. 12-94, and feeds listed in Fig. 12-95, are suggested. The usual procedure is to start with the midrange figure and increase or decrease speeds until the most satisfactory combination, consistent with cutter life and surface quality, is obtained.

In general, speed is REDUCED for hard or abrasive materials, deep cuts, or high alloy content metals. INCREASE speed for soft materials, better finishes, and light cuts.

Refer to the ***rules for determining speeds and feeds*** to calculate the cutting speed and feed for a specific material. See Fig. 12-96.

EXAMPLE PROBLEM: Determine the approximate cutting speed and feed for a 6 in. diameter side cutter (HSS) with 16 teeth, when milling free cutting steel.

INFORMATION AVAILABLE:

Recommended cutting speed for free cutting steel (midpoint in range) . . . . 75 FPM
Recommended feed per tooth (midpoint on range) . . . . 0.008 in.
Cutter diameter . . . . 6 in.
Number of teeth on cutter 16

To determine SPEED SETTING (cutter RPM), the following is given in Fig. 12-96.

RULE: Divide the feet per minute (FPM) by the circumference of the cutter, expressed in feet.

$$\text{FORMULA: RPM} = \frac{\text{FPM} \times 12}{\pi D}$$

$$= \frac{75 \times 12}{3.14 \times 6} = \frac{900}{18.84}$$

$$= 48 \text{ RPM}^*$$

MATERIAL	HIGH SPEED STEEL CUTTER FEET PER MINUTE	METERS PER MINUTE*	CARBIDE CUTTER FEET PER MINUTE	METERS PER MINUTE*
Aluminum	550-1000	170-300	2200-4000	670-1200
Brass	250- 650	75-200	1000-2600	300- 800
Low Carbon Steel	100- 325	30-100	400-1300	120- 400
Free Cutting Steel	150- 250	45- 75	600-1000	180- 300
Alloy Steel	70- 175	20- 50	280- 700	85- 210
Cast Iron	45- 60	15- 20	180- 240	55- 75

Reduce speeds for hard materials, abrasive materials, deep cuts, and high alloy materials. Increase speeds for soft materials, better finishes, light cuts, frail work, and setups. Start at midpoint on the range and increase or decrease speed until best results are obtained.
*Figures rounded off.

Fig. 12-94. Study recommended cutting speeds for milling. Speed is given in surface feet per minute (FPM) and in surface meters per minute (MPM).

TYPE OF CUTTER	MATERIAL ALUMINUM	BRASS	CAST IRON	FREE CUTTING STEEL	ALLOY STEEL
End mill	0.009 (0.22) 0.022 (0.55)	0.007 (0.18) 0.015 (0.38)	0.004 (0.10) 0.009 (0.22)	0.005 (0.13) 0.010 (0.25)	0.003 (0.08) 0.007 (0.18)
Face mill	0.016 (0.40) 0.040 (1.02)	0.012 (0.30) 0.030 (0.75)	0.007 (0.18) 0.018 (0.45)	0.008 (0.20) 0.020 (0.50)	0.005 (0.13) 0.012 (0.30)
Shell end mill	0.012 (0.30) 0.030 (0.75)	0.010 (0.25) 0.022 (0.55)	0.005 (0.13) 0.013 (0.33)	0.007 (0.18) 0.015 (0.38)	0.004 (0.10) 0.009 (0.22)
Slab mill	0.008 (0.20) 0.017 (0.43)	0.006 (0.15) 0.012 (0.30)	0.003 (0.08) 0.007 (0.18)	0.004 (0.10) 0.008 (0.20)	0.001 (0.03) 0.004 (0.10)
Side cutter	0.010 (0.25) 0.020 (0.50)	0.008 (0.20) 0.016 (0.40)	0.004 (0.10) 0.010 (0.25)	0.005 (0.13) 0.011 (0.28)	0.003 (0.08) 0.007 (0.18)
Saw	0.006 (0.15) 0.010 (0.25)	0.004 (0.10) 0.007 (0.18)	0.001 (0.03) 0.003 (0.08)	0.003 (0.08) 0.005 (0.13)	0.001 (0.03) 0.003 (0.08)

Increase or decrease feed until the desired surface finish is obtained.
Feeds may be increased 100 percent or more depending upon the rigidity of the machine and the power available, if carbide tipped cutters are used.

Fig. 12-95. Chart gives recommended feet in inches per tooth and millimeters per tooth for high speed steel (HSS) milling cutters.

To determine FEED SETTING (feed in inches per minute or F)

RULE: Multiply feed per tooth per revolution by number of teeth on cutter and by speed (RPM).

FORMULA: $F = FTR \times T \times RPM$

$= 0.008 \times 16 \times 48$

$= 6.1$*

*The speed and feed are only approximate. Set machine to closest setting, either higher or lower. Increase or reduce speed until satisfactory cutting conditions are achieved.

NOTE: When metric information is given, refer to the appropriate feed and speed charts, and rules for determining speed and feed, Figs. 12-94 through 12-96.

CUTTING FLUIDS

Cutting fluids serve several purposes:

1. They carry away the heat generated during the machining operation.
2. They act as a lubricant.
3. They prevent the chips from sticking or fusing to the cutter teeth.
4. They flush away chips.
5. They influence the finish quality of the machined surface.

As this list shows, it is important that the correct cutting fluid be used for the material being machined. See Fig. 12-97.

MILLING WORK HOLDING ATTACHMENTS

One of the more important features of the milling machine is its adaptability to a large number of work-holding attachments. Each of these attachments increases the usefulness of the milling machine.

Vises

The ***vise*** is probably the most widely employed method of holding work for milling. The jaws are

RULES FOR DETERMINING SPEED AND FEED			
TO FIND	**HAVING**	**RULE**	**FORMULA**
Speed of cutter in feet per minute (FPM)	Diameter of cutter and revolutions per minute	Diameter of cutter (in inches) multiplied by 3.1416 (π) multiplied by revolutions per minute, divided by 12	$FPM = \frac{\pi D \times RPM}{12}$
Speed of cutter in meters per minute	Diameter of cutter and revolutions per minute	Diameter of cutter multiplied by 3.1416 (π) multiplied by revolutions per minute, divided by 1000	$MPM = \frac{D(mm) \times \pi \times RPM}{1000}$
Revolutions per minute (RPM)	Feet per minute and diameter of cutter	Feet per minute, multiplied by 12, divided by circumference of cutter (πD)	$RPM = \frac{FPM \times 12}{\pi D}$
Revolutions per minute (RPM)	Meters per minute and diameter of cutter in millimeters (mm)	Meters per minute multipled by 1000, divided by the circumference of cutter (πD)	$RPM = \frac{MPM \times 1000}{\pi D}$
Feed per revolution (FR)	Feed per minute and revolutions per minute	Feed per minute, divided by revolutions per minute	$FR = \frac{F}{RPM}$
Feed per tooth per revolution (FTR)	Feed per minute and number of teeth in cutter	Feed per minute (in inches or millimeters) divided by number of teeth in cutter × revolutions per minute	$FTR = \frac{F}{T \times RPM}$
Feed per minute (F)	Feed per tooth per revolution, number of teeth in cutter, and RPM	Feed per tooth per revolutions multiplied by number of teeth in cutter, multiplied by revolutions per minute	$F = FTR \times T \times RPM$
Feed per minute (F)	Feed per revolution and revolutions per minute	Feed per revolution multiplied by revolutions per minute	$F = FR \times RPM$
Number of teeth per minute (TM)	Number of teeth in cutter and revolutions per minute	Number of teeth in cutter multiplied by revolutions per minute	$TM = T \times RPM$

RPM = Revolutions per minute
T = Teeth in cutter
D = Diameter of cutter
π = 3.1416 (pi)
FRM = Speed of cutter in feet per minute
TM = Teeth per minute
F = Feed per minute
FR = Feed per revolution
FTR = Feed per tooth per revolution
MPM = Speed of cutter in meters per minute

Fig. 12-96. Study rules for determining speed and feed.

Aluminum and its Alloys	Kerosene, kerosene and lard oil, soluble oil
Plastics	Dry
Brass, Soft	Dry, soluble oil, kerosene and lard oil
Bronze, High Tensile	Soluble oil, lard oil, mineral oil, dry
Cast Iron	Dry, air jet, soluble oil
Copper	Soluble oil, dry, mineral lard oil, kerosene
Magnesium	Low viscosity neutral oils
Malleable Iron	Dry, soda water
Monel Metal	Lard oil, soluble oil
Slate	Dry
Steel, Forging	Soluble oil, sulphurized oil, mineral lard oil
Steel, Manganese	Soluble oil, sulphurized oil, mineral lard oil
Steel, Soft	Soluble oil, mineral lard oil, sulphurized oil, lard oil
Steel, Stainless	Sulphurized mineral oil, soluble oil
Steel, Tool	Soluble oil, mineral lard oil, sulphurized oil
Wrought Iron	Soluble oil, mineral lard oil, sulphurized oil

Fig. 12-97. Read through recommended cutting fluids for various materials.

hardened to resist wear and ground for accuracy. A milling vise, like other work-holding attachments, is keyed to the table slot with ***lugs,*** Fig. 12-98.

A ***flanged vise*** has slotted flanges for fastening the vise to the table, Fig. 12-99. The slots permit the vise to be mounted parallel to, or at right angles to, the spindle on a horizontal milling machine.

The body of a ***swivel vise*** is similar to a flange vise but is fitted with a circular base, graduated in degrees. This permits it to be locked at any angle to the spindle. See Figs. 12-100 and 12-101.

The ***toolmaker's universal vise*** permits compound or double angles to be machined without complex or multiple setups. Refer to Figs. 12-102 and 12-103.

A ***magnetic chuck,*** shown in Fig. 12-104, is ideally suited for many milling operations. The magnet eliminates the need for time consuming hold-down clamps to mount the work to the table. The magnetic chuck can only be used with ferrous metals that can be held magnetically.

A ***rotary table*** can perform a variety of operations, such as: cutting segments of circles, circular slots, cutting irregular shaped slots, and similar opera-

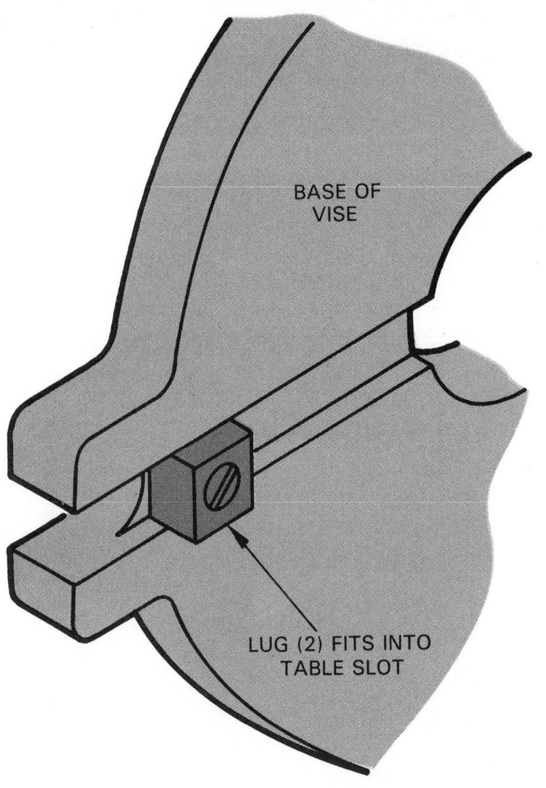

Fig. 12-98. Lugs on base of work holding attachments position device on worktable.

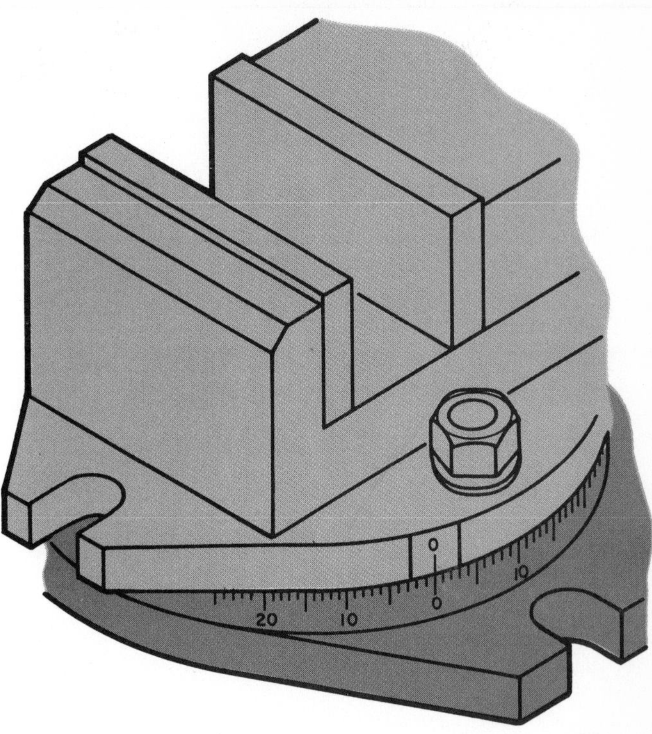

Fig. 12-101. Base of a swivel vise is graduated in degrees. It can be positioned and locked at angle to the machine spindle.

Fig. 12-99. This is a typical flanged vise.

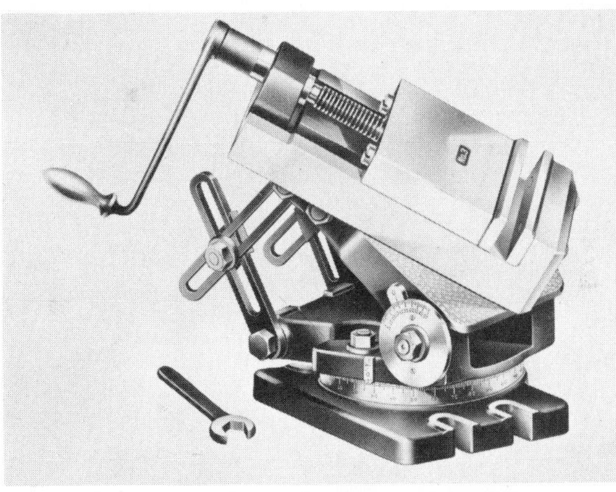

Fig. 12-102. Note construction of universal vise. (Brown & Sharpe Mfg. Co.)

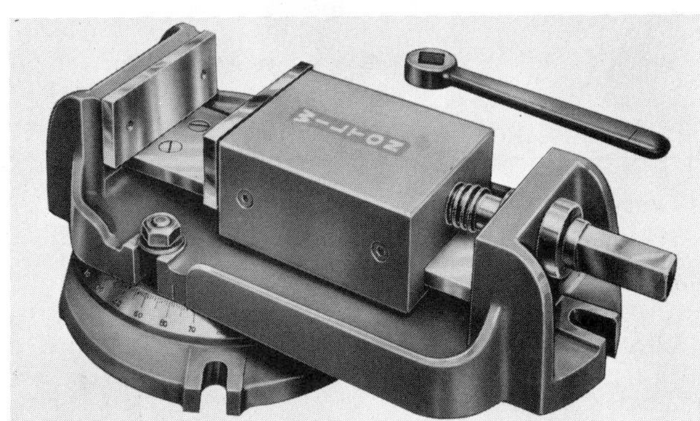

Fig. 12-100. Swivel vise is handy and will rotate to align work easily. (Wilton Tool Mfg. Co.)

tions. A dividing attachment can be fitted to many rotary tables in place of the handwheel, Figs. 12-105 and 12-106.

The table is graduated in degrees around its circumference and adjustments can be made accurately with the handwheel to 1/30 of a degree (2 minutes).

An ***index table*** permits the rapid positioning of work, Fig. 12-107. Indexing is usually in 15 deg. increments. However, a clamping device allows the table to be locked at any setting.

A ***dividing head*** will divide the circumference of circular work into equally spaced divisions. It is one

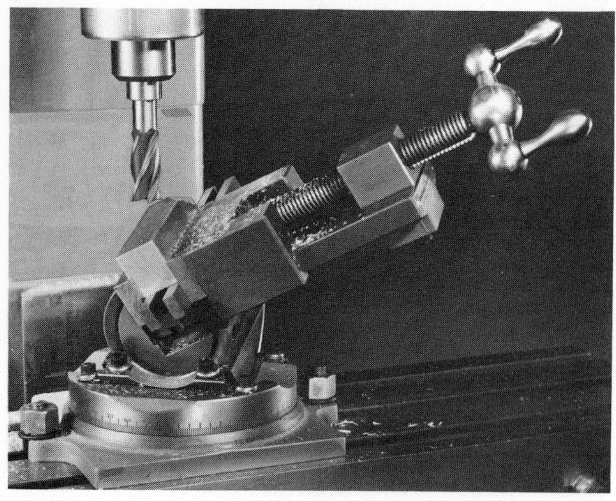

Fig. 12-103. Universal vise is being used on light vertical milling machine to cut a compound angle (double angle). Note how vise is tilted.

Fig. 12-104. This milling machine has a magnetic vise attached. (O.S.Walker Co., Inc.)

Fig. 12-105. Study construction of rotary table. (Troyke Mfg. Co.)

Fig. 12-106. Rotary table is being used to machine part with round and curved shapes.

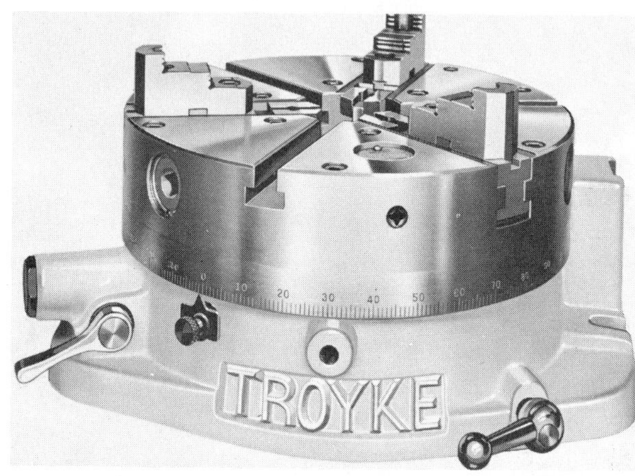

Fig. 12-107. This indexing table is like rotary table with 3-jaw lathe chuck attached.

of the more important milling attachments, Fig. 12-108. This feature makes a dividing head indispensable when milling gear teeth, cutting splines, and spacing holes on a circle. It also makes possible the milling of squares, hexagons, etc., when required.

The dividing head consists of two parts, the dividing unit and the footstock. Work may be mounted between centers, in a chuck, or in a collet, Figs. 12-109 through Fig. 12-111.

An ***index plate*** identified by circles of holes on its face, and the ***index crank,*** which revolves on the index plate, are fundamental in the dividing operation. See Fig. 12-112.

Rotating the index crank causes the dividing head spindle (to which work is mounted) to rotate. The standard ratio for the dividing head is five turns of the index crank for one complete revolution of the spindle (5:1); or 40 turns of the index crank for one revolution of the spindle (40:1).

The ratio between index crank turns and spindle revolution, plus the index plate with its series of

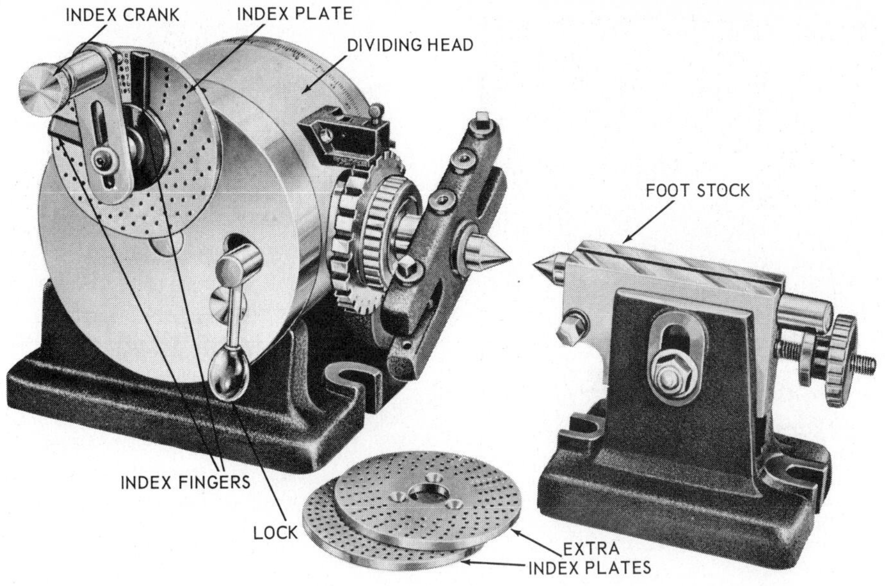

Fig. 12-108. Study parts of dividing head and foot stock. (L.W. Chuck Co.)

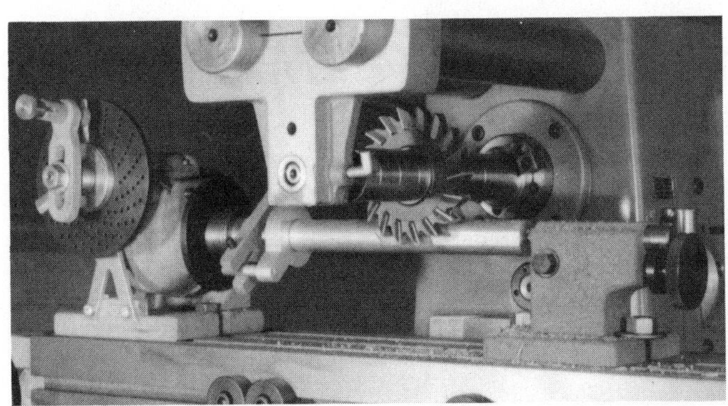

Fig. 12-109. Work has been mounted between centers for a milling operation that requires use of dividing head.

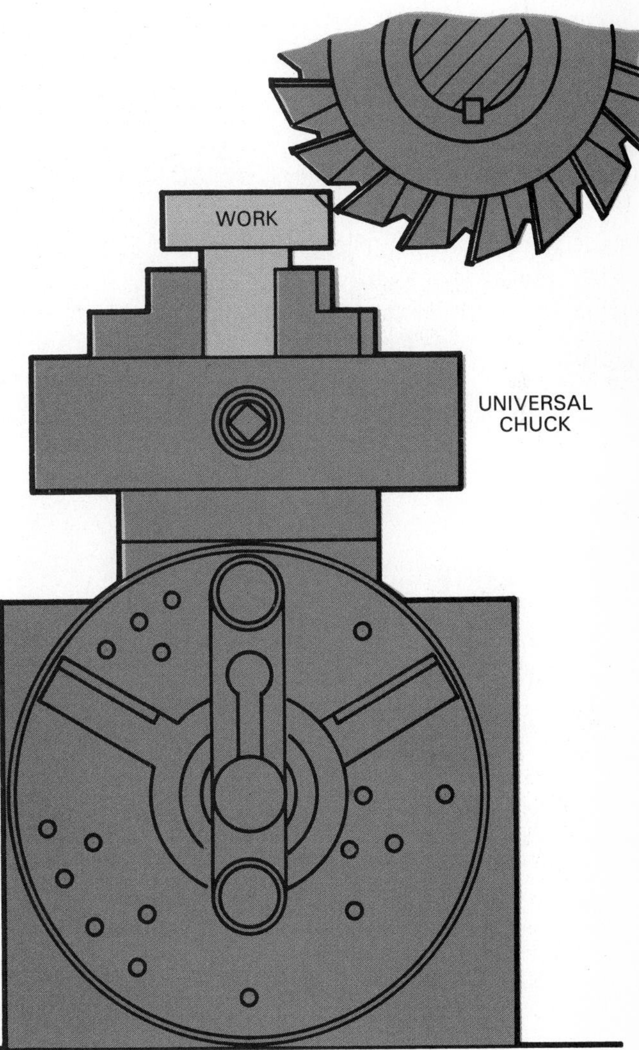

Fig. 12-110. A 3-jaw universal chuck mounted on a dividing head permits short work to be milled.

Fig. 12-111. A cam is being milled while held in a collet. (Cincinnati Lathe & Tool Co.)

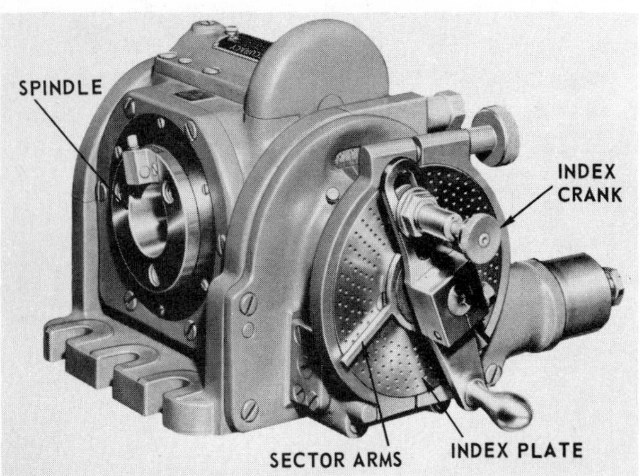

Fig. 12-112. Note index plate, sector arms, and index crank. (Kearney & Trecker Corp.)

equally spaced hole circles, makes it possible to divide the circumference of the work into the required number of equal spaces. Refer to Fig. 12-113.

For example: If 10 teeth were to be cut in a gear, it would require 1/10 of 5 turns (assuming dividing head has a 5:1 ratio), or one-half turn of the index crank for each tooth.

For 25 teeth, the number of crank turns would be 1/25 of 5, or dividing 5 by 25, or 5/25. By further reduction, this becomes 1/5 of a turn of the crank for each tooth. This is where the holes in the index plate come into use, they allow fractional turns to be made accurately.

Select an index plate with a series of holes divisible by 5. On one such plate, the circles have 47, 49, 51, 53, 54, 57, and 60 holes. In this situation,

Fig. 12-113. Internal mechanism of a dividing head use precision gears. Forty turns of index crank rotates spindle (work) one complete revolution. (Kearney & Trecker Corp.)

60 is divisible by 5; then 1/5 of 60 or 12 holes on the 60 hole circle. Indexing would be through 12 holes on the 60 hole circle.

When indexing, it is NOT necessary to count 12 holes each time the work is repositioned after a tooth has been cut. The two arms, called ***sector arms*** or ***index fingers,*** are loosened. One is positioned until it touches the pin on the index crank. The other is moved clockwise until the arms are 12 holes apart. Remember—do NOT count the hole the pin is in, Fig. 12-114.

To index, move the work clear of the cutter. Disengage the crank by withdrawing the pin from the index plate and rotate it through the section taken up by the sector arms. Drop the pin into the hole at this position and lock the dividing head mechanism. Move the sector arms in the same direction as crank rotation to catch up with the pin in the index crank. Repeat the operation for each cut. Crank rotation is clockwise, Fig. 12-115.

The dividing head spindle swivels 5 deg. below horizontal, and 5 deg. beyond perpendicular. Refer to Fig. 12-116.

A precision device, a dividing head has an indexing accuracy of about one minute of arc. This is the equivalent of 1/21,600 part of a circle.

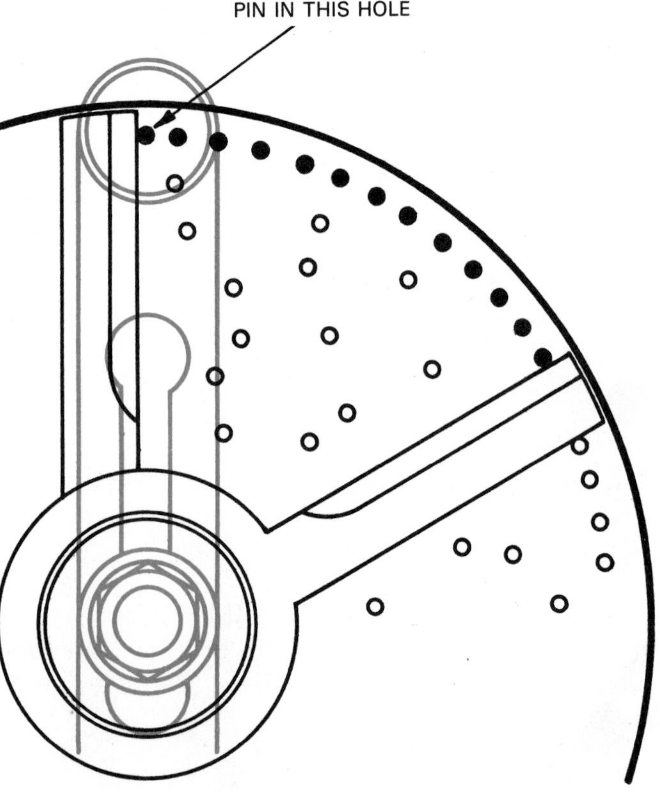

Fig. 12-114. Position sector arms for proper movement. When positioning sector arms, DO NOT COUNT the hole pin on index crank is in.

Fig. 12-115. When cutting gears on milling machine, many machinists make a light cut at each tooth position first. This will check whether dividing head has been set up properly.

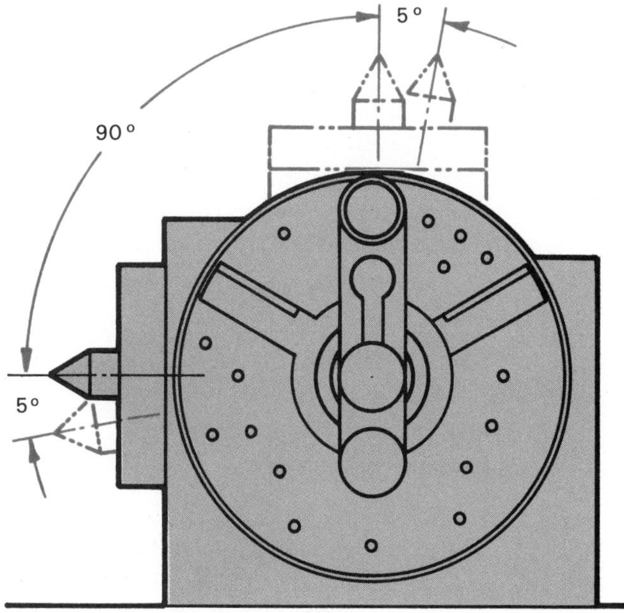

Fig. 12-116. Most dividing heads pivot through an arc of 100 deg.

MILLING OPERATIONS

The versatility of the milling machine family permits many different machining operations to be performed. Because of the number of varied operations that can be performed, it is not possible to cover all of them in a book of this type. Only basic operations will be described.

VERTICAL MILLING MACHINE

The vertical milling machine is capable of performing milling, drilling, boring, and reaming operations, Fig. 12-117. It differs from the horizontal mill in that the spindle is normally in a vertical position.

The ***spindle head*** swivels 90 deg. left or right for machining at any angle, Fig. 12-118. The ***ram,*** on which it is mounted, can be adjusted in and out. On many vertical mills, it revolves 180 deg. on a horizontal plane. Both swivels are graduated in degrees with a vernier scale to assure accurate angular settings, Fig. 12-119 and Fig. 12-120.

Cutters for vertical milling machine

Although adapters are available that permit the use of side and angle cutters, face mills and end mills are the cutters normally used, Fig. 12-121.

Taper shank end mills and drills are fitted in an adapter, Fig. 12-122A. Some machine spindles have a Brown & Sharpe taper. Taper shanks, that are large enough, are mounted directly, Fig. 12-122B. When tapers are too small to fit directly into the spindle, a sleeve must be employed. Straight shank end mills are held in a spring collet, Fig. 12-122C, or in an end mill adapter, Fig. 12-122D. Small drills, reamers, and similar tools are held in a standard Jacobs chuck fitted to the spindle by one of the above methods.

VERTICAL MILLING MACHINE OPERATIONS

In addition to the usual precautions that must be observed when getting a machine tool ready for a job, the spindle head alignment must be checked. Make sure that the spindle head is at right angles

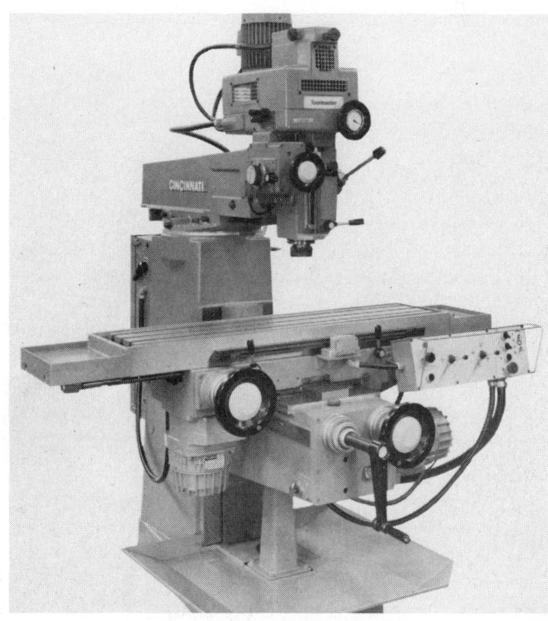

Fig. 12-117. Vertical milling machine is similar in operation to horizontal machine, except for spindle mounting. (Cincinnati Milacron)

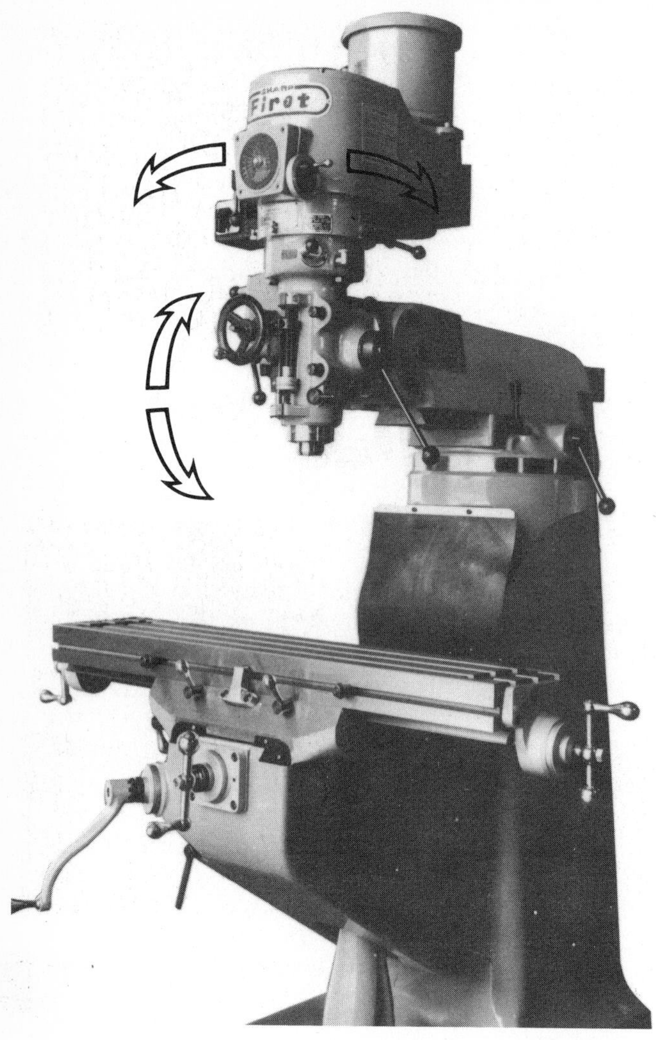

Fig. 12-118. Angular head adjustments are possible on many vertical milling machines. (Sharp Industries, Inc.)

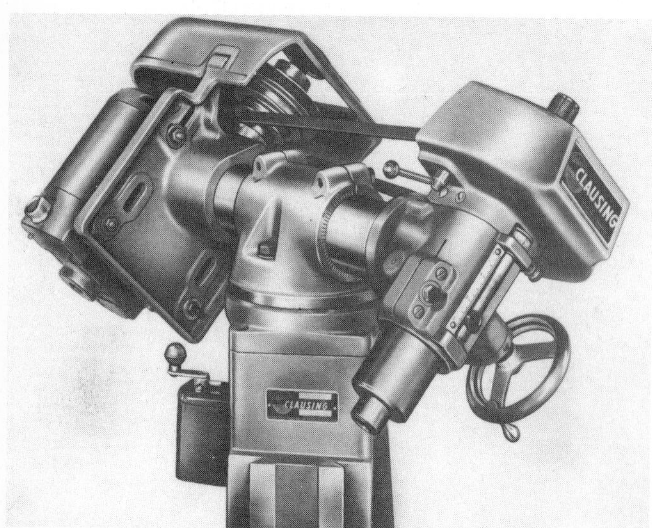

Fig. 12-119. Closeup shows how angular settings are possible with head of this light vertical milling machine. (Clausing Industries Corp.)

Fig. 12-120. Vernier scale on this machine permits spindle head to be set with extreme accuracy. (South Bend Lathe, Inc.)

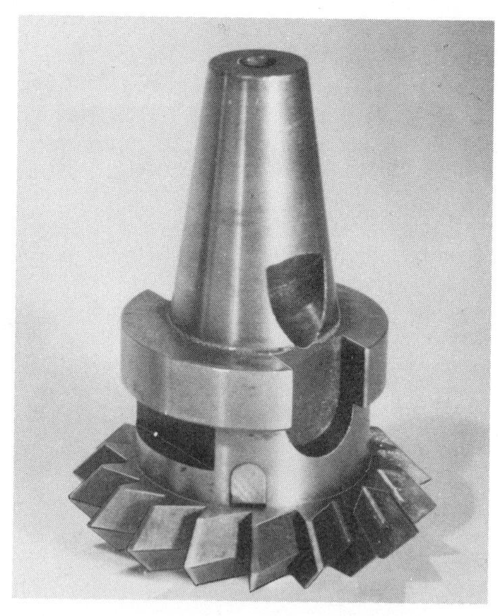

Fig. 12-121. Adapter permits arbor type cutters to be used on a vertical milling machine.

to the work table. Otherwise, it is NOT possible to machine a flat surface, Fig. 12-123.

If a vise is utilized to mount the work, wipe the vise base and work table clean. Inspect for burrs and nicks. They prevent the vise from seating properly on the table. Bolt it firmly to the machine.

The next step is to align the vise with a dial indicator if extreme accuracy is required, Fig. 12-124. However, for many jobs the vise can be aligned with a square, Fig. 12-125. Angular settings can be made with a protractor, Fig. 12-126, or by using the degree divisions on the base of a swivel vise.

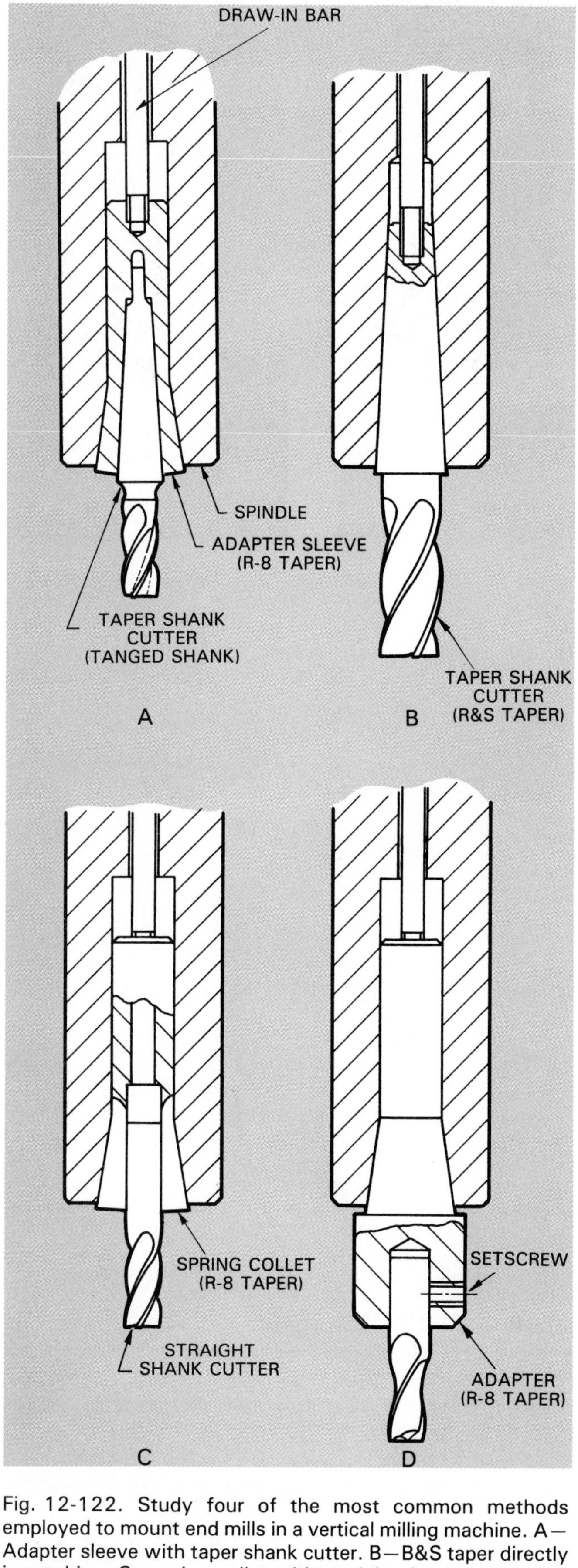

Fig. 12-122. Study four of the most common methods employed to mount end mills in a vertical milling machine. A—Adapter sleeve with taper shank cutter. B—B&S taper directly in machine. C—spring collet with straight shank cutter. D—Adapter with setscrew on straight shank cutter.

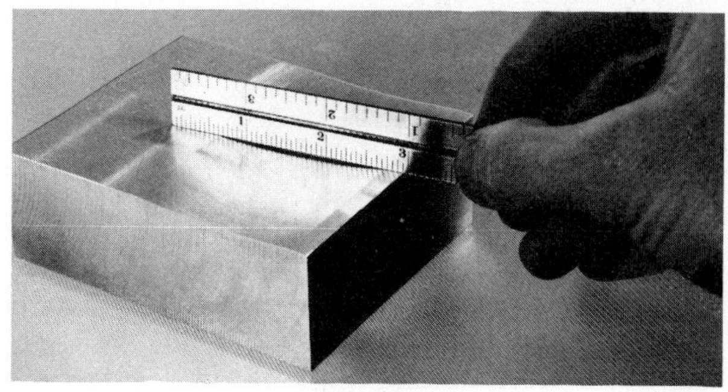

Fig. 12-123. Note irregular machined surface that results when spindle head is NOT at right angles with table.

Fig. 12-124. Use of a dial indicator permits extreme accuracy in aligning solid vise jaw.

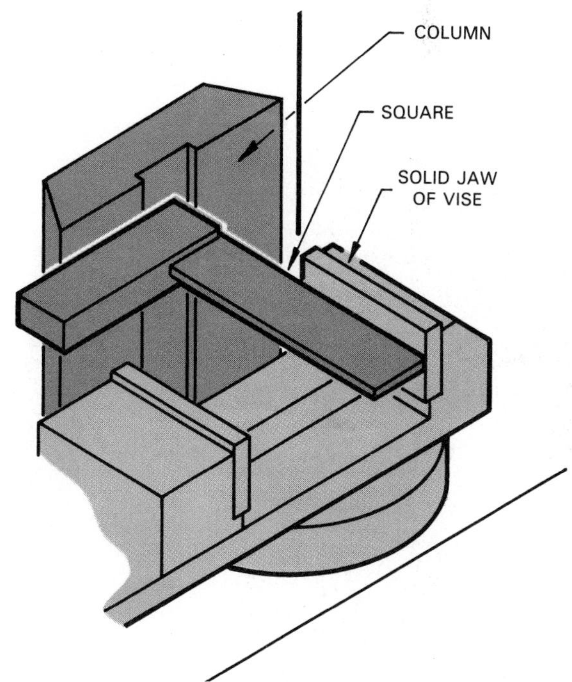

Fig. 12-125. Note procedure for squaring solid vise jaw with a steel square.

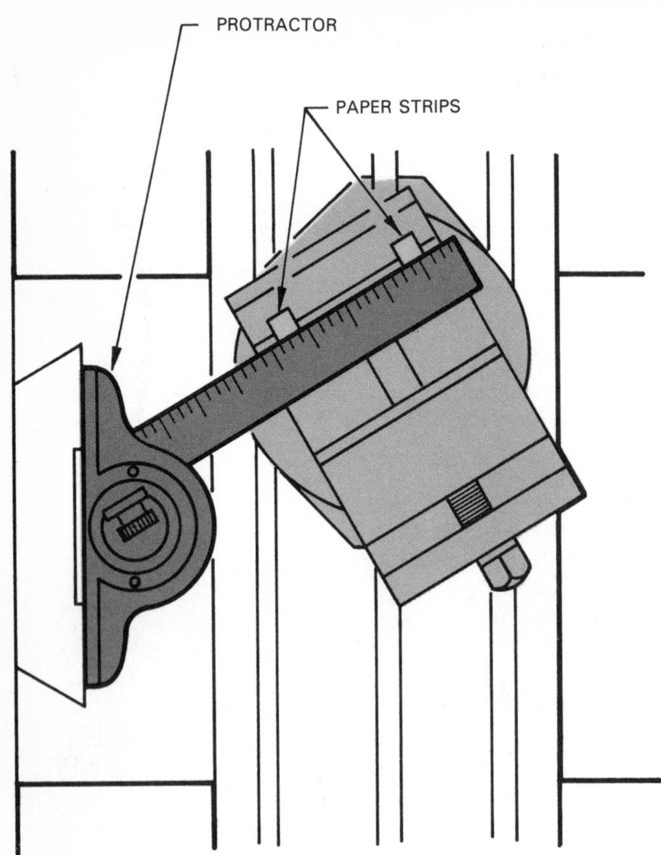

Fig. 12-126. Angular settings can be made with protractor head and steel rule of a combination set. Paper strips aid in determining whether setting is accurate.

Fig. 12-127. Machinist is setting work on parallels with brass hammer.

Wipe the vise jaws and bottom clean of chips and dirt. Place clean parallels in the vise and place the work on them.

Tighten the jaws and tap the work onto the parallels with a mallet or soft face hammer. Thin paper strips can be employed to check whether the work is firmly on the parallels, Fig. 12-127.

Never strike the vise handle with a hammer or mallet to put additional holding pressure on the jaws.

Protect the vise jaws and parallels with soft metal strips if the work is rough.

Squaring stock

A definite sequence must be followed to machine several surfaces or a piece square with one another.

1. Machine the first surface. Remove the burrs and place the first machined surface against the FIXED vise jaw. Insert a section of soft metal rod between the work and MOVABLE jaw if that portion of the work is rough or NOT square. See Fig. 12-128.
2. Machine the second surface.
3. Remove the burrs and reposition the work in the vise, C, Fig. 12-128. Then machine the third side. This side must be machined to dimension.

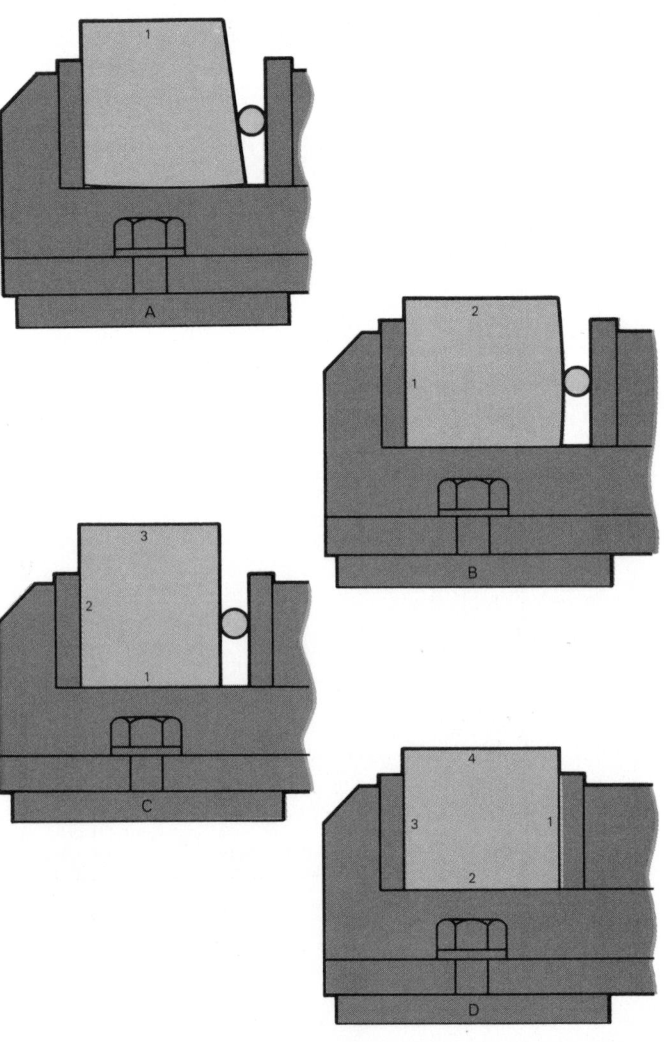

Fig. 12-128. Sequence for squaring work on milling machine. A—Square top. B—Square side. C-Square bottom. D—Square other side.

Take a light cut and "mike" for size. The difference between this measurement and the required thickness is the amount of material that must be removed.

DANGER! Stop the machine before attempting to make measurements.

4. Repeat the above operation to machine the fourth side.
5. If the piece is short enough, the ends may be machined by placing it in a vertical position with the aid of a square, Fig. 12-129. Otherwise, it may be machined as shown in Fig. 12-130.

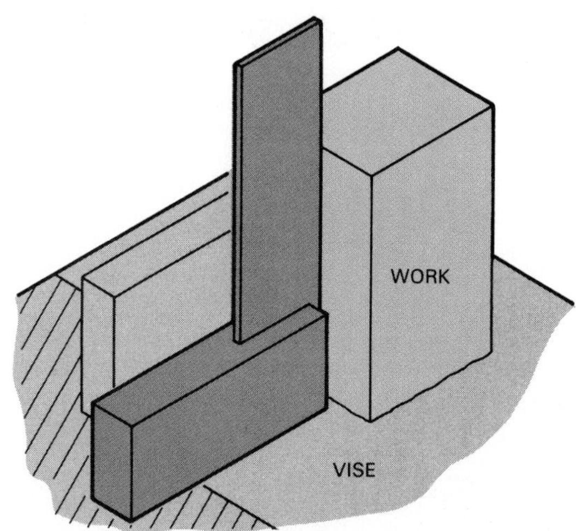

Fig. 12-129. Use a square to position short pieces for machining ends. Movable jaw is NOT shown for clarity.

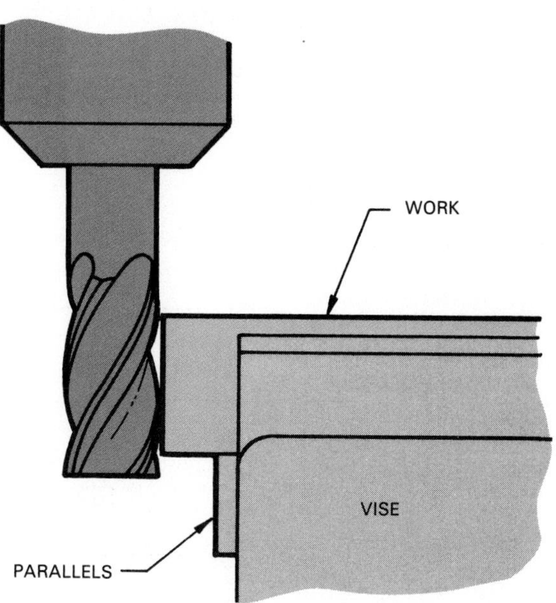

Fig. 12-130. Here is another technique for squaring ends of work. Jaw must be dial indicated to be sure it is at right angles to column.

Machining angular surfaces

Angular surfaces (bevels, chamfers, and tapers) may be milled by tilting the spindle head assembly to the required angle, Fig. 12-131. They may also be made by setting the work at the specified angle in a vise, Fig. 12-132.

A ***fixture*** is often used when many similar pieces must be milled, Fig. 12-133. ***Compound angles*** (angles on two planes) are made in a ***universal vise,*** Fig. 12-134.

When the pivoted spindle head is employed for angular cutting, it is essential that the vise be aligned with the dial indicator. Make a layout of the

Fig. 12-131. Angular surface is being cut with spindle head set to required angle.

Fig. 12-132. Angular surface may also be cut by positioning work at desired angle.

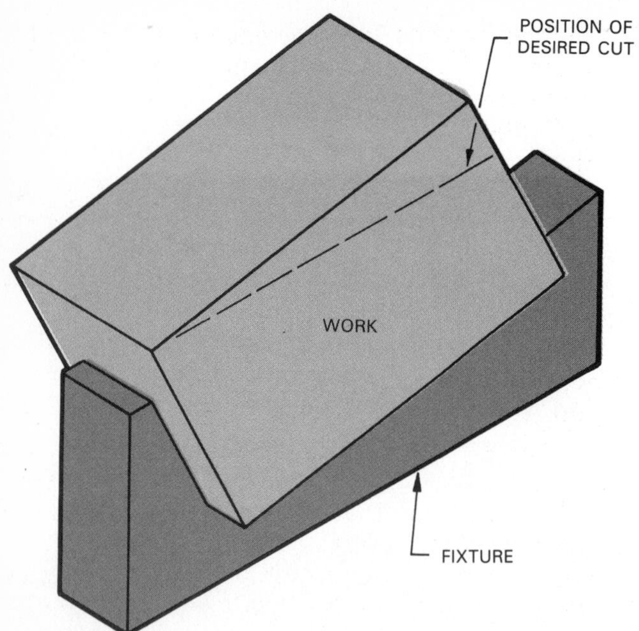

Fig. 12-133. If many pieces are to have angular surfaces machined, much time can be saved by utilizing a fixture.

Fig. 12-135. Work can be set at desired angle with aid of a protractor head.

Fig. 12-134. Angular cut is being made on work held in a universal vise. With this setup, compound angles or angles on two planes are easily cut.

Fig. 12-136. Machinist is positioning angular work with a surface gage.

desired angle on the work and clamp it in the vise. Position the cutter and machine to the line.

Work mounted at an angle in the vise for machining must be set up carefully. Alignment may be made with a protractor head fitted with a spirit level, Fig. 12-135, or with a surface gage, Fig. 12-136.

Milling a keyseat or slot

An end mill may be used to cut a keyseat or slot. After aligning the vise with a dial indicator, the work is clamped in the vise, or to the machine table. If mounted directly to the table, a section of paper between the table and the work will seat the work more solidly and prevent slippage.

A sharp cutter, equal in diameter to the keyseat of slot, is selected. A two-flute end mill is employed when a blind keyseat or slot is to be machined. Otherwise, a multi-flute end mill is used. See Figs. 12-137 and 12-138.

Locating end mill to cut keyseat or slot on round work

After the milling machine has been set up, cutter mounted, and the work secured in a vise, between centers, in V-blocks, or in a fixture, precise

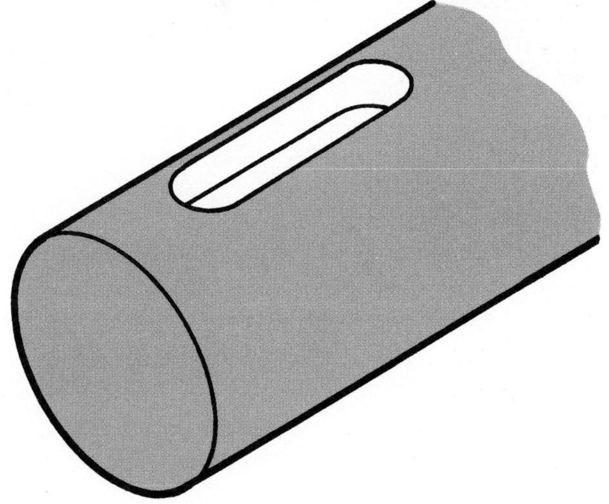

Fig. 12-137. Blind slot or keyseat has been cut in shaft.

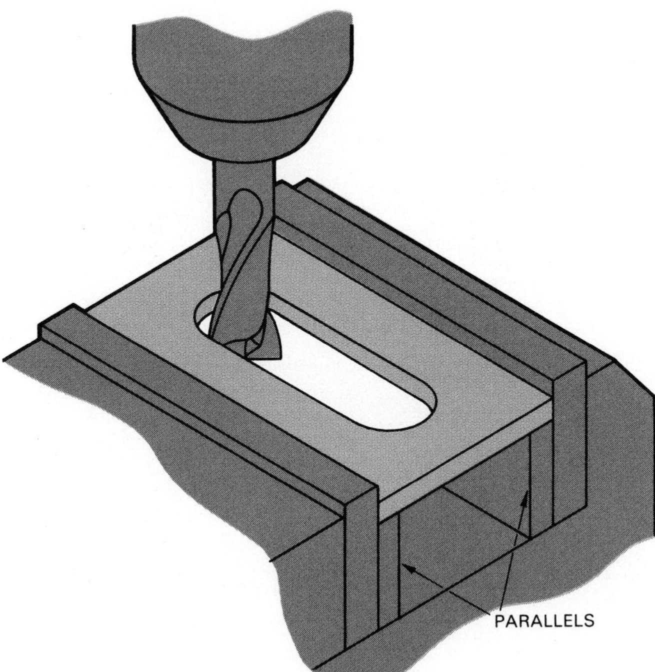

Fig. 12-138. Internal opening is being machined with a two lip end mill.

centering of the end mill is done as follows:

1. Lock the knee to the column.
2. Hold the end of a long, narrow strip of paper between the cutter and the work. Carefully move the cutter towards the work until the paper strip is pulled lightly from your fingers.

WARNING! Pay close attention when using this alignment technique. Use a paper strip long enough to keep your fingers well clear of the cutter. Release the paper as soon as your feel the cutter "grabbing" it.

3. Unlock the knee from the column and lower the work until the cutter is slightly above the work. Move the cutter in half the work diameter, plus half the cutter diameter, plus the paper thickness.
4. Again, using a long, narrow strip of paper, employ the same technique to get the required depth. See Fig. 12-139.

Correct keyseat depth may be obtained from tables in a machinist's handbook.

Machining internal openings

Internal openings are easily machined on a vertical milling machine, Fig. 12-140. A two-flute end mill must be utilized if the cutter must make the initial opening. It can be fed directly into the material in much the same manner as a drill.

When the slot is wider than the cutter diameter, it is important that the direction of feed, in relation to cutter rotation, be observed. Feed direction is normally AGAINST CUTTER ROTATION, Fig. 12-141. This applies ONLY when the cutter is removing metal from one side of the opening.

Machining multi-level surfaces

Milling multi-level surfaces is probably the easiest of the milling operations, Fig. 12-142. A layout of the various levels is made on the work's surface. Cuts are made until the lines are reached. For ac-

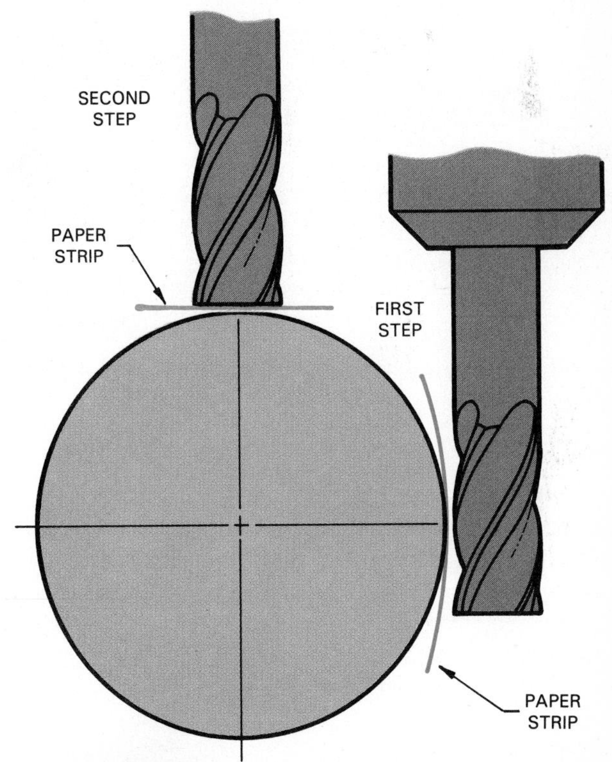

Fig. 12-139. Note use of paper strip to position cutter on exact center of round stock.

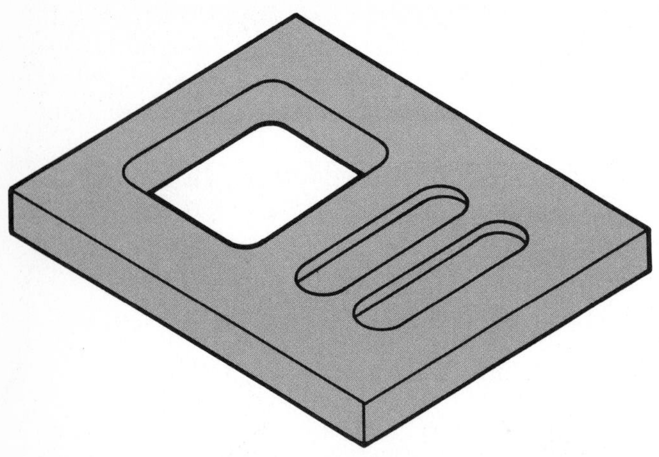

Fig. 12-140. Internal openings have been milled in an aircraft part.

Fig. 12-142. Stepped surface is being milled.

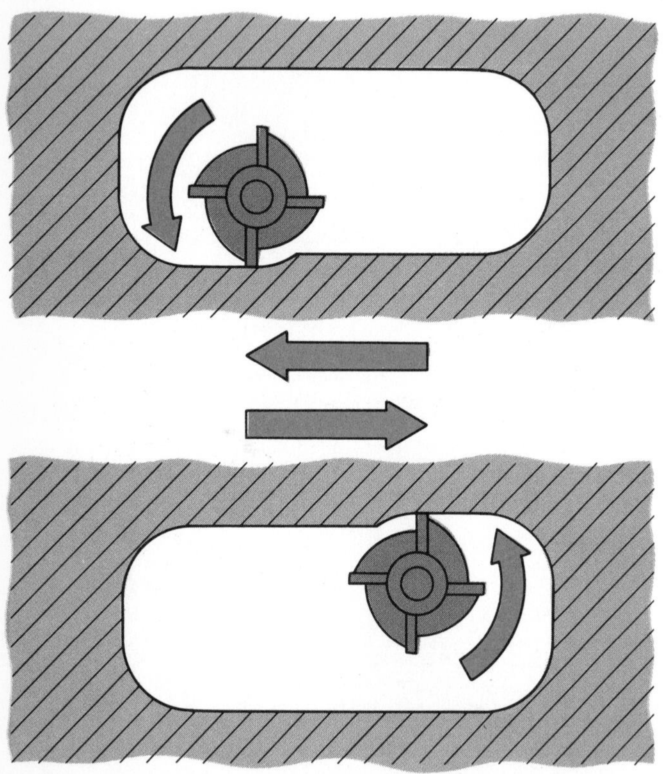

Fig. 12-141. For this type cut, feed direction is always against cutter rotation.

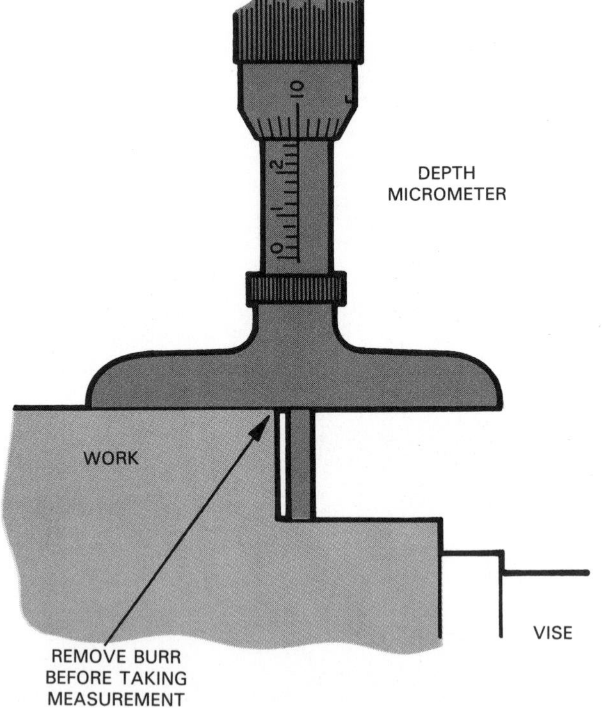

Fig. 12-143. Check cut depth with a depth micrometer before making final cutter adjustment. Remove burrs before making measurement. STOP machine BEFORE attempting to remove burrs and making measurement.

curacy, cut depth must be checked with a depth micrometer (remove burrs before making measurement), Fig. 12-143. Make table adjustments accordingly.

Milling and boring

Holes may be located for drilling, reaming, and boring to very close tolerances on a vertical milling machine. The first hole can be located with a ***wiggler,*** Fig. 12-144, or ***edge finder,*** Fig. 12-145, and machined. Then, it is possible to locate any remaining holes within 0.001 in. (0.025 mm) using the micrometer feed dials, Fig. 12-146. Tolerances of 0.0001 in. (0.0025 mm) are possible when a ***measuring rod and dial indicator*** attachment, Fig. 12-147, or ***digital readout gaging system,*** Fig. 12-148, is fitted to the machine.

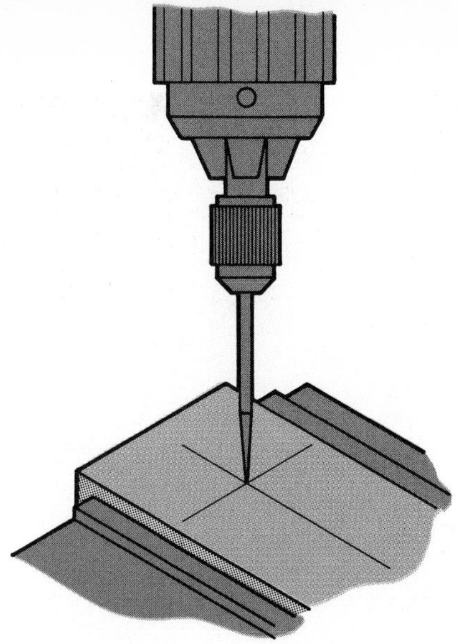

Fig. 12-144. To insure accuracy, align first hole with "wiggler." Remaining holes can be located with micrometer dials on cross and longitudinal feeds of machine.

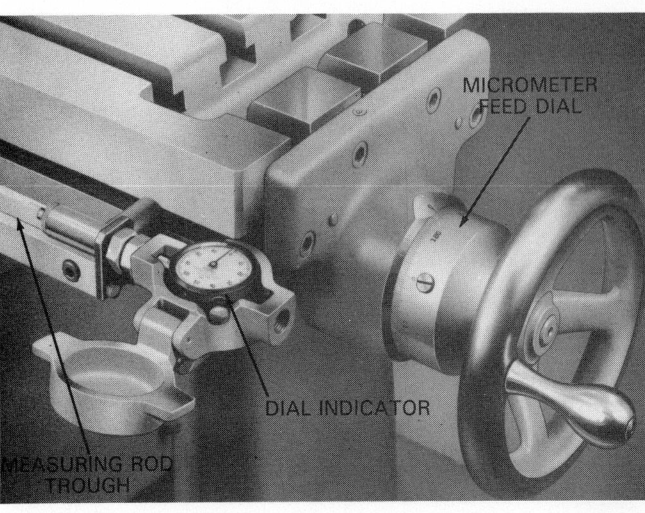

Fig. 12-146. Micrometer feed dials make it possible to maintain very close positional tolerances when drilling a series of holes. (South Bend Lathe, Inc.)

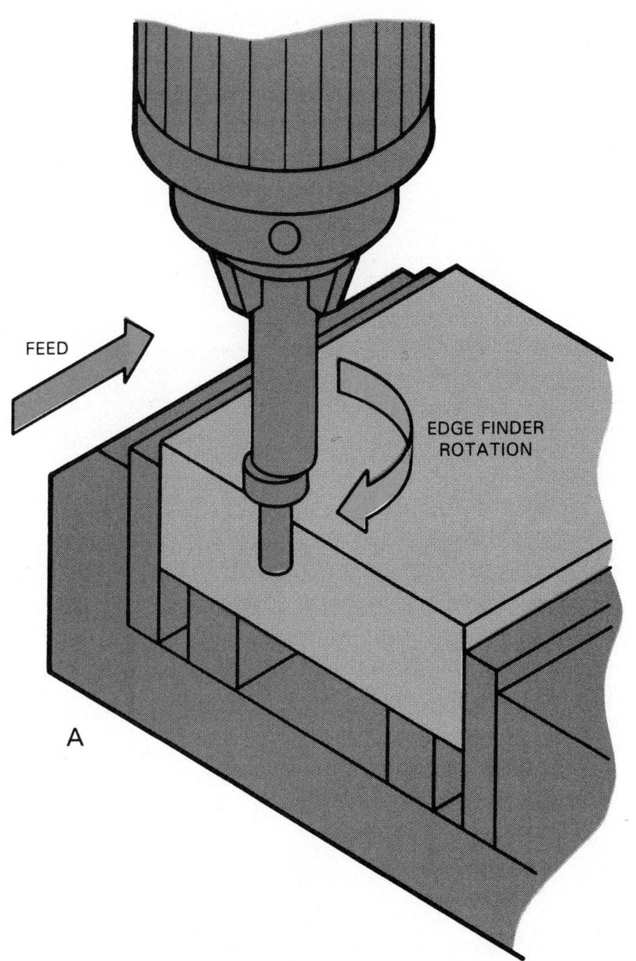

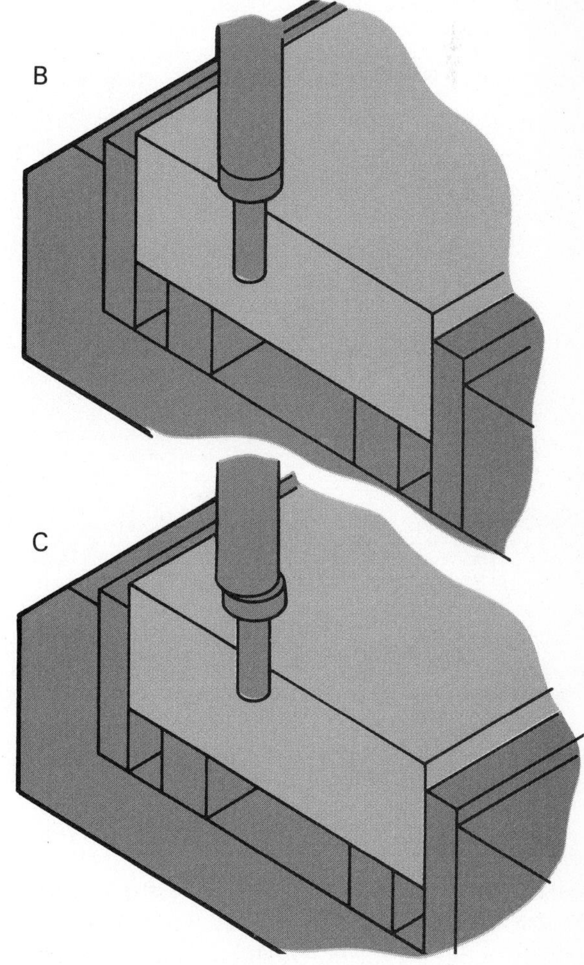

Fig. 12-145. Edge finder is a precision positioning tool that will locate edge of work in relation to center of spindle with 0.0002 in. (0.005 mm) accuracy. A—With spindle rotating at moderate speed, and with edge finder tip as shown, slowly feed tip of tool against work. B—Edge finder tip will gradually become centered with shank. C—When tip becomes exactly centered, tip will abruptly jump sideways about 1/32 in. (0.8 mm). Stop table movement immediately when this occurs. Center of spindle will be exactly one-half tip diameter away from edge of work. Set micrometer dial to "0" and, with edge finder clear of work, move table longitudinally required distance PLUS one-half tip diameter. Same procedure is followed to get traverse measurement.

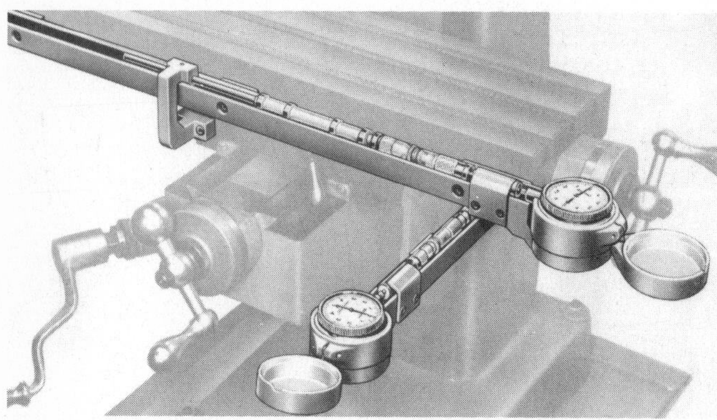

Fig. 12-147. Measuring rods and dial indicator attachments will allow precise location of work. (Clausing Industries, Inc.)

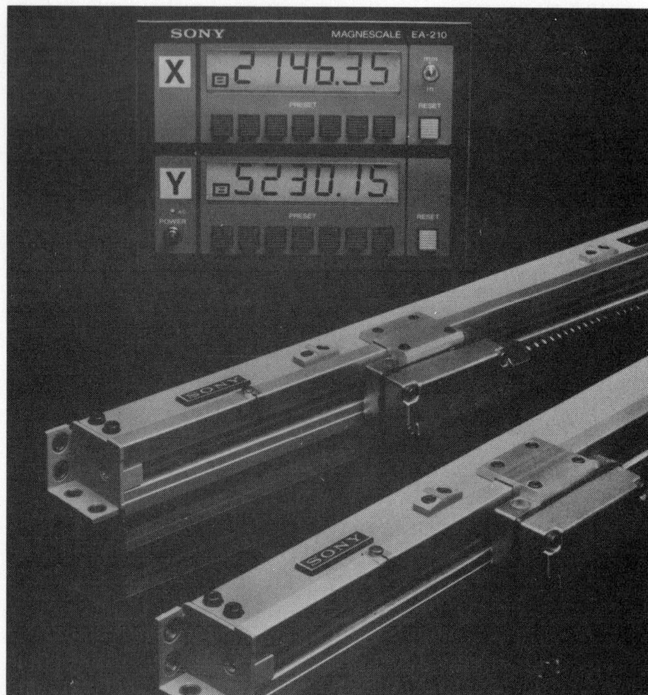

Fig. 12-148. Digital readout gaging systems are linear scales that employ magnetic or photoelectric scanning devices to display absolute distance from a predetermined datum point. Capabilities include inch/metric conversion. (National Machine Systems)

Boring permits holes to be machined accurately with fine surface finishes, Fig. 12-149. A single point tool is fitted to the ***boring head*** which in turn is mounted in the spindle, Fig. 12-150. Hole diameter is obtained by off-setting the tool point from center. The adjustment is graduated for a direct reading, Fig. 12-151. A hole must first be drilled in the work before boring can be started.

Other highly specialized attachments are available for special jobs and long production runs. See Figs. 12-152 through 12-154.

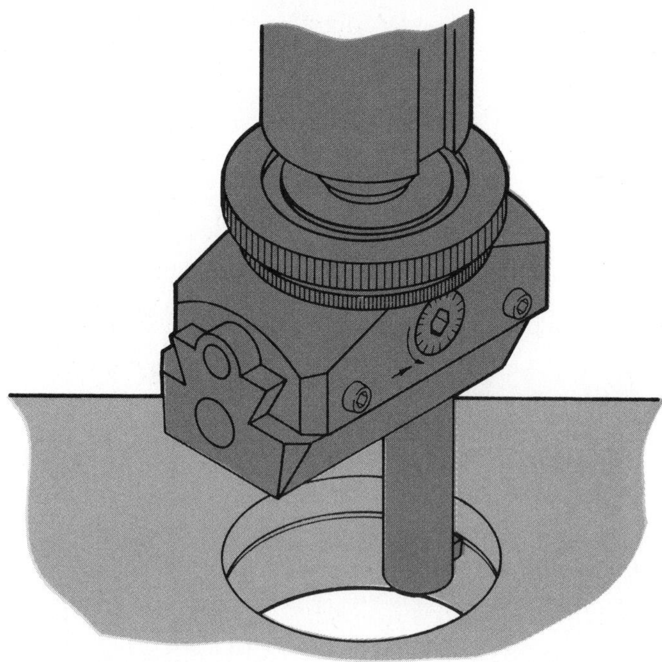

Fig. 12-149. Boring on vertical milling machine produces a very accurately sized hole.

Fig. 12-150. Note offset boring head. (Lido Tools)

Fig. 12-151. Dial on adjusting screw permits accurate cutting tool settings. Note that some boring head dials indicate actual tool movement while others indicate actual material removal.

Fig. 12-152. Slotting attachment is mounted on vertical milling machine. Rotary motion in spindle head is changed to reciprocating or up and down movement for cutting tool. (Bridgeport Machines)

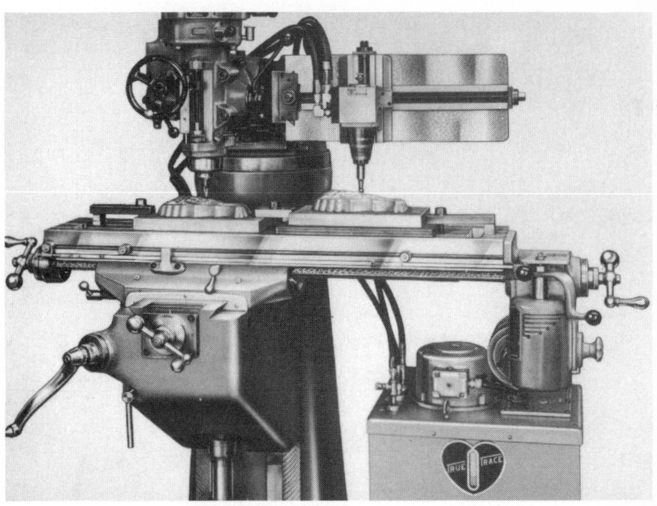

Fig. 12-153. Vertical mill fitted with a hydraulic tracer unit is machining irregular shape. (Bridgeport Machines)

Fig. 12-154. Right angle milling attachment is attached to spindle head. This unit permits machine to be employed for internal and confined area milling.

MILLING MACHINE CARE

1. Check and lubricate the machine with the recommended lubricants.
2. Clean the machine thoroughly after each job. Use a brush to remove chips, Fig. 12-155. Never attempt to clean the machine while it is running.
3. Keep the machine clear of tools.

Fig. 12-155. Use a brush to remove metal chips . . . NEVER your hand!

4. Check each setup for adequate clearance between the work and the various parts of the machine.
5. NEVER force a cutter into a collet or holder. Check to see why it does not fit properly.
6. Use a sharp cutter. Protect your hands when mounting it.
7. Check the machine to determine whether it is level. This should be done at regular intervals.
8. Have ALL guards in place before attempting to operate a milling machine.
9. Check coolant level and condition if the reservoir is built into the milling machine. Change it if it becomes contaminated.
10. Start the machining operation only after you are sure that everything is in satisfactory working condition. It may be necessary to make special fixtures to hold odd shaped or difficult-to-mount work, Fig. 12-156.
11. Use attachments designed for the machine.

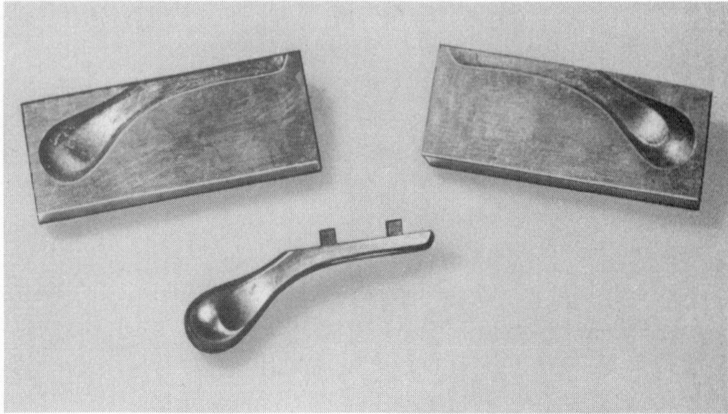

Fig. 12-156. An example of how odd-shaped work can be held for machining on a mill. Special fixtures will hold small part.

HORIZONTAL MILLING MACHINE OPERATIONS

Like the vertical milling machine, the horizontal mill is also a very versatile machine, Fig. 12-157. Many different machining operations can be performed on it.

Milling flat surfaces

A careful study of the drawing will let you determine what is to be done, what cutter is best suited for the job, and the most advantageous way to hold the work. Flat surfaces may be milled with a plain or slab cutter mounted on an arbor (peripheral milling), or with an inserted tooth face or shell milling cutter (face milling). The method employed will be determined by the size and shape of the work.

After the method has been selected, the following sequence of operations is recommended.

1. Check and lubricate machine. Wipe work table and examine it for nicks and burrs. Nicks or burrs will prevent work or holding attachments from seating properly on table.
2. Mount the work directly on the table if possible. Fig. 12-158 illustrates one method for mounting long work. Use a vise if the work cannot be mounted to the table. Clean its base and bolt it firmly in place. Locate the vise as close to the machine column as work shape and arbor support will permit. When possible, pivot the vise so the solid jaw supports the work against cutting rotation, Fig. 12-159.
3. Align the vise with a dial indicator if extreme accuracy is required, Fig. 12-160. Otherwise, a square or machine arbor will do, Figs. 12-125 and 12-161. Angular vise settings can be made using the vise base graduations or with a protractor, Fig. 12-126.
4. Wipe the vise jaws and bottom clean.
5. Place clean parallels in the vise with the work seated on them. Tighten the jaws and tap the work onto the parallels with a mallet or soft face hammer. Thin paper strips can be used to check that the work is seated tightly on the parallels.
6. Select an arbor that is as short as the job will permit. Wipe the taper section of the arbor and spindle opening with a dry cloth. Insert the arbor and draw it tightly into place with the draw-in bar.
7. When possible, the cutter should be wide enough to machine the area in one pass and it should have the smallest diameter possible while large enough to provide adequate clearance. Refer to Figs. 12-162 and 12-163.

DANGER! Use a section of cloth or gloves to protect your hands when mounting a cutter on the arbor, Fig. 12-164.

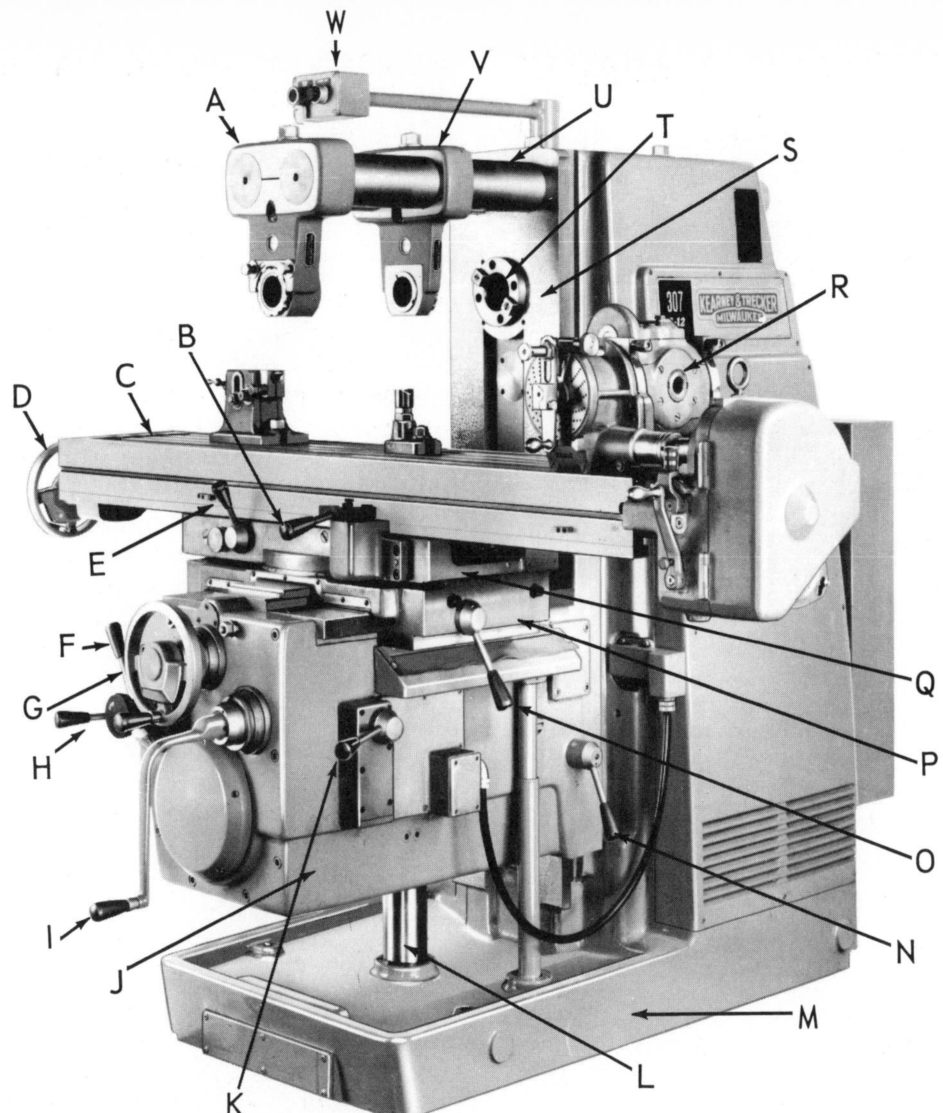

Fig. 12-157. Study parts of horizontal milling machine: A—Arbor support. B—Table feed lever. C—Table. D—Table (longitudinal) handwheel. E—Table clamp lever (lock). F—Saddle power feed lever. G—Saddle (in-out) handwheel. H—Knee power feed lever. I—Knee (up-down) crank. J—Knee. K—Rapid traverse lever. L—Telescopic coolant return and elevating screw. M—Base. N—Knee clamp lever. O—Saddle clamp lever (lock). P—Saddle. Q—Saddle plate. R—Universal dividing head. S—Column. T—Spindle. U—Overarm. V—Inner arbor support. W—Spindle stop-start and master switch (arm also engages clutch). (Kearney & Treacker Corp.)

8. Key the cutter to the arbor. Position it as close to the column as the work will permit. If a helical slab mill is used, mount it so the cutting pressure forces it TOWARDS the column. Refer to Fig. 12-165.
9. Position and lock the arbor support into place, Fig. 12-166. Then, tighten the arbor nut.
10. Adjust the machine to the proper cutting speed and feed.
11. Turn on the machine and check cutter rotation and direction of power feed. If satisfactory, loosen all worktable and knee locks and position the work under the rotating cutter until it just touches the surfaces. Set the micrometer dial to 0. Back the work away from the rotating cutter. Make a light cut with ample cutting fluid flooding the surface. Make a measurement and raise the table the required distance.

Should a previously machined surface require additional machining, it will be best to position the cutter in the following manner. Hold a long, narrow strip of paper with its loose end between the work and the cutter, Fig. 12-167. Raise the table until the paper is pulled lightly from your fingers.

After the cutter has been positioned, it is only necessary to move the cutter clear of the work and raise the table the required distance, plus the thickness of the paper.

Tighten all locks (except longitudinal) and feed the work into the cutter. As soon as cutting starts, turn on the coolant and power feed.

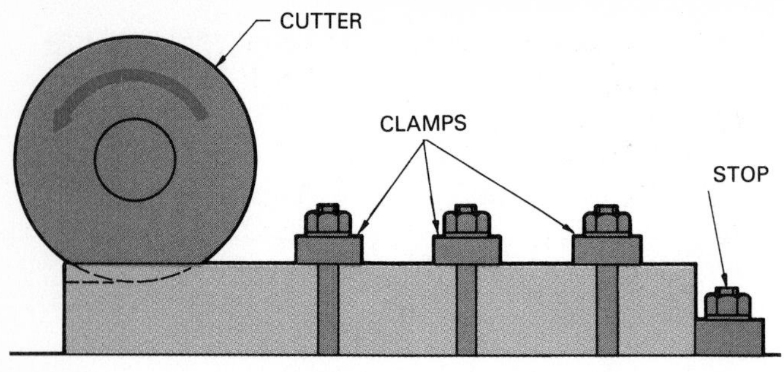

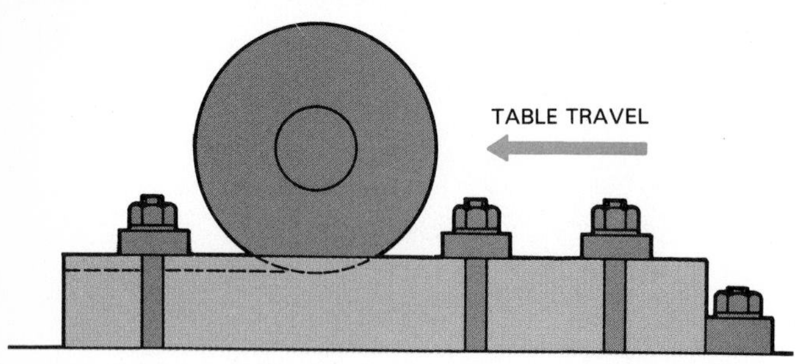

Fig. 12-158. Reposition clamps as the cut progresses across the work.

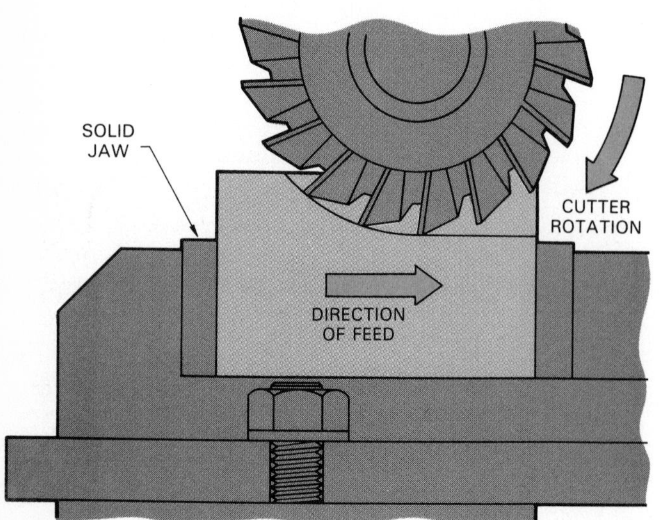

Fig. 12-159. Solid jaw of vise should be in this position whenever setup permits.

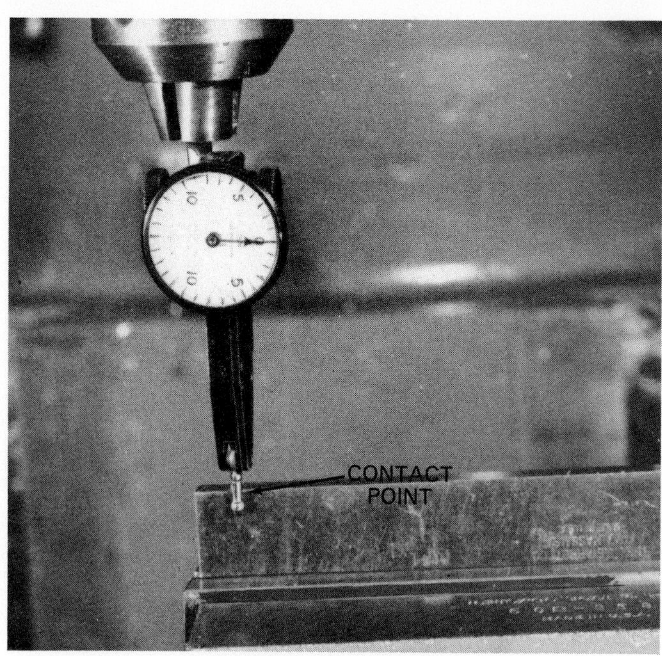

Fig. 12-160. Alignment of vise jaw with dial indicator will insure work accuracy.

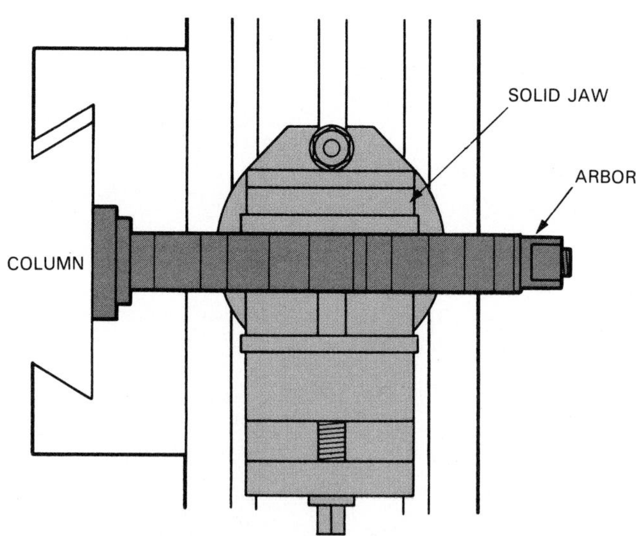

Fig. 12-161. Arbor can also be used to align a vise.

Unless there is an emergency, do NOT stop the work during the machining operation. This will cause a slight depression to be cut in the machined surface, Fig. 12-168.

WARNING! Do NOT attempt to feel the machined surface while the cut is in progress or while the cutter is rotating.

Complete the cut and stop the cutter. Return the work to the starting position.

CAUTION! Avoid feeding the work back to the starting position while the cutter is rotating. This

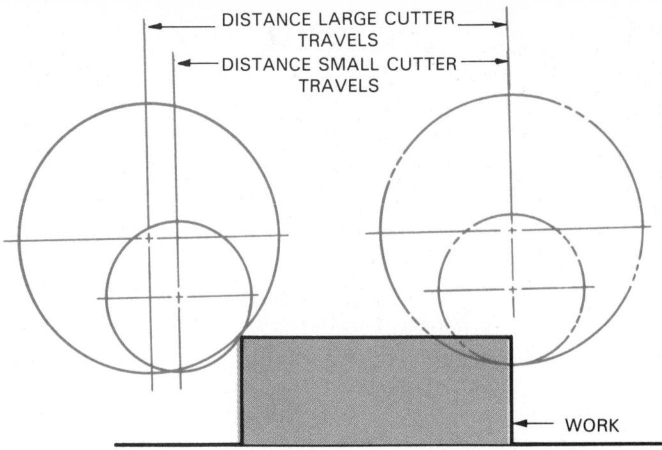

Fig. 12-162. Small diameter cutter is more efficient than a large diameter cutter because it travels less distance while doing same amount of work.

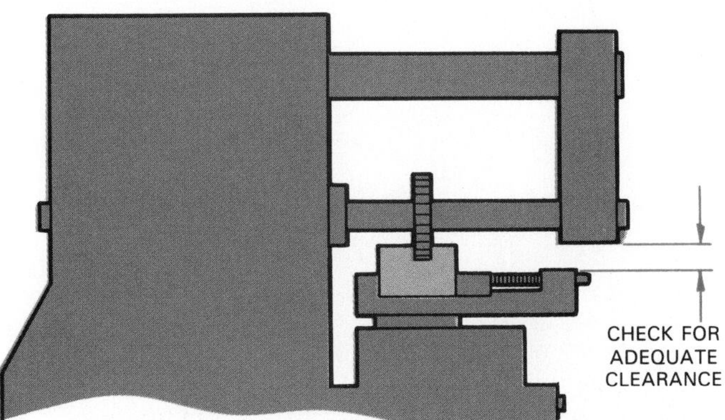

Fig. 12-163. Use smallest cutter diameter possible, but be certain it is large enough for adequate clearance.

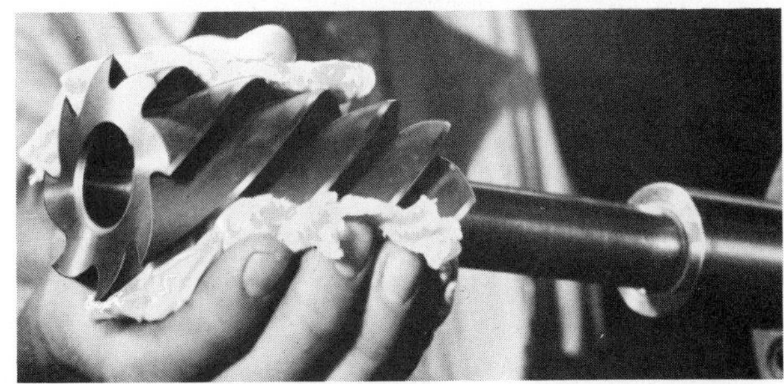

Fig. 12-164. It is advisable to use a cloth to protect hands when mounting or removing cutter from an arbor.

will cause a series of depressions to be made on the newly machined surface.

Repeat the above operations if additional metal must be removed to bring the metal to size.

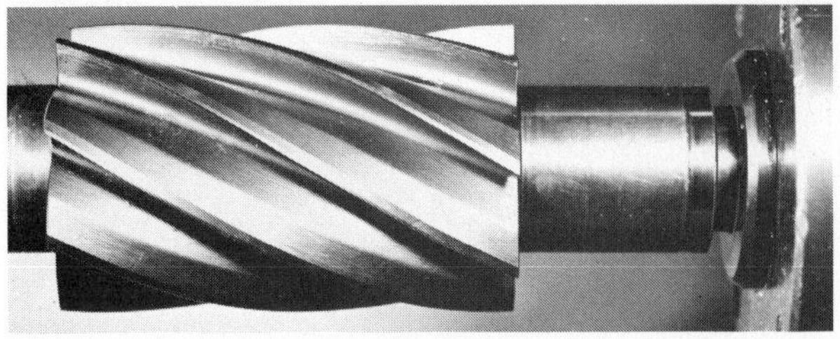

Fig. 12-165. Helical cutters should be mounted so cutting pressures tend to force cutter towards spindle.

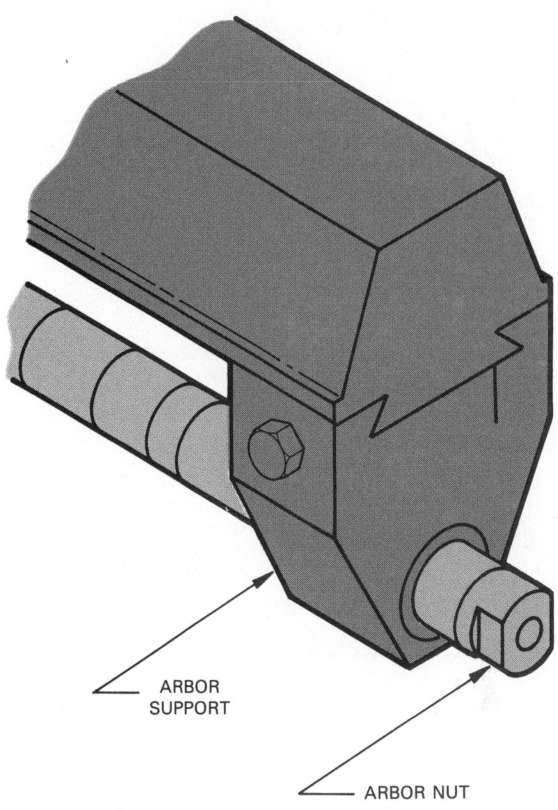

Fig. 12-166. Arbor nut must NOT be tightened until AFTER arbor support has been positioned and locked in place.

Fig. 12-167. Machinist is positioning cutter employing paper strip technique. Pay attentionwhen positioning cutter by this method!

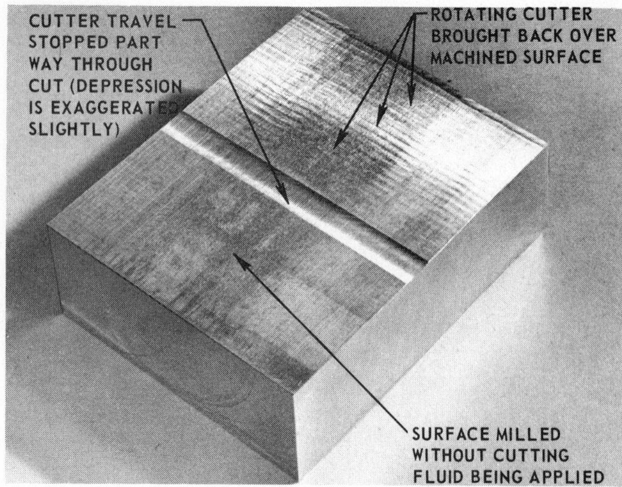

Fig. 12-168. Stopping work movement in middle of cut will cause a slight depression in machined surface. Also, note ridges that were made when revolving cutter was brought back over machined surface.

Squaring stock

The sequence for squaring stock on a horizontal milling machine is the same as that used on a vertical milling machine. Where possible, the cutter should be wide enough to make a full width cut on the material in one pass. If the material is short enough, the ends can be machined by placing it in a vertical position with the aid of a square, Fig. 12-129. If too long for this technique, the ends can be squared as shown in Fig. 12-169.

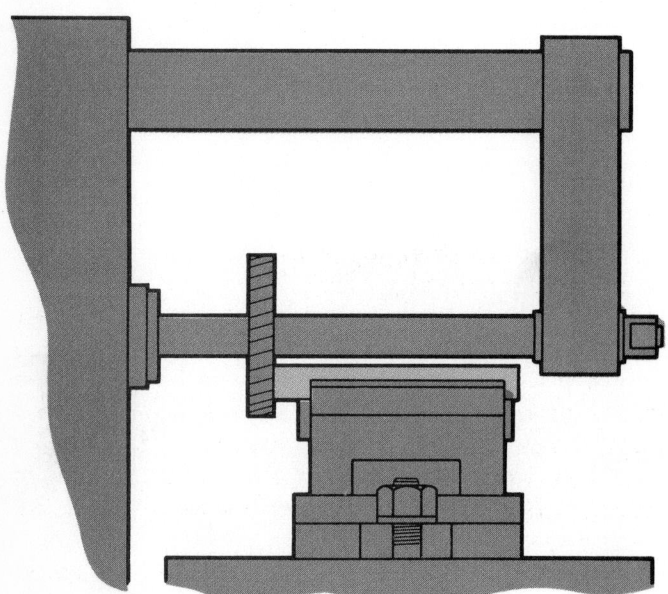

Fig. 12-169. Here is another method of squaring work ends. Be sure there is adequate clearance between work and arbor.

Face milling

Face milling makes use of a cutter that machines a surface at right angles to the spindle axis and parallel to the face of the tool. Face milling cutters over 6 in. (150 mm) in diameter are usually of the inserted tooth type and mount directly to the spindle nose. They are used to mill large, flat surfaces. Face mills smaller than 6 in. (150 mm) are called ***shell mills*** and are held on a short arbor.

1. Select a cutter that is 3/4 in. (20 mm) to 1 in. (25 mm) larger in diameter than the width of the surface to be machined, Fig. 12-170.
2. The work should project about 1 in. (25 mm) beyond the edge of the table to provide adequate clearance, Fig. 12-171. In face milling, it is frequently necessary to mount the work on an angle plate, Fig. 12-172.
3. Adjust the machine for correct speed and feed.
4. Slowly feed the work into the cutter until it starts to remove material. Roughing cuts up to 1/4 in. (6 mm) may be taken.

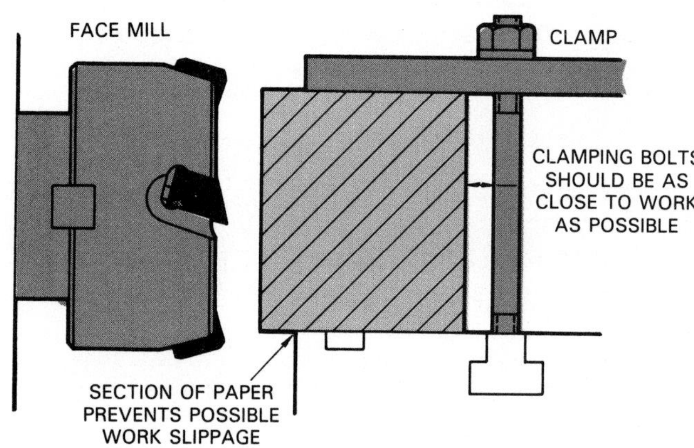

Fig. 12-170. Select cutter that is 3/4 in. (20 mm) to 1 in. (25 mm) larger in diameter than width of surface to be machined.

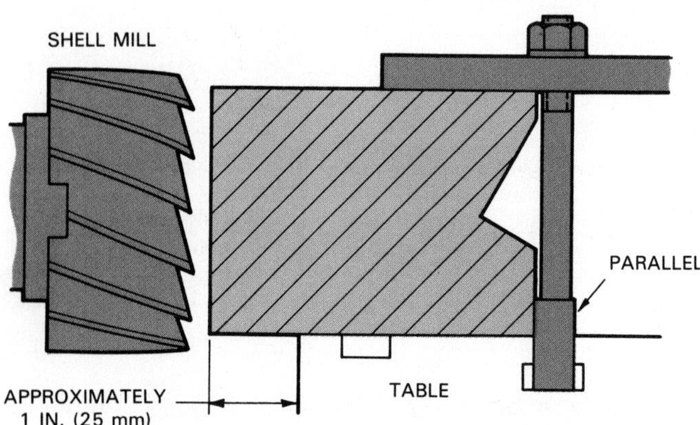

Fig. 12-171. Work should project approximately 1 in. (25 mm) beyond table edge to provide adequate clearance.

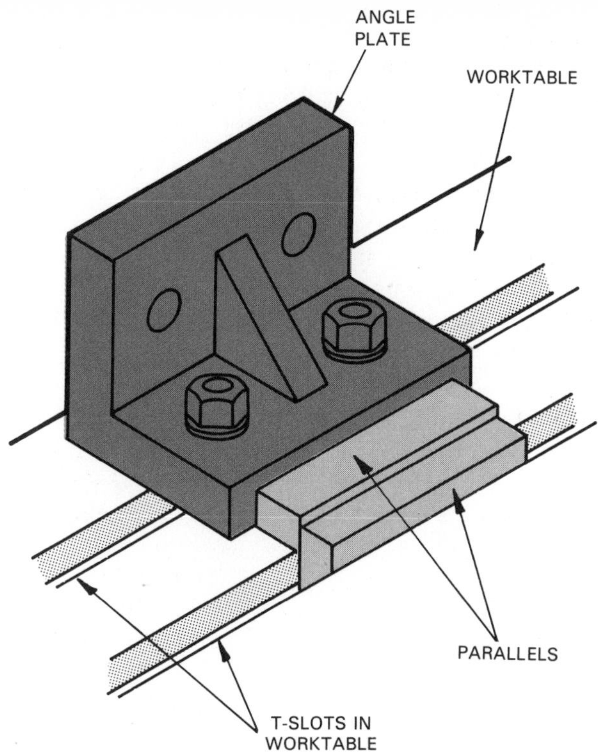

Fig. 12-172. An angle plate is often used to mount work for face milling. Check that mounting clamps clear cutter. Note how parallels are employed to align angle plate.

5. Use adequate cutting fluid.
6. Upon completion of the cut, stop the cutter. Return the work to the starting position for additional machining if needed.
7. Make the finishing cut and tear down the setup. Use a brush to remove chips. Clean and store the cutter.

Side milling

Side milling refers to any milling operation that involves the use of half side and side milling cutters. When employed in pairs to machine opposite sides of a piece at the same time, the setup is called ***straddle milling,*** Fig. 12-173.

Cutters used for this operation should be kept in matched pairs. That is, they should be sharpened at the same time to maintain equal diameters. Shoulder width of the machined surface is determined by the thicknesses of the spacers between the cutters, Figs. 12-174 and 12-175.

Gang milling involves mounting several cutters on the arbor to machine several surfaces in one pass. It is a variation of straddle milling, Fig. 12-176. Gang milling is used when many similar pieces must be made.

A ***side milling cutter*** can also be utilized to machine grooves, keyseats, and, when used with a dividing head or rotary table, squares, hexagons, etc., on round stock.

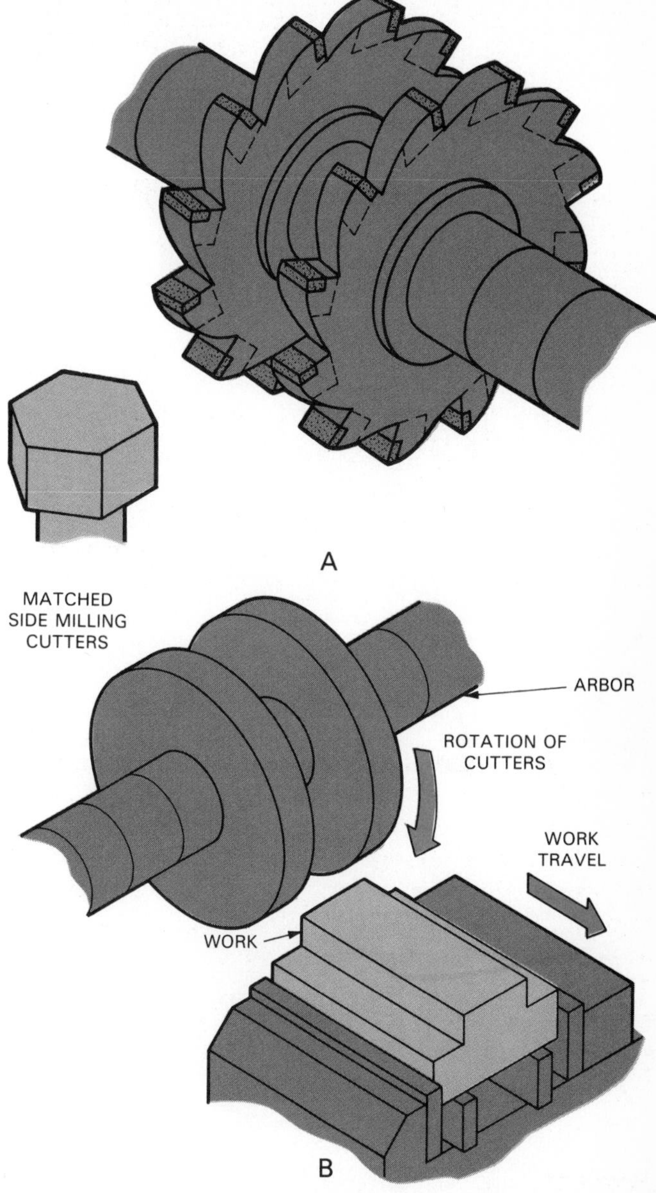

Fig. 12-173. A—Straddle milling has spacer between cutters. Two cutters are machining a hexagon on a machine part. B—Note example of straddle milling flat work.

Locating side cutter for milling slot in square or rectangular work

The machine is set up in much the same manner as it was for milling flat surfaces. Indicate the vise to assure accuracy. Exercising the same care in placing the side cutter on the arbor as was followed with the slab cutter. A plain side milling cutter may be used if the slot is NOT too deep. Otherwise, a ***staggered-tooth side milling cutter*** should be employed.

Position the cutter by one of the following methods:

1. Make a layout on the end of the work, as shown in Fig. 12-177.

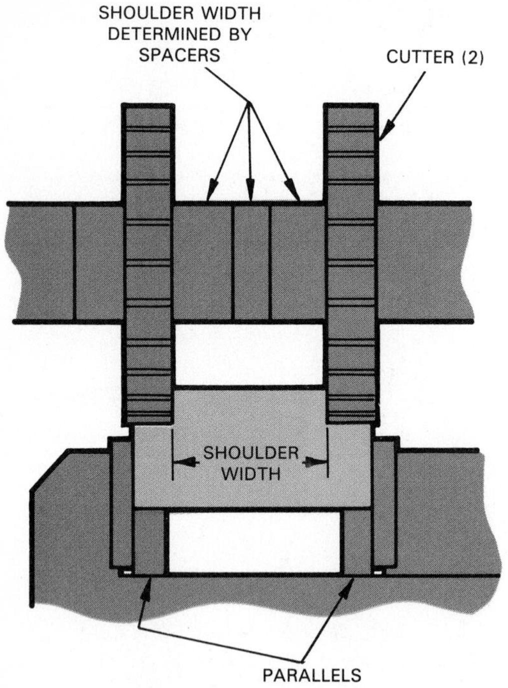

Fig. 12-174. Arbor spacers set distance between cutters. Shim stock washers are used with spacers when other than standard sizes are required.

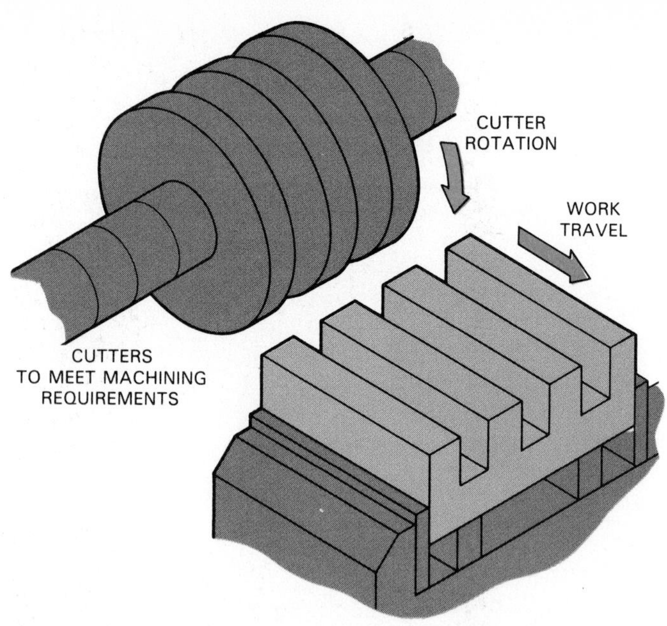

Fig. 12-176. Gang milling requires use of two or more milling cutters mounted on a single arbor.

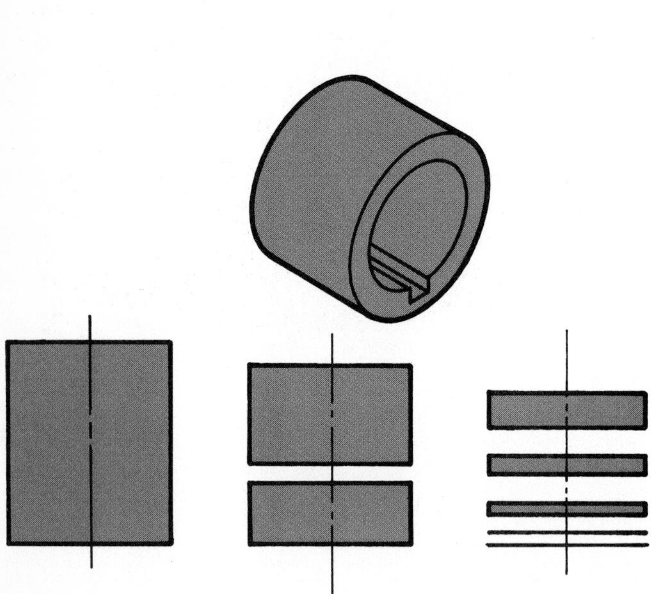

Fig. 12-175. Arbor spacers are available in a large selection of sizes. Special sizes can be made by surface grinding standard sizes to needed dimensions. Shim stock spacers can be used to build up standard size spacers to desired dimension.

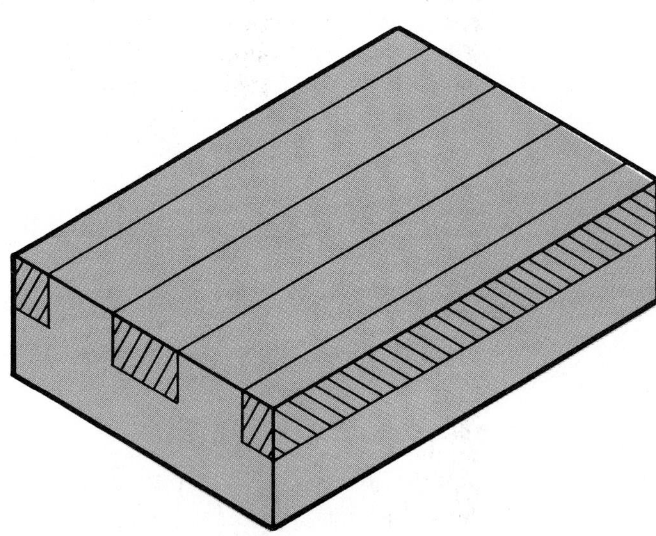

Fig. 12-177. Work has been laid out for milling.

2. Position the cutter with a rule, Fig. 12-178. Make a light cut part way up the piece. Remove the burrs. Measure the cut depth with a depth micrometer. The difference between this measurement and the required depth equals the amount of material that must be removed.

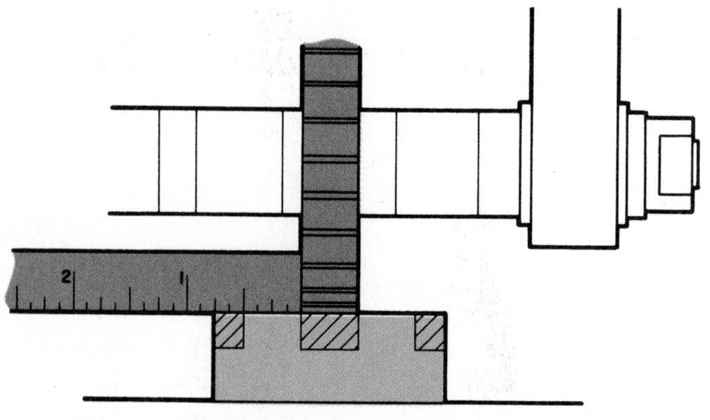

Fig. 12-178. Cutter is being positioned with aid of a steel rule.

3. Use the paper strip technique to bring the side of the cutter against the side of the work, Fig. 12-179. Move the cutter in the required distance, PLUS the thickness of the paper. The paper strip or depth micrometer positioning technique may be used to set the cutter to the desired depth.

SAFETY NOTE! When using the paper strip technique to position a cutter, remember to use a long paper strip and keep your fingers clear of the revolving cutter.

Locating side cutter for milling slot or keyseat in round stock

There are many situations that require keyseats for the standard square key to be cut in round stock. The keyseat must be kept precisely on center if it is to be in alignment with the keyway in the mating piece.

After the milling machine has been set up, cutter mounted and work positioned in a vise, between centers, in V-blocks, or in a fixture, you must center the cutter. Precise centering of the cutter may be accomplished by one of the following methods:

1. Center the cutter visually on the work. With the aid of a steel square and rule, adjust the table until both sides measure the same, Fig. 12-180. Due to the difficulty of obtaining exact measurements with a rule, most machinists prefer to use a depth micrometer in place of the rule.
2. Short pieces cannot always be centered by the above method. For situations of this type, the work is positioned under and brought lightly into contact with the rotating cutter. Traverse (in-out) feed is utilized to pass the work under the cutter. Because the work is circular in shape, an oval shaped cut will result, and the oval will be perfectly centered. To center the cutter, position it on the oval, Fig. 12-181.
3. The previously mentioned narrow paper strip technique may be used to center the cutter. Hold the strip between the work and the cutter. Carefully move the work towards the cutter until it causes the paper to be lightly pulled from between your fingers. Lower the table until the cutter is slightly above the work. Move the cutter in half the diameter of the work, plus half the cutter thickness, plus the paper thickness, Fig. 12-182. The same technique may be employed to center a Woodruff keyseat cutter. Fig. 12-183 illustrates this method.

Lock the saddle to prevent traverse table movement after the cutter has been centered.

Correct keyseat depth can be obtained from tables in one of the many machinist's handbooks. The paper strip technique is used to set the cutter to the required depth. Tighten the knee locks after the depth setting has been made.

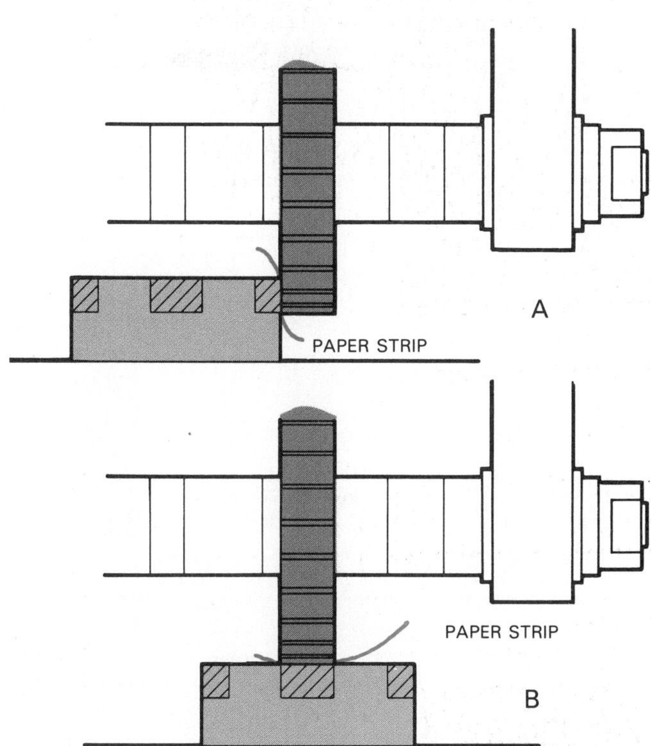

Fig. 12-179. Positioning the cutter. A—Use paper strip to secure internal dimension. Read micrometer to move cutter over work correct distance. B—Use paper strip to position cutter for depth. Then read micrometer dial as cutter is lowered.

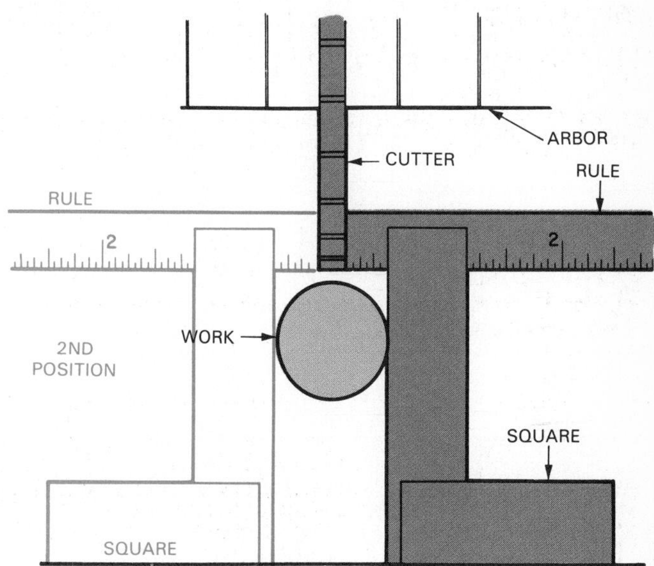

Fig. 12-180. Study how to center cutter on round stock with steel rule and machinist's square.

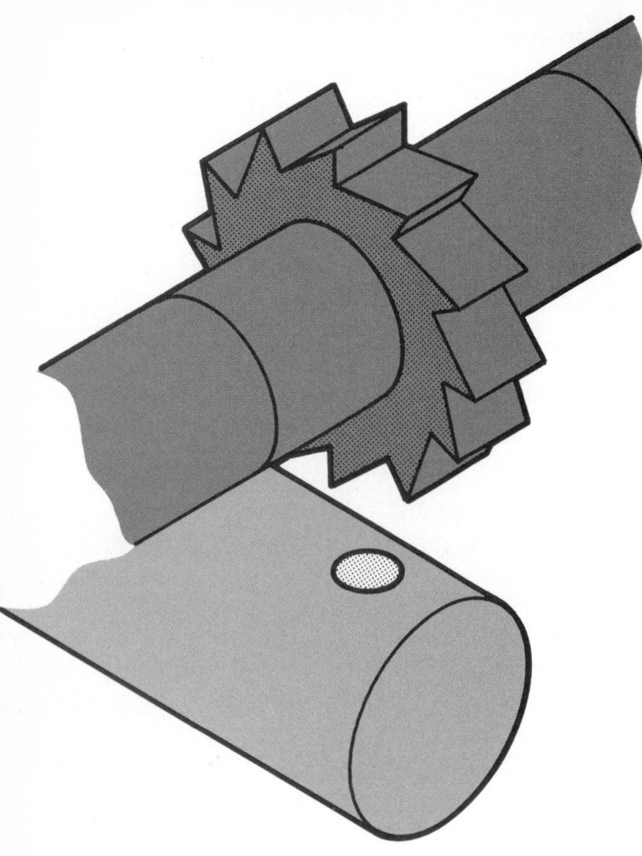

Fig. 12-181. Cutter is being positioned on center using an oval made in work with cutter as a guide.

When the Woodruff keyseat cutter is used, it must also be positioned longitudinally on the work. Slowly fed into the piece until the required depth is attained. This can be checked by placing a key in the cut and ''miking'' the section.

Cutting fluid should be applied liberally during the cutting operation.

SLITTING

Slitting and ***slotting*** thin stock into various widths for the production of flat gages, templates, etc., is a fairly common milling operation, Fig. 12-184. It is performed with a ***slitting saw*** and is likely to give considerable trouble if extreme care is NOT exercised.

A slitting saw of the smallest diameter permitting adequate clearance is used. It must be keyed to the arbor (key should also pass into spacers on either side of cutter)

Best results can be obtained if the cutter is mounted for ***climb milling.*** That is, the work and cutter move in the same direction at the point of contact, Fig. 12-185. Cutting pressure is downward and will tend to press the work onto the table or holding device.

Adjust the table gibs until there is heavy drag felt when the table is moved by hand. This will remove table ''play'' and prevent the cutter from jumping in the cut.

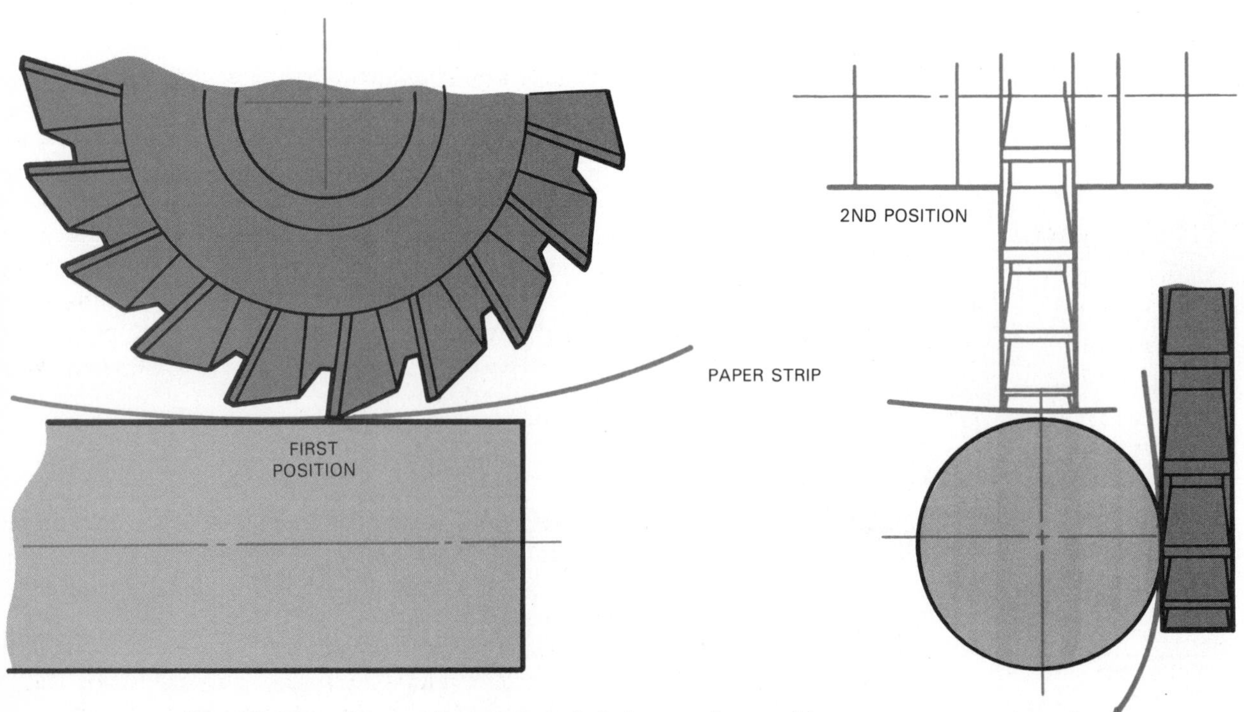

Fig. 12-182. Paper strip technique is being used to position cutter on round stock.

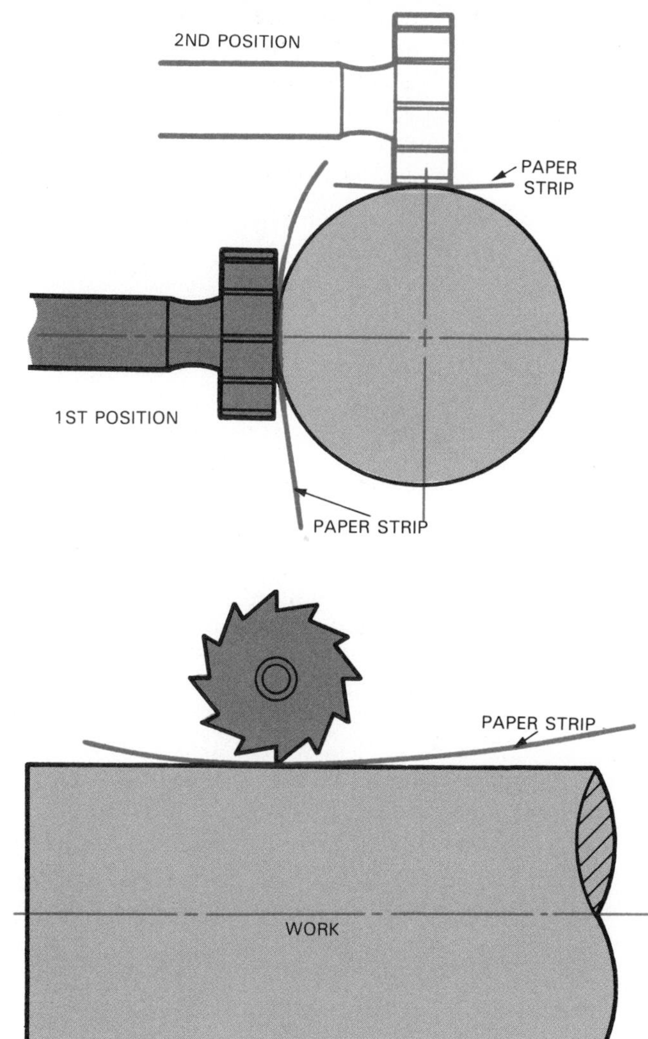

Fig. 12-183. Paper strip technique can also be employed to center a Woodruff keyseat cutter.

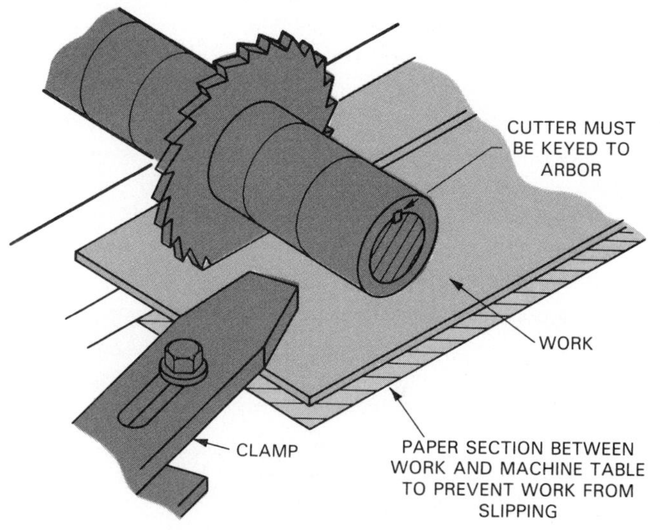

Fig. 12-184. This is a typical slitting or sawing operation setup. Work must be clamped securely.

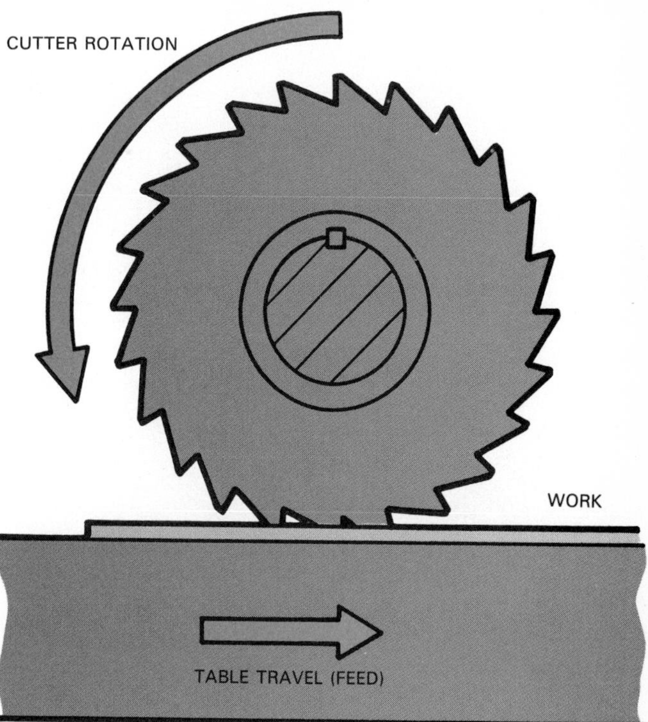

Fig. 12-185. Note cutter setup and feed for best slitting results. Use a slow feed.

If the section is narrow enough, the piece may be clamped in a vise, Fig. 12-186. It should be well supported on parallels.

WARNING! Do NOT permit the parallels to project out into the cutter path.

Long strips must be clamped to the work table. The clamp shown in Fig. 12-187 is made from a section of angle iron and is recommended. The work is aligned with the column face and must be positioned to permit the saw to make the cut over the center of a table T-slot. See Fig. 12-188.

A piece of paper between the work and table will prevent the metal from slipping during the slitting operation. The cutter is set to a depth equal to the work thickness plus 1/16 in. (1.5 mm). Always use a sharp cutter!

SLOTTING

Slotting is similar to slitting except that the cut is made only part way through the work, Fig. 12-189. The slot in a screw head is an example of slotting.

1. Mount the cutter as for conventional milling. Use a sharp cutter of a width suitable for the job. Note the difference between a slotting cutter and a slitting cutter, Fig. 12-190.
2. Set the machine for the correct cutting speed.

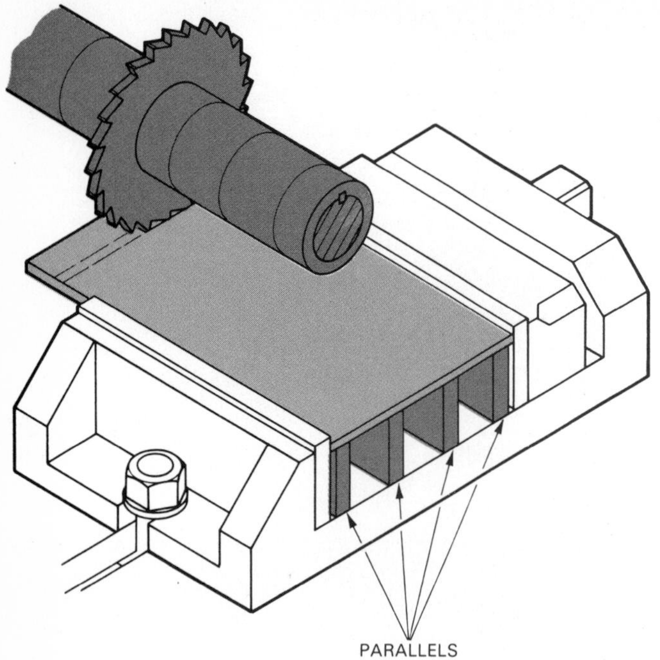

Fig. 12-186. Slitting work is held in vise. Be sure parallels do NOT project into cutter path!

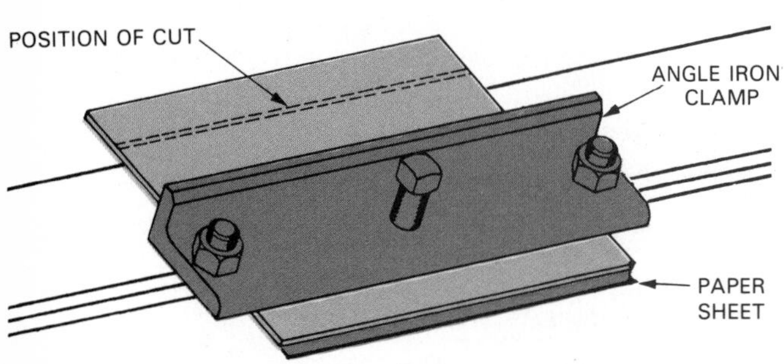

Fig. 12-187. Clamp has been made from angle iron. Paper sheet prevents work movement.

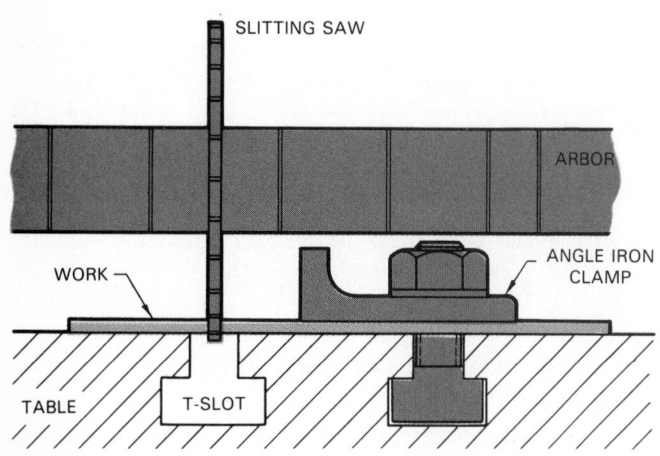

Fig. 12-188. Position work so cut is made over a T-slot.

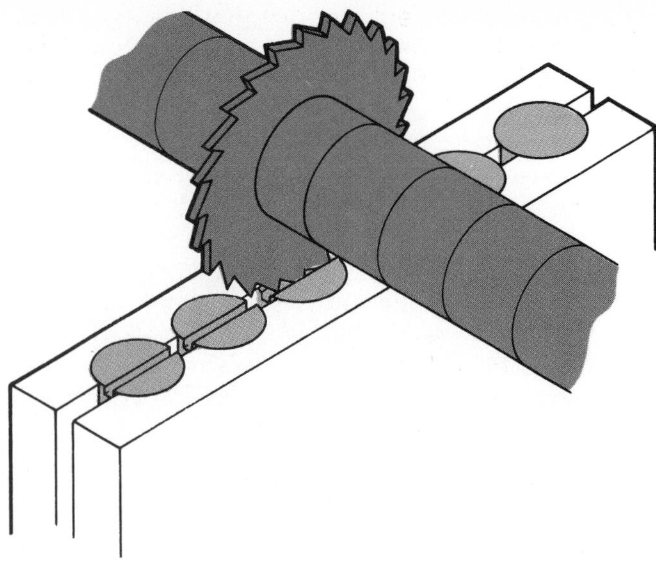

Fig. 12-189. Screw heads are being slotted with a slitting saw. Special slotting saw is available which has many more teeth than a slitting saw of a similar diameter.

Fig. 12-190. Left—Slitting saw. Right—Slotting saw. Note how cutters differ.

Use the slowest feed possible, increasing feed if conditions warrant it.

3. Align the vise and mount the work.
4. Position the cutter and make a light cut. Check the trial cut and make adjustments if necessary.

DANGER! Stop the machine before making measurements and adjustments.

5. Adjust the work for proper cut depth.
6. Apply cutting fluid and make the cut.

WARNING! Avoid standing directly in line with the cutter! Despite all precautions, saws shatter occasionally and can cause serious injury.

DRILLING AND BORING ON HORIZONTAL MILLING MACHINE

The machinist often finds it necessary to produce accurately spaced and drilled holes in work. The milling machine offers a convenient way to make these holes in a specified and precise alignment.

Small drills are held in a standard Jacobs type chuck mounted in the machine spindle, Fig. 12-191. Larger, taper shank drills are fitted into an adapter sleeve, Fig. 12-192.

Boring is done with a single point cutting tool fitted in a ***boring head,*** Fig. 12-193. The boring head may be equipped with a taper shank and mounted directly in the spindle, or straight shank and held in a collet or adapter. See Fig. 12-194.

A wiggler will aid in aligning the machine for drilling the hole prior to boring, or when holes are small enough to be drilled and reamed, Fig. 12-195. A dial indicator must be used to realign previously made holes for boring to final size, Fig. 12-196.

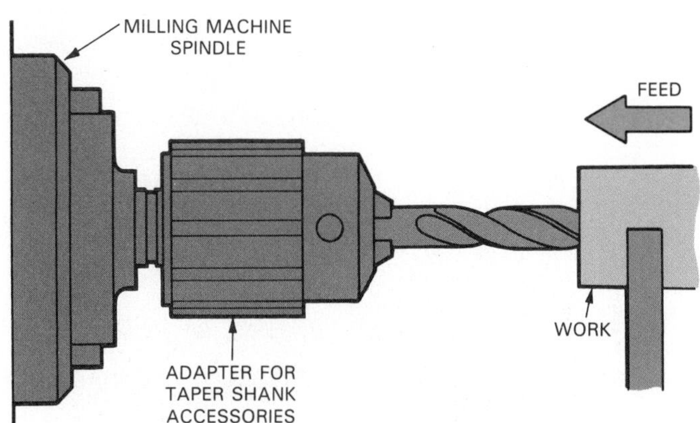

Fig. 12-191. Drilling can be done on a horizontal mill as shown.

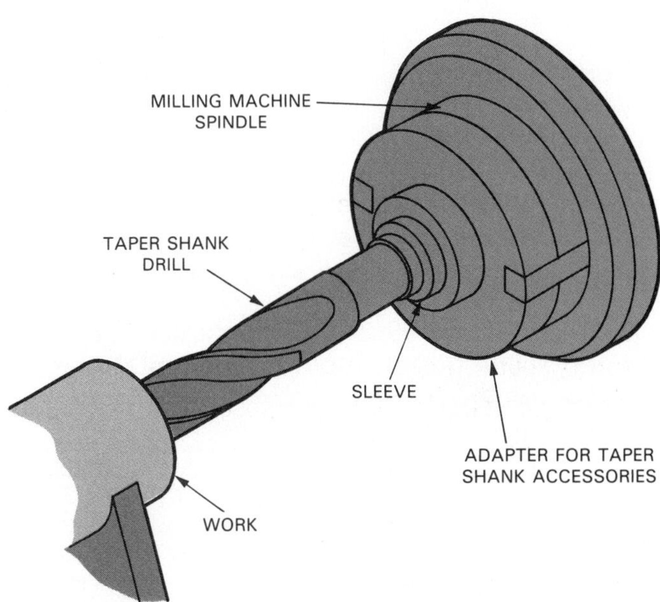

Fig. 12-192. Taper shank drills can be used by fitting an adapter to spindle.

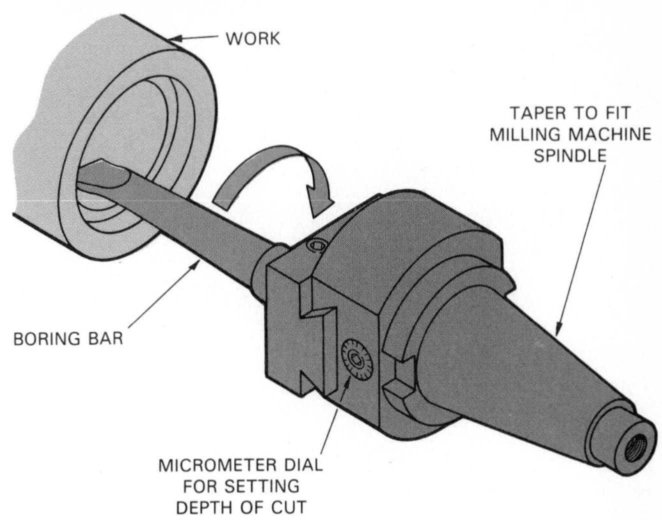

Fig. 12-193. Boring permits large holes to be machined to close tolerances.

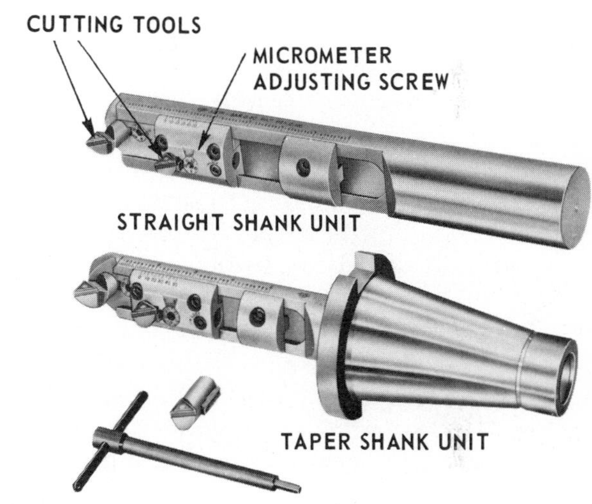

Fig. 12-194. Multipurpose boring bars permit multi-step simultaneous boring. Adjustments can be made without removing boring bar from machine. (Aloris Tool Co., Inc.)

Fig. 12-195. Holes can be located with a wiggler.

Cutting a spur gear

A ***gear*** is a toothed wheel usually fitted to a shaft. It usually engages a similar toothed wheel to smoothly transmit power or motion at a definite

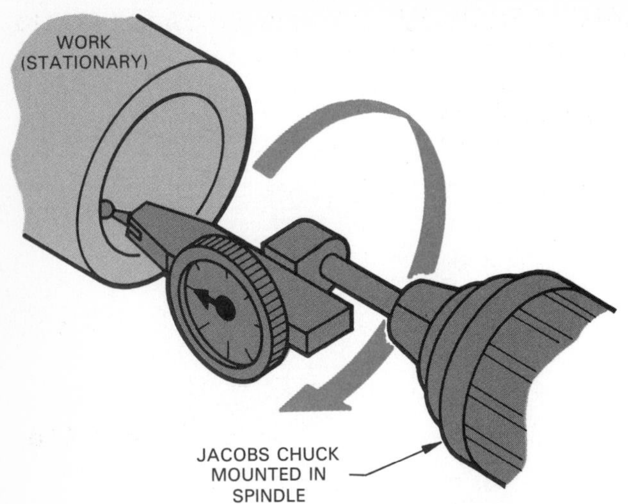

Fig. 12-196. Existing holes can be realigned on a horizontal milling machine with the aid of a dial indicator.

ratio between the shafts. The teeth are shaped so contact between the mating gears is continually maintained while they are in operation.

The ***spur gear*** has teeth that run straight across the face and are perpendicular to the sides. It is the simplest and a widely used gear, Fig. 12-197.

Gear cutting requires a knowledge of ***gear nomenclature*** (gear terminology) to aid in determining the proper gear cutter to use, the depth of the teeth, and the dividing head setup.

Gear nomenclature

The following information is necessary to calculate data needed to machine a simple spur gear (inch base). Inch base gears and metric base gears are NOT interchangeable. The various gear parts are shown in Fig. 12-198.

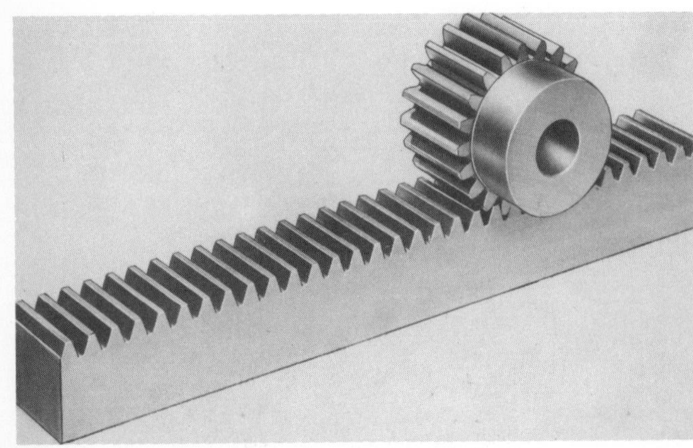

Fig. 12-197. Spur gear meshes in rack gear. Spur gear is simplest of gears. Teeth are cut straight across gear face. Rack is a flat section of metal with teeth cut into it. Combination of spur gear and rack converts rotary motion to linear motion. (Boston Bear)

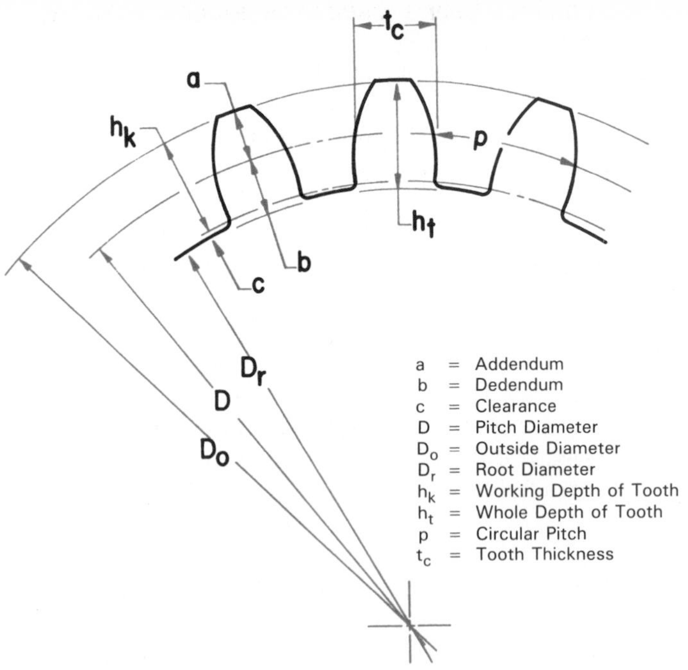

Fig. 12-198. Study various parts of a spur gear.

1. ***Pitch diameter*** (D): The diameter of the pitch circle.

$$D = \frac{N}{P} \text{ or } D = 0.3183pN \text{ or } D = \frac{D_oN}{N + 2}$$

2. ***Diametral pitch*** (P): The number of teeth per inch of pitch diameter.

$$D = \frac{N}{D} \text{ or } P = \frac{N + 2}{D_o} \text{ or } P = \frac{\pi}{p}$$

3. ***Circular pitch*** (p): The distance, measured on the pitch circle, between similar points on adjacent teeth.

$$p = \frac{\pi}{P} \text{ or } p = \frac{\pi D}{N} \text{ or } p = \frac{\pi D_o}{N + 2}$$

4. ***Number of teeth*** (N): The number of teeth on a gear.

$$N = DP \text{ or } N = D_oP - 2 \text{ or } N = \frac{\pi D}{p}$$

5. ***Outside diameter*** (D_o): Diameter or size of the gear blank.

$$D_o = D = 2a \text{ or } D_o = \frac{N}{P} + 2\left(\frac{1}{P}\right) \text{ or } D_o = \frac{N + 2}{P}$$

6. ***Whole depth of tooth*** (h_t): Total depth of a tooth space, equal to the addendum(a) plus dedendum(b), or the depth to which each tooth is cut.

$$h_t = a + b \text{ or } h_t = \frac{2.250}{P} \text{ or } h_t = \frac{2.157}{p}$$

7. ***Working depth*** (h_k): The sum of the addendums of the two mating gears.

$$h_k = a_1 + a_2$$

8. ***Clearance*** (c): The difference between the working depth and the whole depth of a gear tooth. The amount by which the dedendum

on a given gear exceeds the addendum of the mating gear.

$$c = \frac{0.157}{p}$$

9. ***Addendum*** (a): The distance the tooth extends ABOVE the pitch circle.

$$a = \frac{1}{P} \text{ or } a = \frac{D}{N} \text{ or } a = \frac{D_o}{n + 2}$$

10. ***Dedendum*** (b): The distance the tooth extends BELOW the pitch circle.

$$b = \frac{1.157}{p}$$

11. ***Tooth thickness*** (t_c): Thickness of the tooth at the pitch circle. The dimension used in measuring tooth thickness with vernier gear tooth caliper.

$$t_c = \frac{1.5708}{P}$$

12. ***Pitch circle:*** An imaginary circle located approximately half the distance from the roots and tops of the gear teeth. It is tangent to the pitch circle of the mating gear.
13. ***Pressure angle*** (θ): The angle of pressure between contacting teeth of mating gears. It represents the angle at which the forces from the teeth of one gear is transmitted to the mating tooth of another gear. Pressure angles of 14 1/2°, 20°, and 25° are standard. However, the 20° is replacing the older 14 1/2°.
14. ***Distance between centers of two mating gears*** (C): This distance may be calculated by adding the number of teeth of both gears and dividing one-half that sum by the diametral pitch.

$$C = \frac{N_1 + N_2}{2} \div P$$

15. N_1 = Number of teeth on first gear.
16. N_2 = Number of teeth on second gear.

Gear cutters

No one gear cutter can be employed to cut all gears. Gear cutters are made with eight different forms for each diametral pitch (P), depending upon the number of teeth for which the cutter is to be used. Fig. 12-200 illustrates the comparative sizes for gear teeth. The cutter range is as follows:

No. of Cutter	Range of Teeth
1	135 to a rack
2	55 to 134
3	35 to 54
4	26 to 34
5	21 to 25
6	17 to 20
7	14 to 16
8	12 to 13

With the information furnished, it is possible to calculate the data needed to cut a simple spur gear (inch base).

EXAMPLE PROBLEM: Calculate the data needed to cut a 40 tooth, 10 diametral pitch gear.

1. Diameter (D_o) of gear blank needed.
   Having: Diametral pitch (P) = 10
   Number of teeth (N) = 40
   FORMULA: $D_o = \frac{N + 2}{P} = \frac{40 + 2}{10} = \frac{42}{10}$
   $= 4.200$
   $= 4.200$ diameter
2. Whole depth of tooth (h_t) needed. This will be the depth of the cut.
   Having: Diametral pitch (P) = 10
   FORMULA: $h_t = \frac{2.157}{P} = \frac{2.157}{10} = 0.216$
   $= 0.216$ in.
3. The dimension of the addendum (a) is needed to measure the gear tooth for determining whether it is being machined to specifications.
   Having: Diametral pitch (P) = 10
   FORMULA: $a = \frac{1}{P} = \frac{1}{10} = 0.100$
   $= 0.100$ in.
4. Tooth thickness (t_c) is needed to determine whether the gear is being machined to specifications.
   Having: Diametral pitch (P) = 10
   FORMULA: $t_c = \frac{1.5708}{P} = \frac{1.5708}{10} = 0.157$
   $= 0.157$ in.
5. Reference to the gear cutter chart indicates that a No. 3 cutter, with a range of 35 to 54 teeth must be used to cut 40 teeth.
6. Using a 40:1 ratio dividing head for this job means that the index crank must be turned through one complete revolution to position the gear blank for each cut. A 5:1 ratio dividing head would require the use of an index plate that would permit a setting of one-eighth turn for each cut.

Cutting the gear

A few simple precautions, carefully followed, will

Fig. 12-199. This is a typical gear cutter.

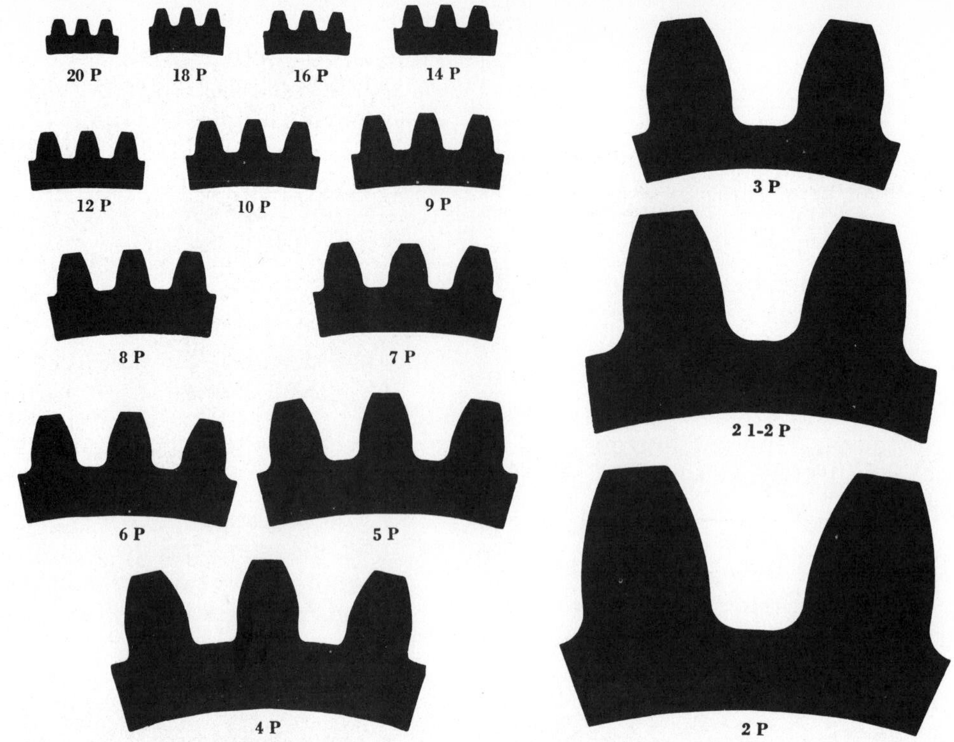

Fig. 12-200. Note comparative sizes of gear teeth. Diametral Pitch is shown.

greatly reduce the possibilities of an inaccurately machined gear.

1. Set up the milling machine as previously described. Check center alignment of the dividing head and foot stock.
2. Press the gear blank on a mandrel and mount the unit to the dividing head. Cutting is done TOWARDS the dividing head, Fig. 12-201.
3. Indicate the gear blank longitudinally and rotate it through one complete revolution. Make adjustments if necessary.
4. Center the cutter on the gear blank. Use a depth micrometer and a steel square, Fig. 12-202. Position the cutter for depth. Use the paper strip technique. Set the micrometer dial to 0; then raise the table to within 0.040 in. of finished depth. For the example problem, it would be raised to make a cut of 0.216 minus 0.040 in., or 0.176 in.
5. Move the work until the cutter just starts to remove metal. Back it away from the cutter. Using the dividing head, bring the next cut into position. Repeat this sequence around the gear blank until you are back at the original cutting position. If there is exact alignment with the first cut, you are ready to cut the gear.
6. Make the roughing cuts. Use liberal quantities of cutting fluid.
7. The finish cut requires more care. Set the vertical scale on the gear tooth vernier caliper to the distance calculated for the addendum (a), Fig. 12-203. Raise the work to within a few thousandths of the calculated whole depth of the tooth (h_t) and make cuts at TWO positions. Make your measurement and adjust until the reading equals as calculated for tooth thickness (t_c). Make the finish cuts. Press the completed gear from the mandrel. Remove all burrs and cut the keyway if required.

Spiral gears, Fig. 12-204, and helical gears, Fig. 12-205, are cut on a universal type milling machine utilizing a ***universal dividing head*** geared to the table feed screw, Fig. 12-206.

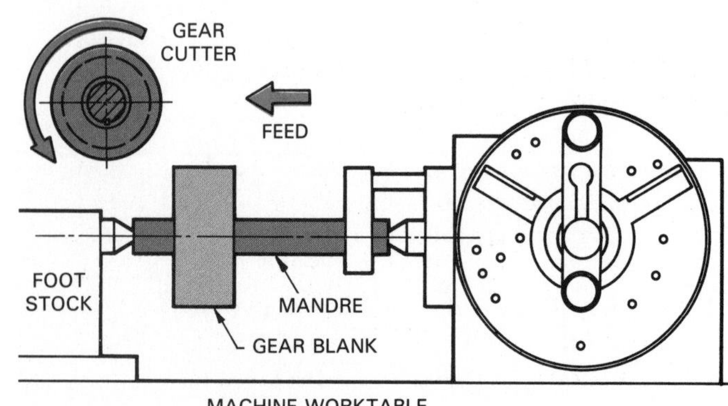

Fig. 12-201. Cutting is done toward dividing head.

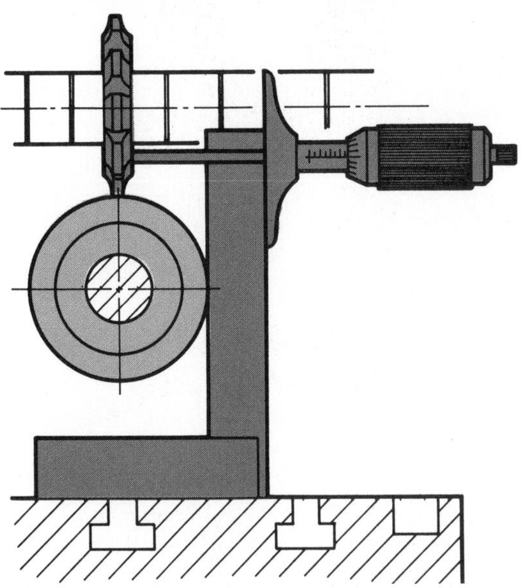

Fig. 12-202. A depth micrometer can be used to center gear cutter on gear blank.

Fig. 12-204. Study setup for cutting a spiral gear. Universal dividing head is coupled to automatic feed mechanism of worktable. Operation is similar to cutting threads on a lathe. Note how table is angled with respect to cutter. (Kearney & Trecker Corp.)

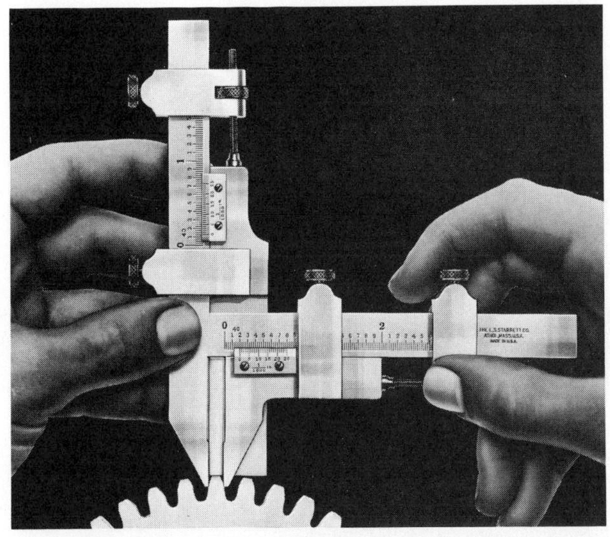

Fig. 12-203. Machinist is measuring gear tooth with gear tooth vernier caliper. Tool is read in same manner as a vernier caliper and vernier height gage. (L.S.Starrett Co.)

Fig. 12-205. Left—Helical gear. Right—Spur gear.

PRECAUTIONS WHEN OPERATING A MILLING MACHINE

1. Avoid performing a machining operation on the milling machine until you are thoroughly familiar with how it should be done.
2. Some materials that are machined produce chips, dust, and fumes that are dangerous to your health. NEVER machine materials that contain asbestos, Fiberglas, beryllium, and beryllium copper unless you are fully aware of the precautions that must be taken.

Fig. 12-206. Universal dividing head and gearing mechanism rotates work at a predetermined rate as work moves into cutter. In this situation, teeth are being machined in a helical milling cutter blank. (Cincinnati Milacron)

3. Maintain cutting fluids properly. Discard them when they become rancid or contaminated.
4. Be sure the cutter rotates in the proper direction. Expensive cutters can be quickly ruined.
5. Carefully store milling cutters, arbors, collets, adapters, etc., after use. They can be damaged if not stored properly.
6. Never start a cut until you are sure there is adequate clearance on all moving parts!
7. Exercise care when handling long sections of metal. Accidentally contacting a light fixture of busbar and cause severe electrical burns and even electrocution!
8. Carefully read instructions when using the new synthetic oils, solvents, and adhesives. Many of them are dangerous if NOT handled correctly.
9. Use adequate ventilation for jobs where dust and fumes are a hazard. Return oils and solvents to proper storage. Wipe up spilled fluids. Do NOT pour used coolants, oil, solvents, etc., down a drain!
10. Should the area where you work be extremely noisy, wear hearing protectors. Take no chances. Protect your sight and hearing at all times in the shop!

INDUSTRIAL APPLICATIONS

The milling machines found in industry operate on the same basic principle as those found in training programs. In many cases, the same equipment is utilized.

Some small job shops, in order to secure maximum use from their equipment, operate multipurpose machines like the vertical/horizontal machine shown in Fig. 12-207.

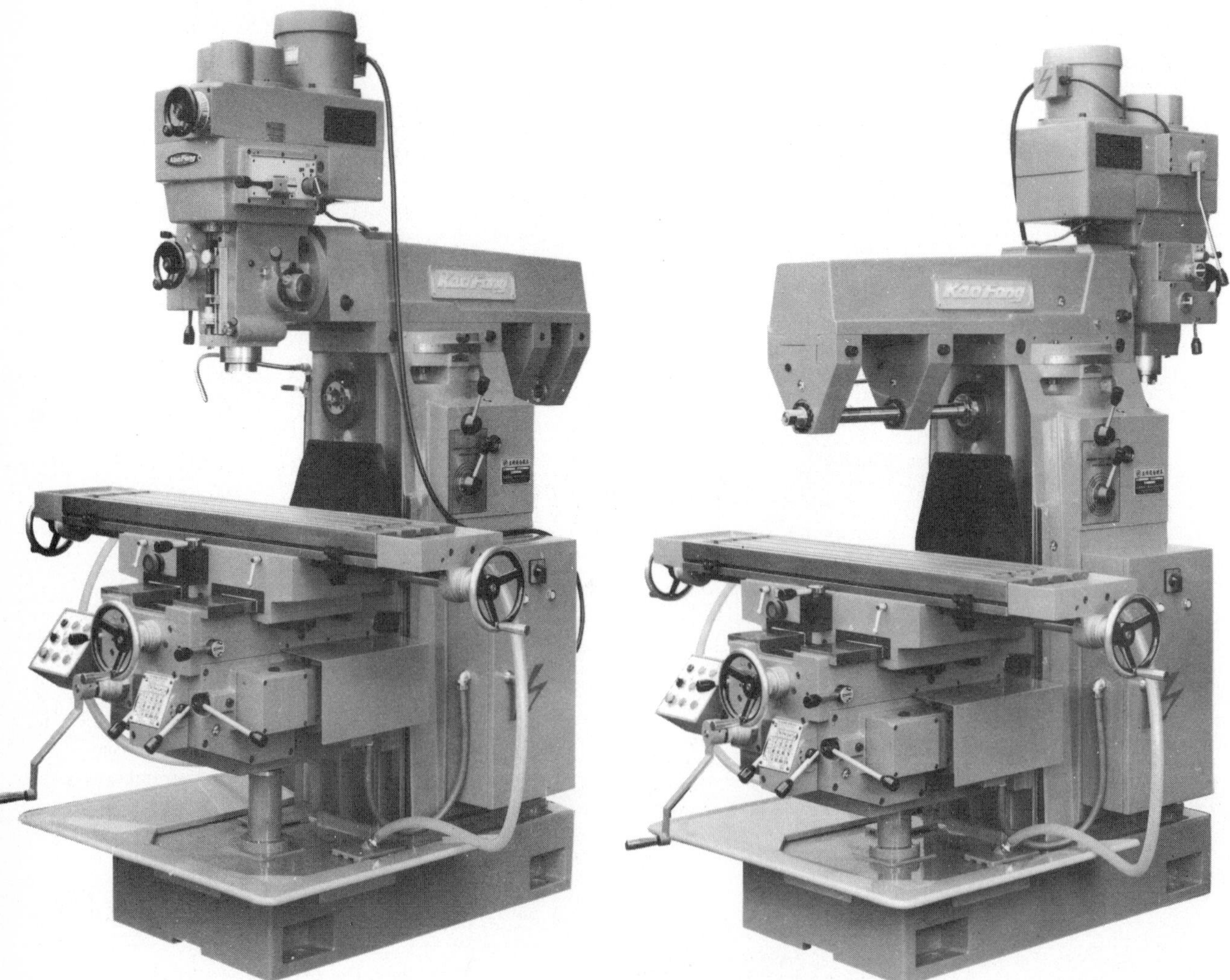

Fig. 12-207. Note multipurpose milling machine has two spindles that swing over table. Left. As a vertical mill. Right. As a horizontal mill. (Sharp Industries, Inc.)

Optical scanners are available for many vertical milling machines. The scanner follows lines on a drawing, Fig. 12-208. This activates a slave mill, usually fitted with an end mill, that duplicates scanner movement, Fig. 12-209. The drawings that the scanner follows are made 10 to 20 times actual machined size. Finished accuracy is within 0.0001 in. (0.002 mm).

Figs. 12-210 through 12-220 illustrate a number of machines utilized by industry. Most are computer controlled.

Fig. 12-210. Machinist is using computer controlled fixed bed milling machine. Note variety of cutting tools. (Portage Machine Co.)

Fig. 12-208. Optical scanning unit follows line drawing of a complex pattern.

Fig. 12-211. Machinist is milling a shoulder radius on a huge steel forging. Note cutter size and steel frame thickness. (General Dynamics—Electro Dynamics Div.)

Fig. 12-209. Slave milling unit follows information received from scanning unit and translates it into machine movement. Automatic compensation is made for tool wear.

Fig. 12-212. An aft head of a solid propellant rocket chamber is being contoured on a template controlled three-dimensional milling machine. The template, located at top of work table, guides cutting tool through a hydraulically controlled system. Three-dimensional milling of this unit eliminates much welding (potential sources of structural weakness) that would otherwise be necessary. (Avco Corp., Lycoming Div.)

Fig. 12-213. Note double housing milling machine. Two additional milling heads can be mounted on crossrail of this fixed bed type machine. (Rockford Machine Tool Co.)

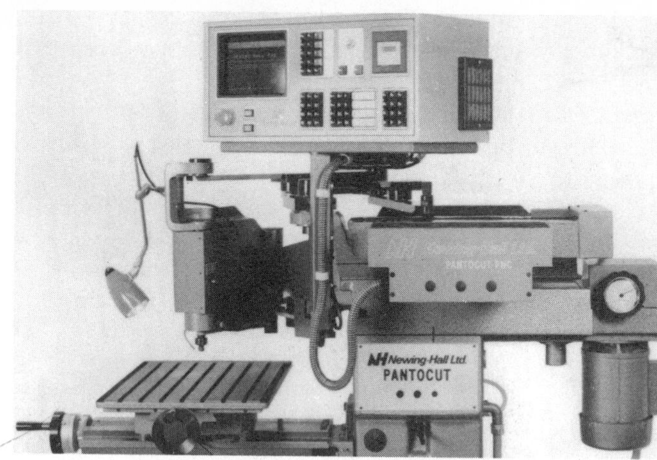

Fig. 12-214. This CNC profile milling and engraving machine has video display. Display provides accurate positional and system status at all times. A digitizing tablet and remote programming module simplify complex programming. (Newing Hall Ltd.)

Fig. 12-215. These are examples of parts produced on profile milling and engraving machine. Note irregular shapes that would be difficult to cut. (Newing Hall Ltd.)

Fig. 12-216. Note examples of engraving done on profile milling and engraving machine. (Newing Hall Ltd.)

Fig. 12-217. CNC vertical milling machine has numerical control of three axes. Input data may be in either cartesian or polar coordinates, with incremental or absolute values in inches or metric. (Bridgeport Machines, Inc.)

TEST YOUR KNOWLEDGE—Chapter 12

Please do not write in the text. Place your answers on a separate sheet of paper.

1. Milling machines fall into two broad classifications: ____ and ____ and ____ types.
2. There are three basic types of ____ and ____ milling machines.
   a. A ____ ____ has a horizontal spindle and the work table has three movements.
   b. A ____ ____ is similar to the above machine but a fourth movement has been added to the work table to permit it to cut helical shapes.
   c. A ____ ____ ____ has the spindle perpendicular or at right angles to the work table.
3. List the four methods of machine control. Briefly describe each of them.
4. Stop the machine before making ____ and ____.
5. Metal chips must never be removed with your ____. Use a ____.
6. Treat all small cuts and skin punctures as potential sources of infection. The following should be done:

Fig. 12-218. This is horizontal machining center with an 80-tool automatic tool changer. Fitted with eight-station pallet loading, machine has simultaneous linear positioning in all three axes as well as two-axis circular interpolation. It is also fitted with fourth axis contouring and indexing system. (Bridgeport Machines, Inc.)

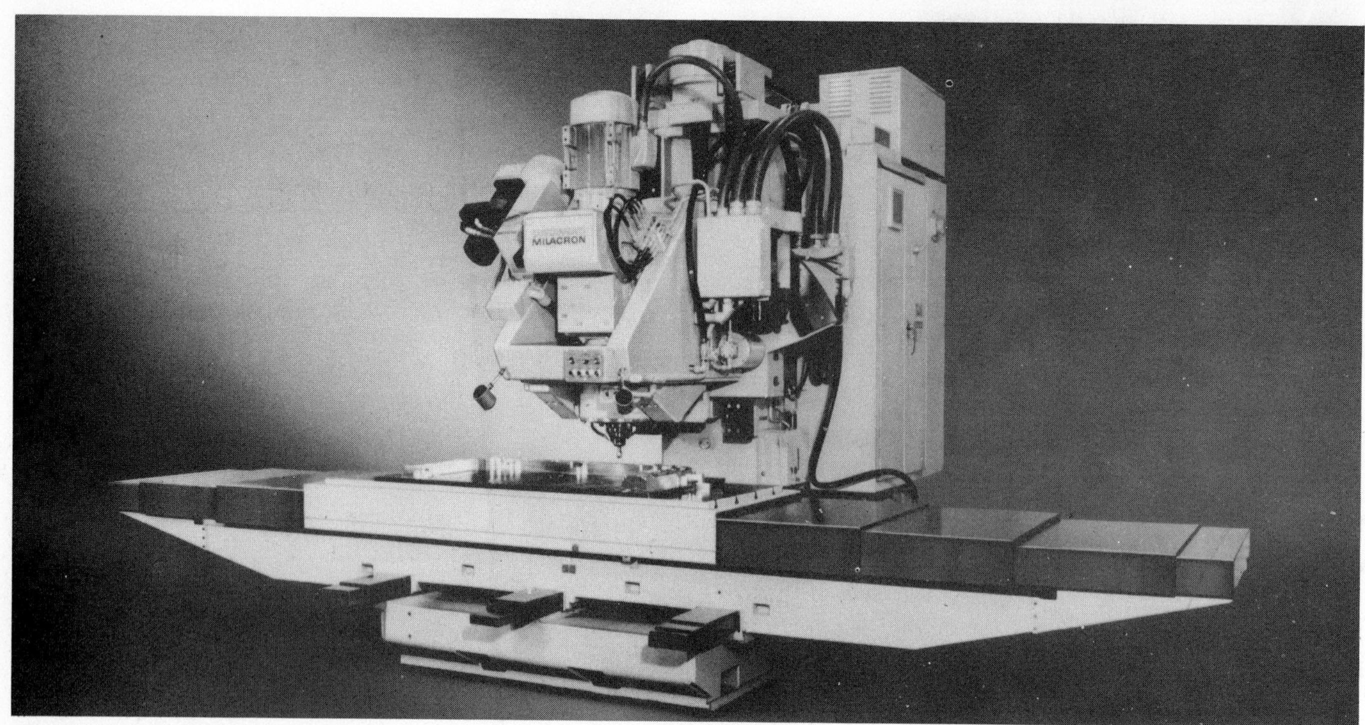

Fig. 12-219. Note 5-axis CNC vertical axis profiling/contouring milling machine. (Cincinnati Milacron)

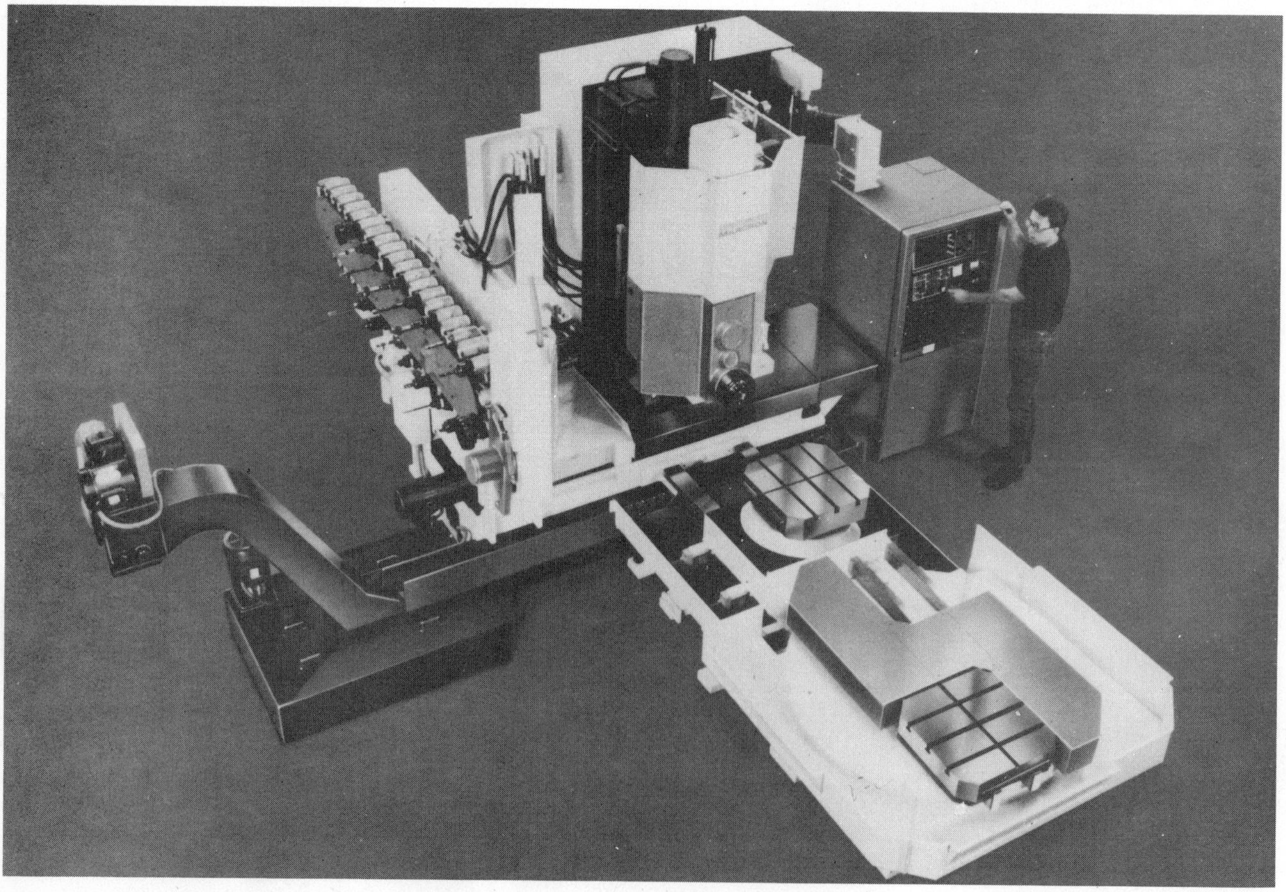

Fig. 12-220. Machinist is using CNC machining center fitted with a two-pallet automatic workchanger. An automatic tool changer and high pressure coolant system that flushes chips into a chip conveyor permits periods of unmanned operation. (Cincinnati Milacron)

a. Clean them thoroughly.
b. Apply antiseptic and cover with a bandage.
c. Promptly report the injury to your instructor.
d. All of the above.
e. None of the above.

7. Milling cutters are sharp. Protect your hands with a ________ of ________ when handling them.
8. Milling operations fall into two main categories:
   a. ________ milling—the surface being machined is parallel with the cutter face.
   b. ________ milling—the surface being machined is parallel with the periphery of the cutter.
9. What are two general types of milling cutters?
10. What is the term "HAND" used to describe in reference to an end mill?

Match each term with the correct sentence below.

11. ____ Two-flute end mill.
12. ____ Multi-flute end mill.
13. ____ Fly cutter.
14. ____ Shell end mill.
15. ____ Face milling cutter.
16. ____ Plain milling cutter.
17. ____ Slab cutter.
18. ____ Side milling cutter.
19. ____ Staggered-tooth side cutter.
20. ____ Metal slitting saw.
   a. Has cutting teeth on the circumference and on both sides.
   b. Cutter with helical teeth designed to cut with a shearing action.
   c. A facing mill with a single point cutting tool.
   d. Can be fed into work like a drill.
   e. Cutter with teeth located around the circumference.
   f. Recommended for conventional milling where plunge cutting (going into work like a drill) is NOT required.
   g. Intended for machining large flat surfaces parallel to the cutter face.
   h. Thin milling cutter designed to machine narrow slots and for cutoff operations.
   i. Mounts on a stub arbor.
   j. Has alternate right-hand and left-hand helical teeth.
21. Flat surfaces are machined with ________ or ________ tooth ________ milling cutters.
22. Make a sketch that illustrates climb milling.
23. Make a sketch illustrating conventional milling.
24. In climb milling:
   a. The work is fed into the rotation of the cutter.
   b. The work moves in the same direction as the rotation of the cutter.
   c. Neither of the above.
25. In conventional milling:
   a. The work is fed into the rotation of the cutter.
   b. The work moves in the same direction as the rotation of the cutter.
   c. Neither of the above.
26. What is a draw-in bar and how is it used?
27. ________ ________ refers to the distance, measured in ________ or ________, a point (tooth) on the circumference of a cutter moves in ________ ________.
28. ________ is the rate at which the work moves into the cutter.
29. Calculate machine speed (RPM) and feed (F) for a 1.5 in. diameter tungsten carbide 5 tooth (T) end mill machining cast iron. Recommended cutting speed is 190 FPM. Feed per tooth (FTR) is 0.004 in. Use the following formulas:

$$RPM = \frac{FPM \times 12}{\pi D}, \text{ and } F = FTR \times T \times RPM$$

30. Determine machine speed (RPM) and feed (F) for a 2 1/2 diameter HSS shell end mill with 8 teeth (T) machining aluminum. Recommended cutting speed is 550 FPM. Feed per tooth (FTR) is 0.010 in. Use the formulas given in problem 29.
31. Calculate machine speed (RPM) for machining aluminum with a 6 in. diameter HSS side milling cutter. Recommended cutting speed is 550 FPM. Use the appropriate formula given in problem 29.
32. Determine machine speed (RPM) and feed (F) for a 4 in. diameter HSS side milling cutter with 16 teeth (T) milling free cutting steel. Recommended cutting speed is 200 FPM. Feed per tooth (FTR) is 0.005 in. Use the formulas given in problem 29.
33. Calculate machine speed (RPM) and feed (F) for a 2.5 in. diameter HSS slab milling cutter with 8 teeth (T) machining brass. Recommended cutting speed is 250 FPM. Feed per tooth (FTR) is 0.006 in. Use the formulas given in problem 29.
34. Cutting fluids serve three purposes. List them.

Match each term with the correct sentence below.

35. ____ Flanged vice. 39. ____ Dividing head.
36. ____ Swivel vise. 40. ____ Indexing table.
37. ____ Universal vise. 41. ____ Magnetic chuck.
38. ____ Rotary table. 42. ____ Vise lug.
   a. Keys vise to a slot in work table.
   b. Can only be mounted parallel or at right angles on work table.
   c. Needed when cutting segments of circles, circular slots, and locating angularly spaced holes and slots.
   d. Can only be used with ferrous metals.
   e. Has circular base graduated in degrees.
   f. Permits rapid positioning of circular work

in 15 deg. increments and can be locked at any angular setting.

g. Used to divide circumference of round work into equally spaced divisions.

h. Permits compound angles (angles on two planes) to be machined without complex or multiple setups.

43. _______ _______ mills and _______ _______ mills are the cutters normally used on a vertical milling machine.

44. The _______ end mill is used when the cutter must be fed into the work like a drill.

45. Blind holes or closed keyseats are made with a _______ end mill.

46. Face milling cutters over 6 in. (150 mm) diameter are usually of the _______ _______ type.

47. A _______ scale on the spindle head of a vertical milling machine assures accurate angular settings.

48. List three methods for machining chamfers, bevels, and tapered sections on a vertical milling machine.

49. A _______ or _______ _______ can be used to locate the first hole of a series to be drilled on a vertical milling machine.

50. The most accurate way to align a vise on a milling machine is with a _______ _______.

51. Explain how to center an end mill on round stock for the purpose of machining a keyseat. Use the paper strip technique.

52. Gang milling means:
    a. Several cutters being used at same time to machine a job.
    b. Two or more cutters straddling job.
    c. Several side cutters being used at same time to machine a job.
    d. All of the above.
    e. None of the above.

53. Why should a milling cutter be keyed to the arbor?

54. When sawing (slitting) thin stock, the _______ diameter cutter that provides adequate clearance should be used.

55. In general, use the _______ arbor possible that permits adequate clearance between the arbor support and the work.

56. Describe how to safely remove or mount a milling cutter on an arbor.

57. Prepare the information necessary to cut a 20 pitch gear with 100 teeth. The dividing head used has 40:1 ratio and the index plate has the following series of holes: 33, 37, 41, 45, 49, 53, and 57.

53. What will be the outside diameter of a gear with 36 teeth-10 pitch?

59. Two gears in mesh have pitch circles of 6.125 in. and 8.500 in. What is their center distance?

60. List five precautions to be observed when operating a milling machine.

RESEARCH AND DEVELOPMENT

1. Prepare a display panel that includes samples of the various types of milling cutters. Include manufacturers' catalogs and price lists.
2. Develop a bulletin board using illustrations of the various types of milling machines. If available, include how much each machine costs.
3. The milling machine and its inventor, Eli Whitney, played an important part in developing mass production techniques. Prepare a term paper on Whitney's project of producing 10,000 muskets, with interchangeable parts, for the federal governmment in 1798, and how it led to the invention of the milling machine.
4. Prepare a video presentation or a series of 35 mm slides illustrating the safety precautions that must be followed when operating a vertical milling machine.
5. Make a series of posters dealing with milling machine safety.
6. Cutting fluids play an important part in any machining operation. Secure samples of cutting fluids used by industry and conduct a series of experiments to show the quality of surfaces machined dry and with the various cutting compounds. Your experiment should include milling aluminum, brass, and steel.
7. Demonstrate how to use a wiggler and edge finder.
8. Demonstrate how boring is done on a vertical milling machine.
9. Demonstrate how to center a cutter on round stock for milling a keyseat. Use the paper strip technique.
10. Overhaul and paint a milling machine in your training facility.
11. Demonstrate how to use a dividing head.
12. Demonstrate how to cut a spur gear.
13. Demonstrate how to machine helical gears.
14. Present a motion picture or video tape on CNC milling machines. Lead the discussion on what was seen.
15. Milling machines were the first machine tools to be automated. Do a research project on automated milling machines. Include samples of the programs, tapes, and specialized drawings used.
16. Develop a simple project to be made by a NC of CNC milling machine. Prepare the program to do the job.

Chapter 13

PRECISION GRINDING

After studying this chapter, you will be able to:

- Explain how precision grinders operate.
- Identify the various types of precision grinding machines.
- Select, dress, and true grinding wheels.
- Safely operate a surface grinder using various work holding devices.
- Solve common surface grinding problems.
- List safety rules related to precision grinding.

Grinding, like milling, drilling, sawing, planing, and turning, is a cutting operation. However, instead of having one, two, or multiple cutting edges working, grinding makes use of an abrasive tool composed of thousands of cutting edges. See Fig. 13-1.

The grinding wheel might be compared to a many toothed milling cutter as each of the abrasive particles is actually a separate cutting edge, Fig. 13-2.

In precision grinding, each abrasive grain removes a relatively small amount of material permitting a smooth, accurate surface to be generated. It is also one of the few machining operations that can produce a smooth, accurate surface on material regardless of its hardness. Grinding is frequently a finishing operation.

SAFETY NOTE! Never attempt to operate these machines while your senses are impaired by medication or other substances.

TYPES OF SURFACE GRINDERS

While all grinding operations might be called surface grinding because all grinding is done on the surface of the material, industry classifies ***surface grinding*** as the grinding of FLAT SURFACES.

Fig. 13-1. Grinding wheel removes material in same manner as a milling cutter, but in much smaller chips of metal.

Fig. 13-2. Closeup shows abrasive grains that make up a typical grinding wheel, magnified about 50x. (Cincinnati Milacron)

There are two basic types of surface grinding machines:

1. ***Planer type surface grinders*** make use of a reciprocating motion to move the work table back and forth under the grinding wheel, Fig. 13-3. Three variations of planer type surface grinding are illustrated in Fig. 13-4.
2. ***Rotary type surface grinders*** have circular work tables that revolve under the rotating grinding wheel, Fig. 13-5. Two variations of the technique are shown in Fig. 13-6.

The planer type surface grinder is frequently found in training centers. It slides the work back and forth under the edge of the grinding wheel. Table movement can be controlled manually and/or by means of a mechanical or hydraulic drive mechanism.

A manually operated machine is shown in Fig. 13-7. All work and grinding wheel movements are made by hand.

The ***traverse handwheel*** (9) controls the left-and-right reciprocating movement of the table. Cross-feed (in-and-out motion) is controlled by the ***cross-feed handwheel*** (8). The ***down-feed handwheel*** (1) controls the up-and-down adjustment of the grinding wheel. It is located on the vertical column.

The machine shown in Fig. 13-8 makes use of hydraulic traverse feed and cross-feed. It is also fitted with a coolant attachment.

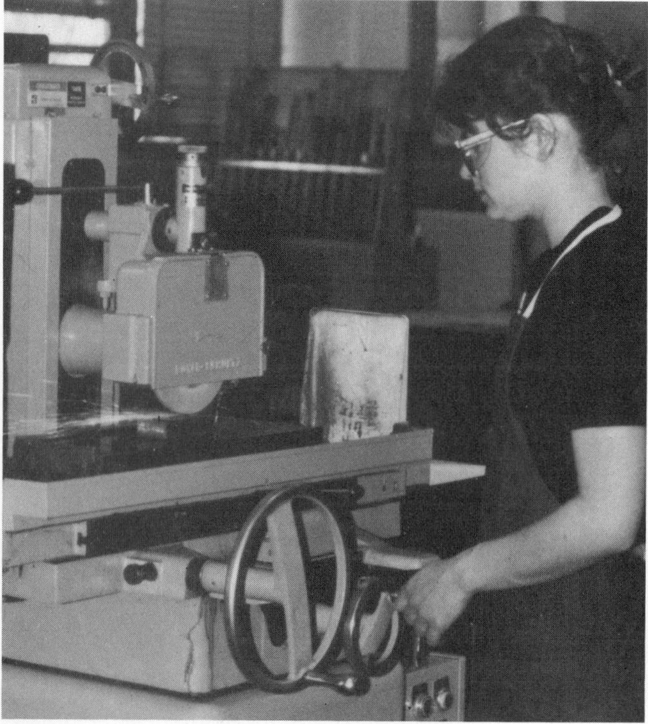

Fig. 13-3. Planer type surface grinder machines flat surfaces. (William L. Schotta, Millersville University)

Both machines operate in much the same manner. However, the operator of a manually operated machine must develop a rhythm to get a smooth, even cutting stroke. Spring stops act as cushions at the end of maximum table travel, Fig. 13-9.

Adjustable table stops on the hydraulically activated traverse feed permit the operator to establish

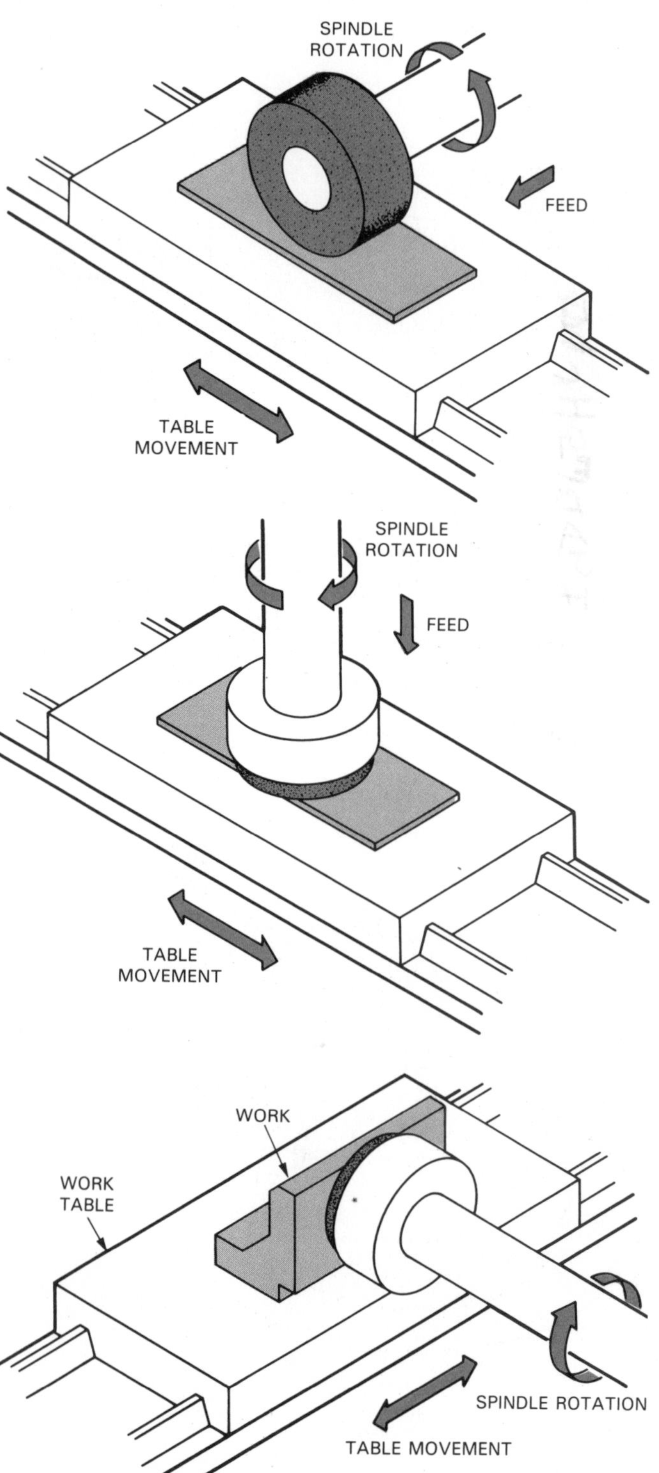

Fig. 13-4. Note three variations of planer type surface grinder.

exact table positioning, Fig. 13-10. At the end of the stroke, table direction is reversed automatically. Automatic cross-feed moves the work in or out a predetermined distance at the completion of each cutting cycle.

A control console is located on the front of the machine, Fig. 13-11. Table travel is started and stopped from this station. Table speed is also controllable from this location. Some grinding machines have another control for ***dwell***—a hydraulic cushion at the end of each stroke.

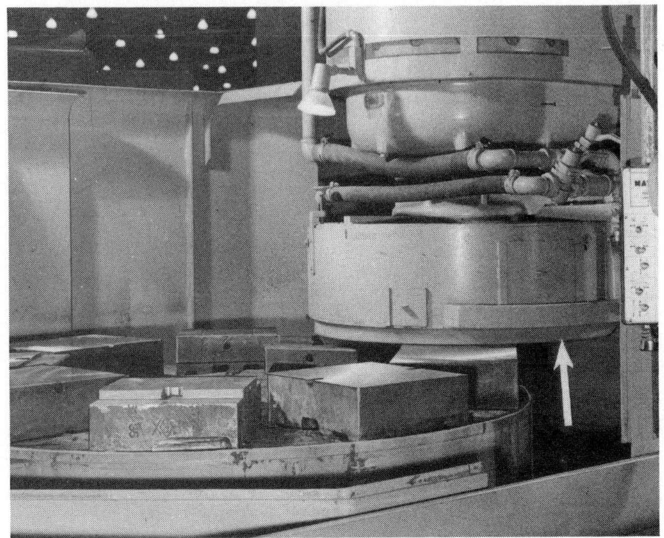

Fig. 13-5. This is a rotary surface grinder. (Norton Co.)

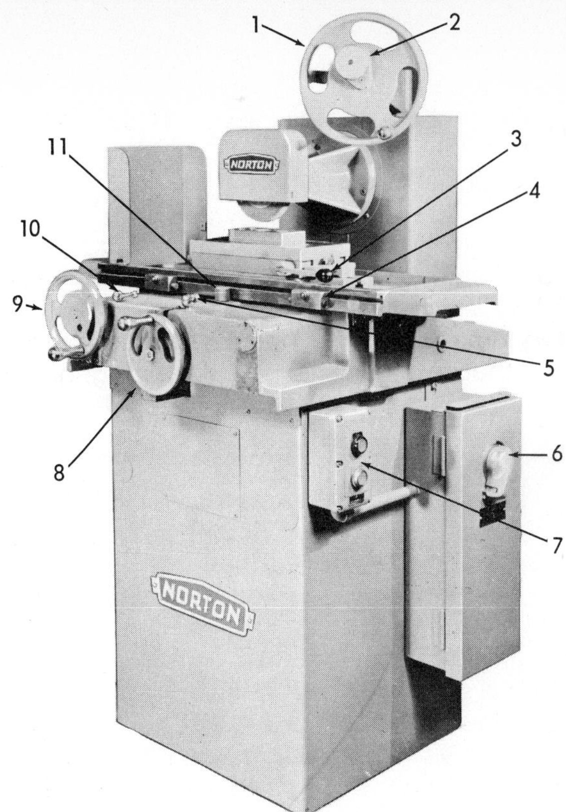

1. Grinding wheel downfeed handwheel... 1 turn equals 0.050 in.
2. Fine downfeed mechanism in increments of 0.0001 in.
3. Control handle for permanent magnetic chuck.
4. Table dog.
5. Cross-feed friction brake for "drag" or locking.
6. Master switch.
7. Electric motor controls.
8. Cross-feed handwheel... one turn equals 0.200 in.
9. Table traverse handwheel.
10. Table traverse friction brake for "drag" or locking.
11. Table traverse stop.

Fig. 13-7. Study parts of manually operated surface grinder. (Norton Co.)

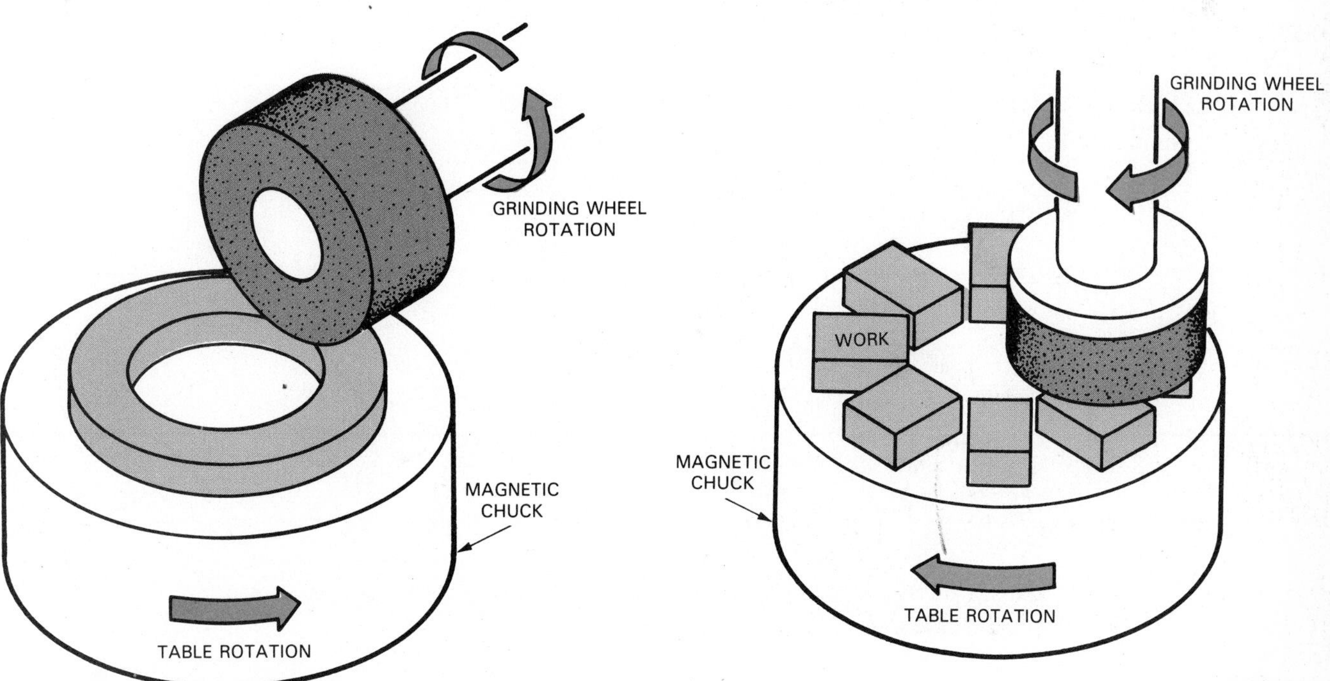

Fig. 13-6. Compare two variations of the rotary type surface grinder.

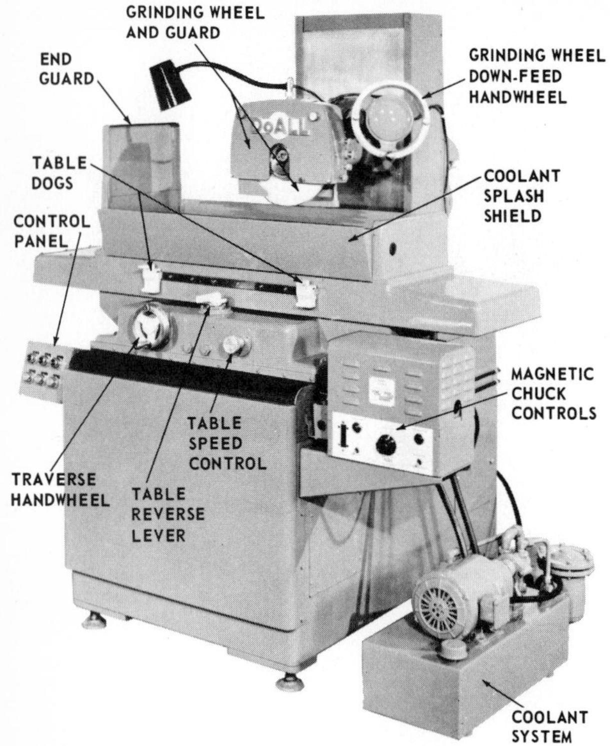

Fig. 13-8. Cross-feed and longitudinal table movements are hydraulically actuated on this surface grinder. (The DoALL Co.)

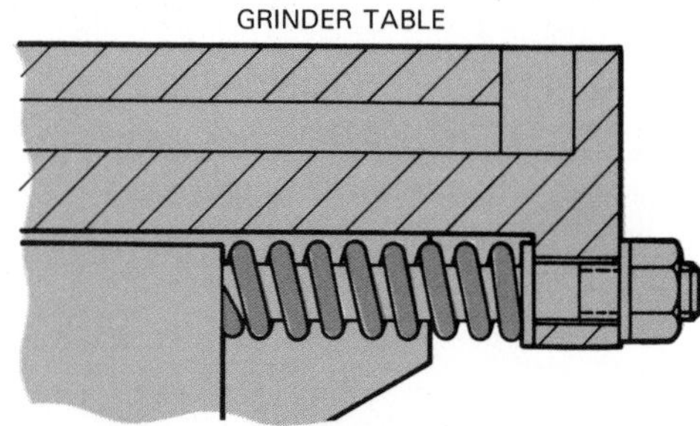

Fig. 13-9. Springs on work table guide are often used to cushion end of stroke on some manually operated surface grinders.

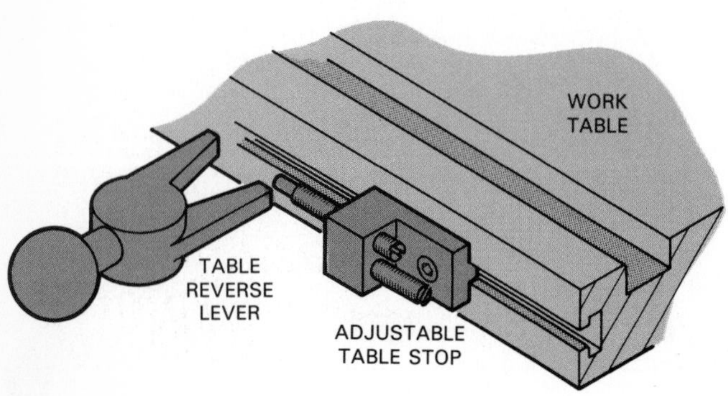

Fig. 13-10. Adjustable table stop is employed to regulate length of worktable stroke.

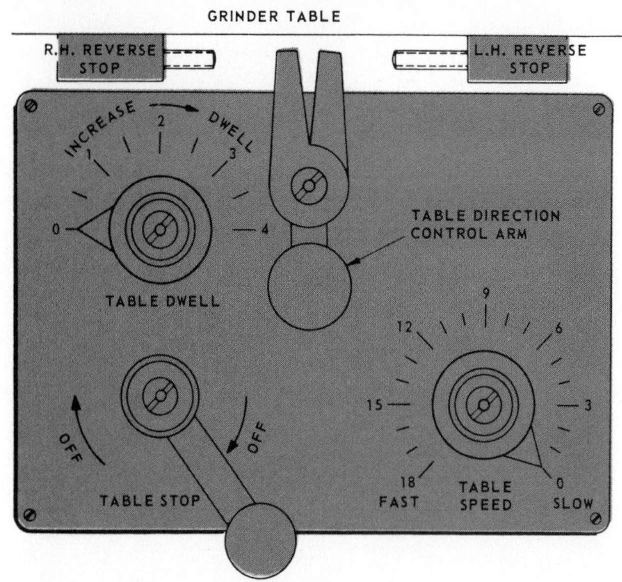

Fig. 13-11. Note typical surface grinder control console. Table dwell sets up a cushioning action at end of each table stroke.

WORK HOLDING DEVICES

Much work done on a surface grinder is held in position by a ***magnetic chuck,*** Fig. 13-12. This holds the work by exerting a magnetic force. Nonmagnetic materials (aluminum, brass, etc.) can be ground by bracing with steel blocks or parallels to prevent movement.

A magnetic chuck makes use of a permanent magnet, Fig. 13-13. This eliminates cords needed for electromagnets and the danger of electrical con-

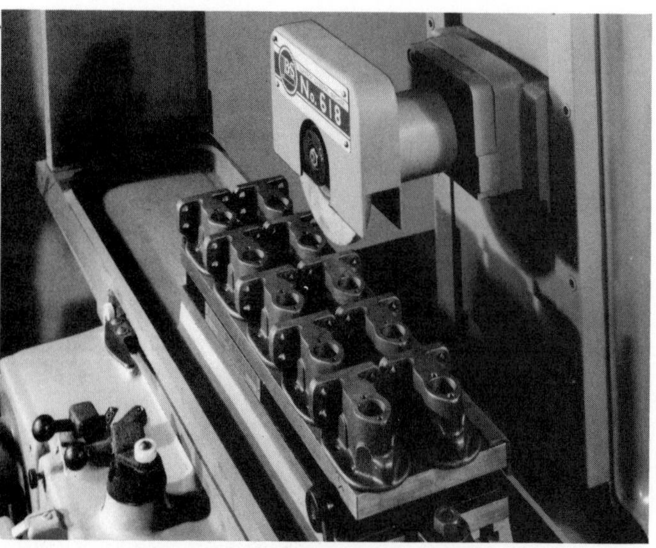

Fig. 13-12. Magnetic chuck is being used to hold multiple pieces for surface grinding. (Brown & Sharpe Mfg. Co.)

Fig. 13-13. Note magnetic chuck with a permanent type magnet. (O.S. Walker Co., Inc.)

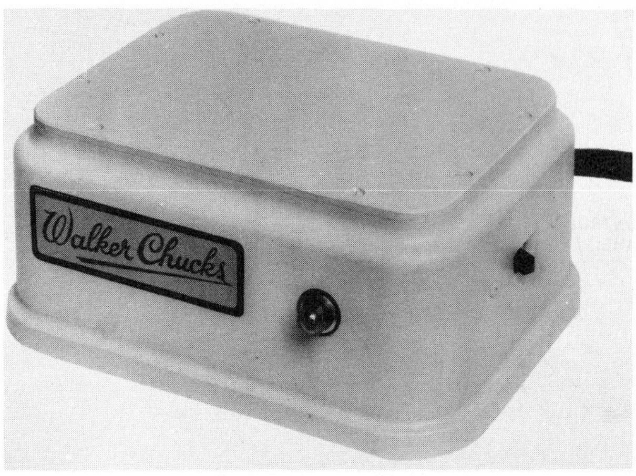

Fig. 13-15. Demagnetizer can be used to neutralize part after being clamped in magnetic chuck. (L-W Chuck Co.)

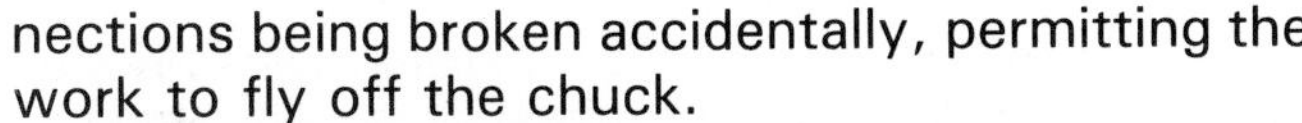
nections being broken accidentally, permitting the work to fly off the chuck.

An ***electromagnetic chuck*** utilizes an electric current to create the magnetic field. See Fig. 13-14.

Frequently, work mounted on a magnetic chuck becomes magnetized and must be demagnetized before it can be used. A ***demagnetizer*** may be employed to neutralize the piece, Fig. 13-15.

Other ways to mount work on a surface grinder are:

1. A ***universal vise,*** Fig. 13-16.
2. An ***indexing head*** with centers, Fig. 13-17.
3. ***Clamps*** to hold the work directly on worktable, Fig. 13-18.
4. ***Double-faced masking tape*** can be used to hold thin sections of nonmagnetic materials. Refer to Fig. 13-19.

Fig. 13-16. A universal vise can be used for grinding operations.

GRINDING WHEELS

As mentioned, each abrasive particle in a grinding wheel is a cutting tooth. As the wheel cuts, the

Fig. 13-14. This magnetic chuck uses electric current to create magnetic field. (O.S. Walker Co., Inc.)

Fig. 13-17. Centers and an indexing head are used for grinding tasks when shape of work permits. Indexing head is utilized in much same manner as dividing head in milling.

Fig. 13-18. Work can also be bolted directly to table for grinding. (Brown & Sharpe Mfg. Co.)

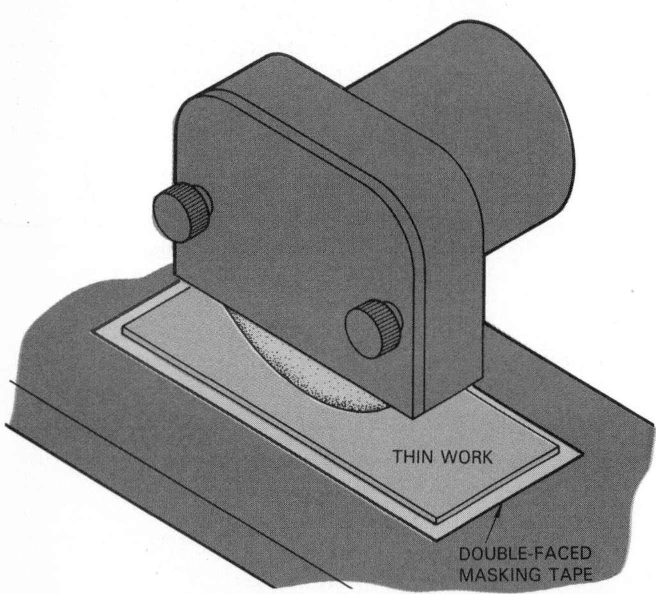

Fig. 13-19. Double-faced masking tape is used to mount thin nonferrous material when it is ground.

metal chips dull the abrasive grains and wear away the ***bonding material*** (material that holds abrasive particles together). The ideal grinding wheel, of course, would be one in which the bonding medium wears away slowly enough to get maximum use from the individual abrasive grains. However, it would also wear rapidly enough to permit dulled abrasive particles to drop off and expose new particles.

Because so many factors govern grinding wheel efficiency, the wheel eventually dulls and must be dressed with a ***diamond dressing tool.*** See Figs. 13-20 and 13-21. Failure to dress the wheel of a precision grinding machine in time, will result in the wheel face becoming loaded or glazed so it cannot cut freely, Fig. 13-22.

Only manufactured abrasives are suitable for modern high-speed grinding wheels. The properties and spacing of abrasive particles and composition of the bonding medium can be controlled to get the desired grinding performance, Fig. 13-23.

To aid in duplicating grinding performance, a standard system of marking grinding wheels has been defined by American National Standards Institute (ANSI). Called ANSI Standard B74.13-1977, it is used by all grinding wheel manufacturers. Five factors were considered:

1. ***Abrasive type*** classifies the abrasive material in the grinding wheel. Manufactured abrasives fall

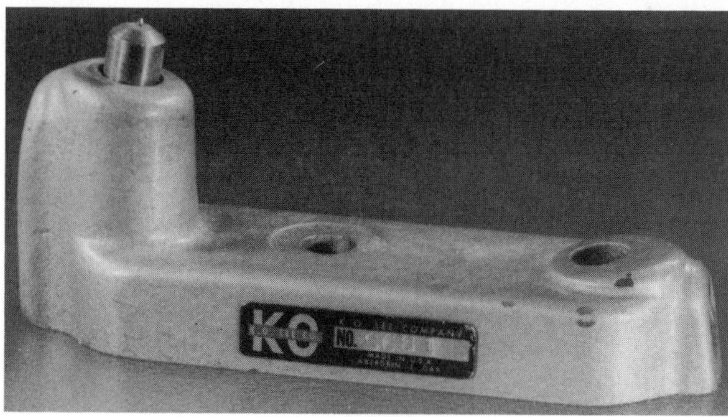

Fig. 13-20. Diamond wheel dressing tool will true and clean wheel face.

Fig. 13-21. Note closeup of diamond chip. Diamond should be rotated a partial turn each time it is used. This will put a new edge of diamond into position.

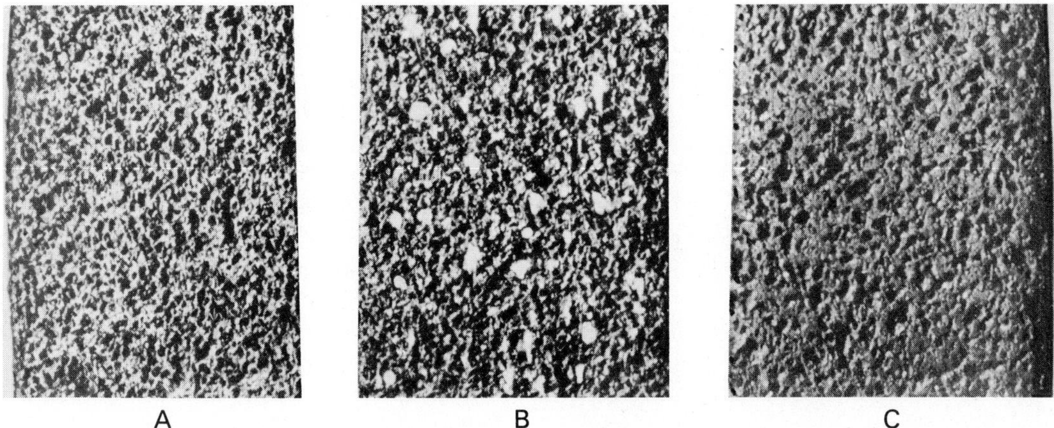

Fig. 13-22. Study grinding wheels in various conditions: A—Properly dressed. B—Loaded. C—Glazed. (Norton Co.)

into two main groups. Letter symbols are used to identify them:

A = Aluminum Oxide
C = Silicon Carbide

An optional prefix number may be employed to designate a particular type of aluminum oxide or silicon carbide abrasive.

2. ***Grain size*** is indicated by a number, usually from 8 (coarse) to 600 (very fine).
3. ***Grade*** is the strength of the bond holding the wheel together ranging from A (soft) to Z (hard).
4. ***Structure*** refers to grain spacing or the manner in which the abrasive grains are distributed throughout the wheel. It is numbered 1 to 16—the higher the number the "more open" the structure (wider grain spacing). The use of this number is optional.
5. ***Bond*** indicates the type of material that holds the abrasive grains (wheel) together. Eight types are used:

B = Resinoid
BF = Resinoid reinforced
E = Shellac
O = Oxychloride
R = Rubber
RF = Rubber reinforced
S = Silicate
V = Vitrified

An additional number or letter(s) is sometimes used as the manufacturer's private marking to identify the grinding wheel. Its application is optional.

The adoption of a standardized grinding wheel marking system has guaranteed, to a reasonable degree, duplication of grinding performance. The wheel marking system is shown in Fig. 13-24.

Grinding wheel shapes

Grinding wheels are made in many standard shapes, Fig. 13-25. While twelve basic face shapes are generally available, the face may be changed

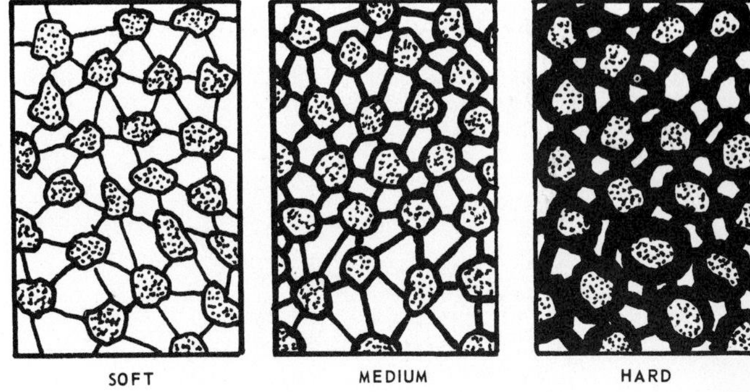

Fig. 13-23. Wheel hardness is determined by type and percentage of bond and grain spacing.

to suit specific job requirements, Fig. 13-26. Wheels used for internal grinding are manufactured in a large selection of shapes and sizes, Fig. 13-27.

How to mount grinding wheels

Select a grinding wheel recommended for the job. Check its soundness by lightly tapping the wheel as shown in Fig. 13-28. A sound wheel will give a clear "metallic ring." If cracked, the tone will be "flat" and NOT a clear "ringing sound."

DANGER! Discard unsound grinding wheels. If possible, break them into several pieces to assure they are not used.

Mount the wheel on the spindle. It should fit snugly. Never force a grinding wheel on a shaft. The blotter rings or compressible washers should be large enough to extend beyond the wheel flanges, Fig. 13-29. It is essential that the wheel be mounted properly or excessive strains will develop during the grinding operation and the wheel could shatter.

WARNING! It is good practice NOT to stand in line with the grinding wheel, especially during the first few passes across the work.

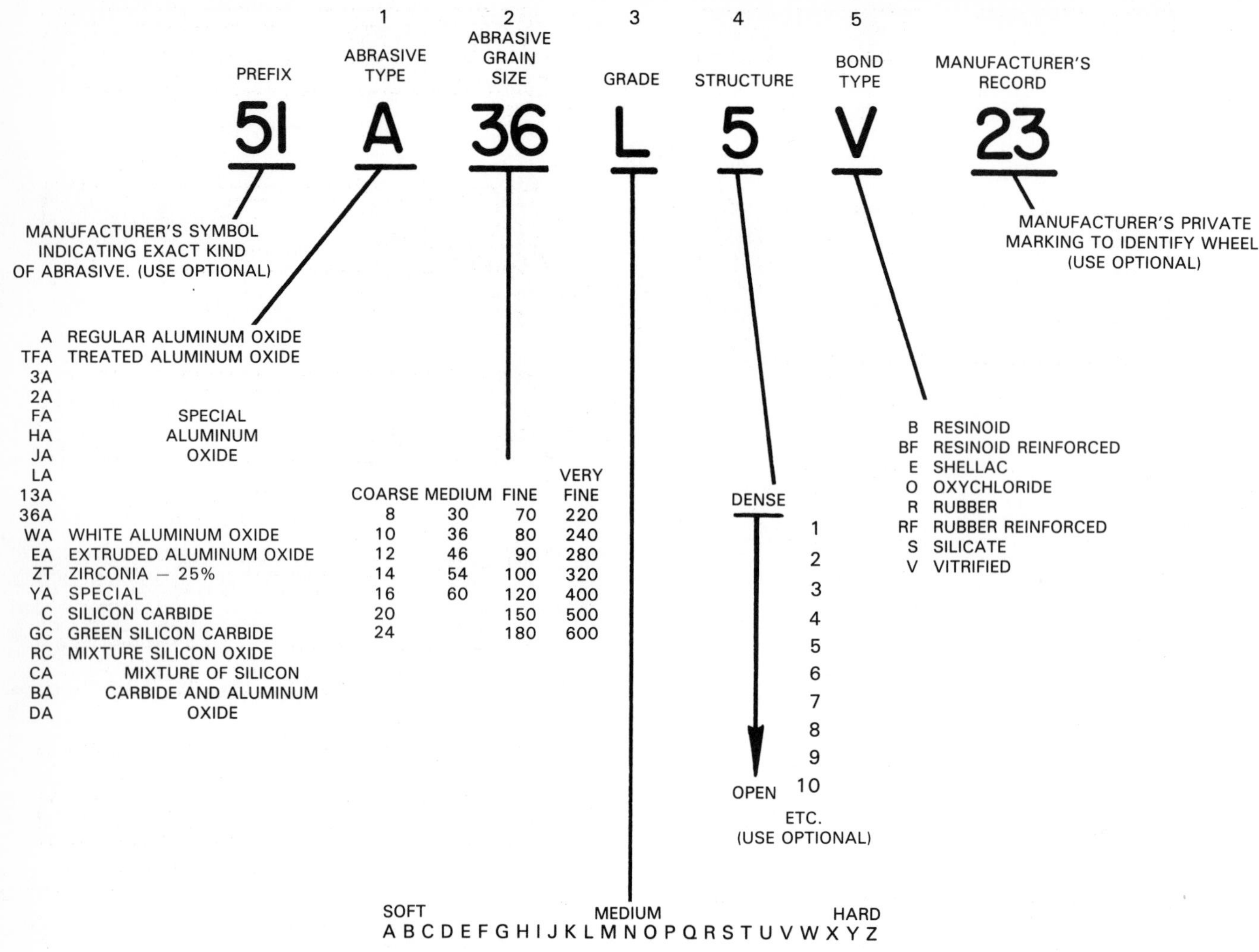

Fig. 13-24. Study standard system for marking grinding wheels.

CUTTING FLUIDS (COOLANTS)

Cutting fluids are an important factor in lessening wear on the grinding wheel. They help to maintain accurate dimensions and are important to the quality of the surface finish produced. As a coolant, it is equally important for it to remove the heat generated during the grinding operation. Heat must be removed as fast as it is generated.

Several types of grinding fluids are utilized:

1. ***Water-soluble chemical fluids*** are solutions that take advantage of the excellent cooling ability of water. They are usually transparent and include a rust inhibitor, water softeners, ***detergents*** to improve the cleaning ability of water, and ***bacteriostasis*** (substance that regulates and controls growth of bacteria).
2. ***Polymers*** are added to improve lubricating qualities.
3. ***Water-soluble oil fluids*** are coolants that take advantage of the excellent cooling qualities of water. It is usually "milky white." It is also less expensive than most chemical type fluids. Bacteriostases are also added to control bacteria growth.

The coolant can be applied by flooding the grinding area, Fig. 13-30. The fluid recirculates by means of a pump and holding tank built into the machine. A mist system forces the coolant over the wheel or applies it to the work surface under pressure (air). It cools by evaporation. A coolant can also be applied manually by pumping the fluid from a pressure type oil pump can, Fig. 13-31.

WARNING! If coolant is applied manually with a hand-held oil pump can, keep the tip a safe distance from the wheel.

For safety, long equipment life, and quality control, a coolant system should be cleaned at regular intervals. Cleaning means removing all dirt and sludge from the holding tank, Fig. 13-32. Discard the fluid when it becomes contaminated.

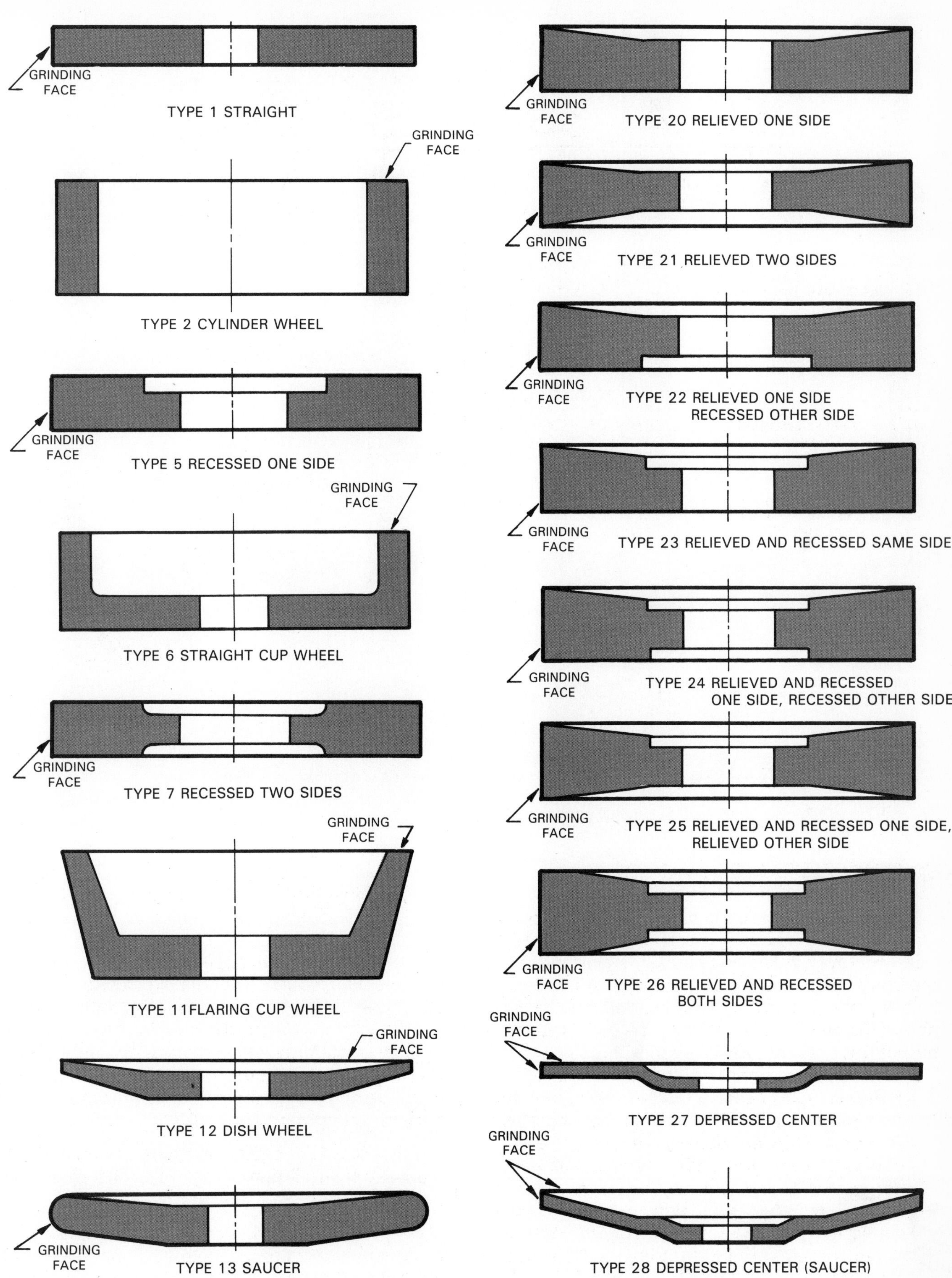

Fig. 13-25. Note standard grinding wheel shapes.

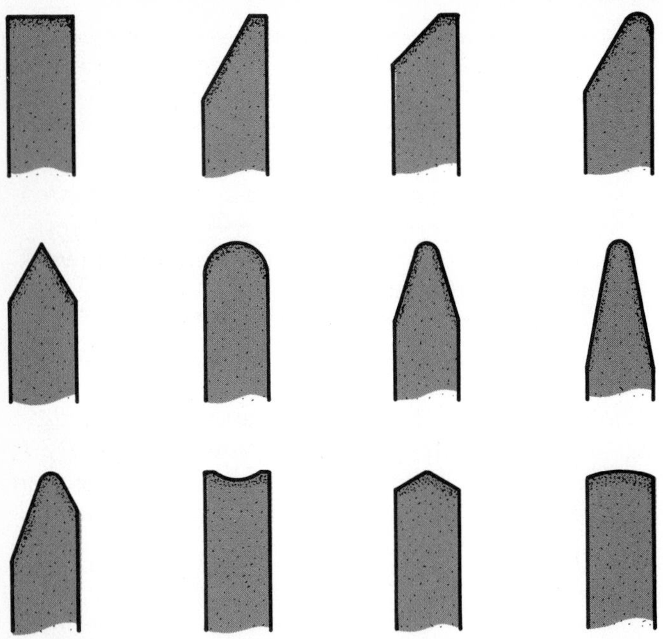

Fig. 13-26. Note standard grinding wheel faces.

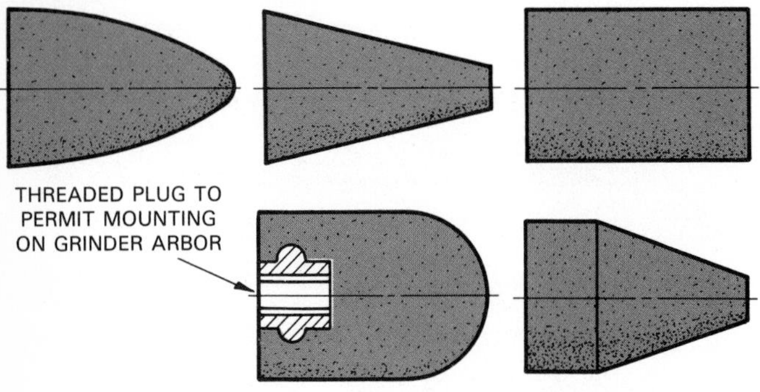

Fig. 13-27. These are a few of the many grinding wheel shapes available for internal grinding.

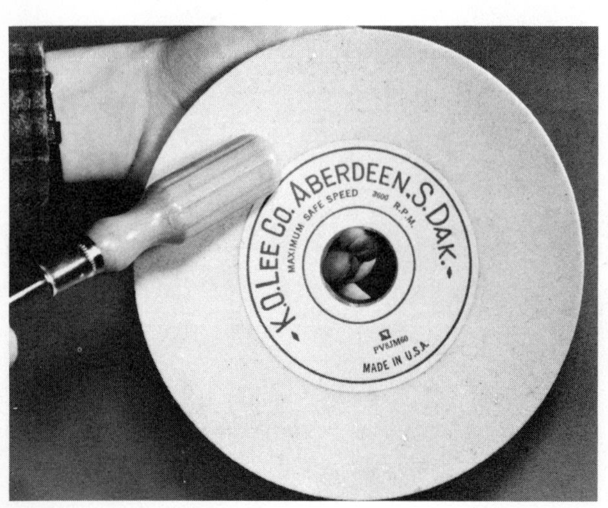

Fig. 13-28. Check soundness of a grinding wheel before mounting it on machine. A sound wheel will give a clear "metallic ring" when tapped as shown.

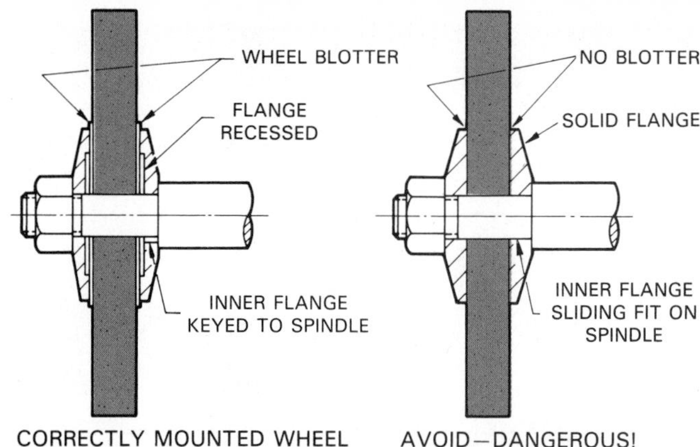

Fig. 13-29. Do not operate a grinder unless wheel is properly mounted. Left. Correct. Right. Incorrect!

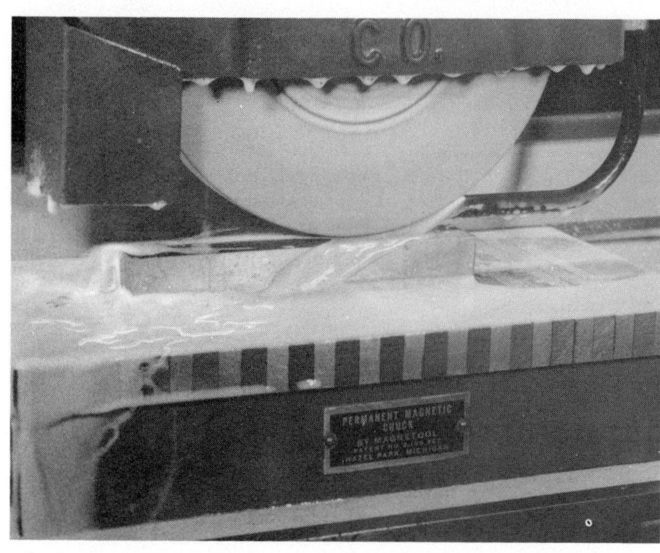

Fig. 13-30. Coolant must flood area being ground.

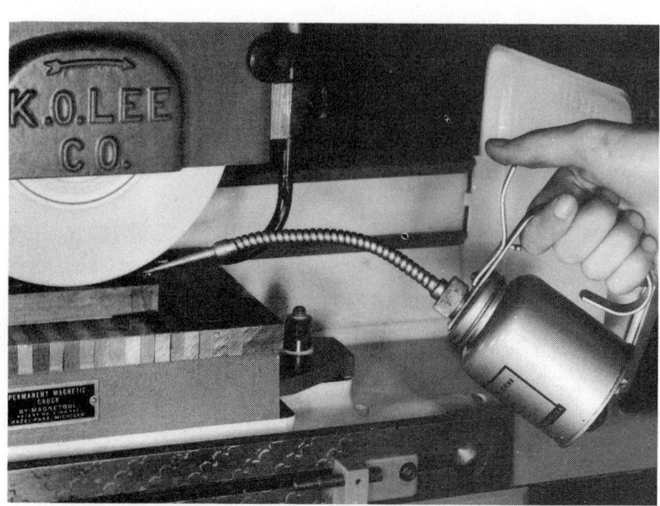

Fig. 13-31. Pressure or pump type oil can be employed to apply coolant. Keep tip away from wheel.

Fig. 13-32. Coolant tank must be cleaned frequently. Chips and grinding wheel residue can mar ground surface.

GRINDING APPLICATIONS

The following procedure is recommended to produce a surface that is flat or free of waviness:

1. Select and mount a suitable grinding wheel.
2. True and dress the wheel with a diamond dressing tool, Fig. 13-33.
3. Mount the work-holding device. If a magnet chuck is used, it should be "ground-in" to assure a surface true and parallel with table travel, Fig. 13-34. Grind off as little material as possible to true the surface. For high precision, this should be done each time the chuck is remounted on the machine.
4. Check the coolant system to be sure it is operating satisfactorily.
5. Locate the work and energize the chuck. If the work is already ground on one surface, protect it and the chuck surface by fitting a piece of oiled paper between them before energizing the chuck, Fig. 13-35.
6. Adjust the table stops, Fig. 13-36.
7. Check the holding power of the magnetic chuck by trying to move the work.
8. Down-feed the wheel until it just touches the highest point on the work surface. The grinding wheel can be set to the approximate position by down-feeding until it just touches a section of paper placed between the wheel and the work surface.
9. Turn on the coolant, spindle, and hydraulic pump motors.
10. Set the cross-feed to move the table in or out about 0.020 in. (0.5 mm) at the end of each cycle.
11. With the wheel clear of the work, down-feed about 0.001 to 0.003 in. (0.025 to 0.075 mm) for average roughing cuts per pass.
12. Use light cuts of 0.0001 in. (0.0025 mm) for finishing the surface. It is wise to redress the wheel for finishing cuts.
13. When the work surface has been ground to the required dimension and finish, use the following procedure to turn off the machine:
    a. Move the grinding wheel clear of the work.
    b. Turn off table travel.
    c. Turn off coolant.
    d. Let the grinding wheel run for a few moments after the coolant has been turned off. This will permit the wheel to free itself of all traces of fluid, Fig. 13-37; otherwise, the wheel can absorb some of the coolant and become out of balance.
    e. Use a squeegee to remove excess coolant from the work. De-energize the chuck and remove the work.

DANGER! Be careful of the sharp edges on newly ground work when removing it from the machine.

    f. Clean the machine. Put a light coating of oil on the chuck's work surface to prevent possible rusting.
    g. Replace all tools to proper storage.

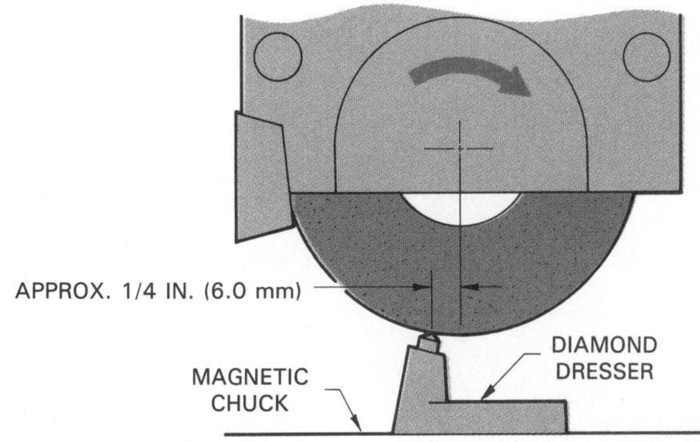

Fig. 13-33. Position diamond wheel dresser, as shown, for best results when cleaning or truing a grinding wheel.

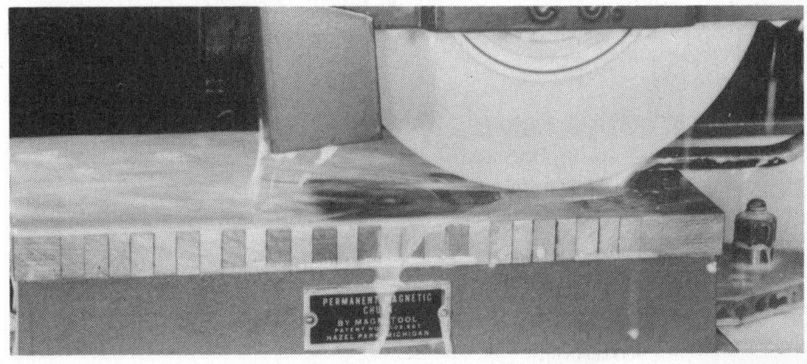

Fig. 13-34. Machine operator is "grinding in" surface of a magnetic chuck to true it before use. Note how surface is flooded with coolant.

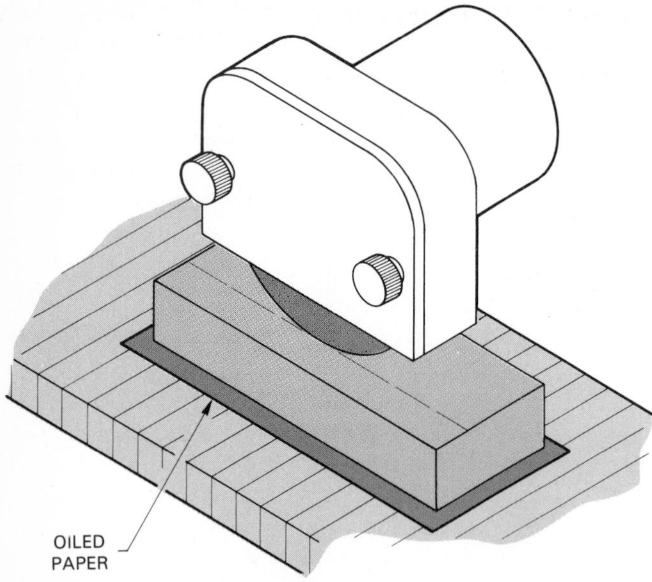

Fig. 13-35. A piece of oiled paper between magnetic chuck and newly ground surface will protect its finish when work is removed from chuck.

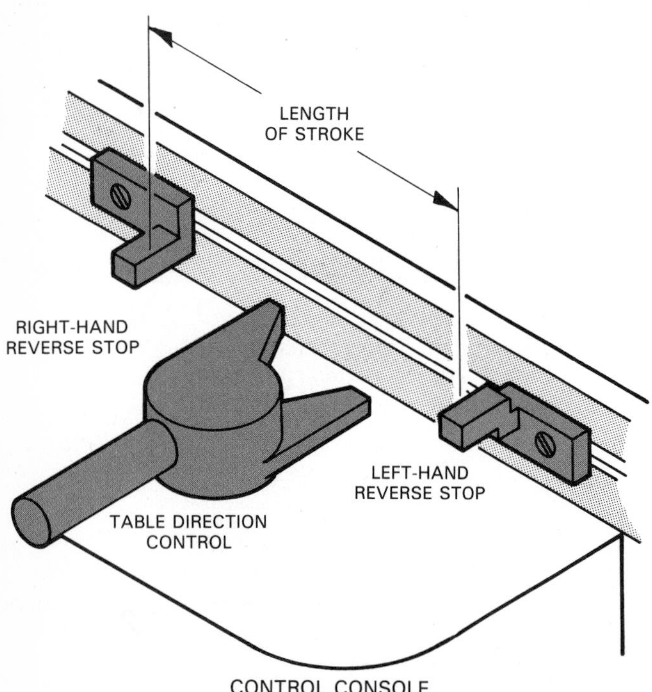

Fig. 13-36. Adjustable stops regulate length of table stroke. Care must be taken to be sure adjustment permits entire work surface to be ground.

Creep grinding

Creep grinding is a relatively new production-type machining technique that makes a deep cut into the work. It is also sometimes known as ***deep grinding.*** Special grinding machines are required for this type of work.

Fig. 13-37. Note sequence for turning surface grinder off.

Creep grinding is a surface grinding operation that is often performed in a single pass with an unusually large depth of cut, Fig. 13-38. In contrast to conventional surface grinding, the depth of cut is increased 1000-10,000 times and the work speed is reduced in the same proportion. MACHINING TIME can be reduced as much as 50-80 percent.

The tools (grinding wheels, work holding devices, etc.) must be designed for this heavy duty work.

GRINDING PROBLEMS

There are many problems peculiar to precision surface grinding. A few of the more common difficulties, with suggestions for their solution, are as follows:

Irregular table movement or ***no table movement*** (hydraulic type machines) may be caused by clogged hydraulic lines, insufficient hydraulic fluid, hydraulic pump not functioning properly, or inadequate table lubrication. A cold hydraulic system may also cause these symptoms. Let the machine warm up for at least fifteen minutes before use. Air in hydraulic lines can cause erratic table movement. Make corrections as recommended by the manufacturer of the machine.

Irregular scratches, of no noticeable pattern, are frequently caused by a dirty coolant system, or by particles becoming loosened in the wheel guard.

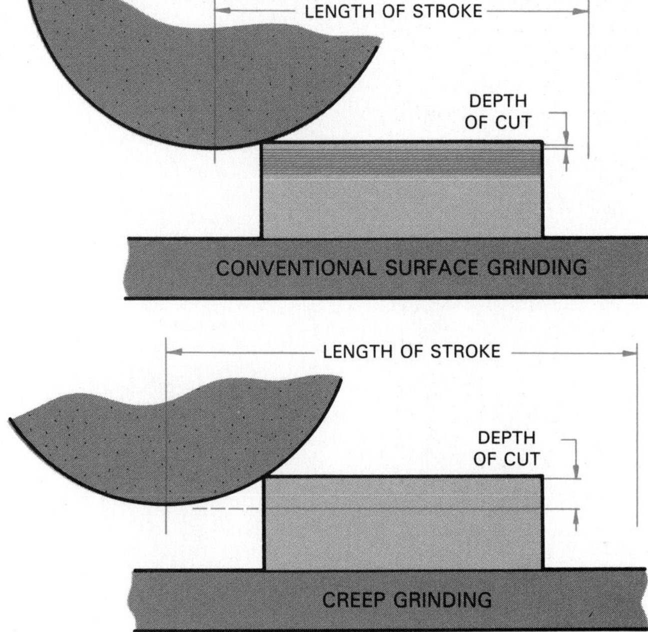

Fig. 13-38. Study difference between conventional and creep grinding. Creep grinding equipment must be especially designed for this heavy duty work.

This could also mean that the grinding wheel is too soft, and the abrasive particles are carried to the wheel by the coolant system.

Work surface waviness can be caused by a wheel being out of round and can be corrected by truing the wheel.

Chatter or ***vibration marks*** may be caused by a glazed or loaded grinding wheel. There is a slipping action between the wheel and the work. The wheel cuts until the glazed section comes into position and slides, rather than cuts, over the work. Correct by redressing the grinding wheel. The same effect can also be caused by a grinding machine that is NOT mounted solidly or by a wheel that is loose on the spindle. Check for these conditions and make corrections if necessary.

Burning or ***work surface checking*** may be the result of too little coolant reaching the work surface, a wheel that is too hard, or a wheel with grain that is too fine. Make corrections as indicated by inspection.

Wheel glazing or ***loading*** often indicates that the wrong coolant is being used. A dull diamond on the wheel dresser can also cause this problem.

Deep, irregular marks on the work surface may be caused by a loose grinding wheel.

Work not flat may be caused by insufficient coolant, a nicked or dirty chuck surface, or a wheel that is too hard. Check and make necessary corrections.

Work that is not parallel is frequently caused by a chuck that has NOT been "ground in" the last time it was mounted on the machine. A nicked or dirty chuck can also cause the same problem. Insufficient coolant may allow the work to expand in the center of the cut permitting more material to be removed in that area than at each end. As the piece cools, the center is depressed, Fig. 13-39. Correct by directing more fluid to the cutting area.

SAFETY

1. Always wear approved type eye protection when performing any grinding operation!
2. Never place a wheel on a grinder before checking it for soundness. Destroy faulty wheels!
3. Do NOT attempt to use a grinder unless all guards and safety devices are in place and securely attached!
4. Change coolant fluid before it become contaminated. It is good practice to set up a schedule for replacing the fluid at regular intervals or adding chemicals to control bacteria growth.
5. Have cuts and grinding "burns" treated promptly! Major infections can occur from untended minor injuries.
6. Wipe up all spilled coolants from the floor around the machine right away.

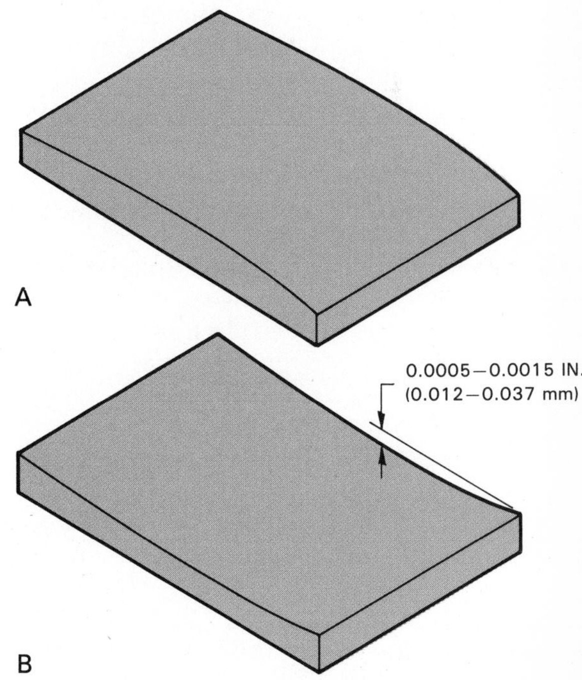

Fig. 13-39. A—Insufficient coolant may allow work to expand in center of cut, causing more material to be removed in center than at each end. B—As piece cools, center becomes depressed.

7. Stop the machine before making measurements or major machine and work adjustments.
8. Never attempt to operate a grinder until you have been instructed in its proper and safe operation. When in doubt, consult your instructor!
9. Make sure the grinding wheel is clear of the work before starting the machine.
10. If a magnetic chuck is used, make sure it is holding the work solidly before starting to grind.
11. If automatic feed is used, run the work through one cycle by hand to be sure there is adequate clearance and that the dogs are adjusted properly.
12. Keep all tools clear of the work table.
13. Check the wheel often to prevent it from becoming glazed or loaded. Dress the wheel when required.
14. Grinding wheels should NOT be operated at speeds higher than that specified by the manufacturer.
15. Remove your watch, rings, etc. before using a magnetic chuck to prevent them from becoming magnetized or pulled toward the machine.

UNIVERSAL TOOL AND CUTTER GRINDER

The ***universal tool and cutter grinder*** is a grinding machine designed to support cutters (primarily milling cutters) while they are sharpened to specified tolerances. Special attachments permit straight, spiral, and helical cutters to be sharpened accurately. See Fig. 13-40.

Other attachments enable the machine to be adapted to all types of internal and external cylindrical grinding. Refer to Figs. 13-41 through 13-43.

TOOL AND CUTTER GRINDING WHEELS

The wheel shapes most frequently employed for tool and cutter grinding are shown in Fig. 13-44. Charts prepared by the various grinding wheel manufacturers are used to determine the correct wheel composition (abrasive, grain size, and bond) for the job at hand.

Keep the grinding wheel clean and sharp by frequent dressing with a diamond tool. Use light cuts to prevent drawing the temper out of the tooth cutting edge.

Crowding the wheel into the cutter is a common mistake because the cutters are made from materials (HSS and cemented carbides) that do NOT give off a brilliant shower of sparks when in contact with a grinding wheel. This creates the illusion that the cut being made is too light.

Fig. 13-40. Note universal tool and cutter grinder. (K.O. Lee Co.)

Fig. 13-41. Cylindrical grinding is being done on this tool and cutter grinder. A powered workhead rotates work. (K.O. Lee Co.)

Fig. 13-42. Tool and cutter grinder can also be employed for internal grinding. (K.O. Lee Co.)

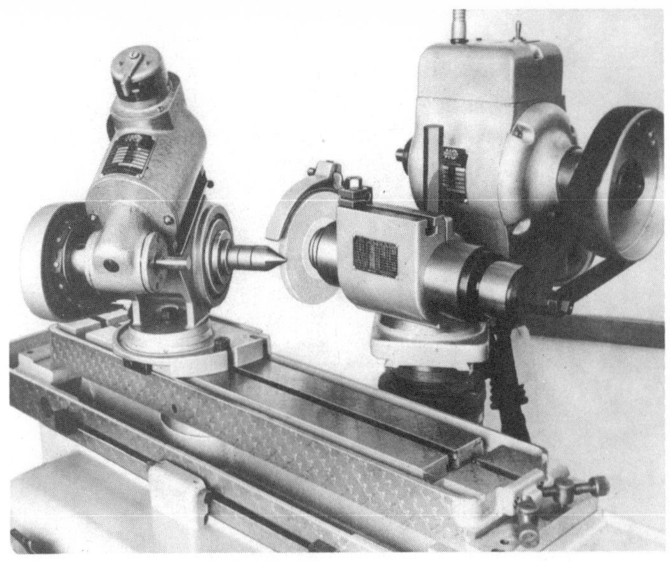

Fig. 13-43. Center is being trued on tool and cutter grinder. Work is rotated by powered workhead. (K.O. Lee Co.)

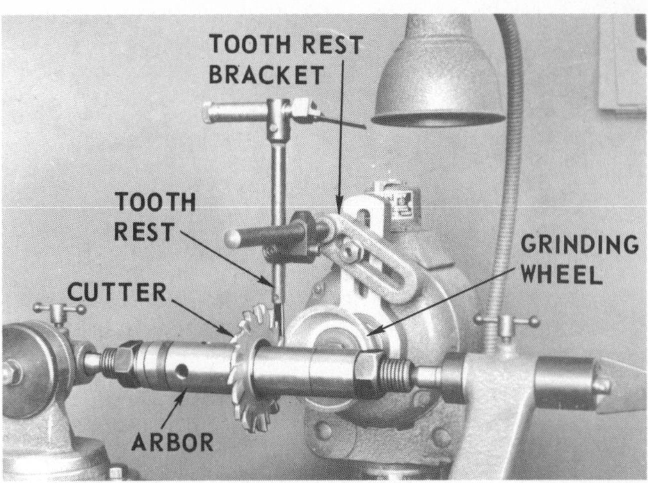

Fig. 13-45. Tooth rest is being used to position teeth of cutter. Note that cup-shaped wheel does grinding. (K.O. Lee Co.)

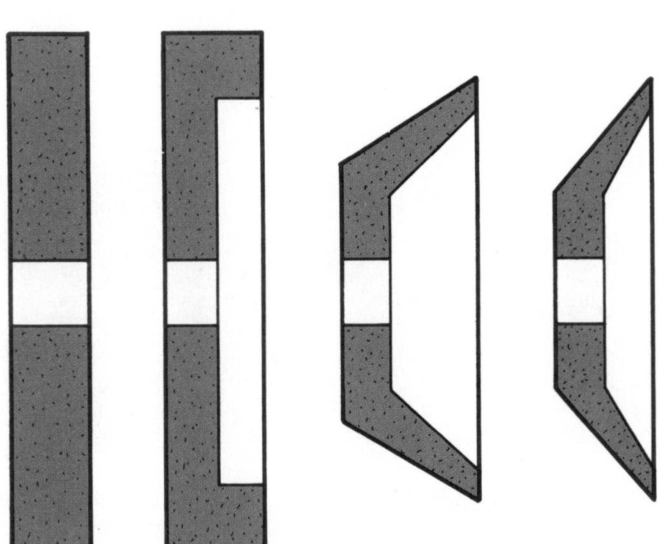

Fig. 13-44. Note wheel shapes used for grinding cutters.

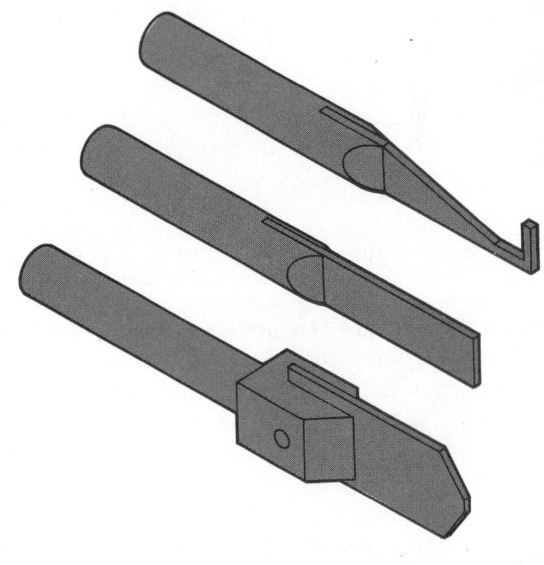

Fig. 13-46. Note several types of tooth rests.

SHARPENING CUTTERS

A ***tooth rest*** locates each tooth quickly and accurately into position, Fig. 13-45. Several types are employed to permit different cutter types to be sharpened, Fig. 13-46. The ***supporting bracket*** can be mounted to the work table or on the grinding wheel housing, Fig. 13-47.

Grinding plain milling cutters

1. Select the correct wheel for the job. True it with a diamond tool.
2. Mount the cutter on a suitable arbor and place the unit between centers, Fig. 13-48.

Fig. 13-47. Note supporting bracket for tooth rests.

Fig. 13-48. Cutters must be mounted on an arbor for sharpening. Approved type eye protection is a must. If machine is not equipped with a vacuuming system, a respirator mask should also be worn. (Norton Co.)

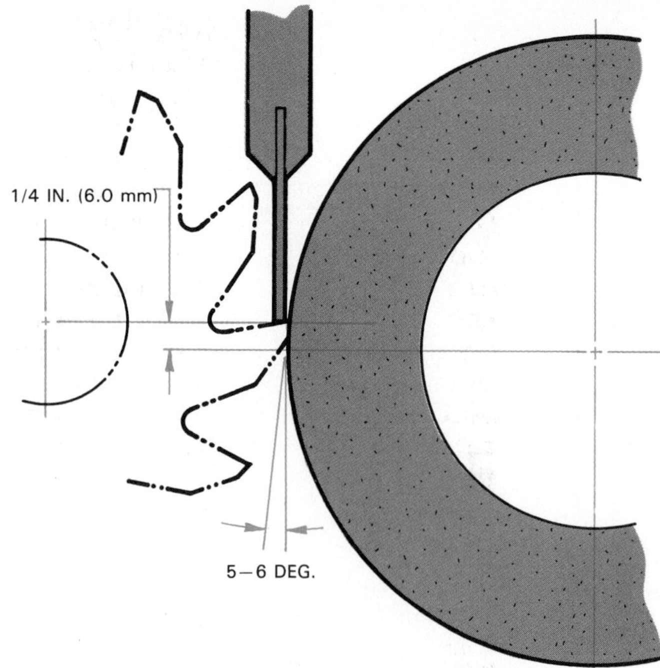

Fig. 13-49. This setup is for grinding a 6 in. (150 mm) diameter cutter.

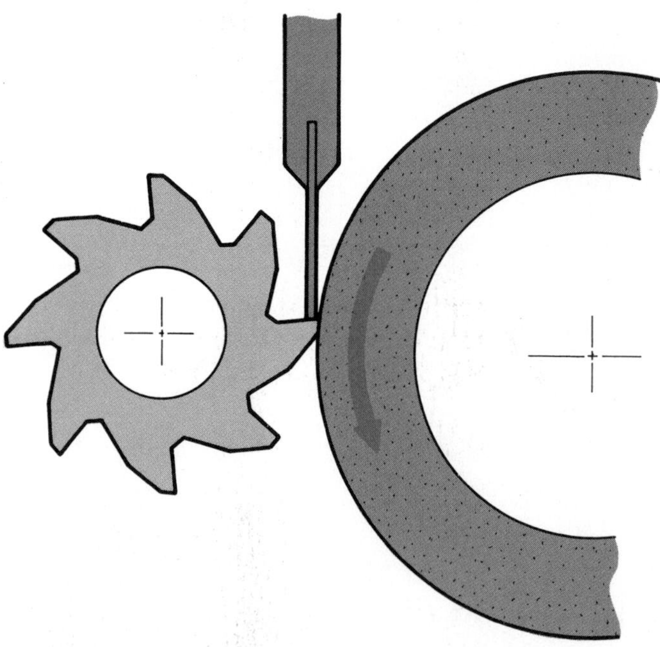

Fig. 13-50. In this setup, wheel grinds AWAY from cutting edge of tooth. With this technique, there is less chance of drawing temper out of tooth and no burr is formed.

3. Mount the tooth rest to the wheel head. Position the edge about 1/4 in. (6.0 mm) above the center line of the grinding wheel, Fig. 13-49. This will produce a 5 to 6 deg. clearance angle on the tooth cutting edge of a 6 in. (1.50.0 mm) diameter cutter. Adjust to suit the cutter being ground.
4. The setup should permit the wheel to grind away from the tooth cutting edge, Fig. 13-50. While requiring more machining care than grinding into a cutting edge of the tooth there is less chance of drawing the temper. Also, no burr is formed that must be oilstoned off to secure a sharp edge, Fig. 13-51.
5. Flaring cup wheels are also used for cutter and tool grinding. How they are setup is shown in Fig. 13-52. Since there is a greater area of contact when using flare cup wheels, LIGHTER CUTS should be taken than with straight grinding wheels.
6. Start the machine and feed the cutter into the wheel. Take a light cut. A bit of thinned layout bluing should be applied to the back of the tooth. This will allow a visual check of how the grinding operation is progressing and whether the setup is producing the proper clearance angle.
7. When satisfied with the setup, bring the next tooth into position on the tooth rest and grind that tooth.
8. Repeat the operation until all of the teeth are

sharpened. Make necessary adjustments to assure tooth concentricity (cutting surfaces of all teeth are same distance from arbor hole center line).

After a cutter has been sharpened several times, the clearance angle ***flat*** (land) will become too wide. Then, it becomes necessary to grind in a secondary clearance angle, Fig. 13-53.

If it becomes apparent that more material is being removed from some teeth than others, a quick check must be made to determine the cause:

1. The grinding wheel may be too soft and wear down too rapidly. As the wheel wears, less material is removed from the cutter tooth.
2. The tooth rest may NOT be mounted solidly and moves during the grinding operation.
3. The arbor may NOT be running true on the centers. Test arbor runout with an indicator as it is rotated.

When the trouble has been ''pin pointed,'' make the necessary corrections and continue the operation.

An ***indexing disc*** may also be used to position each tooth for sharpening, Fig. 13-54. It is mounted on the arbor. The divisions are normal to each other, plus or minus 4 minutes (1/15 degree). They are available in a range of graduations.

Side teeth on a side-milling cutter must also be sharpened. This is done by mounting the cutter on a stub arbor and fitting the unit into a ***workhead*** rather than between centers. Facing mills are sharpened in the same manner. See Fig. 13-55.

Grinding helical teeth cutters

Slabbing cutters and other cutters having helical teeth are sharpened in much the same manner as plain milling cutters, Fig. 13-56. However, these

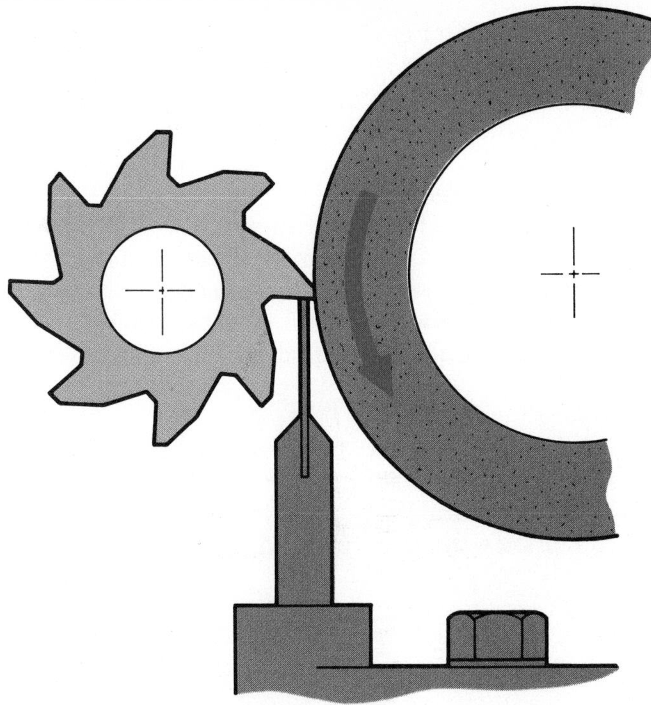

Fig. 13-51. When wheel grinds INTO cutting edge of tooth, there is some danger that a burr will be formed or temper drawn.

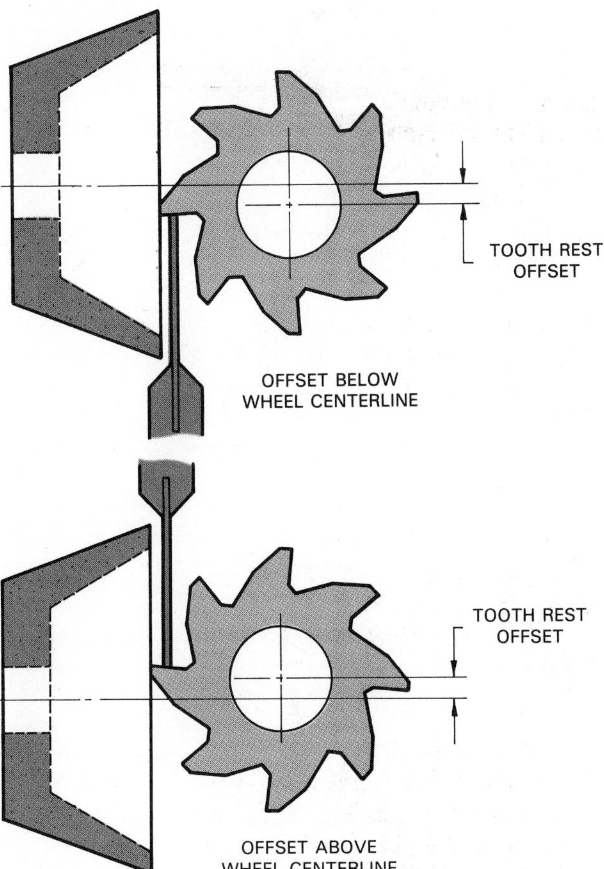

Fig. 13-52. Milling cutter is being ground with a flare cup wheel.

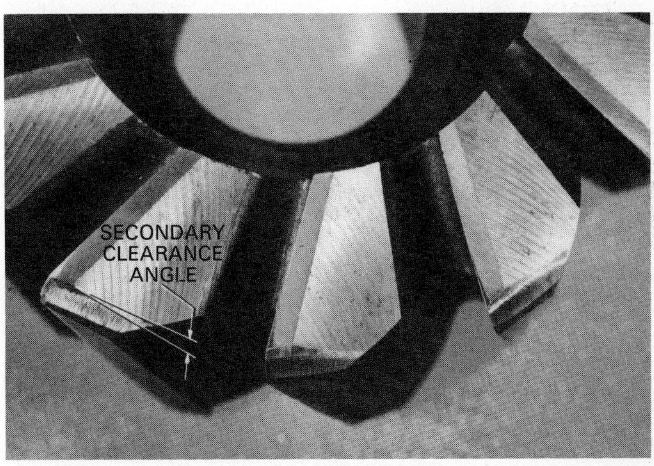

Fig. 13-53. This cutter has been sharpened so many times that a secondary clearance angle is needed to permit tool to cut properly.

Fig. 13-54. An indexing disc is often used to position each tooth for sharpening. (K.O. Lee Co.)

cutters must be held against the tooth rest as the table is traversed. This will impart a twisting motion to keep the tooth correctly located against the grinding wheel.

Grinding end mills

End mills are sharpened in much the same way as helical teeth cutters with the end mill mounted in a workhead rather than between centers. The end teeth are sharpened with the same technique used to sharpen the side teeth on a side milling cutter.

Grinding form cutters

Form tooth cutters must be ground radially to preserve the tooth shape, Figs. 13-57 and 13-58. An index disc may be employed or a special form cutter grinder may be utilized.

Fig. 13-55. Grinder is sharpening side of teeth of a face milling cutter.

Fig. 13-57. A convex cutter is typical of form tooth cutters. (Standard Tool Co.)

Fig. 13-56. Note setup for sharpening a cutter with helical teeth. (K.O. Lee Co.)

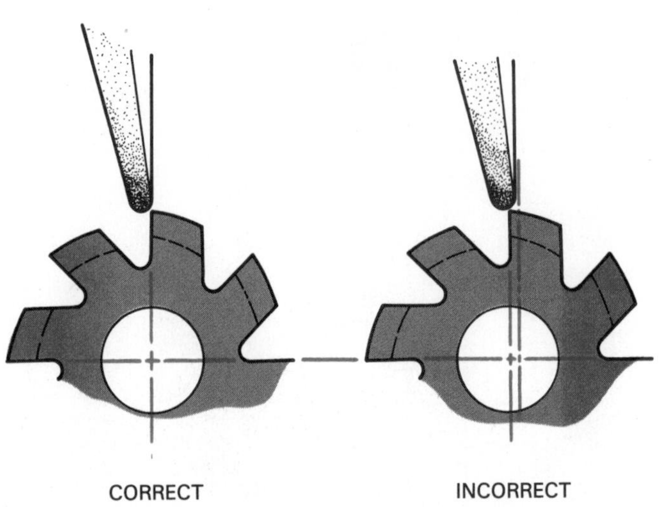

Fig. 13-58. Form tooth cutters must be ground radially; otherwise, form or shape cutter was designed to machine will be altered.

Grinding taps

A universal tool and cutter grinder may also be used to resharpen taps. Normally a tap becomes dull when the leading edges of the starting chamfer become worn. The chamfer can be reground by mounting the tap in a workhead, Fig. 13-59. Flutes are reground with a straight wheel with an edge that has been shaped to fit the flutes, Fig. 13-60.

Grinding reamers

The cutting action of a machine reamer takes place at the front end of the teeth, Fig. 13-61. The worktable is pivoted at a 45 deg. angle. Using a cup wheel, adjust the tooth rest and/or grinding head to give the correct clearance. Sharpen the reamer in the same manner employed to sharpen a face milling cutter.

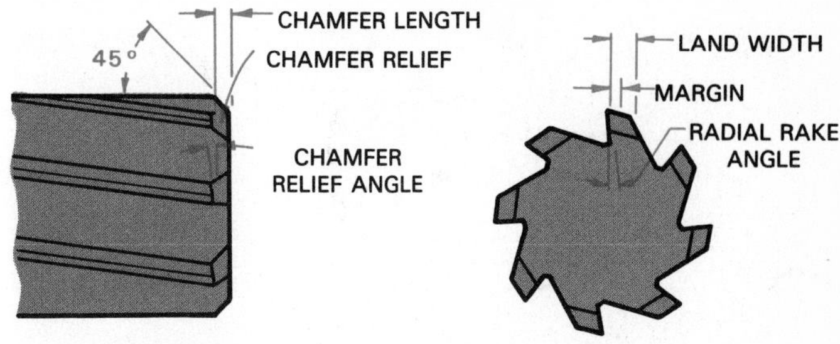

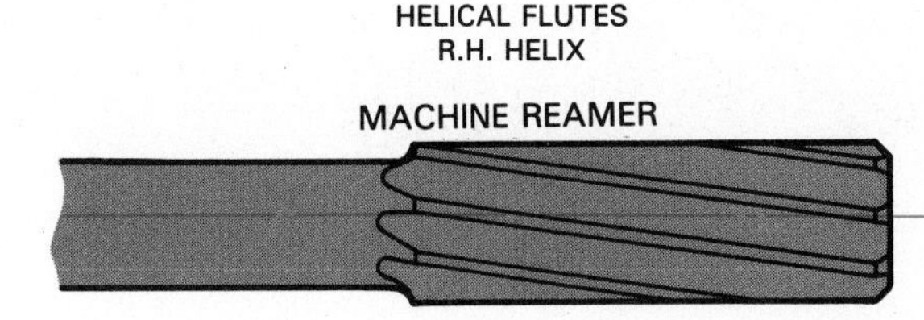

Fig. 13-61. Study cutting edges of a machine reamer.

CYLINDRICAL GRINDING

With a ***cylindrical grinder,*** it is economically feasible to machine hardened steel to tolerances of 0.00001 in. (0.0002 mm) with extremely fine surface finishes. See Figs. 13-62 and 13-63.

Fig. 13-59. Note setup for regrinding chamfer on a tap. (K.O. Lee Co.)

On this machine, work is mounted between centers and rotates while in contact with the grinding wheel, Fig. 13-64. Straight, taper, and form grinding is possible with this technique. Two variations of cylindrical grinding are:

1. With ***traverse grinding*** a fixed amount of material is removed from the rotating work as it moves past the revolving grinding wheel. Work wider than the face of the grinding wheel can be ground. See Fig. 13-65.
2. With ***plunge grinding,*** the work still rotates; however, it is NOT necessary to move the grinding wheel across the work surface. The area being ground is no wider than the wheel face.

Fig. 13-60. Flutes are being reground on tap to renew cutting edges of teeth. (K.O. Lee Co.)

Fig. 13-62. Universal grinding machine can be used for internal and cylindrical grinding as well as cutter grinding. (Landis Tool Co.)

Fig. 13-63. Heavy-duty 60 in. x 228 in. (1500 mm x 5700 mm) cylindrical grinder is being used for reconditioning forming rolls used by steel mills. (Simmons Machine Tool Corp.)

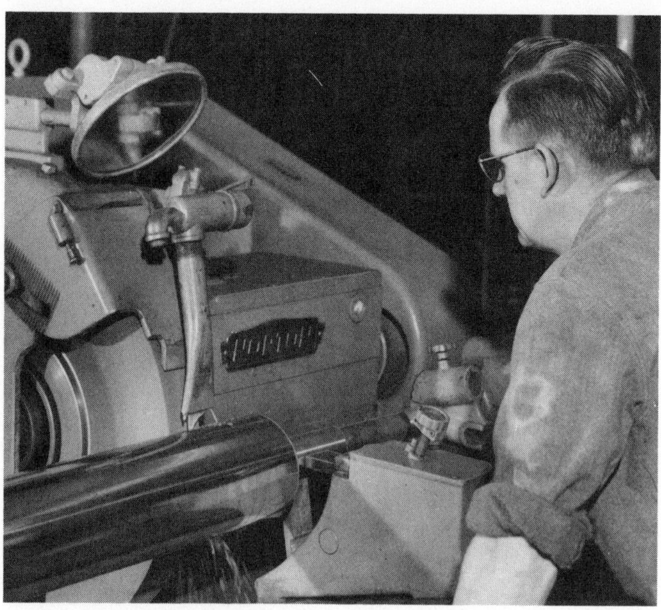

Fig. 13-65. With traverse grinding, work moves past revolving grinding wheel. (Norton Co.)

Fig. 13-64. Note closeup of a cylindrical grinding operation. (Cincinnati Milacron)

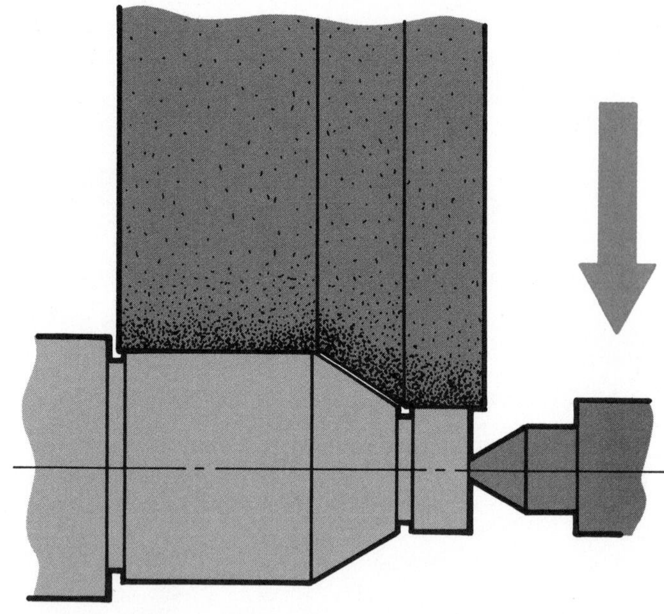

Fig. 13-66. With plunge grinding, grinding wheel is fed into rotating work. Since work is no wider than grinding wheel, reciprocating motion is not needed.

Grinding wheel infeed is continuous rather than incremental (minute changes at end of each cut), Fig. 13-66.

Grinding to a shoulder by both techniques is shown in Fig. 13-67.

Holding and driving the work

As the work rotates on centers, it is extremely important that the centers be free of dirt and nicks. They must also run absolutely true. If possible, the head center should be ground in place. The center holes must also be clean, of the correct shape and depth, and well lubricated.

Long work is best supported by ***work rests,*** Fig. 13-68. They support from the back and bottom and are adjustable to compensate for material removed in the grinding operation.

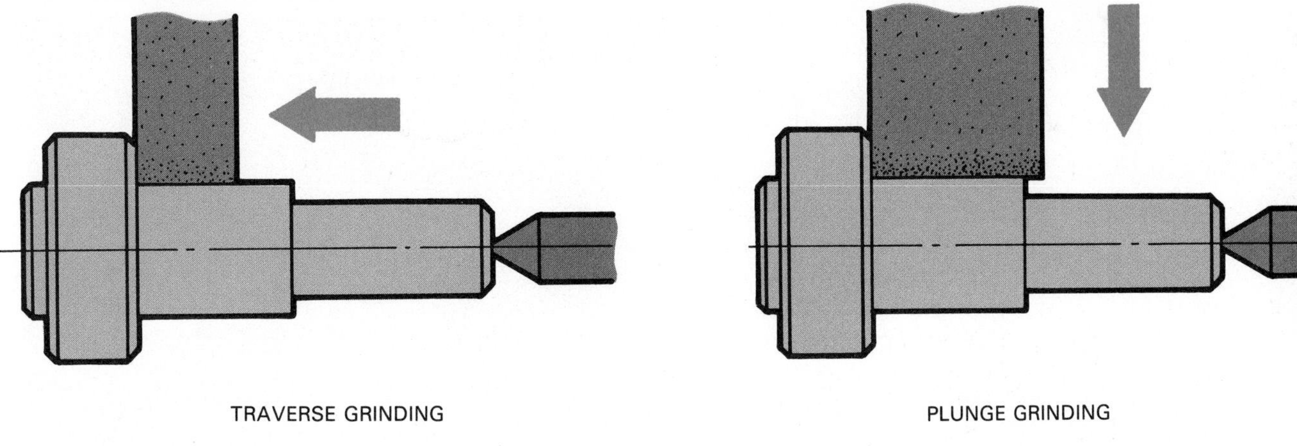

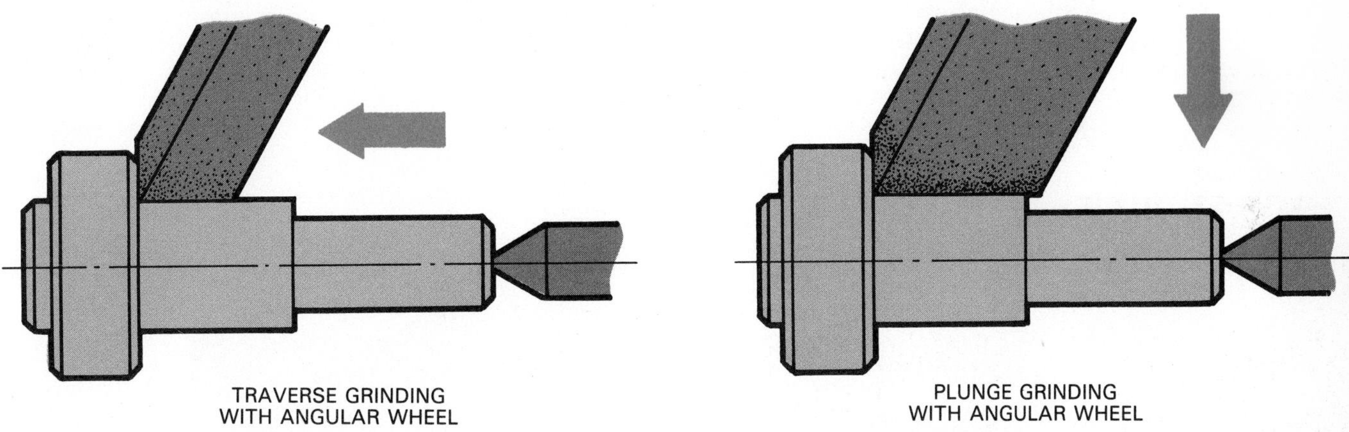

Fig. 13-67. Study grinding to a shoulder by traverse and plunge grinding techniques.

Work rotation is accomplished through a drive plate that revolves around the head stock center and an adjustable drive pin and dog, Fig. 13-69. Work may also be mounted in a chuck.

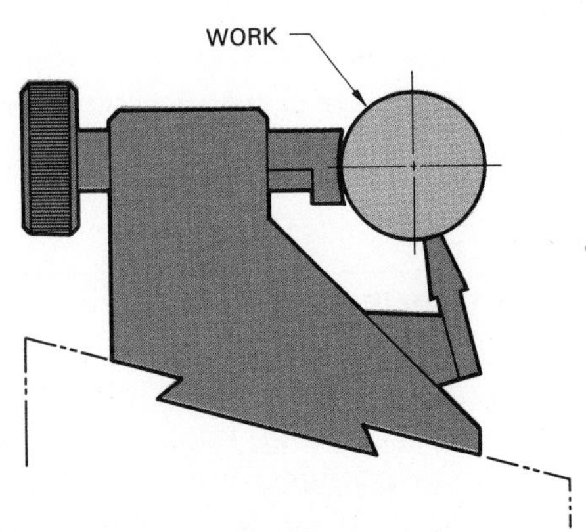

Fig. 13-68. Work rest is placed every four or five diameters along work for support. It must be adjusted after each grinding pass.

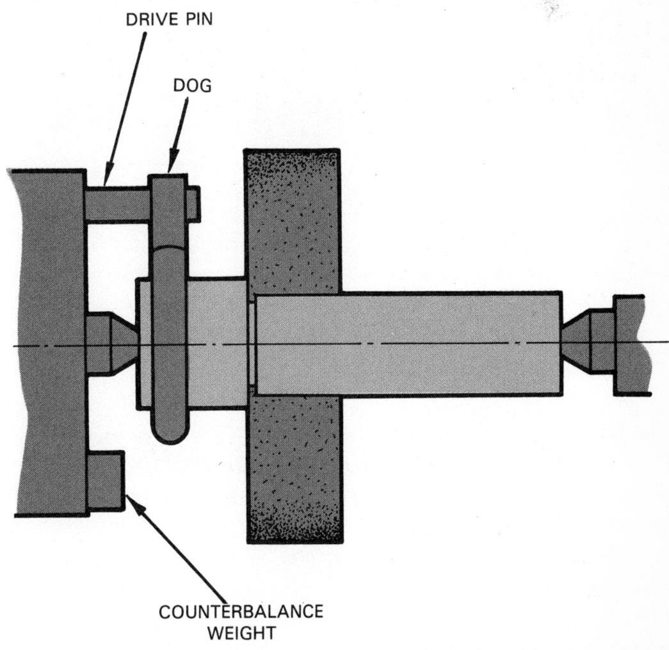

Fig. 13-69. Note one method used to rotate work on a cylindrical grinder.

Machine operation

To assure a good finish and size accuracy, it is vital that work rotation and traverse table movement (back-and-forth on front of wheel) be smooth and steady.

Table movement should be such that the wheel will OVER RUN the work end by about one-third the width of the wheel face, Fig. 13-70. This permits the grinding wheel to do a more accurate grinding job. ***Insufficient run off*** will result in work that is oversize. ***Complete run off*** of the grinding wheel will cause the piece to be undersize.

The grinding wheel must be true and balanced. Otherwise, vibration will cause chatter marks on the work and may cause it to be out of round.

Cutting speeds and feeds can be determined from information available on charts furnished by the grinding machine and grinding wheel manufacturers. They will also specify what coolant will give the best results.

Fig. 13-71. Internal grinding operation is being performed on a universal grinding machine. Note how extended workpiece is supported. (Norton Co.)

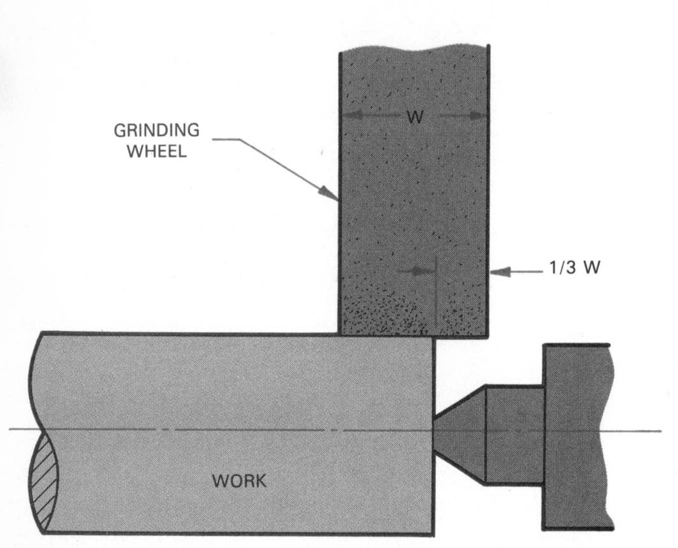

Fig. 13-70. Table movement should permit about one-third of wheel's width to run beyond end of work. This permits wheel to do a more accurate job.

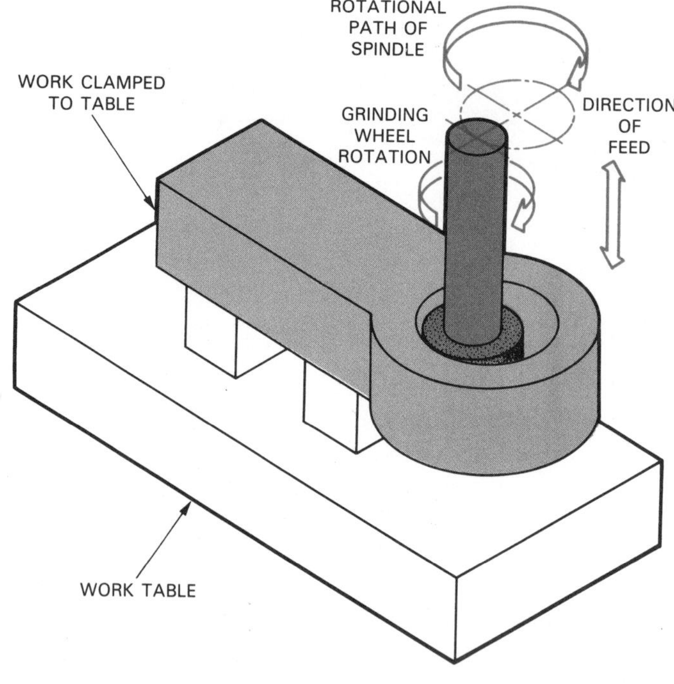

Fig. 13-72. Study internal grinding of work too large or odd-shaped to be rotated.

INTERNAL GRINDING

Internal grinding is done to secure a fine surface finish and accuracy on inside diameters, Fig. 13-71. Work is mounted in a chuck and rotates. During the grinding operation, the revolving grinding wheel moves in and out of the hole.

A special grinding machine which finishes holes in pieces too large to be rotated by the conventional machine is shown in Fig. 13-72. Hole diameter is controlled by regulating the diameter of the circle in which the grinding head moves.

CENTERLESS GRINDING

The work does not have to be supported between centers in centerless grinding because the work is rotated against the grinding wheel. Instead, the piece is positioned on a work support blade. The workpiece is fed automatically between a regulating or feed wheel and a grinding wheel. Primarily, the regulating wheel causes the piece to rotate, and the

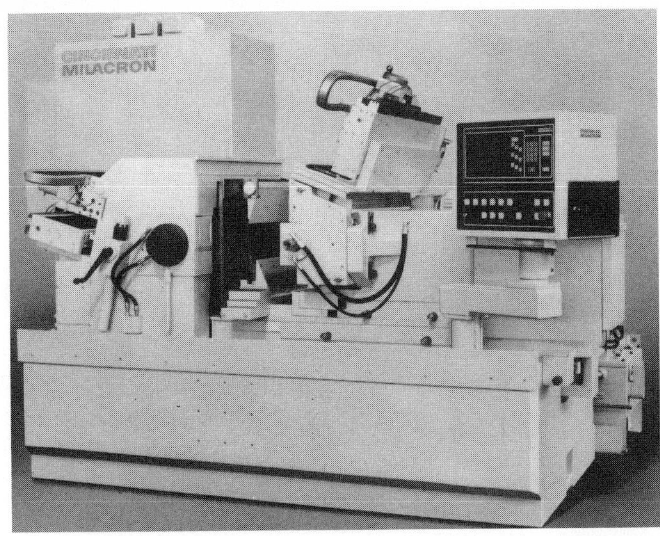

Fig. 13-73. This is a computer controlled (CNC) centerless grinding machine. (Cincinnati Milacron)

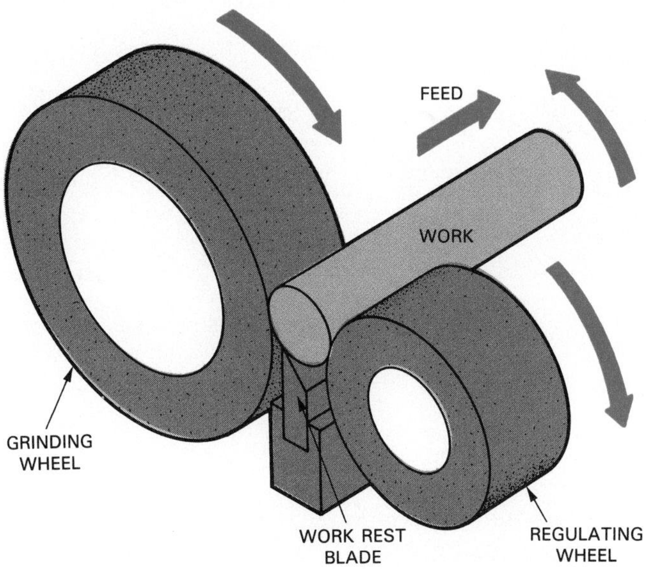

Fig. 13-74. Note how centerless grinding is done.

grinding wheel does the cutting. Feed through the wheels is obtained by setting the regulating wheel at a slight angle.

There are four variations of centerless grinding:

1. ***Through feed grinding*** can only be employed to produce simple cylindrical shapes. Work is fed continuously by hand, or from a feed hopper, into the gap between the grinding wheel and the regulating wheel. The finished pieces DROP OFF the work support blade, Fig. 13-75.
2. ***Infeed grinding*** is a centerless grinding technique that feeds the work into the wheel gap until it reaches a STOP. See Fig. 13-76. The piece is ejected at the completion of the grinding operation. Work diameter is controlled by regulating the width of the gap between the regulating wheel and grinding wheel. Work with a shoulder can be ground using this technique.
3. ***End feed grinding*** is a form of centerless grinding ideally suited for grinding short tapers and spherical shapes. Both wheels are dressed to the required shape and work is fed in from the side of the wheel to an end stop. The finished piece is ejected automatically.
4. ***Internal centerless grinding*** minimizes distortion in finishing thin-wall work and eliminates reproduction of hole-size errors and waviness in the finish. Refer to Fig. 13-78.

Centerless grinding is utilized when large quantities of the same part are required. Production is high and costs are relatively low because there is no need to drill center holes nor to mount work in a holding device. Almost any material can be ground by this technique.

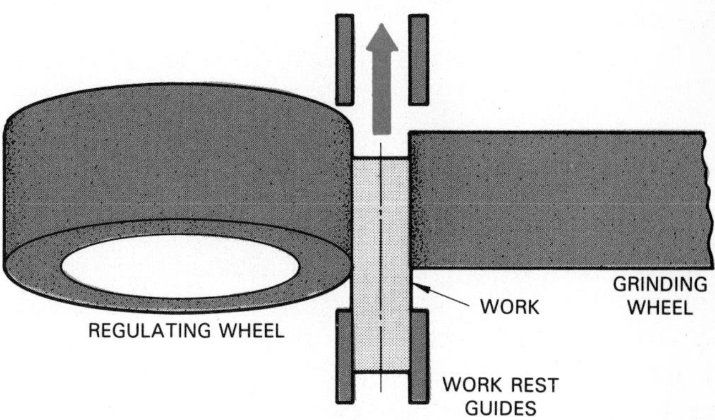

Fig. 13-75. Angle of regulating wheel pulls work over grinding wheel with through feed centerless grinding.

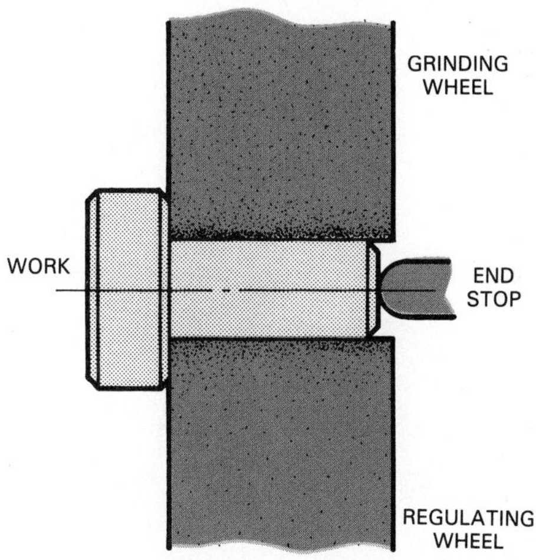

Fig. 13-76. Note infeed centerless grinding.

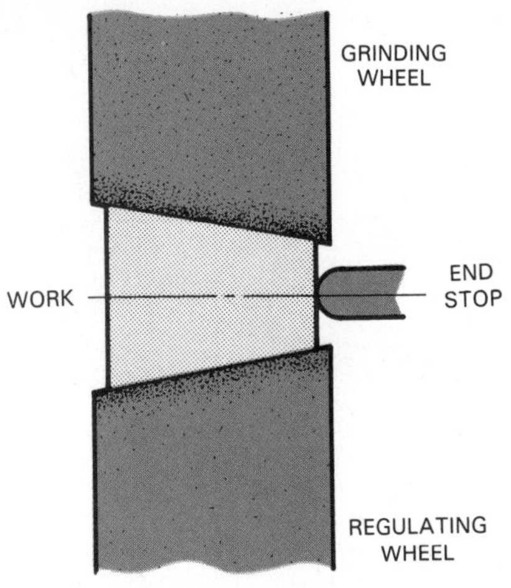

Fig. 13-77. Note end feed centerless grinding.

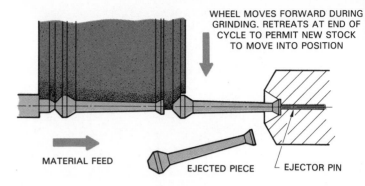

Fig. 13-79. Form grinding of this engine part is done at rate of 200 pieces an hour. Material was heat treated before grinding.

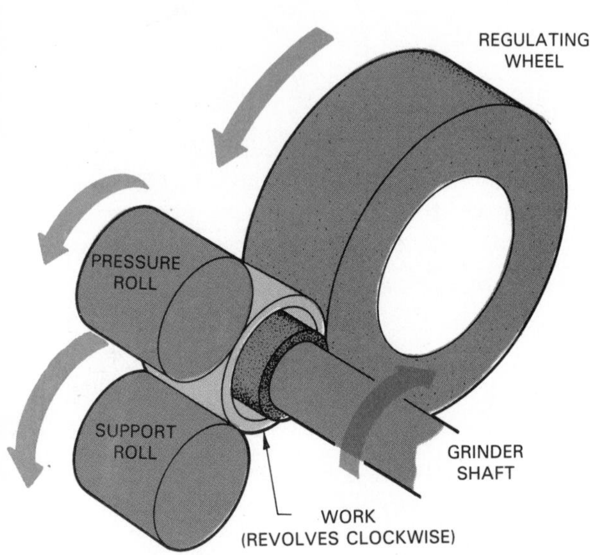

Fig. 13-78. Study setup for internal centerless grinding.

Fig. 13-80. This CNC thread grinding machine has electronic controls that provide thread accuracy and cost-efficient operation. (Reishauer Elgin)

FORM GRINDING

In ***form grinding,*** the grinding wheel is shaped to produce the required contour on the work. Fig. 13-79 shows this principle.

Thread grinding is an example of form grinding, Figs. 13-80 and 13-81. A form or template guides the diamond particle wheel that dresses the wheel that grinds the required thread shape. There is automatic compensation on the grinding machine for the material removed from the grinding wheel when it is dressed.

OTHER GRINDING TECHNIQUES

In addition to the grinding techniques already described, industry makes considerable use of other abrasive type processes.

Abrasive belt machining

Abrasive belt grinding was first employed for light stock removal and polishing operations, Fig. 13-82. However, the capability of the technique has advanced to the stage where high rate metal removal to close tolerances is possible. This is primarily due to tougher and sharper abrasive grains, improved adhesives, and stronger backings.

A major advantage of abrasive belt grinding is its versatility. A machine can be converted quickly from heavy stock removal to finishing operations, or for grinding a different material, by simply changing the abrasive belt.

Abrasive belts, because of their length, run cool and require light contact pressure, thus reducing the possibility of metal distortion caused by heat. Soft contact wheels and flexible belts conform to irregular shapes. Belts may be used dry or with a coolant. Several abrasive grinding machine applications are shown in Fig. 13-83.

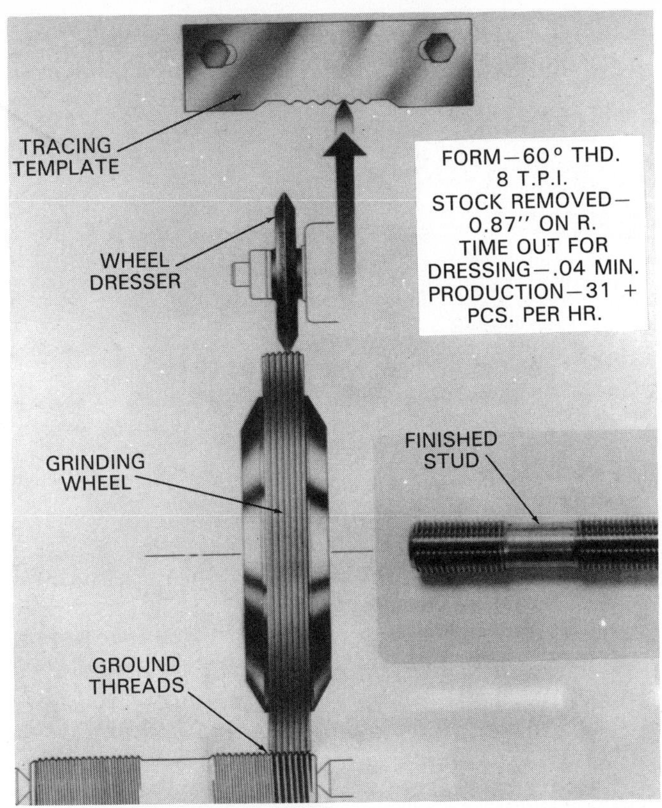

Fig. 13-81. Precision threads are being form ground on a special stud. (Jones & Lamson Machine Co.)

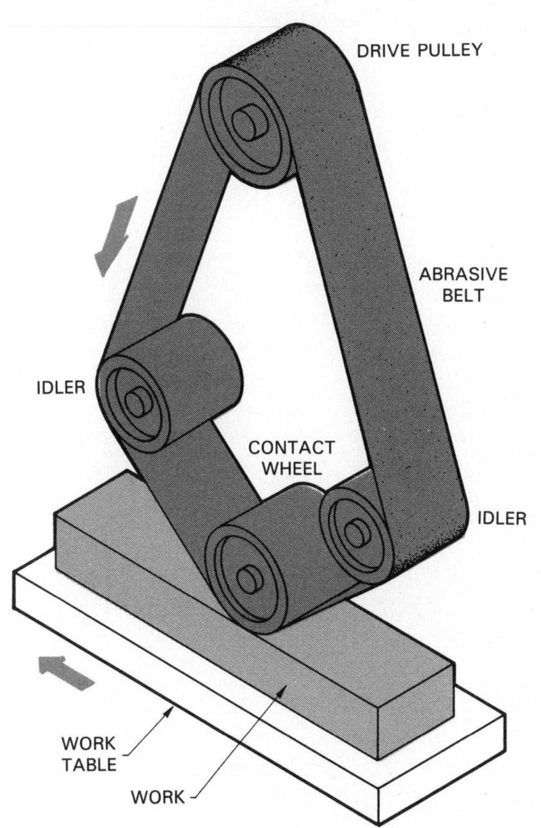

Fig. 13-82. Study abrasive belt grinding.

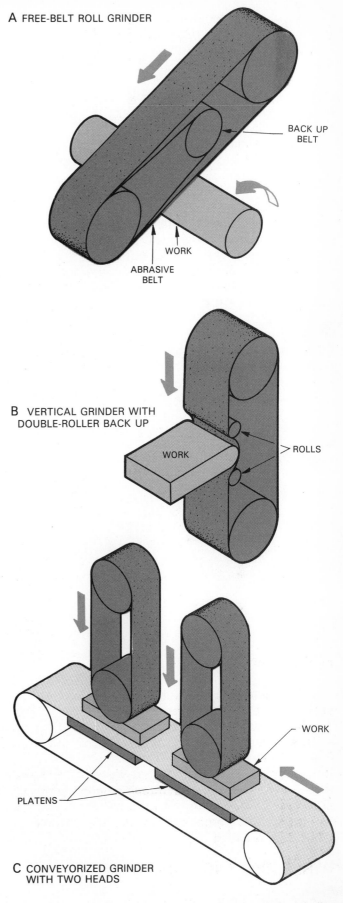

A FREE-BELT ROLL GRINDER

BACK UP BELT

WORK

ABRASIVE BELT

B VERTICAL GRINDER WITH DOUBLE-ROLLER BACK UP

WORK

ROLLS

WORK

PLATENS

C CONVEYORIZED GRINDER WITH TWO HEADS

Fig. 13-83. Note a few of many abrasive belt grinding techniques.

The most satisfactory belt speed for grinding ferrous and nonferrous metals is between 5000-9000 ***sfm*** (surface feet per minute). Slower speeds of 1500-3000 sfm are required for tougher materials like titanium.

Abrasive belt grinding usually requires support behind the belt. This may be in the form of contact wheels, rolls, or platens.

Contact wheels are usually made of cloth or rubber. Hardness and/or density of the contact wheel affects stock removal and finish. Serrated or slotted wheels improve cutting action and prolong abrasive belt life.

Platens are made of metal (some have cemented carbide inserts) and are usually NOT as effective as contact wheels. They are flat but can be shaped to conform to the contour required on the work. Jets of air or water may be applied between the belt and platen to reduce friction.

Electrolytic grinding

Electrolytic grinding is actually an electrochemical machining process, Fig. 13-84. Applications of the technique include rapid removal of stock from alloy steel parts, sharpening carbide tools, and machining heat-sensitive work.

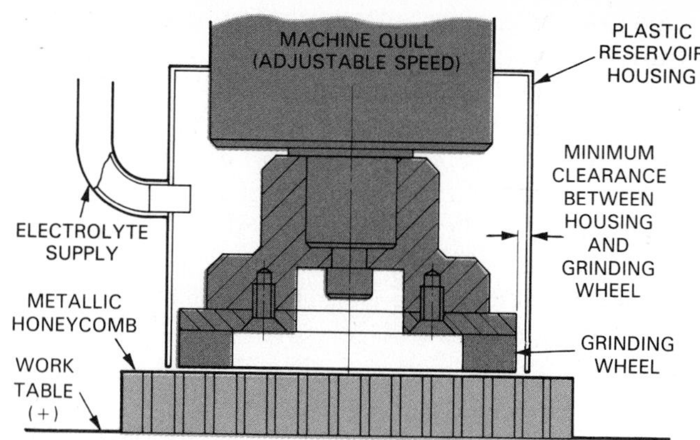

Fig. 13-84. This is an electrolytic grinding process (ELG).

An electric current is passed between a metal bonded grinding wheel (cathode) and the work (anode) through a conductive electrolyte, Fig. 13-85. The work face is attacked electrochemically by the current (similar to electroplating in reverse).

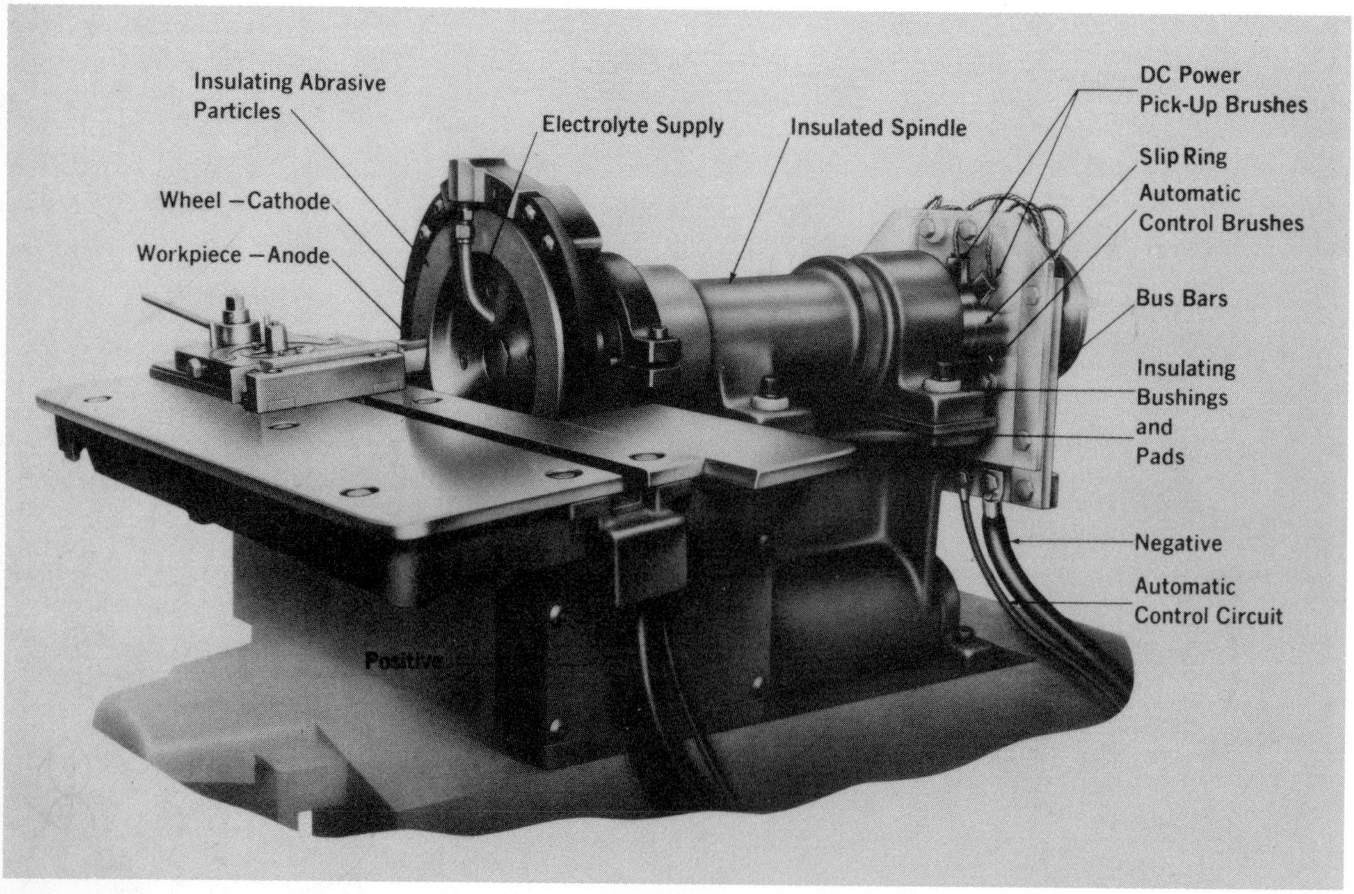

Fig. 13-85. Study nomenclature (parts) of an electrochemical grinding machine designed to sharpen carbide lathe tools. (Hammond Machinery Builders)

The dissolved material is removed by the wheel. No burr is developed making it possible to machine fragile work like stainless steel and exotic metal honeycomb sections. Neither is heat generated nor is there a metallurgical change in the metal.

An electrochemical grinding machine for sharpening carbide tools is shown in Fig. 13-86.

Computer controlled (CNC) grinders

Many types of computer controlled (NC and CNC) grinders are available. See Figs. 13-87 through 13-89. They are designed to operate automatically. Functions, such as: tool positioning, spindle start and stop, vertical feed motion, and linear feed rates, are programmed into a machine's computer. The operator is then relieved from the responsibility of controlling and monitoring the numerous coordinate settings and related machine functions.

The grinding wheel path representing the contours of the part (at a selected distance from part edge) is programmed directly to its dimensional specifications. Grinding wheel wear (wheel diameter decreases each time it is dressed) is compensated for automatically.

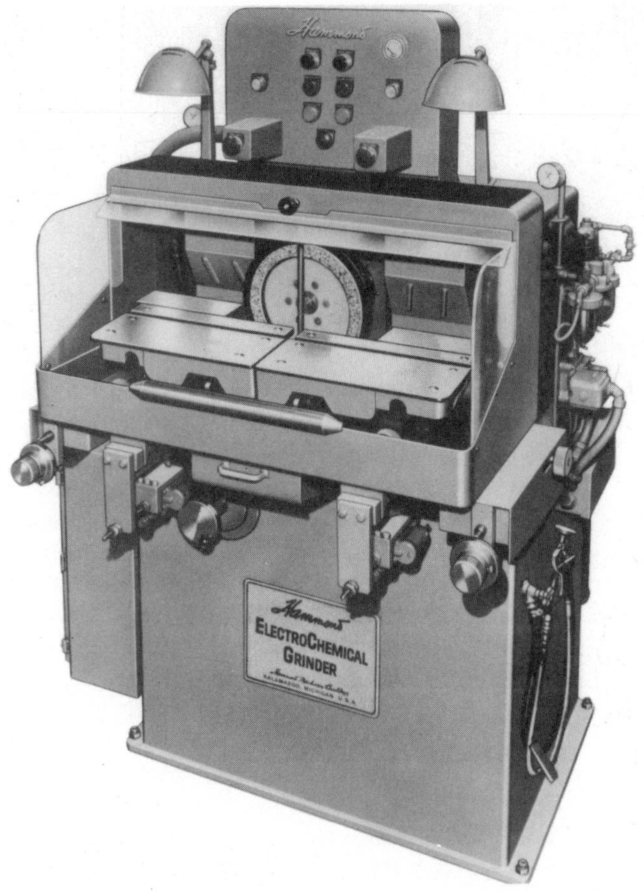

Fig. 13-86. This is an electrochemical grinding machine. (Hammond Machinery Builders)

Fig. 13-87. CNC gear grinding machine is solid enough for creep-feed grinding which is now considered most cost-efficient method of grinding gears. (Reishauer Elgin)

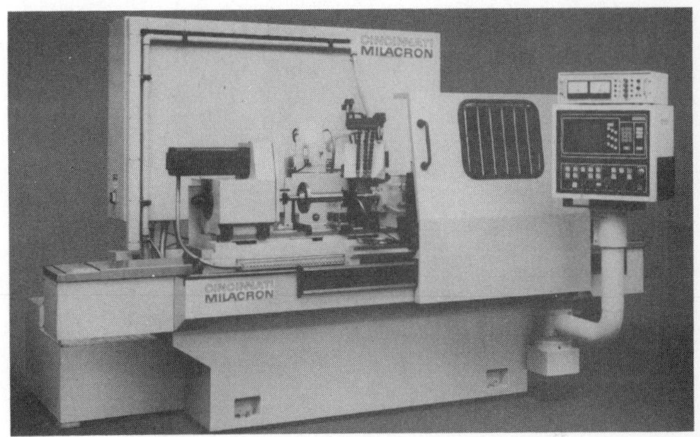

Fig. 13-88. This is a CNC cylindrical grinder. (Cincinnati Milacron)

Fig. 13-89. This CNC engine camshaft grinder is equipped with an in-process gaging system as well as automatic loader and unloader. (Cincinnati Milacron)

The part's contour is generated in a continuous motion by the X- and Y-axes slides of the machine. See Fig. 13-90.

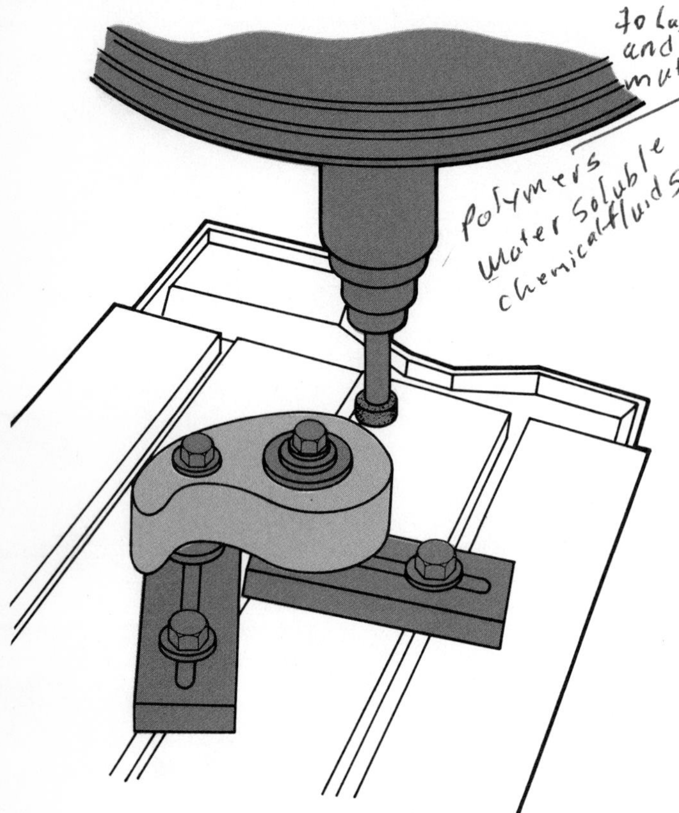

Fig. 13-90. With CNC grinding of this cam, for example, computer frees operator from responsibility of controlling and monitoring grinding wheel during grinding operation.

TEST YOUR KNOWLEDGE—Chapter 13

Please do not write in the text. Place your answers on a separate sheet of paper.

1. Industry classifies surface grinding as the grinding of ________ surfaces.
2. Surface grinding operations fall into two categories. List them.
3. Three work holding devices are frequently used to hold work for surface grinding. What are they?
4. List five (5) factors that are distinguishing characteristics of a grinding wheel.
5. The ideal grinding wheel will:
   a. Wear way as the abrasive particles become dull.
   b. Wear away at a predetermined rate.
   c. Wear away slowly to save money.
   d. All of the above.
   e. None of the above.
6. A solid grinding wheel will give off a ________ when struck lightly with a metal rod.
7. List the two conditions that commonly prevent a grinding wheel from cutting efficiently.
8. Why are cutting fluids or coolants necessary for grinding operations?
9. List the two basic types of cutting fluids.
10. A ________ wheel dressing tool is usually used to true and dress the wheels used for precision grinding.
11. Chatter and vibration marks are caused on the work when the grinding wheel is ________ or ________.
12. The problems in question 11 can be corrected by ________ the ________.
13. Irregular scratches on the work are usually caused by a ________ system. How can this problem be corrected?
14. What is the difference between conventional grinding and creep grinding?
15. A ________ and ________ is a grinding machine designed to support cutters (usually milling cutters) while they are being sharpened.
16. List the two variations of cylindrical grinding.
17. With ________ grinding, it is NOT necessary to support work between centers or mount work in a chuck while being rotated against grinding wheel.
18. Make sketches of nine (9) standard grinding wheel shapes.
19. The grinding technique that employs a belt on which abrasive particles are bonded for stock removal, finishing, and polishing operations is known as ________.
20. ________ or ________ grinding is actually an electrochemical machining operation. How is it done?

RESEARCH AND DEVELOPMENT

1. Secure samples of work produced by precision grinding. Compare them with a surface roughness comparison standard and determine the degree of roughness of each sample.
2. Prepare a specimen board with surfaces finished by the various precision grinding techniques. Use illustrations to indicate the type of machine used to produce each surface.
3. Check all of the grinding wheels in the shop or lab. Discard the ones that would be dangerous to use. Design a storage rack so the good wheels can be stowed safely with little danger of them becoming damaged.
4. Inspect the coolant system on the grinders in the shop or lab. Clean and make necessary repairs.

5. Prepare a list of recommendations that will improve precision grinding operations in the shop.
6. Contact grinding wheel manufacturers and secure photos that show how grinding wheels are manufactured. Design a bulletin board display around the material.
7. Demonstrate how to sharpen a milling cutter.
8. Demonstrate the correct way to true and dress a grinding wheel.
9. Research the various types of coolants and the material on which they are used. Make a poster on your findings and mount it near the grinding machines.
10. Prepare a poster that lists the problems encountered with precision grinding and how they can be corrected. Mount the poster near the grinding machines.
11. Prepare a handbook on how to safely operate precision grinding machines. Duplicate it for each member in the class.

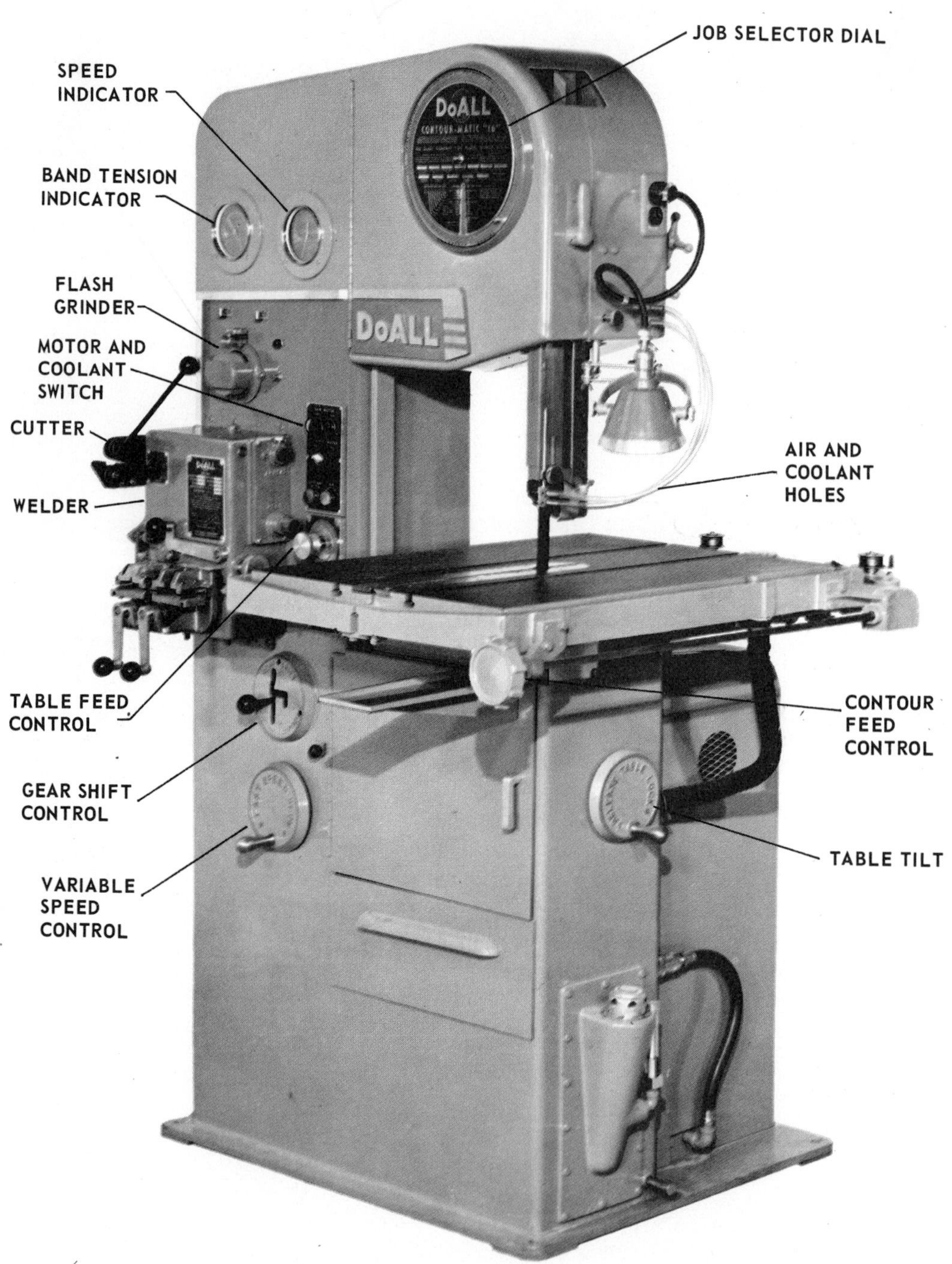

Fig. 14-1. Study parts of vertical band saw for metal machining. (DoALL Co.)

Chapter 14

BAND MACHINING

After studying this chapter, you will be able to:
Describe how a band machine operates.
Explain the advantages of band machining.
Select the proper blade for the job to be done.
Weld a blade and mount it on a band machine.
Safely operate a band machine.

Band machining is a widely employed machining technique that makes use of a continuous saw blade. Chip removal is rapid, Figs. 14-1 and 14-2. Each tooth is a precision cutting tool and accuracy can be held to close tolerances, eliminating or minimizing many secondary machining operations.

DANGER! Never attempt to operate this machine while your senses are impaired by medication or other substances.

BAND MACHINING OPERATIONS

Band machining offers several major advantages over other machining techniques:

1. Band machining MAINTAINS SHARPNESS, Fig. 14-3. Wear is distributed over many teeth. Chip load is uniform and constant on each tooth, minimizing tool wear.
2. Band machining provides UNRESTRICTED CUTTING GEOMETRY, Figs. 14-4 and 14-5. Cutting can be done at any angle, in any direction, and the length of the cut is unlimited.
3. Band machining is EFFICIENT, Fig. 14-6. Excess chip production wastes power. Band machining produces the desired shape with a minimum of chips.

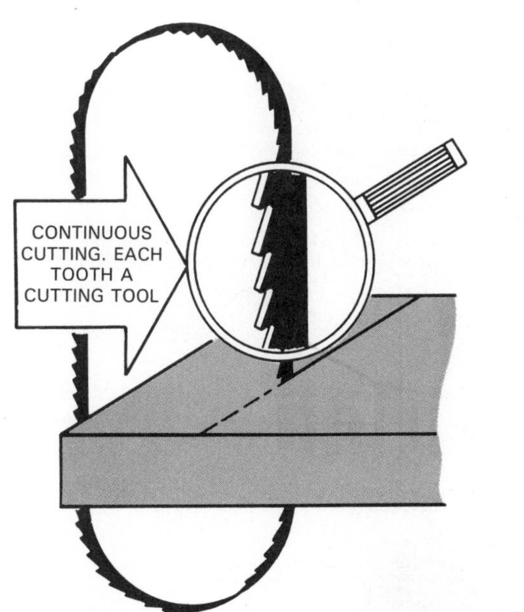

Fig. 14-2. Band machining makes use of a continuous saw blade. Each tooth is a cutting tool.

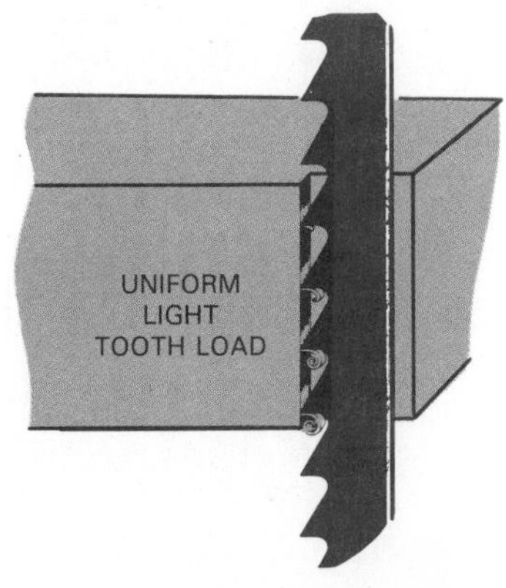

Fig. 14-3. Wear is distributed over many cutting edges (teeth) with band machining.

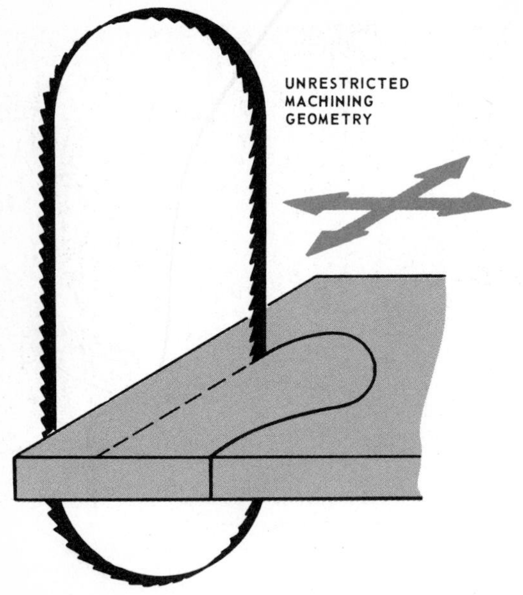

Fig. 14-4. Band machining permits machining at any angle or direction. Cut length is almost unlimited.

Fig. 14-5. Machining geometry in band machining is also unrestricted. (DoALL Co.)

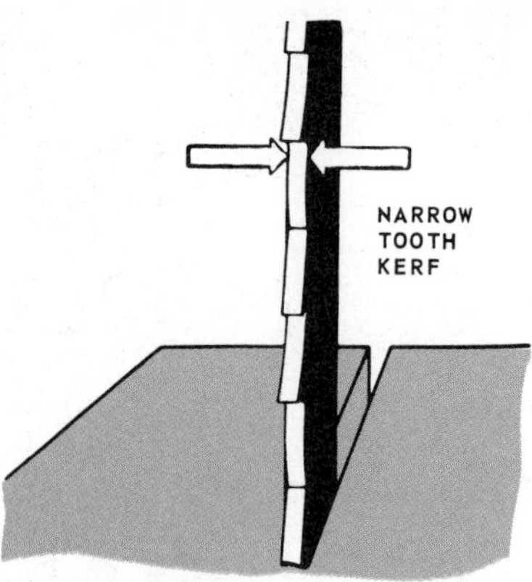

Fig. 14-6. Band machining is very efficient, with little waste.

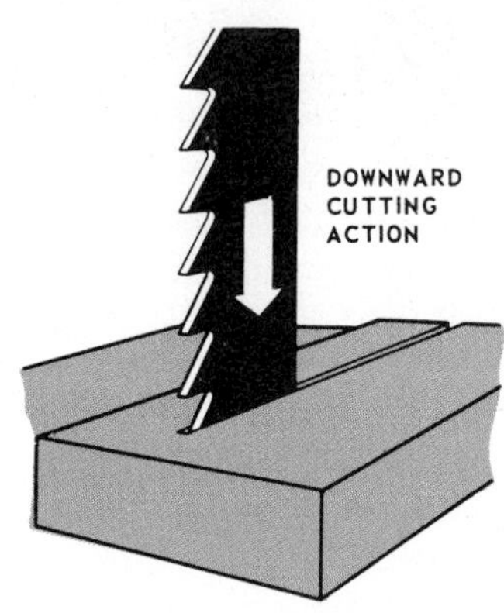

Fig. 14-7. Cutting action helps hold work on table.

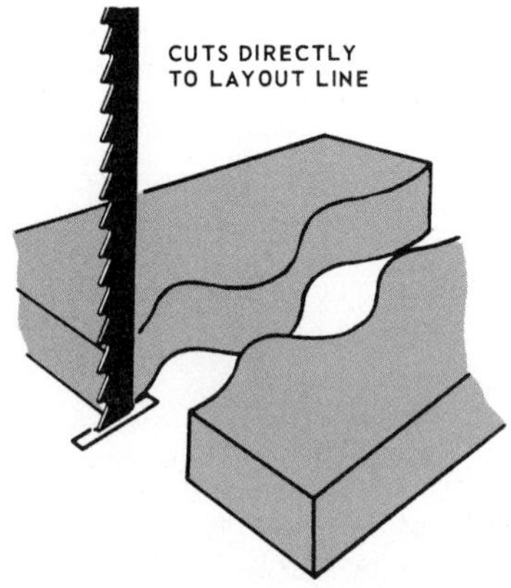

Fig. 14-8. Band machining produces little waste. Unwanted material is removed in solid sections.

4. Band machining provides a BUILT-IN TOOLHOLDER, Fig. 14-7. Cutting action is downward and cutting forces hold the work to the table. In most situations, work need NOT be clamped.
5. Band machining has LITTLE WASTE, Fig. 14-8. Band machining cuts directly to shape. Unwanted material is removed in solid sections rather than in chips, Figs. 14-9 and 14-10.

BAND BLADE SELECTION

Some blade manufacturers list more than 500 different band saw blades. Points that must be considered by the machinist when selecting the correct

Fig. 14-9. A band machine has capacity to make three-dimensional cuts. Limited production items can be machined from solid piece quickly and economically. (DoALL Co.)

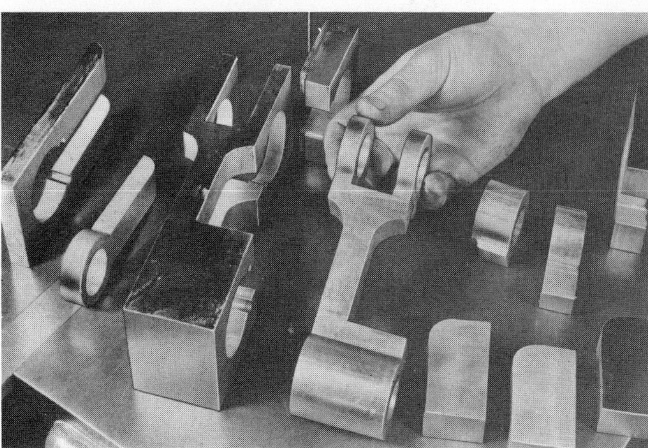

Fig. 14-10. Unwanted material is removed in solid sections rather than by being reduced to chips.

blade for a specific job are:

1. ***Blade type***—There are six basic kinds of band saw blades, Fig. 14-11:
   A. Tungsten carbide.
   B. Bimetal (high-speed cutting edge with flexible carbon steel back).

MATCH THE TOOLS TO THE JOB		
TYPE OF BLADE	APPLICATIONS	BAND MACHINE
T/C Inserted tungsten carbide teeth on fatigue-resistant blade.	Heavy production and slabbing operations in tough materials.	Horizontal cutoff machines over 5 hp with positive feed. Vertical contour machines over 5 hp with positive feed.
IMPERIAL BIMETAL HSS cutting edge with flex-resistant carbon-alloy back.	Mild to tough production and cutoff applications.	Horizontal cutoff machines over 1 1/2 hp with controlled feed, generally with variable-speed drives and with coolant system. Vertical contour machines over 1 1/2 hp with coolant system.
DEMON M-2 HSS blade	Heavy-duty toolroom and maintenance shop work. Full-time production applications.	Horizontal cutoff machines over 1 1/2 hp with controlled feed, generally with variable-speed drives and with coolant system. Vertical contour machines over 1 1/2 hp with coolant system.
DEMON SHOCK-RESISTANT M-2 HSS blade specially processed for greater shock resistance.	Structurals, tubing, materials of varying cross section.	Horizontal cutoff machines over 1 1/2 hp with controlled feed, generally with variable-speed drives and with coolant system. Vertical contour machines with 1 1/2 hp with coolant system.
DART Carbon-alloy, hard-edge, spring-tempered-back blade.	Superior accuracy for light toolroom and maintenance shop applications as well as light manufacturing.	Horizontal cutoff machines under 1 1/2 hp with coolant system. Vertical contour machines under 1 1/2 hp with coolant system.
STANDARD CARBON All-purpose, hard-edge, flexible-back blade.	Light toolroom and maintenance shop applications.	Horizontal cutoff machines under 1 1/2 hp with weight feed and without coolant. Generally step speeds. Vertical contour machines under 1 1/2 hp without coolant.

Fig. 14-11. Study recommendations for using basic blade types. (DoALL Co.)

C. High-speed steel.
D. Shock resistant.
E. Hard edge with spring tempered back.
F. Carbon steel flexible back.

2. ***Blade characteristics*** include width, pitch, set, gage, and tooth form.
   A. ***Width*** of the blade is important—the wider the blade, the greater its strength and the more accurate it will cut, Fig. 14-12. Use the WIDEST blade the machine will accommodate when making STRAIGHT cuts. CONTOUR cutting should utilize the widest blade that will cut the required radii, Fig. 14-13. Widths form 1/16 to 2 in. (1.5 to 50.0 mm) are available.
   B. ***Pitch*** refers to the number of teeth per inch or the distance between each tooth measured in millimeters, Fig. 14-14. The THICKNESS of the material to be cut determines the pitch. At least THREE teeth should be in contact with the work for best performance. Pitches from 2 to 32 per inch are manufactured.

Fig. 14-12. Blade width is from tooth tip to other edge or back.

WIDTH OF BLADE	SMALLEST RADIUS	THE WIDTH OF THE BLADE IS DETERMINED BY THE SMALLEST RADIUS TO BE CUT
1/16	1/16	
3/32	1/8	
1/8	7/32	
3/16	3/8	
1/4	5/8	
5/16	7/8	
3/8	1 1/4	
1/2	3	

Fig. 14-13. Note how blade width affects size of radius that can be cut.

MATERIAL THICKNESS	BAND PITCH
Less than 1 in. (25 mm)	10 or 14
1 to 3 in. (25 to 75 mm)	6 or 8
3 to 6 in. (75 to 150 mm)	4 to 6
6 to 12 in. (150 to 300 mm)	2 or 3

Fig. 14-14. Note recommended band pitches to saw various thicknesses of material.

   C. ***Set*** provides clearance for the blade back, Figs. 14-15 and 14-16. ***Raker set*** is recommended for cutting LARGE solids or THICK plate and bar stock. ***Wavy set*** should be used for work with VARYING THICKNESSES such as pipe, tubing, and structural materials. ***Straight set*** is specified for FREE CUTTING materials, like aluminum and magnesium.
   D. ***Gage*** refers to blade thickness, Fig. 14-17. Extra strength can be obtained when using narrow blades by securing a blade of a heavier gage.
   E. ***Tooth form*** is the shape of the tooth, Fig. 14-18. There are three basic forms. Each has its specific application.
      a. ***Standard tooth blades,*** with well rounded gullets, are usually best for most ferrous metals, hard bronze, and brass.
      b. ***Skip tooth blades*** provide more gullet and better chip clearance without weakening the blade body. They are recom-

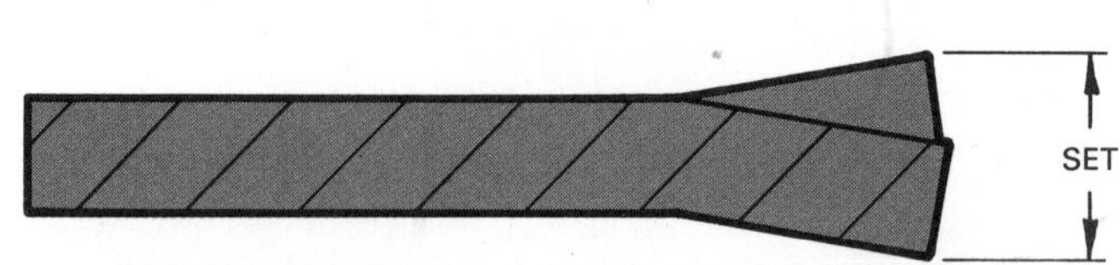

Fig. 14-15. Blade set refers to side angle of teeth.

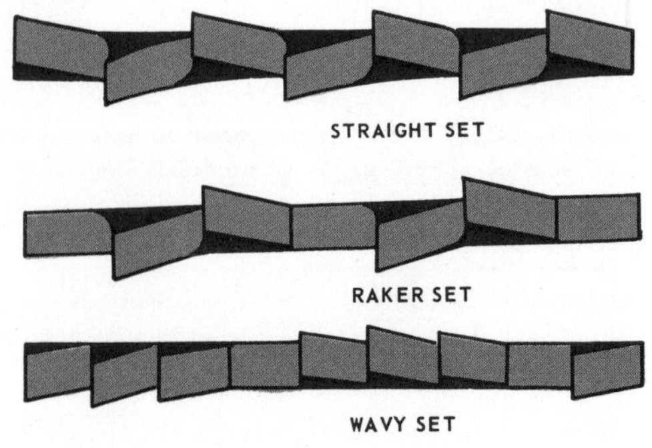

Fig. 14-16. Study different types of blade set.

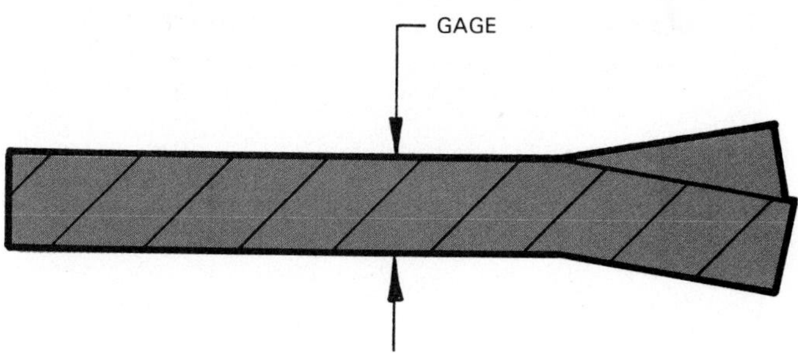

Fig. 14-17. Blade thickness is referred to as gage.

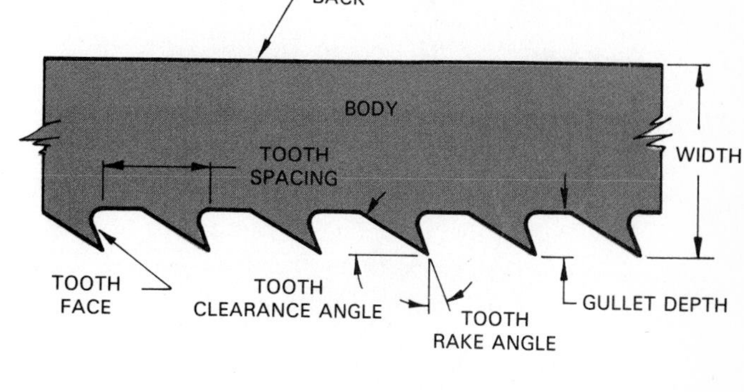

Fig. 14-19. Study terminology commonly used with saw blades.

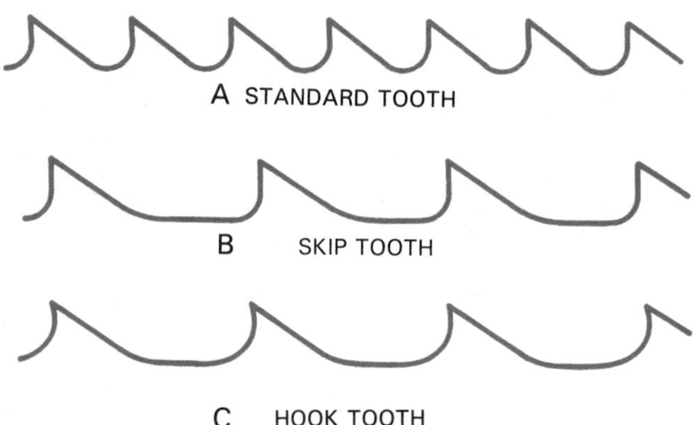

Fig. 14-18. A—Standard tooth blades, with their well rounded gullets, are usually best for most ferrous metals, hard bronzes, and brasses. B—Skip tooth blades provide more gullet and better chip clearance without weakening blade body. This type blade is best suited for aluminum, copper, magnesium, and soft brasses. C—Hook tooth blades offer two advantages over skip tooth blade: blade design makes it feed easier and its chip breaker design prevents it from gumming up.

Fig. 14-20. This device is used to determine best blade and cutting speed for a specific job. (DoALL Co.)

mended for most aluminum, magnesium, and brass alloys.

c. ***Hook tooth blades*** offer two advantages over the skip tooth blade. Blade design makes it feed easier and its chip breaker design prevents gumming up.

The parts of a blade are shown in Fig. 14-19.

Many band machines have built-in blade selection devices, Fig. 14-20. With them, it is a simple matter to dial in the various bits of information necessary to determine the best blade for the job. Following this recommendation will result in the job being done faster and with a better finish.

WELDING BLADES

Band saw blade stock can be purchased as ready to use ***welded bands.*** However, it is more economical to buy it in 100, 300, or 500 ft. (30, 90, or 150 meter) ***strip-out containers.*** The desired length of blade material is withdrawn from the container, ends squared, and the blade welded. Extreme care must be taken to make a good weld (as strong as blade).

WARNING! Wear leather gloves and eye protection when handling band saw blade stock or welded blades.

Preparing blade for welding

Use snips or blade cut-off shears to cut the blade to length. Cuts must be square or the problems shown in Fig. 14-21 will result.

After squaring the blade ends, it will be necessary to remove several teeth (depending upon blade pitch) by grinding. As about 1/4 in. (6.5 mm) of the blade is consumed in the welding process, the teeth

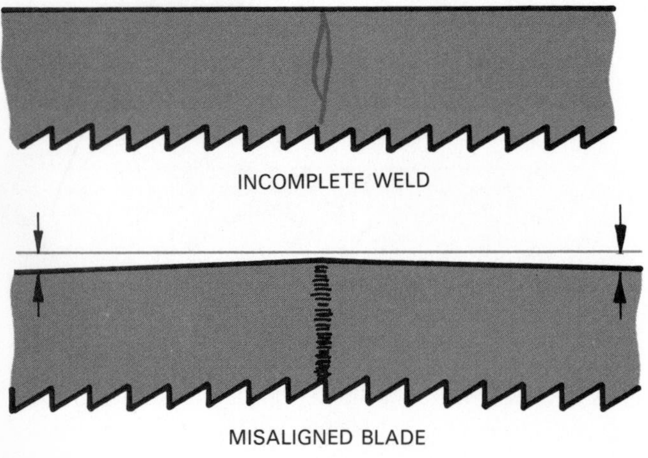

Fig. 14-21. Note common problems encountered when welding band saw blades; they are usually result of poorly squared blade ends or dirty blade material.

Fig. 14-23. Heavy-duty "flash" welder in operation. It is evident how welder got its name. Eye protection is a must!

must be ground off to assure uniform tooth spacing after the weld has been made. Remove only the teeth. Do NOT grind into the back of the blade.

Making the blade weld

For convenience, most band machines have the welder built-in. This is a resistance-type butt welder, Fig. 14-22. Blades up to 1/2 in. (12.5 mm) wide can be welded in a light-duty welder. A heavy-duty "flash" butt welder is required for heavier blades. Refer to Fig. 14-23.

The blade ends are butted together and clamped in the welder's jaws. Clean the jaws before and after making a weld. The teeth should be placed against the aligning plates. Pressure (determined by width and thickness of blade being joined) is applied to the band ends. Check the blade to be sure the ends are touching across their entire width and in the center of the gap between the welder jaws.

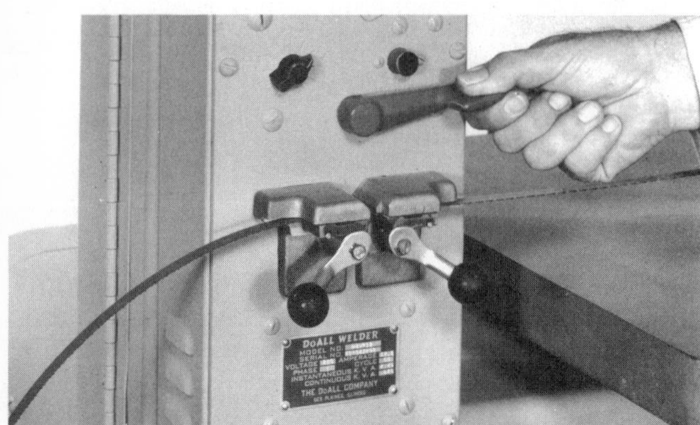

Fig. 14-22. Resistance butt welder is used for blades up to 1/2 in. (12.5 mm) in width. (DoALL Co.)

After checking, stand to one side to be clear of any "flash" that might result. Press the welder switch in as far as it will go; then release it immediately. The weld is made automatically and the resulting weld should look like Fig. 14-24.

DANGER! Wear approved tinted glass type eye protection when welding band saw blades. The bright flash can cause eye injury.

The weld, at this point, is brittle and must be ***annealed*** before it can be used. Follow the recommendations for annealing furnished by the manufacturer of the machine being employed. Avoid overheating the blade or it will remain brittle. Let it cool slowly.

Remove the "flash" formed during welding on the grinder built into the welder. A finished weld should look like Fig. 14-25. Use care when grinding to prevent dulling the teeth. The blade will also be weakened if it becomes "dished" during the grinding, Fig. 14-26.

Fig. 14-24. Flash buildup at point of weld should be uniform across blade. Do NOT flex blade at weld until section has been annealed. It is very brittle and will break! (DoALL Co.)

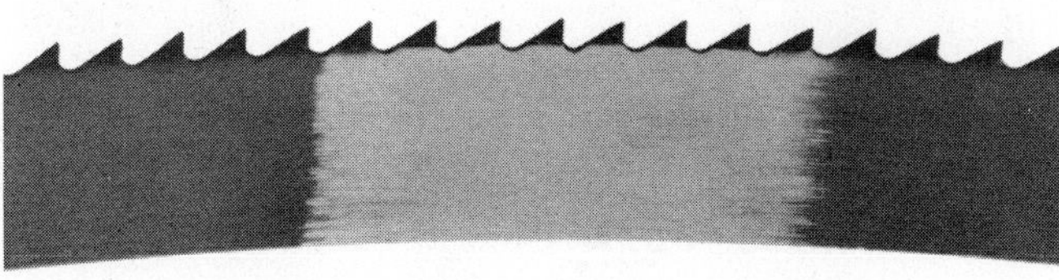

Fig. 14-25. This is a properly made and cleaned weld. Weld, if done properly, is as strong as blade itself.

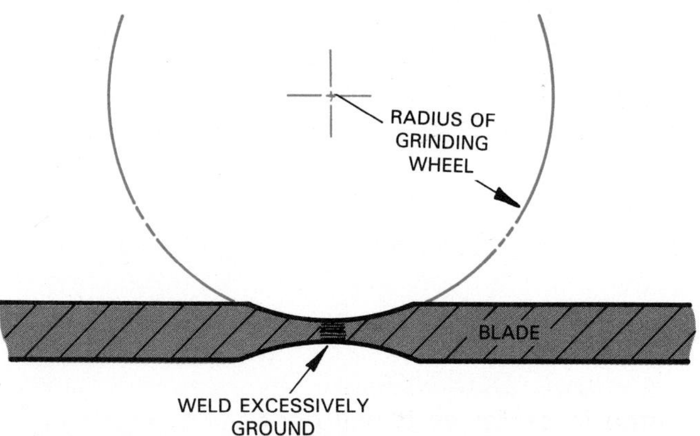

Fig. 14-26. A blade is seriously weakened if it becomes "dished" during grinding operation that removes weld flash.

BAND MACHINE PREPARATION

As with all other machine tools, a band machine must be made ready with care if the tool is to operate at maximum efficiency.

Band machine lubrication

Use the grades of lubricants as specified in the manufacturer's manual on the machine. Develop a definite lubrication sequence to reduce the possibility of missing a vital point.

Band blade guides

Select and install blade guides suitable for the job at hand. Use blade guide inserts for light sawing, Fig. 14-27. Roller guides are recommended for continuous high-speed sawing, Fig. 14-28.

Guides must be the proper width, Fig. 14-29. If TOO WIDE, the saw teeth will be damaged. If TOO NARROW, the blade will have a tendency to twist in the work and make it difficult to follow the desired path or cut.

Inspect and clean the upper and lower guide units. Make sure the back-up bearings are NOT clogged with chips.

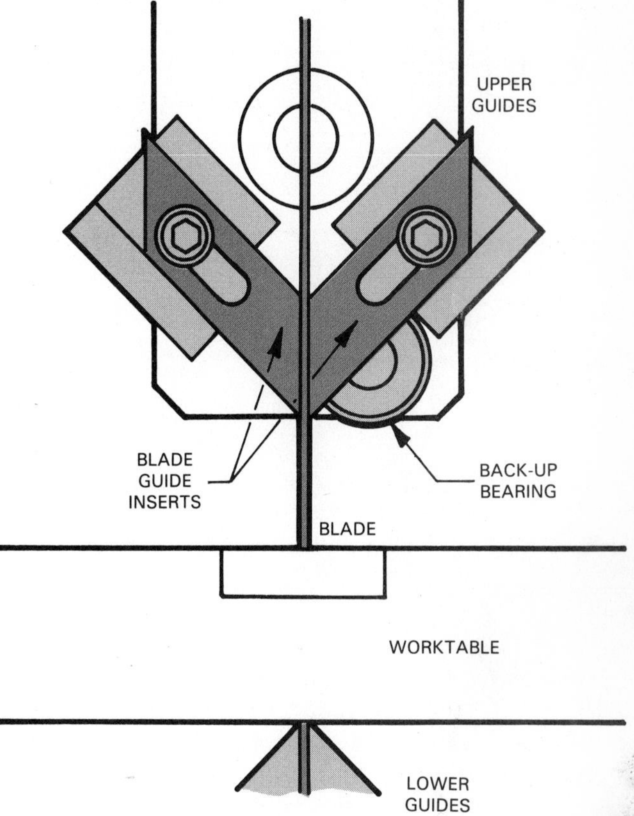

Fig. 14-27. Note blade guide inserts.

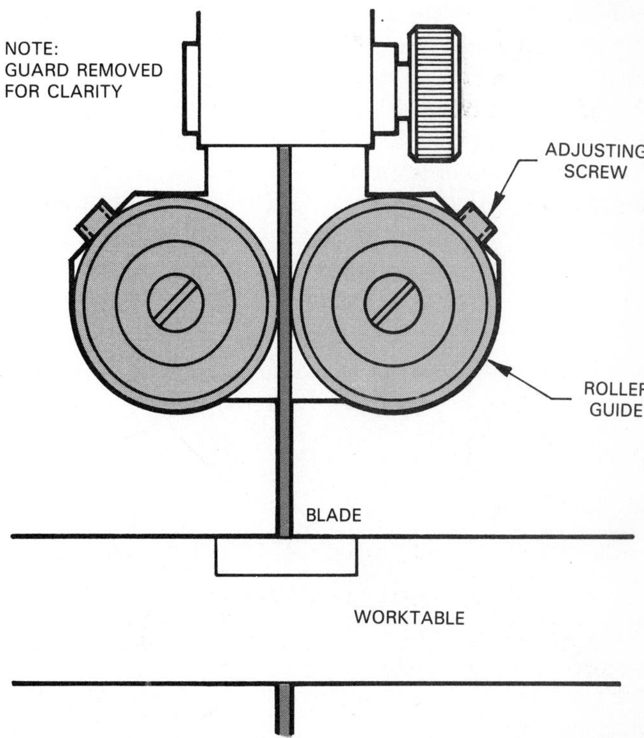

Fig. 14-28. Roller guides are recommended for continuous high-speed cutting.

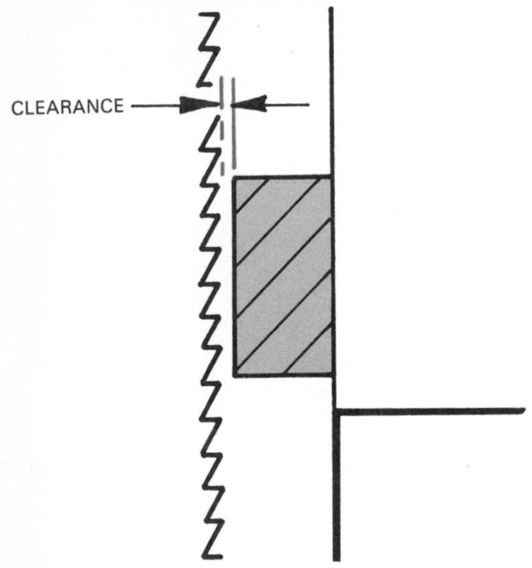

Fig. 14-29. Blade guides of proper width must be employed or band will be ruined.

Typically, there should be 0.001 - 0.002 in. (0.025 - 0.050 mm) clearance between the guide and blade for best results, Fig. 14-30. Follow the manufacturer's recommendations.

Blade tracking

Adjust the band carrier wheels so the blade tracks correctly. Again, it is important that the manufacturer's recommendations be carefully followed. However, if they are NOT available, the following will usually permit satisfactory operation on most band machines:

1. The center of the band should ride directly over the center of the wheel crown on the rubber tire, Fig. 14-31. Replace the tire if it has become frayed or damaged.

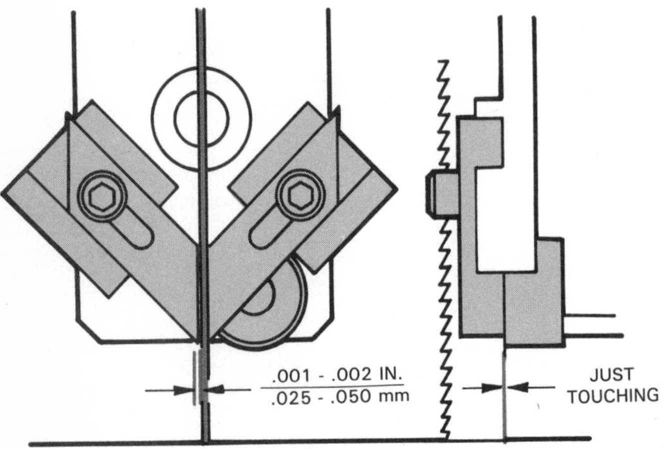

Fig. 14-30. Note how blade guides are adjusted to produce long blade life.

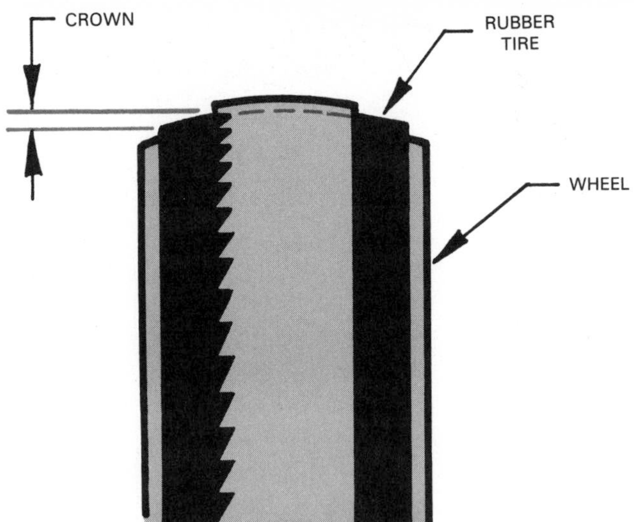

Fig. 14-31. Blade should ride directly on centerline of wheel crown.

2. There should be no noticeable gap between the back of the band and the back-up bearings of the saw guides. See Fig. 14-30.
3. Remember that the blade must be installed with the teeth DOWNWARD.

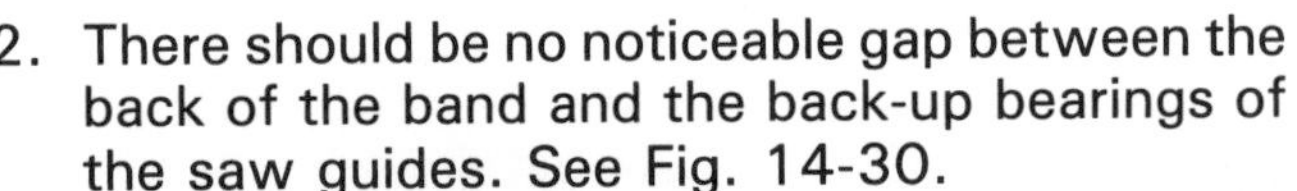
Observe extreme caution when handling saw blades. They are sharp and can cause serious injury.

DANGER! Be sure that the power to the band machine has been turned off at the master switch before attempting to install a blade.

Band blade tension

Blade tension refers to the pressure put on the saw band to keep it taut and tracking properly. On small machines, it is usually applied by means of a hand crank, Fig. 14-32. On large heavy-duty machines, tension is applied hydraulically.

The amount of blade tension is determined by the WIDTH and PITCH of the blade. Use the band tension chart furnished with the machine.

Many band machines have built-in ***tension meters*** that make it easy to adjust and maintain proper blade tension. This is especially important when a new blade is installed. A new blade has a tendency to stretch slightly when first used. This can create a safety problem if tension is NOT readjusted before it falls off too far.

Band cutting speed

As with other machining operations, best results will be obtained if recommended cutting speeds are maintained. See Fig. 14-33 for band speeds of a few selected materials.

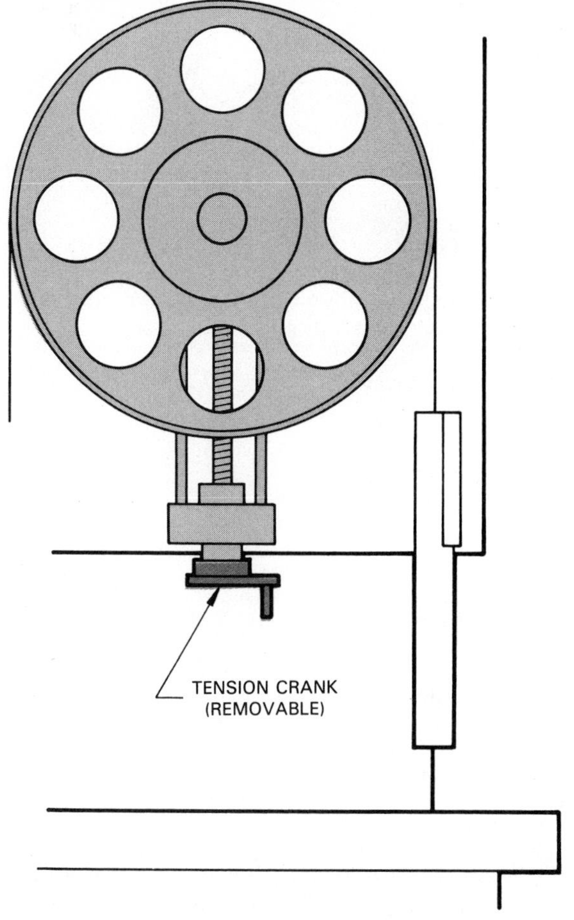

Fig. 14-32. This is a typical location of blade tension crank. Crank is removable. On some band machines, tension crank is located on front of head.

Band cutting fluids

Cutting fluid can be applied on a band machine by flooding, in the form of a mist, or as a solid type lubricant.

Flooding is recommended for heavy-duty band machining.

Mist coolant is used for high-speed sawing of free-machining nonferrous metals. It is also employed when tough hard-to-machine materials are cut.

Solid lubricants are applied when the machine does NOT have a built-in coolant system.

Coolant manufacturers will furnish a coolant chart with recommendations for band machining operations upon request.

BAND MACHINING OPERATIONS

The vertical band machine is designed to do the following sawing operations.

Straight sawing

Straight, two-dimensional sawing is band machining in its simplest form, Fig. 14-34. The operator just follows a straight layout line.

Splitting, shown in Fig. 14-35, is another job that can be done rapidly on a band machine.

Contour sawing

Contour sawing is possible on a vertical band machine, Fig. 14-36. Machine size is the limiting factor on the work dimensions that can be cut.

MATERIAL	THICKNESS INCHES (millimeters)	BAND SPEED SURFACE FEET PER MINUTE (meters per minute)
Low-to-medium carbon steels	Under 1 in. (25 mm)	345 - 360 sfm (105 - 110) mpm
	1 to 6 in. (25 to 150 mm)	295 - 345 sfm (90 - 105) mpm
Medium-to-high carbon steels	Under 1 in. (25 mm)	225 - 250 sfm (70 - 75) mpm
	1 to 6 in. (25 to 150 mm)	200 - 225 sfm (60 - 70) mpm
Free machining steels	Under 1 in. (25 mm)	260 - 395 sfm (80 - 120) mpm
	1 to 6 in. (25 to 150 mm)	260 - 345 sfm (80 - 105) mpm
Titanium, pure and alloys	Under 1 in. (25 mm)	100 - 115 sfm (30 - 35) mpm
	1 to 6 in. (25 to 150 mm)	90 - 110 sfm (30 - 35) mpm

Fig. 14-33. Note recommended cutting speeds for a few metals and alloys.

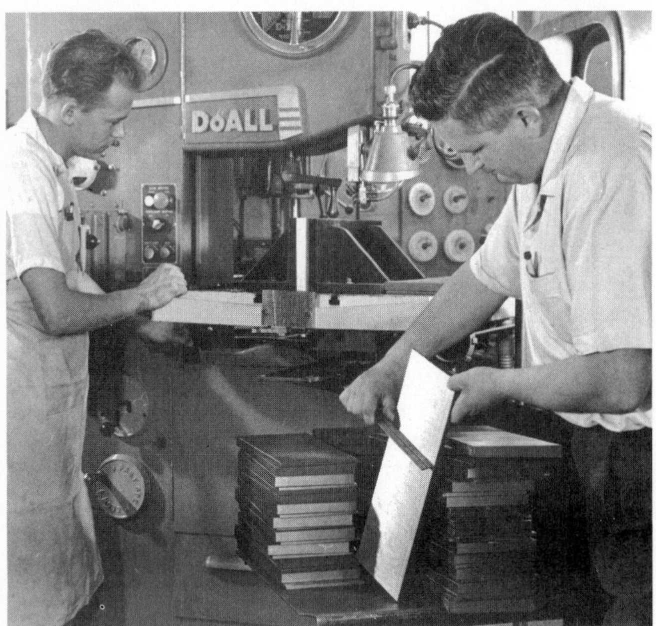

Fig. 14-34. This is an example of two-dimensional band machining. Two pieces 16 in. long, 6 in. wide, and tapering from 3/4 in. at one end to 1/4 in. at the other, are cut quickly and accurately from a 16 in. x 6 in. x 1 in. section of steel. (DoALL Co.)

Fig. 14-35. During this slitting operation, note how this bearing is held in place for machining. (DoALL Co.)

Angular sawing

The table on the vertical band machine is usually mounted on trunnions permitting it to be tilted. This makes it possible to machine compound angular cuts, Fig. 14-37.

Internal cuts

Precision internal cuts can be made on a vertical band machine, Fig. 14-38. The band is threaded

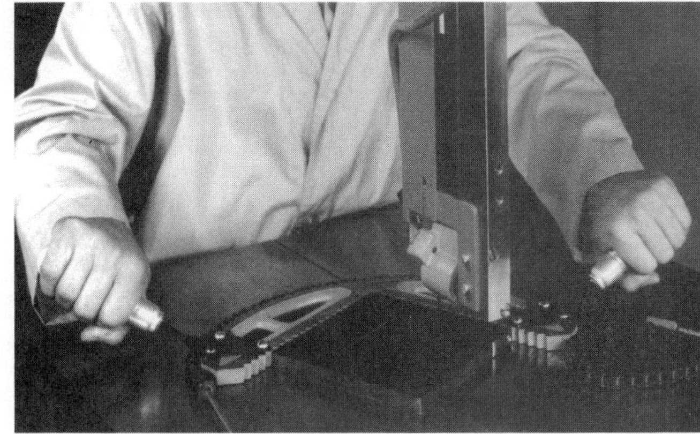

Fig. 14-36. Machinist is using contour sawing. (DoALL Co.)

Fig. 14-37. Compound angular cuts can be made by tilting the worktable. (DoALL Co.)

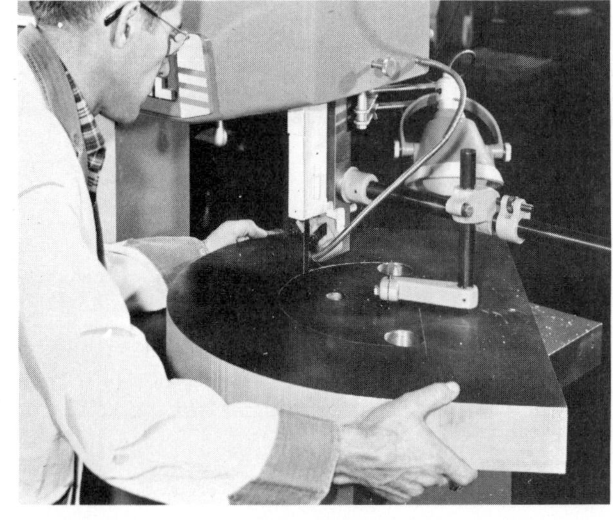

Fig. 14-38. Precision internal cuts can be made on band machine. Blade is threaded through holes drilled in piece, welded, and work guided along prescribed lines. (DoALL Co.)

through a hole drilled in the piece, welded, and the work maneuvered along the prescribed line. Additional holes must be drilled if sharp corners are specified, Fig. 14-39. It may be necessary to use the blade as a file to get the work into cutting position, Fig. 14-40.

After completing the internal cut, cut the band close to the weld so the entire weld can be cut away prior to rewelding. It is recommended that there be no more than ONE WELD in the band.

WARNING! Exercise extreme care when the blade breaks through the work at the completion of the cut. If possible, a section of back-up metal should be between your hand and where the band will break through the work. The moving blade can cause serious injury! See Fig. 14-41.

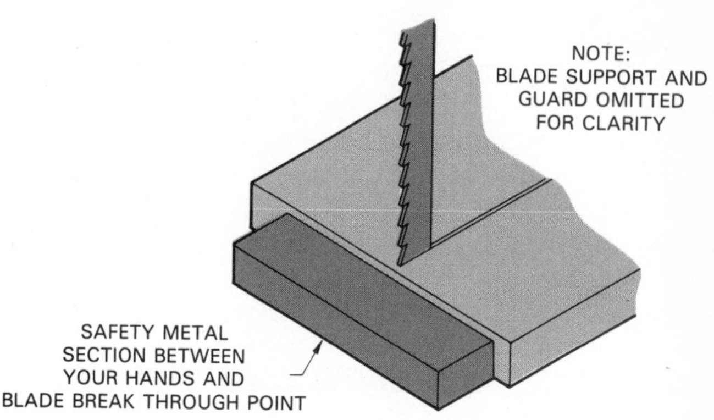

Fig. 14-41. CAUTION! Never have your hands directly in front of blade. Use a back-up metal section where blade breaks through work.

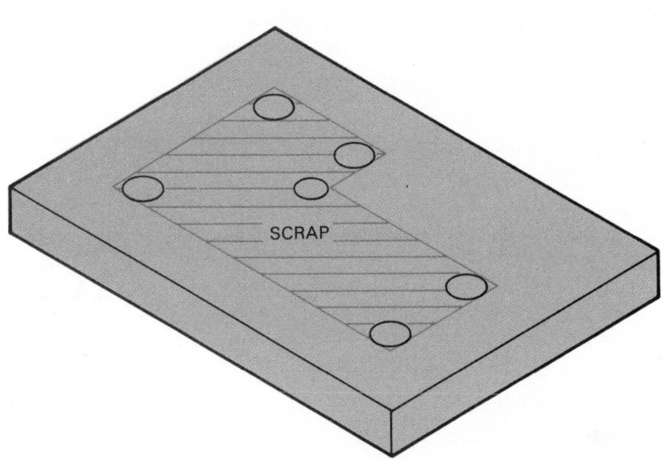

Fig. 14-39. Typical layout of drilled holes when sharp corners are specified on an internal cut.

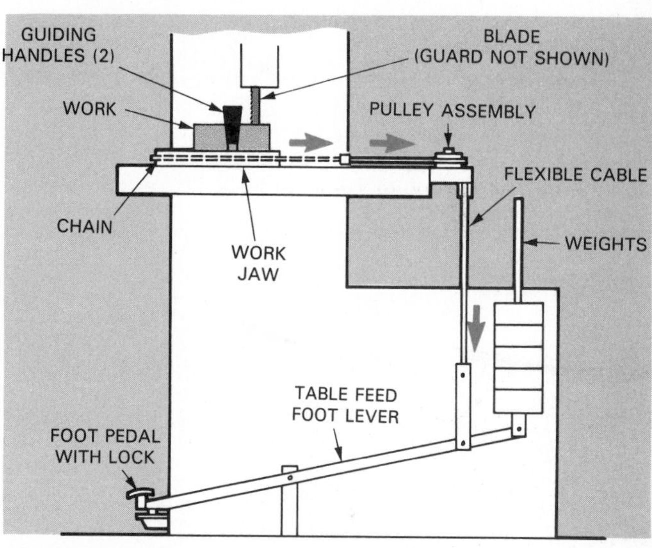

Fig. 14-42. This power feed device makes use of weights to pull work into blade.

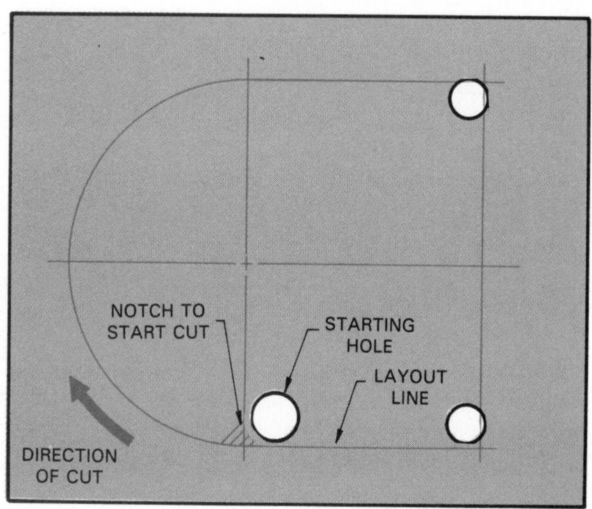

Fig. 14-40. Start internal cuts by using blade as a file, and notch work until blade can be positioned to start cut.

BAND MACHINE POWER FEED

Power feed or mechanical pressure attachments are available for band machines. The simplest attachment makes use of weights to pull work into the blade, Fig. 14-42. Both hands of the operator are free to guide the work.

Several types of hydraulic power feed attachments have been devised. On some vertical band machines, the worktable is hydraulically actuated and feeds the work into the blade at a constant rate. Accidental overfeeding is eliminated greatly extending band life. See Fig. 14-43.

A handwheel, connected to the work by a sprocket and chain, guides work along the layout line. The servomechanism on the lower blade guide senses changes in feed force on the work and automatically counteracts them. To maintain a constant feeding force on the work, the device advances, slows, stops, or reverses the worktable. To leave the operator's second hand free, a foot switch permits them to move the table by remote control.

Another type feed mechanism is a self-contained unit attached to the machine. A hydraulic cylinder applies and maintains constant pressure on the work through a sprocket and chain system. Fig. 14-44 shows the parts of this type system.

OTHER BAND MACHINING APPLICATIONS

The great versatility of the band machine is further utilized by the addition of accessories and/or minor tool modifications.

Band filing

A smooth, uniformly finished surface may be obtained rapidly and with considerable accuracy on a band machine fitted for filing, Fig. 14-45.

A series of small file segments make up the ***file band.*** The individual units interlock and form a continuous file. The segments are fitted to a flexible back. File guides replace the regular saw guides when a band file is used, Fig. 14-46.

A variety of file shapes and cuts are available.

Band polishing

Parts can be polished on a band machine with a polishing attachment, Fig. 14-47. A continuous ***band abrasive cloth*** replaces the saw blade. Best

TROUBLESHOOTING BAND MACHINES

PROBLEM	***CORRECTION***
1. Teeth dull prematurely.	a. Use slower cutting speed. b. Replace blade with a finer pitch band. c. Be sure proper type cutting fluid is used. d. Increase feed pressure. e. Check to be sure band is installed with teeth pointing down.
2. Band teeth breaking out.	a. Reduce feed pressuure. b. Use finer pitch band if thin material is being cut. c. Be sure work is held solidly as it is fed into band. d. Use a heavier-duty cutting fluid.
3. Band breaks.	a. Change to a heavier band. b. Reduce cutting speed. c. Check wheels for damage. d. If blade breaks at weld, use longer annealing time. Reduce heat gradually. e. Use finer pitch blade. f. Reduce feed pressure. g. Decrease band tension. h. Check blade guides for proper adjustment. i. Use cutting fluid.
4. Cutting rate too slow.	a. Increase band speed. b. Use coarser pitch blade. c. Increase feed pressure. d. Use cutting fluid.
5. Band makes "belly-shaped" cut.	a. Increase blade tension. b. Adjust guides close to work. c. Use coarser pitch band. d. Increase feed pressure.
6. Band does NOT run true against saw guide backup bearing.	a. Remove burr on back of band where joined. b. If hunting back and forth against backup bearing on guide, reweld blade with back of band in true alignment. c. Check alignment of wheels. d. Check backup bearing. Replace if worn.
7. Premature loss of set.	a. Band too wide for radius being cut. b. Reduce cutting speed. c. Apply cutting fluid.

(Courtesy of DoALL Co.)

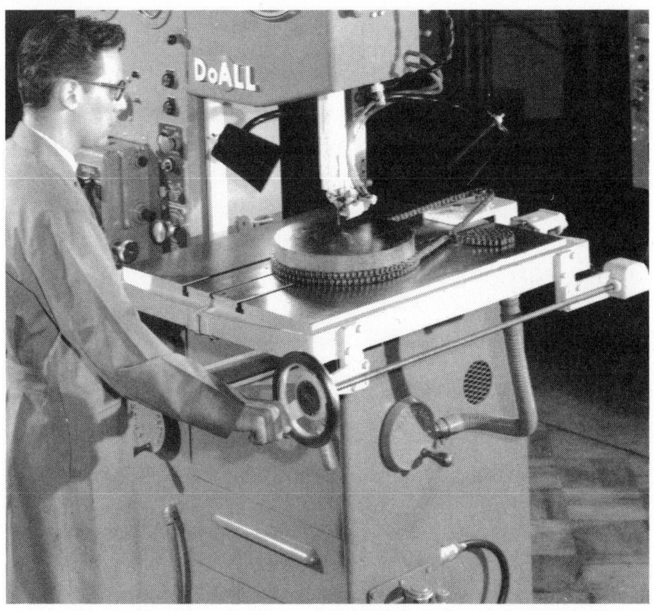

Fig. 14-43. Band machine is fitted with a servo-contour feed attachment. Only one hand is needed to guide cutting operation. Table movement is controlled by a foot switch. (DoALL Co.)

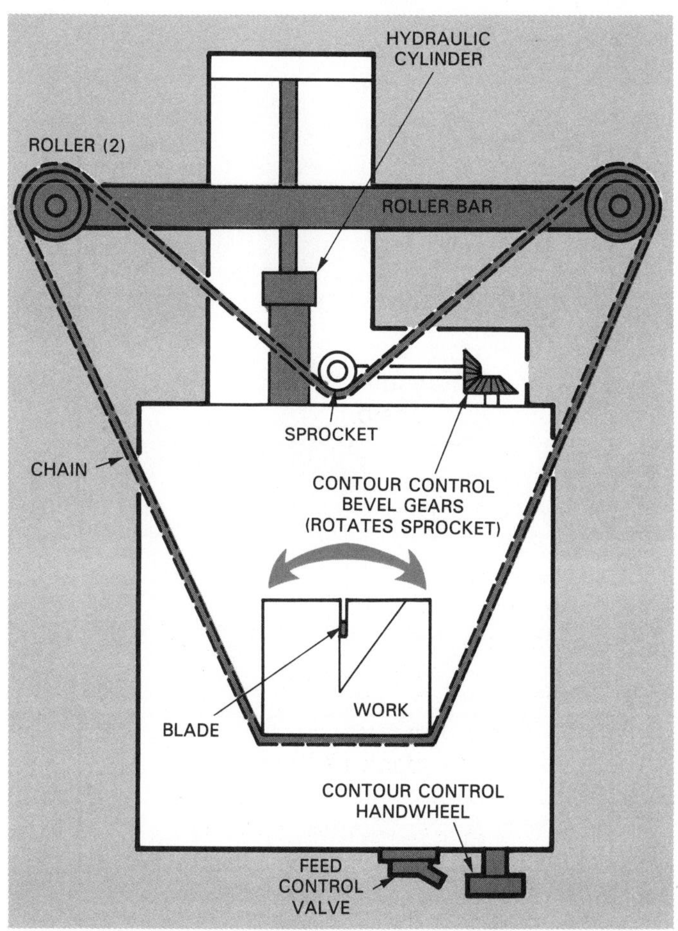

Fig. 14-44. Study parts of hydraulically actuated power feed unit.

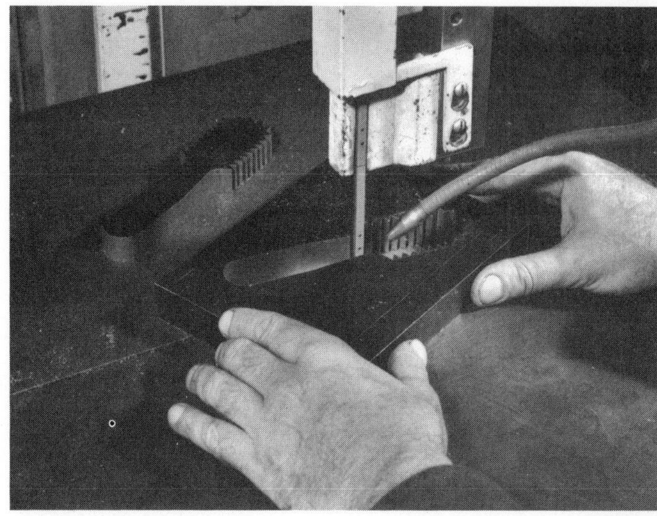

Fig. 14-45. Filing is being done on this band machine. (DoALL Co.)

results can be obtained if the back of the abrasive band is lubricated with graphite powder. Abrasive band life will also be greatly extended.

Friction sawing

Friction sawing makes use of extremely high cutting speeds of between 6000 and 15,000 feet per minute (fpm) or 1800 to 4500 meters per minute (mpm) and heavy pressure to cut ferrous metals. Cutting friction generates intensive heat and softens the metal.

The blade teeth do NOT actually cut, they are used to scoop out the softened metal. As a matter of fact, dull teeth are superior as they generate heat better. Friction is a spectacular operation as can be testified to by the shower of sparks produced.

The technique is a rapid way to cut ferrous metals under 1 in. (25 mm) in thickness. Hardness is NOT a deterring factor as long as the metal has a low melting point range.

Most band machines are NOT adaptable for friction sawing.

SAFETY NOTE! Friction sawing requires a full face shield, leather gloves, and a transparent shield fitted around the cut area.

Other band tools

Tooth-type bands are most commonly used on the vertical band machine. However, other types of blades have been developed for special work.

The ***knife-edge blade*** is employed to cut material that would tear or fray when machined by a conventional blade, Fig. 14-48. For example: sponge rubber, cork, cloth, corrugated cardboard, and rubber would tear easily.

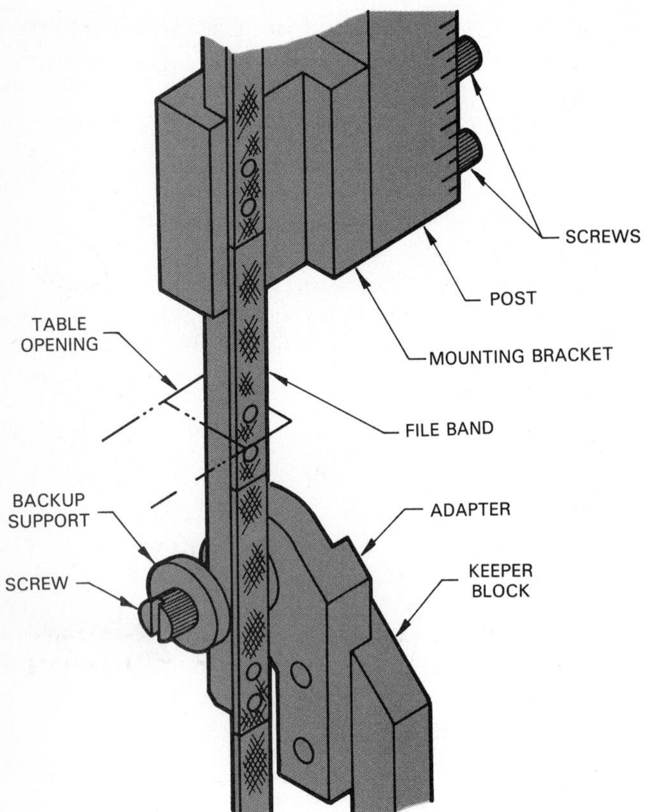

Fig. 14-46. Guides are setup for band filing. Worktable has been removed for clarity.

The ***diamond-edge band*** is specially designed to cut material that is difficult or impossible to cut with a conventional toothed blade, Fig. 14-49. The diamonds are only on the front edge of the band where the cutting is accomplished! On a ***wire band,*** Fig. 14-50, diamonds are fused around the circumference on the band permitting it to cut in any direction, Fig. 14-50.

Friable materials (materials that are easily crumbled or reduced to powder) of extreme hardness or with abrasive qualities can be cut economically with these bands.

In addition to the bands mentioned, blades with unusual characteristics are available to meet almost any band machining requirement.

Specialized vertical band machines

The vertical band machine is manufactured in a wide range of sizes, and has been adapted to do many kinds of band machining. Some large machines have been fitted with closed circuit TV and a remote control console to permit the operator to contour machine large sections of material or to perform hazardous or dangerous work. For example, this type machine is employed to machine toxic and/or radioactive materials, Fig. 14-51.

Computer controlled band machines (NC and CNC) are now available. They have "X" and "Y" table movement capability plus ***circular interpolation*** (can cut circular and elliptical shaped contours). Input is by computer, by tape, or manually done. See Fig. 14-52.

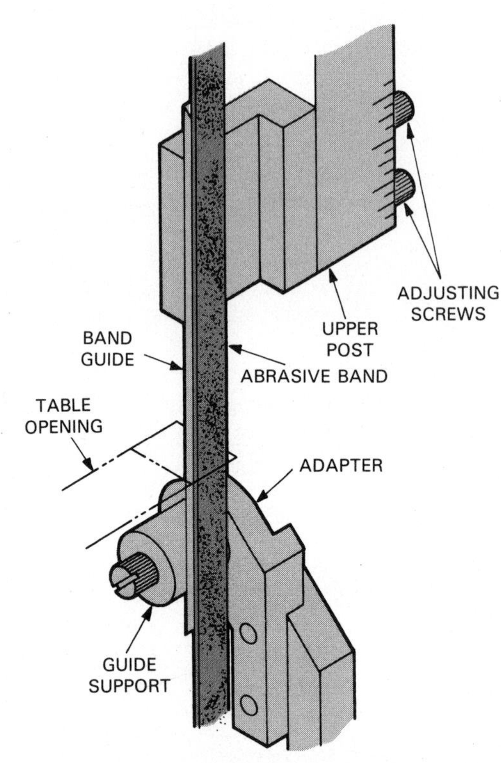

Fig. 14-47. Polishing can be done on a band machine by replacing saw blade with an abrasive band, and utilizing guide and support shown. Worktable has been removed for clarity.

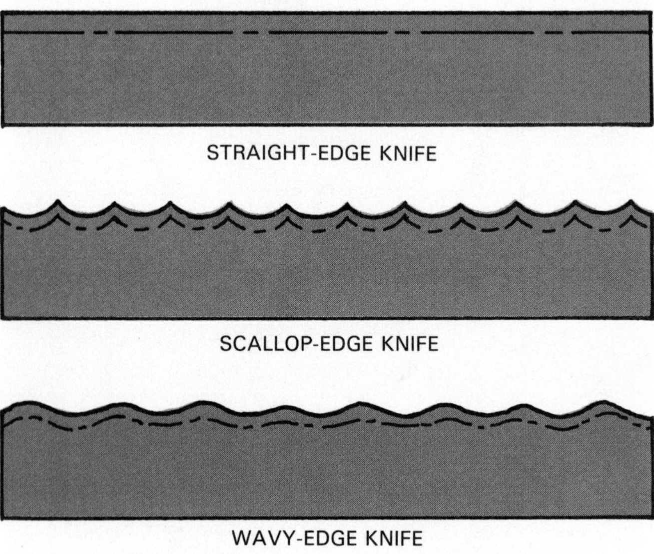

Fig. 14-48. Study types of knife edge blades.

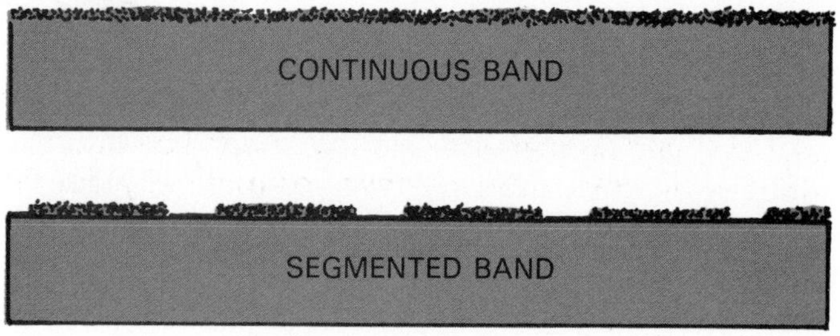

CONTINUOUS BAND

SEGMENTED BAND

Fig. 14-49. Cutting edge of diamond band is impregnated with diamond dust.

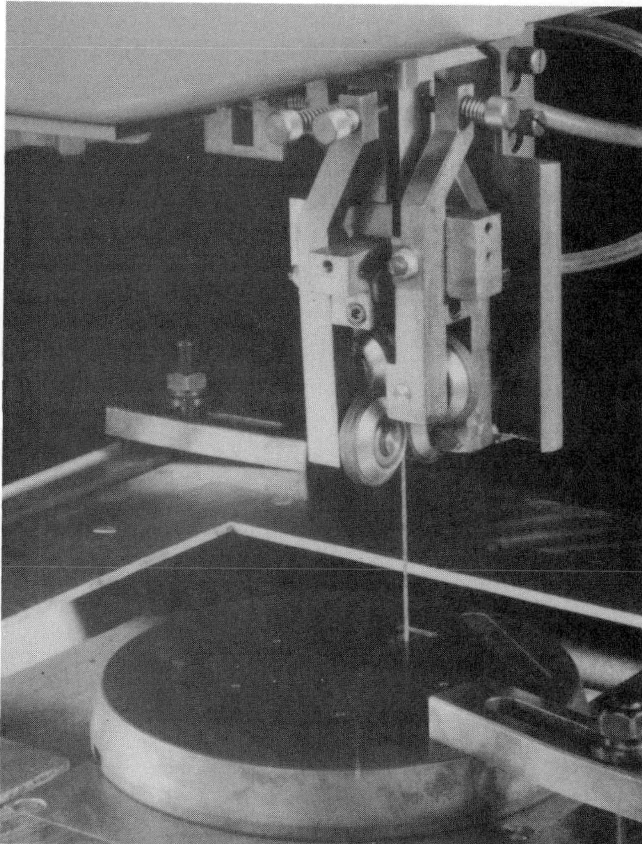

Fig. 14-50. Abrasive wire (diamond impregnated) is cutting an extrusion die opening. Note guides used to allow unidirectional cutting. (DoALL Co.)

BAND MACHINING SAFETY

1. Wear eye protection and leather gloves when handling band blades or blade material!
2. Get help when handling heavy material.
3. Have cuts and bruises treated immediately! Report all injuries!
4. Remove burrs and sharp edges from the work as soon as possible. They can cause serious cuts!
5. Do not clean chips from the machine with your hands! Use a brush. Stop the machine before attempting to clean it!
6. Keep your hands away from the moving blade! Use a push stick or section of metal for additional safety. Never have your hands in line with the cutting edge of the band!
7. Other than changing blade speeds (some) machines require band to be running when speed changes are made), make no adjustments until the blade has come to a complete stop!
8. Be sure all guards are in place before starting to operate a band machine.
9. Do not start a cut until the guides have been set properly! Guides should be positioned as close to the work as the job will permit.
10. Do NOT attempt to operate the machine until you have received instructions on its safe operation!

TEST YOUR KNOWLEDGE—Chapter 14

Please do not write in the text. Place your answers on a separate sheet of paper.

1. Band machining makes use of a ________ ________ ________ blade.
2. The cutting tool must be installed with the ________ facing ________.

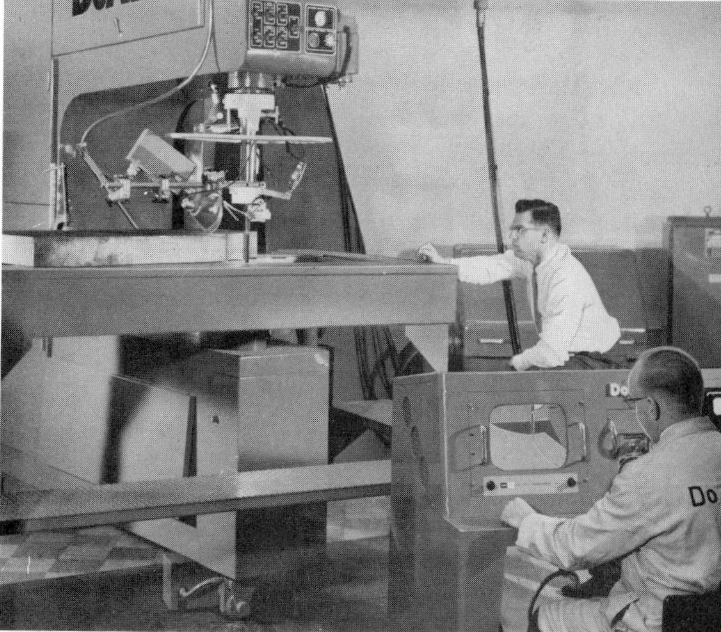

Fig. 14-51. This band-type machine is fitted with a closed circuit TV camera and remote control console. This type machine allows very large work or toxic material to be machined. (DoALL Co.)

3. List three (3) advantages band machining has over other machining techniques.
4. What two points must be considered when selecting a blade for a specific job?
5. Blade pitch refers to:
   a. Width of the blade in inches or millimeters.
   b. Thickness of the blade in inches or millimeters.
   c. Number of teeth per inch of blade or tooth spacing in millimeters.
   d. All of the above.
   e. None of the above.
6. Tooth form is the:
   a. Shape of the tooth.
   b. Thickness of the blade.
   c. Number of teeth on blade.
   d. All of the above.
   e. None of the above.
7. When making straight cuts, use the ______ blade the machine can accommodate.
8. The joint of a properly welded blade should be ______________________.
9. The blade must be ______ after welding because the joint is extremely ______ and cannot be used in this condition.
10. A ______________ welder is used to weld band machine blades.
11. Blade tension is the pressure put on the saw band to:
    a. Cut the metal more rapidly.
    b. Keep it taut and tracking properly.
    c. Reduce the power needed to do the cutting.
    d. All of the above.
    e. None of the above.

Fig. 14-52. Computer controlled band machine has "X" and "Y" table movement. Controller has circular interpolation and it can cut circular and elliptical shapes. Input commands are made by computer, tape, or manually. (DoALL Co.)

12. Cutting fluids are applied on the band machine in the form of:
    a. ________ ________ is recommended for heavy-duty sawing.
    b. ________ ________ is used for high-speed sawing of free machining nonferrous metals.
    c. ________ ________ is applied when the machine is NOT fitted with a built-in coolant system.
13. What is the simplest form of band machining?
14. The worktable on many vertical band machines can be tilted to make ________ cuts.
15. How can internal cuts be made on a band machine?
16. Smooth, uniformly finished surfaces are possible when the machine is fitted for ________ ________.
17. Describe friction sawing.
18. Of what use is a knife-edge blade on a band machine?
19. When are diamond-edge bands used?
20. What is unique about a diamond impregnated wire band?

RESEARCH AND DEVELOPMENT

1. Overhaul a worn band machine in your training area. Follow procedures outlined by your instructor.
2. Collect literature on the various types of vertical band machines. Bind and catalog the collection for a technical library.
3. Secure samples of the various types of blades used on band machines. Mount and label them on a suitable display panel.
4. Visit a local industry that uses band machines. Note the type of work being done on them. After getting permission, talk with machine operators to find out what their reactions are to operating a band machine. What type work is the most difficult to perform on the machine?
5. Secure a motion picture or video tape on band machining. Prepare a list of items to be discussed with the class after the tape or film has been seen.
6. Demonstrate the proper way to weld a band saw blade.
7. Prepare a paper on the history of band machining.
8. Borrow samples of work produced on a band machine. Develop a display around them and explain how they were made.
9. Machine a three-dimensional job similar to the one illustrated in Fig. 14-9 and Fig. 14-10.

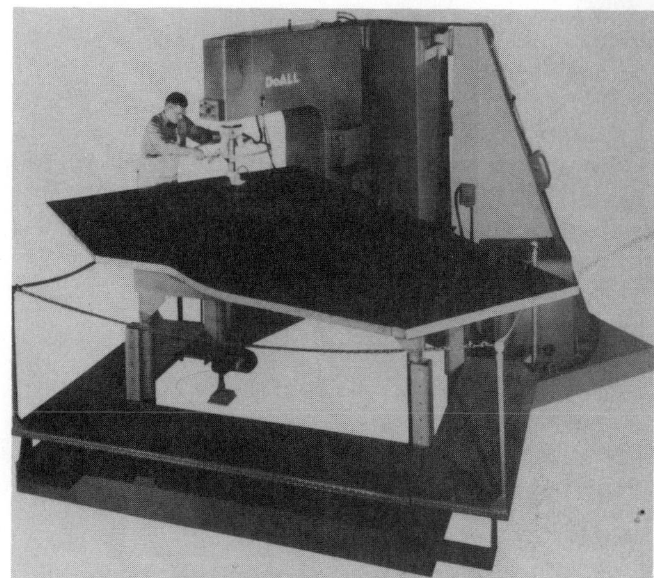

Band machining is very versatile. In this example, machinist is cutting huge plate. (DoALL Co.)

Chapter 15

QUALITY CONTROL

After studying this chapter, you will be able to:

- Explain the need for quality control.
- Point out the difference between the two basic quality control techniques.
- Describe how some nondestructive testing methods operate.
- Explain advanced methods for assuring quality control during and after machining.

The primary purpose of ***quality control*** is to seek out and prevent potential product defects in the manufacturing process before they can cause injuries or damage and substandard products. The eventual goal of quality control is NOT to detect imperfect products, but to PREVENT them from ever being made.

Quality control is one of the most important segments of industry. It plays a vital role in improving the competitive position of a manufacturer.

HISTORY OF QUALITY CONTROL

The development of the science of modern quality control closely parallels that of the airplane. Early aircraft were made ''by-guess-and-by-golly'' with little concern for quality control, Fig. 15-1.

As the theory of flight became more refined, more care was taken in selecting materials that went into the planes, Fig. 15-2. Tests were made to determine material strengths. Engines were made to specific standards and were checked many times during their manufacture. Aircraft dependability increased greatly. In the automotive field, Henry Ford started the mass production of the Model-T using similar quality control methods.

The volume production of all-metal aircraft in the early 1930s introduced many new quality control techniques, Fig. 15-3. ***Jigs*** and ***fixtures,*** which held the parts while they were machined or fabricated, were aligned with optical tools. The use of iron

Fig. 15-1. In producing this 1910 airplane, which only one was made, builder used little or no quality control in its construction. (U.S. Air Force)

Fig. 15-2. De Haviland aircraft of 1917 was constructed according to established specifications. Materials were inspected and some were tested before use.

Fig. 15-3. The Martin B-10 was one of the first quantity, all-metal aircraft. Many quality control techniques were developed and employed during its manufacture. (U.S. Air Force)

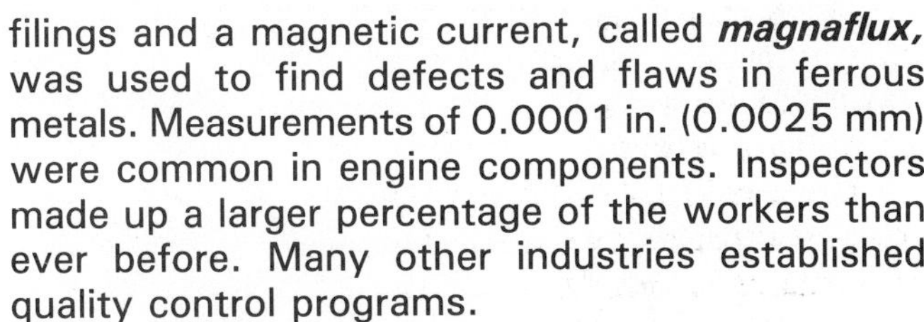

filings and a magnetic current, called ***magnaflux,*** was used to find defects and flaws in ferrous metals. Measurements of 0.0001 in. (0.0025 mm) were common in engine components. Inspectors made up a larger percentage of the workers than ever before. Many other industries established quality control programs.

The advent of World War II and the need for thousands of high performance aircraft led to the introduction of many of the quality control techniques in use today, Fig. 15-4. The inspector became a vital part of the manufacturing team.

Aerospace vehicles require quality control programs of great scope and magnitude, Fig. 15-5. Because they are subjected to pressures, stresses, and temperatures seldom found on earth, a breakdown or malfunction of any of the thousands of critical parts would be disasterous.

At the same time, other industries were placing increasing emphasis on the quality control techniques developed by the aerospace industry. As products became more complex and sophisticated, and with a growing demand for reliability, an increased portion of industry's budget had to be spent on quality control.

Fig. 15-4. The B-17 Flying Fortress was backbone of Air Force during 1940s. Mass production was made possible only because of modern quality control procedures.

Fig. 15-5. Artist's illustration of one of NASA's designs for a proposed manned space station. Construction of space station and launch vehicle will require a quality control program second to none. Why is quality control so important for such an undertaking? (Rockwell International)

CLASSIFICATIONS OF QUALITY CONTROL

Quality control techniques fall into two basic classifications:

1. ***Destructive testing,*** results in the part being destroyed during the quality control testing program.
2. ***Nondestructive testing,*** is done in such a manner that the usefulness of the product is NOT impaired.

Destructive testing

Destructive testing is a costly and time consuming quality control technique. A specimen is selected at random from a great number of pieces, Fig. 15-6. Statistically, at least, it indicates the characteristics of the undestroyed, and untested, remaining pieces. It gives no assurance, however, of perfection because many of the untested parts could be defective and be used in the manufacture of a product.

Nondestructive testing

Nondestructive testing is a basic tool of industry, Fig. 15-7. It is well suited for testing electronic and aerospace products where the performance of EACH part is critical. Each piece can be tested individually AND as part of a completed assembly.

Fig. 15-6. Automobile bodies, selected at random, are cut apart so quality control inspectors can check placement and size of welds struck on plant's highly-automated framing line. Disassembly is carried out to assure that structural integrity of bodies is maintained. (American Motors Corp.)

Fig. 15-7. Quality control engineer is checking dimensional accuracy of tire mold tread ring with dial indicators. This is a nondestructive quality control technique because part is not damaged in any way. (Aluminum Co. of America)

NONDESTRUCTIVE TESTING TECHNIQUES

You are familiar with many methods of nondestructive testing: measuring, weighing, and visual inspection. While satisfactory for many products, they leave much to be desired for others. To assure effective quality control in all areas of manufacturing, industry has developed more sophisticated testing techniques.

Measuring

Micrometers, vernier tools, dial indicators, gages, etc., fall into the ***measuring*** category of quality control testing. To guarantee their accuracy, these tools must be checked at frequent intervals against known ***standards,*** Fig. 15-8. The calibrating is done in a ***precision tool calibration laboratory,*** Fig. 15-9.

The shape of some products is so complex that conventional measuring tools cannot be used to give accurate measurements. Tools have been devised that measure electronically and check thousands of individual reference points on the object against specifications. Refer to Fig. 15-10.

Measuring can be done visually on an ***optical comparator,*** which is a gaging system for inspection and

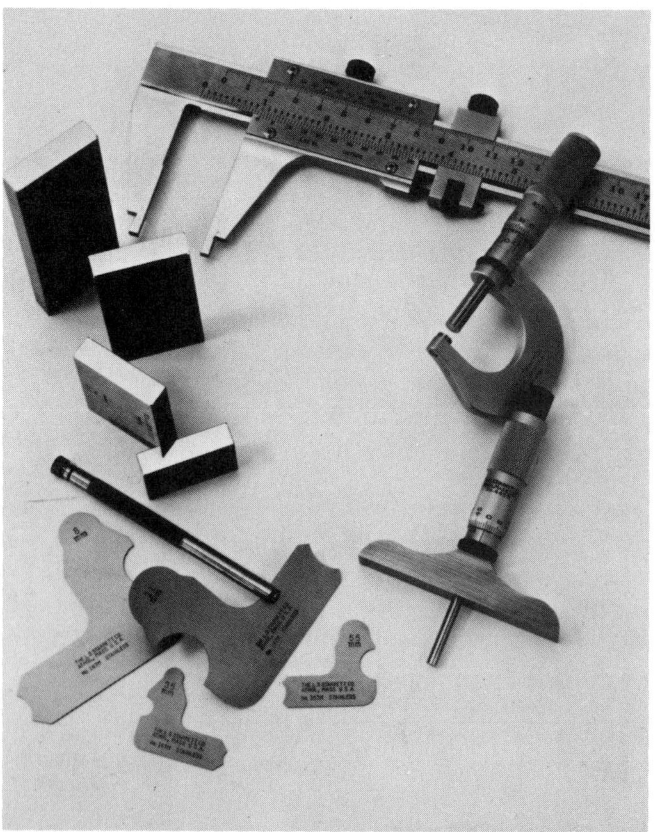

Fig. 15-8. Some quality control techniques are based on measurement. To guarantee accuracy of measuring tools, they must be checked against known standards at frequent intervals. (L.S. Starret Co.)

Fig. 15-9. Personnel in the precision tool calibration laboratory test and keep an accurate set of records on all production measuring tools used in the manufacturing process.

precise measurement of small parts and sections of larger parts. See Fig. 15-11. A part can be magnified up to 500X with the absence of distortion to permit accurate measurement. An enlarged

Fig. 15-11. Parts can be magnified up to 500X on an optical comparitor. Functions on this optical device are virtually indispensable in today's technology. (Bridgeport Machines, Inc.)

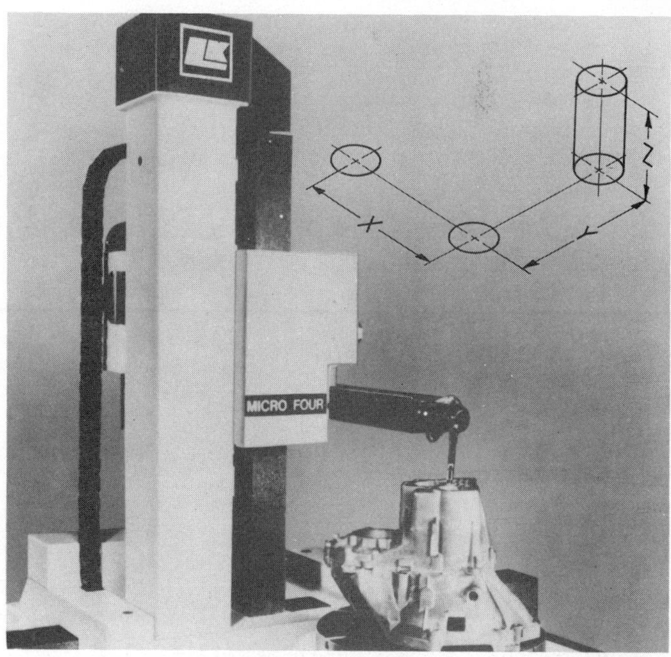

Fig. 15-10. Left. Complex parts, like these turbine wheels, cannot be checked for accuracy with conventional measuring tools. Why do you think this is so? Right. Computer aided, direct-reading measuring device is required for inspection of complex parts, like turbine wheels or precision transaxle housings. Measurements can be made in three planes (X, Y, and Z axes) without error. Measurements are to 0.0002 in. (0.005 mm). English to metric conversion is accomplished with the push of a button. (Howmet Corp. and Cincinnati Milacron)

image of the part being inspected is projected on a screen where it is superimposed upon an accurate drawing overlay of the part. Variations as small as 0.0005 in. (0.013 mm) can be noted by a skilled operator, Fig. 15-12.

Quality control can also be assured by ***statistical means*** by measuring a number of parts in a production run, Fig. 15-13. This is carefully worked out by a mathematical approach to quality control. A wide variation in the accuracy of the parts being inspected will make it necessary to increase the inspection rate to include more or all of the units manufactured.

Special gaging and inspection tools can be designed for almost any application, Fig. 15-14. Combining precision tools with electronic devices permits accurate inspection to be made by semi-skilled workers, Fig. 15-15.

Radiography (X-ray) inspection

Inspection by ***radiography*** involves passing gamma rays through a part and onto light sensitive film to detect flaws (crack, pores, etc.). It has become routine in the acceptance or inspection of critical parts and materials, Fig. 15-16. The technique involves the use of X-rays and ***gamma radiation*** (highly energetic, penetrating radiation found

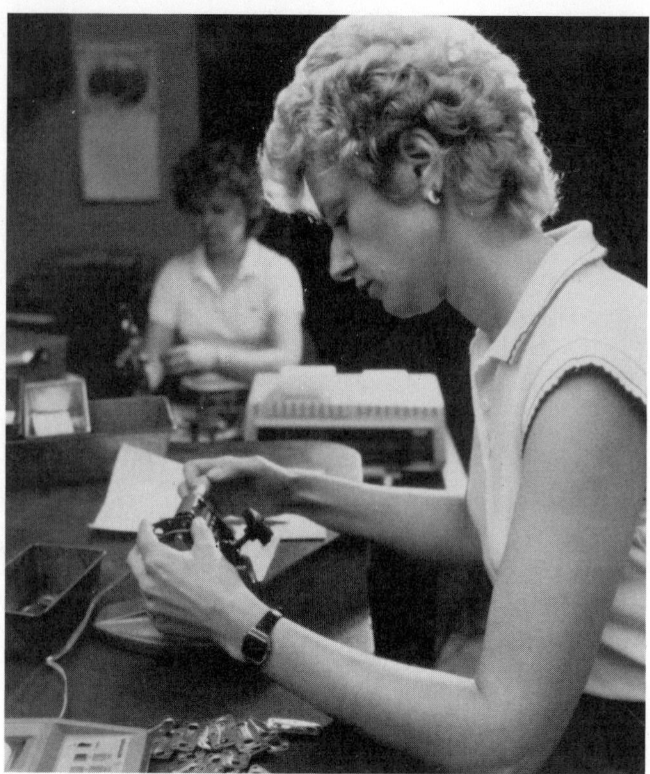

Fig. 15-13. Quality control inspector is performing frequency distribution on plate lamination thickness, using a digimatic micrometer and a mini-processor. (Master Lock Co.)

Fig. 15-12. Quality control engineer is checking hole location on a plate lamination of an automatic lock body assembly, using an optical comparitor with a part overlay transparency. (Master Lock Co.)

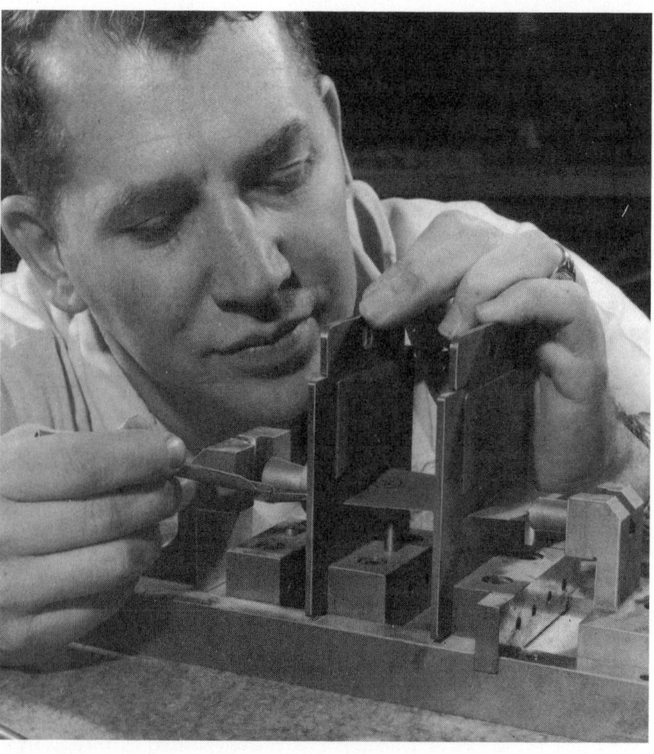

Fig. 15-14. Engineer is checking shape of a turbine blade for a jet engine with guillotine type gage. (Westinghouse Electric Co.)

Fig. 15-15. Electronic tools, like this stroboscopic light are also employed in quality control. Stroboscopic light is similar to timing light used to check ignition timing on automobile engines. Here strobe light is checking lathe spindle speed against indicated speed on control dial.
(Clausing Industrial Corp.)

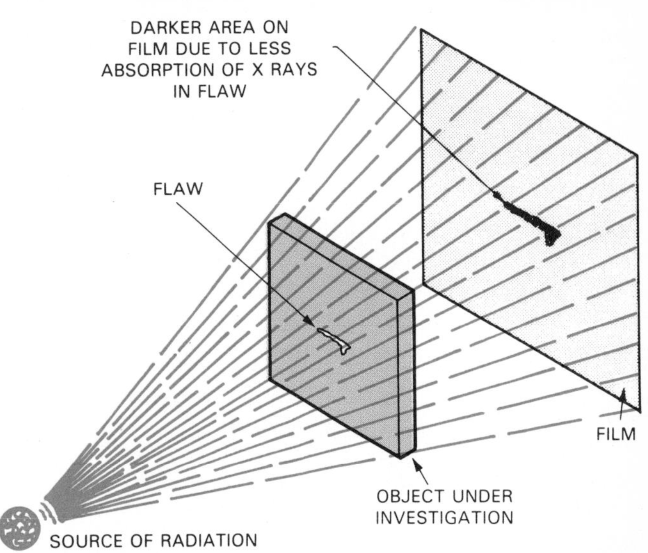

Fig. 15-17. Note how radiographic inspection works.

Fig. 15-16. Laser alignment device is utilized, with an X-ray generator, to precisely locate X-ray radiation on an electron beam welded, titanium structure. Highly accurate alignment is required due to vary narrow heat affected zone characteristic of EB welds. (Grumman Aerospace Corp.)

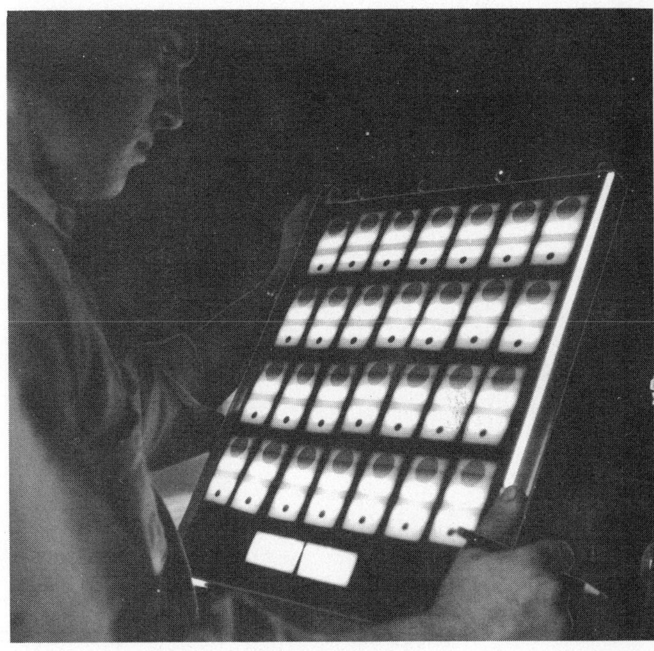

Fig. 15-18. X-ray examination is being used to check a number of similar parts. Hidden flaws are easily detected.
(Westinghouse Electric Corp.)

in certain radioactive elements) projected through the object under inspection and onto a section of photographic film, Fig. 15-17. The developed film has an image of the internal structure of the part or assembly, Fig. 15-18.

Many kinds of ***peripheral*** (outside circumference) inspection operations can be made because of the ***omnidirectional*** (all directions) characteristics of the rays, Fig. 15-19.

Radiography offers many advantages:

1. Inspection sensitivity is high.
2. Projected image is geometrically correct.
3. A permanent record is produced.
4. Image interpretation is highly accurate.

In addition, the internal and hidden parts of complex assemblies can be inspected without fear of

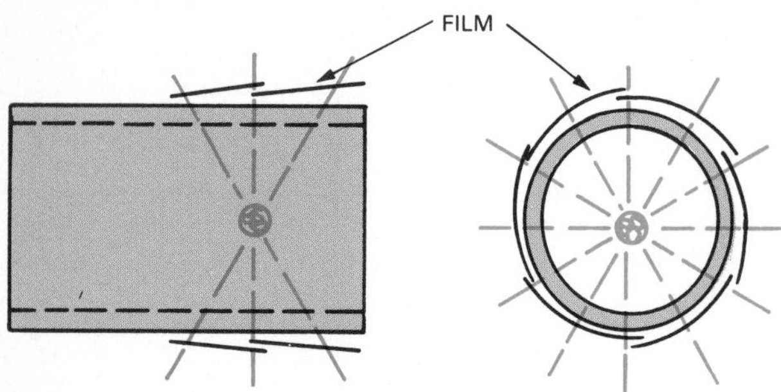

Fig. 15-19. Note technique used to inspect cylindrical objects by radiographic means. Flaw would cause more exposure of film and resulting image of flaw on film after development.

damage. Objects can also be inspected without taking them out of use. For example: wheels and axles on locomotives and railway cars are inspected at regular intervals without having to remove them and take them to special facilities.

CAUTION! Technicians making radiographic tests must observe strict safety precautions when handling radioactive materials. Overexposure to X-rays can be harmful.

Magnetic particle inspection

Magnetic particle inspection, more commonly known as ***magnafluxing*** (trademark: Magnaflux Corp.) is a nondestructive testing technique employed to detect flaws on or near the surface of ferromagnetic materials (iron based metals). The technique is rapid, but shows only serious defects, NOT scratches nor minor visual defects, Fig. 15-20.

The magnetic particle inspection technique is based on the theory that every conductor of electricity is surrounded by a circular magnetic field. If the part is made of ferromagnetic material, these lines of force will, to a large extent, be contained within the piece. A circular magnetic field, if NOT interrupted, has no poles. However, because a flaw or other imperfection present in the piece is oriented (positioned) to cut through these magnetic lines of force, poles will be formed at each edge of the flaw. These poles will hold the finely divided magnetic particles, thus outlining the flaw.

The limitations of this technique are apparent when the flaw is parallel to the line of magnetic force. The flaw will not interrupt the force, so no indication of it will appear when magnetic particles are applied, Fig. 15-21.

A magnetic field is introduced into the part and fine particles of iron are blown (dry method) or flowed in liquid suspension (wet method) over the part. Because the flaw will distrub or distort the magnetic field, it will have different magnetic properties than the surrounding metal. Many of the iron particles will be attracted to the area and form a definite indication of the flaw (its exact location, shape, and extent), Fig. 15-22.

Fluorescent penetrant inspection

The theory of ***fluorescent penetrant inspection*** is based on capillary action to show flaws in parts, Fig. 15-23. A penetrant solution is applied to the part's surface by dipping, spraying, or brushing. Capillary action literally pulls the solution into the defect. The surface is rinsed clean, and before or after drying, a wet or dry developer is applied. This

Fig. 15-20. Large Magnaflux machine is checking huge machined section. (U.S. Army)

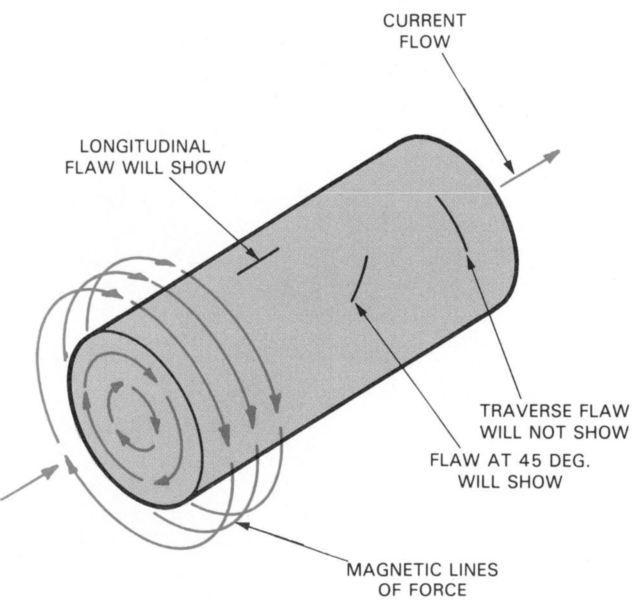

Fig. 15-21. Note theory, scope, and limitations of magnetic particle inspection procedure. (Magnaflux Corp.)

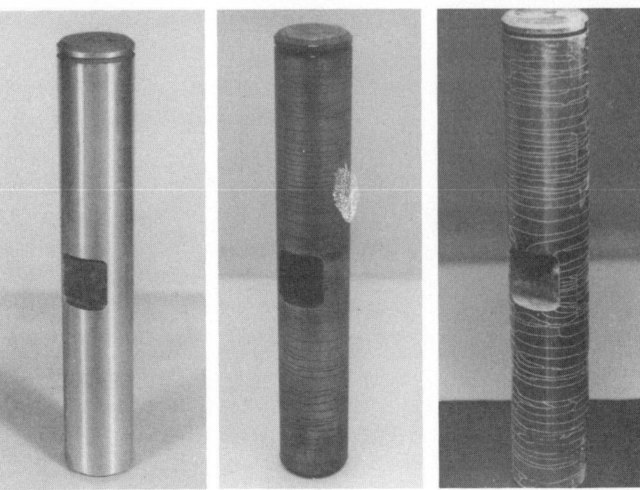

Fig. 15-23. Three ways to inspect kingpin for a truck front axle. Left. Under visual or "eye-balling" inspection, it appears sound and safe for service. Center. Inspection with Magnaflux shows that excess heat generated during grinding operation has caused dangerous cracks. Right. Same kingpin and same cracks. This time is was treated with a fluorescent penetrant testing material and photographed under black light. (Magnaflux Corp.)

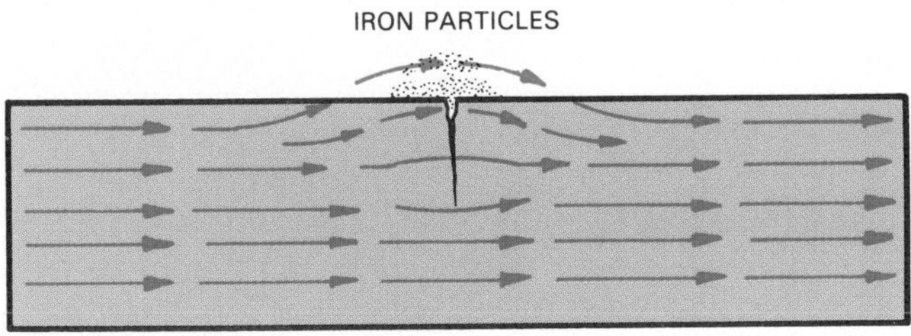

Fig. 15-22. Crack in steel bar generates magnetic field outside part to hold iron particles. Build up of iron particles makes even tiny flaws visible.

acts as a blotter, and draws the penetrant back to the surface.

When inspected under a BLACK LIGHT, any defects glow with fluorescent brilliance, marking the defect. This is because dye penetrant has flowed into and still remains in the flaw.

Spotcheck (Trademark: Magnaflux Corp.)

Spotcheck is another of the penetrant inspection tools, Fig. 15-24. It is easy to use, accurate,

Fig. 15-24. Study spotcheck penetrant inspection technique. 1—Apply cleaner to loosen grease and dirt. 2—Remove grease and dirt loosened by cleaner. 3—(Left photo) Spray on penetrant. 4—Apply developer. 5—(Right photo) Inspect for flaws. Flaws stand out sufficiently to be seen visually. (Magnaflux Corp.)

economical, and does NOT require a black light to bring out the flaws. Application is similar to that described for the fluorescent type penetrant.

The specimen being inspected is coated with a red liquid dye which soaks into the surface crack or flaw. The liquid is washed off and the part dried. A developer is then dusted or sprayed on the part. Flaws and cracks show up red against the white background of the developer.

Ultrasonic inspection

Ultrasonic testing techniques utilize sound waves, above the audible range, to detect cracks and flaws in almost any kind of material that is capable of conducting sound. It may also be employed to measure the thickness of the same materials from one side.

Ultrasonic test equipment is shown in Figs. 15-25 through 15-27.

The human ear can hear sound waves whose frequencies range from about 20 to 20,000 cycles per second. Sound waves oscillating (vibrating) with a frequency greater than 20,000 cycles per second are ***inaudible*** (cannot be heard) and are known as ***ultrasound.***

Ultrasonic testing equipment utilizes waves of millions of cycles per second. The term ***"megahertz"*** (written MHz), meaning millions of cycles per second.

Sound waves are employed to obtain information about the interior structure of a material by observing the echos that are reflected from within the material, Fig. 15-28. It is possible to judge distances by the length of time required to receive an echo from an obstruction (flaw), Fig. 15-29.

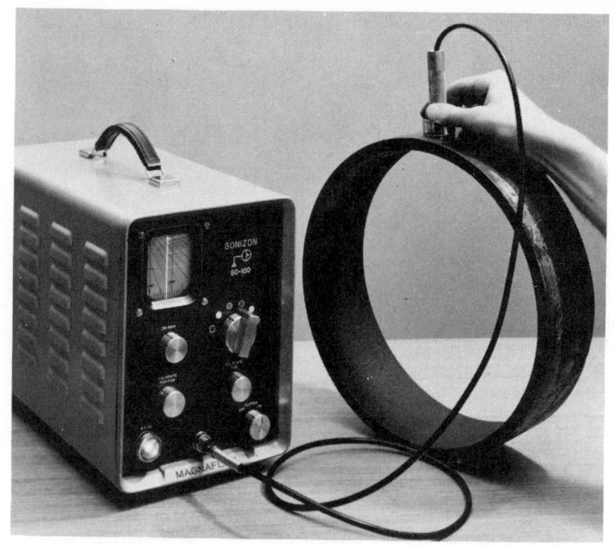

Fig. 15-27. Note portable contact type ultrasonic testing device. (Magnaflux Corp.)

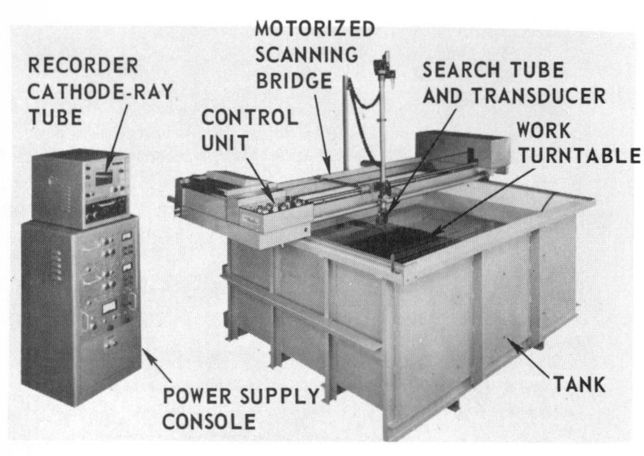

Fig. 15-25. This is a complete automatic ultrasonic immersion testing system.

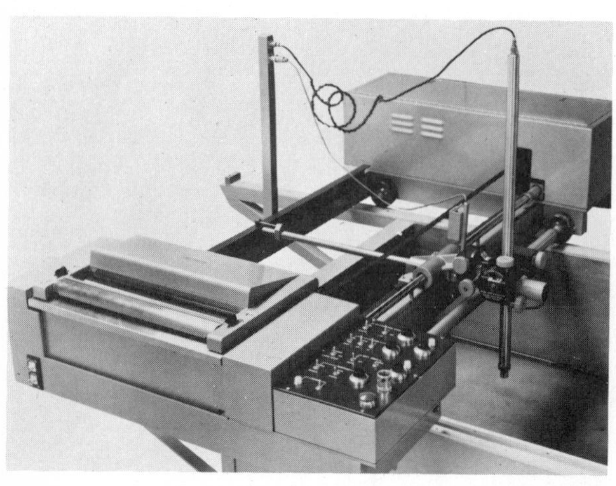

Fig. 15-26. Close-up shows ultrasonic scanning bridge with manipulator, search tube, and recorder.

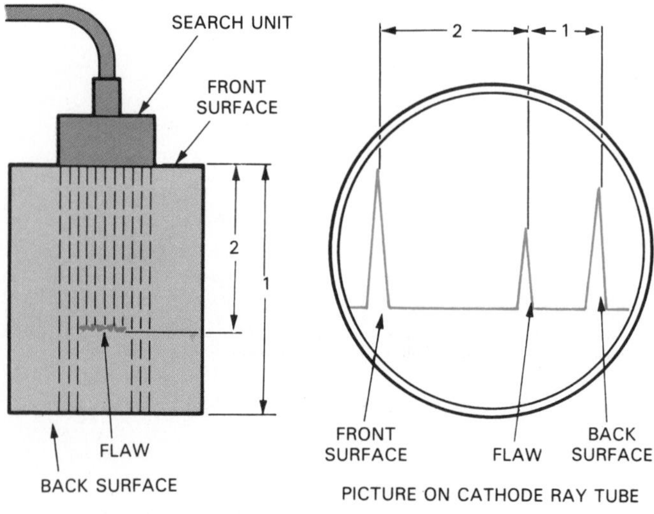

Fig. 15-28. Note how ultrasonic sound waves are used to detect and locate flaw in a test piece.

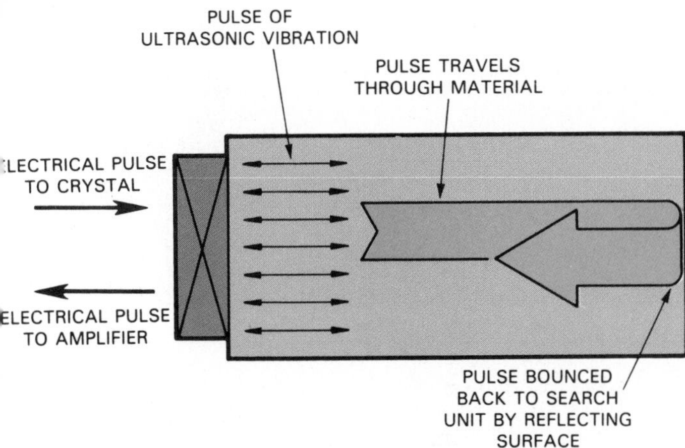

Fig. 15-29. Study how sound waves travel through part and bounce back to locate flaws in material.

The obstruction, from which the echoes are received, it either front surface, back surface, or internal. The ultrasonic testing equipment is partly a timing device for measuring the relative length of elasped time between the sending of the sound waves and the return of the various echoes.

High frequency sound is produced by a piezo-electric transducer (crystal) which is electrically pulsed and then vibrates at its own natural frequency, Fig. 15-30.

In order to transmit the sound waves from the transducer to the metal and return the echoes to the transducer, it is necessary to provide a liquid coupling. This is accomplished with a film of oil, glycerine, or water between the transducer and the test piece, Fig. 15-31. The same results can be achieved by immersing both the test piece and transducer in water, Fig. 15-32.

Immersed testing is ideal for production type testing as there is no contact between the

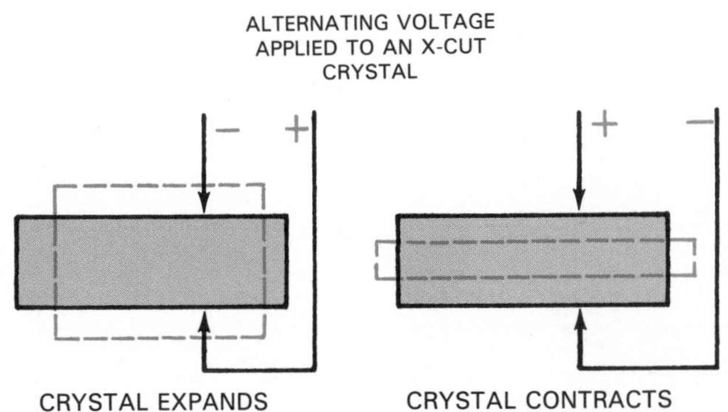

Fig. 15-30. A piezoelectric transducer is a device that receives energy from one system and retransmits it to another system, often in a different form. For inspection, it is pulsed electrically to generate ultrasonic sound waves.

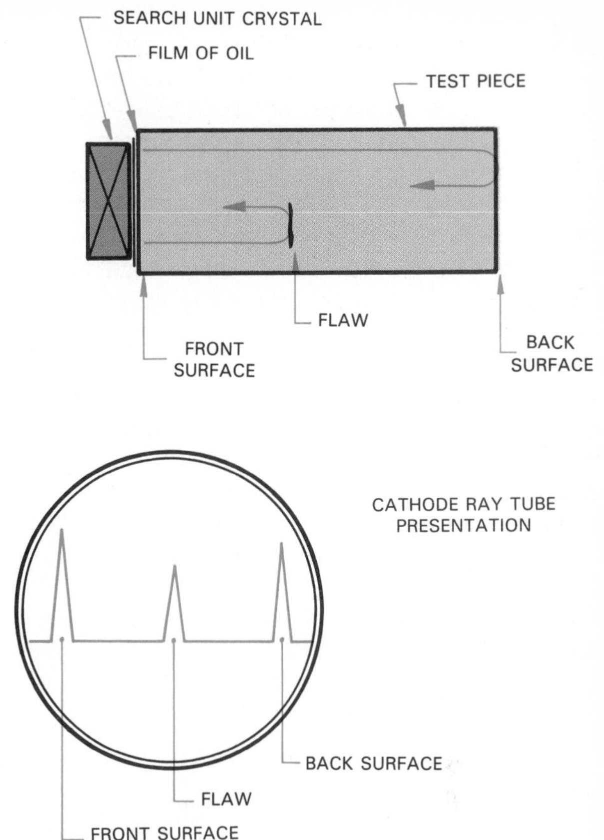

Fig. 15-31. Study action of contact type ultrasonic inspection. Transducer is placed in contact with test piece. A film of oil, water, or glycerine is employed to make a positive contact. Note how CRT records extra voltage spike indicating flaw.

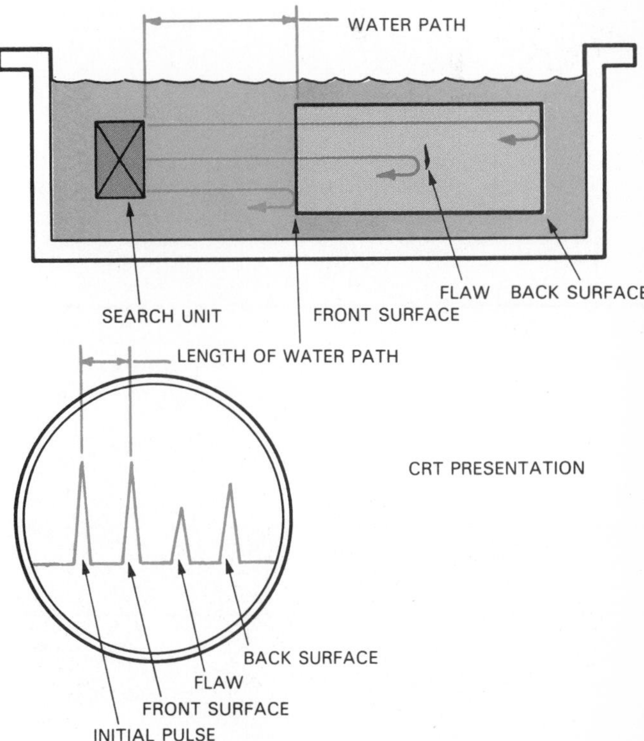

Fig. 15-32. Study immersion type ultrasonic testing. Note the extra "pip" or spike on CRT.

transducer and the work, and thus no transducer wear occurs.

The transducer vibrates for about two-millionths of a second. The result is a very short burst of sound waves that travel through the liquid to the surface of the test material. A portion of the sound is immediately reflected from the surface of the metal as a very large echo. Part of the sound will NOT be reflected and will continue into the test material. If this portion of the sound encounters no interference, in the form of a discontinuity in the material (flaw), it will continue until it is partially reflected from the back surface as a second echo or back reflection. If there is a flaw in the interior, a portion of the sound will be reflected from the flaw and will return to the receiver as a separate echo between the echoes received from the front and back surfaces.

After the transducer has given off its short burst of sound waves, it stops vibrating and "listens" for the returning echoes. When the echoes are received, they cause the transducer to vibrate and to generate an electric current which can be visually displayed on a trace of a ***cathode ray tube*** (CRT), which is a television type pictuure tube, Fig. 15-33.

Information transmitted to the CRT can be expanded or condensed to improve readability. This is shown in Fig. 15-34.

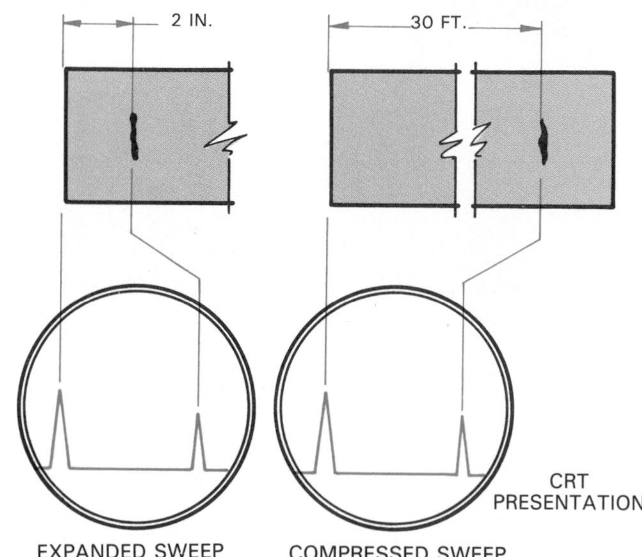

Fig. 15-34. CRT (scope) presentation can be expanded or compressed for easier reading or to compensate for different material thicknesses.

There are two basic categories of ultrasonic testing:

1. ***Pulse echo*** uses sound waves generated by a transducer that travel through the test piece, Fig. 15-35. The reflected sound waves (echoes) locate the flaws. One crystal is used to transmit the sound and receive the echoes.
2. ***Through inspection*** has one crystal that transmits the wave through the piece, and another crystal(s) that picks up the signal on the opposite end of the piece(s). See Fig. 15-36.

The beam is partially blocked by a flaw. The reduced intensity of the beam activates a signaling device (flashes a light, rings a bell, etc.) to

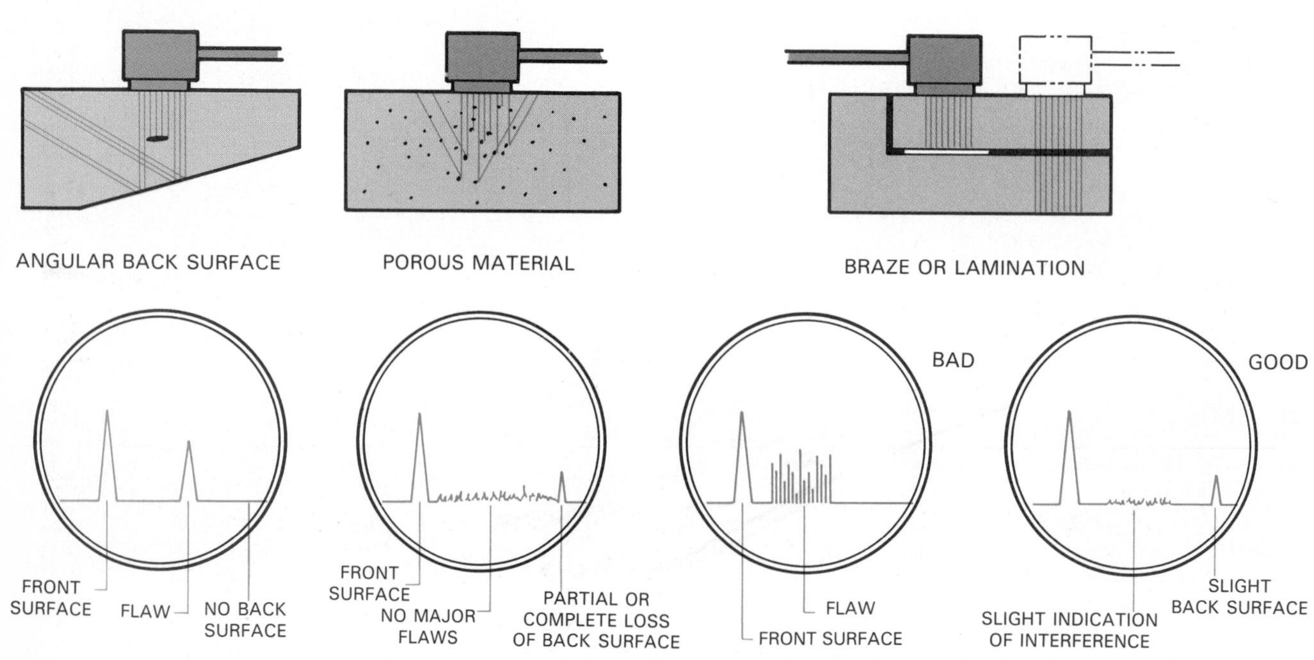

Fig. 15-33. Note various flaws as they appear on cathode-ray tube (CRT).

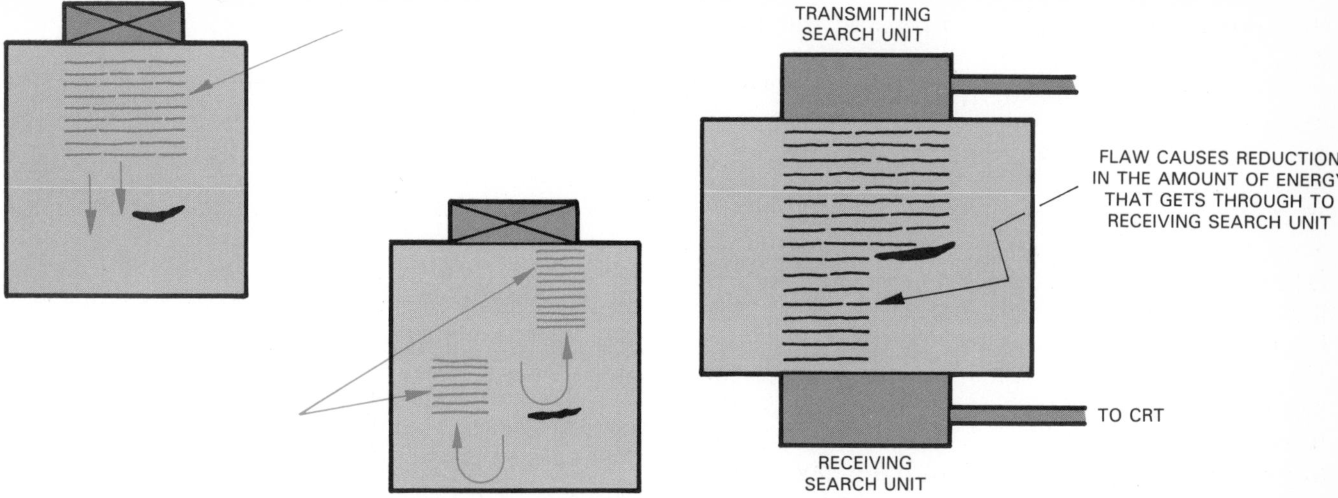
TRANSMITTING
SEARCH UNIT
FLAW CAUSES REDUCTION
IN THE AMOUNT OF ENERGY
THAT GETS THROUGH TO
RECEIVING SEARCH UNIT
TO CRT
RECEIVING
SEARCH UNIT

Fig. 15-35. With pulse echo ultrasonic inspection, reflected wave will return sooner when it bounces off of flaw.

Fig. 15-36. Through type ultrasonic inspection uses sensor to detect waves on opposite side of piece.

alert the operator to the flaw. The technique is frequently employed to check the integrity of helicopter rotor blades for example. The technique has been automated to do this work.

Apparently, there is no size limitation on work that can be tested by ultrasonic techniques. Fig. 15-37 shows one application.

Inspection by laser

The laser has also been adapted for quality control, Fig. 15-38. In addition to being able to check the fit of components on an assembly line, the laser under computer control, can be made sensitive enough to detect tool wear and compensate for it automatically, Fig. 15-39. This helps prevent imperfect parts from ever being made.

Eddy-current inspection

The ***eddy-current test*** of metal products is based on the fact that flaws in a conductor will cause different impedance (resistance) changes in a coil brought near it. A slightly different eddy-current will

Fig. 15-37. This wing section is longest adhesively bonded panel ever produced. Bonding integrity is being checked with ultrasonic test equipment. Assembly is made up of 30 individual parts. (AVCO Aerostructures Div., AVCO Corp.)

Fig. 15-38. Laser is being used to inspect part used in an automatic transmission. Inspection gives an immediate indication whether machine adjustments are required.
(Ford Motor Company)

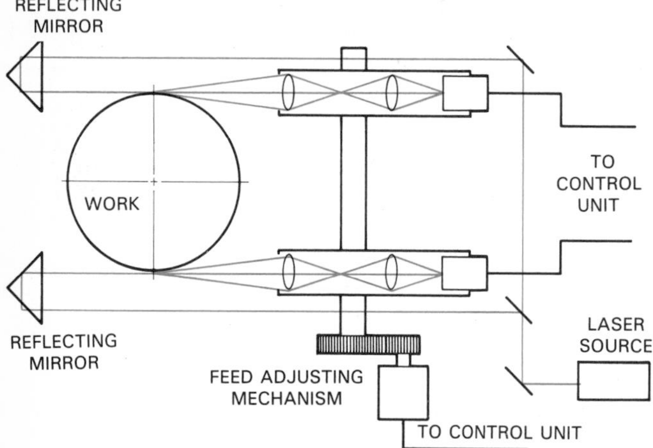

Fig. 15-39. Study one type of laser inspection device. With it, tool wear can be continuously monitored. Note how laser light could detect work or part oversize or undersize.

result in a test coil placed next to a metal part WITHOUT flaws. This difference will show which parts pass and fail the inspection.

The application for eddy-current testing methods can be divided into two general categories:

1. The ***eddy-current differential system*** is used to detect flaws (cracks, seams, holes, etc.) in metal parts, such as: wire, tubing, and bar stock, as they move off of the production line. The test equipment must be sensitive to rapid change.
2. The ***eddy-current absolute system*** is employed for the detection of variations in dimension, composition, and other physical properties of a metal product. Instrumentation for this type operation must be responsive to comparatively small changes.

Eddy-current instrumentation is designed to detect these changes and convert them into a form that allows them to be monitored by the operator. See Figs. 15-40 and 15-41.

Work being tested can be passed through encircling detection coils at speeds up to 500 fpm (150 mpm), Fig. 15-42. When a flaw in the test piece is detected, the eddy-current unit does one or more of the following:

1. Flashes a warning light to alert the operator.
2. Sounds a tone alarm.
3. An automatic rejection device, plugged into the unit, ejects the part that does NOT meet standards.
4. A device marks the section that contains the flaw so it can be removed.
5. Provides an electronic record of the section under test.

Fig. 15-40. This is a control panel for an eddy-current flow detector.

Fig. 15-41. These are various sizes and types of encircling test coils for checking nonferromagnetic tubing, bar stock, or wire with eddy-current system.

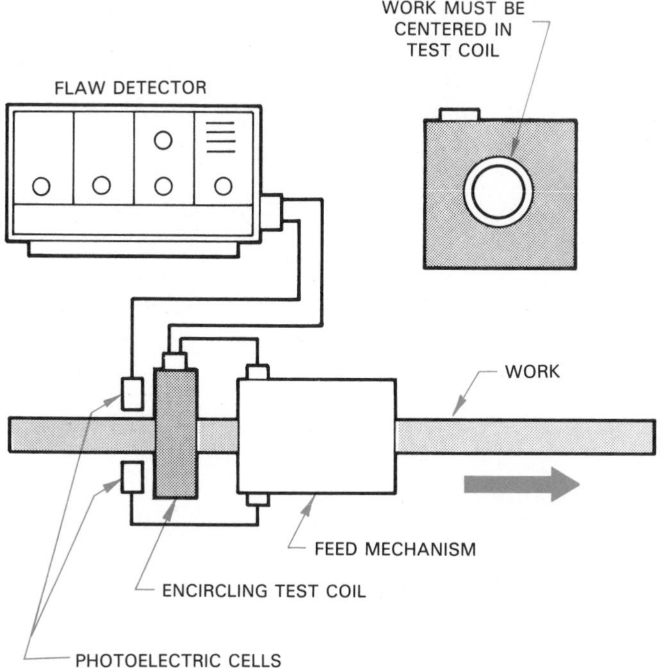

Fig. 15-42. Study operation of eddy-current flaw detection. Photoelectric cells turn off alarm system when end of a test piece passes into test coil. Any flaw would change small current in test coil.

OTHER QUALITY CONTROL TECHNIQUES

In addition to the quality control techniques described, industry makes wide use of highly specialized testing devices. For example, the quality of a machined surface may be highly critical. The surface finish can be inspected VISUALLY with a ***comparative checking unit,*** Fig. 15-43, or ELECTRONICALLY with a ***profilimeter,*** Fig. 15-44.

Because of the versatility of the computer, many new quality control techniques are being developed as their need arises. It will also make 100 percent inspection possible, Fig. 15-45.

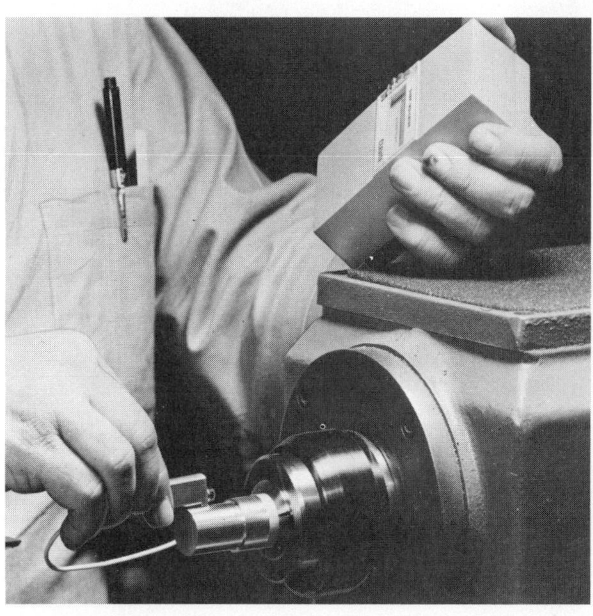

Fig. 15-44. Surface roughness can also be checked electronically with a profilometer. (Clevite Corp.)

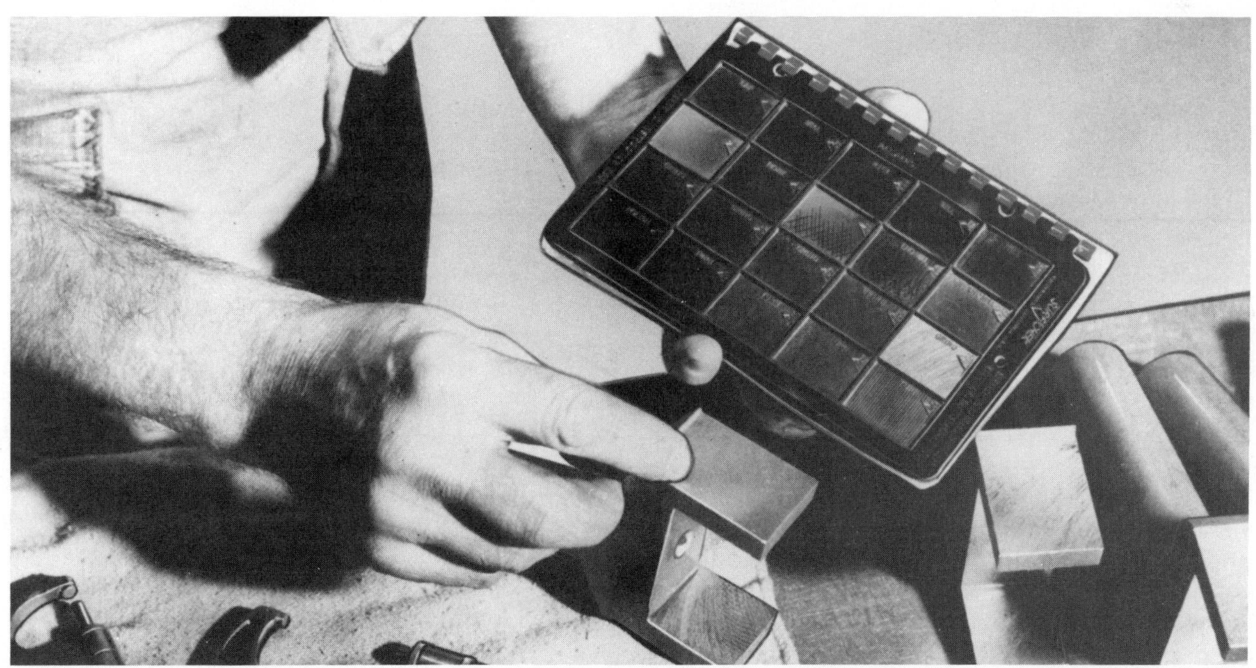

Fig. 15-43. Machinist is using surface finish comparative type checking panel. (Surf-Chek)

Fig. 15-45. New quality control devices make 100 percent inspection possible. Here lasers scan front end, door frames, and tail assembly to an accuracy of 3.0 mm or about 1/9 in. (Pontiac Div., General Motors Corp.)

Remember! Quality control should NOT be done so imperfect parts will be detected and discarded but to PREVENT THEM from ever being manufactured.

TEST YOUR KNOWLEDGE—Chapter 15

Please do not write in the text. Place your answers on a separate sheet of paper.

1. Quality control is an important segment of industry. Its purpose is to:
   a. Improve product quality.
   b. Maintain quality.
   c. Help to reduce costs.
   d. All of the above.
   e. None of the above.
2. Quality control falls into two basic classifications. Name and explain each.
3. Precision measuring tools, like micrometers, vernier tools, dial indicators, etc., are inspected and calibrated in a ________ ________ ________ laboratory.
4. Inspection by radiography involves the use of ________ or gamma rays.
5. List four (4) advantages of radiographic inspection.
6. Magnetic particle inspection is more commonly known as ________.
7. Describe the fluorescent penetrant inspection process.
8. Quality control is an important industrial tool because:
   a. It can be done easily.
   b. It guarantees that the parts being produced meet predetermined standards and specifications.
   c. It can be done by unskilled labor.
   d. All of the above.
   e. None of the above.
9. The ________ ________ is an optical gaging instrument designed for the inspection of small parts and sections of larger parts.
10. Explain how the magnetic particle inspection technique operates.
11. Only ________ metals can be inspected by the magnetic particle technique.
12. Ultrasonic inspection makes use of:
    a. Accurately made measuring fixtures.
    b. High frequency sound beam.
    c. X-rays.
    d. All of the above.
    e. None of the above.

13. Make a sketch showing the two methods of liquid coupling used for ultrasonic testing.
14. List the two basic categories of ultrasonic testing. Briefly describe each.
15. Make a sketch showing how ultrasonic testing is done.

RESEARCH AND DEVELOPMENT

1. Devise a way to use an overhead projector to demonstrate the optical comparitor. comparitor.
2. Develop simple measuring fixtures to check a simple project against the plans.
3. Penetrants described in the text are easy to use. The cost is within reach of most budgets. Carefully analyze the needs of your shop. If needed, present your analysis to request that penetrant materials be purchased.
4. Select a machine part and examine it carefully. What points on the piece come under the quality control program in the plant where it was manufactured? What points must be checked against specifications if the part is to be interchangeable with other components of the assembly?
5. Devise a quality control program for your training area.
6. Ask for the demonstration of the Magnaflux technique the next time you take a field trip to a plant that uses the system.
7. Check the micrometers and vernier measuring tools in your instructional area. Make any needed repairs and adjustments.
8. Show a film on quality control. Preview it before showing and prepare an outline and quiz on the film's more significant points of interest.

Fig. 16-1. Four different automotive transmission cases are produced on this complex machine. Each case is drilled, reamed, bored, faced, chamfered, and tapped. A fully machined case is turned out at the rate of one every 51 seconds. (The Cross Co.)

Chapter 16

AUTOMATION

After studying this chapter, you will be able to:

- Define the term "automation."
- Describe several automated production systems.
- Explain the operation of N/C (numerical control) and CNC (computer-assisted numerical control) systems.
- Point out how manual and computer-assisted programming is done.
- Discuss the use of robotics in automated production systems.

Automation is a system for the continuous automatic production of a product. See Figs. 16-1 and 16-2. A machine, or group of machines, is activated electronically, hydraulically, mechanically, pneumatically, or in combination, to automatically perform one or more of the five basic manufacturing processes:

1. Making.
2. Inspecting.
3. Assembling.
4. Testing.
5. Packaging.

The computer, integrated with specially designed machines, has revolutionized automation technology while improving product quality and reducing manufacturing costs, Fig. 16-3. Human intervention is also reduced to an absolute minimum.

One of the newer automated production techniques is known as ***flexible manufacturing system*** (FMS). However, many other terms are used to describe this general category of machining/manufacturing equipment: ***computer integrated manufacturing system*** (CIMS), ***computer integrated and manufacturing system*** (CIAM), and ***flexible manufacturing system complex*** (FMC).

FMS is a range of systems, similar in concept, that bring together work stations (machine tools), automated material handling and/or transfer, and computer control in an integrated manner, Fig. 16-4.

FMS is capable of producing a selected range of work configurations randomly and simultaneously. Work is transferred to and from member machines by automated fixture carts or conveyors. Robots may also be employed in some operation, Fig. 16-5.

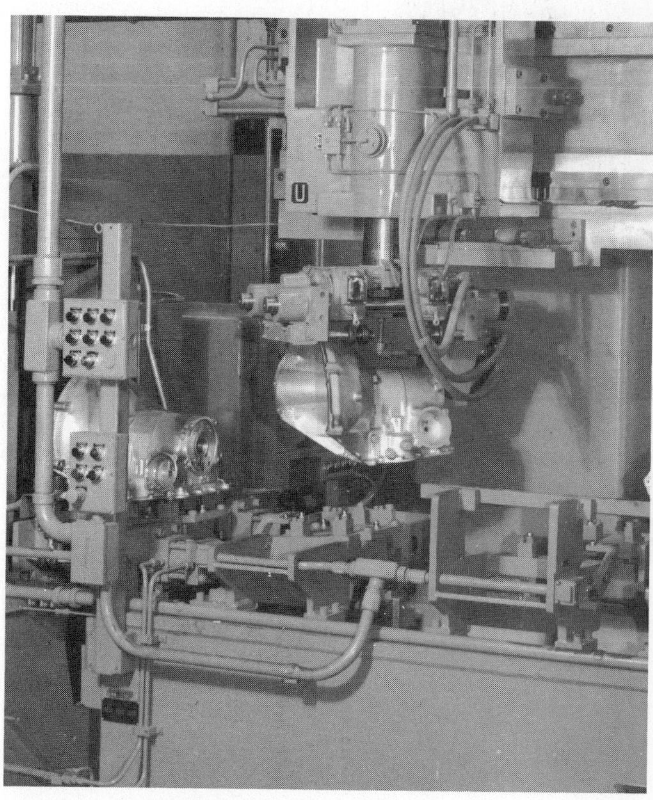

Fig. 16-2. Note section of machine shown in Fig. 16-1. Transmission case is pivoted into a new position for operation at this station. (The Cross Co.)

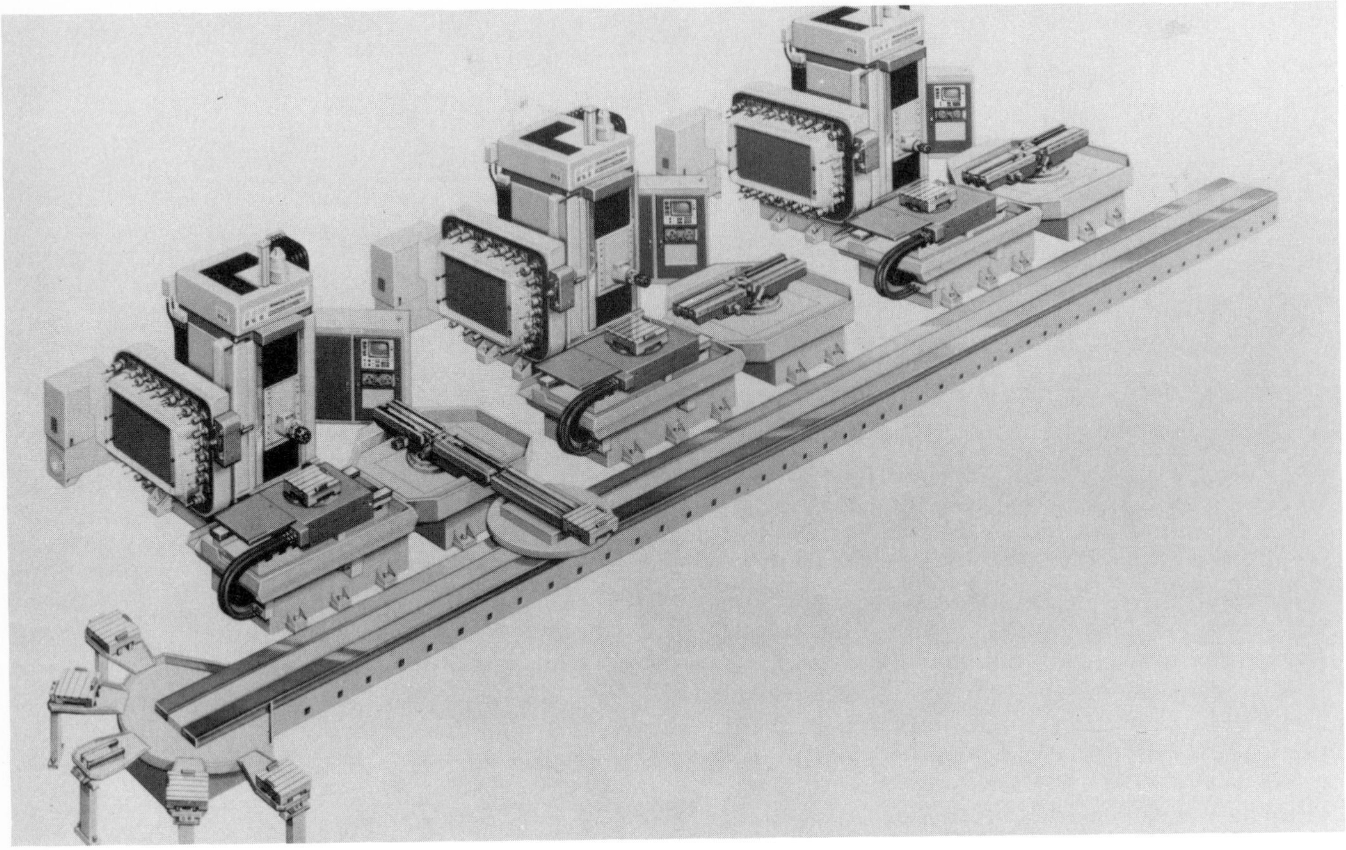

Fig. 16-3. This group of machines, called a palletized cell, combines general purpose machine tools, using a common pallet designed to hold parts for machining. Each machine has a microcomputer. Note how work moves from station to station. (Kearney & Trecker Corp.)

Fig. 16-4. Flexible manufacturing system (FMS) consists of several machine tools, fed by automated materials handling equipment and robots, all under the direction and control of a central computer. Parts move from station to station on automated carts shown in foreground. (Kearney & Trecker Corp.)

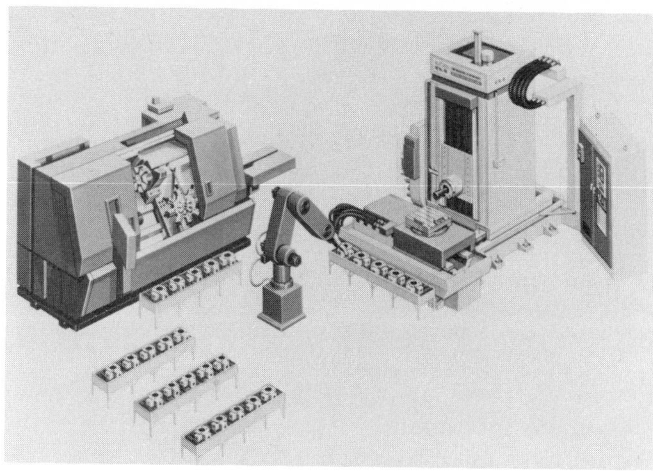

Fig. 16-5. Robot, or automated cell, consists of several machine tools with robotic material handling. Cell becomes a fully automatic process through application of robotics, power clamping, and other forms of automation. (Kearney & Trecker Corp.)

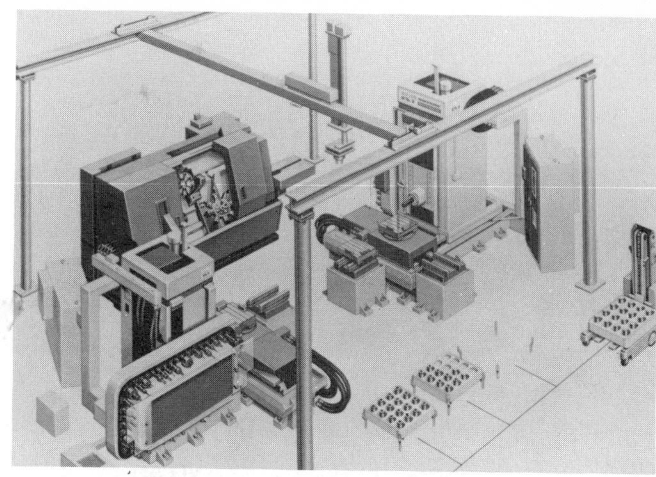

Fig. 16-6. With FMS, each step in manufacturing process is linked with succeeding one. There is an automated flow of raw material, total machining of part across machines, and then removal and storage of finished part. (Kearney & Trecker Corp.)

Machine tools employed in FMS are CNC (computer-assisted numerical control) machining centers, but other CNC machine tools and automatic gaging equipment may be included.

Each step in the manufacturing process is computer controlled and linked with the succeeding one, Fig. 16-6. The measuring sensors, often utilizing lasers, are sensitive enough to detect tool wear as it occurs and compensate for it automatically. Every part is inspected, Fig. 16-7. Problems such as tool malfunction, tool breakage or damage, etc., are immediately identified so corrections can be made before additional parts are produced that do not meet specifications.

Fig. 16-7. Lasers scan front end, door frames, and tail assembly of automobile body. Inspection is done to prevent imperfect parts from being moved to next stage of production. (Pontiac Div., General Motors Corp.)

Flexible manufacturing systems require fewer but more highly skilled machinists, and the system can be designed to operate on second and third shifts with very little human supervision. Advancements in numerical control (NC or N/C) and computer technology have made equipment and controls for FMS possible.

AN INTRODUCTION TO NUMERICAL CONTROL

In machining, when cutting operations are to be performed on conventinal machine tools, the machinist first studies the print. After determining the machining sequence, the cutter is installed and the work mounted and positioned on the machine.

Machining is done by moving one or more of the machine's lead and feed screws. After selecting the cutting speed and feed rate, the cutter is fed into the material and MANUALLY guided through the various machining operations that will produce the part specified on the print.

Numerical control (NC) is NOT a machining process; it is the operation of a machine tool by a series of coded instructions. The code consists of ***alphanumeric data*** (numbers, letters of alphabet, and other symbols) that are translated into pulses of electric current. The pulses of current activate motors and other devices to run the machine through the specified machining cycle, Fig. 16-8.

The motors, called ***servos,*** are connected to the machine's lead and feed screws. They provide the power that positions the work and feeds the cutter. The coded instructions are on punched paper tape, punched plastic tape, magnetic tape, floppy disk, or they are sent directly from a computer. These instructions control the electronic impulses that instruct the servos when to start, in what direction to move, and how far to move.

Feed rate, cutter speed, and on some NC machines, tool changes are controlled by the same set of instructions or ***program.*** See Fig. 16-9.

NC machine tool positioning systems

Some NC machine tools are fitted with a ***closed loop system*** to control the positioning of the work. This system uses an electronic feed-back device, called a ***transducer,*** to continually monitor tool movement. It "tells" the servo(s) when the instructed distance (tool position) has been reached. The feed-back unit also serves as a check on the accuracy of the desired tool movement.

Other NC machine tools are fitted with an ***open loop system*** to control work positioning. It has no feed-back for monitoring or comparing purposes. The system relies on the integrity of the control unit for accuracy.

Cartesian coordinate system

Since the ***NC program*** operating instructions must instruct the servos in what direction the work and/or tool must move, there has to be some way to define this movement.

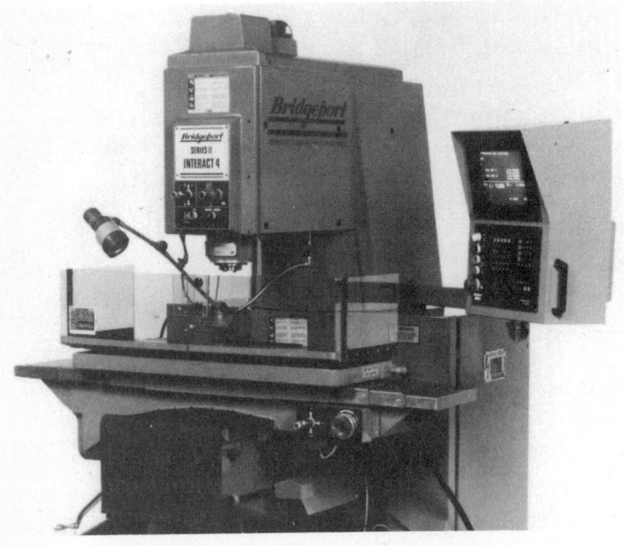

Fig. 16-8. Numerical control is NOT a machining process. Rather, it is operation of a machine by a series of coded instructions. Code consists of alphanumeric data that are translated into pulses of electric current that activate motors and other devices to run machine through machining operation. Note that this NC machine has no conventional hand controls and has an on-board computer to control machine. (Bridgeport Machines, Inc.)

Fig. 16-9. Automatic tool changes are also controlled by NC program on machines like this machining center. Note tools contained in magazine on left side of machine. (Kearney & Trecker Corp.)

The ***Cartesian coordinate system*** is the basis of all NC programming, Fig. 16-10. Programs in either inch or metric units specify the destination of a particular movement. With it, the ***axis of movement*** (X, Y, or Z) and the direction of movement (+ or −) can be identified. To determine whether the movement is positive (+) or negative (−), the program is written as though the tool, rather than the work, is doing the moving.

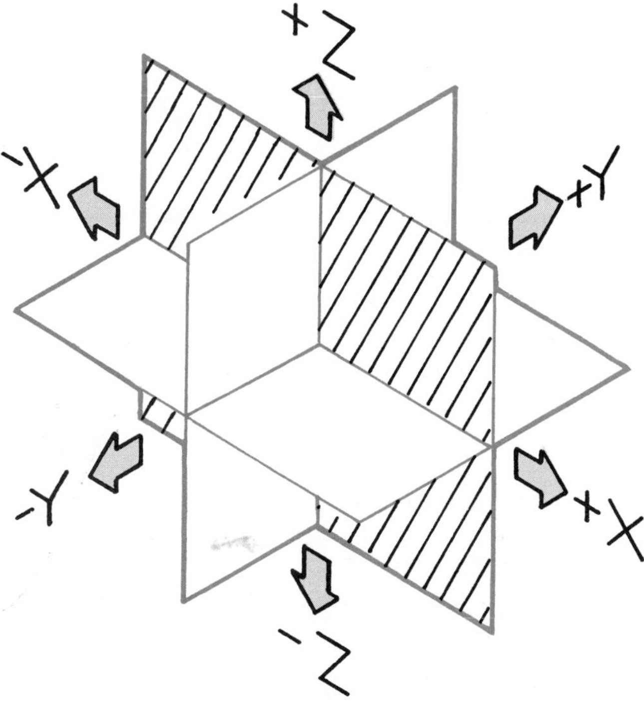

Fig. 16-10. Cartesian coordinate system is basis of all NC programming.

Spindle motion is assigned the Z axis. This means that for a drill press or vertical milling machine, for example, the Z axis is vertical. For such machines as a lathe or horizontal milling machine, the Z axis is horizontal. See Fig. 16-11.

The system of coordinates used for machine axis designation is specified according to the ***Right-Hand Rule*** of Cartesian coordinates, Fig. 16-12.

NOTE: Most NC machine tools manufactured today are of the ***CNC*** (computer-assisted numerical control) type. That is, they are equipped with an onboard minicomputer control system. Machining instructions can be programmed directly into computer memory and NOT require input via punched tapes. However, since large numbers of tape controlled machines are in use, tape information will be included in this chapter.

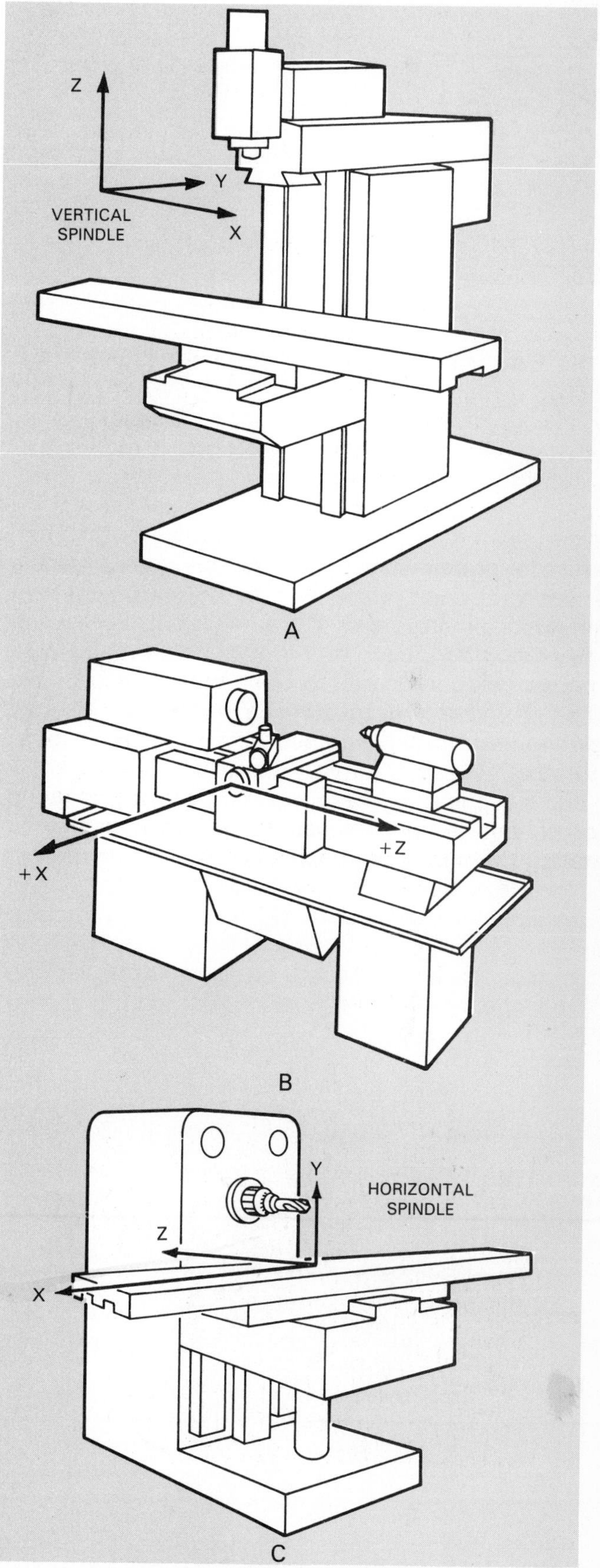

Fig. 16-11. Axes of machine tool movements. A—Vertical milling machine. B—Lathe. C—Horizontal milling machine. Spindle motion is assigned Z axis. Note how it differs between machine with vertical spindle and machines with horizontal spindle.

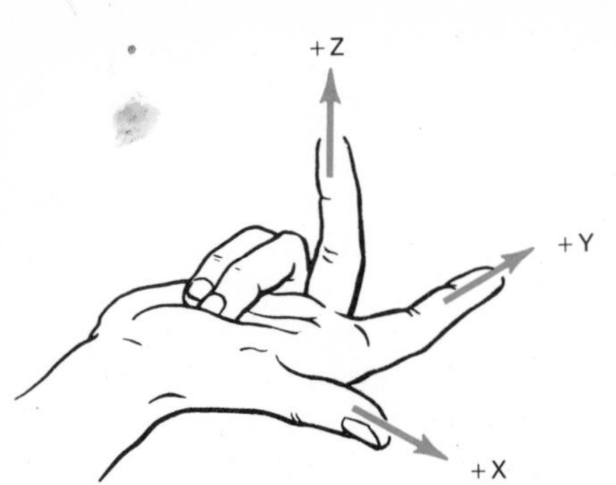

Fig. 16-12. Machine tool axes are specified according to right-handed system of Cartesian coordinates. Hold right hand as shown. Sequence of thumb, forefinger, and next finger point in positive (+) directions of X, Y, and Z axes of right-handed coordinate system. (IBM)

NC tool positioning

NC tool positioning may be by incremental or absolute dimensioning, Fig. 16-13. With ***incremental positioning,*** each tool movement is made with reference to the prior or last tool position, Fig. 16-14. ***Absolute positioning*** measures all tool movement from a fixed point, origin, or zero point. See Fig. 16-15.

Where possible, absolute dimensioning should be employed because a mistake in dimensioning an individual point does NOT affect remaining dimensions. It is also easier to check for errors.

A ***machine control unit*** (MCU) is a microcomputer or electronic control circuit for controlling the machine. Most will accept program information in either the incremental or absolute mode. It is also possible to switch back and forth from one mode to the other in the same program.

Either inch or metric dimensions may be used. The choice can be made by a switch or by a specific code in the part program.

BASIC NC SYSTEMS

There are two basic NC systems: point-to-point and contour or continuous path.

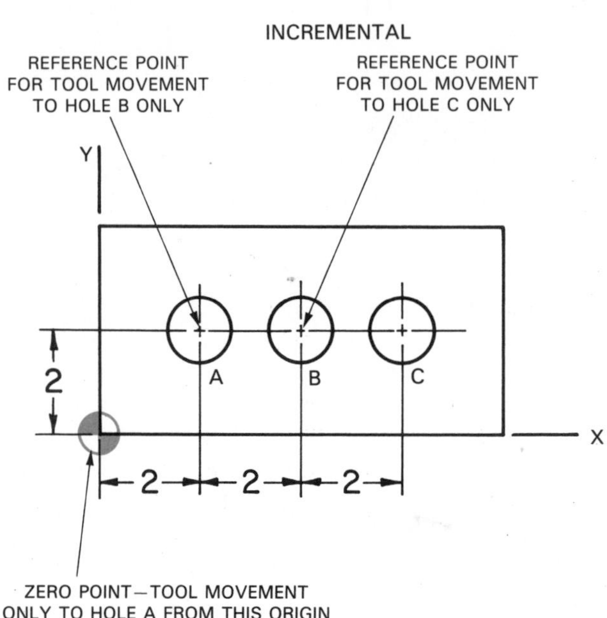

Fig. 16-14. Incremental positioning system.

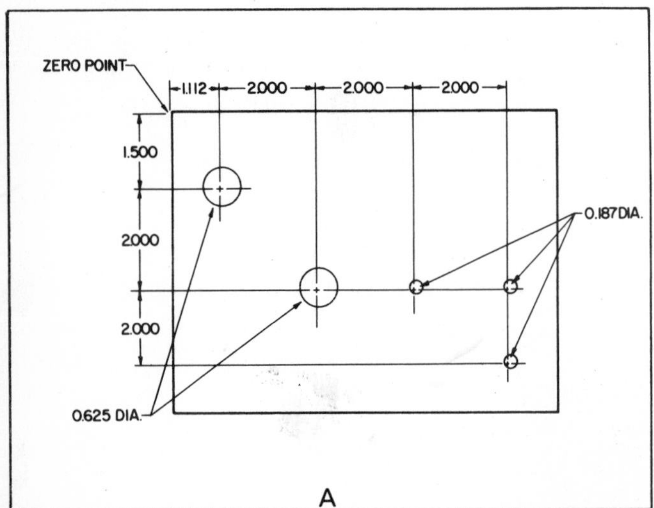

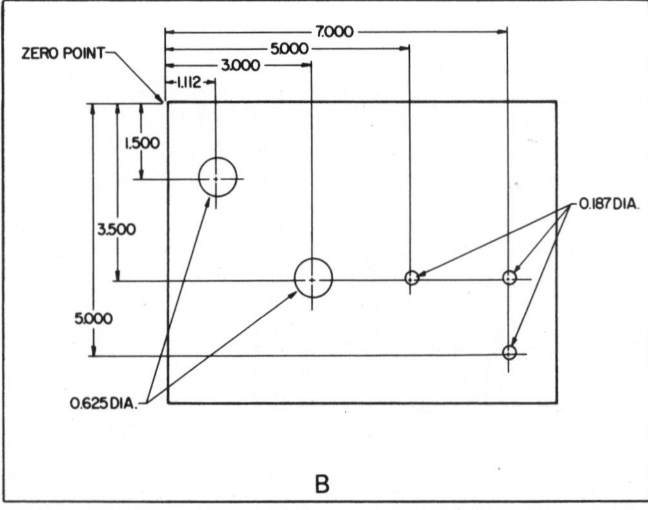

Fig. 16-13. Spindle positioning. A—Incremental measuring. In this system, each set of coordinates has its point of origin from last point established. B—Absolute measuring. In this system, all coordinates are measured from fixed point (zero point) of origin.

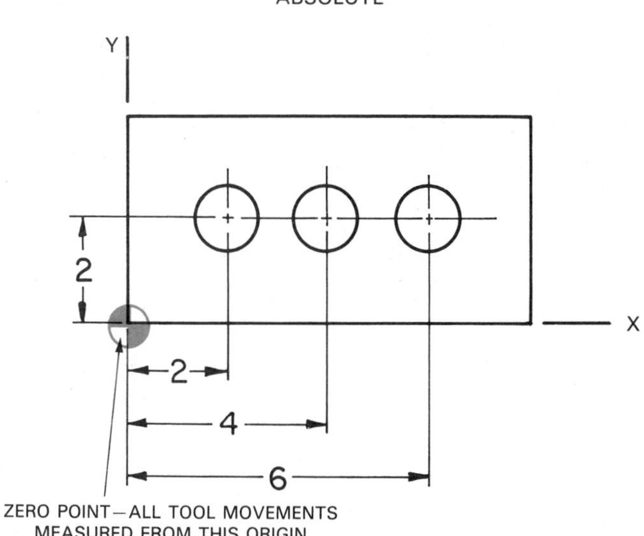

Fig. 16-15. Absolute positioning system.

Point-to-point system

The ***point-to-point system*** is usually employed for drilling, punching, spot welding, straight line milling, or any operation performed at a fixed location in terms of a two-axis coordinate position. While moving from one work position to the next, the cutting tool is NOT in contact with the work and a rapid traverse feed can be used.

Tool movement from one point to the next does NOT have to follow a specific path, except in straight line milling. See Fig. 16-16.

Some point-to-point NC machines have the capacity of straight line milling along either the X or Y axis, or at a 45 deg. angle between two points with no change of depth along the line the tool travels.

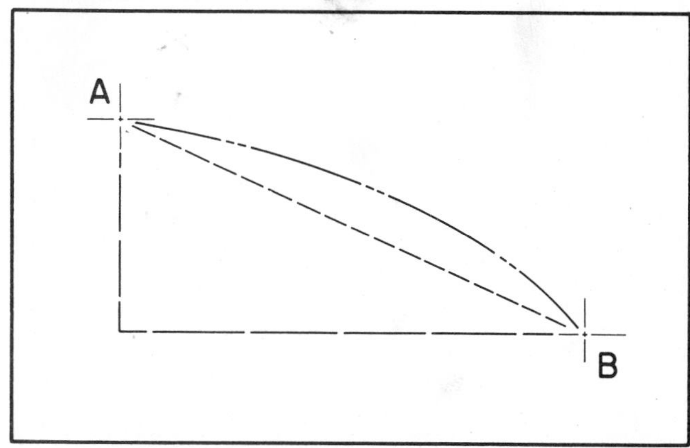

Fig. 16-16. In point-to-point NC systems, there is no concern what path is taken when tool moves from Point A to Point B, except in straight line milling.

Point-to-point systems generally do NOT require computers for program preparation. Anyone having a basic knowledge of machine practices and the ability to read engineering drawings can prepare a program with a minumum of instructions.

Contour (continuous path) system

A ***contour*** or ***continuous path system*** precisely controls machine and tool movement at all times, in all planes, as the cutter moves along the program path, Fig. 16-17. Cutting is continuous and can be on up to six axes simultaneously, Fig. 16-18.

Cutter location is monitored continuously through a feed-back mechanism to the machine control unit (MCU). This is required to maintain the correct direction of cutter movement and proper cutting speed and feed.

Cutter size and other variables must also be considered when programming, Figs. 16-19 and 16-20.

Geometrical complexity makes a computer mandatory for preparing programs to machine two- and three-dimensional shapes, Fig. 16-21. This is necessary for making circular, elliptical, or similar complex cuts. The tool must be fed a constantly changing series of instructions. Far too many to be calculated manually.

In a way, continuous path machining might be described as a very sophisticated point-to-point system because tool movement is only a series of straight lines. However, the straight line movements are so small (0.0001 in. or 0.003 mm) and blend so well, that they appear to be a continuous smooth curved cut, Fig. 16-22.

To reduce the number of calculations required to program a curved surface, most newer NC machine tools have a MCU encoded for ***circular interpolation.*** All that has to be programmed are:

1. The coordinate locations of the end points of the arc.
2. The radius of the circle of which the arc is a component.
3. The coordinate location of the circle.

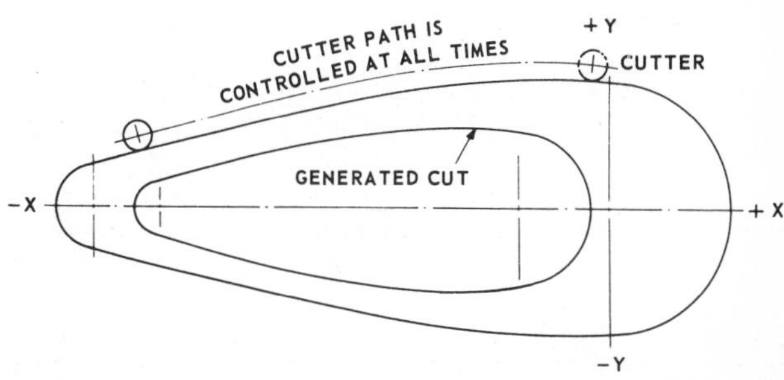

Fig. 16-17. With contour or continuous path system, cutter path is controlled at all times. By adding a Z axis, three-dimensional machining is possible.

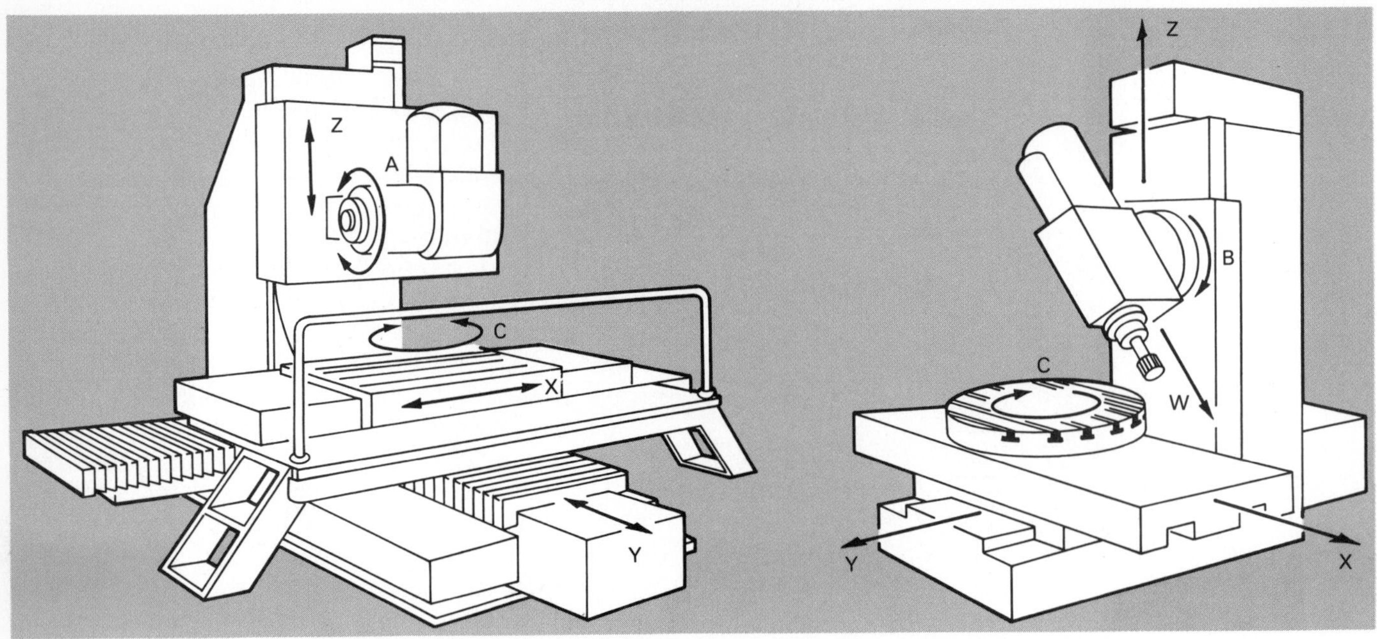

Fig. 16-18. Left—5-axes NC milling machine. Right—6-axes machining center.

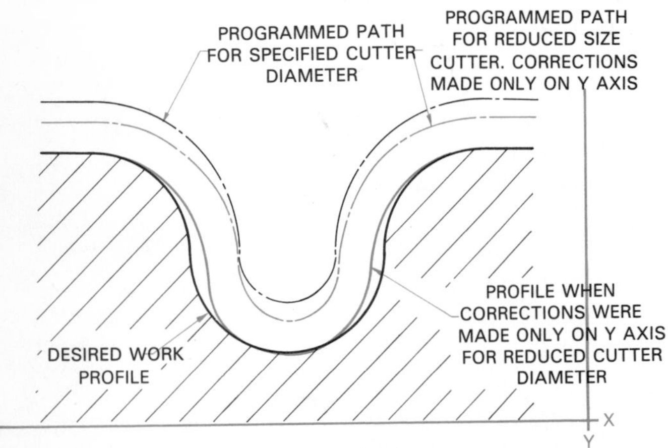

Fig. 16-19. On contour program, adjustments must be made if a cutter is smaller or larger than one specified; otherwise, contour will deviate from required part outline.

4. The direction the cutter is to proceed, Fig. 16-23.

The circular interpolator in the MCU automatically computes the necessary number of intermediate points to describe the circular cut. It also generates the electronic signals that will run the servos and guide the cutting tool in making the cut.

Very complex parts can be produced economically in small numbers, without the need for expensive jigs, fixtures, and templates by the continuous path system, Fig. 16-24. On some NC machine tools, ***mirror image parts*** (right-hand and left-hand units) can be machined using the same program. See Fig. 16-25.

PROGRAMMING NC MACHINES

An ***NC machining program*** is a sequence of instructions that ''tells'' the machine what operations to perform, and where on the material they are to be done. A program is also referred to as ***software.***

Each line of an NC program is called a ***block.*** It consists of a sequence number, preparatory codes for setting up the machine, coordinates for the destination of the tool, and additional commands (feed rate, coolant on, coolant off, etc.), Fig. 16-26.

NC equipment can be programmed manually or by computer assist.

Manual programming

Manual programming may be done if the part is NOT too complex. It can be accomplished by anyone who can interpret engineering drawings and has a working knowledge of machine tool operations, Fig. 16-27.

A program is developed by converting each machining sequence and machine function into a coded block of information that the MCU can understand, Fig. 16-28. The code consists of alphanumeric date or letters, numbers, punctuation marks,and special characters. Each identifies a different machine function.

Each block of information is a line on the ***program sheet*** or ***program manuscript*** and is identified by a sequence number, Fig. 16-29. Included in this information are the coordinate dimensions (X, Y, and Z movement) or location where the operation (drilling, punching, spot welding, etc.) is to take

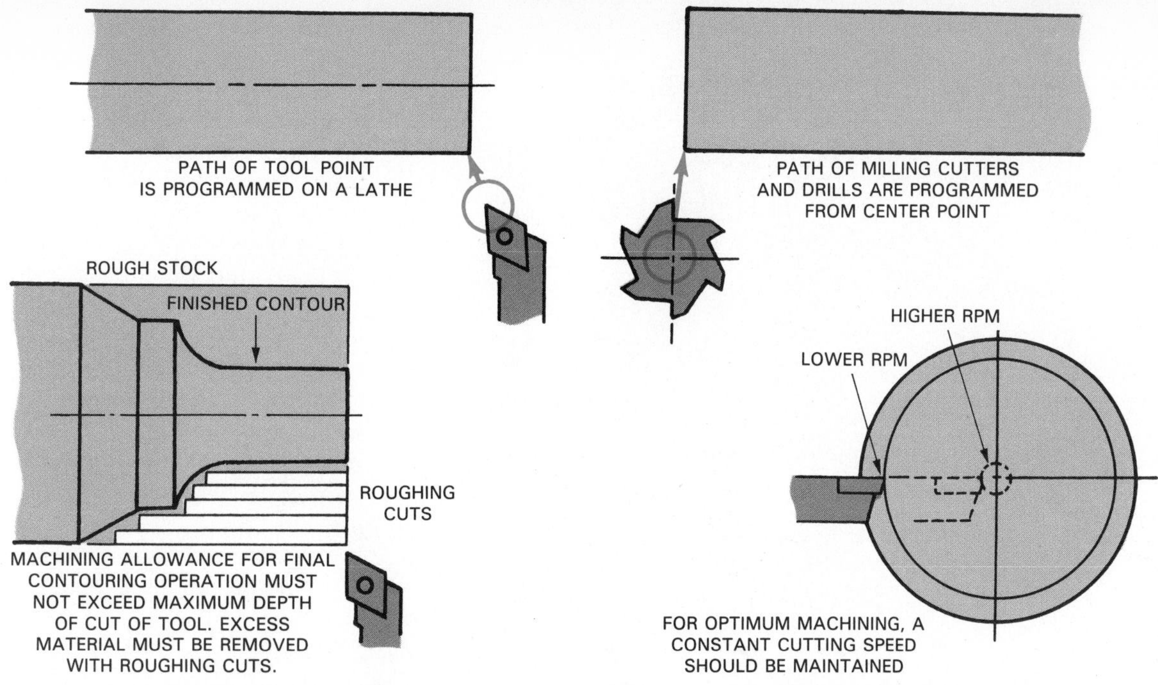

Fig. 16-20. These are variables that must be considered when designing an NC program.

Fig. 16-21. This small aluminum alloy impeller was machined on a 5-axes CNC milling machine. Accuracy is imperative as impeller spins at 300,000 RPM in a compressor. (Rigid Machine Tool Inc.)

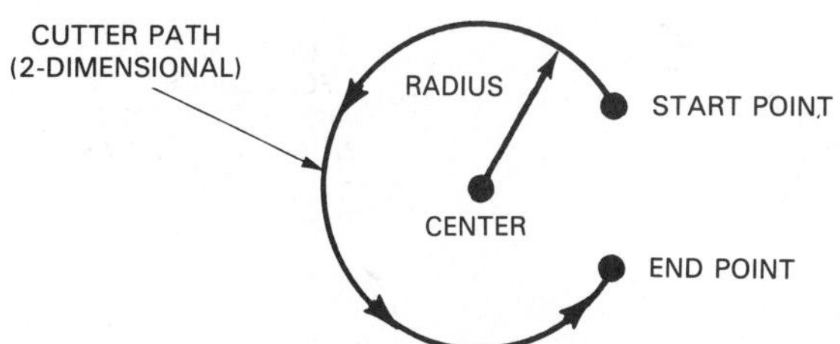

Fig. 16-23. On NC units encoded with circular interpolation, only coordinate location of end points of arc, radius of circle, coordinate location of circle, and direction of cut need to be programmed. Circular interpolator in MCU automatically computes necessary number of intermediate points required to describe circular cut. On some machines, circle is divided into four 90 deg. segments and must be programmed accordingly.

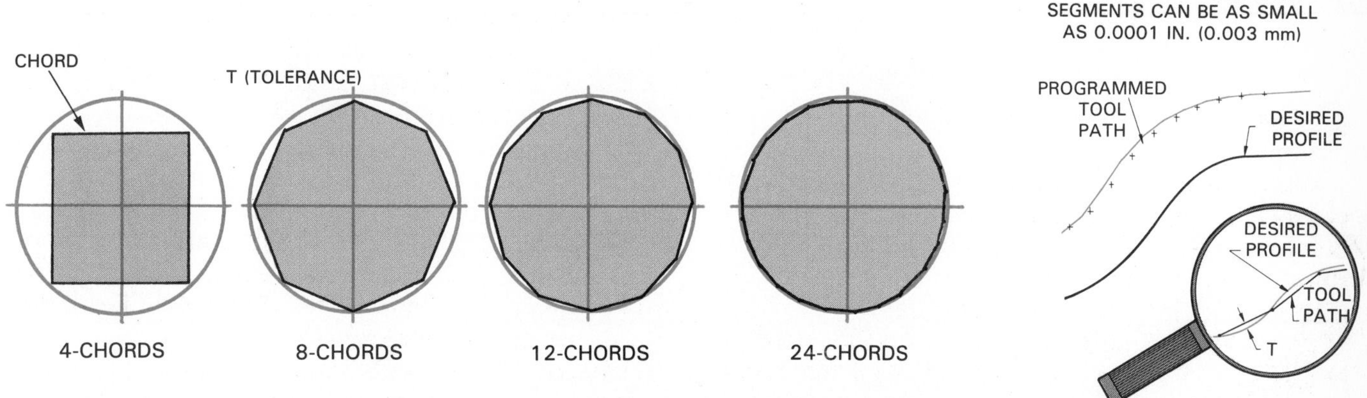

Fig. 16-22. Contour obtained from contour or continuous path machining are result of a series of straight line movements. Degree to which contour corresponds with specified curve depends upon how many movements or chords are used. Note how, as number of chords increase, the closer it is to a perfect circle. Actual number of lines or points needed is determined by tolerance allowed between design or curved surface and one machined.

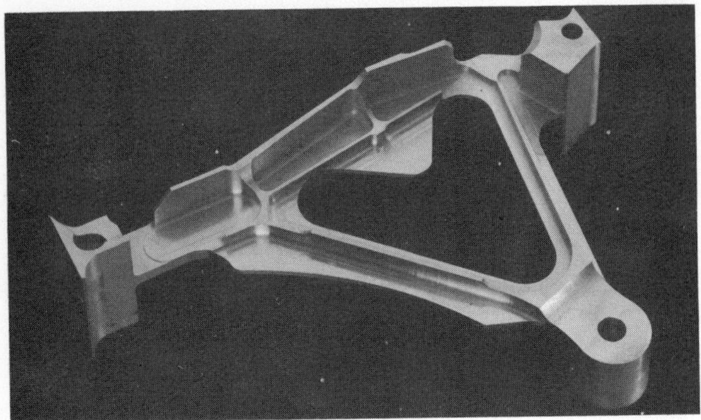

Fig. 16-24. Complex parts, like this aircraft component, are produced quickly and economically by continuous path NC system. (Cincinnati Milacron)

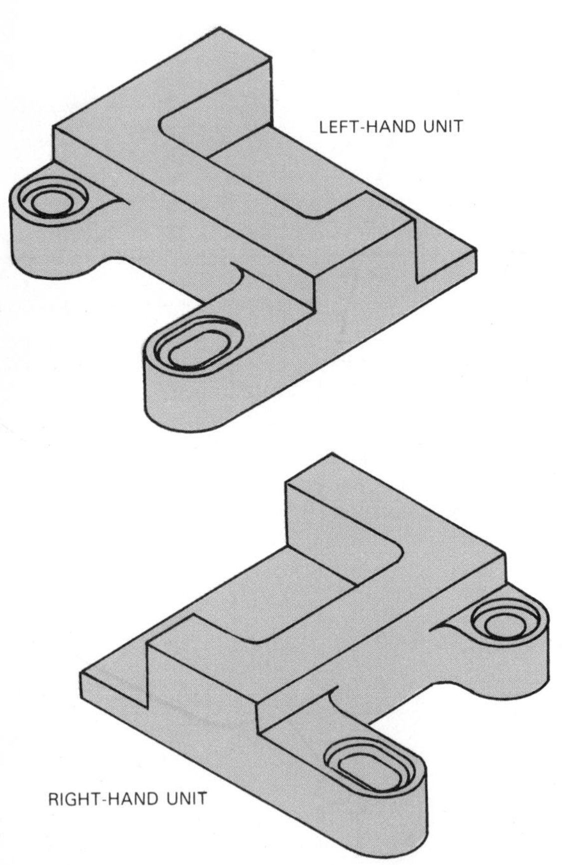

Fig. 16-25. On many NC machine tools, mirror image parts can be machined using the same program.

place, along with miscellaneous functions (spindle on, spindle off, tool change, etc.).

Every block of information must be separated with an ***end of block*** (EOB) code. An ***end of program*** code completes the program. A final EOB rewinds the tape. It will then be ready to repeat the machining cycle.

The coded information is sometimes punched into a paper or mylar tape on a ***tape punch/reader unit,*** Fig. 16-30. The information can also be recorded on magnetic tape. However, with today's equipment, the information is often placed directly into the on-board computer as electronic data. The input information is proofed for omissions and/or errors and corrected before being released to the production area.

Computer-assisted NC programming

Computer-assisted programming reduces and simplifies numerical calculations that the programmer must perform when programming the machining or move complex parts, Fig. 16-31.

PREPARATORY FUNCTIONS (G-CODES)*	
CODE	**FUNCTION**
G00	Rapid traverse (slides move only at rapid traverse speed).
G01	Linear interpolation (slides move at right angles and/or at programmed angles).
G02	Circular interpolation CW (tool follows a quarter part of circumference in a clockwise direction).
G03	Circular interpolation CCW (tool follows a quarter part of circumference in a counterclockwise direction).
G04	Dwell (timed delay of established duration. Length is expressed in X or F word).
G33	Thread cutting.
G70	Inch programming.
G71	Metric programming.
G81	Drill.
G90	Absolute coordinates.
G91	Incremental coordinates.

*G-CODES may vary on different N/C machines.

MISCELLANEOUS FUNCTIONS (M-CODES)*	
CODE	**FUNCTION**
M00	Stop machine until operator restart.
M02	End of program.
M03	Start spindle—CW.
M04	Start spindle—CCW.
M05	Stop spindle.
M06	Tool change.
M07	Coolant on.
M09	Coolant off.
M30	End program and rewind tape.
M52	Advance spindle.
M53	Retract spindle.
M56	Tool inhibit.

*M-CODES may vary on different N/C machines.

Fig. 16-26. Examples of codes of NC prepared and miscellaneous functions.

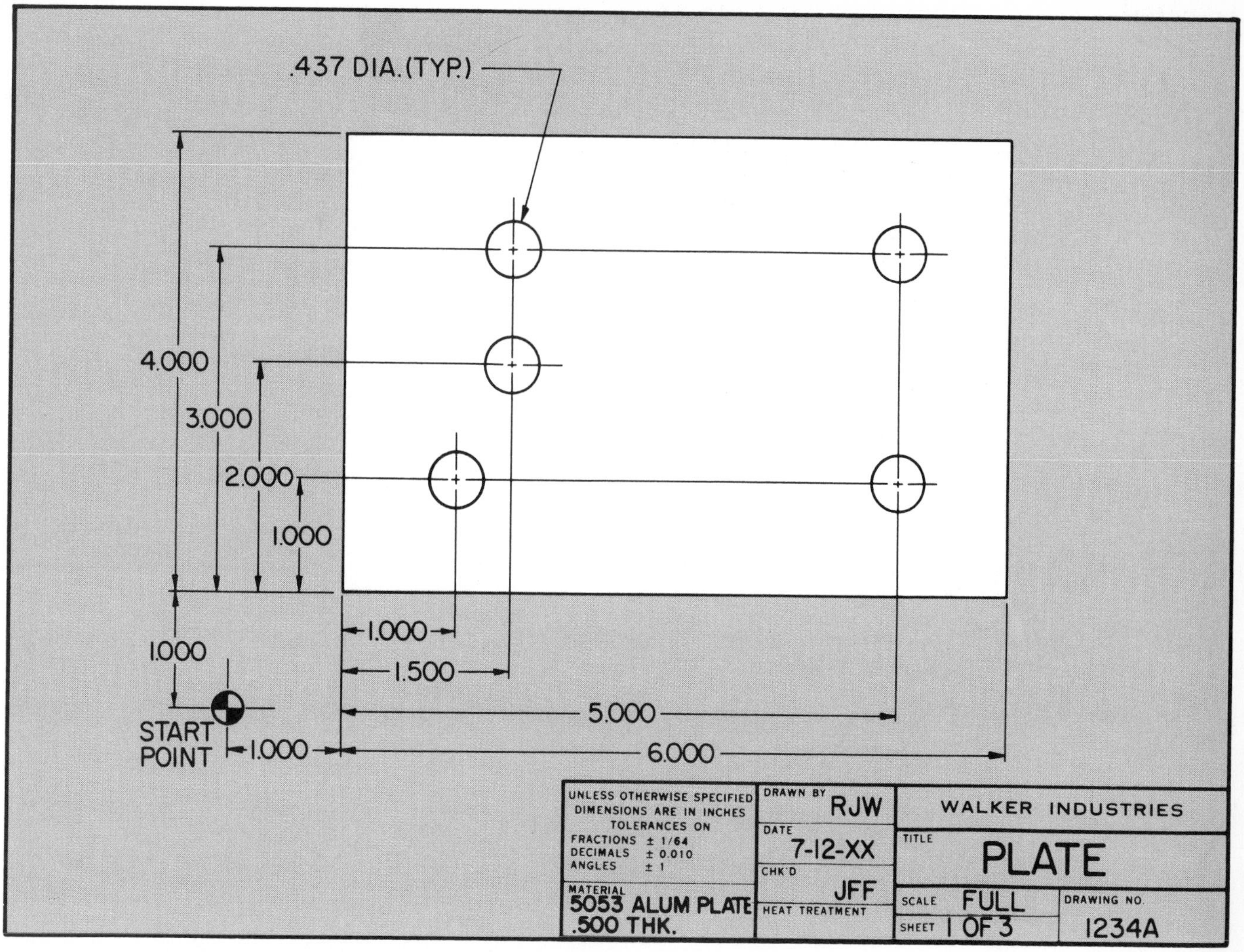

Fig. 16-27. Point-to-point programming can be done manually by anyone who can interpret engineering drawings and has a working knowledge of machine tool operations.

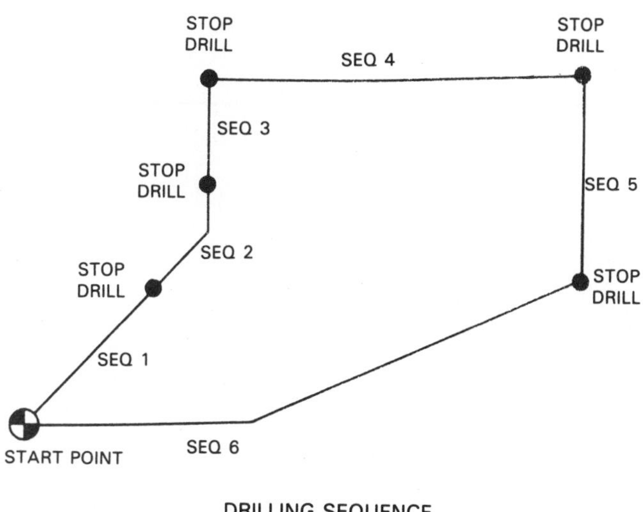

Fig. 16-28. Machining operations must be converted into individual coded blocks of information that MCU can understand. This is planned drilling sequence for part shown in Fig. 16-27.

PART NO. 1234A	MACHINE 2-AXIS DRILLING WITH FIXED FEED RATE	REMARKS SET POINT (0,0) M56 IS TOOL INHIBIT
PART NAME PLATE		
TAPE NO. 1234A	TOOLING 7/16 DIA. DRILL	
DATE 7-12-XX PAGE 1 OF 1		PROCESSOR JFF APPROVED LJ

N	G	X	Y	Z	F	EOB	M	INSTRUCTIONS
000	90					EOB		7/16 DIA. DRILL
00						EOB		SET DEPTH STOP
0		0	0			EOB	03	
1		2000	2000			EOB	07	
2		2500	4000			EOB		
3		2500	4000			EOB		
4		6000	4000			EOB		
5		6000	2000			EOB		
6		0	0			EOB	0956	
7						EOB	05	
8						EOB	30	CHANGE PART

Fig. 16-29. Information given on print shown in Fig. 16-27 was developed into this program.

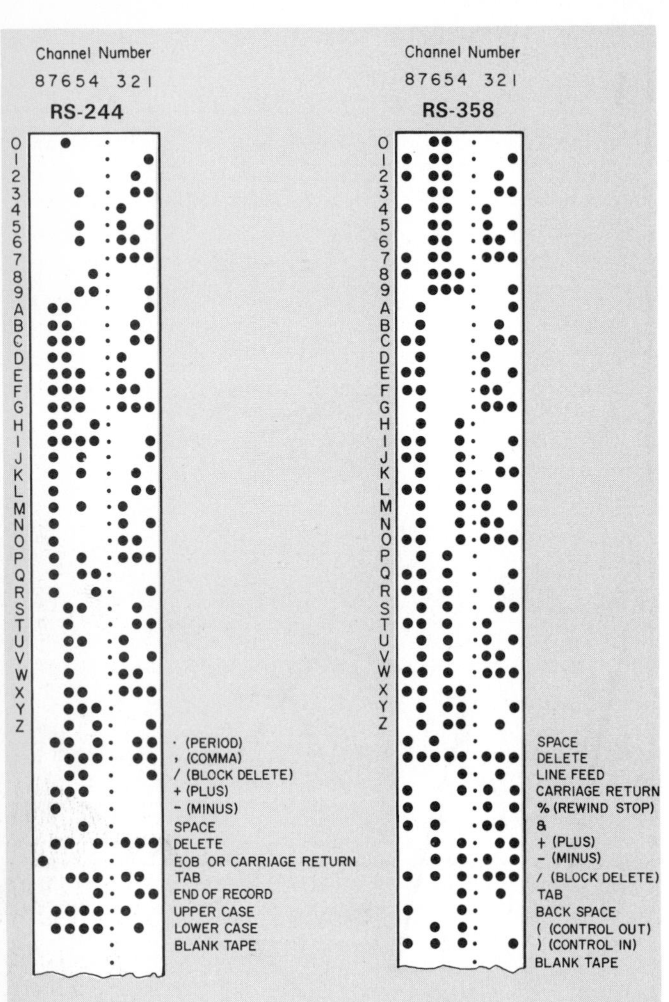

A

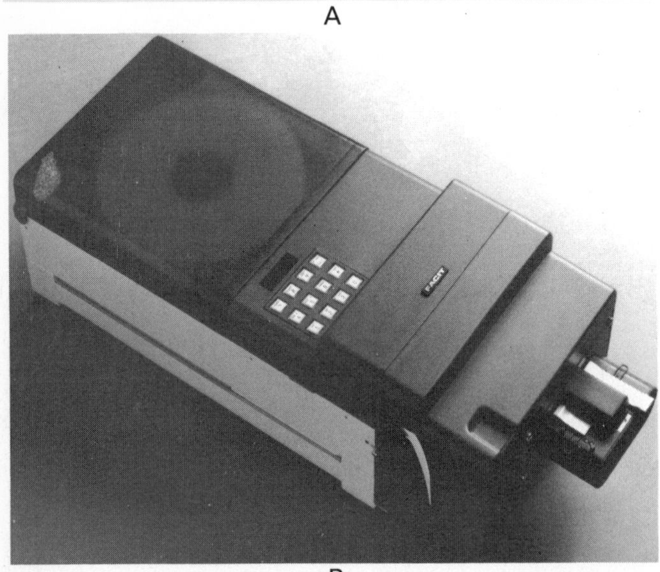

B

Fig. 16-30. A—Tape code. Note that every level of RS-244 tape has an odd number of punches for parity check. RS-358 tape has an even number. Parity check is a method of automatically checking to reduce possibility of tape errors by malfunctioning tape punch. Both tapes are in current use. B—Tape punch/reader unit. Program information is transferred to tape using this unit. It can also be used to read tape prepared elsewhere. (Facit)

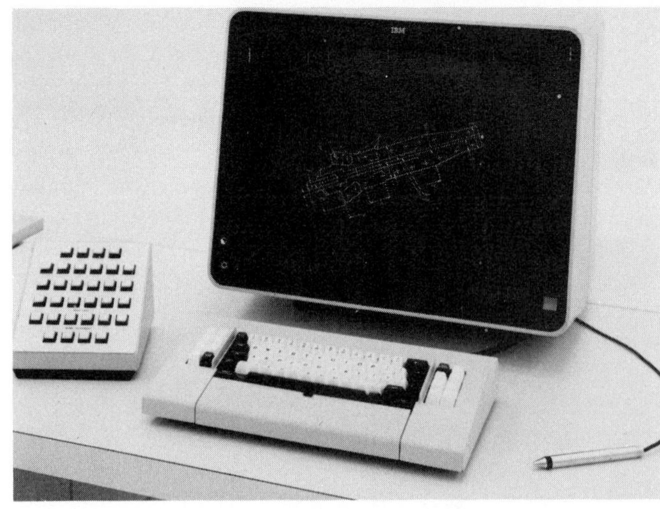

Fig. 16-31. Computer-assisted programming reduces and simplifies numerical calculations that programmer must perform on more complex contour machining operations. (IBM)

Most NC machine tools manufactured today have control units of the CNC (Computer-assisted numerical control) type. That is, CNC machines have a self-contained NC system with a computer dedicated exclusively to that particular machine, Fig. 16-32. They can also be controlled by a larger, more powerful host computer.

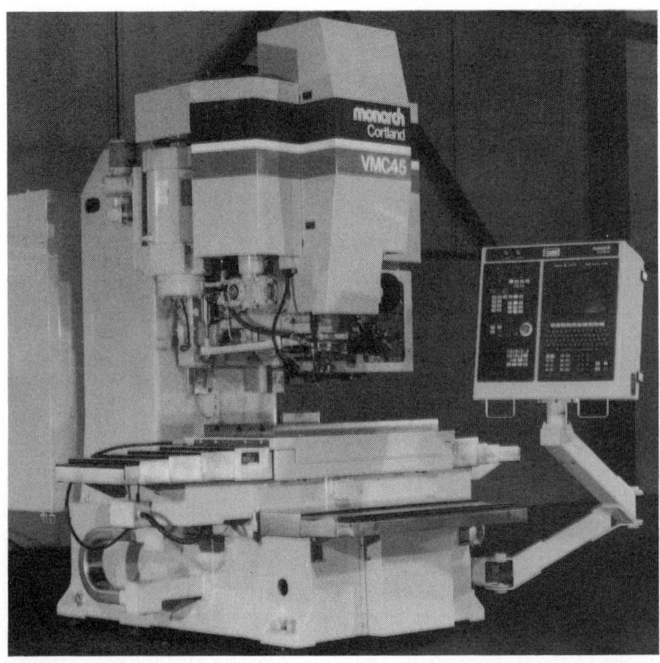

Fig. 16-32. Most NC machine tools manufactured today are of CNC (computer numerical control) type. That is, they have a self-contained NC system with an on-board computer dedicated exclusively to that particular machine. (Monarch Machine Tool Co.)

A ***host computer*** is a main-frame computer that monitors and controls other NC and CNC machines.

Programming, in one of the many computer languages, that describes the geometry of the part to be made can be entered into the system in a number of ways. The display screen (cathode ray tube or CRT) can show the full operational data as the part is being machined, Fig. 16-33. Graphic capability may include a picture of the part and tools, and a simulation of cutter path, often in vivid color. See Fig. 16-34.

Computer memory capacity is given as an equivalent tape capacity in feet and/or meters. After use, the program can be moved to another, larger computer, or stored on magnetic tape or disk.

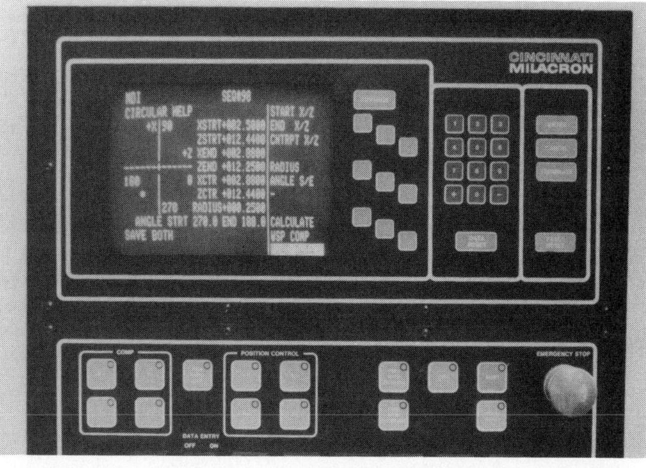

Fig. 16-33. Display screen (cathode ray tube or CRT) can show full informational data as part is being machined. (Cincinnati-Milacron)

COMPUTER LANGUAGES

There are many different computer-assisted languages: APT (Automatic Programmed Tools), ADAPT (Adaptation of APT), COMPACT II, AUTOMAP (Automatic Machining Programming), etc. Some are very powerful and provide programming for milling complex curved sections that require up to five and sometimes six axes of motion by machine tools, Fig. 16-35. Other programming languages are designed to be used by a specific machine or for a specific machining application.

Fig. 16-35. Controllable pitch propeller blade for ship is being machined on 5-axis milling machine. Reflections from overhead lights show how geometry of machined surface changes from hub to tip. Blade is about 8 ft. or 2.5 m long. (Bird-Johnson Company)

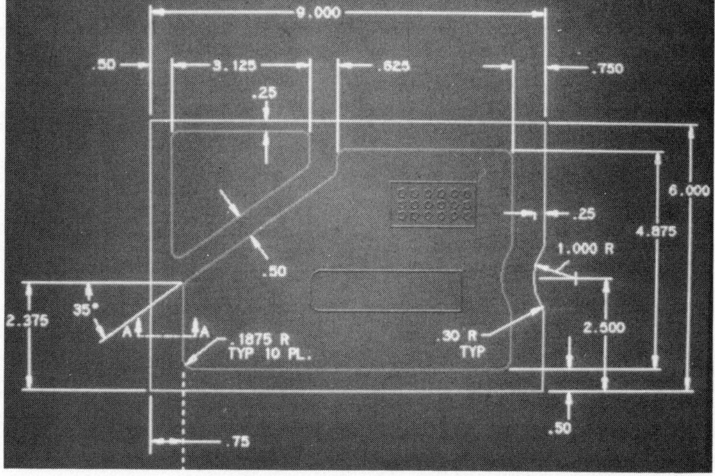

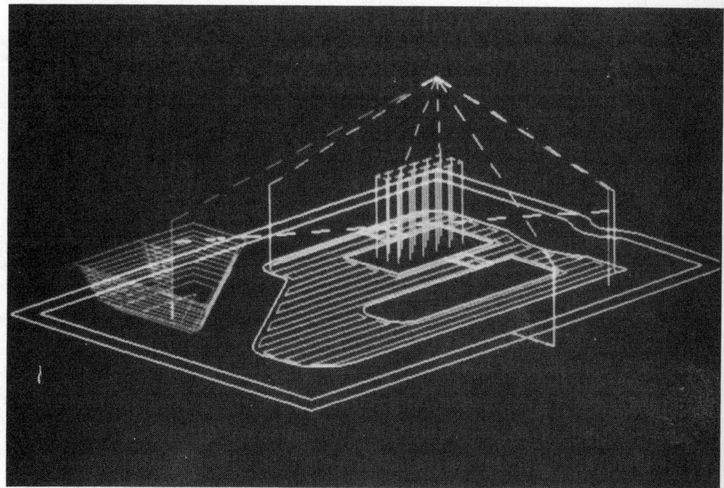

Fig. 16-34. Left. Top view of job with dimensions. Right. An isometric view of tool paths relative to part design. Both views were created using computer graphics. (Computervision)

Part programming languages consist of a vocabulary of words, numbers, and other symbols. When combined according to certain rules, called ***syntax,*** the languages are capable of producing sets of instructions for machining parts.

APT is the most powerful of the computer-assisted programming languages in common usage. It is a system that defines, in a series of statements, part geometry, cutter operations, and machine tool capabilities.

The spelling of an APT word used in a programming statement must be spelled exactly as it is spelled in the ***APT system dictionary.*** The dictionary specifies the only spelling the computer will recognize. For example:

1. CØØLNT = ***Coolant.*** CØØLNT/ØN means turn the coolant on. It will continue on until getting a CØØLNT/ØFF or STØP command.
2. FEDRAT = ***Feed rate.*** FEDRAT/6, IPM means feed rate in all directions will be 6 in. per minute.
3. SPINDL = ***SPINDLE.*** SPINDL/1000 RPM, CLW means the spindle is to rotate 1000 revolutions per minute in a clockwise direction. Spindle stays on until SPINDL/ØFF or STØP command.
4. GØRGT = Tool movement is to the right. See Fig. 16-36.
5. GØLFT = Tool movement is to the left.
6. FINI = Part program is completed.

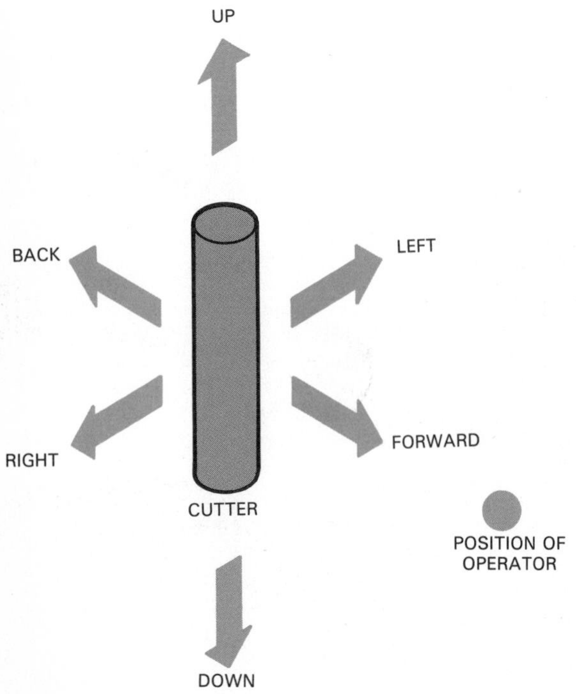

APT programming, as with most computer-assisted programming languages, is done in three parts:

1. ***Geometry***—part features are defined in geometric figures.
2. ***Machining statements***—this is used to direct the cutter around the geometry of the part in a predetermined sequence.
3. ***Auxiliary function statements***—this includes tool changes, cutting speeds, feed rate, coolant on, coolant off, etc.

NC AND THE FUTURE

In reality, computer technology has reached the stage where an engineer can use computer graphics to design a part, Fig. 16-37. The computer is used to place the ***design elements*** (lines, curves, circles, etc.) and dimensions on the computer display screen, Fig. 16-38. Design changes can be made easily using the computer keyboard. When the design has been edited and proofed, the computer can be instructed to analyze the geometry of the part and calculate the tool paths that will be required when machining the part.

The tool path is translated into a detailed sequence of machine axes movement commands that will enable an appropriate CNC machine tool to produce the part, Fig. 16-39. No engineering drawing or program manuscript is required. However, the data can be used to produce ***hard copy*** (detail and assembly drawings) of the part and to determine how it fits into the overall product assembly. Refer to Fig. 16-40.

UP
BACK
LEFT
RIGHT
FORWARD
CUTTER
POSITION OF OPERATOR
DOWN

Fig. 16-36. Note APT (automatic programmed tools) motion commands.

Fig. 16-37. Computer technology has reached stage where an engineer can, using computer graphics, design a part. He or she can instruct computer to analyze geometry of part and calculate tool paths and program that will be required to machine part. (Computervision)

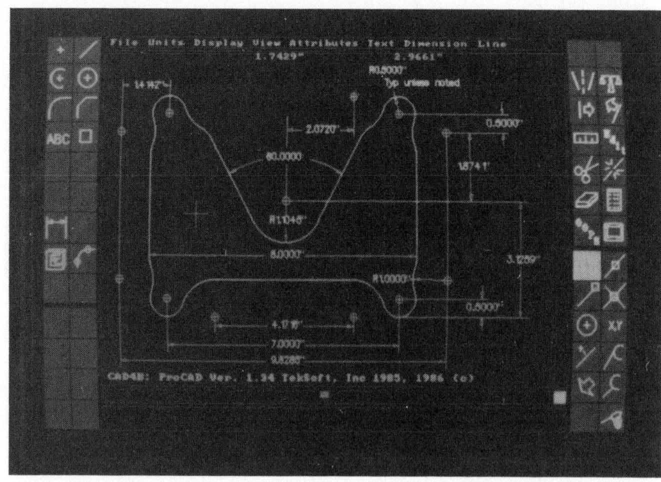

Fig. 16-38. This is a dimensioned view of part designed using computer graphics. Design changes can be made easily. (TekSoft, CAD/CAM Systems)

Fig. 16-39. Tool path data developed on a computer can be translated into a detailed sequence of machine axes movement commands that will enable an appropriate CNC machine to produce part. (Maho Machine Tool Corp.)

The computer generated instructions can be stored in a central master computer for direct transfer to a CNC machine tool for parts manufacture. This is known as ***DNC*** (direct numerical control).

The data can also be stored for future use on a punched tape, magnetic tape (similar to an audio tape cassette), or a disk. However, before being released for production, the program is verified by machining a sample part from some inexpensive material, like plastic or wax. See Fig. 16-41.

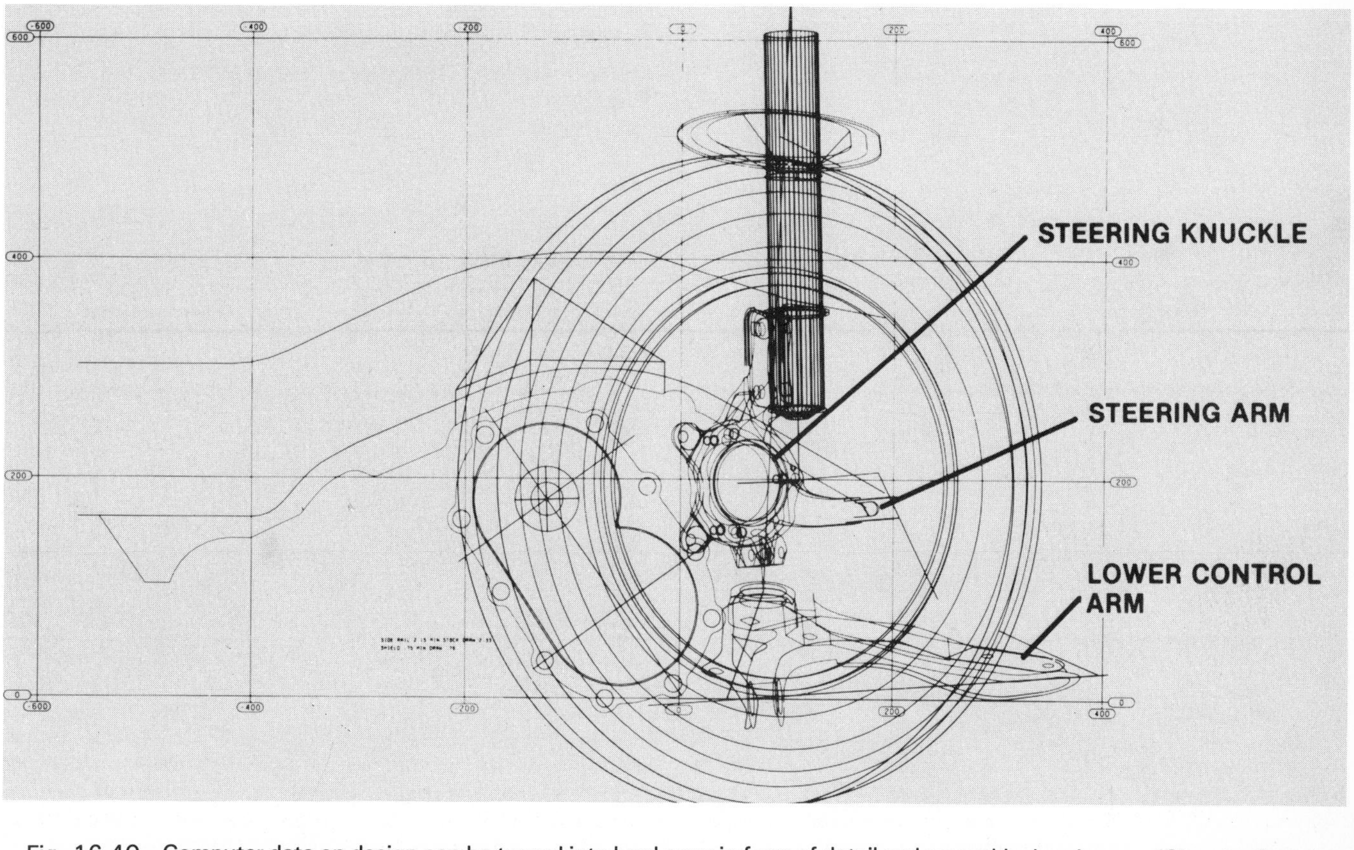

Fig. 16-40. Computer data on design can be turned into hard copy in form of detail and assembly drawings. (Chrysler Corp.)

Fig. 16-41. NC and CNC programs are usually checked out by using them to produce part in an inexpensive material, like plastic or wax. (Maho Machine Tool Corp.)

The system that makes all of this possible is called ***CAD/CAM*** (computer aided design/computer aided manufacturing). See Fig. 16-42.

ADVANTAGES AND DISADVANTAGES OF NC

NC can offer many advantages over traditional machining techniques. A few of the more important advantages include:

1. Increased productivity.
2. Reduced tooling costs.
3. Improved quality control.
4. Economical production of difficult to manufacture parts.
5. Versatility. NC machines can be programmed to produce a single piece or a large scale production run.
6. Elimination of many jigs and fixtures.
7. All, or nearly all, of the machining operations can often be performed on a single machine.
8. Direct savings in labor.
9. Easy inch/metric conversion.
10. Increased machining capabilities.

NC does have some disadvantages. Most can be overcome by careful planning.

1. High initial cost of equipment.
2. Shortage of skilled technicians to service NC equipment.
3. Increased maintenance costs over traditional machine tools.
4. Machine capabilities must be fully utilized.

NC FOR TRAINING PROGRAMS

Many NC and CNC machine tools have been developed especially for training programs, Figs. 16-43 and 16-44. The complexity of the work they are capable of performing is limited by machine size and on-board computer capacity.

When using these machines, students and trainees are able to develop programs, introduce the program to the machine computer, and see the resulting finished part within a reasonable time. Fig. 16-45 shows this type CNC machine.

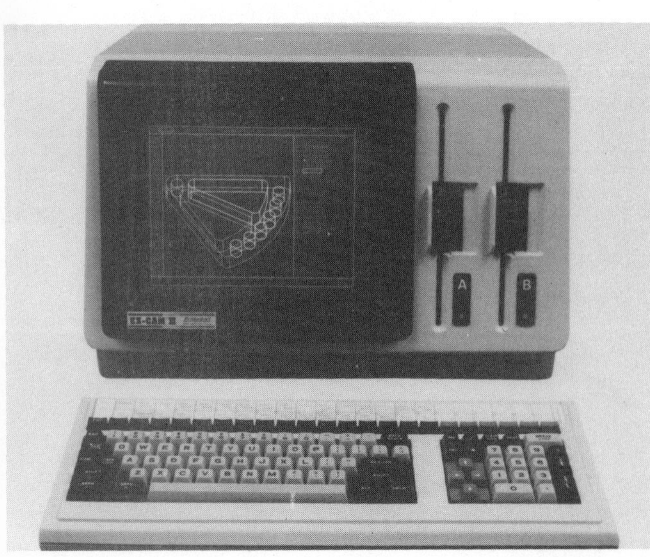

Fig. 16-42. Computers make CAD/CAM possible. (Bridgeport Machines, Inc.)

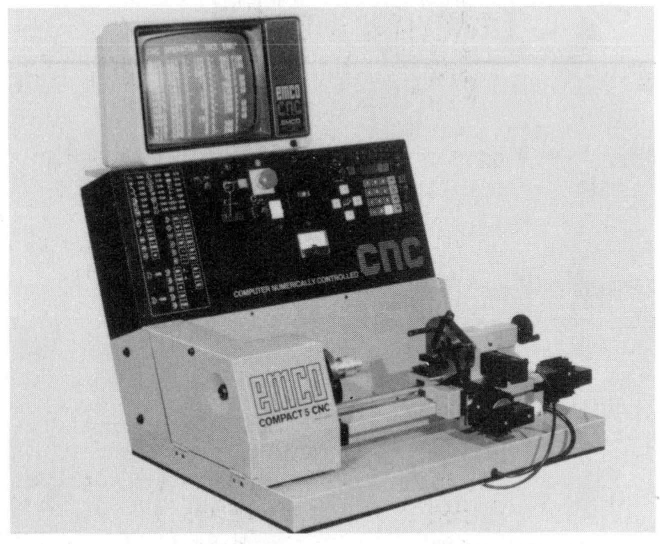

Fig. 16-43. This CNC lathe is designed especially for training purposes. (Emco Maier)

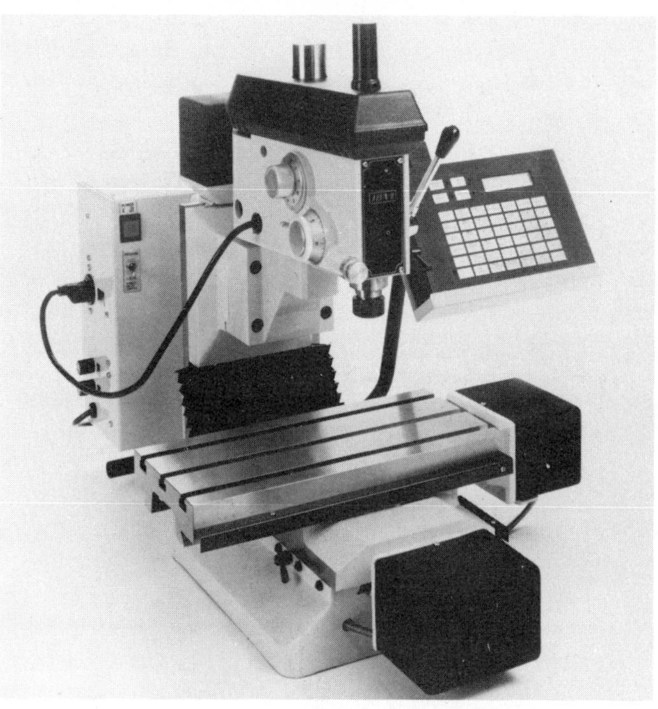

Fig. 16-44. CNC vertical milling/drilling machine is designed for both training and industrial use. (Dyna Electronics Inc.)

Fig. 16-45. Industry and many schools use relatively inexpensive CNC machine tools for educational purposes. They simulate actual industrial practices on how programs are developed. Students and trainees can see resulting finished part within a reasonable time.
(Millersville University, Dept. of Industry and Technology)

OTHER NC APPLICATIONS

NC and CNC systems have been adapted to a broad range of metalworking related equipment. Spot welding, riveting, and punching were the first to use NC systems because they are basically point-to-point operations. See Fig. 16-46.

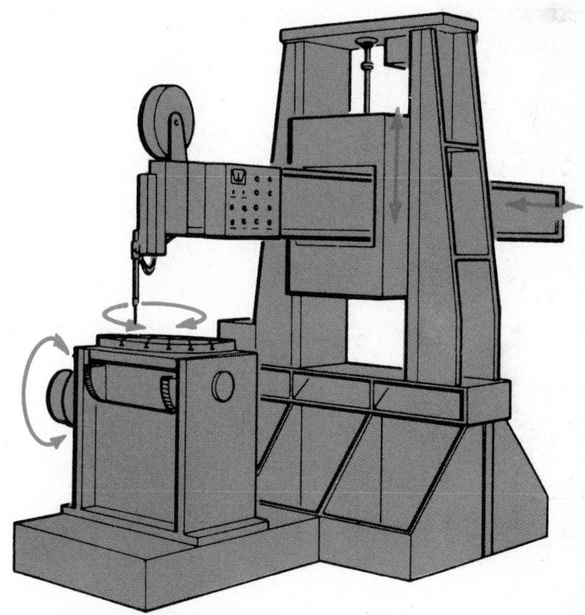

Fig. 16-46. Drawing shows welding machine with four axes of controlled movement.

Multi-operation equipment, like that employed to machine automobile engine blocks, are computer controlled. The machine shown in Fig. 16-47 has the rough castings loaded into one end. The castings are then transferred from station-to-station, through several machining operations. The system then unloads the finished blocks automatically at the other end.

Coordinates for machining auto body dies (shaped steel blocks that form sheet metal body sections) can be scanned or ''lifted'' from carefully made clay and wood models, Fig. 16-48. The coordinates are stored in computer memory and mathematically define body geometry. The information is then used to control the machine tools when the three-dimensional body dies are machined, Fig. 16-49.

ROBOTICS IN AUTOMATION

Automated systems have been adapted to do tasks other than to remove or machine metal. Programmable industrial robots have found considerable use in loading and unloading metal cutting machine tools and other industrial operations. See Fig. 16-50.

The Robot Institute of America (RIA) has adopted the following definition for industrial robots:

> ''A ***robot*** is a programmable, multifunctional manipulator designed to move material, parts, tools, or specialized devices through variable programmed motions for the performance of a variety of tasks.''

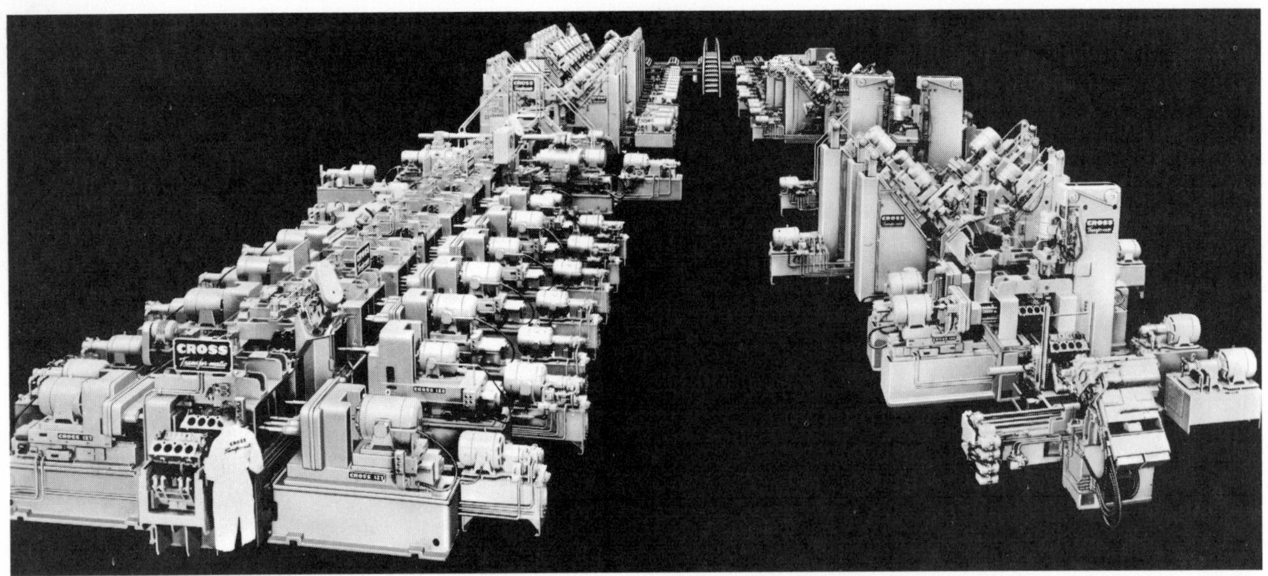

Fig. 16-47. This is a large machining complex employed to machine four- and eight-cylinder automobile engine blocks. At any one time, 104 blocks are having some machining or inspection operation performed on them. Machine is almost two city blocks long and performs 555 operations, including: 265 drilling, 6 milling, 56 reaming, 101 countersinking, 106 tapping, and 133 inspection operations. It produces 100 pieces an hour. (Cross Co.)

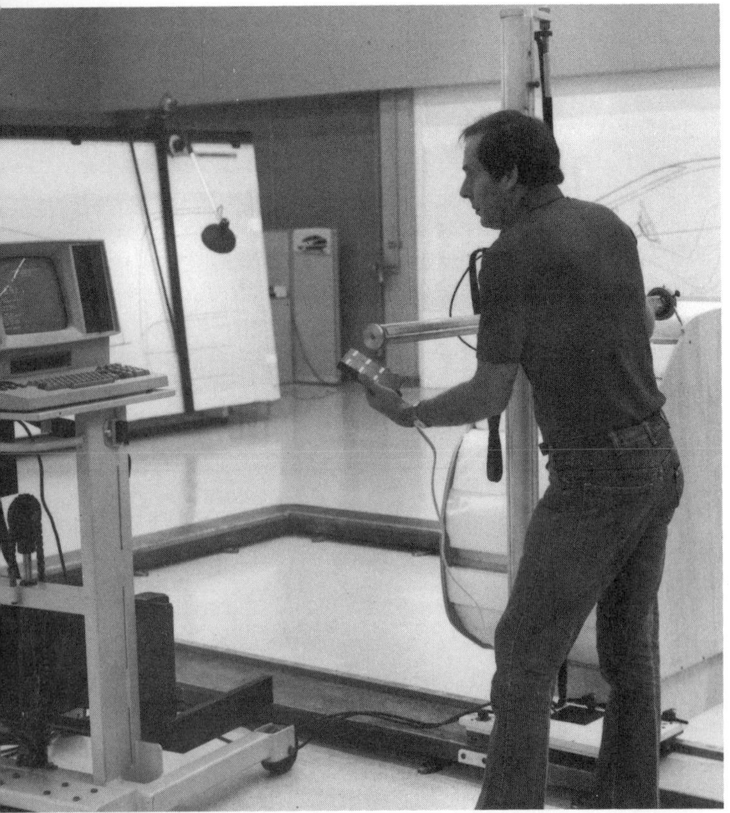

Fig. 16-48. Coordinates for machining auto body dies (shaped steel blocks used to form sheet metal body sections) can be scanned or ''lifted'' from carefully made clay and/or wood models. Coordinates mathematically define body geometry and can be used to control cutting tools when three-dimensional body dies are machined.
(Advanced Concept Center, General Motors Corp.)

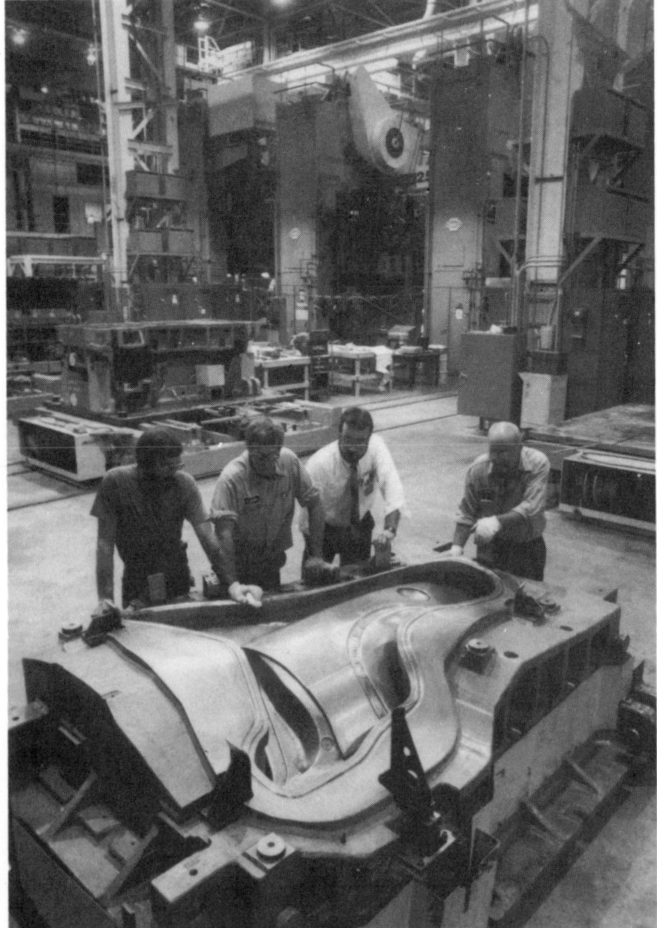

Fig. 16-49. Body dies machined using technique shown in Fig. 16-48. (Buick Div., General Motors Corp.)

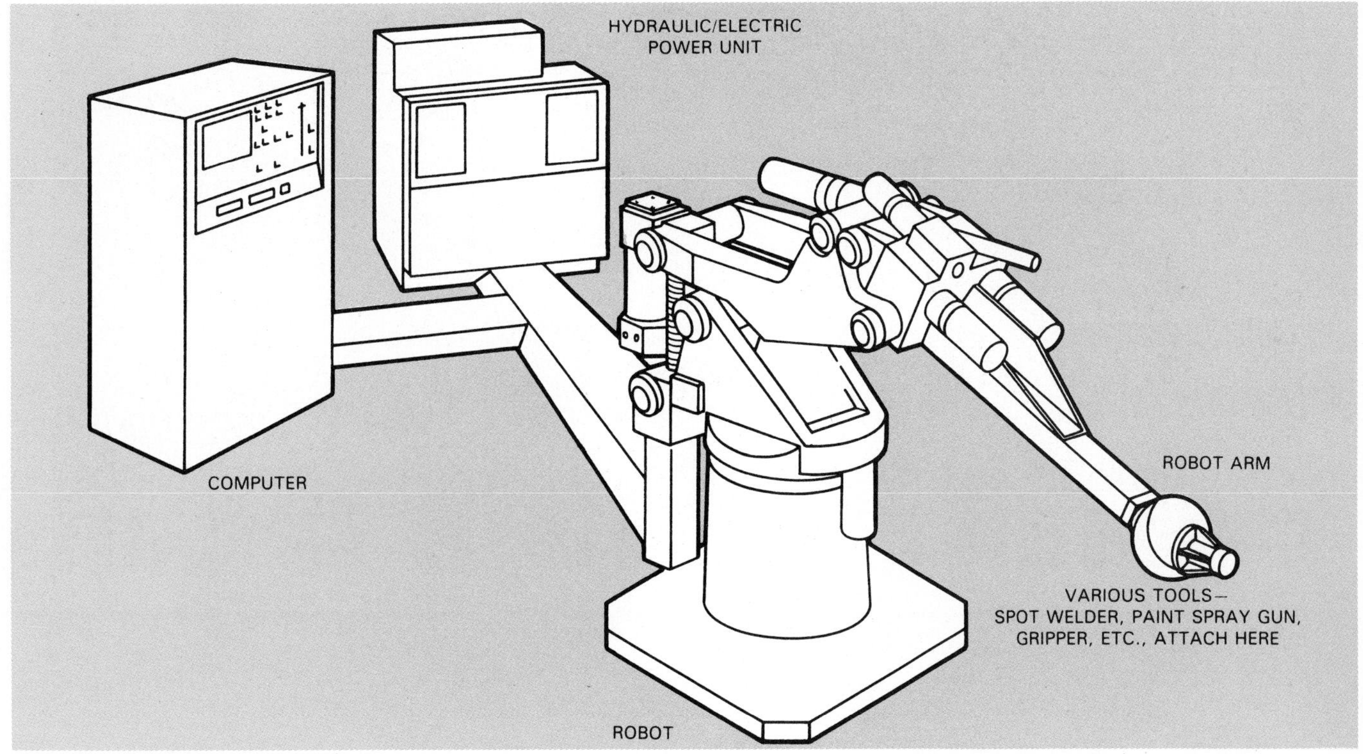

Fig. 16-50. Robots can be programmed to do many types of jobs—spot welding, paint spraying, drilling, deburring, inspecting, etc. Most of the jobs they perform are hazardous or very monotonous for human operators.

There are two implied requirements in the definition:

1. The robot must be a multiaxis machine that is capable of moving parts or tools to a specific location as well as positioning them at any attitude.
2. There must be a control system that can be programmed to drive the ***manipulator*** (device that grasps part or tool) through a series of specified motions and be capable of interacting with other machines.

Many types of robots have been developed, Fig. 16-51. Some are capable of selecting and positioning specifically shaped parts for machining or storage, Fig. 16-52. A laser is utilized to ''see'' and define the outline so the correct part will be selected.

THE FUTURE AUTOMATION

Only time will tell what impact the computer and new automation techniques will have on our society. Some workers, mostly unskilled and semiskilled, will lose their jobs, much like the home artisans did during the Industrial Revolution. The same thing happened to the carriage maker, blacksmith, buggy whip manufacturer, and feed dealer when Henry Ford started to mass produce the automobile, Fig. 16-53.

These developments were condemned then, just as automation and robots are condemned today. However, the machines eventually helped employ, directly or indirectly, many more people than the number they originally replaced. Better jobs, with higher pay and improved working conditions, were created. These changes also demanded that workers and technicians be better educated.

The key to the future, then, will be men and women who are well versed in industrial technology.

AUTOMATION SAFETY

1. Observe the safe operating procedures that are used with traditional machine tools and machining.
2. Wear approved eye protection and snug fitting apron or shop coat.
3. Be sure work holding devices are positioned correctly and securely fastened to the work table.
4. Know how to safely stop machining operations in case of an emergency.
5. Carefully check tool clearance to be sure the cutter will clear the work, work holding devices, clamps, etc., when manually positioning the work and during rapid traverse.

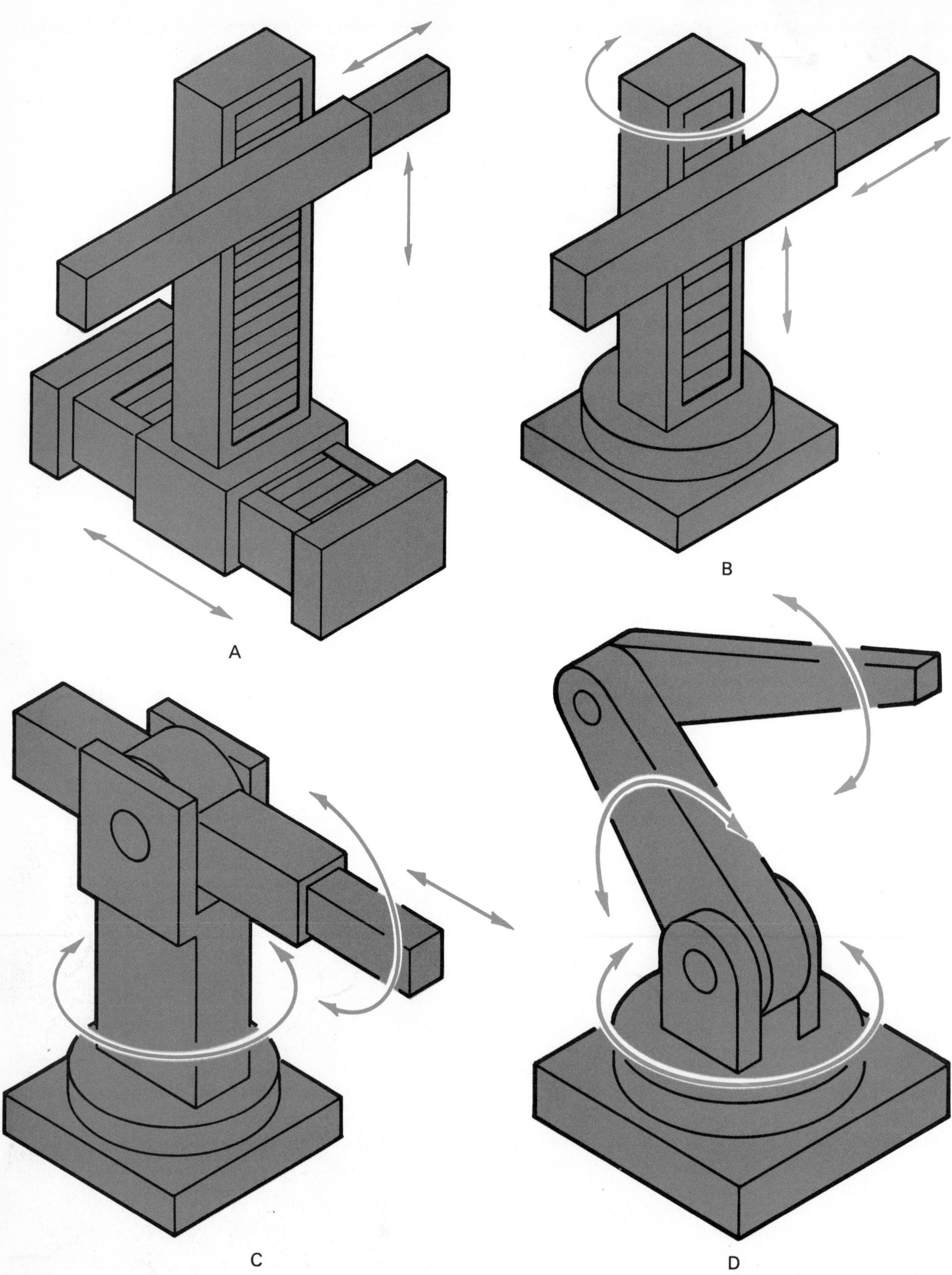

16-51. Basic geometric configurations of robots. All provide three articulations (specific arm movements). A—Cartesian coordinate. B—Cylindrical coordinates. C—Polar coorindates. D—Revolute coordinates.

Fig. 16-52. Some robots can be programmed to "see" with lasers. They define the shape of parts, pick them out of a group of parts, and properly orient them (set in a definite position) on machine tool for cutting operations. When machining operation is complete, robot will remove finished part and place it on materials handling system for next operation. (MTS Systems Corp.)

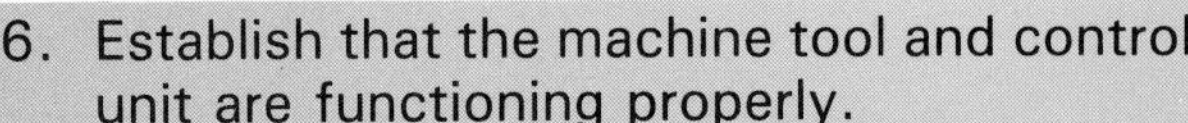

6. Establish that the machine tool and control unit are functioning properly.
7. Initiate a "dry run" for a safety check of tool positioning for each machining operation.
8. Continually monitor machining operation to be sure each tool is cutting properly.
9. Remove burrs and sharp edges before inspecting finished parts for accuracy, finish of machined surfaces, etc.

TEST YOUR KNOWLEDGE—Chapter 16

Please do not write in the text. Place your answers on a separate sheet of paper.

1. What is automation?
2. How are the machines in automation activated?
3. List the five (5) basic manufacturing processes that automated machines can perform.
4. What do the following acronyms stand for in their relation to automation? (An acronym is a word formed using the initial letters of words in a phrase. For example: RPM stands for revolutions per minute.)
   a. FMS.
   b. CIMS.
   c. CIAM.
   d. NC.
   e. CNC.
   f. CAD/CAM.

Fig. 16-53. Then and now in the manufacture of motor cars. When Henry Ford opened the first production line, he paid the workers $5 per day, a very good salary for the time. (Ford Motor Company)

5. Describe the differences between manual machining techniques and NC methods.
6. Prepare a sketch showing the Cartesian coordinate system.
7. Prepare two similar sketches. Show incremental dimensioning on the first sketch and absolute dimensioning on the second sketch.
8. What is an NC program?
9. List the two basic NC systems.
10. Using sketches, show how the two NC systems differ.
11. Which of the two NC methods require the use of a computer? Why is a computer required?
12. What do the following terms mean?
    a. MCU.
    b. Alphanumeric data.
    c. Program sheet.
13. How is an NC program often proofed?

RESEARCH AND DEVELOPMENT

1. The advent of automation is frequently thought of as the "Second Industrial Revolution." Develop a research paper on the FIRST Industrial Revolution with emphasis on working conditions, how people lived, wages, etc. Compare them with the same conditions that exist today.
2. Design and construct a simple machine that will illustrate how NC works.
3. Review up-to-date technical magazines that have articles on NC, CNC, robotics, etc. Prepare a brief outline of each article for class discussion.
4. Arrange to visit a plant that utilizes automated equipment. If such a visit is NOT possible, show a video tape or motion picture that illustrates automation.
5. If your shop/lab is fortunate enough to have an NC or CNC machine tool, ask your instructor to assign a programming problem. Prepare the program, edit and proof it, and follow through to the finished machined part.
6. Design a bulletin board on automation.

Chapter 17

FASTENERS

After studying this chapter, you will be able to:

- Identify several types of fasteners.
- Explain why inch-based fasteners are not interchangeable with metric-based fasteners.
- Describe how some fasteners are used.
- Select the proper fastening technique for a specific job.
- Describe chemical fastening techniques.

A ***fastener*** is any device used to hold two objects or parts together. This would include bolts, nuts, screws, pins, keys, rivets, and even chemical bonding agents or adhesives. This chapter will explain and illustrate common fasteners.

It is critical to choose the proper fasteners for each job, Fig. 17-1. A poorly selected fastener can greatly reduce the safety and dependability originally designed into a product. Improper fasteners could increase assembly costs, and result in an inferior or faulty product. To improve quality, several fastening techniques are frequently employed in the same assembly. In fact, one auto manufacturer uses more than 11,000 kinds and sizes of fasteners.

THREADED FASTENERS

Threaded fasteners utilize the wedging action of the screw thread to clamp parts together. To achieve maximum strength, a threaded fastener should screw into its mating part at least a distance equal to one and one-half times the thread diameter. See Fig. 17-2.

Threaded fasteners vary in cost from thousands of dollars for special bolts that attach the wings to the fuselage of large aircraft, to a fraction of a penny for small machine screws. See Fig. 17-3.

Most threaded fasteners are available in metric sizes. Since many American manufacturers have started to use metric fasteners in their products, some problems have arisen. Metric threads, while having the same ***basic profile*** (shape) as unified threads, are NOT interchangeable, Fig. 17-4.

Until a full changeover to metric sizes is made, and products already made with unified threads

Fig. 17-1. Note huge Rolls-Royce turbofan engine on test stand. Many types of fasteners had to be designed or selected for this engine. Some operate at high temperatures, others at sub-zero temperatures. A poorly selected fastener can greatly reduce safety factor designed into engine.

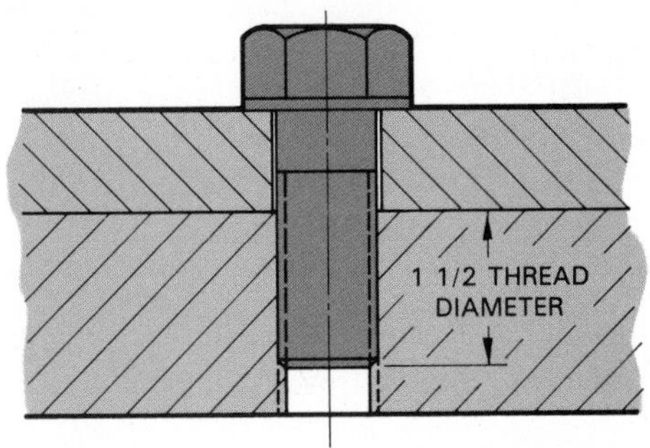

Fig. 17-2. For maximum strength, a threaded fastener must screw into mating part a distance equal to 1 1/2 times diameter of thread.

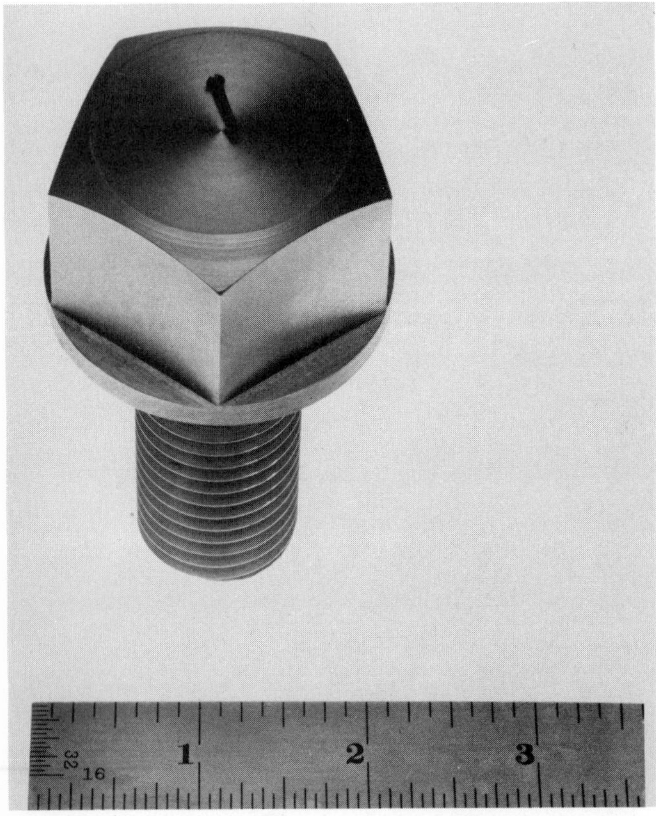

Fig. 17-3. There is a wide range of threaded fasteners available to industry. Do you notice anything different about the head of the large threaded fastener?

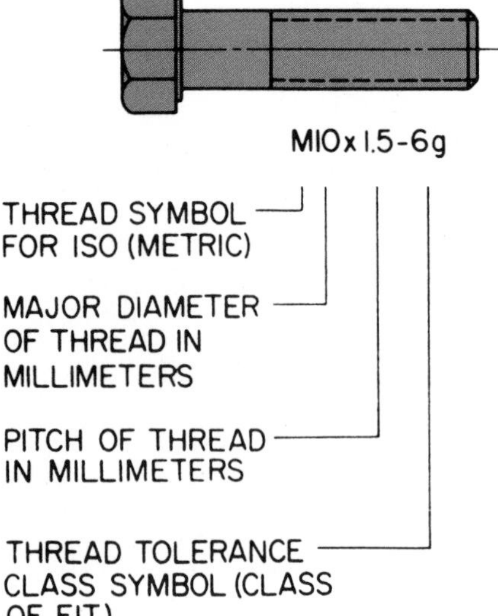

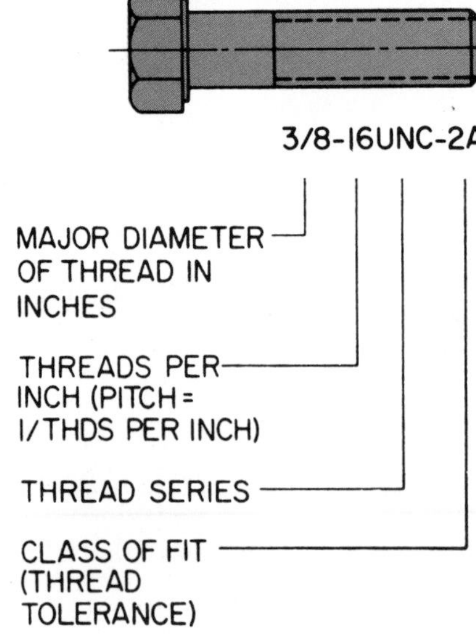

Fig. 17-4. Metric threads have same basic profile (shape) as unified thread series; however, unified and metric threads are NOT interchangeable.

wear out or are discarded, some method will have to be devised to easily identify metric threaded fasteners from inch-based fasteners. While no foolproof method has yet been contrived, Figs. 17-5 and 17-6 illustrate two possible solutions.

Machine screws

Machine screws are widely used in general assembly work. They have slotted or recessed heads. They are also made in a number of head styles, Fig. 17-7.

Machine screws are available in body diameters from #0000 (0.021 in.) to 3/4 in. (0.750 in.) and in lengths from 1/8 (0.125) to 3 in. Metric sizes are also manufactured. Nuts, in either square or hexagonal types, are purchased separately.

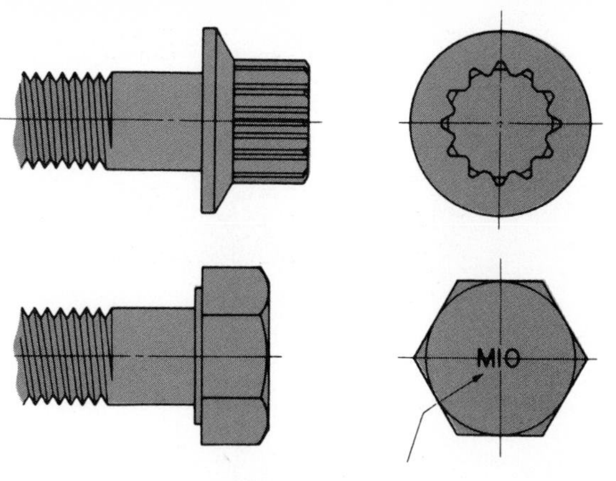

Fig. 17-5. Metric fasteners are manufactured in same variety of head shapes. However, there is a problem in finding an easy way to identify metric-threaded fasteners from inch-threaded fasteners. Some larger size hex headed metric fasteners have size stamped on head. A twelve-spline flange head is under consideration for eight sizes of metric threads: 5, 6.3, 8, 10, 13, 14, 16, and 20 mm.

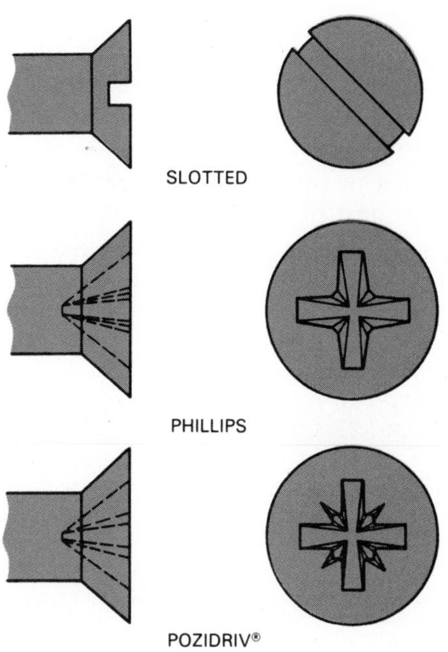

Fig. 17-6. The Pozidriv® cross-recessed head has been suggested as a means of identifying metric screws. It would be used in place of Phillips cross-recessed head, which would only be used with inch-based screws.

Machine bolts

Machine bolts are employed to assemble parts that do NOT require close tolerances, Fig. 17-8. They are manufactured with square and hexagonal heads, in a range of diameters from 1/2 in. to 30

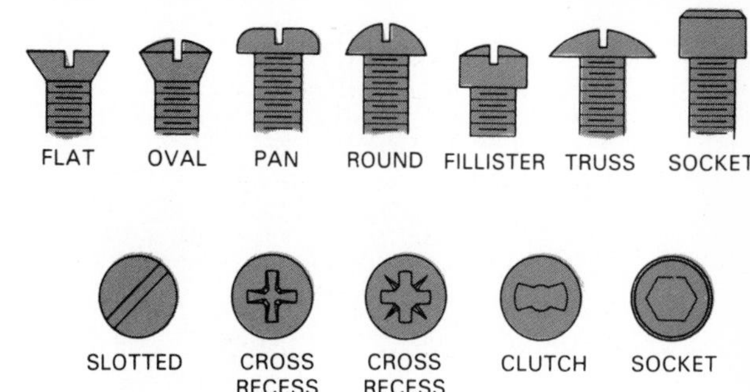

Fig. 17-7. Study a sampling of many types of machine screws available.

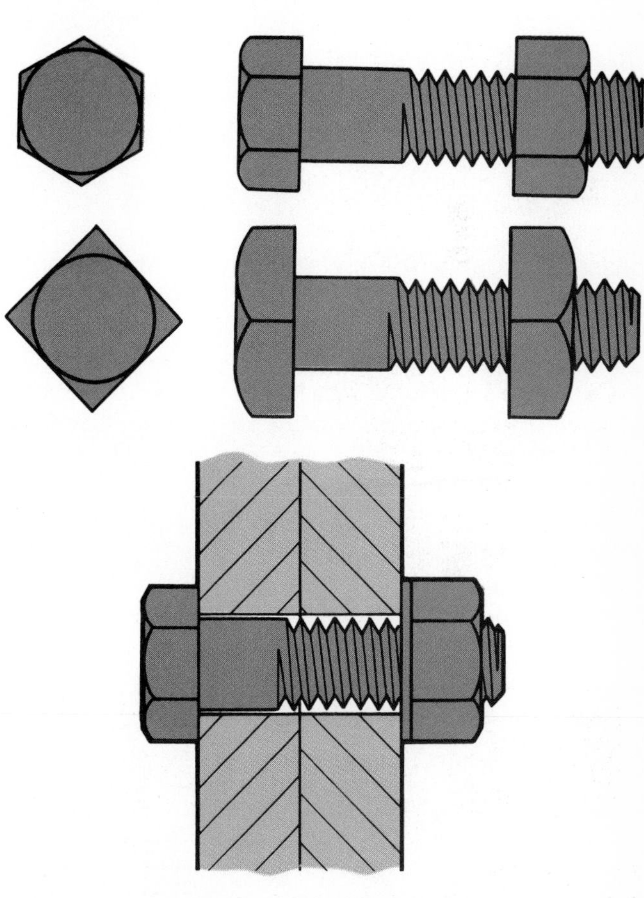

Fig. 17-8. Machine bolts use nut to produce clamping force.

in. The nuts are similar in shape to the bolt head. They are usually furnished with the machine bolts. Tightening the nut produces the clamping action.

Cap screws

Cap screws are found in assemblies requiring a higher quality and a more finished appearance, Fig. 17-9. Instead of tightening a nut to develop clamping action as with the machine bolt, the cap screw

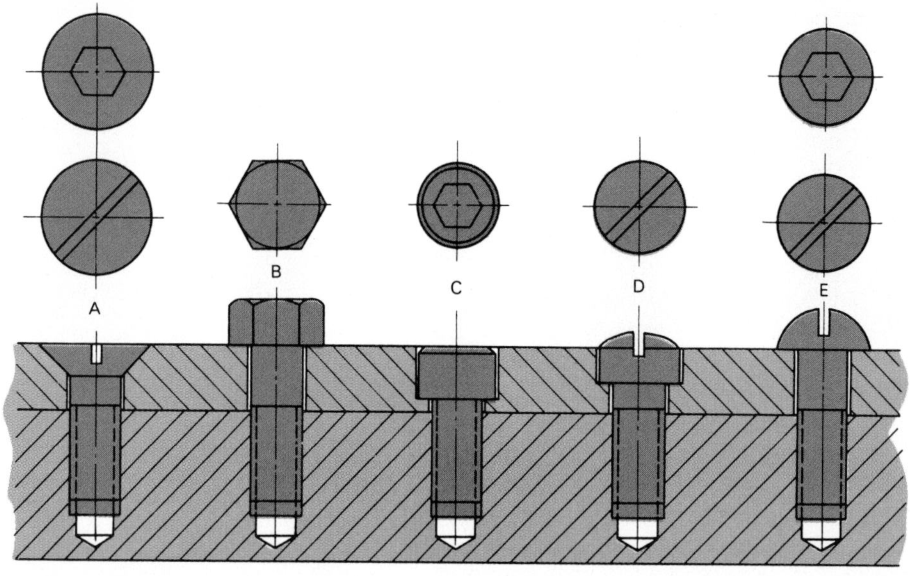

Fig. 17-9. Many types of cap screws are manufactured. A—Flat head. B—Hexagonal head. C—Socket head. D—Fillister head. E—Button or round head.

passes through a clearance hole in one of the pieces and screws into a threaded hole in the other part. Clamping action is accomplished by tightening the bolt into the threaded part.

Cap screws are held to much closer tolerances in their manufacture. They are provided with a machined or semifinished bearing surface under the head. Some types of cap screws are heat treated.

Cap screws are stocked in coarse and fine thread series and in diameters from 1/4 to 2 in. Lengths from 3/8 to 10 in. are available. Metric sizes can also be supplied.

Setscrews

Setscrews are semipermanent fasteners that prevent pulleys from slipping on shafts, hold collars in place on assemblies, and position shafts on assemblies. They are usually made of heat treated steels.

Setscrews are classified in two ways, by their head style, and by their point style, Fig. 17-11.

The ***thumbscrew*** can be turned by hand. It is sometimes employed in place of a setscrew in assemblies that require rapid or frequent dissembly, Fig. 17-12. They are available with points similar to those on setscrews.

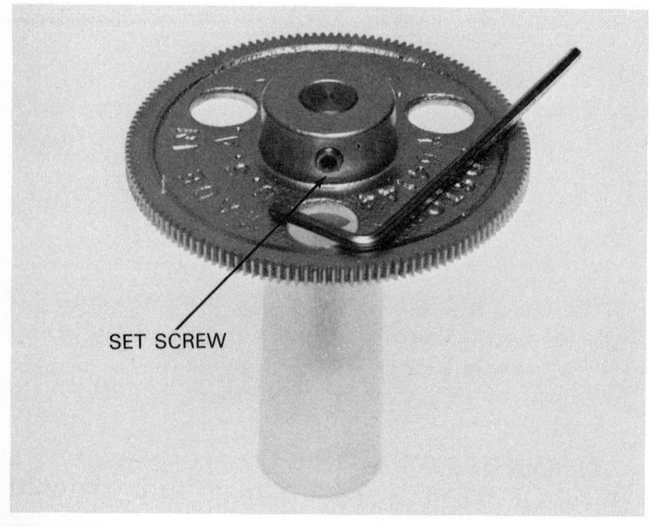

Fig. 17-10. Note this typical application of a setscrew.

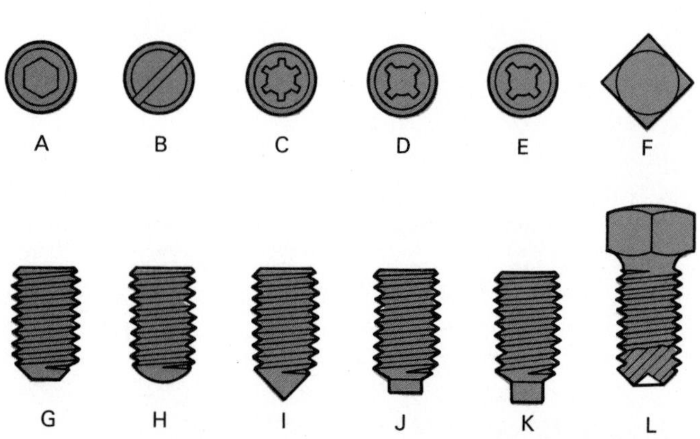

Fig. 17-11. Study setscrew head and point designs. A—Socket head. B—Slotted head. C, D, and E—Fluted head. F—Square head. G—Flat point. H—Oval point. I—Cone point. J—Half dog point. K—Full dog point. L—Cut point.

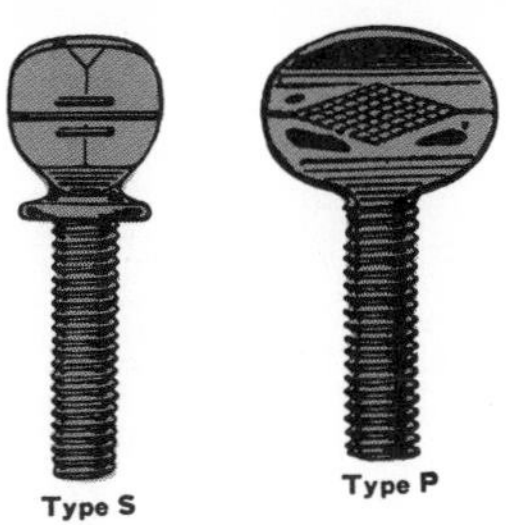

Fig. 17-12. Thumb screws can be removed or installed by hand. (Parker-Kalon)

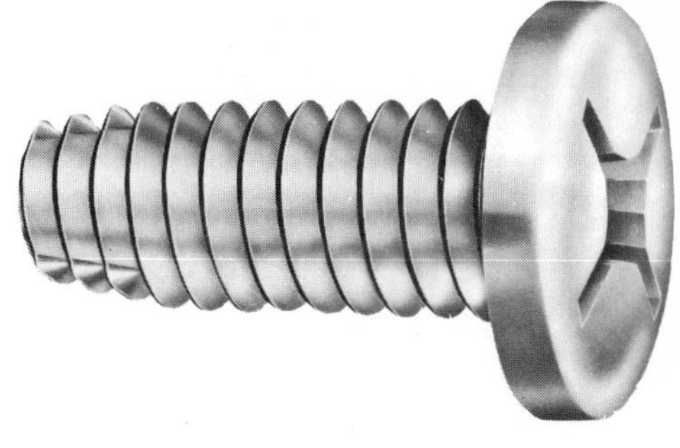

Fig. 17-14. Note thread forming screw.

Stud bolts

Stud bolts are headless bolts that are threaded the entire length, or, more commonly, on both ends, Fig. 17-13. One end is designed for semipermanent installation in a tapped hole while the other end is threaded for standard nut assembly to clamp the pieces together.

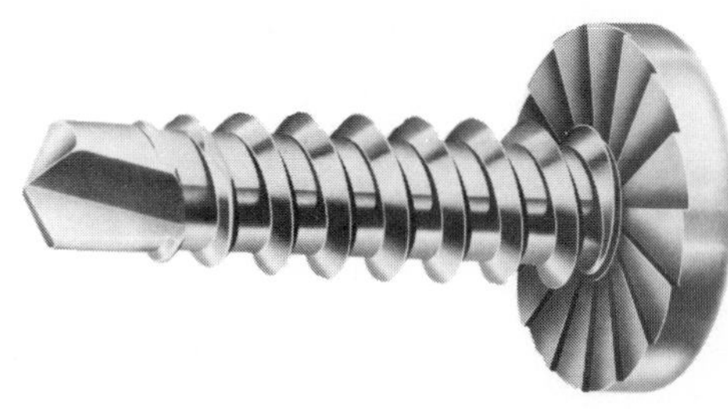

Fig. 17-15. This is a self-drilling screw. It is also known as TEKS®. (USM Corp., Fastener Group)

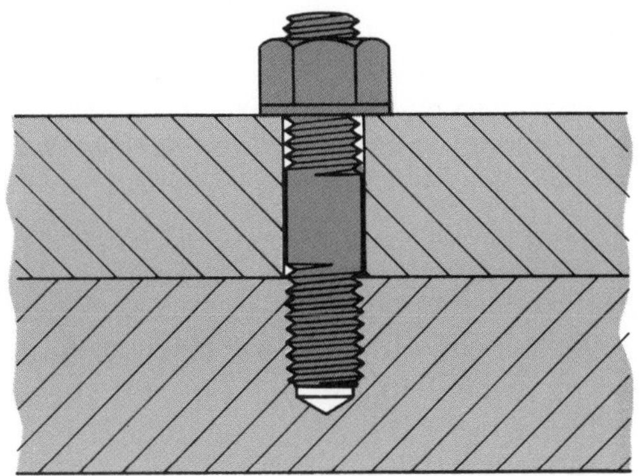

Fig. 17-13. One end of stud bolt usually threads into part; other end accepts a nut.

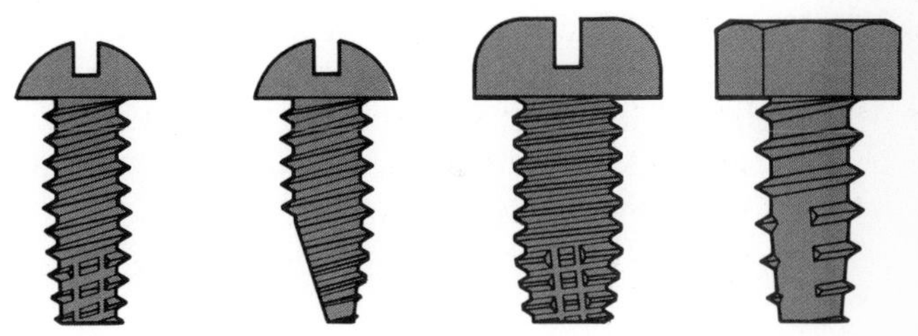

Fig. 17-16. Study variations of thread cutting screws.

Thread forming screws

Thread forming screws produce a thread in the part as they are driven, Fig. 17-14. This feature eliminates a costly tapping operation. A variation of the thread forming screw eliminates expensive drilling or punching, and aligning operations because the screw drills its own hole as it is driven into place. See Fig. 17-15.

Thread cutting screws

Thread cutting screws differ from thread forming screws because they actually cut threads into the material when driven. Refer to Fig. 17-16. Thread cutting screws are hardened and are employed to join heavy gage sheet metal, and to thread into nonferrous metal assemblies.

Drive screws

Drive screws are simply hammered into a drilled or punched hole of the proper size. A permanent assembly results, Fig. 17-17.

Nuts

Nuts, for most threaded fasteners, have external hexagonal or square shapes, and are used with bolts having the same shaped head. They are available in various degrees of finish.

A ***regular nut*** is ***unfinished*** or NOT MACHINED, except on the threads, Fig. 17-18.

A ***regular semifinished nut*** is MACHINED on the bearing face to provide a truer surface for the washer, Fig. 17-19.

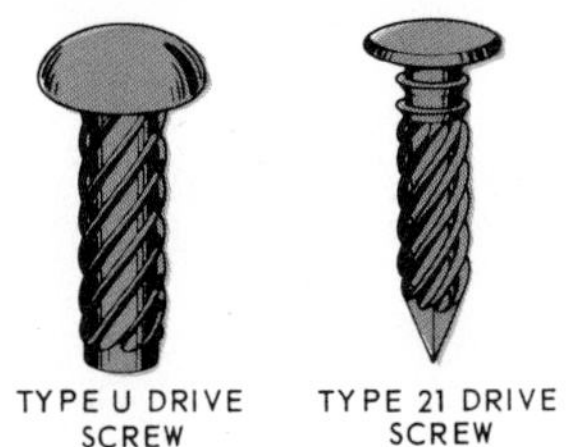

Fig. 17-17. Drive screws are hammered or forced into place in presized hole. (Parker-Kalon)

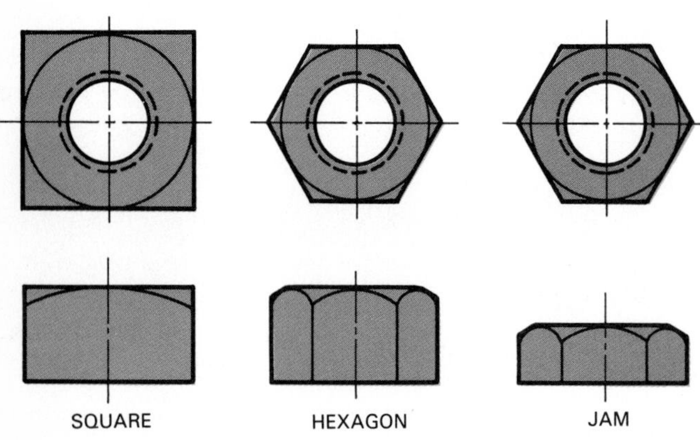

Fig. 17-18. Regular nuts are not machined, except for threads.

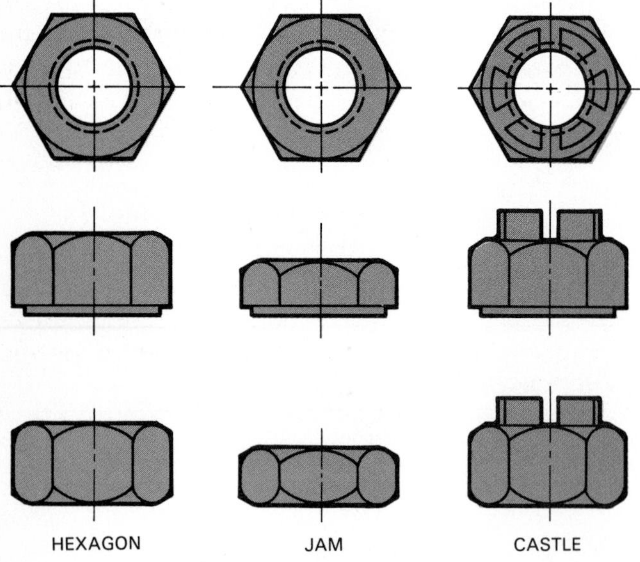

Fig. 17-19. Regular semifinished nuts have a machined bearing surface.

A ***heavy semifinished nut*** is identical in finish to the regular semifinished nut. However, the body is thicker for additional strength, Fig. 17-20.

The ***jam nut*** is thinner than the standard cut. It is frequently used to lock a full size nut in place.

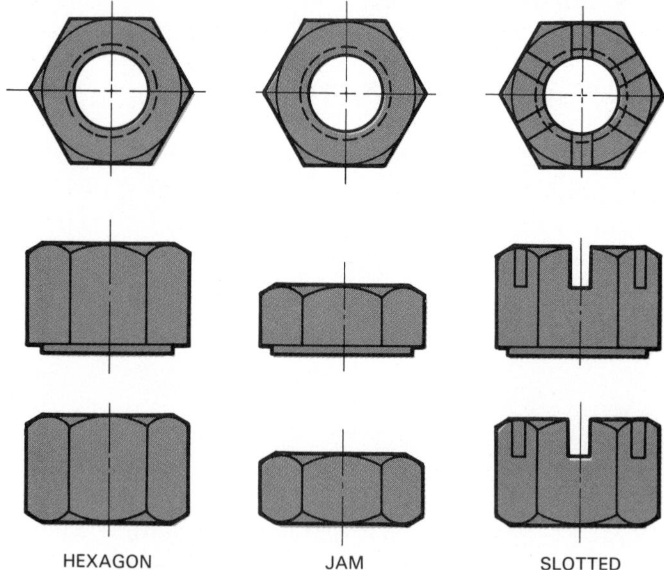

Fig. 17-20. Heavy semifinished nuts are thicker than regular nuts.

Castellated and ***slotted nuts*** have slots across the flats so they can be locked in place with a cotter pin or safety wire. The cotter pin or wire is inserted through the slot and a hole drilled in the bolt or stud to prevent the nut from turning loose.

These nut types are being replaced on many applications by self-locking nuts, Fig. 17-21. ***Self-locking nuts*** have nylon inserts, or are slightly deformed to produce a friction fit, so they cannot vibrate loose. No hole is required when self-locking nuts are employed in an assembly.

In critical assemblies, use a NEW self-locking nut to replace a used self-locking nut that has been removed for any reason. The used nut may not have adequate locking action remaining and may loosen in service.

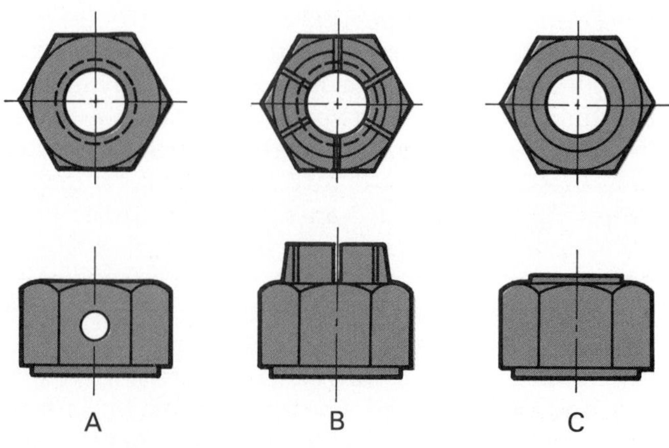

Fig. 17-21. These are self-locking nuts.

Acorn nuts are used when appearance is of primary importance, or where projecting threads must be protected. They are available in high or low crown styles. See Fig. 17-22.

The ***wing nut*** is found where frequent adjustment or frequent removal is necessary. It can be loosened and tightened rapidly without the need of a wrench. Refer to Fig. 17-23.

Nuts are usually manufactured of the same materials as their mating bolts.

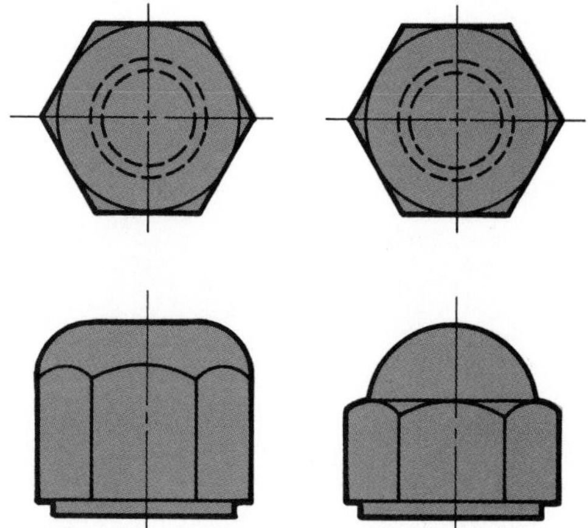

Fig. 17-22. Acorn or cap nuts look good and will protect threads.

Fig. 17-23. Wing nut looks as if it has ''wings.'' Like thumbscrew, it is turned by hand. (Parker-Kalon)

Inserts

Inserts are a special form of nut or internal thread. They are designed to provide higher strength threads in soft metals and plastics. The type shown in Fig. 17-24 is frequently used to replace damaged or stripped threads. The threaded hole is drilled and tapped. The insert is then screwed into the hole. Its internal thread is standard size and form. For optimum results, inserts must be installed according to the manufacturer's instructions.

Washers

Washers provide an increased bearing surface for bolt heads and nuts, distributing the load over a larger area. They also prevent surface marring.

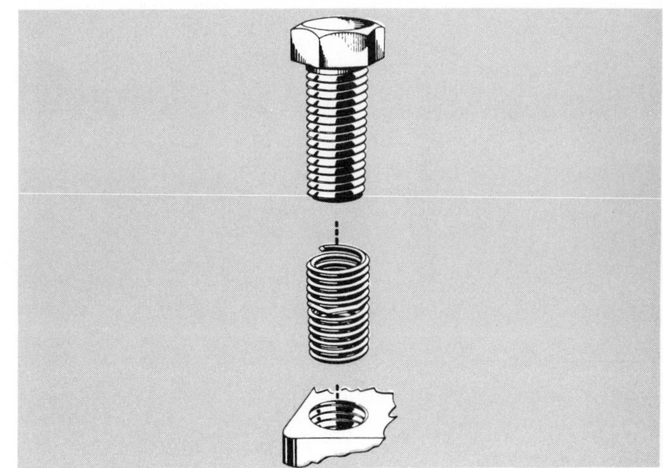

Fig. 17-24. Insert for threaded hole is frequently used to replace damaged or stripped threads in a part. (Heli-Coil Corp.)

The ***standard washer*** is produced in light, medium, heavy-duty, and extra heavy-duty series. See Fig. 17-25.

Lock washers

The application of a ***lock washer*** will prevent a bolt or nut from loosening under vibration.

The ***split-ring lock wahser*** is rapidly being replaced by the ***tooth-type lock washer*** which has greater holding power on most applications, Fig. 17-26.

Preassembled lock washer and screw nuts and ***lock washer and nut units*** have a washer mounted on the nut. They are employed to lower assembly time and reduce waste in the mass-assembly market, Fig. 17-27.

NONTHREADED FASTENING DEVICES

Nonthreaded fasteners comprise a large group of holding devices. These include dowel pins, cotter pins, retaining rings, rivets, keys, and adhesives. Each has its advantages.

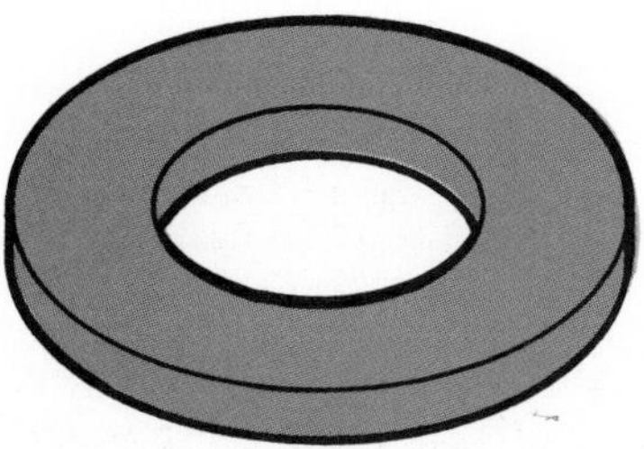

Fig. 17-25. Standard flat washer provided bearing surface for fastener.

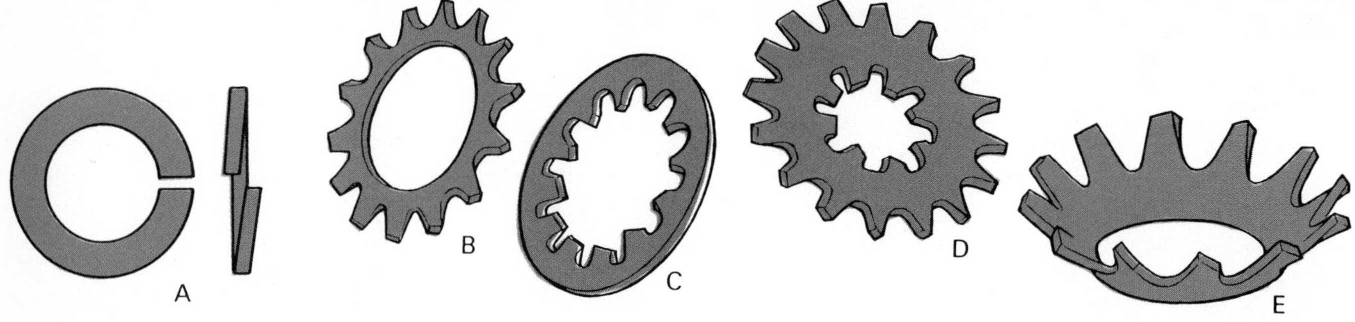

Fig. 17-26. Study lock washer variations. A—Split-ring type. B—External type, this type should be used whenever possible because it provides greatest resistance to turning. C—Internal type, it is used with small head screws and to hide teeth either for appearance or to prevent snagging. D—Internal-external type, it is employed when mounting holes are oversize. E—Countersunk type, it is for applications with flat or oval head screws.

Fig. 17-27. Lock washer and screw units and lock washer and nut units are frequently used to simplify assembly. (Shakeproof Div., Illinois Tool Works, Inc.)

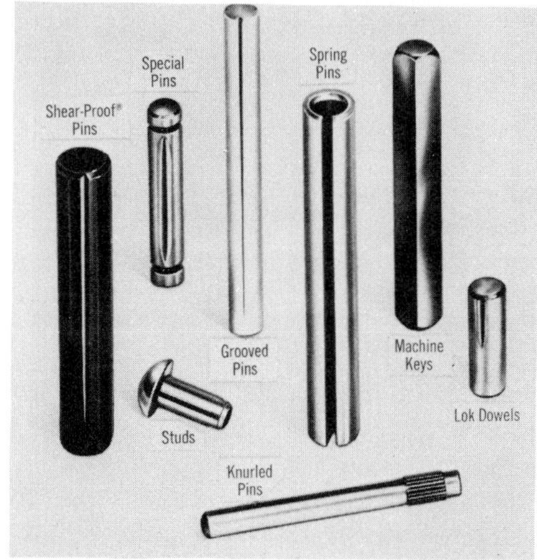

Fig. 17-28. Note types of dowel pins. They are made in a wide range of sizes. (Driv-lok Inc.)

Dowel pins

Dowel pins are made of heat treated alloy steel and are found in assemblies where parts must be accurately positioned and held in absolute relation to one another See Figs. 17-18 and 17-29. They assure perfect alignment and facilitate quicker disassembly and reassembly of parts in exact relationship to other parts. They are fitted into reamed holes and are available in diameters from 1/16 to 1 in. They are also available in metric sizes.

Dowel pins are normally 0.0002 in. (0.005 mm) oversize (plain steel finish) but are available in 0.001 in. (0.025 mm) oversize (black finish) for repairs.

Taper pins are made with a uniform taper of 1/4 in. per foot in lengths up to 6 in., with diameters as small as 5/32 in. at the large diameter.

Cotter pins

The ***cotter pin*** is fitted into a hole drilled crosswise in a shaft, Fig. 17-30. It prevents parts from slip-

Fig. 17-29. This knob is attached to its shaft with a taper pin.

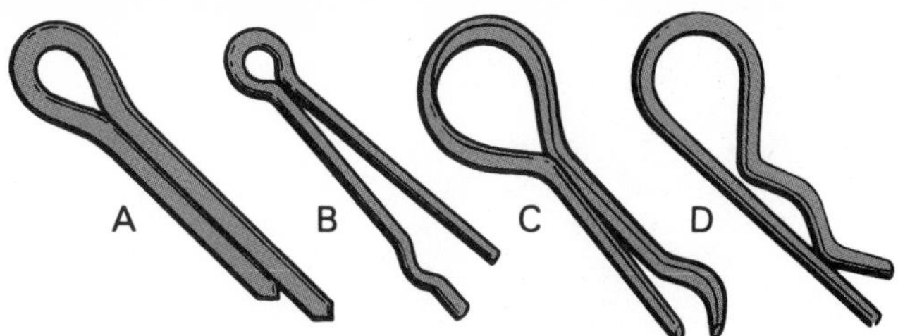

Fig. 17-30. Study cotter pin types. A—Standard. B—Humped. C—Clinch. D—Hitch.

ping or turning off. Other types of retaining devices are replacing the cotter pin.

Retaining rings

The ***retaining ring*** has been developed for both internal and external applications, Fig. 17-31. While most retaining rings must be seated in grooves, a self-locking type does NOT require this special recess. See Fig. 17-32.

Special pliers are needed for rapid assembly and disassembly of the retaining rings, Fig. 17-33.

Retaining rings reduce both cost and weight of the product on which they are employed.

Rivets

Permanent assemblies can be made with ***rivets,*** Fig. 17-34. Solid rivets can be ***set*** (deformed larger on one end) by hand or machine.

Blind rivets are mechanical fasteners that have been developed for applications where the joint is NOT accessible from one side. They require special tools for installation, Fig. 17-35. Blind rivet types are shown in Fig. 17-36.

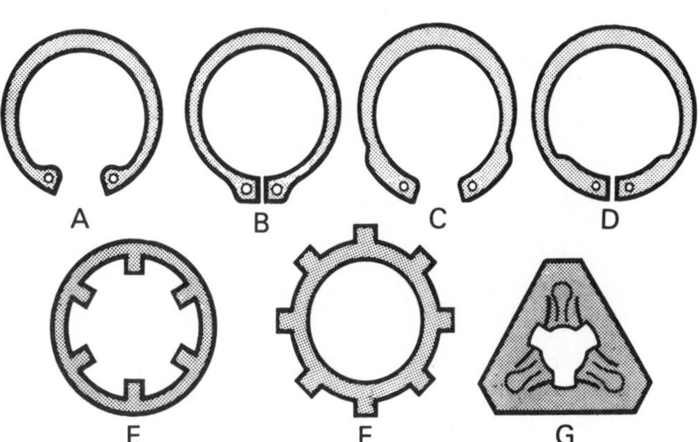

Fig. 17-31. Study retaining ring types. A—Basic internal ring. B—Basic external ring. C—Inverted internal ring. D—Inverted external ring. E—External self-locking ring. F—Internal self-locking ring. G—Triangular self-locking ring.

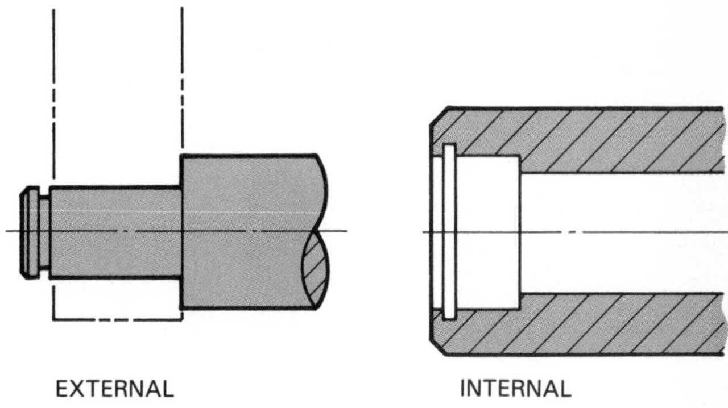

Fig. 17-32. Grooves are machined to receive retaining rings. They eliminate many other expensive machining operations.

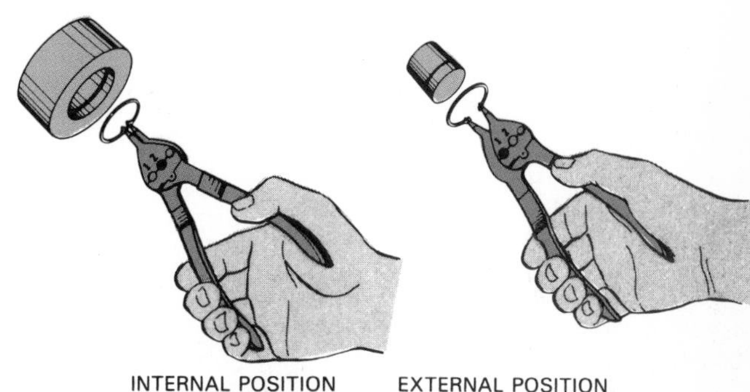

Fig. 17-33. Special pliers are used to install Truarc® retaining rings. (Waldes Kohinoor Inc.)

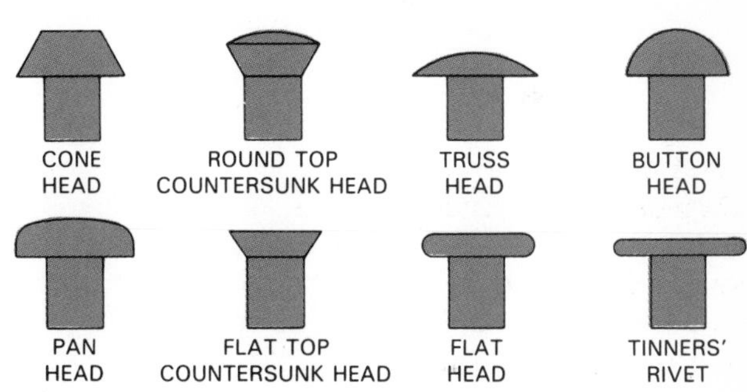

Fig. 17-34. Study rivet head styles.

Keys

A ***key*** is a small section of metal that prevents a gear or pulley from rotating on its shaft. One-half of the key fits into a ***keyseat*** on the shaft while the other half of the key fits into a ***keyway*** in the hub of the gear or pulley, Fig. 17-37. Fig. 17-38 shows commonly used keys.

A ***square key*** is usually one-fourth the shaft diameter. It may be slightly tapered on the top to make it easier to install.

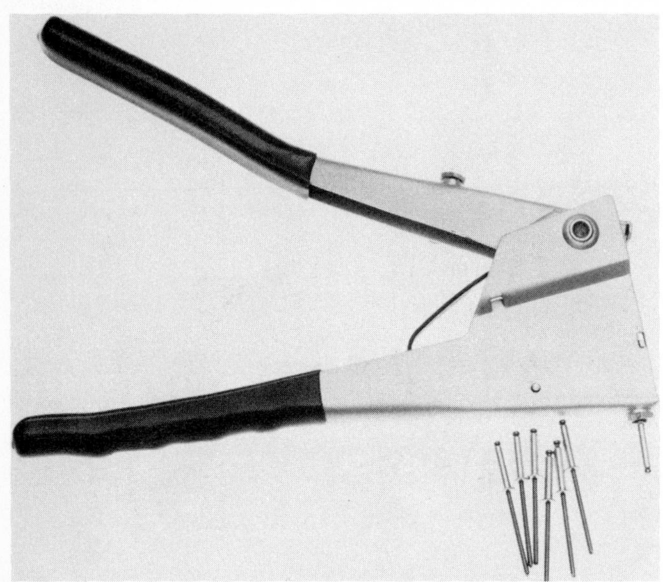

Fig. 17-35. Rivet gun or pliers are used to insert one type of blind rivet.

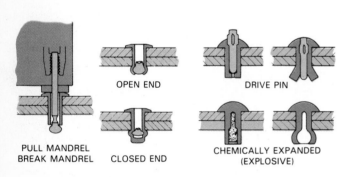

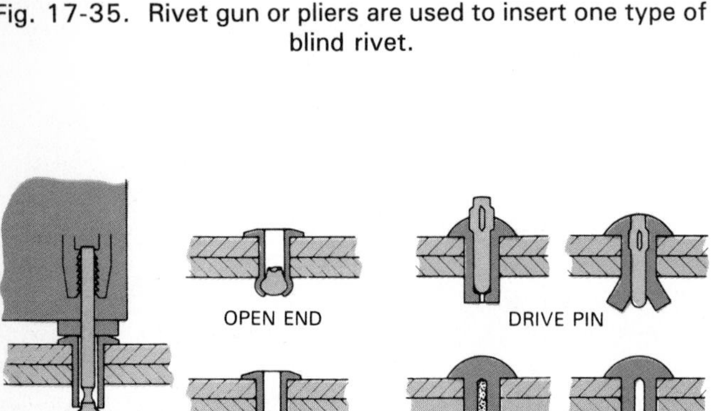

Fig. 17-36. Note types of blind rivets.

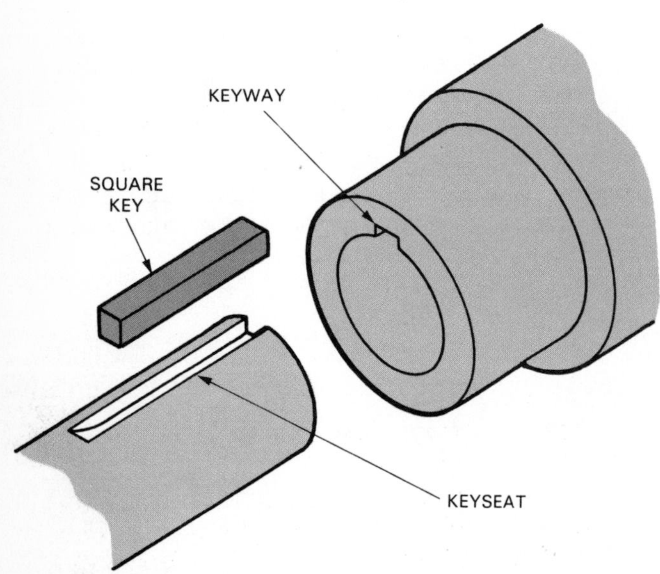

Fig. 17-37. This square key is employed to prevent pulley or gear from rotating on shaft.

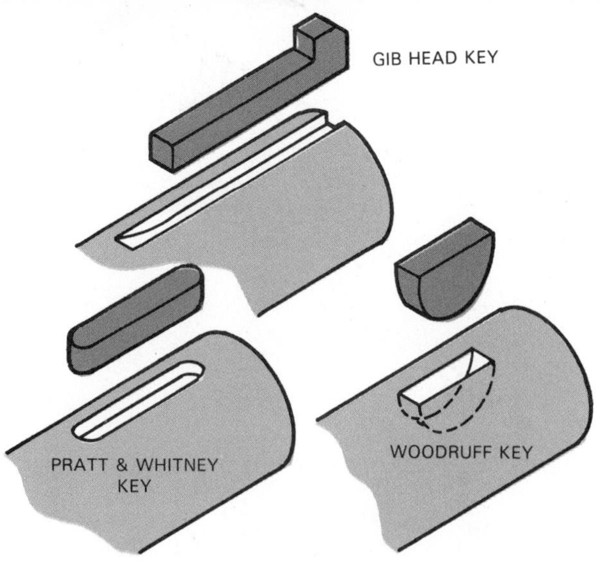

Fig. 17-38. Compare three types of keys.

The Pratt And Whitney Key is similar to the square key, except it is rounded at both ends. It fits into a keyseat of the same shape.

The ***gib head key*** is interchangeable with the square key. The head design permits easy removal from the assembly.

A ***Woodruff key*** is semicircular and fits into a keyseat of the same shape. The top of the key fits into the keyway of the mating part.

ADHESIVES

Adhesives provide one of the newer ways to join metals and to keep threaded fasteners from vibrating loose. In some applications, the resulting joints are stronger than the metal itself. Adhesive bonded joints do NOT require costly and time consuming operations such as drilling, countersinking, riveting, etc.

The major drawback of adhesives is heat. While some adhesives retain their strength at temperatures as high as 700 °F (371 °C), most of them should NOT be used above temperatures of 150-200 °F (66-93 °C).

Adhesives for locking threaded fasteners in place are made in a number of chemical formulations. The desired permanence of the threaded joint will determine the type of adhesive to be employed.

Adhesive bonded assemblies offer many advantages over conventional fastening techniques.

1. The load is distributed evenly over the entire joined area, Fig. 17-39.
2. There is continuous contact between the mating surfaces, Fig. 17-40.
3. Full strength of the mating parts is maintained. Holes for the insertion of fasteners are

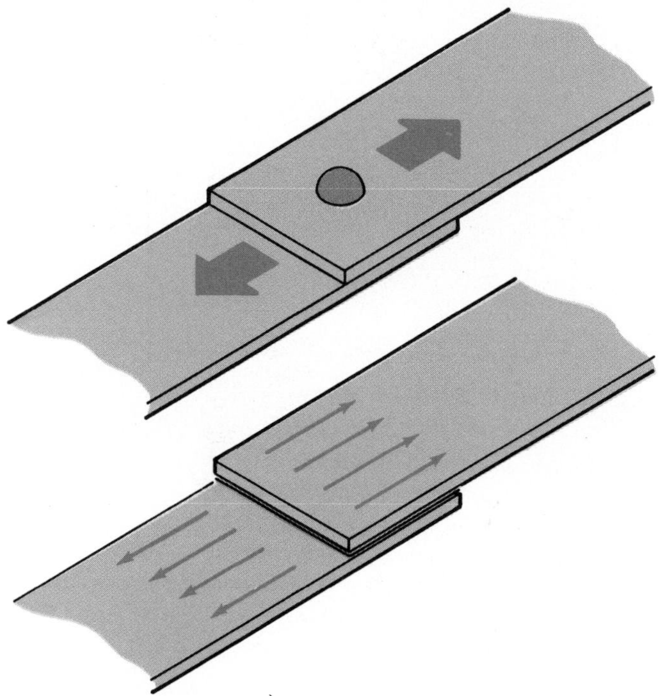

Fig. 17-39. When an adhesive is used to join metal, load is distributed evenly over entire joint. Rivet or conventional threaded fastener localizes load in a small area.

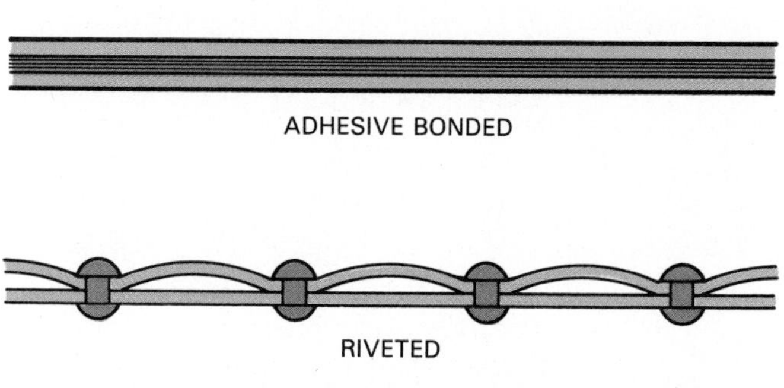

Fig. 17-40. On surfaces joined with an adhesive, mating surfaces are in continuous contact.

eliminated. Extreme heat such as required in welding is not needed so there is no danger of the work becoming distorted or having its head treatment affected.

4. Smooth surfaces result because there are no external projections as with rivets and bolts, nor is the surface marred due to the heat and pressure used when the sections are spot welded.

Many commercial adhesives are sold in small quantities. They are suitable for use in training areas and the home, Fig. 17-41.

Adhesives are available in liquid, paste, or solid form. Many can be applied directly from the container. Others must be mixed with a catalyst or

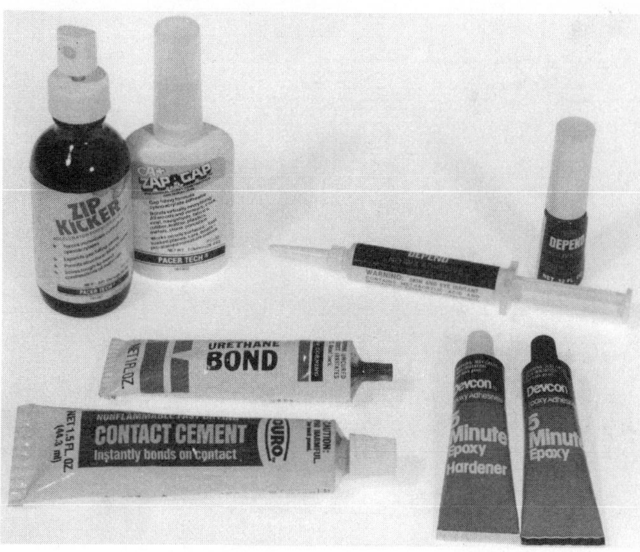

Fig. 17-41. Adhesives for joining metal to metal and metal to other materials are available in good hardware stores. They are similar to those found in industry.

hardener. A few pressure sensitive adhesives are manufactured in sheet form.

Using adhesives

Most adhesives require the following five steps to produce solid bonded joints:

1. ***Surface preparation*** is critical! All adhesives require CLEAN surfaces to produce full strength bonds. Preparation may range from simply wiping the surfaces with a solvent to multistage cleaning and chemical treatment.
2. ***Adhesive preparation*** must be done properly. Mixing, delivery to work area, setting up equipment, etc. must all meet recommendations.
3. ***Adhesive application*** can be done with a brush, roller, spray, dipping, etc., Fig. 17-42. Follow the instructions for the type adhesive.
4. ***Assembly*** involves positioning of the materials to be joined, using jigs to align materials, etc.
5. ***Bond development*** is the curing of adhesives, evaporation of solvents, application of pressure and/or heat, etc., Fig. 17-43.

FASTENER SAFETY

1. Wear approved eye protection when making openings, drilling, punching, countersinking, etc., for fasteners.
2. Do NOT use your hands to remove metal chips from holes for fasteners! Use a brush. Burrs are usually raised around the opening and can cause nasty cuts.
3. If compressed air is employed to clean drilled or tapped holes, wear approved eye

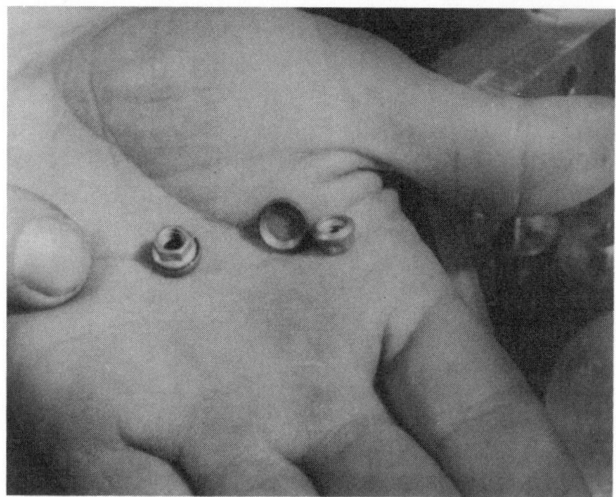

Fig. 17-42. Self-aligning nuts used in aircraft are assembled with a cyanoacrylate adhesive. Two part fasteners are assembled automatically and drop onto an indexing table. A needle-tip applicator (left photo) applies a precise amount of adhesive; then parts are brought together. Setup produes 3600 assemblies an hour. Parts are shown before and after assembly (right photo). (Loctite Corp.)

Fig. 17-43. Autoclave is for bonding large assemblies. Steam applies heat and pressure simultaneously. Assembly is wrapped in rubber blanket to protect bond from moisture. (3-M Corp.)

protection and protect the work in such a manner that there is no danger of flying chips injuring nearby workers.

4. Remove all burrs! Avoid checking whether burrs have been removed with your fingers.
5. Some adhesives use solvents that are highly flammable and/or toxic. Do NOT apply such adhesives near areas where there are open flames. Apply them only in well-ventilated areas. Use a suitable respirator when using such adhesives.
6. Follow the instructions on the adhesive container! Remove any adhesive from your skin promptly by washing in water.
7. Cyanoacrylate adnesives cure in a very short time (5 to 15 seconds). Do NOT allow any of this adhesive to get onto your fingers as it will cause them to adhere together. Unless a suitable solvent is available, it could require a painful surgical operation to separate the joined fingers. To be safe, wear disposable plastic gloves when preparing or applying adhesives. Handle adhesives with care!
8. CAUTION! The chemicals in adhesives for joining metals, or metal to some other material, can cause severe skin irritation. All of them can impair or cause loss of your sight.
9. Adhesive fumes can also be dangerous. Work in a well ventilated area. Above all, follow the instructions on the container when mixing and using adhesives. Only mix the amount you will need.

TEST YOUR KNOWLEDGE—Chapter 17

Please do not write in the text. Place your answers on a separate sheet of paper.

1. For best results, a threaded fastener should screw into it mating part a distance equal to ________ times the diameter of the thread.
2. There are many ways of joining material. List four types of threaded fasteners. Describe how each is used.
3. ________ screws are used for general assembly work.
4. To prevent a pulley from slipping on a shaft, a ________ screw is often employed.
5. The ________ bolt is threaded at both ends.

6. ________ or ________ are employed when the parts are to be joined permanently.
7. Why are lock washers used?
8. While most ________ ________ must be seated in grooves, a self-locking type does NOT require the special recess. They have been developed for both internal and external applications.
9. When is a jam nut employed?
10. The shape of the ________ nut permits it to be loosened and tightened without a wrench.

Match each work in the left column with the most correct sentence in the right column. Place the appropriate letter in the blank.

11. ____ Rivet.
12. ____ Jam nut.
13. ____ Drive screw.
14. ____ Thread cutting screw.
15. ____ Acorn nut.
16. ____ Dowel pin.
17. ____ Blind rivet.
18. ____ Keyway.
19. ____ Keyseat.
20. ____ Key.

a. Developed for applications where it is used in confined area.
b. Used where parts must be aligned accurately and held in absolute relation with another part.
c. Prevents a pulley or gear from slipping on a shaft.
d. Protects projecting threads.
e. Are hammered into a drilled or punched hole.
f. Used to make permanent assemblies.
g. Slot cut in gear or pulley to receive "c."
h. Locks a regular nut in place.
i. Eliminate costly tapping operations.
j. Slot cut in shaft to receive "c."

21. List the steps, in their proper sequence, when using adhesives to join metals.
22. List five (5) safety precautions that must be observed when using fasteners.

RESEARCH AND DEVELOPMENT

1. Prepare samples of work making use of the various threaded and nonthreaded fasteners. Mount them on a panel or arrange them on a table for the class to examine.
2. Secure samples of fasteners NOT described in this chapter. Classify them according to the material on which they are used and their recommended applications.
3. Develop a paper on how early fasteners were made. Use drawings to show how they looked.
4. Contact a manufacturer of fasteners and request samples of a machine bolt or cap screw in the various stages it must pass through until becoming a finished product. If it is not possible to secure actual samples, make a drawing showing the various stages of bolt manufacture.
5. Make a display of the various fasteners explained and described in this chapter. Mount and label them on a display panel.
6. Collect catalogs on fasteners for the school's technical library.
7. Secure samples of several adhesives suitable for joining metal to metal. Demonstrate the proper and safe way to use these materials.
8. Devise a test method that will determine the strength of the various adhesives.
9. Organize and label the storage of fasteners in your training area. Inventory them and determine which fasteners will have to be reordered.

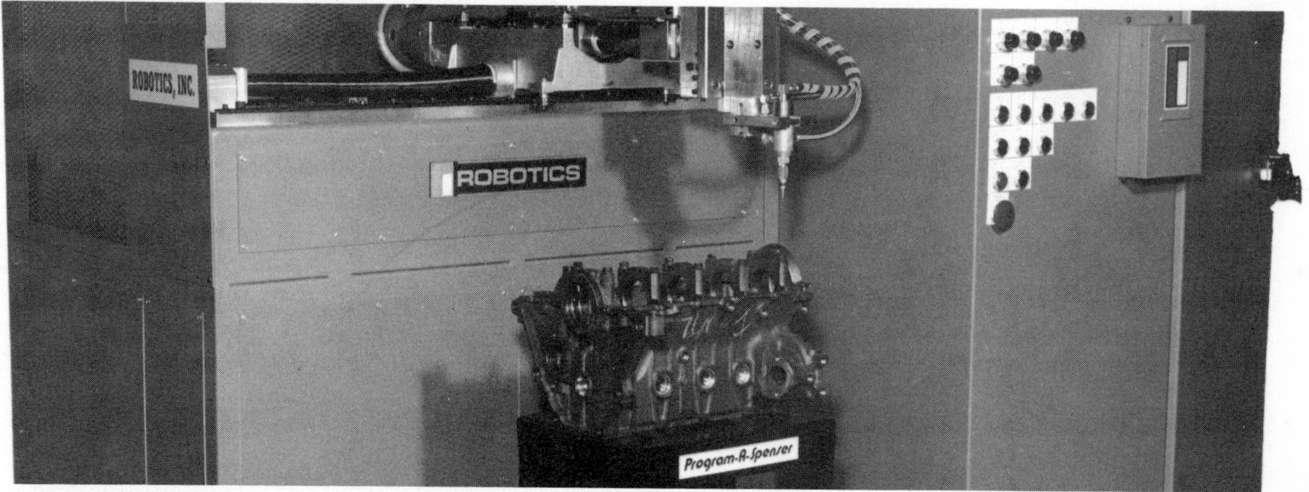

Fig. 17-44. Robotic device is used to apply formed-in-place gasket of silicon adhesive onto automotive engine block oil pan mating surface. (Dow Corning)

Fig. 18-1. Industry uses more than a thousand different metals, nonmetals, and composites of metals and nonmetals. Several hundreds will be utilized in aircraft like this advanced X-Wing helicopter. The X-Wing is a concept for VTOL (Vertical Takeoff and Landing) aircraft which uses a four-bladed helicopter-like rotor system that spins for hover and low speed flight and stops at approximately 200 knots to become a fixed wing for high speed flight. Each metal and composite will be selected for what it can contribute to aircraft safety and reliability. The United States Aerospace is the world leader in the development of new metals and composites and advanced production techniques. (United Technologies Sikorsky Aircraft)

Chapter 18

METAL CHARACTERISTICS

After studying this chapter, you will be able to:

- Explain how metals are classified.
- Describe the characteristics of metals.
- Recognize the hazards that are posed when certain metals are machined.
- Explain the characteristics of some reinforced composite materials.

More than a thousand different metals and alloys are used by the metalworking industry. Most of them are worked in the machine shop. However, it must be noted that there is an apparent "new materials revolution," especially in the aerospace industries. Many nonmetals and composites of metals and nonmetals are rapidly emerging into what once was almost the "exclusive domain" of metals. Refer to Fig. 18-1.

Can you tell by looking at a piece of metal whether it is a ***ferrous metal*** (metal containing iron), or a ***nonferrous metal*** (metal containing no iron)? It is an ***alloy*** (mixture of two or more metals)? Could it be a ***base metal,*** like tin, copper, or zinc? When you consider it from a practical point of view, it is almost impossible to find out much about a piece of metal by just looking at it.

A machinist is NOT expected to have a complete understanding of all the technicalities of metals. However, a working knowledge of the various materials and the common terms associated with them is essential.

In addition to the conventional metals worked in a modern machine shop, several of the newer materials will be described in this chapter. Plastics are described in Chapter 21.

CLASSIFYING METALS

Modern industrial metals may be classified as:

1. Ferrous metals.
2. Nonferrous metals.
3. High temperature metals.
4. Rare metals.

Because of space exploration and national defense, great strides have been made in the development of the latter two groups.

NOTE: Refer to the back of this text for tables that show physical properties of metals, dimensional tolerances, feeds, and speeds for machining.

FERROUS METALS

Ferrous metals comprise the family of irons and steels, and their alloys. For simplicity and a better understanding, they can be further subdivided into various categories.

Cast irons

Cast irons are iron alloys that contain 2.0 to 5.0 percent carbon with small quantities of silicon and manganese. There may also be traces of other elements.

Of the cast irons, ***gray iron*** and ***malleable iron*** are most widely used in industry. They can be found in great quantities in automotive, railroad, farm equipment, and machine tool bodies. See Fig. 18-2.

Malleable iron can be hammered into shape without fracturing. Most cast irons can be readily machined once the hard surface scale has been penetrated. Carbide cutting tools are recommended because of the abrasive nature of the iron scale. NO cutting fluids should be used on cast iron. Compressed air is recommended if a coolant is needed.

Steels

Steel is often considered the "backbone" of the metalworking industry, Fig. 18-3. ***Carbon steel*** is very common; it is an alloy of iron and carbon and/or other alloying elements but carbon is the major

Fig. 18-2. Many quality gray and malleable iron castings are used in modern machine tool construction because of its rigidity, stability with wide changes in temperture, and ease of machining to close tolerances. Many parts in this grinder are made of quality iron castings. (Bridgeport Machines, Inc.)

Fig. 18-3. Molten iron from a blast furnace is being charged into a basic oxygen furnace vessel at a Bethlehem Steel plant. After charge has been completed, vessel will return to its upright position for oxygen "blow." Blast furnace iron, mixed with scrap and selected additives, will then be refined into steel.

alloying agent. The alloying elements impart to iron (basic ingredient of all ferrous metals) the desired characteristics needed to perform a specific job.

The physical properties of steel are unique. Steel can be made soft enough to be easily machined. By careful ***heat treatment,*** the soft steel can be transformed into a "glass hard" material. By varying the heat treatment procedure slightly, steel can be given a hard, wear resistant surface while retaining a soft, tough core to resist breaking. Its magnetic qualities also make steel ideal for many electrical applications.

Carbon steels are classified according to the amount of carbon they contain. The ***carbon content*** is measured in percentage or in points (100 points equal 1 percent). They are available in all standard mill forms, Fig. 18-4.

Low carbon steels do NOT contain enough carbon to be hardened (less than 0.30 percent or 30 points). They are easy to work and can be case hardened. Low carbon steel is often called ***mild steel*** or ***machine steel.*** They are used for nuts, bolts, screws, gun parts, precision shafting, tie rods, tool cylinders, and similar applications.

Hot rolled steel is steel that has been rolled to finished size while hot. It is easily identified by a black oxide surface scale, Fig. 18-5.

Cold finished steel is steel that has been ***"pickled"*** or treated with a dilute acid solution to remove the oxide coating. After pickling, the steel is drawn or rolled to finish size and shape while cold. Cold finished steel is characterized by a smooth bright finish. The process improves the machinability of the steel.

Medium carbon steel contains 0.30 to 0.60 percent carbon (30 to 60 points). The carbon content is sufficient to allow partial hardening with proper heat treatment. The heat treating process improves the strength of the steel.

Available in all standard mill forms, medium carbon steels are used for many machine parts, automotive gears, camshafts, crankshafts, cap screws, precision shafting, etc.

High carbon steels contain 0.060 to 1.50 percent carbon (60 to 150 points). They are available in hot rolled form. However, some high carbon steel shapes may be purchased with ground surfaces. Drill rod and ground flat stock are examples.

High carbon steels are found in products that must be heat treated, Fig. 18-6. Applications include heavy machinery parts, control rods, wrenches, hammers, screwdrivers, pliers, springs, and a large variety of agricultural equipment.

Adding controlled amounts of sulphur and/or lead to carbon steels result in improved machinability without greatly affecting the metal's mechanical properties. Usually, machining involves the removal of considerable metal. The machine tool must be

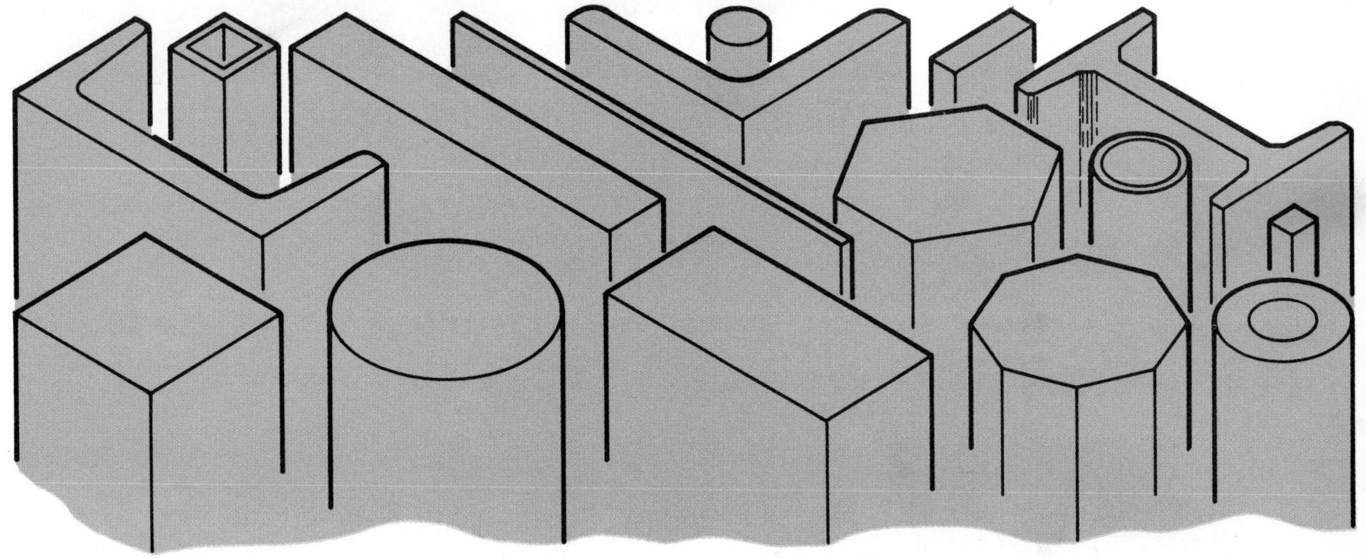

Fig. 18-4. These are a small portion of the hundreds of different shapes and sizes of steel available.

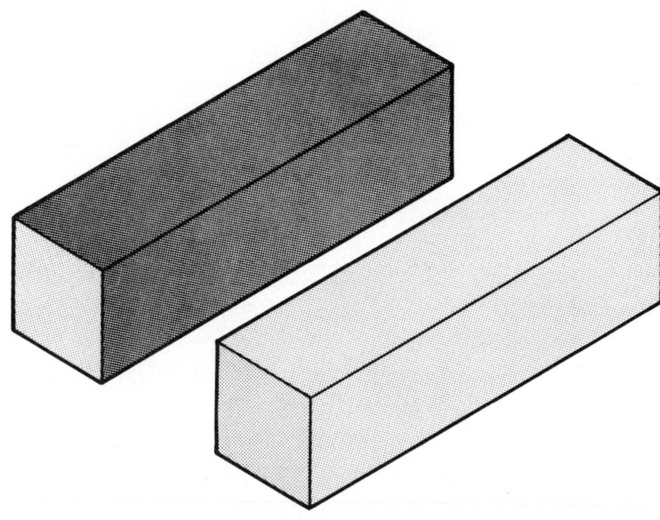

Fig. 18-5. Left. Hot rolled steel is characterized by a black oxide coating. Right. Cold finished steel has a smooth shiny surface.

Fig. 18-6. Teeth of giant helical gear are being hardened by induction method. Only faces of teeth are hardened to minimize their wear. (Philadelphia Gear Corp.)

capable of increased cutting speeds before the machining of high carbon steels will prove economical.

Alloy steels

Alloy steels have other metal elements added to change the metals' characteristics. Alloy steels are more costly to produce than carbon steels because of the increased number of special operations that must be performed in their manufacture, Fig. 18-7.

Elements such as nickel, chromium, molybdenum, vanadium, manganese, and tungsten are used to make alloy steel harder, stronger, and tougher. A combination of two or more of the above elements usually imparts some of the characteristic properties of each one added.

Chromium-nickel steels, for example, develop good hardening properties with good ***ductility*** (property of metal that permits permanent deformation by hammering, rolling, and drawing without breaking or fracturing). Chromium-molybdenum combinations develop excellent hardenability with satisfactory ductility and a certain amount of heat resistance.

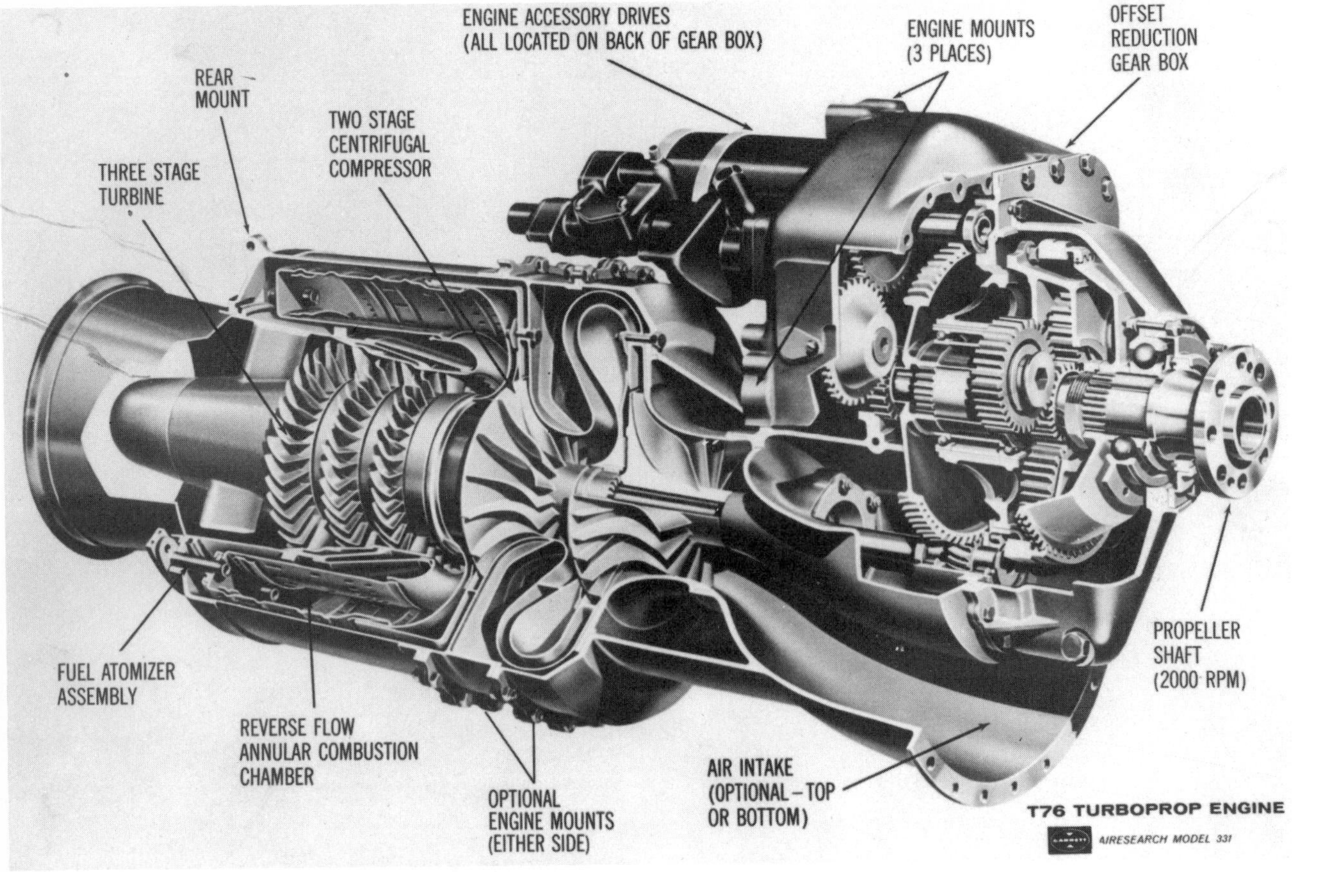

Fig. 18-7. Many parts of this turboprop engine are made from alloy and high temperature steels. (Airsearch Mfg. Co.)

Alloy metallic elements

Metallic elements added to alloy steels, and properties they impart to steel, are:

Nickel imparts toughness and strength to steel, particularly at low temperatures. Nickel steels permit more economical heat treatment and have improved resistance to corrosion. They are especially suitable for the case-hardening process and are used for applications such as armor plate, roller bearings, and aircraft engine parts.

Chromium is added when toughness, hardness, and wear resistance are desired. It is the basis of ***stainless steel.*** Chromium steel is found extensively in automotive and aircraft parts, Fig. 18-8.

Molybdenum is employed as an alloying agent when the steel must remain tough at high temperatures.

Vanadium, when added as an alloying element, produces a steel that has a fine grain structure and increased toughness at high temperatures.

Manganese purifies steel and adds strength and toughness. Manganese steel is used for parts that must withstand shock and hard wear, Fig. 18-9.

Tungsten, when added in the proper amount, makes steel that has a fine, dense structure with improved heat treatment qualities. It is one of the

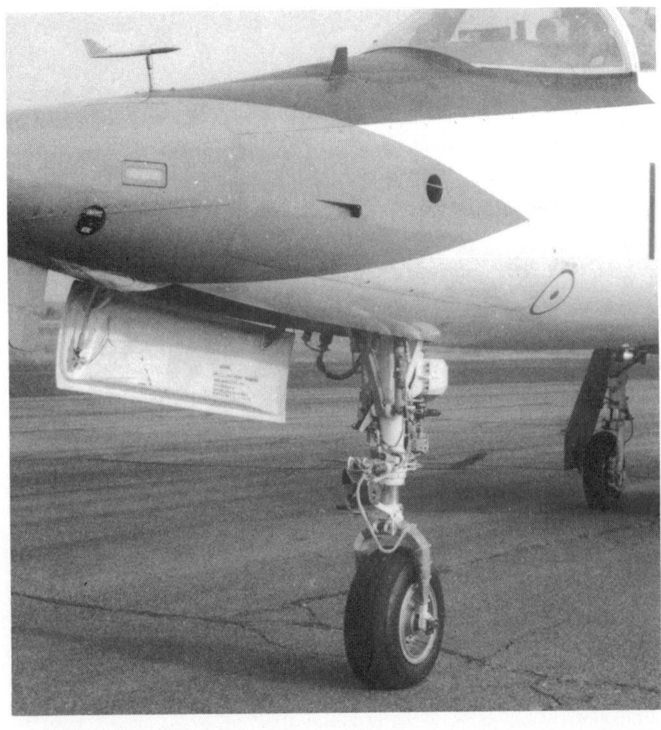

Fig. 18-8. Chromium steel is used extensively in landing gear assembly of this aircraft. It has ability to withstand the shock of landing this multi-ton aircraft. (Grumman Aerospace Corp.)

Fig. 18-9. Many components on this large earth moving vehicle are made from manganese steel. This type steel is strong and tough enough to withstand constant abrasive wear.

principle alloying agents in many tool steels. Tools made with these steels retain their strength and hardness at high temperatures.

Cobalt is the chief element in high speed steels because it improves the ***red hardness*** (still hard when red hot) of cutting tool materials. Wear resistance is also improved.

Some alloy steels possess relatively high strength at moderatly elevated temperatures. They are finding many applications in air frame structures of aerospace vehicles.

Tool steel

Tool steel is the term usually applied to steels found in devices that cut, shear, or form materials. They may be either carbon or alloy steels. Steels in the lower carbon content range (0.70 to 0.90 percent or 70 to 90 points) are used for tools subject to SHOCK. Higher carbon content tool steels (1.10 to 1.30 percent or 110 to 130 points) are utilized when tools with KEEN CUTTING EDGES are required.

Drills, reamers, milling cutters, punches, and dies are made from alloy tool steels. Although several tool steels can be hardened using water as a quenching medium, most of them must be hardened in oil or in air. The latter are often known as ***oil hardened*** or ***air hardened steels.***

Some alloy steels are also classified as ***high speed steels*** because they are capable of making deeper cuts at higher cutting speeds than regular tool steels. They possess red hardness, or the ability to retain their hardness at high temperatures, Fig. 18-10. They also possess high abrasion resistance. In spite of the development and wide spread use of cemented carbides and ceramics, high speed steels remain a major cutting tool material.

Fig. 18-10. Two high-speed milling cutters are being used to cut slots in an aluminum plate.

Tungsten carbide

Tungsten carbide is the hardest human-made metal. It is almost as hard as a diamond. The metal is shaped by molding tungsten, carbon, and cobalt powders under heat and pressure in a process known as ***sintering.*** The metals fuse together without melting.

Tungsten carbide, while NOT a true steel, is usually classified with the steels. Tools made from this family of materials can cut many times faster than high speed cutting tools, Fig. 18-11.

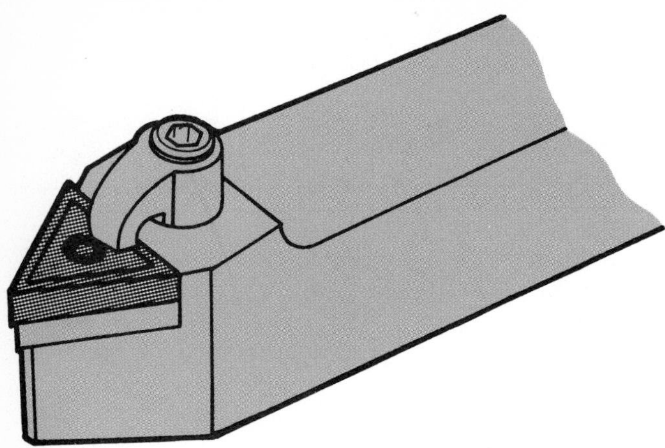

Fig. 18-11. Cutting tools made from tungsten carbide can cut many times faster than high-speed cutters.

Stainless steels

There are more than a hundred different stainless steels. However, one characteristic common to all of them is that they contain enough chromium to render them corrosion resistant. ***Stainless steels*** may be divided into three basic groups:

The ***austenitic*** classification includes the chromium-nickel and chromium-nickel-manganese stainless steels. Generally, they are hardenable only by cold working. The American Iron and Steel Institute (AISI) 300 series of stainless steels are in this category.

The ***martensitic*** stainless steel alloys of iron, carbon, and chromium are characteristically magnetic in nature and obtain their hardness through normal heat treating processes.

The ***ferritic*** stainless steel have more than 18 percent chromium. They are nonhardenable and all of them are magnetic.

Stainless steels may be machined with techniques normal for mild steels. However, some precautions must be observed with stainless steel:

1. Feeds must be high enough to insure that the cutting edge(s) get under the previous cuts and thus avoid the hardened portions.
2. Tools must be as large as possible, because the life of the cutting edge(s) depends on good heat dissipation into the body of the cutting tool.
3. Use finishing cuts when working to close tolerances.
4. The machine should be adjusted so there is minimum "play." Otherwise, the cutting tool may "ride" the work and glaze and/or harden the surface.

Identifying steels

Because different kinds of steels look alike, several methods of identification have been devised. They include identification by chemical composition, mechanical properties, their ability to meet a standard specification or industry accepted practice, or their ability to be fabricated. The shape of the ***mill form*** (rod, bar, structural shape, etc.), and the intended use for the metal, can also determine the method of identification.

The American Iron and Steel Institute (AISI) and the Society of Automotive Engineers (SAE) have devised almost identical standards that are widely used for identifying steel. Both systems use an identical ***four-number code*** (some steels require a fifth digit) that describes the physical characteristics of the various steels.

The AISI system also makes use of a ***prefix letter*** (A, B, C, etc.) that indicates the steel manufacturing process.

Note! Refer to the tables in the back of this text for more information on this subject.

The four-numeral code works as follows: The FIRST DIGIT classifies the steel. The SECOND DIGIT indicates the approximate percentage of the alloying element in the steel. The LAST TWO DIGITS show the approximate carbon content of the steel in points or hundredths of one percent.

For example, steel designated SAE 1020 is a carbon steel with approximately 20 points or 0.20 percent carbon. The AISI and SAE four-digit code applies primarily to bar, rod, and wire products.

Color coding is another method of identifying the many kinds of steel, Fig. 18-12. Each commonly used steel is designated by a specific color. The color coding is painted on the ends of bars 1 in. (25 mm) or larger. On bars smaller than 1 in. (25 mm), the color code may be applied to the end of the bar or on an attached tag.

The ***spark test*** is also employed at times to determine grades of steel, Fig. 18-13. The metal is

Fig. 18-12. Note color coding of steel rods.

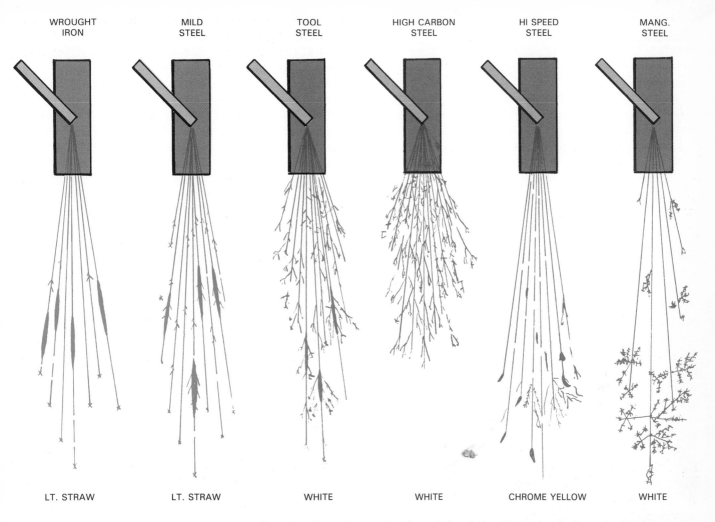

WROUGHT IRON | MILD STEEL | TOOL STEEL | HIGH CARBON STEEL | HI SPEED STEEL | MANG. STEEL

LT. STRAW | LT. STRAW | WHITE | WHITE | CHROME YELLOW | WHITE

Fig. 18-13. Spark test is sometimes employed to determine grade of steel. Steel should touch grinding wheel lightly and resulting sparks are then observed.

touched to the grinding wheel lightly and the resulting sparks are carefully observed.

SAFETY NOTE! During a spark test, wear approved eye protection! The grinder eye shield should be clean and in place. The tool rest must also be properly adjusted!

NONFERROUS METALS

There are many metals that do NOT have iron as their basic ingredient. Known as nonferrous metals, they offer specific properties, or combination of properties, that make them ideal for tasks where ferrous metals are not suitable, Fig. 18-14.

Aluminum

Aluminum has come to mean a large family of aluminum alloys, NOT just a single metal. As first produced, aluminum is 99.5 to 99.76 percent pure. It is somewhat soft and NOT very strong.

Fig. 18-14. Aluminum alloys are utilized in construction of this high-speed train designed to travel at 200 mph (320 km/h). Metal is light, strong, and corrosion resistant. (French National Railroads)

The strength of aluminum can be greatly increased by adding small amounts of alloying elements, heat treating, or cold working. A combination of the three techniques has produced aluminum alloys that, pound for pound, are stronger than structural steel. In addition to increasing strength, alloying elements can be selected that will improve welding characteristics, corrosion resistance, machinability, etc.

Many new aluminum alloys are being developed. ALCOA (Aluminum Company of America) is now producing Alithalite®, aluminum-lithium alloys. If used on a commercial aircraft like the Boeing 767, it would reduce the weight of the aircraft by about seven tons. No new tools or manufacturing techniques are needed to work this new aluminum alloy.

There are two main classes of aluminum alloys: wrought alloys and cast alloys. The shape of ***wrought alloys*** is changed by mechanically working them: forging, rolling, extruding, hammering, etc. ***Cast alloys*** are shaped by pouring metal into a mold and allowing it to solidify, Fig. 18-15.

Each alloy is given an identifying number. Known as the ***Aluminum Association Designation System,*** it is a four-digit code, plus a temper designation. ***Temper designation*** indicates the degree of hardness of the alloy. It follows the alloy identification number and is separated from it by a dash.

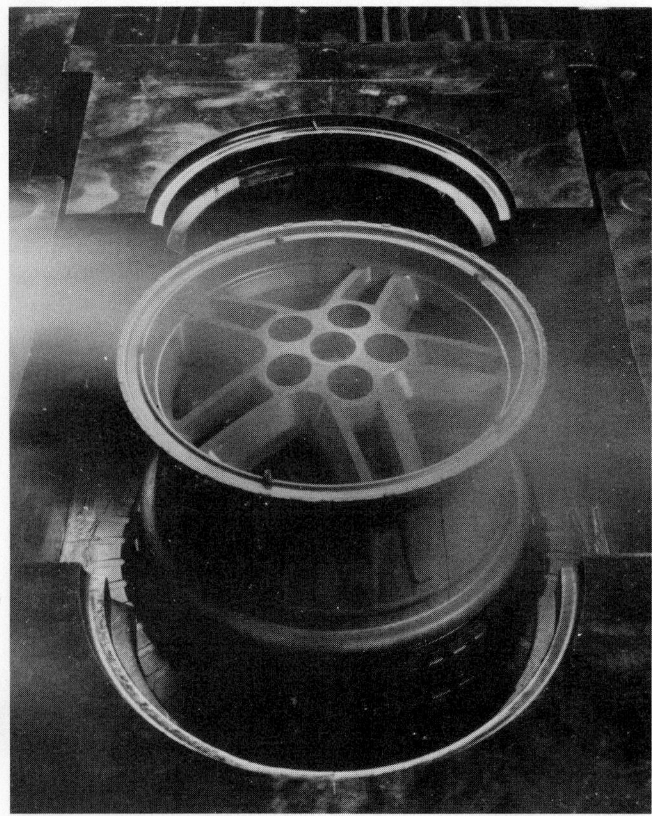

Fig. 18-15. Die cast aluminum wheel is emerging from die (mold) of a casting machine. Wheel is lighter, stronger, and more attractive than conventional pressed steel wheels. (Kelsey-Hayes)

Note! See the table in the rear of this text for the Aluminum Association Designation System.

Aluminum alloys possess many desirable qualities. They are extremely strong and corrosion resistant under most conditions. The alloys are lighter than most commercially available metals. They can be shaped and formed easily, and are readily available in a multitude of sizes, shapes, and alloys.

Machining aluminum

Most of the wrought aluminum alloys possess excellent machining characteristics. They are capable of being machined to intricate shapes at high cutting speeds. However, the makeup of an aluminum alloy is a factor that can affect machinability. Some aluminum alloys, being of a ***nonabrasive nature*** (those containing copper, magnesium, or zinc), have improved machinability. Other alloys with ***abrasive constituents*** (such as silicon) reduce tool life and machined surfaces may have a slightly gray finish with little lustre.

Most aluminum alloys are easier to machine to a good finish when in full hard temper than when in an annealed state. Machining characteristics of more commonly used aluminum alloys are:

1. Number 1100 and 3003 ***aluminum*** alloys have good machinability but are gummy in nature. Turnings are long and stringy, causing difficulty in chip disposal. Good results can be obtained if the cutting tools have large top and side rake angles, with keen, smooth cutting edges.
2. Number 5052 ***aluminum*** has turnings that are long and stringy and the machined surface is NOT as good as on 3003. Machinability is good, however.
3. Number 5056 ***aluminum*** alloy has good machinability with the advantage of fairly easy chip disposal.
4. Number 2017-T4 and 2014T6 ***aluminum*** alloys machine to an excellent finish. Of the two, 2014-T6 has better machinability because of the heat treating method employed. It causes greater tool wear.
5. Number 2024-T3 ***aluminum*** alloy has good machining characteristics with properly sharpened and honed tools. Surface finishes are excellent.
6. Number 6061-T6 ***aluminum*** alloy contains silicon and magnesium. It is more difficult to machine than the 2000 series alloys. Properly sharpened cutting tools and coolants with good lubricating qualities are essential. Fine finishes are obtainable with moderately heavy cuts.

7. Number 7075-T6 ***aluminum*** is the highest strength aluminum alloy that is commercially available. Machining qualities are good.

High speed cutting tools will produce satisfactory results when machining most aluminum alloys. However, optimum results will require the use of carbide or ceramic tools. Recommended tool geometry for aluminum is shown in Fig. 18-16.

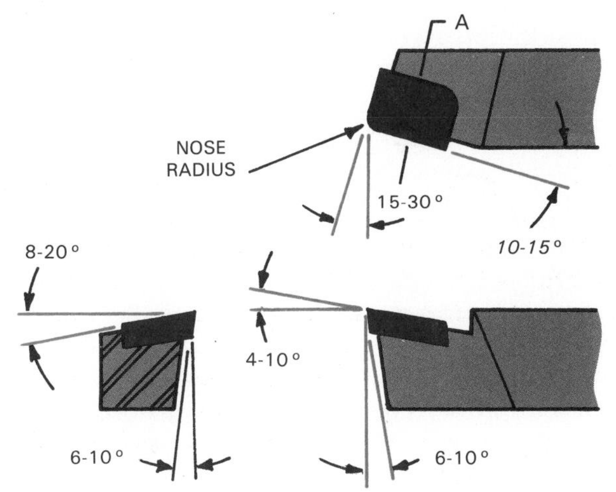

Fig. 18-16. Study angles of carbide tool for turning aluminum alloys.

Magnesium

Magnesium alloys are the lightest of the structural metals. They have a high strength to ratio weight. They have excellent machining properties, and can be machined by all common metalworking techniques.

In spite of their many advantages, magnesium alloys must be worked with extreme care. Several of the alloys developed for use at elevated temperatures in aerospace vehicles contain thorium, a low-level RADIOACTIVE material. They must be handled according to strict safety precautions devices for radioactive materials.

Another concern when machining magnesium is the possibility that the chips, not the solid metal, may ignite from the heat generated during the cutting operation. This is especially true where fine magnesium particles are formed when machined without adequate dissipation, such as in grinding.

To guard against magnesium fires, use a straight mineral oil cutting fluid in adequate quantities to flood the work. Water-base coolants must be AVOIDED, as they react with the chips and actually intensify a magnesium fire once it gets started.

Do NOT allow magnesium chips to accumulate on or around the machine. Use sharp tools because the tool rubbing against the work, without a definite cutting action, generates heat. This heat may cause easily ignited fine particles to be formed.

DANGER! Extreme care must be taken when machining magnesium because the chips or particles are highly flammable. Burning magnesium chips are so intensely hot (600 °F or 315 °C), they cannot be extinguished by conventional fire fighting techniques. Water or a commercial extinguishing agent actually INTENSIFIES the fire!

Because of the relatively high thermal conductivity of magnesium, there is normally no fire hazard when a solid section of the metal is exposed to fire.

The lathe tool recommended for magnesium is shown in Fig. 18-17.

Titanium

Titanium is a metal as strong as steel but only half as heavy. It bridges the gap between aluminum and steel. It is silvery in appearance, and extremely resistant to corrosion. Most titanium alloys are capable of continuous operation at temperatures up to about 800 °F (427 °C). This makes titanium ideal for use in high speed aircraft components. Aluminum fails rapidly at temperatures above 250 °F (121 °C).

Machining titanium

Titanium can be machined with conventional tools if the following practices are observed:

1. Tool and work setups must be rigid.
2. Tools must be kept sharp.
3. Good coolants must be applied in adequate quantities.
4. Cutting speeds must be slower with heavier feeds than those used for steel.

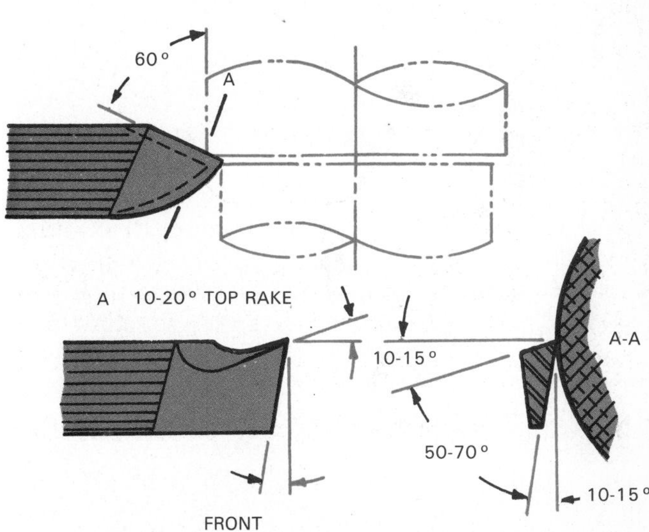

Fig. 18-17. Study typical lathe tool for turning magnesium.

Turning titanium that is commercially pure is very similar to turning 18-8 stainless steel. The alloys are somewhat more difficult to machine. Tungsten carbide and some types of ceramic tools produce the best results.

Milling titanium is more difficult than turning because the chips tend to weld to the cutter teeth. Climb milling may alleviate the problem to a great extent. Cast alloy tools often prove more economical to use than carbide tools. A water-base coolant is recommended.

Drilling titanium with conventional, high-speed steel drills will produce satisfactory work. Drills should be no longer than necessary to produce the required depth hole and still allow the chips to flow unhampered.

Tapping titanium is one of the more difficult machining operations. The tap has a tendency to freeze or bind in the hole. Careful selection of cutting fluid will minimize this problem.

Sawing titanium requires a slow speed of about 50 fpm (15 m/min) with heavy, constant pressure. The tooth geometry of the blade must be designed for sawing titanium.

COPPER BASE ALLOYS

Brass and bronze are the most familiar of the ***copper base alloys.*** However, lesser known heat treatable alloys are available. The newer alloys include copper and exotic metals like zirconium and beryllium. Most copper base alloys are available in rod, bar, tube, wire, strip, and sheet forms.

Copper

Copper is a ***base metal;*** that is, a pure metallic element. It is probably the oldest known metal. It can be shaped easily but becomes hard when worked and must be annealed or softened. Copper is difficult to machine because of its toughness and softness.

With copper, keep tools honed sharp and make as deep a cut as possible. Cutting fluids are NOT ordinarily needed, except for tapping copper.

Brass

Brass is an alloy of copper and zinc. Its color is determined by the percentage of zinc it contains. It ranges in color from reddish yellow to a silvery yellow. Most brasses can be readily machined.

Bronze

Bronze is an alloy of copper and tin. It is harder than brass and is much more expensive. Many special bronze alloys include additional alloying elements such as aluminum, nickel, silicon, and phosphorous. Most bronzes are relatively easy to machine with sharp tools.

Beryllium copper

Beryllium is one of the newer copper base alloys. Its machining qualities are similar to those of copper.

WARNING! Machining beryllium copper can pose a definite health hazard if precautions are NOT observed. The fine dust generated by machining and filing can cause severe respiratory damage. A respirator-type face mask must be worn. Special procedures must also be followed when cleaning machines used to machine beryllium copper.

In addition, a vacuum system should be employed to remove the beryllium copper dust, or the work should be liberally flooded with cutting fluid. Do NOT permit cutting tools to become dull; dull tools generate more dust than sharp tools.

Beryllium copper can be heat treated. It should NOT be machined in the annealed state. Recommended tool geometry for lathe tools is shown in Fig. 18-18.

When machining beryllium, employ a mineral oil-base cutting fluid. The fluid should be selected for its cooling properties, rather than for lubrication.

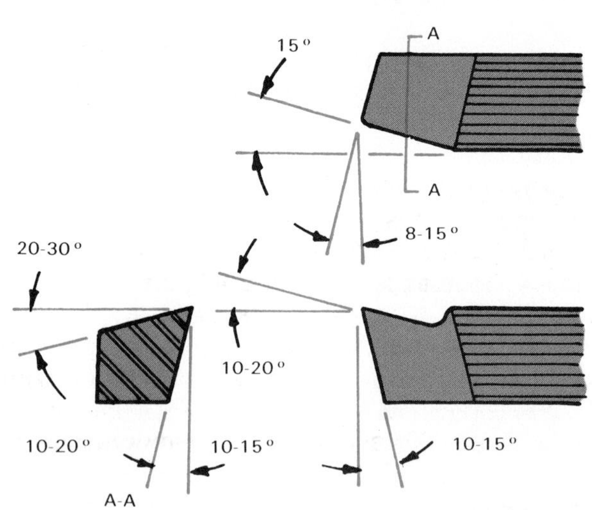

Fig. 18-18. Note tool shape for turning beryllium copper.

HIGH TEMPERATURE METALS

The nuclear and aerospace industries are chiefly responsible for the development of a number of high temperature metals, Fig. 18-19.

High temperature metals have the unique properties of high strength for extended periods at elevated temperatures. They are sometimes called ***superalloys.***

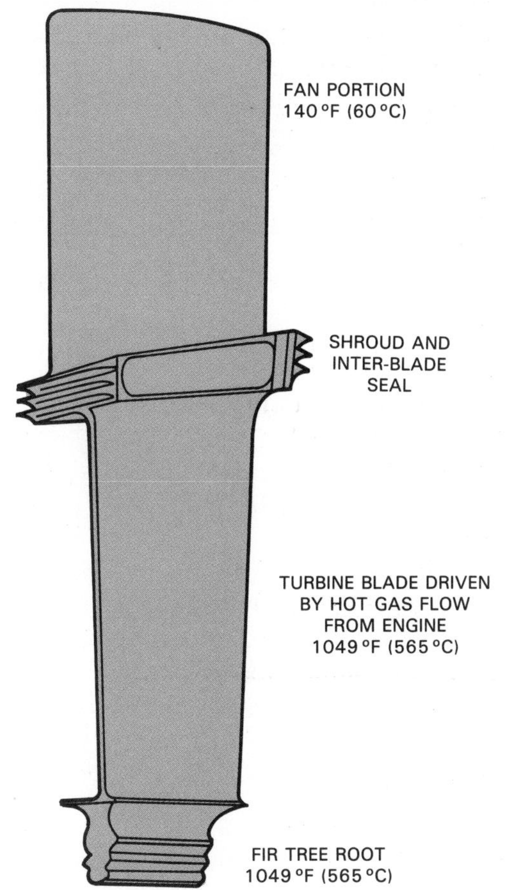

Fig. 18-19. Metal used for manufacture this fan jet turbine engine must be able to withstand high temperatures for long periods of time without failing.

Nickel base alloys

Nickel base alloys are known commercially as Iconel-X, Hastealloy-X, Rene 41, etc. They have many uses in jet engines, rocket engines, and electric heat treating furnaces. These applications require metals that can operate at temperatures of 1200 to 1900 °F (649 to 1038 °C). These metals are NOT easy to machine by conventional methods.

Molybdenum

Molybdenum has excellent elevated temperature strength in the 1900 to 2500 °F (1038 to 1372 °C) range and has found many applications in modern technology. It also has great resistance to corrosion by acids, molten glass, and metals.

Contrary to what might be implied, molybdenum machines similar to cast iron if both the work and cutting tool are mounted rigidly. For most work, tungsten carbide and ceramic tools are preferred over high-speed tools.

Tantalum

Tantalum alloys are specified where dependability at temperatures above 2000 °F (1094 °C) is required. Tantalum is used for rocket nozzles, heat exchangers in nuclear reactors and missiles and space structures.

Tantalum is NOT an easy metal to machine. It is gummy and has a tendency to tear. High speed steel tools are usually recommended. Extreme cutting angles are used to keep the tool and chip clear of the work. The tool should be well supported, with little overhang.

Tungsten

Tungsten melts at a higher temperature 6200 °F (3429 °C) than other known metals. However, tungsten is NOT resistant to oxidation at high temperatures 930 °F (499 °F) and must be protected with a suitable coating like silicides.

It has many uses in rocket engines, welding electrodes, high temperature furnaces, and is an ideal metal for breaker points in electrical devices.

Machining is quite difficult, but it can be done with carbide and ceramic tools if the work is preheated to about 400 °F (204 °C). The final shaping of a tungsten part is frequently by grinding. Adequate cooling of the grinding wheel with an oil-base compound is recommended.

RARE METALS

Name just about any rare metal known, and the odds are that someone is attempting to find a way to use it for aerospace applications. Most metals in this category are only available in small quantities for experimental purposes. Many of them cost considerably more than gold.

Included in the rare metal group are such metals as: scandium, yttrium (prounounced itream), cerium, europium, lanthanum, and holmium. While they may seem strange and almost unknown at the present time, it has NOT been too long ago the uranium, titanium, and beryllium were in the same category. Today, these three metals are frequently found in a modern machine shop.

Beryllium is the latest of these metals to be introduced in more than experimental quantities. It has only been slightly more than 100 years ago that aluminum was considered a rare metal and worth many times more than gold.

Other Materials

In addition to conventional metals and plastics, the modern machine shop is also expected to work other types of materials.

Honeycomb

Many ways have been devised to give existing metals greater strength and rigidity while reducing weight. ***Honeycomb*** sandwich structures is one such material, Fig. 18-20. Sections of thin metal (aluminum, stainless steel, titanium, and nonmetals like Nomex®) are bonded together in such a manner that when expanded, a structure is formed that is similar in appearance to the wax comb honeybees use to store honey.

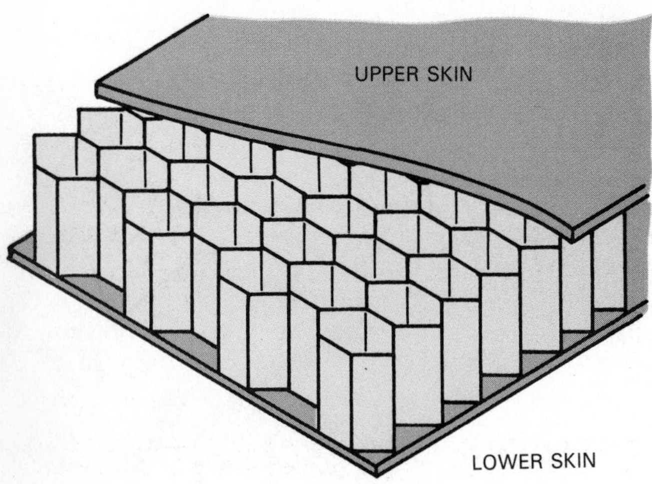

Fig. 18-20. Honeycomb has great strength and rigidity for its weight. It has many applications in aerospace industries.

When honeycomb is rigidly bonded between two metal or composite sheets to form a sandwich panel, it becomes a strong structure. It has a very high strength-to-weight ratio and rigidity-to-weight ratio. The bonding is done with an adhesive, or fused by brazing or resistant welding.

Because of its fragile nature (before shaping and bonding into rigid units), the material can cause problems in machining, Fig. 18-21. In addition to special tools that literally pare the material off, electrolytic grinding is the most rapid method for machining honeycomb. It does NOT leave a burr that would create problems in its removal.

The space shuttle, and late model aerospace vehicles utilize large quantities of aluminum, stainless steel, and titanium honeycomb in their structures. See Fig. 18-22.

Composites

Composites are a relatively new development that utilize fibers of conventional materials, and some NOT so common materials, in both pure and alloy forms. Fibers such as pure iron, graphite, boron, and fiberglass are bonded together with special ***plastic matrixes*** (binding substances like epoxies). These materials are generally lighter, stronger, and more rigid than many conventional metals.

Present usage of composites is mainly concerned with aerospace and some automotive applications, Fig. 18-23. However, some composites are finding applications in such things as fishing poles, skis, golf clubs, tennis rackets, bicycle frames, etc., Fig. 18-24. Much research is being done to reduce the cost of composites so they can be employed to make lighter and safer automobile bodies and other components.

TEST YOUR KNOWLEDGE—Chapter 18

Please do not write in the text. Place your answers on a separate sheet of paper.

1. What are the four main categories of metals?
2. Iron and steel are classified as ________ metals.
3. Carbide cutting tools are recommended for machining cast iron because:
   a. Cast iron is hard and brittle.
   b. Of the hard surface scale.
   c. Cast iron is difficult to machine.
   d. All of the above.
   e. None of the above.
4. Carbon steel is an alloy of ________ and ________.
5. How are carbon steels classified?
6. Hot rolled steel is characterized by the ________ on its surface.
7. The machinability of carbon steel is improved if ________ or ________ is added as an alloying element.
8. Nickel, chromium, molybdenum, vanadium, and tungsten are used to make steel ________, ________, and ________.
9. Drills, reamers, milling cutters, etc., are usually made from ________ steel.
10. The chief characteristic of stainless steel as its resistance to ________.
11. List the three basic groups of stainless steel.
12. Aluminum, magnesium, and titanium are ________ metals.
13. Magnesium is the ________ of the structural metals.
14. Titanium is a metal that is as ________ as steel but only ________ as ________.
15. Brass and bronze are ________ base alloys.
16. Brass is an alloy of copper and ________.
17. Bronze is an alloy of copper and ________.
18. Why must a machinist take special precautions when working beryllium copper?
19. Nickel base alloys, molybdenum, tantalum, and tungsten are classified as ________ ________ metals.
20. What is the structural material known as honeycomb?
21. What are composites?

Fig. 18-21. Machining honeycomb can be a delicate operation. Three-dimensional milling machine is carving aluminum honeycomb to airfoil shape for tail section of aircraft. (Hexcel Corp.)

RESEARCH AND DEVELOPMENT

1. Prepare a display board showing specimens of ferrous, nonferrous, high temperature, and rare metals. Label them according to type, classification, and use.
2. Many terms are used to describe the various properties of metals. Research their meanings as they pertain to metals.

Machinability	Ductility
Malleability	Elasticity
Brittleness	Hardness
Toughness	Tensile strength
Yield point	Elongation
Stress	Plasticity

3. Secure samples of honeycomb. Prepare a report explaining why is is difficult to machine by conventional methods. Include some of the techniques employed to machine the material.

Fig. 18-22. Large quantities of aluminum, stainless steel, titanium and nonmetal honeycomb are utilized in construction of space shuttle and its Boeing 747 carrier aircraft. Honeycomb was only material light and strong enough to do the job. (NASA)

Fig. 18-23. Until advent of composites, it was not possible to take advantage of aerodynamic characteristics of forward swept aircraft wing in high-speed flight. Metal was not rigid enough. The X-29 is first of many such aircraft that will explore use of this type wing. (Grumman Aerospace Corp.)

Fig. 18-24. This is one of the first commercial applications of a composite. Bicycle frame is fabricated from a graphite composite called MAGNAMITE* which is stiffer than steel and lighter than titanium. (O.F. Mossberg & Sons, Inc.)

*Registered trademark of Hercules,Inc.

4. Secure samples of composites. Prepare a report on how they are made and shaped.
5. Contact firms manufacturing high temperature metals and request pamphlets on them for your shop technical library.
6. Show a film or video on how aluminum is made. Discuss the details of the film with the class.
7. Show a film or video on steel making. Prepare a quiz to be given after the presentation and discussion.
8. Secure copies of trade magazines dealing with metals and other materials.

Chapter 19

HEAT TREATMENT OF METALS

After studying this chapter, you will be able to:

- Explain why some metals are heat treated.
- List some of the metals that can be heat treated.
- Describe some types of heat treating techniques and how they are performed.
- Case harden low carbon steel.
- Harden and temper some carbon steels.
- Compare hardness testing techniques.
- Point out the safety precautions that must be observed when heat treating metals.

Since many parts produced in the machine shop must be heat treated before use, it is important that the machinist be familiar with the basic science of heat treating metals.

Heat treatment involves the controlled heating and cooling of a metal or alloy to obtain certain desirable changes in its physical characteristics, Fig. 19-1. These changes include improving resistance to shock, obtaining toughness, and increasing wear resistance and hardness, Fig. 19-2. Heat treatment involves a number of processes.

Similar techniques can be employed to ***anneal*** (soften) metals to make them easier to machine, or to ***case harden*** (produce a hard exterior surface) steel for better resistance to wear.

SAFETY NOTE! Never attempt to heat treat metals while your senses are impaired by medication or other substances.

HEAT TREATABLE METALS

Steel and most of its alloys are hardenable. However, when heat treating carbon steel, it must be remembered that the carbon content of the metal is an important consideration. Carbon steels are classified by the percentage of carbon they contain in "points" or hundredths of 1 percent. For example: 60 point carbon steel would contain 60/100 (0.60) of 1 percent carbon. Carbon steel with less than 50 points carbon CANNOT be hardened.

Many aluminum alloys, magnesium, copper, beryllium, and titanium are also capable of being heat treated.

Fig. 19-1. Computerized material testing system is used to determine tensile strength of metals. Metal is subjected to a slowly applied force that pulls metal apart. Material fractures when ultimate tensile strength of specimen is exceeded. (MTS Systems Corp.)

Fig. 19-2. Many parts of this huge ore carrying truck are heat treated. Without heat treated parts (wheels, drive shaft, gears, axles, etc.), vehicle would not be able to maintain its grueling work load for long without part wear and failure. (Euclid)

Fig. 19-3. Shown is cryogenic quenching area in a modern heat treating facility. Characteristics of several aluminum alloys and some space age metals are greatly improved by heating them to a predetermined temperature and quickly cooling (quenching) them in liquid nitrogen at about −300 °F (−185 °C). Grumman Aerospace Corp.)

GUIDE TO HEAT TREATMENT

Heat treatment is done by heating the metal to a predetermined temperature, then ***quenching*** it (cooling it rapidly) in water, brine, oil, blasts of cold air, or liquid nitrogen. Refer to Fig. 19-3.

Desired qualities do NOT always prevail after quenching. Stresses can develop that, under certain conditions, may cause some steels to shatter. Therefore, the metal may have to be reheated to a lower temperature, followed by another cooling cycle to develop the proper degree of hardness and toughness.

Types of heat treatment

The heat treatment of metals may be divided into two major categories. One deals with FERROUS metals, the other with NONFERROUS metals. Because each area is so broad, it is beyond the scope of this text to include more than basic information on the heat treatment of metals.

Changes in the physical characteristics of steel and its alloys can be affected by six basic types of heat treatment.

Stress relieving

Stress relieving is done to remove internal stresses that have developed in parts that have been cold worked, machined, or welded, Fig. 19-4. To stress relieve, steel parts are heated to 1000-1200 °F (547-660 °C), held at this temperature one hour or more per inch of thickness, and then slowly air or furnace cooled. The technique is sometimes called ***process annealing.***

Annealing

Annealing is a process that reduces the hardness of a metal to make it easier to machine or work, Fig. 19-5. It involves heating the metal to slightly above its critical temperature but never more than 50-75 °F (10-24 °C) above this point, Fig. 19-6. The time it is held at this temperature depends upon

Fig. 19-4. Stresses that develop in metals when they are machined or welded are removed by stress relieving.

Fig. 19-5. Annealing reduces hardness of metal and improves its machinability. When done in a controlled atmosphere, oxidation does not form and part remains bright. (Lindberg Steel Treating Co.)

the shape and thickness of the part. After holding at this temperature, the piece is allowed to cool slowly in the furnace or some insulating material.

For some steels, it may be necessary to use the box annealing method or a controlled atmosphere furnace to prevent the work from scaling or ***decarbonizing*** (loss of carbon on surface). With box annealing, the part is placed in a metal box and the entire unit is heated and allowed to cool slowly in the sealed furnace.

In reducing the hardness of the metal, the machinability is improved. Many nonferrous metals can also be softened by annealing.

Normalizing

Normalizing is a process where the metal is heated to slightly above its upper critical temperature and allowed to cool to room temperature. Normalizing is employed to refine the grain structure of some steels and thereby improve machinability. It is a process closely related to annealing.

Case hardening

Low carbon steel CANNOT be hardened to any great degree by conventional heat treatment. However, a hard shell can be put on the surface, while the inner portion remains relatively soft and tough, Fig. 19-7.

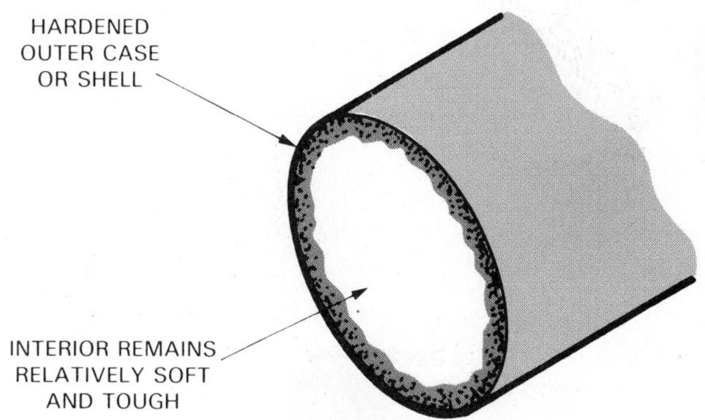

Fig. 19-7. Cross section of a case hardened piece shows that interior of section remains relatively soft and tough.

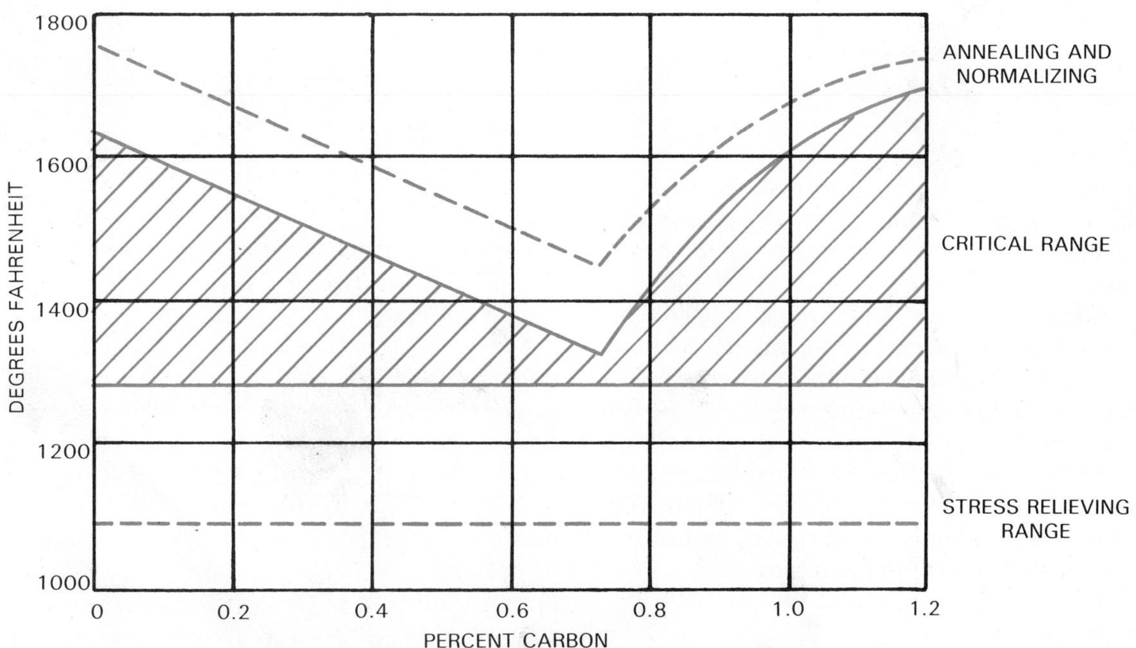

Fig. 19-6. Note critical range diagram for plain carbon steels.

Case hardening is accomplished by heating the piece to a red heat and introducing small quantities of carbon or nitrogen to the metal's surface. This can be done by one of the following methods: carburizing, cyaniding, or nitriding.

1. During ***carburizing,*** sometimes termed the ***pack method,*** the steel is buried in a dry ***carnonaceous material*** (material rich in carbon) and heated to just above its transformation range, Fig. 19-8. In the ***transformation range*** (1350-1650 °F or 745-915 °C), steels undergo internal atomic changes that radically affect the properties of the material. It is held at this temperature (15 minutes to 1 hour) until the desired case is attained. The part is removed from the furnace and quenched. Deep cases can be obtained by this method.

Fig. 19-9. Operator is carefully removing die block from cyanide salt pot. (Master Lock Co.)

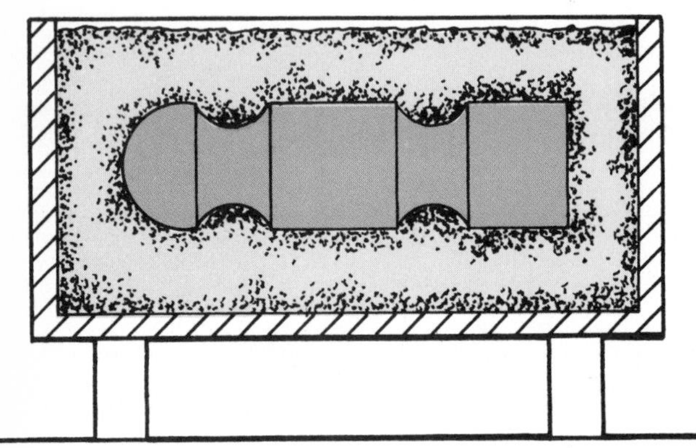

Fig. 19-8. Part is packed in container of carbonaceous material and is ready to be case hardened.

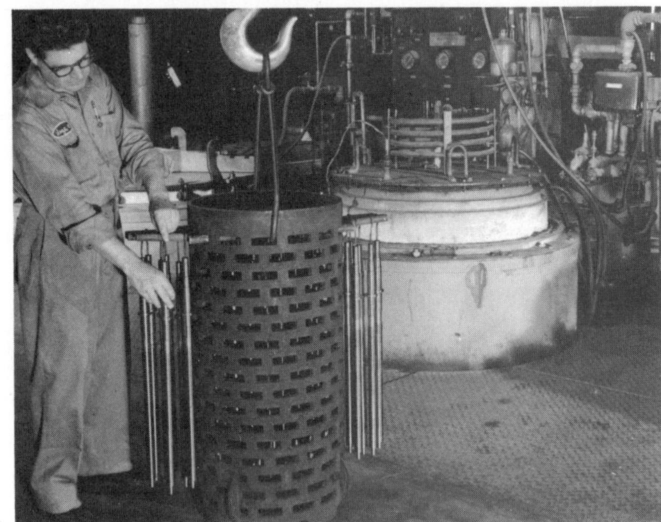

Fig. 19-10. Nitriding operation develops high hardness without quenching, and distortion is virtually nonexistant. (Lindberg Steel Treating Co.)

2. During the ***liquid salt method*** of case hardening, also known as ***cyaniding,*** the part is heated in a molten cyanide salt bath and then quenched, Fig. 19-9. The immersion period is usually less than one hour. A high hardness is imparted to the work and the parts treated have good wear resistance.
3. During the ***gas method*** or ***nitriding method*** of case hardening, the parts are placed in a special airtight heating chamber where ammonia gas is introduced at high temperature, Fig. 19-10. The ammonia decomposes into nitrogen and hydrogen. The nitrogen enters the steel to form ***nitrides*** which give an extreme hardness to the metal's surface. Wear resistance and high temperature hardness are greatly increased.

Surface hardening

Surface hardening is often used when only a medium hard surface is required on high carbon or alloy steels, Fig. 19-11. The internal structure of the metal is NOT affected. Flame hardening and induction hardening are used to attain these characteristics.

Flame hardening involves the rapid heating of the surface with an acetylene torch, and immediately quenching the heated surface, Fig. 19-12. The flame must be moved constantly to prevent burning or hardening the metal too deeply.

Induction hardening makes use of a high frequency electrical induction current to heat the metal, Fig. 19-13. The quenching medium follows the induction coil. The technique is rapid and tends to minimize distortion in the piece being heat treated. Induction hardening is ideal for production hardening.

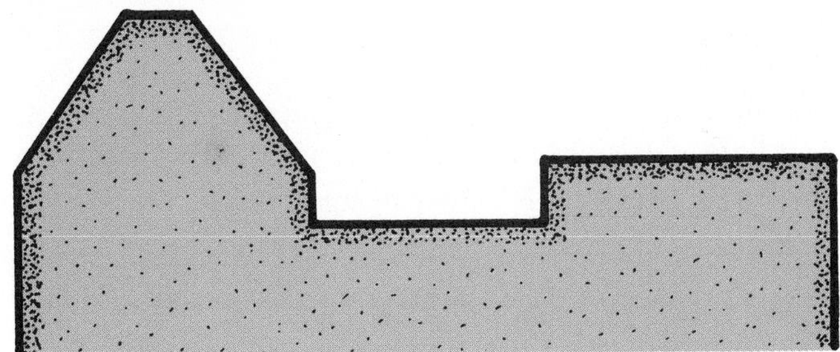

Fig. 19-11. Lathe ways are frequently surface hardened for improved wear resistance.

Fig. 19-12. Lathe bed is being flame hardened.

Fig. 19-13. Induction hardening is being done on a shaft used on a piece of farm machinery. A cold water spray hardens shaft after it has been heated to proper temperature with an induction coil. (Lindberg Steel Treating Co.)

Hardening

Hardening is a technique normally employed to obtain optimum physical qualities in steel, Fig. 19-14. It is accomplished by heating the metal to a predetermined temperature for a specified period of time.

The temperature which steel will harden is called ***critical temperature*** and ranges from 1400 to 2400 °F (760 to 1316 °C), depending upon the alloy and carbon content. For a "rule-of-thumb" range of hardening temperatures for carbon steel. See Fig. 19-15.

After heating, the part is quenched in water, brine, oil, liquid nitrogen, or blasts of cold air. Water or brine is employed to quench plain carbon steel. Oil is usually used to quench alloy steels. Blasts of cold air or liquid nitrogen is used for the high alloy steels.

Fig. 19-14. Molds for plastic bowling pins are being placed in a furnace for hardening. (Lindberg Steel Treating Co.)

CARBON CONTENT	HARDENING TEMPERATURE RANGE
0.65 to 0.80%	1450 to 1550 °F (788 to 843 °C)
0.80 to 0.95%	1410 to 1460 °F (766 to 793 °C)
0.95 to 1.10%	1400 to 1450 °F (760 to 788 °C)
Over 1.10%	1380 to 1430 °F (749 to 777 °C)

Fig. 19-15. Chart shows "rule-of-thumb" or rough practical method to determine range of hardening temperature for carbon steel. Metal must be quenched in water, brine, light oil, or blasts of cold air.

Quenching leaves the steel hard and brittle. It may fracture if exposed to sudden changes in temperature. For most purposes, this brittleness and hardness must be reduced by a tempering or drawing operation.

Tempering

Tempering or ***drawing*** is used to lower a metal's brittleness or hardness. It involves heating the steel to below critical range. The exact temperature will depend upon the type of steel used and its application. This information can be found in steel producer's catalogs and the various machinist's handbooks. Hold the temperature until complete penetration is achieved and then quench.

With the internal stresses released, the toughness and impact resistance increases. As the temperature is raised, ductility is improved but there is a decrease in hardness and strength.

HEAT TREATMENT OF OTHER METALS

Many other metals and their alloys are potentially heat treatable. You should have a basic understanding of these processes.

Heat treating aluminum

Aluminum is a general term applied to the base metal and its alloys. When heat treating the metal, it is imperative that the EXACT ALLOY be known or the part to be heat treated may be ruined.

Aluminum alloys are heat treatable in much the same manner as steel. That is, the metal is heated to a predetermined temperature, quenched, and then reheated to a lower temperature. However, some aluminum alloys age harden at room temperature and must be kept refrigerated to keep them soft and ***ductile*** (property of metal which permits it to be drawn out or hammered thin) while they are being worked.

Heat treating temperatures for several of the more commonly used aluminum alloys are shown in Fig. 19-16. Information on other alloys, each of which requires special treatment to bring out optimum physical qualities, can be obtained from handbooks available from the various producers of aluminum.

Heat treating brass

Brass can be annealed after cold working by heating to 1100 °F (593 °C) and cooling. The rate of cooling has no appreciable effect on the metal.

Heat treating copper

Copper is annealed in much the same manner as brass. The metal may be quenched or allowed to cool slowly at room temperature.

Heat treating titanium

Most titanium alloys are heat treatable. However, special facilities are necessary because it is a ***reactive metal;*** it readily absorbs oxygen, carbon, and nitrogen. These elements greatly affect the strength of titanium and its resistance to fatigue and corrosion.

HEAT TREATING EQUIPMENT

To obtain maximum results when heat treating, it is important that the equipment be in good condition and be suitable for the job to be done.

Quenching media

The main problem in heat treating is to cool the heated metal at a uniform rate over its entire area. Water, oil, air, and liquid nitrogen is each a standard ***quenching media*** used to draw the heat from the heat treated part.

Water has the most severe cooling features. It is employed mainly when treating plain carbon steel for maximum hardness. Water has the disadvantage

HEAT TREATABLE ALUMINUM ALLOYS

SOLUTION				PRECIPTATION (Aging)		
Aluminum Alloy	Heat to F° (°C)	Quench	Resulting Temper	Heat to F° (C°)	Hold for Hours	Resulting Temper
2024-0	910-930 °F (488-499 °C)	In cold water as quickly as posible after removal from furnace	—	Room	48-96	2024-T4
6061-0	960-980 °F (515-527 °C)		6061-W	315-325 °F (157-162 °C)	16-20	6061-T6
				345-355 °F (174-179 °C)	6-10	
7075-0	860-930 °F (460-499 °C)		7075-W	245-255 °F (118-124 °C)	20-26	7075-T6
NOTE: When heat treating clad 2024 and clad 7075 aluminum, hold temperature for shortest possible time.						

Fig. 19-16. Note heat treating temperatures for several types of aluminum alloys.

of forming gases when the hot metal is immersed in it. The gas bubbles adhere to the metal, retard cooling, and cause soft spots in the treated piece.

Brine, 5 to 10 percent salt in the quench water, prevents the formation of gases and gives better cooling results.

Mineral oils cool more slowly and produce less distortion in the treated part than does water. Special quenching oils have been developed. They have high ***flash points*** (lowest temperature at which vapor of oil can be made to ignite in air) and do NOT have a disagreeable odor. In production heat treatment using oil, the quenching bath must be filtered and cooled down to room temperature. Oils are used to harden alloy steels.

WARNING! Quenching heated metal in oil should only be done in a well ventilated area. Avoid inhaling any of the fumes!

Freely ***circulated air*** is used to cool some highly alloyed steels. The air, used as a cooling media, must be dry because any moisture may cause the steel to fracture.

Liquid nitrogen a super-cold cooling medium, is used with several aluminum and space age alloys. Special facilities are required.

Furnaces

The ***furnace*** must be capable of reaching and maintaining the temperatures needed for heat treating. The furnaces are fired by gas, oil, or electricity.

There are two types of gas furnaces: the ***box*** or ***muffle type,*** Fig. 19-17, and the ***pot type,*** Fig. 19-18. Industrial applications make use of many variations of these two variations.

SAFETY NOTE! Gas fired furnaces are noisy. Hearing protectors should be worn when working near them.

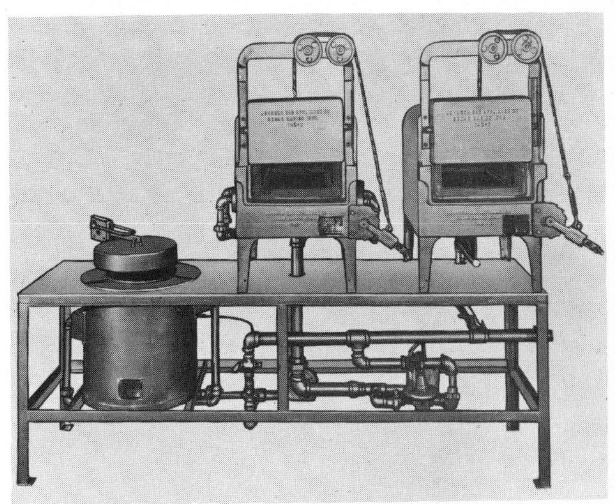

Fig. 19-18. This unit consists of two muffle furnaces and a pot type heat treating furnace. (Johnson Gas Appliance Co.)

Electric furnaces are quiet, require no elaborate venting, heats to temperature quickly, and the temperature can be controlled with accuracy, Fig. 19-19. It is also possible to automatically time the heating and cooling cycles.

Fig. 19-17. This is a box of muffle type, gas fired heat treating furnace. (McEnglevan Heat Treating & Mfg. Co.)

Fig. 19-19. This is a two-chamber, electric heat treating furnace. The upper high temperature chamber is fitted for controlled atmosphere operation. The lower chamber is a recirculating type draw or tempering furnace. (Lucifer Furnaces, Inc.)

Electric pot-type furnaces are also used by industry. See Figs. 19-20 and 19-21.

Some industrial heat treating furnaces are fitted with continuous belts for high production rates, Fig. 19-22. Others may be several stories high.

Some furnaces can be sealed and flooded with ***inert gases*** (gas that does not oxidize metal's surface not is absorbed by part being heat treated). The furnace is sealed and then a vacuum is drawn to keep atmospheric gases from contaminating the inert gas and the metal during heat treatment operations. See Fig. 19-23.

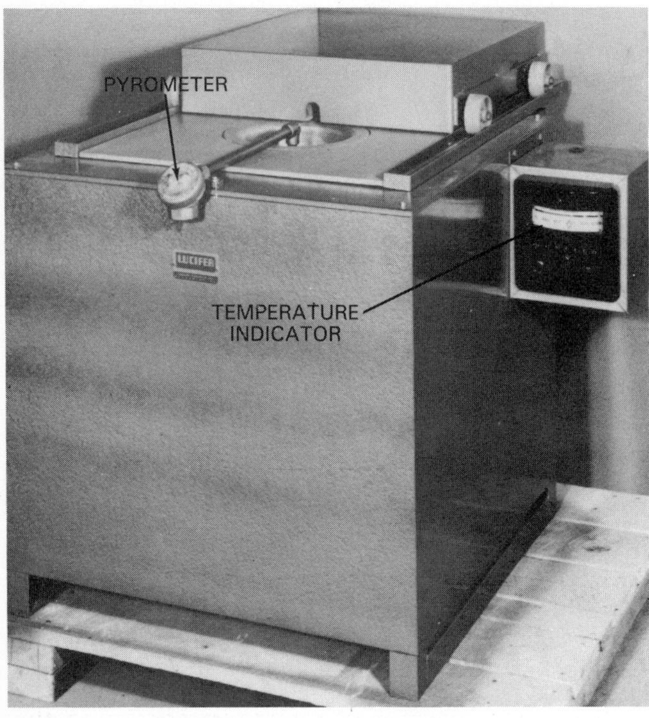

Fig. 19-20. This is an electric pot-type furnace for salt bath, lead, or cyanide processing of steel. (Lucifer Furnaces, Inc.)

Fig. 19-22. Completely self-contained heat treating furnace has continuous belt that feeds parts into furnace and delivers treated parts at far end. (Lindberg Steel Treating Co.)

HARDENING CARBON STEEL

Heat treating any metal requires maintaining accurate temperatures. A ***pyrometer*** is an instrument that accurately measures furnace temperature, Fig. 19-24. On electric furnaces, the pyrometer can be set to the desired temperature and can be used to maintain this temperature once it is reached.

If the furnace is NOT equipped with a pyrometer, it will be necessary to judge the temperature by the color of the metal as it heats. Color charts are available from the steel companies. See Fig. 19-25.

Fig. 19-21. Note two automatic, liquid pot-type furnaces. One at right has a water quench tank with a conveyor. Furnace on far left incorporates an automatic oil quench cycle. (Lindberg Steel Treating Co.)

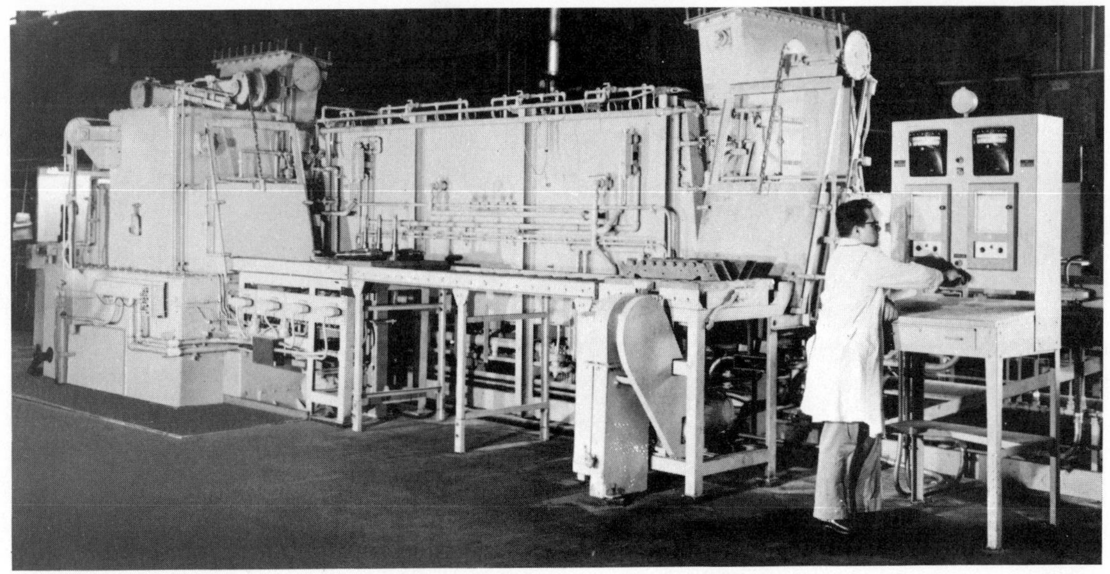

Fig. 19-23. This is a production type, controlled atmosphere heat treating furnace. (Lindberg Steel Treating Co.)

Fig. 19-24. Pyrometer is used to measure temperatures in a furnace. This pyrometer is used on gas fired furnaces. (Johnson Gas Appliance Co.)

TEMPERING STEEL

TEMPERING DEGREES	TEMPER COLOR
380 °F (193 °C)	Straw
420 °F (216 °C)	Dark straw
460 °F (238 °C)	Yellow-brown
500 °F (260 °C)	Spotted red-brown
540 °F (282 °C)	Purple
580 °F (308 °C)	Dark blue
620 °F (327 °C)	Pale blue
660 °F (349 °C)	Light blue-gray

Fig. 19-25. Before metal becomes incandescent (glowing red), steel will pass through above colors. Colors are also useful for tempering steel if no pyrometer is fitted to the furance.

The following procedure is recommended when hardening carbon steels. Oxidation of the metal during heat treating can be avoided by wrapping the part in stainless steel foil, Fig. 19-26.

1. Place the metal in the furnace and bring the metal to the desired temperature. If it is a gas furnace, light it according to the manufacturer's instructions.

DANGER! When lighting a gas furnace, stand to one side and do NOT look into the fire box.

When the gas has ignited, adjust the air and gas valves for the best operation.

2. Heat the metal to its critical temperature 1300-1600 °F or (715-871 °C). Avoid placing the part being heat treated directly in the gas flames. If it HAS not been placed in a section of stainless steel foil, position it in a section of iron pipe, as shown in Fig. 19-27. Allow it to ''soak'' in the furnace until it is heated evenly throughout.
3. Preheat the tong jaws and remove the piece from the furnace.

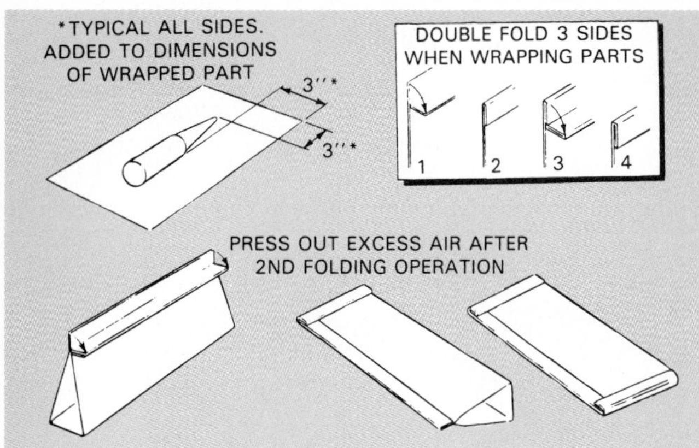
*TYPICAL ALL SIDES.
ADDED TO DIMENSIONS
OF WRAPPED PART
3''*
3''*
DOUBLE FOLD 3 SIDES
WHEN WRAPPING PARTS
1
2
3
4
PRESS OUT EXCESS AIR AFTER
2ND FOLDING OPERATION

Fig. 19-26. Study procedure for wrapping parts to be heat treated in stainless steel foil as protection against oxidation and decarburization. (Lindberg, Div. Solar Industries)

Fig. 19-27. Work being heat treated must be protected from direct flames in a gas furnace by inserting it in a section of pipe. Elevate unit from furnace floor to permit uniform heating.

WARNING! Dress properly for this operation, Fig. 19-28. The metal is very hot and serious burns can result from relatively minor accidents.

4. Quench the piece in water, brine, or oil depending upon the type of steel used. Steels are usually classed, except for some alloy steels, as water hardening or oil hardening types, according to the quenching medium to be employed on them. The quenching technique is critical.
5. To secure even hardness throughout the piece, dip long slender sections straight down into the quenching fluid with an up and down motion. Avoid a circular motion as this may cause the piece to warp. Other shaped parts should be moved around in such a manner to permit them to cool quickly and evenly.

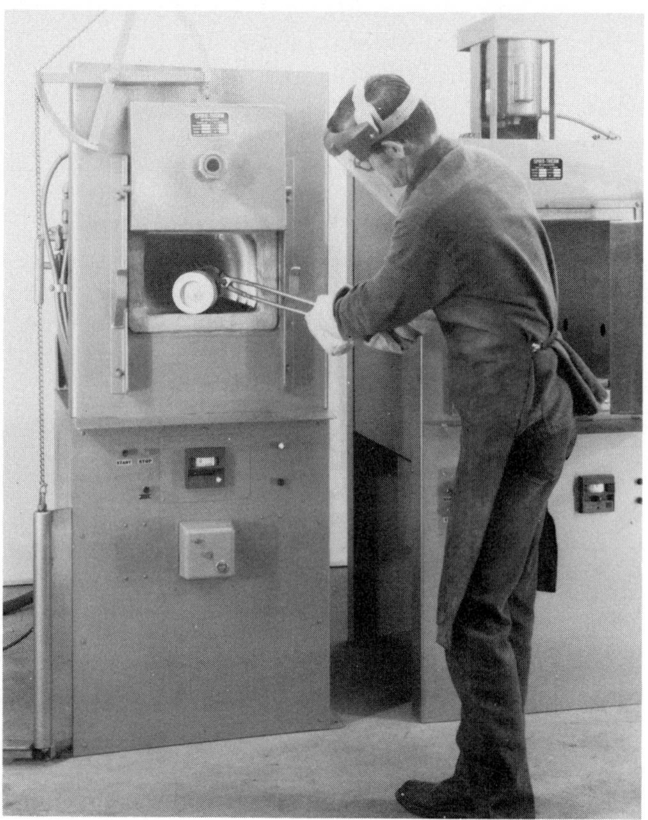

Fig. 19-28. Heat treating temperatures are very hot. Dress properly for job and keep area around furnace clean so there is no danger of slipping or stumbling. Also preheat tongs before grasping heated part. (McEnglevan Heat Treating & Mfg. Co.)

Steel, that has been hardened properly, will be ''glass hard'' and too brittle for most purposes. Hardness may be checked by trying to file the work surface. A file will NOT cut the surface if the piece has been hardened properly. Do NOT use a NEW FILE for testing hardness!

TEMPERING CARBON STEEL

As mentioned, tempering or drawing is employed to relieve the stresses and strains that developed in the metal during hardening. Until tempering is done, the ''glass hard'' steel may crack or shatter from shock (dropping, striking, etc.) or from sudden changes in temperature.

Tempering is done as follows:

1. Polish the hardened piece with abrasive cloth.
2. Reheat the piece to the correct tempering temperature. This is determined by the type of steel and the job the finished part is to do. It will

range from about 380 °F or 193 °C (metal will turn a silvery yellow or straw color) to 700 °F or 371 °C (metal will turn light gray or blue color). Color charts available from steel companies can be employed as a guide for determining when the desired temperature is attained if no pyrometer is available on the furnace. Quench immediately upon reaching the required temperature.

3. Small tools are best tempered by placing them on a steel plate that has been heated red hot. Have the point of the tool extend beyond the edge of the plate, Fig. 19-29. Watch the temper color as it heats up. Quench when the proper color has reached the tool point.

NOTE! Hot liquid baths of oil, molten salts, and lead are often employed in place of a furnace for heating parts to their proper tempering temperature. The pieces are held in the bath until the heat permeates them. They are then removed from the bath and allowed to cool in still air.

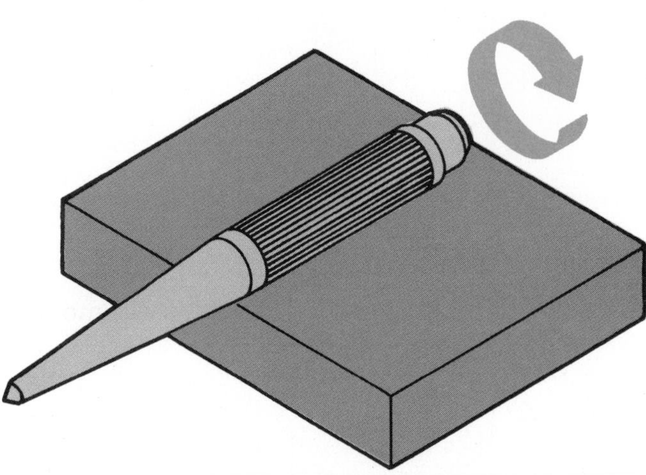

Fig. 19-29. Use a heated steel plate when tempering small tools. Have point of tool extend beyond edge of plate.

CASE HARDENING LOW CARBON STEEL

Of the several case hardening techniques, the simplest is ***carburizing,*** requiring a minimum of equipment. It uses a nonpoisonous commercial compound such as ***Kasenit.***

DANGER! Cyaniding is NOT recommended because cyanide is a deadly poison and VERY DANGEROUS to use under any but ideal conditions.

WARNING! Work in a well-ventilated area and wear full face protection, leather apron, and heat resistant gloves (avoid asbestos gloves) when using Kasenit.

There are two methods recommended for using Kasenit to harden low carbon steel. The first method is as follows:

1. Bring the furnace to temperature.
2. Bring the work to a bright red (1650-1700 °F or 900-930 °C). Use a pyrometer or thermocouple to monitor the temperature, Fig. 19-30.
3. Dip, roll, or spinkle Kasenit on the piece, Fig. 19-31. The powder will melt and adhere to the surface forming a shell.
4. Reheat to a bright red and hold at this temperature for a few minutes.
5. Quench in cold water.

The second method for case hardening low carbon steel is to:

1. Secure a container large enough to hold the work. A tin can will do if care is taken to burn off the tin coating before use.
2. Completely cover the job with Kasenit, Fig. 19-8.
3. Place the entire unit in the furnace and heat to a red heat. Hold the temperature for 5 to 30 minutes depending upon the depth of case required.
4. Quench the job in clean, cool water using DRY tongs to remove the piece from the container.

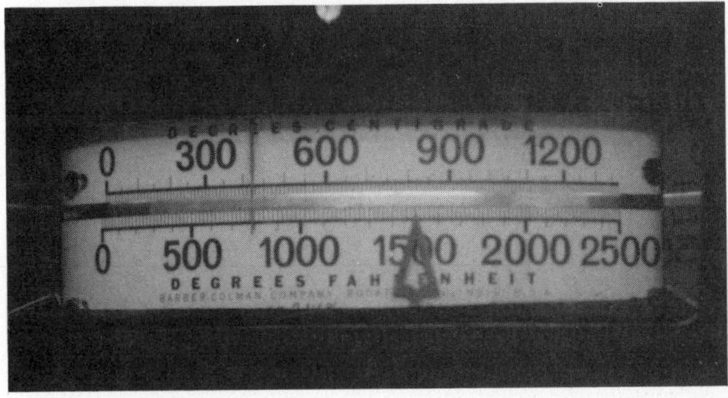

Fig. 19-30. A pyrometer or thermocouple (electronic temperature measuring device) should be used to monitor, and on electric furnaces, maintain desired temperature. Thermocouple has been set for 1500 °F and furnace has already attained a temperature of 800 °F. Approximately what would these temperatures be on the upper Centigrade scale?

HARDNESS TESTING

Hardness testing will make sure that the metal has been brought to the proper degree of heat treatment or cold work, or a combination of the two, for its particular use. With hardness testing, it is also

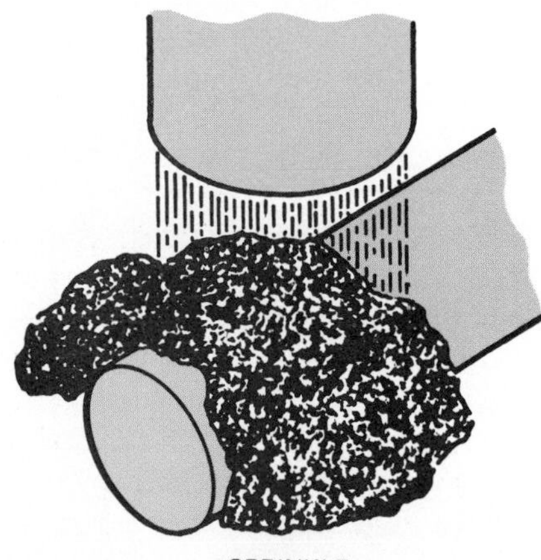

Fig. 19-31. Dip, roll, or sprinkle Kasenit, or other safe commercial case hardening compound, on work until a shell of compound has been formed.

possible to establish standards for hardness that can be cited on drawings and specifications.

The more commonly utilized technique is concerned with the distance a steel ball or special shaped diamond penetrates into the metal under a specific load.

Brinell and Rockwell testing machines, known as ***indention hardness testers,*** are used. The ***hardness number*** indicates the degree of material hardness. Refer to Fig. 19-32.

HARDNESS CONVERSION TABLE
(APPROXIMATE)

Values vary depending on grades and conditions of material involved. Rockwell "B" Scale should not be used over B-100. The "C" Scale should not be used under C-20.

Brinell	Rockwell		Shore Sclero-scope	Tensile Lbs. Sq. In.	Brinell	Rockwell	Shore Sclero-scope	Tensile Lbs. Sq. In.
Hard No.	B Scale	C Scale	Hard No.	In 1000 Lbs.	Hard No.	B Scale	Hard No.	In 1000 Lbs.
782	...	72	107	383	163	84	25	84
744	...	69	100	365	159	83	25	82
713	...	67	96	350	156	82	24	80
683	...	65	92	334	153	81	24	79
652	...	63	88	318	149	80	23	78
627	...	61	85	307	146	78	23	77
600	...	59	81	294	143	77	22	76
578	...	58	78	284	140	76	..	74
555	...	56	75	271	137	75	..	73
532	...	54	72	260	134	74	..	71
512	...	52	70	251	131	72	..	70
495	...	51	68	242	128	71	..	69
477	...	49	66	233	126	70	..	67
460	...	48	64	226	124	69	..	66
444	...	47	61	217	121	67	..	65
430	...	45	59	210	118	66	..	63
418	...	44	57	205	116	65	..	62
402	...	43	55	197	114	64	..	61
387	...	41	53	189	112	62	..	60
375	...	40	52	183	109	61	..	59
364	...	39	50	178	107	59	..	58
351	*(110)*	38	49	172	105	58	..	57
340	*(109)*	37	47	167	103	57	..	56
332	*(108.5)*	36	46	162	101	56	..	55
321	*(108)*	35	45	157	99	54	..	54
311	*(107.5)*	34	44	152	97	53	..	53
302	*(107)*	33	42	148	96	52	..	53
293	*(106)*	31	41	144	95	51	..	52
286	*(105.5)*	30	40	140	93	50	..	52
277	*(104.5)*	29	39	136	92	49	..	51
269	*(104)*	28	38	132	90	48	..	50
262	*(103)*	27	37	128	88	47	..	49
255	*(102)*	26	36	125	87	46	..	48
248	*(101)*	25	36	121	86	45	..	48
241	100	24	35	118	85	44	..	47
235	99	*(22)*	34	115	83	43	..	47
228	98	*(21)*	33	113	82	42	..	46
223	97	*(20)*	33	109	81	41	..	46
217	96	*(19)*	32	106	80	40	..	45
212	95	*(18)*	31	104	79	39	..	45
207	94	*(17)*	30	101	78	38	..	44
202	93	*(15)*	30	99	77	37	..	44
196	92	*(13)*	29	96	76	36	..	43
192	91	*(12)*	29	94	75	35	..	43
187	90	*(10)*	28	91	74	33	..	42
183	89	*(9)*	28	90	73	31	..	42
179	88	..	27	89	72	30	..	41
174	87	..	27	88	71	29	..	41
170	86	..	26	86	70	27	..	40
166	85	..	26	85	69	26	..	40

Fig. 19-32. Study hardness conversion table.

Brinell hardness tester

The ***Brinell Hardness Tester*** is employed extensively in the laboratory and in production, Fig. 19-33. The Brinell test is a measure of resistance of the material and is an excellent index of such factors as machinability, uniformity of grade, temper after heat treatment, and body hardness of the metal.

The Brinell test for determining the hardness of a metallic material consists of applying a known load to the surface of the material being tested through a hardened steel ball of known diameter. The diameter, or depth, of the resulting permanent impression in the metal is measured. The Brinell hardness is taken as the quotient of the applied load, divided by the area of the surface of the impression which is assumed to be spherical:

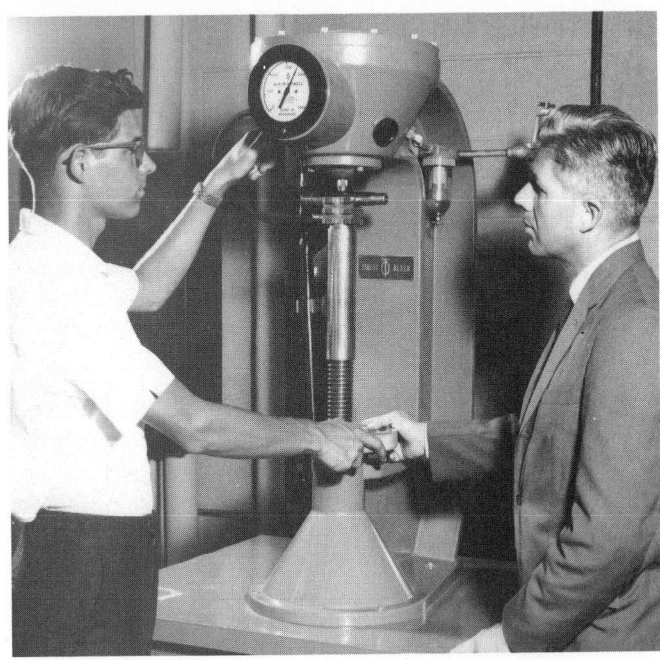

Fig. 19-33. Engineers are preparing to use Brinell Hardness Tester. (Tinius Olsen Testing Machine Co.)

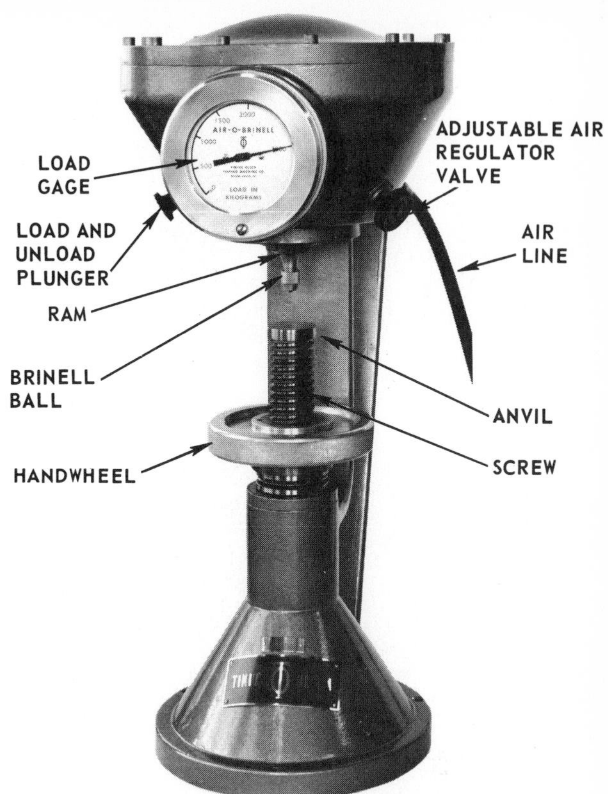

Fig. 19-34. Study parts of compressed air type Brinell Hardness Tester. (Tinius Olsen Testing Machine Co.)

$$BHN = \frac{P}{\frac{D}{2}\left(D - \sqrt{D^2 - d^2}\right)}$$

Where:

BHN = Brinell Hardness Number in kilograms per square millimeter.

D = Diameter of steel ball in millimeters.

d = Diameter of impression in millimeters.

To perform the Brinell test on a compressed air type hardness tester, Fig. 19-34, the following instructions are suggested:

1. Gradually turn the adjustable air regulator valve in a clockwise direction until the desired load is indicated on the load gauge. Any Brinell load from 500 to 3000 kilograms can be selected by adjusting this regulator. Check the load reading when making the initial test of a series, and adjust the air regulator valve, if necessary, so that the desired load is indicated on the dial when a specimen is actually under load.
2. Insert the test specimen on the anvil, then turn the handwheel until the gap allows for insertion of the specimen into the machine. This distance between the surface of the specimen and the Brinell ball should be kept to a minimum prior to applying the load.
3. Pull out the load and unload plunger on the left side of the machine. The load will be instantly released and the test completed when the plunger is pushed in.
4. Read the impression with the special microscope and obtain the Brinell hardness number from the hardness table.

   NOTE! Do NOT apply the test load when the anvil is within travel range of the ram and Brinell ball unless the test specimen is in place. Otherwise, a Brinell impression will be made on the machine's anvil.
5. The thickness of the test piece must be such that no bulge, or other marking showing the effects of the load, appears on the side opposite the impression.
6. When necessary, the surface on which the impression is to be made should be filed, ground, machined, or polished with abrasive material so that the edge of the impression shall be clearly enough defined to permit measurement of the diameter to the specified accuracy.

Rockwell hardness tester

The Rockwell testing technique is the most widely used of all hardness testing.

When using the ***Rockwell Hardness Tester,*** BOTH a steel ball and specially designed diamond cone penetrator are utilized, Fig. 19-35. A minor load of 10 kg is applied first, after which the dial gage is set to zero. The major load is then added and removed.

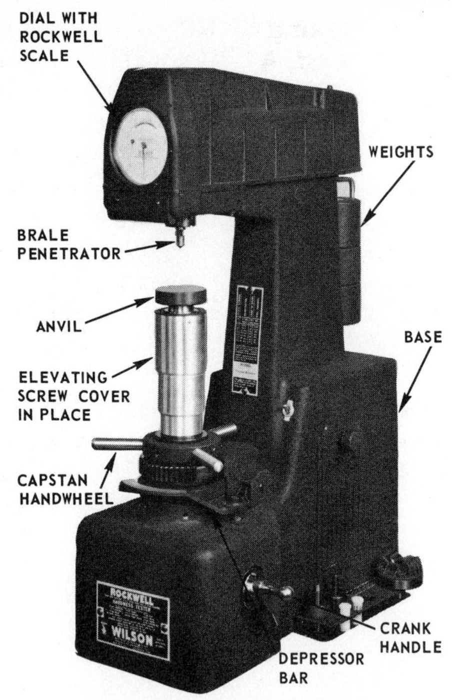
DIAL WITH ROCKWELL SCALE
WEIGHTS
BRALE PENETRATOR
ANVIL
BASE
ELEVATING SCREW COVER IN PLACE
CAPSTAN HANDWHEEL
ROCKWELL
WILSON
CRANK HANDLE
DEPRESSOR BAR

Fig. 19-35. Study components of Rockwell Hardness Tester. (Wilson Mechanical Instrument Div., American Chain and Cable Co., Inc.)

The hardness number represents the additional depth to which the test ball, Fig. 19-36, or conical diamond penetrator, Fig. 19-37, is driven by the major load BEYOND the depth of the previously applied light load. The hardness number is automatically indicated on the dial gage.

Typically, a 1/16 in. steel ball is employed in conjunction with the 100 kg load, for testing such metals as brass, bronze, and soft steel. All readings made with the 1/16 in. ball and 100 kg load are B readings and the letter B must be prefixed or placed before the number.

NOTE! There is no Rockwell hardness designated by a figure alone. It must ALWAYS be prefixed by the proper scale letter. See Fig. 19-38.

The conical diamond test point known as a ***Brale® Penetrator,*** is used with the 150 kg load for testing hardened steel or any hard metals, Fig. 19-39. All readings with the Brale® Penetrator and 150 kg load are C readings and the letter C must be prefixed to the hardness number.

Special penetrators are available for testing soft materials. The scale designation depends upon the ball size used to make the test. See Fig. 19-38.

Two weights are normally supplied with the Rockwell Tester. One of them has a red marking and

DIAL IS NOW IDLE
DIAL NOW SET TO ZERO
DIAL READS B-C PLUS CONSTANT AMOUNT DUE TO ADDED SPRING OF MACHINE UNDER MAJOR LOAD, BUT WHOSE VALUE WILL DISAPPEAR FROM DIAL READING WHEN MAJOR LOAD IS WITHDRAWN
ROCKWELL HARDNESS NUMBER IS REGISTERED ON DIAL
DIAL IS NOW IDLE
MAJOR LOAD (Not Applied)
MINOR LOAD (Not applied)
1/16" DIA. STEEL BAL.
TEST SPECIMEN
ELEVATING SCREW
MAJOR LOAD (Applied)
MINOR LOAD (Applied)
MAJOR LOAD Released)
MINOR LOAD (Released
WORK PLACED IN TESTER
TEXT SPECIMEN RAISED TO APPLY MINOR LOAD
DEPRESSOR BAR IS RELEASED. MAJOR LOAD IS APPLIED
CRANK IS TURNED TO REMOVE MAJOR LOAD. MINOR LOAD STILL APPLIED
TEST SPECIMEN HAS BEEN LOWERED. IT CAN NOW BE REMOVED FROM TESTER

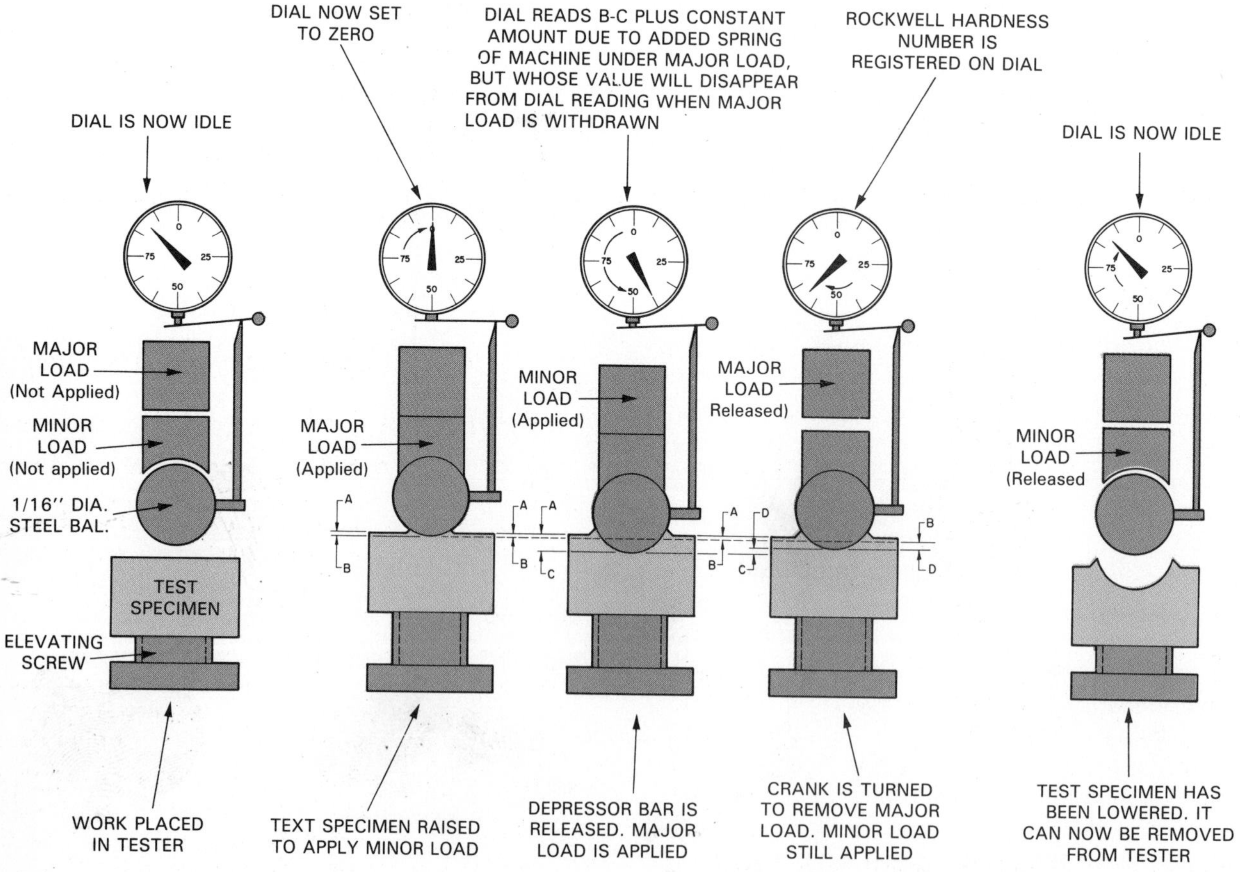

Fig. 19-36. Diagram shows how 1/16 in. steel ball penetrator is used to make a Rockwell Hardness reading. Size of ball has been greatly exaggerated for clarity.

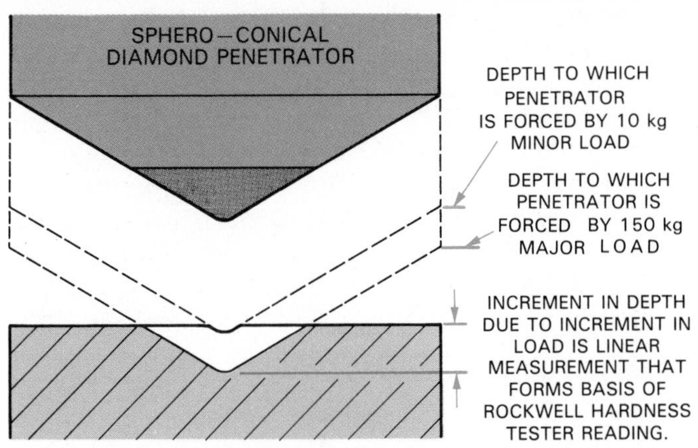

Fig. 19-37. Note how conical diamond (Brale®) penetrator is employed to determine hardness.

SCALE SYMBOL	PENETRATOR	MAJOR LOAD kg	DIAL FIGURES	TYPICAL APPLICATIONS OF SCALES
B	1/16 in. ball	100	Red	Copper alloys, soft steels, aluminum alloys, malleable iron, etc.
C	Diamond cone	150	Black	Steel, hard cast iron, titanium, deep case hardened steel, etc.
A	Diamond cone	60	Black	Cemented carbides, thin, steel, and shallow case hardened steel.
D	Diamond cone	100	Black	Thin steel, medium case hardened steel, and pearlite malleable iron.
E	1/8 in. ball	100	Red	Cast iron, aluminum and magnesium alloys, and bearing materials.
F	1/16 in. ball	60	Red	Annealed copper alloys, thin soft sheet metals.
G	1/16 in. ball	150	Red	Phosphor bronze, beryllium copper, malleable iron, etc.
H	1/8 in. ball	60	Red	Aluminum, zinc, lead.

Fig. 19-38. Symbol is used as a prefix to value read from dial. It depends upon load, type of penetrator, and scale from which dial readings are taken. They are shown above.

Fig. 19-39. This is a closeup of a conical diamond or Brale® Penetrator. (Wilson Mech. Inst. Div., American Chain and Cable Co., Inc.)

the other black. However, three different loads can be applied. The weight arm together with the link and weight pan apply a load of 60 kg. The weight with the red marking is placed on the weight pan for the 100 kg load. A 150 kg load is applied when the weight with the black marking is added. The black weight is NEVER used alone.

When making tests with the 100 kg load, the dial scale with red figures is used. The dial scale with black figures apply with the 150 kg load.

The 60 kg load (weight of weight pan alone) is employed with the Brale® Penetrator for testing extremely hard metals, such as tungsten carbide alloys. The 1/16 in. ball is extensively used for testing sheet brass.

To operate the Rockwell Hardness Tester, select the proper penetrator point and mount it. Check that the weight for the desired test load is in position. Place the correct anvil in the elevating screw with extreme care or the penetrator might be damaged. See Fig. 19-40.

Inspect the test specimen and remove any scale or burr that would flatten under the test and give a false reading.

The following operations are considered to give the most precise hardness readings. Note that the numbers in Fig. 19-41 correspond to the following sequence numbers.

1. Place the test specimen on the anvil.
2. Gently raise the piece until it comes into contact with the penetrator. Continue to turn the capstan handwheel slowly until the small pointer (8) is nearly vertical and slightly to the right of the dot. Continue raising the work until the long pointer is approximately upright (within five divisions plus or minus). The minor load (10 kg) has now been applied.
3. Set the dial to ''zero'' (line marked ''set'') by

Fig. 19-40. Protect penetrator with your finger when removing and replacing an anvil. Hitting hard, but brittle, diamond with anvil may fracture diamond. A steel ball penetrator may be deformed if struct with anvil. (Wilson Mech. Inst. Div., American Chain and Cable Co., Inc.)

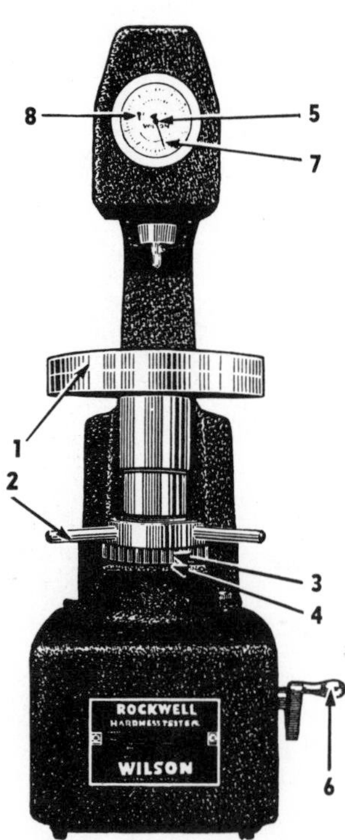

Fig. 19-41. Study Rockwell Tester operating procedures.

turning the knurled ring that is located below the handwheel, Fig. 19-42.

4. Carefully push down on the depressor bar to apply the major load. The penetrator is forced into the work. The depth it penetrates depends upon the metal's hardness.

Fig. 19-42. Thumb is used to "zero-in" dial pointer before applying major load. Same thumb is used to push down depressor bar to apply major load.
(Wilson Mech. Inst. Div., American Chain and Cable Co., Inc.)

5. Watch the pointer until it comes to rest.
6. Pull the crank handle forward to lift the major load, but leave the minor load still applied.
7. Read the Rockwell Hardness Number. If the test has been made with the 1/16 in. ball, and the load is 100 kg, the reading is taken from the red scale and the letter B is prefixed to the number to signify the condition of the test. The letter C is prefixed to the number if the Brale® Penetrator and 150 kg load were employed. The reading is then made from the black scale.

After completing the test, lower the work from the penetrator and remove it from the testing machine.

Like other precision tools, the Rockwell Tester must be handled with care if it is to maintain its accuracy. There are a few precausions that must be observed:

1. When moving the tester, it should only be grasped by the cast iron base and NEVER by any parts that are attached to the base.
2. The tester should be leveled on a solid bench, in a location free from grit and vibration. Keep the machine covered with NOT in use!
3. False readings will result if the shoulder on the Brale® Penetrator is NOT kept clean, Fig. 19-43. Carefully wipe it with a clean soft cloth before installation in the tester.
4. Use only lubricants specified by the manufacturer.
5. Long work must be supported.
6. A smooth surface is necessary for accurate readings. Clean the work with an abrasive cloth to remove scale and roughness. Castings and forgings should have a spot ground or machined where the test is to be made so that

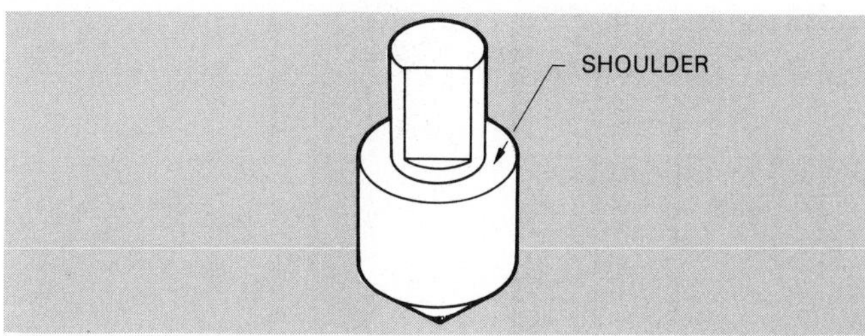

Fig. 19-43. Shoulder of Brale® Penetrator must be free of dirt and burrs before it is mounted in tester. Otherwise, incorrect readings will result.

the penetrator will test the true metal underneath.

7. The specimen must be thick enough so that the undersurface does NOT show the slightest indication of the test.
8. Corrections must be added to readings made on round stock if it is NOT possible to file or grind a flat spot in the test area.
9. Be careful NOT to damage the penetrator or the anvil by forcing them together when a test piece is NOT in the machine.
10. If used on case hardened steel, accurate readings cannot be made unless the ''case'' is several times as thick as the indentation depth.

Scleroscope hardness testing machine

A ***Scleroscope*** drops a hammer onto the test piece and the resulting bounch or rebound of the hammer is used to determine hardness.

Two styles of Scleroscopes are in use. One is fitted with a vertical scale, Fig. 19-44; the other is a dial recording instrument, Fig. 19-45. They may be employed for testing the hardness of all metals, ferrous and nonferrous, polished or unpolished, with virtually no limitation in size or shape, Fig. 19-46.

Hardness testing with the Scleroscope is essentially a nonmarring test. No craters that may require refinishing of the test area are produced.

The theory of the Scleroscope hardness test involves a diamond hammer that is dropped from a fixed height and makes a minute indentation in the metal. The hammer rebounds, but NOT to its original height, because some of the energy in the falling hammer is dissipated in producing the tiny indentation. The rebound of the hammer varies in proportion to the hardness of the metal — the harder the metal, the higher the rebound.

The tester scale consists of units which are determined by dividing the average rebound of the hammer from quenched tool steel of ultimate hardness into 100 parts. These rebounds will range from 95 to 105. The scale is carried higher than 100 to cover super-hard metals, Fig. 19-46.

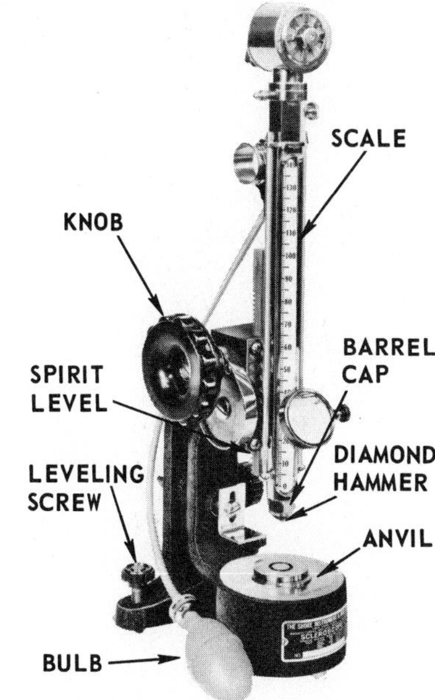

Fig. 19-44. Study parts of Shore Scleroscope. (Shore Inst. and Mfg. Co., Inc.)

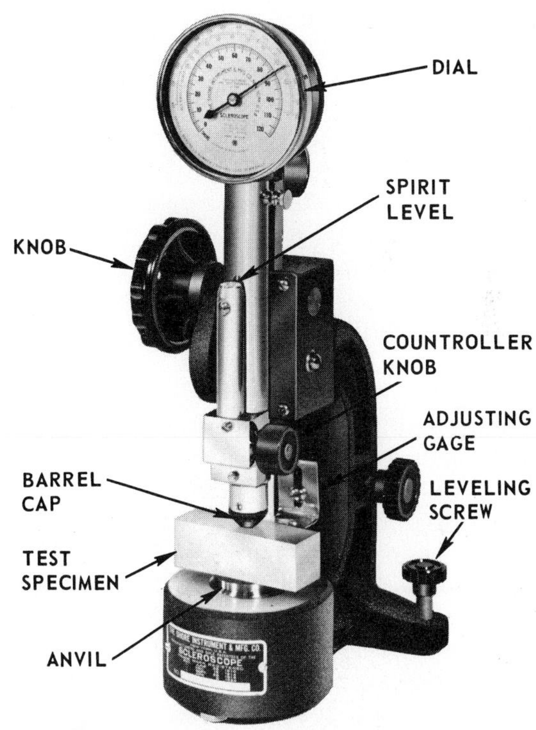

Fig. 19-45. This is a direct reading Scleroscope. (Shore Inst. and Mfg. Co., Inc.)

The Scleroscope is capable of yielding accurate hardness readings on the softest or the hardest metals without changing the scale or diamond hammer.

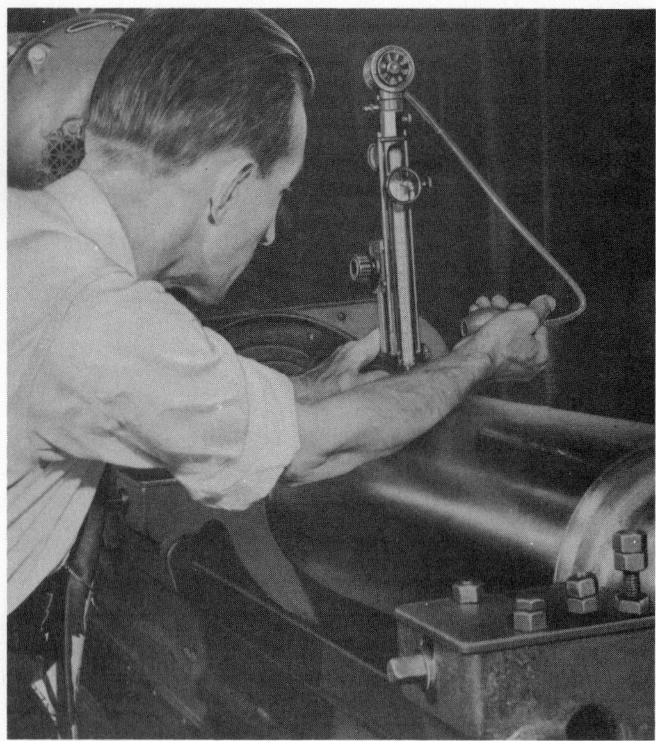

Fig. 19-46. Vertical scale Scleroscope is being employed to test hardness of a forming roll without making substantial dent in its surface. (Shore Inst. and Mfg. Co., Inc.)

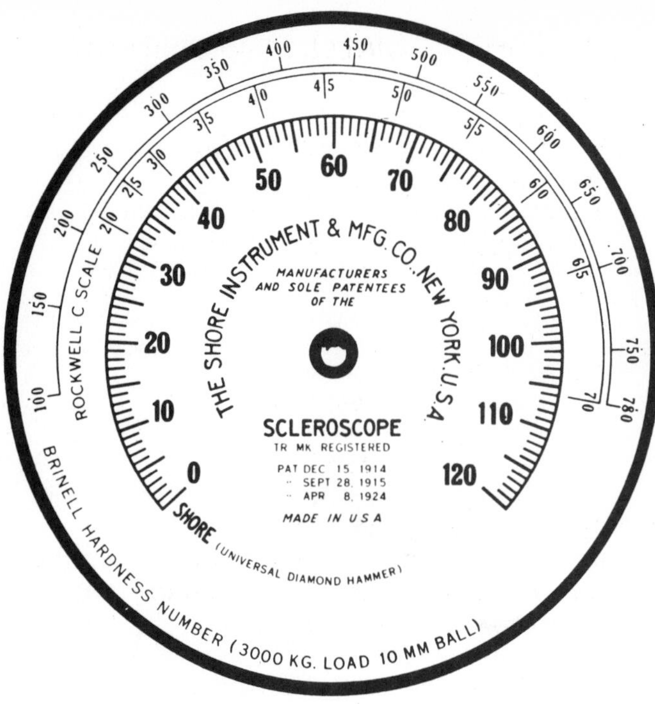

Fig. 19-47. Study scales on Scleroscope dial.

When testing objects within the capacity of the clamping stand of the vertical scale. Scleroscope, the specimen must be mounted on the anvil, Fig. 19-43. The unit should be leveled by turning the leveling screws while observing the spirit level. The instrument is operated pneumatically by means of the rubber bulb.

To perform a test, revolve the knob to bring the barrel cap firmly into contact with the test specimen. It is essential that a firm pressure be maintained during the test. Squeeze and release the rubber bulb to draw the hammer to the up position. As torque is maintained on the knob, again squeeze and release the rubber bulb and observe the reading on the scale.

The height to which the hammer rebounds on the first bounce indicates the hardness of the specimen. The average of several tests is the correct hardness of the piece. Do NOT make more than one test at a given spot or false readings will result.

While the method may sound unorthodox, the results are very close to those obtained with Brinell and Rockwell testers.

The dial recording Scleroscope works on the same rebound principle but is direct reading, Fig. 19-44. Like the vertical scale instrument, the average of several tests is the correct hardness of the test piece.

HEAT TREATING SAFETY

1. Make sure that the furnace is in good operating condition before attempting to use it. Avoid lighting a furnace until you have been instructed in its safe operation. NEVER stand in front of a gas furnace nor look into it when it is being ignited.
2. Heat treating involves metal being raised to a very high temperatures. Handle the hot metal with the appropriate tools!
3. Wear an approved full face safety shield and the proper protective clothing! Wear heat resistant gloves (not asbestos gloves) and a leather apron (NEVER a cloth apron, especially not a greasy nor oil soaked apron).
4. Never look into the furnace unless you are wearing tinted goggles or glasses under your face shield.
5. Work only in areas that are well ventilated.
6. Do NOT stand over the quenching bath when immersing hot work.
7. The use of potassium cyanide in a school shop or lab as a case hardening medium must be avoided. If you work in a situation that permits its use, do NOT breath any of the resulting fumes. Wash thoroughly after completing the heat treating operations.
8. Have any burn, cut or bruise treated immediately no matter how minor it may appear.

TEST YOUR KNOWLEDGE—Chapter 19

Please do not write in the text. Place your answers on a separate sheet of paper.

1. Heat treating is done to:
   a. Obtain certain desirable changes in the metal's physical characteristics.
   b. Increase the hardness of the metal.
   c. Soften (anneal) the metal.
   d. All of the above.
   e. None of the above.
2. What does heat treating involve?
3. Carbon steels are classified by the percentage of carbon in "points" or hundredths of 1 percent they contain. If this statement is true, briefly explain what it means.
4. In addition to steel, list four other metals that area capable of being heat treated.
5. In addition to water, ________, ________, ________ of cold ________ and ________ ________ are used as quenching mediums.

The following are matching questions. Each word in the left column matches one of the sentences. On your work sheet, place the letter next to the appropriate number to match the words and statements.

6. Stress relieving.
7. Annealing.
8. Normalizing.
9. Case hardening.
10. Surface hardening.
11. Hardening.

a. Involves heating metal to slightly above its upper critical temperature and then permitting it to cool slowly in insulating material. Hardness of metal is reduced.
b. Used to refine grain structure of steel and to improve its machinability.
c. Done to reduce stress that has developed in parts that have been welded, machined, or cold worked during process.
d. Used when only a medium hard surface is required on high carbon or alloy steels.
e. Only outer surface of low carbon steel is hardened while inner portion remains relatively soft and tough.
f. Accomplished by heating metal to its critical range and cooling rapidly.

12. Tempering a section of hardened steel makes it:
    a. Brittle.
    b. Soft.
    c. Tough.
    d. All of the above.
    e. None of the above.
13. What advantages does an electric heat treating furnace have over a gas fired heat treating furnace?
14. The ________ is used to measure and monitor the high temperatures needed in heat treating.
15. What is hardness testing?
16. List three types of commonly used hardness testers.
17. What safety precautions must be observed when lighting a gas-fired heat treating furnace?
18. List five (5) safety precautions that must be observed when heat treating metal.

RESEARCH AND DEVELOPMENT

1. Prepare a glossary of heat treating terms. Duplicate it and make copies available to member of the class.
2. When heat treating aluminum alloys, the terms solution heat treatment and precipitation heat treatment are used. What do they mean?
3. The Metcalf Experiment is one method employed to show the grain structure of heat treated steel and the effects caused by over heating. How is it performed? Perform the experiment and mount the pieces that show the results on a panel for observation.
4. Demonstrate the proper way to harden and temper a piece of tool steel.
5. Demonstrate the proper way to case harden low carbon steel by carburizing. Use Kasenit as the carbon source.
6. Secure samples of work that have been heat treated by various techniques.
7. The Moh Scale was the first hardness testing technique. Research the Moh Scale and explain how it was used?
8. Secure handbooks from the various steel manufacturers and/or distributors for inclusion in the shop's technical library.
9. Arrange a field trip to a local industry that has a heat treating area. Ask to have the various hardness testers demonstrated.
10. Secure literate on various hardness testers for inclusion in shop's technical library. library.

Chapter 20

METAL FINISHES

After studying this chapter, you will be able to:

- Describe how the quality of a machined surface is determined.
- Explain why the quality of a machined surface has a direct bearing on production costs.
- Describe some metal finishing techniques.

The term ***metal finish*** refers to the degree of smoothness or roughness remaining on a part surface. A machined surface has geometric irregularities that are produced by the cutting action of the tool. Each type of cutting tool leaves its own characteristic surface marking. See Fig. 20-1.

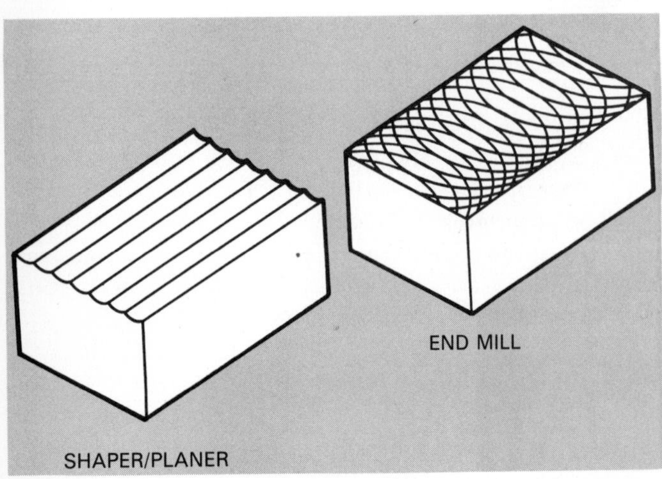

Fig. 20-1. Each type cutting tool leaves its characteristic markings.

QUALITY OF MACHINED SURFACES

At one time, the quality of a machined surface was noted by the symbol "𝑓." This was NOT based on specific standards. Therefore, the engineer or drafter included explanatory notes such as rough grind, smooth turn, surface grind, etc., on the drawing. These notes indicate the general surface finish quality desired.

The technique left much to be desired because each machinist interpreted the specifications differently. Often, the piece was better finished than it had to be, which increased its production cost. The problem reached such proportions that in the early 1940s the Standard Associations of Canada, Great Britain, and the United States developed tentative surface roughness or texture standards.

The terms and ratings of surface roughness or texture standards relate to surfaces produced by machining, grinding, casting, molding, forging, etc. They are NOT concerned with luster, appearance, color, corrosion resistance, wear resistance, hardness, and the many other characteristics that may be governing considerations in specific applications.

The standards also did NOT define the different degrees of surface roughness and waviness suitable for specific purposes, nor do they specify the means by which any degree of such irregularities may be obtained or produced.

The standards deal only with the height, width, and direction of surface irregularities, since these are of practical importance in specifications.

The present surface finish system averages, arithmetically, the irregular contours on a surface in ***microinches*** (millionths of an inch and shown as XX μ in.) or ***micrometers*** (millionths of a meter and shown as XX μ m). With established standards, a universal set of numbers and symbols indicating surface roughness or texture are now used on drawings and in specifications, Figs. 20-2 and 20-3.

In addition to surface roughness, other surface conditions are considered and values given for them.

SYMBOL	DESCRIPTION
 √	Basic Surface Roughness/Texture Symbol. Surface may be produced by any method.
 √ (with horizontal bar)	Material Removal by Machining Required. Horizontal bar indicates that material removal by machining is required. Material must be provided for that purpose.
.187  √ (with horizontal bar)	Material Removal Allowance. Number indicates amount of material that must be removed in inches/millimeters. Tolerances may be added.
 √ (with circle in vee)	Material Removal Prohibited. Circle in vee indicates that surface must be produced by processes such as casting, forging, hot finishing, cold finishing, powder metallurgy, or injection molding without subsequent removal of material.
 √‾	Surface Texture Symbol. To be used when any surface characteristics are specified above horizontal line or to right of symbol. Surface may be produced by any method.

Fig. 20-2. Study surface roughness or texture symbols.

ROUGHNESS HEIGHT RATING		SURFACE DESCRIPTION	PROCESS
Microinch	Micrometer		
1000√	25.2√	Very rough	Saw and torch cutting, forging, or sand casting.
500√	12.5√	Rough machining	Heavy cuts, coarse feeds in turning, milling, and boring.
250√	6.3√	Coarse	Very coarse surface grind, rapid feeds in turning, planing, milling, boring, and filing.
125√	3.2√	Medium	Machining operations with sharp tools, high speeds, fine feeds, and light cuts.
63√	1.6√	Good machine finish	Sharp tools, high speeds, extra fine feeds, and cuts.
32√	0.8√	High grade machine finish	Extremely fine feeds and cuts on lathe, mill, and shaper required. Easily produced by centerless, cylindrical, and surface grinder.
16√	0.4√	High quality machine finish	Very smooth reaming or fine cylindrical or surface grinding, or coarse hone or lapping of surface.
8√	0.2√	Very fine machine finish	Fine honing and lapping of surface.
2 - 4√	0.05 0.1√	Extremely smooth machine finish	Extra fine honing and lapping of surface.

Fig. 20-3. Study chart giving description of roughness values. When used on a drawing, number indicates roughest surface in microinches/micrometers that is acceptable for that specific application.

Waviness ordinarily takes the form of smoothly rounded waves and are caused by tool and machine vibration and chatter, Fig. 20-4. This is measured with reference to a nominal or geometrically perfect surface. Waviness is of GREATER magnitude than roughness.

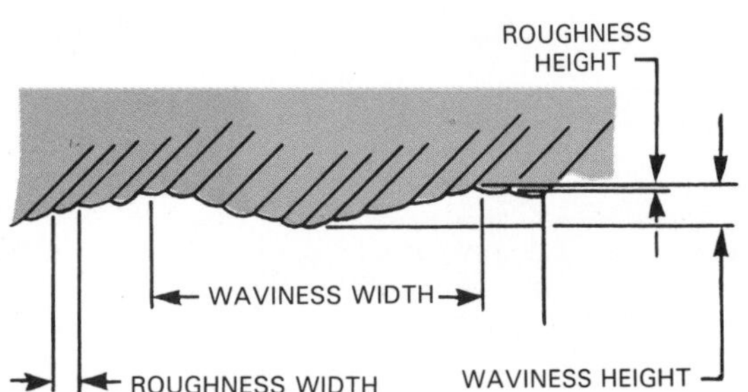

Fig. 20-4. Note how surface waviness is measured.

Waviness is specified in inches (millimeters) as the maximum allowable peak-to-valley height. It is measured with a sensitive dial indicator with an 0.06 in. (1.5 mm) diameter ball contact. Fig. 20-5 shows how acceptable waviness tolerances are specified on drawings, in reports, and as specifications.

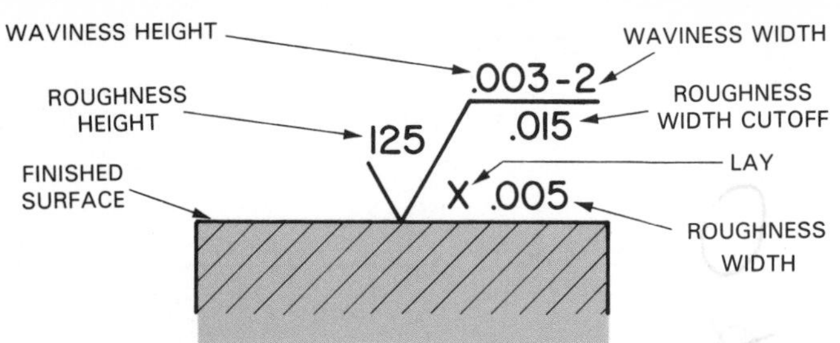

Fig. 20-5. On drawings, symbols show roughness, waviness, and lay. They specify finishes required on a surface.

Lay is the term that describes the direction of the predominant tool marks, grain, or pattern of surface roughness. See Fig. 20-6.

Description of degrees of surface roughness

Milling and turning can produce surface finishes in the order of 125 to 8 μ inches (3.2 to 0.2 μ meters). Grinding, depending on the coarseness of the wheel and feed rate, has a range of 63 to 4 μ inches (1.6 to 0.1 μ meters).

Lapping produces the SMOOTHEST finish on a production basis, Figs. 20-7 and 20-8. It is used by automotive and other industries to produce mating surfaces flat and smooth enough to form a gasketless oil tight seal in automatic transmissions and other applications. Surfaces are as fine as 2 to 3 μ inches (0.05 to 0.07 μ meters).

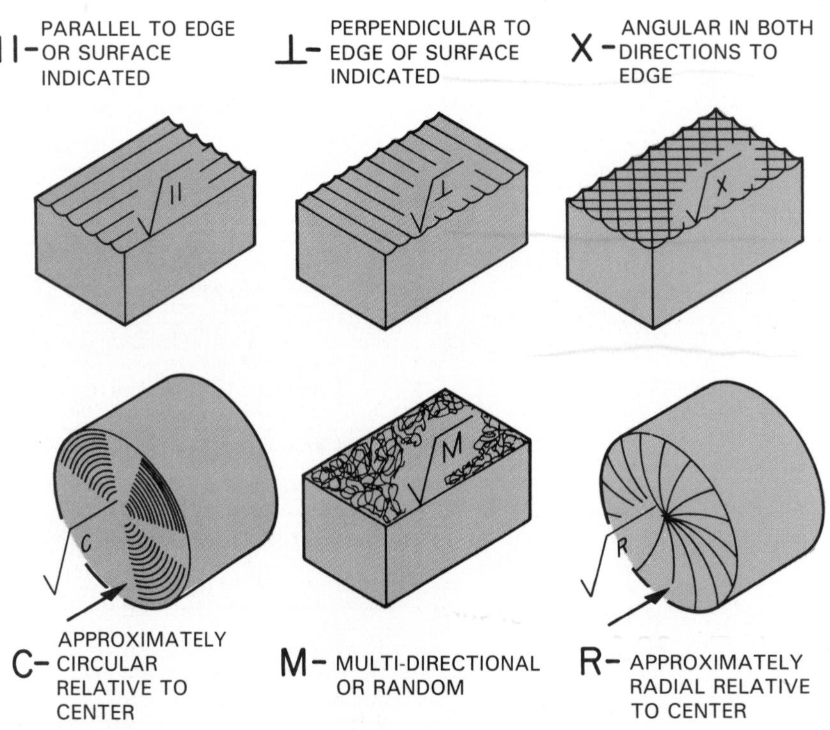

Fig. 20-6. Compare how direction of predominant tool marks (lay) is specified.

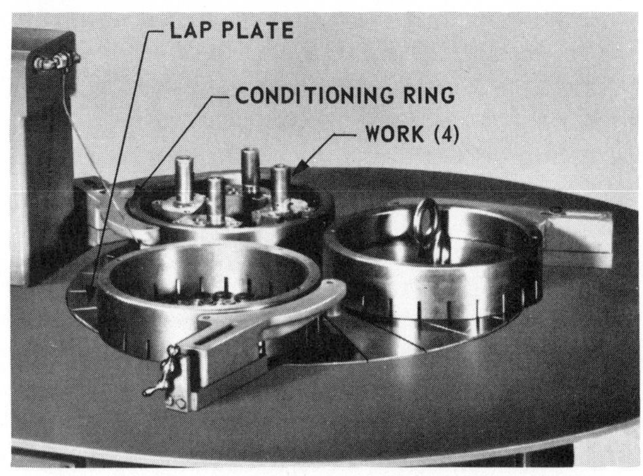

Fig. 20-7. This is a machine for lapping flat surfaces. A heavy cast alloy plate revolves slowly under power. Large cast alloy adjustable conditioning rings are held in position and rotate freely on lap plate to keep plate flat and true. Work is placed inside conditioning rings where they also rotate on lap to create a cutting action which forms a true flat surface. Abrasive grains, suspended in a suitable vehicle, are continually fed on lap plate and uniformly distributed under work during lapping action. (LAPMASTER, Div. John Crane)

Fig. 20-8. High production lapping machine will handle work up to 14.5 in. (368 mm) in diameter and produce microinch finishes of 2 to 3 RMS with absolute uniformity. (LAPMASTER, Div. John Crane)

A brief description of the various degrees of surface roughness is given in Fig. 20-3.

Several tools have been developed to measure surface quality. The most accurate is the ***profilometer*** which measures and amplifies surface roughness electronically, Fig. 20-9. Surface

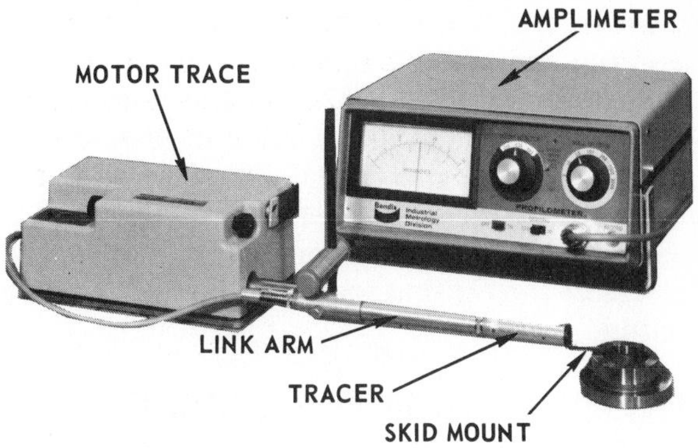

Fig. 20-9. The profilometer measures surface roughness electronically in microinches and/or micrometers. Skid mount must always be in moving contact with surface of material before there will be a reading. (Micrometrical Div., Bendix Corp.)

roughness can be read on a meter or a hard copy graphic readout can be provided by some models.

The ***surface roughness gage*** is a visual or comparison tool, Fig. 20-10. It contains sample specimens of the various degrees of surfaces that conform to American National Standard (ANSI Y14.36-1978) values.

Economics of machined surfaces

The quality of a machined surface has a direct bearing on production costs—the FINER the finish,

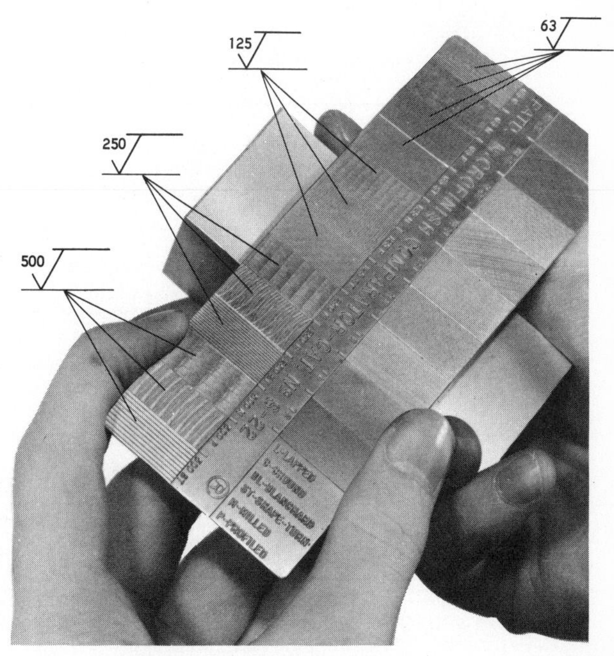

Fig. 20-10. A surface roughness gage is being used to visually compare surface of a milled aluminum block. As it is often difficult to check machined surfaces visually, a test by feel is also used. More obvious roughness standards are identified by appropriate symbols in microinches.

the HIGHER the cost. Care must be taken to meet the required specifications but NOT to exceed them if costs are to be kept within acceptable limits.

The chart in Fig. 20-11 illustrates the range of surface quality normally attainable by the various machining processes. Values are relative and will vary depending upon the condition of the machine, sharpness of the cutting tool, and the material being machined.

OTHER METAL FINISHING TECHNIQUES

While the quality of the machined surface is of paramount importance in the machining of metal, other finishing techniques are also employed for one or more of the following reasons:

1. ***Appearance*** affects product salability and is more important than often realized. A product is much more attractive with a proper finish than when the metal is left unfinished. A car would be drab if left unfinished (unpainted), Fig. 20-12.
2. ***Protection*** is important because all metals are affected to some degree by contaminants in the atmosphere and by abrasion. For example, metal sheet is often textured as a means of protection. The textured surface will reduce reflected light so that surface defects (scratches and small dents that occur during product use) are ''hidden'' within the irregularities of the surface treatment.
3. ***Identification*** makes the product stand out over its competition. Finishes are also applied to make the product blend into the surroundings. See Fig. 20-13.

MACHINE PROCESS	MACHING FINISHES/MICROINCHES								
	500	250	125	63	43	16	8	4	2
ABRASIVE CUTOFF									
AUTOMATIC SCREW MACHINE									
BORE									
BROACH									
COUNTERBORE									
COUNTERSINK									
DRILL									
DRILL (CENTER)									
FACE									
FILE									
GRIND, CYLINDRICAL									
GRIND, SURFACE									
HONE, CYLINDRICAL									
HONE, FLAT									
LAP									
MILL, FINISH									
MILL, ROUGH									
REAM									
SAW									
SHAPE									
SPOTFACE									
SUPER FINISH									
TURN, SMOOTH									
TURN, DIAMOND									
TURN, ROUGH									

Fig. 20-11. Study typical surface finishes: left of heavy line equals practical commercial finish and right of heavy line are obtainable at higher cost.

Fig. 20-12. An automobile would be quite drab if a color finish were not applied. Automotive finishes are planned in design stage as with this four-passenger sports car. (Pontiac Div., General Motors Corp.)

4. A form of ***cost reduction*** can result by a surface finish. For example, an expensive metal can be coated onto a less expensive metal, or other material. The finished part will be less costly than if it were made from the more costly metal. The part will retain the properties of the higher priced metal. Silver and gold, for example, are often applied to steel to improve electrical conduction, heat distribution properties, solderability, and appearance.

Regardless of the finish that is to be employed, the surfaces must be cleaned of all contaminants before the finish can be applied, Fig. 20-15. Oxidation can be removed mechanically (sand blasting or burnishing), or chemically (etching with an acid, etc.). Solvents are often employed to remove oil and grease.

Fig. 20-14 shows the many finishes that can be applied to aluminum.

Fig. 20-13. A computer designed camoflage pattern for this aircraft that will be employed in desert areas. It has been developed to make aircraft blend into surroundings, both in the air and on the ground. (Evans & Sutherland)

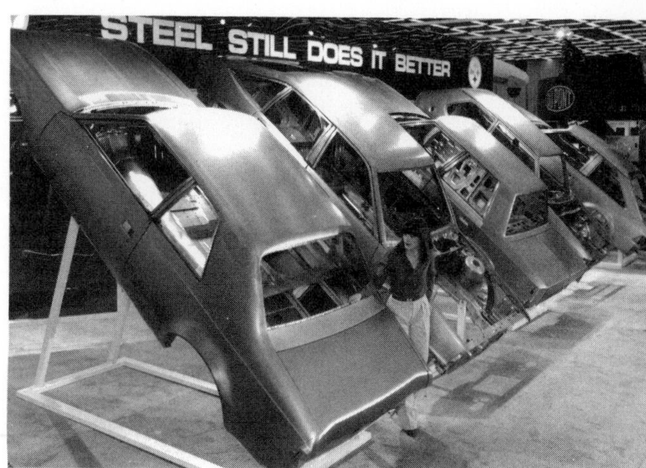

Fig. 20-15. Before colorful protective finish can be applied to these auto bodies, steel surfaces must be cleaned of all dirt, oil, and grease. Otherwise, finish will not adhere properly and allow rust to form quickly, requiring expensive repairs.

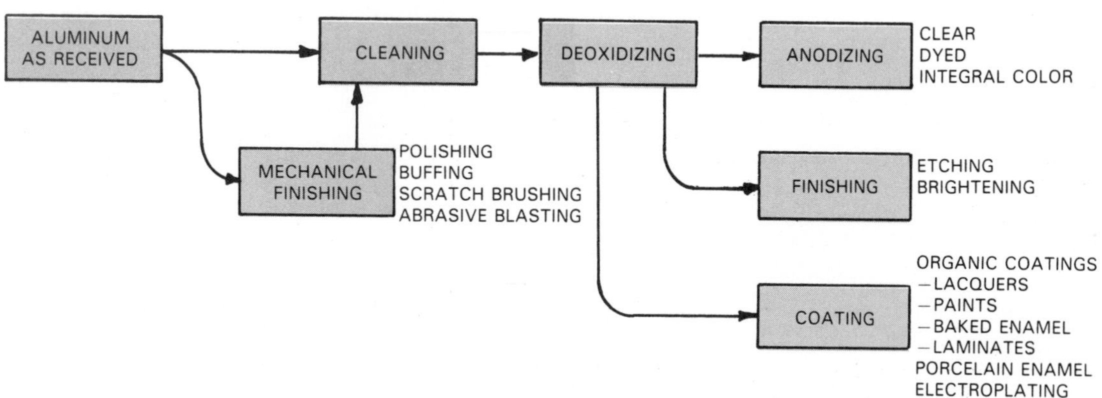

Fig. 20-14. Note various types of finishes that can be applied to aluminum.

Special cleaning methods have been devised to clean complex machined castings. They are needed to remove loose casting sand, metal chips, and other foreign matter entrapped in recesses and passages during manufacturing operations, Fig. 20-16.

Organic coatings

A wide range of finishes fall into the category of ***organic coatings:*** paints, varnishes, lacquers, enamels, various plastic-base materials, and epoxies in both clear and ***pigmented*** (color) formulas. With the exception of the epoxies, they are set by the evaporation of their solvents. This may be accomplished by air drying or baking. Epoxies require the addition of a ***catalyst*** or hardner to set.

A primer is often required to secure satisfactory bonding between the metal and the finish. Some types of castings may need a filler to smooth out the rough cast surface, Fig. 20-17.

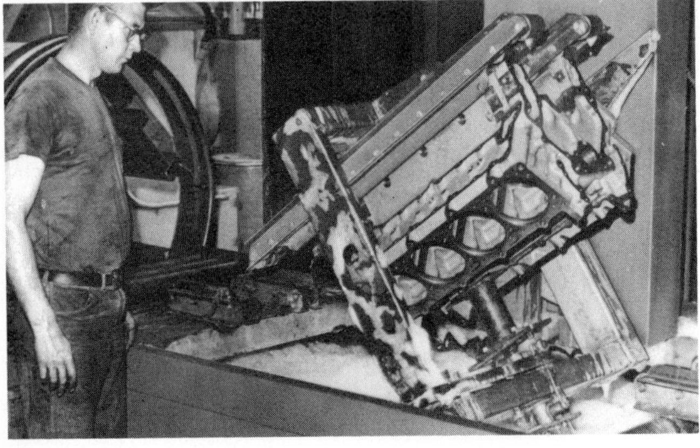

Fig. 20-16. Truck engine block is being cleaned after machining operations. Chemicals and part agitation removes foreign matter from deep cavities, holes, and recesses of casting. The final operation of cleaning cycle is application of rust inhibiting chemicals. (Magnus Chemical)

Fig. 20-17. A filler is used to smooth surface irregularities on a lathe bed casting. Here filler is smoothed with an air sander before finish is applied. (Clausing Industrial, Inc.)

Note that finishes are SELDOM applied to machined surfaces!

Organic coatings are applied in the following ways:

1. ***Brushing,*** at one time, was the only way of applying finishes.
2. With ***spraying,*** the finish is atomized and carried to the work surface by air pressure, Fig. 20-18. Small pressurized spray cans, offered in a wide range of colors, are available. Spraying is easily adapted to mass-production techniques, Fig. 20-19.

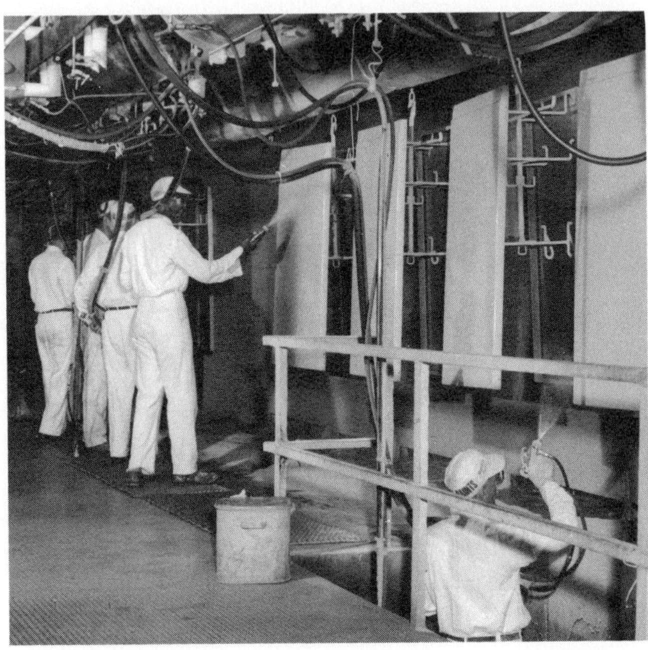

Fig. 20-18. Panels for machine bases are being spray painted. Men in pit are spraying edges. (The Devilbiss Co.)

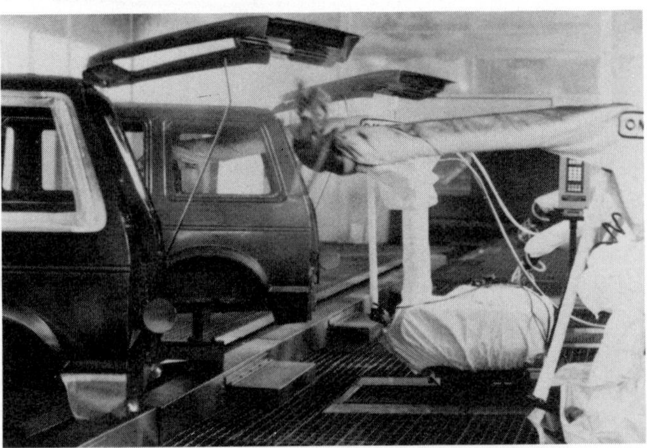

Fig. 20-19. Robot painting is commonly used by automotive industry. Robot movement is controlled by computer. (Chrysler Corp.)

3. ***Roller coating*** can be used only on flat surfaces. It is a low cost technique that can be mechanized.
4. With ***dipping,*** the part is submersed into the finish, removed, and allowed to dry, Fig. 20-20. Dipping is widely used today by the automobile industry to apply body primer and rust proofing. The coating is dried by warm air.
5. During ***flow coating,*** the part is flooded with the finish and allowed to drain while held in an atmosphere saturated with solvent vapor. Drying is then delayed until draining is complete.

Inorganic coatings

Several well-known finishing materials fall into the ***inorganic coatings*** category: anodizing, glass coating, chemical blackening, etc.

Anodizing is the best known of this type finish. This process forms a protective layer of aluminum oxide on aluminum parts.

There are three classes of anodizing: ordinary anodizing, hard coat anodizing, and electrobrightening. The basic procedure for producing all three is the same, Fig. 20-21.

The anodized coating of ***aluminum oxide*** forms by reaction of the aluminum with an electrolyte when the aluminum is used as the anode. Oxygen is liberated at the surface of the aluminum and an oxide forms. Ordinary anodizing leaves a layer of 0.0001 to 0.0006 in. (0.003 to 0.015 mm) in thickness. Hard coating produces a layer about 10 times thicker. It has superior abrasion, erosion, and corrosion resistance; however, the strength of the material is slightly reduced.

Electrobrightening occurs when the aluminum is the anode in an electrolyte that dissolves the oxide film at about the same rate that it is formed. The process leaves a smooth, bright, mirror-like finish.

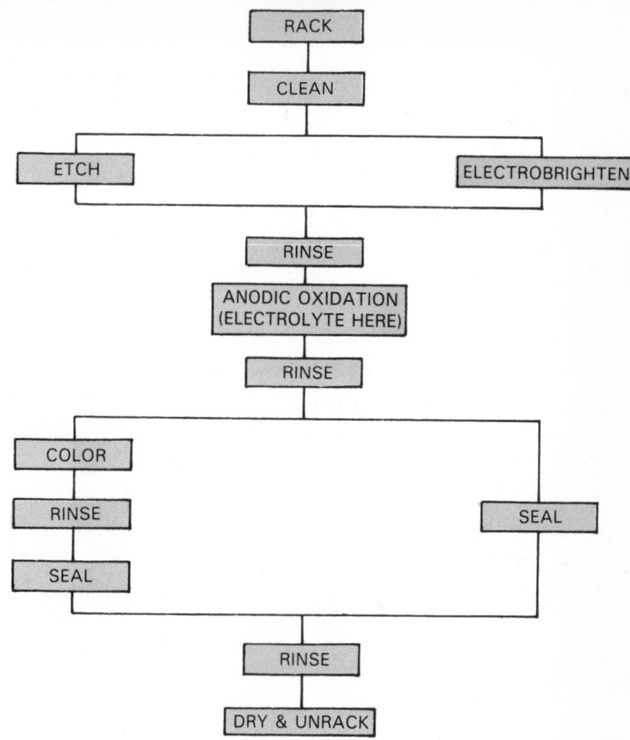

Fig. 20-21. Note sequence of operations when anodizing aluminum.

The anodized coating can be dyed in a wide range of colors. The color becomes part of the surface of the metal.

Vitreous or ***porcelain enamel*** is a glass coating that has been fused to sheet or cast iron surfaces. They form an extremely hard coating that is smooth and easy to clean, Fig. 20-22. Available in many colors, they can be applied to metals that remain

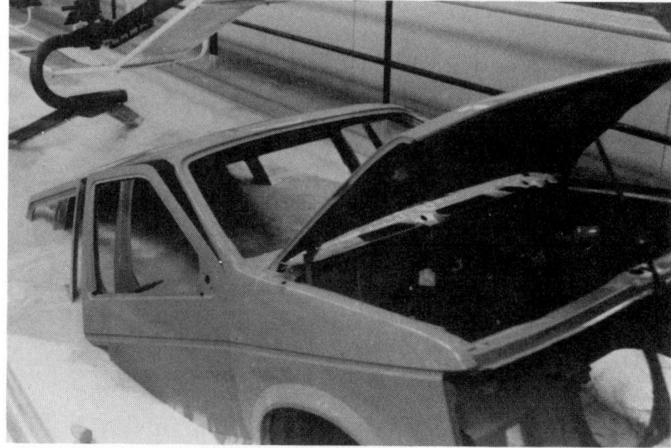

Fig. 20-20. Electrically charged particles of paint evenly coat and protect all bare surfaces of this van body as it is submerged in tank containing tens of thousands of gallons of primer. (Chrysler Corp.)

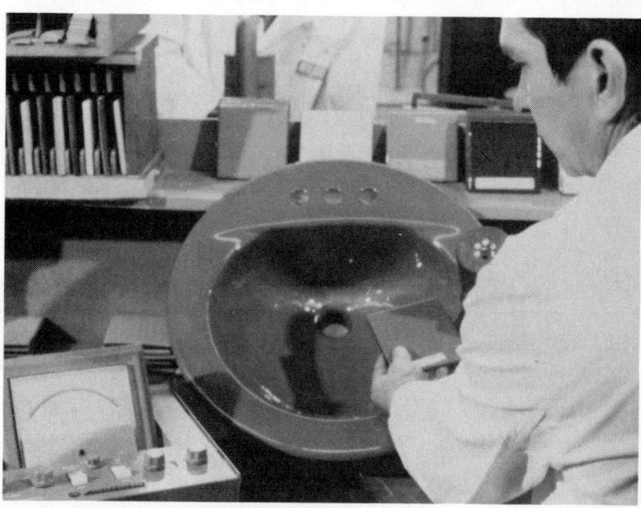

Fig. 20-22. Vitreous or porcelain enamel is a glass coating fused to steel or cast surfaces. It is an extremely hard coating that is smooth and easy to clean. (Eljer Plumbingware)

solid at the firing (melting) temperatures of the coating material. The finish is applied as a powder or as a thin slurry known as "slip." After drying, it is fired at about 1500 °F (815 °C) until it fuses to the metal surface, Fig. 20-23.

Fig. 20-23. Porcelain is applied as a powder, called "frit," or as a thin slurry, known as "slip." After application, unit is fired at 1500 °F (815 °C) until porcelain has fused to metal surface. (Eljer Plumbingware)

Chemical blackening is a surface finishing technique for steel that is often used. Because blackening changes the metal's surface instead of simply adhering to it, there is no surface build-up. Critical dimensions remain unchanged. The process also offers several advantages:

1. The black finish enhances the appearance of many items. Because the cost is so low, it is possible to finish parts that were previously left unfinished.
2. Corrosion resistance is improved.
3. Light glare is greatly reduced. When employed on moving tools and machine parts, safety factors are improved by reducing eye fatigue.
4. Abrasion resistance is improved.
5. Adhesion qualities are improved. Paint and other finishes take a better hold faster, and last longer.

Metal coatings

Metal coatings, with the exception of electroplating, are primarily applied to steel. They adhere to the surface tightly enough to offer protection from corrision as well as improving wear resistance and/or improving the appearance of the item. For example, cutting tools are now available with a titanium oxide coating that improves tool life up to 800 percent.

The more commonly employed metal coatings include the following:

Electroplating refers to when metal coating is deposited on a metal surface by the use of an electrical current. Practically any metal can be used as a coating.

Coating thickness can be controlled closely and, unlike many other metal coating processes, can deposit wear resistant coatings as well as adding a coating that is attractive in appearance. Allowances must be made on machined surfaces to permit the additional metal deposited by the plating process. Chrome plating is one example of electroplating.

Metal spraying utilizes metal wire or powder heated to its melting point and sprayed by air pressure onto the work surface to produce the desired coating. See Fig. 20-24.

In general, most inorganic materials that can be melted without decomposition can be applied by spraying. Flame sprayed coatings can be applied to build up worn or scored surfaces so they can be remachined to the required size, Fig. 20-25. Super hard coatings can be sprayed when abrasion resistant surfaces are needed.

The ***detonation gun coating process*** (D-Gun®) is another technique for depositing a metallic coating on a workpiece. Almost any material that can be melted without decomposing can be sprayed. The process was invented and developed by the Union Carbide Corporation.

The D-Gun is essentially a water cooled barrel several feet long and about one inch in diameter. It is fitted with valving for introducing gases and material to be sprayed. See Fig. 20-26.

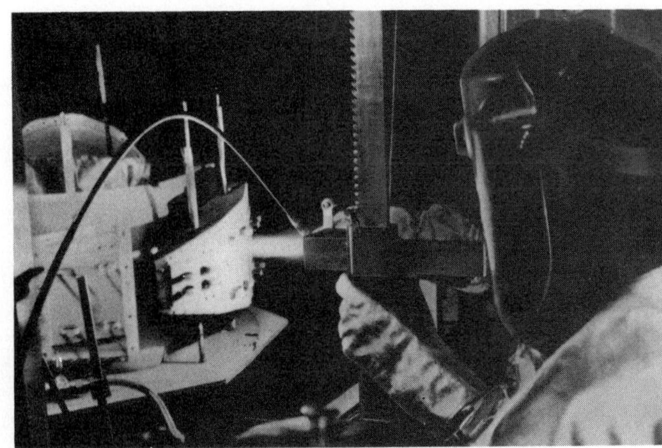

Fig. 20-24. Plasma flame process is one of several metal spraying techniques. It is capable of temperatures up to 30,000 °F (16 649 °C) and can spray any material that will melt without decomposing. Note how operator is dressed for protection against high heat generated by process. (Metco Inc.)

Fig. 20-25. Metallizing gun is applying stainless steel onto a roller surface. (Metco Inc.)

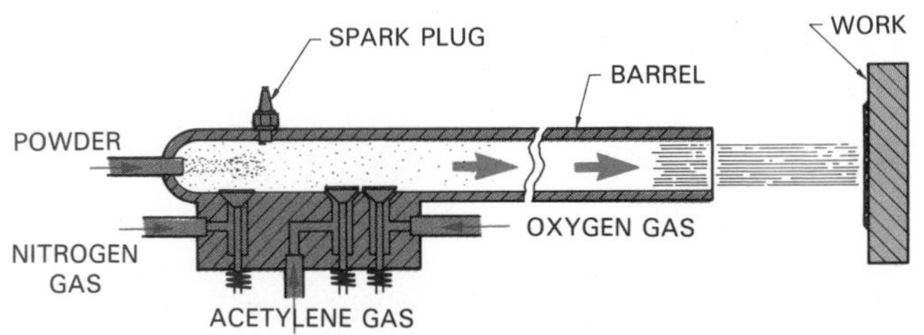

Fig. 20-26. This is a cross section of a D-Gun® or detonation gun used for applying metal coatings. (Union Carbide Corp.)

A carefully measured mixture of gases, usually oxygen and acetylene, is fed into the barrel along with a charge of coating material in powder form. The powder has a particle size of less than 100 microns. A spark ignites the gases. The powder is heated and accelerated in a high temperature, high velocity gas stream. This forces the molten material against the surface to be coated. A pulse of nitrogen purges the barrel after each detonation and the process is repeated many times a second.

Each detonation, called a "pop," results in a circle of coating a few microns thick and about an inch in diameter. The molten or semimolten coating droplets quickly solidify on the work surface. The complete coating is made of many overlapping "pops" until the surface is built up to the required thickness. The process can be fully automated.

The process has the ability to apply coatings with very high melting points to fully heat treated parts WITHOUT danger of changing the metallurgical properties or strength of the part and WITHOUT danger of thermal distortion.

Coatings include pure metals and metallic alloys such as nickel and nichrome, tungsten carbide, and ceramics. These coatings are used on many applications, especially to combat wear (abrasive, erosive, and adhesive) often in very corrosive environments. Refer to Fig. 20-27.

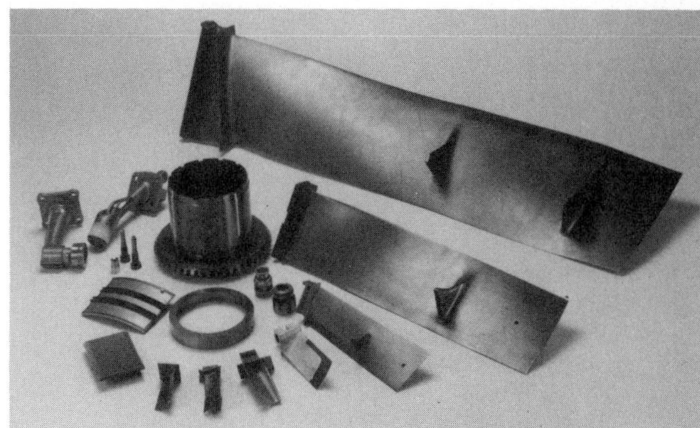

Fig. 20-27. These are examples of metal and ceramic coating application on gas turbine engines that were applied by detonation gun coating process. (Union Carbide Corp.)

Mechanical finishes

Buffing is a power polishing operation. Buffing wheels are attached to a buffing lathe, Fig. 20-28. The wheels are charged with different grades of abrasives that remove scratches and polish the metal's surface to a high luster.

> WARNING! Unless the buffing lathe is fitted with a high efficiency dust collecting system, an approved filter mask must be worn.

Diamond dust and air powered, hand held polishing units are utilized to polish the hardened steel dies used for die casting or the injection molding of plastics. See Fig. 20-29.

Deburring techniques

Power brushing is frequently employed to remove burrs from machined surfaces, Figs. 20-30 and 20-31. Wire or fiber brushes replace the buffing wheels on the buffing lathe. In removing burrs, the wheels produce a satin sheen on the brushed surface.

Hand deburring of small holes and intricate parts is tedious and expensive so other methods are desirable, Fig. 20-32. One technique, known as

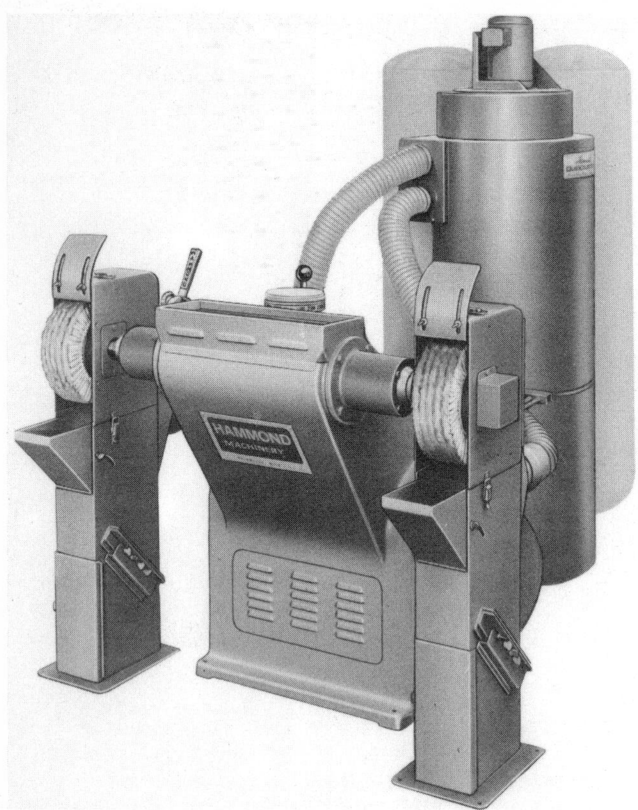

Fig. 20-28. This buffing lathe equipped has wheel guards and dust collection system. (Hammond Machinery Inc.)

Fig. 20-29. Worker is polishing a steel die with a diamond dust to produce a mirror like finish. (Diamond Tool Div., Engis Corp.)

extrude-hone, makes use of a ***silicon putty*** (very similar to "Silly Putty" found in toy stores) permeated with finely divided abrasive particles.

The silicon plastic is forced into and through the part to be deburred. As it flows through the openings to be deburred, the abrasive grains remove the machine burrs—chip, wire, or hair in the part.

Fig. 20-30. Buffing lathe is fitted with wire wheels to remove burrs. Wheel guards were removed for clarity. (Osborn Mfg. Co.)

Fig. 20-31. Top gear has burrs. Lower gear has had burrs removed using wire brushes. (Osborn Mfg. Co.)

Fig. 20-33 illustrates several difficult jobs deburred on a production basis by this process.

TEST YOUR KNOWLEDGE—Chapter 20

Please do not write in the text. Place your answers on a separate sheet of paper.

1. The symbol ________ was used at one time on drawings to designate a machined surface.

Fig. 20-32. Block shown, a production item, has 80 intersecting holes; a few are shown in cutaway sections. Access to them for hand deburring was virtually impossible. Abrasive charged silicon putty removed all burrs rapidly and economically. (South Shore Tool and Development Corp.)

Fig. 20-33. High cost of burr removal by hand was eliminated and production rate substantially increased by silicon plastic process. (South Shore Tool and Development Corp.)

2. Why was its use discontinued?
3. Surface roughness is now measured in _______ and _______. What does each equal?
4. In addition to surface roughness, other surface conditions were given values. Waviness was one such condition. It means:
   a. Very rough surfaces.
   b. Smoothly rounded undulations caused by tool and machine conditions.
   c. Scratches on the machined surface.
   d. All of the above.
   e. None of the above.
5. Lay is another surface finish condition. What does it mean?
6. A 500/12.5 surface finish is _______ than a 125/3.2 surface finish.
7. While the quality of a machined surface is of paramount importance in the machining of metal, other finishing methods are used in the machine shop. They are employed for one or more of the following reasons: (Explain each.)
8. Regardless of the finishing method utilized, the surface to be finished must be thoroughly _______ of all contaminants.
9. Paints, lacquers, and enamels help make up the family of _______ coatings.
10. What are the five (5) ways employed to apply the finishes in question number nine?
11. List three (3) types of anodizing.
12. Describe electroplating.

RESEARCH AND DEVELOPMENT

1. Secure a copy of the publication SURFACE TEXTURE—ANSI B46.1 for the shop technical library. This ANSI paper is on the measurement of surface roughness.
2. Make a collection of brochures advertising the various types of surface roughness/texture checking equipment. Develop a bulletin board display around them.
3. Secure or make samples of machined surfaces that match the various degrees of surface roughness. Mount them on a display panel. Identify each sample with the method employed to machine it and the correct symbol of roughness.
4. Prepare a paper that will explain the techniques used to develop average roughness values. They are explained in many drafting books, machinists' handbooks, in addition to ANSI B46.1. The term RMS is often used in the formulas. What does it mean?
5. Demonstrate electroplating. Secure the necessary equipment from the Science Department.
6. Prepare a demonstration of the anodizing process.
7. Devise and construct a safe method to clean work made in the shop.
8. Secure examples of work that have been electroplated and anodized.
9. Contact a local machine shop and find what equipment they employ to check for surface quality.
10. Prepare a term paper on flame spraying. Report your findings to the class.

Chapter 21

MACHINING PLASTICS

After studying this chapter, you will be able to:

- Explain the general characteristics of some plastics.
- Describe the hazards associated with machining certain plastics.
- Safely machine several types of plastics.
- Sharpen cutting tools to machine plastics.

Plastics, while relative newcomers to industry compared to metals, are being used in increasingly larger quantities each year. No longer are plastics rarities in the machine shop. For this reason, it is important that the machinist be familiar with the machining characteristics of the more common plastics, such as nylon, Teflon® , Delrin® , acrylics, and laminated plastics.

Note! Teflon® is the Du Pont trademark for fluorocarbon resins. Delrin® is the Du Pont trademark for acetal resin.

> DANGER! Special care must be taken when working plastics. The dust and fumes given off by some plastics may be irritating to the skin, eyes, and respiratory system. Other plastics have fillers such as asbestos, glass, etc. that can be harmful to your health. Be sure you are aware of the safety precautions that must be observed before you attempt to machine any plastic!

NYLON

Nylon is the name for a group of polyamide resins. It is tough and has a high tensile (pull), impact, and flexural (bending) strength. Nylon is highly resistant to abrasion and is NOT affected by most common chemicals, greases, and solvents.

Nylon is excellent for bearings, gears, cams, and rollers. Another application for parts machined from this plastic is in the preparation of experimental parts. Research and development projects also make considerable use of machined nylon pieces.

General machining precautions for nylon

Most types of nylon can be machined using techniques normally employed for machining soft brass. Coolants, while permitting higher cutting speeds, are NOT necessary to produce good quality machined surfaces. When employed, coolants should be of the soluble-oil type.

To assure accuracy, parts machined from nylon should be brought to room temperature before checking dimensions.

Since nylon is NOT as rigid as metal, it must be well supported during machining operations. Otherwise, deflection of the unsupported stock will result in inaccuracies.

Turning nylon

Nylon can be turned on a standard metalworking lathe. While tool bits sharpened to machine soft brass will prove satisfactory, best results are obtained using the tool bit shapes shown in Fig. 21-1. Tools must be kept very sharp!

The best finish on nylon is obtained with a high speed and fine feed.

Milling nylon

Conventional milling cutters, providing they are kept very sharp, can be utilized for milling nylon, Fig. 21-2. Climb milling will minimize burring. Surface speeds in excess of 100 fpm (30 m/m) can be employed.

Vertical milling is practical using fly cutters or two-lip end mills. Cutters must be kept very sharp to prevent plastic from melting or becoming gummy. Feeds of 10 ipm (25 mm/m) or higher have proven satisfactory. Smoother surface finishes can be acquired using lighter feeds.

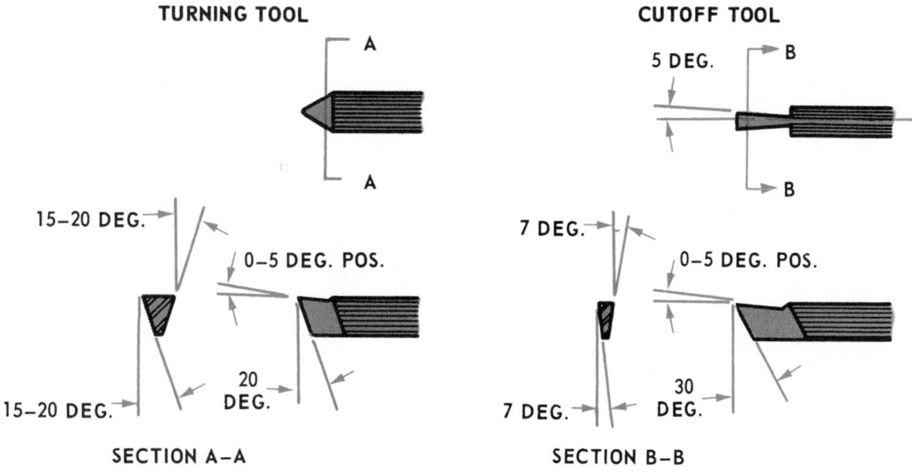

Fig. 21-1. Note typical lathe cutting tools for nylon. (Du Pont)

Fig. 21-2. Nylon and most other plastics can be readily milled providing cutters are kept very sharp.

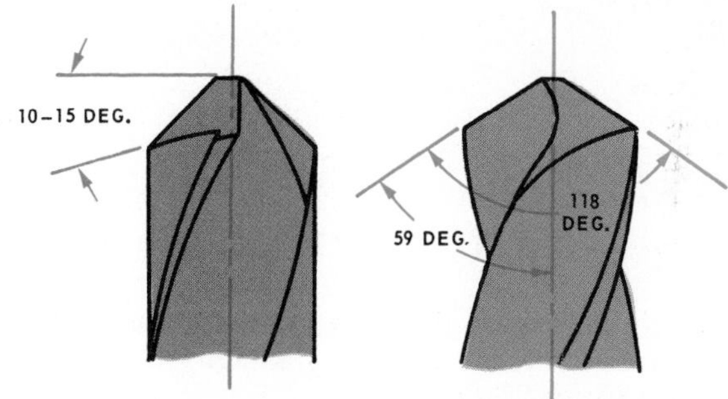

Fig. 21-3. This is the drill point recommended for drilling nylon. (Du Pont)

Drilling nylon

Drilling requires extra care because this operation produces considerable heat. Standard twist drills, sharpened as shown in Fig. 21-3, will produce acceptable results. However, best results are obtained utilizing drills designed specifically for plastics. Drills for plastics have flutes that are highly polished and have a long lead. These drills should be sharpened as shown in Fig. 21-3.

Use heavy feeds to prevent the excessive heat that results when the drill scrapes the plastic rather than cutting it. When coolants are NOT employed, the drill must be withdrawn from the hole frequently to clean out chips and to prevent overheating. Holes will be to size if the drill is kept cool.

Nylon can be reamed using an expansion reamer, adjusted to a few thousands oversize. Holes finished with solid reamers tend to be undersize.

Threading and tapping nylon

Nylon can be threaded and tapped with conventional equipment. But before tapping, the hole should be chamfered to reduce the chance of the first few threads tearing. See Fig. 21-4.

Production tapping requires a tap 0.005 in. (0.125 mm) oversize unless a self-locking thread is desired. The tap can be made oversize by chrome plating.

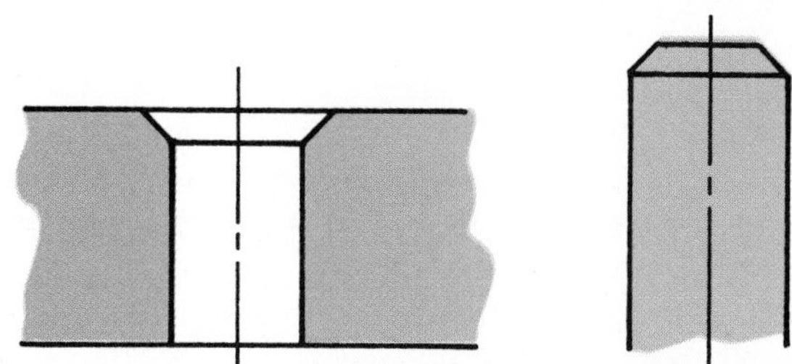

Fig. 21-4. When hand-threading nylon, and many other plastics, first few threads may tear if hole or rod is not chamfered first.

Threading can be done in nylon with a regular single point cutting tool, Fig. 21-5. Use the same procedure as with metal. However, because of nylon's resiliency, the finish cut should NOT be less than 0.005 in. (0.125 mm). Support long work with a ball bearing follow rest.

Fig. 21-5. Threading plastics on a lathe can be done using same procedure recommended for metal. However, cutting tool must be kept very sharp and have plenty of clearance.

Sawing nylon

In sawing nylon and other plastics, good results can be obtained by employing a band saw. A band saw blade quickly dissipates the heat. Dry cutting is best accomplished with a skip tooth, metal cutting blade with 4 to 6 teeth per inch. However, the blade must be sharp to prevent gumming of the nylon, which usually freezes the blade in the cut.

Also used extensively in cutting plastics are hollow-ground, plastic cutting, circular saw blades. Blades with a slight set are available where quantities of both extra thick or thin plastics are sawed.

Annealing nylon

Like metal, machined nylon parts require ***annealing*** to insure against dimensional changes. It is recommended that annealing be carried out in the absence of air, preferably by immersion in a suitable liquid. High temperature boiling hydrocarbons, like waxes and oils, are recommended for annealing nylon. A temperature of 300 °F (148 °C) is often employed for general annealing.

An annealing time of 15 minutes per 0.125 in. (3.175 mm) of cross section is normally required. Allow the part to cool slowly in a draft free area. Placing the heated piece in a cardboard container is a simple way of insuring slow, even cooling.

DELRIN® ACETAL RESIN

Parts manufactured from ***Delrin acetal resin*** have an unusual combination of physical properties that bridge the gap between metals and plastics. These properties include excellent dimensional stability, high strength, and rigidity. The plastic is utilized to make parts in business machines (gears, cams, bearings, and printing wheels). Delrin is also replacing brass and zinc for many applications in the automotive and plumbing industries.

Delrin has low friction, requires a minimum use of lubricants and is very quiet in operation.

Its machining characteristics are very similar to nylon. Recommended machining, cutting, and finishing operations are shown in Fig. 21-6.

TEFLON

Teflon is filling a wide range of needs in the electronic, electrical, chemical, and processing industries. It has a very low friction coefficient. Two flat sections of Teflon rubbed together generate about the same friction as do two ice cubes rubbed together. This makes Teflon ideal for bearings and seals in food processing equipment where lubricating oil would contaminate the food.

Teflon works well in both high temperature and ***cryogenic*** (very low temperature) applications.

Teflon is costly, but will do things that no other material can do well.

Teflon has a tendency to pick up metal shavings and chips. No machining should be attempted until the machine is totally clean of all metal particles.

Teflon general machining characteristics

Teflon has a high thermal expansion rate. It expands at a rate about 10 times that of steel. When tolerances are critical, measurements should be made at room temperature or, where applicable, at the temperature at which the part will be used.

It is recommended that the plastic be stored at 74 °F (21 °C) or above, for at least 48 hours before and during the machining operations.

Turning Teflon

Teflon is more flexible than the other plastics described. Care must be taken to support the work properly to prevent deflection of the material away from the cutting tool.

Tools must be sharp and have generous clearances so the cutting edge of the tool does not rub, Fig. 21-7. Chips must NOT be allowed to accumulate around the work because they prevent the heat from dissipating. See Fig. 21-8.

Cutting fluids are needed if tolerances are critical. Large amounts of water-base coolant will deter thermal expansion in Teflon.

TABLE I
MACHINING, CUTTING, AND FINISHING OPERATIONS WITH DELRIN® ACETAL RESIN

	Equipment		Cutting Speed			
Operation	Machines	Tools	RPM/FPM	Feed	Coolant Use	Remarks
Sawing	Std.	Std. 14 T.P.I. Slight Set	100 300 FPM (30.5-91.5 M/min)	Med.		Coolant improves finish of cut
Drilling	Std.	Std. Twist Drills 118°	1500 RPM for 0.500" (1.27 cm) drill	Med.	At med. and high speeds	On-Size holes drilled without coolant
Turning	Std.	Std.	690-840 FPM (210-256 M/min)	.002"-.005" (0.051-0.13 mm)	At high speeds	Depth of cut—.016"-.200" (0.41-5.08 mm); support long lengths
Milling	Std.	Std. cutters; single fluted end mills	Similar to brass	Similar to brass	Not required	—
Shaping	Std.	Std.	Max.	Similar to brass	Not required	—
Reaming	Std.	Expansion type preferred	Similar to brass	Med.	At med. and high speeds	—
Threading and Tapping	Std.	Std.	Similar to brass	Similar to brass	At med. and high speeds	Coolant facilitates cutting to dimensions
Blanking and Punching	Std.	Std.	—	—	Not required	Primarily for 1/16" (0.16 cm) thick stock
Filing and Sanding	Std.	Std. File, Std. Abrasive Paper and Discs	—	—	Wet sanding	Finish will vary with type of file
Finishing	Std.	6-12" (15.2-30.4 cm) dia. muslin pumice and water polishing compound	1000-2000 RPM	—	Not required	Use light pressure and rotate part

Fig. 21-6. Study recommended machining, cutting, and finishing operations for Delrin® acetal resins. (Du Pont)

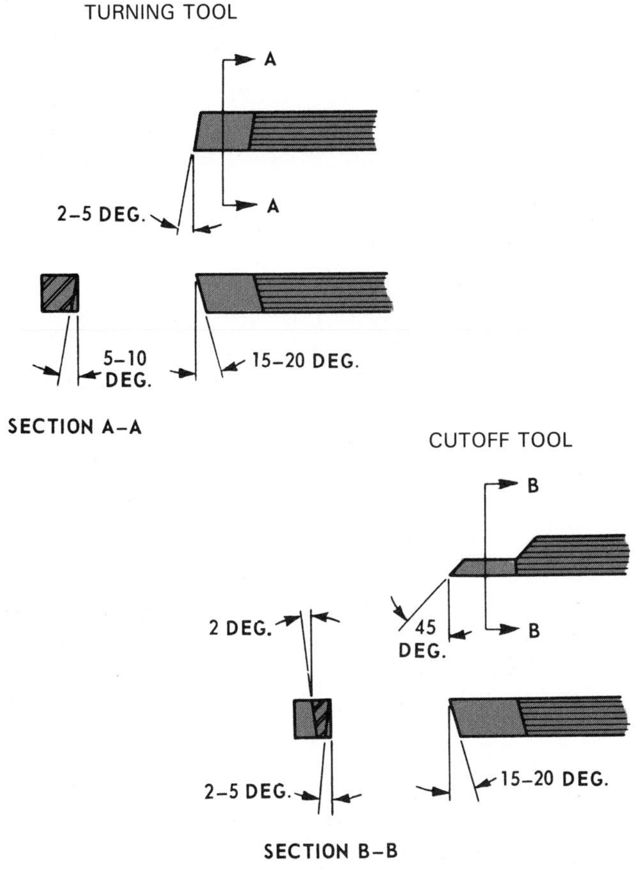

Fig. 21-7. Note typical lathe cutting tools for machining Teflon. (Du Pont)

Fig. 21-8. When turning plastics, do not allow chips to accumulate around work. Chips prevent heat from being dissipated and work may become distorted.

Best surface finishes are achieved at cutting speeds of 200-500 fpm (60-150 mpm) with feeds of 0.0002 to 0.010 in. (0.005 to 0.25 mm).

Drilling Teflon

Drills sharpened as shown in Fig. 21-9 will provide satisfactory cutting action in Teflon.

Teflon tends to swell slightly during the drilling operation and results in a hole smaller than the drill. To compensate for this swelling, the machinist must

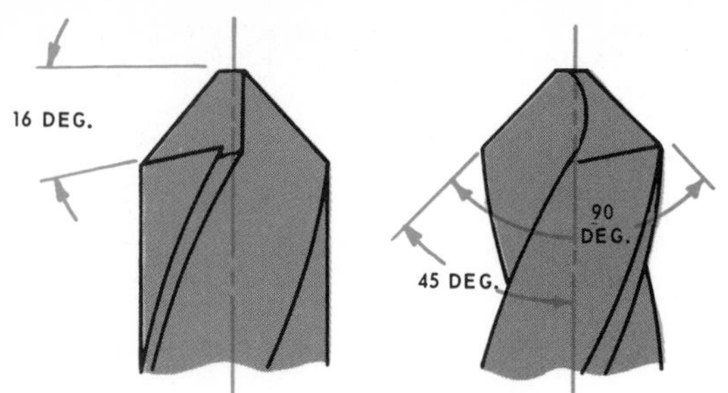

Fig. 21-9. This drill point is recommended for Teflon. (Du Pont)

use a drill slightly oversize. You should test drill on a piece of scrap material to determine the exact drill size needed for the conditions at hand.

For close tolerance drilling in Teflon, feeds of 0.004-0.006 in. (0.10-0.15 mm) are suggested.

Milling Teflon

Teflon is milled in much the same manner as the other plastics. Only newly sharpened and honed cutters should be employed. The work must be solidly supported.

With shell mills, it is recommended that the milling head of the machine be tilted slightly into the cut, Fig. 21-10. This will eliminate cutter drag marks on the material.

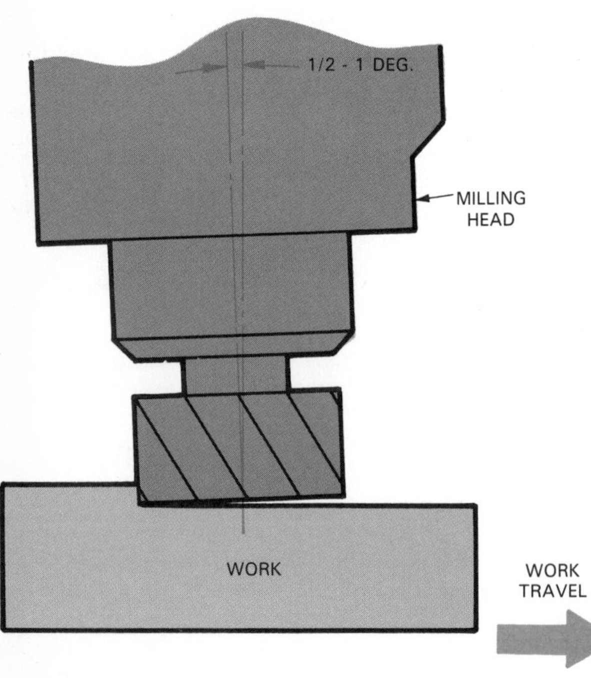

Fig. 21-10. When shell milling, it is recommended that cutter be tilted slightly (1/2 to 1 deg.) into cut to machine with periphery of cutter. This will eliminate cutter "drag" marks on machined surface.

Reaming Teflon

Reaming of Teflon is NOT advised. Holes should be bored with a single point tool if close tolerances are specified.

Threading and tapping Teflon

Teflon is threaded and tapped with the same general techniques as those suggested for nylon.

Sawing Teflon

A rigid machine, with first class saw guides, is essential if a square cut is to be made in Teflon. No coolant is needed. Maximum machine speeds can be utilized with a skip tooth blade having 4 to 6 teeth per inch.

Annealing Teflon

In order to maintain dimensional stability, Teflon should be annealed. Normally, the plastic is heated to a temperature above that which the finished part will be exposed, but below 621 °F (333 °C). Above this temperature, Teflon gels. One hour of annealing for each 1 in. (25 mm) of thickness is adequate. Allow the part to cool slowly. Heating is usually done in an oven.

The following oven annealing procedure for Teflon is recommended:

1. Anneal the rod or tubing.
2. Rough machine the part to within 0.06 in. (1.5 mm) of finished size.
3. Reanneal again, but at temperatures slightly lower than that of the initial annealing.
4. Finish machine after the Teflon has reached room temperature.

ACRYLICS

The ***acrylic plastics*** (Lucite® and Plexiglas®) have an unusual combination of desirable characteristics. They have good dimensional stability and high impact strength to temperatures as high as 200 °F (94 °C). Acrylics are easy to machine, form, and polish.

Note Lucite® is a Du Pont trademark for acrylic plastics. Plexiglas® is a Rhom & Haas Co. trademark for acrylic plastics.

Because of their unusual "light piping" (they transmit light as a hose carries water) and edge lighting qualities, acrylics find considerable use in light control and optical applications.

General machining characteristics for acrylics

Acrylics are machined in much the same manner as the other plastics, Fig. 21-11. Generally, little difficulty will be encountered if sharp tools with adequate clearance are used. Drills should be sharpened as shown in Fig. 21-12.

Fig. 21-11. Transparent acrylics present a unique machining experience. You can see tool cutting inside work. Poor cutting action can be spotted immediately and corrections made.

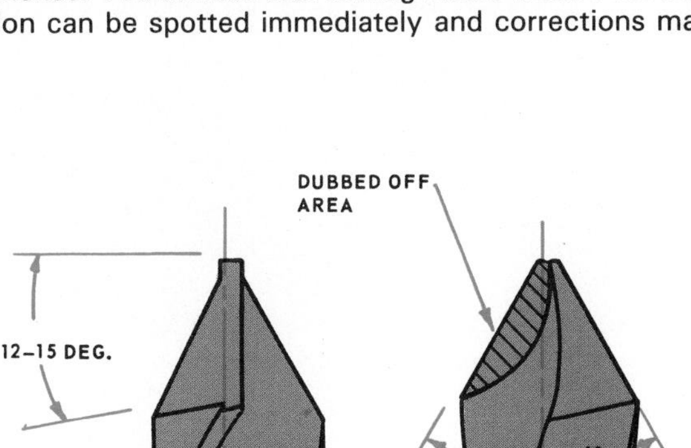

Fig. 21-12. This is recommended drill point for making through holes in acrylics. Tip angle should be increased to 118 deg. for blind holes. Smoothly finished holes can be produced by first drilling a pilot hole and filling it with wax. Wax will allow chips to move up flutes without sticking to drill. (Rohm and Haas)

LAMINATED PLASTICS

Laminated plastics consist of layers of reinforcing materials (cotton fabric, paper, asbestos, glass fiber, etc.) that have been impregnated with synthetic resins. The resins are cured under heat and pressure.

General machining characteristics for laminated plastics

Machining of laminated plastics can be accomplished with conventional machine tools, Fig. 21-13. However, a dust collector system and filtered dust mask must be employed for machine operator safety.

Some laminated plastics, like those containing glass fiber and asbestos, are highly abrasive. Carbide tools are recommended!

Fig. 21-13. Laminated plastics, especially those containing asbestos and glass fiber material, cause rapid tool wear. When possible, carbide tools should be used.

Turning laminated plastics

A round nose tool produces the best surface finish. Speeds up to 4000 fpm (1220 m/m) can be used. Lathe work is usually done dry. However, internal threading may sometimes require a lubricant.

Drilling laminated plastics

Drilling operations are similar to those used with nylon. However, drilling parallel with the laminations should be avoided whenever possible because the material may split along the laminations when drilled. See Fig. 21-14.

Milling laminated plastics

Speeds up to 1000 fpm (305 m/m) are possible with good results. Feeds up to 20 in. (500 mm) per minute have been used. For cotton fabric-base laminates, best results are obtained by operating at the highest spindle speed the cutter will stand, with the maximum feed, that will produce an acceptable surface finish.

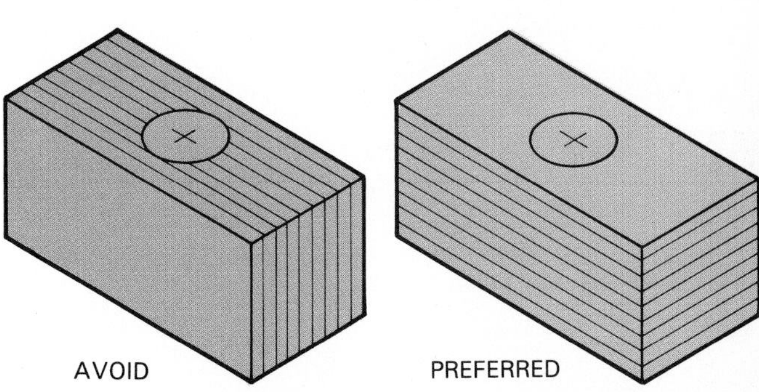

Fig. 21-14. Whenever possible, holes should be drilled at right angles to laminations. Holes drilled parallel to laminations have a tendency to split some types of material.

Climb milling is recommended to keep the work held tightly in the holding device and to prevent an edge from being raised.

Sawing laminated plastics

Band sawing is advised for curved or straight cuts in laminated plastics when smooth edges and close tolerances are NOT specified.

Blades with 5 to 8 teeth per inch and medium to high set can be operated at speeds up to 8000 fpm (2400 m/m). Feed the work as fast as it will cut without forcing the blade.

Threading and tapping laminated plastics

Hand threading of laminated plastics can be done with standard taps and dies. It is normally done dry. High speed steel taps that are slightly oversize 0.002-0.005 in. (0.05-0.13 mm) should be used if available.

A slight chamfer on the hole to be tapped, or rod to be threaded, will improve the work quality by preventing the first few threads from tearing.

TEST YOUR KNOWLEDGE—Chapter 21

Please do not write in the text. Place your answers on a separate sheet of paper.

1. Give a common trade name for each of the following groups of plastics:
   a. Polyamide resins.
   b. Acetal resins.
   c. Fluorocarbon resins.
   d. Acrylic resins.
2. If plastics are to be machined with any degree of accuracy, the cutting tools must be ________.
3. When hand threading plastics, why is it recommended that the hole or rod be chamfered?
4. Like metal, many plastics require ________ to insure against dimensional changes.
5. When turning many plastics on a lathe, care must be taken to prevent the ________ from accumulating around the ________. If this is NOT done, ________ will build up and cause the plastic to become ________.
6. What is unique about Teflon?
7. Machining plastics can create health problems for the machinist if precautions are NOT taken. What are these problems and how can they best be handled?
8. What are laminated plastics?
9. When drilling laminated plastics, what should be avoided?

RESEARCH AND DEVELOPMENT

1. Contact manufacturers of plastics and request pamphlets on recommended machining techniques and safety precautions. Place the accumulated material in the technical library.
2. Secure samples of various plastics and demonstrate recommended machining techniques. Point out the differences between machining plastics and metal.
3. Develop a safety program to be followed when machining plastics. It can be in the form of a bulletin board display, pamphlet, or series of safety posters.
4. Review the various metalworking technical magazines and make photocopies of the many uses of plastics in the machine shop. Prepare a term paper on your findings.
5. Visit a machine shop that works plastics. Prepare a term paper on your observations. Secure samples, if possible, of the products produced.

Chapter 22

HIGH ENERGY RATE FORMING (HERF)

After studying this chapter, you will be able to:

- Describe several HERF techniques.
- Compare the advantages and disadvantages of HERF.
- Explain some industrial applications of HERF.

The introduction of super-tough alloys for aerospace vehicles and the need for shaping thin, brittle metal has been responsible for the development of new ways to do the work. One of these new techniques is known as ***high energy rate forming*** or ***HERF.*** There is little similarity between it and conventional metalworking processes like turning, drilling, milling, etc.

Shaping metals by conventional presses and drop hammers parallels HERF. However, problems develop when shaping the new alloys by conventional means. There is ***''spring-back''*** where the metal tends to try to regain its original shape, Fig. 22-1. It is difficult and costly to shape these metals to acceptable tolerances by conventional means.

In HERF, the metal is shaped in microseconds with pressures generated by the sudden application of large amounts of energy. The metal, in most cases, is slammed against the die and shaped so rapidly that there is no tendency for the material to try to return to its original shape.

The great pressures are generated by detonating explosives, releasing compressed gases, discharging powerful electrical sparks, or electomagnetic energy.

HERF offers many advantages. Tool costs are reduced. There is usually no need for expensive machinery and apparently there is no limitation to the size of the sections that can be formed.

EXPLOSIVE FORMING

Explosive forming uses the high pressure wave of an explosive charge to form the metal. It is an older technique adapted to the missile age. It originated in the late 1800s to shape ornate door knobs and similar products.

The sheer size of many aerospace and marine components, up to 144 in. by 230 in. by 1 in. thick (3657 mm by 5842 mm by 25 mm thick), makes it impossible to form them in existing presses. The presses are either too small or are NOT powerful enough to provide the pressures required to shape the high strength alloys. See Figs. 22-2 and 22-3.

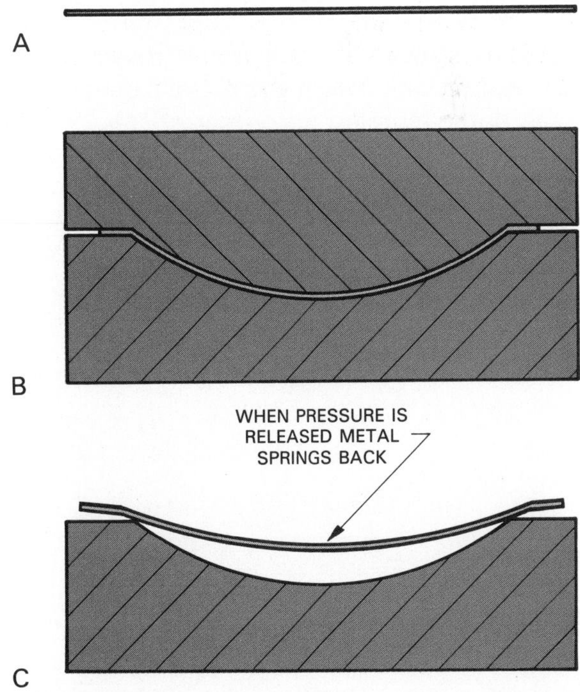

Fig. 22-1. Many metals tend to ''spring-back'' to near their original shape after being formed by conventional means. A—Flat sheet metal blank ready for forming. B—Formed between dies. C—Metal tries to return to original shape when male die is removed or pressure released.

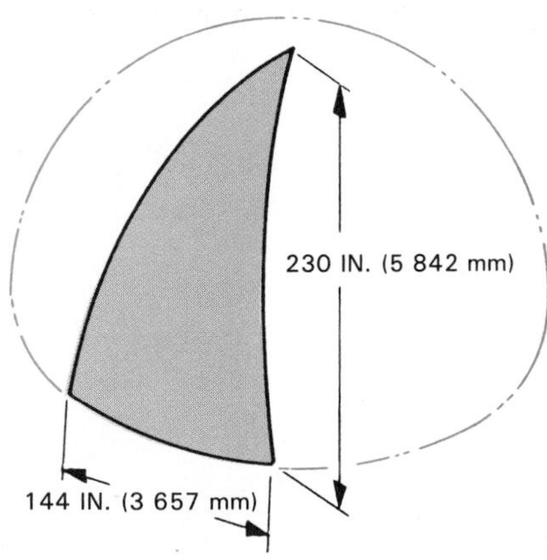

Fig. 22-2. Explosives forming was used to shape segments for top and bottom domes of giant aerospace vehicle fuel tanks. Segments were welded after forming.

Explosive forming utilizes the pressure wave generated by an explosion in a fluid to force the material against the walls of the die, Figs. 22-4 through 22-6. The fluid has the effect of rounding off the pressure pulse generated by the detonation. The metal is cut or fabricated to a shape determined by the contours of the finished part, Fig. 22-7. The preform is placed in the die, filled with water and an explosive charge suspended in it.

A large holding ring, clamped over the outer edge of the work, assures the necessary seal for drawing a vacuum in the die. A ***vacuum*** is necessary between the work and the die; otherwise, an air cushion would develop as the metal is forced into the die. This would prevent the metal from seating in the die and assuming its proper shape.

When detonated, the explosion slams the material against the die walls, Fig. 22-8. This is accomplished in microseconds.

Placement and quantity of the high explosive is critical. The charge can range from a few ounces

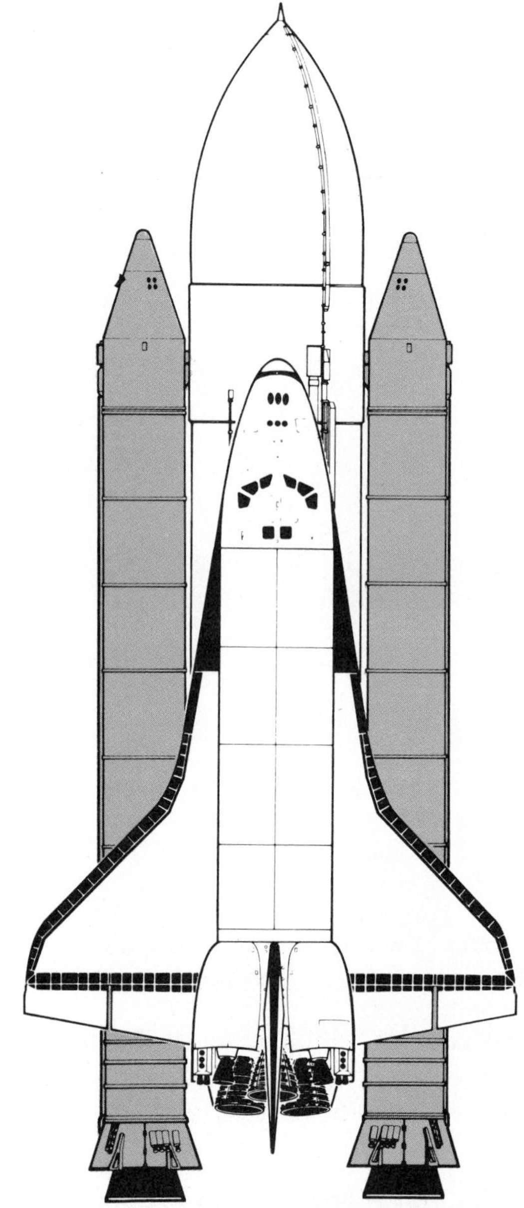

Fig. 22-3. High energy rate forming (HERF) shaped end sections of external fuel tanks of Space Shuttle. (NASA)

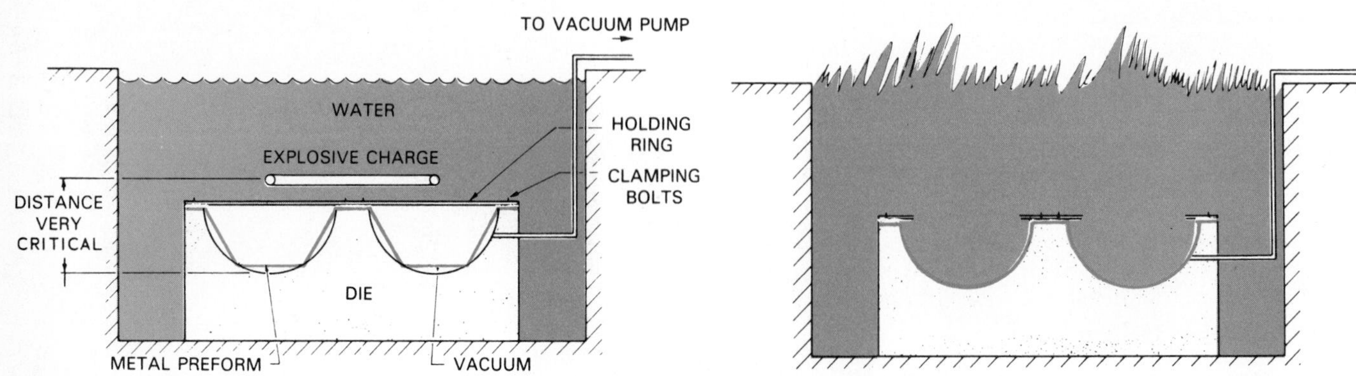

Fig. 22-4. Diagram shows principle of explosive forming process. In some applications, a few dollars worth of explosives will do work of a press that may cost a million or more dollars.

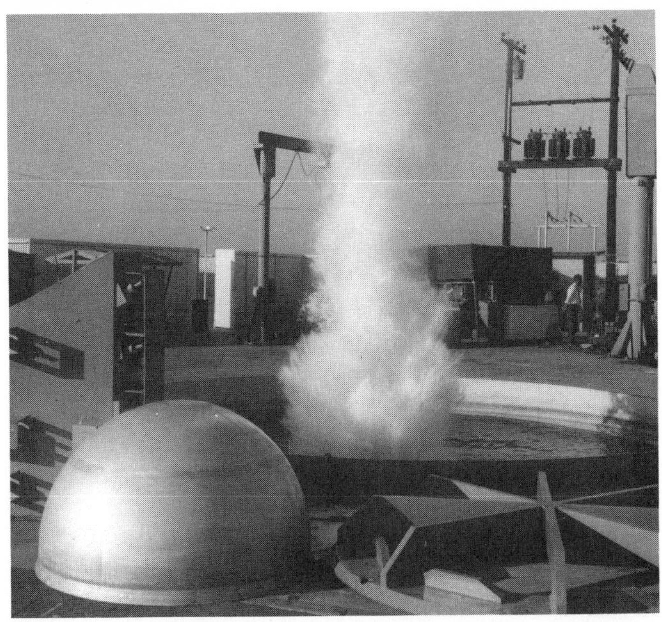

Fig. 22-5. Explosive forming is being used to shape dome of a fuel tank. Water has effect of rounding off pressure pulse generated by detonation causing forces to be equalized over entire surface of material being formed. (The Ryan Company)

Fig. 22-7. Worker is welding preform of a torus (donut) shaped fuel tank for explosive forming. This is first step in process. Formed conventionally, it would require forty sections rather than twelve and would require 60 percent more welding footage.

Fig. 22-6. Note a few of many aircraft and missile parts shaped by explosive forming. (The Ryan Company)

Fig. 22-8. Once resembling a cake pan, exploded metal now is torus, or donut shaped. Engineers examine inner contour of formed part, removed from die underneath. Metal lip clamps around edge of die and is secured by a holding ring. This assures necessary tight seal for drawing a vacuum into die during forming operation.

Fig. 22-9. Many parts on nuclear submarines are shaped by HERF. (General Dynamics Corp., Electric Boat Div.)

to form small parts, to the many pounds needed to form large sections of aluminum and steel up to 4 in. (100 mm) thick. For example, the heavy steel missile hatches on the Navy's submarines are formed by this technique, Fig. 22-9. Many forms of explosives are utilized: rod, sheet, granules, liquid, stick, cord, and plastic.

Depending upon placement of the explosive, most operations fall into two categories: stand-off operations and contact operations.

Stand-off operations

During ***stand-off HERF,*** the charge is located some distance from the work. Its energy is transmitted through a fluid medium such as water. This technique is used to form and size parts.

Contact operations

During ***contact HERF,*** the charge is touching the work and the explosive energy acts directly on the metal. Welding, hardening, compacting powdered metals, and controlled cutting are done with this technique.

Explosive forming—advantages and disadvantages

While explosive forming offers many advantages, there are some drawbacks associated with the process:

1. The technique has NOT been developed to the stage where a part can always be formed properly on the first shot.
2. Since the operation utilizes the "big-bang" principle (an explosion) to do forming, the noise can be a problem. Also, strict laws prohibit the use of explosives in populated areas. These factors usually make it necessary to locate the facility away from populated areas. This increases transportation and handling costs.
3. Personnel must be highly skilled in the safe handling of high explosives. Insurance rates are high!

ELECTROHYDRAULIC FORMING

Electrohydraulic forming, also called ***capacitor discharge forming*** or ***spark forming,*** is a variation of explosive forming, Fig. 22-10. High voltage electrical energy is discharged from a ***capacitor bank*** (device used to store electrical energy) into a thin wire or foil suspended between two electrodes. The unit is immersed in water, as with explosive forming. Many of the titanium metal (tough, light metal that is difficult to form) parts on aerospace vehicles are formed using this technique, Fig. 22-11.

As the wire or foil is vaporized by the electric current discharge, the vapor products expand, converting the electrical energy to hydraulic energy. The shock wave forms the metal against the die.

As the energy is less than that associated with explosives, it is usually necessary to repeat the operation several times to achieve the desired results.

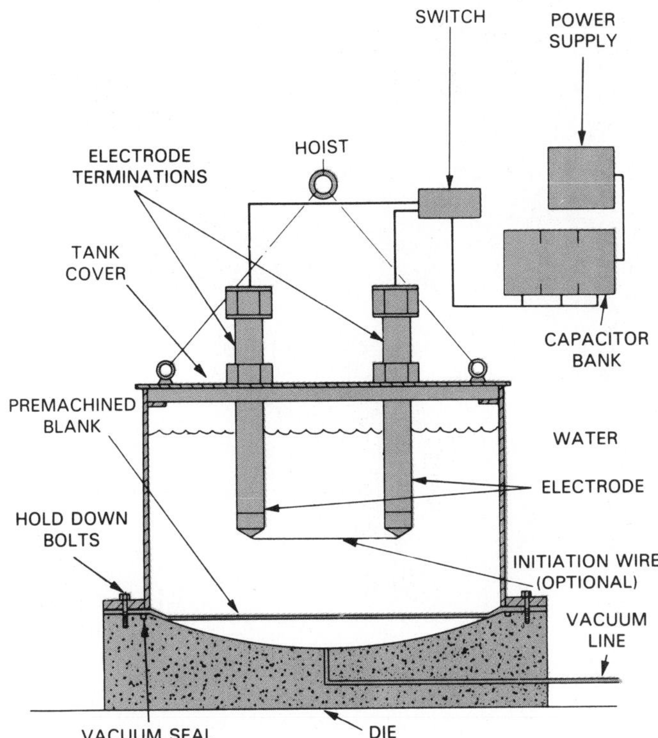

Fig. 22-10. This diagram shows setup for using electrical energy as a source of power for HERF operations. (NASA)

Fig. 22-11. Mach 3 aircraft, like this proposed design, will have to be manufactured from hard to form heat resistant metals. HERF techniques will be required to form many complex parts. (Grumman Aerospace Corp.)

A well designed electrohydraulic forming facility can be adapted to automation. The capability of generating high pressures without a "bang" permits the unit to be employed in conventional industrial facilities.

MAGNETIC FORMING

Magnetic forming, also termed ***electromagnetic forming*** or ***magnetic pulse forming,*** utilizes an insulated induction coil wrapped around or placed within the work. Coil location depends upon whether the metal is to be squeezed inward or bulged outward. The coil is shaped to produce the desired shape in the work. Refer to Figs. 22-12 through 22-15.

As very high momentary currents are passed through the coil, an intense magnetic field is

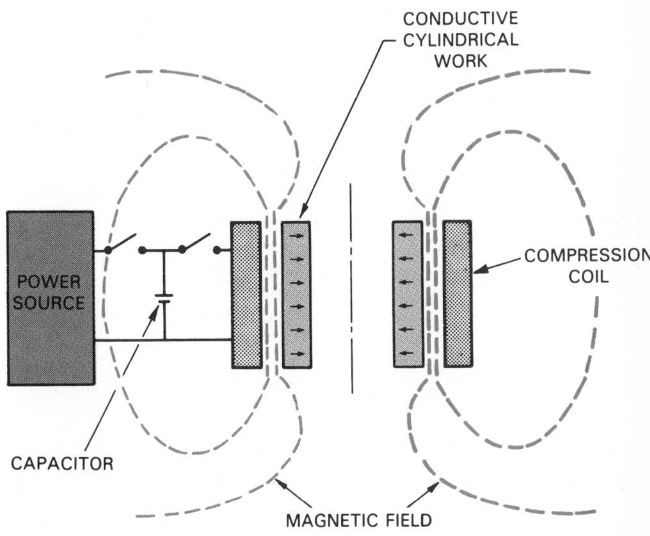

Fig. 22-12. Note how magnetic pulse metal forming works. A heavy magnetic field is produced by discharging a capacitor through a coil. During brief impulse, eddy currents in work restrict magnetic field to surface of work, creating a uniform force to form metal. Process can be employed directly on highly conductive metals. Low conductive metals can be shaped by using aluminum (highly conductive) layer between work and coils.

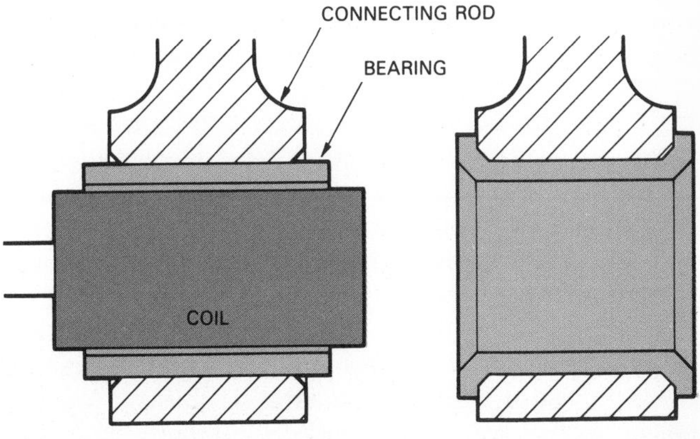

Fig. 22-13. Magnetic pulse is being used to expand a bearing sleeve into a connecting rod.

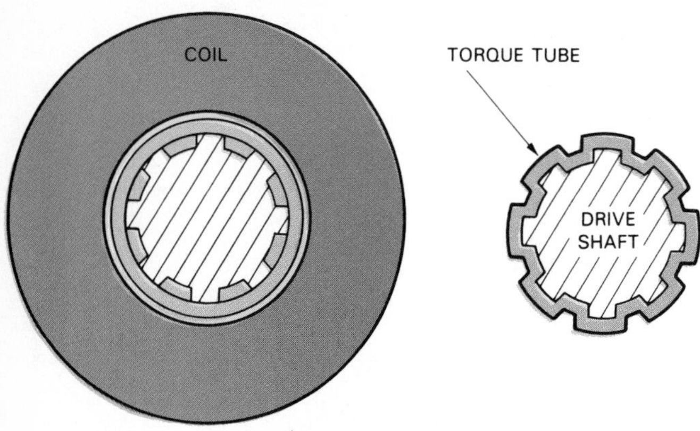

Fig. 22-14. Torque tube is being compressed on automotive drive shaft.

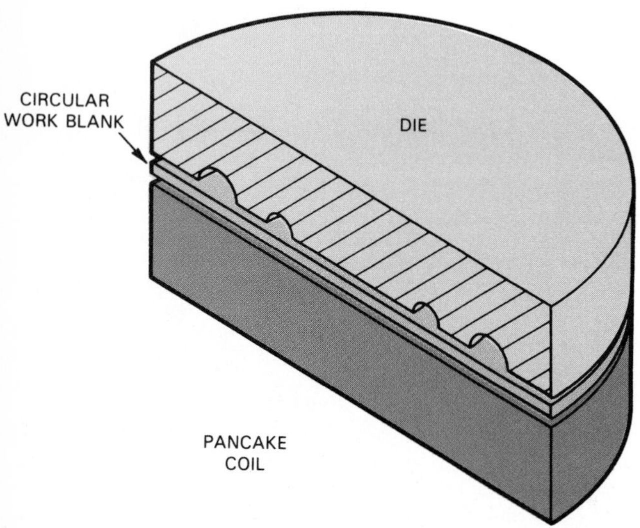

Fig. 22-15. Flat forming applications or forcing sheet metal into a die, requires a pancake coil to provide uniform magnetic pressure.

developed. This causes the work to collapse, compress, shrink, or expand depending on the design of the coil. See Fig. 22-16.

The power source is basically the same as that used for electrohydraulic forming—a capacitor bank. The spark gap is replaced by a coil. Energy is obtained by capacitor discharge through the coil. In fact, properly designed equipment can be used to perform both operations.

Energy storage capacity and ability to utilize that energy determines the size of the work that can be formed. Highly conductive metals can be formed easily. Nonconductive or low-conductivity materials can be formed if they are wrapped or coated with a high-conductivity auxiliary material.

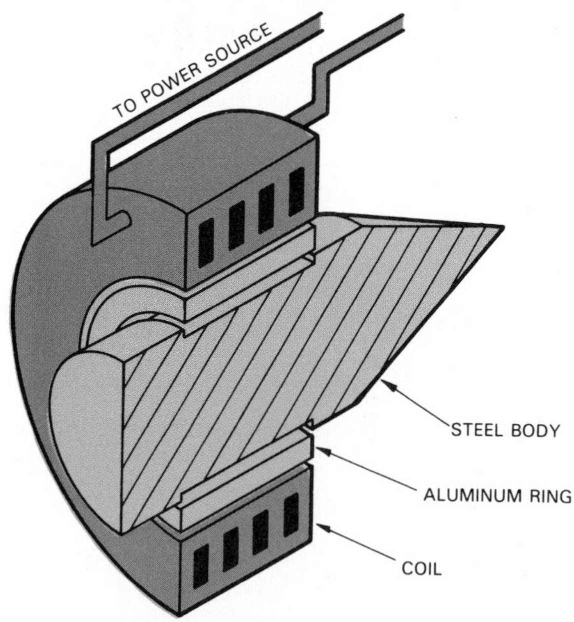

Fig. 22-16. A magnetic field can also be employed to shrink or squeeze parts together.

PNEUMATIC-MECHANICAL FORMING

Pneumatic-mechanical forming uses a punch and die operated by high-pressure gas. It was the first of the HERF techniques to become a standard production tool. The operation has much in common with conventional forging since a punch and die are employed, Figs. 22-17 and 22-18. However, the forces developed are many times more powerful and are sufficient to shape hard-to-work materials. The machine also requires less space than the conventional forging press.

High-pressure gas is used to accelerate a punch into a die in pneumatic-mechanical forming. The punch and die are mounted on opposed rams that meet with equal force, thus taking most of the strain off the frame of the machine. Some machines make use of a recoil mechanism similar to that used on an artillary piece. The metal blank is preheated prior to the forming operation.

Pneumatic-mechanical forming can be accomplished by several means. One makes use of a blank cartridge to power the ram for shaping small parts. However, the most widely employed method utilizes a two part cylinder, Fig. 22-19. Gas is stored in one part of the cylinder at approximately 2000 psi (13 800 kPa). Gas is also stored in the second section of the cylinder but at a much lower pressure of about 200 psi (1380 kPa). A plate with an orifice separates the two sections of the cylinder. The piston (ram), with a special seal, closes off the orifice. In this way, the high-pressure gas acts only on a small section of the piston (area of orifice) while

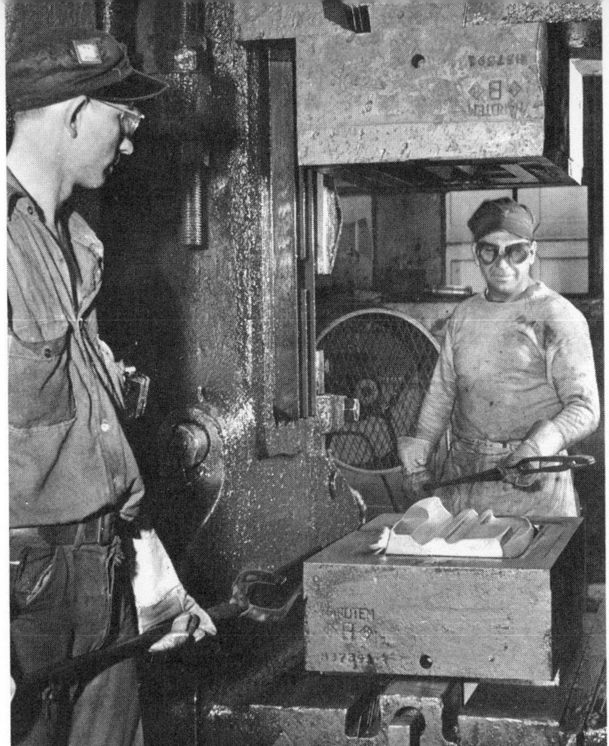

Fig. 22-17. A conventional drop forge is shaping an aluminum blank. This technique may require a dozen or more strokes to complete forming job. Pneumatic-mechanical forming also uses punch and die to do same thing. However, machine is capable of exerting pressures many times greater than conventional forging press and takes up considerably less space. (Mueller Brass Co.)

Fig. 22-18. This is a DYNAPAK (Trademark-General Dynamics Corp.) pneumatic-mechanical press. Because of unique self-reacting framework system, machine imparts no shock load to floor. This makes it possible to use it in close proximity to conventional chip-making machines. (General Dynamics Corp.)

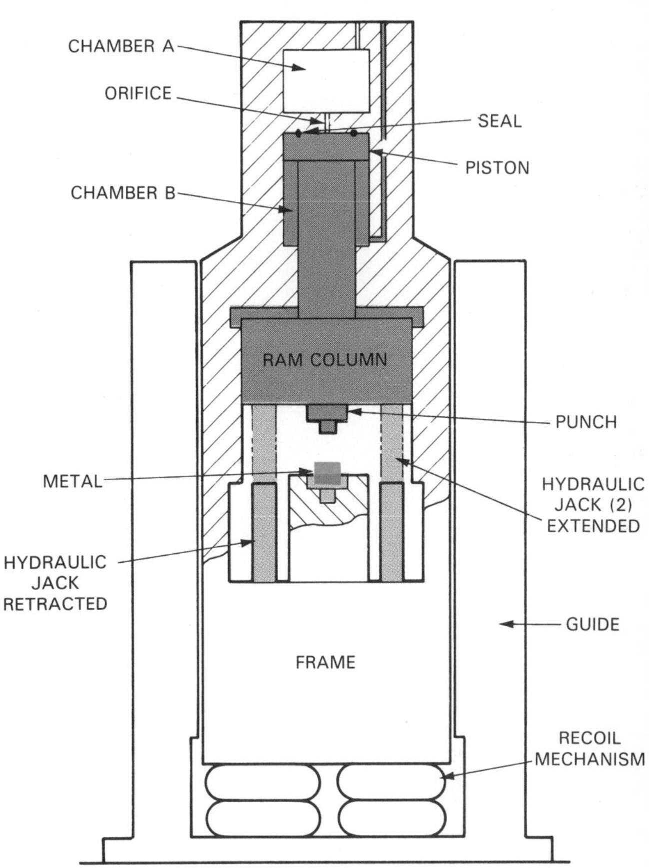

Fig. 22-19. Note cross section drawing of a pneumatic-mechanical forming press. (General Dynamics Corp.)

the low pressure acts on the entire area of the piston. This maintains a stable balance between the two sections, Fig. 22-20.

To make the machine operate, the pressure is increased slightly in the high-pressure section. This upsets the balance and the piston starts to move. When the seal is disengaged, the high pressure acts on the entire surface area of the piston driving it forward at tremendous speed, Fig. 22-21. Hard to

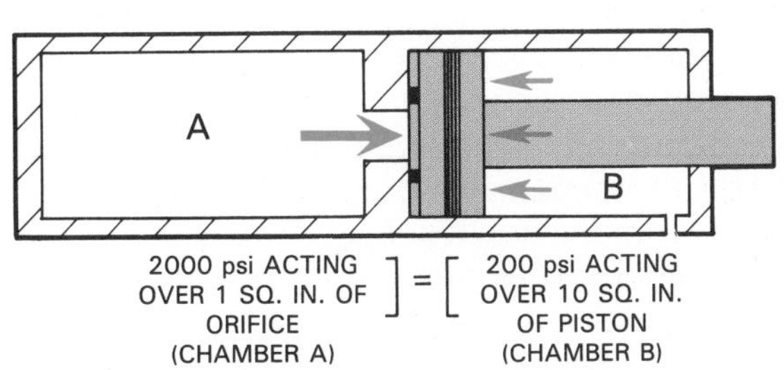

Fig. 22-20. High-pressure gas acts on only a small area of piston (area of orifice) while low pressure gas acts on entire area of piston. This maintains a balanced situation between two.

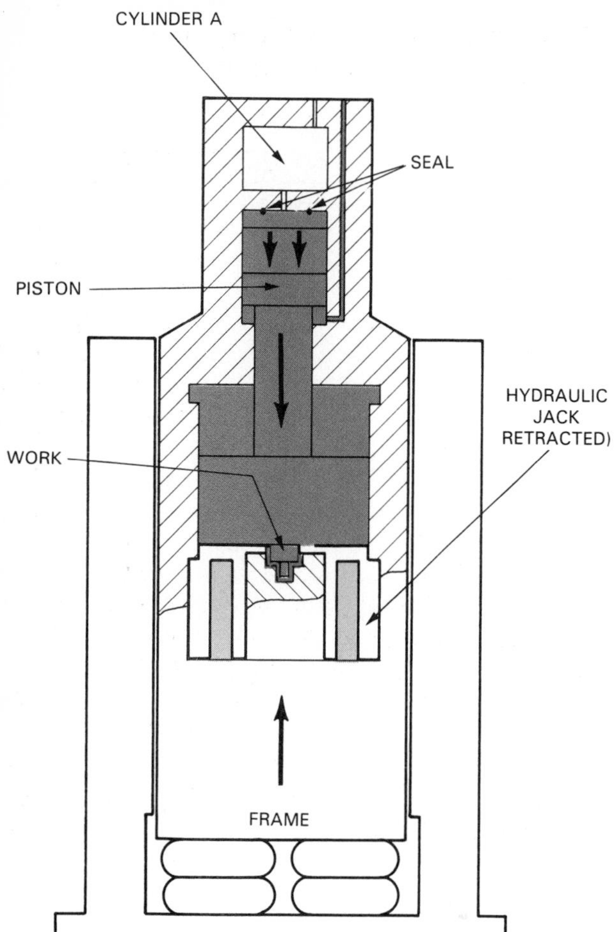

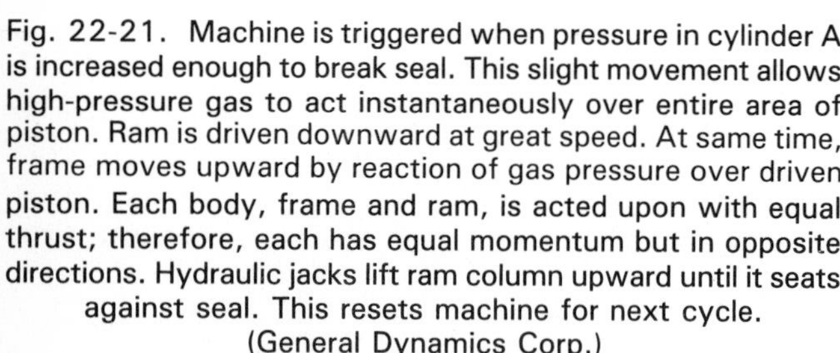

Fig. 22-21. Machine is triggered when pressure in cylinder A is increased enough to break seal. This slight movement allows high-pressure gas to act instantaneously over entire area of piston. Ram is driven downward at great speed. At same time, frame moves upward by reaction of gas pressure over driven piston. Each body, frame and ram, is acted upon with equal thrust; therefore, each has equal momentum but in opposite directions. Hydraulic jacks lift ram column upward until it seats against seal. This resets machine for next cycle. (General Dynamics Corp.)

shape metals are usually formed with one blow, Fig. 22-22. Some space age materials can best be shaped by this technique.

In this method of forging, precise control is possible. Heavy, expensive equipment is eliminated. Fewer operations are necessary because most parts can be formed with one blow. The parts are produced to close tolerances with smoothly finished surfaces that require a minimum of machining, Fig. 22-23. Production is rapid. Substantial savings are possible because less material is used, fewer operations must be performed to finish the part, and higher production rates are obtained.

Fig. 22-22. Front-wheel spindle for an automobile was formed from alloy steel in one blow on a pneumatic-hydraulic press. (General Dynamic Corp.)

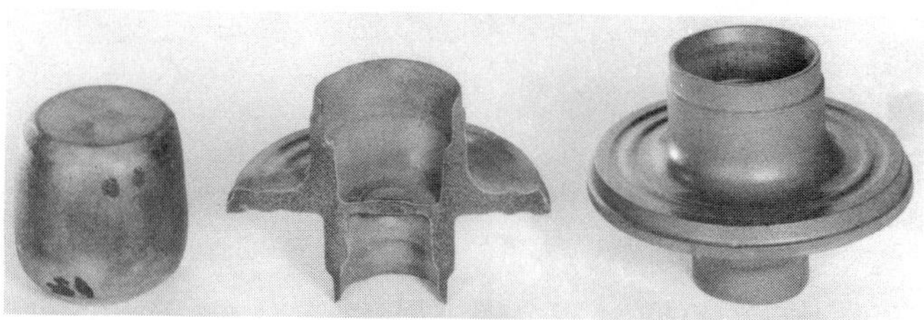

Fig. 22-23. Wheel hub was formed on a conventional press (center) and a hub was formed on HERF machine (right). Note better finish on HERF part. Section of metal (left) is blank from which wheel hub is formed. (General Dynamics Corp.)

TEST YOUR KNOWLEDGE—Chapter 22

Please do not write in the text. Place your answers on a separate sheet of paper.

1. The abbreviation HERF means ________ ________ ________ ________.
2. In HERF, metal is shaped:
   a. By the slow application of great pressure.
   b. In microseconds with pressure generated by the sudden application of large amounts of energy.
   c. By conventional forging methods.
   d. All of the above.
   e. None of the above.
3. The pressures needed in HERF are generated by:
   a. Detonating explosives.
   b. Releasing compressed gases.
   c. Electomagnetic energy.
   d. All of the above.
   e. None of the above.

4. Many metals tend to _______ _______ to their original shape after being formed by conventional means. This problem is greatly reduced or entirely eliminated when _______ is used to shape the metal.
5. Of all the HERF techniques _______ _______ is the oldest having been developed in the late 1800s.
6. What is explosive forming?
7. Why must a vacuum be pulled in the die when explosive forming?
8. Depending upon the placement of the explosive, most operations fall into two categories. List them.
9. _______ _______ is a variation of explosive forming. However, _______ is used in place of the explosive charge to generate the required energy.
10. What HERF technique employs a very high electric current passing through an induction coil shaped to produce the required configuration in the work?
11. The technique in question 10 can be used to _______ _______ or _______ _______ on the work to produce the desired shape, depending upon placement of the coil.
12. In pneumatic-mechanical forming, _______ _______ is used to accelerate the _______ into the _______.

RESEARCH AND DEVELOPMENT

1. Develop a series of slides for the overhead projector, showing step-by-step, how the various HERF techniques work. Use them to illustrate a talk on HERF to be given to the class.
2. Secure information on the various HERF techniques from trade journals and companies making use of HERF. Make the material into a bulletin board display.
3. Get samples of work produced by HERF. If there are no such companies in your area, use photos from trade magazines and brochures to develop a display panel. Employ a sketch of the HERF process employed to produce the particular pieces displayed or pictured.
4. Contact a company using HERF techniques. Request the loan of a film, video tape, or slides that could be used to illustrate HERF.

DANGER! Students should NOT experiment with explosive forming!

Fig. 23-1. Series of dies replace usual cutting tools with chipless machining. Work is transferred from station to station to permit various forming operations. Scrap is kept to an absolute minimum. (National Machinery Co.)

Chapter 23

CHIPLESS MACHINING

After studying this chapter, you will be able to:

- Explain how chipless machining is done.
- Recognize the five basic operations of chipless machining and their variations.
- Describe the intraform process and how it differs from other chipless machining techniques.

Chipless machining forms a metal wire or rod into the desired shape using a series of dies. It is another of the metalworking processes finding increased usage in modern industry. The technique will NOT replace conventional machining. However, for some jobs, chipless machining does make substantial savings possible. It can reduce the amount of scrap metal that results and can increase production speed. The process is sometimes called ***cold heading*** or ***cold forming.***

In chipless machining, a series of dies replace the usual cutting tools on the lathe, drill press, or milling machine. See Figs. 23-1 and 23-2.

Material used in chipless machining is usually in coil form and is referred to as ***wire.*** This material is turned into needed, and often complex, shapes. Refer to Fig. 23-3.

Accuracy can be held to tolerances of 0.002 in. (0.05 mm) and closer if required. However, costs increase in proportion to the precision wanted. In most cases, waste is totally eliminated.

The technique is an economical and efficient way to make bolts, nuts, screws, and other fasteners, Fig. 23-4. Another example, almost all spark plug bases are made by chipless machining, Fig. 23-5.

There are five basic operations done by machines using this process, Fig. 23-6. Combinations and

Fig.23-2. Action shot shows fingers transferring part from station to station. (National Machinery Co.)

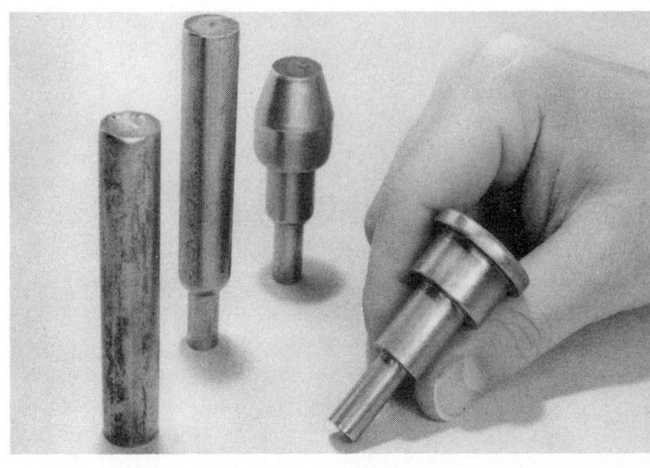
Fig. 23-3. Parts can be manufactured quickly and inexpensively using chipless machining techniques. No scrap resulted from this job and tolerances were held to within 0.002 in. (0.05 mm). (National Machinery Co.)

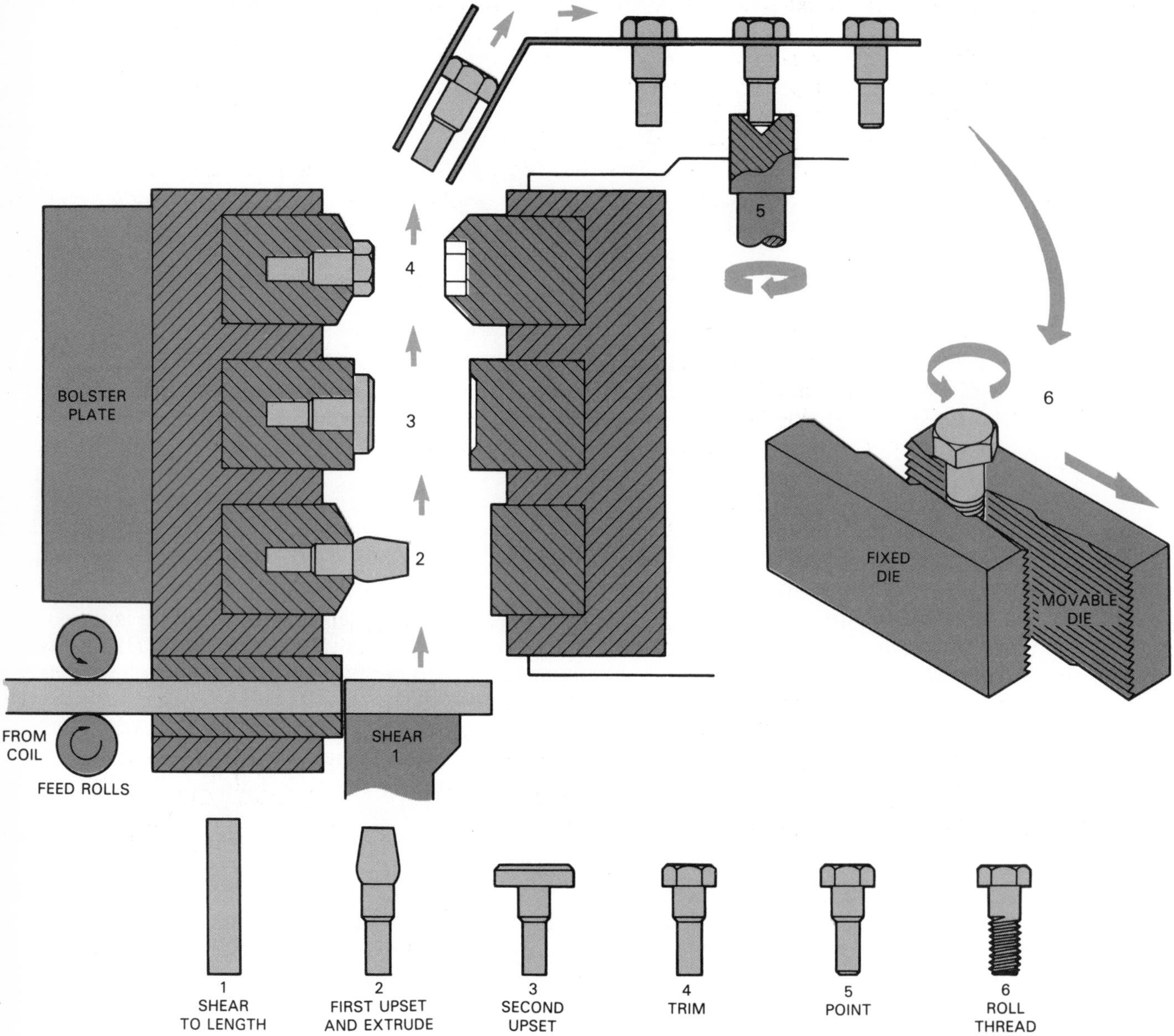

BOLSTER PLATE
4
3
2
5
6
FIXED DIE
MOVABLE DIE
FROM COIL
FEED ROLLS
SHEAR 1
1 SHEAR TO LENGTH
2 FIRST UPSET AND EXTRUDE
3 SECOND UPSET
4 TRIM
5 POINT
6 ROLL THREAD

Fig. 23-4. Heavy arrows and numbers indicate sequence involved in producing bolts by chipless machining. Trace part flow through process.

BLANK
WASTE
FINISHED

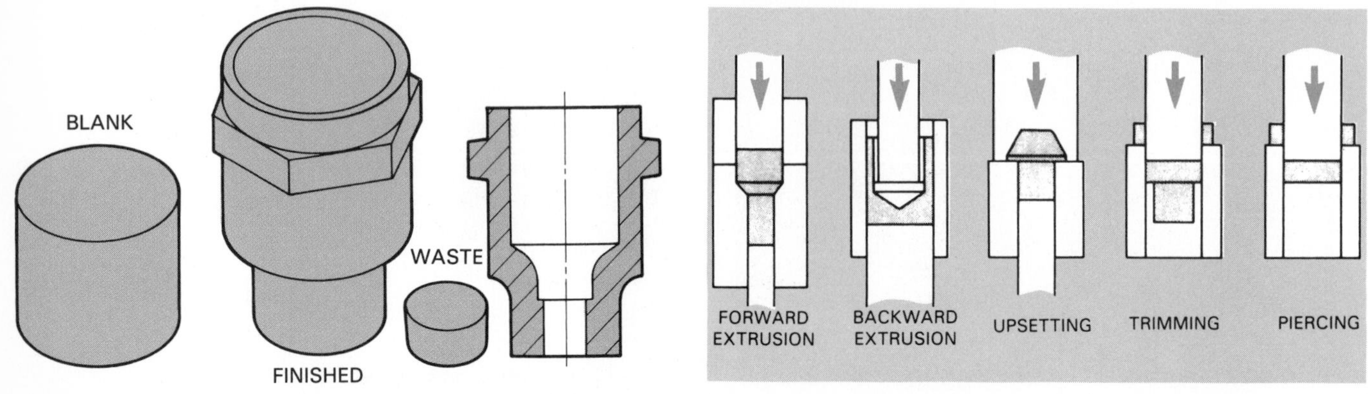

Fig. 23-5. Almost all spark plug bases are made by chipless machining. Note small amount of scrap of less than 2 percent.

FORWARD EXTRUSION
BACKWARD EXTRUSION
UPSETTING
TRIMMING
PIERCING

Fig. 23-6. Study five (5) basic operations performed by machines designed for cold forming process.

variations of these operations make possible a wide range of applications, Fig. 23-7.

Metals ranging from aluminum alloys to medium carbon steel can be shaped using this technique. Stainless steel, copper, and nickel alloys can be cold formed but the ease in which they are shaped depends upon the part design. Material up to 1.5 in. (37.5 mm) diameter can be formed on some machines.

INTRAFORM MACHINING

Intraform® (Cincinnati Milacron trademark) is a development of chipless machining that makes it possible to form profiles on the inside diameter(s) of cylindrical pieces. Forming inside profiles would be extremely difficult and expensive to do by conventional machining techniques.

In this process, a section of hollow cylindrical stock is placed over a steel mandrel, Fig. 23-8. It is them squeezed by rapidly pulsating dies, Fig. 23-9. At the completion of the operation, the mandrel's profile is produced on the inside diameter of the part, Fig. 23-10.

As Fig. 23-11 shows, fixed rolls cause the four dies to pulsate rapidly around the outside diameter of the work. The tops of the cams are shaped to permit a smooth continuous squeezing action of the dies. Even though the work is being squeezed by the dies more than 1000 times per minute, noise and vibration are NOT a problem. A cross section of the Intraform machine is shown in Fig. 23-12.

The technique has proven to be a practical way to produce rifle barrels for example, Fig. 23-13. Predrilled steel blanks are fed into the machine which forms the chamber and rifling. In addition to

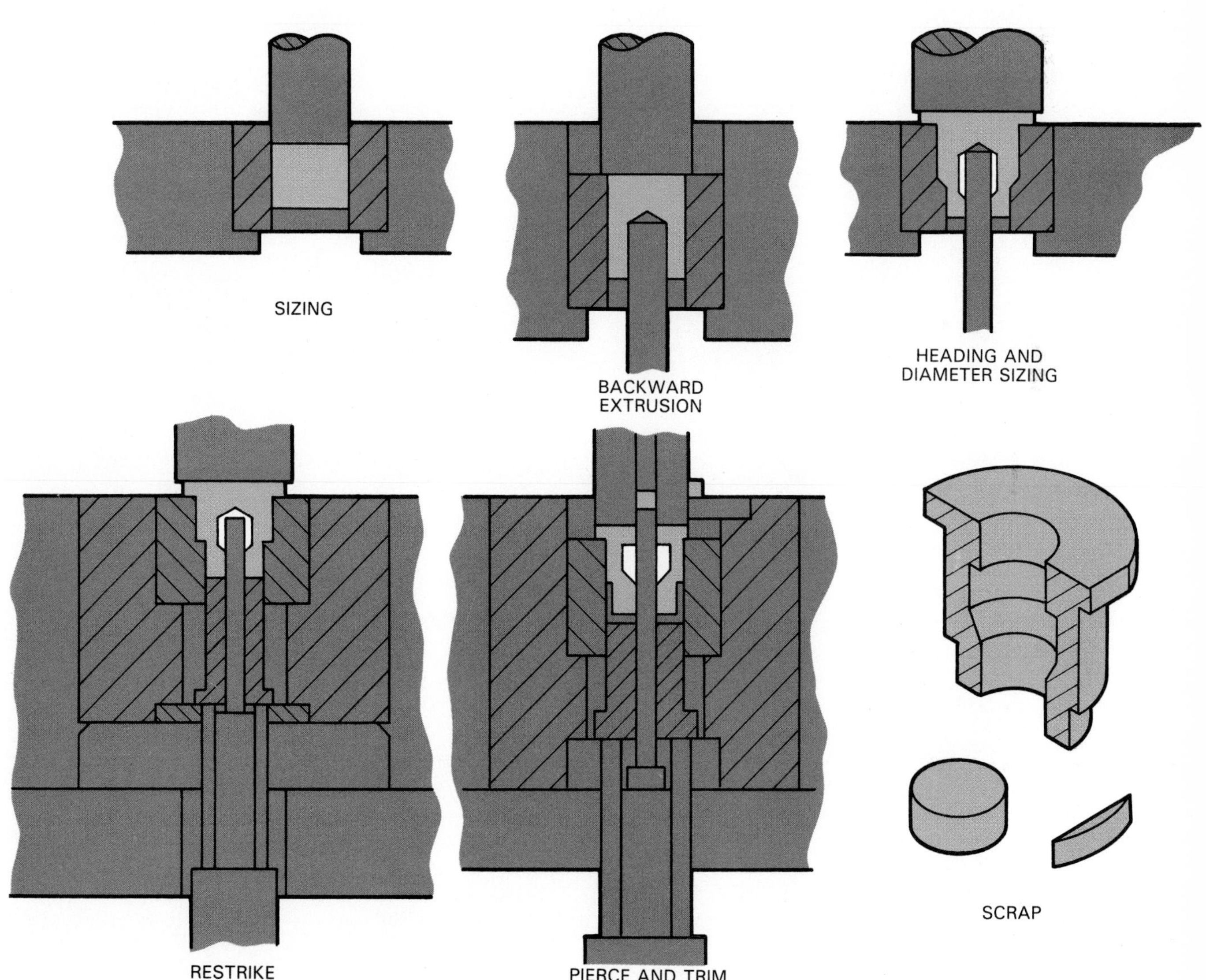

Fig. 23-7. Note combinations and variations of five basic chipless machining operations.

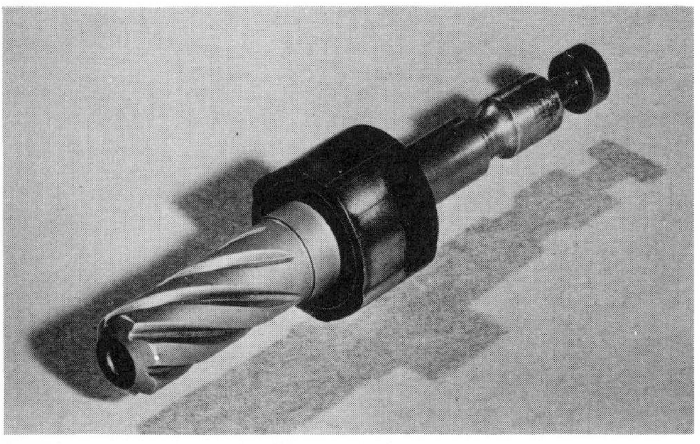

Fig. 23-8. Two-piece Intraform mandrel is employed in production of this automotive starter clutch housing. (Cincinnati Milacron)

Fig. 23-10. Clutch housing and sectioned part show helical spline and cam clutch profile, which was formed in one operation, at a production rate of 220 parts per hour. (Cincinnati Milacron)

Fig. 23-9. Note dies used in production of automotive starter clutch housing. (Cincinnati Milacron)

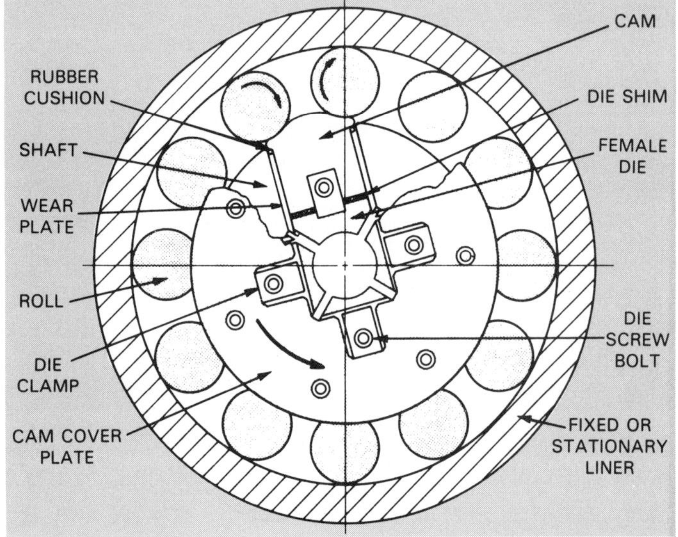

Fig. 23-11. Drawing illustrates Intraform machine die head in open position.

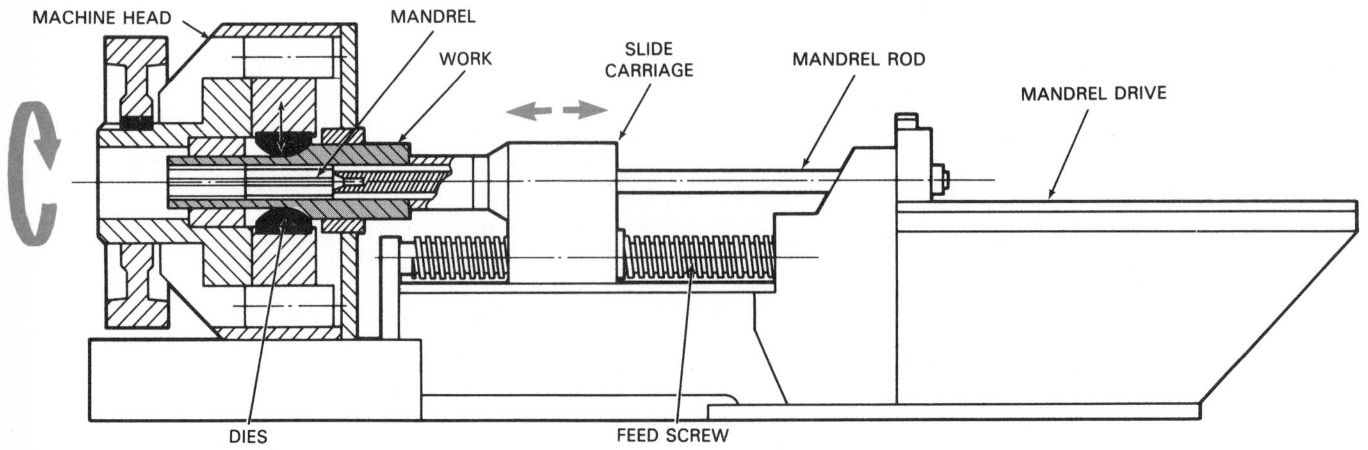

Fig. 23-12. Study cross-sectional view of Intraform machine.

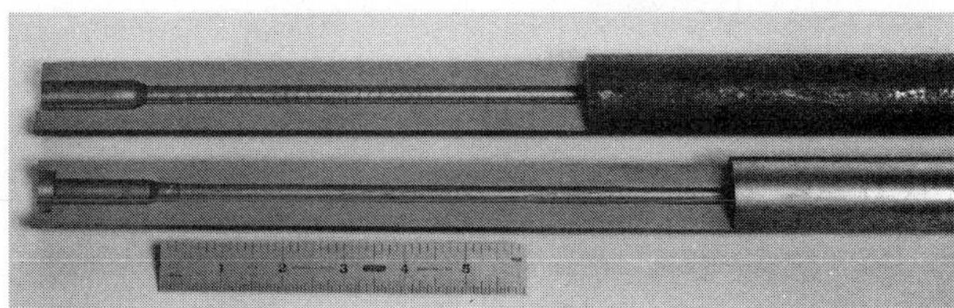

Fig. 23-13. This shows predrilled, rolled steel blank and Intraformed chamber and rifling for a 30 caliber rifle barrel. About 60 percent of all barrels produced by Intraform process are target rifle quality, as compared to about 10 percent by conventional methods. (Cincinnati Milacron)

improving the surface finish of the bore, the operation also improves the physical characteristics of the metal.

A typical Intraform sequence is shown in Figs. 23-14 through 23-16.

TEST YOUR KNOWLEDGE—Chapter 23

Please do not write in the text. Place your answers on a separate sheet of paper.

1. Chipless machining is also known as _______ _______ or _______ _______.
2. How does chipless machining make substantial savings possible?
3. A series of _______ replace the usual cutting tools of the lathe, drill press, and milling machine.
4. Chipless machining is still the most economical way to make _______, _______, _______, and other types of _______.

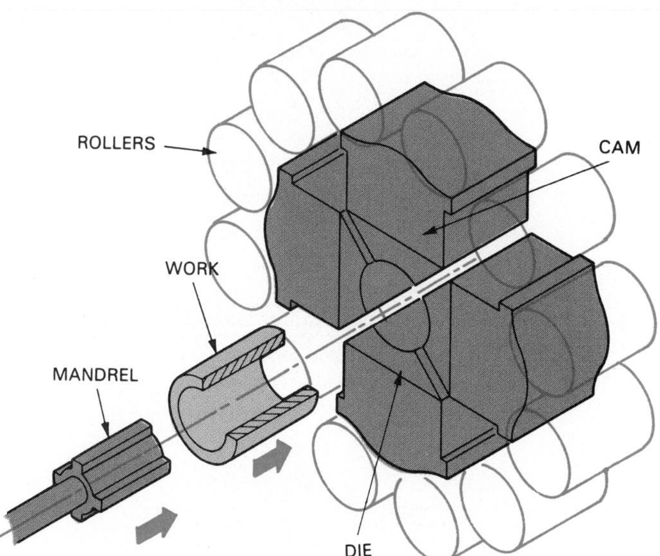

Fig. 23-14. Work is ready to be formed by Intraform process.

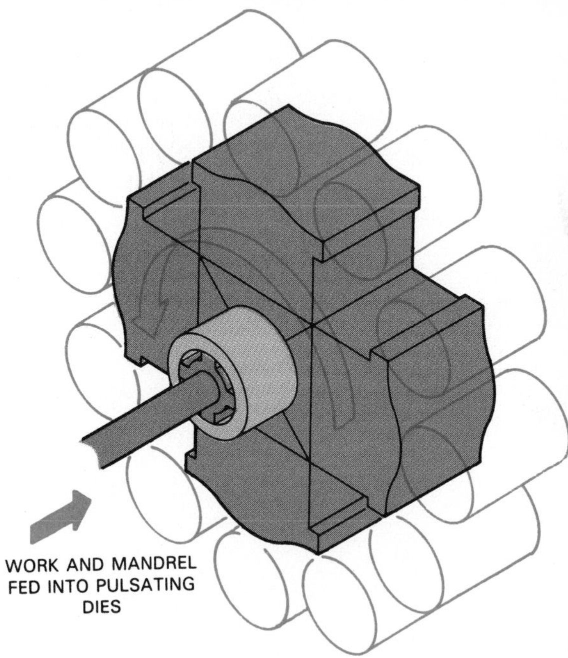

Fig. 23-15. Work and mandrel are between Intraform dies. Contact with rotating dies causes freewheeling work, and mandrel, to rotate at about 80 percent of die rpm. Work feeds over mandrel.

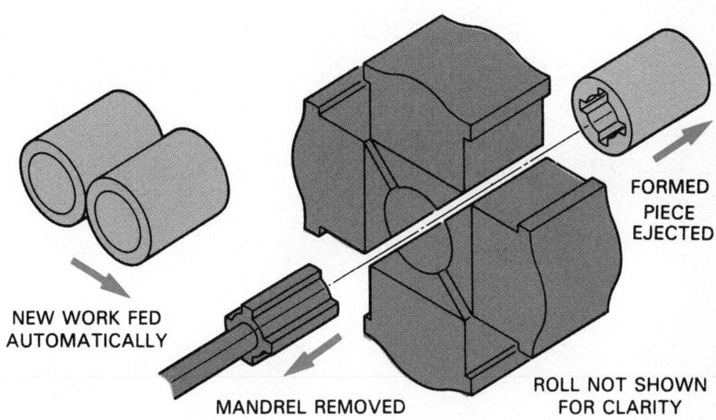

Fig. 23-16. When operation is completed, mandrel is retracted. Next piece feeds into position and formed piece is ejected.

5. List the five (5) basic operations performed by machines making use of the chipless machining process.
6. Intraform® is a chipless machining technique that can form profiles on the _______ _______ of _______ pieces.
7. The Intraform technique has proven to be a practical way to produce:
   a. Socket wrenches.
   b. Rifle barrels.
   c. Automotive starter clutch housings.
   d. All of the above.
   e. None of the above.

RESEARCH AND DEVELOPMENT

1. Secure samples of work produced by chipless machining. Mount them on a display panel with an illustrated explanation of the process.
2. What does the term "plasticity," as it relates to metal and chipless machining mean?
3. Secure material from companies that use the chipless machining process for the shop technical library. Prepare a display.
4. Contact a company making use of chipless machining and request samples of a product in various stages of manufacture. Prepare a display panel of the samples.

Chapter 24

ELECTRO-MACHINING PROCESSES

After studying this chapter, you will be able to:

- Explain the advantages and disadvantages of the electro-machining processes.
- Describe electrical discharge machining.
- Explain electrical discharge wire cutting.
- Describe electro-chemical machining.

The most notable advantages of ***electro-machining*** is that mechanical forces have no influence on the processes. Material is NOT removed by ''brute force'' as with conventional machining techniques. The electrical energy is applied directly to remove metal by erosion. The electrical energy is NOT just a source of power for the motor that drives a cutting tool.

Electrical discharge machining and Electro-chemical machining are two electro-machining techniques that have made a great impact on the field of metalworking and machining. Neither technique produces a chip as metal is removed. The particles are disposed of completely by vaporization or reduced to microscopic particles. The metals must be conductors of electricity in order to be machined by either of these processes.

NOTE! Never attempt to operate electro-machining equipment while your senses are impaired by medication or other subtances.

ELECTRICAL DISCHARGE MACHINING (EDM)

Electrical discharge machining (EDM) is a process where hard, tough, fragile, and/or heat sensitive metals, that are difficult to machine by conventional ''chip-making'' techniques, can be worked to close tolerances.

Die blanks, for example, can be worked in heat treated condition, Fig. 24-1. This eliminates warping and distortion that frequently occurs when a finished die is heat treated. Superhard metals are readily worked to tolerances as close as 0.0002 in. (0.005 mm) with very fine surface finishes. ''Washed-out'' (worn) dies can also be worked in the heat treated state.

EDM principle

In a gasoline engine, sparking or arcing takes place at the spark plug gap when the ignition coil fires to ignite the fuel mixture. After prolonged operation, the spark plug electrodes will be eroded by the action of the electric arcs. This is the basis of EDM. See Figs. 24-2 and 24-3.

An ***electric discharge machine,*** Fig. 24-4, is composed of the following parts.

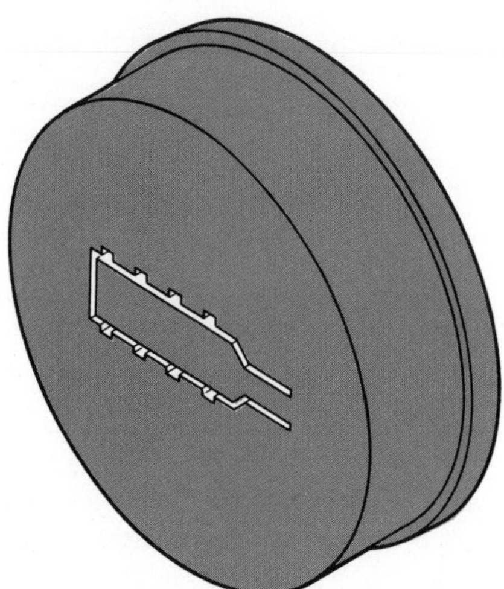

Fig. 24-1. Die planks, like this extrusion die used to produce storm window frames, can be machined by EDM while in heat-treating stage. Job can be done quickly, with no posibility of die warping, and at a considerable savings over traditional machining techniques.

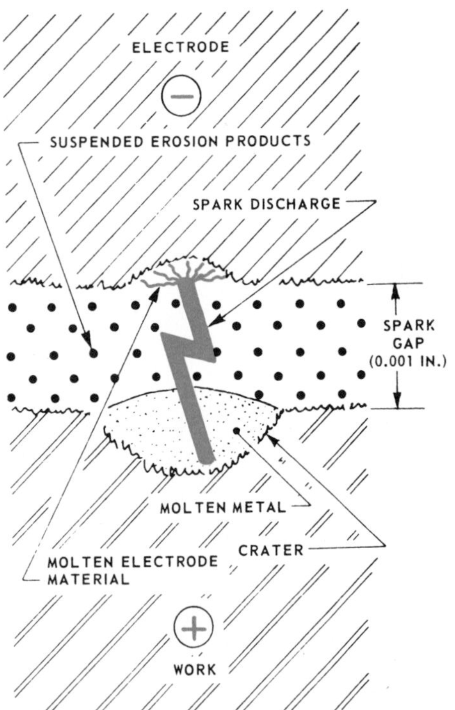

Fig. 24-2. Study how EDM process works. Spark or arc from electrode causes work to erode.

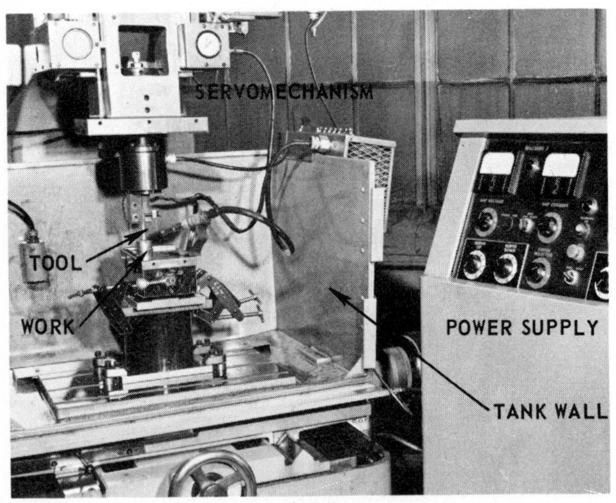

Fig. 24-3. Note typical setup for EDM. Front panel of dielectric tank has been removed for clarity.

1. A ***power supply*** is needed to provide direct current and a method to control voltage and frequency.
2. The ***electrode*** can be compared to the cutting tool of conventional machining. Copper, tungsten, and graphite alloys are the most effective materials for the electrode.
3. A ***servomechanism*** (drive unit) is used to accurately control electrode movement and to maintain the correct distance between the work and electrode as machining progresses.
4. A ***coolant,*** usually a light mineral oil, is used to form a dielectric (nonconducting) barrier between the electrode and work at the arc gap.

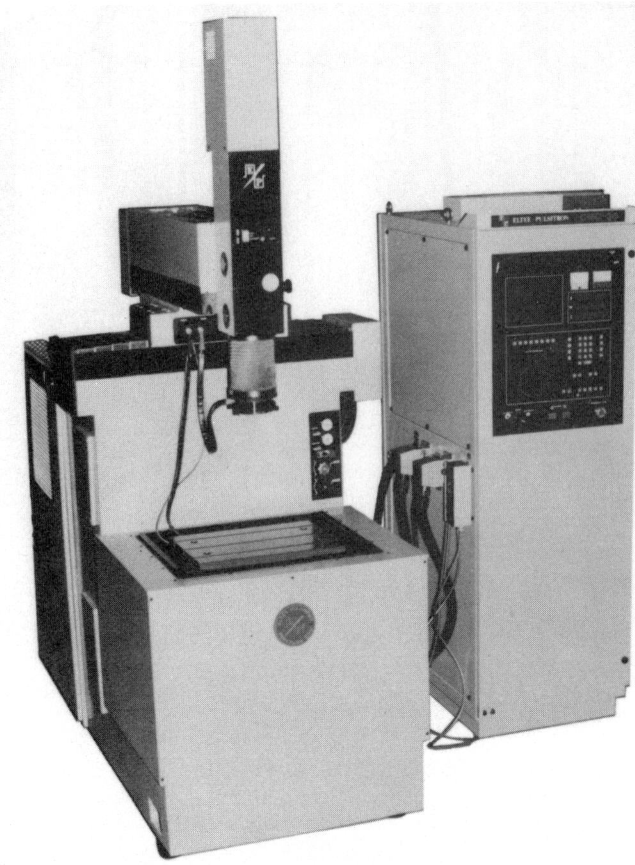

Fig. 24-4. This is a CNC ram-type EDM machine. (Eltree Pulsitron)

The servomechanism maintains a thin gap of about 0.001 in. (0.025 mm) between the electrode and work. The electrode and work are submerged in the fluid, Fig. 24-5. When the voltage across the gap reaches a point sufficient to cause the dielectric (fluid) to break down, a spark occurs. Each spark erodes a tiny particle of metal, but since the sparking occurs 20,000 to 30,000 times per second, appreciable quantities of metal are removed.

Besides providing a nonconductive barrier, the dielectric fluid also serves to flush particles from the gap, keep the electrode and work cool, and prevent fusion of the electrode with the work. A filter removes particles from the fluid.

Roughing cuts are made at low voltage and low frequency, with high amperage and high capacitance (opposition to any change in voltage.) Finishing cuts require high voltage and high frequency, with low amperage and low capacitance.

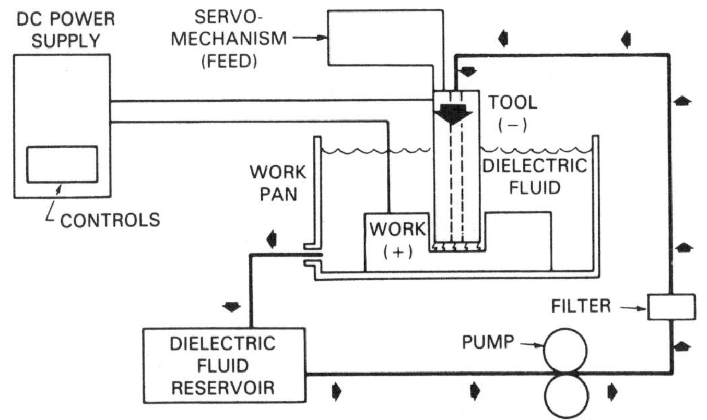

Fig. 24-5. Study diagram showing principle of EDM process.

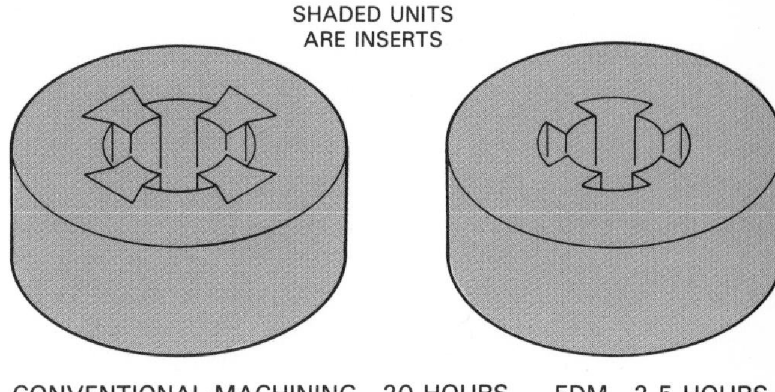

Fig. 24-7. This guide would be quite expensive to manufacture by conventional machining techniques.

Hard metals erode at a much slower rate than soft metals. As the electrode is also consumed, but at a much slower rate than the work, considerable savings can be effected by making interchangeable electrodes for roughing, sizing, and finishing. Long runs may require several sets of electrodes.

EDM applications

The EDM process is employed to:

1. Shape carbide tools and dies.
2. Machine complex shapes in hard, tough metals. Refer to Fig. 24-6.
3. Machine applications where the physical characteristics of the metal or its use makes it impractical or very expensive to machine by conventional methods, Fig. 24-7.
4. Eliminate tedious and expensive hand work in die making because cavity produced in metal is a ''mirror image'' of the electrode, Fig. 24-8.

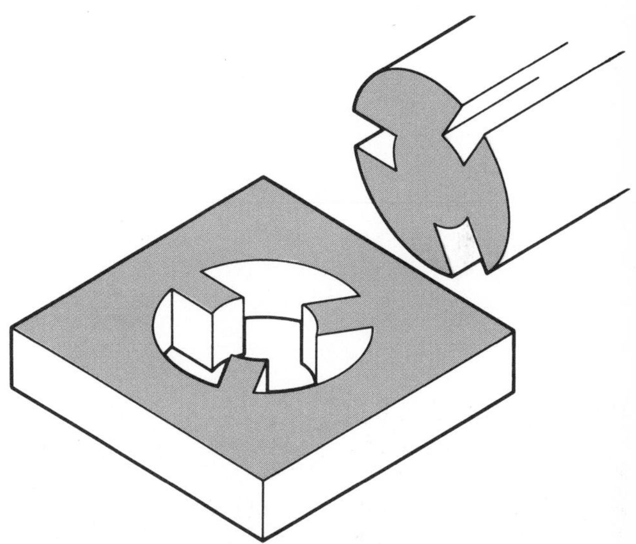

Fig. 24-8. An EDM electrode must be an exact reversal or mirror image of cut to be made. Many electrodes are made from graphite.

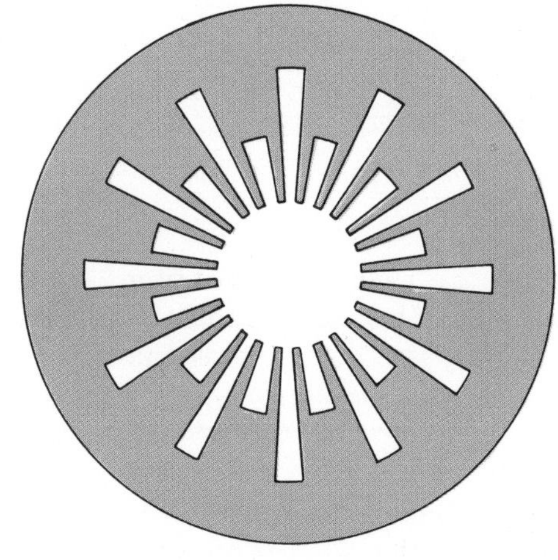

Fig. 24-6. EDM makes it possible to machine complex shapes to close tolerances in hard, tough materials. This die is employed to make electronic parts.

ELECTRICAL DISCHARGE WIRE CUTTING (EDWC)

Similar to band machining, ***electrical discharge wire cutting*** (EDWC) makes use of a small diameter wire electrode in place of the saw. Similar to EDM, material is removed by spark erosion as the wire electrode is fed through the workpiece, Fig. 24-9. The wire electrode is fed from a spool over sapphire or diamond guides. The wire electrode is used only ONCE because it becomes warped or distorted in one pass through the work.

A steady stream of deionized water cools the electrode and work. A starter or threading hole is also required.

The EDWC technique is well suited for CNC applications. This makes it possible to produce dies, shaped carbide cutting tools, and punches in less than one third the time required by conventional methods. When employed for other types of work,

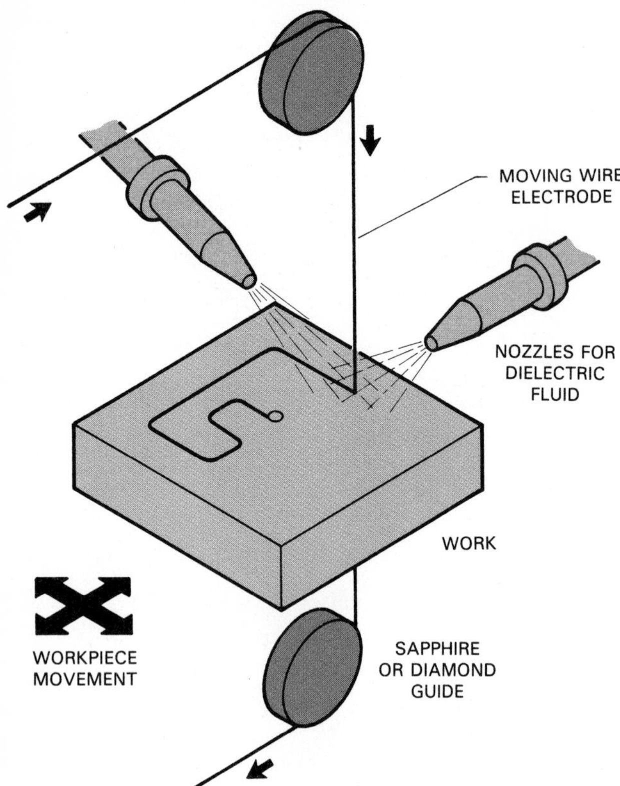

Fig. 24-9. Study how electrical discharge wire cutting (EDWC) works. It differs from EDM in that a fine, moving wire electrode is used to make cut instead of a solid electrode. This technique is ideal for CNC operations.

layers of sheet metal stacked up to 6 in. (150 mm) thick can be gang cut to produce a number of parts in one pass.

Examples of EDWC work are in Fig. 24-10.

ELECTRO-CHEMICAL MACHINING (ECM)

Electro-chemical machining, more commonly known as ECM, might be classified as electroplating in reverse. As in electro-plating, the process requires DC electricity and a suitable ***electrolyte*** (an electrically conductive fluid). However, with ECM, the metal is REMOVED from the work rather than being deposited onto it, Fig. 24-11.

The electrolyte for ECM is usually common salt (NaCl) mixed with water. A stream of electrolyte is pumped at high pressures through a gap between the positively (+) charged work and the negatively (−)charged tool (electrode). The current passing through the gap removes material from the work by electrolysis, duplicating the shape of the electrode tool as it advances into the metal. In some applications, tolerances as close as 0.0004 in. (0.010 mm) can be maintained.

The work is NOT touched by the tool, there is no friction, no heat, no sparking, and no tool wear. The machined surface is burr free and in some instances, is highly polished. The operation of the machine is most unique because the only sound heard is the rush of liquid. Refer to Figs. 24-12 and 24-13.

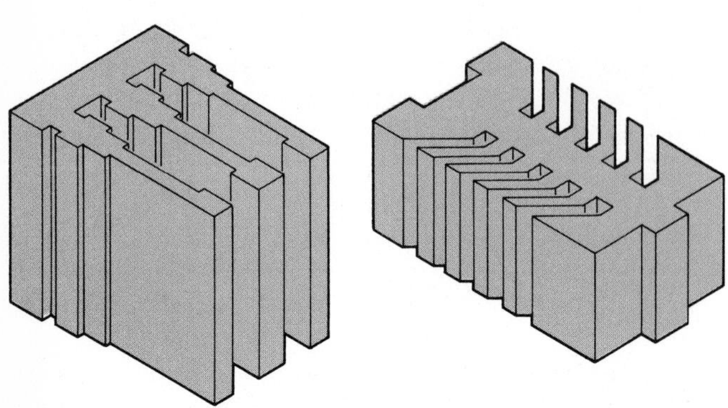

Fig. 24-10. These are examples of work done by EDWC technique.

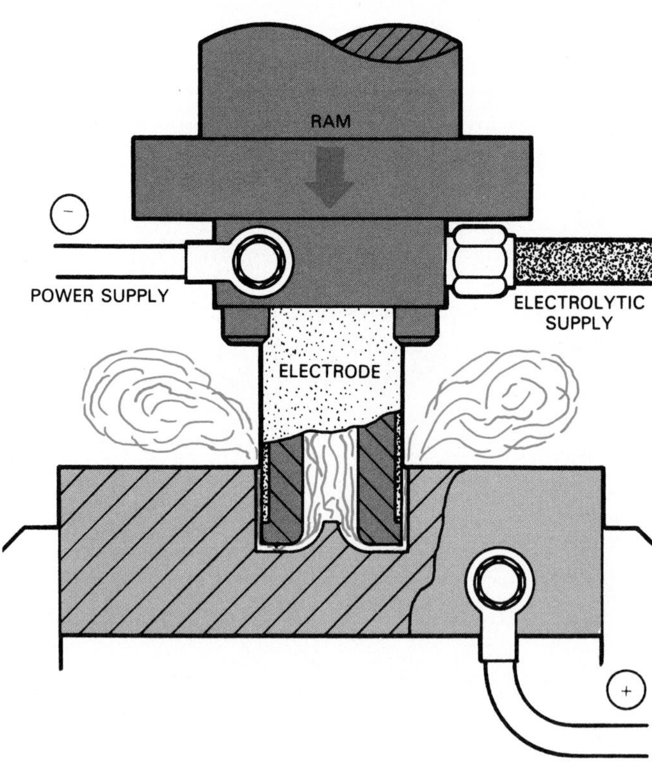

Fig. 24-11. Study how electro-chemical machining (ECM) works.

ECM advantages

ECM offers many advantages:

1. Metal is removed rapidly—up to 1 cu. in. per minute for each 10,000 amps of machining current.

Fig. 24-12. This is a vertical ECM machine. Size of work enclosure and fixed open height impose limits on size of work and fixtures that can be accomodated under ram. (ANOCUT, Inc.)

Fig. 24-13. Rush of fluid is only noise heard when ECM machine is in operation.

2. The kind of metal or its hardness does NOT affect the speed of material removal. Cast iron is about the only metal that offers problems and it is machined by other techniques.
3. It is accurate. Difficult shapes can be machined easily, Fig. 24-14.
4. The machined metal is stress free and will NOT warp or spring out of shape when removed from the machine.

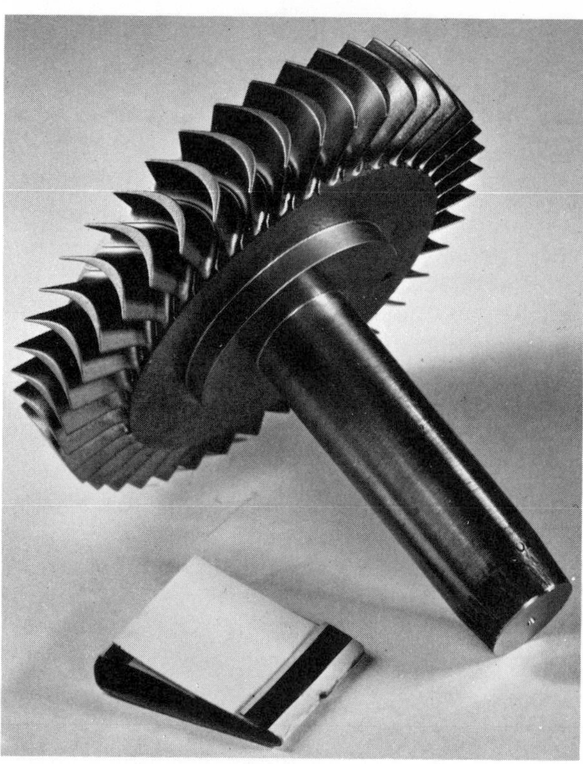

Fig. 24-14. Blades of this titanium turbine wheel are machined in metal blank. One blade is machined in two wheels simultaneously, at a feed rate of approximately 0.250 in. (6.5 mm) per minute.

5. There is no tool wear.
6. Several operations (milling, grinding, deburring, and polishing) can often be eliminated with ECM. See Fig. 24-15.

Note that there are continuing advances being made in ECM. The ability to produce even more complex shapes with simple tooling widens the range of application for ECM.

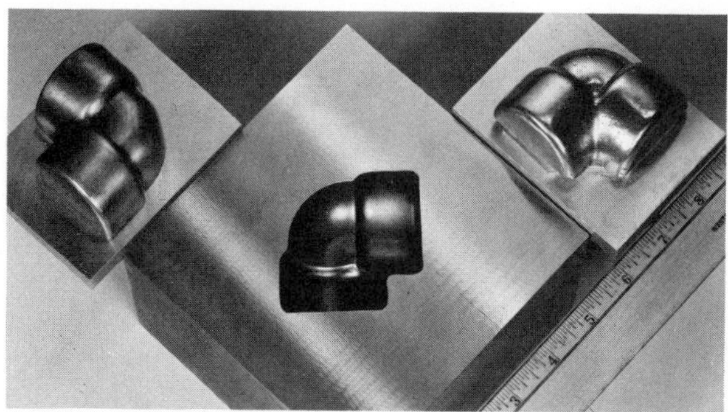

Fig. 24-15. This pipe elbow forging die was machined into hardened steel in 20 minutes to shape and finish shown. Conventional machining and polishing would require several hours to produce same form. There would also be danger of warping when machined die was heat treated.

TEST YOUR KNOWLEDGE—Chapter 24

Please do not write in the text. Place your answers on a separate sheet of paper.

1. EDM stands for _______ _______ _______.
2. EDWC stands for _______ _______ _______ _______.
3. ECM stands for _______ _______ _______.
4. Metals to be machined by EDM, EDWC, and ECM must be _______ or _______; otherwise, none of the processes can be used.
5. None of the techniques produce a _______ as metal is removed. The particles are disposed of completely by _______ or reduced to _______ _______ of metal.
6. EDM is a process that permits metals that are:
   a. Hard, brittle, or tough to be machined.
   b. Difficult or impossible to work by conventional means to be machined to close tolerances.
   c. Fragile or heat sensitive to be machined.
   d. All of the above.
   e. None of the above.
7. Explain the functions of the dielectric fluid used in EDM.
8. The EDM machine _______ maintains a very thin gap of about 0.001 in. (0.025 mm) between the electrode and the work.
9. The electrode in EDM produces a cavity in work that is an exact _______ _______ of tool.
10. How does EDWC differ from EDM?
11. ECM might well be classified as _______ in reverse.
12. In ECM:
    a. The work is not touched by the tool.
    b. There is no friction or heat generated.
    c. There is no tool wear.
    d. All of the above.
    e. None of the above.
13. In ECM, metal is removed _______.
14. What are five advantages that ECM offers?

RESEARCH AND DEVELOPMENT

1. Some of the larger tool and metalworking machinery supply houses have demonstration models of electro-machining units. Contact such firms that are nearby to borrow samples of metals worked during demonstrations along with the electrodes used.
2. Make a collection of literature on electro-machining processes for the shop library.
3. Develop and produce a working model of either process. Make photos as you develop the machine and prepare a paper on the project.
4. Secure samples of work produced by EDM, EDWC, and ECM and prepare a display.

Chapter 25

NONTRADITIONAL MACHINING TECHNIQUES

After studying this chapter, you will be able to:

- Describe several nontraditional machining techniques
- Explain how nontraditional machining techniques differ from traditional machining processes.
- Summarize how to do several nontraditional machining techniques.
- List the advantages and disadvantages of several nontraditional machining techniques.

The metalworking industry is responsible for cutting, shaping, and fabricating both metals and nonmetals. Since the range of material is so broad, and with additional materials being developed rapidly, new and different machining techniques have been devised. This chapter will describe a few of these new systems that differ from the tratitional chip removal methods of the lathe, drill press, milling machine, grinder, and saw.

CHEMICAL MACHINING

Chemical machining techniques are a refinement of the process used by photoengravers to prepare printing cuts and plates. Chemicals, usually in an aqueous solution (water and chemicals), are employed to etch away selected portions of the metal to produce an accurately contoured part.

In general, chemical machining falls into two categories: chemical milling and chemical blanking.

Chemical milling

Chemical milling, also called chem-mill or contour etching, is a recognized and accepted technique for machining metal to exacting tolerances through chemical action. The process makes possible the selective removal of metal from relatively large surface areas, Fig. 25-1. For example, this can be done for weight reduction of sheet metal parts, a critical factor in aerospace vehicle performance.

Refinements in the process make it possible to remove metal to form shapes or microscopic parts that would be difficult or impossible to do by conventional machining techniques, Fig. 25-2. Tapers and multiple depth cuts can be produced by chem-milling, Fig. 25-3.

Chemical milling is NOT a rival to conventional milling; both processes should be considered complimentary.

Basically, chem-milling is a process in which the prepared part is immersed in an ***etchant*** (usually a strong alkaline solution) where the resulting chemical action removes the desired metal. The im-

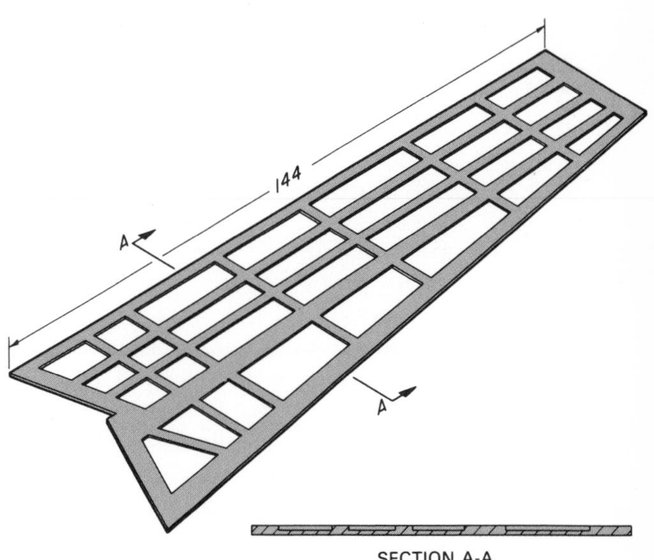

Fig. 25-1. Chemical milling is employed to remove metal to close tolerances. This aircraft wing panel has been chemically reduced in sections where spars are not attached. Considerable weight reduction is accomplished with no sacrifice in structural strength.

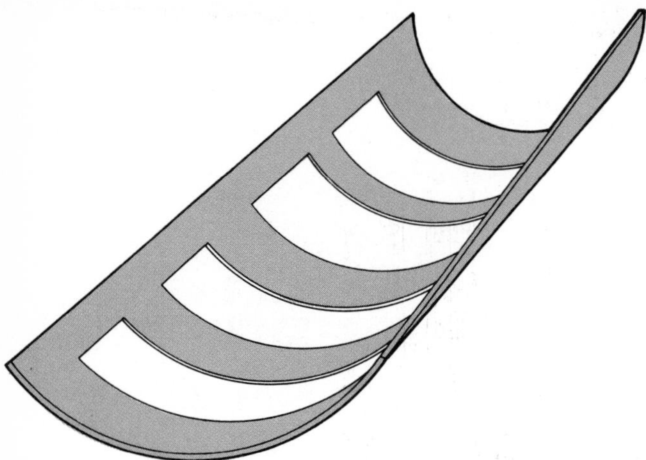

Fig. 25-2. Outer skin of an aircraft engine housing is integrally stiffened. Aluminum sheet is approximately 48 in. by 72 in. (1219 mm by 1829 mm) and is chem-milled after being formed.

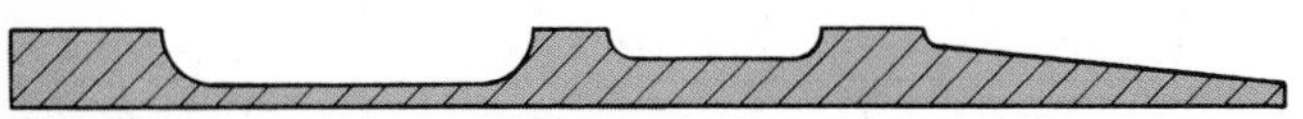

Fig. 25-3. Multi-depth cuts are possible with chem-milling by masking shallower sections when correct depth has been reached. Tapers are produced by withdrawing metal from etchant at a predetermined rate.

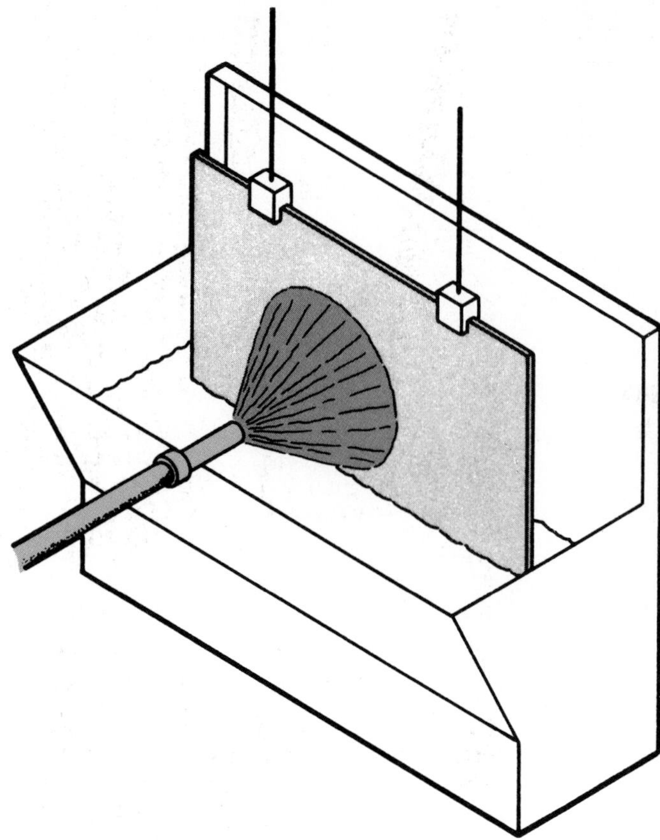

Fig. 25-4. Cleaning involves removal of all grease and dirt that might affect etching process.

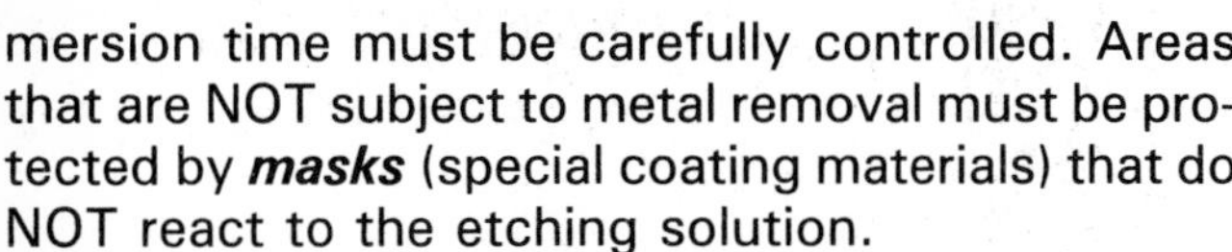

mersion time must be carefully controlled. Areas that are NOT subject to metal removal must be protected by ***masks*** (special coating materials) that do NOT react to the etching solution.

Steps in chem-milling

The principle steps in chem-milling are shown in Figs. 25-4 to 25-9. Study them closely.

Advantages of chem-milling

Chem-milling offers many advantages:

1. Tooling costs are low.
2. Tolerances of plus or minus 0.003 in. (0.07 mm) are obtainable on cuts up to 0.50 in. (12.5 mm) deep.
3. The only size limitations imposed are by the size of the immersion tanks available.
4. Warping and distortion of formed sections is negligible.
5. Contoured or shaped parts can be chem-milled after they are formed, Fig. 25-10.
6. Many parts can be produced simultaneously.
7. Unsupported pieces as thin as 0.015 in. (0.4 mm) can be machined without danger of buckling.
8. Both sides of the metal can be milled at the same time.

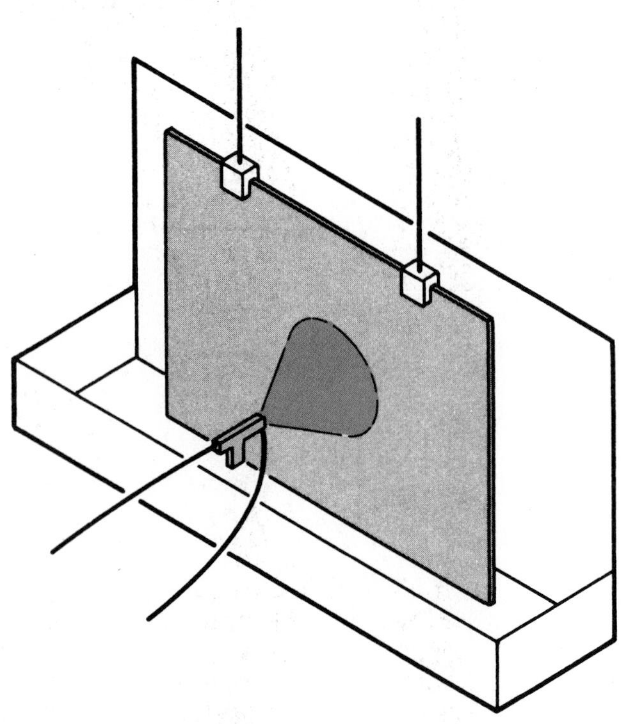

Fig. 25-5. During masking, entire part is coated with a masking material, applied by brushing, dipping, spraying, or roller coating. Masked metal sheet is then baked to remove all solvents.

Fig. 25-6. While scribing and stripping, a template is placed over entire part. Then areas to be exposed are circumscribed and masking material stripped away. (Grumman Aerospace Corp.)

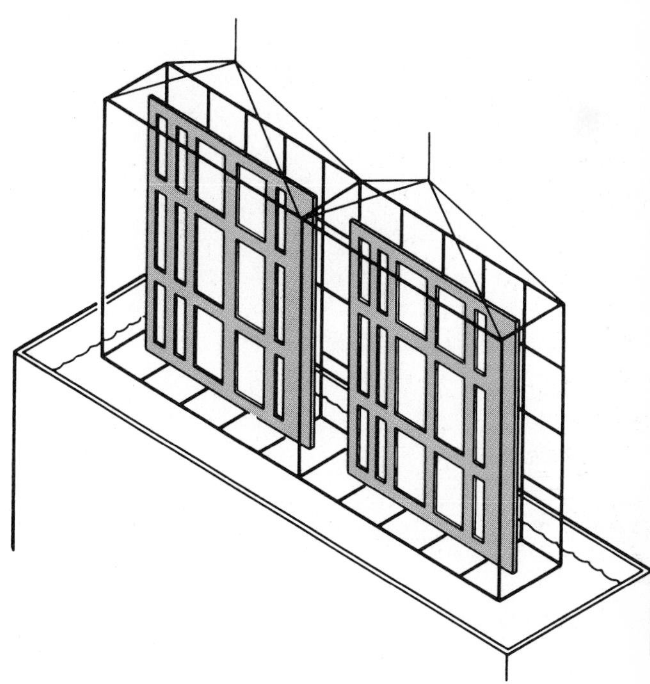

Fig. 25-8. After rinsing, in rinsing and solvent strip phase, parts are lowered into solvent tank which releases maskant bond and residue of maskant is stripped from part.

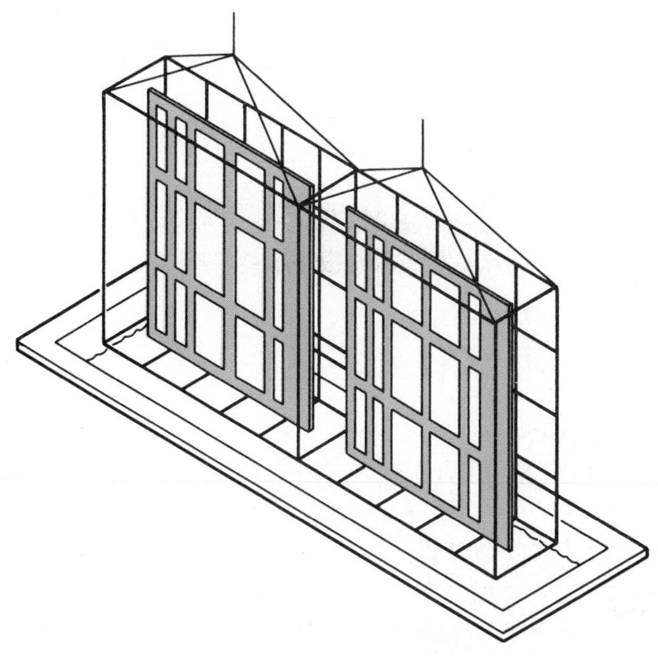

Fig. 25-7. During etching, parts are racked and lowered into etchant for milling operation.

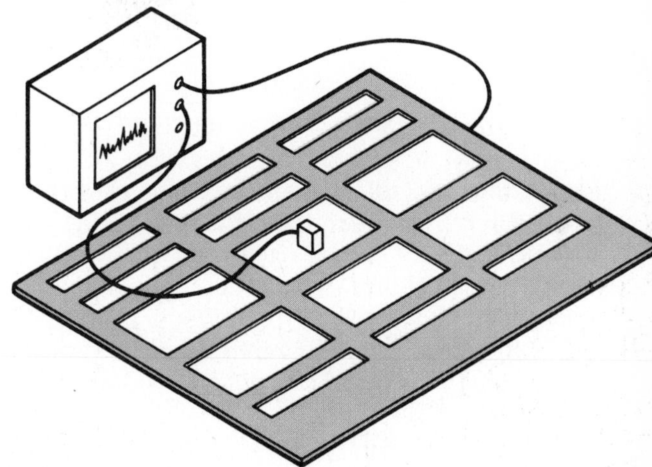

Fig. 25-9. Inspection includes measuring accuracy of chemical milling etch with aid of an ultrasonic thickness gage.

9. Any metal, regardless of its state of heat treatment, can be machined chemically.
10. No burrs are produced in the area machined.

Disadvantages of chem-milling

Chem-milling does offer some disadvantages:

1. The process is slow and it takes considerable time to remove large quantities of metal.
2. All surface imperfections must be removed before etching. These areas will etch at a faster rate and are amplified on the finished surface.
3. The technique is NOT recommended for etching holes.
4. Surface finishes on deep etches are NOT comparable with conventionally machined surfaces.

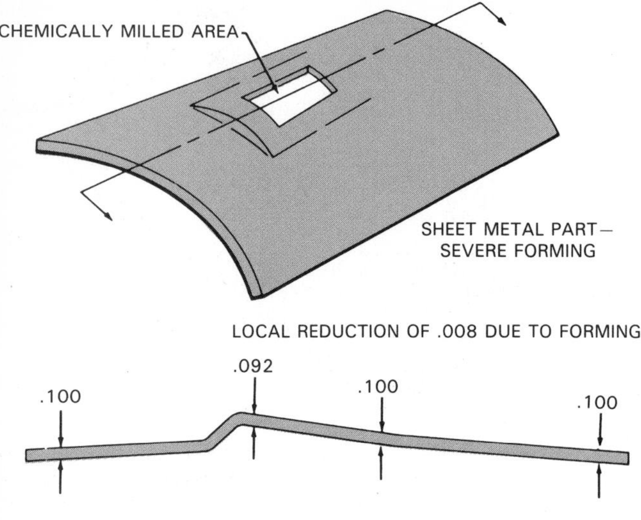

VIEW A-A PRIOR TO CHEMICAL MILLING

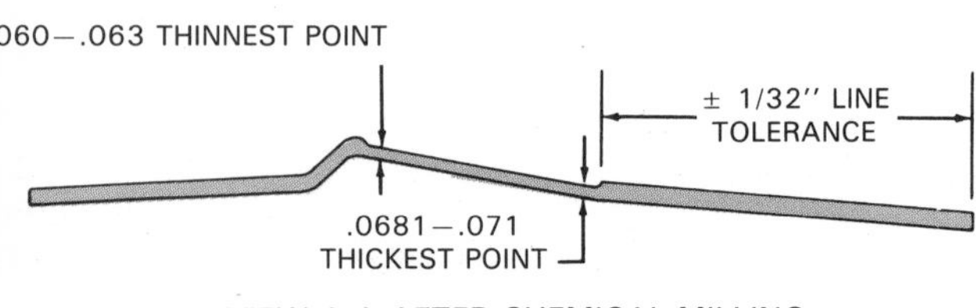

VIEW A-A AFTER CHEMICAL MILLING

Fig. 25-10. This aircraft part was chem-milled after it was formed. Pieces even more severely formed than this part can be chem-milled economically. (Grumman Aerospace Corp.)

5. Lateral dimensions are difficult to hold because the etchant works sideways as well as in depth. Typical lateral tolerances work on an etch factor of 3:1. This means that for every 0.003 in. (0.007 mm) of etched depth, 0.001 in. (0.03 mm) of UNDERCUT will occur. The top edges of the cavity will be sharp; however, the inside edges and corners will have a radius approximately equal to the cut depth, Fig. 25-11.

CHEMICAL BLANKING

Chemical blanking, also termed ***chem-blanking, photoforming,*** or ***photoetching*** involves TOTAL REMOVAL of metal from certain areas by chemical action. It is a variation of chemical milling. Chem-blanking is employed by the aerospace and electronics industries to produce small, intricate, ultrathin parts. See Figs. 25-12 and 25-13.

Metal foil as thin as 0.00008 in. (0.002 mm) can be worked by chem-blanking. A cigarette paper is about 0.001 in. (0.025 mm) thick. This ultrathin metal must be laminated to plastic backing to protect it from damage during shipping, Fig. 25-14.

The process is NOT recommended for metals thicker than 0.090 in. (3.00 mm). However, almost any metal can be chem-blanked.

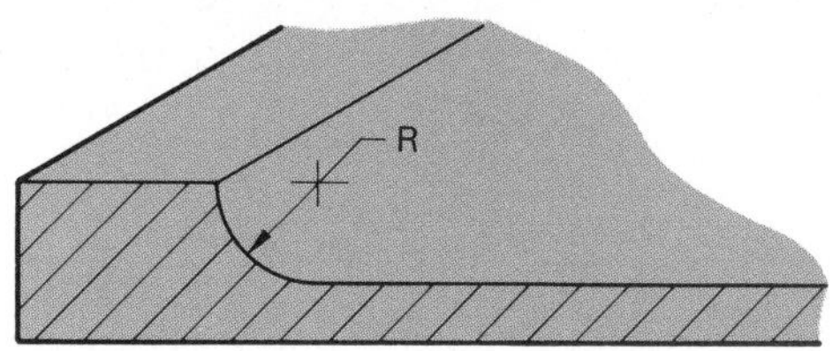

Fig. 25-11. Inside edges of a chem-milled section will have a radius equal to depth of etch. This is an advantage in many applications.

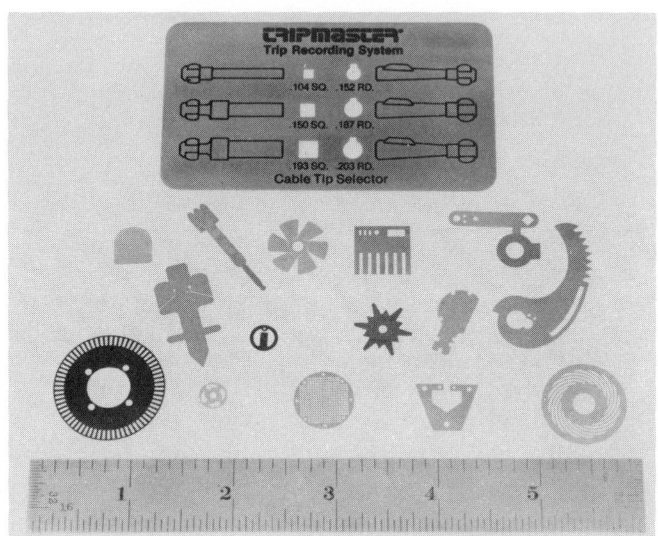

Fig. 25-12. This is a sampling of parts made by chemical blanking. (Microphoto, Inc.)

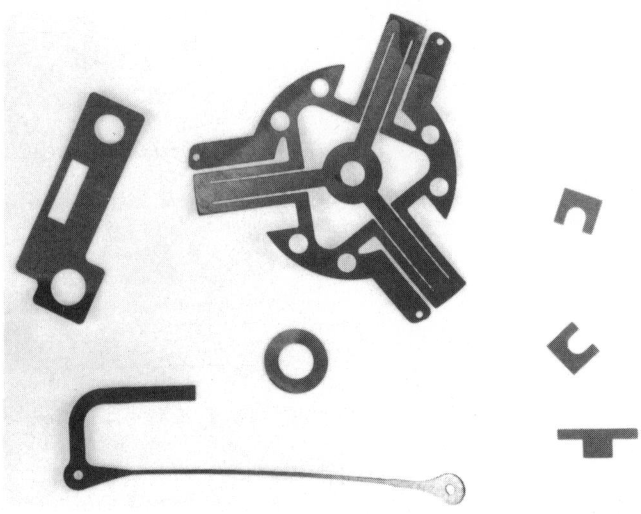

Fig. 25-13. These parts range in thickness from 0.0001 in. to 0.002 in. (0.0025 to 0.05 mm) and are for electronic applications. (Hamilton Watch Co.)

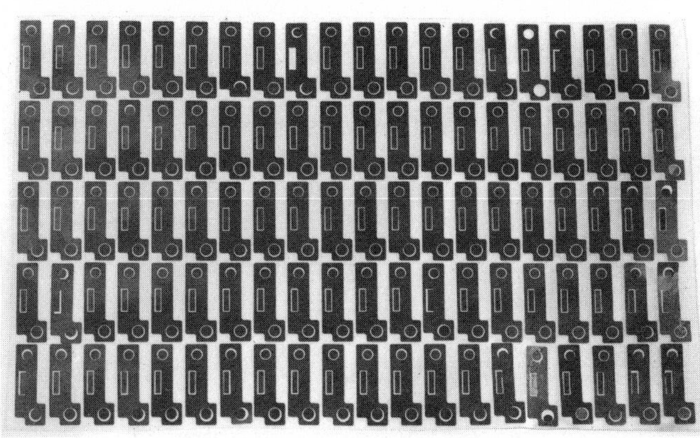

Fig. 25-14. Plastic sheet holds quantity of 0.001 in. (0.025 mm) thick computer components. Computer parts have been laminated to plastic to assure flat, crease-free condition.

Fig. 25-16. Master drawing is photographically reduced to required size. Since many identical parts must be made, it is more economical to etch several at a time. (Microphoto, Inc.)

Steps in chem-blanking

There are several major steps in the chem-blanking process:

1. A master drawing of the part is made, Fig. 25-15. Depending upon the tolerances that must be met, it may be drawn 50 times the size of the required part.
2. The drawing is reduced photographically. This produces a film or glass master the exact size of the required part. Since many identical parts must usually be made, a multiple negative, with several images, is often produced on a photorepeating machine, Fig. 25-16.
3. Metal sheet is thoroughly cleaned and coated with a photosensitive resist. The multiple image negative is contact printed on the metal blank's surface. After exposure, the images are developed, Fig. 25-17. The resist in unexposed areas is dissolved during the developing process and exposes bare metal, Fig. 25-18.
4. The processed metal blank is placed in a spray etcher, Fig. 25-19. Spraying is often preferred over immersion or splashing because it offers a higher etching rate and better tolerance control. Fig. 25-20 shows one type of horizontal conveyorized spray etcher used for chem-blanking. The spray nozzles move back and forth and the holding tray oscillates to assure maximum exposure of the work to the etchant. Etching time varies from a bit over three minutes for metal 0.0001 in. (0.002 mm) thick, to as much as one hour for 0.010 in. (0.25 mm) material. Etching removes all of the metal not protected by the resist coating.
5. After etching, the photoresist is removed with solvents and the metal flushed with warm water and dried. Visual inspection of each chem-blanked unit assures strict adherence to specifications, Fig. 25-21.

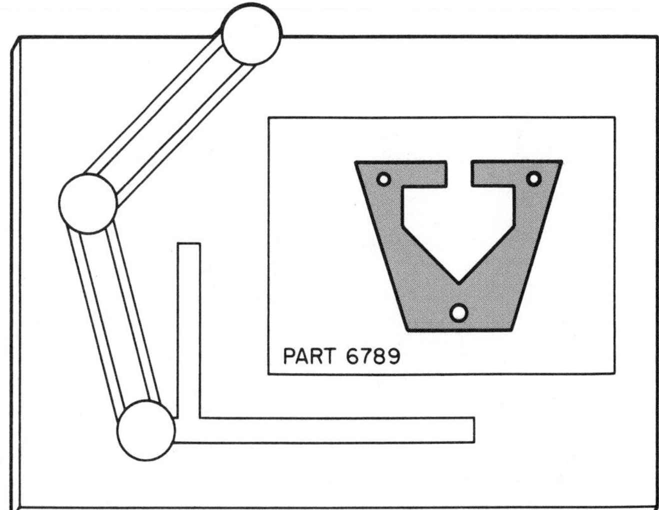

Fig. 25-15. Master drawing may be up to 50 times actual size of required part, depending on tolerances that must be met. (Microphoto, Inc.)

Advantages of chemical blanking

Like all nontraditional machining techniques, chemical blanking can offer certain advantages over conventional machining techniques:

1. Tooling costs are low because the job consists of an inked drawing on dimensinally stable plastic film. Almost any pattern that can be drawn can be produced. The art work for small parts can be drawn oversize. A good drafter can make these large drawings to within plus or minus 0.005 in. (0.13 mm) on Mylar film. The possible error diminishes when the drawing is reduced to actual size.

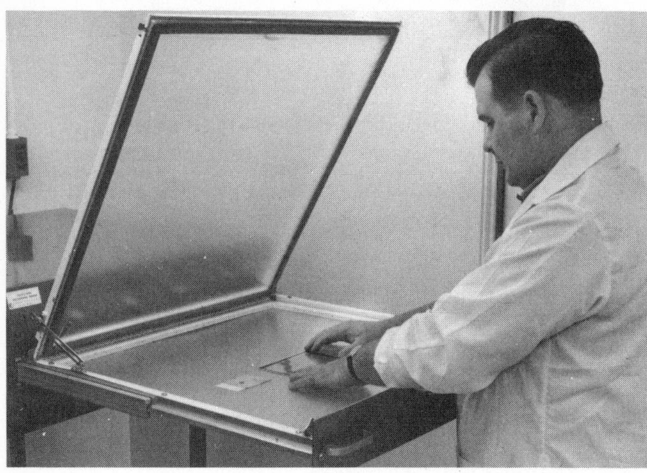

Fig. 25-17. Technician is using a double-sided printing unit. In double-sided printing, metal must be placed between two identical photographic plates, which must be perfectly aligned. Light source deposits part image on photographic plates. (Hamilton Watch Co.)

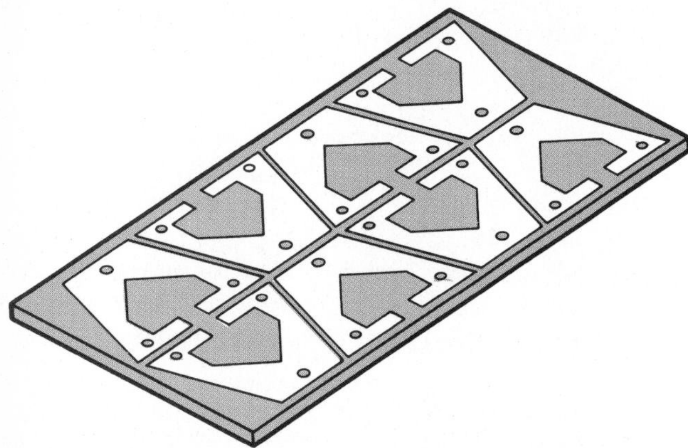

Fig. 25-18. Metal sheet, with a photosensitive coating, has image of part contact printed on its surface using multiple part negative. During development, photosensitive material in unexposed areas is dissolved, exposing bare metal. (Microphoto, Inc.)

Fig. 25-19. After designs have been developed on metal, strips are placed in spray etcher. These are stainless steel computer parts. (Hamilton Watch Co.)

2. There are no burrs on the milled part.
3. Initial quantities of newly designed parts can often be produced in a matter of hours.
4. Design changes are made easily by changing the existing art work.
5. Ultrathin metal foils can be worked with no fear of distortion.
6. Accuracy increases as metal thickness decreases.
7. Metal characteristics (brittleness, hardness, etc.) have no significant effect on the process.

Disadvantages of chemical blanking

The disadvantages of chemical blanking are:

1. The process is relatively slow. Metal removal seldom exceeds 0.001 in. (0.02 mm) per minute.
2. Highly skilled workers are required.
3. Chemicals are highly corrosive and etching equipment must be isolated from other plant equipment. Considerable care must be employed when disposing of expended etchant and rinse water to prevent environmental damage.
4. Good photographic facilities are required, but they are NOT always available.
5. Maximum metal thickness that can be worked is limited. Practical limit for production purposes is 0.090 in. (3.00 mm).
6. Tolerances increase with metal thickness

HYDRODYNAMIC MACHINING (HDM)

The advent of composites caused new problems for the metalworking areas that had to shape, form, and fabricate them. Most composites are made of ***"layups"*** (several layers of a fabric like material) bonded together into three dimensional shapes. The many sections that make up the layups have to be accurately cut to outline shape, Fig. 25-22. However, the nature of composites quickly dulled conventional cutting tools.

Hydrodynamic machining (HDM) or ***water-jet cutting,*** was developed to shape composites, thin metal, plastics, etc., quickly and accurately. Computer controlled, the technique typically uses a 55,000 psi (379 000 kPa) water jet to cut complex shapes with minimal waste, Fig. 25-23. Depending upon the material cut, tolerances can usually be held to ±0.004 in. (0.01 mm). No heat is produced to damage the material and there is little ***particulate*** (fragments) generated.

ULTRASONIC MACHINING

The science of ***ultrasonics*** (silent sound waves) has found applications in many areas—machining, welding, quality control, and cleaning, to name a

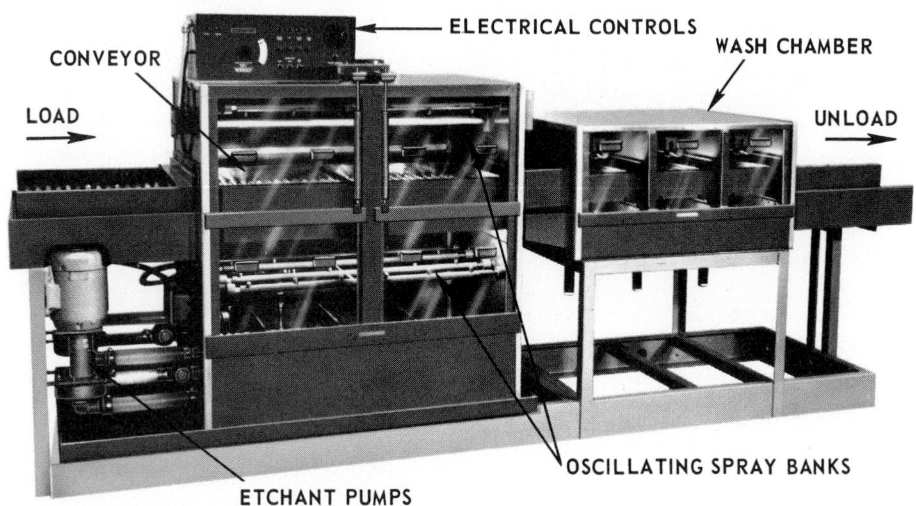

Fig. 25-20. This continuous spray etching machine was developed for conveyorized operation. It also includes rinsing compartment at exit end. (Chemcut)

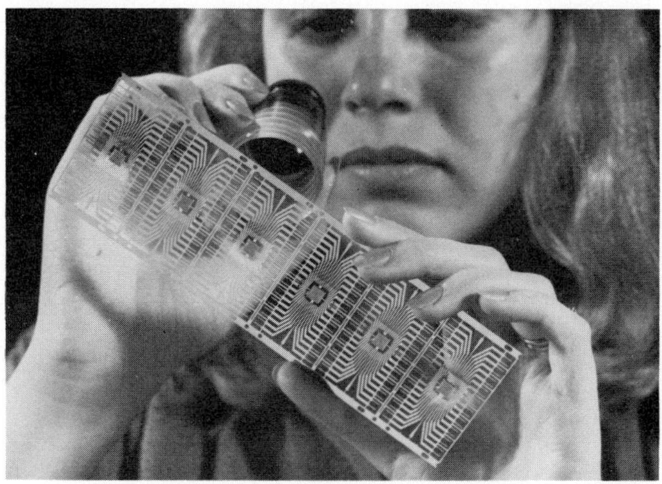

Fig. 25-21. Inspector is visually checking microchip section that was chem-blanked. It will become part of electronic control module of an automotive emission/fuel control system. (Delco Electronics Div. of General Motors Corp.)

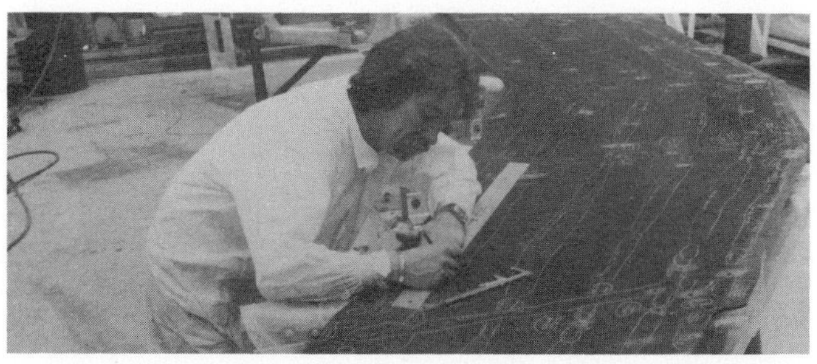

Fig. 25-22. After being cut to outline shape by a computer-driven water jet or laser, each layer of boron/epoxy composite is transferred to a forming mold. Here wing skins for X-29 experimental aircraft are shown in mold. (Grumman Aerospace Corp.)

Fig. 25-23. Operating from an overhead gantry, this 6-axis robot can cut complex shapes from a sheet of raw material using a high pressure water jet. (GCA/Industrial Systems Group)

few. Considerable research is being done to develop new uses and to improve existing techniques.

The average person can hear sounds that vibrate between 20 to 20,000 cycles (times) per second. Below 20 cycles per second, sound waves are called ***infrasonic.*** Above 20,000 cycles per second, sound waves are known as ***ultrasonic.*** Industrial applications make use of energy up to 100,000 cycles per second.

Ultrasonic waves are created by passing an electric current (usually 60 cycle AC) into a suitable generator (transducer) to produce the desired frequency. The sound waves may be utilized in conjunction with a fluid as in quality control and cleaning applications, or applied directly to the cutting tool or metal as it is being machined, welded, or formed.

Ultrasonic assisted machining

Ultrasonic machining applies soundwaves to the tool or metal as it is cutting or being cut. Several variations are in use. A fairly recent development is an ultrasonic assist for conventional metal cutting processes. The transducer is fitted to a standard machine tool to make the tool vibrate at high frequency, Fig. 25-24. The ultrasonic assist makes possible a 10 to 50 percent reduction in tool forces and there is almost complete elimination of tool chatter. Chatter reduction is a distinct advantage when boring operations require long slender boring tools. Tool wear is reduced and more cutting can be done between sharpenings. Surface finishes are also improved, Fig. 25-25.

Similar applications have been made in grinding. Ultrasonic waves are passed through the grinding wheel. Particles and chips, that normally become imbedded in the wheel, are vibrated loose and washed away by the coolant. Grinding temperatures are also reduced. Wheel life is extended. Material can be removed more rapidly and surface finishes are improved for the same expenditure of power. Ultrasonic vibrations imparted to grinder coolant have been found to produce similar results.

Drilling, reaming, honing, milling, and EDM (electrical discharge machining) techniques employing ultrasonic assist are in general use.

Impact machining

Impact or ***slurry machining*** uses ultrasonics and a special tool to force an abrasive against the work. It was an early application of ultrasonics to machining, Fig. 25-26. With the exception of diamond tools, it is the only commercially feasible way to machine extremely hard, brittle, and frangible (breakable) materials. The technique works best on hard, brittle materials (glass, quartz, silicon, and carbides), but it is ineffective on soft materials, such as aluminum and copper.

Fig. 25-24. Ultrasonic unit is being used to assist machining on a lathe. Unit is mounted directly to cutting tool and requires no modification on lathe. However, conventional surface or cylindrical grinder cannot accept such a unit without considerable modification to machine. (Sonobond Ultrasonics Corp.)

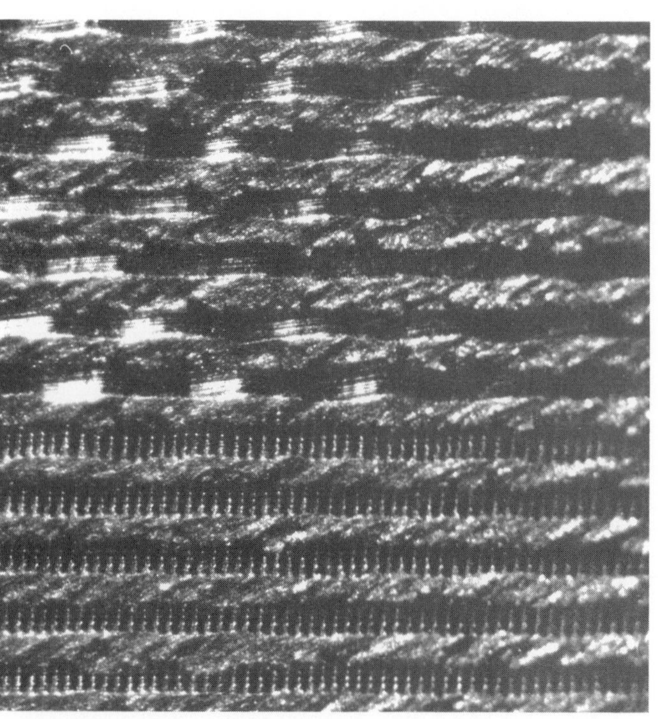

Fig. 25-25. Above. Microphotograph showing surface appearance of aluminum without ultrasonic power assist. Below. With ultrasonic power assist. (Sonobond Ultrasonics Corp.)

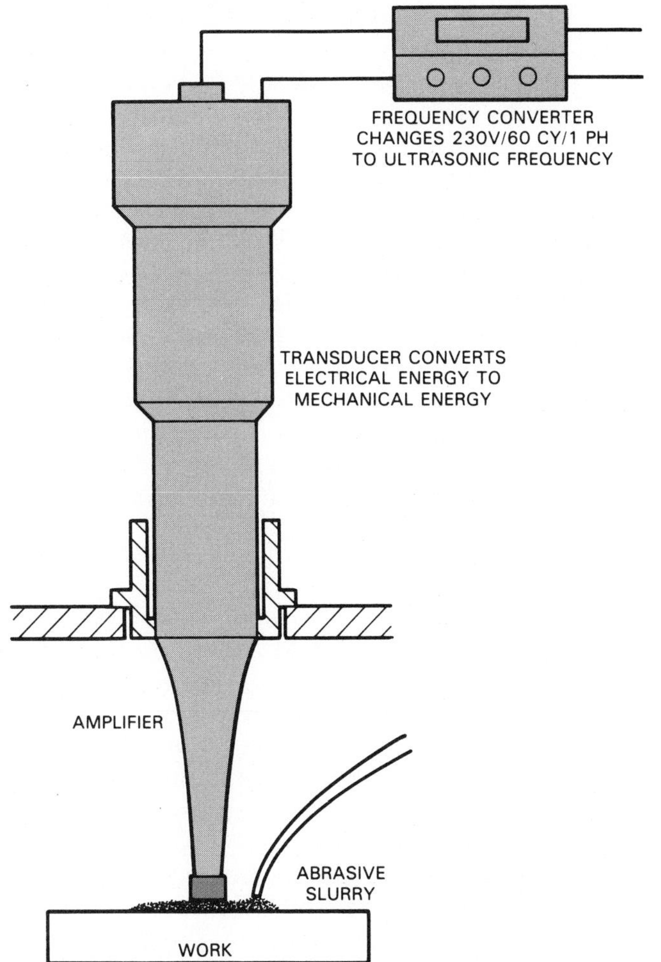

Fig. 25-26. Study ultrasonic machining setup.

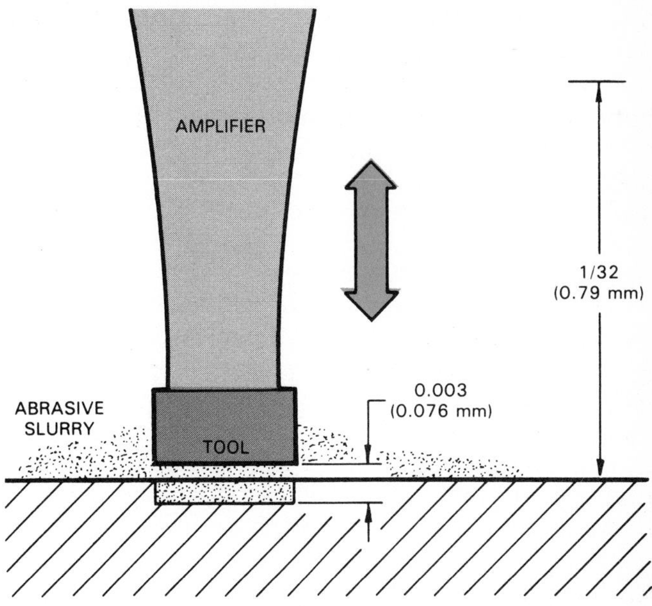

Fig. 25-27. Tool motion in ultrasonic (impact) machining is slight, only 0.003 in. (0.076 mm). The 1/32 in. (0.79 mm) is for reference purposes only.

Machining is done by a shaped cutting tool oscillating about 25,000 times per second. This rapid oscillation pounds a slurry of fine abrasive particles against the work. The tool stroke, at the vibrating end, is only 0.003 in. (0.076 mm), Fig. 25-27. The tool does NOT touch the work. No heat is generated, eliminating any possibility of the work becoming distorted. Microscopic portions of the work are chipped away to produce the shape desired. The machined section is a mirror image of the cutting tool, Fig. 25-28.

A solid funnel-shaped horn must be employed to amplify and transmit the vibrations. Ultrasonic vibrations directly from the transducer do NOT produce enough motion to produce the required tool movement.

Industrial applications of impact machining include:

1. Slicing and cutting germanium and silicon wafers into tiny "chips" for transistors, diodes, and rectifiers for the electronics industry.
2. Machining complex shapes in nonconductive and semiconductive materials that cannot be satisfactorily handled by EDM and ECM.

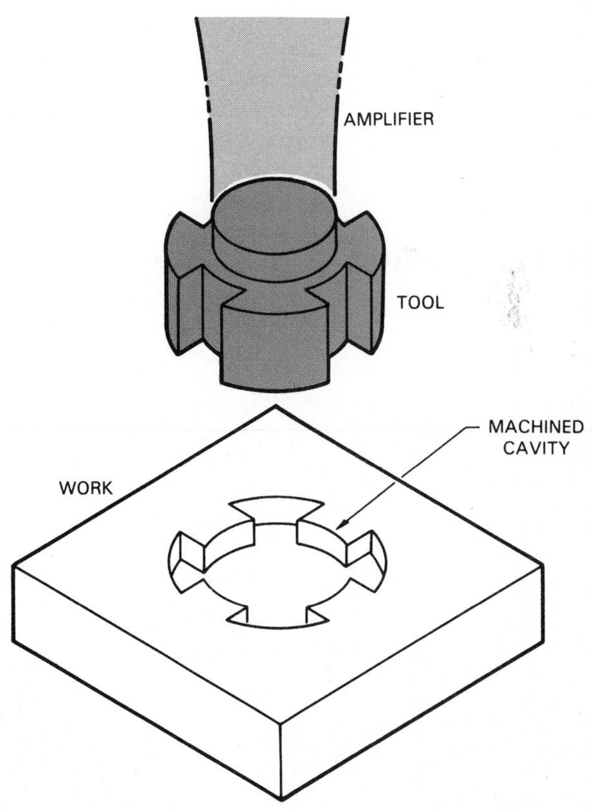

Fig. 25-28. Machined section is a mirror image of tool.

3. Shaping virtually unmachinable space age materials.

The process is slow and the surface finish is dependent upon the abrasive grit size used. One inch or 25 mm is about the deepest cut possible with existing equipment. Tolerances of 0.001 in. (0.025 mm) can be maintained on hole size and geometry in most materials. Equipment cost is moderate and the training period for machine operators is short and does NOT require special skills.

Other ultrasonic applications

Industrial applications of ultrasonics are varied. A principal use is to improve the cleaning power of chemical solvents. As the waves pass through the cleaning solution, microscopic vapor pockets are created in the fluid. These ''bubbles'' (really areas of very high vacuum) form and collapse about 20,000 times per second. This creates local pressures as high as 10,000 psi and heat. The pressure and heat smash against the work to be cleaned and literally tear away the dirt, oil, grease, chips, flux, and other contaminants on the work surfaces.

The cleaning action is so thorough that the technique is employed to decontaminate work that has been exposed to radioactive solutions and gases.

Nondestructive testing utilizing ultrasonics has been developed to the point where the technique can be fully automated. Flaws as small as 0.001 in. by 0.250 in. (0.025 mm by 6.25 mm) can be detected on a continuous basis.

Sound waves are also used to weld metals to nonmetals. Precision welding of dissimilar materials to form a solid bond has been developed successfully. For example, ultrasonic waves can weld aluminum wires to special glass on some electronic circuits.

ELECTRON BEAM MACHINING (EBM)

The ***electron beam micro cutter-welder*** uses a high energy, highly focused beam of electrons to weld or cut materials. It is the direct result of the special needs of the atomic energy, electronics, and aerospace industries. In some aspects, it is the MOST PRECISE and versatile of the new machining techniques. See Fig. 25-29.

The first application of the electron beam technique was concerned with welding ''impossible'' jobs. The development of refined focusing systems made it possible to control the cutting action with a high degree of precision. As the electron beam can cut any known metal or nonmetal that can exist in a high vacuum, its micro cutting capabilities are almost unlimited.

Electron beam cutting action can be controlled so precisely that it is possible to drill holes as small as 0.0002 in. (0.005 mm) in diameter and mill slots having widths of 0.0005 in. (0.0125 mm), Fig. 25-30. The finish of the completed work is similar to a very fine machined edge.

The electron beam machine is similar in some respects to a television set. In the television set, a heated tungsten filament emits a beam of electrons. The beam is concentrated by an electron optic system into a small diameter beam. The beam is moved so rapidly by a deflection system that a glowing picture is produced on a fluorescent screen (TV picture tube).

The electron beam machine tool also makes use of a beam of electrons. However, the electron beam is several times more intense than the one used to produce a television picture. Controls similar to

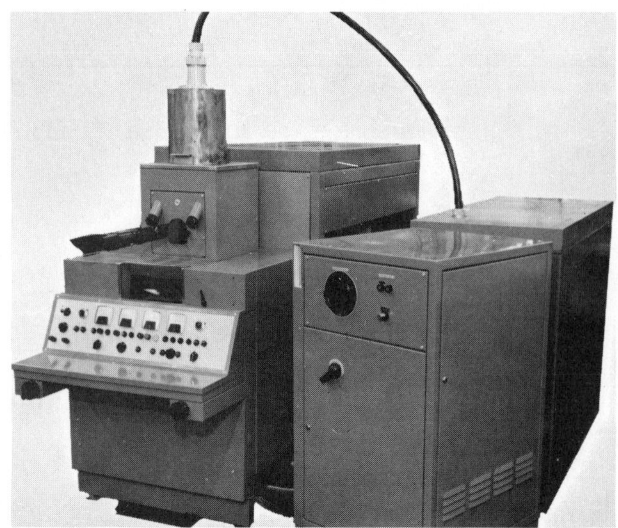

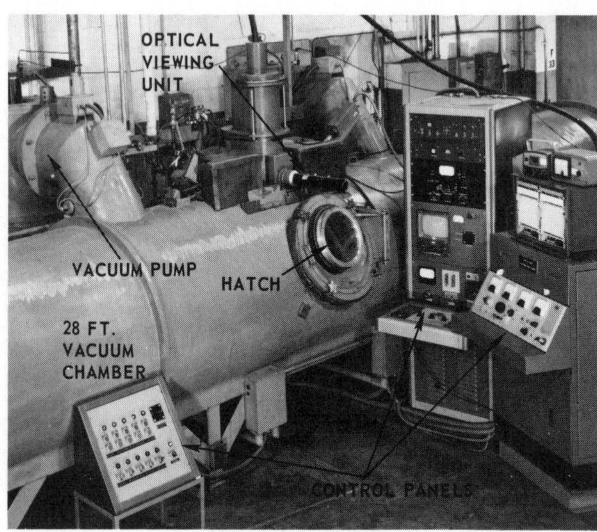

Fig. 25-29. Left—This is an electron beam micro cutter-welder. (Hamilton Standard Div., United Technologies) Right—This is a partial view of a 28 ft. (8 534 m) electron beam cutting and welding machine. (Westinghouse Electric Corp.)

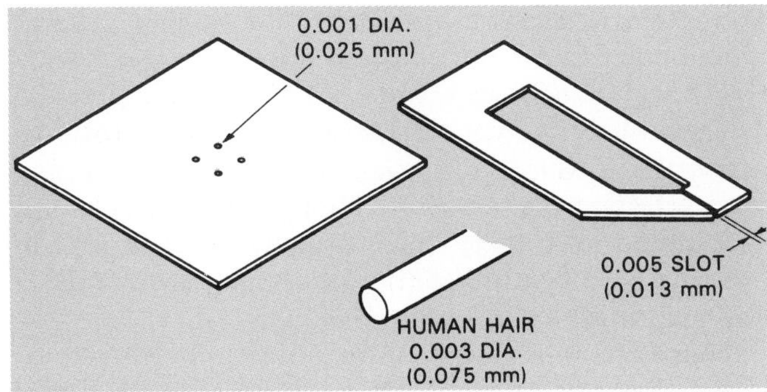

Fig. 25-30. Note size of work done by electron beam technique. Parts machined and human hair are drawn to same scale.

those on the TV set focus and adjust the electron beam for cutting or welding, Fig. 25-31.

The electron beam machine is basically a source of thermal energy, Fig. 25-32. The beam of electrons can be focused to a very sharp point. Cutting is achieved by alternately heating and cooling the area to be cut. The heating and cooling must be carefully controlled so the material at the point of focus is heated to a high enough temperature to vaporize it yet not cause the surrounding area to melt. This is accomplished by employing a pulsating technique, Fig. 25-33.

The beam is on for only a few milliseconds and is off for a considerably longer period of time. Temperatuures of 12,000 °F (6 649 °C) can be achieved at the focal point of the electron beam.

Cut geometry (shape of cut) is controlled by movement of the worktable in the vacuum chamber and by employing the deflection coil to bend the beam of electrons to the desired cutting path. Initial hole diameter or cut width is controlled by the amount of power applied and the duration of the cutting time. See Figs. 25-34 and 25-35.

Fig. 25-31. Adjustments and position of electron beam are controlled by dials similar in function to those found on a TV set. (Hamilton Standard Div., United Technologies)

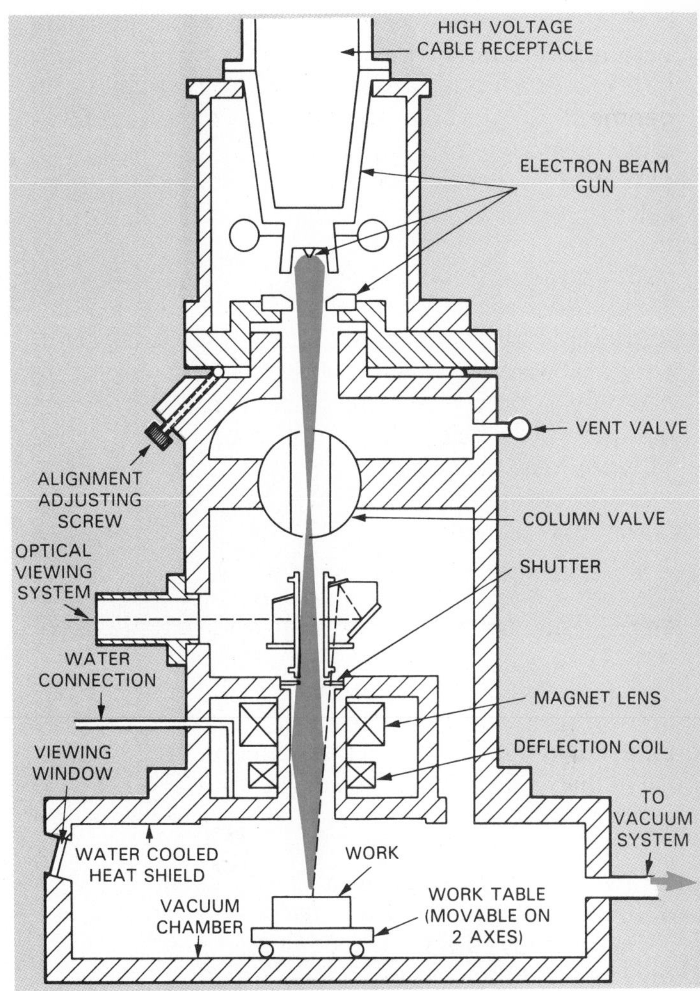

Fig. 25-32. Study cross section of electron beam micro cutter-welder.

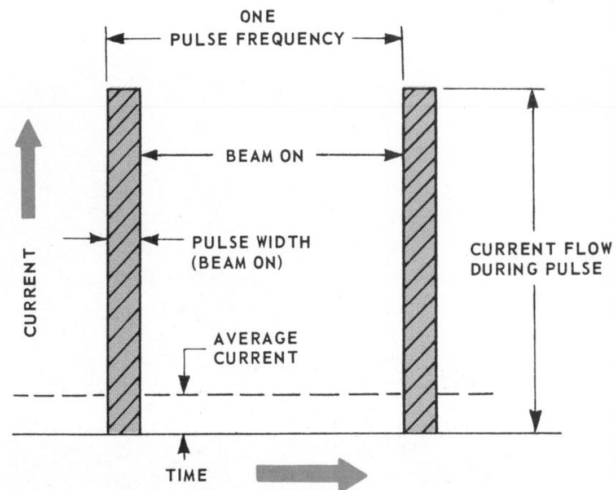

Fig. 25-33. Note schematic of beam pulsing. (Hamilton Standard Div., United Technologies)

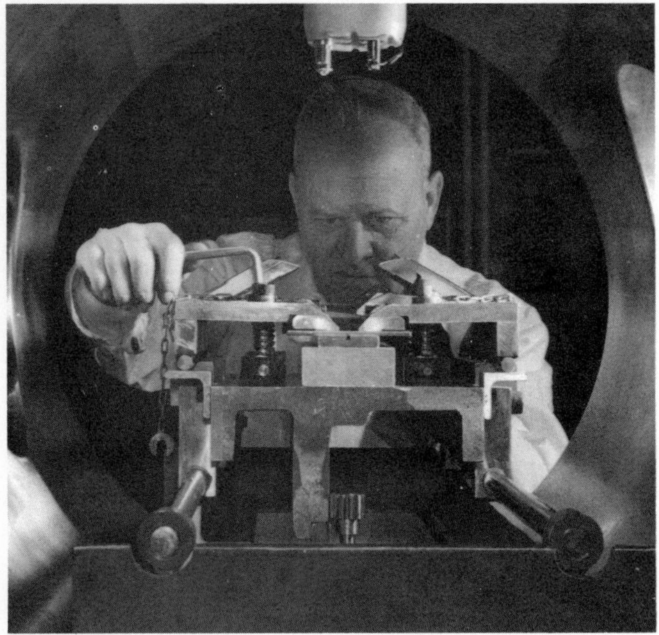

Fig. 25-34. Engineer is preparing a job for machining in an electron beam machine.

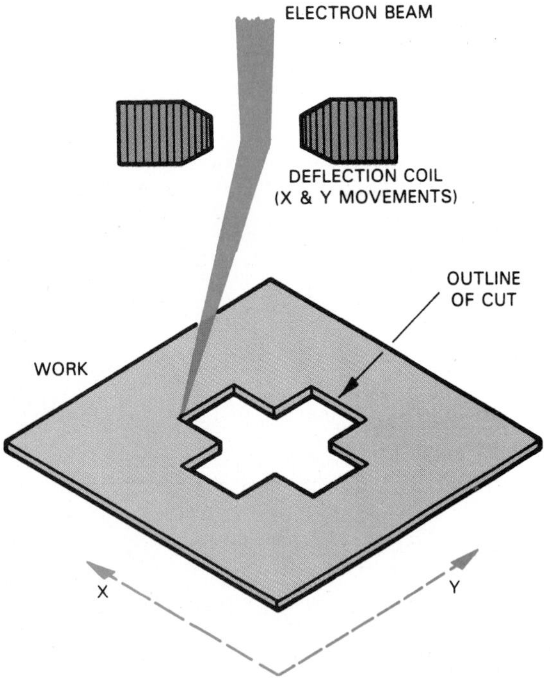

ELECTRON BEAM

DEFLECTION COIL
(X & Y MOVEMENTS)

OUTLINE
OF CUT

WORK

X

Y

Fig. 25-35. Path of cut is controlled by deflecting electron beam or by movement of worktable. Cutting must be done in a vacuum.

LASER BEAM MACHINING

The term ***LASER*** (pronounced LAY-zer) is an abbreviation for ***L***ight ***A***mplification by ***S***timulated ***E***mission of ***R***adiation. The laser produces a narrow and intense beam of light that can be focused optically onto an area only a few microns in diameter. Refer to Fig. 25-36.

Depending upon the initial energy source used to activate the laser, it is possible to instantaneously create temperatures up to 75,000 °F (41 649 °C) at the point of focus. This is almost seven times the average temperature of the sun. No known material can withstand such heat, Fig. 25-37.

There has been a dramatic increase in the use of lasers in part manufacturing, Fig. 25-38. They are used for cutting, drilling, slotting, scribing, heat treating, and welding, Fig. 25-39.

The energy output of a laser is usually NOT continuous. It lasts only a fraction of a second. When employed for machining and cutting, it operates at 1 to 5 cycles per second, Fig. 25-40. The cycle can be controlled manually or electronically. The laser operates on the principle shown in Fig. 25-41.

The laser unit in Fig. 25-42 can operate in either continuous or pulsed modes. With most materials, the edge quality of the cut is better with continous beam. Cutting speed is also much greater. However, until recently, pulsing was essential for cutting parts with intricate details. This is because starting, stopping, and turning corners concentrates heat in localized areas on the work. This overheating causes the part to burn away from the program path, affecting cut quality and dimensional accuracy.

Fig. 25-36. Laser produces a narrow and intense beam of light that is almost seven times average temperature of sun. Here a laser is shown drilling small holes in difficult to machine space age metal. (Perkins-Elmer)

The laser unit shown in Fig. 25-42 uses a water assisted cutting system (WACS) which extracts the heat produced by continuous cutting. This allows a continuous beam to be used for cutting many parts, where pulsating, with its lower cutting speeds, was previously required.

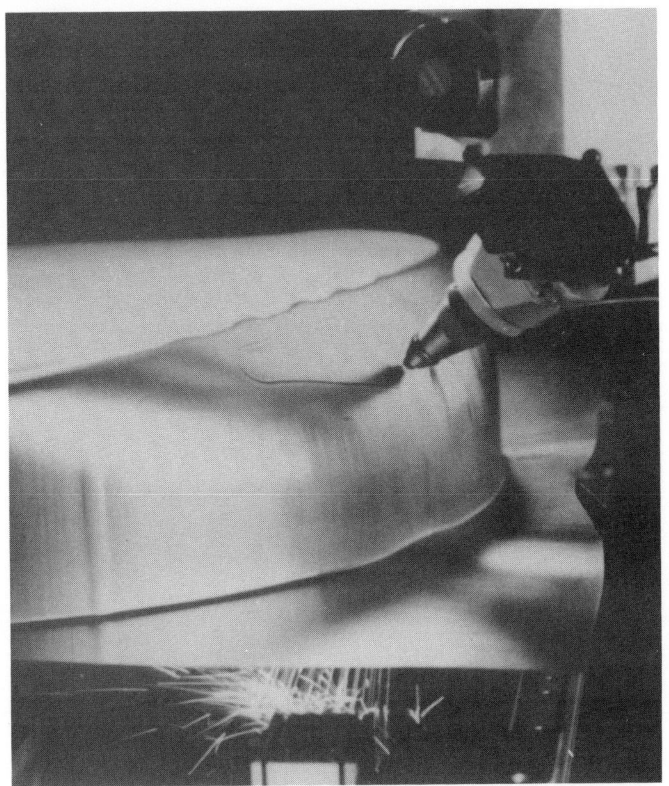

Fig. 25-37. When used with CNC, laser is capable of cutting 3-D workpieces such as this automobile tire well. (Trumpf Inc.)

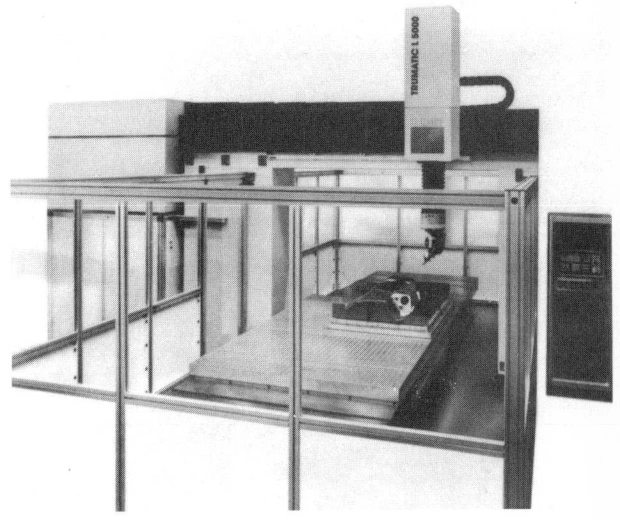

Fig. 25-38. This is a 5-axes CNC laser cutting center. (Trumpf Inc.)

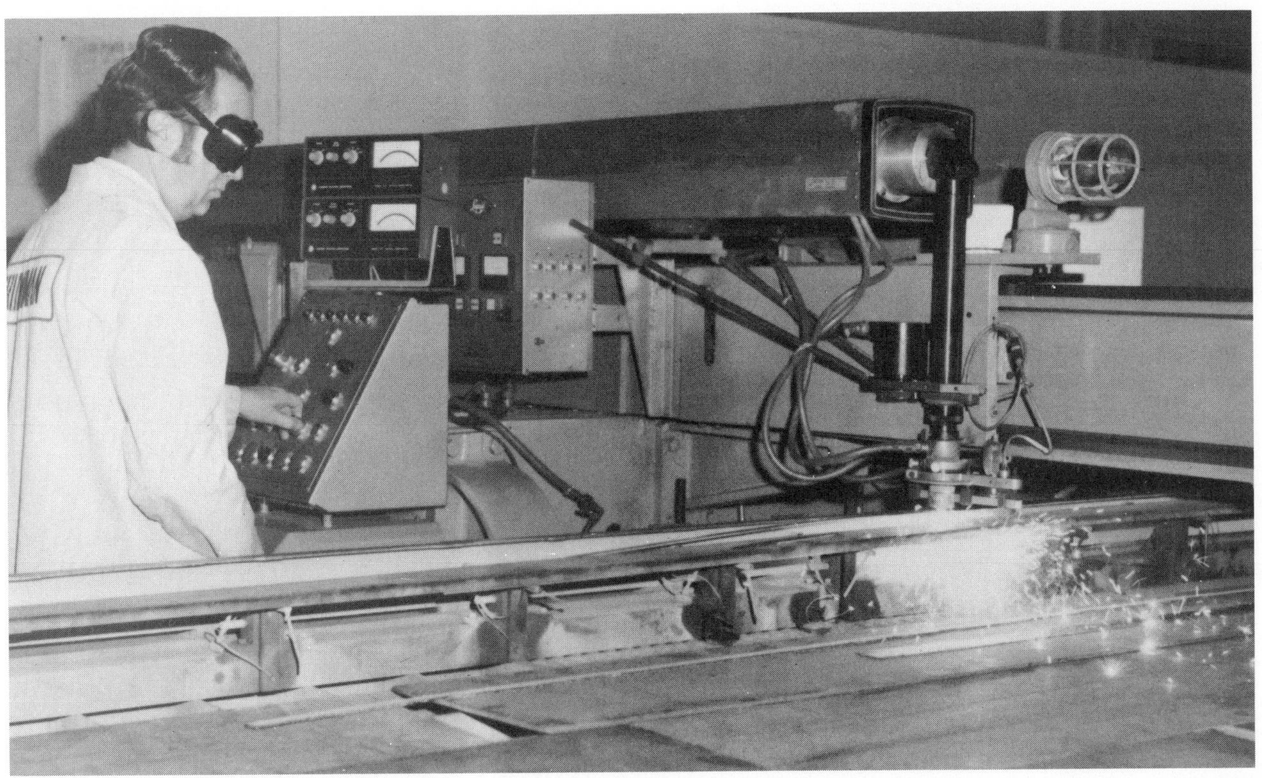

Fig. 25-39. In laser welding, a light beam is focused on and used to vaporize sections being welded. Molten metal surrounds point of vaporization as beam is moved along joint. (Grumman Aerospace Corp.)

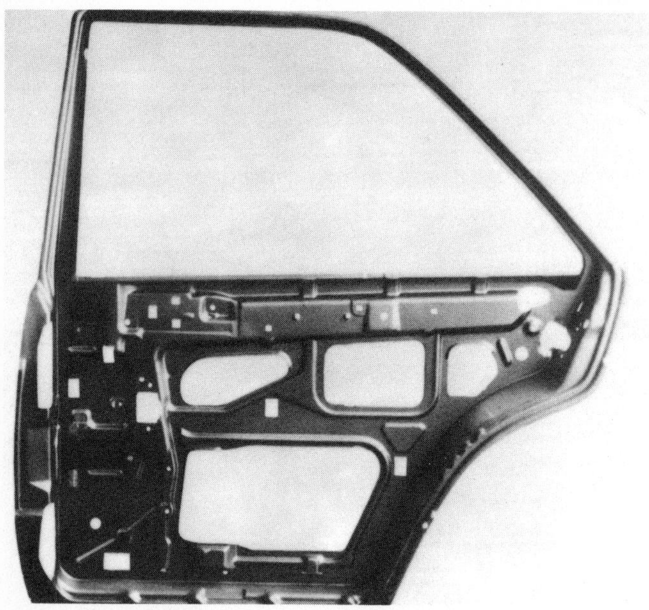

Fig. 25-40. Openings in this auto car door were accurately and quickly cut by a laser. An autofocus device in cutting nozzle compensated for any workpiece-to-workpiece variations. (Trumpf Inc.)

Fig. 25-42. This modern laser is cutting sheet metal sections using a continuous beam. Water assist removes heat to prevent cutting problems. (Amada Laser Systems)

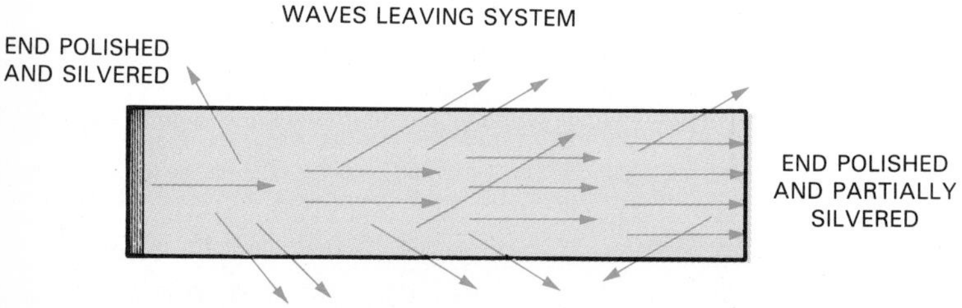

WAVES LEAVING SYSTEM

END POLISHED AND SILVERED

END POLISHED AND PARTIALLY SILVERED

A—Ends of a ruby rod are flattened so they are parallel, and silvered to form mirrors. Mirrors at one end is made to reflect only part of light so that when there is a buildup in energy between mirrors, beam can escape.

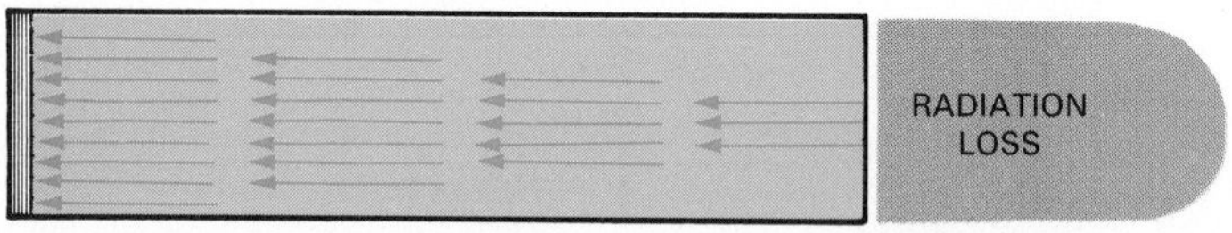

RADIATION LOSS

B—Soon after chromium atoms in ruby crystal are pumped by a flash lamp to a higher energy level, they drop to another level and stimulated emission takes place. Waves moving at angles to crystal's axis leave system, but those traveling along axis grow by stimulated emission of photons.

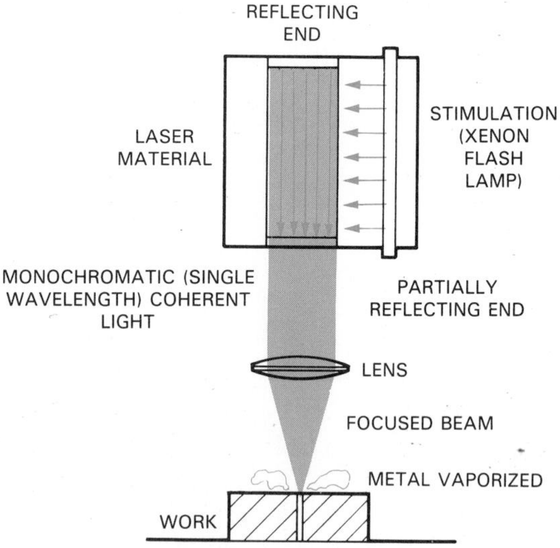

LASER BEAM

LENS

WORK

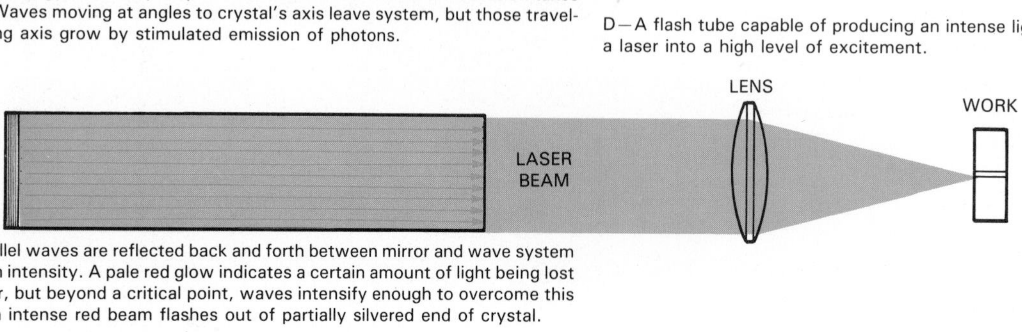

C—Parallel waves are reflected back and forth between mirror and wave system grows in intensity. A pale red glow indicates a certain amount of light being lost at mirror, but beyond a critical point, waves intensify enough to overcome this loss. An intense red beam flashes out of partially silvered end of crystal.

REFLECTING END

LASER MATERIAL

STIMULATION (XENON FLASH LAMP)

MONOCHROMATIC (SINGLE WAVELENGTH) COHERENT LIGHT

PARTIALLY REFLECTING END

LENS

FOCUSED BEAM

METAL VAPORIZED

WORK

D—A flash tube capable of producing an intense light is employed to "pump" a laser into a high level of excitement.

Fig. 25-41. Study operation of laser.

Until recently, aluminum, stainless steel, and titanium parts cut to shape by a laser required a time-consuming secondary operation to remove the oxide and dross that formed on the edge and surface of the cut. Titanium had a tendency to become embrittled by the oxygen that was absorbed during the cutting operation.

The CLEAN CUT™ process, shown in Fig. 25-43, virtually eliminates these problems by flooding the work surface with argon or nitrogen during the cutting operation. Since no oxygen can contaminate the metal, there is no oxide formation or embrittlement. Parts can be welded immediately after cutting because they require no cleaning.

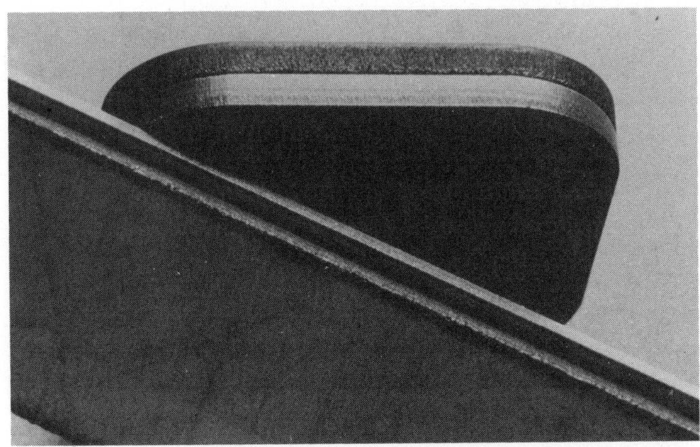

Fig. 25-43. This precision cut was made with laser using argon or nitrogen gas barrier to protect cut from oxygen. (Amada Laser Systems)

TEST YOUR KNOWLEDGE—Chapter 25

Please do not write in the text. Place your answers on a separate sheet of paper.

1. Chemical machining falls into two categories. Briefly describe each of them.
2. Chemical milling is also known as ________ ________ or ________ ________.
3. In proper order, list the principal steps followed in chemical milling.
4. A mask must be used to protect the portion of a chem-milled job that is:
   a. Not to be etched.
   b. To be etched.
   c. To be cleaned.
   d. All of the above.
   e. None of the above.
5. In proper order, list the principal steps followed in chemical blanking.
6. Briefly describe water jet machining.
7. What is ultrasonics?
8. Sound waves below 20 cycles per second are called ________.
9. Sound waves above 20,000 cycles per second are called ________.
10. List five (5) areas where the science of ultrasonics has found industrial applications.
11. Impact machining makes use of a ________ tool that pounds ________ ________ ________ against the work to do the cutting.
12. Impact machining is one of the very few commercially feasible methods for machining which of the following materials?
    a. Hard materials.
    b. Brittle materials.
    c. Frangible materials.
    d. All of the above.
    e. None of the above.
13. Impact machining has several disadvantages. It is ________ and the surface finish is dependent upon the ________ ________ of the ________ used. Also, ________ (________) is the deepest cut possible with existing equipment.
14. With impact machining, tolerances of ________ can be maintained on hole size and geometry in most materials.
15. The development of electron beam machine was direct result of special needs of the:
    a. Electronics industries.
    b. Atomic energy industries.
    c. Aerospace industries.
    d. All of the above.
    e. None of the above.
16. The first application of the electron beam technique was concerned with ________ ________ ________.
17. Holes as small as ________ in diameter can be drilled using the electron beam technique.
18. The electron beam machine is basically a source of:
    a. Thermal energy.
    b. Sonic energy.
    c. Fluid energy.
    d. All of the above.
    e. None of the above.
19. The electron beam technique cuts material by:
    a. Alternately heating and cooling the area to be cut.
    b. Vaporizing the material.
    c. Making use of a pulsing technique.
    d. All of the above.
    e. None of the above.
20. List two methods employed to control the shape of the cut with EBM.
21. What does the term LASER mean?
22. Describe how the laser operates.
23. The laser is capable of machining holes as small as ________ in diameter.

RESEARCH AND DEVELOPMENT

1. Prepare a file for the shop technical library on chemical milling and chemical blanking techniques. Secure literature from manufacturers of chem-milling and chem-blanking equipment and clippings from the various technical magazines.
2. Secure samples of work produced by the chemical machining techniques.
3. Develop and produce equipment that will permit you to demonstrate chemical milling. Prepare a technical paper on the process with photographs and submit it to one of the professional industrial education magazines.
4. Conduct a series of chemical milling experiments. Use different metals; however, use the etchant for an equal time on each. Prepare a technical paper on your experiment. List the depth of etch, and what effect heat and cold have on etching rate. Develop a table showing times required to achieve equal etch depths on various metals, quality of surface finish, amount of undercut, and how it can be controlled.
5. Secure information on the use of water jet machining.
6. Secure samples of work that have been machined using ultrasonic techniques. Mount the samples, if they are small enough, on a display panel. Include a sketch showing the machining technique employed.
7. Gather information on other uses of ultrasonics. Prepare a bulletin board display using the material.
8. Demonstrate how ultrasonic sound waves can be measured. Borrow a transducer and oscilloscope from the science department.
9. Construct an ultrasonic assist and experiment with it on the lathe.
10. Design and construct an impact machining device. Demonstrate it on various materials. Prepare an evaluation of your work.
11. Prepare a bulletin board display featuring electron beam machining. Use material from technical magazines and manufacturer's literature, brochures, etc.
12. Prepare a research paper on electron beam machining and welding techniques. Include the history of its development and how the atomic energy, electronics, and aerospace industries utilize its unique characteristics.
13. Prepare a researach paper on use of the laser by industry. Use illustrations from magazines.

SAFETY NOTE! Because of the inherent dangers of using the laser, it is NOT recommended that an attempt be made to design and construct a laser capable of cutting metal.

Chapter 26

POWDER METALLURGY

After studying this chapter, you will be able to:

- Explain several applications for parts made by the powder metallurgy process.
- Describe how powder metallurgy parts are produced.
- Relate how powder metallurgy parts can be machined.

Powder metallurgy, abbreviated P/M, is a technique used to shape parts from metal powders, Fig. 26-1. Sometimes called ***sintering,*** it was developed in the late 1920s by the automotive industry to make electric motor bearings. See Fig. 26-1.

POWDER METALLURGY APPLICATIONS

The P/M process is widely employed by the industry for such applications as:

1. Self-lubricating bearings and bearing materials.
2. Precision finished machine parts, gears, cams, ratchets, etc., with tolerances as close as ±0.0005 in. (0.0127 mm). See Fig. 26-2.
3. Permanent metal filters (sintered bronze fuel filters for example).
4. Fabricating materials that are difficult to work:
   a. Tough cutting tools (tungsten carbide).
   b. Supermagnets (aluminum-nickel alloys or Alnico).
   c. Mixtures of metals and ceramics for jet and rocket applications that require the heat resistance of ceramics as well as the heat transfer qualities of metals. This is also used for special cutting tools (cermets).
   d. High density counterweights for aerospace instruments that require maximum weight concentration in a minimum space.
   e. Storage battery elements (nickel-cadmium or Nicads®).

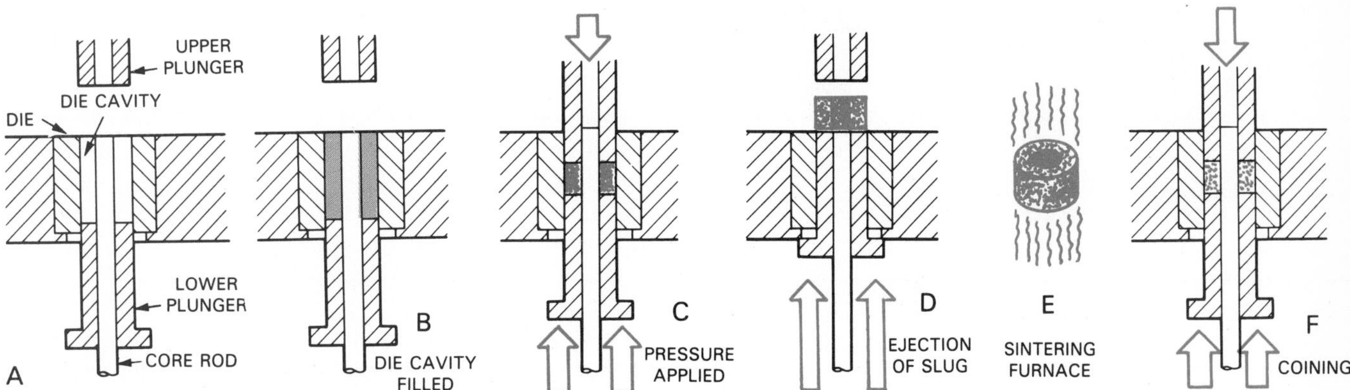

Fig. 26-1. Study steps in fabricating by powder metallurgy process: A—Note cross section of die and die cavity. Depth of cavity is determined by thickness of required part, and amount of pressure that will be applied. B—Die cavity is filled with proper metal powder mixture. C—Pressure as high as 50 tons per square inch is applied. D—Briquette is pushed from die cavity. E—Pieces are then passed through a sintering furnace to convert them into a strong useful product. F—Some pieces can be used as they come from furnace. Others may require a coining or sizing operation to bring them to exact size and to improve their surface finish.

Fig. 26-2. Powder metallurgy parts are produced by mixing carefully selected metal powders. Automotive parts shown are made from ferrous base metal powders. They are as strong as similar parts machined from solid metal but can be made by powder metallurgy process much more quickly and economically. (Metal Powder Industries Federation)

Fig. 26-3. Metal powders are mixed and blended to meet physical requirements of part. (Burgess-Norton Mfg. Co.)

POWDER METALLURGY PROCESS

The first phase in the manufacture of powder metal products in the careful mixing of high purity metal powders, Fig. 26-3. The are carefully

Fig. 26-4. Metal powder is compacted at pressures up to 50 tons per square inch. (Burgess-Norton Mfg. Co.)

weighed and thoroughly mixed into a blend of correct proportions. Many materials can be used: iron, steel, stainless steel, brass, bronze, nickel, chromium, etc. Combinations of these metals and nonmetals can also be used.

Briquetting

The powder blend is then fed into a precision die. The die cavity has the shape of the desired part, but it is several times deeper than the thickness of the part. The powder is compressed by an upper and lower punch, Fig. 26-4. Pressures applied range from 15 to 50 tons per square inch. This portion of the operation is known as ***briquetting.***

The piece, as ejected from the die, appears to be solid metal. However, it is quite brittle and fragile. It will crumble if NOT handled carefully, Fig. 26-5.

Sintering

To transfer the briquette or ''green compact'' into a strong, useful unit, it must be ***''sintered''*** at 1500 to 2300 °F (815 to 1260 °C) for 30 minutes to 2 hours, depending upon the powder mixture. This is done in a controlled atmosphere furnace, Fig. 26-6.

Forging

Many parts can be utilized as they come from the furnace, Fig. 26-7. Because of shrinkage and distortion caused by the heating operation, the pieces may be subject to a ***sizing, coining,*** or ***forging*** operation, Fig. 26-8. This consists of pressing the sintered pieces into accurate size dies to obtain precise finished dimensions, higher densities, and smoother surface finishes, Fig. 26-9.

Fig. 26-5. This is a close-up of briquetting press showing "green compacts" moving from press. (Delco Moraine, Div., General Motors Corp.)

Fig. 26-7. Finished pieces are emerging from furnace. They may be used "as is," or they may require additional operations to make them usable.

When the part is to be formed, it must be reheated just prior to the forging operation, Fig. 26-10.

Fig. 26-6. This sintering furnace permits continuous operation. Atmosphere within furnace is carefully controlled to prevent oxidation or contamination of sinterings. (Delco Moraine, Div., General Motors Corp.)

Fig. 26-8. Hot forging takes place in closed dies with compact at red heat. (Burgess-Norton Mfg. Co.)

Fig. 26-9. Quality control manager is gaging finished P/M parts to determine whether they meet design specifications. (Burgess-Norton Mfg. Co.)

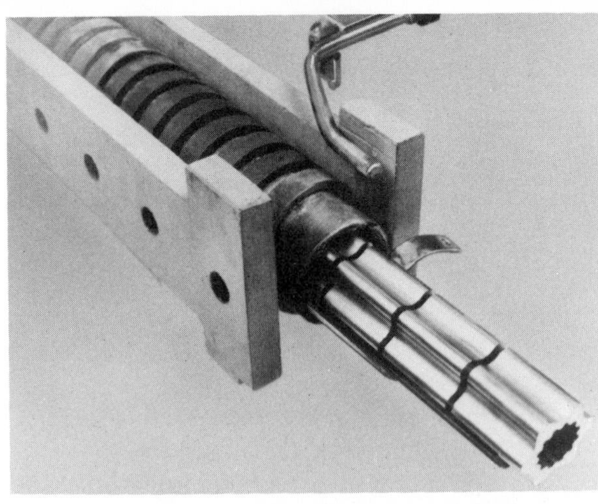

Fig. 26-10. Induction heating brings P/M compacts up to hot forging temperature rapidly. After forging, part is ready for additional machining, if needed. (Burgess-Norton Mfg. Co.)

Fig. 26-11. Flat lapping is a grinding operation that generates a micro finish. It is being performed on these gear faces. (Burgess-Norton Mfg. Co.)

Powder metallurgy costs

The tool cost is moderate and the process is best suited to quantity production. If production quantities exceed several thousand units, the finished piece can be produced at less than the cost of rough sand castings.

Powder metal parts can be drilled, tapped, plated, heat treated, machined, and ground, Fig. 26-11.

TEST YOUR KNOWLEDGE—Chapter 26

Please do not write in the text. Place your answers on a separate sheet of paper.

1. Powder metallurgy, abbreviated ________, is the technique of ________ parts from ________ ________.
2. The powder metallurgy process is used to make:
   a. Self-lubricating bearings.
   b. Precision machine parts.
   c. Permanent metal filters.
   d. All of the above.
   e. None of the above.
3. List the steps in making a part by the powder metallurgy technique.
4. Parts made from metal powder can be:
   a. Drilled.
   b. Heat treated.
   c. Turned on a lathe.
   d. All of the above.
   e. None of the above.

5. What is a briquette or "green compact?"
6. Why is it often necessary to size, coin, or forge parts made from metal powders after they have been sintered?
7. What do the above operations do to the sintered piece?

RESEARCH AND DEVELOPMENT

1. Secure a bearing and a fuel filter made using powder metallurgy technology. Examine the units under a microscope and:
   a. Make a sketch with exaggerated details that shows the structure of each example.
   b. Use the sketch to prepare a projectual for the overhead projector. Use the projected image to explain your findings to the class.
   c. Have a micro-photograph made of the grain structure of each part. Use this to illustrate a presentation on powder metallurgy.
2. Contact a firm that manufactures products using powder metallurgy and request samples of units in the various stages of the manufacturing process. Prepare a bulletin board display.
3. Secure samples of different products made by the powder metallurgy process. Prepare a display panel showing these products and how they are used. For example, fuel filters made by the process have the ability to separate water from gasoline.

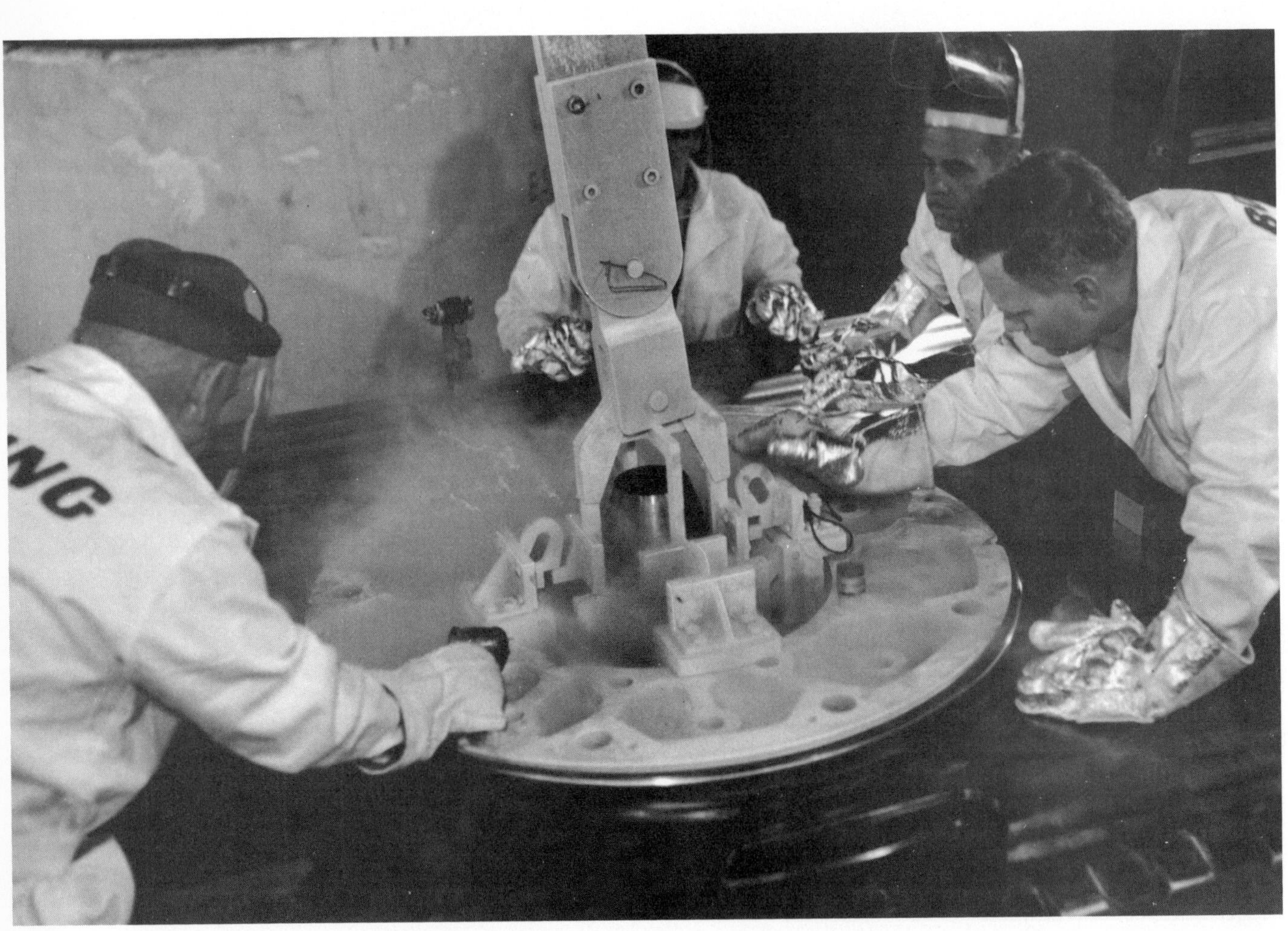

Fig. 27-1. Wing pivot on a proposed Super Sonic Transport (SST) cooled to −300 °F (−184 °C) is being mated to wing clevis joint. Note special protection gloves worn by technicians. (Boeing Company)

Chapter 27

CRYOGENIC APPLICATIONS

After studying this chapter, you will be able to:

- Explain how the science of cryogenics is used in industry.
- Describe some applications for cryogenics.

The science of cryogenics is one of the more recent additions to the technology of machining or metalworking. The technique is NOT a new nor different way to work metal, but rather it is employed to improve or reinforce other metalworking techniques. The term ''cryogenic'' is taken from the Greek word ''kryos,'' meaning ''icy cold'' and the Latin ''generatus,'' meaning to make or create. ***Cryogenic*** means, literally, to make ''icy cold.''

Cryogenics deals with temperatures starting at the point where oxygen liquefies (about −300 °F or −184 °C), and goes down to just about absolute zero (−459 °F or −293 °C). At this point, everything but helium freezes. Oxygen and nitrogen look something like ''white beach sand'' or ''table salt.'' Metal also acts strangely. Lead coils act like steel springs, some metals increase tremendously in strength, while others become ''super conductors'' of electricity.

Wide use is made of cryogenics in annealing and heat treating metals. The characteristics of several aluminum alloys and some space age metals are greatly improved by heating them and then quenching them in liquid nitrogen.

Another application of ultracold is to shrink-fit metal parts together. As most metals shrink in size as they become cold, one part of the assembly is made slightly oversize and is immersed in liquid nitrogen. The exact amount of oversize is determined by part size and type of metal. The diameter is reduced (shrunk) by the extreme temperature drop until it fits easily into its mating part. The cooled part expands as it returns to room temperature and thus is held or locked in place, Fig. 27-1. The parts do NOT become distorted as they would if the parts were pressed together or heated (expanded) until they could be fitted together.

TEST YOUR KNOWLEDGE—Chapter 27

Please do not write in the text. Place your answers on a separate sheet of paper.

1. What does the term ''cryogenic'' mean?
2. The science of cryogenics deals with temperature starting at ________ or ________ and goes down to temperatures near ________ or ________.
3. What does shrink fitting mean?
4. Why is it better to use the super-low temperatures of cryogenics to shrink fit parts together, rather than heat?

RESEARCH AND DEVELOPMENT

1. Demonstrate shrink fitting two metal parts together. Use ''dry ice'' to cool the part.

> DANGER! Handle ''dry ice'' with insulated gloves and wear protective eye wear and clothing. ''Dry ice'' can cause severe burns if NOT handled with caution!

2. Examine technical, scientific, and popular mazagines for examples showing super cold applications. HINT: The aerospace and electronics industries make use of cryogenic applications.

Chapter 28

JIGS AND FIXTURES

After studying this chapter, you will be able to:

- Explain why jigs and fixtures are used.
- Describe a jig.
- Describe a fixture.
- Elaborate on the classifications of jigs and fixtures.

Jigs and ***fixtures*** are devices widely employed in production machine shops to hold work while machining operations are performed. They position the work and guide the cutting tool(s) so that all of the parts produced are uniform and within specifications. Manufacturing costs are reduced when large numbers of identical and interchangeable parts must be produced. Jigs and fixtures are often justified when limited production is required because relatively unskilled workers may operate the machines.

Jigs and fixtures are also employed in assembly operations (welding, riveting, etc.) to position and hold work while fabricating standarized parts.

JIGS

A ***jig*** is a device that holds the work in place and guides the cutting tool during the machining operation (drilling, reaming, tapping, etc.). Hardened steel bushings guide the drill or cutting tool, Fig. 28-1.

The jig is seldom mounted solidly to the drill press table. However, for safety, it is usually nested between guide bars that are mounted solidly to the table, Fig. 28-2.

Fig. 28-1. This is a lift type drill jig. Left. Jig is open to receive work shown in foreground. Right. Jig is closed, with work in place ready for drilling. (Ex-Cell-O Corp.)

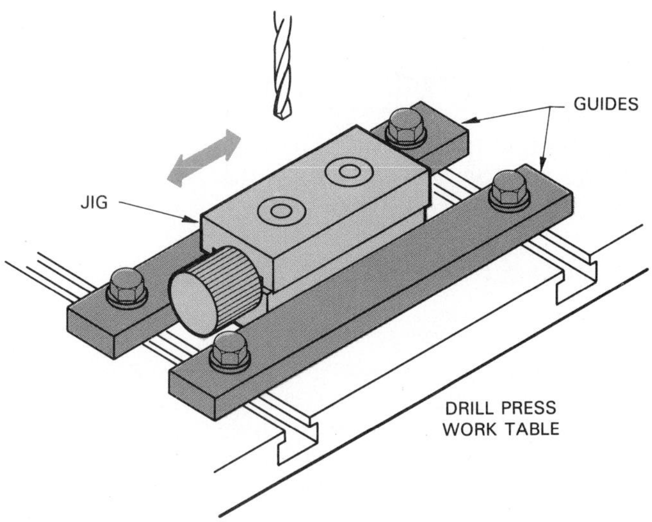

Fig. 28-2. Drill jig is nested between guide bars to prevent "merry-go-round."

Fig. 28-4. This is a two-piece drill jig. Part drilled in jig is shown in foreground.

Jig types

Drill jigs fall into two general types: open jigs and boxed (closed) jigs.

Open jigs—the drill template or plate jig is the simplest form of the open type drill jig. It consists of a plate with holes to guide the drill. It fits over the work, Fig. 28-3.

In its more elaborate form, clamps hold the work in place and the unit is fitted with a base plate to provide clearance for the drill as it breaks through the work, Fig. 28-4. The circular drill jig is a variation of the plate jig, Fig. 28-5.

Box or ***closed jigs***—the box jig encloses the work and is used when holes are to be drilled in several directions, Fig. 28-6. These are more complex than open jigs. The work is fitted into place through a hinged or swinging cover. The clamps that hold the work in place are premanently mounted to the jig. Box jigs are more costly to make than open jigs.

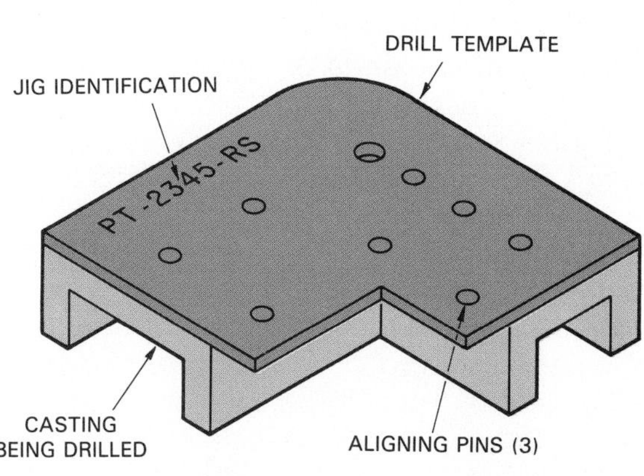

Fig. 28-3. Note simple drill template.

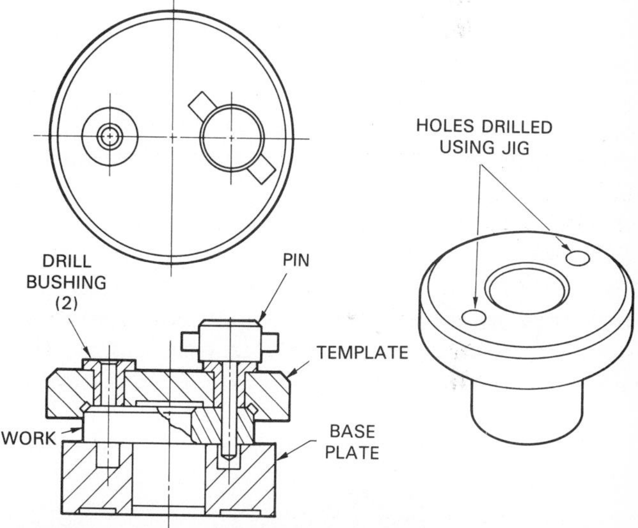

Fig. 28-5. With this circular type drill jig, pin is placed in first hole drilled to hold work in position when drilling second hole.

Fig. 28-6. These are box type drill jigs. Three stations are permanently set up for drilling, reaming, and tapping drill press head castings. Drilling machine in background drills holes for pinion shaft (quill feed mechanism) and column locks. (Clausing Industries Inc.)

A combination of the open and box jig is often used when several different operations must be performed on a job. Slip bushings are utilized to guide the drills and are removed for subsequent operations such as reaming, tapping, countersinking, counterboring, or spot facing.

FIXTURES

Fixtures are employed to position and hold work rigidly while machining operations are performed on it. The fixture does NOT guide the cutting tool(s).

Fixtures fall into many classifications. The class is determined by the type of machine tool on which it is used: milling machine (vertical or horizontal), lathe, band saw, grinder, etc. Fixture design ranges from a simple modification to vise jaws, Fig. 28-7, to very large and complex fixtures used by the aerospace industry, Fig. 28-8.

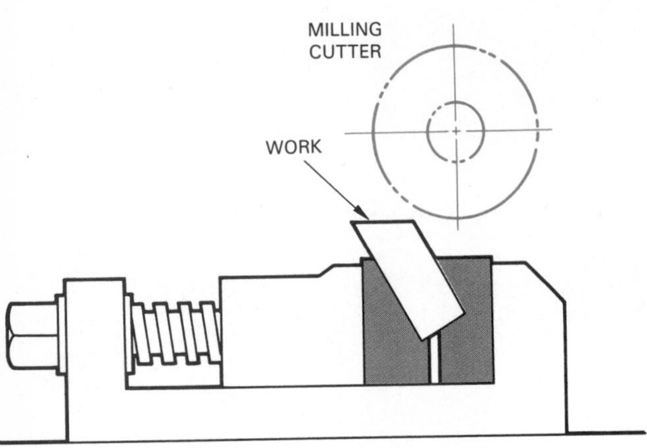

Fig. 28-7. Modified vise jaws (shaded portion) position work so that an angular cut is made on work.

Refer to Figs. 28-9 through 28-14 for different fixture applications.

JIG AND FIXTURE CONSTRUCTION

Jigs and fixtures are designed for specific jobs. Their complexity is determined by the number of pieces to be produced, the degree of accuracy required, and the kind of machining operations that must be performed.

The body of a jig or fixture may be built-up, welded, or cast. Commercial components are available in a wide range of sizes, types, and shapes. See Figs. 28-15 through 28-18.

TEST YOUR KNOWLEDGE—Chapter 28

Please do not write in the text. Place your answers on a separate sheet of paper.

1. Jigs and fixtures are devices used in ________ ________ ________ to ________ ________ while machining operations are performed.
2. When are jigs and fixtures used?
3. What is a jig?
4. Jigs fall into two general types. List and briefly describe each type.
5. A combination of the two jig types listed in question four is often used when ________.
6. What is a fixture?

RESEARCH AND DEVELOPMENT

1. Contact local industry and borrow examples of small jigs and fixtures they no longer use. Explain them to the class. If possible, include jobs produced on them.
2. Secure samples of products that have components produced with a jig and/or fixture.
3. Make a bulletin board display of magazine illustrations, drawings, and/or photographs showing various kinds of jigs and fixtures.
4. Design and manufacture a simple template jig for a job to be produced in the training area.
5. Design and manufacture a fixture for a training area product to be machined on a lathe, grinder, drill press, etc. Work in close cooperation with the drafting department in designing the project and producing the prints for the product.
6. Seek permission to visit a machine shop utilizing jigs and fixtures. Take 35 mm slides of the jigs and fixtures in use. Make a presentation to the class and employ the slides in your talk.

Fig. 28-8. Many fixtures are used in aircraft fuselage assembly. In fixture construction, extreme accuracy is critical, requiring use of lasers to assure precise alignment. (British Aerospace)

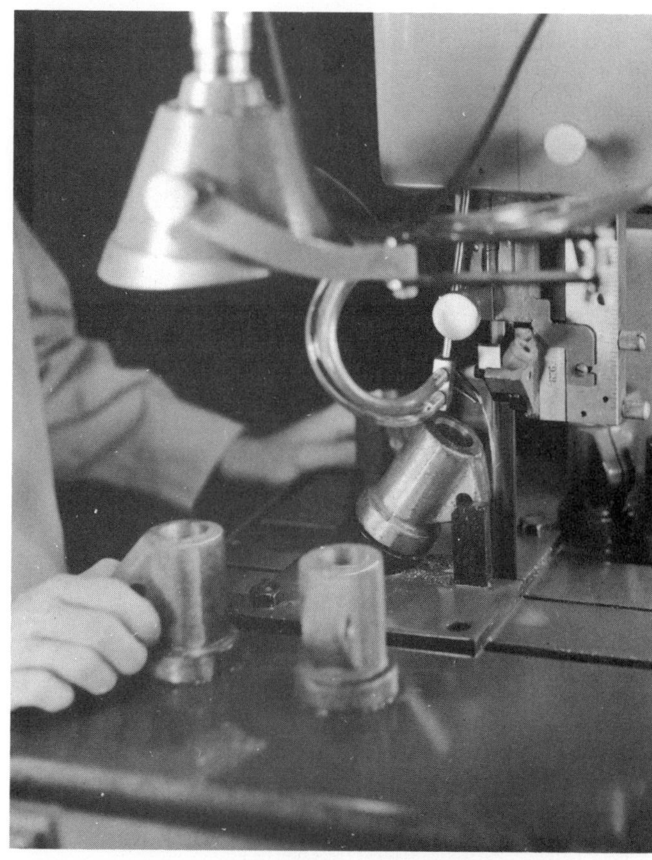

Fig. 28-9. Fixture is used to hold clamp device while being slotted on a band saw. (The DoALL Co.)

Fig. 28-10. Many jigs and fixtures are required in manufacture of autos. Fixtures hold body sections in accurate alignment while robots weld them together. (Pontiac Motor Div., GMC)

Fig. 28-11. This fixture is equipped with two dial indicators and three air gages. It is used when inspecting five critical dimensions on aluminum automatic transmission torque converter stators. Unit being inspected is shown at left foreground. (Ex-Cell-O Corp.)

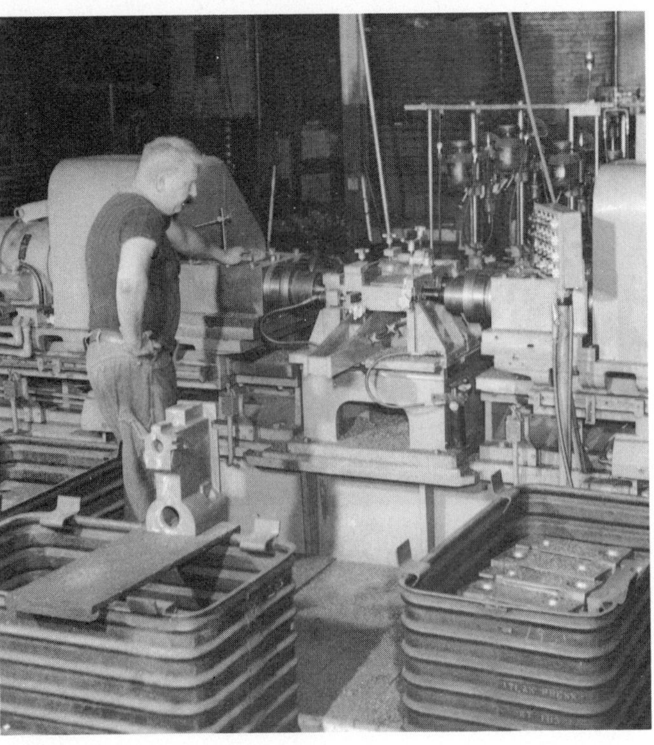

Fig. 28-12. Bearing surfaces for drill head column and quill are bored simultaneously in this fixture to assure accurate alignment. (Clausing Industries Inc.)

Fig. 28-13. Drill head casting is being held in position for machining. (Clausing Industries Inc.)

Fig. 28-14. Machinist is placing casting for body of a horizontal milling machine in a fixture for machining. Fixture permits machining to very close tolerances on a mass production basis. (Clausing Industries Inc.)

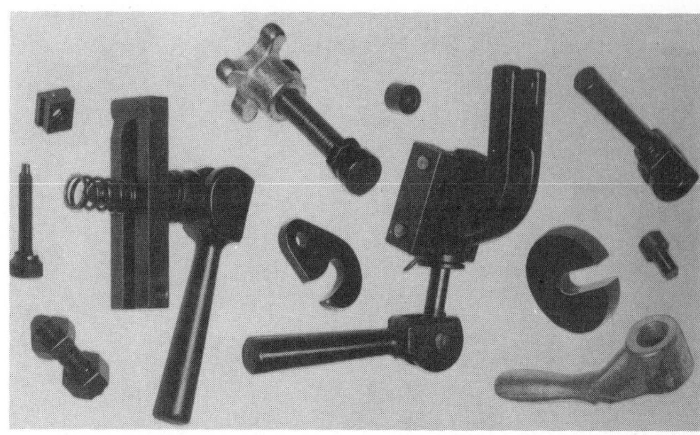

Fig. 28-17. Jig and fixture clamping components are employed to hold various types and shapes of work in a jig or fixture or to machine worktable. (Ex-Cell-O Corp.)

Fig. 28-15. These are samples of drill jigs manufactured from commercially available components. Drill bushings are used to guide drill. (Ex-Cell-O Corp.)

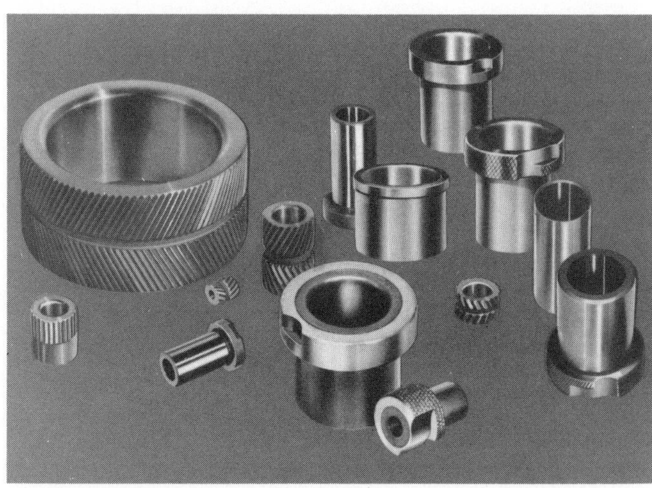

Fig. 28-16. Note a few of the many types of drill bushings available commercially. They assure accuracy and are more economical than shop-made bushings. (Ex-Cell-O Corp.)

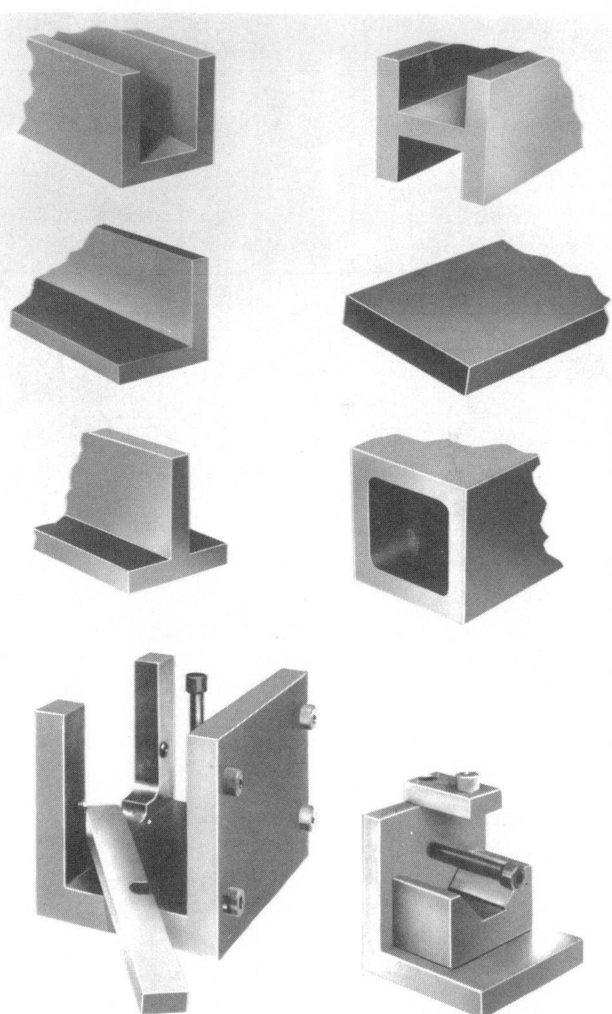

Fig. 28-18. Standard cast iron shapes are machined parallel and square to save time and money in both designing and building jigs and fixtures. Sections of different shapes can be bolted together to form complex holding devices. Two completed units are shown at bottom of illustration. (Ex-Cell-O Corp.)

Chapter 29

HYDRAULIC POWER TRANSMISSION

After studying this chapter, you will be able to:

- Define the term ''hydraulics.''
- Explain the advantages of hydraulics as a means of power transmission on machine tools.
- Describe some of the physical properties of fluids.
- Explain how hydraulic systems multiply forces.
- Define the term ''fluidics.''

Hydraulic power is applied to all types of machine tools to control and guide them through machining operations. Its simplicity, smoothness of operation, flexiblity and ease of control, makes hydraulic power transmission ideal for many machine tool applications.

In this chapter, we will examine the basic of hydraulics, its application to some machine tools, and also study fluidics or the technology of moving fluids.

BASIC HYDRAULICS

The term ***''hydraulics''*** is based on the Greek word for WATER. Originally, it was concerned only with the physical behavior of water at rest and in motion. It has since been broadened to include the study of ALL LIQUIDS.

Advantage of hydraulic power

1. ***Ease of control***—simple hydraulic valving permits accurate control of fluid pressure and flow. This makes possible an infinite number of machine speeds and feeds. Large forces can be controlled by small forces.
2. ***No mechanical parts***—there is no need for a complicated system of belts, gears, cams, and levers. Lubrication is also inherent in the unit.
3. ***Simple installation***—installations are simple and power can be transmitted up, down, around corners, and over relatively long distances with little loss of efficiency.
4. ***Separate controls***—speed and feed units are separate and can be controlled independently.
5. ***Smooth aplications of power***—transmitted motion is remarkably smooth and constant. Maximum pressures are generated almost instantly at the start of the cut and remain constant during the entire cutting stroke.
6. ***Rapid directional changes***—with hydraulics, rapid reversals at the end of the cutting stroke are cushioned and shock-free because there is no metal to metal contact. The cushioning action, known as ***''dwell,''*** can be increased or decreased as needed.
7. ***Flexibility***—a great number of motions, locating, clamping, feeding, driving, etc., are possible and can be built into a single machine.
8. ***Contour machining possible***—precision contour machining is possible for some applications without the expense of a CNC machine tool nor the need for a complicated NC program. This is done through the use of special valving systems. Tool feed is controlled automatically by a stylus, mounted on a hydraulic piston, as it traces over a master template or pattern. One, two, and three-dimensional cutting is possible. Several cutting units can be controlled by one tracing unit. See Figs. 29-1 and 29-2.

Physical properties of fluids

To better under stand the operation of hydraulically activated units, it is necessary to learn about the physical properties of liquids.

1. ***Shapelessness of fluids***—fluids have NO outer shape of their own as do solids. They quickly conform to the shape of their containers. This makes it possible to transmit fluids easily through pipes, tubing, and hoses. This can be done by gravity or by applying pressure.
2. ***Relative incompressiblity of liquids***—the relatively slight decrease in volume when liquids are

Fig. 29-1. Close-up shows tracer unit and cutting head. Note that cutter is same shape as tracer finger. (Bridgeport Machines Inc.)

Fig. 29-2. This vertical milling machine is fitted with two cutting heads controlled hydraulically by tracer unit at right. (Bridgeport Machines Inc.)

initially pressurized acts like a ''shock absorber'' and cushions machine movement. However, when constant force is applied to a confined liquid, the liquid quickly exhibits the same characteristics as a solid.

3. ***Transmission of forces through liquids***—the directionof a blow on a SOLID almost entirely determines the direction of transmitted force. When force is applied to a column of LIQUID, the force is not only transmitted straight through to the other end, but also equally in every direction throughout the column, Fig. 29-3.
4. ***Pascal's Law***—pressure set up in a liquid acts equally in all directions. The shape of the container in no way alters this pressure relation. Pascal's Law is the foundation of modern hydraulics.

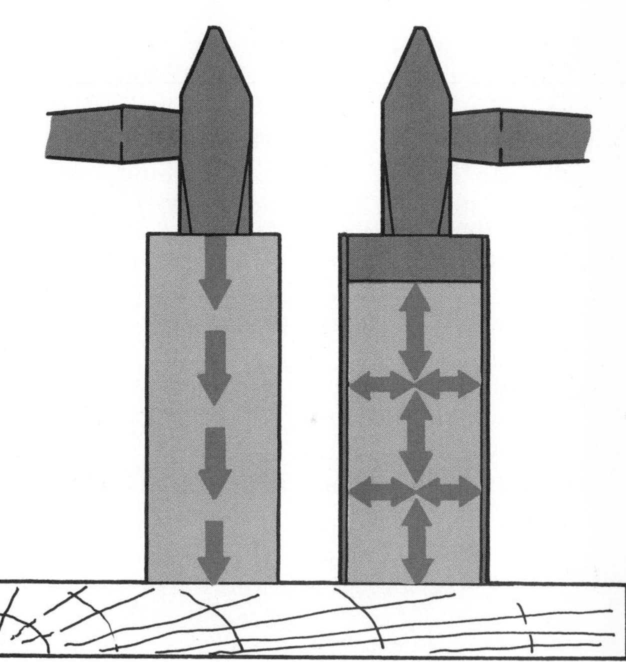

Fig. 29-3. Pressure in a confined liquid is equal on all surfaces.

Transmission of forces in hydraulic systems

Forces applies to a confined liquid is transmitted equally throughout the liquid regardless of the container shape. Force applied to Piston #1, Fig. 29-4, will be transmitted to Piston #2.

Pressure is defined as the force divided by the area over which it is distributed. For example, a force of 100 lbs. applied to a piston of 10 sq. in. will create a pressure in the liquid of 10 psi (pounds per square inch).

The shape of the container has no bearing on the degree of pressure, providing an unobstructed passageway is available. Therefore, a realtively

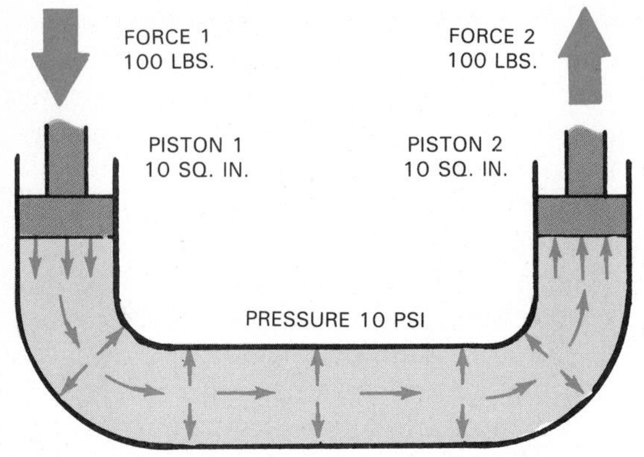

Fig. 29-4. Note pressures and forces in basic hydraulic system.

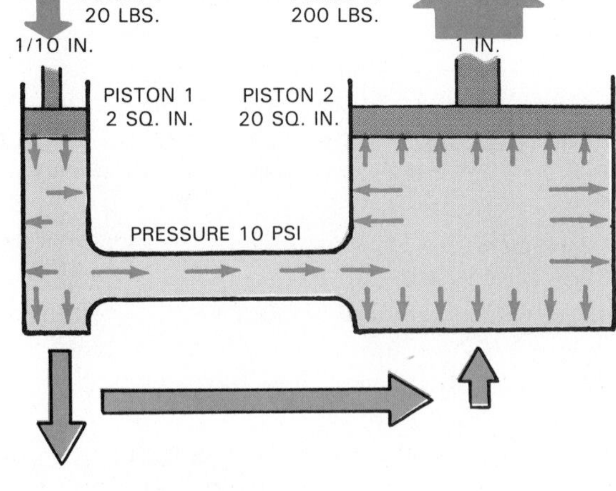

Fig. 29-6. Study principle of multiplying forces hydraulically.

small pipe can be employed to connect the two cylinders and work well. See Fig. 29-5.

Multiplication of forces

The output force was equal to the input force in the previous examples. By using pistons of different sizes, forces can be multiplied, Fig. 29-6. However, note that the piston travel is no longer equal.

The distance factor can be explained in this manner. Piston #1 has an area of 2 sq. in. When it is pushed down 1 in., 2 cu. in. of liquid will be displaced. In order to accommodate the displaced fluid (2 cu. in.), Piston #2 will have to move only 1/10 in., because its area is 10 times LARGER than Piston #1.

Because of Pascal's Law, the general rule is if two pistons are used in an hydraulic system, the force acting on each piston will be directly proportional to its area. The magnitude of each force will be the product of pressure and piston area. The distances moved will be inversely proportional to their areas.

The hydraulic press operates on the two piston configuration. See Fig. 29-7.

Fig. 29-7. This hydraulic press is capable of generating many tons of pressure where needed on work. It can be used to straighten bent shafts or to force press-fit parts together. (Dake Corp.)

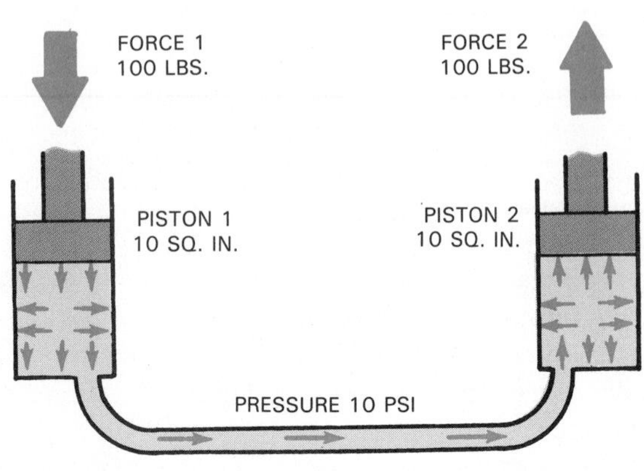

Fig. 29-5. Small diameter pipe or tubing can be used to connect two cylinders with no reduction in pressure or force.

HYDRAULIC APPLICATIONS ON MACHINE TOOLS

Pascal's Law is utilized in hydraulic applications on machine tools. However, the pressure is created by a pump rather than by a small piston and weights, Fig. 29-8.

An interesting application of hydraulics to a machine tool is a 3-dimensional tracer mechanism, as shown in Fig. 29-9. The cutter duplicates the movement of the tracer finger as it moves over a master template.

When under the control of the tracer finger, each cutter will follow the same path as that of the tracer finger along the template. The tracer fingers must match the cutter or cutters if the work is to duplicate the form of the template. As some deflection (movement) of the tracer finger must take place before the valve mechanism will respond, the tracer finger must be made slightly larger than the cutter diameter to compensate for this "lag."

A cross section of a tracer unit is in Fig. 29-10.

The tracer unit controls the movements of the table and cross slide to duplicate profile shapes. Vertical movement is accomplished by cutter travel.

Fig. 29-9. Close-up of universal tracer unit on a Hydro-Tel milling machine. Tracer finger is fitted in hydraulically actuated tracer unit to right. Cutter on left duplicates movement of tracer finger. (Cincinnati Milacron)

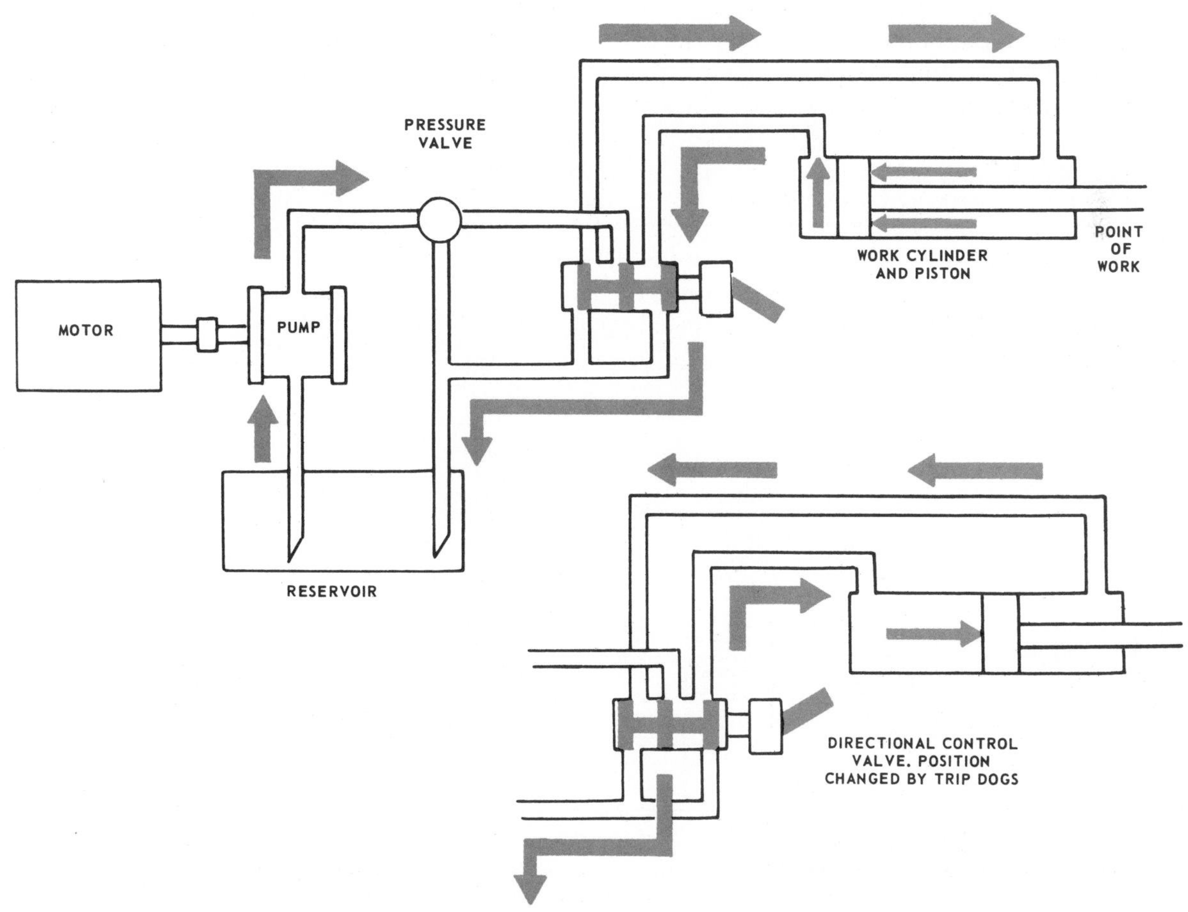

Fig. 29-8. Basic hydraulic circuit is employed on most machine tool applications.

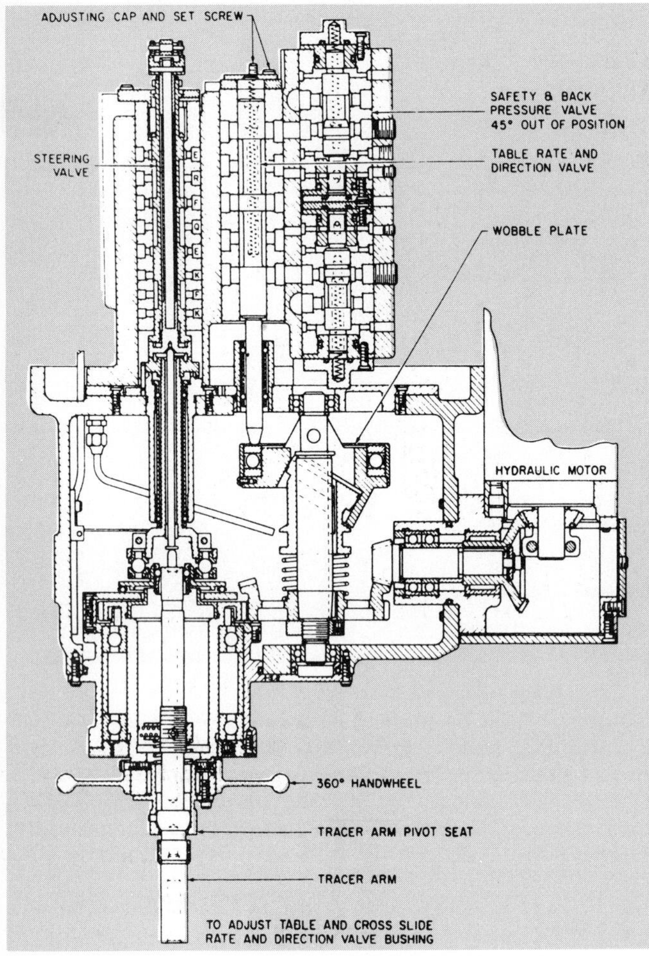

Fig. 29-10. Note hydraulic valving mechanism of Hydro-Tel tracer unit. (Cincinnati Milacron)

Machine tool application

Hydraulic units are employed on machine tools because of their simplicity, dependability, and the ease by which speeds and feeds can be controlled through an almost infinite range.

A schematic drawing of the hydraulic system of one type of surface grinder is shown in Fig. 29-11. A schematic drawing makes use of simple symbols for the various hydraulic components rather than complicated drawings.

Hydraulics are also utilized to actuate the worktable and tool travel on many planers. A diagram of how one such system operates is shown in Fig. 29-12.

The shaper employs hydraulic pressure to impart movement to the ram, Fig. 29-13. Because of hydraulic system flexibility, ram speed can be changed and stroke length varied while the machine is in operation. Power on the cutting stroke is uniform through the entire stroke.

Many machine tool variable speed units are activated hydraulically. Pressure in the hydraulic system causes a piston to spread or close a two-piece U-pulley. The pulley variable speed unit can be opened to reduce speed or closed to increase speed.

Fig. 29-14 shows a typical system.

DANGER! Relatively small hydraulic systems can exert tons of force. To prevent serious injury, use extreme care when working around hydraulic systems!

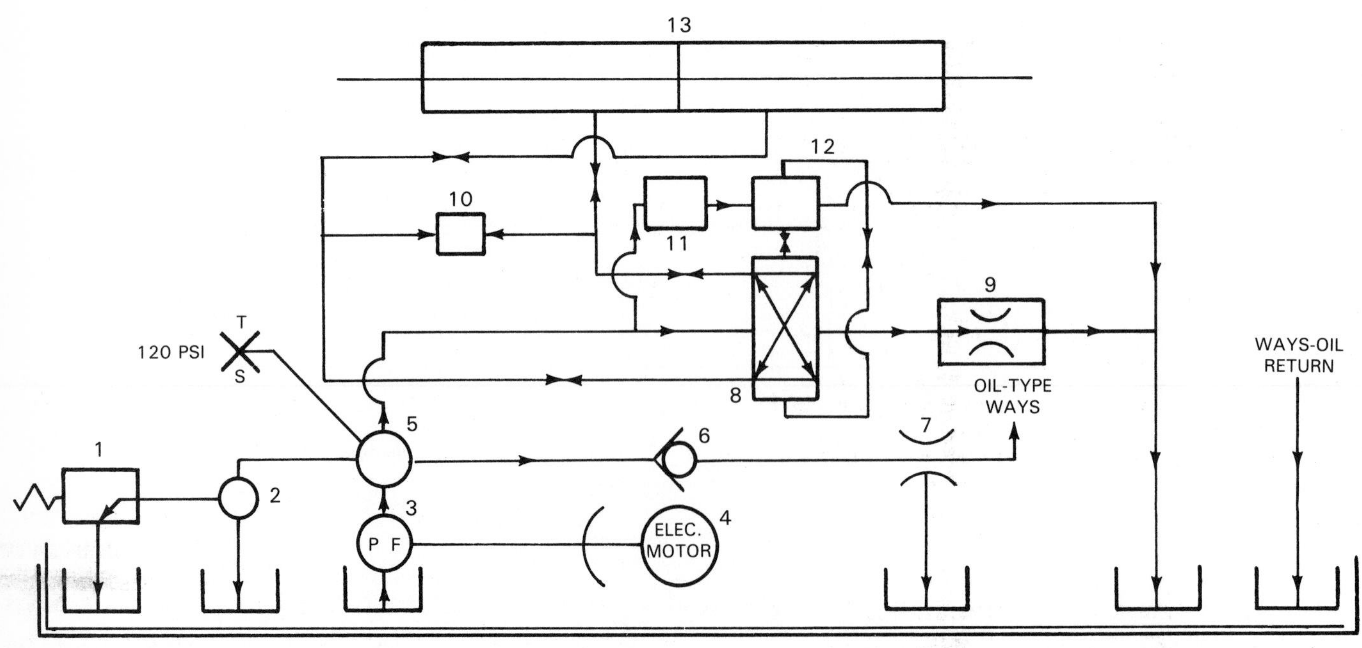

Fig. 29-11. Study drawing of hydraulic system on a surface grinder. 1—Relief valve. 2—Filter assembly. 3—Constant flow rotary gear pump. 4—Electric motor. 5—Special elbow. 6—Oil filter and check valve. 7—Pressure relief-oil ways. 8—Four-way valve. 9—Table control. 10—Table stop. 11—Dwell. 12—Directional valve. 13—Table drive cylinder. (K.O. Lee Co.)

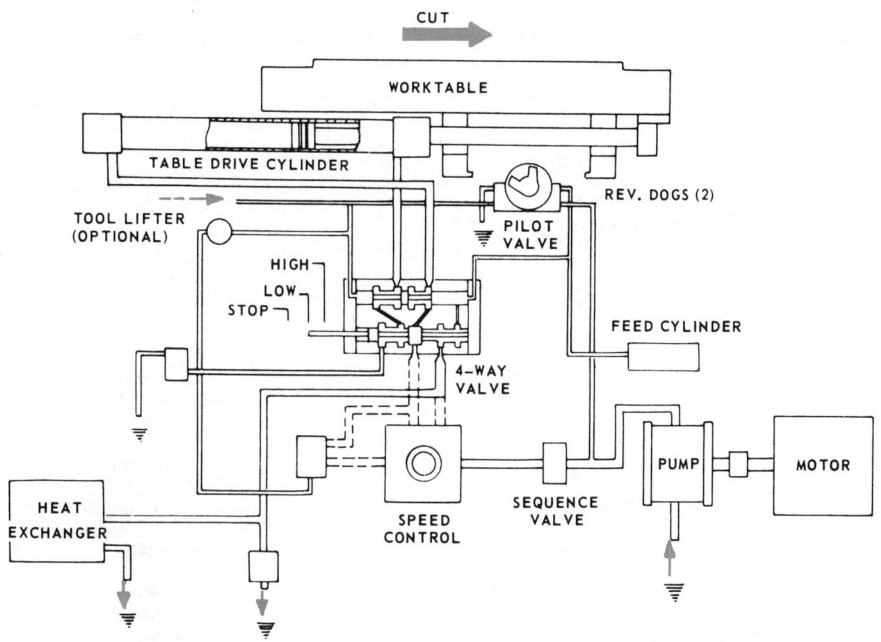

Fig. 29-12. A hydraulic cylinder is employed to impart movement to worktable of this planer.

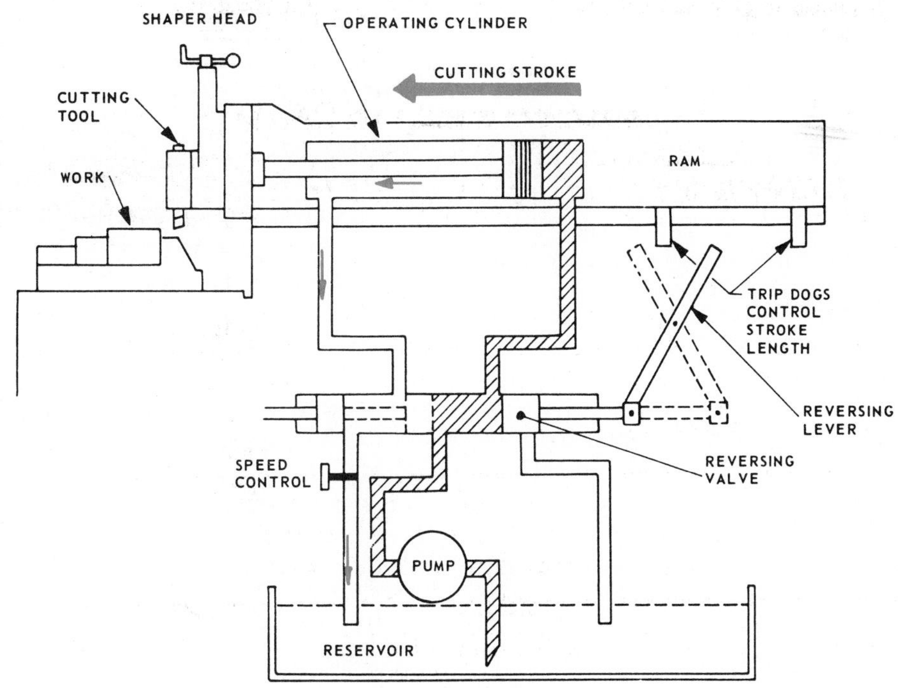

Fig. 29-13. Study operation of hydraulically activated shaper.

FLUIDICS

Fluidics (pronounced flew-id-iks) is a relatively new development in the technology of fluid power. The term ***fluid*** is employed to denote a continuous medium, either gaseous or liquid.

Fluidics is a contraction of the words ''fluid'' and ''logic.'' It is the technology that utilizes the dynamic forces of a fluid in motion to duplicate and often improve upon many of the tasks normally accomplished by electronic, electromechanical, and mechanical systems. Refer to Fig. 29-15.

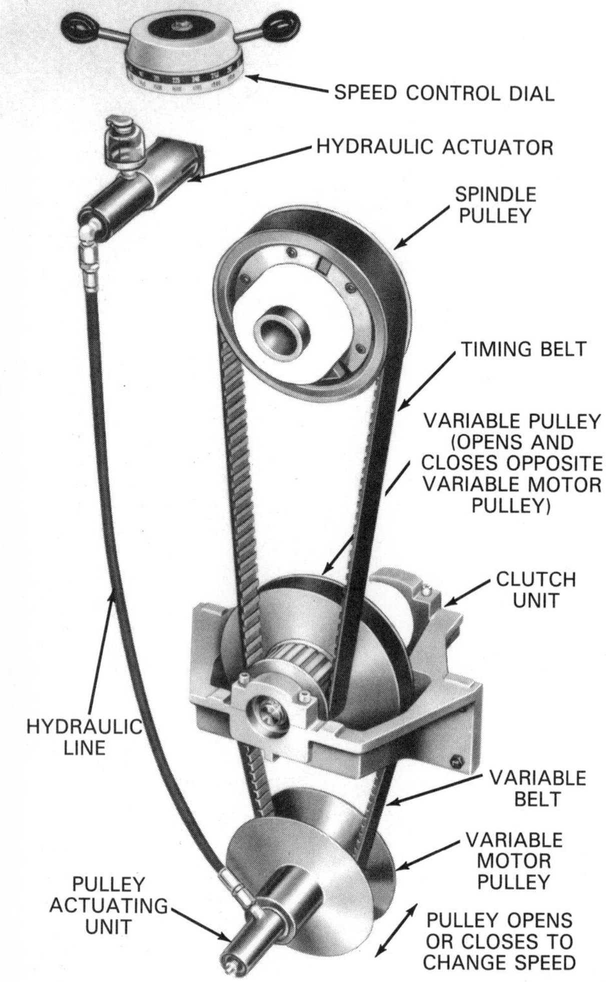

Fig. 29-14. This split pulley is hydraulically actuated from top of machine. It is used to control spindle speed on many lathes. (Clausing Industries Inc.)

Fig. 29-15. A fluidic unit, used to control a glass press, is being checked out by its designer. It replaces an electromechanical control system that required manual adjustment. Production has to be stopped while adjustments were made. Fluidic controls can be adjusted with press in operation. (Corning Glass Works)

Low-pressure streams of fluid are used to control, guide, and amplify the delivery of larger amounts of fluid energy which perform the required work, Fig. 29-16. The technique is roughly similar to the way large electric currents are controlled by smaller impulses of electric current inside of transistors, integrated circuits, relays, etc.

Advantages of fluidics

Fluidics can offer specific advantages over some electronic and mechanical type controls in many applications:

1. ***Simplicity***—fluidic devices have NO moving parts to wear out, jam, or break down. Refer to Fig. 29-17.
2. ***Ruggedness and reliability***—the physical ruggedness of fluidic devices assure their reliability under conditions of extreme vibration or shock. One machine tool application has already completed 500,000,000 cycles without failure.
3. ***Hazardous situations***—any explosive-proofing requirement can be satisfied. Fluidic devices can be employed in close proximity to explosive materials and highly combustible gases and liquids where electronic controls might cause dangerous sparks.
4. ***Nuclear applications***—electronic controls are affected by radiation factors. Fluidic devices function with reliability under high levels of pulsed or constant radiation.
5. ***Extremes in temperature***—the control units are reliable at temperatures as high as 1400 °F (760 °C). Electronic and other type controls fail at high temperatures.
6. ***Atmospheric conditions***—fluid devices are made of materials that are NOT affected by dust, dirt, humidity, or corrosive atmospheric conditions. Devices are usually made of glass and sometimes plastics or other material.

Fluidic controls have limitations. In some cases, they are more expensive than conventional electronic and hydraulic applications. Also, they do NOT act as fast as electronic systems. Fluid action is limited to the speed of sound. Electronic signals move at the speed of light.

How fluidics work

Fluidic control systems are those in which a fluid (liquid or a gas) flows through intricate and precisely made channels within a solid component or circuit, Fig. 29-18. The ***fluidic control system*** is used to perform logic, computation, and control functions

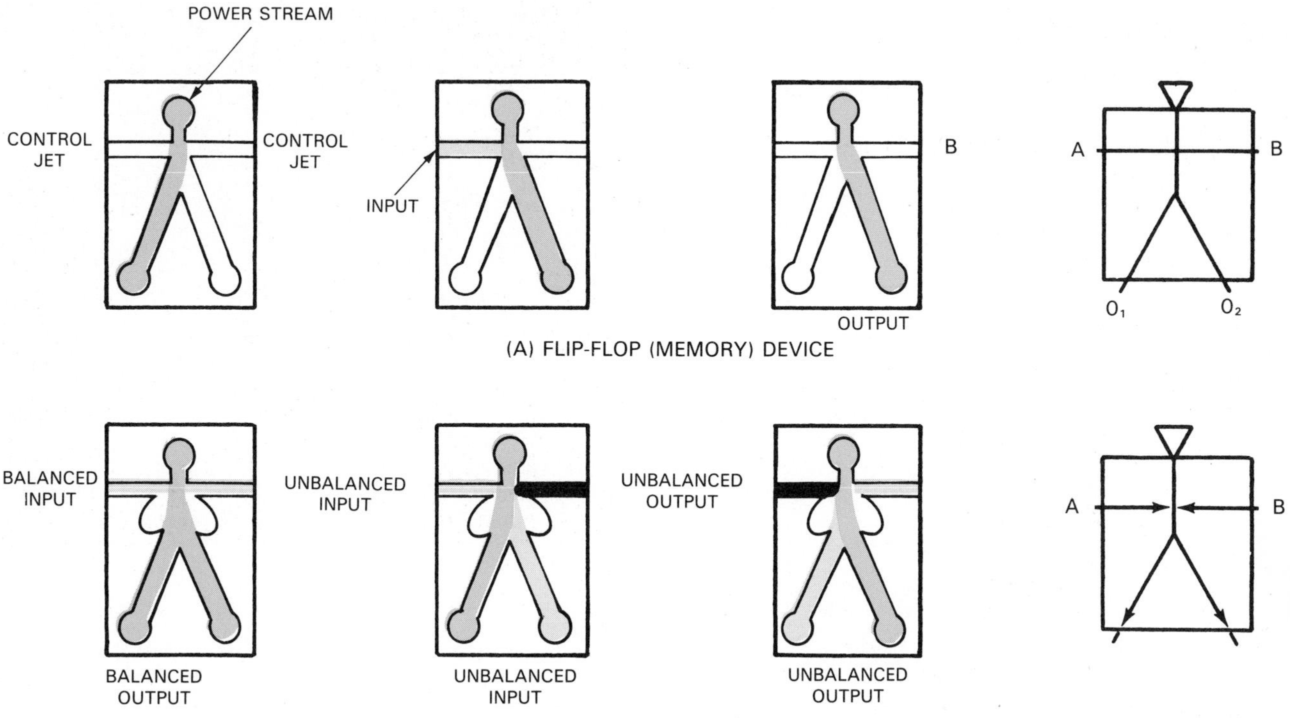

Fig. 29-16. Note action of two of the many types of fluidic controls used in machine control units. A—Flip-flop device. It works very much like an electrical switch used to turn machine on or off. A small input pressure flips power stream from one output leg to other. It will remain this way (memory) until a small input pressure is applied. B—Proportional amplifier. Power streams will split evenly if control streams are of equal pressure. However, if control jet streams are unbalanced, forces bend power stream and outputs are unequal. Output is proportional to control pressure applied. Device can be compared to accelerator on an automobile. When pressure is applied, machine speeds up in proportion to force applied to accelerator. Release pressure on accelerator and car slows down. Again, deceleration will be in proportion to forces removed.

for hydraulic and pneumatic equipment. All fluidic elements are classified as either active or passive. The ***active devices*** require energy to operate; the ***passive systems*** do NOT.

The different types of fluidic devices can be organized and connected together to perform complex tasks or steps in sequence, Fig. 29-19. This is similar to how electronic components can be connected into a circuit to complete a task, amplify radio waves to operate a radio speaker for example. Integrated fluidic circuits can also be used to do complex tasks, like control machine tool operation. See Fig. 29-20.

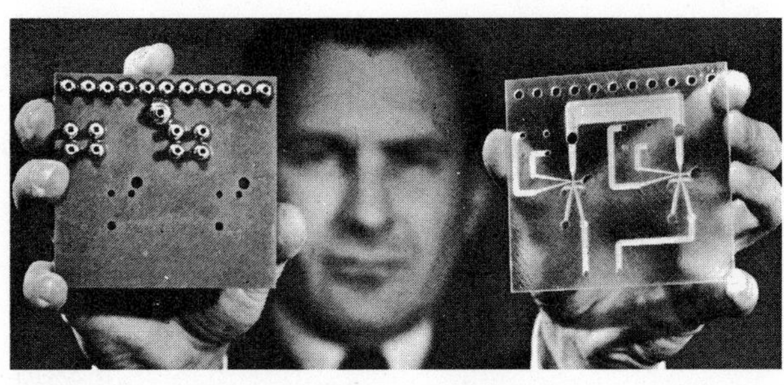

Fig. 29-17. Fluidic devices have no moving parts to fail. Glass section is bonded to back plate. Plastic hoses are attached to fittings and connected to other units for desired control functions. (Corning Glass Works)

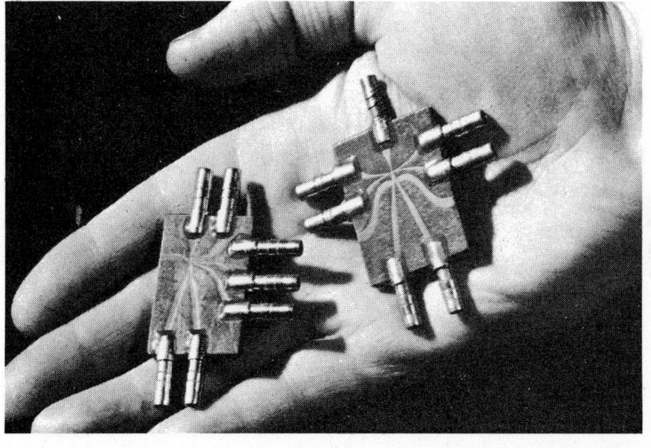

Fig. 29-18. Size of these fluidic devices can be determined by comparison with human hand. Note small intricate channels. (Corning Glass Works)

Fig. 29-19. This integrated fluid amplifier circuit is part of a machine control. It was made by stacking individual components and fusing them together. A similar block is shown broken open to illustrate various units that make up circuit. (Corning Glass Works)

Fig. 29-20. Fluidics are employed to control this combination flash and drilling machine. It removes flash from aluminum die castings and drills an internal passageway. Machine has three cylinders driving punches, and a high-speed drill with a built-in drive cylinder. (C.A. Norgren Co.)

TEST YOUR KNOWLEDGE—Chapter 29

Please do not write in the text. Place your answer on a separate sheet of paper.

1. Hydraulic power is applied to machine tools:
   a. Because of its rapid speed.
   b. Because of its smoothness of operation.
   c. Because of its quiet operation.
   d. All of the above.
   e. None of the above.
2. What does the term ''hydraulic'' mean?
3. List four advantages of hydraulic power.
4. Of what use is the slight decrease in volume when a liquid is compressed?
5. Pascal's Law states that ________ set up in a liquid acts ________ in ________ ________.
6. Prepare a sketch showing how the input force on a hydraulic mechanism can be increased.
7. Fluidics is the technology that makes use of ________________________________.
8. The word ''fluidics'' is a contraction of fluid and logic. The term fluid, as defined for this technology, means:
   a. A liquid.
   b. A gas.
   c. Either gaseous or liquid.
   d. All of the above.
   e. None of the above.
9. How does a schematic drawing of a hydraulic circuit differ from a conventional drawing of the same circuit?

RESEARCH AND DEVELOPMENT

1. Write to the manufacturers of hydraulically actuated machines used in your training area and request schematics and parts lists. Check the machines to determine whether the hydraulic fluid is as specified. Also check that all joints and packing meet specifications for safe and economic operation.
2. Demonstrate Pascal's Law. Use equipment borrowed from the science department.
3. Construct a ''bread-board'' installation of a typical hydraulic installation.
4. Prepare information on the science of fluidics. Prepare a bulletin board display around them.
5. Construct a hydraulic arbor press. Utilize a hydraulically activated automobile jack for the power source.

Proposed orbiting space station is presenting many challenges in its design and manufacture. No one is sure how materials will hold up for long periods of time in vacuum of space. Extreme cold will greatly affect strength of metals. Also, how will components machined at room temperature on earth fit when assembled in space where gravity does not act upon them? What other potential problems can you identify when the station is assembled? (NASA)

GLOSSARY OF TERMS

A AXIS: An angle defining rotary motion of a machine tool or slide around the X axis, such that a right-handed screw advanced in positive (+) A direction would be advanced in the positive (+) X direction.

ABRASIVE: A material that penetrates and cuts a material that is softer than itself. It may be natural (emery, corundum and diamonds) or artificial, (silicon carbide, aluminum oxide).

ABRASIVE ACCURACY: Accuracy as measured from a specified reference zero.

ABSOLUTE COORDINATE: Values of X, Y, and Z coordinates used in designating a point in space.

ABSOLUTE DIMENSION: A dimension expressed with regard to origin of a coordinate axis, but not necessarily coinciding with absolute zero point.

ABSOLUTE SYSTEM: A system in which all coordinate locations are measured from a fixed location on machine table or from an absolute zero point established by a programmer or machine designer.

ABSOLUTE ZERO: −459.69° Fahrenheit or −273.16° Centigrade.

ACCANDEC: A means of accelerating and decelerating feed rates to smooth starts and stops when an operation is under numerical control and when feed rate values are changed.

ACCEPTANCE TEST: A test to determine performance, capability, and conformity of software or hardware to design specifications.

ACCURACY: Conformity of an indicated value to a value accepted as a standard.

ACCURATE: Made within the tolerances allowed.

ACME THREAD: Similar in form to square thread in that top and bottom of thread is flat; however, sides have a 29 deg. included angle. The Acme thread is used for feed and adjusting screws on machine tools.

ACQUISITION: A function which obtains information from memory locations or data files for use in data manipulation or handling.

ACTIVE STORAGE: Data storage locations which hold data being transformed into motion.

ACUTE ANGLE: An angle of less than 90 deg.

ADAPTIVE CONTROL (AC): A system that automatically and continuously monitors on-line performance of a manufacturing operation by measuring one or more variables of activity, comparing measured quantities with other measured quantities, and adjusting or modifying activity to meet production standards.

ADDENDUM: That portion of gear tooth that projects above or outside of pitch circle.

ADDRESS: Means of identifying information or a location in a control system.

ALIGN: Adjusting to given points.

ALLOWANCE: Limits permitted for satisfactory performance of machined parts.

ALLOY: A mixture of two or more metals fused or melted together to form a new metal.

ALPHANUMERIC CODE: A code system consisting of characters, including numbers, letters, punctuation marks, and such signs as $, @, and #. Also referred to as alphameric code.

AMPLIFIER: A device or system which provides an output that is different in magnitude from control output.

ANALOG: Presentation of a variable by physical quantity in a defined relationship; for example, a dial indication of pressure.

ANGLE PLATE: A precisely made tool of cast iron, steel, or granite that is used to hold work in a vertical position for layout or machining. Faces are at right angles (90 deg.) and may have slotted openings for easier mounting of the work or clamping to the machine tool table.

ANNEALING: The process of heating metal to a given temperature (exact temperature and period temperature is held depends upon composition of metal being annealed) and cooling it slowly to remove stresses and induce softness.

ANODIZING: A process for applying an oxide coating to aluminum. It is done electrolytically in an acid solution with equipment similar to that used for electroplating. Technique can be varied to produce a light colored, porous coating that can be dyed in a variety of colors to a harder and nonporous coating for protection against corrosion.

ANSI: American National Standards Institute that helps standardize industry.

APRON: A covering plate or casting that encloses and protects a mechanism. Portion of lathe carriage that contains gears, clutches, and levers for moving carriage by hand and power feed.

ARBOR: A shaft or spindle for holding cutting tools.

ASSEMBLY: A unit fitted together from manufactured parts. A machine tool may comprise several assemblies.

AUTOMATICALLY PROGRAMMED TOOLS (APT): A computer-assisted program system describing parts illustrated on a design, in a sequence of statements, part geometry, cutter operations, and machine tool capabilties. Used for turning, point-to-point, and multiaxis milling.

AUTOMATIC SYSTEM FOR POSITIONING OF TOOLS (AUTOSPOT): A general-purpose computer program used in preparing instructions for NC positioning and straight-cut systems.

AUTOMATION: An industrial technique whereby mechanical labor and mechanical control are substituted for human labor and human control. Basically an extension and a refinement of mass production.

AXIS: Center line, real or imaginary, passing through an object about which it could rotate. A point of reference.

B AXIS: The angle defining rotating motion of a machine tool member or slide about the Y axis.

BACK GEARS: Gears fitted to belt driven machine tools to increase number of spindle speeds. Used to slow spindle speed of lathe for cutting threads, knurling, and for making heavy roughing cuts.

BACKLASH: Lost motion (play) in moving parts, such as thread in a nut or in teeth of meshing gears.

BATCH PROCESSING: A manufacturing operation in which a designated quantity of material is manufactured. Also, a method of processing a job so that each is completed before next job is started.

BED: One of the principal parts of a machine tool. It contains ways or bearing surfaces that support and guide work or cutting tool.

BERYLLIUM: A metal that weighs almost 80 percent less than steel, yet offers virtually equal strength characteristics. It is easy to machine but is brittle. Used in missiles and aircraft where weight is critical and in nuclear reactors. One of the "exotic" metals.

BEVEL: Angle formed by a line or a surface that is not at right angles to another line or surface.

BINARY CODE: A code in which each allowable position has one of two possible states. It can be expressed as 1 (one) or 0 (zero), on or off, etc.

BISTABLE: Elements which have two output possibilities, and will hold a given condition until switched.

BIT: An abbreviation of "binary digit." A single character of a language employing exactly two distinct kinds of characters.

BLANKING: A stamping operation in which a die is used to shear or cut a desired shape from flat sheets or strips of metal.

BLOCK: Words grouped together as a unit to provide complete information for a cutting operation. One or more rows of punched holes in a tape separated from other words by an end of block character.

BLOWHOLE: A hole produced in a casting when gases are entrapped during the pouring operation.

BOMB OUT: Complete failure of a computer routine resulting in need to restart or reprogram computer.

BRAZING: Joining metals by fusion of nonferrous alloys that have melting temperatures above 800 °F but lower than the metals being joined.

BRITTLENESS: In some respects, opposite of toughness. Characteristics that cause metal to break easily.

BUFFING: Process of bringing out luster of metal. Buffing is accomplished by using cloth wheels, usually cotton or muslin discs sewed together, and a tripoli compound. Proper wheel speed depends on size of wheel.

BUG: A flaw or defect in a program code or in design of computer rendering program incapable of performing objectives for which it was written.

BURNISHING: Process of finishing a metal surface by compressing its surface. Often done by tumbling work with steel balls.

BURR: Sharp edge remaining on metal after cutting, stamping, or machining. Burr can be dangerous if not removed.

BUSHING: A bearing for a revolving shaft. A hardened steel tube used on jigs to guide drills and reamers.

BYTE: A sequence of binary digits operated upon as a single unit. A byte may be composed of 8, 12, or 16 binary digits depending upon system.

C AXIS: An angle defining rotary motion of a machine tool part or slide around the Z axis.

CALIBRATION: Adjustment of a device so that output is within a designated tolerance for specific input values.

CAM: A rotating or sliding element that, because of curvature of its driving surface, imparts complicated motions to followers or driven elements of machine tool.

CANNED CYCLE: In NC, a set of operations preset in hardware or software and initiated by a single command. Several operations are performed in a predetermined sequence; function ends with a return to beginning condition.

CARBURIZING: A process that introduces carbon to surface of steel by heating metal below its melting temperature in contact with carbonaceous solids, liquids, or gases and holding at that temperature for a predetermined time. Piece is then quenched.

CASE HARDENING: A process of surface hardening iron base alloys so that surface layer or case is made substantially harder than interior or core. Typical case hardening processes are carburizing, cyaniding, and nitriding.

CASTING: An object made by pouring molten metal into a mold.

CAT HEAD: A sleeve or collar which fits over out-of-round or irregular shaped work permitting it to be supported in a steady rest. The work is centered in cat head by using adjusting screws located around its circumference.

CATHODE RAY TUBE (CRT): A screen resembling a television display upon which computer generated and textual data is displayed.

CELL: Minimum unit in flexible manufacturing concept. Usually, it is comprised of one or two machines with computer control and automated work handling equipment.

CENTER, DEAD: A stationary center.

CENTER LINE: A line used to indicate an axis of a symmetrical part. Center line consists of a series of long and short dashes.

CENTER, LIVE: A rotating center.

CENTRAL PROCESSING UNIT (CPU): Heart of computer that handles memory and computational functions.

CENTRIFUGAL CASTING: A casting technique in which mold is rotated during pouring and solidification of metal. It produces a casting with certain desirable characteristics.

CERMETS: A combination of ceramics and metals that is finding increased use for high temperature applications. Resistance to high temperatures and wear indicates great promise as a super high-speed cutting tool.

CHAD: Pieces of material that are removed when punching holes in tape.

CHANNEL: Paths parallel to edge of punch tape in which holes are located. Standard NC tape has eight channels. They are also known as tracks or levels.

CHARACTER: One of a set of elementary marks or events which may be combined to express information.

CHASER: A thread cutting tool that fits into a die head used on a turret lathe or screw machine. Usually a hardened steel plate with several teeth of correct pitch cut into it. Three or four chasers are used in a die head.

CHASING THREADS: Cutting threads on a machine tool.

CHATTER: Vibrations caused by cutting tool springing away from work. It produces small ridges on machined surface.

CHEMICAL MILLING: Controlled removal of metal by chemicals rather than by conventional machining methods.

CHIP: A small piece of semiconductor material on which electronic components are formed. Integrated circuits are formed on chips.

CHIP BREAKER: A small groove ground on top of cutting tool, near cutting edge, to break chips into small sections.

CHUCK: A device to hold work or cutting tools on a machine tool.

CIRCULAR PITCH: Distance from center of one gear tooth to center of next tooth measured on pitch circle.

CLEARANCE: Distance by which one object clears another object.

CLIMB MILLING: Feeding work into milling cutter in same direction it rotates.

CLOCKWISE: From left to right, in a circular motion. The direction clock hands move.

CLOSED LOOP SYSTEM: A system in which a reference signal is compared with a position signal generated by a monitoring unit on machine tool (feedback), difference is used to adjust machine tool to reduce difference to zero. An imaginary loop is formed by data flow.

CODE: A system of organized symbols (bits) representing information in a language that can be understood and handled by a control system.

COINING: Process that impresses image or characters on die and punch onto a plain metal surface.

COLD HEADING: An operation in which metal is worked cold.

COLOR HARDEN: A hardening technique usually done for appearance only.

COLOR TEMPER: Using color range steel passes through when heated to determine proper degree of hardness.

COMMAND: A signal from a machine control unit (MCU) initiating one step in a complete program.

COMPUTER AIDED DESIGN (CAD): Use of computers to aid in designing a product.

COMPUTER AIDED MANUFACTURING (CAM): Use of computers to aid in various phases of manufacturing.

COMPUTER GRAPHICS: Graphs, charts, and/or drawings generated by a computer. They are displayed on a video screen or printed by a plotter or printer.

COMPUTER INTEGRATED MANUFACTURING SYSTEM (CIMS): A multimachine manufacturing complex linked by a material handling system and includes such features as toolchangers and load/unload stations. Under control of a computer, various workpieces are introduced into system, then randomly and simultaneously transported to NC machine tools and other processing stations.

CONCAVE SURFACE: A curved depression in surface of an object.

CONCENTRIC: Having a common center.

CONE PULLEY: A one-piece pulley having two or more diameters.

CONTINUOUS CASTING: A casting technique in which ingot is continuously solidified while it is being poured. Length of casting is not determined by mold dimensions.

CONTINUOUS PATH OPERATION: An operation in which rate and direction of relative movement of machine members is under continuous control so that machine travels through designated path at a specified rate without pausing.

CONTOUR: Outline of an object.

CONTOUR CONTROL SYSTEM: A system in which cutting path can result from coordinated, simultaneous motion of two or more axes.

CONTOURING CONTROL SYSTEM: An NC system that generates a contour by controlling a machine or cutting tool in a path resulting from coordinated, simultaneous motion of two or more axes.

CONTROL: A signal received at system input; used as intelligence to produce a modification in output.

CONTROLLER: A device through which commands are introduced and manipulated to compute, encode, and store data, produce readouts and process computation and output. In NC, also known as a machine control unit (MCU).

CONVENTIONAL: Not original, customary, or traditional.

CONVEX SURFACE: A rounded surface on an object.

COOLANT: A fluid or gas used to cool cutting edge of a tool to prevent it from burning up during machining operation.

CORE: A body of sand or other material that is formed to a desired shape and placed in a mold to produce a cavity or opening in a casting.

COUNTERBORE: Enlarging a hole to a given depth and diameter.

COUNTERCLOCKWISE: From right to left in a circular motion.

COUNTERSINK: Chamfering a hole to receive a flathead screw.

CRYOGENICS: Study and development of extremely low temperature processes, techniques, and equipment.

CURSOR: Movable pointer used by a CRT operator to indicate where entries or actions are to take place.

CUTTER COMPENSATION: A feature on certain NC machines enabling operator to incrementally adjust in a direction normal to programmed tool path for changes in cutter radius, length, or deflection.

CUTTER OFFSET: 1. Difference between a part surface and axial center of a cutter or cutter path during a machining operation. 2. An NC feature enabling a machine operator to use an oversize or undersize cutter.

CUTTING FLUID: A liquid used to cool and lubricate cutting tool to improve quality of surface finish.

CYANIDING: A process of case hardening a ferrous alloy by heating in molten cyanide causing metal to absorb carbon.

CYCLE: 1. A sequence of operations repeated regularly. 2. Time necessary for one sequence of operations to occur.

DATA: Information that is input to a computer system and is then processed so that it can be output in a sensible form. It usually consists of numbers, letters, or symbols that refer to or describe an object, idea, condition, situation, or other type of information.

DEAD BAND: Range through which an input can be varied without initiating response.

DEBUG: In NC, process of detecting, locating, and removing software errors and hardware problems causing malfunctions in a computer.

DECARBURIZING: Process of removing carbon from metals.

DECELERATION DISTANCE: Calculated distance for decreasing speed of an axis of motion to avoid overshooting a position.

DEDENDUM: Portion of gear tooth between pitch circle and root circle, and is equal to addendum plus clearance.

DEMAGNETIZING: Removal of magnetism from a piece held in a magnet chuck.

DIE: A tool used to cut external threads. Also, a tool used to impart a desired shape to a piece of metal.

DIE CASTING: A method of casting metal under pressure by injecting it into metal dies of a die casting machine.

DIE CAVITY: A hollow space inside a die where metal solidifies to form a casting.

DIE CHASERS: See CHASER.

DIE STOCK: Handle for holding a threading die while rotating die over workpiece.

DIGITIZING: Process of converting a scaled, but non-mathematical, drawing into coordinate NC locations for programming purposes.

DIRECT NUMERICAL CONTROL (DNC): Use of a shared computer to program, service, and log a process such as a machine tool cutting operation. Part program data is distributed via data lines to machine tools.

DISK: A random access storage component of a computer to store NC programs or other data.

DIVIDING HEAD: A machine tool attachment for accurate spacing of holes, slots, gear teeth, and flutes. When geared to table lead screw, it can be used to machine spirals.

DOG: A projecting piece on side of a machine tool worktable to trip automatic feed mechanism off or for reverse travel.

DOG, LATHE: A device for clamping work so that it can be machined between centers.

DRAFT: Clearance on a pattern that allows easy withdrawal of pattern from mold.

DRIFT: A tapered piece of flat steel used to separate tapered shank tools from sleeves, sockets, or machine tool spindles.

DRILLING: Cutting round holes by use of a cutting tool sharpened on its point.

DRILL ROD: A carbon steel rod accurately and smoothly ground to size. Available in a large range of sizes.

DRIVE FIT: Using force or pressure to fit two pieces together. One of several classes of fits.

DROP FORGING: A forming operation, usually done under impact, that compresses metal in dies designed to produce desired shape.

DUCTILITY: Property of a metal that permits permanent deformation by hammering, rolling, and drawing without breaking or fracturing.

DUMP: To remove all or part of contents of a computer storage device. Also, printed copy resulting from operation.

DWELL: A timed delay of programmed duration.

ECCENTRIC: Not on a common center. A device that converts rotary motion into a reciprocating (back and forth) motion.

EDIT: To modify format of a program or to alter data output or input by inserting or deleting characters.

ELECTROPLATING: A plating process accomplished by passing an electric current from an anode (usually made of plating material) to work, through an electrolyte containing salts of plating metal in solution.

EMERY: A natural abrasive.

END OF BLOCK (EOB): A character that represents end of a line or block of information in an NC program.

EXPANSION FIT: Reverse of shrink fit. Piece to be fitted is placed in liquid nitrogen or dry ice until it shrinks enough to fit into mating piece. Interference develops between fitted pieces as cooled piece expands.

EZY-OUT: A tool for removing broken bolts and studs from a hole. It is made in several sizes.

FACE: To make a flat surface by machining.

FACEPLATE: A circular plate that fits to headstock spindle and drives or carries work to be machined.

FAMILY OF PARTS: Grouping of workpieces into logical families so they can be produced by same group of machines, tooling, and people with only minor changes of procedure or setup.

FATIGUE: Tendency for metal to break or fracture under repeated or fluctuating stresses.

FEEDBACK: Information returned from output of a machine or process for use as input in subsequent operations.

FERROUS: Denotes family of metals in which iron is major ingredient.

FILLET: Curved surface that connects two surfaces that form an angle.

FIT: Clearance or interference between two mating parts. There are several classes of fits.

FIXTURE: A device for holding work in a machine tool. IT DOES NOT GUIDE CUTTING TOOL.

FLAME HARDEN: A method of surface hardening steel by rapidly heating surface with flame of an oxyacetylene torch, and quenching.

FLASH: A thin fin of metal formed at parting line of a forging or casting where a small portion of metal is forced out between edges of die.

FLASK: A wooden or metal form consisting of a cope (top portion) and a drag (bottom portion) used to hold sand that forms mold.

FLEXIBLE MANUFACTURING SYSTEM (FMS): Comprised of a series of work stations and storage places linked by an automated material handling system to provide a manufacturing capability for a selected range of parts under computer control.

FLIP-FLOP: A bistable subsystem that will alternate from one output to another upon receipt of correctly phased input signals.

FLOPPY DISK: A device similar to a 45 rpm record and made of a flexible material. It is used for storing computer data.

FLOTURN PROCESS: Another term for shear spinning.

FLOW: A quantity passing a point per unit of time. In pneumatics, this is commonly represented by CFM, or cubic feet per minute.

FLUIDICS: Control systems wherein a liquid or gas flows through intricate and precise channels, within a solid component or circuit, to perform logic, computation and control functions for hydraulic and pneumatic equipment.

FLUORESCENT PENETRANT INSPECTION: A nondestructive testing technique. An oil base penetrant is sprayed on work and is drawn into every crack and flaw. Surface is rinsed with solvent to remove excess penetrant. After developing, the surface is viewed under a ''black light.'' Defects glow with fluorescent brilliance.

FLUTE: A groove machined in a cutting tool to facilitate easy chip removal and to permit cutting fluid to reach the cutting point.

FLUX: Fusible material used in brazing and welding to dissolve and facilitate removal of oxides and other undesirable substances.

FLY CUTTER: A single point tool fitted in an arbor. Inexpensive to make, but is relatively inefficient because only one point does the cutting.

FORCE FIT: Interference between to mating parts is sufficient to require force to press pieces together. Parts are considered permanently assembled.

FORGE: To form metal with heat and/or pressure.

FORMING: Operations or steps necessary to shape metal to a desired form. Change does not intentionally change thickness of metal.

FORTRAN: Acronym for Formula Translation, a universal computing language in which many NC programs are written.

FREE FIT: Used when tolerances are liberal. Clearance is sufficient to permit a shaft to run freely without binding or overheating when properly lubricated.

G FUNCTION: A preparatory code on a program tape to indicate a special function. The letter ''g'' is actually a lower case letter when shown on a printout. For example: g90 indicates that all following coordinate locations are expressed in absolute dimensions. Function can only be used on machines having the capabilties called for by function.

GAGE: A tool used for checking metal parts to determine whether they are within specified limits.

GANG MILLING: Using two or more milling cutters to machine several surfaces at one time.

GARBAGE: Erroneous, faulty, unwanted, or extraneous data in a computer or NC program.

GATE: 1. Point where molten metal enters mold cavity. 2. In FLUIDICS, a device or circuit which allows passage of a signal only if certain control requirements have been satisfied.

GEARS: Toothed wheels that transmit rotary motion from one shaft to another shaft without slippage.

GIB: A wedge-shaped strip that can be adjusted to maintain a proper fit of movable surface of a machine tool.

GRADUATE: To divide into equal parts by engraving or cutting lines or graduations into metal.

GRADUATIONS: Lines that indicate points of measurement on measuring tools and machine dials.

GUERIN PROCESS: A method of forming metal sheet in which metal is forced to conform to shape of a male die by application of force to a confined rubber pad.

H CODE: A code designated by prefix H followed by two digits. It calls out a specific fixture offset value which may be stated for one, two, or three axes.

HALF NUTS: Mechanism that locks lathe carriage to lead screw for purpose of cutting threads.

HARDENING: Heating and quenching of certain iron-base alloys for purpose of producing a hardness superior to that of untreated material.

HARDNESS TESTING: Techniques used to determine degree of hardness of heat-treated material.

HEAT TREATMENT: Careful application of a combination of heating and cooling cycles to a metal or alloy in solid state to bring about certain desirable conditions, such as hardness and toughness.

HELICAL GEARS: Gears with teeth cut at some angle, other than at right angles, to gear face permitting two or more teeth to be engaged at all times. Their operation is smoother and not as noisy as operation of spur gears.

HELIX: Path a point generates as it moves at a fixed rate of advance on surface of a cylinder, such as screw threads or flutes on a twist drill.

HIGH ENERGY RATE FORMING: A metal forming technique involving release of a source of high energy such as explosives, electrical or pneumatic-mechanical.

HOB: A special type gear cutter designed to cut gear teeth on a continuous basis.

HOBBING: Cutting gear teeth with a hob. Gear blank and hob rotate together as in mesh during cutting operation.

HOME POSITION: Fixed location in basic coordinate axes of machine tool. Usually, point in work process in which tools are fully retracted permitting any necessary changes.

HONING: A process used to produce an extremely fine surface finish on an object after grinding operation. Honing permits a closer fit on critical parts. Abrasive blocks are forced against work surface under very light spring pressure in a rotary motion and at same time moved back and forth. The area is flooded with cutting fluid. Honing is an expensive operation.

HYDROFORM: A method of forming parts in rubber under accurately controlled fluid pressure. Metal is formed over a movable male die in a flexible diaphragm.

HYDROSPIN: Another name for shear spinning.

ID: Abbreviation for inside diameter.

IDLER GEAR: A gear or gears placed between two other gears to transfer motion from one to other without changing direction of rotation or ratio.

IMPEDANCE: A total opposition to circuit flow, represented by resistance, capacitance, and inductance, combined to a resultant.

INCREMENTAL SYSTEM: Programming where each coordinate location is given in terms of distance and direction along rectangular axes from previous location and not from a fixed zero location.

INDEPENDENT CHUCK: A chuck in which each jaw can be moved independently of other jaws.

INDEXING: Term used to describe correct spacing of holes, slots, etc., on periphery of a cylindrical piece using a dividing or indexing head.

INDICATOR: A sensitive instrument capable of measuring slight variations when testing trueness of work, machines or machine attachments.

INPUT: The actual transfer by an appropriate medium of information into a computer or machine control unit.

INSERTED TOOTH CUTTER: A milling cutter with teeth that can be replaced when they become damaged or worn rather than replacing entire cutter.

INSPECTION: Measuring and checking of finished parts to determine whether they have been made to specifications.

INTEGRATED CIRCUITS (ICs): A very small single structure assembly of electronic components containing many circuits and functions.

INTERCHANGEABLE: Refers to a part that has been made to specific dimensions and tolerances and is capable of being fitted in a mechanism in place of a similarly made part.

INTERFACE: Functioning connection between two elements or components within a system.

INTERPOLATION: Method of determining moves between two or three coordinate points.

INVESTMENT CASTING: A process that involves making a wax, plastic, or even a frozen mercury pattern, surrounding it with a wet refractory material, melting or burning pattern after investment material has dried and set, and finally pouring metal (usually under air or centrifugal pressure) into cavity.

JARNO TAPER: A standard taper of 0.600 in. per foot. Used on machine tools.

JIG: A device that holds work in position and positions and guides cutting tool.

JO-BLOCK: Precisely made steel blocks used by industry as a standard of measurement. They are made in a range of sizes and with a dimensional accuracy of ±0.000002 (two millionths) inch, with a flatness and parallelism of ±0.000003 (three millionths) inch.

KEY: A small piece of metal imbedded partially in shaft and partially in hub to prevent rotation of gear or pulley on shaft.

KEYWAY: Slot or recess in shaft that holds key.

KNEE: Unit that supports saddle and table of a column and knee-type milling machine.

KNURLING: Operation that presses grooved, hardened steel wheels (knurls) into surface of cylindrical work, usually rotating on a lathe, to product rows of uniformly spaced serrations by deforming material which provide a better grip, or for decorative purposes.

LAND: Metal left between flutes or grooves in drills, reamers, taps, and other cutting tools.

LAPPING: Process of finishing surfaces with a very fine abrasive, like diamond dust or abrasive flours.

LARD OIL: A cutting oil made from animal fats. It is often mixed with mineral oils that improve lubricating qualities.

LAY OUT: To locate and scribe points for machining and forming operations.

LEAD: Distance a nut will advance on a screw in one revolution.

LEAD SCREW: Long precision screw on front of lathe bed that is geared to spindle to transmit motion to carriage for thread cutting.

LONGITUDINAL MOVEMENT: Lengthwise movement.

M FUNCTION: A miscellaneous function designated by letter M followed by a two digit number that calls out functions such as coolant on and off, spindle rotation direction, etc., in an NC workpiece program.

MACHINABILITY: Characteristic of a material that describes ease or difficulty of machining it.

MACHINABILITY INDEX: Table that indicates degree of ease or difficulty of machining a material. It is based on the machining characteristics of a common steel (ANSI B1112 = 100). Magnesium alloy (Machinability Index = 500 to 2000) is relatively easy to machine. Tool steel (Machinability Index = 34) is difficult to machine.

MACHINE CONTROL UNIT (MCU): Arrangement of electronics, input and output devices that comprise control which reads a workpiece program and generates signal necessary to direct an NC machine tool.

MACHINE TOOL: Name given to that class of machines which, taken as a group, can reproduce themselves.

MACHINIST: A person who is skilled in use of machine tools and is capable of making complex machine setups.

MAGNAFLUX: A nondestructive inspection technique that makes use of a magnetic field and magnetic particles to locate flaws in materials.

MAGNAGLO: See FLUORESCENT PENETRANT INSPECTION. For use on magnetic materials only.

MAGNETIC CHUCK: A work-holding device that uses magnetic fields to hold work for machining (grinding).

MAJOR DIAMETER: Largest diameter of a thread measured perpendicular to axis.

MALLEABILITY: Property of metal that determines its ease in being shaped when subjected to mechanical working (forging, rolling, etc.).

MANDREL: A slightly tapered, hardened steel shaft that supports work that cannot be held by any other method for machining between centers.

MANUAL DATA INPUT (MDI): Feature of a machine control unit that allows a programmer or operator to enter data directly to unit rather than through outside storage medium such as punch tape, disk, or computer memory.

MARFORM: A drawing process that forms metal sheet by using a movable steel punch and a rubber headed ram.

MATCH PLATE: Production type of pattern equipment usually made of metal. Consists of a plate on each side of which is mounted matching halves of pattern.

MENU: An arrangement of options presented to user of design or controlling equipment from which one chooses possibilities to complete a task.

MESH: To engage gears to a working contact.

MICROSECOND: One millionth of a second.

MILL: To remove metal with a rotating cutter on a milling machine.

MILLING MACHINE: A machine that removes metal from work by means of a rotary cutter.

MILLISECOND: One thousandth of a second.

MINOR DIAMETER: Smallest diameter of a screw thread measured across roots and perpendicular to axis. Also known as "root diameter."

MITER GEARS: Right angle bevel gears having same number of teeth. Used to transmit power through shafts at right angles to each other.

MORSE TAPER: A standard taper of approximately 5/8 in. per foot. Used on lathe centers, drill shanks, etc.

MUSIC WIRE: A carbon steel wire used to manufacture springs.

NANOSECOND: One billionth of a second.

NC: Abbreviation for National Coarse series of screw threads.

NECKING: Machining a groove around a cylindrical shaft.

NF: Abbreviation for the National Fine series of screw threads.

NITRIDING: A case hardening technique in which a ferrous alloy is heated in an atmosphere of ammonia or in contact with a nitrogenous material to produce surface hardness by absorption of nitrogen. Quenching is not necessary.

NONFERROUS: Metals containing no iron.

NORMALIZING: A process in which ferrous alloys are heated to approximately 100 °F above critical temperature range and cooled slowly in still air at room temperature to relieve stresses that may have developed during machining, welding or forming operations.

NUMERICAL CONTROL (NC): A machine control system is which actions are controlled by direct insertion of numerical data. Data is automatically interpreted.

NUMERICAL CONTROL SYSTEM: A system in which actions are controlled by direct insertion of NUMERICAL DATA at some point. The system must automatically interpret this data.

NUMERICAL DATA: Data in which information is expressed by a set of numbers or symbols.

OBTUSE ANGLE: An angle of more than 90 deg.

OD: Abbreviation for outside diameter.

OFF-CENTER: Eccentric, not accurate.

OIL HARDENING: Using a mineral oil as a quenching medium in heat treatment or surface hardening of certain alloys.

OPEN LOOP SYSTEM: A control system that has no means for comparing output with input for machine control purposes.

OUT-OF-TRUE: Not on center, eccentric, out of alignment.

PEENING: An operation that involves mechanical working of metal by means of hammer-like blows.

PERMANENT MOLD: A mold ordinarily made of metal that is used for repeated production of similar castings.

PICKLING: A technique employed in removal of stains and oxide scales from metal surfaces by immersion in acid baths.

PINION: Smaller of two mating gears.

PITCH: Distance from a point on one thread to a corresponding point on next thread.

PITCH DIAMETER: Diameter of an imaginary cylinder that would pass through threads at such points as to make width of thread and width of space equal at point where they are cut by cylinder. It is equal to major diameter of thread minus depth of one thread. In gearing, diameter of pitch circle, which is an imaginary circle located at about midpoint on the teeth, where teeth of both gears contact each other.

PLASTER MOLD CASTING: A casting process which uses plaster molds in place of sand molds. Castings produced in a plaster mold have a much better surface finish than those cast in sand. Used primarily with aluminum.

POSITION CONTROL SYSTEM: A positioning system in which controlled motion is required only to reach a given end point, with no path control during the transition from one point to the next. Also known as POINT-TO-POINT SYSTEM.

POSITION SENSOR, also POSITION TRANSDUCER: A device for measuring a position, and converting this measurement into a form convenient for transmission.

PRECISION: Closeness of agreement among repeated measurements of same characteristic by same method under same conditions. (ASA C85)

PREPARATORY FUNCTION: A command changing mode of operation of control such as from positioning to contouring or calling for a repeat cycle of machine.

PRESS FIT: A class of fit where interference between mating parts is sufficient to require force to press pieces together. Assembly is considered permanent.

PRINTOUT: A printed sheet giving all data of a program that has either been manually or computer processed. Printout is used for reference and visual checking.

PROGRAMMER: In NC work, there are two important types of programmers. A workpiece programmer writes instructions for computer to act upon to develop a specific program tape. A computer programmer develops routines that give computer basic intelligence to act upon instructions that have been prepared by a workpiece programmer.

PYROMETER: A device for measuring high temperatures. Temperatures are determined by measuring electric current generated in thermocouple as it heats up.

QUANTUM: Numerical value of smallest unit of measure used in a numerical control system.

QUENCHING: Process of rapid cooling from an elevated temperature by contact with fluids or gases.

QUICK RETURN: Mechanism on some machine tools that can be engaged to return worktable rapidly to its starting point during noncutting cycle.

QUILL: Steel tube in head of some machine tools that encloses bearings and rotating spindle on which are mounted cutting tools. It is geared to a handwheel and/or lever that is used to raise or lower rotating cutting tool on work surface. Quill can be locked in position.

RACK: A flat strip with teeth designed to mesh with teeth on a gear. Used to change rotary motion to reciprocating motion.

RAM: Part of shaper that moves back and forth and carries cutting tool.

RAPID TRAVERSE: Used to bring work on milling table rapidly into cutting position and return table quickly to starting position for next cut.

RAW STORAGE: Storage media, such as drawings, containing all information (geometry of part, cutter size, feeds and speeds, etc.) necessary for preparing tape.

READOUT: A visual display of data as it is taking place.

REAM: To finish a drilled hole to exact size with a reamer.

REAMER: A cutting tool used to produce a smooth, accurate hole by removing a small amount of metal from a drilled hole.

RELIEF: An undercut of offset surface to provide clearance.

RIGHT ANGLE: An angle of 90 deg.

RISER: A reservoir of molten metal provided to compensate for contraction of cast metals as they solidify.

ROLLING: A process of forming and shaping metal by passing it through a series of driven rollers.

ROOT DIAMETER: Smallest or "minor diameter" of a thread.

ROTARY TABLE: A milling attachment that gives a rotary motion to piece. It consists of a circular worktable rotated by a handwheel through a worm and worm gear. Hub of handwheel is graduated in degrees permitting precise spacing of holes, slots, grooves, etc., around piece.

ROUGHING: Rapid removal of stock without regard for quality of surface finish.

ROW: A path perpendicular to edge of a tape along which information may be stored by presence or absence of holes or magnetized areas.

RUNNER: Channel of a gating system through which molten metal flows from sprue to casting and risers.

SAE: Abbreviation for Society of Automotive Engineers.

SAFE EDGE: Edge of a file on which no teeth have been cut.

SANDBLAST: Process of cleaning castings and metals by blowing sand at them under very high air pressure.

SAND MOLD CASTING: A process which involves pouring molten metal into a cavity that has been formed in a sand mold.

SCALE: Surface oxidation caused on metals by heating them in air.

SCRAPING: Process of removing an exceedingly small portion of wearing surfaces of machinery by means of scrapers to bring surfaces to a precision fit and finish not attainable by ordinary filing techniques.

SCRIBE: To draw a line with a scriber or other sharp pointed tool.

SEQUENCE NUMBER: A number identifying relative location of blocks or groups of blocks on a tape.

SERVOMECHANISM: An automatic reset device for controlling large amounts of power by means of very small amounts of power and automatically correcting performance of a mechanism.

SETOVER: Distance a lathe tailstock has been offset from normal center line of machine. Used in one method of taper turning.

SET UP: Term used to describe positioning of workpiece, attachments and cutting tools on machine tool.

SHEAR SPINNING: A process whereby a metal blank, which may be flat or preformed to some shape, is clamped between tailstock of shear spinning machine and a power driven spinning mandrel. Mandrel is same shape and size as inside of finished part, and metal is forced to flow onto mandrel by action of forming rollers.

SHIM: Pieces of sheet metal, available in many thicknesses, that are used between mating parts to provide proper clearance.

SHRINK FIT: A fit in which outer member is expanded by heating to permit insertion of inner member and a tight fit is obtained as outer member shrinks as it cools. A very tight fit is made and must be considered permanent.

SINE SPINNING: Another name for shear spinning.

SINTERING: A method of bonding metal powders that have been compacted by heating them to a predetermined temperature.

SOFTWARE: All of the program manuscripts, tapes, decks of tabulating cards, methods sheets, flow charts, and other programming documentation associated with computers and NC.

SOLDERING: A method of joining metals by means of a nonferrous filler metal, without fusion of base metals, and is normally carried out at temperatures lower than 800 °F (427 °C).

SPINNING: A process that makes a sheet metal disc into hollow shape by pressing tool against it and forcing it against a rotating form (chuck).

SPLINE: A series of grooves, cut lengthwise, around a shaft or hole.

SPOTFACE: To machine a circular spot on surface of a part to furnish a flat bearing surface for head of a bolt or nut.

SPRUE HOLE: Opening in a mold into which molten metal is poured.

SPUR GEAR: A gear having straight teeth that are cut parallel to direction of rotation. Most commonly used gears.

STAKING: Joining of two parts by upsetting metal at their junction.

STANDARD: An accepted base for a uniform system of measurement and quality.

STELLITE: An alloy of cobalt, chromium, and tungsten used to make high-speed cutting tools.

STOPS: Projections on side of worktable to disengage automatic power feed. Also known as a dog.

STRADDLE MILL: Using two or more milling cutters to perform several milling operations simultaneously.

STRAIGHT-CUT CONTROL SYSTEM: System in which controlled cutting action occurs only along a path parallel to linear, circular, or other machine ways.

STRAIGHTEDGE: A precision tool for checking accuracy of flat surfaces.

STRAIN: A measure of change in shape or size of a body, compared to original shape or size.

STRESS: Intensity, at a point in a body, of internal forces.

STRETCH FORM: A process of forming parts and shapes with large curvatures by stretching sheet over a form of desired shape.

SUPER FINISH: A surface finish where surface irregularities have been reduced to a few millionths of an inch to produce an exceptionally smooth and long-wearing surface.

SURFACE PLATE: A plate of iron or granite that has been ground, and sometimes lapped, to a smooth flat surface. It is used to give a base for layout measurements and inspection.

SURFACE ROUGNESS SCALE: A series of small plates visualizing degree of roughness for a particular surface. They establish a standard permitting a machinist or an inspector to compare specified finishes visually and by feel.

T FUNCTION: A code identifying a tool select command on a program tape. As with m and g codes, the t also appears as a lower case letter on a printout.

TANG: Flats or tongue machined on end of tapered shanks. Tang fits into a slot in mating part and prevents taper from rotating in mating part.

TANTALUM: A metal that is capable of withstanding temperatures in 2500-4000 °F (1 372-2 206 °C) range. A metal finding more and more use in space age.

TAP: Tool used to cut internal threads.

TAPER: Piece that increases or decreases in size at uniform rate to assume wedge or conical shape.

TAPPING: Operation of producing internal threads with a tap. May be done by hand or machine. Also, process of removing molten metal from a furnace.

TEMPERING: A sequence in heat-treating consisting of reheating quench hardened or normalized parts to a temperature below transformation range and holding it for a sufficient time to produce desired properties.

TEMPLATE: A pattern or guide.

TENSILE STRENGTH: Maximum load a piece can support in tension without breaking or failing.

TENSION: Stress due to forces that tend to make a body longer.

THREAD: Act of cutting a screw thread.

THREAD ROLLING: A technique for applying a thread to a bolt or screw by rolling it between two grooved die plates, one of which is in motion, or between rotating circular rolls.

TITANIUM: A metal used for applications that require properties of lightweight, high strength, and good temperature and corrosion resistance. It weighs only about half as much as steel yet is almost as strong as some commonly used steels.

TOLERANCE: Permissible deviation from a basic dimension.

TOOL CRIB: A room or area in a machine shop where tools and supplies are stored and dispensed as needed.

TOOL FUNCTION: A command identifying a tool and calling for its selection either automatically or manually. Actual changing of tool may be initiated by a separate tool change command.

TOOLROOM: Area or department where tools, jigs, fixtures, and dies are manufactured.

TRACK: A path parallel to edge of a tape along which information may be stored by presence or absence of holes or magnetized areas.

TRAIN: A series of meshed gears.

TRUE: On center.

TUMBLER GEARS: Gears in gear train that can be adjusted to reverse direction of rotation of driven gear.

UNIFIED THREADS: A series of screw threads that have been adopted by United States, Canada, and Great Britain to attain interchangeability of certain screw threads. Revised standard provides greater strength, easier assembly, and longer tool life.

UNIVERSAL CHUCK: A chuck on which all jaws move simultaneously at a uniform rate to center round or hexagonal stock automatically.

V-BLOCK: Square or rectangular steel blocks with a 90 deg. V accurately machined through its center. It is provided with a clamp for holding round stock for drilling, milling, and laying out operations. Blocks are furnished as pairs and are frequently hardened and ground for additional accuracy.

VENTS: Narrow openings in molds that permit gases, generated during pouring, to escape.

VERTICAL MILLING ATTACHMENT: A mechanism that can be attached to some milling machines to convert them into vertical milling machines.

V-WAYS: Raised portion on machine tool beds that act as bearing surfaces. They guide and align movable portion of machine that rides on them. They are shaped like an inverted V.

WAYS: Flat or V-shaped bearing surfaces on a machine that aligns and guides movable part of machine that rides on them.

WHEEL DRESSER: A device to true face of a grinding wheel.

WORK HARDNESS: Name applied to increase in hardness that develops in metal as a result of cold forming.

WORKING DRAWING: A drawing or drawings that give machinist necessary information to make and assemble a mechanism.

WRINGING FIT: A fit that is practically metal to metal. It is selective rather than interchangeable and requires a twisting motion to assemble.

X-RAY: A nondestructive inspection technique that has become a routine step in acceptance of parts and materials.

ZERO: Point from which all dimensions are referenced in an absolute positioning system.

ZYGLO: A fluorescent penetrant inspection technique for detecting flaws in nonmagnetic metals and solids.

USEFUL TABLES

METALS WE USE

SHAPES	LENGTH	HOW MEASURED	*HOW PURCHASED
Sheet less than 1/4 in. thick	to 144 in.	Thickness x width, widths to 72 in.	Weight, foot or piece
Plate more than 1/4 in. thick	to 20 ft.	Thickness x width	Weight, foot or piece
Band	to 20 ft.	Thickness x width	Weight, or piece
Rod	12 to 20 ft.	Diameter	Weight, foot or piece
Square	12 to 20 ft.	Width	Weight, foot or piece
Flats	Hot rolled 20–22 ft. Cold finished	Thickness x width	Weight, foot or piece
Hexagon	12 to 20 ft.	Distance across flats	Weight, foot or piece
Octagon	12 to 20 ft.	Distance across flats	Weight, foot or piece
Angle	Lengths to 40 ft.	Leg length x leg length x thickness of legs	Weight, foot or piece
Channel	Lengths to 60 ft.	Depth x web thickness x flange width	Weight, foot or piece
I–beam	Lengths to 60 ft.	Height x web thickness x flange width	Weight, foot or piece

* Charge made for cutting to other than standard lengths.

PHYSICAL PROPERTIES OF METALS

METAL	SYMBOL	SPECIFIC GRAVITY	SPECIFIC HEAT	MELTING POINT*		LBS. PER CUBIC INCH
				DEG. C	DEG. F.	
Aluminum (Cast)	Al	2.56	.2185	658	1217	.0924
Aluminum (Rolled).	Al	2.71	–	–	–	.0978
Antimony	Sb	6.71	.051	630	1166	.2424
Bismuth	Bi	9.80	.031	271	520	.3540
Boron.	B	2.30	.3091	2300	4172	.0831
Brass.	–	8.51	.094	–	–	.3075
Cadmium.	Cd	8.60	.057	321	610	.3107
Calcium	Ca	1.57	.170	810	1490	.0567
Carbon	C	2.22	.165	–	–	.0802
Chromium	Cr	6.80	.120	1510	2750	.2457
Cobalt	Co	8.50	.110	1490	2714	.3071
Copper.	Cu	8.89	.094	1083	1982	.3212
Columbium . . .	Cb	8.57	–	1950	3542	.3096
Gold	Au	19.32	.032	1063	1945	.6979
Iridium.	Ir	22.42	.033	2300	4170	.8099
Iron.	Fe	7.86	.110	1520	2768	.2634
Iron (Cast) . . .	Fe	7.218	.1298	1375	2507	.2605
Iron (Wrought) .	Fe	7.70	.1138	1500–1600	2732–2912	.2779
Lead	Pb	11.37	.031	327	621	.4108
Lithium	Li	.057	.941	186	367	.0213
Magnesium . . .	Mg	1.74	.250	651	1204	.0629
Manganese . . .	Mn	8.00	.120	1225	2237	.2890
Mercury	Hg	13.59	.032	38.7	37.7	.4909
Molybdenum. . .	Mo	10.2	.0647	2620	4748	.368
Monel Metal. . .	–	8.87	.127	1360	2480	.320
Nickel	Ni	8.80	.130	1452	2646	.319
Phosphorus . . .	P	1.82	.177	43	111.4	.0657
Platinum.	Pt	21.50	.033	1755	3191	.7767
Potassium. . . .	K	0.87	.170	62	144	.0314
Selenium.	Se	4.81	.084	220	428	.174
Silicon.	Si	2.40	.1762	1427	2600	.087
Silver.	Ag	10.53	.056	961	1761	.3805
Sodium.	Na	0.97	.290	97	207	.0350
Steel	–	7.858	.1175	1330–1378	2372–2532	.2839
Strontium	Sr	2.54	.074	–	–	.0918
Sulphur.	S	2.07	.175	115	235.4	.075
Tantalum	Ta	10.80	–	2850	5160	.3902
Tin	Sn	7.29	.056	232	450	.2634
Titanium.	Ti	5.3	.130	1900	3450	.1915
Tungsten	W	19.10	.033	3000	5432	.6900
Uranium	U	18.70	–	–	–	.6755
Vanadium	V	5.50	–	1730	3146	.1987
Zinc	Zn	7.19	.094	419	786	.2598

* Circular of the Bureau of Standards No. 35, Department of Commerce and Labor.

RECOMMENDED TURNING RATES USING HIGH-SPEED TOOLS
STAINLESS STEELS

NATURE OF STOCK	TYPE NO.	SPEED SFPM	FEED* INCHES PER REVOLUTION
The free machining grades	430 F 416 303	100–140 90–135 80–120	0.003–0.005 for finish cuts and up to 0.015 for roughing cuts
High carbon grades which are slowed down on account of their abrasive action on tools	410 430 420 431 440 302 304 316	75–115 75–115 45–85 45–85 30–60 45–80	0.003–0.008 0.003–0.008 0.003–0.008 0.003–0.008 0.003–0.008 0.004–0.008

DECIMAL EQUIVALENTS NUMBER SIZE DRILLS

NO.	SIZE OF DRILL IN INCHES	NO.	SIZE OF DRILL IN INCHES	NO.	SIZE OF DRILL IN INCHES	NO.	SIZE OF DRILL IN INCHES
1	.2280	21	.1590	41	.0960	61	.0390
2	.2210	22	.1570	42	.0935	62	.0380
3	.2130	23	.1540	43	.0890	63	.0370
4	.2090	24	.1520	44	.0860	64	.0360
5	.2055	25	.1495	45	.0820	65	.0350
6	.2040	26	.1470	46	.0810	66	.0330
7	.2010	27	.1440	47	.0785	67	.0320
8	.1990	28	.1405	48	.0760	68	.0310
9	.1960	29	.1360	49	.0730	69	.0292
10	.1935	30	.1285	50	.0700	70	.0280
11	.1910	31	.1200	51	.0670	71	.0260
12	.1890	32	.1160	52	.0635	72	.0250
13	.1850	33	.1130	53	.0595	73	.0240
14	.1820	34	.1110	54	.0550	74	.0225
15	.1800	35	.1100	55	.0520	75	.0210
16	.1770	36	.1065	56	.0465	76	.0200
17	.1730	37	.1040	57	.0430	77	.0180
18	.1695	38	.1015	58	.0420	78	.0160
19	.1660	39	.0995	59	.0410	79	.0145
20	.1610	40	.0980	60	.0400	80	.0135

LETTER SIZE DRILLS

A	0.234	J	0.277	S	0.348
B	0.238	K	0.281	T	0.358
C	0.242	L	0.290	U	0.368
D	0.246	M	0.295	V	0.377
E	0.250	N	0.302	W	0.386
F	0.257	O	0.316	X	0.397
G	0.261	P	0.323	Y	0.404
H	0.266	Q	0.332	Z	0.413
I	0.272	R	0.339		

CONVERSION TABLE
METRIC TO ENGLISH

WHEN YOU KNOW	MULTIPLY BY: * = Exact VERY ACCURATE	APPROXIMATE	TO FIND
	LENGTH		
millimeters	0.0393701	0.04	inches
centimeters	0.3937008	0.4	inches
meters	3.280840	3.3	feet
meters	1.093613	1.1	yards
kilometers	0.621371	0.6	miles
	WEIGHT		
grains	0.00228571	0.0023	ounces
grams	0.03527396	0.035	ounces
kilograms	2.204623	2.2	pounds
tonnes	1.1023113	1.1	short tons
	VOLUME		
milliliters		0.2	teaspoons
milliliters	0.06667	0.067	tablespoons
milliliters	0.03381402	0.03	fluid ounces
liters	61.02374	61.024	cubic inches
liters	2.113376	2.1	pints
liters	1.056688	1.06	quarts
liters	0.26417205	0.26	gallons
liters	0.03531467	0.035	cubic feet
cubic meters	61023.74	61023.7	cubic inches
cubic meters	35.31467	35.0	cubic feet
cubic meters	1.3079506	1.3	cubic yards
cubic meters	264.17205	264.0	gallons
	AREA		
square centimeters	0.1550003	0.16	square inches
square centimeters	0.00107639	0.001	square feet
square meters	10.76391	10.8	square feet
square meters	1.195990	1.2	square yards
square kilometers		0.4	square miles
hectares	2.471054	2.5	acres
	TEMPERATURE		
Celsius	*9/5 (then add 32)		Fahrenheit

CONVERSION TABLE
ENGLISH TO METRIC

WHEN YOU KNOW	MULTIPLY BY: * = Exact VERY ACCURATE	APPROXIMATE	TO FIND
	LENGTH		
inches	*25.4		millimeters
inches	* 2.54		centimeters
feet	* 0.3048		meters
feet	*30.48		centimeters
yards	* 0.9144	0.9	meters
miles	* 1.609344	1.6	kilometers
	WEIGHT		
grains	15.43236	15.4	grams
ounces	*28.349523125	28.0	grams
ounces	* 0.028349523125	.028	kilograms
pounds	* 0.45359237	0.45	kilograms
short ton	* 0.90718474	0.9	tonnes
	VOLUME		
teaspoons		5.0	milliliters
tablespoons		15.0	milliliters
fluid ounces	29.57353	30.0	milliliters
cups		0.24	liters
pints	* 0.473176473	0.47	liters
quarts	* 0.946352946	0.95	liters
gallons	* 3.785411784	3.8	liters
cubic inches	* 0.016387064	0.02	liters
cubic feet	* 0.028316846592	0.03	cubic meters
cubic yards	* 0.764554857984	0.76	cubic meters
	AREA		
square inches	* 6.4516	6.5	square centimeters
square feet	* 0.09290304	0.09	square meters
square yards	* 0.83612736	0.8	square meters
square miles		2.6	square kilometers
acres	* 0.40468564224	0.4	hectares
	TEMPERATURE		
Fahrenheit	*5/9 (after subtracting 32)		Celsius

TABLE OF CUTTING SPEEDS

MATERIAL	ROUGHING CUT	FINISHING CUT	THREADING
1/8 IN. DIAMETER			
Machine Steel and Bronze	2880 rpm	3200 rpm	1020 rpm
Cast Iron	1920	2560	800
Tool Steel (Annealed)	1600	2400	640
Brass	4800	6400	1600
Aluminum	6400	9600	1600
3/16 IN. DIAMETER			
Machine Steel and Bronze	2880	3200	1120
Tool Steel	1600	2400	640
Brass	4800	6400	1600
Aluminum	6400	9600	1600
1/4 IN. DIAMETER			
Machine Steel and Bronze	1440	1600	560
Tool Steel	800	1200	320
Brass	2400	3200	800
Aluminum	3200	4800	800
3/8 IN. DIAMETER			
Machine Steel and Bronze	960	1066	270
Tool Steel	540	800	220
Brass	1700	2100	530
Aluminum	2130	3200	540
1/2 IN. DIAMETER			
Machine Steel and Bronze	720	800	280
Tool Steel	400	600	160
Brass	1200	1600	400
Aluminum	1600	2400	400
5/8 IN. DIAMETER			
Machine Steel and Bronze	576	640	160
Tool Steel	320	480	200
Brass	960	1280	320
Aluminum	1280	1920	320
3/4 IN. DIAMETER			
Machine Steel and Bronze	500	550	176
Tool Steel	266	400	106
Brass	800	1066	266
Aluminum	1066	1600	266
1 IN. DIAMETER			
Machine Steel and Bronze	360	400	140
Tool Steel	200	300	80
Brass	600	800	200
Aluminum	800	1200	200
1 1/4 IN. DIAMETER			
Machine Steel	288	320	112
Tool Steel	160	240	64
Brass	480	640	160
Aluminum	640	960	160
1 1/2 IN. DIAMETER			
Machine Steel and Bronze	240	270	67
Tool Steel	134	200	53
Brass	400	534	134
Aluminum	534	800	134
1 3/4 IN. DIAMETER			
Machine Steel and Bronze	205	230	80
Tool Steel	115	170	50
Brass	340	450	115
Aluminum	456	680	115
2 IN. DIAMETER			
Machine Steel	180	200	50
Tool Steel	100	150	40
Brass	300	400	100
Aluminum	400	600	100
2 1/2 IN. DIAMETER			
Machine Steel	141	160	56
Tool Steel	80	120	32
Brass	240	320	80
Aluminum	320	480	80
3 IN. DIAMETER			
Machine Steel	120	140	40
Tool Steel	65	100	40
Brass	200	270	65
Aluminum	270	400	65
3 1/2 IN. DIAMETER			
Machine Steel	103	115	40
Tool Steel	60	85	23
Brass	171	228	57
Aluminum	228	342	57
4 IN. DIAMETER			
Machine Steel	90	100	35
Tool Steel	50	75	20
Brass	150	200	50
Aluminum	200	300	50
4 1/2 IN. DIAMETER			
Machine Steel	80	90	31
Tool Steel	45	67	18
Brass	133	178	45
Aluminum	178	267	45
5 IN. DIAMETER			
Machine Steel	72	80	28
Tool Steel	40	58	16
Brass	120	160	40
Aluminum	160	240	40

60° V-TYPE THREAD DIMENSIONS

WITH SIZES OF TAP DRILL AND CLEARANCE DRILL

FRACTIONAL SIZES

NATIONAL SPECIAL THREAD SERIES

Nominal Size	Threads per Inch	Major Diameter Inches	Minor Diameter Inches	Pitch Diameter Inches	Tap Drill for 75% Thread †	Clearance Drill Size *
1/16"	64	.0625	.0422	.0524	3/64"	51
5/64"	60	.0781	.0563	.0673	1/16"	45
3/32"	48	.0938	.0667	.0803	49	40
7/64"	48	.1094	.0823	.0959	43	32
1/8"	32	.1250	.0844	.1047	3/32"	29
9/64"	40	.1406	.1081	.1244	32	24
5/32"	32	.1563	.1157	.1360	1/8"	19
5/32"	36	.1563	.1202	.1382	30	19
11/64"	32	.1719	.1313	.1516	9/64"	14
3/16"	24	.1875	.1334	.1604	26	8
3/16"	32	.1875	.1469	.1672	22	8
13/64"	24	.2031	.1490	.1760	20	3
7/32"	24	.2188	.1646	.1917	16	1
7/32"	32	.2188	.1782	.1985	12	1
15/64"	24	.2344	.1806	.2073	10	1/4"
1/4"	24	.2500	.1959	.2229	4	17/64"
1/4"	27	.2500	.2019	.2260	3	17/64"
1/4"	32	.2500	.2094	.2297	7/32"	17/64"
5/16"	20	.3125	.2476	.2800	17/64"	21/64"
5/16"	27	.3125	.2644	.2884	J	21/64"
5/16"	32	.3125	.2719	.2922	9/32"	21/64"
3/8"	20	.3750	.3100	.3425	21/64"	25/64"
3/8"	27	.3750	.3269	.3509	R	25/64"
7/16"	24	.4375	.3834	.4104	X	29/64"
7/16"	27	.4375	.3894	.4134	Y	29/64"
1/2"	12	.5000	.3918	.4459	27/64"	33/64"
1/2"	24	.5000	.4459	.4729	29/64"	33/64"
1/2"	27	.5000	.4519	.4759	15/32"	33/64"
9/16"	27	.5625	.5144	.5384	17/32"	37/64"
5/8"	12	.6250	.5168	.5709	35/64"	41/64"
5/8"	27	.6250	.5769	.6009	19/32"	41/64"
11/16"	11	.6875	.5694	.6285	19/32"	45/64"
11/16"	16	.6875	.6063	.6469	5/8"	45/64"
3/4"	12	.7500	.6418	.6959	43/64"	49/64"
3/4"	27	.7500	.7019	.7259	23/32"	49/64"
13/16"	10	.8125	.6826	.7476	23/32"	53/64"
7/8"	12	.8750	.7668	.8209	51/64"	57/64"
7/8"	18**	.8750	.8028	.8389	53/64"	57/64"
7/8"	27	.8750	.8269	.8509	27/32"	57/64"
15/16"	9	.9375	.7932	.8654	53/64"	61/64"
1"	12	1.0000	.8918	.9459	59/64"	1 1/64"
1"	27	1.0000	.9519	.9759	31/32"	1 1/64"
1 5/8"	5 1/2	1.6250	1.3888	1.5069	1 29/64"	1 41/64"
1 7/8"	5	1.8750	1.6152	1.7451	1 11/16"	1 57/64"
2 1/8"	4 1/2	2.1250	1.8363	1.9807	1 29/32"	2 5/32"
2 3/8"	4	2.3750	2.0502	2.2126	2 1/8"	2 13/32"

† Refer to tables of "DIAMETERS OF NUMBERED DRILLS" and "DIAMETERS OF LETTERED DRILLS" for sizes.

* Clearance drill makes hole with standard clearance for diameter of nominal size.

** Standard spark plug size.

60° V-TYPE THREAD DIMENSIONS
WITH SIZES OF TAP DRILL AND CLEARANCE DRILL
INTERNATIONAL STANDARD—METRIC

Major Diameter m/m	Pitch m/m	Minor Diameter m/m	Pitch Diameter m/m	Tap Drill for 75% Thread m/m	Tap Drill for 75% Thread † No. or Inches	Clearance Drill Size *
2.0	.40	1.48	1.740	1.6	1/16"	41
2.3	.40	1.78	2.040	1.9	48	36
2.6	.45	2.02	2.308	2.1	45	31
3.0	.50	2.35	2.675	2.5	40	29
3.5	.60	2.72	3.110	2.9	33	23
4.0	.70	3.09	3.545	3.3	30	16
4.5	.75	3.53	4.013	3.75	26	10
5.0	.80	3.96	4.480	4.2	19	3
5.5	.90	4.33	4.915	4.6	14	15/64"
6.0	1.00	4.70	5.350	5.0	9	1/4"
7.0	1.00	5.70	6.350	6.0	15/64"	19/64"
8.0	1.25	6.38	7.188	6.8	H	11/32"
9.0	1.25	7.38	8.188	7.8	5/16"	3/8"
10.0	1.50	8.05	9.026	8.6	R	27/64"
11.0	1.50	9.05	10.026	9.6	V	29/64"
12.0	1.75	9.73	10.863	10.5	Z	1/2"
14.0**	1.25	12.38	13.188	13.0	33/64"	9/16"
14.0	2.00	11.40	12.701	12.0	15/32"	9/16"
16.0	2.00	13.40	14.701	14.0	35/64"	21/32"
18.0**	1.50	16.05	17.026	16.5	41/64"	47/64"
18.0	2.50	14.75	16.376	15.5	39/64"	47/64"
20.0	2.50	16.75	18.376	17.5	11/16"	13/16"
22.0	2.50	18.75	20.376	19.5	49/64"	57/64"
24.0	3.00	20.10	22.051	21.0	53/64"	31/32"
27.0	3.00	23.10	25.051	24.0	15/16"	1 3/32"
30.0	3.50	25.45	27.727	26.5	1 3/64"	1 13/64"
33.0	3.50	28.45	30.727	29.5	1 11/64"	1 21/64"
36.0	4.00	30.80	33.402	32.0	1 17/64"	1 7/16"
39.0	4.0	33.80	36.402	35.0	1 3/8"	1 9/16"
42.0	4.50	36.15	39.077	37.0	1 29/64"	1 43/64"
45.0	4.50	39.15	42.077	40.0	1 37/64"	1 13/16"
48.0	5.00	41.50	44.752	43.0	1 11/16"	1 29/32"

† Refer to tables of "DIAMETERS OF NUMBERED DRILLS" and "DIAMETERS OF LETTERED DRILLS" for sizes.
* Clearance drill makes hole with standard clearance for diameter of nominal size.
** Standard spark plug size.

FRACTIONAL INCHES INTO DECIMALS AND MILLIMETERS

INCH	DECIMAL INCH	MILLIMETER	INCH	DECIMAL INCH	MILLIMETER
1/64	0.0156	0.3967	33/64	0.5162	13.0968
1/32	0.0312	0.7937	17/32	0.5312	13.4937
3/64	0.0468	1.1906	35/64	0.5468	13.8906
1/16	0.0625	1.5875	9/16	0.5625	14.2875
5/64	0.0781	1.9843	37/64	0.5781	14.6843
3/32	0.0937	2.3812	19/32	0.5937	15.0812
7/64	0.1093	2.7781	39/64	0.6093	15.4781
1/8	0.125	3.175	5/8	0.625	15.875
9/64	0.1406	3.5718	41/64	0.6406	16.2718
5/32	0.1562	3.9687	21/32	0.6562	16.6687
11/64	0.1718	4.3656	43/64	0.6718	17.0656
3/16	0.1875	4.7625	11/16	0.6875	17.4625
13/64	0.2031	5.1593	45/64	0.7031	17.8593
7/32	0.2187	5.5562	23/32	0.7187	18.2562
15/64	0.2343	5.9531	47/64	0.7343	18.6531
1/4	0.25	6.5	3/4	0.75	19.05
17/64	0.2656	6.7468	49/64	0.7656	19.4468
9/32	0.2812	7.1437	25/32	0.7812	19.8437
19/64	0.2968	7.5406	51/64	0.7968	20.2406
5/16	0.3125	7.9375	13/16	0.8125	20.6375
21/64	0.3281	8,3343	53/64	0.8281	21.0343
11/32	0.3437	8.7312	27/32	0.8437	21.4312
23/64	0.3593	9.1281	55/64	0.8593	21.8281
3/8	0.375	9.525	7/8	0.875	22.225
25/64	0.3906	9.9218	57/64	0.8906	22.6218
13/32	0.4062	10.3187	29/32	0.9062	23.0187
27/64	0.4218	10.7156	59/64	0.9218	23.4156
7/16	0.4375	11.1125	15/16	0.9375	23.8125
29/64	0.4531	11.5093	61/64	0.9531	24.2093
15/32	0.4687	11.9062	31/32	0.9687	24.6062
31/64	0.4843	12.3031	63/64	0.9843	25.0031
1/2	0.50	12.7	1	1.0000	25.4

Order of Ductility of Metals

1. Gold
2. Platinum
3. Silver
4. Iron
5. Copper
6. Aluminum
7. Nickel
8. Zinc
9. Tin
10. Lead

FEEDS AND SPEEDS FOR HIGH-SPEED DRILLS

Drill Diameter, Inches	Cast Iron		Bronze or Brass		Drop Forgings Alloy Steel Tool Steel Annealed		Drop Forgings Alloy Steel Heat-Treated		Steel Castings		Mild Steel	
	Feed	Speed	Feed	Speed	Feed	Speed	Feed	Speed	Feed	Speed	Feed	Speed
1/16	.002	4550	.002	9150	.002	3650	.002	2750	.002	3650	.002	4250
	.004	6700	.004	12000	.003	4550	.003	3650	.003	4550	.003	5600
1/8	.002	2550	.002	4550	.002	1800	.002	1225	.002	1800	.002	2100
	.004	3350	.004	5600	.003	2250	.003	1800	.003	2250	.003	2800
3/16	.004	1500	.004	3100	.003	1200	.003	900	.003	1200	.003	1400
	.006	2200	.007	5600	.004	1500	.004	1200	.005	1500	.005	1900
1/4	.004	1150	.004	2300	.003	925	.003	750	.003	925	.003	1050
	.006	1650	.007	2750	.004	1150	.004	925	.005	1150	.005	1500
5/16	.006	925	.007	1825	.004	725	.004	500	.004	725	.005	850
	.009	1325	.010	2200	.006	925	.005	725	.006	925	.007	1200
3/8	.006	750	.007	1525	.004	600	.004	400	.004	600	.005	700
	.009	1100	.010	1850	.006	750	.005	600	.006	750	.007	925
7/16	.009	650	.010	1300	.006	525	.005	350	.006	525	.006	600
	.012	950	.014	1525	.009	650	.006	525	.010	650	.010	800
1/2	.008	575	.010	1150	.006	375	.005	300	.006	375	.006	525
	.012	850	.014	1375	.009	575	.006	375	.010	575	.010	700
9/16	.012	500	.014	1000	.008	350	.007	275	.010	350	.010	575
	.016	750	.018	1200	.012	500	.010	350	.014	500	.014	625
5/8	.012	450	.014	900	.008	300	.007	250	.010	300	.010	425
	.016	675	.018	1100	.012	450	.010	300	.014	450	.014	565
11/16	.012	410	.014	800	.008	275	.007	225	.010	275	.010	375
	.016	625	.018	1000	.012	410	.010	275	.014	410	.014	525
3/4	.012	375	.014	750	.008	250	.007	200	.010	250	.010	350
	.016	550	.018	900	.012	375	.010	250	.014	375	.014	475
13/16	.014	350	.016	700	.010	240	.009	190	.014	240	.014	325
	.020	525	.022	850	.014	350	.012	240	.016	350	.016	450
7/8	.014	325	.016	650	.010	225	.009	175	.014	225	.014	300
	.020	475	.022	800	.014	325	.012	225	.016	325	.016	400
15/16	.014	300	.016	625	.010	200	.009	160	.014	200	.014	275
	.020	450	.022	725	.014	300	.012	200	.016	300	.016	375
1	.014	280	.016	575	.010	185	.009	150	.014	185	.014	265
	.020	425	.022	675	.014	280	.012	185	.016	280	.016	350

Speeds and feeds shown apply to average working conditions and materials. They are recommended with regard to conserving DRILLS and avoiding excessive MACHINE TOOL wear.

Under many conditions, these speeds and feeds may be considerably increased; under others they must be decreased. This is dependent on judgment of operator, and performance obtained.

Excessive speeds and feeds will show up by action of machine and drill. Same applies to lower speeds and feeds. Operator will notice whether he/she is getting proper performance by experience, and will advance or retard as case may justify.

Feeds and speeds should be changed in proper proportions and a liberal use of cooling compound will increase life of tools.

REMEMBER! NEVER dip drill into water to cool while grinding. This will cause tiny checks, or cracks at cutting edge which will cause drill to dull quickly.

To determine feed and speed according to the above chart, we illustrate as follows:

You are going to drill Drop Forgings heat-treated. We suppose you will use a 1/2 in. Drill. Follow column down to where the 1/2 in. drill meets it; there you will find that a feed from .005 to .006 and a speed of from 300 to 375 rpm are recommended. Start by using .005 feed and 300 rpm. If drill and machine seem to turn smoothly without strain, then both feed and speed can be advanced. Operator will soon find which is best.

Do not leave a drill in after it shows signs of dulling or laboring; then is the time to regrind, proper grinding is essential. (Chicago-Latrobe)

Probable Percentage of Full Thread Produced in Tapped Hole Using Stock Sizes of Drill

Tap	Tap Drill	Decimal Equiv. of Tap Drill	Theoretical % of Thread	Probable Oversize (Mean)	Probable Hole Size	Percentage of Thread	Tap	Tap Drill	Decimal Equiv. of Tap Drill	Theoretical % of Thread	Probable Oversize (Mean)	Probable Hole Size	Percentage of Thread
0-80	56	.0465	83	.0015	.0480	74	8-32	29	.1360	69	.0029	.1389	62
	3/64	.0469	81	.0015	.0484	71		28	.1405	58	.0029	.1434	51
1-64	54	.0550	89	.0015	.0565	81	8-36	29	.1360	78	.0029	.1389	70
	53	.0595	67	.0015	.0610	59		28	.1405	68	.0029	.1434	57
1-72	53	.0595	75	.0015	.0610	67		9/64	.1406	68	.0029	.1435	57
	1/16	.0625	58	.0015	.0640	50	10-24	27	.1440	85	.0032	.1472	79
2-56	51	.0670	82	.0017	.0687	74		26	.1470	79	.0032	.1502	74
	50	.0700	69	.0017	.0717	62		25	.1495	75	.0032	.1527	69
	49	.0730	56	.0017	.0747	49		24	.1520	70	.0032	.1552	64
2-64	50	.0700	79	.0017	.0717	70		23	.1540	67	.0032	.1572	61
	49	.0730	64	.0017	.0747	56		5/32	.1563	62	.0032	.1595	56
3-48	48	.0760	85	.0019	.0779	78		22	.1570	61	.0032	.1602	55
	5/64	.0781	77	.0019	.0800	70	10-32	5/32	.1563	83	.0032	.1595	75
	47	.0785	76	.0019	.0804	69		22	.1570	81	.0032	.1602	73
	46	.0810	67	.0019	.0829	60		21	.1590	76	.0032	.1622	68
	45	.0820	63	.0019	.0839	56		20	.1610	71	.0032	.1642	64
3-56	46	.0810	78	.0019	.0829	69		19	.1660	59	.0032	.1692	51
	45	.0820	73	.0019	.0839	65	12-24	11/64	.1719	82	.0035	.1754	75
	44	.0860	56	.0019	.0879	48		17	.1730	79	.0035	.1765	73
4-40	44	.0860	80	.0020	.0880	74		16	.1770	72	.0035	.1805	66
	43	.0890	71	.0020	.0910	65		15	.1800	67	.0035	.1835	60
	42	.0935	57	.0020	.0955	51		14	.1820	63	.0035	.1855	56
	3/32	.0938	56	.0020	.0958	50	12-28	16	.1770	84	.0035	.1805	77
4-48	42	.0935	68	.0020	.0955	61		15	.1800	78	.0035	.1835	70
	3/32	.0938	68	.0020	.0958	60		14	.1820	73	.0035	.1855	66
	41	.0960	59	.0020	.0980	52		13	.1850	67	.0035	.1885	59
5-40	40	.0980	83	.0023	.1003	76		3/16	.1875	61	.0035	.1910	54
	39	.0995	79	.0023	.1018	71	1/4-20	9	.1960	83	.0038	.1998	77
	38	.1015	72	.0023	.1038	65		8	.1990	79	.0038	.2028	73
	37	.1040	65	.0023	.1063	58		7	.2010	75	.0038	.2048	70
5-44	38	.1015	79	.0023	.1038	72		13/64	.2031	72	.0038	.2069	66
	37	.1040	71	.0023	.1063	63		6	.2040	71	.0038	.2078	65
	36	.1065	63	.0023	.1088	55		5	.2055	69	.0038	.2093	63
6-32	37	.1040	84	.0023	.1063	78		4	.2090	63	.0038	.2128	57
	36	.1065	78	.0026	.1091	71	1/4-28	3	.2130	80	.0038	.2168	72
	7/64	.1094	70	.0026	.1120	64		7/32	.2188	67	.0038	.2226	59
	35	.1100	69	.0026	.1126	63		2	.2210	63	.0038	.2248	55
	34	.1110	67	.0026	.1136	60	5/16-18	F	.2570	77	.0038	.2608	72
	33	.1130	62	.0026	.1156	55		G	.2610	71	.0041	.2651	66
6-40	34	.1110	83	.0026	.1136	75		17/64	.2656	65	.0041	.2697	59
	33	.1130	77	.0026	.1156	69		H	.2660	64	.0041	.2701	59
	32	.1160	68	.0026	.1186	60							

(Concluded on following page) *(Standard Tool Co.)*

Probable Percentage of Full Thread Produced in Tapped Hole Using Stock Sizes of Drill

Tap	Tap Drill	Decimal Equiv. of Tap Drill	Theoretical % of Thread	Probable Oversize (Mean)	Probable Hole Size	Percentage of Thread
5/16-24	H	.2660	86	.0041	.2701	78
	I	.2720	75	.0041	.2761	67
	J	.2770	66	.0041	.2811	58
3/8-16	5/16	.3125	77	.0044	.3169	72
	O	.3160	73	.0044	.3204	68
	P	.3230	64	.0044	.3274	59
3/8-24	21/64	.3281	87	.0044	.3325	79
	Q	.3320	79	.0044	.3364	71
	R	.3390	67	.0044	.3434	58
7/16-14	T	.3580	86	.0046	.3626	81
	23/64	.3594	84	.0046	.3640	79
	U	.3680	75	.0046	.3726	70
	3/8	.3750	67	.0046	.3796	62
	V	.3770	65	.0046	.3816	60
7/16-20	W	.3860	79	.0046	.3906	72
	25/64	.3906	72	.0046	.3952	65
	X	.3970	62	.0046	.4016	55
1/2-13	27/64	.4219	78	.0047	.4266	73
	7/16	.4375	63	.0047	.4422	58
1/2-20	29/64	.4531	72	.0047	.4578	65
9/16-12	15/32	.4688	87	.0048	.4736	82
	31/64	.4844	72	.0048	.4892	68
9/16-18	1/2	.5000	87	.0048	.5048	80
	33/64	.5156	65	.0048	.5204	58
5/8-11	17/32	.5313	79	.0049	.5362	75
	35/64	.5469	66	.0049	.5518	62
5/8-18	9/16	.5625	87	.0049	.5674	80
	37/64	.5781	65	.0049	.5831	58
3/4-10	41/64	.6406	84	.0050	.6456	80
	21/32	.6563	72	.0050	.6613	68
3/4-16	11/16	.6875	77	.0050	.6925	71
7/8-9	49/64	.7656	76	.0052	.7708	72
	25/32	.7812	65	.0052	.7864	61
7/8-14	51/64	.7969	84	.0052	.8021	79
	13/16	.8125	67	.0052	.8177	62
1"-8	55/64	.8594	87	.0059	.8653	83
	7/8	.8750	77	.0059	.8809	73
	57/64	.8906	67	.0059	.8965	64
	29/32	.9063	58	.0059	.9122	54
1"-12	29/32	.9063	87	.0060	.9123	81
	59/64	.9219	72	.0060	.9279	67
	15/16	.9375	58	.0060	.9435	52
1"-14	59/64	.9219	84	.0060	.9279	78
	15/16	.9375	67	.0060	.9435	61
1 1/8-7	31/32	.9688	84	.0062	.9750	81
	63/64	.9844	76	.0067	.9911	72
	1"	1.0000	67	.0070	1.0070	64
	1 1/64	1.0156	59	.0070	1.0226	55
1 1/8-12	1 1/32	1.0313	87	.0071	1.0384	80
	1 3/64	1.0469	72	.0072	1.0541	66
1 1/4-7	1 3/32	1.0938	84			
	1 7/64	1.1094	76			
	1 1/8	1.1250	67			
1 1/4-12	1 5/32	1.1563	87			
	1 11/64	1.1719	72			
1 3/8-6	1 3/16	1.1875	87			
	1 13/64	1.2031	79	No Test Results Available		
	1 7/32	1.2188	72			
	1 15/64	1.2344	65			
1 3/8-12	1 9/32	1.2813	87			
	1 19/64	1.2969	72	Reaming Recommended		
1 1/2-6	1 5/16	1.3125	87			
	1 21/64	1.3281	79			
	1 11/32	1.3438	72			
	1 23/64	1.3594	65			
1 1/2-12	1 13/32	1.4063	87			
	1 27/64	1.4219	72			

Taper Pipe		Straight Pipe	
Thread	Drill	Thread	Drill
1/8-27	R	1/8-27	S
1/4-18	7/16	1/4-18	29/64
3/8-18	37/64	3/8-18	19/32
1/2-14	23/32	1/2-14	47/64
3/4-14	59/64	3/4-14	15/16
1-11 1/2	1 5/32	1-11 1/2	1 3/16
1 1/4-11 1/2	1 1/2	1 1/4-11 1/2	1 33/64
1 1/2-11 1/2	1 47/64	1 1/2-11 1/2	1 3/4
2-11 1/2	2 7/32	2-11 1/2	2 7/32
2 1/2-8	2 5/8	2 1/2-8	2 21/32
3-8	3 1/4	3-8	3 9/32
3 1/2-8	3 3/4	3 1/2-8	3 25/32
4-8	4 1/4	4-8	4 9/32

FEEDS AND SPEEDS FOR HIGH-SPEED STEEL DRILLS, REAMERS AND TAPS

		DRILLS			REAMERS		TAPS – S.F.M.			
							THREADS PER INCH			
MATERIAL	BRINELL	S.F.M.	POINT	FEED	S.F.M.	FEED	3–7 1/2	8–15	16–24	25–UP
Aluminum	99–101	200–250	118 deg.	M	150–160	M	50	100	150	200
Aluminum bronze	170–187	60	118 deg.	M	40–45	M	12	25	45	60
Bakelite	. . . .	80	60–90 deg.	M	50–60	M	50	100	150	200
Brass	192–202	200–250	118 deg.	H	150–160	H	50	100	150	200
Bronze, common	166–183	200–250	118 deg.	H	150–160	H	40	80	100	150
Bronze, phosphor, 1/2 hard	187–202	175–180	118 deg.	M	130–140	M	25	40	50	80
Bronze, phosphor, soft	149–163	200–250	118 deg.	H	150–160	H	40	80	100	150
Cast iron, soft	126	140–150	90 deg.	H	100–110	H	30	60	90	140
Cast iron, medium soft	196	80–110	118 deg.	M	50–65	M	25	40	50	80
Cast iron, hard	293–302	45–50	118 deg.	L	67–75	L	10	20	30	40
Cast iron, chilled *	402	15	150 deg.	L	8–10	L	5	5	10	10
Cast steel	286–302	40–50 *	118 deg.	L	70–75	L	20	30	40	50
Celluloid	. . . .	100	90 deg.	M	75–80	M	50	100	150	200
Copper	80–85	70	100 deg.	L	45–55	L	40	80	100	150
Drop forgings (steel)	170–196	60	118 deg.	M	40–45	M	12	25	45	60
Duralumin	90–104	200	118 deg.	M	150–160	M	50	100	150	200
Everdur	179–207	60	118 deg.	L	40–45	L	20	30	40	50
Machinery steel	170–196	110	118 deg.	H	67–75	H	35	50	60	85
Magnet steel, soft	241–302	35–40	118 deg.	M	20–25	M	20	40	50	75
Magnet steel, hard *	321–512	15	150 deg.	L	10	L	5	10	15	25
Manganese steel, 7–13 percent	187–217	15	150 deg.	L	10	L	15	20	25	30
Manganese copper, 30 percent Mn. *	134	15	150 deg.	L	10– 12	L	. .	. .	. .	. .
Malleable iron	112–126	85–90	118 deg.	H	. . .	H	20	30	40	50
Mild steel, .20–.30 C	170–202	110–120	118 deg.	H	75–85	H	40	55	70	90
Molybdenum steel	196–235	55	125 deg.	M	35–45	M	20	30	35	45
Monel metal	149–170	50	118 deg.	M	35–38	M	8	10	15	20
Nickel, pure *	187–202	75	118 deg.	L	40	L	25	40	50	80
Nickel steel, 3 1/2 percent	196–241	60	118 deg.	L	40–45	L	8	10	15	20
Rubber, hard	. . . .	100	60–90 deg.	L	70–80	L	50	100	150	200
Screw stock, C.R.	170–196	110	118 deg.	H	75	H	20	30	40	50
Spring steel	402	20	150 deg.	L	12–15	L	10	10	15	15
Stainless steel	146–149	50	118 deg.	M	30	M	8	10	15	20
Stainless steel, C.R. *	460–477	20	118 deg.	L	15	L	8	10	15	20
Steel, .40 to .50 C	170–196	80	118 deg.	M	8–10	M	20	30	40	50
Tool, S.A.E., and Forging steel	149	75	118 deg.	H	35–40	H	25	35	45	55
Tool, S.A.E., and Forging steel	241	50	125 deg.	M	12	M	15	15	25	25
Tool, S.A.E., and Forging steel *	402	15	150 deg.	L	10	L	8	10	15	20
Zinc alloy	112–126	200–250	118 deg.	M	150–175	M	50	100	150	200

* Use specially constructed heavy-duty drills.
Carbon Steel Tools should be run at speeds 40 to 50 percent of those recommended for High Speed.
Spiral Point Taps may be run at speeds 15 to 20 percent faster than Regular Taps.

Length of Pins	TAPER PIN AND REAMER SIZE*															
	6/0	5/0	4/0	3/0	2/0	0	1	2	3	4	5	6	7	8	9	10
* 3/8	50	44	38	32	29											
* 1/2	51	45	39	33	30	27										
* 5/8	52	46	41	34	30	27	21									
* 3/4	1/16	47	42	7/64	1/8	9/64	5/32	16	13/64	15/64	I	P				
*1		49	44	37	31	29	25	11/64	9	1	H	O	W			
*1 1/4					32	30	26	19	10	2	G	5/16	V	29/64		
*1 1/2						*1/8	*9/64	20	3/16	7/32	F	N	V	29/64	35/64	43/64
*1 3/4						*31	*29	*5/32	14	3	E	N	U	29/64	35/64	21/32
*2						*33	*30	*25	*16	4	D	M	U	7/16	35/64	21/32
*2 1/4						*7/64	*1/8	*26	*11/64	*13/64	C	M	23/64	7/16	17/32	21/32
*2 1/2						*37	*31	*28	*19	* 8	*B	L	T	7/16	17/32	41/64
*2 3/4						*40	*33	*29	*21	*11	*1	9/32	S	27/64	17/32	41/64
*3						*42	*35	*30	*5/32	*3/16	*2	J	11/32	27/64	33/64	41/64
*3 1/4									*24	*14	*2	*I	R	27/64	33/64	5/8
*3 1/2									*26	*16	*3	*H	Q	Z	33/64	5/8
*3 3/4										*17	*4	*G	*Q	Z	1/2	5/8
*4										*19	*5	*F	*P	13/32	1/2	39/64
*4 1/4												*1/4	*O	Y	1/2	39/64
*4 1/2												*D	*O	X	31/32	39/64
*4 3/4												*C	*5/16	*25/64	31/64	19/32
*5												*B	*N	*W	31/64	19/32
*5 1/4															15/32	19/32
*5 1/2															*15/32	37/64
*5 3/4															*15/32	37/64
*6															*29/64	37/64

* Hole sizes too small to admit taper pin reamers of standard length. Special, extra length reamers are required for these cases.

SCREWS THREAD ELEMENTS FOR UNIFIED AND NATIONAL FORM OF THREAD

THREADS PER INCH (n)	PITCH (p) $p = \frac{1}{n}$	SINGLE HEIGHT — SUBTRACT FROM BASIC MAJOR DIAMETER TO GET BASIC PITCH DIAMETER	DOUBLE HEIGHT — SUBTRACT FROM BASIC MAJOR DIAMETER TO GET BASIC MINOR DIAMETER	83 1/3 PERCENT DOUBLE HEIGHT — SUBTRACT FROM BASIC MAJOR DIAMETER TO GET MINOR DIAMETER OF RING GAGE	BASIC WIDTH OF CREST AND ROOT FLAT $\frac{p}{8}$	CONSTANT FOR BEST SIZE WIRE ALSO SINGLE HEIGHT OF 60 DEG. V-THREAD	DIAMETER OF BEST SIZE WIRE
3	.333333	.216506	.43301	.36084	.0417	.28868	.19245
3 1/4	.307692	.199852	.39970	.33309	.0385	.26647	.17765
3 1/2	.285714	.185577	.37115	.30929	.0357	.24744	.16496
4	.250000	.162379	.32476	.27063	.0312	.21651	.14434
4 1/2	.222222	.144337	.28867	.24056	.0278	.19245	.12830
5	.200000	.129903	.25981	.21650	.0250	.17321	.11547
5 1/2	.181818	.118093	.23619	.19682	.0227	.15746	.10497
6	.166666	.108253	.21651	.18042	.0208	.14434	.09623
7	.142857	.092788	.18558	.15465	.0179	.12372	.08248
8	.125000	.081189	.16238	.13531	.0156	.10825	.07217
9	.111111	.072168	.14434	.12028	.0139	.09623	.06415
10	.100000	.064952	.12990	.10825	.0125	.08660	.05774
11	.090909	.059046	.11809	.09841	.0114	.07873	.05249
11 1/2	.086956	.056480	.11296	.09413	.0109	.07531	.05020
12	.083333	.054127	.10826	.09021	.0104	.07217	.04811
13	.076923	.049963	.09993	.08327	.0096	.06662	.04441
14	.071428	.046394	.09279	.07732	.0089	.06186	.04124
16	.062500	.040595	.08119	.06766	.0078	.05413	.03608
18	.055555	.036086	.07217	.06014	.0069	.04811	.03208
20	.050000	.032475	.06495	.05412	.0062	.04330	.02887
22	.045454	.029523	.05905	.04920	.0057	.03936	.02624
24	.041666	.027063	.05413	.04510	.0052	.03608	.02406
27	.037037	.024056	.04811	.04009	.0046	.03208	.02138
28	.035714	.023197	.04639	.03866	.0045	.03093	.02062
30	.033333	.021651	.04330	.03608	.0042	.02887	.01925
32	.031250	.020297	.04059	.03383	.0039	.02706	.01804
36	.027777	.018042	.03608	.03007	.0035	.02406	.01604
40	.025000	.016237	.03247	.02706	.0031	.02165	.01443
44	.022727	.014761	.02952	.02460	.0028	.01968	.01312
48	.020833	.013531	.02706	.02255	.0026	.01804	.01203
50	.020000	.012990	.02598	.02165	.0025	.01732	.01155
56	.017857	.011598	.02320	.01933	.0022	.01546	.01031
60	.016666	.010825	.02165	.01804	.0021	.01443	.00962
64	.015625	.010148	.02030	.01691	.0020	.01353	.00902
72	.013888	.009021	.01804	.01503	.0017	.01203	.00802
80	.012500	.008118	.01624	.01353	.0016	.01083	.00722
90	.011111	.007217	.01443	.01202	.0014	.00962	.00642
96	.010417	.006766	.01353	.01127	.0013	.00902	.00601
100	.010000	.006495	.01299	.01082	.0012	.00866	.00577
120	.008333	.005413	.01083	.00902	.0010	.00722	.00481

Using the Best Size Wires, measurement over three wires minus Constant for Best Size Wire equals Pitch Diameter.

MACHINE SCREW AND CAP SCREW HEADS

	SIZE	A	B	C	D
FILLISTER HEAD	#8	.260	.141	.042	.060
	#10	.302	.164	.048	.072
	1/4	3/8	.205	.064	.087
	5/16	7/16	.242	.077	.102
	3/8	9/16	.300	.086	.125
	1/2	3/4	.394	.102	.168
	5/8	7/8	.500	.128	.215
	3/4	1	.590	.144	.258
	1	1 5/16	.774	.182	.352
FLAT HEAD	#8	.320	.092	.043	.037
	#10	.372	.107	.048	.044
	1/4	1/2	.146	.064	.063
	5/16	5/8	.183	.072	.078
	3/8	3/4	.220	.081	.095
	1/2	7/8	.220	.102	.090
	5/8	1 1/8	.293	.128	.125
	3/4	1 3/8	.366	.144	.153
ROUND HEAD	#8	.297	.113	.044	.067
	#10	.346	.130	.048	.073
	1/4	7/16	.1831	.064	.107
	5/16	9/16	.236	.072	.150
	3/8	5/8	.262	.081	.160
	1/2	13/16	.340	.102	.200
	5/8	1	.422	.128	.255
	3/4	1 1/4	.526	.144	.320
HEXAGON HEAD	1/4	.494	.170	7/16	
	5/16	.564	.215	1/2	
	3/8	.635	.246	9/16	
	1/2	.846	.333	3/4	
	5/8	1.058	.411	15/16	
	3/4	1.270	.490	1 1/8	
	7/8	1.482	.566	1 5/16	
	1	1.693	.640	1 1/2	
SOCKET HEAD	#8	.265	.164	1/8	
	#10	5/16	.190	5/32	
	1/4	3/8	1/4	3/16	
	5/16	7/16	5/16	7/32	
	3/8	9/16	3/8	5/16	
	7/16	5/8	7/16	5/16	
	1/2	3/4	1/2	3/8	
	5/8	7/8	5/8	1/2	
	3/4	1	3/4	9/16	
	7/8	1 1/8	7/8	9/16	
	1	1 5/16	1	5/8	

STANDARD SYSTEM OF MARKING

GENERAL

Taps, dies, and other threading tools will be marked with the nominal size, number of threads per inch, and the proper symbol to identify the thread form. These symbols are in agreement with the A.S.A. B1-7-1949 Standard on Nomenclature, Definitions, and Letter Symbols for Screw Threads.

The marking for the British Threads are the fully abbreviated form based on the data in the British Standard Institute, Specification No. 84-1940.

SYMBOLS USED FOR AMERICAN THREADS ARE:

Symbol	Reference
NC	American National Coarse Thread Series
NF	American National Fine Thread Series
NEF	American National Extra Fine Thread Series
N	American National 8, 12 and 16 Thread Series (8N, 12N, 16N)
NH	American (National) Hose Coupling and Fire Hose Coupling Threads
NM	National Miniature Screw Thread
NGO	American (National) Gas Outlet Thread
NS	American Special Thread (60° Thread Form)
NPT	American (National) Taper Pipe Thread
NPTF	Dryseal American (National) Taper Pipe Thread
PTF	Dryseal SAE Short Internal Taper Pipe Thread
ANPT	Military Aeronautical Pipe Thread Specification MIL-P-7105
NPS	American (National) Straight Pipe Thread
NPSC	American (National) Straight Pipe Thread in Pipe Couplings
NPSF	Dryseal American (National) Fuel Internal Straight Pipe Thread
NPSH	American (Standard) Straight Pipe Thread for Hose Couplings and Nipples
NPSI	Dryseal American (National) Intermediate Internal Straight Pipe Thread
NPSL	American (National) Internal Straight Pipe Thread for Locknut Connections (Loose Fitting Mechanical Joints)
NPSM	American (National) Internal Straight Pipe Thread for Mechanical Joints (Free Fitting)
NPTR	American (National) Internal Taper Pipe Thread for Railing Joints (Mechanical Joints)
AMO	American Standard Microscope Objective Thread
ACME C	Acme Screw Thread — Centralizing Type
ACME G	Acme Screw Thread — General Purpose Type
STUB ACME	Stub Acme Threads
N. BUTT	National Buttress Screw Thread
V	A 60° "V" Thread with Truncated Crests and Roots. The Theoretical "V" Form is usually flatted several thousandths of an inch to the user's specifications.
SB	Manufacturers Stovebolt Standard Thread
STI	Special Threads for Helical Coil Wire Screw Thread Inserts

SYMBOLS USED FOR BRITISH THREADS ARE:

Symbol	Reference
BSW	British Standard Whitworth Coarse Thread Series
BSF	British Standard Fine Thread Series
BSP	British Standard Taper Pipe Thread
BSPP	British Standard Pipe (Parallel) Thread
WHIT	Whitworth Standard Special Thread
BA	British Association Standard Thread

BENT SHANK TAPPER TAPS

In addition to the regular marking, bent shank tapper taps will be marked with the table number to which they are made.

GROUND THREAD TAPS — LIMIT NUMBERS

All standard Ground Thread Taps made to Tables 327 and 329 will be marked with the letter G to designate Ground Thread. The letter G will be followed by the letter H to designate above basic (L below basic) and a numeral to designate the Pitch Diameter limits.

Example: G H3 Indicates a Ground Thread Tap with Pitch Diameter limits .0010 to .0015 in. over basic.

Pitch Diameter limits for Taps to 1 in. diameter inclusive.

L1 = Basis to Basic minus .0005
H1 = Basic to Basic plus .0005
H2 = Basic plus .0005 to Basic plus .0010
H3 = Basic plus .0010 to Basic plus .0015
H4 = Basic plus .0015 to Basic plus .0020
H5 = Basic plus .0020 to Basic plus .0025
H6 = Basic plus .0025 to Basic plus .0030

(Morse Twist Drill & Machine Co.)

HARDNESS CONVERSION TABLE

BRINELL IDENTATION DIAMETER, mm	BRINELL HARDNESS NUMBER		ROCKWELL HARDNESS NUMBER		ROCKWELL SUPERFICIAL HARDNESS NUMBER SUPERFICIAL DIAMOND PENETRATOR			TENSILE STRENGTH (APPROXIMATE) 1000 psi
	STANDARD BALL	TUNGSTEN-CARBIDE BALL	B SCALE	C SCALE	15 N SCALE	30 N SCALE	45 N SCALE	
2.45	...	627	...	58.7	89.6	76.3	65.1	347
2.50	...	601	...	57.3	89.0	75.1	63.5	328
2.55	...	578	...	56.0	88.4	73.9	62.1	313
2.60	...	555	...	54.7	87.8	72.7	60.6	298
2.65	...	534	...	53.5	87.2	71.6	59.2	288
2.70	...	514	...	52.1	86.5	70.3	57.6	274
2.75	...	495	...	51.0	85.9	69.4	56.1	264
2.80	...	477	...	49.6	85.3	68.2	54.5	252
2.85	...	461	...	48.5	84.7	67.2	53.2	242
2.90	...	444	...	47.1	84.0	65.8	51.5	230
2.95	429	429	...	45.7	83.4	64.6	49.9	219
3.00	415	415	...	44.5	82.8	63.5	48.4	212
3.05	401	401	...	43.1	82.0	62.3	46.9	202
3.10	388	388	...	41.8	81.4	61.1	45.3	193
3.15	375	375	...	40.4	80.6	59.9	43.6	184
3.20	363	363	...	39.1	80.0	58.7	42.0	177
3.25	352	352	...	37.9	79.3	57.6	40.5	170
3.30	341	341	...	36.6	78.6	56.4	39.1	163
3.35	331	331	...	35.5	78.0	55.4	37.8	158
3.40	321	321	...	34.3	77.3	54.3	36.4	152
3.45	311	311	...	33.1	76.7	53.3	34.4	147
3.50	302	302	...	32.1	76.1	52.2	33.8	143
3.55	293	293	...	30.9	75.5	51.2	32.4	139
3.60	285	285	...	29.9	75.0	50.3	31.2	136
3.65	277	277	...	28.8	74.4	49.3	29.9	131
3.70	269	269	...	27.6	73.7	48.3	28.5	128
3.75	262	262	...	26.6	73.1	47.3	27.3	125
3.80	255	255	...	25.4	72.5	46.2	26.0	121
3.85	248	248	...	24.2	71.7	45.1	24.5	118
3.90	241	241	100.0	22.8	70.9	43.9	22.8	114
3.95	235	235	99.0	21.7	70.3	42.9	21.5	111
4.00	229	229	98.2	20.5	69.7	41.9	20.1	109
4.05	223	223	97.3	...	...	...	...	104
4.10	217	217	96.4	...	...	...	...	103
4.15	212	212	95.5	...	...	...	...	100
4.20	207	207	94.6	...	...	...	...	99
4.25	201	201	93.8	...	...	...	...	97
4.30	197	197	92.8	...	...	...	...	94
4.35	192	192	91.9	...	...	...	...	92
4.40	187	187	90.7	...	...	...	...	90
4.45	183	183	90.0	...	...	...	...	89
4.50	179	179	89.0	...	...	...	...	88
4.55	174	174	87.8	...	...	...	...	86
4.60	170	170	86.8	...	...	...	...	84
4.65	167	167	86.0	...	...	...	...	83
4.70	163	163	85.0	...	...	...	...	82
4.80	156	156	82.9	...	...	...	...	80
4.90	149	149	80.8	...	...	...	...	73
5.00	143	143	78.7	...	...	...	...	71
5.10	137	137	76.4	...	...	...	...	67
5.20	131	131	74.0	...	...	...	...	65
5.30	126	126	72.0	...	...	...	...	63
5.40	121	121	69.0	...	...	...	...	60
5.50	116	116	67.6	...	...	...	...	58
5.60	111	111	65.7	...	...	...	...	56

(Carpenter Steel Co.)

WIRE GAGES

NUMBER OF WIRE GAGE	AMERICAN OR BROWN & SHARPE, INCHES	WASHBURN & MOEN MFG. CO., A. S. & W. ROEBLING, INCHES	IMPERIAL WIRE GAGE, INCHES	STUBS' STEEL WIRE, INCHES	BIRMINGHAM OR STUBS' IRON WIRE, INCHES
0000000	....	.4900	.5000	...	...
000000	.5800	.4615	.4640	...	...
00000	.5165	.4305	.4320	...	.500
0000	.460	.3938	.4000	...	.454
000	.40964	.3625	.3720	...	.425
00	.3648	.3310	.3480	...	.380
0	.32486	.3065	.3240	...	.340
1	.2893	.2830	.3000	.227	.300
2	.25763	.2625	.2760	.219	.284
3	.22942	.2437	.2520	.212	.259
4	.20431	.2253	.2320	.207	.238
5	.18194	.2070	.2120	.204	.220
6	.16202	.1920	.1920	.201	.203
7	.14428	.1770	.1760	.199	.180
8	.12849	.1620	.1600	.197	.165
9	.11443	.1483	.1440	.194	.148
10	.10189	.1350	.1280	.191	.134
11	.090742	.1205	.1160	.188	.120
12	.080808	.1055	.1040	.185	.109
13	.071961	.0915	.0920	.182	.095
14	.064084	.0800	.0800	.180	.083
15	.057068	.0720	.0720	.178	.072
16	.05082	.0625	.0640	.175	.065
17	.045257	.0540	.0560	.172	.058
18	.040303	.0475	.0480	.168	.049
19	.03589	.0410	.0400	.164	.042
20	.031961	.0348	.0360	.161	.035
21	.028462	.0317	.0320	.157	.032
22	.025347	.0286	.0280	.155	.028
23	.022571	.0258	.0240	.153	.025
24	.0201	.0230	.0220	.151	.022
25	.0179	.0204	.0200	.148	.020
26	.01594	.0181	.0180	.146	.018
27	.014195	.0173	.0164	.143	.016
28	.012641	.0162	.0148	.139	.014
29	.011257	.0150	.0136	.134	.013
30	.010025	.0140	.0124	.127	.012
31	.008928	.0132	.0116	.120	.010
32	.00795	.0128	.0108	.115	.009
33	.00708	.0118	.0100	.112	.008
34	.006304	.0104	.0092	.110	.007
35	.005614	.0095	.0084	.108	.005
36	.005	.0090	.0076	.106	.004
37	.004453	.0085	.0068	.103	...
38	.003965	.0080	.0060	.101	...
39	.003531	.0075	.0052	.099	...
40	.003144	.0070	.0048	.097	...

DIMENSIONAL TOLERANCES

Allowances for Turning Machine-Straightened Bars

When ordering bars that are to be turned, it is recommended that allowances be made for finishing from hot rolled diameters not less than amounts shown in following table, and specify hot rolled sizes accordingly:

NOMINAL DIAMETER OF HOT ROLLED BAR, INCHES	MINIMUM ALLOWANCE ON DIAMETER FOR TURNING, INCHES
1 1/2 to 3, inclusive.	1/8
Over 3 to 6, inclusive	1/4
Over 6 to 8, inclusive	3/8

STANDARD DIMENSIONAL TOLERANCES

Hot Rolled Bars: Rounds and Squares

SPECIFIED SIZE, INCHES	VARIATIONS FROM SIZE INCHES		OUT-OF-ROUND (1) OR SQUARE (2) INCHES
	OVER	UNDER	
1/4 to 5/16 incl. (3).	(4)	(4)	(4)
Over 5/16 to 7/16 incl. (3). . .	0.006	0.006	0.009
Over 7/16 to 5/8 incl. (3) . . .	0.007	0.007	0.010
Over 5/8 to 7/8 incl.	0.008	0.008	0.012
Over 7/8 to 1 incl.	0.009	0.009	0.013
Over 1 to 1 1/8 incl.	0.010	0.010	0.015
Over 1 1/8 to 1 1/4 incl. . . .	0.011	0.011	0.016
Over 1 1/4 to 1 3/8 incl. . . .	0.012	0.012	0.018
Over 1 3/8 to 1 1/2 incl. . . .	0.014	0.014	0.021
Over 1 1/2 to 2 incl.	1/64	1/64	0.023
Over 2 to 2 1/2 incl.	1/32	0	0.023
Over 2 1/2 to 3 1/2 incl. . . .	3/64	0	0.035
Over 3 1/2 to 4 1/2 incl. . . .	1/16	0	0.046
Over 4 1/2 to 5 1/2 incl. . . .	5/64	0	0.058
Over 5 1/2 to 6 1/2 incl. . . .	1/8	0	0.070
Over 6 1/2 to 8 incl.	5/32	0	0.085

(1) Out-of-round is difference between maximum and minimum diameters of bar, measured at same cross section.

(2) Out-of-square is difference in two dimensions at same cross section of a square bar, each dimension being distance between opposite faces.

(3) Round sections in size range of 1/4 in. to approximately 5/8 in. diameter are commonly produced on rod mills in coils. Tolerances on product made this way have not been established; for such tolerances producer should be consulted.

Variations in size of coiled product made on rod mills are greater than size tolerances for product made on bar mills.

(4) Squares in this size are not commonly produced as hot rolled product.

Hot Rolled Bars: Hexagons and Octagons

SPECIFIED SIZES BETWEEN OPPOSITE SIDES, INCHES	VARIATIONS FROM SIZE INCHES		MAXIMUM DIFFERENCE 3 MEASUREMENTS FOR HEXAGONS ONLY INCHES
	UNDER	OVER	
1/4 to 1/2 incl.	0.007	0.007	0.011
Over 1/2 to 1 incl.	0.010	0.010	0.015
Over 1 to 1 1/2 incl.	0.021	0.021	0.025
Over 1 1/2 to 2 incl.	1/32	1/32	1/32
Over 2 to 2 1/2 incl.	3/64	3/64	3/64
Over 2 1/2 to 3 1/2 incl.	1/16	1/16	1/16

Hot Rolled Bars: Flats

	VARIATIONS FROM THICKNESS FOR THICKNESSES GIVEN OVER OR UNDER INCHES			VARIATIONS FROM WIDTH INCHES	
SPECIFIED WIDTHS, INCHES	1/8 TO 1/2 INCL.	OVER 1/2 TO 1 INCL.	OVER 1 TO 2 INCL.	OVER	UNDER
To 1 incl.	0.008	0.010	. .	1/64	1/64
Over 1 to 2 incl. . . .	0.012	0.015	1/32	1/32	1/32
Over 2 to 4 incl. . . .	0.015	0.020	1/32	1/16	1/32
Over 4 to 6 incl. . . .	0.015	0.020	1/32	3/32	1/16
Over 6 to 8 incl. . . .	0.016	0.025	1/32	1/8	5/32
Over 8 to 10 incl. . .	0.021	0.031	1/32	5/32	3/16

Cold Finished Bars: Rounds

SPECIFIED SIZE, INCHES	VARIATIONS FROM SIZE INCHES (1)	
	OVER	UNDER
Over 1/2 to 1 excl.	0.002	0.002
1 to 1 1/2 excl.	0.0025	0.0025
1 1/2 to 4 incl. (2)	0.003	0.003

(1) When it is necessary to heat treat, or heat treat and pickle after cold finishing, because of special hardness or mechanical property requirements, tolerances are double those shown in table.
(2) Size tolerances for sizes over 4 in. are availale upon request.

Cold Finished Bars:*
Hexagons, Octagons and Squares

SPECIFIED SIZE, INCHES	VARIATIONS FROM SIZE INCHES (1)	
	OVER	UNDER
Over 1/2 to 1 incl.	0	0.004
Over 1 to 2 incl.	0	0.006
Over 2 to 3 incl.	0	0.008
Over 3 (2)	0	0.010

* Bars with plus or minus tolerances to meet other special needs are available upon request.

(1) When it is necessary to heat treat, or heat treat and pickle after cold finishing, because of special hardness or mechanical property requirements, tolerances are double those shown in the table.
(2) Size tolerances for sizes over 4 in. are available upon request.

Cold Finished Bars: Flats

SPECIFIED WIDTH SIZE OR THICKNESS, INCHES	VARIATIONS FROM WIDTH OVER OR UNDER, INCHES (1)		VARIATIONS FROM THICK-NESSES OVER OR UNDER, INCH
	FOR THICK-NESSES 1/4 IN. AND UNDER	FOR THICK-NESSES OVER 1/4 IN.	
Over 1/8 to 1 incl.	. . .	. . .	0.002
Over 3/8 to 1 incl.	0.004	0.002	0.002
Over 1 to 2 incl.	0.006	0.003	0.003
Over 2 to 3 incl.	0.008	0.004	0.004
Over 3 to 4 1/2 incl. (2)	0.010	0.005	0.005

(1) When it is necessary to heat treat, or heat treat and pickle after cold finishing, because of special hardness or mechanical property requirements, tolerances are double those shown in the table.
(2) Width tolerances over 4 1/2 in. and thickness tolerances over 4 1/2 in. are available upon request.

Machine-Cut Bars
Machine-Cut after Machine Straightening

SPECIFIED SIZES AS THEY APPLY TO ROUNDS, SQUARES, HEXAGONS, OCTAGONS AND WIDTH OF FLATS, INCHES	VARIATIONS FROM SPECIFIED LENGTHS, INCHES			
	TO 12 FT. INCL.		OVER 12 FT. TO 25 FT. INCL.	
	OVER	UNDER	OVER	UNDER
To 3 incl.	1/8	0	3/16	0
Over 3 to 6 incl.	3/16	0	1/4	0
Over 6 to 9 incl.	1/4	0	5/16	0
Over 9 to 12 incl.	1/2	0	1/2	0

Hot or Cold Cutting
Length Tolerances

SPECIFIED SIZES AS THEY APPLY TO ROUNDS, SQUARES, HEXAGONS, OCTAGONS AND WIDTHS OF FLATS, INCHES	VARIATIONS FROM SPECIFIED LENGTHS, INCHES			
	TO 12 FT. INCL.		OVER 12 FT. TO 25 FT. INCL.	
	OVER	UNDER	OVER	UNDER
To 2 incl.	1/2	0	3/4	0
Over 2 to 4 incl.	3/4	0	1	0
Over 4 to 6 incl.	1	0	1 1/4	0
Over 6 to 9 incl.	1 1/4	0	1 1/2	0
Over 9 to 12 incl.	1 1/2	0	2	0

Hot Finished and Cold Finished Bars for Machining

Camber is greatest deviation of a side from a straight line. Measurement is taken on concave side of bar with a straightedge. Unless otherwise specified, hot finished and cold finished bars for machining purposes are furnished machine-straightened to following tolerances:
Hot finished:

1/8 in. in any 5 feet; but may not exceed 1/8 in. x $\frac{\text{No. of feet in length}}{5}$

Cold finished:

1/16 in. in any 5 feet; but may not exceed 1/16 in. x $\frac{\text{No. of feet in length}}{5}$

Wire
Drawn, centerless ground, centerless ground and polished round and square wire

SPECIFIED SIZE, INCHES	TOLERANCE, INCHES	
		UNDER
1/2	0.002	0.002
Under 1/2 to 5/16 incl.	0.0015	0.0015
Under 5/16 to 0.050 incl.	0.001	0.001

The maximum out-of-round tolerance for round wire is one-half of total size tolerance shown in above table.

Tolerance for wire for which final operation is a surface treatment for purpose of removing scale or drawing lubricant.

SPECIFIED SIZE, INCHES	TOLERANCE, INCHES	
	OVER	UNDER
1/2	0.004	0.004
Under 1/2 to 5/16 incl.	0.003	0.003
Under 5/16 to 0.050 incl.	0.002	0.002

Drawn Wire in Hexagons and Octagons — Size Tolerances.

SPECIFIED SIZE,* INCHES	TOLERANCE, INCHES	
	OVER	UNDER
1/2	0	0.004
Under 1/2 to 5/16 incl.	0	0.003
Under 5/16 to 1/8 incl.	0	0.002

* Distance across flats.
Producer should be consulted for all tolerances for half round, oval and half oval wires. (Carpenter Steel Co.)

Surface speed – feet/minute	
Round bars	$\frac{\text{Diameter} \times 3.1416 \times \text{RPM}}{12}$
Hexagon bars (distance across corners)	$\frac{\text{Size} \times 3.1416 \times \text{RPM} \times 1.155}{12}$
	$\frac{\text{Distance} \times 3.1416 \times \text{RPM}}{12}$
Square bars (distance across corners)	$\frac{\text{Size} \times 3.1416 \times \text{RPM} \times 1.414}{12}$
	$\frac{\text{Distance} \times 3.1416 \times \text{RPM}}{12}$
Revolutions – number/minute	
Round bars	$\frac{\text{SFM} \times 12}{\text{Diameter} \times 3.1416}$
Hexagon bars	$\frac{\text{SFM} \times 12}{\text{Size} \times 3.1416 \times 1.155}$
	$\frac{\text{SFM} \times 12}{\text{Distance} \times 3.1416}$
Square bars	$\frac{\text{SFM} \times 12}{\text{Size} \times 3.1416 \times 1.414}$
	$\frac{\text{SFM} \times 12}{\text{Distance} \times 3.1416}$
Feed – inches/revolution	$\frac{\text{Feed inches per minute}}{\text{RPM}}$
	$\frac{\text{Diameter} \times 3.1416 \times \text{Feed}}{\text{SFM} \times 12}$
Feed – inches/tooth	$\frac{\text{Feed}}{\text{No. of teeth}}$
Time for actual machining – seconds	$\frac{\text{Revolutions required} \times 60 \text{ seconds}}{\text{RPM}}$
Machine time	Time for machining + idle time
Tapping or threading time – seconds	$\frac{\text{No. of threads} \times 60 \text{ (seconds)}}{\text{Actual threading speed in RPM}}$

NOTE: All sizes are in inches.

COLOR CODES FOR MARKING STEELS

S.A.E. Number	Code Color	S.A.E. Number	Code Color	S.A.E. Number	Code Color	S.A.E. Number	Code Color
	CARBON STEELS	2115	Red and bronze	T1340	Orange and green	3450	Black and bronze
1010	White	2315	Red and blue	T1345	Orange and red	4820	Green and purple
1015	White	2320	Red and blue	T1350	Orange and red		CHROMIUM STEELS
X1015	White	2330	Red and white		NICKEL-CHROMIUM STEELS	5120	Black
1020	Brown	2335	Red and white	3115	Blue and black	5140	Black and white
X1020	Brown	2340	Red and green	3120	Blue and black	5150	Black and white
1025	Red	2345	Red and green	3125	Pink	52100	Black and brown
X1025	Red	2350	Red and aluminum	3130	Blue and green		CHROMIUM-VANADIUM STEELS
1030	Blue	2515	Red and black	3135	Blue and green	6115	White and brown
1035	Blue		MOLYBDENUM STEELS	3140	Blue and white	6120	White and brown
1040	Green	4130	Green and white	X3140	Blue and white	6125	White and aluminum
X1040	Green	X4130	Green and bronze	3145	Blue and white	6130	White and yellow
1045	Orange	4135	Green and yellow	3150	Blue and brown	6135	White and yellow
X1045	Orange	4140	Green and brown	3215	Blue and purple	6140	White and bronze
1050	Bronze	4150	Green and brown	3220	Blue and purple	6145	White and orange
1095	Aluminum	4340	Green and aluminum	3230	Blue and purple	6150	White and orange
	FREE CUTTING STEELS	4345	Green and aluminum	3240	Blue and aluminum	6195	White and purple
1112	Yellow	4615	Green and black	3245	Blue and aluminum		TUNGSTEN STEELS
X1112	Yellow	4620	Green and black	3250	Blue and bronze	71360	Brown and orange
1120	Yellow and brown	4640	Green and pink	3312	Orange and black	71660	Brown and bronze
X1314	Yellow and blue	4815	Green and purple	3325	Orange and black	7260	Brown and aluminum
X1315	Yellow and red	X1340	Yellow and black	3335	Blue and orange		SILICON-MANGANESE STEELS
X1335	Yellow and black		MANGANESE STEELS	3340	Blue and orange	9255	Bronze and aluminum
	NICKEL STEELS	T1330	Orange and green	3415	Blue and pink	9260	Bronze and aluminum
2015	Red and brown	T1335	Orange and green	3435	Orange and aluminum		

GRINDING WHEEL MARKINGS

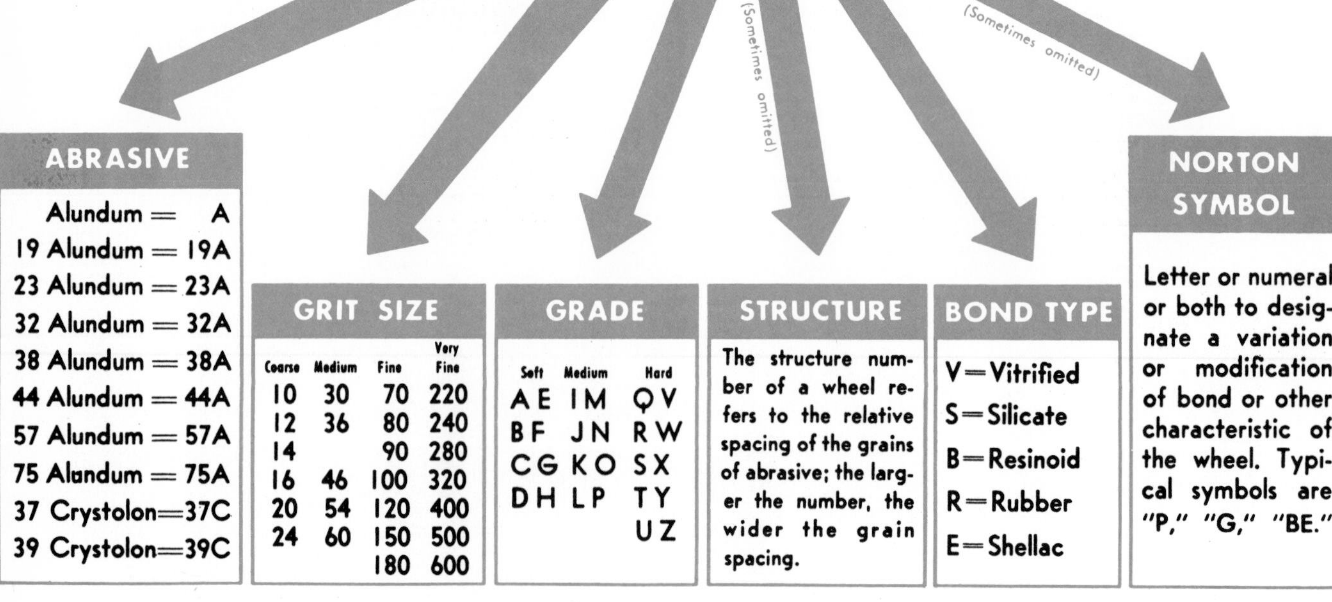

STANDARD SYMBOLS USED IN DIMENSIONING

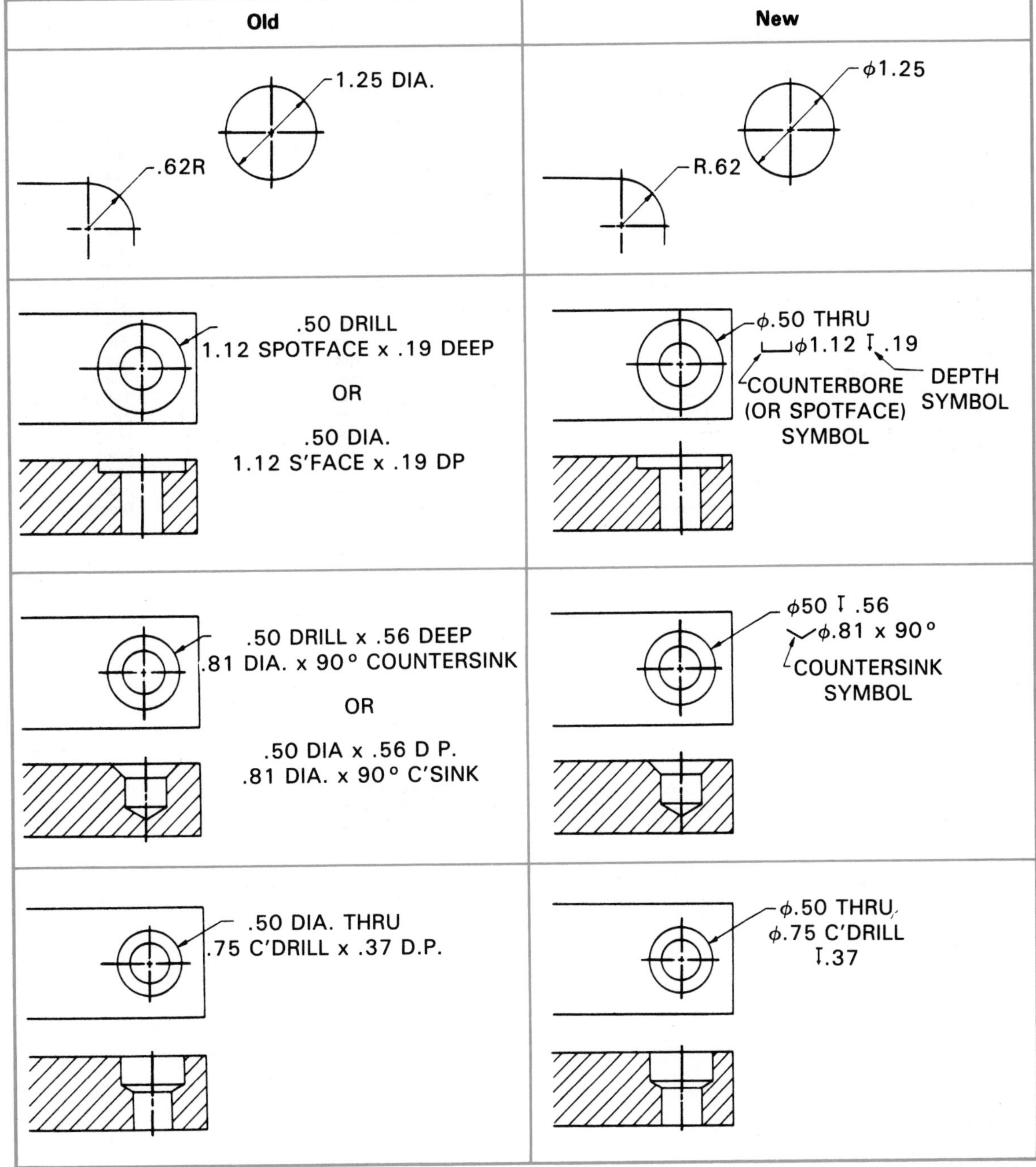

GENERAL MATERIAL CHART

MATERIAL Sheet (less than 1/4" thick)	HOW MEASURED	HOW PURCHASED	CHARACTERISTICS
Copper	Gauge number (Brown & Sharpe & Amer. Std.)	24 x 96" sheet or 12 or 18" by lineal feet on roll	Pure metal
Brass	Gauge number (B & S and Amer. Std.)	24 x 76" sheet or 12 or 18" by lineal feet on roll	Alloy of copper & zinc
Aluminum	Decimal	24 x 72" sheet or 12 or 18" by lineal feet on roll	Available as commercially pure metal or alloyed for strength, hardness, & ductility
Galvanized steel	Gauge number (U.S. Std.)	24 x 96" sheet	Mild steel sheet with zinc plating, also available with zinc coating that is part of sheet
Black annealed steel sheet	Gauge number (U.S. Std.)	24 x 96" sheet	Mild steel with oxide coating-hot rolled
Cold rolled steel sheet	Gauge number (U.S. Std.)	24 x 96" sheet	Oxide removed and cold rolled to final thickness
Tin plate	Gauge number (U.S. Std.)	20 x 28" sheet 56 or 112 to pkg	Mild steel with tin coating
Nickel silver	Gauge number (Brown & Sharp)	6 or 12" wide by lineal sheet	Copper 50%, Zinc 30%, nickel 20%
Expanded	Gauge number (U.S. Std.)	36 x 96"	Metal is pierced and expanded (stretched) to diamond shape; also available rolled to thickness after it has been expanded
Perforated	Gauge number (U.S. Std.)	30 x 36" 36 x 48"	Design is cut in sheet; many designs available

RULES FOR DETERMINING SPEED AND FEED

TO FIND	HAVING	RULE	FORMULA
Speed of cutter in feet per minute (FPM)	Diameter of cutter and revolutions per minute	Diameter of cutter (in inches) multiplied by 3.1416 (π) multiplied by revolutions per minute, divided by 12	$FPM = \frac{\pi D \times RPM}{12}$
Speed of cutter in meters per minute	Diameter of cutter and revolutions per minute	Diameter of cutter multiplied by 3.1416 (π) multiplied by revolutions per minute, divided by 1000	$MPM = \frac{D(mm) \times \pi \times RPM}{1000}$
Revolutions per minute (RPM)	Feet per minute and diameter of cutter	Feet per minute, multiplied by 12, divided by circumference of cutter (πD)	$RPM = \frac{FPM \times 12}{\pi D}$
Revolutions per minute (RPM)	Meters per minute and diameter of cutter in millimeters (mm)	Meters per minute multipled by 1000, divided by the circumference of cutter (πD)	$RPM = \frac{MPM \times 1000}{\pi D}$
Feed per revolution (FR)	Feed per minute and revolutions per minute	Feed per minute, divided by revolutions per minute	$FR = \frac{F}{RPM}$
Feed per tooth per revolution (FTR)	Feed per minute and number of teeth in cutter	Feed per minute (in inches or millimeters) divided by number of teeth in cutter × revolutions per minute	$FTR = \frac{F}{T \times RPM}$
Feed per minute (F)	Feed per tooth per revolution, number of teeth in cutter, and RPM	Feed per tooth per revolutions multiplied by number of teeth in cutter, multiplied by revolutions per minute	$F = FTR \times T \times RPM$
Feed per minute (F)	Feed per revolution and revolutions per minute	Feed per revolution multiplied by revolutions per minute	$F = FR \times RPM$
Number of teeth per minute (TM)	Number of teeth in cutter and revolutions per minute	Number of teeth in cutter multiplied by revolutions per minute	$TM = T \times RPM$

RPM = Revolutions per minute
T = Teeth in cutter
D = Diameter of cutter
π = 3.1416 (pi)
FRM = Speed of cutter in feet per minute

TM = Teeth per minute
F = Feed per minute
FR = Feed per revolution
FTR = Feed per tooth per revolution
MPM = Speed of cutter in meters per minute

CUTTING FLUID CHART

Material	Cutting Fluid
Aluminum and its Alloys	Kerosene, kerosene and lard oil, soluble oil
Plastics	Dry
Brass, Soft	Dry, soluble oil, kerosene and lard oil
Bronze, High Tensile	Soluble oil, lard oil, mineral oil, dry
Cast Iron	Dry, air jet, soluble oil
Copper	Soluble oil, dry, mineral lard oil, kerosene
Magnesium	Low viscosity neutral oils
Malleable Iron	Dry, soda water
Monel Metal	Lard oil, soluble oil
Slate	Dry
Steel, Forging	Soluble oil, sulphurized oil, mineral lard oil
Steel, Manganese	Soluble oil, sulphurized oil, mineral lard oil
Steel, Soft	Soluble oil, mineral lard oil, sulphurized oil, lard oil
Steel, Stainless	Sulphurized mineral oil, soluble oil
Steel, Tool	Soluble oil, mineral lard oil, sulphurized oil
Wrought Iron	Soluble oil, mineral lard oil, sulphurized oil

BOLT TORQUING CHART

METRIC STANDARD						
GRADE OF BOLT		5D	8G	10K	12K	
MIN. TENSILE STRENGTH		71,160 P.S.I	113,800 P.S.I.	142,200 P.S.I.	170,679 P.S.I.	
GRADE MARKINGS ON HEAD		5D	8G	10K	12K	SIZE OF SOCKET OR WRENCH OPENING
METRIC		FOOT POUNDS				METRIC
BOLT DIA.	U.S. DEC EQUIV.					BOLT HEAD
6mm	**.2362**	**5**	**6**	**8**	**10**	**10mm**
8mm	**.3150**	**10**	**16**	**22**	**27**	**14mm**
10mm	**.3937**	**19**	**31**	**40**	**49**	**17mm**
12mm	**.4720**	**34**	**54**	**70**	**86**	**19mm**
14mm	**.5512**	**55**	**89**	**117**	**137**	**22mm**
16mm	**.6299**	**83**	**132**	**175**	**208**	**24mm**
18mm	**.709**	**111**	**182**	**236**	**283**	**27mm**
22mm	**.8661**	**182**	**284**	**394**	**464**	**32mm**

SAE STANDARD / FOOT POUNDS						
GRADE OF BOLT	SAE 1 & 2	SAE 5	SAE 6	SAE 8		
MIN. TEN STRENGTH	64,000 P.S.I.	105,000 P.S.I.	133,000 P.S.I	150,000 P.S.I.		
MARKINGS ON HEAD					SIZE OF SOCKET OR WRENCH OPENING	
U.S. STANDARD	FOOT POUNDS				U.S. REGULAR	
BOLT DIA.					BOLT HEAD	NUT
1/4	**5**	**7**	**10**	**10.5**	**3/8**	**7/16**
5/16	**9**	**14**	**19**	**22**	**1/2**	**9/16**
3/8	**15**	**25**	**34**	**37**	**9/16**	**5/8**
7/16	**24**	**40**	**55**	**60**	**5/8**	**3/4**
1/2	**37**	**60**	**85**	**92**	**3/4**	**13/16**
9/16	**53**	**88**	**120**	**132**	**7/8**	**7/8**
5/8	**74**	**120**	**167**	**180**	**15/16**	**1.**
3/4	**120**	**200**	**280**	**296**	**1-1/8**	**1-1/8**

FRACTION	INCHES	M/M
1/64	.01563	.397
1/32	.03125	.794
3/64	.04688	1.191
1/16	.06250	1.588
5/64	.07813	1.984
3/32	.09375	2.381
7/64	.10938	2.778
1/8	.12500	3.175
9/64	.14063	3.572
5/32	.15625	3.969
11/64	.17188	4.366
3/16	.18750	4.763
13/64	.20313	5.159
7/32	.21875	5.556
15/64	.23438	5.953
1/4	.25000	6.350
17/64	.26563	6.747
9/32	.28125	7.144
19/64	.29688	7.541
5/16	.31250	7.938
21/64	.32813	8.334
11/32	.34375	8.731
23/64	.35938	9.128
3/8	.37500	9.525
25/64	.39063	9.922
13/32	.40625	10.319
27/64	.42188	10.716
7/16	.43750	11.113
29/64	.45313	11.509
15/32	.46875	11.906
31/64	.48438	12.303
1/2	.50000	12.700

FRACTION	INCHES	M/M
33/64	.51563	13.097
17/32	.53125	13.494
35/64	.54688	13.891
9/16	.56250	14.288
37/64	.57813	14.684
19/32	.59375	15.081
39/64	.60938	15.478
5/8	.62500	15.875
41/64	.64063	16.272
21/32	.65625	16.669
43/64	.67188	17.066
11/16	.68750	17.463
45/64	.70313	17.859
23/32	.71875	18.256
47/64	.73438	18.653
3/4	.75000	19.050
49/64	.76563	19.447
25/32	.78125	19.844
51/64	.79688	20.241
13/16	.81250	20.638
53/64	.82813	21.034
27/32	.84375	21.431
55/64	.85938	21.828
7/8	.87500	22.225
57/64	.89063	22.622
29/32	.90625	23.019
59/64	.92188	23.416
15/16	.93750	23.813
61/64	.95313	24.209
31/32	.96875	24.606
63/64	.98438	25.003
1	1.00000	25.400

THE POWDER METALLURGY PROCESSES

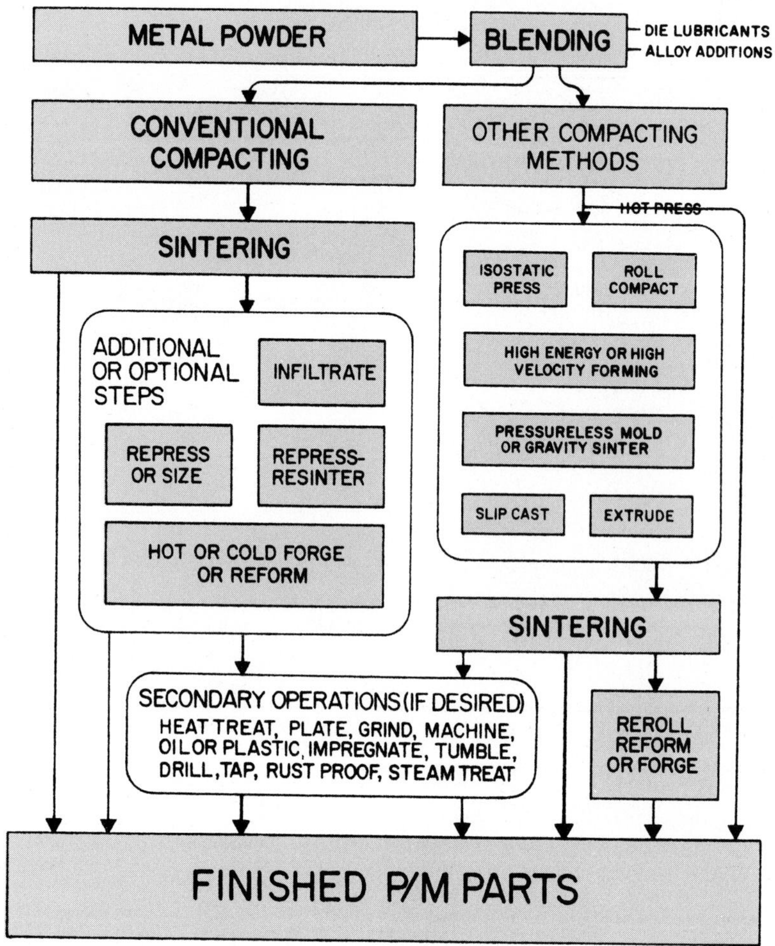

SINTERING	Sintering is a solid state phenomenon in which powdered metal particles become metallurgically bonded below the melting point of the metal. No adhesives or cements are used.
INFILTRATION	Pores of P/M parts are filled with a lower melting point metal such as copper base alloy. When the part is sintered, the infiltrant material melts and penetrates into the P/M part by capillary action.
COINING AND SIZING	Basically, this operation involves repressing sintered parts in a die similar to the original compacting die.
IMPREGNATION	The pores of the P/M part are filled with a lubricant or other non-metallic such as plastic resin. This may be done by means of a vacuum or by soaking. The part then becomes self-lubricating or pressure-tight if resin is used.

WELDING PROCESSES

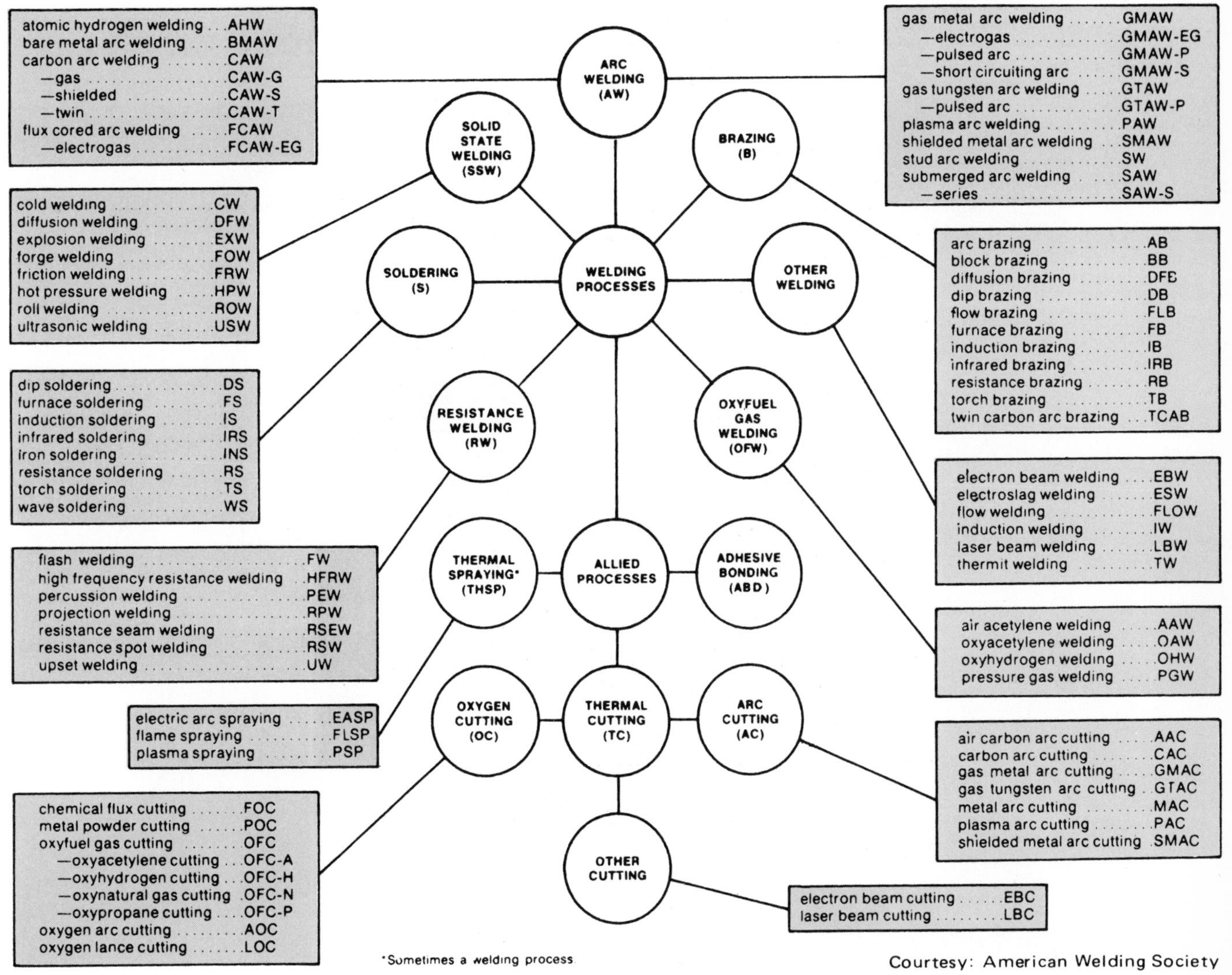

*Sometimes a welding process

Courtesy: American Welding Society

ACKNOWLEDGEMENTS

The author expresses his sincere thanks to the many organizations and manufacturers who cooperated so generously in supplying the technical information and many of the photographs used in this textbook.

We have endeavored to give them credit where due. Any omission was purely accidental.

John R. Walker
Bel Air, Maryland

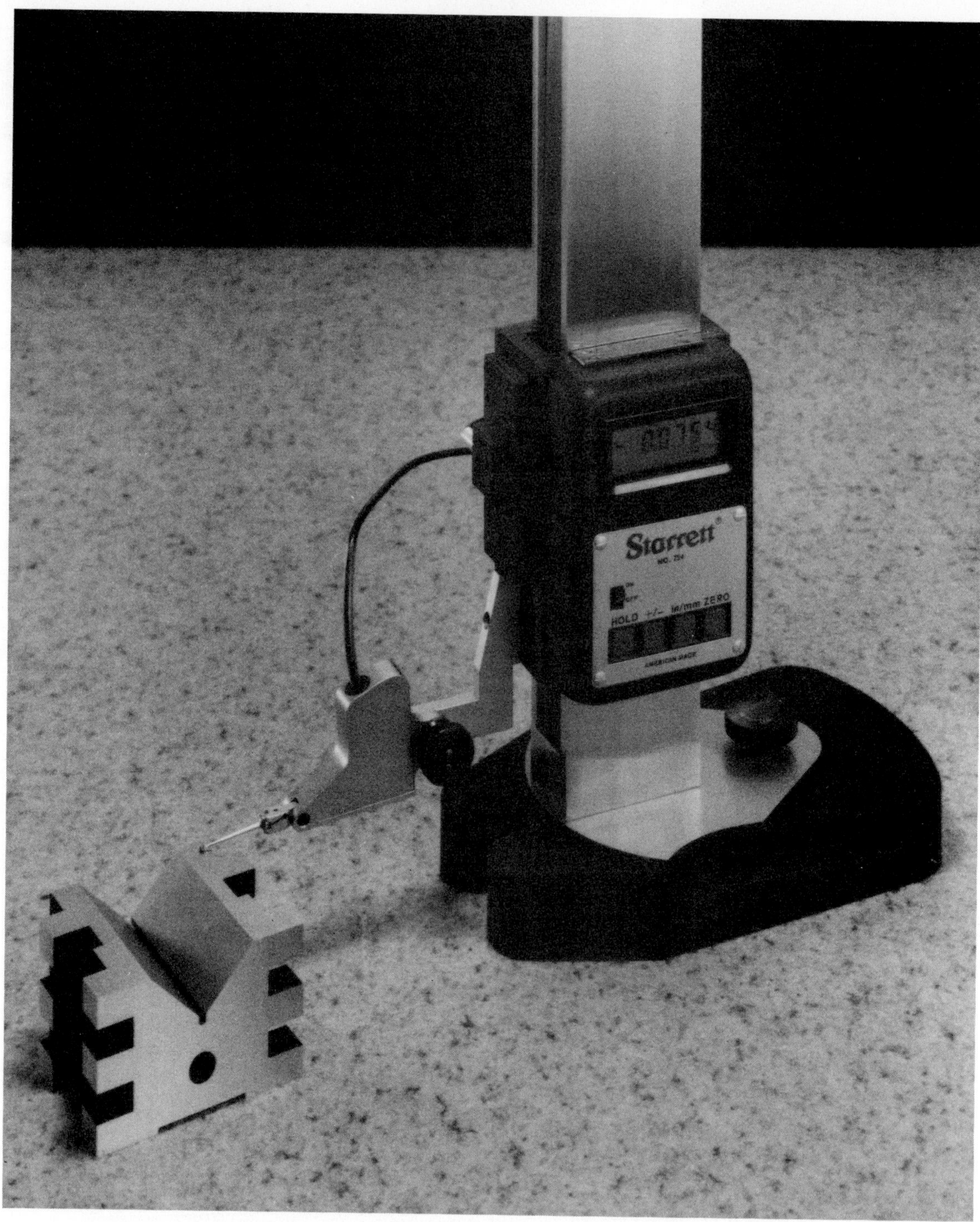

Digital readout height gauge. Note tool options along bottom of measuring unit. Measurements can be made in either inches or millimeters. (The L.S. Starrett Co.)

Giant presses, some of them exerting forces of 1,600 tons, stamp out more than 60 different parts at Volkswagen South Charleston, Volkswagen of America's metal stamping plant. For safety, dual controls have to be activated before the presses will operate. Control panel at bottom is on a decoiling blanker which cuts coiled steel to varying size for components formed by the huge stamping presses.

INDEX

B

C

D

E

F

G

H

I

M

N

O

P

Q

R

T

U

V

W